# McGraw-Hill's Compilation of

# DATA COMMUNICATIONS STANDARDS

**Edited by**

**Harold C. Folts**  Data Communications
Standards Consultant

**Harry R. Karp**  Editor-in-Chief
Data Communications

McGraw-Hill Publications Co.
1221 Avenue of the Americas
New York, New York 10020

Library of Congress Cataloging in Publication Data

Main entry under title:
McGraw-Hill's compilation of
data communications standards.

    1. Data transmission systems—Standards. 2. Com-
puter networks—Standards. I. Folts, Harold C., 1935-
II. Karp, Harry R., 1922- III. Title: Compilation of data
communications standards.
TK5105.M3  621.38'021  78-17191
ISBN 0-07-099782-9

**Second Printing April 1979**

# Foreword

Progress in data communications is ultimately dependent on the ability of diverse components in complex networks to trade information back and forth. In this light, it is a short step to the realization that data communications is impossible without standardization since no information transfer can take place until a common basis for communications has been established. All U.S. and international organizations dedicated to the establishment of a common basis—that is, standards—for information transfer in the data communications field are represented in this book, as are all of the data communications standards they have promulgated.

The editors extend their thanks for receiving reprint permission from the following national and international groups: The Consultative Committee for International Telegraph and Telephone (CCITT) of the International Telecommunications Union; the International Organization for Standardization (ISO); the American National Standards Institute (ANSI); the Electronic Industries Association (EIA); and the Specification and Consumer Information Branch, General Services Administration (Fed. Std.).

The editors also thank Glenn Hartwig, Bill Left, Janet Eyler, and Ruby Katayama for their efforts in the production of this volume.

# Preface

Modern data communications represents an advancing and complex technology with many facets. Data communications, on a worldwide basis, necessitates a vast assemblage of equipment, transmission systems, and user applications—all interworking in harmony. Growing at an astronomical rate, data communications soon will represent a substantial portion of the world's total telecommunications services.

Around 1960, the urgent need for standards in the data communications field was recognized and primary development activities were established by the International Telegraph and Telephone Consultative Committee (CCITT), the International Organization for Standardization (ISO), the American National Standards Institute (ANSI), the Electronic Industries Association (EIA), and many other organizations throughout the world. International activity through the present by these organizations has led to a high level of international cooperation transparent to national ideologies.

The most often used data communications standards that have been put into effect by CCITT, ISO, ANSI, EIA, and the U.S. Federal Government are presented in their entirety in this book. This one volume serves as a single source for this wealth of information. Each section is preceded by an introduction giving the background of the responsible organization that developed the collection of standards. Additionally, a cross reference is provided to identify corresponding standards published by other activities. Many of these standards were developed through joint effort and are thus published by each participating organization. All the different versions may not be identical, however, because of differences in application. Therefore, each of the cross-referenced standards should be carefully studied to determine which is appropriate for the application under consideration.

Each standard has been developed through the participation of many industry and government experts who devoted long and hard hours preparing proposals and negotiating agreements. With a wide diversity of inputs, particularly on the international scene, achieving an optimal point of technical compromise requires intensive cooperation of the individual participants as well as intimate understanding of the influencing factors *in toto*. Much well-deserved praise is given by the editors to the individuals responsible for achieving the level of agreement needed in producing these universally acceptable standards. The final standards are the result of long hours of intensive work and personal sacrifice.

This book has been assembled in an attempt to bring an orderly, usable consolidation of current data communications standards to implementers and users. In this way, the outstanding work of these organizations and the individual participants can be widely and effectively applied.

# Table of Contents

vi

# Part 1

# CCITT Data Communications Standards

In the very early days when the initial telecommunications technology began to evolve into practical implementation, the need for standards to establish international services was quickly recognized. As a result, the Union Telegraphique was established in 1865 by a treaty among participating nations. This activity finally evolved into today's organization known as the International Telecommunications Union (ITU). In 1947, the ITU was made a specialized agency of the United Nations Organization and moved to its present headquarters in Geneva, Switzerland.

There are now over 160 member countries in the ITU. Originally, membership consisted of the more technologically advanced countries, but today it is rapidly expanding to include the many emerging nations. Membership is broken into five basic categories as follows:

A. Administration—The member countries governmental telecommunications administration. In the case of the USA, this is the Department of State because provision of telecommunication services is left to private industry. The Administration of each country is the only voting member.

B. Recognized Private Operating Agencies (RPOA)—The private or government controlled corporations that provide public telecommunication services. For example, AT&T is an RPOA from the United States. RPOAs serve as principal advisers to the national administration of their countries.

C. Scientific and Industrial Organizations—Commercial organizations that have an interest in telecommunications are allowed to participate as observers and contribute to the work.

D. International Organizations—Any private or commercial organization that has an interest can participate as an observer and contribute to the work. Such organizations include the International Organization for Standardization (ISO) and the International Air Transport Association (IATA).

E. Specialized Treaty Agencies—Any recognized treaty agency in related fields is also allowed to participate. For example, the World Meteorological Organization.

The technical issues and development of standards (called Recommendations) are dealt with in the ITU by two committees. The first is the International Radio Consultative Committee (CCIR) and the other is the International Telegraph and Telephone Consultative Committee (CCITT). The CCITT, which is the significant committee from the data communications point of view, is broken into approximately 18 Study Groups each of which is assigned a specific area of responsibility. Study Groups VII and XVII have direct responsibilities for data communications interfaces, services, and transmission work. There is related activity in Study Group I, which is considering a new Teletex service; Study Group III, tariff principles: Study Group VIII, terminals; Study Group IX, telegraph practices (TELEX); and Study Group XIV, facsimile.

Study Group VII, Public Data Networks, was officially formed by the Vth Plenary Assembly in December, 1972. The Study Group was given the responsibility of developing Recommendations for implementing new digital networks specifically tailored for data communications services. The X-series of Recommendations represent the results of Study Group VII's work. This has, by far, been the most active Study Group in the history of CCITT. There are often as many as 200 participants in a meeting dealing with at least 100 technical contributions each time and generating up to 100 additional documents, reports, and draft Recommendations. Interest is very keen in furthering the state-of-the-art in this new technology.

Study Group XVII was originally formed as Special Study Group A by the Plenary Assembly in 1960 to deal with the then new technology of data transmission over the telephone network. This group has primarily dealt with modern modulation techniques and associated interfaces with data terminal equipment, and has produced the V-series Recommendations. Its work is still very important because it will be necessary to transmit data over telephone (analog) type circuits for many years ahead, even with the emergence of the advanced technology of public data networks.

The recommendations provided in this section include a selection from the X-series and V-series which are applicable to interconnection of users with telecommunications networks. These were approved by the Sixth Plenary Assembly Sept-Oct 1976. Additionally, there are three Provisional Recommendations X.3, X.28, and X.29 subsequently approved under the Accelerated Procedures as well as a number of Provisional Amendments also approved for X.25, V.10, V.11, V.21, V.23, V.24, V.26, V.26bis, V.27, V.27bis, V.27 ter, V.29, and V.54. The work of CCITT in data communications is rapidly advancing to keep in step with technology.

(Editor's note: Text in bold-face type represents provisional revisions approved September 1977.)

# CCITT CROSS REFERENCE

The majority of CCITT Recommendations are unique in themselves and are not duplicated elsewhere. The few that can be cross referenced are as follows:

| NUMBER | TITLE | RELATED CROSS REFERENCE STANDARDS |
|---|---|---|
| X.4 | General structure of signals of International Alphabet No 5 code for data transmission over public data networks | ISO 1155, 1177; ANSI X3.15, X3.16 |
| X.21 | General purpose interface between data terminal equipment (DTE) and data circuit-terminating equipment (DCE) for synchronous operation on public data networks | ANSI DRAFT BSR X3.69 |
| X.24 | List of definitions of interchange circuits between data terminal equipment (DTE) and data circuit-terminating equipment (DCE) on public data networks | ANSI DRAFT BSR X3.69 |
| X.25 | Interface between data terminal equipment (DTE) and data circuit-terminating equipment (DCE) for terminals operating in the packet mode on public data networks | ANSI DRAFT under development |
| X.26 | (V.10 electrical characteristics applied to public data network interfaces) | V.10 |
| X.27 | (V.11 electrical characteristics applied to public data network interfaces) | V.11 |
| V.3 | International Alphabet No. 5 | ISO 646; ANSI X3.4 |
| V.4 | General structure of signals of International Alphabet No. 5 for data transmission over public telephone networks | ISO 1155, 1177; ANSI X3.15, X3.16 FED STD 1010, 1011 |
| V.5 | Standardization of data-signalling rates for synchronous data transmission in the general switched telephone network | ANSI X3.1; EIA RS-269B FED STD 1013 |
| V.6 | Standardization of data-signalling rates for synchronous data transmission of leased telephone-type circuits | ANSI X3.1; EIA RS-269B FED STD 1013 |
| V.10 | Electrical characteristics for unbalanced double-current interchange circuits for general use with integrated circuit equipment in the field of data communications | CCITT X.26; EIA RS-423; FED-STD-1030 |
| V.11 | Electrical characteristics for balanced double-current interchange circuits for general use with integrated circuit equipment in the field of data communications | CCITT X.27; EIA RS-422; FED-STD-1020; |
| V.24 | List of definitions for interchange circuits between data terminal equipment and data circuit-terminating equipment | EIA RS-232C, RS-449 |
| V.26bis | 2400/1200 bits per second modem standardized for use in general switched telephone network | FED-STD-1005 |
| V.27bis | 4800 bits per second modem with automatic equalizer standardized for use on leased telephone-type circuits | FED-STD-1006 |
| V.27ter | 4800/2400 bits per second modem standardized for use in the general switched telephone network | FED-STD-1006 |
| V.28 | Electrical characteristics for unbalanced double-current interchange circuits | EIA RS-232C |
| V.54 | Loop test devices for modems | EIA RS-449 |

THE INTERNATIONAL TELEGRAPH AND TELEPHONE CONSULTATIVE COMMITTEE

# CCITT

# SIXTH PLENARY ASSEMBLY

GENEVA, 27 SEPTEMBER - 8 OCTOBER 1976

## ORANGE BOOK

## VOLUME VIII.1

3

# DATA TRANSMISSION
# OVER THE TELEPHONE NETWORK

Published by the
INTERNATIONAL TELECOMMUNICATION UNION
GENEVA, 1977

ISBN 92-61-00441-5

# PART I

## Series V Recommendations

## DATA TRANSMISSION OVER THE TELEPHONE NETWORK

### PRELIMINARY NOTES

1.    For principles governing the collaboration between the International Telecommunication Union and other international organizations in the study of data transmission, reference should be made to Recommendation A.20, Volume I.

2.    In this Volume, the expression "Administration" is used for shortness to indicate both a telecommunication Administration and a recognized private operating agency.

### PRELIMINARY NOTE

This document contains amendments, adopted under the accelerated procedure for the provisional approval of Recommendations, Resolution No. 2 of the Sixth Plenary Assembly, for the Recommendations V.10, V.11, V.21, V.23, V.24, V.26, V.26 *bis*, V.27, V.27 *bis*, V.27 *ter*, V.29 and V.54 published in the Orange Book, Volume VIII.1, Geneva, 1976.

# GENERAL

**Recommendation V.1**

### EQUIVALENCE BETWEEN BINARY NOTATION SYMBOLS
### AND THE SIGNIFICANT CONDITIONS OF A TWO-CONDITION CODE

*(New Delhi, 1960, amended at Geneva, 1964 and 1972)*

Binary numbering expresses numbers by means of two digits normally represented by the symbols 0 and 1. Transmission channels are especially well suited to the transmission of signals by a modulation having two significant conditions (two-condition modulation). These two significant conditions are sometimes called "space" and "mark" or "start" and "stop", or they may be called condition A or condition Z [1].

It is very useful to make the two conditions of a two-condition modulation correspond to the binary digits 0 and 1. Such equivalence will facilitate the transmission of numbers resulting from binary calculation, the conversion of codes for binary numbers and of codes for decimal numbers, maintenance operations and relations between transmission personnel and the personnel in charge of data-processing machines.

At first sight, it does not seem to matter whether the symbol 0 corresponds in transmission to condition A or condition Z, the symbol 1 then corresponding to condition Z or condition A or vice versa.

In telegraphy, however, when a telegraphic communication is set up and the sending of signals is stopped (called the idle condition of the line), the signal sent over the line consists of condition Z throughout the suspension of transmission.

It is logical (and for certain VF telegraph systems also essential) to use the same rule in data transmission. During the "idle periods" of transmission, condition Z should be applied to the circuit input.

Data transmission on a circuit is often controlled by perforated tape. On perforated tapes used for telegraphy, condition Z is represented by perforation. When binary numbers are represented by means of perforations, it is customary to represent the symbol 1 by a perforation. It is therefore logical to make this symbol 1 correspond to condition Z.

For these reasons, the CCITT *unanimously declares the following view:*

1.  In transmitting data by two-condition code, in which the digits are formed using binary notation, the symbol 1 of the binary notation will be equivalent to condition Z of the modulation, and the symbol 0 of the binary notation will be equivalent to condition A of the modulation.

2.  During periods when there is no signal sent to the input of the circuit, the circuit input condition is condition Z.

3.  If perforation is used, one perforation corresponds to one unit interval under condition Z.

---

[1] Definitions of condition A and condition Z: *List of definitions of essential telecommunication terms*, Part I, No. 31.38.

4. In accordance with Recommendation R.31, the sending of symbol 1 (condition Z) corresponds to the tone being sent on a channel using amplitude modulation.

5. In accordance with Recommendation R.35, when frequency modulation is used, the sending of symbol 0 corresponds to the higher frequency, while the sending of symbol 1 corresponds to the lower frequency.

6. a) for phase modulation with reference phase:

the symbol 1 corresponds to a phase equal to the reference phase;

the symbol 0 corresponds to a phase opposed to the reference phase.

b) for differential two-phase modulation where the alternative phase changes are 0 degree or 180 degrees:

the symbol 1 corresponds to a phase inversion from the previous element;

the symbol 0 corresponds to a no-phase inversion from the previous element.

7. A summary of equivalence is shown in Table 1/V.1.

TABLE 1/V.1 – **Summary of equivalence** (see Note 1)

| | Digit 0<br><br>"Start" signal in start-stop code<br>Line available condition<br>in telex switching<br>"Space" element of start-stop code<br>Condition A | Digit 1<br><br>"Stop" signal in start-stop code<br>Line idle condition<br>in telex switching (Note 2)<br>"Mark" element of start-stop code<br>Condition Z |
|---|---|---|
| Amplitude modulation | Tone-off | Tone-on |
| Frequency modulation | High frequency | Low frequency |
| Phase modulation with reference phase | Opposite phase to the reference phase | Reference phase |
| Differential two-phase modulation where the alternative phase changes are 0 degree or 180 degrees | No-phase inversion | Inversion of the phase |
| Perforations | No perforation | Perforation |

*Note 1.* – The standardization described in this Recommendation is general, whether over telegraph-type circuits or over circuits of the telephone type, making use of electromechanical or electronic devices.
*Note 2* – It primarily applies to anisochronous use.

**Recommendation V.2**

## POWER LEVELS FOR DATA TRANSMISSION OVER TELEPHONE LINES

*(New Delhi, 1960, amended at Geneva, 1964)*

The objectives in specifying data signal levels are as follows:

a) to ensure satisfactory transmission and to permit coordination with devices such as signalling receivers or echo suppressors, the data signal levels on international circuits should be controlled as closely as possible;

b) to ensure correct performance of multichannel carrier systems from the point of view of loading and noise, the mean power of data circuits should not differ much from the conventional value of channel loading (−15 dBm0 for each direction of transmission: see Note below). This conventional value makes allowance for a reasonable proportion (under 5%) of the channels in a multichannel system being used for non-speech applications at fixed power levels at about −10 dBm0 simultaneously in both directions of transmission.

If the proportion of non-speech applications (including data) does not exceed the above figure of 5% then the mean power of −10 dBm0 simultaneously in both directions of transmission would be allowable for data transmission also.

However, assuming an appreciably higher (e.g. 10 to 20%) proportion of non-speech circuits (due to the development of data transmission) on an international carrier system, a reduction of this power by 3 dB might be reasonable. In this way the sum of the mean powers in both directions of transmission in a duplex (i.e. transmitting tones in both directions simultaneously) system would be −10 dBm0 (i.e. −13 dBm0 for each direction). The power transmitted on the channel of a simplex (i.e. transmitting in one direction only) system or on either channel of a half duplex (i.e. transmitting in opposite directions consecutively) would be −10 dBm0 (assuming that there were no echoes);

*Note.* − The distribution of long-term mean power among the channels in a multichannel carrier telephone system (conventional mean value of −15 dBm0), probably has a standard deviation in the neighbourhood of 4 dB (*Green Book*, Volume III, Supplement No. 5).

c) it is probable that Administrations will wish to fix specific values for the signal power level of data modulators either at the subscriber's line terminals or at the local exchanges. The relation between these values and the power levels on international circuits depends on the particular national transmission plan; in any case, a wide range of losses among the possible connections between the subscriber and the input to international circuits must be expected;

d) considerations a) to c) suggest that specification of the maximum data signal level only is not the most useful form. One alternative proposal would be to specify the nominal power at the input to the international circuit. The nominal power would be the statistically estimated mean power obtained from measurement on many data transmission circuits.

For these reasons, the CCITT *unanimously declares the following view:*

## A. DATA TRANSMISSION OVER LEASED TELEPHONE CIRCUITS (PRIVATE WIRES) SET UP ON CARRIER SYSTEMS

1. The maximum power output of the subscriber's apparatus into the line shall not exceed 1 mW.

2. For systems transmitting tones continuously, for example frequency-modulation systems, the maximum power level at the zero relative level point shall be −10 dBm0. When transmission of data is discontinued for any appreciable time, the power level should preferably be reduced to −20 dBm0 or lower.

3. For systems not transmitting tones continuously, for example amplitude-modulation systems, a higher level up to −6 dBm0 at the zero relative level point may be used provided that the sum of the mean powers during the busy hour on both directions of transmission does not exceed 64 microwatts (corresponding to a mean level of −15 dBm0 on each direction of transmission simultaneously). Further, the level of any tones above 2400 Hz should not be so high as to cause interference on adjacent channels on carrier-telephone systems (see Recommendation G.224).

*Note1.* − In suggesting these limits, the CCITT has in mind that the recommended maximum level of −5 dBm0 for private circuits for alternate telephony and telegraphy may no longer be acceptable having regard to the recommendation that "to avoid overloading carrier systems, the mean power should be limited to 32 μW if such systems are subject to considerable extension".

*Note 2.* — The proposed limit of − 10 dBm0 for continuous tone systems is in line with the existing Recommendation H.41 (T.11) for frequency-modulation phototelegraph transmissions.

*Note 3.* — It is not possible to give any firm estimate of the proportion of international circuits which will at any time be carrying data transmissions. If the proportion should reach a high level, the provisional limits now proposed would need to be reconsidered.

## B. DATA TRANSMISSION OVER THE SWITCHED TELEPHONE SYSTEM

The maximum power output of the subscriber's equipment into the line shall not exceed 1 mW at any frequency.

In systems continuously transmitting tone, such as frequency or phase-modulation systems, the power level of the subscriber's equipment should be adjusted to make allowance for loss between his equipment and the point of entry to an international circuit; so that the corresponding nominal level of the signal at the international circuit input shall not exceed − 10 dBm0 (simplex systems) or − 13 dBm0 (duplex systems).

In systems not transmitting tones continuously, such as amplitude-modulation systems and multifrequency systems, higher levels may be used, provided always that the mean power of all the signals at the international circuit input during any one hour in both directions of transmission does not exceed 64 microwatts (representing a mean level of − 15 dBm0 in each direction of transmission simultaneously).

Furthermore, the frequency level in carrier telephone systems which are part of a circuit should not be so high that it might cause interference in adjacent channels. Recommendation G.224 could be referred to with a view to providing adequate levels.

*Note 1.* — In practice, it is no easy matter to assess the loss between a subscriber's equipment and the international circuit, so that this part of the present Recommendation should be taken as providing general planning guidance. As mean level at the international circuit input, the mean figure obtained from measurement or calculation (on numerous transmission data) may be adopted.

*Note 2.* — In switched connections the loss between subscribers' telephones may be high: 30 to 40 dB. The level of the signals received will then be very low, and these signals may suffer disturbance from the dialling pulses sent over other circuits. Hence the transmission level should be as high as possible.

If there is likely to be a heavy demand for data-transmission international connections over the switched network, some Administrations might want to provide special 4-wire subscribers' lines. If so, the levels to be used might be those proposed for leased circuits.

**Recommendation V.3**

### INTERNATIONAL ALPHABET No. 5

*(Mar del plata, 1968, amended at Geneva, 1972)*

*Introduction*

A seven-unit alphabet capable of meeting the requirements of private users on leased circuits and of users of data transmission by means of connections set up by switching on the general telephone network or on telegraph networks has been established jointly by the CCITT and the International Organization for Standardization (ISO).

This alphabet — Alphabet No. 5 — is not intended to replace Alphabet No. 2. It is a supplementary alphabet for the use of those who might not be satisfied with the more limited possibilities of Alphabet No. 2. In such cases it is considered as the alphabet to be used as common basic language for data transmission and for elaborated message systems.

Alphabet No. 5 does not exclude the use of any other alphabet that might be better adapted to special needs.

## 1. *Scope and field of application*

1.1      This Recommendation contains a set of 128 characters (control characters and graphic characters such as letters, digits and symbols) with their coded representation. Most of these characters are mandatory and unchangeable, but provision is made for some flexibility to accommodate special national and other requirements.

1.2      The need for graphics and controls in data processing and in data transmission has been taken into account in determining this character set.

1.3      This Recommendation consists of a general table with a number of options, notes, a legend and explanatory notes. It also contains a specific International Reference Version, guidance on the exercise of the options to define specific national versions and application oriented versions.

1.4      This character set is primarily intended for the interchange of information within message transmission systems and between data processing systems and associated equipment.

1.5      This character set is applicable to all Latin alphabets.

1.6      The character set includes facilities for extension where its 128 characters are insufficient for particular applications.

1.7      The definition of some control characters in this Recommendation assumes that data associated with them is to be processed serially in a forward direction. Their effect when included in strings of data which are processed other than serially in a forward direction or included in data formated for fixed record processing may have undesirable effects or may require additional special treatment to ensure that the control characters have their desired effect.

## 2. *Implementation*

2.1      This set should be regarded as a basic alphabet in an abstract sense. Its practical use requires definitions of its implementation in various media. For example, this could include punched tapes, punched cards, magnetic tapes and transmission channels, thus permitting interchange of data to take place either indirectly by means of an intermediate recording in a physical medium, or by local electrical connection of various units (such as input and output devices and computers) or by means of data transmission equipment.

2.2      The implementation of this coded character set in physical media and for transmission, taking also into account the need for error checking, is the subject of ISO publications.

3.     *Basic code table*

TABLE 1/V.3 – **Basic code**

| b7 | 0 | 0 | 0 | 0 | 1 | 1 | 1 | 1 |
|----|---|---|---|---|---|---|---|---|
| b6 | 0 | 0 | 1 | 1 | 0 | 0 | 1 | 1 |
| b5 | 0 | 1 | 0 | 1 | 0 | 1 | 0 | 1 |

| b4 | b3 | b2 | b1 | | 0 | 1 | 2 | 3 | 4 | 5 | 6 | 7 |
|----|----|----|----|----|---|---|---|---|---|---|---|---|
| 0 | 0 | 0 | 0 | 0 | NUL | TC₇ (DLE) | SP | 0 | @ | P | ` | p |
| 0 | 0 | 0 | 1 | 1 | TC₁ (SOH) | DC₁ | ! | 1 | A | Q | a | q |
| 0 | 0 | 1 | 0 | 2 | TC₂ (STX) | DC₂ | " | 2 | B | R | b | r |
| 0 | 0 | 1 | 1 | 3 | TC₃ (ETX) | DC₃ | £(#) | 3 | C | S | c | s |
| 0 | 1 | 0 | 0 | 4 | TC₄ (EOT) | DC₄ | $(¤) | 4 | D | T | d | t |
| 0 | 1 | 0 | 1 | 5 | TC₅ (ENQ) | TC₈ (NAK) | % | 5 | E | U | e | u |
| 0 | 1 | 1 | 0 | 6 | TC₆ (ACK) | TC₉ (SYN) | & | 6 | F | V | f | v |
| 0 | 1 | 1 | 1 | 7 | BEL | TC₁₀ (ETB) | ' | 7 | G | W | g | w |
| 1 | 0 | 0 | 0 | 8 | FE₀ (BS) | CAN | ( | 8 | H | X | h | x |
| 1 | 0 | 0 | 1 | 9 | FE₁ (HT) | EM | ) | 9 | I | Y | i | y |
| 1 | 0 | 1 | 0 | 10 | FE₂ (LF) | SUB | * | : | J | Z | j | z |
| 1 | 0 | 1 | 1 | 11 | FE₃ (VT) | ESC | + | ; | K | | k | |
| 1 | 1 | 0 | 0 | 12 | FE₄ (FF) | IS₄ (FS) | , | < | L | | l | |
| 1 | 1 | 0 | 1 | 13 | FE₅ (CR) | IS₃ (GS) | – | = | M | | m | |
| 1 | 1 | 1 | 0 | 14 | SO | IS₂ (RS) | . | > | N | ^ | n | ‾ |
| 1 | 1 | 1 | 1 | 15 | SI | IS₁ (US) | / | ? | O | _ | o | DEL |

CCITT - 5610

*Notes about Table 1/V.3*

1    The Format Effectors are intended for equipment in which horizontal and vertical movements are effected separately. If equipment requires the action of CARRIAGE RETURN to be combined with a vertical movement, the Format Effector for that vertical movement may be used to effect the combined movement. For example, if NEW LINE (symbol NL, equivalent to CR + LF) is required, $FE_2$ shall be used to represent it. This substitution requires agreement between the sender and the recipient of the data.

The use of these combined functions may be restricted for international transmission on general switched telecommunication networks (telegraph and telephone networks).

2    The symbol £ is assigned to position 2/3 and the symbol $ is assigned to position 2/4. In a situation where there is no requirement for the symbol £ the symbol # (number sign) may be used in position 2/3. Where there is no requirement for the symbol $ the symbol ¤ (currency sign) may be used in position 2/4. The chosen allocations for symbols to these positions for international information interchange should be agreed between the interested parties. It should be noted that, unless otherwise agreed between sender and recipient, the symbols £, $ or ¤ do not designate the currency of a specific country.

3    National use positions. The allocations of characters to these positions lies within the responsibility of national standardization bodies. These positions are primarily intended for alphabet extensions. If they are not required for that purpose, they may be used for symbols.

4    Positions 5/14, 6/0 and 7/14 are provided for the symbols UPWARD ARROW HEAD, GRAVE ACCENT and OVERLINE. However, these positions may be used for other graphical characters when it is necessary to have 8, 9 or 10 positions for national use.

5    Position 7/14 is used for the graphic character ‾ (OVERLINE), the graphical representation of which may vary according to national use to represent (TILDE) or another diacritical sign provided that there is no risk of confusion with another graphic character included in the table.

6    The graphic characters in positions 2/2, 2/7, 2/12 and 5/14 have respectively the significance of QUOTATION MARK, APOSTROPHE, COMMA and UPWARD ARROW HEAD ; however, these characters take on the significance of the diacritical signs DIAERESIS, ACUTE ACCENT, CEDILLA and CIRCUMFLEX ACCENT when they are preceded or followed by the BACKSPACE character (0/8).

## 4. *Legend*

### 4.1 *Control characters*

| Abbreviation | Note of Table 1/V.3 | Meaning | Position in the code table |
|---|---|---|---|
| ACK | | Acknowledge | 0/6 |
| BEL | | Bell | 0/7 |
| BS | | Backspace | 0/8 |
| CAN | | Cancel | 1/8 |
| CR | 1 | Carriage return | 0/13 |
| DC | | Device control | – |
| DEL | | Delete | 7/15 |
| DLE | | Data link escape | 1/0 |
| EM | | End of medium | 1/9 |
| ENQ | | Enquiry | 0/5 |
| EOT | | End of transmission | 0/4 |
| ESC | | Escape | 1/11 |
| ETB | | End of transmission block | 1/7 |
| ETX | | End of text | 0/3 |
| FE | | Format effector | – |
| FF | 1 | Form feed | 0/12 |
| FS | | File separator | 1/12 |
| GS | | Group separator | 1/13 |
| HT | | Horizontal tabulation | 0/9 |
| IS | | Information separator | – |
| LF | 1 | Line feed | 0/10 |
| NAK | | Negative acknowledge | 1/5 |
| NUL | | Null | 0/0 |
| RS | | Record separator | 1/14 |
| SI | | Shift-in | 0/15 |
| SO | | Shift-out | 0/14 |
| SOH | | Start of heading | 0/1 |
| SP | | Space (see 7.2) | 2/0 |
| STX | | Start of text | 0/2 |
| SUB | | Substitute character | 1/10 |
| SYN | | Synchronous idle | 1/6 |
| TC | | Transmission control | – |
| US | | Unit separator | 1/15 |
| VT | 1 | Vertical tabulation | 0/11 |

## 4.2  *Graphic characters*

| Graphic | Notes of Table 1/V.3 | Name | Position in the code table |
|---|---|---|---|
| (space) | | Space (see 7.2) | 2/0 |
| ! | | Exclamation mark | 2/1 |
| " | 6 | Quotation mark, Diaeresis | 2/2 |
| £ | 2 | Pound sign | 2/3 |
| # | 2 | Number sign | 2/3 |
| $ | 2 | Dollar sign | 2/4 |
| ¤ | 2 | Currency sign | 2/4 |
| % | | Percent sign | 2/5 |
| & | | Ampersand | 2/6 |
| ' | 6 | Apostrophe, Acute accent | 2/7 |
| ( | | Left parenthesis | 2/8 |
| ) | | Right parenthesis | 2/9 |
| * | | Asterisk | 2/10 |
| + | | Plus sign | 2/11 |
| , | 6 | Comma, Cedilla | 2/12 |
| — | | Hyphen, Minus sign | 2/13 |
| . | | Full stop (period) | 2/14 |
| / | | Solidus | 2/15 |
| : | | Colon | 3/10 |
| ; | | Semi-colon | 3/11 |
| < | | Less-than sign | 3/12 |
| = | | Equal sign | 3/13 |
| > | | Greater-than sign | 3/14 |
| ? | | Question mark | 3/15 |
| ^ | 4,6 | Upward arrow head, Circumflex accent | 5/14 |
| _ | | Underline | 5/15 |
| ` | 4 | Grave accent | 6/0 |
| — | 4,5 | Overline, Tilde | 7/14 |

## 5.  *Explanatory notes*

### 5.1  *Numbering of the positions in Table 1/V.3*

Within any one character the bits are identified by $b_7$, $b_6$ ... $b_1$, where $b_7$ is the highest order, or most significant bit, and $b_1$ is the lowest order, or least significant bit. If desired these may be given a numerical significance in the binary system, thus:

Bit identification:  $b_7$  $b_6$  $b_5$  $b_4$  $b_3$  $b_2$  $b_1$

Significance:        64  32  16   8   4   2   1

In the table the columns and rows are identified by numbers written in binary and decimal notations.

Any one position in the table may be identified either by its bit pattern, or by its column and row numbers. For instance, the position containing the digit 1 may be identified:

—  by its bit pattern in order of decreasing significance, e.g. 011 0001 [2];

—  by its column and row numbers, e.g. 3/1.

The column number is derived from bits $b_7$, $b_6$ and $b_5$ giving them weights of 4, 2 and 1 respectively. The row number is derived from bits $b_4$, $b_3$, $b_2$ and $b_1$ giving them weights of 8, 4, 2 and 1 respectively.

_____

[2] Order of transmitting bits is not necessarily the same as shown here. For the order of the transmission of bits, see I in Recommendation V.4 or X.4.

### 5.2    *Diacritical signs*

In the character set, some printing symbols may be designed to permit their use for the composition of accented letters when necessary for general interchange of information. A sequence of three characters, comprising a letter, BACKSPACE and one of these symbols, is needed for this composition, and the symbol is then regarded as a diacritical sign. It should be noted that these symbols take on their diacritical significance when they are preceded or followed by one BACKSPACE character; for example, the symbol corresponding to the code combination 2/7 normally has the significance of APOSTROPHE, but becomes the diacritical sign ACUTE ACCENT when it precedes or follows a BACKSPACE character.

In order to increase efficiency, it is possible to introduce accented letters (as single characters) in the positions marked by Note 3 in Table 1/V.3. According to national requirements, these positions may contain special diacritical signs.

### 5.3    *Names, meanings and fonts of graphic characters*

This Recommendation assigns at least one name to denote each of the graphic characters displayed in Tables 1/V.3 and 2/V.3. The names chosen to denote graphic characters are intended to reflect their customary meanings. However, this Recommendation does not define and does not restrict the meanings of graphic characters. Nor does it specify a particular style or font design for the graphic characters.

Under the provision of Note 3 of Table 1/V.3 graphic characters which are different from the characters of the international reference version may be assigned to the national use positions. When such assignments are made, the graphic characters should have distinct forms and be given distinctive names which are not in conflict with any of the forms or the names of any of the graphic characters in the international reference version.

### 5.4    *Uniqueness of character allocation*

A character allocated to a position in Table 1/V.3 may not be placed elsewhere in the table. For example, in the case of position 2/3 the character not used cannot be placed elsewhere. In particular the POUND sign (£) can never be represented by the bit combination of position 2/4.

### 6.    *Versions of Table 1/V.3*

### 6.1    *General*

6.1.1    In order to use Table 1/V.3 for information interchange, it is necessary to exercise the options left open, i.e. those affected by Notes 2 to 5. A single character must be allocated to each of the positions for which this freedom exists or it must be declared to be unused. A code table completed in this way is called a *version*.

6.1.2    The Notes to Table 1/V.3, the Explanatory Notes and the Legend apply in full to any version.

### 6.2    *International reference version*

This version is available for use when there is no requirement to use a national or an application-oriented version. In international information processing interchange the international reference version (Table 2/V.3) is assumed unless a particular agreement exists between sender and recipient of the data.

TABLE 2/V.3 – **International reference version**

| | | | | | b7 | 0 | 0 | 0 | 0 | 1 | 1 | 1 | 1 |
|---|---|---|---|---|---|---|---|---|---|---|---|---|---|
| | | | | | b6 | 0 | 0 | 1 | 1 | 0 | 0 | 1 | 1 |
| | | | | | b5 | 0 | 1 | 0 | 1 | 0 | 1 | 0 | 1 |
| b4 | b3 | b2 | b1 | | | 0 | 1 | 2 | 3 | 4 | 5 | 6 | 7 |
| 0 | 0 | 0 | 0 | 0 | | NUL | TC7 (DLE) | SP | 0 | @ | P | ` | p |
| 0 | 0 | 0 | 1 | 1 | | TC1 (SOH) | DC1 | ! | 1 | A | Q | a | q |
| 0 | 0 | 1 | 0 | 2 | | TC2 (STX) | DC2 | " | 2 | B | R | b | r |
| 0 | 0 | 1 | 1 | 3 | | TC3 (ETX) | DC3 | # | 3 | C | S | c | s |
| 0 | 1 | 0 | 0 | 4 | | TC4 (EOT) | DC4 | ¤ | 4 | D | T | d | t |
| 0 | 1 | 0 | 1 | 5 | | TC5 (ENQ) | TC8 (NAK) | % | 5 | E | U | e | u |
| 0 | 1 | 1 | 0 | 6 | | TC6 (ACK) | TC9 (SYN) | & | 6 | F | V | f | v |
| 0 | 1 | 1 | 1 | 7 | | BEL | TC10 (ETB) | ' | 7 | G | W | g | w |
| 1 | 0 | 0 | 0 | 8 | | FE0 (BS) | CAN | ( | 8 | H | X | h | x |
| 1 | 0 | 0 | 1 | 9 | | FE1 (HT) | EM | ) | 9 | I | Y | i | y |
| 1 | 0 | 1 | 0 | 10 | | FE2 (LF) | SUB | * | : | J | Z | j | z |
| 1 | 0 | 1 | 1 | 11 | | FE3 (VT) | ESC | + | ; | K | [ | k | { |
| 1 | 1 | 0 | 0 | 12 | | FE4 (FF) | IS4 (FS) | , | < | L | \ | l | \| |
| 1 | 1 | 0 | 1 | 13 | | FE5 (CR) | IS3 (GS) | – | = | M | ] | m | } |
| 1 | 1 | 1 | 0 | 14 | | SO | IS2 (RS) | . | > | N | ^ | n | ‾ |
| 1 | 1 | 1 | 1 | 15 | | SI | IS1 (US) | / | ? | O | _ | o | DEL |

CCITT-11540

The following characters are allocated to the optional positions of Table 1/V.3:

| | | |
|---|---|---|
| # | Number sign | 2/3 |
| ¤ | Currency sign | 2/4 |
| @ | Commercial at | 4/0 |
| [ | Left square bracket | 5/11 |
| \ | Reverse solidus | 5/12 |
| ] | Right square bracket | 5/13 |
| { | Left curly bracket | 7/11 |
| \| | Vertical line | 7/12 |
| } | Right curly bracket | 7/13 |

CCITT-4929

It should be noted that no substitution is allowed when using the international reference version.

### 6.3   *National versions*

6.3.1   The responsibility for defining national versions lies with the national standardization bodies. These bodies shall exercise the options available and make the required selection.

6.3.2   If so required, more than one national version can be defined within a country. The different versions shall be separately identified. In particular when for a given national use position, e.g. 5/12 or 6/0, alternative characters are required, two different versions shall be identified, even if they differ only by this single character.

6.3.3   If there is in a country no special demand for specific characters, it is strongly recommended that the characters of the international reference version be allocated to the same national use positions.

### 6.4   *Application-oriented versions*

Within national or international industries, organizations or professional groups, application-oriented versions can be used. They require precise agreement among the parties concerned, who will have to exercise the options available and to make the required selection.

### 7.   *Functional characteristics related to control characters*

Some definitions given below are stated in general terms and more explicit definitions of use may be needed for specific implementation of the code table on recording media or on transmission channels. These more explicit definitions and the use of these characters are the subject of ISO publications.

### 7.1   *General designations and control characters*

The general designation of control characters involves a specific class name followed by a subscript number.

They are defined as follows:

TC   — *Transmission control characters*

Control characters intended to control or facilitate transmission of information over telecommunication networks.

The use of the TC characters on the general telecommunication networks is the subject of ISO publications.

The transmission control characters are:

ACK, DLE, ENQ, EOT, ETB, ETX, NAK, SOH, STX and SYN.

**FE**   —   *Format effectors*

Control characters mainly intended for the control of the layout and positioning of information on printing and/or display devices. In the definitions of specific format effectors, any reference to printing devices should be interpreted as including display devices. The definitions of format effectors use the following concept:

*a)*   a page is composed of a number of lines of characters;

*b)*   the characters forming a line occupy a number of positions called character positions;

*c)*   the active position is that character position in which the character about to be processed would appear if it were to be printed. The active position normally advances one character position at a time.

The format effector characters are:

BS, CR, FF, HT, LF and VT (see also Note 1 to to Table 1/V.3).

**DC**   —   *Device control characters*

Control characters for the control of a local or remote ancillary device (or devices) connected to a data processing and/or telecommunication system. These control characters are not intended to control telecommunication systems; this should be achieved by the use of TCs.

Certain preferred uses of the individual DCs are given in 7.2 below.

**IS**   —   *Information separators*

Control characters that are used to separate and qualify data logically. There are four such characters. They may be used either in hierarchical order or non-hierarchically; in the latter case their specific meanings depend on their applications.

When they are used hierarchically, the ascending order is:

US, RS, GS, FS.

In this case data normally delimited by a particular separator cannot be split by a higher order separator but will be considered as delimited by any higher order separator.

7.2   *Specific control characters*

Individual members of the classes of controls are sometimes referred to by their abbreviated class name and a subscript number (e.g. $TC_5$) and sometimes by a specific name indicative of their use (e.g. ENQ).

Different but related meanings may be associated with some of the control characters but in an interchange of data this normally requires agreement between the sender and the recipient.

**ACK**   —   *Acknowledge*

A transmission control character transmitted by a receiver as an affirmative response to the sender.

**BEL**   —   *Bell*

A control character that is used when there is a need to call for attention; it may control alarm or attention devices.

**BS**   —   *Backspace*

A format effector which moves the active position one character position backwards on the same line.

CAN  –  *Cancel*

A character, or the first character of a sequence, indicating that the data preceding it is in error. As a result this data is to be ignored. The specific meaning of this character must be defined for each application and/or between sender and recipient.

CR   –  *Carriage return*

A format effector which moves the active position to the first character position on the same line.

*Device controls*

DC1  –  A device control character which is primarily intended for turning on or starting an ancillary device. If it is not required for this purpose, it may be used to restore a device to the basic mode of operation (see also DC2 and DC3), or for any other device control function not provided by other DCs.

DC2  –  A device control character which is primarily intended for turning on or starting an ancillary device. If it is not required for this purpose, it may be used to set a device to a special mode of operation (in which case DC1 is used to restore the device to the basic mode), or for any other device control function not provided by other DCs.

DC3  –  A device control character which is primarily intended for turning off or stopping an ancillary device. This function may be a secondary level stop, e.g. wait, pause, stand-by or halt (in which case DC1 is used to restore normal operation). If it is not required for this purpose, it may be used for any other device control function not provided by other DCs.

DC4  –  A device control character which is primarily intended for turning off, stopping or interrupting an ancillary device. If it is not required for this purpose, it may be used for any other device control function not provided by other DCs.

*Examples of use of the device controls:*

1)  One switching

    on  –  DC2        off  –  DC4

2)  Two independent switchings

    First one    on  –  DC2        off  –  DC4
    Second one   on  –  DC1        off  –  DC3

3)  Two dependent switchings

    General      on  –  DC2        off  –  DC4
    Particular   on  –  DC1        off  –  DC3

4)  Input and output switching

    Output       on  –  DC2        off  –  DC4
    Input        on  –  DC1        off  –  DC3

DEL  –  *Delete*

A character used primarily to erase or obliterate an erroneous or unwanted character in punched tape. DEL characters may also serve to accomplish media-fill or time-fill. They may be inserted into or removed from a stream of data without affecting the information content of that stream, but then the addition or removal of these characters may affect the information layout and/or the control of equipment.

DLE   —  *Data link escape*

A transmission control character which will change the meaning of a limited number of contiguously following characters. It is used exclusively to provide supplementary data transmission control functions. Only graphic characters and transmission control characters can be used in DLE sequences.

EM   —  *End of medium*

A control character thay may be used to identify the physical end of a medium, or the end of the used portion of a medium, or the end of the wanted portion of data recorded on a medium. The position of this character does not necessarily correspond to the physical end of the medium.

ENQ   —  *Enquiry*

A transmission control character used as a request for a response from a remote station — the response may include station identification and/or station status. When a "Who are you?" function is required on the general switched transmission network, the firse use of ENQ after the connection is established shall have the meaning "Who are you?" (station identification). Subsequent use of ENQ may, or may not, include the function "Who are you?", as determined by agreement.

EOT   —  *End of transmission*

A transmission control character used to indicate the conclusion of the transmission of one or more texts.

ESC   —  *Escape*

A control character which is used to provide an additional control function. It alters the meaning of a limited number of contiguously following bit combinations which constitute the escape sequence. Escape sequences are used to obtain additional control functions which may provide among other things graphic sets outside the standard set. Such control functions must not be used as additional transmission controls.
The use of the character ESC and of the escape sequences in conjunction with code extension techniques is the subject of an ISO Standard.

ETB   —  *End of transmission block*

A transmission control character used to indicate the end of a transmission block of data where data is divided into such blocks for transmission purposes.

ETX   —  *End of text*

A transmission control character which terminates a text.

FF   —  *Form feed*

A format effector which advances the active position to the same character position on a pre-determined line of the next form or page.

HT   —  *Horizontal tabulation*

A format effector which advances the active position to the next pre-determined character position on the same line.

*Information separators*

IS$_1$ (US) — A control character used to separate and qualify data logically; its specific meaning has to be defined for each application. If this character is used in hierarchical order as specified in the general definition of IS, it delimits a data item called a UNIT.

IS$_2$ (RS) — A control character used to separate and qualify data logically; its specific meaning has to be defined for each application. If this character is used in hierarchical order as specified in the general definition of IS, it delimits a data item called a RECORD.

IS$_3$ (GS) — A control character used to separate and qualify data logically; its specific meaning has to be defined for each application. If this character is used in hierarchical order as specified in the general definition of IS, it delimits a data item called a GROUP.

IS$_4$ (FS) — A control character used to separate and qualify data logically; its specific meaning has to be defined for each application. If this character is used in hierarchical order as specified in the general definition of IS, it delimits a data item called a FILE.

LF       — *Line feed*

A format effector which advances the active position to the same character position of the next line.

NAK    — *Negative acknowledge*

A transmission control character transmitted by a receiver as a negative response to the sender.

NUL    — *Null*

A control character used to accomplish media-fill or time-fill. NUL characters may be inserted into or removed from a stream of data without affecting the information content of that stream, but then the addition or removal of these characters may affect the information layout and/or the control of equipment.

SI       — *Shift-in*

A control character which is used in conjunction with SHIFT-OUT and ESCAPE to extend the graphic character set of the code. It may reinstate the standard meanings of the bit combinations which follow it. The effect of this character when using code extension techniques is described in an ISO Standard.

SO       — *Shift-out*

A control character which is used in conjunction with SHIFT-IN and ESCAPE to extend the graphic character set of the code. It may alter the meaning of the bit combinations of columns 2 to 7 which follow it until a SHIFT-IN character is reached. However, the characters SPACE (2/0) and DELETE (7/15) are unaffected by SHIFT-OUT. The effect of this character when using code extension techniques is described in an ISO Standard.

SOH    — *Start of heading*

A transmission control character used as the first character of a heading of an information message.

SP — *Space*

 A character which advances the active position one character position on the same line. This character is also regarded as a non-printing graphic.

STX — *Start of text*

 A transmission control character which precedes a text and which is used to terminate a heading.

SUB — *Substitute character*

 A control character used in the place of a character that has been found to be invalid or in error. SUB is intended to be introduced by automatic means.

SYN — *Synchronous idle*

 A transmission control character used by a synchronous transmission system in the absence of any other character (idle condition) to provide a signal from which synchronism may be achieved or retained between data-terminal equipment.

VT — *Vertical tabulation*

 A format effector which advances the active position to the same character position on the next pre-determined line.

**Recommendation V.4**

## GENERAL STRUCTURE OF SIGNALS OF INTERNATIONAL ALPHABET No. 5 CODE FOR DATA TRANSMISSION OVER PUBLIC TELEPHONE NETWORKS [3]

*(Mar del Plata, 1968, amended at Geneva, 1976)*

The CCITT,

 I. *considering, firstly,*

 the agreement between the International Organization for Standardization (ISO) and the CCITT on the main characteristics of a seven-unit alphabet (International Alphabet No. 5) to be used for data transmission and for telecommunications requirements that cannot be met by the existing five-unit International Alphabet No. 2;

 the interest, both to the users and to the telecommunication services, of an agreement concerning the chronological order of transmission of bits in serial working.

 *declares the view*

 that the agreed rank number of the unit in the alphabetical table of combinations should correspond to the chronological order of transmission in serial working on telecommunication circuits;

 that, when this rank in the combination represents the order of the bit in binary numbering, the bits should be transmitted in serial working with the low order bit first;

---

[3] See Recommendation X.4 for data transmission over public data networks.

that the numerical meaning corresponding to each information unit considered in isolation is that of the digit:

0 for a unit corresponding to condition A (travail = space), and
1 for a unit corresponding to condition Z (repos = mark),

in accordance with the definitions of these conditions for a two-condition transmission system;

II. *considering, moreover,*

that it is often desirable, in data transmission, to add an extra "parity" unit to allow for the detection of errors in received signals;

the possibility offered by this addition for the detection of faults in data terminal equipment;

the need to reserve the possibility of making this addition during the transmission itself, after the seven information units proper have been sent;

*declares the view*

that signals of the International Alphabet No. 5 code for data transmission should, in general, include an additional "parity" unit;

that the rank of this unit and, hence, the chronological order of the transmission in serial working should be the eighth of the combination thus completed;

III. *considering*

that, in start-stop systems working with electromechanical equipment, the margin of such equipment and the reliability of the connection are considerably increased by the use of a stop element corresponding to the duration of two unit intervals of the modulation;

that for transmissions over telephone circuits via modems installed on the user's premises, the latter must be able to use the connections at the highest possible practical rate in characters per second, and that in such a case a single stop unit leads to a gain of about 10% as regards this practical rate;

that, however, it does not appear that the production of electronic devices capable of working at will with start-stop signals having a stop element equal to one- or two-unit intervals should lead to costly complications and that such an arrangement can have the advantage of appreciably limiting the error rate without greatly reducing the practical efficiency of the connection,

*declares the view*

that in start-stop systems using combinations of the seven-unit alphabet normally followed by a parity unit, the first information unit of the transmitted combination should be preceded by a start element corresponding to condition A (space);

that the duration of this start element should be one-unit interval for the modulation rate under consideration, at transmitter output;

that the combination of seven information units, normally completed by its parity unit, should be followed by a stop element corresponding to condition Z (mark);

that for start-stop systems using the seven-unit code on switched telephone networks, a two-unit stop element should be used with electromechanical data terminal equipments operating at modulation rates up to and including 200 bauds. In other cases, the use of a one-unit stop element is preferable. However, this is subject to a mutual agreement between Administrations concerned;

that similar situations when a one-unit stop element can be used may apply to leased circuits;

that the start-stop receivers should be capable of correctly receiving start-stop signals comprising a single-unit stop element, whose duration will be reduced by a time interval equal to the deviation corresponding to the degree of gross start-stop distortion permitted at receiver input. However, for electromechanical equipment which must use a two-unit stop element (eleven-unit code signal) with a modulation rate of 200 bauds or less, receivers should be capable of correctly receiving signals with a stop element reduced to one unit;

IV.  *considering, finally,*

that the direction of the parity unit can only be that of the even parity on the perforated tapes, particularly owing to the possibility of deletion (combination 7/15 of the alphabet) which causes a hole to appear in all tracks;

that, on the other hand, the odd parity is considered essential in the equipment which depends on transitions in the signals to maintain Ysynchronism (in cases where combination 1/6 (SYNC) of the alphabet does not permit of an economical solution),

*declares the view*

that the parity unit of the signal should correspond to the even parity in links or connections operated on the principle of the start-stop system,

that this parity should be odd on links or connections using end-to-end synchronous operation,

that arrangements should be made when necessary to reverse the direction of the parity unit at the input and output of the synchronous equipment connected either to apparatus working on the start-stop principle or receiving characters on perforated tape.

**Recommendation V.5**

## STANDARDIZATION OF DATA-SIGNALLING RATES FOR
## SYNCHRONOUS DATA TRANSMISSION IN THE GENERAL SWITCHED TELEPHONE NETWORK

*(former Recommendation V.22, Geneva, 1964; amended at Mar del Plata, 1968, at Geneva, 1972 and 1976)*

1.      Data transmission by international communications carried on the general switched telephone network using a synchronous transmission procedure will be done with a specific mode of modulation, two- or multi-condition, and serial transmission (see Note 1). For synchronous data transmission on leased telephone-type circuits see Recommendation V.6.

2.      The data signalling rates for synchronous transmission in the general switched telephone network will be:

600, 1200, 2400 or 4800 bits/s (see Note 2)

The users will choose among these rates, in accordance with their needs and the facilities afforded by the connection.

3.      Data signalling rates should in no case deviate from the nominal value by more than $\pm$ 0.01%.

*Note 1.* — The application of parallel data transmission is a subject of other Recommendations.

*Note 2.* — Modems for use in the general switched telephone network at these data signalling rates; see Recommendations V.23, V.26 *bis* and V.27 *ter*, respectively.

*Note 3.* — For data transmission at 200 (300) bit/s, see Recommendation V.21.

**Recommendation V.6**

## STANDARDIZATION OF DATA SIGNALLING RATES FOR
## SYNCHRONOUS DATA TRANSMISSION ON LEASED TELEPHONE-TYPE CIRCUITS

*(former Recommendation V.22 bis, Geneva, 1972; amended at Geneva, 1976)*

1.    Data transmission by international communications carried on leased telephone-type circuits (either normal quality or special quality circuits) using a synchronous transmission procedure will be done with a specific mode of modulation, two- or multi-condition, and serial transmission (see Note 1). For synchronous data transmission in the general switched telephone network see Recommendation V.5.

2.    It is recommended that for synchronous transmission the data signalling rates should be divided into two distinct classes to be known as "preferred" and "supplementary". The union of these classes is defined to be the "permitted" data signalling rates.

   a)    *Preferred range of data signalling rates (bits per second)*

   |  |  |
   |---|---|
   | 600 (see Note 2) | 4800 (see Note 2) |
   | 1200 (see Note 2) | 7200 |
   | 2400 (see Note 2) | 9600 (see Note 2) |
   | 3600 | |

   b)    *Supplementary range of data signalling rates (bits per second)*

   |  |  |
   |---|---|
   | 1800 | 6600 |
   | 3000 | 7800 |
   | 4200 | 8400 |
   | 5400 | 9000 |
   | 6000 | 10 200 |
   | | 10 800 |

   c)    *Permitted range of data signalling rates (bits per second)*

   The range is defined as 600 times "$N$" bits per second where $1 \leqslant N \leqslant 18$    $N$: a positive integer.

This algorithm, with the addition of 2000 bits per second (see Note 3), yields the total range of data signalling rates which is the union of the preferred and supplementary data signalling rates.

In determining the permitted range, the CCITT has in mind the need to restrict the number of data signalling rates (and hence modem design required), yet at the same time to allow the best use to be made of technical progress in both modem development and improvement in the telephone plant. It is considered that a geometric progression in standard rates provides the most satisfactory basis of development.

3.    Data signalling rates should in no case deviate from the nominal value by more than $\pm$ 0.01%.

*Note 1.* —  The application of parallel data transmission is a subject of other Recommendations.

*Note 2.* —  Modems for use on leased telephone-type circuits at these data signalling rates; see Recommendations V.23, V.26, V.27, V.27 *bis* and V.29, respectively.

*Note 3.* —  It is recognized that there is substantial usage of a 2000 bit/s data signalling rate in some countries and that this usage will continue.

## INTERFACES AND VOICE-BAND MODEMS [1]

**Recommendation V.10**

## ELECTRICAL CHARACTERISTICS FOR UNBALANCED DOUBLE-CURRENT INTERCHANGE CIRCUITS FOR GENERAL USE WITH INTEGRATED CIRCUIT EQUIPMENT IN THE FIELD OF DATA COMMUNICATIONS [2]

*(Geneva, 1976)*

1. *Introduction*

This Recommendation deals with the electrical characteristics of the generator, receiver and interconnecting leads of an unbalanced interchange circuit employing a differential receiver.

In the context of this Recommendation an unbalanced interchange circuit is defined as consisting of an unbalanced generator connected to a receiver by an interconnecting lead and a common return lead.

Annexes are provided to give guidance on a number of application aspects as follows:

*Annex 1*   Waveshaping examples

*Annex 2*   Compatibility with other interfaces

*Annex 3*   Operational constraints

*Annex 4*   Multipoint operation

*Annex 5*   Special considerations for coaxial cable applications

*Note.* — Generator and load devices meeting the electrical characteristics of this Recommendation need not operate over the entire signalling rate range specified. They may be designed to operate over narrower ranges to satisfy specific requirements more economically, particularly at lower signalling rates.

The interconnecting cable is normally not terminated, but the matter of terminating coaxial interconnecting cable is dealt with in Annex 5 of this Recommendation. Where the interchange circuit incorporates the special provisions for coaxial applications with cable termination this shall be referred to as "complying with Recommendation V.10 (SPECIAL)".

Reference measurements are described which may be used to verify certain of the recommended parameters but it is a matter for individual manufacturers to decide what tests are necessary to ensure compliance with the Recommendation.

---

[1] Former Recommendations V.10, V.11 and V.13 for data transmission over the telex network have been renumbered as S.15, S.16 and S.17 respectively and published in Volume VII of the *Orange Book*.

[2] This Recommendation is also designated as X.26 in the Series X Recommendations.

2.    *Field of application*

The electrical characteristics specified in this Recommendation apply to interchange circuits operating with data signalling rates up to 100 kbit/s [3], and are intended to be used primarily in Data Terminal Equipment (DTE) and Data Circuit-terminating Equipment (DCE) implemented in integrated circuit technology.

This Recommendation applies to new work and is not intended to apply to DCE implemented in discrete component technology, for which the electrical characteristics covered by Recommendation V.28 are more appropriate.

Typical points of application are illustrated in Figure 1/V.10.

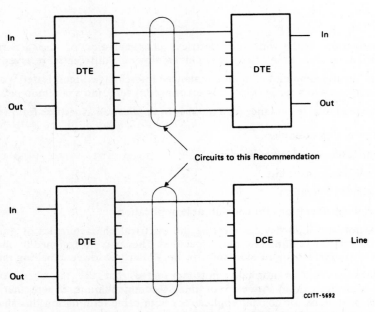

FIGURE 1/V.10 — Typical applications of unbalanced interchange circuits

Whilst the unbalanced interchange circuit is primarily intended for use at the lower signalling rates, its use should be avoided in the following cases:

1)    where the interconnecting cable is too long for proper unbalanced circuit operation;

2)    where extraneous noise sources make unbalanced circuit operation impossible;

3)    where it is necessary to minimize interference with other signals.

Whilst a restriction on maximum cable length is not specified, guidelines are given with respect to conservative operating distance as a function of signalling rates (see Annex 3).

_____

[3] Signalling rates above the suggested 100 kbit/s may also be employed, but the maximum suggested operating distances should be shortened accordingly (see Figure 10/V.10).

3.    *Symbolic representation of an interchange circuit* (Figure 2/V.10)

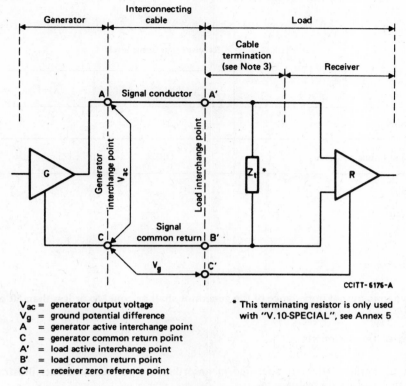

$V_{ac}$ = generator output voltage
$V_g$ = ground potential difference
A  = generator active interchange point
C  = generator common return point
A' = load active interchange point
B' = load common return point
C' = receiver zero reference point

\* This terminating resistor is only used with "V.10-SPECIAL", see Annex 5

FIGURE 2/V.10 – Symbolic representation of an unbalanced interchange circuit

*Note 1.* – Two interchange points are shown above. The output characteristics of the generator, excluding any interconnecting cable, are defined at the "generator interchange point". The electrical characteristics to which the receiver must respond are defined at the "load interchange point".

*Note 2.* – The connection of the signal common return is dealt with in 9. below. Points C and C' may be connected to protective ground if required by national regulations.

*Note 3.* – The interconnecting cable is normally not terminated. The termination of coaxial interconnecting cable is dealt with in Annex 5.

4.    *Generator polarities and receiver significant levels*

4.1    *Generator*

The signal conditions for the generator are specified in terms of the voltage between output points A and C shown in Figure 2/V.10.

When the signal condition 0 (space) for data circuits, or ON for control and timing circuits, is transmitted the output point A is positive with respect to point C. When the signal condition 1 (mark) for data circuits, or OFF for control and timing circuits, is transmitted the output point A is negative with respect to point C.

4.2    *Receiver*

The receiver significant levels are shown in Table A/V.10, where $V_{A'}$ and $V_{B'}$ are respectively the voltage at points A' and B' relative to point C':

TABLE A/V.10 – **Receiver significant levels**

|  | $V_{A'} - V_{B'} < -0.3$ V | $V_{A'} - V_{B'} > +0.3$ V |
|---|---|---|
| Data circuits | 1 | 0 |
| Control and timing circuits | OFF | ON |

5.     *Generator*

5.1    *Output impedance*

The total dynamic output impedance of the generator shall be equal to or less than 50 ohms.

5.2    *Static reference measurements*

The generator characteristics are specified in accordance with measurements illustrated in Figure 3/V.10 and described in 5.2.1 to 5.2.4 below.

5.2.1  *Open circuit measurement* (Figure 3a)/V.10)

The open circuit voltage measurement is made with a 3900 ohm resistor connected between points A and C. For either binary state, the magnitude of the signal voltage ($V_0$) shall be equal to or greater than 4.0 volts but not greater than 6.0 volts.

*Note.* – The figure of 4 volts is required to allow limited compatibility with Recommendation V.28, but a value of 3 volts may be used where Recommendation V.28 compatibility is not required. This is subject to further study.

5.2.2  *Test termination measurement* (Figure 3b)/V.10)

With a test load of 450 ohms connected between output points A and C, the magnitude of the output voltage $(V_t)$ shall be equal to or greater than 0.9 of the magnitude of $V_0$.

5.2.3  *Short-circuit measurement* (Figure 3c)/V.10)

With the output points A and C short-circuited the current flowing through point A shall not exceed 150 milliamperes for either logical condition.

5.2.4  *Power-off measurements* (Figure 3d)/V.10)

Under power-off condition, with voltage ranging between $+0.25$ volt and $-0.25$ volt applied between the output point A and point C, the magnitude of the output leakage current $(I_x)$ shall not exceed 100 microamperes.

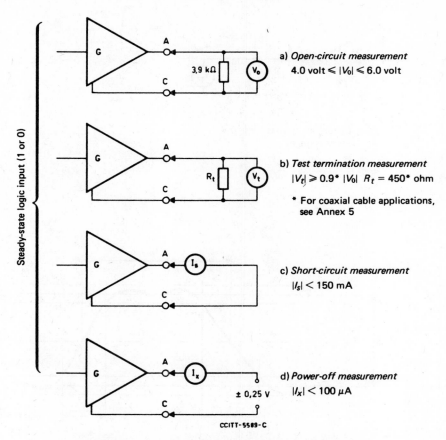

**Steady-state logic input (1 or 0)**

**a)** *Open-circuit measurement*
4.0 volt $\leqslant |V_0| \leqslant$ 6.0 volt

**b)** *Test termination measurement*
$|V_t| \geqslant 0.9^* |V_0|$　$R_t = 450^*$ ohm

\* For coaxial cable applications,
see Annex 5

**c)** *Short-circuit measurement*
$|I_s| <$ 150 mA

**d)** *Power-off measurement*
$|I_x| <$ 100 μA

± 0,25 V

CCITT-5589-C

FIGURE 3/V.10 – Generator parameter reference measurements

## 5.3　*Generator output rise-time measurement* (Figure 4/V.10)

### 5.3.1　*Waveform*

The measurement will be made with a resistor of 450 ohms connected between points A and C. A test signal, with a nominal signal element duration $t_b$ and composed of alternate ones and zeros, shall be applied to the input. The change in amplitude of the output signal during transitions from one binary state to the other shall be monotonic between 0.1 and 0.9 of $V_{ss}$.

### 5.3.2　*Waveshaping*

Waveshaping of the generator output signal shall be employed to control the level of interference (near-end cross-talk) which may be coupled to adjacent circuits in an interconnection. The rise time $(t_r)$ of the output signal shall be controlled to ensure the signal reaches 0.9 $V_{ss}$ between 0.1 and 0.3 of the duration of the unit interval $(t_b)$ at signalling rates greater than 1 kbit/s, and between 100 and 300 microseconds at signalling rates of 1 kbit/s or less. The method of waveshaping is not specified but examples are given in Annex 1.

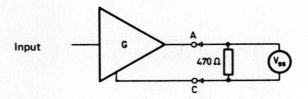

$V_{ss}$ = difference between steady
state voltages

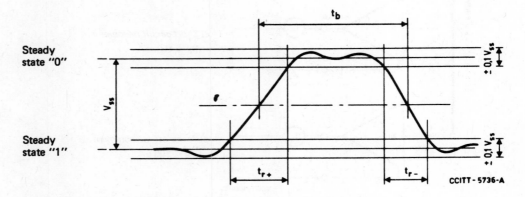

$t_b$ = nominal duration of the test signal element
$100 \mu s \leqslant t_r \leqslant 300 \mu s$ when $t_b \geqslant 1$ ms
$0.1 t_b \leqslant t_r \leqslant 0.3 t_b$ when $t_b < 1$ ms

FIGURE 4/V.10 — Generator output rise-time measurement

## 6.    Load

### 6.1    Characteristics

The load consists of a receiver (R) as shown in Figure 2/V.10. The electrical characteristics of the receiver are specified in terms of the measurements illustrated in Figures 5/V.10, 6/V.10 and 7/V.10 and described in 6.2, 6.3 and 6.4 below. A circuit meeting these requirements results in a differential receiver having a high input impedance, a small input threshold transition region between −0.3 and +0.3 volts differential, and allowance for an internal bias voltage not to exceed 3 volts in magnitude. The receiver is electrically identical to that specified for the balanced receiver in Recommendation V.11.

### 6.2    Receiver input voltage  −  current measurements (Figure 5/V.10)

With the voltage $V_{ia}$ (or $V_{ib}$) ranging between −10 volts and +10 volts, while $V_{ib}$ (or $V_{ia}$) is held at 0 volt, the resultant input current $I_{ia}$ (or $I_{ib}$) shall remain within the shaded range shown in Figure 5/V.10. These measurements apply with the power supply of the receiver in both the power-on and power-off conditions.

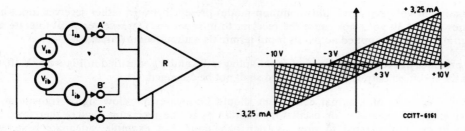

FIGURE 5/V.10 – Receiver input voltage-current measurements

## 6.3 *DC input sensitivity measurements* (Figure 6/V.10)

Over the entire common-mode voltage $(V_{cm})$ range of $+7$ volts to $-7$ volts, the receiver shall not require a differential input voltage $(V_i)$ of more than 300 millivolts to assume correctly the intended binary state. Reversing the polarity of $V_i$ shall cause the receiver to assume the opposite binary state.

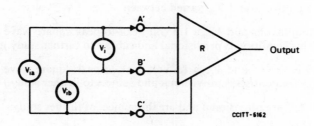

CCITT-6162

| Applied voltages | | Resulting input voltage $V_i$ | Output binary state | Purpose of measurement |
|---|---|---|---|---|
| $V_{ia}$ | $V_{ib}$ | | | |
| $-12$ V | 0 V | $-12$ V | | |
| 0 V | $-12$ V | $+12$ V | (not specified) | To ensure no damage to receiver inputs |
| $+12$ V | 0 V | $+12$ V | | |
| 0 V | $+12$ V | $-12$ V | | |
| $+10$ V | $+4$ V | $+6$ V | 0 | To guarantee correct operation at $V_i = 6$ V (maintain correct logic state) |
| $+4$ V | $+10$ V | $-6$ V | 1 | |
| $-10$ V | $-4$ V | $-6$ V | 1 | |
| $-4$ V | $-10$ V | $+6$ V | 0 | |
| | | | | 300 mV threshold measurement |
| $+0.30$ V | 0 V | $+0.3$ V | 0 | } $V_{cm} = 0$ V |
| 0 V | $+0.30$ V | $-0.3$ V | 1 | |
| $+7.15$ V | $+6.85$ V | $+0.3$ V | 0 | } $V_{cm} = +7$ V |
| $+6.85$ V | $+7.15$ V | $-0.3$ V | 1 | |
| $-7.15$ V | $-6.85$ V | $-0.3$ V | 1 | } $V_{cm} = -7$ V |
| $-6.85$ V | $-7.15$ V | $+0.3$ V | 0 | |

FIGURE 6/V.10 – Receiver input sensitivity measurement

The maximum voltage (signal plus common-mode) present between either receiver input and receiver ground shall not exceed 10 volts nor cause the receiver to misoperate. The receiver shall tolerate a maximum differential signal of 12 volts applied across its input terminals without being damaged.

In the presence of the combinations of input voltages $V_{ia}$ and $V_{ib}$ specified in Figure 6/V.10, the receiver shall maintain the specified output binary state and shall not be damaged.

*Note.* – Designers of terminal equipment should be aware that slow signal transitions with noise present may give rise to instability or oscillatory conditions in the receiving device; therefore, appropriate techniques should be implemented to prevent such behaviour. For example, adequate hysteresis may be incorporated into the receiver to prevent such conditions.

6.4     *Input balance test* (Figure 7/V.10)

The balance of the receiver input resistances and bias voltages shall be such that the receiver shall remain in the intended binary state under the conditions shown in Figure 7/V.10 and described as follows:

a)     with $V_i$ = +720 millivolts and $V_{cm}$ varied between −7 and +7 volts;

b)     with $V_i$ = −720 millivolts and $V_{cm}$ varied between −7 and +7 volts;

c)     with $V_i$ = +300 millivolts and $V_{cm}$ a 1.5 volt peak-to-peak square wave at the highest applicable signalling rate. (This condition is provisional and subject to further study.);

d)     with $V_i$ = −300 millivolts and $V_{cm}$ a 1.5 volt peak-to-peak square wave at the highest applicable signalling rate. (This condition is provisional and subject to further study.)

*Note.* – The values of $V_i$ are provisional and are the subject of further study.

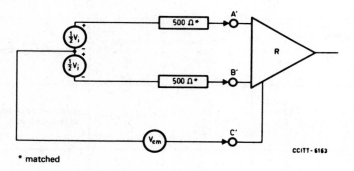

FIGURE 7/V.10 – Receiver input balance test

7.     *Environmental constraints*

In order to operate an unbalanced interchange circuit at signalling rates ranging between 0 and 100 kbit/s, the following conditions apply:

1)     The total peak differential noise measured between the points A' and B' at the load interchange point (with the generator interchange point connected to a 50 ohm resistor substituted for the generator) shall not exceed the expected amplitude of the received signal minus 0.3 volts (provisional value).

2)     The worst-case combination of generator-receiver ground potential difference ($V_g$, Figure 2/V.10) and longitudinally induced peak random noise voltage measured between the receiver points A' or B' and C' with the generator ends of the cable A and C joined together shall not exceed 4 volts.

## 8.    *Circuit protection*

Unbalanced generator and load devices complying with this Recommendation shall not be damaged under the following conditions:

1)    generator open circuit;

2)    short-circuit between the conductors of the interconnecting cable;

3)    short-circuit between the conductors and Point C or C'.

The above faults 2) and 3) might cause power dissipation in the interchange circuit devices to approach the maximum power dissipation that may be tolerated by a typical integrated circuit (IC) package. The user is therefore cautioned that where multiple generators and receivers are implemented in a single IC package, only one such fault per package might be tolerable at any one time without damage occuring.

The user is also cautioned that generator and receiver devices complying with this Recommendation might be damaged by spurious voltages applied between their input or output points and points C or C' (Figure 2/V.10). In those applications where the interconnecting cable may be inadvertently connected to other circuits or where it may be exposed to a severe electromagnetic environment, protection should be employed.

## 9.   Category 1 and Category 2 receivers

In order to provide flexibility in the choice of generator (V.10 or V.11), two categories of receiver are defined as follows:

Category 1—receivers shall have both input terminals A' and B' connected to individual terminals at the load interchange point, independent of all other receivers, as shown in Figure 8A/V.10, and as applied in Annex 2, Figure 9bis/V.10.

Category 2—receivers shall have one terminal connection for each A' input terminal at the load interchange point, and all B' input terminals shall be connected together within the DCE or DTE and shall be brought to one common B' input terminal as shown in Figure 8B/V.10.

The specification of the category to be used in any application is part of the appropriate DCE Recommendation, using this type of interface electrical characteristics.

## 10.  Signal common return

The interconnection between the generator and the load interchange points in Figure 2/V.10 shall consist of a signal conductor for each circuit and one signal common return for each direction as shown in Figures 8A/V.10 and 8B/V.10. Signal common return may be implemented by more than one lead, where required to accomplish interworking, as described in Annex 2, section 2, and as shown in Figure 9bis/V.10.

To minimize the effects of ground potential difference $V_g$ and longitudinally-coupled noise on the signal at the load interchange point, the signal common return shall be connected to ground only at the C terminal of the generator interchange point. For example, the B' terminal of all the receivers in DTE which interconnect with unbalanced generators in DCE shall connect to signal common return circuit 102b, which is connected to ground only in DCE. Signal common return circuit 102a is used to interconnect terminal B' of the receivers in DCE with the grounded terminal C of the unbalanced generators in DTE, as in Figures 8A/V.10 and 8B1V.10.

## 11.  Detection of generator power-off or circuit failure

Certain applications require detection of various fault conditions in the interchange circuits, e.g.

1)    generator in power-off condition;

2)    receiver not interconnected with a generator;

3)    open-circuited interconnecting cable;

4)  short-circuited interconnecting cable;

5)  input signal to the load remaining within the transition region ($\pm$ 300 millivolts) for an abnormal period of time.

When detection of one or more fault conditions is required by specific applications, additional provisions are required in the load and the following items must be determined:

a)  which interchange circuits require fault detection;

b)  what faults must be detected;

c)  what action must be taken when a fault is detected, e.g. which binary state must the receiver assume?

The method of detection of fault conditions is application-dependent and is therefore not further specified.


Where electrical characteristics conforming to Recommendation V.10 are used, the following interchange circuits of Recommendation V.24, where implemented, shall be used to detect either a power-off condition in the equipment connected through the interface or disconnection of the interconnecting cable:

Circuit 105 (Request to send)
Circuit 107 (Data set ready)
Circuit 108.1/108.2 (Connect data set to line/Data terminal ready)
Circuit 120 (Transmit backward channel line signal)
Circuit 202 (Call request)
Circuit 213 (Power indication)

The receivers of these circuits shall interpret a circuit fault as an OFF condition.

The interchange circuits monitoring circuit-fault conditions in data network interfaces are indicated in Recommendation X.24.

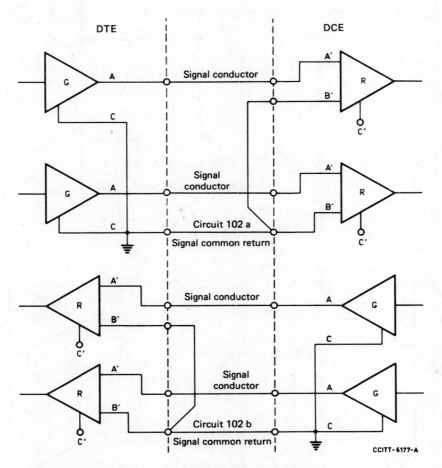

FIGURE 8/V.10 – Interconnection (example) of signal common return

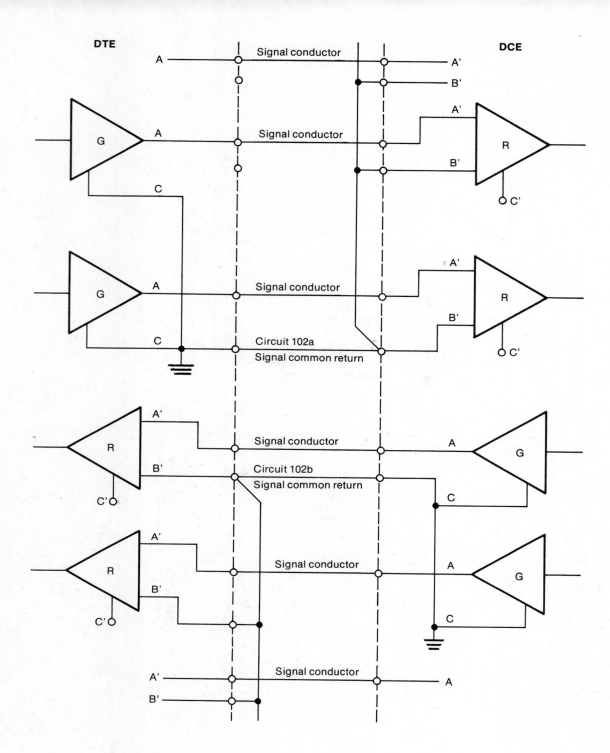

**Figure 8A/V.10—Interconnection of signal common return for category 1 receivers**

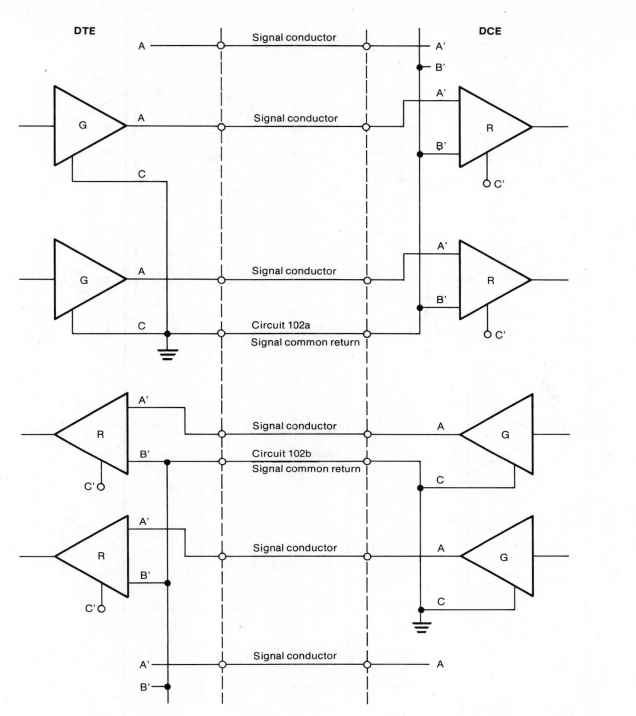

**Figure 8B/V.10 — Interconnection of signal common return for category 2 receivers**

37

**Figure 9bis/V.10—Interconnection of signal common return by more then one conductor in order to accomplish inter-operation of V.10 generators with category 1 receivers**

ANNEX 1

(to Recommendation V.10)

### Waveshaping examples

The required waveshaping may be accomplished either by providing a slew-rate control in the generator or by inserting an RC filter at the generator interchange point. A combination of these methods may also be employed. An example of the RC filter method is shown in Figure 9/V.10. Typical values of capacitance $C_w$, with the value of $R_w$ selected so that $R_w + R_d$ is approximately 50 ohms, are given for typical cable with an interconductor shunt capacitance of approximately 0.05 microfarads per kilometre.

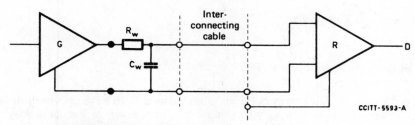

$R_d$ = generator internal resistance
$R_w$ = 50 ohms − $R_d$

| $C_w$ (microfarads) | Data signalling rate range (kbit/s) |
|---|---|
| 1.0 | 0 - 2.5 |
| 0.47 | 2.5 - 5.0 |
| 0.22 | 5.0 - 10 |
| 0.1 | 10 - 25 |
| 0.047 | 25 - 50 |
| 0.022 | 50 - 100 |

FIGURE 9/V.10 − Example method for waveshaping

ANNEX 2

(to Recommendation V.10)

### Compatibility with other interfaces

1. *Compatibility of Recommendation V.10 and Recommendation V.11 interchange circuits in the same interface*

The electrical characteristics of Recommendation V.10 are designed to allow the use of balanced (see Recommendation V.11) and unbalanced circuits within the same interface. For example, the balanced circuits may be used for data and timing whilst the unbalanced circuits may be used for associated control circuit functions.

2.      *Recommendation V.10 interworking with Recommendation V.11*

The basic differential receiver specifications of Recommendations V.10 and V.11 are electrically identical. It is therefore possible to interconnect an equipment using Recommendation V.10 receivers and generators on one side of the interface with an equipment using Recommendation V.11 generators and receivers on the other side of the interface. Such interconnection would result in interchange circuits according to Recommendation V.11 in one direction and interchange circuits according to Recommendation V.10 in the other direction. Where such interworking is contemplated, the following technical considerations must be taken into account.

2.1     Interconnecting cable lengths are limited by performance of the circuits working to the Recommendation V.10 side of the interface.

2.2     The optional cable termination resistance $(Z_t)$, if implemented, in the equipment using Recommendation V.11 must be removed.

**2.3 V.10-type receivers shall be of Category 1.**

3.      *Recommendation V.10 interworking with Recommendation V.28*

The unbalanced electrical characteristics of Recommendation V.10 have also been designed to permit limited interworking, under certain conditions, with generators and receivers to Recommendation V.28. Where such interworking is contemplated, the following technical limitations must be considered:

3.1     Separate DTE and DCE signal return paths will not be available at the Recommendation V.28 side of the interface.

3.2     Data signalling-rate limitations according to Recommendation V.28 shall apply.

3.3     Interconnecting cable lengths are limited by the Recommendation V.28 performance restrictions.

3.4     Probability of satisfactory operation will be enhanced by providing the maximum generator voltage possible on the Recommendation V.10 side of the interface within the limitations according to Recommendation V.10.

3.5     Whilst Recommendation V.28 type generators may use potentials in excess of 12 volts, many existing equipments are designed to operate with power supplies of 12 volts or less. Where this is the case, no further protection of Recommendation V.10 receivers is required; however, in the general case, protection against excessively high voltages from Recommendation V.28 generators must be provided for the Recommendation V.10 receivers.

3.6     Power-off detectors in Recommendation V.28 receivers may not necessarily work with Recommendation V.10 generators.

ANNEX 3

(to Recommendation V.10)

**Operational constraints**

No electrical characteristics of the interconnection cable are specified in this Recommendation. However, guidance is given herein concerning operational constraints imposed by cable length and near-end cross-talk.

The maximum operating distance for the unbalanced interchange circuit is primarily a function of the amount of interference (near-end cross-talk) coupled to adjacent circuits in the equipment interconnection. Additionally the unbalanced circuit is susceptible to exposure to differential noise resulting from any imbalance between the signal conductor and signal common return at the load interchange point. Increasing the physical separation and interconnection cable length between the generator and load interchange points might increase the exposure to common-mode noise and the degree of near-end cross-talk. Accordingly, users are advised to restrict the cable length to a minimum consistent with the generator-load physical separation requirements.

The curve of cable length versus signalling rate given in Figure 10/V.10 may be used as a conservative guide. This curve is based upon calculations and empirical data using twisted-pair telephone cable with a shunt capacitance of 0.052 microfarads per kilometre, a 50 ohm source impedance, a 6 volt source signal and maximum near-end cross-talk of 1 volt peak. The rise time $(t_r)$ of the source signal at signalling rates below 1000 bit/s is 100 microseconds and above 1000 bit/s is 0.1 $t_b$ (see Figure 4/V.10).

The user is cautioned that the curve given in Figure 10/V.10 does not account for common-mode noise or near-end cross-talk levels beyond the limits specified, that may be introduced between the generator and load by exceptionally long cables. On the other hand operation within the signalling-rate and distance bounds of Figure 10/V.10 will usually ensure that the distortion of the signal appearing at the receiver input will be acceptable. Many applications, however, can tolerate greater levels of signal distortion, and correspondingly greater cable lengths can be employed. The generation of near-end cross-talk can be reduced by using lower source resistances and increasing the amount of waveshaping employed.

Experience has shown that in most practical cases the operating distance at the lower signalling rates may be extended to several kilometres.

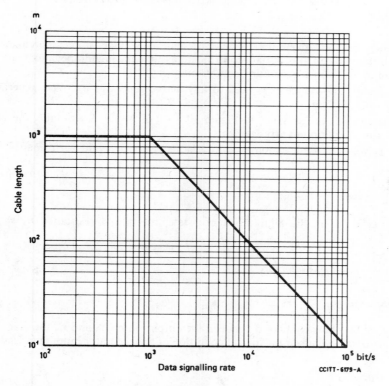

FIGURE 10/V.10 – **Data signalling rate vs cable length for unbalanced interchange circuit**

ANNEX 4

(to Recommendation V.10)

**Multipoint operation**

It is considered that further study is required before parameters for this application can be defined, and this Annex is intended as a guideline for this study. The values quoted for the high-impedance state are tentative and are provided only for guidance.

1.    *Multipoint interchange circuit arrangements*

The point-to-point interchange circuit arrangement of one generator and one load might be expanded to a multipoint arrangement by adding generators, receivers or both at interchange points along the interconnecting cable, as shown in Figure 11/V.10.

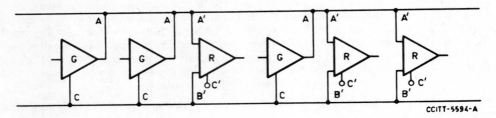

FIGURE 11/V.10 – **Multipoint interchange circuit arrangement**

Only one generator at a given time would present its voltage at its interchange point. All other generators would be isolated by an appropriate control and assume the high impedance state defined below. All receivers would be continuously in an operating condition.

The combined load impedance presented to any active generator by other generators, receivers and cable must not be less than 50 ohms.

The guidelines on cable length given in Figure 10/V.10 of Annex 3, assuming the same environmental conditions, apply equally to multipoint arrangements.

If multiple receivers are used, extreme caution is necessary to avoid performance degradation due to reflection.

2.    *High-impedance state*

When the generator is in the high-impedance state, the output resistance $R_h$ shall be greater than 10 000 ohms.

3.    *High impedance state measurements*

3.1    *Static measurements*

With test load of 50 ohms connected between the output points A and C, the magnitude of the output voltage shall not exceed 2 mV whatever the logical condition of the generator input data lead.

3.2    *Dynamic measurements*

During transitions of the generator output between the low-impedance state and the high-impedance state, the signal measured across a 50 ohm test load connected between the output points A and C shall be such that the change in amplitude goes from 10% to 90% of the steady state voltage in less than 10 microseconds.

ANNEX 5

(to Recommendation V.10)

**Special considerations for coaxial cable applications**

It is recognized that where coaxial cables are used for interconnecting purposes it may be desirable to include a terminating resistance at the receiver end of the cable. This is considered to be a special case for which special generator characteristics are required. The terminating resistance shall in no case be less than 50 ohms and the reference measurements under 5.2.2 and 5.3 shall be made with a 50 ohm test termination. Use of this special application will require appropriate agreement with the proper authority.

The alternative set of electrical characteristics applied in the coaxial cable case is the following (this shall be referred to as "V.10 SPECIAL") [4]:

### 5.2.2 *bis*    *Test termination measurement* (Figure 3b)/V.10

With a test load $(R_t)$ of 50 ohms connected between output points A and C, the magnitude of the output voltage $(V_t)$ shall be equal to or greater than 0.5 of the magnitude of $V_0$.

### 5.3.1 *bis*    *Waveform* (Figure 4/V.10)

The measurement will be made with a resistor of 50 ohms connected between points A and C. A test signal, with a nominal signal element duration $t_b$ and composed of alternate ones and zeros, shall be applied to the input. The change in amplitude of the output signal during transitions from one binary state to the other shall be monotonic between 0.1 and 0.9 of $V_{ss}$.

### 5.3.2 *bis*    *Waveshaping*

Waveshaping is not normally required for coaxial cable applications.

### 9 *bis*    *Signal common return*

In applications where coaxial cables are used, the screen of the coaxial cable shall be connected to ground only at point C at the generator end as shown in Figure 12/V.10.

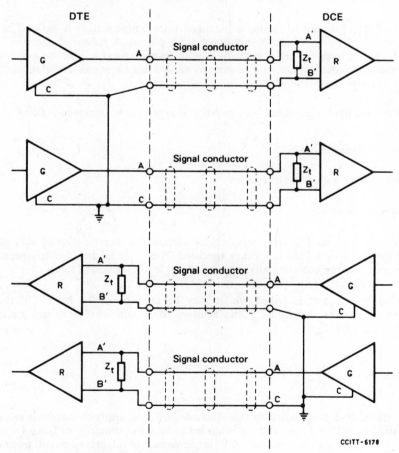

$Z_t$ = optional termination impedance

FIGURE 12/V.10 – **Interconnection with coaxial cable**

---

[4] All the electrical characteristics specified in Recommendation V.10 other than those set down in this Annex are applicable to the coaxial cable case with a cable termination.

**Recommendation V.11**

## ELECTRICAL CHARACTERISTICS FOR BALANCED DOUBLE-CURRENT INTERCHANGE CIRCUITS FOR GENERAL USE WITH INTEGRATED CIRCUIT EQUIPMENT IN THE FIELD OF DATA COMMUNICATIONS [5]

*(Geneva, 1976)*

### 1.    *Introduction*

This Recommendation deals with the electrical characteristics of the generator, receiver and interconnecting leads of a differential signalling (balanced) interchange circuit with an optional d.c. offset.

The balanced generator and load components are designed to cause minimum mutual interference with adjacent balanced or unbalanced interchange circuits (see Recommendation V.10) provided that waveshaping is employed on the unbalanced circuits.

In the context of this Recommendation, a balanced interchange circuit is defined as consisting of a balanced generator connected by a balanced interconnecting pair to a balanced receiver. For a balanced generator the algebraic sum of both the outlet potentials, with respect to earth, shall be constant for all signals transmitted; the impedances of the outlets with respect to earth shall be equal. The degree of balance of the interconnecting pair is a matter for further study.

Annexes are provided to give guidance on a number of application aspects as follows:

*Annex 1* Cables and terminations

*Annex 2* Compatibility with other interfaces

*Annex 3* Multipoint operation

*Note.* –  Generator and load devices meeting the electrical characteristics of this Recommendation need not operate over the entire signalling rate range specified. They may be designed to operate over narrower ranges to satisfy requirements more economically, particularly at lower signalling rates.

Reference measurements are described which may be used to verify certain of the recommended parameters but it is a matter for individual manufacturers to decide what tests are necessary to ensure compliance with the Recommendation.

### 2.    *Field of application*

The electrical characteristics specified in this Recommendation apply to interchange circuits operating with data signalling rates up to 10 Mbit/s, and are intended to be used primarily in Data Terminal Equipment (DTE) and Data Circuit-terminating Equipment (DCE) implemented in integrated-circuit technology.

This Recommendation applies to new work and is not intended to apply to DCE implemented in discrete component technology, for which the electrical characteristics covered by Recommendation V.28 are more appropriate.

---

[5] This Recommendation is also designated as X.27 in the Series X Recommendations.

Typical points of application are illustrated in Figure 1/V.11.

Whilst the balanced interchange circuit is primarily intended for use at the higher signalling rates, its use at the lower rates may be necessary in the following cases:

1) where the interconnecting cable is too long for proper unbalanced circuit operation;

2) where extraneous noise sources make unbalanced circuit operation impossible;

3) where it is necessary to minimize interference with other signals.

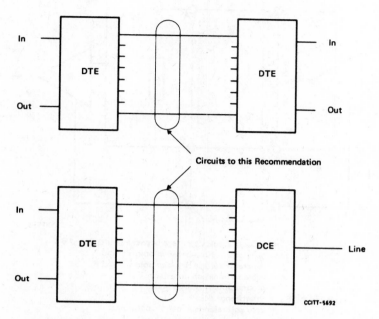

FIGURE 1/V.11 – Typical applications of balanced interchange circuits

3.    *Symbolic representation of interchange circuit* (Figure 2/V.11)

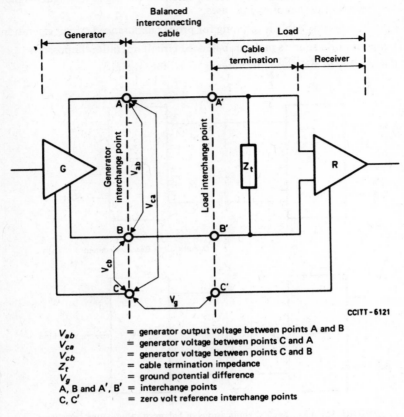

| | |
|---|---|
| $V_{ab}$ | = generator output voltage between points A and B |
| $V_{ca}$ | = generator voltage between points C and A |
| $V_{cb}$ | = generator voltage between points C and B |
| $Z_t$ | = cable termination impedance |
| $V_g$ | = ground potential difference |
| A, B and A', B' | = interchange points |
| C, C' | = zero volt reference interchange points |

FIGURE 2/V.11 – Symbolic representation of a balanced interchange circuit

*Note 1.* – Two interchange points are shown above. The output characteristics of the generator, excluding any interconnecting cable, are defined at the "generator interchange point". The electrical characteristics to which the receiver must respond are defined at the "load interchange point".

*Note 2.* – Point C may be connected to C' by CCITT Recommendation V.24 Circuit 102 and to protective ground if required by national regulations.

4.    *Generator polarities and receiver significant levels*

4.1    *Generator*

The signal conditions for the generator are specified in terms of the voltage between output points A and B shown in Figure 2/V.11.

When the signal condition 0 (space) for data circuits or ON for control and timing circuits is transmitted, the output point A is positive with respect to point B. When the signal condition 1 (mark) for data circuits or OFF for control and timing circuits is transmitted, the output point A is negative with respect to point B.

## 4.2    *Receiver*

The receiver differential significant levels are shown in Table A/V.11 below, where $V_{A'}$ and $V_{B'}$ are respectively the voltages at points A' and B' relative to point C':

TABLE A/V.11 – **Receiver differential significant levels**

|                             | $V_{A'} - V_{B'} < -0.3\ V$ | $V_{A'} - V_{B'} > +0.3\ V$ |
| --------------------------- | --------------------------- | --------------------------- |
| Data circuits               | 1                           | 0                           |
| Control and timing circuits | OFF                         | ON                          |

## 5.    *Generator*

### 5.1    *Resistance and offset voltage*

1.    The total generator resistance between points A and B shall be equal to or less than 100 ohms and adequately balanced with respect to point C. (It is left for further study as to the degree of balance required both statically and dynamically.)

2.    The magnitude of the generator d.c. offset voltage (see 5.2.2 below) shall not exceed 3 V under all operating conditions.

### 5.2    *Static reference measurements*

The generator characteristics are specified in accordance with measurements illustrated in Figure 3/V.11 and described in 5.2.1 to 5.2.4 below.

### 5.2.1    *Open-circuit measurement* (Figure 3a)/V.11)

The open-circuit voltage measurements are made with a 3900 ohm resistor connected between points A and B. For either binary state, the magnitude of the differential voltage $(V_0)$ shall be not more than 6.0 volts, nor shall the magnitude of $V_{0a}$ and $V_{0b}$ be more than 6.0 volts.

### 5.2.2    *Test-termination measurement* (Figure 3b)/V.11)

With a test load of two resistors, each 50 ohms, connected in series between the output points A and B, the differential voltage $(V_t)$ shall not be less than 2.0 volts or 50% of the magnitude of $V_0$, whichever is greater. For the opposite binary state the polarity of $V_t$ shall be reversed $(V_t)$. The difference in the magnitudes of $V_t$ and $V_t$ shall be less than 0.4 volts. The magnitude of the generator offset voltage $V_{0s}$ measured between the centre of the test load and point C shall not be greater than 3.0 volts. The magnitude of the difference in the values of $V_{0s}$ for one binary state and the opposite binary state shall be less than 0.4 volts.

*Note.* – Under some conditions this measurement does not determine the degree of balance of the internal generator impedances to point C. It is left for further study whether additional measurements are necessary to ensure adequate balance in generator output impedances.

### 5.2.3    *Short-circuit measurement* (Figure 3c)/V.11)

With the output points A and B short-circuited to point C, the currents flowing through each of output points A or B shall not exceed 150 milliamperes for either logical condition.

### 5.2.4   *Power-off measurements* (Figure 3d)/V.11

Under power-off condition with voltages ranging between $+0.25$ volt and $-0.25$ volt applied between each output point and point C, as indicated in Figure 3d)/V.11, the magnitude of the output leakage currents ($I_{xa}$ and $I_{xb}$) shall not exceed 100 microamperes.

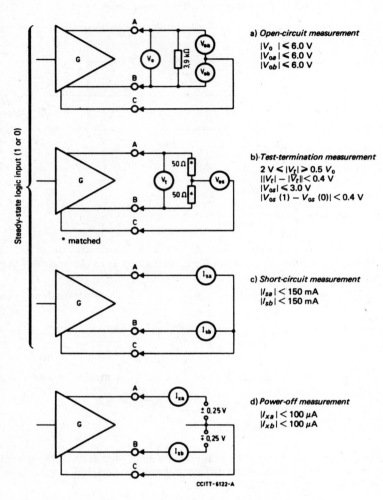

a) *Open-circuit measurement*

$|V_o| \leqslant 6.0$ V
$|V_{oa}| \leqslant 6.0$ V
$|V_{ob}| \leqslant 6.0$ V

b) *Test-termination measurement*

$2$ V $< |V_t| \geqslant 0.5 \, V_o$
$||V_t| - |\bar{V}_t|| < 0.4$ V
$|V_{os}| \leqslant 3.0$ V
$|V_{os}(1) - V_{os}(0)| < 0.4$ V

c) *Short-circuit measurement*

$|I_{sa}| < 150$ mA
$|I_{sb}| < 150$ mA

d) *Power-off measurement*

$|I_{xa}| < 100 \, \mu$A
$|I_{xb}| < 100 \, \mu$A

FIGURE 3/V.11 – Generator-parameter reference measurements

**5.3** *Dynamic voltage balance and rise time* (Figure 4/V.11)

With the measurement configuration shown in Figure 4/V.11, a test signal with a nominal signal element duration $t_b$ and composed of alternate ones and zeros, shall be applied to the input. The change in amplitude of the output signal during transitions from one binary state to the other shall be monotonic between 0.1 and 0.9 $V_{ss}$ within 0.1 of $t_b$ or 20 nanoseconds, whichever is greater. Thereafter the signal voltage shall not vary more than 10% of $V_{ss}$ from the steady state value.

The resultant voltage due to imbalance ($V_E$) shall not exceed 0.4 V peak-to-peak (the value of $V_E$ is provisional and is subject to further study to determine whether voltage peaks of very short duration should be included).

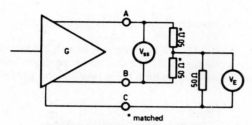

$V_E < 0.4$ V peak-to-peak (provisional)
$V_{ss}$ = difference between signal steady-state voltages

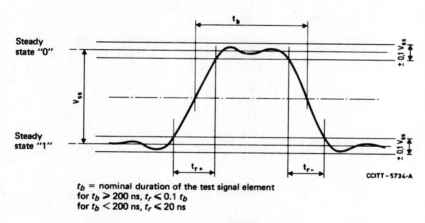

$t_b$ = nominal duration of the test signal element
for $t_b \geqslant 200$ ns, $t_r \leqslant 0.1\ t_b$
for $t_b < 200$ ns, $t_r \leqslant 20$ ns

CCITT - 5734-A

FIGURE 4/V.11 – Generator dynamic balance and rise-time measurement

6.      *Load*

6.1     *Characteristics*

The load consists of a receiver (R) and an optional cable termination resistance $(Z_t)$ as shown in Figure 2/V.11. The electrical characteristics of the receiver are specified in terms of the measurements illustrated in Figures 5/V.11, 6/V.11 and 7/V.11 and described in 6.2, 6.3 and 6.4 below. A circuit meeting these requirements results in a differential receiver having a high input impedance, a small input threshold transition region between $-0.3$ and $+0.3$ volts differential, and allowance for an internal bias voltage not to exceed 3 volts in magnitude. The receiver is electrically identical to that specified for the unbalanced receiver in Recommendation V.10.

6.2     *Receiver input voltage  —  current measurements* (Figure 5/V.11)

With the voltage $V_{ia}$ (or $V_{ib}$) ranging between $-10$ volts and $+10$ volts, while $V_{ib}$ (or $V_{ia}$) is held at 0 volt, the resultant input current $I_{ia}$ (or $I_{ib}$) shall remain within the shaded range shown in Figure 5/V.11. These measurements apply with the power supply of the receiver in both the power-on and power-off conditions.

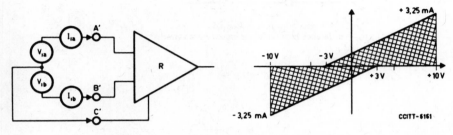

FIGURE 5/V.11 – Receiver input voltage-current measurements

**6.3**     *D.c. input sensitivity measurements* (Figure 6/V.11)

Over the entire common mode voltage $(V_{cm})$ range of $+7$ volts to $-7$ volts, the receiver shall not require a differential input voltage $(V_i)$ of more than 300 millivolts to assume correctly the intended binary state. Reversing the polarity of $V_i$ shall cause the receiver to assume the opposite binary state.

The maximum voltage (signal plus common mode) present between either receiver input and receiver signal ground shall not exceed 10 volts nor cause the receiver to malfunction. The receiver shall tolerate a maximum differential signal of 12 volts applied across its input terminals without being damaged.

In the presence of the combinations of input voltages $V_{ia}$ and $V_{ib}$ specified in Figure 6/V.11, the receiver shall maintain the specified output binary state and shall not be damaged.

*Note.* — Designers of terminal equipment should be aware that slow signal transitions with noise present may give rise to instability or oscillatory conditions in the receiving device; therefore, appropriate techniques should be implemented to prevent such behaviour. For example, adequate hysteresis may be incorporated into the receiver to prevent such conditions.

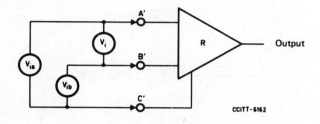

| Applied voltages | | Resulting input voltage $V_i$ | Output binary state | Purpose of measurement |
|---|---|---|---|---|
| $V_{ia}$ | $V_{ib}$ | | | |
| −12 V | 0 V | −12 V | | |
| 0 V | −12 V | +12 V | (not specified) | To ensure no damage |
| +12 V | 0 V | +12 V | | to receiver inputs |
| 0 V | +12 V | −12 V | | |
| +10 V | + 4 V | + 6 V | 0 | To guarantee correct |
| + 4 V | +10 V | − 6 V | 1 | operation at $V_i = 6$ V |
| −10 V | − 4 V | − 6 V | 1 | (maintain correct |
| − 4 V | −10 V | + 6 V | 0 | logic state) |
| | | | | 300 mV threshold measurement |
| +0.30 V | 0   V | +0.3 V | 0 | $\}$ $V_{cm} = 0$ V |
| 0   V | +0.30 V | −0.3 V | 1 | |
| +7.15 V | +6.85 V | +0.3 V | 0 | $\}$ $V_{cm} = +7$ V |
| +6.85 V | +7.15 V | −0.3 V | 1 | |
| −7.15 V | −6.85 V | −0.3 V | 1 | $\}$ $V_{cm} = −7$ V |
| −6.85 V | −7.15 V | +0.3 V | 0 | |

FIGURE 6/V.11 − Receiver input sensitivity measurement

### 6.4    *Input balance test* (Figure 7/V.11)

The balance of the receiver input resistance and bias voltages shall be such that the receiver shall remain in the intended binary state under the conditions shown in Figure 7/V.11 and described as follows:

a)   with $V_i$ = +720 millivolts and $V_{cm}$ varied between −7 and +7 volts;

b)   with $V_i$ = −720 millivolts and $V_{cm}$ varied between −7 and +7 volts;

c)   with $V_i$ = +300 millivolts and $V_{cm}$ a 1.5 volt peak-to-peak square wave at the highest applicable signalling rate. (This condition is provisional and subject to further study.);

d)   with $V_i$ = −300 millivolts and $V_{cm}$ a 1.5 volt peak-to-peak square wave at the highest applicable signalling rate (This condition is provisional and subject to further study).

*Note.* −  The values of $V_i$ are provisional and are the subject of further study.

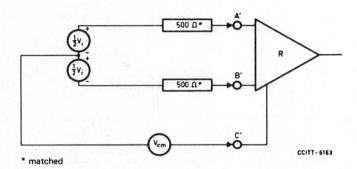

FIGURE 7/V.11 − Receiver input balance test

### 6.5    *Terminator*

The use of a terminator is optional depending upon the specific environment in which the interchange circuit is employed (see Annex 1). In no case shall the total load resistance be less than 100 ohms.

### 7.    *Environmental constraints*

In order to operate a balanced interchange circuit at signalling rates ranging between 0 and 10 Mbit/s, the following conditions apply:

1)   For each interchange circuit a balanced interconnecting pair is required.

2)   Each interchange circuit must be appropriately terminated (see Annex 1).

3)   The total common-mode voltage at the receiver must be less than 7 volts peak. This value is provisional and is subject to further study.

The common mode voltage at the receiver is the worst case combination of:

a)   generator-receiver ground-potential difference ($V_g$, Figure 2/V.11);

b)   longitudinally induced random noise voltage measured between the receiver points A' or B' and C' with the generator ends of the cable A, B and C joined together; and

c)   generator bias voltage, if any.

Unless the generator is of a type which generates no bias voltage, the sum of a) and b) above, which is the element of the common mode voltage due to the environment of the interchange circuit, must be less than 4 volts peak.

8.    *Circuit protection*

Balanced generator and load devices complying with this Recommendation shall not be damaged under the following conditions:

1)    generator open circuit;

2)    short-circuit between the conductors of the interconnecting cable;

3)    short-circuit between either or both conductors and point C and C'.

The above faults 2) and 3) might cause power dissipation in the interchange circuit devices to approach the maximum power dissipation that may be tolerable by a typical integrated circuit (IC) package. The user is therefore cautioned that where multiple generators and receivers are implemented in a single IC package, only one such fault per package might be tolerable at any one time without damage occurring.

The user is also cautioned that the generator and receiver devices complying with this Recommendation might be damaged by spurious voltages applied between their input or output points and points C and C' (Figure 2/V.11). In those applications where the interconnecting cable may be inadvertently connected to other circuits, or where it may be exposed to a severe electromagnetic environment, protection should be employed.

9.    *Detection of generator power-off or circuit failure*

Certain applications require detection of various fault conditions in the interchange circuits, e.g.:

1)    generator in power-off condition;

2)    receiver not interconnected with a generator;

3)    open-circuited interconnecting cable;

4)    short-circuited interconnecting cable;

5)    input signal to the load remaining within the transition region ($\pm$ 300 millivolts) for an abnormal period of time.

When detection of one or more fault conditions is required by specific applications, additional provisions are required in the load and the following items must be determined:

a)    which interchange circuits require fault detection;

b)    what faults must be detected;

c)    what action must be taken when a fault is detected, e.g. which binary state must the receiver assume?

The method of detection of fault conditions is application-dependent and is therefore not further specified.

Where electrical characteristics conforming to Recommendation V.11 are used, the following interchange circuits of Recommendation V.24, where implemented, shall be used to detect either a power-off condition in the equipment connected through the interface or disconnection of the interconnecting cable:

Circuit 105 (Request to send)
Circuit 107 (Data set ready)
Circuit 108.1/108.2 (Connect data set to line/Data terminal ready)
Circuit 120 (Transmit backward channel line signal)
Circuit 202 (Call request)
Circuit 213 (Power indication)
The receivers of these circuits shall interpret a circuit fault as an OFF condition.
The interchange circuits monitoring circuit-fault conditions in data network interfaces are indicated in Recommendation X.24.

# ANNEX 1

## (to Recommendation V.11)

### Cable and terminations

No electrical characteristics of the interconnecting cable are specified in this Recommendation. Guidance is given herein concerning operational constraints imposed by the length, balance and terminating resistance of the cable.

1. *Cable*

Over the length of the cable, the two conductors should have essentially the same values of:

1) capacitance to ground;

2) longitudinal resistance and inductance;

3) coupling to adjacent cables and circuits.

## 2.    *Cable length*

The maximum permissible length of cable separating the generator and the load in a point-to-point application is a function of the data signalling rate. It is further influenced by the tolerable signal distortion and the environmental constraints such as ground potential difference and longitudinal noise. Increasing the distance between generator and load might increase the exposure to ground potential difference.

As an illustration of the above conditions, the curves of cable length versus data signalling rate in Figure 8/V.11 may be used for guidance.

These curves are based upon empirical data using twisted pair telephone cable (0.51 mm wire diameter) both unterminated and terminated in a 100 ohm resistive load. The cable length restrictions shown by the curves are based upon the following assumed signal quality requirements at the load:

1)    signal rise and fall time equal to, or less than, one-half the duration of the signal element;

2)    a maximum voltage loss between generator and load of 6 dB.

At the higher signalling rates (see Figure 8/V.11) the sloping portion of the curves shows the cable length limitation established by the assumed signal rise and fall time requirements. The cable length has been arbitrarily limited to 1000 metres by the assumed maximum allowable loss of 6 dB.

These curves assume that the environmental limits specified in this Recommendation have been achieved. At the higher signalling rates these conditions are more difficult to attain due to cable imperfections and common-mode noise. Operation within the signalling rate and distance bounds of Figure 8/V.11 will usually ensure that distortion of the signal appearing at the receiver input will be acceptable. Many applications, however, can tolerate much greater levels of signal distortion and in these cases correspondingly greater cable lengths may be employed.

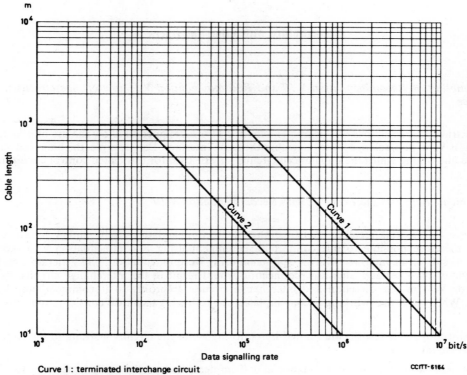

Curve 1 : terminated interchange circuit
Curve 2 : unterminated interchange circuit

FIGURE 8/V.11 – Data signalling rate vs cable length for balanced interchange circuit

Experience has shown that in many practical cases the operating distance at lower signalling rates may extend to several kilometres.

For synchronous transmission where the data and signal element timing are transmitted in opposite directions, the phase relationship between the two may need to be adjusted to ensure conformity with the relevant requirements of signal quality at the interchange point.

3.      *Cable termination*

The use of a cable termination is optional and dependent on the specific application. At the higher signalling rates (above 200 kbit/s) or at any signalling rate where the cable propagation delay is of the order of half the signal element duration a termination should be used to preserve the signal rise time and minimize reflections. The terminating impedance should match as closely as possible the cable characteristic impedance in the signal spectrum.

Generally, a resistance in the range of 100 to 150 ohms will be satisfactory, the higher values leading to lower power dissipation.

At the lower signalling rates, where distortion and rise-time are not critical, it may be desirable to omit the termination in order to minimize power dissipation in the generator.

ANNEX 2

(to Recommendation V.11)

**Compatibility with other interfaces**

1.      *Compatibility of Recommendation V.10 and Recommendation V.11 interchange circuits in the same interface*

The electrical characteristics of Recommendation V.11 are designed to allow the use of unbalanced (see Recommendation V.10) and balanced circuits within the same interface. For example, the balanced circuits may be used for data and timing whilst the unbalanced circuits may be used for associated control circuit functions.

2.      *Recommendation V.11 interworking with Recommendation V.10*

The basic differential receiver specifications of Recommendations V.10 and V.11 are electrically identical. It is therefore possible to interconnect an equipment using Recommendation V.10 receivers and generators on one side of the interface with an equipment using Recommendation V.11 generators and receivers on the other side of the interface. Such interconnection would result in the interchange circuits according to Recommendation V.11 in one direction and interchange circuits according to Recommendation V.10 in the other direction. Where such interworking is contemplated, the following technical considerations must be taken into account:

2.1     Interconnecting cable lengths are limited by performance of the circuits working to the Recommendation V.10 side of the interface.

2.2     The optional cable termination resistance $(Z_t)$, if implemented, in the equipment using Recommendation V.11 must be removed.

3.      *Recommendation V.11 interworking with Recommendation V.35*

Equipment having interchange circuits according to Recommendation V.11 is not intended for interworking with equipment having interchange circuits according to electrical characteristics of Recommendation V.35.

## ANNEX 3

### (to Recommendation V.11)

### Multipoint operation

It is considered that further study is required before parameters for this application can be precisely defined and this Annex, giving provisional figures, is intended as a guideline for this study.

1.  *General*

The point-to-point interchange circuit arrangement of one generator and one load might be expanded to a multipoint arrangement by adding generators, receivers or both, at interchange points along the interconnecting cable, as shown in Figures 9/V.11 and 10/V.11.

Only one generator at a given time would present its differential voltage at its interchange point. All other generators would be isolated by an appropriate control, and assume the high impedance state defined below. All receivers would be continuously in an operating condition.

In multipoint configuration, one or more terminators might be required at the cable end or at interchange points, depending on the application. The combined load impedance presented to any active generator by other generators, receivers, cable and terminators, must not be less than 100 ohms.

The operation of a multipoint arrangement must not be perturbed by any of its components when they are either in a high impedance state or a power-off state [6]. The generators and receivers must tolerate without damage the transmitted signals with their maximum amplitude within the specified limits.

Generators on the same multipoint line must have the same nominal d.c. offset voltage in order to operate correctly. However, generators with different d.c. offsets could be used on the same line provided that these differences be compensated at the common reference point.

2.  *Configurations*

Several topological arrangements are to be considered:

—  cluster of circuits at the end of a line;

—  multidrop line;

—  star configuration.

Figure 9/V.11 illustrates a cluster configuration. Each line should be correctly terminated at the receiving end in order to avoid reflections and leads from that point to the receivers kept as short as possible.

Figure 10/V.11 illustrates a multidrop line. The presence of several generators along the line imposes the need to terminate the line correctly at both ends to avoid reflections. The length of the tapping connections along the line must be short enough to avoid mismatching of the main line. A length corresponding to a propagation delay of 1% of the unit interval of the signals seems to be an acceptable limit for the length of the tapping connection.

This tolerable propagation delay and other configurations are to be studied further.

The guidelines on cable length given in Figure 8/V.11 of Annex 1, assuming the same environmental conditions, apply equally to multipoint arrangements. Above 1 or 2 Mbit/s, however, the environmental conditions may be more difficult to control.

---

[6] In the power-off state of any device it is assumed that the supply collapses to zero and is replaced by a very low impedance or short circuit.

### 3. High impedance state

#### 3.1 Static measurements

When in the high impedance state and with test loads of 50 ohms connected between each generator output point and point C, the magnitude of the voltage $V_h$ measured between points A and B shall not exceed 4 mV whatever the logical condition of the generator input data lead (Figure 11/V.11).

When the generator is in the high impedance state, with voltages ranging between $-6$ V and $+6$ V applied between each output point and point C, as indicated in Figure 12/V.11, the magnitude of the output leakage currents $I_{xa}$ and $I_{xb}$ shall not exceed 150 µA.

The same condition applies under power-off condition.

#### 3.2 Dynamic measurements

During transitions of the generator output between the low impedance state and the high impedance state, the differential signal measured across a 100 ohm test load connected between the generator points A and B shall be such that the change in amplitude goes from 10% to 90% of the steady state voltage in less than 10µs.

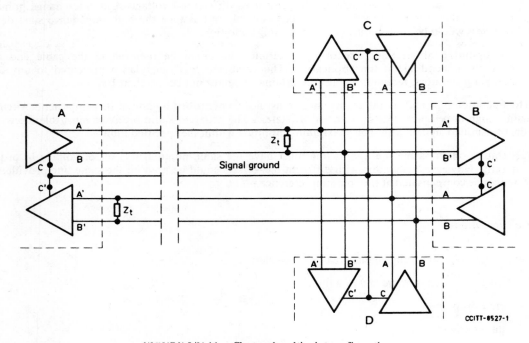

FIGURE 9/V.11 – Clustered multipoint configuration

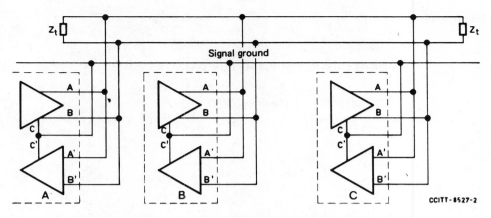

FIGURE 10/V.11 – **Distributed multipoint configuration**

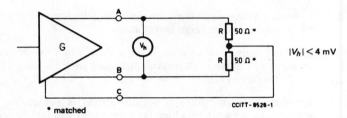

FIGURE 11/V.11 – **High impedance state static measurement**

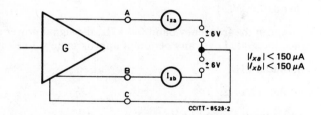

FIGURE 12/V.11 – **Generator output leakage current measurement**

**Recommendation V.15**

## USE OF ACOUSTIC COUPLING FOR DATA TRANSMISSION

*(Geneva, 1972)*

The CCITT,

*considering*

that there is a wide variety of telephone instruments in existence and that the acoustic path involved in the use of any coupling device cannot be accurately prescribed for all cases, and hence it will be difficult to ensure satisfactory transmission in all situations,

*recommends*

that acoustic coupling of data transmission equipment via telephone instruments to the telephone transmission network should not be used for permanent installations.

It is, however, recognized that there may be a need for a means to provide temporary connection of portable data transmission equipment to the network in circumstances where it may not be possible to obtain convenient access to the subscriber's line terminals.

The use of acoustic coupling for temporary communications is subject to the agreement of the Administration in charge of the telephone network to which the equipment will be connected.

If an Administration decides to permit acoustic coupling for temporary data transmission stations, the acoustic coupling equipment conform to the following:

1.     The maximum power output of the subscriber's equipment into the line shall not exceed 1 mW at any frequency.

The mean permitted telephone line signal power shall not exceed $-13$ dBm0 for duplex operation and $-10$ dBm0 for simplex operation when integrated over any period of approximately 3 seconds. [See b) and c) of the introduction of Recommendation V.2.]

2.     If $p$ is the signal power in the frequency band 0-4 kHz, the signal power outside this band shall not exceed the following values when integrated over any period of approximately 3 seconds:

$p$ — 20 dB in the band 4 to 8 kHz,

$p$ — 40 dB in the band 8 to 12 kHz,

$p$ — 60 dB in each 4- kHz band above 12 kHz.

3.     The frequencies emitted by the transducer shall be such as not to interfere with national and international telephone signalling systems and pilot signals involved in the telephone connection envisaged.

4.     Adequate protection shall be provided in the transducer to avoid causing any dangerous electric potential and currents to the telephone system.

5.     It shall not be possible to cause acoustic shock to telephone users under any normal condition or when the acoustic coupler develops any single fault.

6.     The mechanical arrangements of the transducer shall not cause mechanical damage to the telephone instrument.

7.     In addition to the contents of this Recommendation, the regulations of the national Administration must also be complied with.

# MODEMS FOR PARALLEL DATA TRANSMISSION
## USING TELEPHONE SIGNALLING FREQUENCIES

*(Geneva, 1976)*

Systems for parallel data transmission can be used economically when the transmitting sets (outstations) use the signalling frequencies of push-button telephone sets to transmit data to a central receiving set (instation) via the switched telephone network.

## 1.    *Scope*

In many networks, the introduction of keyboard telephone sets allows simple, one-way data transmission at speeds up to about 10 characters per second to be made from a large number of push-button telephone sets serving as outstations to a common instation, via the general switched telephone network. Transmissions in the instation-to-outstation direction are generally confined to simple acoustic signals and voice replies.

The CCITT therefore *unanimously recommends*

that the modems to be used for stations operating in the general switched telephone network should meet the specifications shown below.

## 2.    *General characteristics*

### 2.1    *Data channel*

The transmission system uses two sets of frequencies in accordance with Recommendation Q.23. Each character is transmitted in the form of two simultaneously transmitted frequencies. These two frequencies belong to two separate sub-assemblies. Each of these two assemblies consists of four frequencies ["2 (1/4)" code]. This coding can thus be used to transmit 16 different character combinations and perhaps more (see Note).

The actual transmission consists in sending a frequency pair for a time greater than 30 ms, followed by a silence period of not less than 25 ms.

*Note.* — In order to stretch the set of characters, several frequency pairs may be transmitted before the silence period. It should be noted that in this case character coding and decoding will not be effected by the DCE but by the DTE.

61

## 2.2 *Backward channel*

The following possibilities might be considered:

*a)* a telephone channel not simultaneous with data transmission in the forward direction;

*b)* a backward channel for audible signalling;

*c)* a backward channel for electrical signalling.

Possibilities *b)* and *c)* are provided on a basis of non-simultaneity or, optionally, simultaneity with the data transmission channels in the forward direction.

A loudspeaker will be built into the outstation modem. Optionally, a continuous signalling output may be provided. If the national regulations permit, an output for response to the channel may also be provided as an option.

## 3. *Frequency assignments*

## 3.1 *Data transmission channel*

The 2 groups of 4 frequencies specified in Recommendation Q.23 are defined as follows:

— low group frequencies: 697, 770, 852, 941 Hz;

— high group frequencies: 1209, 1336, 1477, 1633 Hz.

The frequency pairs are assigned to the different digits as shown in Table 1/V.19.

TABLE 1/V.19

|  | $B_1 = 1209$ Hz | $B_2 = 1336$ Hz | $B_3 = 1477$ Hz | $B_4 = 1633$ Hz |
|---|---|---|---|---|
| $A_1 = 697$ Hz | 1 | 2 | 3 | A |
| $A_2 = 770$ Hz | 4 | 5 | 6 | B |
| $A_3 = 852$ Hz | 7 | 8 | 9 | C |
| $A_4 = 941$ Hz | * | 0 | # | D |

## 3.2 *Backward channel*

For audible signals and electrical signalling, the backward channel frequency will be 420 Hz. This frequency may be amplitude-modulated at a rate of up to 5 bauds.

Use may also be made of an FM backward channel similar to that of the Recommendation V.23 type modem, or of the No. 2 transmission channel of a Recommendation V.21 type modem (if the frequency 1633 Hz is not used). These two types of backward channel may be used at the same time as the data frequencies in the forward direction; the use of these backward channels is optional.

## 4. *Tolerances*

## 4.1 *Data frequency tolerances*

The data frequency tolerances are defined in Recommendation Q.23; the difference between each frequency and its nominal frequency must not exceed $\pm 1.8\%$ of the nominal frequency. Apart from this tolerance of $\pm 1.8\%$ on transmission, the instation receiver should be able to accept a difference of $\pm 6$ Hz due to the carrier systems.

### 4.2　Frequency tolerance on backward channel

The tolerance of 420 Hz on the backward channel should be $\pm$ 4 Hz; the receiver of the outstation should also be able to accept a difference of $\pm$ 6 Hz due to the carrier systems.

### 5.　Line power levels

On the basis of Recommendation V.2, the following maximum power levels are recommended for each frequency transmitted, measured at the relative zero point:

　　— 13 dBm0 for the data transmission channel without the simultaneous backward channel;

　　— 16 dBm0 for the data transmission channel with the simultaneous backward channel;

　　— 10 dBm0 for the non-simultaneous backward channel;

　　— 16 dBm0 for the simultaneous backward channel.

### 6.　Power levels on reception

In view of the provision of Recommendation V.2 and the statistical values of the maximum transmission loss between subscribers, it is recommended that the instation receiver should be able to detect frequency pairs received at $-45$ dBm.

*Note.* — Studies should be continued with a view to permitting levels on reception below $-45$ dBm.

### 7.　Character reception

A character will be detected and delivered to the DTE interface if, and only if, the two frequencies corresponding to the character are detected and are stable for at least 10 ms.

The silent period will be detected and delivered to the DTE interface if no frequency belonging to the code appears for at least 10 ms.

*Note.* — During silent periods, the microphone of the telephone set is connected to the telephone line, so that interfering signals (ambient noise, speech) may be received. The receiver must be fitted with devices capable of distinguishing between these interfering signals and data signals (speech protection). It would be advisable to study further the method of assessing receiver response to the simulation of data signals by interfering signals. A reproducible test signal should be defined, so that comparable measurements can be made.·

### 8.　Detection of line signal received on the data channel

Circuit 109 must be in the ON position when a character is received; the circuit may be switched from ON to OFF:

　　1)　on detecion of the silent period;

　　2)　after a time-out of 60 $\pm$ 10 ms following detection of the silent period.

### 9.　Timing for characters received

By its very principle, the system is asynchronous; however, it may be useful to provide the DTE, on an optional basis, with a signal which indicates the sampling times of the data wires. In this case, it is advisable to use circuit 131, which will switch from OFF to ON when the character reaches the interface, and then back to OFF after a time $T$. This time will be chosen in such a way that the data are stable at the DTE interface.

The value $T = 15$ ms may be recommended by way of example.

This clock may optionally be disabled on reception of a silent period.

10.   *Interface of instation modem*

The functional characteristics of the interchange circuits concerned are as defined in Recommendation V.24 (see Note 1).

*List of interchange circuits concerned*

102       Signal ground or common return

104       Received data (8 circuits. These circuits are designated $A_1$, $A_2 \ldots B_4$ according to their correspondence with the relevant frequency in Table 1/V.19 (see Note 2 below)

105       Request to send (see Note 3 below)

107       Data set ready

108/1     Connect data set to line (see Note 4 below)

108/2     Data terminal ready (see Note 4 below)

109       Data channel received line signal detector

125       Calling indicator

130       Transmit backward tone

191       Transmitted voice answer (see Note 3 below)

The following interchange circuits are optional:

110       Data signal quality detector

131       Received character timing

*Note 1.* — Manufacturers who marketed a modem of this type prior to the publication of this Recommendation may regard the interface defined in this paragraph as optional.

*Note 2.* — To make the interface compatible with the relevant specifications of Recommendation V.30, the combination $A_4$, $B_4$ may be transmitted on circuit 104 instead of a pause ("1" on all circuits), provided circuit 107 is in the ON position and circuit 105 is in the OFF position. This simulated idle combination is optional.

*Note 3.* — These circuits are required if the "telephone channel" facility is provided in the modem. The electrical characteristics of interchange circuit 191 are still under study.

*Note 4.* — Circuit 108 must be available either as circuit 108/1 — *Connect data set to line*, or as circuit 108/2 — *Data terminal ready.* For automatic calling, this circuit is used exclusively as circuit 108/2.

The electrical characteristics of these interchange circuits are as defined in Recommendation V.28, using the 25-pin connector and the pin arrangement defined in ISO DIS 2110-2, column D. Data circuits: when the frequency corresponding to a circuit is emitted, the corresponding interchange circuit will be negative. When the frequency is interrupted, the corresponding interchange circuit will be positive.

11.   *Interface of outstation modems*

In view of the purpose of these modems, which are or will be more or less integrated in economic terminals, the specification of the interface is liable to result in a much higher equipment cost. Hence no interface is recommended.

**Recommendation V.20**

## PARALLEL DATA TRANSMISSION MODEMS STANDARDIZED FOR UNIVERSAL USE IN THE GENERAL SWITCHED TELEPHONE NETWORK

*(former Recommendation V.30, Mar del Plata, 1968 amended at Geneva, 1972)*

There is a need for one-way data transmission systems where a large number of low-cost sending stations (outstations) transmit to a central receiving station (instation) over the switched telephone network.

The following systems are desired:

*a)* transmitting 16-character combinations;

*b)* transmitting 64-character combinations;

*c)* transmitting 256-character combinations.

In most cases a character signalling rate of 20 characters per second will be sufficient; 40 characters per second may be required for some applications of the 16 character combination system.

The transmission from the instation to the outstations is limited either to simple acknowledgement signals (data collection systems) or to analogue signals (voice-answering systems).

The use of normal push-button telephone sets in the outstation for some of these applications may be of advantage for the user. However, it is recognized that for the time being on some telephone systems there exist certain limitations in the frequency band 600 to 900 Hz. This is due to the characteristics of the telecommunication path, such as signalling frequencies and metering pulses. Therefore, for a universal system the frequency band of the data channel is 900 to 2000 Hz, which excludes the use of the normal push-button telephone set.

A so-called parallel data-transmission system using two or three times one out of four frequencies can fulfil the above requirements.

For these reasons, the CCITT *unanimously declares the following view:*

1. Parallel data-transmission systems can be used economically when a large number of low-cost sending stations (outstations) wish to transmit to a central receiving station (instation) over the switched telephone network (or on leased telephone circuits).

Apart from the possibility of the use, on a restricted scale, of a system that is compatible with multifrequency push-button telephone signalling devices, the following system is recommended as a universally applicable system for the switched telephone circuits.

2. *Facilities*

2.1 *Data channel*

The basic system has a maximum of 16-character combinations and a modulation rate of up to 40 bauds. This permits a character signalling rate of up to 20 characters per second when an inter-character rest condition is used, or up to 40 characters per second with the use of a binary timing channel. This basic system consists of two groups of four frequencies, one frequency from each group being transmitted simultaneously (two times one out of four).

The basic system includes provision for expansion up to 64-character combinations by the addition of a third four-frequency group (three times one out of four). No use is foreseen for the system with 64-character combinations at character signalling rates above 20 characters per second, within this class of inexpensive parallel transmission equipment.

An expansion of the basic system to cater for 256 characters (up to 20 characters per second) is achieved by using only two groups for the conveyance of data, each character being transmitted in two sequential parts. The two half characters are positively identified by the two different conditions of a binary channel. The timing channel mentioned above is recommended to be used for this purpose.

Where an inter-character rest condition is required the full number of frequency combinations in the modem will not be available to the user as character combinations:

a)   with the 16-frequency combination system, only 15 characters will be available unless a timing channel is used from frequency group B;

b)   with the 64-frequency combination system only 63 characters are available.

These recommended systems have an inherent transmission error-detecting capability.

2.2   *Backward channel*

Provision is made for the following facilities:

a)   a speech channel non-simultaneous with forward data;

b)   a backward channel for audible signalling;

c)   a backward channel for electrical signalling purposes.

Facilities b) and c) are provided, either non-simultaneous or optionally simultaneous with the forward data channels.

A loudspeaker will be provided in the outstation modem. On an optional basis a d.c. signalling output will be provided. If national regulations permit, a voice-answering output will also be provided on an optional basis.

3.   *Frequency allocations*

3.1   *Data channels*

Frequency allocations and designations as shown in Table 1/V.20 are recommended.

TABLE 1/V.20

| Group \ Channel No. | 1 | 2 | 3 | 4 |
|---|---|---|---|---|
| A | 920 Hz | 1000 Hz | 1080 Hz | 1160 Hz |
| B | 1320 Hz | 1400 Hz | 1480 Hz | 1560 Hz |
| C | 1720 Hz | 1800 Hz | 1880 Hz | 1960 Hz |

For the basic 16-character system only groups A and C are used.

If an inter-character rest condition is used, during the time no input data circuits are operated, rest frequencies are sent to line. The highest frequency in each group is recommended to be the rest frequency.

### 3.2  *Timing channel*

If a timing channel is provided in the 16-character system this should consist of a selected pair of group B frequencies. The recommended frequencies are $F_{B2} = 1400$ Hz and $F_{B3} = 1480$ Hz.

In the case where this timing channel will be used to identify the two halves of the character in the 256-character system, the higher frequency is transmitted simultaneously with the first half of the character.

No timing channel is provided in the 64-character combination system.

### 3.3  *Backward channel*

The frequency of the backward channel for audible and electrical signalling shall be 420 Hz. This tone may be amplitude modulated at rates up to, say, 5 bauds.

A frequency modulated backward channel which is similar to that of a Recommendation V.23 modem can also be used simultaneously with the forward data frequencies, use of this channel being optional.

### 3.4  *Tolerances*

The tolerances on both data and backward frequencies should be ± 4 Hz.

The receiver should cater for ± 6 Hz difference due to carrier systems in addition to the transmitter tolerance of ± 4 Hz.

### 4.  *Power levels*

Based on Recommendation V.2 the following maximum power levels measured at the zero relative level point are recommended for each transmitted frequency:

### 4.1  *Data and timing channels*

4.1.1  16-character system without timing channel and with a non-simultaneous backward channel: −13 dBm0.

4.1.2  All other cases: −16 dBm0.

### 4.2  *Backward channel*

4.2.1  Non-simultaneous: −10 dBm0.

4.2.2  Simultaneous: −16 dBm0.

In systems where either the simultaneous or the non-simultaneous backward channel is used, all power levels should be −16 dBm0.

The maximum difference between any data tone at the transmitter terminal should be 1 dB.

### 5.  *Threshold levels of the data channel received signal detector*

When the level of the received signal in group C exceeds −49 dBm, circuit 109 shall be ON. When the level of this received signal is less than −54 dBm, circuit 109 shall be OFF. The detector circuit which causes circuit 109 to turn ON or OFF shall exhibit hysteresis action such that the level at which the OFF to ON transition occurs shall be at least 2 dB greater than that for the ON to OFF transition.

Group C was chosen for this purpose because it is the most critical from a received level point of view.

6. *Minimum level of received signal on the backward channel*

The expected minimum level is −45 dBm for the 420-Hz tone. This information is provided to assist equipment manufacturers.

7. *Instation modem interface*

The functional characteristics of interchange circuits comply with Recommendation V.24.

7.1 *List of essential interchange circuits:*

| | |
|---|---|
| 102 | Signal ground or common return |
| 104 | Received data [12 or 8 circuits depending on whether Group B is provided or not. These received data circuits are designated A1, A2 ... C4, each corresponding to its relevant frequency (see Table 1/V.20)] |
| 105 | Request to send (see Note 2) |
| 107 | Data set ready |
| 108/1 | Connect data set to line |
| 108/2 | Data terminal ready } (see Note 1) |
| 109 | Data channel received line signal detector |
| 125 | Calling indicator |
| 130 | Transmit backward tone |
| 191 | Transmitted voice answer (see Note 2) |

The following optional interchange circuits may be provided:

| | |
|---|---|
| 110 | Data signal quality detector |
| 124 | Select frequency groups |
| 131 | Received character timing. |

*Note 1.* − This circuit shall be capable of use as circuit 108/1 − *Connect data set to line* or circuit 108/2 − *Data terminal ready*, depending upon its use. For automatic calling it shall be used as 108/2 only.

*Note 2.* − These circuits are required if the speech channel facility is provided in the modem. The electrical characteristics of interchange circuits 191 and 192 are left for further study.

7.2 The electrical characteristics of the interchange circuits comply with Recommendation V.28.

Data circuits: when the frequency corresponding to the circuit is ON, the appropriate interchange circuit will be negative. When the frequency in this channel is OFF, the interchange circuit will be positive.

For timing purposes in the 256-character system, a single interchange circuit is selected from Group B so that positive polarity indicates the first half of the character period and a negative polarity indicates the second half of the character.

8. *Outstation modem interface*

The functional characteristics of interchange circuits comply with Recommendation V.24.

8.1 *List of essential interchange circuits:*

| | |
|---|---|
| 102 | Signal ground or common return (see Note 2) |
| 103 | Transmitted data (nine or six circuits depending on whether Group B is provided or not). |
| | These circuits are designated A1, A2 ... C3, each corresponding to its relevant frequency (see Table 1/V.20) |
| 105 | Request to send |
| 129 | Request to receive |

8.2    The following optional interchange circuits may be provided:

    107      Data set ready

    119      Received backward channel data

    192      Received voice answer (see Note 1)

When the optional timing channel is used then the appropriate data circuits are operated.

*Note 1.* — See 7.1, Note 2 above.

*Note 2.* — The transmitted data circuits (103) will all use the same common return (102). The control circuits may operate each on their own return circuit.

8.3    *Electrical characteristics*

The data and control interchange circuits at the outstation will be operated by the opening or closing of contacts carrying only direct current. The electrical characteristics of interchange circuits comply with Recommendation V.31.

9.    *Correspondence for each group* (Table 2/V.20)

TABLE 2/V.20

| At outstation closing of circuit | Number of the channel on line | At instation negative polarity on circuit |
|---|---|---|
| 1 | 1 | 1 |
| 2 | 2 | 2 |
| 3 | 3 | 3 |
| None | 4 | 4 |

Not more than one circuit per group may be closed at a time.

10.    *Character set*

This Recommendation includes the allocation of transmission frequencies to the interchange circuits.

The allocation of interchange circuits to the code combinations to be transmitted, i.e. definition of a character set, must conform to the conditions defined in this Recommendation and must take into account the application requirements and the type of input media (paper tape, punched cards, keyboards, etc.).

For this reason the recommendation for a character set is primarily for ISO in collaboration with CCITT.

*Note.* — Examples of alphabets and coding methods are given in Supplements Nos. 20 and 21 to the *White Book*, Volume VIII, and in Supplements Nos. 56 and 57 to the *Blue Book*, Volume VIII.

**Recommendation V.21**

### 200-BAUD MODEM STANDARDIZED FOR USE IN THE GENERAL SWITCHED TELEPHONE NETWORK [7]

*(Geneva, 1964, amended at Mar del Plata, 1968, and at Geneva, 1972 and 1976)*

*Note.* — The modem, designed for use on connections set up by switching in the general public network, can obviously be used on leased lines.

A system of data transmission at a low modulation rate, such that data could be transmitted over a telephone circuit operated alternatively for telephone calls and data transmissions, using simple input/output equipment and easy operating procedures, would be economical.

The modulation rate must be such as to allow the use of current types of data sources and sinks, especially electromechanical devices.

The system for data transmission will be duplex, either for simultaneous two-way data transmission or for the transmission of signals sent in the backward direction for error-control purposes. The transmission must be such that use can be made of normal telephone circuits, and this applies both to the bandwidth available and to the restrictions imposed by signalling in the telephone networks.

The two correspondents are brought into contact by a telephone call, and the circuit is put into the data-transmission position:

*a)* manually by agreement between the operators, or

*b)* automatically.

For these reasons, the CCITT *unanimously declares the following view:*

1. Data transmission may take place at low modulation rates on telephone calls set up on switched telephone circuits (or on leased telephone circuits).

2. The communication circuit for data transmission is a duplex circuit whereby data transmission in both directions simultaneously is possible at 200 bauds or less.

The modulation is a binary modulation obtained by frequency shift.

*Note 1.* — Modems to this Recommendation may operate also at modulation rates of up to 300 bauds. For the moment, however, reliable 300 bauds transmission cannot be guaranteed in all cases. It may, therefore, be necessary to carry out tests in order to verify whether operation at rates up to 300 bauds is possible.

*Note 2.* — In view of the constraints mentioned in Note 1, future modem designs should take account of the need to achieve satisfactory operation at all modulation rates up to and including 300 bauds.

3. For channel No. 1, the nominal mean frequency is 1080 Hz.

For channel No. 2, it is 1750 Hz.

The frequency deviation is ± 100 Hz. In each channel, the higher characteristic frequency $(F_A)$ corresponds to the symbol 0.

---

[7] See Notes 1 and 2 under 2. of this Recommendation.

The characteristic frequencies [8] as measured at the modulator output must not differ by more than ± 6 Hz from the nominal figures.

A maximum drift frequency of ± 6 Hz is assumed for the line. Hence the demodulation equipment must tolerate drifts of ± 12 Hz between the frequencies received and their nominal values.

4.      Data may be transmitted by synchronous or asynchronous procedures. With synchronous transmission, the modem will not have to provide the signals which would be necessary to maintain synchronism when transmission is not proceeding.

5.      It will be for the user to decide whether, in view of the connections he makes with this system, he will have to request that the data circuit-terminating equipment be equipped with facilities for disabling echo suppressors. The international characteristics of the echo suppressor tone disabler have been standardized by the CCITT (Recommendation G.161, C) and the disabling tone should have the following characteristics:

—      disabling tone transmitted: 2100 ± 15 Hz at a level of − 12 ± 6 dBm0;

—      the disabling tone to last at least 400 ms, the tone disabler should hold in the disabled mode for any single-frequency sinusoid in the band from 390-700 Hz having a level of − 27 dBm0 or greater, and from 700-3000 Hz having a level of − 31 dBm0 or greater. The tone disabler should release for any signal in the band from 200-3400 Hz having a level of − 36 dBm0 or less;

—      the tolerable interruptions by the data signal to last not more than 100 ms.

6.      The maximum power output of the subscriber's equipment into the line shall not exceed 1 mW at any frequency.

The power level of the subscriber's equipment should be adjusted to make allowance for loss between his equipment and the point of entry to an international circuit, so that the corresponding nominal level of the signal at the international circuit input shall not exceed − 13 dBm0. (See Recommendation V.2, B.)

7.      a)      When both channels are used for simultaneous both-way data transmission, channel No. 1 is used for transmission of the caller's data (i.e. the person making the telephone call) towards the called station, while channel No. 2 is used for transmission in the other direction.

b)      When one channel is used for data transmission and the other is used for transmission of check signals, service signals, etc., only, it is channel No. 1 which is used for transmission from the calling to the called station regardless of the direction in which the data are transmitted.

c)      The procedure for the assignment of the channels described under a) and b) above applies in the case of the general service of data transmission, making it possible to transmit data or check signal, service signal, etc., bilaterally between any two subscribers. In special cases which do not come under this rule, the procedure of assignment of the channels is determined by the prior agreement between the correspondents, bearing in mind the requirement proper to each service.

8.      *Interchange circuits*

a)      *List of interchange circuits essential for the modems when used on the general switched telephone network or non-switched leased telephone circuits* (see Table 1/V.21)

The configurations of interchange circuits are those essential for the particular switched network or leased circuit requirement indicated. Where one or more of such requirements are provided in a modem, then all of the appropriate interchange circuit facilities should be provided.

---

[8] The nominal characteristic frequencies:
Channel No. 1 ($F_A$ = 1180 Hz and $F_z$ = 980 Hz);
Channel No. 2 ($F_A$ = 1850 Hz and $F_z$ = 1650 Hz).

| | | General switched telephone network including terminals equipped for manual calling, manual answering, automatic calling automatic answering (Note 1) | Non-switched leased telephone circuits (Note 1) | |
|---|---|---|---|---|
| | Interchange circuit | | | |
| Number | Designation | | point-to-point | Multipoint |
| 102 | Signal ground or common return | X | X | X |
| 102a (Note 6) | DTE common return | X | X | X |
| 102b | DCE common return | X | X | X |
| 103 | Transmitted data | X | X | X |
| 104 | Received data | X | X | X |
| 105 | Request to end | | X (Note 2) | X |
| 106 | Ready for sending | X | X | X |
| 107 | Data set ready | X | X | X |
| 108/1 | Connect data set to line | X (Note 3) | X | X |
| 108/2 | Data terminal ready | X (Note 3) | X (Note 4) | |
| 109 | Data channel received line signal detector | X | X | X |
| 125 | Calling indicator | X | | |
| 126 | Select transmit frequency | | | X (Note 5) |

*Note 1.* – Interchange circuits indicated by X must be properly terminated according to Recommendation V.24 in the data terminal equipment and data circuit-terminating equipment.

*Note 2.* – Circuit 105 is not required when alternate voice/data service is used on non-switched leased point-to-point circuits.

*Note 3.* – This circuit shall be capable of operation as circuit 108/1 – *Connect data set to line* or circuit 108/2 – *Data terminal ready* depending on its use. For automatic calling it shall be used as 108/2 only.

*Note 4.* – In the leased point-to-point case, where alternate voice/data service is to be provided, circuit 108/2 may be used optionally.

*Note 5.* – Circuit 126 controls the functions of circuits 126 and 127 as defined in Recommendation V.24.

**Note 6.— Interchange circuits 102a and 102b are required where the electrical characteristics defined in Recommendation V.10 are used.**

b)        *Response times of circuits 106 and 109*

*Definitions*

i)        Circuit 109 response times are the times that elapse between the connection or removal of a tone to or from the modem receive line terminals and the appearance of the corresponding ON or OFF condition on circuit 109.

The test tone should have a frequency corresponding to the characteristic frequency of binary 1 and be derived from a source with an impedance equal to the nominal input impedance of the modem under test:

The level of the test tone should fall into the level range between 1 dB above the actual threshold of the received line signal detector and the maximum admissible level of the received signal. At all levels within this range the measured response times shall be within the specified limits.

ii)     Circuit 106 response times are the times from the connection of an ON or OFF condition on:

—   circuit 105 (where it is provided) to the appearance of the corresponding OFF or ON condition on circuit 106;

—   circuit 109 (where circuit 105 is not provided) to the appearance of the corresponding ON or OFF condition on circuit 106.

c)      *Response times*

| | | |
|---|---|---|
| *Circuit 106* | | |
| OFF to ON | 20-50 ms (see Note 1) | 400-1000 ms (see Note 2) |
| ON to OFF | | ⩽ 2 ms |
| *Circuit 109* | | |
| OFF to ON | ⩽ 20 ms (see Note 1) | 300-700 ms (see Note 2) |
| ON to OFF | 20-80 ms (see Note 1) | 20-80 ms (see Note 2) |

*Note 1.* – These times are used on leased point-to-point networks without alternate voice-data facilities and on leased multipoint facilities.

*Note 2.* – These times are used on general switch network service and on leased point-to-point circuits with alternate voice-data.

d)      *Threshold of data channel received line signal detector*

Level of received line signal at received line signal terminals of modem for all types of connection, i.e. general switched telephone network or non-switched leased telephone circuit:

greater than $-43$ dBm          circuit 109 ON

less than        $-48$ dBm          circuit 109 OFF

The condition of circuit 109 for levels between $-43$ dBm and $-48$ dBm is not specified except that the signal detector shall exhibit a hysteresis action such that the level at which the OFF to ON transition occurs shall be at least 2 dB greater than for the ON to OFF transition.

Where transmission conditions are known on switched or leased circuits, Administrations should be permitted at the time of modem installation to change these response levels of the received line signal detector to less sensitive values (e.g. $-33$ dBm and $-38$ dBm respectively).

e)      *Clamping to binary 1 condition of circuit 104*

Two options shall be provided in the modem:

i)      When clamping is not used there is no inhibition of the signals on circuit 104. There is no protection against noise, supervisory and control tones, switching transients, etc., appearing on circuit 104.

ii)     When clamping is used, circuit 104 is held in a marking condition (binary 1) when circuit 109 is in the OFF condition. When circuit 109 is ON the clamp is removed and circuit 104 can respond to the input signals of the modem.

## 9. *Electrical characteristics of interchange circuits*

a) Use of electrical characteristics conforming to Recommendation V.28 is recommended with the connector pin assignment plan specified by ISO DIS 2110.

b) Application of electrical characteristics conforming to Recommendation V.10 and V.11 is recognized as an alternative together with the use of the connector and pin assignment plan specified by ISO DIS 4902.

    i) Concerning circuits 103, 104, 105 (where used), 106, 107, 108, and 109, the receivers shall be in accordance with Recommendation V.11 or, alternatively, Recommendation V.10, Category 1. Either V.10 or V.11 generators may be utilized.

    ii) Where circuits 125 and/or 126 are used, Recommendation V.10 applies with receivers configured as specified by Recommendation V.10 for Category 2.

    iii) Interworking between equipment applying Recommendation V.10 and/or V.11 and equipment applying Recommendation V.28 is allowed on a noninterference basis. The onus for adaptation to V.28 equipment rests solely with the alternative V.10/V.11 equipment.

*Note.*—Manufacturers may wish to note that the long-term objective is to replace electrical characteristics specified in Recommendation V.28, and Study Group XVII has agreed that the work shall proceed to develop a more efficient all balanced interface for the V-Series application which minimizes the number of interchange circuits. It is expected that this work would be based upon the alternative application given in 9 b) above utilizing the V.11 electrical characteristics.

10.  The following information is provided to assist equipment manufacturers:

    a) The nominal range of attenuations in subscriber-to-subscriber connections is from 5 to 30 dB at the reference frequency (800 or 1000 Hz), assuming up to 35 dB attenuation at the frequency 1750 Hz.

    b) The data modem should have no adjustment for send level or receive sensitivity under the control of the operator.

11. In case of interruption of a leased circuit, the use of a non-standardized modem over the switched connection established as a substitute for the leased circuit is not recommended.

**600/1200-BAUD MODEM STANDARDIZED
FOR USE IN THE GENERAL SWITCHED TELEPHONE NETWORK**

*(Geneva, 1964, amended at Mar del Plata, 1968, and at Geneva, 1972 and 1976)*

*Note.* — The modem, designed for use on connections set up by switching in the general public network, can obviously be used on leased lines.

1. The principal characteristics recommended for a modem to transmit data at medium speed in the general switched telephone network are as follows:

— use of modulation rates up to 600/1200 bauds on the communication channel (see Recommendation V.5);

— frequency modulation with synchronous or asynchronous mode of operation;

— inclusion of a backward channel at modulation rates up to 75 bauds for error control, use of this channel being optional.

2. *Modulation rates and characteristic frequencies for the forward data-transmission channel*

|  | $F_0$ | $F_Z$ (symbol 1, mark) | $F_A$ symbol 0, space) |
|---|---|---|---|
| Mode 1: up to 600 bauds | 1500 Hz | 1300 Hz | 1700 Hz |
| Mode 2: up to 1200 bauds | 1700 Hz | 1300 Hz | 2100 Hz |

It is understood that the modem would be used in mode 1 when the presence of long loaded cables and/or the presence on some connections of signalling receivers operating close to 2000 Hz would prevent satisfactory transmission in mode 2. The modem could be used in mode 2 on suitable connections.

3. *Tolerances on the characteristic frequencies for the forward channel*

It should be possible with all rates of modulation to permit a tolerance, at the transmitter, of $\pm$ 10 Hz on both the $F_A$ and $F_Z$ frequencies. This tolerance should be considered as a limit.

Acceptance of these tolerances would give a tolerance of $\pm$ 10 Hz for the mean-frequency $F_0 = F_A + F_Z/2$.

The tolerance on the frequency difference $F_A - F_Z$ with regard to the nominal value would be $\pm$ 20 Hz.

A maximum frequency drift of $\pm$ 6 Hz has been assumed in the connection between the modems which might consist of several carrier circuits connected in tandem. This would make the tolerances on the mark and space frequencies at the receiving modem $\pm$ 16 Hz.

4.     *Modulation rate and characteristic frequencies for the backward channel*

The modulation rate and characteristic frequencies for the backward channel are as follows:

|  | $F_Z$ (symbol 1, mark) | $F_A$ (Symbol 0, space) |
|---|---|---|
| Modulation rate up to 75 bauds | 390 Hz | 450 Hz |

In the absence of any signal on the backward channel interface, the Condition Z signal is to be transmitted.

5.     *Tolerances on the characteristic frequencies of the backward channel*

As the backward channel is a VF telegraph-type channel, the frequency tolerances should be as recommended in Recommendation R.35 for frequency-shift voice-frequency telegraphy.

The ± 6 Hz frequency drift in the connection between the modems postulated in 3. above would produce additional distortion in the backward channel. This should be taken into account in the design.

6.     *Division of power between the forward and backward channels*

Considering the following table which shows the levels of power for total power remaining equal to 1 mW.

| Forward channel level (dBm) | Backward channel level (dBm) |
|---|---|
| 0 | $-\infty$ |
| −1 | −7 |
| −2 | −4 |
| −3 | −3 |

equal division of power between the forward and backward channels could be recommended provisionally.

7.     The following information is provided to assist equipment manufacturers:

    *a)*    The nominal range of attenuations in subscriber-to- subscriber connections is from 5 to 30 dB at the reference frequency (800 or 1000 Hz), assuming up to 35 dB attenuation at the recommended mean frequency $(F_0)$ of the forward channel.

    *b)*    A convenient range of sensitivity at the mean frequency $F_0$ for data receivers has been found to be − 40 to 0 dBm for the forward channel at the subscribers' terminals.

    *c)*    The data modem should have no adjustment for send level or receive sensitivity under the control of the operator.

8.     *Interchange circuits*

The configurations of interchange circuits are those essential for the particular switched network or leased circuit requirement as indicated in Tables 1/V.23 and 2/V.23. Where one or more of such requirements are provided in a modem, then all the appropriate interchange circuits should be provided.

a)   *List of interchange circuits essential for the modems when used on the general switched telephone network, including terminals equipped for manual calling or answering or automatic calling or answering (see Table 1/V.23).*

**TABLE 1/V.23**

| Interchange circuit | | Forward (data) channel one-way system (Note 1) | | | | Forward (data) channel either-way system (Note 1) | |
| --- | --- | --- | --- | --- | --- | --- | --- |
| | | Without backward channel | | With backward channel | | Without backward channel | With backward channel |
| No. | Designation | Transmit end | Receive end | Transmit end | Receive end | | |
| 102 | Signal ground or common return | X | X | X | X | X | X |
| 102a (Note 4) | DTE common return | X | X | X | X | X | X |
| 102b (Note 4) | DCE common return | X | X | X | X | X | X |
| 103 | Transmitted data | X | | X | | X | X |
| 104 | Received data | | X | | X | X | X |
| 105 | Request to send | | | | | X | X |
| 106 | Ready for sending | X | | X | | X | X |
| 107 | Data set ready | X | X | X | X | X | X |
| 108/1 or 108/2 (Note 2) | Connect data set to line Data terminal ready | X | X | X | X | X | X |
| 109 | Data channel received line signal detector | | X | | X | X | X |
| 111 | Data signalling rate selector (DTE) | X | X | X | X | X | X |
| 114 (Note 3) | Transmitter signal element timing (DCE) | X | | X | | X | X |
| 115 (Note 3) | Receiver signal element timing (DCE) | | X | | X | X | X |
| 118 | Transmitted backward channel data | | | | X | | X |
| 119 | Received backward channel data | | | X | | | X |
| 120 | Transmit backward channel line signal | | | | | | X |
| 121 | Backward channel ready | | | | X | | X |
| 122 | Backward channel received line signal detector | | | X | | | X |
| 125 | Calling indicator | X | X | X | X | X | X |

*Notes applicable to Tables 1/V.23 and 2/V.23*

*Note 1.* – Interchange circuits indicated by X must be properly terminated according to Recommendation V.24 in the data terminal equipment and data circuit-terminating equipment.

*Note 2.* – This circuit shall be capable of operation as circuit 108/1 – *Connect data set to line* or circuit 108/2 – *Data terminal ready* depending on its use. For automatic calling it shall be used as 108/2 only.

*Note 3.* – These circuits are required when the optional clock is implemented in the modem.

**Note 4.**—Interchange circuits 102a and 102b are required where the electrical characteristics defined in Recommendation V.10 are used.

b)     *List of interchange circuits essential for the modems when used on non-switched leased telephone circuits* (see Table 2/V.23)

## TABLE 2/V.23

| No. | Designation | Forward (data) channel one-way system (Note 1) | | | | Forward (data) channel either way or both ways simultaneously system (Note 1) | |
|---|---|---|---|---|---|---|---|
| | | Without backward channel | | With backward channel | | Without backward channel | With backward channel |
| | | Transmit end | Receive end | Transmit end | Receive end | | |
| 102 | Signal ground or common return | X | X | X | X | X | X |
| 102a (Note 4) | DTE common return | X | X | X | X | X | X |
| 102b (Note 4) | DCE common return | X | X | X | X | X | X |
| 103 | Transmitted data | X | | X | | X | X |
| 104 | Received data | | X | | X | X | X |
| 105 | Request to send | X | | X | | X | X |
| 106 | Ready for sending | X | | X | | X | X |
| 107 | Data set ready | X | X | X | X | X | X |
| 108/1 | Connect data set to line | X | X | X | X | X | X |
| 109 | Data channel received line signal detector | | X | | X | X | X |
| 111 | Data signalling rate selector (DTE) | X | X | X | X | X | X |
| 114 (Note 3) | Transmitter signal element timing (DCE) | X | | X | | X | X |
| 115 (Note 3) | Receiver signal element timing (DCE) | | X | | | X | X |
| 118 | Transmitted backward channel data | | | | X | | X |
| 119 | Received backward channel data | | | X | | | X |
| 120 | Transmit backward channel line signal | | | | X | | X |
| 121 | Backward channel ready | | | | X | | X |
| 122 | Backward channel received line signal detector | | | X | | | X |

c)      *Response times of circuits 106 and 109, 121 and 122*

*Definitions*

i)      Circuits 109 and 122 response times are the times that elapse between the connection or removal of a tone to or from the modem receive line terminals and the appearance of the corresponding ON or OFF condition on circuits 109 and 122.

The test tone should have a frequency corresponding to the characteristic frequency of binary 1 and be derived from a source with an impedance equal to the nominal input impedance of the modem.

The level of the test tone should fall within the level range between 3 dB above the actual threshold of the received line signal detector and the maximum admissible level of the received signal. At all levels within this range the measured response times shall be within the specified limits.

ii)     Circuit 106 response times are from the connection of an ON or OFF condition on:

—   circuit 105 (where it is provided) to the appearance of the corresponding ON or OFF condition on circuit 106;

—   circuit 107 (where circuit 105 is not provided) to the appearance of the corresponding ON or OFF condition on circuit 106.

iii)    Circuit 121 response times are from the connection of an ON or OFF condition on:

—   circuit 120 (where it is provided) to the appearance of the corresponding ON or OFF condition on circuit 121;

—   circuit 109 (where circuit 120 is not provided) to the appearance of the corresponding ON or OFF condition on circuit 121.

d)      *Response times*

| Circuit 106 | | |
|---|---|---|
| OFF to ON | 750 ms to 1400 ms (see Note 1) | a) 20 ms to 40 ms (see Note 2)<br>b) 200 ms to 275 ms (see Note 2) |
| ON to OFF | ≤ 2 ms | |
| Circuit 109 | | |
| OFF to ON | 300 ms to 700 ms (see Note 1) | 10 ms to 20 ms (see Note 2) |
| ON to OFF | 5 ms to 15 ms (see Note 1) | 5 ms to 15 ms (see Note 2) |
| Circuit 121 | | |
| OFF to ON | 80 ms to 160 ms | |
| ON to OFF | ≤ 2 ms | |
| Circuit 122 | | |
| OFF to ON | < 80 ms | |
| ON to OFF | 15 ms to 80 ms | |

*Note 1.* — For automatic calling and answering, the longer response times of circuits 106 and 109 are to be used during call establishment only.

*Note 2.* — The choice of response times depends upon the system application:

*a)* no protection given against line echoes;

*b)* protection given against line echoes.

*Note 3.* — The above parameters are provisional and are the subject of further study.

e)        *Threshold of data channel and backward channel received line signal detectors*

Level of received line signal at receive line terminals of modem for all types of connections, i.e. general switched telephone network or non-switched leased telephone circuits:

greater than  $-43$ dBm          circuits 109/122  ON

less than      $-48$ dBm          circuits 109/122  OFF

The condition of circuits 109 and 122 for levels between $-43$ dBm and $-48$ dBm is not specified except that the signal detectors shall exhibit a hysteresis action such that the level at which the OFF to ON transition occurs is at least 2 dB greater than that for the ON to OFF transition.

Where transmission conditions are known on switched or leased circuits, Administrations should be permitted at the time of modem installation to change these response levels of the received line signal detectors to less sensitive values (e.g. $-33$ dBm and $-38$ dBm respectively).

f)        *Clamping to binary condition 1 of circuit 104 (Received data) and circuit 119 (Received backward channel data)*

Two options shall be provided in the modem:

i)      When clamping is not used there is no inhibition of the signals on circuits 104 and 119. There is no protection against noise, supervisory and control tones, switching transients etc. from appearing on circuits 104 and 119.

ii)     When clamping is used, circuit 104 is held in a marking condition (binary 1) under the conditions defined below. When these conditions do not exist the clamp is removed and circuit 104 can respond to the input signals of the modem:

  —     when circuit 109 is in the OFF condition;

  —     when circuit 105 is in the ON condition and the modem is used in half duplex mode (turn-around systems). To protect circuit 104 from false signals a delay device shall be provided to maintain circuit 109 in the OFF condition for a period of 150 $\pm$ 25 ms after circuit 105 has been turned from ON to OFF. The use of this additional delay is optional.

iii)    When clamping is used, circuit 119 is held in a marking condition (binary 1) under the conditions defined below. When these conditions do not exist the clamp is removed and circuit 119 can respond to the input signals of the modem:

  —     when circuit 122 is in the OFF condition.

## 9. *Electrical characteristics of interchange circuits*
a) Use of electrical characteristics conforming to Recommendation V.28 is recommended with the connector pin assignment plan specified by ISO DIS 2110.
b) Application of electrical characteristics conforming to Recommendation V.10 and V.11 is recognized as an alternative together with the use of the connectors and pin assignment plan specified by ISO DIS 4902.
  i) Concerning circuits 103, 104, 105 (where used), 106, 107, 108, 109, and, where the optional clock is implemented in the modem, circuits 114 and 115, the receivers shall be in accordance with Recommendation V.11 or alternatively Recommendation V.10, Category 1. Either V.10 or V.11 generators may be utilized.
  ii) In the case of circuits 111, 118, 119, 120, 121, 122 and 125, Recommendation V.10 applies with receivers configured as specified by Recommendation V.10 for Category 2.
  iii) It is preferred that backward channel circuits appear on a separate connector and comprise circuits 118, 119, 120, 121, 122 (Category 2) and 102, 102a and 102b.
  iv) Interworking between equipment applying Recommendation V.10 and/or V.11 and equipment applying Recommendation V.28 is allowed on a noninterference basis. The onus for adaptation to V.28 equipment rests solely with the alternative V.10/V.11 equipment.

*Note.*—Manufacturers may wish to note that the long-term objective is to replace electrical characteristics specified in Recommendation V.28, and Study Group XVII has agreed that the work shall proceed to develop a more efficient all balanced interface for the V-Series application which minimizes the number of interchange circuits. It is expected that this work would be based upon the alternative application given in 9 b) above utilizing the V.11 electrical characteristics.

## 10. *Equipment for the disablement of echo suppressors*

(See 5. of Recommendation V.21.)

## 11. *Inclusion of a clock in the modem*

A clock is not an essential item in the standardized modem. However, the modem may conveniently include a clock when used primarily for synchronous transmission.

If such a clock is included in the modem, a synchronizing pattern consisting of alternate binary 0 and binary 1 at clock rate should be transmitted for the whole interval between the OFF to ON transitions of interchange circuits 105 and 106. Users should note that part of this synchronizing pattern may appear at the distant receiver on circuit 104 after the OFF to ON transition of circuit 109. The data terminal equipment should make provision to differentiate between these false signals and true data.

**Recommendation V.24**

## LIST OF DEFINITIONS FOR INTERCHANGE CIRCUITS BETWEEN
## DATA-TERMINAL EQUIPMENT AND DATA CIRCUIT-TERMINATING EQUIPMENT [9]

*(Geneva, 1964, amended at Mar del Plata, 1968, and at Geneva, 1972 and 1976)*

### CONTENTS

I  — Scope

II  — Line of demarcation

III  — Definitions of interchange circuits

    1. 100-series.   General application

    2. 200-series.   Specifically for automatic calling

IV  — Operational requirements

---

### I.  SCOPE

I.1      This Recommendation applies to the interconnecting circuits being called interchange circuits at the interface between DTE and DCE for the transfer of binary data, control and timing signals and analogue signals as appropriate. This Recommendation also applies to both sides of separate intermediate equipment, which may be inserted between these two classes of equipment (see Figure 1/V.24).

Electrical characteristics for interchange circuits are detailed in appropriate Recommendations for electrical characteristics, or in certain special cases, in Recommendations for DCE.

In any type of practical equipment a selection will be made from the range of interchange circuits defined in this Recommendation, as appropriate. When by mutual arrangement other circuits are to be used, these additional circuits should conform to the electrical characteristics specified in the appropriate Recommendation.

The actual interchange circuits to be used in a particular DCE are those indicated in the appropriate Recommendation.

The usage and operational requirements of the interchange circuits and the interaction between them are recommended in IV. of this Recommendation. For proper operation of the DCE it is important that the guidelines in IV. of this Recommendation are observed.

I.2      The DCE may include signal converters, timing generators, pulse regenerators, and control circuitry, together with equipment to provide other functions such as error control, automatic calling and automatic answering. Some of this equipment may be separate intermediate equipment or it may be located in the DCE.

I.3      The range of interchange circuits defined in this Recommendation is applicable, for example:

    *a)*    to synchronous and asynchronous data communications,

    *b)*    to data communication on leased line service, either 2-wire or 4-wire, either point-to-point or multipoint operation,

    *c)*    to data communication on switched network service, either 2-wire or 4-wire,

    *d)*    where short interconnecting cables are used between DTE and DCE. An explanation of short cables is given in II. below.

---

[9] In this Recommendation the terms "data terminal equipment" and "data circuit-terminating equipment" are indicated by DTE and DCE respectively.

## II.　LINE OF DEMARCATION

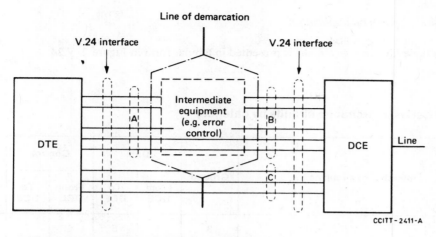

Without intermediate equipment the selections A and B are identical.
Selection C may be a selection specifically for automatic calling.

FIGURE 1/V.24 — **Illustration of general layout of communication equipment**

The interface between DTE and DCE is located at a connector, which is the interchange point between these two classes of equipment. Separate connectors may be provided for the interchange circuits associated with the signal-conversion or similar equipment and those associated with the automatic calling equipment.

The connector(s) will not necessarily be physically attached to the DCE and may be mounted in a fixed position near the DTE.

An interconnecting cable or cables will normally be provided with the DTE. The use of short cables is recommended. Their length should be limited solely by the load capacitance and other electrical characteristics specified in the relevant Recommendation on electrical characteristics.

## III. DEFINITIONS OF INTERCHANGE CIRCUITS

III.1 *100 series — General application*

A list of these interchange circuits is presented in tabular form in Figure 2/V.24.

**Figure 2/V.24—100-series interchange circuits by category**

| Interchange circuit number | Interchange circuit name | Ground | Data | | Control | | Timing | |
|---|---|---|---|---|---|---|---|---|
| | | | From DCE | To DCE | From DCE | To DCE | From DCE | To DCE |
| 1 | 2 | 3 | 4 | 5 | 6 | 7 | 8 | 9 |
| 102 | Signal ground or common return | X | | | | | | |
| 102a | DTE common return | X | | | | | | |
| 102b | DCE common return | X | | | | | | |
| 103 | Transmitted data | | | X | | | | |
| 104 | Received data | | X | | | | | |
| 105 | Request to send | | | | | X | | |
| 106 | Ready for sending | | | | X | | | |
| 107 | Data set ready | | | | X | | | |
| 108/1 | Connect data set to line | | | | | X | | |
| 108/2 | Data terminal ready | | | | | X | | |
| 109 | Data channel received line signal detector | | | | X | | | |
| 110 | Data signal quality detector | | | | X | | | |
| 111 | Data signalling rate selector (DTE) | | | | | X | | |
| 112 | Data signalling rate selector (DCE) | | | | X | | | |
| 113 | Transmitter signal element timing (DTE) | | | | | | | X |
| 114 | Transmitter signal element timing (DCE) | | | | | | X | |
| 116 | Select standby | | | | | X | | |
| 117 | Standby indicator | | | | X | | | |
| 118 | Transmitted backward channel data | | | X | | | | |
| 119 | Received backward channel data | | X | | | | | |
| 120 | Transmit backward channel line isgnal | | | | | X | | |
| 121 | Backward channel ready | | | | X | | | |
| 122 | Backward channel received line signal detector | | | | X | | | |
| 123 | Backward channel signal quality detector | | | | X | | | |
| 124 | Select frequency groups | | | | | X | | |
| 125 | Calling indicator | | | | X | | | |
| 126 | Select transmit frequency | | | | | X | | |
| 127 | Select receive frequency | | | | | X | | |
| 128 | Receiver signal element timing (DTE) | | | | | | | X |
| 129 | Request to receive | | | | | X | | |
| 130 | Transmit backward tone | | | | | X | | |
| 131 | Received character timing | | | | | | X | |
| 132 | Return to non-data mode | | | | | X | | |
| 133 | Ready for receiving | | | | | X | | |
| 134 | Received data present | | | | X | | | |
| 140 | Remote loopback for point-to-point circuits | | | | | X | | |
| 141 | Local loopback | | | | | X | | |
| 142 | Test indicator | | | | X | | | |
| 191 | Transmitted voice answer | | | | | X | | |
| 192 | Received voice answer | | | | X | | | |

84

*Circuit 102 — Signal ground or common return*

This conductor established the signal common return for unbalanced interchange circuits with electrical characteristics according to Recommendation V.28 and the d.c. reference potential for balanced circuits according to Recommendations V.11 and V.35.

Within the DCE, this circuit shall be brought to one point, and it shall be possible to connect this point to protective ground or earth by means of a metallic strap within the equipment. This metallic strap can be connected or removed at installation, as may be required to meet applicable regulations or to minimize the introduction of noise into electronic circuitry.

*Circuit 102a — DTE common return*

This conductor is connected to the DTE circuit common and is used as the reference potential for the unbalanced Recommendation V.10 type interchange circuit receivers within the DCE.

*Circuit 102b — DCE common return*

This conductor is connected to the DCE circuit common and is used as the reference potential for the unbalanced Recommendation V.10 type interchange circuit receivers within the DTE.

*Note.* — Where a mixture of Recommendation V.10 and V.11 circuits is used in the same interface, separate provision must be made for the Recommendation V.10 common return circuits 102a and 102b, and for a Recommendation V.11 d.c. reference potential conductor circuit 102, or protective ground connection, as required.

*Circuit 103 — Transmitted data*

Direction: To DCE

The data signals originated by the DTE, to be transmitted via the data channel to one or more remote data stations, are transferred on this circuit to the DCE.

*Circuit 104 — Received data*

Direction: From DCE

The data signals generated by the DCE, in response to data channel line signals received from a remote data station, are transferred on this circuit to the DTE.

*Circuit 105 — Request to send*

Direction: To DCE

Signals on this circuit control the data channel transmit function of the DCE.

The ON condition causes the DCE to assume the data channel transmit mode.

The OFF condition causes the DCE to assume the data channel non-transmit mode, when all data transferred on circuit 103 have been transmitted.

*Circuit 106 — Ready for sending*

Direction: From DCE

Signals on this circuit indicate whether the DCE is conditioned to transmit data on the data channel.

The ON condition indicates that the DCE is conditioned to transmit data on the data channel.

The OFF condition indicates that the DCE is not prepared to transmit data on the data channel.

*Circuit 107 — Data set ready*

Direction: From DCE

Signals on this circuit indicate whether the DCE is ready to operate.

The ON condition indicates that the signal-conversion or similar equipment is connected to the line and that the DCE is ready to exchange further control signals with the DTE to initiate the exchange of data.

The OFF condition indicates that the DCE is not ready to operate.

*Circuit 108/1 — Connect data set to line*

Direction: To DCE

Signals on this circuit control switching of the signal-conversion or similar equipment to or from the line.

The ON condition causes the DCE to connect the signal-conversion or similar equipment to the line.

The OFF condition causes the DCE to remove the signal-conversion or similar equipment from the line, when the transmission to line of all data previously transferred on circuit 103 and/or circuit 118 has been completed.

*Circuit 108/2 — Data terminal ready*

Direction: To DCE

Signals on this circuit control switching of the signal-conversion or similar equipment to or from the line.

The ON condition, indicating that the DTE is ready to operate, prepares the DCE to connect the signal-conversion or similar equipment to the line and maintains this connection after it has been established by supplementary means.

The DTE is permitted to present the ON condition on circuit 108/2 whenever it is ready to transmit or receive data.

The OFF condition causes the DCE to remove the signal-conversion or similar equipment from the line, when the transmission to line of all data previously transferred on circuit 103 and/or circuit 118 has been completed.

*Circuit 109 — Data channel received line signal detector*

Direction: From DCE

Signals on this circuit indicate whether the received data channel line signal is within appropriate limits, as specified in the relevant Recommendation for DCE.

The ON condition indicates that the received signal is within appropriate limits.

The OFF condition indicates that the received signal is not within appropriate limits.

*Circuit 110 — Data signal quality detector*

Direction: From DCE

Signals on this circuit indicate whether there is a reasonable probability of an error in the data received on the data channel. The signal quality indicated conforms to the relevant DCE Recommendation.

The ON condition indicates that there is no reason to believe that an error has occurred.

The OFF condition indicates that there is a reasonable probability of an error.

*Circuit 111 — Data signalling rate selector* (DTE source)

Direction: To DCE

Signals on this circuit are used to select one of the two data signalling rates of a dual rate synchronous DCE, or to select one of the two ranges of data signalling rates of a dual range asynchronous DCE.

The ON condition selects the higher rate or range or rates.

The OFF condition selects the lower rate or range of rates.

*Circuit 112 — Data signalling rate selector* (DCE source)

Direction: From DCE

Signals on this circuit are used to select one of the two data signalling rates or ranges of rates in the DTE to coincide with the data signalling rate or range of rates in use in a dual rate synchronous or dual range asynchronous DCE.

The ON condition selects the higher rate or range of rates.

The OFF condition selects the lower rate or range of rates.

*Circuit 113 — Transmitter signal element timing* (DTE source)

Direction: To DCE

Signals on this circuit provide the DCE with signal element timing information.

The condition on this circuit shall be ON and OFF for nominally equal periods of time and the transition from ON to OFF condition shall nominally indicate the centre of each signal element on circuit 103.

*Circuit 114 — Transmitter signal element timing* (DCE source)

Direction: From DCE

Signals on this circuit provide the DTE with signal element timing information.

The condition on this circuit shall be ON and OFF for nominally equal periods of time. The DTE shall present a data signal on circuit 103 in which the transitions between signal elements nominally occur at the time of the transitions from OFF to ON condition of circuit 114.

*Circuit 115 — Receiver signal element timing* (DCE source)

Direction: From DCE

Signals on this circuit provide the DTE with signal element timing information.

The condition of this circuit shall be ON and OFF for nominally equal periods of time, and a transition from ON to OFF condition shall nominally indicate the centre of each signal element on circuit 104.

*Circuit 116 — Select standby*

Direction: To DCE

Signals on this circuit are used to select the normal or standby facilities, such as signal converters and communication channels.

The ON condition selects the standby mode of operation, causing the DCE to replace predetermined facilities by their reserves.

The OFF condition causes the DCE to replace the standby facilities by the normal. The OFF condition on this circuit shall be maintained whenever the standby facilities are not required for use.

*Circuit 117 — Standby indicator*

Direction: From DCE

Signals on this circuit indicate whether the DCE is conditioned to operate in its standby mode with the predetermined facilities replaced by their reserves.

The ON condition indicates that the DCE is conditioned to operate in its standby mode.

The OFF condition indicates that the DCE is conditioned to operate in its normal mode.

*Circuit 118 — Transmitted backward channel data*

Direction: To DCE

This circuit is equivalent to circuit 103, except that it is used to transmit data via the backward channel.

*Circuit 119 — Received backward channel data*

Direction: From DCE

This circuit is equivalent to circuit 104, except that it is used for data received on the backward channel.

*Circuit 120 — Transmit backward channel line signal*

Direction: To DCE

This circuit is equivalent to circuit 105, except that it is used to control the backward channel transmit function of the DCE.

The ON condition causes the DCE to assume the backward channel transmit mode.

The OFF condition causes the DCE to assume the backward channel non-transmit mode, when all data transferred on circuit 118 have been transmitted to line.

*Circuit 121 — Backward channel ready*

Direction: From DCE

This circuit is equivalent to circuit 106, except that it is used to indicate whether the DCE is conditioned to transmit data on the backward channel.

The ON condition indicates that the DCE is conditioned to transmit data on the backward channel.

The OFF condition indicates that the DCE is not conditioned to transmit data on the backward channel.

*Circuit 122 — Backward channel received line signal detector*

Direction: From DCE

This circuit is equivalent to circuit 109, except that it is used to indicate whether the received backward channel line signal is within appropriate limits, as specified in the relevant Recommendation for DCE.

*Circuit 123 — Backward channel signal quality detector*

Direction: From DCE

This circuit is equivalent to circuit 110, except that it is used to indicate the signal quality of the received backward channel line signal.

*Circuit 124 — Select frequency groups*

　　Direction: To DCE

　　Signals on this circuit are used to select the desired frequency groups available in the DCE.

　　The ON condition causes the DCE to use all frequency groups to represent data signals.

　　The OFF condition causes the DCE to use a specified reduced number of frequency groups to represent data signals.

*Circuit 125 — Calling indicator*

　　Direction: From DCE

　　Signals on this circuit indicate whether a calling signal is being received by the DCE.

　　The ON condition indicates that a calling signal is being received.

　　The OFF condition indicates that no calling signal is being received, and this condition may also appear during interruptions of a pulse-modulated calling signal.

*Circuit 126 — Select transmit frequency*

　　Direction: To DCE

　　Signals on this circuit are used to select the required transmit frequency of the DCE.

　　The ON condition selects the higher transmit frequency.

　　The OFF condition selects the lower transmit frequency.

*Circuit 127 — Select receive frequency*

　　Direction: To DCE

　　Signals on this circuit are used to select the required receive frequency of the DCE.

　　The ON condition selects the lower receive frequency.

　　The OFF condition selects the higher receive frequency.

*Circuit 128 — Receiver signal element timing* (DTE source)

　　Direction: To DCE

　　Signals on this circuit provide the DCE with signal element timing information.

　　The condition on this circuit shall be ON and OFF for nominally equal periods of time. The DCE shall present a data signal on circuit 104 in which the transitions between signal elements nominally occur at the time of the transitions from OFF to ON condition of the signal on circuit 128.

*Circuit 129 — Request to receive*

　　Direction: To DCE

　　Signals on this circuit are used to control the receive function of the DCE.

　　The ON condition causes the DCE to assume the receive mode.

　　The OFF condition causes the DCE to assume the non-receive mode.

*Circuit 130 — Transmit backward tone*

Direction: To DCE

Signals on this circuit control the transmission of a backward channel tone.

The ON condition causes the DCE to transmit a backward channel tone.

The OFF condition causes the DCE to stop the transmission of a backward channel tone.

*Circuit 131 — Received character timing*

Direction: From DCE

Signals on this circuit provide the DTE with character timing information, as specified in the relevant Recommendation for DCE.

*Circuit 132 — Return to non-data mode*

Direction: To DCE

Signals on this circuit are used to restore the non-data mode provided with the DCE, without releasing the line connection to the remote station.

The ON condition causes the DCE to restore the non-data mode. When the non-data mode has been established, this circuit must be turned OFF.

*Circuit 133 — Ready for receiving*

Direction: To DCE

Signals on this circuit control the transfer of data on circuit 104, indicating whether the DTE is capable of accepting a given amount of data (e.g. a block of data), specified in the appropriate Recommendation for intermediate equipment, for example, error control equipment.

The ON condition must be maintained whenever the DTE is capable of accepting data, and causes the intermediate equipment to transfer the received data to the DTE.

The OFF condition indicates that the DTE is not able to accept data, and causes the intermediate equipment to retain the data.

*Circuit 134 — Received data present*

Direction: From DCE

Signals on this circuit are used to separate information messages from supervisory messages, transferred on circuit 104, as specified in the appropriate Recommendation for intermediate equipment, e.g. error control equipment.

The ON condition indicates the data which represent information messages.

The OFF condition shall be maintained at all other times.

**Circuit 140—Remote loopback for point-to-point circuits**

**Direction: to DCE**

**Signals on this circuit are used to control the loop 2 test condition in a remote DCE.**

**The ON condition of circuit 140 causes the local DCE to command the establishment of the loop 2 test condition in the remote DCE.**

**The OFF condition of circuit 140 causes the local DCE to command the release of the loop 2 test condition in the remote DCE.**

*Circuit 141—Local loopback*
   **Direction: to DCE**

   Signals on this circuit are used to control the loop 3 test condition in the local DCE.

   The ON condition of circuit 141 causes the establishment of the loop 3 test condition in the local DCE.

   The OFF condition of circuit 141 causes the release of the loop 3 test condition in the local DCE.

*Circuit 142 — Test indicator*

   Direction: From DCE

   This circuit is used to indicate to the DTE the test mode status of the DCE.

   The ON condition indicates that the DCE is in the test mode, precluding transmission of data to a remote DTE.

   The OFF condition indicates that the DCE is in the non-test mode, with no test in progress.

*Circuit 191 — Transmitted voice answer*

Direction: To DCE

Signals generated by a voice answer unit in the DTE are transferred on this circuit to the DCE.

The electrical characteristics of this analogue interchange circuit are part of the appropriate DCE Recommendation.

*Circuit 192 — Received voice answer*

Direction: From DCE

Received voice signals, generated by a voice answering unit at the remote data terminal, are transferred on this circuit to the DTE.

The electrical characteristics of this analogue interchange circuit are part of the appropriate DCE Recommendation.

III.2    *200-series — Specifically for automatic calling*

A list of these interchange circuits is presented in tabular form in Figure 3/V.24.

For the proper procedures, refer to the relevant Recommendation for automatic calling procedures.

| Interchange circuit number | Interchange circuit name | From DCE | To DCE |
|:---:|---|:---:|:---:|
| 201 | Signal ground or common return | X | X |
| 202 | Call request | | X |
| 203 | Data line occupied | X | |
| 204 | Distant station connected | X | |
| 205 | Abandon call | X | |
| 206 | Digit signal ($2^0$) | | X |
| 207 | Digit signal ($2^1$) | | X |
| 208 | Digit signal ($2^2$) | | X |
| 209 | Digit signal ($2^3$) | | X |
| 210 | Present next digit | X | |
| 211 | Digit present | | X |
| 213 | Power indication | X | |

FIGURE 3/V.24 — 200-series interchange circuits specifically for automatic calling

*Circuit 201 — Signal ground or common return*

This conductor establishes the signal common reference potential for all 200-series interchange circuits. Within the automatic calling equipment this circuit shall be brought to one point, and it shall be possible to connect this point to protective ground or earth by means of a metallic strap within the equipment. This metallic strap can be connected or removed at installation as may be required to meet applicable regulations or to minimize the introduction of noise into electronic circuitry.

*Circuit 202 — Call request*

Direction: To DCE

Signals on this circuit are used to condition the automatic calling equipment to originate a call and to switch the automatic calling equipment to or from the line.

The ON condition causes the DCE to condition the automatic calling equipment to originate a call and to connect this equipment to the line.

The OFF condition causes the automatic calling equipment to be removed from the line and indicates that the DTE has released the automatic calling equipment.

*Circuit 203 — Data line occupied*

Direction: From DCE

Signals on this circuit indicate whether or not the associated communication channel is in use (e.g. for automatic calling, data or voice communication, test procedures).

The ON condition indicates that the communication channel is in use.

The OFF condition indicates that the communication channel is not in use, and that the DTE may originate a call.

*Circuit 204 — Distant station connected*

Direction: From DCE

Signals on this circuit indicate whether a connection has been established to a remote data station.

The ON condition indicates the receipt of a signal from a remote DCE signalling that a connection to that equipment has been established.

The OFF condition shall be maintained at all other times.

*Circuit 205 — Abandon call*

Direction: From DCE

Signals on this circuit indicate whether a preset time has elapsed between successive events in the calling procedure.

The ON condition indicates that the call should be abandoned.

The OFF condition indicates that call origination can proceed.

*Digit signal circuits:*

Circuit 206 — Digit signal $(2^0)$
Circuit 207 — Digit signal $(2^1)$
Circuit 208 — Digit signal $(2^2)$
Circuit 209 — Digit signal $(2^3)$

Direction: To DCE

On these circuits the DTE presents the following code combinations, being the digits to be called and associated control characters.

| Information | Binary states | | | |
|---|---|---|---|---|
| | 209 | 208 | 207 | 206 |
| Digit 1 | 0 | 0 | 0 | 1 |
| Digit 2 | 0 | 0 | 1 | 0 |
| Digit 3 | 0 | 0 | 1 | 1 |
| Digit 4 | 0 | 1 | 0 | 0 |
| Digit 5 | 0 | 1 | 0 | 1 |
| Digit 6 | 0 | 1 | 1 | 0 |
| Digit 7 | 0 | 1 | 1 | 1 |
| Digit 8 | 1 | 0 | 0 | 0 |
| Digit 9 | 1 | 0 | 0 | 1 |
| Digit 0 | 0 | 0 | 0 | 0 |
| Control character EON | 1 | 1 | 0 | 0 |
| Control character SEP | 1 | 1 | 0 | 1 |

The control character EON (end of number) causes the DCE to take the appropriate actions to await an answer from the called data station.

The control character SEP (separation) indicates the need for a pause between successive digits, and causes the automatic calling equipment to insert the appropriate time interval.

### Circuit 210 — *Present next digit*

Direction: From DCE

Signals on this circuit indicate whether the automatic calling equipment is ready to accept the next code combination.

The ON condition indicates that the automatic calling equipment is ready to accept the next code combination.

The OFF condition indicates that the automatic calling equipment is not ready to accept signals on the digit signal circuits.

### Circuit 211 — *Digit present*

Direction: To DCE

Signals on this circuit control the reading of the code combination presented on the digit signal circuits.

The ON condition causes the automatic calling equipment to read the code combination presented on the digit signal circuits.

The OFF condition on this circuit prevents the automatic calling equipment from reading a code combination on the digit signal circuits.

### Circuit 213 — *Power indication*

Direction: From DCE

Signals on this circuit indicate whether power is available within the automatic calling equipment.

The ON condition indicates that power is available within the automatic calling equipment.

The OFF condition indicates that power is not available within the automatic calling equipment.

## IV. OPERATIONAL REQUIREMENTS

In the following, operational requirements are given for the usage of interchange circuits. It also explains in further detail the required correlation between interchange circuits, where implemented. These guidelines can also be used for the selection of interchange circuits for DCE which are not currently covered by a CCITT Recommendation.

### IV.1    *Data circuits*

It is evident that proper data transmission may be impaired when the required condition is not present on an implemented control interchange circuit. Therefore, the DTE shall not transfer data on circuit 103 unless an ON condition is present on all of the following four circuits, where implemented: circuit 105, circuit 106, circuit 107 and circuit 108.1/108.2.

All data transferred on circuit 103 during the time an ON condition is present on all of the above four circuits, where implemented, shall be transmitted by the DCE.

Refer also to IV.4 and IV.5 below for further explanation.

The DTE shall not transfer data on circuit 118 unless an ON condition is present on all of the following four circuits, where implemented: circuit 120, circuit 121, circuit 107 and circuit 108.1/108.2.

All data transferred on circuit 118 during the time an ON condition is present on all of the above four circuits, where implemented, shall be transmitted by the DCE.

IV.2  *Idle periods*

During intervals when circuit 105 and circuit 106 are in the ON condition and no data are available for transmission, the DTE may transmit binary 1 condition, reversals or other sequences to maintain timing synchronizing, e.g. SYN coded characters, idle characters according to the code used, etc.

Specific requirements, where applicable, are stated in the appropriate DCE Recommendations.

IV.3  *Clamping*

When clamping is used the following clamping conditions shall be provided by the DCE.

1.  In all applications the DCE shall hold, where implemented:

    *a)*  circuit 104 in the binary 1 condition when circuit 109 is in the OFF condition, and

    *b)*  circuit 109 in the binary 1 condition when circuit 122 is in the OFF condition.

2.  In addition a DCE arranged for half-duplex (CCITT definition: simplex) operation (turn-around system), shall also hold, where implemented:

    *a)*  circuit 104 in the binary 1 condition and circuit 109 in the OFF condition when circuit 105 is in the ON condition, and for a short interval (to be specified in Recommendations for DCE) following the ON to OFF transition on circuit 105, and

    *b)*  circuit 119 in the binary 1 condition and circuit 122 in the OFF condition, when circuit 120 is in the ON condition, and for a short interval (to be specified in Recommendations for DCE) following the ON to OFF transition on circuit 120.

Without these clamping conditions, there is no inhibition of signals due to excessive noise, supervisory and control signals, switching transients, feedback from the local transmitter, etc., from appearing on circuit 104, circuit 119, circuit 109 and circuit 122.

IV.4  *Operation of circuits 107 and 108/1 and 108/2*

Signals on circuit 107 are to be considered as responses to signals which initiate connection to line, e.g. circuit 108/1. However, the conditioning of a data channel, such as equalization and clamp removal, cannot be expected to occur before circuit 107 is turned ON.

When circuit 108/1 or 108/2 is turned OFF, it shall not be turned ON again until circuit 107 is turned OFF by the DCE.

A wiring option shall be provided within the DCE to select either circuit 108/1 or circuit 108/2 operation.

When the DCE is conditioned for automatic answering of calls, connection to the line occurs only in response to a combination of the calling signal and an ON condition on circuit 108/2.

In certain special dedicated circuit (lease line) applications, circuit 108/1 might not be implemented, in which case the condition on this circuit is assumed to be permanently ON.

Under certain test conditions, both the DTE and the DCE may exercise some of the interchange circuits. It is then to be understood that when circuit 107 is OFF, the DTE is to ignore the conditions on any interchange circuit from the DCE except those on circuit 125 and the timing circuits. Additionally, when circuit 108/1 or 108/2 is OFF the DCE is to ignore the conditions on any interchange circuit from the DTE. The ON conditions on circuits 107 and 108/1 or 108/2 are therefore prerequisite conditions for accepting as valid the signals on interchange circuits from the DCE or DTE respectively, other than circuit 125. The OFF condition on circuit 108/1 or 108/2 shall not disable the operation of circuit 125.

IV.5     *Interrelationship of circuits 103, 105 and 106*

The DTE signals its intent to transmit data by turning ON circuit 105. It is then the responsibility of the DCE to enter the transmit mode, i.e. be prepared to transmit data, and also to alert the remote DCE and condition it to receive data. The means by which a DCE enters the transmit mode and alerts and conditions the remote DCE are described in the appropriate modem Recommendation.

When the transmitting DCE turns circuit 106 ON, the DTE is permitted to transfer data across the interface on circuit 103. By turning ON circuit 106 it is implied that all data transferred across the interface prior to the time that any one of the four circuits: 105, 106, 107, 108.1/108.2 is again turned OFF, will be transferred to the telecommunication channels; however, the ON condition of circuit 106 is not necessarily a guarantee that the remote DCE is in the receive mode. (Depending on the complexity and sophistication of the transmitting signal converter, there may be a delay ranging from less than a millisecond up to several seconds between the time a bit is transferred across the interface until the time a signal element representing this bit is transmitted on the telecommunication channel.)

The DTE shall not turn circuit 105 OFF before the end of the last bit (data bit or stop element) transferred across the interface on circuit 103. Similarly, in certain full duplex switched network applications where circuit 105 is not implemented (see specific DCE Recommendations), this requirement applies equally when circuit 108/1 or 108/2 is turned OFF to terminate a switched network call.

Where circuit 105 is provided, the ON and OFF conditions on circuit 106 shall be responses to the ON and OFF conditions on circuit 105. For the appropriate response times of circuit 106, and for the operation of circuit 106 when circuit 105 is not provided, refer to the relevant Recommendation for DCE.

When circuit 105 or circuit 106 or both are OFF, the DTE shall maintain a binary 1 condition on circuit 103. When circuit 105 is turned OFF it shall not be turned ON again until curcuit 106 is turned OFF by the DCE.

*Note.* — These conditions also apply to the relationship between circuits 120, 121 and 118.

IV.6     *Timing circuits*

It is desirable that the transfer of timing information across the interface shall not be restricted to periods when actual transmission of data is in progress; however, during intervals when timing information is not transferred across the interface, the circuit involved should be held in the OFF condition. The following conditions apply:

a)     *Circuit 113 — Transmitter signal element timing* (DTE source)

Where circuit 113 is used, the DTE shall transfer timing information across the interface on this circuit at all times that the timing source in the DTE is capable of generating this information, e.g. when the DTE is in a power-on condition.

b)     *Circuit 114 — Transmitter signal element timing* (DCE source)

Where circuit 114 is used, the DCE shall transfer timing information across the interface on this circuit at all times that the timing source in the DCE is capable of generating this information, e.g. when the DCE is in a power-on condition. It is recognized that a DCE which derives power from the central office battery over the local telephone loop is in a power-off condition when disconnected from the loop, i.e. on-hook.

c)     *Circuit 115 — Receiver signal element timing* (DCE source)

Where circuit 115 is used, the DCE shall transfer timing information across the interface on this circuit at all times that the timing source is capable of generating this information.

It is recognized that a DCE which derives power from the serving central office via the local telephone loop, is in a power-off condition with timing sources stopped, when the DCE is disconnected from the line. It is also recognized that some timing sources will not continue to run indefinitely without a driving (external synchronization) signal.

Accuracy and stability of this signal as defined in the DCE Recommendations is required only when circuit 109 is ON. Drift during the OFF condition of circuit 109 is acceptable; however, resynchronization of the signal on circuit 115 must be accomplished as rapidly as possible following the turning ON of circuit 109 for the next transmission as indicated in the relevant DCE Recommendation.

d)     *Circuit 128 — Receiver signal element timing* (DTE source)

Where circuit 115 is used, the DTE may also provide timing information on circuit 128 under conditions established by mutual agreement (e.g. when a synchronous DCE has asynchronous standby facilities).

IV.7     *Circuit 125 — Calling indicator*

The operation of circuit 125 shall not be impaired or disabled by any condition on any other interchange circuit.

IV.8     *Usage of circuits 126 and 127*

Originally, these circuits were defined for operational control of a 2-wire, frequency-divided duplex DCE, such as the Recommendation V.21 type modem. Transmitter and receiver control were separated, so that local testing of both data channels might be performed as national Administrations required.

The modem according to Recommendation V.21 does not require separate operational control by the DTE of circuits 126 and 127. Some modems, however, select the transmit and receive frequencies according to the condition of circuit 125. In these cases external operation of circuit 126 may not be required except to override the condition selected according to circuit 125.

The use of circit 127 may become necessary in certain types of 4-wire operation on multidrop circuits.

IV.9     *Interrelationship of circuits 202 to 211*

*Circuit 202*

Circuit 202 must be turned OFF between calls or call attempts and shall not be turned ON before circuit 203 is turned OFF.

*Circuit 204*

The ON condition of this circuit must be maintained until the DTE has released the automatic calling equipment, i.e. until circuit 202 is turned OFF.

*Circuit 205*

The OFF condition shall be maintained on this circuit after circuit 204 comes ON.

The initial time interval starts when circuit 202 comes on. Subsequent time intervals start each time circuit 210 is turned OFF.

*Circuits 206, 207, 208 and 209*

The conditions on these four circuits shall not change whilst circuit 211 is ON.

*Circuit 210*

When circuit 210 is turned OFF, it shall not be turned ON again before circuit 211 is turned OFF.

*Circuit 211*

Circuit 211 shall neither be turned ON when circuit 210 is in the OFF condition, nor before the DTE has presented the required code combination on the digit signal circuits.

Circuit 211 shall not be turned OFF before circuit 210 is turned OFF.

**Recommendation V.25**

### AUTOMATIC CALLING AND/OR ANSWERING EQUIPMENT ON THE GENERAL SWITCHED TELEPHONE NETWORK, INCLUDING DISABLING OF ECHO SUPPRESSORS ON MANUALLY ESTABLISHED CALLS

*(Mar del Plata, 1968, amended at Geneva, 1972 and 1976)*

1.    *Scope*

1.1    This Recommendation is concerned with the setting up of a data connection when automatic calling and/or answering equipment is used over international circuits.

Automatic calling and answering equipment used within any single Administration's area or between two Administrations by bilateral agreement are not necessarily constrained by these proposals. In particular, the use of 2100 Hz answering tone, as described in the text, could be substituted by another tone when the equipment is used over circuits not equipped with echo suppressors. Similarly, the calling tone could be omitted by bilateral agreements but attention is drawn to 7. and 8. below.

1.2    This Recommendation describes the sequences of events involved in establishing a connection between an automatic calling data station [10] and an automatic answering data station for Series V Recommendations modems specified for general switched network operations. The system configuration proposed is shown in Figure 1/V.25.

Consideration is given only to:

a)    the events which affect the interfaces between the data terminal equipment and the data circuit-terminating equipment, and

b)    the events on the line during establishment of a data call.

Interactions within the data circuit-terminating equipment are not considered, since such consideration is unnecessary for purposes of international standardization.

---

[10] In this Recommendation the term "data station" is used as synonymous with the term "terminal installation for data transmission" which is defined in Definition 53.05.

1.3     The proposed procedures are intended to be suitable for the four types of calls, namely:

*a)*     automatic calling data station to automatic answering data station;

*b)*     manual data station to automatic answering data station;

*c)*     automatic calling data station to manual data station;

*d)*     disabling of echo suppressors in the case of manual data stations.

1.4     The data terminal equipment is responsible for:

*a)*     during call establishment:

    i)      ensuring that the data circuit-terminating equipment is available for operation,

    ii)     providing the telephone number,

    iii)    deciding to abandon the call if it is unsuccessfully completed;

*b)*     after call is established:

    i)      establishing identities,

    ii)     exchanging such traffic as is appropriate,

    iii)    initiating disconnect at calling and answering data station.

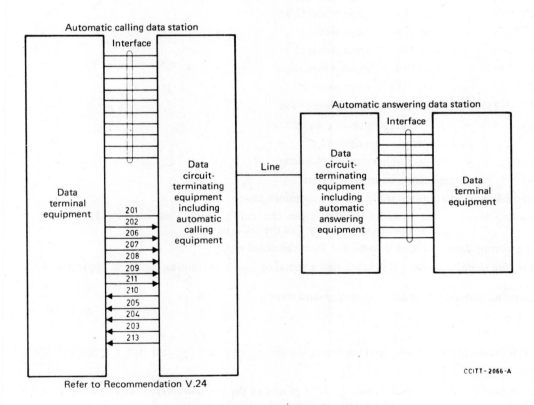

FIGURE 1/V.25 – **System configuration**

2.      *Abbreviations and definitions*

The following abbreviations are used in this Recommendation:

CT 104   = Circuit 104  –  Received data

CT 105   = Circuit 105  –  Request to send

CT 106   = Circuit 106  –  Ready for sending

CT 107   = Circuit 107 — Data set ready

CT 108/1 = Circuit 108/1 — Connect data set to line

CT 108/2 = Circuit 108/2 — Data terminal ready

CT 109   = Circuit 109 — Data channel received line signal detector

CT 119   = Circuit 119 — Received backward channel data

CT 120   = Circuit 120 — Transmit backward channel line signal

CT 121   = Circuit 121 — Backward channel ready

CT 122   = Circuit 122 — Backward channel received line signal detector

CT 125   = Circuit 125 — Calling indicator

CT 201   = Circuit 201 — Signal ground or common return

CT 202   = Circuit 202 — Call request

CT 203   = Circuit 203 — Data line occupied

CT 204   = Circuit 204 — Distant station connected

CT 205   = Circuit 205 — Abandon call

CT 206   = Circuit 206 — Digit signal ($2^0$)

CT 207   = Circuit 207 — Digit signal ($2^1$)

CT 208   = Circuit 208 — Digit signal ($2^2$)

CT 209   = Circuit 209 — Digit signal ($2^3$)

CT 210   = Circuit 210 — Present next digit

CT 211   = Circuit 211 — Digit present

CT 213   = Circuit 213 — Power indication

DCE    = Data circuit-terminating equipment

DTE    = Data terminal equipment

EON    = End-of-number control character

SEP    = Separation control character

The following definitions apply to this Recommendation:

*Calling tone:*   the tone transmitted from the calling end. This may be 1300 Hz or any tone corresponding to binary 1 of the DCE in use.

*Answering tone:* the tone transmitted from the called end.

*Starting signal:*  binary 1, synchronizing signal or equalizer training signal, as appropriate.

3.     *Interface procedures at call-originating data station*

*Event*

3.1    DTE checks if CT 213 ON, and the following circuits OFF: CT 202, CT 210, CT 205, CT 204, CT 203.

3.2    DTE puts CT 202 ON.

3.3    DTE puts CT 108/2 ON (CT 108/2 can be placed in the ON condition at any time up to and including event 3.16).

3.4    For half-duplex modems, DTE puts CT 105 ON if the calling end wishes to transmit first. CT 105 can be placed ON at any time up to and including event 3.20.

3.5    Line goes "off hook".

3.6    DCE puts CT 203 ON.

3.7    Telephone system puts dial tone on line [11].

3.8    DCE puts CT 210 ON.

3.9    DTE presents the first or appropriate digit on CT 206, CT 207, CT 208 and CT 209.

---

[11] Some countries apply the second dial tone to the line after the initial digit is transferred.

3.10    DTE puts CT 211 ON after digit signals have been presented.

3.11    DCE dials first digit; then takes CT 210 OFF.

3.12    DTE takes CT 211 OFF.

3.13    Events 3.8 to 3.12 are repeated (but this process may be interrupted by SEP) until the last digit signal is presented and transferred. Event 3.8 is then repeated but event 3.14 follows.

3.14    DTE presents EON on CT 206, CT 207, CT 208 and CT 209; it then puts CT 211 ON.

3.15    DCE takes CT 210 OFF.

3.16    DTE takes CT 211 OFF and puts CT 108/2 ON, if not previously ON.

3.17    The interrupted calling tone, as shown in Figure 2/V.25, is transmitted to line from the calling DCE.

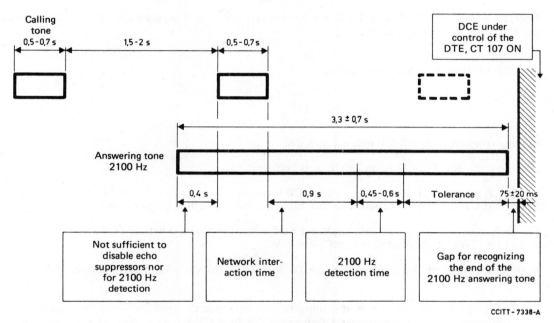

FIGURE 2/V.25 — **Timing of line signals**

3.18    a)            If the call is answered by a data station, then 2100 Hz tone is received by the calling DCE. Echo suppressors are disabled during coincidence of a silent period in the interrupted calling tone (event 3.17) with 2100 Hz answering tone. The 2100 Hz answering tone must not activate CT 104 and CT 109.

        b)    If the call is not answered, or is answered by a non-data station, then no 2100 Hz is received at the calling data station. If no answering tone is received after an elapsed time, CT 205 comes ON. This time is measured from event 3.15 and selectable in the range of 10-40 seconds. The DTE must respond by turning CT 202 OFF.

3.19    When 2100 Hz has been recognized by the DCE for a period of 450 to 600 ms, the interrupted calling tone is discontinued by the DCE as shown in Figure 2/V.25. The DCE transfers control of the connection to the telephone line from CT 202 to CT 108/2.

3.20    The DCE examines the line to determine the end of the 2100 Hz answering tone. The DCE detects an absence of the 2100 Hz tone for 75 ± 20 ms, and then puts CT 107 ON:

    i)    If CT 105 is ON, the starting signal is put on the line. After its delay as specified in the appropriate Series V Recommendation, CT 106 comes ON and the DTE can then transmit data.

ii)  If CT 105 is OFF, the incoming starting signal is recognized and after its delay as specified in the appropriate Series V Recommendation, the DCE puts CT 109 ON to allow the examination of CT 104 by the DTE.

iii)  For the duplex modem case, where CT 105 is not used, the starting signal is put on the line after CT 107 is put ON. The DCE then puts ON CT 109 and CT 106 after a delay as specified in the appropriate Series V Recommendation.

*Note.* — There may be an interim period during which certain existing V.21 modems may not be able to provide the silent period between the end of the answering tone and the application of the starting signal. In this case, the use of a selective answering tone detector (see 11. below) will be essential.

3.21    The DCE turns ON CT 204. The DTE then may turn OFF CT 202 without disconnecting the call.

*Note 1.* — After event 3.19, both CT 202 and CT 108/2 must be turned OFF to disconnect. The ON of CT 205 is an indication to DTE to disconnect.

*Note 2.* — Where CT 105 or CT 120 are not implemented, the timing of CT 106 or CT 121 shall be related to CT 107 and CT 109 respectively.

4.       *Interface procedure at called data station*

*Event*

4.1      Ringing received on line. DCE puts CT 125 ON.

4.2      a)               If CT 108/2 is ON, DCE goes "off hook".

b)   If CT 108/1 or CT 108/2 is OFF, the DCE waits for CT 108/1 or CT 108/2 to come ON, and then goes "off hook". If CT 108/1 or CT 108/2 does not turn ON, then the call is not answered.

4.3      The DCE goes "off hook", maintains silence on the line for a period between 1.8 and 2.5 seconds then transmits 2100 Hz [12] for a period, as shown in Figure 2/V.25.

4.4      At the end of the 2100 Hz transmission, the DCE puts CT 107 ON after a silent period of 75 ± 20 ms (see Figure 2/V.25).

i)  If CT 105 is ON, the DCE transmits the starting signal. After its delay as specified in the appropriate Series V Recommendation, the DCE puts CT 106 ON. The DTE can then transmit data.

ii)  If CT 105 is OFF, the DCE receives the starting signal and after its delay as specified in the appropriate Series V Recommendation, puts CT 109 ON in expectation of receiving data.

iii)  As 3.20 iii).

5.       *Proposed line procedures*

The line procedures outlined consider the half duplex case of the Series V Recommendations modems. For reasons of simplicity, the same timing of line signals will be used for duplex modems (including modems with backward channel).

Systems which operate in the half-duplex mode and which employ automatic calling equipment shall determine by pre-arrangement which of the two data stations — calling or answering — shall first transmit to the other upon the establishment of the data connection. As indicated in 3. above, the DTE at the data station which is to transmit first must put CT 105 ON, at the appropriate point in the call establishment sequence. For correct operation, it is necessary that the longer response times of CT 106 and CT 109 as specified in the appropriate Series V Recommendation are used during call establishment.

---

[12] The 2100 Hz tolerance will be ± 15 Hz in accordance with Recommendation G.161.

Figure 2/V.25 shows the timings of line signals when automatic calling and automatic answering are employed. The sequence of operation is as follows:

After the DCE has dialled the digits of the directory number for the automatic answering data station, followed by the EON character, the DCE sends the calling tone to the answering data station. The calling tone consists of a series of interrupted bursts of binary 1 signal or 1300 Hz ON for a duration of not less than 0.5 second and not more than 0.7 second and OFF for a duration of not less than 1.5 second and not more than 2.0 seconds.

1.8 to 2.5 seconds after the called data station is connected to the line (i.e., CT 125 and CT 108 are ON), it sends a continuous 2100 Hz answering tone for a duration of not less than 2.6 seconds and not more than 4.0 seconds.

The answering tone propagates towards the calling data station and, during the course of one or two interruptions between bursts of calling tone, causes any echo suppressors in the circuit to disable. The answering tone is recognized by the calling data station for a period of between 0.45 and 0.60 second after its arrival. The calling data station terminates the calling tone burst sequence and recognizes the end of the answering tone for a period of 75 ± 20 ms after its arrival at the calling data station. At the end of this delay, the DCE puts CT 107 ON. Similarly, the answering data station delays for a period of 75 ± 20 ms after terminating the answering tone before putting CT 107 ON.

To keep the echo suppressors disabled, it is necessary to ensure that following the 75 ± 20 ms silent period after the transmission of the 2100 Hz answering tone from the called data station, which serves to disable the echo suppressor during the silent period in the calling tone, energy is maintained as specified in Recommendation G.161.

The DCE at the data station at which CT 105 has been turned ON (by pre-arrangement) commences to send the starting signal. Data communication can commence after CT 106 is put ON at that data station.

During the automatic calling and answering procedures, the echo suppressors will be disabled. If signal gaps exceed 100 ms at any time, e.g., during modem turn-around, they may become re-enabled.

6.    *Manual data station calling automatic answering data station*

The procedure for establishment of a call from a manual data station to an automatic answering data station is similar to that from an automatic calling data station, except that no tone is transmitted from the calling data station until the called data station has answered. The manual operator dials the required number, hears 2100 Hz returned from the automatic answering data station and then presses his data button to connect the data circuit-terminating equipment to the line during the period that 2100 Hz is being received. CT 107 comes ON at the time specified in event 3.20.

Satisfactory disabling of echo suppressors by the answering tone, however, will require that no speech signals from the microphone at the calling data station enter the telecommunications circuit for a period of at least 400 ms during the receipt of answering tone. This may be accomplished by a handset switch or other appropriate means.

7.    *Automatic calling data station calling manual data station*

An operator answering a call from an automatic calling equipment hears an interrupted calling tone of 0.5 to 0.7 second ON and 1.5 to 2.0 seconds OFF. The data button must be depressed to connect the modem to line. A period of about 2.6 to 4.0 seconds of 2100 Hz tone is transmitted to the calling data station to disable echo suppressors and notify the calling data station that the connection is being established. This sequence is followed by data transmission, as required.

8.      *Disabling of echo suppressors in the case of manual data stations*

The procedures as described in 6. and 7. above with regard to the manually operated data stations, can obviously be used for disabling echo suppressors when manual switching from voice conversation to data is required, which is the preferred principle of operation. Considering the type of DCE designed to be used in conjunction with manual connection set-up, it will be necessary to equip the DCE with a 2100 Hz answering tone generator. To avoid modifying existing equipment at the data station which receives the answering tone, the following procedure may replace the operation principle of 6. above. The manual operator operates his data key after the end of the 2100 Hz answering tone. The data station which is to transmit the answering tone is to be agreed between the operators while still in the voice mode.

Care must be exercised in cases of half-duplex modems where transmission of data is started from the data station which transmits the answering tone, to avoid mutilation of the initial data.

*Note.* — Where disabling of echo suppressors is not required in the half-duplex modem case, the 2100 Hz answering tone need not be transmitted. However, the delay between CT 105 to CT 106 ON conditions should be longer than 100 ms in consideration of the echo suppressor suppression hangover time.

9.      *Protection of ordinary telephone users*

As both automatic calling and automatic answering data stations transmit tones to line during call establishment, a normal telephone user who becomes inadvertently connected to one will receive tone signals for a period of sufficient duration to indicate clearly to him that he is incorrectly connected.

10.     *Manual selection of automatic answering, data mode and voice mode*

It is recognized that, at the data station, means should be provided to allow the operator to select between automatic and manual answering of calls. If a call is manually answered, voice mode shall be established. Subsequent switching to the data mode shall be performed by the procedure as specified in 7. above.

Selection of manual or automatic answering of subsequent calls shall be possible after entering the data mode. As an option, automatic answering may be arranged for all subsequent incoming calls. In this case, manual answering may still be achieved by keeping CT 108/2 OFF to cause an audible signal to occur at the telephone instrument.

The DCE shall be disconnected from the line whenever CT 108/1 or CT 108/2 is turned OFF, irrespective of the means employed in establishing the connection.

Procedures for switching to the voice mode between data transmission within the same call shall ensure that CT 107 is turned OFF while in the voice mode.

11.     *2100 Hz tone recognition*

To protect the 2100 Hz tone detector against faulty operation resulting from interference generated by the interrupted calling tone, the detector should be inhibited during the ON periods of the calling tone.

Additionally, in cases where automatic calling equipment is used to set up the call, the 2100 Hz detector must not respond to spurious tones which may arise from speech or service signals during call establishment. It is suggested that the answering tone detection be prevented when the 2100 Hz signal is accompanied by any other signal of comparable level within the ranges 350 Hz to 1800 Hz and 2500 Hz to 3400 Hz.

*Note.* — The relative inhibiting signal levels recommended for the echo suppressor disabling tone detector of Recommendation G.161 are a useful guide for 2100 Hz tone detector inhibiting levels.

**Recommendation V.26**

### 2400 BITS PER SECOND MODEM STANDARDIZED FOR USE ON 4-WIRE LEASED TELEPHONE-TYPE CIRCUITS

*(Mar del Plata, 1968, amended at Geneva, 1972 and 1976)*

On leased circuits, considering that there exist and will come into being many modems with features designed to meet the requirements of the Administrations and users, this Recommendation in no way restricts the use of any other modems.

1.      The principal characteristics for this recommended modem for transmitting data at 2400 bits per second on 4-wire leased point-to-point and multipoint circuits conforming to Recommendation M.1020 are as follows:

   *a)*   it is capable of operating in a full-duplex mode;

   *b)*   four-phase modulation with synchronous mode of operation;

   *c)*   inclusion of a backward (supervisory) channel at modulation rates up to 75 bauds in each direction of transmission, the use of these channels being optional.

2.      *Line signals*

2.1      The carrier frequency is to be 1800 ± 1 Hz. No separate pilot frequencies are provided. The power levels used will conform to Recommendation V.2.

2.2      *Division of power between the forward and backward channels*

If simultaneous transmission of the forward and backward channels occurs in the same direction, a backward channel shall be 6 dB lower in power level than the data channel.

2.3      The data stream to be transmitted is divided into pairs of consecutive bits (dibits). Each dibit is encoded as a phase change relative to the phase of the immediately preceding signal element. At the receiver the dibits are decoded and the bits are reassembled in correct order. Two alternative arrangements of coding are listed in Table 1/V.26. The left-hand digit of the dibit is the one occurring first in the data stream.

TABLE 1/V.26

| Dibit | Phase change (see Note) | |
|:---:|:---:|:---:|
| | Alternative A | Alternative B |
| 00 | 0° | +45° |
| 01 | +90° | +135° |
| 11 | +180° | +225° |
| 10 | +270° | +315° |

*Note.* – The phase change is the actual on-line phase shift in the transition region from the end of one signalling element to the beginning of the following signalling element.

The meaning of phase change for alternatives A and B is illustrated by the line signal diagram in Figure 1/V.26.

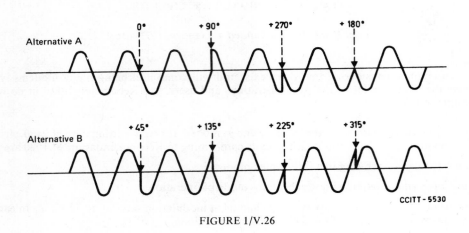

FIGURE 1/V.26

### 2.4    *Synchronizing signal*

For the whole duration of the interval between the OFF to ON transitions of circuits 105 and 106, the line signal shall be that corresponding to the continuous transmission of dibit 11. This shall be known as the synchronizing signal.

*Note.* — Owing to several causes, the stability of timing recovery at the receiver is liable to be data-pattern sensitive. The presence of dibit 11 provides a stabilizing influence irrespective of the cause of lack of stability. Users are advised to include sufficient binary 1s in the data which will ensure that the dibit 11 will occur frequently. In certain cases, the use of a scrambling method may also facilitate timing recovery problems. However, prior agreement is required between users of a circuit.

### 3.    *Data signalling and modulation rates*

The data signalling rate shall be 2400 bits per second ± 0.01%, i.e. the modulation rate is 1200 bauds ± 0.01%.

### 4.    *Received signal frequency tolerance*

Noting that the carrier frequency tolerance allowance at the transmitter is ± 1 Hz and assuming a maximum frequency drift of ± 6 Hz in the connection between the modems, then the receiver must be able to accept errors of at least ± 7 Hz in the received frequencies.

### 5.    *Backward channel*

The modulation rate, characteristic frequencies, tolerances, etc., to be as recommended for backward channel in Recommendation V.23.

6. *Interchange circuits*

6.1 *List of interchange circuits concerned* (see Table 2/V.26)

**TABLE 2/V.26**

| Interchange circuit | | Forward (data) channel half-duplex or full duplex | |
|---|---|---|---|
| No. | Designation | Without backward channel | With backward channel |
| 102 | Signal ground or common return | X | X |
| 102a (Note 1) | DTE common return | X | X |
| 102b (Note 1) | DCE common return | X | X |
| 103 | Transmitted data | X | X |
| 104 | Received data | X | X |
| 105 | Request to send | X | X |
| 106 | Ready for sending | X | X |
| 107 | Data set ready | X | X |
| 108/1 | Connect data set to line | X | X |
| 109 | Data channel received line signal detector | X | X |
| 113 | Transmitter signal element timing (DTE source) | X | X |
| 114 | Transmitter signal element timing (DCE source) | X | X |
| 115 | Receiver signal element timing (DCE source) | X | X |
| 118 | Transmitted backward channel data | | X |
| 119 | Received backward channel data | | X |
| 120 | Transmit backward channel line signal | | X |
| 121 | Backward channel ready | | X |
| 122 | Backward channel received line signal detector | | X |

*Note 1.*—Interchange circuits 102a and 102b are required where the electrical characteristics defined in Recommendation V.10 are used.

6.2 *Threshold and response times of circuit 109*

A fall in level of the incoming line signal to $-31$ dBm or lower for more than $10 \pm 5$ ms will cause circuit 109 to be turned OFF. An increase in level to $-26$ dBm or higher will, within $10 \pm 5$ ms, turn this circuit ON. The condition of circuit 109 for levels between $-26$ dBm and $-31$ dBm is not specified except that the signal level detector shall exhibit a hysteresis action such that the level at which the OFF to ON transition occurs is at least 2 dB greater than that for the ON to OFF transition. These values shall be measured when the synchronizing signal as defined in 2.4 above is being transmitted. It should be noted that the aforementioned times relate only to the defined function of circuit 109 and do not necessarily include the time for the modem to achieve bit synchronism.

*Note.* — The signal levels specified above shall apply unless completion of Recommendation M.1020 indicates otherwise.

### 6.3    *Response times of circuits 106, 121 and 122*

| | | |
|---|---|---|
| *Circuit 106*<br>OFF to ON | 65-100 ms (see Note 1)<br>(Provisional) | 25-45 ms (see Note 2)<br>(Provisional) |
| ON to OFF | $\leqslant$ 2 ms | |
| *Circuit 121*<br>OFF to ON | 80 ms to 160 ms | |
| ON to OFF | $\leqslant$ 2 ms | |
| *Circuit 122*<br>OFF to ON | < 80 ms | |
| ON to OFF | 15 ms to 80 ms | |

*Note 1.* – These times shall be used when infrequent operation of circuit 105 is required, e.g. as in many cases of point-to-point usage. Further study is required to verify the range quoted.

*Note 2.* – These times shall be used when frequent operation of circuit 105 is required, e.g. in many cases of multipoint usage. Further study is required with a view to reducing these times.

### 6.4    *Threshold of circuit 122*

— greater than $-34$ dBm:          circuit 122 ON

— less than $-39$ dBm:          circuit 122 OFF

The condition of circuit 122 for levels between $-34$ dBm and $-39$ dBm is not specified except that the signal detector shall exhibit a hysteresis action such that the level at which the OFF to ON transition occurs is at least 2 dB greater than that for the ON to OFF transition.

### 6.5    *Clamping of circuit 104*

Two options shall be provided in the modem.

i)    When clamping is not used there is no inhibition of the signals on circuit 104. There is no protection against noise, line transients, etc. from appearing on circuit 104.

ii)    When clamping is used, circuit 104 is held in the marking condition (binary 1) when circuit 109 is in the OFF condition. When this condition does not exist the clamp is removed and circuit 104 can respond to the input signals of the modem.

### 6.6    *Clamping of circuit 109*

Two options shall be provided in the modem.

i)    When clamping is not used there is no inhibition of the signals on circuit 119. There is no protection against noise, line transients, etc. from appearing on circuit 119.

ii)    When clamping is used, circuit 119 is held in the marking condition (binary 1) when circuit 122 is in the OFF condition. When this condition does not exist, the clamp is removed and circuit 119 can respond to the input signals of the modem.

## 7.    *Timing arrangements*

Clocks should be included in the modem to provide the data terminal equipment with transmitter signal element timing (Recommendation V.24, circuit 114) and receiver signal element timing (Recommendation V.24, circuit 115). Alternatively, the transmitter signal element timing may be originated in the data terminal equipment instead of in the data circuit-terminating equipment and be transferred to the modem via the appropriate interchange circuit (Recommendation V.24, circuit 113).

## 8. *Electrical characteristics at interchange circuits*

**a)** Use of electrical characteristics conforming to Recommendation V.28 is recommended with the connector pin assignment plan specified by ISO DIS 2110.

**b)** Application of electrical characteristics conforming to Recommendation V.10 and V.11 is recognized as an alternative together with the use of the connectors and pin assignment plan specified by ISO DIS 4902.

**i)** Concerning circuits 103, 104, 105 (where used), 106, 107, 108, 109, 113, 114, and 115, the receivers shall be in accordance with Recommendation V.11 or, alternatively, Recommendation V.10 Category 1. Either V.10 or V.11 generators may be utilized.

**ii)** In the case of circuits 118, 119, 120, 121 and 122, Recommendation V.10 applies with receivers configured as specified by Recommendation V.10 for Category 2.

**iii)** It is preferred that backward channel circuits appear on a separate connector and comprise circuits 118, 119, 120, 121, 122 (Category 2) and 102, 102a and 102b.

**iv)** Interworking between equipment applying Recomendation V.10 and/or V.11 and equipment applying Recommendation V.28 is allowed on a noninterference basis. The onus for adaptation to V.28 equipment rests soley with the alternative V.10/V.11 equipment.

*Note.—* Manufacturers may wish to note that the long-term objective is to replace electrical characteristics specified in Recommendation V.28, and Study Group XVII has agreed that the work shall proceed to develop a more efficient all balanced interface for the V-Series application which minimize the number of interchange circuits. It is expected that this work would be based upon the alternative application given in 8 b) above utilizing the V.11 electrical characteristics.

## 9. The following information is provided to assist equipment manufacturers:

The data modem should have no adjustment for send level or receive sensitivity under the control of the operator.

**2400/1200 BITS PER SECOND MODEM STANDARDIZED FOR USE IN THE
GENERAL SWITCHED TELEPHONE NETWORK**

*(Geneva, 1972, amended at Geneva, 1976)*

The CCITT,

*considering*

a)   that there is a demand for data tramsmission at 2400 bit/s over the generai switched telephone network;

b)   that a majority of connections over the general switched telephone network within some countries are capable of carrying data at 2400 bit/s;

c)   that a much lower proportion of international connections in the general switched telephone service are capable of carrying data at 2400 bit/s,

*unanumously declares the view*

A.   that transmission at 2400 bit/s should be allowed on the general switched telephone network. Reliable transmission cannot be guaranteed on every connection or routing and tests should be made between the most probable terminal points before a service is provided.

The CCITT expects that developments during the next few years in modern technology will bring about modems of more advanced design enabling reliable transmission to be given on a much higher proportion of connections.

*Note.* —  The provisions of this Recommendation are to be regarded as provisional in order to provide service where it is urgently required and between locations where it is expected that a reasonably satisfactory service can be given. The study of improved methods of transmission at 2400 bits/s or above over the general switched telephone network will be urgently continued with the aim to recommend a method of transmission which will enable a more reliable service to be given over a high proportion of the connections encountered in normal service.

B.   that the characteristics of the modems for this service shall provisionally be the following:

1.   *Principal characteristics*

—   use of a data signalling rate of 2400 bit/s with carrier frequency, modulation and coding according to Recommendation V.26, Alternative B (see Note below) on the communication channel. Administrations and users should note that the performance of this modem on international connections may not always be suitable for this service without prior testing and conditioning if required;

— reduced rate capability at 1200 bit/s;

— inclusion of a backward channel at modulation rates up to 75 bauds, use of this channel being optional.

*Note.* — Attention is drawn to the fact that there are some old-type modems currently in operation for which the coding method in accordance with Recommendation V.26, Alternative A, is used.

### 2.        *Line signals at 2400 and 1200 bit/s*

2.1        The carrier frequency is to be 1800 ± 1 Hz. No separate pilot frequencies are provided. The power levels used will conform to Recommendation V.2.

### 2.2        *Phase distortion limits*

The transmitted line signal spectrum should have linear phase characteristics (to be obtained by means of filters or equalizers or digital means). The deviation of the phase distortion characteristic should not exceed the limits specified in Figure 1/V.26 *bis.*

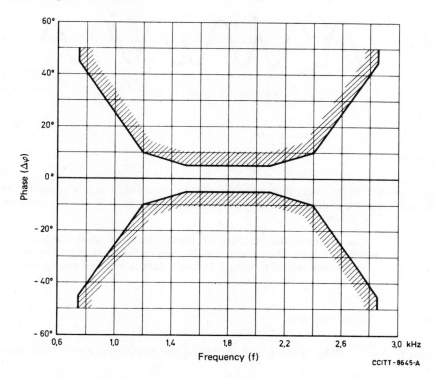

FIGURE 1/V.26 *bis* – Tolerance limit for phase distortion of the signal transmitted to the line

### 2.3        *Division of power between forward and backward channels*

Equal division of power between the forward and backward channels is recommended provisionally.

### 2.4        *Operation at 2400 bit/s*

2.4.1        The data stream to be transmitted is divided into pairs of consecutive bits (dibits). Each dibit is encoded as a phase change relative to the phase of the immediately preceding signal element (see Table 1/V.26 *bis*). At the receiver the dibits are decoded and the bits are reassembled in correct order. The left-hand digit of the dibit is the one occurring first in the data stream.

TABLE 1/V.26 *bis*

| Dibit | Phase change (see Note) |
|-------|-------------------------|
| 00 | +45° |
| 01 | +135° |
| 11 | +225° |
| 10 | +315° |

*Note.* – The phase change is the actual on-line phase shift in the transition region from the end of one signalling element to the beginning of the following signalling element.

The meaning of phase change is illustrated by the line signal diagram given in Figure 2/V.26 *bis*.

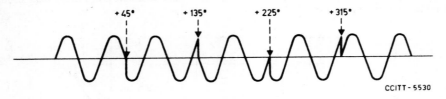

FIGURE 2/V.26 *bis*

### 2.4.2  *Synchronizing signal*

For the whole duration of the interval between the OFF to ON transitions of circuits 105 or 107 and 106, the line signal shall be that corresponding to the continuous transmission of dibit 11. This shall be known as the synchronizing signal [see 5.2 ii) below].

*Note.* – Owing to several causes the stability of timing recovery at the receiver is liable to be data-pattern sensitive. The presence of dibit 11 provides a stabilizing influence irrespective of the cause of lack of stability. Users are advised to include sufficient binary 1s in the data which will ensure that the dibit 11 will occur frequently.

### 2.4.3  *Data signalling and modulation rates*

The data signalling rate shall be 2400 bit/s $\pm$ 0.01%, i.e. the modulation rate is 1200 bauds $\pm$ 0.01%.

### 2.5  *Operation at 1200 bit/s*

2.5.1   Coding and modulation used are 2-phase differential modulation with binary 0 for +90° and binary 1 for +270°.

2.5.2   The data signalling rate shall be 1200 bit/s $\pm$ 0.01%, the modulation rate remains at 1200 bauds $\pm$ 0.01%.

### 3.  *Received signal frequency tolerance*

Noting that the carrier frequency tolerance allowance at the transmitter is $\pm$ 1 Hz and assuming a maximum frequency drift of $\pm$ 6 Hz in the connection between the modems, then the receiver must be able to accept errors of at least $\pm$ 7 Hz in the received frequencies.

4.    *Backward channel*

4.1    *Modulation rate and characteristic frequencies for the backward channel*

The modulation rate and characteristic frequencies for the backward channel are as follows:

|  | $F_Z$ (symbol 1, mark) | $F_A$ symbol 0, space) |
|---|---|---|
| Modulation rate up to 75 bauds | 390 Hz | 450 Hz |

In the absence of any signal on the backward channel interface, the condition Z signal is to be transmitted.

4.2    *Tolerances on the characteristic frequencies of the backward channel*

As the backward channel is a VF telegraph-type channel, the frequency tolerances should be as recommended in Recommendation R.35 for frequency-shift voice-frequency telegraphy.

The $\pm$ 6 Hz frequency drift in the connection between the modems postulated in 3. above would produce additional distortion in the backward channel. This should be taken into account in the design.

5.    *Interchange circuits*

5.1    The list of interchange circuits essential for the modems when used on the general switched telephone network, including terminals equipped for manual calling or answering or automatic calling or answering is given in Table 2/V.26 *bis*.

**TABLE 2/V.26** *bis*

| Interchange circuit | | Forward (data) channel one-way system (see Note 1) | | | | Forward (data) channel either-way system (see Note 1) | |
| --- | --- | --- | --- | --- | --- | --- | --- |
| | | Without backward channel | | With backward channel | | Without backward channel | With backward channel |
| No. | Designation | Transmit end | Receive end | Transmit end | Receive end | | |
| 102 | Signal ground or common return | X | X | X | X | X | X |
| 102a (Note 3) | DTE common return | X | X | X | X | X | X |
| 102b (Note 3) | DCE common return | X | X | X | X | X | X |
| 103 | Transmitted data | X | | X | | X | X |
| 104 | Received data | | X | | X | X | X |
| 105 | Request to send | X | | X | | X | X |
| 106 | Ready for sending | X | | X | | X | X |
| 107 | Data set ready | X | X | X | X | X | X |
| 108/1 or 108/2 (see Note 2) | Connect data set to line / Data terminal ready | X | X | X | X | X | X |
| 109 | Data channel received line signal detector | | X | | X | X | X |
| 111 | Data signalling rate selector (DTE source) | X | X | X | X | X | X |
| 113 | Transmitter signal element timing (DTE source) | X | | X | | X | X |
| 114 | Transmitter signal element timing (DCE source) | X | | X | | X | X |
| 115 | Receiver signal element timing (DCE source) | | X | | X | X | X |
| 118 | Transmitted backward channel data | | | | X | | X |
| 119 | Received backward channel data | | | X | | | X |
| 120 | Transmit backward channel line signal | | | | | | X |
| 121 | Backward channel ready | | | | X | | X |
| 122 | Backward channel received line signal detector | | | X | | | X |
| 125 | Calling indicator | X | X | X | X | X | X |

*Note 1.* — Interchange circuits indicated by X must be properly terminated according to Recommendation V.24 in the data terminal equipment and data circuit-terminating equipment.

*Note 2.* — This circuit shall be capable of operation as circuit 108/1 or circuit 108/2 depending on its use. For automatic calling it shall be used as 108/2 only.

*Note 3.*—Interchange circuits 102a and 102b are required where the electrical characteristics defined in Recommendation V.10 are used.

5.2     *Response times of circuits 106, 109, 121 and 122* (see Table 3/V.26 *bis*).

*Definition*

i)      Circuit 109 response times are the times that elapse between the connection or removal of the test synchronizing signal to or from the modem receive line terminals and the appearance of the corresponding ON and OFF condition on circuit 109.

The level of the test synchronizing signal should fall within the level range between 3 dB above the actual OFF to ON threshold of the received line signal detector and the maximum admissible level of the received signal. At all levels within this range, the measured response times shall be within the specified limits.

ii)     Circuit 106 response times are from the connection to an ON or OFF condition on:

—       circuit 105 to the appearance of the corresponding ON or OFF condition on circuit 106; or

—       circuit 107 (where circuit 105 is not required to initiate the synchronizing signal) to the appearance of the corresponding ON or OFF condition on circuit 106.

TABLE 3/V.26 *bis* — Response times

| Circuit 106 OFF to ON | 750 ms to 1400 ms (see Note 1) | a)  65 ms to 100 ms (see Note 2)<br>b) 200 ms to 275 ms (see Note 2) |
|---|---|---|
| ON to OFF | ≤ 2 ms | |
| Circuit 109 OFF to ON | 300 ms to 700 ms (see Note 1) | 5 ms to 15 ms (see Note 2) |
| ON to OFF | 5 ms to 15 ms (see Note 1) | 5 ms to 15 ms (see Note 1) |
| Circuit 121 OFF to ON | 80 ms to 160 ms | |
| ON to OFF | ≤ 2 ms | |
| Circuit 122 OFF to ON | < 80 ms | |
| ON to OFF | 15 ms to 80 ms | |

*Note 1.* — For automatic calling and answering, the longer response times of circuits 106 and 109 are to be used during call establishment only.

*Note 2.* — The choice of response times depends upon the system application: *a)* limited protection given against line echoes; *b)* protection given against line echoes.

*Note 3.* — The above parameters and procedures, particularly in the case of automatic calling and answering are provisional and are the subject of further study. Especially the shorter response times for circuit 109 may need revision to prevent remnants of the synchronizing signal from appearing on circuit 104.

5.3     *Threshold of data channel and backward channel received line signal detectors*

Level of received line signal at receive line terminals of modem for all types of connections, i.e. general switched telephone network or non-switched leased telephone circuits:

— greater than $-43$ dBm:          circuits 109/122 ON

— less than $-48$ dBm             circuits 109/122 OFF

The condition of circuits 109 and 122 for levels between $-43$ dBm and $-48$ dBm is not specified except that the signal detectors shall exhibit a hysteresis action such that the level at which the OFF to ON transition occurs is at least 2 dB greater than that for the ON to OFF transition.

Where transmission conditions are known and allowed, it may be desirable at the time of modem installation to change these response levels of the received line signal detector to less sensitive values (e.g. $-33$ dBm and $-38$ dBm respectively).

5.4     *Clamping to binary condition 1 of circuit 104*

Two options shall be provided in the modem:

i)     When clamping is not used there is no inhibition of the signals on circuit 104. There is no protection against noise, supervisory and control tones, switching transients, etc. from appearing on circuit 104.

ii)    When clamping is used, circuit 104 is held in a marking condition (binary 1) under the conditions defined below. When these conditions do not exist the clamp is removed and circuit 104 can respond to the input signals of the modem:

— when circuit 109 is in the OFF condition;

— when circuit 105 in the the ON condition and the modem is used in half duplex mode (turn-around systems). To protect circuit 104 from false signals a delay device shall be provided to maintain circuit 109 in the OFF condition for a period of 150 $\pm$ 25 ms after circuit 105 has been turned from ON to OFF. The use of this additional delay is optional.

6.     *Timing arrangements*

Clocks should be included in the modem to provide the data terminal equipment with transmitter signal element timing (Recommendation V.24, circuit 114) and receiver signal element timing (Recommendation V.24, circuit 115). Alternatively, the transmitter signal element timing may be originated in the data terminal equipment instead of in the data circuit-terminating equipment and be transferred to the modem via the appropriate interchange circuit (Recommendation V.24, circuit 113).

**7. Electrical characteristics of interchange circuits**

**a) Use of electrical characteristics conforming to Recommendation V.28 is recommended with the connector pin assignment plan specified by ISO DIS 2110.**

**b) Application of electrical characteristics conforming to Recommendation V.10 and V.11 is recognized as an alternative together with the use of the connectors and pin assignment plan specified by ISO DIS 4902.**

**i) Concerning circuits 103, 104, 105, 106, 107, 108, 109, 113, 114 and 115, the receivers shall be in accordance with Recommendation V.11 or, alternatively, Recommendation V.10, Category 1. Either V.10 or V.11 generators may be utilized.**

**ii) In the case of circuits 111, 118, 119, 120, 121, 122 and 123, Recommendation V.10 applies with receivers configured as specified by Recommendation V.10 for Category 2.**

**iii) It is preferred that backward channel circuits appear on a separate connector and comprise circuits 118, 119, 120, 121, 122 (Category 2) and 102, 102a and 102b.**

**iv) Interworking between equipment applying Recommendation V.10 and/or V.11 and equipment applying Recommendation V.28 is allowed on a noninterference basis. The onus for adaptation to V.28 equipment rests solely with the alternative V.10/V.11 equipment.**

*Note.*—Manufacturers may wish to note that the long-term objective is to replace electrical characteristics specified in Recommendation V.28, and Study Group XVII has agreed that the work shall proceed to develop a more efficient all balanced interface for the V-Series application which minimizes the number of interchange circuits. It is expected that this work would be based upon the alternative application given in 7 b) above utilizing the V.11 electrical characteristics.

8.     The following information is provided to assist equipment manufacturers:

The data modem should have no adjustment for send level or receive sensitivity under the control of the operator.

9.     It will be for the user to decide whether, in view of the connections he makes with this system, he will have to request that the data circuit-terminating equipment be equipped with facilities for disabling echo suppressors. The international characteristics of the echo suppressor tone disabler have been standardized by the CCITT (C. of Recommendation G.161) and the disabling tone should have the following characteristics:

— disabling tone transmitted: 2100 ± 15 Hz at a level of −12 ± 6 dBm0,

— the disabling tone to last at least 400 ms, the tone disabler should hold in the disabled mode for any single-frequency sinusoid in the band from 390-700 Hz having a level of −27 dBm0 or greater, and

from 700-3000 Hz having a level of −31 dBm0 or greater. The tone disabler should release for any signal in the band from 200-3400 Hz having a level of −36 dBm0 or less,

— the tolerable interruptions by the data signal to last not more than 100 ms.

10.    *Fixed compromise equalizer*

A fixed compromise equalizer shall be incorporated into the receiver. The characteristics of this equalizer may be selected by Administrations but this should be the matter for further study.

**4800 BITS PER SECOND MODEM WITH MANUAL EQUALIZER
STANDARDIZED FOR USE ON LEASED TELEPHONE-TYPE CIRCUITS**

*(Geneva, 1972, amended at Geneva, 1976)*

1. *Introduction*

This modem is intended to be used primarily on Recommendation M.1020 circuits but this does not preclude the use of this modem over circuits of lower quality at the discretion of the concerned Administration.

On leased circuits, considering that there exist and will come into being many modems with features designed to meet the requirements of the Administrations and users, this Recommendation in no way restricts the use of any other modems.

The principal characteristics for this recommended modem for transmitting data at 4800 bits per second on leased circuits are as follows:

a)  it is capable of operating in a full-duplex mode or half-duplex mode;

b)  differential eight-phase modulation with synchronous mode of operation;

c)  possibility of a backward (supervisory) channel at modulation rates up to 75 bauds in each direction of transmission, the use of these channels being optional;

d)  inclusion of a manually adjustable equalizer.

2. *Line signals*

2.1    The carrier frequency is to be 1800 ± 1 Hz. No separate pilot frequencies are provided. The power levels used will conform to Recommendation V.2.

2.2    *Division of power between the forward and backward channels*

If simultaneous transmission of the forward and backward channels occurs in the same direction, a backward channel should be 6 dB lower in power level than the forward (data) channel.

2.3    The data stream to be transmitted is divided into groups of three consecutive bits (tribits). Each is encoded as a phase change relative to the phase of the immediately preceding signal tribits element (see Table 1/V.27). At the receiver the tribits are decoded and the bits are reassembled in correct order. The left-hand digit of the tribit is the one occurring first in the data stream as it enters the modulator portion of the modem after the scrambler.

118

TABLE 1/V.27

| Tribit values | | | Phase change (see Note) |
|---|---|---|---|
| 0 | 0 | 1 | 0° |
| 0 | 0 | 0 | 45° |
| 0 | 1 | 0 | 90° |
| 0 | 1 | 1 | 135° |
| 1 | 1 | 1 | 180° |
| 1 | 1 | 0 | 225° |
| 1 | 0 | 0 | 270° |
| 1 | 0 | 1 | 315° |

*Note.* – The phase change is the actual on-line phase shift in the transition region from the end of one signalling element to the beginning of the following signalling element.

3. *Data signalling and modulation rates*

The data signalling rate shall be 4800 bits per second $\pm$ 0.01%, i.e. the modulation rate is 1600 bauds $\pm$ 0.01%.

4. *Received signal frequency tolerance*

The carrier frequency tolerance allowance at the transmitter is $\pm$ 1 Hz and assuming a maximum frequency drift of $\pm$ 6 Hz in the connection between the modems, then the receiver must be able to accept errors of at least $\pm$ 7 Hz in the received frequencies.

5. *Backward channel*

The modulation rate, characteristic frequencies, tolerances, etc. to be as recommended for backward channel in Recommendation V.23. This does not preclude the use of a higher speed backward channel with operational capability of 75 bauds or higher, bearing the same characteristic frequencies as the V.23 backward channel.

6.     *List of essential interchange circuits* (see Table 2/V.27)

**TABLE 2/V.27**

| Interchange circuit (see Note 1) | | Forward (data) channel half-duplex or full duplex | |
|---|---|---|---|
| **No.** | **Designation** | **Without backward channel** | **With backward channel** |
| 102 | Signal ground or common return | X | X |
| 102a (Note 1) | DTE common return | X | X |
| 102b (Note 1) | DCE common return | X | X |
| 103 | Transmitted data | X | X |
| 104 | Received data | X | X |
| 105 (see Note 2) | Request to send | X | X |
| 106 | Ready for sending | X | X |
| 107 | Data set ready | X | X |
| 108/1 | Connect data set to line | X | X |
| 109 | Dwa channel received line signal detector | X | X |
| 113 | Transmitter signal element timing (DCE source) | X | X |
| 114 | Transmitter signal element timing (DTE source) | X | X |
| 115 | Receiver signal element timing (DCE source) | X | X |
| 118 | Transmitted backward channel data | | X |
| 119 | Received backward channel data | | X |
| 120 | Transmit backward channel line signal | | X |
| 121 | Backward channel ready | | X |
| 122 | Backward channel received line signal detector | | X |

*Note 1.*—**Interchange circuits 102a and 102b are required where the electrical characteristics defined in Recommendation V.10 are used.**

*Note 2.* – Not essential for 4-wire full-duplex continuous carrier operation.

7.     *Threshold and response times of circuit 109*

A fall in level of the incoming line signal to $-31$ dBm or lower for more than $10 \pm 5$ ms will cause circuit 109 to be turned OFF. An increase in level to $-26 \pm 1$ dBm or higher will turn this circuit ON after a delay of

*a)*    $13 \pm 3$ ms for fast operations,

*b)*    100 ms to 1200 ms for slow operation,

where the choice of the delay for slow operation depends upon the application. Delays within the range of *b)* may be provided for 4-wire full-duplex continuous carrier operation.

8.     *Timing arrangements*

Clocks should be included in the modem to provide the data terminal equipment with transmitter signal element timing (Recommendation V.24, circuit 114) and receiver signal element timing (Recommendation V.24, circuit 115). Alternatively, the transmitter signal element timing may be originated in the data terminal equipment and be transferred to the modem via the appropriate interchange circuit (Recommendation V.24, circuit 113).

### 9. Electrical characteristics of interchange circuits

**a)** Use of electrical characteristics conforming to Recommendation V.28 is recommended with the connector pin assignment plan specified by ISO DIS 2110.

**b)** Application of electrical characteristics conforming to Recommendation V.10 and V.11 is recognized as an alternative together with the use of the connectors and pin assignment plan specified by ISC DIS 4902.

**i)** Concerning circuits 103, 104, 105 (where used), 106, 107, 108, 109, 113, 114 and 115, the receivers shall be in accordance with Recommendation V.11 or, alternatively, Recommendation V.10, Category 1. Either V.10 or V.11 generators may be utilized.

**ii)** In the case of circuits 118, 119, 120, 121 and 122, Recommendation V.10 applies with receivers configured as specified by Recommendation V.10 for Category 2.

**iii)** It is preferred that backward channel circuits appear on a separate connector and comprise circuits 118, 119, 120, 121, 122 (Category 2) and 102, 102a and 102b.

**iv)** Interworking between equipment applying Recommendation V.10 and/or V.11 and equipment applying Recommendation V.28 is allowed on a noninterference basis. The onus for adaptation to V.28 equipment rests solely with the alternative V.10/V.11 equipment.

*Note*—Manufacturers may wish to note that the long-term objective is to replace electrical characteristics specified in Recommendation V.28, and Study Group XVII has agreed that the work shall proceed to develop a more efficient all balanced interface for the V-Series application which minimizes the number of interchange circuits. It is expected that this work would be based upon the alternative application given in 9 b) above utilizing the V.11 electrical characteristics.

**10.** The following information is provided to assist equipment manufacturers:

- the data modem should have no adjustment for send level or receive sensitivity under the control of the operator;

- no fall-back rate has been included because the convenient rate would be 3200 bit/s, not a permitted rate;

- circuit 108/2 has not been included in the list of interchange circuits because it was considered that the modem would not be suitable for switched network use until an automatic equalizer had been recommended.

**11.** *Synchronizing signal*

During the interval between the OFF to ON transition of circuit 105 and the OFF to ON transition of circuit 106, synchronizing signals for properly conditioning the receiving modem must be generated by the transmitting modem. These signals are defined as

*a)* signals to establish basic demodulator requirements;

*b)* signals to establish scrambler synchronization.

The actual composition of the synchronization signals is continuous 180 degrees phase reversals on line for $9 \pm 1$ ms followed by continuous 1s at the input to the transmit scrambler for *b)*. Condition *b)* shall be *sustained until the OFF to ON transition of circuit 106.*

**12.** *Response time for circuit 106*

The time between the OFF to ON transition of circuit 105 and the OFF to ON transition of circuit 106 shall be optionally 20 ms $\pm$ 3 ms or 50 ms $\pm$ 20 ms.

**13.** *Line signal characteristics*

A 50% raised cosine energy spectrum shaping is equally divided between the receiver and transmitter.

## 14. *Scrambler*

A self-synchronizing scrambler/descrambler having the generating polynomial $1 + x^{-6} + x^{-7}$, with additional guards against repeating patterns of 1, 2, 3, 4, 6, 9 and 12 bits, shall be included in the modem. The Appendix shows a suitable logical arrangement.

At the transmitter the scrambler shall effectively divide the message polynomial, of which the input data sequence represents the coefficients in descending order, by the scrambler generating polynomial, to generate the transmitted sequence, and at the receiver the received polynomial, of which the received data sequence represents the coefficients in descending order, shall be multiplied by the scrambler generating polynomial to recover the message sequence.

The detailed scrambling and descrambling processes are described in the Appendix.

## 15. *Equalizer*

A manually adjustable equalizer with the capability of compensating for the amplitude and group delay distortion within the limits of Recommendation M.1020 shall be provided in the receiver. The transmitter shall be able to send an equalization pattern while the receiver shall incorporate a means of indicating correct adjustment of the equalizer controls. The equalizer pattern is generated by applying continuous 1s to the input of the transmitter scrambler defined above.

## 16. *Alternative equalization and scrambler techniques*

This Recommendation does not preclude the use of alternative equalization techniques, for example manually adjustable transmit equalizers for use in multipoint polled networks and for point-to-point networks with an unattended location.

These techniques, and their incorporation in the modem, and a new scrambler, should be the subject of further study.

*Note.* — For modems with automatic adaptive equalizers, see Recommendation V.27 *bis*.

# Appendix

## (to Recommendation V.27)

### Detailed scrambling and descrambling processes

### 1. *Scrambling*

The message polynomial is divided by the generating polynomial $1 + x^{-6} + x^{-7}$. (See Figure 1/V.27.) The coefficients of the quotient of this division taken in descending order from the data sequence to be transmitted.

The transmitted bit sequence is continuously searched over a span of 45 bits for sequences of the form

$$p(x) = \sum_{i=0}^{32} a_i x^i$$

where $a_i = 1$ or $0$ and $a_i = a_{i+9}$ or $a_{i+12}$

If such a sequence occurs, the bit immediately following the sequence is inverted before transmission.

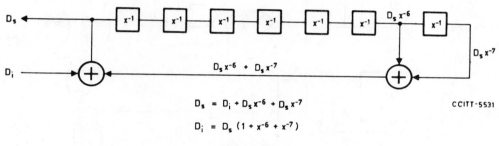

$$D_s = D_i + D_s x^{-6} + D_s x^{-7}$$

$$D_i = D_s (1 + x^{-6} + x^{-7})$$

CCITT-5531

FIGURE 1/V.27

### 2. *Descrambling*

At the receiver the incoming bit sequence is continuously searched over a span of 45 bits for sequences of the form $p(x)$. If such a sequence occurs, the bit immediately following the sequence is inverted. The polynomial represented by the resultant sequence is then multiplied by the generating polynomial $1 + x^{-6} + x^{-7}$ to form the recovered message polynomial. The coefficients of the recovered polynomial, taken in descending order, form the output data sequence.

### 3. *Elements of scrambling process*

The factor $1 + x^{-6} + x^{-7}$ randomizes the transmitted data over a sequence length of 127 bits.

The equality $a_i = a_{i+9}$ in the guard polynomial $p(x)$ prevents repeated patterns of 1, 3 and 9 bits from occurring for more than 42 successive bits.

The equality $a_i = a_{i+12}$ in $p(x)$ prevents repeated patterns of 2, 4, 6 and 12 bits from occurring for more than 45 successive bits.

4. Figure 2/V.27 is given as an indication only, since with another technique this logical arrangement might take another form.

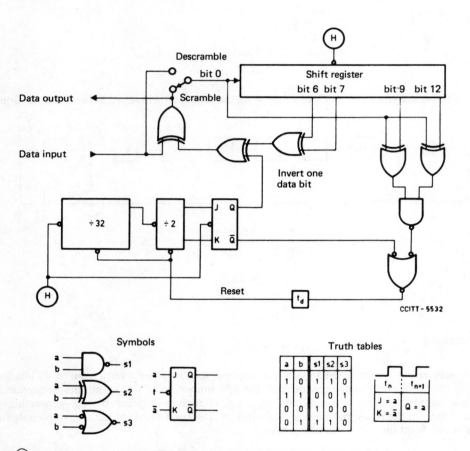

124

*Note 1.* – Ⓗ represents the clock signal. The negative going transition is the active transition.

*Note 2.* – There is a delay time, due to physical circuits, between a negative going transition of Ⓗ and the end of the "0" state represented by $t_d$ on the non-RESET wire; therefore the first coincidence between bit 0 and bit 9 or bit 12 is not taken into account by the counter.

*Note 3.* – The same voltage convention is used for data signals and logical circuits in the diagram.

FIGURE 2/V.27 – **An example of scrambler and descrambler circuitry**

**Recommendation V.27** *bis*

**4800 BITS PER SECOND MODEM WITH AUTOMATIC EQUALIZER**
**STANDARDIZED FOR USE ON LEASED TELEPHONE-TYPE CIRCUITS**

*(Geneva, 1976)*

*Introduction*

This modem is intended to be used over any general leased circuits not necessarily conforming to Recommendation M.1020. A provision for a fast start-up sequence is made to allow the use of this modem for multipoint polling applications if the circuits used conform to Recommendation M.1020.

On leased circuits, considering that there exist and will come into being many modems with features designed to meet the requirements of the Administrations and users, this Recommendation in no way restricts the use of any other modems. This Recommendation does not eliminate the need for manually equalized modems according to Recommendation V.27 or application of other automatically equalized 4800 bits per second modems.

The provisions of this Recommendation are to be regarded as provisional in order to provide service where it is urgently required and between locations where it is expected that a reasonably satisfactory service can be given.

1.      *Principal characteristics*

The principal characteristics for this recommended modem are very similar to the characteristics of a modem conforming to Recommendation V.27 with the exception of the equalizer used and these characteristics are as follows:

*a)*   operates in a full-duplex or half-duplex mode over 4-wire leased circuits or in a half-duplex mode over 2-wire leased circuits;

*b)*   at 4800 bits per second operation, modulation is 8-phase differentially encoded as described in Recommendation V.27;

*c)*   reduced rate capability at 2400 bits per second with 4-phase differentially encoded modulation scheme as described in Recommendation V.26, Alternative A;

*d)*   possibility of a backward (supervisory) channel at modulation rates up to 75 bauds in each direction of transmission, the provision and the use of these channels being optional;

*e)*   inclusion of an automatic adaptive equalizer with a specific start-up sequence for Recommendation M.1020 lines and an alternate start-up sequence for much lower grade lines.

2.      *Line signals at 4800 and 2400 bits per second operation*

2.1     *Carrier frequency*

The carrier frequency is to be 1800 ± 1 Hz. No separate pilot tones are provided. The power levels used will conform to Recommendation V.2.

2.1.1   *Spectrum at 4800 bits per second*

A 50% raised cosine energy spectrum shaping is equally divided between the receiver and transmitter. The energy density at 1000 Hz and 2600 Hz shall be attenuated 3.0 dB ± 2.0 dB with respect to the maximum energy density between 1000 Hz and 2600 Hz.

### 2.1.2 Spectrum at 2400 bits per second

A minimum of 50% raised cosine energy spectrum shaping is equally divided between the receiver and transmitter. The energy density at 1200 Hz and 2400 Hz shall be attenuated 3.0 dB ± 2.0 dB with respect to the maximum energy density between 1200 Hz and 2400 Hz.

### 2.2 Division of power between the forward and backward channel

If simultaneous transmission of the forward and backward channels occur in the same direction, a backward channel should be 6 dB lower in power level than the forward (data) channel.

### 2.3 Operation at 4800 bits per second

### 2.3.1 Data signalling and modulation rate

The data signalling rate shall be 4800 bits per second ± 0.01%, i.e. the modulation rate is 1600 bauds ± 0.01%.

### 2.3.2 Encoding data bits

The data stream to be transmitted is divided into groups of three consecutive bits (tribits). Each is encoded as a phase change relative to the phase of the preceding signal tribit element (see Table 1/V.27 bis). At the receiver, the tribits are decoded and the bits are reassembled in correct order. The left-hand digit of the tribit is the one occurring first in the data stream as it enters the modulator portion of the modem after the scrambler.

TABLE 1/V.27 bis

| Tribit values | | | Phase change (see Note) |
|---|---|---|---|
| 0 | 0 | 1 | 0° |
| 0 | 0 | 0 | 45° |
| 0 | 1 | 0 | 90° |
| 0 | 1 | 1 | 135° |
| 1 | 1 | 1 | 180° |
| 1 | 1 | 0 | 225° |
| 1 | 0 | 0 | 270° |
| 1 | 0 | 1 | 315° |

*Note.* -- The phase change is the actual on-line phase shift in the transition region from the end of one signalling element to the beginning of the following signalling element.

### 2.4 Operation at 2400 bits per second

### 2.4.1 Data signalling and modulation rate

The data signalling rate shall be 2400 bits per second ± 0.01%, i.e. the modulation rate is 1200 bauds ± 0.01%.

## 2.4.2   *Encoding data bits*

At 2400 bits per second the data stream is divided into groups of two bits (dibits). Each dibit is encoded as a phase change relative to the phase of the immediately preceding signal element (see Table 2/V.27 *bis*). At the receiver, the dibits are decoded and reassembled in the correct order. The left-hand digit of the dibit is the one occurring first in the data stream as it enters the modulator portion of the modem after the scrambler.

TABLE 2/V.27 *bis*

| Dibit value | Phase change (see Note) |
|---|---|
| 00 | 0° |
| 01 | 90° |
| 11 | 180° |
| 10 | 270° |

*Note.* – The phase change is the actual on-line phase shift in the transition region from the end of one signalling element to the beginning of the following signalling element.

## 2.5   *Operating sequences*

### 2.5.1   *"Turn-on" sequence*

During the interval between the OFF to ON transition of circuit 105 and the OFF to ON transition of circuit 106, synchronizing signals for proper conditioning of the receiving modem must be generated by the transmitting modem. These are signals to establish carrier detection, AGC if required, timing synchronization, equalizer convergence and descrambler synchronization.

Two sequences are defined, i.e.,

*a)*   a short one for 4-wire circuits conforming to Recommendation M.1020 operation,

*b)*   a long one for 4-wire circuits which are much worse than Recommendation M.1020 and for 2-wire circuits.

The sequences, for both data rates, are divided into three segments as in Table 3/V.27 *bis*.

TABLE 3/V.27 *bis*

| Type of line signal | Segment 1 Continuous 180° phase reversals | Segment 2 0°-180° 2-phase equalizer conditioning pattern | Segment 3 Continuous scrambled ONEs | Total of segments 1, 2 and 3 Total "Turn-on" sequence time | |
|---|---|---|---|---|---|
| | | | | 4800 | 2400 |
| Number of symbol intervals | *a)* 14 SI* *b)* 50 SI | *a)* 58 SI *b)* 1074 SI | 8 SI | *a)* 50 ms *b)* 708 ms | *a)* 67 ms *b)* 943 ms |

\* SI = symbol intervals. The duration of segments 1, 2 and 3 are expressed in number of symbol intervals, these numbers being the same in fallback operation.

2.5.1.1    The composition of Segment 1 is continuous 180° phase reversals on line for 14 symbol intervals in the case of sequence *a)*, for 50 symbol intervals in the case of sequence *b)*.

2.5.1.2    Segment 2 is composed of an equalizer conditioning pattern which is a pseudo-random sequence generated by the polynomial:

$$1 + x^{-6} + x^{-7}$$

When the pseudo-random sequence contains a ZERO, 0° phase change is transmitted; when it contains a ONE, 180° phase change is transmitted. Segment 2 begins with the sequence 0°, 180°, 180°, 180°, 180°, 180°, 0°, ... according to the pseudo-random sequence and continues for 58 symbol intervals in the case of sequence *a)* and 1074 symbol intervals in the case of sequence *b)*. The detailed pseudo-random generation is described in the Appendix to this Recommendation.

2.5.1.3    Segment 3 commences transmission according to the encoding described in 2.3 and 2.4 above with continuous data ONEs applied to the input of the data scrambler. Segment 3 is 8 symbol intervals. At the end of Segment 3, circuit 106 is turned ON and user data are applied to the input of the data scrambler.

### 2.5.2    *"Turn-off" sequence*

The line signal emitted after the ON to OFF transition of circuit 105 is divided into two segments as shown in Table 4/V.27 *bis*.

TABLE 4/V.27 *bis*

| | Segment A | Segment B | Total of segments A and B |
|---|---|---|---|
| Type of line signal | Remaining data followed by continuous scrambled ONEs | No transmitted energy | Total "Turn-off" time |
| Duration | 5 ms to 10 ms | 20 ms | 25 to 30 ms |

If an OFF to ON transition of circuit 105 occurs during the turn-off sequence, it will not be taken into account until the end of the turn-off sequence.

### 3.    *Received signal frequency tolerance*

Noting that the carrier frequency tolerance allowance of the transmitter is ± 1 Hz and assuming a maximum drift of ± 6 Hz in the connection between the modems, then the receiver must be able to accept errors of at least ± 7 Hz in the received frequencies.

### 4.    *Backward channel*

The modulation rate, characteristic frequencies, tolerances, etc. to be as recommended for the backward channel in Recommendation V.23. This does not preclude the use of a higher speed backward channel with operational capability of 75 bauds or higher, bearing the same characteristic frequencies as the V.23 backward channel.

## 5.    Interchange circuits

### 5.1    List of essential interchange circuits (Table 5/V.27 bis)

**TABLE 5/V.27** bis

| Interchange circuit (see Note) | | Forward (data) channel half-duplex or full duplex | |
|---|---|---|---|
| **No.** | **Designation** | **Without backward channel** | **With backward channel** |
| 102 | Signal ground or common return | X | X |
| 102a (Note 1) | DTE Common return | X | X |
| 102b (Note 1) | DCE common return | X | X |
| 103 | Transmitted data | X | X |
| 104 | Received data | X | X |
| 105 | Request to send | X | X |
| 106 | Ready for sending | X | X |
| 107 | Data set ready | X | X |
| 108/1 | Connect data set to line | X | X |
| 109 | Data channel received line signal detector | X | X |
| 111 | Data signal rate selector (DTE source) | X | X |
| 113 | Transmitter signal element timing (DCE source) | X | X |
| 114 | Transmitter signal element timing (DCE source) | X | X |
| 115 | Receiver signal element timing (DCE source) | X | X |
| 118 | Transmitted backward channel data | | X |
| 119 | Received backward channel data | | X |
| 120 | Transmit backward channel line signal | | X |
| 121 | Backward channel ready | | X |
| 122 | Backward channel received line signal detector | | X |

*Note 1.*—Interchange circuits 102a and 102b are required where the electrical characteristics defined in Recommendation V.10 are used.

### 5.2    Response times of circuits 106, 109, 121 and 122 (Table 6/V.27 bis)

TABLE 6/V.27 *bis* – **Response times**

| Circuit 106 | 4800 bits per second | 2400 bits per second |
|---|---|---|
| OFF to ON | *a)*   50 ms <br> *b)* 708 ms | *a)*   67 ms <br> *b)* 944 ms |
| ON to OFF | ≤ 2 ms | |
| **Circuit 109** | | |
| OFF to ON | See 5.2.1 | |
| ON to OFF | 5 to 15 ms | |
| **Circuit 121** | | |
| OFF to ON | 80 to 160 ms | |
| ON to OFF | ≤ 2 ms | |
| **Circuit 122** | | |
| OFF to ON | < 80 ms | |
| ON to OFF | 15 to 80 ms | |

*Note. − a)* and *b)* refer to sequence *a)* and sequence *b)* as defined in 2.5.

5.2.1    *Circuit 109*

Circuit 109 must run ON after synchronizing is completed and prior to user data appearing on circuit 104.

5.2.2    Circuit 106 response times are from the connection of an ON or OFF condition on circuit 105 to the appearance of the corresponding ON or OFF condition on circuit 106.

5.3    *Threshold of data channel and backward channel received line signal detectors*

Levels of received line signal at receiver line terminals:

— *For use over ordinary quality leased circuits* (ref. Recommendation M.1040)

Threshold for circuits 109/122:

— greater than −43 dBm:        OFF to ON

— less than −48 dBm:        ON to OFF

— *For use over special quality leased circuits* (ref. Recommendation M.1020)

Threshold for circuit 109:

— greater than −26 dBm:        OFF to ON

— less than −31 dBm:        ON to OFF

Threshold for circuit 122:

— greater than −34 dBm:        OFF to ON

— less than −39 dBm:        ON to OFF

The condition of circuits 109 and 122 for levels between the above levels is not specified except that the signal detectors shall exhibit a hysteresis action such that the level at which the OFF to ON transition occurs is at least 2 dB greater than that for the ON to OFF transition.

5.4    *Clamping to binary condition 1 of circuit 104 (Received data)*

Two options shall be provided in the modem:

*a)*    When clamping is not used there is no inhibition of the signals on circuit 104. There is no protection against noise, supervisory and control tones, switching transients, etc. from appearing on circuit 104.

*b)*    When clamping is used, circuit 104 is held in a marking condition (binary 1) under the conditions defined below. When these conditions do not exist, the clamp is removed and circuit 104 can respond to the input signals of the modem:

— when circuit 109 is in the OFF condition;

— when circuit 105 is in the ON condition and the modem is used in half-duplex mode (turn-around systems).

6.    *Timing arrangement*

Clocks should be included in the modem to provide the data terminal equipment with transmitter element timing (Recommendation V.24, circuit 114) and receiver signal element timing (Recommendation V.24, circuit 115). The transmitter element timing may be originated in the data terminal equipment and be transferred to the modem via the appropriate interchange circuit (Recommendation V.24, circuit 113.).

## 7. *Electrical characteristics of interchange circuits*

**a)** Use of electrical characteristics conforming to Recommendation V.28 is recommended with the connector pin assignment plan specified by ISO DIS 2110.

**b)** Application of electrical characteristics conforming to Recommendation V.10 and V.11 is recognized as an alternative together with the use of the connectors and pin assignment plan specified by ISC DIS 4902.

i) Concerning circuits 103, 104, 105, 106, 107, 108, 109, 113, 114 and 115, the receivers shall be in accordance with Recommendation V.11 or, alternatively, Recommendation V.10, Category 1. Either V.10 or V.11 generators may be utilized.

ii) In the case of circuits 111, 118, 119, 120, 121 and 122, Recommendation V.10 applies with receivers configured as specified by Recommendation V.10 for Category 2.

iii) It is preferred that backward channel circuits appear on a separate connector and comprise circuits 118, 119, 120, 121, 122 (Category 2) and 102, 102a and 102b.

iv) Interworking between equipment applying Recommendation V.10 and/or V.11 and equipment applying Recommendation V.28 is allowed on a noninterference basis. The onus for adaptation to V.28 equipment rests solely with the alternative V.10/V.11 equipment.

*Note.* — Manufacturers may wish to note that the long-term objective is to replace electrical characteristics specified in Recommendation V.28, and Study Group XVII has agreed that the work shall proceed to develop a more efficient all balanced interface for the V-Series application which minimizes the number of interchange circuits. It is expected that this work would be based upon the alternative application given in 7 b) above utilizing the V.11 electrical characteristics.

**8.**     The following information is provided to assist equipment manufacturers:

The data modem should have no adjustment for send level or receive sensitivity under the control of the operator.

At 4800 bits per second operation, the transmitter energy spectrum shall be shaped in such a way that when continuous data ONEs are applied to the input of the scrambler, the resulting transmitted spectrum shall have a substantially linear phase characteristic over the band of 1100 Hz to 2500 Hz.

At 2400 bits per second operation, the transmitter energy spectrum shall be shaped in such a way that when continuous data ONEs are applied to the input of the scrambler, the resulting transmitted spectrum shall have a substantially linear phase-characteristic over the band of 1300 Hz to 2300 Hz.

**9.**     *Equalizer*

An automatic adaptive equalizer shall be provided in the receiver. The receiver shall incorporate a means of detecting loss of equalization and be able to recover equalization from the normal data-modulated received line signal without initiating a new synchronizing signal from the distant transmitter.

**10.**     *Scrambler*

A self-synchronizing scrambler/descrambler having the generating polynomial:

$$1 + x^{-6} + x^{-7}$$

with additional guards against repeating pattern of 1, 2, 3, 4, 6, 8, 9 and 12 bits, shall be included in this modem. In the Appendix, Figure 2/V.27 *bis* shows a suitable logical arrangement (see Note). The scrambler/descrambler is the same as that in Recommendation V.27 with the addition of circuitry to guard against repeating patterns of 8 bits.

*Note.* — Figures 1/V.27 *bis* and 2/V.27 *bis* in the Appendix are given as an indication only, since with another technique these logical arrangements might take another form.

At the transmitter the scrambler shall effectively divide the message polynomial, of which the input data sequence represents the coefficients in descending order, by the scrambler generating polynomial to generate the transmitted sequence, and at the receiver the received polynomial, of which the received data sequence represents the coefficients in descending order, shall be multiplied by the scrambler generating polynomial to recover the message sequence.

## 11. *Options*

Since this modem is equipped with an automatic adaptive equalizer, and can operate on 2-wire lines, operation over the general switched network is possible. Thus, in the event of failure of the leased line, the general switched network may serve as a stand-by facility.

Options can be added to this modem in order to allow the use of the general switched network when the leased line fails. These options can also be added for use on 2-wire leased lines where echo protection is required.

Additional information for these options can be found in Recommendation V.27 *ter*.

Appendix

(to Recommendation V.27 *bis*)

**A two phase equalizer training generator for 4800 bit/s**

Rapid convergence for the equalizer with the least amount of circuitry is more readily accomplished by sending only an in-phase or out-of-phase carrier during training. This implies that the only tribits sent to the modulator will be 001 (0° phase) or 111 (180° phase). Refer to Figure 1/V.27 *bis* for circuitry to generate the sequence and Figure 3/V.27 *bis* for timing the sequence.

Let T1 be a timing signal equal to 1600 Hz (symbol clock), that is true (high) for one 4800 Hz period, and low for two 4800 Hz clock periods. T2 is the inversion of T1.

During T1 select the input to the scrambler, during T2 select the first stage of the scrambler. During the period when T2 is high, C forces the output high. This may be accomplished by circuitry shown in Figure 2/V.27 *bis*.

If T1 is forced continually high and T2 is forced continually low, normal operation is restored.

In order to ensure consistent training, the same pattern should always be sent. To accomplish this, the data input to the scrambler should be in mark hold during the training, and the first seven stages of the scrambler should be loaded with 0011110 (right-hand-most first in time) on the first coincidence on T1 and the signal that will cause the mute should be removed from the transmitter output. [Generally this signal will be *Request to send* (RTS)].

This particular starting point was chosen in order to ensure a pattern that has continuous 180° phase reversals at the beginning in order to ensure rapid clock acquisition, followed by a pattern that will ensure rapid equalizer convergence.

Within eight symbol intervals prior to the ON condition of *Ready for sending* (RFS), the scrambler should be switched to normal operation, keeping the scrambler in mark hold until RFS, to synchronize the descrambler.

*Note.* — At 2400 bits per second, a similar technique may be used with appropriate clocking changes.

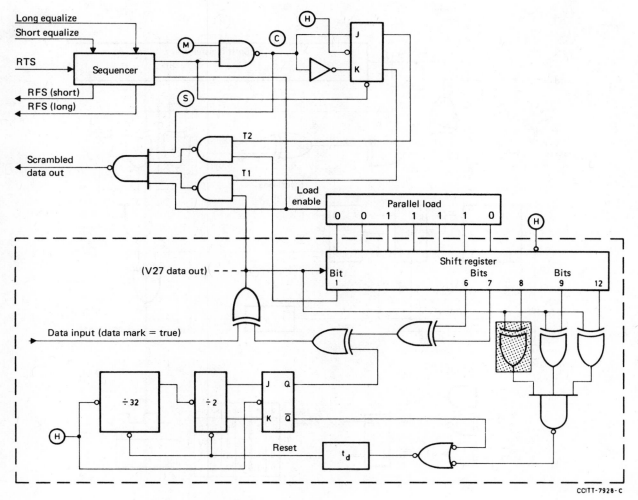

CCITT-7928-C

*Note 1.* – The dotted line encloses the V27 scrambler.

*Note 2.* – Shaded rectangle is for guarding against 8-bit repeating pattern.

*Note 3.* – Ⓗ is 3 times baud rate clock.

*Note 4.* – Ⓜ is baud rate clock (1600 Hz).

*Note 5.* – Diagrams shown with positive logic.

*Note 6.* – Signals Ⓒ and Ⓢ are identified only to correlate with Figure 3/V.27 *bis*.

FIGURE 1/V.27 *bis* – **An example of sequence generator and scrambler circuitry**

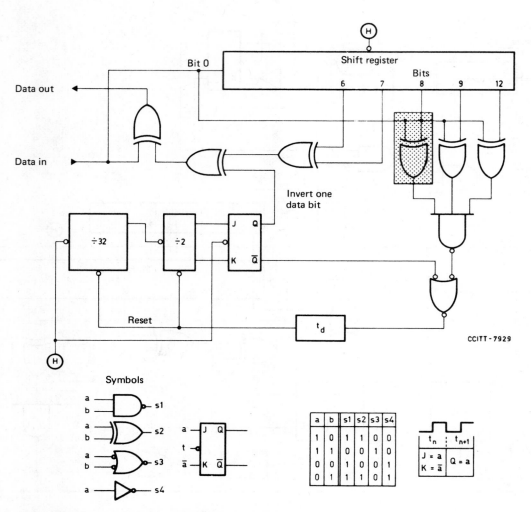

*Note 1.* – Shaded rectangle is for guarding against 8-bit repeating pattern.
*Note 2.* – Ⓗ represents clock signal. The negative going transition is the active transition.
*Note 3.* – There is a delay time due to physical circuits between a negative going transition of Ⓗ and the end of the "0" state represented by $t_d$ on the non-reset wire; therefore the first coincidence between bit 0 and bit 8 or bit 9 or bit 12 is not taken into account by the counter.

FIGURE 2/V.27 *bis* – **An example of descrambler circuitry**

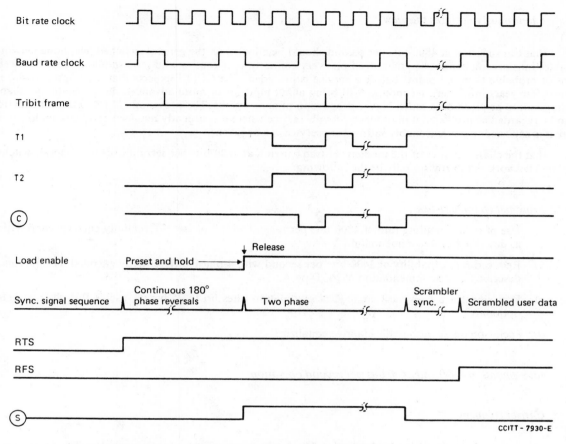

Bit rate clock

Baud rate clock

Tribit frame

T1

T2

C

Load enable    Preset and hold ——→ ↓ Release

Sync. signal sequence | Continuous 180° phase reversals | Two phase | Scrambler sync. | Scrambled user data

RTS

RFS

S

CCITT - 7930-E

FIGURE 3/V.27 *bis* — **Synchronizing signal sequence for 4800 bit/s** (see Figure 1/V.27 *bis*)

**Recommendation V.27 *ter***

### 4800/2400 BITS PER SECOND MODEM STANDARDIZED FOR USE IN THE GENERAL SWITCHED TELEPHONE NETWORK

*(Geneva, 1976)*

The CCITT,

*considering*

a)   that there is a demand for data transmission at 4800 bits per second over the general switched telephone network;

b)   that a majority of connections over the general switched telephone network within some countries are capable of carrying data at 4800 bits per second;

c)   that a lower proportion of international connections in the general switched telephone network are capable of carrying data at 4800 bits per second;

d)   that other international connections in the general switched telephone network may still support operations at 2400 bits per second using a built-in fallback capability, *unanimously declares the view*

that transmission at 4800 bits per second should be allowed on the general switched telephone network. Reliable transmission cannot be guaranteed on every connection or routing and tests should be made between the most probable terminal points before a service is provided. The CCITT expects that developments during the next few years in modern technology will bring about modems of more advanced design enabling reliable transmission to be given on a much higher proportion of connections. The provisions of this Recommendation are to be regarded as provisional in order to provide service where it is urgently required and between locations where it is expected that a reasonably satisfactory service can be given;

that the characteristics of the modem for transmission at 4800 bits per second over the general switched telephone network shall provisionally be the following:

1. *Principal characteristics*

   a) Use of data signalling rate of 4800 bits per second with 8-phase differentially encoded modulation as described in Recommendation V.27.

   b) Reduced rate capability at 2400 bits per second with 4-phase differentially encoded modulation as described in Recommendation V.26, Type A.

   c) Provision for a backward channel at modulation rates up to 75 bauds, use of this channel being optional.

   d) Inclusion on an automatic adaptive equalizer.

2. *Line signals at 4800 and 2400 bits per second operation*

2.1 *Carrier frequency*

The carrier frequency is to be $1800 \pm 1$ Hz. No separate pilot tones are provided. The power levels used will conform to Recommendation V.2.

2.1.1 *Spectrum at 4800 bits per second*

The 50% raised cosine energy spectrum shaping is equally divided between the receiver and transmitter. The energy density at 1000 Hz and 2600 Hz shall be attenuated $3.0$ dB $\pm$ $2.0$ dB with respect to the maximum energy density between 1000 Hz and 2600 Hz.

2.1.2 *Spectrum at 2400 bits per second*

A minimum of 50% raised cosine energy spectrum shaping is equally divided between the receiver and transmitter. The energy density at 1200 Hz and 2400 Hz shall be attenuated $3.0$ dB $\pm$ $2.0$ dB with respect to the maximum energy denisty between 1200 Hz and 2400 Hz.

2.2 *Division of power between the forward and backward channel*

Equal division between the forward and backward channels is recommended (if provided).

2.3 *Operation at 4800 bits per second*

2.3.1 *Data signalling and modulation rate*

The data signalling rate shall be 4800 bits per second $\pm$ 0.01%, i.e. the modulation rate is 1600 bauds $\pm$ 0.01%.

### 2.3.2    *Encoding data bits*

The data stream to be transmitted is divided into groups of three consecutive bits (tribits). Each is encoded as a phase change relative to the phase of the preceding signal tribits element (see Table 1/V.27 *ter*). At the receiver, the tribits are decoded and the bits are reassembled in correct order. The left-hand digit of the tribit is the one occurring first in the data stream as it enters the modulator portion of the modem after the scrambler.

TABLE 1/V.27 *ter*

| Tribit values | | | Phase change (see Note) |
|---|---|---|---|
| 0 | 0 | 1 | 0° |
| 0 | 0 | 0 | 45° |
| 0 | 1 | 0 | 90° |
| 0 | 1 | 1 | 135° |
| 1 | 1 | 1 | 180° |
| 1 | 1 | 0 | 225° |
| 1 | 0 | 0 | 270° |
| 1 | 0 | 1 | 315° |

*Note.* – The phase change is the actual on-line phase shift in the transition region from the end of one signalling element to the beginning of the following signalling element.

### 2.4    *Operation at 2400 bits per second*

### 2.4.1    *Data signalling and modulation rate*

The data signalling rate shall be 2400 bits per second ± 0.01% i.e. the modulation rate is 1200 bauds ± 0.01%.

### 2.4.2    *Encoding data bits*

At 2400 bits per second the data stream is divided into groups of two bits (dibits). Each dibit is encoded as a phase change relative to the phase of the immediately preceding signal element (see Table 2/V.27 *ter*). At the receiver, the dibits are decoded and reassembled in the correct order. The left-hand digit of the dibit is the one occurring first in the data stream as it enters the modulator portion of the modem after the scrambler.

TABLE 2/V.27 *ter*

| Dibit values | Phase change (see Note) |
|---|---|
| 00 | 0° |
| 01 | 90° |
| 11 | 180° |
| 10 | 270° |

*Note.* – The phase change is the actual on-line phase shift in the transition region from the end of one signalling element to the beginning of the following signalling element.

## 2.5    *Operating sequences*

### 2.5.1    *"Turn-on" sequence*

During the interval between the OFF to ON transition of circuit 105 and the OFF to ON transition of circuit 106, synchronizing signals for proper conditioning of the receiving modem must be generated by the transmitting modem. These are signals to establish carrier detect, AGC if required, timing synchronization, equalizer convergence and descrambler synchronization.

The synchronizing signals are defined in two separate sequences with the long sequence used once at the beginning of the established connection and the short sequence used for subsequent turn-around in which the equalizer training pattern is used to update and refine equalizer convergence.

Two sequences are defined, i.e.,

a)    a short one for turn-around operation,

b)    a longer one for initial establishment of connection.

The sequence *b)* is only used after the first OFF to ON transition of circuit 105 following the OFF to ON transition of circuit 107, or at the OFF to ON transition of circuit 107 if the circuit 105 is already ON. After every subsequent OFF to ON transition of circuit 105, the sequence *a)* is used.

The sequences, for both data rates, are divided into five segments as in Table 3/V.27 *ter*.

TABLE 3/V.27 *ter*

| Type of line signal | Segment 1 | Segment 2 | Segment 3 | Segment 4 | Segment 5 | Total of segments 1, 2, 3, 4 and 5 | |
|---|---|---|---|---|---|---|---|
| | Unmodulated carrier | No trans-mitted energy | Continuous 180° phase reversals | 0° - 180° 2-phase equalizer conditioning pattern | Continuous scrambled ONEs | Nominal total "Turn-on" sequence time | |
| | | | | | | 4800 | 2400 |
| Protection against talker echo | 185 ms to 200 ms | 20 ms to 25 ms | a) 14 SI* b) 50 SI | a)   58 SI b) 1074 SI | 8 SI | a) 265 ms b) 923 ms | a)   281 ms b) 1158 ms |
| Without any protection | 0 ms | 0 ms | a) 14 SI b) 50 SI | a)   58 SI b) 1074 SI | 8 SI | a)   50 ms b) 708 ms | a)   66 ms b)  943 ms |

\* SI = symbol intervals. The duration of segments 3, 4 and 5 are expressed in number of symbol intervals, these numbers being the same in the fallback operation.

2.5.1.1    The composition of Segment 3 is continuous 180° phase reversals on line for 14 symbol intervals in the case of sequence *a)*, for 50 symbol intervals in the case of sequence *b)*.

2.5.1.2    Segment 4 is composed of an equalizer conditioning pattern which is a pseudo-random sequence generated by the polynomial:

$$1 + x^{-6} + x^{-7}$$

When the pseudo-random sequence contains a ZERO, 0° phase change is transmitted; when it contains a ONE, 180° phase change is transmitted. Segment 4 begins with the sequence 0°, 180°, 180°, 180°, 180°, 180°, 0°, . . . according to the pseudo-random sequence and continues for 58 symbol intervals in the case of sequence *a)* and 1074 symbol intervals in the case of sequence *b)*. The detailed pseudo-random generation is described in the Appendix to this Recommendation.

2.5.1.3     Segment 5 commences transmission according to the encoding described in 2.3 and 2.4 above with continuous data ONEs applied to the input of the data scrambler. Segment 5 is 8 symbol intervals. At the end of Segmnt 5, circuit 106 is turned ON and user data are applied to the input of the data scrambler.

### 2.5.2     *"Turn-off" sequence*

The line signal emitted after the ON to OFF transition of circuit 105 is divided into two segments as in Table 4/V.27 *ter*.

TABLE 4/V.27 *ter*

|  | Segment A | Segment B | Total "Turn-off" time |
|---|---|---|---|
| Type of line signals | Remaining data followed by continuous scrambled ONEs | No transmitted energy | Total of segments A and B |
| With or without protection against talker echo | 5 ms to 10 ms | 20 ms | 25 to 30 ms |

If an OFF to ON transition of circuit 105 occurs during the turn-off sequence, it will not be taken into account until the end of the turn-off sequence.

### 3.     *Received signal frequency tolerance*

Noting that the carrier frequency tolerance allowance of the transmitter is $\pm 1$ Hz and assuming a maximum drift of $\pm 6$ Hz in the connection between the modems, then the receiver must be able to accept errors of at least $\pm 7$ Hz in the received frequencies.

### 4.     *Backward channel*

The modulation rate, characteristic frequencies, tolerances, etc. to be as recommended for backward channel in Recommendation V.23. This does not preclude the use of a higher speed backward channel with operational capability of 75 bauds or higher, bearing the same characteristic frequencies as the V.23 backward channel.

### 5.     *Interchange circuits*

### 5.1     *Table of interchange circuits* (see notes 1 and 2 to Table 5/V.27 *ter*)

Interchange circuits essential for the modem when used on the general switched telephone network, including terminals equipped for manual calling or automatic calling or answering are as in Table 5/V.27 *ter*.

**VOLUME VIII.1  —  Rec. V.27 ter**

**TABLE 5/V.27** *ter*

| No. | Designation | Forward (data) channel one-way system | | | | Forward (data) channel either-way system | |
|---|---|---|---|---|---|---|---|
| | | Without backward channel | | With backward channel | | Without backward channel | With backward channel |
| | | Transmit end | Receive end | Transmit end | Receive end | | |
| 102 | Signal ground or common return | X | X | X | X | X | X |
| 102a (Note 1) | DTE common return | X | X | X | X | X | X |
| 102b (Note 2) | DCE common return | X | X | X | X | X | X |
| 103 | Transmitted data | X | | X | | X | X |
| 104 | Received data | | X | | X | X | X |
| 105 | Request to send | X | | X | | X | X |
| 106 | Ready for sending | X | | X | | X | X |
| 107 | Data set ready | X | X | X | X | X | X |
| 108/1 or 108/2 (see Note 3) | Connect data set to line / Data terminal ready | X | X | X | X | X | X |
| 109 | Data channel received line signal detector | | X | | X | X | X |
| 111 | Data signalling rate selector (DTE source) | X | X | X | X | X | X |
| 113 | Transmitter signal element timing (DTE source) | X | | X | | X | X |
| 114 | Transmitter signal element timing (DCE source) | X | | X | | X | X |
| 115 | Receiver signal element timing (DCE source) | | X | | X | X | X |
| 118 | Transmitted backward channel data | | | | X | | X |
| 119 | Received backward channel data | | | X | | | X |
| 120 | Transmit backward channel line signal | | | | | | X |
| 121 | Backward channel ready | | | | X | | X |
| 122 | Backward channel received line signal detector | | | X | | | X |
| 125 | Calling indicator | X | X | X | X | X | X |

*Note 1.*—Interchange circuits 102a and 102b are required where the electrical characteristics defined in Recommendation V.10 are used.

*Note 2.* – Interchange circuits indicated by X must be properly terminated according to Recommendation V.24 in the data terminal equipment and data circuit-terminating equipment.

*Note 3.* – This circuit shall be capable of operation as circuit 108/1 – *Connect data set to line* or circuit 108/2 – *Data terminal ready* depending on its use. For automatic calling it shall be used as 108/2 only.

5.2     *Response times of circuits 106, 109, 121 and 122* (see Tables 6/V.27 *ter* and 7/V.27 *ter*)

TABLE 6/V.27 *ter*

| Response times for operation at 4800 bits per second | | |
|---|---|---|
| Circuit 106 | With protection against talker echo | Without protection against talker echo |
| OFF to ON | a) 215 ± 10 ms + 50 ms<br>b) 215 ± 10 ms + 708 ms | a) 50 ms<br>b) 708 ms |
| ON to OFF | ⩽ 2 ms | ⩽ 2 ms |
| Circuit 109<br>OFF to ON | See 5.2.1 | See 5.2.1 |
| ON to OFF | 5 ms to 15 ms | 5 ms to 15 ms |
| Circuit 121<br>OFF to ON | 80 to 160 ms | 80 to 160 ms |
| ON to OFF | ⩽ 2 ms | ⩽ 2 ms |
| Circuit 122<br>OFF to ON | < 80 ms | < 80 ms |
| ON to OFF | 15 to 80 ms | 15 to 80 ms |

TABLE 7/V27 *ter*

| Response times for operation of 2400 bits per second | | |
|---|---|---|
| Circuit 106 | With protection against talker echo | Without protection against talker echo |
| OFF to ON | a) 215 ± 10 ms + 67 ms<br>b) 215 ± 10 ms + 944 ms | a) 67 ms<br>b) 944 ms |
| ON to OFF | ⩽ 2 ms | ⩽ 2 ms |
| Circuit 109<br>OFF to ON | See 5.2.1 | See 5.2.1 |
| ON to OFF | 5 to 15 ms | 5 to 15 ms |

*Note 1.* – *a)* and *b)* refer to sequence *a)* and sequence *b)* as defined in 2.5.

*Note 2.* – The parameters and procedures, particularly in the case of automatic calling and answering are provisional and are the subject of further study.

### 5.2.1    *Circuit 109*

Circuit 109 must turn ON after synchronizing is completed and prior to user data appearing on circuit 104. Circuit 109 is prevented from turning ON during reception of unmodulated carrier when the optional protection against talker echo is used.

5.2.2   *Circuit 106*

Circuit 106 response times are from the connection of an ON or OFF condition on:

— circuit 105 to the appearance of the corresponding ON or OFF condition on circuit 106; or,

— circuit 107 (where circuit 105 is already on) to the appearance of the corresponding ON or OFF condition on circuit 106.

5.3   *Threshold of data channel and backward channel received line signal detectors*

Level of received line signal at the receive line terminals of the modem for all types of connections, i.e. the general switched telephone network or non-switched 2-wire leased telephone circuits:

— greater than −43 dBm:        circuits 109/122 ON

— less than −48 dBm:           circuits 109/122 OFF

The condition of circuits 109 and 122 for levels between −43 dBm and −48 dBm is not specified except that the signal detectors shall exhibit a hysteresis action, such that the level at which the OFF to ON transition occurs is at least 2 dB greater than that for the ON to OFF transition.

Where transmission conditions are known and allowed, it may be desirable at the time of modem installation to change these response levels of the received line signal detector to less sensitive values (e.g. −33 dBm and −38 dBm respectively).

5.4   *Clamping to binary condition 1 of circuit 104 (received data)*

Two options shall be provided in the modem:

*a)*   When clamping is not used there is no inhibition of the signals on circuit 104. There is no protection against noise, supervisory and control tones, switching transients, etc. from appearing on circuit 104.

*b)*   When a clamping is used, circuit 104 is held in a marking condition (binary 1) under the conditions defined below. When these conditions do not exist, the clamp is removed and circuit 104 can respond to the input signals of the modem:

— when circuit 109 is in the OFF condition;

— when circuit 105 is in the ON condition and the modem is used in half-duplex mode (turn-around systems). To protect circuit 104 from false signals, a delay device shall be provided to maintain circuit 109 in the OFF condition for a period of 150 ± 25 ms after circuit 105 has been turned from ON to OFF. The use of this additional delay is optional.

6.   *Timing arrangement*

Clocks should be included in the modem to provide the data terminal equipment with transmitter element timing (Recommendation V.24, circuit 114) and receiver signal element timing (Recommendation V.24, circuit 115). The transmitter element timing may be originated in the data terminal equipment and be transferred to the modem via the appropriate interchange circuit (Recommendation V.24, circuit 113).

### 7. Electrical characteristics of interchange circuits

a) Use of electrical characteristics conforming to Recommendation V.28 is recommended with the connector pin assignment plan specified by ISO DIS 2110.

b) Application of electrical characteristics conforming to Recommendation V.10 and V.11 is recognized as an alternative together with the use of the connectors and pin assignment plan specified by ISO DIS 4902.

i) Concerning circuits 103, 104, 105, 106, 107, 108, 109, 113, 114 and 115, the receivers shall be in accordance with Recommendation V.11 or, alternatively, Recommendation V.10, Category 1. Either V.10 or V.11 generators may be utilized.

ii) In the case of circuits 111, 118, 119, 120, 121, 122 and 125, Recommendation V.10 applies with receivers configured as specified by Recommendation V.10 for Category 2.

iii) It is preferred that backward channel circuits appear on a separate connector and comprise circuits 118, 119, 120, 121, 122 (Category 2) and 102, 102a and 102b.

iv) Interworking between equipment applying Recommendation V.10 and/or V.11 and equipment applying Recommendation V.28 is allowed on a noninterference basis. The onus for adaptation to V.28 equipment rests solely with the alternative V.10/V.11 equipment.

*Note.*—Manufacturers may wish to note that the long-term objective is to replace electrical characteristics specified in Recommendation V.28, and Study Group XVII has agreed that the work shall proceed to develop a more efficient all balanced interface for the V-Series application which minimizes the number of interchange circuits. It is expected that this work would be based upon the alternative application given in 7 b) above utilizing the V.11 electrical characteristics.

**8.** The following information is provided to assist equipment manufacturers.

The data modem should have no adjustment for send level or receive sensitivity under the control of the operator.

At 4800 bits per second operation, the transmitter energy spectrum shall be shaped in such a way that when continuous data ONEs are applied to the input of the scrambler, the resulting transmitted spectrum shall have a substantially linear phase characteristic over the band of 1100 Hz to 2500 Hz.

At 2400 bits per second operation, the transmitter energy spectrum shall be shaped in such a way that when continuous data ONEs are applied to the input of the scrambler, the resulting transmitted spectrum shall have a substantially linear phase characteristic over the band of 1300 Hz to 2300 Hz.

**9.** It will be up to the user to decide whether, in view of the connection he makes with this system, he will have to request that the data circuit-terminating equipment be equipped with facilities for disabling echo suppressor. The international characteristics of the echo suppressor tone disabler have been standardized by the CCITT (C. of Recommendation G. 161) and the disabling tone should have the following characteristics:

— disabling tone transmitted: $2100 \pm 15$ Hz at a level of $-12 \pm 6$ dBm0;

— the disabling tone to last at least 400 ms; the tone disabler should hold in the disabled mode for any signal frequency sinusoid in the band from 390-700 Hz having a level of $-27$ dBm0 or greater, and from the band 700-3000 Hz having a level of $-31$ dBm0 or greater. The tone disabler should release for any signal in the band from 200-3400 Hz having a level of $-36$ dBm0 or less;

— the tolerable interruptions by the data signal to last not more than 100 ms.

**10.** *Equalizer*

An automatic adaptive equalizer shall be provided in the receiver.

## 11. *Scrambler*

A self-synchronizing scrambler/descrambler having the generating polynomial:

$$1 + x^{-6} + x^{-7}$$

with additional guards against repeating pattern of 1, 2, 3, 4, 6, 8, 9, and 12 bits, shall be included in this modem. In the Appendix, Figure 2/V.27 *ter* shows a suitable logical arrangement (see Note). The scrambler/descrambler is the same as that in Recommendation V.27 with the addition of circuitry to guard against repeating patterns of 8 bits.

*Note.* — Figures 1/V.27 *ter* and 2/V.27 *ter* in the Appendix are given as an indication only, since with another technique these logical arrangements might take another form.

At the transmitter the scrambler shall effectively divide the message polynomial, of which the input data sequence represents the coefficients in descending order, by the scrambler generating polynomial to generate the transmitted sequence, and at the receiver the received polynomial, of which the received data sequence represents the coefficients in descending order, shall be multiplied by the scrambler generating polynomial to recover the message sequence.

### Appendix

(to Recommendation V.27 *ter*)

### A two-phase equalizer training generator for 4800 bit/s

Rapid convergence for the equalizer with the least amount of circuitry is more readily accomplished by sending only an in-phase or out-of-phase carrier during training. This implies that the only tribits sent to the modulator will be 001 (0° phase) or 111 (180° phase). Refer to Figure 1/V.27 *ter* for circuitry to generate the sequence and Figure 3/V.27 *ter* for timing the sequence.

Let T1 be a timing signal equal to 1600 Hz (symbol clock), that is true (high) for one 4800 Hz period, and low for two 4800 Hz clock periods. T2 is the inversion of T1.

During T1, select the input to the scrambler, during T2 select the first stage of the scrambler. During the period when T2 is high, C forces the output high. This may be accomplished by circuitry shown in Figure 2/V.27 *ter*.

If T1 is forced continually high and T2 is forced continually low, normal operation is restored.

144

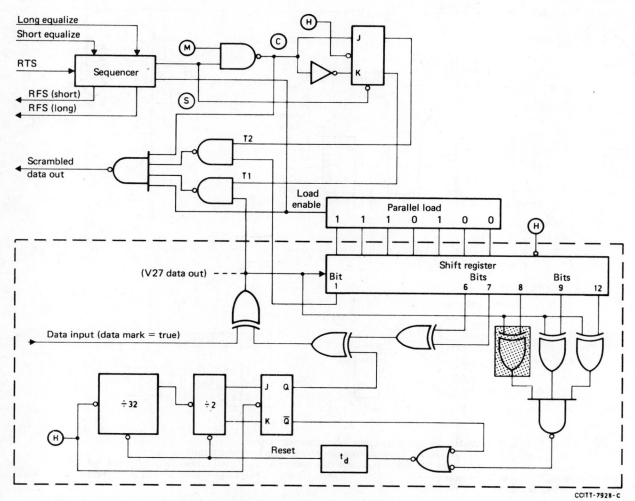

CCITT-7928-C

*Note 1.* – The dotted line encloses the V27 scrambler.

*Note 2.* – Shaded rectangle is for guarding against 8-bit repeating pattern.

*Note 3.* – (H) is 3 times baud rate clock.

*Note 4.* – (M) is baud rate clock (1600 Hz).

*Note 5.* – Diagrams shown with positive logic.

*Note 6.* – Signals (C) and (S) are identified only to correlate with Figure 3/V.27 *ter*.

FIGURE 1/V.27 *ter* – **An example of sequence generator and scrambler circuitry**

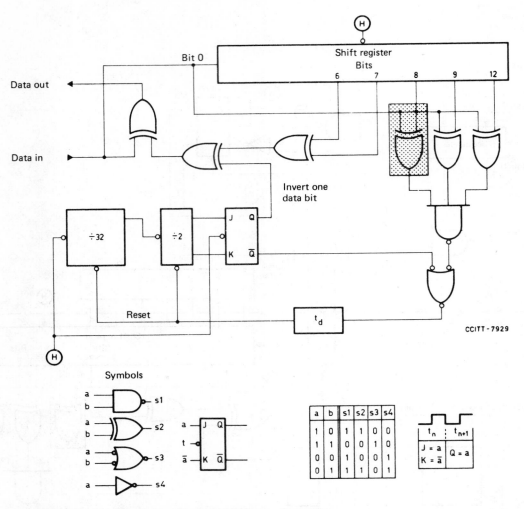

*Note 1.* – Shaded rectangle is for guarding against 8-bit repeating pattern.

*Note 2.* – Ⓗ represents clock signal. The negative going transition is the active transition.

*Note 3.* – There is a delay time due to physical circuits between a negative going transition of a Ⓗ and the end of the "0" state represented by $t_d$ on the non-reset wire; therefore the first coincidence between bit 0 and bit 8 or bit 9 or bit 12 is not taken into account by the counter.

FIGURE 2/V.27 *ter* – **An example of descrambler circuitry**

In order to insure consistent training, the same pattern should always be sent. To accomplish this, the data input to the scrambler should be in mark hold during the training, and the first seven stages of the scrambler should be loaded with 0011110 (right-hand-most first in time) on the first coincidence on T1 and the signal that will cause the mute should be removed from the transmitter output. [Generally this signal will be *Request to send* (RTS).]

This particular starting point was chosen in order to ensure a pattern that has continuous 180° phase reversals at the beginning in order to ensure rapid clock acquisition, followed by a pattern that will ensure rapid equalizer convergence.

Within eight symbol intervals prior to the ON condition of *Ready for sending* (RFS), the scrambler should be switched to normal operation, keeping the scrambler in mark hold until RFS to synchronize the descrambler.

*Note.* — At 2400 bits per second, a similar technique may be used with appropriate clocking changes.

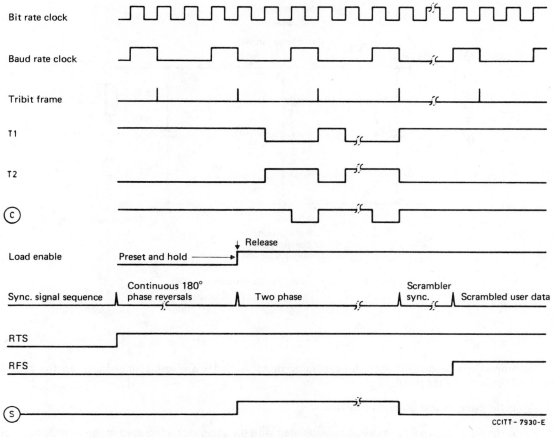

FIGURE 3/V.27 *ter* -- **Synchronizing signal sequence for 4800 bit/s (see Figure 1/V.27 *ter*)**

**Recommendation V.28**

## ELECTRICAL CHARACTERISTICS FOR UNBALANCED DOUBLE-CURRENT INTERCHANGE CIRCUITS

*(Geneva, 1972)*

1.  *Scope*

The electrical characteristics specified in this Recommendation apply generally to interchange circuits operating with data signalling rates below the limit of 20 000 bits per second.

2.  *Interchange equivalent circuit*

Figure 1/V.28 shows the interchange equivalent circuit with the electrical parameters, which are defined below.

This equivalent circuit is independent of whether the generator is located in the data circuit-terminating equipment and the load in the data terminal equipment or vice versa.

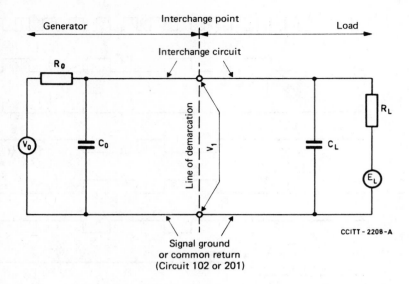

FIGURE 1/V.28 – **Interchange equivalent circuit**

The impedance associated with the generator (load) includes any cable impedance on the generator (load) side of the interchange point.

$V_0$    is the open-circuit generator voltage.

$R_0$    is the total effective d.c. resistance associated with the generator, measured at the interchange point.

$C_0$    is the total effective capacitance associated with the generator, measured at the interchange point.

$V_1$    is the voltage at the interchange point with respect to signal ground or common return.

$C_L$    is the total effective capacitance associated with the load, measured at the interchange point.

$R_L$    is the total effective d.c. resistance associated with the load, measured at the interchange point.

$E_L$    is the open-circuit load voltage (bias).

3.      *Load*

The test conditions for measuring the load impedance are shown in Figure 2/V.28.

The impedance on the load side of an interchange circuit shall have a d.c. resistance ($R_L$) neither less than 3000 ohms nor more than 7000 ohms. With an applied voltage ($E_m$), 3 to 15 volts in magnitude, the measured input current ($I$) shall be within the following limits:

$$I_{\text{min., max.}} = \left| \frac{E_m \pm E_{L\,\text{max.}}}{R_{L\,\text{max., min.}}} \right|$$

The open-circuit load voltage ($E_L$) shall not exceed 2 volts.

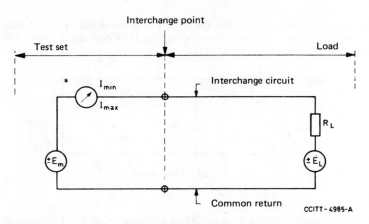

* *Note.* – The internal resistance of the ammeter shall be much less than the load resistance ($R_L$).

FIGURE 2/V.28 – **Equivalent test circuit**

The effective shunt capacitance ($C_L$) of the load, measured at the interchange point, shall not exceed 2500 picofarads.

To avoid inducing voltage surges on interchange circuits the reactive component of the load impedance shall not be inductive.

*Note.* – This is subject to further study.

The load on an interchange circuit shall not prejudice continuous operation with any input signals within the voltage limits specified in 4. below.

## 4.     *Generator*

The generator on an interchange circuit shall withstand an open circuit and a short circuit between itself and any other interchange circuit (including generators and loads) without sustaining damage to itself or its associated equipment.

The open circuit generator voltage ($V_0$) on any interchange circuit shall not exceed 25 volts in magnitude. The impedance ($R_0$ and $C_0$) on the generator side of an interchange circuit is not specified; however, the combination of $V_0$ and $R_0$ shall be selected so that a short circuit between any two interchange circuits shall not result in any case in a current in excess of one-half ampere.

Additionally, when the load open-circuit voltage ($E_L$) is zero, the voltage ($V_1$) at the interchange point shall not be less than 5 volts and not more than 15 volts in magnitude (either positive or negative polarity), for any load resistance ($R_L$) in the range between 3000 ohms and 7000 ohms.

The effective shunt capacitance ($C_0$) at the generator side of an interchange circuit is not specified. However, the generator shall be capable of driving all of the capacitance at the generator side ($C_0$), plus a load capacitance ($C_L$) of 2500 picofarads.

*Note.* – Relay or switch contacts may be used to generate signals on an interchange circuit, with appropriate measures to ensure that signals so generated comply with the applicable clauses of 6. below.

## 5.     *Significant levels* ($V_1$)

For data interchange circuits, the signal shall be considered in the binary 1 condition when the voltage ($V_1$) on the interchange circuit measured at the interchange point is more negative than minus 3 volts. The signal shall be considered in the binary 0 condition when the voltage ($V_1$) is more positive than plus 3 volts.

For control and timing interchange circuits, the circuit shall be considered ON when the voltage ($V_1$) on the interchange circuit is more positive than plus 3 volts, and shall be considered OFF when the voltage ($V_1$) is more negative than minus 3 volts.

| $V_1 < -3$ volts | $V_1 > +3$ volts |
|:---:|:---:|
| 1 | 0 |
| OFF | ON |

FIGURE 3/V.28 – Correlation table

*Note.* – In certain countries, in the case of direct connection to d.c. telegraph-type circuits only, the voltage polarities in Figure 3/V.28 may be reversed.

The region between plus 3 volts and minus 3 volts is defined as the transition region. For an exception to this, see 7. below.

6.    *Signal characteristics*

The following limitations to the characteristics of signals transmitted across the interchange point, exclusive of external interference, shall be met at the interchange point when the interchange circuit is loaded with any receiving circuit which meets the characteristics specified in 3. above.

These limitations apply to all (data, control and timing) interchange signals unless otherwise specified.

1)    All interchange signals entering into the transition region shall proceed through this region to the opposite signal state and shall not re-enter this region until the next significant change of signal condition, except as indicated in 6) below.

2)    There shall be no reversal of the direction of voltage change while the signal is in the transition region, except as indicated in 6) below.

3)    For control interchange circuits, the time required for the signal to pass through the transition region during a change in state shall not exceed one millisecond.

4)    For data and timing interchange circuits, the time required for the signal to pass through the transition region during a change in state shall not exceed 1 millisecond or 3 per cent of the nominal element period on the interchange circuit, whichever is the less.

5)    To reduce crosstalk between interchange circuits the maximum instantaneous rate of voltage change will be limited. A provisional limit will be 30 volts per microsecond.

6)    When electromechanical devices are used on interchange circuits, points 1) and 2) above do not apply to data interchange circuits.

7.    *Circuit failures*

The following interchange circuits, where implemented, shall be used to detect either a power-off condition in the equipment connected through the interface or the disconnection of the interconnecting cable:

Circuit 105          – *Request to send*

Circuit 107          – *Data set ready*

Circuit 108.1/108.2 – *Connect data set to line/Data terminal ready*

Circuit 120          — *Transmit backward channel line signal*

Circuit 202          - *Call request*

Circuit 213          — *Power indication.*

The power-off impedance of the generator side of these circuits shall not be less than 300 ohms when measured with an applied voltage (either positive or negative polarity) not greater than 2 volts in magnitude referenced to signal ground or common return.

The load for these circuits shall interpret the power-off condition or the disconnection of the interconnecting cable as an OFF condition on these circuits.

**Recommendation V.29**

**9600 BITS PER SECOND MODEM STANDARDIZED FOR USE ON
LEASED TELEPHONE-TYPE CIRCUITS**

*(Geneva, 1976)*

### 1.    Introduction

This modem is intended to be used primarily on Recommendation M.1020 circuits but this does not preclude the use of this modem over circuits of lower quality at the discretion of the concerned Administration.

On leased circuits, considering that there exist and will come into being many modems with features designed to meet the requirements of the Administrations and users, this Recommendation in no way restricts the use of any other modems.

The principal characteristics of this recommended modem for transmitting data at 9600 bits per second on leased circuits are as follows:

   *a)*   fallback rates of 7200 and 4800 bits per second;

   *b)*   capable of operating in a full-duplex or half-duplex mode;

   *c)*   combined amplitude- and phase-modulation with synchronous mode of operation;

   *d)*   inclusion of an automatic adaptive equalizer;

   *e)*   optional inclusion of a multiplexer for combining data rates of 7200, 4800 and 2400 bits per second.

*Note.* — The principal use of this recommended modem is on leased circuits. Other applications, such as stand-by operation on the switched network, should be points for further study.

### 2.    Line signals

2.1    The carrier frequency is to be 1700 ± 1 Hz. No separate pilot frequencies are provided. The power levels used will conform to Recommendation V.2.

### 2.2    Signal space coding

2.2.1    At 9600 bits per second, the scrambled data stream to be transmitted is divided into groups of four consecutive data bits (quadbits). The first bit (Q1) in time of each quadbit is used to determine the signal element amplitude to be transmitted. The second (Q2), third (Q3) and fourth (Q4) bits are encoded as a phase change relative to the phase of the immediately preceding element (see Table 1/V.29). The phase encoding is identical to Recommendation V.27.

TABLE 1/V.29

| Q2 | Q3 | Q4 | Phase change (see Note) |
|----|----|----|-------------------------|
| 0 | 0 | 1 | 0° |
| 0 | 0 | 0 | 45° |
| 0 | 1 | 0 | 90° |
| 0 | 1 | 1 | 135° |
| 1 | 1 | 1 | 180° |
| 1 | 1 | 0 | 225° |
| 1 | 0 | 0 | 270° |
| 1 | 0 | 1 | 315° |

*Note.* — The phase change is the actual on-line phase shift in the transition region from the end of one signalling element to the beginning of the following signalling element.

The relative amplitude of the transmitted signal element is determined by the first bit (Q1) of the quadbit and the absolute phase of the signal element (see Table V.29). The absolute phase is initially established by the synchronizing signal as explained in 9. below.

TABLE 2/V.29

| Absolute phase | Q1 | Relative signal element amplitude |
|----------------|----|-----------------------------------|
| 0°, 90°, 180°, 270° | 0 | 3 |
| | 1 | 5 |
| 45°, 135°, 225°, 315° | 0 | $\sqrt{2}$ |
| | 1 | $3\sqrt{2}$ |

Figure 1/V.29 shows the absolute phase diagram of transmitted signal elements at 9600 bits per second.

At the receiver the quadbits are decoded and the data bits are reassembled in correct order.

2.2.2    At the fallback rate of 7200 bits per second, the scrambled data stream to be transmitted is divided into groups of three consecutive data bits. The first data bit in time determines Q2 of the modulator quadbit. The second and third data bits determine Q3 and Q4 respectively of the modulator quadbit. Q1 of the modulator quadbit is a data ZERO for each signal element. Signal elements are determined in accordance with 2.2.1 above. Figure 2/V.29 shows the absolute phase diagram of the transmitted signal elements at 7200 bits per second.

2.2.3    At the fallback rate of 4800 bits per second (see Table 3/V.29), the scrambled data stream to be transmitted is divided into groups of two consecutive data bits. The first data bit in time determines Q2 of the modulator quadbit and the second data bit determines Q3 of the modulator quadbit. Q1 of the modulator quadbit is a data ZERO for each signal element. Q4 is determined by inverting the modulo 2 sum of Q2 + Q3. The signal element is then determined in accordance with 2.2.1 above. Figure 3/V.29 shows the absolute phase diagram of transmitted signal elements at 4800 bits per second.

TABLE 3/V.29

| Data bits | | Quabits | | | | Phase change |
|---|---|---|---|---|---|---|
| | | Q1 | Q2 | Q3 | Q4 | |
| 0 | 0 | 0 | 0 | 0 | 1 | 0° |
| 0 | 1 | 0 | 0 | 1 | 0 | 90° |
| 1 | 1 | 0 | 1 | 1 | 1 | 180° |
| 1 | 0 | 0 | 1 | 0 | 0 | 270° |

The phase changes are identical with Recommendation V.26 (alternative A) and the amplitude is constant with a relative value of 3.

## 3.    Data signalling and modulation rates

The data signalling rates shall be 9600, 7200 and 4800 bits per second ± 0.01%. The modulation rate is 2400 bauds ± 0.01%.

## 4.    Received signal frequency tolerance

The carrier frequency tolerance allowance at the transmitter is ± 1 Hz. Assuming a maximum frequency drift of ± 6 Hz in the connection between the modems, then the receiver must be able to accept errors of at least ± 7 Hz in the received signal frequency.

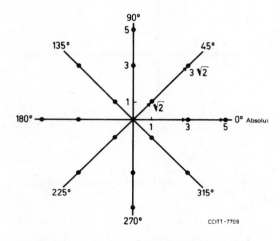

FIGURE 1/V.29 – **Signal space diagram at 9600 bit/s**

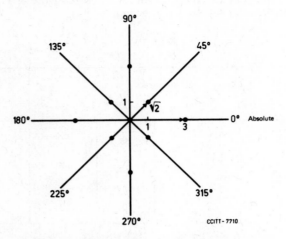

FIGURE 2/V.29 — **Signal space diagram at 7200 bit/s**

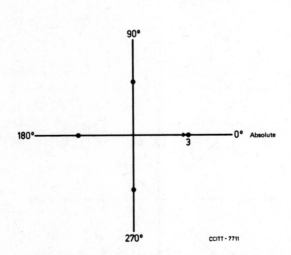

FIGURE 3/V.29 — **Signal space diagram at 4800 bit/s**

5.      *List of essential interchange circuits* (Table 4/V.29)

**TABLE 4/V.29**

| No. | Designation (See Notes 1, 2 and 3) |
|---|---|
| 102 | Signal ground or common return |
| 102a | DTE common return |
| (Note 1) | |
| 102b | DCE common return |
| (Note 1) | |
| 103 | Transmitted data |
| 104 | Received data |
| 015 | Request to send |
| (see Note 3) | |
| 106 | Ready for sending |
| 107 | Data set ready |
| 109 | Data channel received line signal detector |
| 111 | Data signalling rate selector (DTE source) |
| 113 | Transmitter signal element timing (DTE source) |
| 114 | Transmitter signal element timing (DCE source) |
| 115 | Receiver signal element timing (DCE source) |

*Note 1.*—Interchange circuits 102a and 102b are required where the electrical characteristics defined in Recommendation V.10 are used.

*Note 2.* – A manual selector shall be implemented which determines the two signalling rates selected by circuit 111. The manual selector positions shall be designated 9600/7200, 9600/4800 and 7200/4800. The ON condition of circuit 111 selects the higher signalling rate and the OFF condition of circuit 111 selects the lower signalling rate.

*Note 3.* – Not essential for 4-wire full duplex continuous carrier operation.

6.      *Threshold and response times of circuit 109*

6.1     *Threshold*

—    greater than $-26$ dBm:          circuit 109 ON.

—    less than $-31$ dBm:          circuit 109 OFF.

The condition of circuit 109 for levels between $-26$ dBm and $-31$ dBm is not specified except that the signal detector shall exhibit a hysteresis action, such that the level at which the OFF to ON transition occurs is at least 2 dB greater than that for the ON to OFF transition.

6.2     *Response times*

—    ON to OFF: $10 \pm 5$ ms;

—    OFF to ON:

    *a)*    when no new equalization is needed, $15 \pm 10$ ms;

    *b)*    when a new equalization is needed, the circuit 109 will be maintained in OFF condition until synchronization is achieved.

Response times of circuit 109 are the times that elapse between the connection or removal of a line signal to or from the modem receive line terminals and the appearance of the corresponding ON or OFF condition on circuit 109.

*Note.* – Circuit 109 ON to OFF response time should be suitably chosen within the specified limits to ensure that all valid data bits have appeared on circuit 104.

7.    *Timing arrangements*

Clocks should be included in the modem to provide the data terminal equipment with transmitter signal element timing (Recommendation V.24, circuit 114) and receiver signal element timing (Recommendation V.24, circuit 115). Alternatively, the transmitter signal element timing may be originated in the data terminal equipment and be transferred to the modem via the appropriate interchange circuit (Recommendation V.24, circuit 113).

**8. *Electrical characteristics of interchange circuits***

**a) Use of electrical characteristics conforming to Recommendation V.28 is recommended with the connector pin assignment plan specified by ISO DIS 2110.**

**b) Application of electrical characteristics conforming to Recommendation V.10 and V.11 is recognized as an alternative together with the use of the connectors and pin assignment plan specified by ISO DIS 4902.**

**i) Concerning circuits 103, 104, 105 (where used), 106, 107, 109, 113, 114 and 115, the receivers shall be in accordance with Recommendation V.11 or, alternatively, Recommendation V.10, Category 1. Either V.10 or V.11 generators may be utilized.**

**ii) In the case of circuit 111, Recommendation V.10 applies with receivers configured as specified by Recommendation V.10 for Category 2.**

**iii) Interworking between equipment applying Recommendation V.10 and/or V.11 and equipment applying Recommendation V.28 is allowed on a noninterference basis. The onus for adaptation to V.28 equipment rests solely with the alternative V.10/V.11 equipment.**

*Note.*—**Manufacturers may wish to note that the long-term objective is to replace electrical characteristics specified in Recommendation V.28, and Study Group XVII has agreed that the work shall proceed to develop a more efficient all balanced interface for the V-Series application which minimizes the number of interchange circuits. It is expected that this work would be based upon the alternative application given in 8 b) above utilizing the V.11 electrical characteristics.**

9.    The following information is provided to assist equipment manufacturers:

—    The data modem should have no adjustment for send level or receive sensitivity under the control of the operator.

—    The transmitter energy spectrum shall be shaped in such a way that with continuous data ONEs applied to the input of the scrambler the resulting transmitted spectrum shall have a substantially linear phase characteristic over the band of 700 Hz to 2700 Hz and the energy density at 500 Hz and 2900 Hz shall be attenuated 4.5 dB $\pm$ 2.5 dB with respect to the maximum energy density between 500 Hz and 2900 Hz.

10.    *Synchronizing signals*

Transmission of synchronizing signals may be initiated by the modem or by the associated data terminal equipment. When circuit 105 is used to control the transmitter carrier, the synchronizing signals are generated during the interval between the OFF to ON transition of circuit 105 and the OFF to ON transition of circuit 106. When the receiving modem detects a circuit condition which requires resynchronizing, it shall turn circuit 106 OFF and generate a synchronizing signal.

The synchronizing signals for all data rates are divided into four segments as in Table 5/V.29.

TABLE 5/V.29

|  | Segment 1 | Segment 2 | Segment 3 | Segment 4 | Total of segments 1, 2, 3 and 4 |
|---|---|---|---|---|---|
| Type of line signal | No transmitted energy | Alternations | Equalizer conditioning pattern | Scrambled all data ONEs | Total synchronizing signal |
| Number of symbol intervals | 48 | 128 | 384 | 48 | 608 |
| Approximate time in ms* | 20 | 53 | 160 | 20 | 253 |

* Approximate times are provided for information only. The segment duration is determined by the exact number of symbol intervals.

**10.1** Segment 2 of the synchronizing signal consists of alterations between two signal elements. The first signal element (A) transmitted has a relative amplitude of 3 and defines the absolute phase reference of 180°. The second signal element (B) transmitted depends on the data rate. Figure 4/V.29 shows the B signal element at each of the data rates. Segment 2 alternates ABAB....ABAB for 128 symbol intervals.

**10.2** Segment 3 of the synchronizing signals transmits two signal elements according to an equalizer conditioning pattern. The first signal element (C) has a relative amplitude of 3 and absolute phase of 0°. The second signal element (D) transmitted depends on the data rate. Figure 4/V.29 shows the D signal element at each of the data rates. The equalizer conditioning pattern is a pseudo-random sequence generated by the polynomial.

$$1 + x^{-6} + x^{-7}.$$

Each time the pseudo-random sequence contains a ZERO, point C is transmitted. Each time the pseudo-random sequence contains a ONE, the point D is transmitted. Segment 3 begins with the sequence CDCDCDC.... according to the pseudo-random sequence and continues for 384 symbol intervals. The detailed pseudo-random sequence generation is described in Appendix 1.

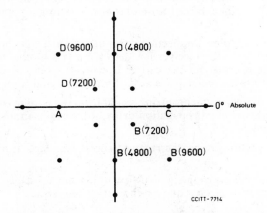

FIGURE 4/V.29 – Signal space diagram showing synchronizing signal points

**10.3** Segment 4 commences transmission according to the encoding described in 2.2 above with continuous data ONEs applied to the input of the data scrambler. Segment 4 duration is 48 symbol intervals. At the end of Segment 4, circuit 106 is turned ON and user data are applied to the input of the data scrambler.

**11.** *Response time for circuit 106*

The time between the OFF to ON transition of circuit 105 and the OFF to ON transition of circuit 106 shall be optionally 15 ms ± 5 ms or 253.5 ms ± 0.5 ms.

The short delay is used when circuit 105 does not control the transmitter carrier. The long delay is used when circuit 105 controls transmitter carrier and a synchronizing signal is initiated by the OFF to ON transition of circuit 105.

The time between the ON to OFF transition of circuit 105 and the ON to OFF transition of circuit 106 shall be suitably chosen to ensure that all valid signal elements have been transmitted.

157

## 12. *Scrambler*

A self-synchronizing scrambler/descrambler having the generating polynomial $1 + x^{-18} + x^{-23}$, shall be included in the modem.

At the transmitter the scrambler shall effectively divide the message polynomial, of which the input data sequence represents the coefficients in descending order, by the scrambler generating polynomial to generate the transmitted sequence. At the receiver the received polynomial, of which the received data sequence represents the coefficients in descending order, shall be multiplied by the scrambler generating polynomial to recover the message sequence.

The detailed scrambling and descrambling processes are described in Appendix 2.

## 13. *Equalizer*

An automatic adaptive equalizer shall be provided in the receiver.

The receiver shall incorporate a means of detecting loss of equalization and initiating a synchronizing signal sequence in its associated local transmitter. The receiver shall incorporate a means of detecting a synchronizing signal sequence from the remote transmitter and initiating a synchronizing signal sequence in its associated local transmitter.

Either modem of a full-duplex connection can initiate the synchronizing signal sequence. The synchronizing signal is initiated when the receiver has detected a loss of equalization or when the transmitter circuit 105 OFF to ON transition occurs in the carrier controlled mode, as described in 10. above. Having initiated a synchronizing signal, the modem expects a synchronizing signal from the remote transmitter.

If the modem does not receive a synchronizing signal from the remote transmitter within a time interval equal to the maximum expected two-way propagation delay, it transmits another synchronizing signal. A time interval of 1.2 seconds is recommended.

If the modem fails to synchronize on the received signal sequence, it transmits another synchronizing signal.

If a modem receives a synchronizing signal when it had not initiated a synchronizing signal and the receiver properly synchronizes, it returns only one synchronizing sequence.

## 14. *Multiplexing* (Table 6/V.29)

A multiplexing option may be included to combine 7200, 4800 and 2400 bits per second data subchannels into a single aggregate bit stream for transmission. Identification of the individual data subchannels is accomplished by assignment to the modulator quadbit as defined in 2.2 above.

TABLE 6/V.29

| Aggregate data rate | Multiplex configuration | Sub-channel data rate | Multiplex channel | Modulator bits | | | |
|---|---|---|---|---|---|---|---|
| | | | | Q1 | Q2 | Q3 | Q4 |
| 9600 bit/s | 1 | 9600 | A | X | X | X | X |
| | 2 | 7200 | A | X | X | X | |
| | | 2400 | B | | | | X |
| | 3 | 4800 | A | X | | X | |
| | | 4800 | B | | X | | X |
| | 4 | 4800 | A | X | | X | |
| | | 2400 | B | | X | | |
| | | 2400 | C | | | | X |
| | 5 | 2400 | A | X | | | |
| | | 2400 | B | | X | | |
| | | 2400 | C | | | X | |
| | | 2400 | D | | | | X |
| 7200 bit/s | 6 | 7200 | A | | X | X | X |
| | 7 | 4800 | A | | X | X | |
| | | 2400 | B | | | | X |
| | 8 | 2400 | A | | X | | |
| | | 2400 | B | | | X | |
| | | 2400 | C | | | | X |
| 4800 bit/s | 9 | 4800 | A | | X | X | |
| | 10 | 2400 | A | | X | | |
| | | 2400 | B | | | X | |

*Note.* – When more than one modulator bit is assigned to a sub-channel, the first bit in time of the sub-channel is assigned to the first bit in time (Q1) of the modulator.

Appendix 1

(to Recommendation V.29)

**Details of the pseudo-random sequence generator**

The equalizer conditioning pattern is determined by a pseudo-random sequence generated by the polynomial $1 + x^{-6} + x^{-7}$. Figure 5/V.29 shows a suitable implementation.

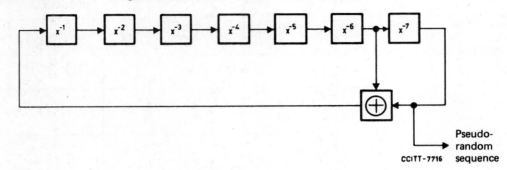

FIGURE 5/V.29

The initial condition of the generator is 0101010. The generator clock is at the symbol rate (2400 symbols per second). The first four conditions of the generator are:

—  initial condition:       0101010
—  after first shift:        1010101
—  after second shift:   1101010
—  after third shift:       1110101

Appendix 2

(to Recommendation V.29)

**Detailed scrambling and descrambling process**

1.    *Scrambling*

The message polynomial is divided by the generating polynomial $1 + x^{-18} + x^{-23}$ (see Figure 6/V.29). The coefficients of the quotient of this division taken in descending order form the data sequence to be transmitted. In order to ensure that proper starting sequence is generated, the shift register is fed with "0" during segments 1, 2, and 3. During segment 4 and normal data transmission it is fed with scrambled data $D_s$ (input data $D_i$ being "1" during segment 4).

$$D_s = D_i + D_s x^{-18} + D_s x^{-23}$$

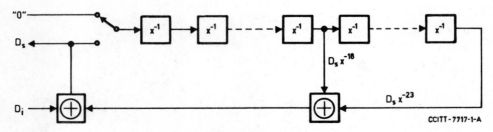

FIGURE 6/V.29

## 2.    *Descrambling*

The polynomial represented by the received sequence is multiplied by the generating polynomial (Figure 7/V.29) to form the recovered message polynomial. The coefficients of the recovered polynomial, taken in descending order, form the output data sequence $D_0$.

$$D_0 = D_i = D_s (1 + x^{-18} + x^{-23})$$

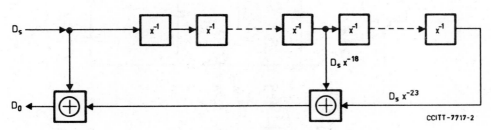

FIGURE 7/V.29

## 3.    *Elements of the scrambling process*

The polynomial $1 + x^{-18} + x^{-23}$ generates a pseudo-random sequence of length $2^{23} - 1 = 8,388,607$. This long sequence does not require the use of a guard polynomial to prevent the occurrence of repeat patterns and is particularly simple to implement with integrated circuits.

**Recommendation V.31**

## ELECTRICAL CHARACTERISTICS FOR SINGLE-CURRENT INTERCHANGE
## CIRCUITS CONTROLLED BY CONTACT CLOSURE

*(Geneva, 1972)*

## 1.    *General*

In general, the electrical characteristics specified in this Recommendation apply to interchange circuits operating at data signalling rates up to 75 bit/s.

Each interchange circuit consists of two conductors (go and return leads) which are electrically insulated from each other and from all other interchange circuits. A common return lead can be assigned to several interchange circuits of a group.

## 2.    *Equivalent circuit of interface*

Figure 1/V.31 shows the equivalent interchange circuit, together with the electrical characteristics laid down in this Recommendation. Some electrical characteristics vary depending upon whether the signal receive side is located in the data circuit-terminating equipment or in the data terminal equipment. This fact is specially referred to below.

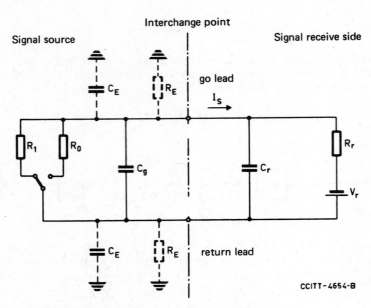

$R_1$ = internal resistance of the signal source in the closed contact condition
$R_0$ = internal resistance of the signal source in the open contact condition
$C_g$ = capacitance of signal source
$C_r$ = capacitance of signal receive side
$V_r$ = open circuit voltage of signal receive side
$I_s$ = current in interchange circuit
$R_r$ = internal resistance of signal receive side
$R_E$ = insulation resistance of signal source if the latter is in the data terminal equipment
$C_E$ = ground capacitance of signal source if the latter is in the data terminal equipment

FIGURE 1/V.31 – **Equivalent circuit of interface**

3.    *Signal source*

The signal source must be isolated from ground or earth irrespective of whether it is located within the data circuit-terminating equipment or within the data terminal equipment.

If the signal receive side is in the data circuit-terminating equipment, the open-contact insulation resistance measured from either leg to ground or to any other interchange circuit shall not fall below 5 megohms and the capacitance measured between the same points shall not exceed 1000 picofardas.

Irrespective of the above, the following specifications apply to the signal source.

3.1    *Internal resistance of signal source $R_1$, $R_0$*

The d.c. resistance of the closed contact $R_1$, including the resistance of the interface cable, measured at the interface (see Figure 1/V.31), should not exceed 10 ohms within the current and voltage ranges of the signal receive side.

The d.c. resistance of the open contact $R_0$ including the insulation resistance of the interface cable should not fall below 250 kilohms when measured at the interface (see Figure 1/V.31) within the voltage range of the signal receive side.

3.2    *Capacitance of signal source $C_g$*

The capacitance of the signal source $C_g$ including that of the interface cable, measured at the interface (see Figure 1/V.31), should not exceed 2500 picofards.

4.       *Signal receive side*

4.1      *Signal receive side in the data circuit-terminating equipment*

The signal receive side in the data circuit-terminating equipment can be floating or connected to ground at any single point.

4.1.1    *Open-circuit voltage of the signal receive side $V_r$*

The open circuit voltage $V_r$ on the signal receive side of the data circuit-terminating equipment, measured at the interface (see Figure 1/V.31), should not fall below 3 volts and should not exceed 12 volts.

4.1.2    *Current at interface $I_S$*

The current $I_S$ supplied by the signal receive side in the data circuit-terminating equipment should not fall below 0.1 milliamp and should not exceed 15 milliamps, when measured at the interface (see Figure 1/V.31) in the closed contact condition, i.e. with an internal resistance of the signal source of $R_1 \leqslant 10$ ohms.

*Note.* –  Irrespective of the current $I_s$ in the closed contact conditions, i.e. with an internal resistance of the signal source of $R_1 \leqslant 10$ ohms, the voltage at the interface should not exceed 150 millivolts, when measured between go and return leads.

4.1.3    *Internal resistance of signal receive side $R_r$*

The internal resistance $R_r$ of the signal receive side of the data circuit-terminating equipment results from the limits for the open-circuit voltage $V_r$ of the signal receive side and the current $I_S$ at the interface, which are specified under 4.1.1 and 4.1.2 above.

Even if $R_r$ has an inductive component, the voltage at the interface should not exceed the maximum of 12 volts specified under 4.1.1 above.

*Note.* –  This item is subject to further study.

4.1.4    *Capacitance of signal receive side $C_r$*

The capacitance of $C_r$ of the signal receive side in the data circuit-terminating equipment, including the capacitance of the cable up to the interface (see Figure 1/V.31), is not specified. It is only necessary to ensure that the signal receive side works satisfactorily, allowing for the capacitance of the signal source $C_g$.

4.2      *Signal receive side in the data terminal equipment*

The signal receive side in the data terminal equipment can be connected to ground at any single point.

4.2.1    *Open circuit voltage of the signal receive side $V_r$*

The open-circuit voltage $V_r$ of the signal receive side of the data terminal equipment, measured at the interface (see Figure 1/V.31), should not fall below 3 volts and should not exceed 52.8 volts.

4.2.2    *Current at the interface $I_S$*

The current $I_S$, supplied by the signal receive side in the data terminal equipment, should not fall below 10 milliamps and not exceed 50 milliamps, when measured at the interface (see Figure 1/V.31) in the closed contact condition, i.e. with an internal resistance of the signal source of $R_1 \leqslant 10$ ohms.

### 4.2.3    *Internal resistance of signal receive side $R_r$*

The internal resistance $R_r$ of the signal receive side in the data terminal equipment is obtained from the limits for the open-circuit voltage $V_r$ of the signal receive side and the current $I_S$ at the interface, which are specified under 4.2.1 and 4.2.2 above.

Even if $R_r$ has an inductive component, the voltage at the interface should not exceed the maximum of 52.8 volts, specified under 4.2.1.

*Note.* — This item is subject to further study.

### 4.2.4    *Capacitance of signal receive side $C_r$*

The capacitance of $C_r$ of the signal receive side in the data terminal equipment including the capacitance of the cable is not specified. It is only necessary to ensure that the signal receive side works satisfactorily, allowing for the capacitance of the signal source $C_g$.

### 5.    *Signal allocation*

Table 1/V.31 shows allocations of digital signals for data, control and timing circuits.

TABLE 1/V.31

|  | Closed contact $R_1 \leqslant 10\ \Omega$ | Open contact $R_0 \geqslant 250\ \text{k}\Omega$ |
|---|---|---|
| Data circuits | "1" condition | "0" condition |
| Control and timing circuits | ON condition | OFF condition |

## WIDEBAND MODEMS

**Recommendation V.35**

### DATA TRANSMISSION AT 48 KILOBITS PER SECOND
### USING 60-108 kHz GROUP BAND CIRCUITS

*(Mar del Plata, 1968, amended at Geneva, 1972 and 1976)*

On leased circuits, considering that there exist and will come into being other modems with features designed to meet the requirements of the Administrations and users, this Recommendation in no way restricts the use of any other modems.

This is a particular system using a group reference pilot at 104.080 kHz.

Principal recommended characteristics to be used for simultaneous both-way operation are the following:

1. *Input/output*

Rectangular polar serial binary data.

2. *Transmission rates*

Preferred mode is synchronous at 48 000 $\pm$ 1 bit/s, with the following exceptions permissible:

a) Synchronous at 40 800 $\pm$ 1 bit/s when it is an operational necessity, or

b) Non-synchronous transmission of essentially random binary facsimile with element durations in the range 21 microseconds to 200 milliseconds.

*Note.* — Operation at half data signalling rate shall be possible when the line characteristics do not permit the above data signalling rates.

3. *Scrambler/descrambler*

Synchronous data should be scrambled to avoid restrictions on the data input format. Such restrictions would be imposed by the need to have sufficient transitions for receiver clock stability, without short repetitive sequences of data signals which would result in high level discrete frequency components in the line signal. Synchronous data should be scrambled and descrambled by means of the logical arrangements described in Appendix 1.

## 4.    *Modulation technique*

The baseband signal (see 5. below) should be translated to the 60-104 kHz band as an asymmetric sideband suppressed carrier AM signal with a carrier frequency of 100 kHz. A pilot carrier will be necessary to permit homochronous demodulation. To simplify the problem of recovery of the pilot carrier for demodulation the serial binary data signal should be modified as stated in 5. below. The transmitted signal should correspond with the following:

a)    The data carrier frequency should be 100 000 $\pm$ 2 Hz.

b)    The nominal level of a frequency translated suppressed carrier 48 kbit/s encoded data baseband signal in the 60-104 kHz band should be equivalent to $-5$ dBm0.

c)    A pilot carrier at $-9 \pm 0.5$ dB relative to the nominal level of the signal in b) should be added such that the pilot carrier would be in phase, to within $\pm$ 0.04 radian, with a frequency translated continuous binary 1 input to the modulator.

d)    The modulator should be linear, and the characteristics of the transmit bandpass filter should be such that the relative attenuation distortion and the relative envelope delay distortion in the range 64 to 101.5 kHz are less than 0.2 dB and 4 microseconds respectively.

## 5.    *Baseband signal*

a)    The scrambled synchronous or random non-synchronous serial binary data signal should be modified by the following transform:

$$\frac{pT_1}{1 + pT_1} \text{ to remove the low-frequency components,}$$

where $p$ is the complex frequency operator, and

$T_1$ is $25/2\pi$ times the minimum binary element duration, i.e. 83 microseconds.

The value of $T_1$ shall have an accuracy of $\pm$ 2%.

In this form the signal is referred to as the baseband signal.

b)    The baseband signal resulting from the transformation should not suffer impairment greater than that resulting from relative attenuation distortion or relative envelope delay distortion of 1.5 dB or 4 microseconds respectively, *and*

i)    distortion due to modification of the baseband signal by the transform

$$\frac{pT_2}{1 + pT_2}$$

where $T_2$ is 3.18 milliseconds; *or*

ii)    distortion due to modification of the baseband signal by the transform

$$\left[ \frac{pT_3}{1 + pT_3} \right]^2$$

where $T_3$ is 6.36 milliseconds.

c)    The frequency range for a) and b) is 0 to 36 kHz.

6.      *Voice channel*

        A service speech channel provided as an integral part of this system should correspond to channel 1 of a 12-channel system, i.e. as a lower sideband SSB signal in the 104-108 kHz band.

   *a)*    The characteristics of this channel may be less stringent than those of a telephone circuit in accordance with Recommendation G.232.

   *b)*    This voice channel is optional.

7.      *Group reference pilots*

7.1     Provision should be made for facilitating the injection of a group reference pilot of 104.08 kHz from a source external to the modem.

7.2     The protection of the group reference pilot should conform to Recommendation H.52.

8.      *Adjacent channel interference*

a)      When transmitting scrambled synchronous serial binary data at 48 kbit/s on the data channel, the out-of-band energy in a 3-kHz band centered at any frequency in the range 1.5 to 58.5 kHz or 105.5 to 178.5 kHz should not exceed −60 dBm0.

b)      When a signal at 0 dBm0 at any frequency in the range 0 to 60 or 104 to 180 kHz is applied to the carrier input terminals, the resulting crosstalk measured in the demodulated data baseband should not exceed a level equivalent to −40 dBm0.

9.      *Line characteristics*

        The characteristics of a channel over which this equipment can be expected to operate satisfactorily should be as given in Recommendation H.14, B.

10.     *Interface*

a)      The interchange circuits should be as shown in Table 1/V.35.


TABLE 1/V.35

| Number | Function |
|--------|----------|
| 102 | Signal ground or common return |
| 103 $\phi$ | Transmitted data |
| 104 $\phi$ | Received data |
| 105 | Request to send |
| 106 | Ready for sending |
| 107 | Data set ready |
| 109 | Data channel receive line signal detector |
| 114 $\phi$ | Transmitter signal element timing |
| 115 $\phi$ | Receiver signal element timing |


b)      The electrical characteristics of the interchange circuits marked ∅ should be as described in Appendix 2; the circuits not marked should conform to Recommendation V.28.

Appendix 1

(to Recommendation V.35)

### Scrambling process

1.   *Definitions*

i)      **applied data bit**

The data bit which has been applied to the scrambler but has not affected the transmission at the time of consideration.

ii)     **next transmitted bit**

The bit which will be transmitted as a result of scrambling the applied data bit.

iii)    **earlier transmitted bits**

Those bits which have been transmitted earlier than the next transmitted bit. They are numbered sequentially in reverse time order, i.e. the first earlier transmitted bit is that immediately preceding the next transmitted bit.

iv)     **adverse state**

The presence of any one of certain repetitive patterns in the earlier transmitted bits.

2.   *Scrambling process*

The binary value of the next transmitted bit shall be such as to produce odd parity when considered together with the twentieth and third earlier transmitted bits and the applied data bit unless an adverse state is apparent, in which case the binary value of the next transmitted bit shall be such as to produce even instead of odd parity.

An adverse state shall be apparent only if the binary values of the $p^{th}$ and $(p + 8)^{th}$ earlier transmitted bits have not differed from one another when $p$ represents all the integers from 1 to $q$ inclusive. The value of $q$ shall be such that, for $p = (q + 1)$, the $p^{th}$ and $(p + 8)^{th}$ earlier transmitted bits had opposite binary values and $q = (31 + 32\, r)$, $r$ being 0 or any positive integer.

At the time of commencement, i.e. when no earlier bits have been transmitted, an arbitrary 20-bit pattern may be assumed to represent the earlier transmitted bits. At this time also it may be assumed that the $p^{th}$ and $(p + 8)^{th}$ earlier transmitted bits have had the same binary value when $p$ represents all the integers up to any arbitrary value. Similar assumptions may be made for the descrambling process at commencement.

*Note 1.* — From this it can be seen that received data cannot necessarily be descrambled correctly until at least 20 bits have been correctly received and any pair of these bits, separated from each other by seven other bits, have differed in binary value from one another.

*Note 2.* — It is not possible to devise a satisfactory test pattern to check the operation of the Adverse State Detector (ASD) because of the large number of possible states in which the 20-stage shift register can be at the commencement of testing. For those modems in which it is possible to bypass the scrambler and the descrambler and to strap the scrambler to function as a descrambler, the following method may be used. A 1 : 1 test pattern is transmitted with the ASD of the scrambler bypassed. If the ASD of the descrambler is functioning correctly the descrambled test pattern will contain a single element error every 32 bits, i.e.

90 000 errors per minute for a modem operating at 48 kbit/s indicates that the descrambler is functioning correctly. The operation of the ASD of the scrambler may be checked in a similar manner with the scrambler strapped as a descrambler and the descrambler bypassed.

3.        Figure 1/V.35 is given as an indication only, since with another technique this logical arrangement might take another form.

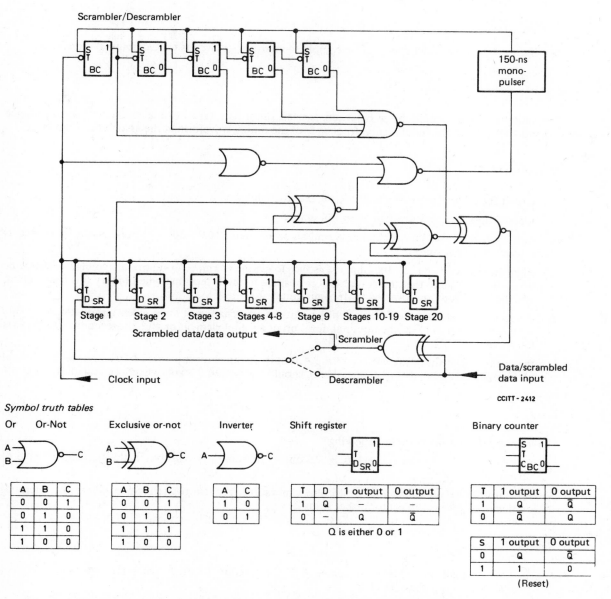

*Note.* – Negative-going transitions of clocks (i.e. 1 to 0 transitions) coincide with data transitions. This is self-synchronizing.

FIGURE 1/V.35 – **An example of scrambler and descrambler circuitry**

## Appendix 2

## (to Recommendation V.35)

## Electrical characteristics for balanced double-current interchange circuits

1.    *Scope*

The electrical characteristics specified here apply to interchange circuits to Recommendation V.35.

2.    *Cable*

The interface cable should be a balanced twisted multi-pair type with a characteristic impedance between 80 and 120 ohms at the fundamental frequency of the timing waveform at the associated terminator.

3.    *Generator*

This circuit should comply with the following requirements:

a)   source impedance in the range 50 to 150 ohms;

b)   resistance between short-circuited terminals and circuit 102: 150 $\pm$ 15 ohms (the tolerance is subject to further study);

c)   when terminated by a 100-ohm resistive load the terminal-to-terminal voltage should be 0.55 volt $\pm$ 20% so that the A terminal is positive to the B terminal when binary 0 is transmitted, and the conditions are reversed to transmit binary 1;

d)   the rise time between the 10% and 90% points of any change of state when terminated as in 3 *c)* should be less than 1% of the nominal duration of a signal element or 40 nanoseconds, whichever is the greater;

e)   the arithmetic mean of the voltage of the A terminal with respect to circuit 102, and the B terminal with respect to circuit 102 (d.c. line offset), should not exceed 0.6 volt when terminated as in *c)*.

4.    *Load*

The load should comply with the following:

a)   input impedance in the range 100 $\pm$ 10 ohms, substantially resistive in the frequency range of operation;

b)   resistance to circuit 102 of 150 $\pm$ 15 ohms, measured from short-circuited terminals (the tolerance on this resistance is subject to further study).

5.    *Electrical safety*

A generator or load should not be damaged by connection to earth potential, short-circuiting, or cross-connection to other interchange circuits.

6.    *Performance in the presence of noise*

A generator, as in 3. above, connected via a cable as in 2. above to a load, as in 4. above, should operate without error in the presence of longitudinal noise or d.c. common return potential differences (circuit 102 offset) as follows:

a)   with $\pm$ 2 volts (peak) noise present longitudinally, i.e. algebraically added to both load input terminals simultaneously with respect to the common return, *or*

b)   with $\pm$ 4 volts circuit 102 offset;

c) if circuit 102 offset and longitudinal noise are present simultaneously, satisfactory operation should be achieved when:

$$\frac{\text{circuit 102 offset}}{2} + \text{longitudinal noise (peak)} = 2 \text{ volts or less.}$$

*Note.* — It has been proposed to perform a test under inclusion of a cable length corresponding to the actual operation. This point is for further study.

**Recommendation V.36**

## MODEMS FOR SYNCHRONOUS DATA TRANSMISSION USING 60-108 kHz GROUP BAND CIRCUITS

*(Geneva, 1976)*

On leased circuits, considering that there exists and will come into being other modems with features designed to meet the requirements of the Administrations and users, this Recommendation in no way restricts the use of any other modems.

The only group reference pilot frequency which can be used in conjunction with this modem is 104.08 kHz.

1. *Scope*

The family of modems covered by this Recommendation should be applicable to the following uses:

a) transmission of data between customers on leased circuits;

b) transmission of a multiplex aggregate bit stream for public data networks;

c) extension of a PCM channel at 64 kbit/s over analogue facilities;

d) transmission of a common channel signalling system for telephony and/or public data networks;

e) extension of single-channel-per-carrier (SCPC) circuit from satellite earth station;

f) transmission of a multiplex aggregate bit stream for telegraph and data signals.

Principal recommended characteristics to be used for simultaneous both-way synchronous operation are the following:

2. *Data signalling rates*

2.1 *Application a)*

The recommended data signalling rate (equals the customer signalling rate) for international use is synchronous at 48 kbit/s. For certain national applications or with bilateral agreement between Administrations the following data signalling rates are applicable: 56, 64 and 72 kbit/s.

2.2 *Applications b), c) and d)*

For these applications the recommended data signalling rate is synchronous at 64 kbit/s.

For those synchronous networks requiring the end-to-end transmission of both the 8 kHz and 64 kHz timing together with the data at 64 kbit/s, a data signalling rate of 72 kbit/s on the line is suggested.

The corresponding data format shall be obtained by inserting one extra bit just before the first bit of each octet of the 64 kbit/s data stream. The alignment pattern, the alignment procedure and the possible use of housekeeping bits in the data format are left for further study.

When the transmission of the 8 kHz timing is not required, the data signalling rate on the line may be 64 kbit/s.

2.3     *Application e)*

The recommended data signalling rate (equals the customer signalling rate) for international use is synchronous at 48 kbit/s. For certain national applications or with bilateral agreement between Administrations the data signalling rate of 56 kbit/s is applicable.

2.4     *Application f)*

The recommended data signalling rate is synchronous at 64 kbit/s.

2.5     The permitted tolerance for all the data signalling rates mentioned above is $\pm$ 1 bit/s.

3.      *Scrambler/descrambler*

In order to be bit sequence independent and to avoid high amplitude spectral components on the line, the data should be scrambled and descrambled by means of the logical arrangements described in the Appendix.

4.      *Baseband signal*

The equivalent baseband signal shaping process is based upon the binary coded partial response pulse, often referred to as class IV, whose time and spectral function are defined by:

$$g(t) = \frac{2}{\pi} \cdot \frac{\mathrm{Sin}\,\frac{\pi}{T}\,t}{\left(\frac{t}{T}\right)^2 - 1}$$

and

$$G(f) = \begin{cases} 2\,Tj\,\mathrm{Sin}\,2\,\pi\,Tf, & |f| < \dfrac{1}{2T} \\[2ex] 0 & , |f| > \dfrac{1}{2T} \end{cases}$$

respectively, where $1/T$ denotes the data signalling rate.

This shaping process should be effected in such a way that the decoding can be achieved by full wave rectification of the demodulated line signal.

The reference to equivalent baseband signals recognizes that the modem implementation may be such that the binary signal at the input and output of the modem is converted to and from the line signal without appearing as an actual baseband signal.

5.      *Line signal in 60-108 kHz band* (at the line output of the modem)

5.1     In the 60-108 kHz band the line signal should correspond to a single sideband signal with its carrier frequency at 100 kHz ± 2 Hz.

5.2     The relationship between the binary signals at the real or hypothetical output of the scrambler and the transmitted line signal states shall be in accordance with the amplitude modulation case of Recommendation V.1, i.e., tone ON for binary 1 and tone OFF for binary 0.

In a practical case this means that the voltage or no voltage conditions which will result from the full wave rectification of the demodulated line signal will correspond with the binary 1 and binary 0 signals respectively at the output of the scrambler.

5.3     The amplitude of the theoretical line signal spectrum, corresponding to binary symbol 1 appearing at the output of the scrambler, is to be sinusoidal, with zeros and maxima at the frequencies listed below:

| Data signalling rate (kbit/s) | Zeros at (kHz) | Maxima at (kHz) |
|---|---|---|
| 64 | 68 and 100 | 84 |
| 48 | 76 and 100 | 88 |
| 56 | 72 and 100 | 86 |
| 72 | 64 and 100 | 82 |

5.4     In the 60-108 kHz band, amplitude distortion of the real spectrum relative to the theoretical spectrum as defined under 5.3 above is not to exceed ± 1 dB; the group delay distortion is not to exceed 8 microseconds. These two requirements are to be met for each frequency band centred on one of the maxima mentioned in 5.3 and whose width is equal to 80% of the frequency band used.

5.5     The nominal level of the line data signal should be − 6 dBm0. The actual level should be within ± 1 dB of the nominal level.

5.6     A pilot carrier at the same frequency as the modulated carrier at the transmitter and with a level of − 9 ± 0.5 dB relative to the actual level mentioned under 5.5 above, should be added to the line signal. The relative phase between the modulated carrier and the pilot carrier at the transmitter should be time invariant.

6.      *Group reference pilot*

6.1     Provision should be made for facilitating the injection of a group reference pilot of 104.08 kHz from a source external to the modem.

6.2     The protection of the group reference pilot should conform to Recommendation H.52.

7.      *Voice channel*

7.1     The service speech channel is an integral part of the applications a) and e) of this system and is used on an optional basis. The channel corresponds to channel 1 of a 12 channel SSB-AM system in the 104-108 kHz band (virtual carrier at 108 kHz). It can transmit continuous voice at a mean level of maximum − 15 dBm0 or pulsed signalling tones according to the individual specifications.

To avoid overloading of the system by peak signals a limiter shall be used with cut-off levels above +3 dBm0.

To avoid stability problems the channel shall be connected to 4-wire equipment only.

For operator-to-operator signalling Recommendation Q.1 shall be followed, but instead of 500/20 Hz a non-interrupted tone of 2280 Hz at a level of −10 dBm0 shall be used.

For other signalling purposes [application e)] the R1 or R2 inband signalling, described in Recommendations Q.322, Q.323 and Q.364, Q.365 respectively, is preferred.

The transmit filter shall be such that any frequency applied to the transmit input terminals at a level of −15 dBm0 will not cause a level exceeding:

    *a)*    −73 dBm0p in the adjacent group,

    *b)*    −61 dBm0 in the vicinity (± 25 Hz) of the pilot 104.08 kHz,

    *c)*    −55 dBm0 in the data band between 64 and 101 kHz,

    *d)*    the values specified in Recommendation V.354 to protect the nearest low level signalling path.

The voiceband is sufficiently protected if the same filter is used in the receive direction of the channel. The attenuation/frequency characteristic, measured between the voice-frequency input and the groupband output or the groupband input and the voice-frequency output, with respect to the value at 800 Hz is limited by:

    −1 dB over the 300-340 Hz band, +2 dB between 540 and 2280 Hz.

7.2    The voice channel is inapplicable to applications b), c), d) and f). It is used on an optional basis for applications a) and e).

*Note.* − When the modem is installed at the repeater station, the voice channel should be extended to the renter's premises.

8.    *Adjacent channel interference*

The adjacent channel interference should conform to Recommendation H.52.

9.    *Line characteristics*

The characteristics of a channel over which this equipment can be expected to operate satisfactorily are given in Recommendation H.14, B. This includes the sectionwise group delay equalization for operation at rates of 48 up to 64 kbit/s for circuits with a maximum of 4 through-group filters.

For circuits with a larger number of sections or where a signalling rate of 72 kbit/s is required the characteristics given in Recommendation H.14, B are not adequate.

Further study is required to decide how best to correct for the group-delay distortion in these cases.

10.    *Interface*

10.1    For applications a), e) and f) the interchange circuits should be as shown in Table 1/V.36.

TABLE 1/V.36

| Interchange circuit | | Remark |
|---|---|---|
| 102 | Signal ground or common return | Note 1 |
| 102a | DTE common return | Note 2 |
| 102b | DCE common return | Note 2 |
| 103 | Transmitted data | Note 3 |
| 104 | Received data | Note 3 |
| 105 | Request to send | Note 4 |
| 106 | Ready for sending | Note 4 |
| 107 | Data set ready | Note 4 |
| 109 | Data channel received line signal detector | Note 4 |
| 113 | Transmitter signal element timing (DTE source) | Note 3 |
| 114 | Transmitter signal element timing (DCE source) | Note 3 |
| 115 | Receiver signal element timing (DCE source) | Note 3 |

*Note 1.* – The provision of this conductor is optional.

*Note 2.* – These conductors are used in conjunction with interchange circuits 105, 106, 107 and 109, if these circuits use the electrical characteristics according to Recommendation V.10.

*Note 3.* – The interchange circuits 103, 104, 113, 114 and 115 use electrical characteristics of Recommendation V.11

*Note 4.* – The use of V.11 electrical characteristics on circuits 105, 106, 107 and 109 is optional. When the V.10 electrical characteristics are used, separate provision must be made for the common return circuits 102a, 102b and 102, if provided. Each V.10 interchange circuit receiver should be implemented with its two leads at the interchange point.

*Note 5.* – The need for an interim period during which the electrical interface characteristics of Recommendation V.35 may be optionally used needs further study.

*Note 6.* – When the modem is installed at the repeater station, this interface should appear at the customer's premises without restrictions regarding the data signalling rate and the provision of the voice channel. The method to achieve this is subject to national regulations.

10.2    For applications b), c) and d) the interfaces may comply with the functional requirements given in Recommendation G.703 for the 64 kbit/s interface. The electrical characteristics of the interface circuits are under study.

## 11.    *Error performance*

11.1    For a hypothetical reference circuit, 2500 km in length, with characteristics in accordance with Recommendation H.14, and with not more than two through-group connection equipments, the performance objective in terms of error rate should be not worse than 1 error per $10^7$ bits transmitted. This is based on an assumed Gaussian noise power of 4 pW per km/per 4 kHz band psophometrically weighted (This figure corresponds to 4 pW0p/km).

11.2    The measuring technique and error performance criteria in a back-to-back modem measurement configuration require further study.

## 12.    *Additional information for the designer*

## 12.1    *Input level variation*

The step-change in the input level is, under normal conditions, smaller than ± 0.1 dB. The gradual input level change is smaller than ± 6 dB, this includes the tolerance of the transmitter output level.

12.2    *Interference from adjacent group bands*

A sinusoidal signal of +10 dBm0 in the frequency bands of 36-60 kHz and 108-132 kHz can appear together with the data line signal at the input of the receiver.

Appendix

(to Recommendation V.36)

**Scrambling process**

1.    *Definitions*

i)      **applied data bit**

The data bit which has been applied to the scrambler but has not affected the transmission at the time of consideration.

ii)     **next transmitted bit**

The bit which will be transmitted as a result of scrambling the applied data bit.

iii)    **earlier transmitted bits**

Those bits which have been transmitted earlier than the next transmitted bit. They are numbered sequentially in reverse time order, i.e. the first earlier transmitted bit is that immediately preceding the next transmitted bit.

iv)     **adverse state**

The presence of any one of certain repetitive patterns in the earlier transmitted bits.

2.    *Scrambling process*

The binary value of the next transmitted bit shall be such as to produce odd parity when considered together with the twentieth and third earlier transmitted bits and the applied data bit unless an adverse state is apparent, in which case the binary value of the next transmitted bit shall be such as to produce even instead of odd parity.

An adverse state shall be apparent only if the binary values of the $p^{th}$ and $(p + 8)^{th}$ earlier transmitted bits have not differed from one another when $p$ represents all the integers from 1 to $q$ inclusive. The value of $q$ shall be such that, for $p = (q + 1)$, the $p^{th}$ and $(p + 8)^{th}$ earlier transmitted bits had opposite binary values and $q = (31 + 32 r)$, $r$ being 0 or any positive integer.

At the time of commencement, i.e. when no earlier bits have been transmitted, an arbitrary 20-bit pattern may be assumed to represent the earlier transmitted bits. At this time also it may be assumed that the $p^{th}$ and $(p + 8)^{th}$ earlier transmitted bits have had the same binary value when $p$ represents all the integers up to any arbitrary value. Similar assumptions may be made for the descrambling process at commencement.

*Note 1.* – From this it can be seen that received data cannot necessarily be descrambled correctly until at least 20 bits have been correctly received and any pair of these bits, separated from each other by seven other bits, have differed in binary value from one another.

*Note 2.* — It is not possible to devise a satisfactory test pattern to check the operation of the Adverse State Detector (ASD) because of the large number of possible states in which the 20 state shift register can be at the commencement of testing. For those modems in which it is possible to bypass the scrambler and the descrambler and to strap the scrambler to function as a descrambler, the following method may be used. A 1 : 1 test pattern is transmitted with the ASD of the scrambler bypassed. If the ASD of the descrambler is functioning correctly the descrambled test pattern will contain a single element error every 32 bits, i.e. 90 000 errors per minute for a modem operating at 48 kbit/s indicates that the descrambler is functioning correctly. The operation of the ASD of the scrambler may be checked in a similar manner with the scrambler strapped as a descrambler and the descrambler bypassed.

3.      Figure 1/V.36 is given as an indication only, since with another technique this logical arrangement might take another form.

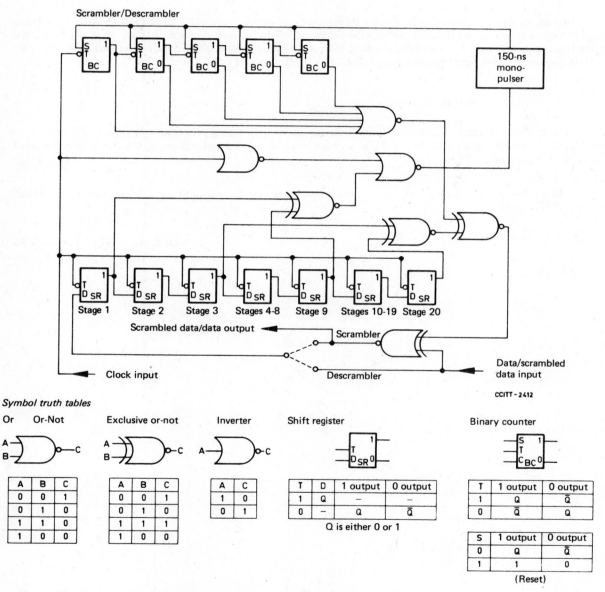

*Symbol truth tables*

*Note.* — Negative-going transitions of clocks (i.e. 1 to 0 transitions) coincide with data transitions. This is self-synchronizing.

FIGURE 1/V.36 – **An example of scrambler and descrambler circuitry**

**LOOP TEST DEVICES FOR MODEMS** [1]

*(Geneva, 1976)*

1.    *Introduction*

The CCITT,

   *considering*

the increasing use being made of data transmission systems, the volume of the information circulating on data transmission networks, the savings to be made by reducing interruption time on such links, the importance of being able to determine responsibilities in maintenance questions for networks, of necessity involving several parties, and the advantages of standardization in this field,

   *unanimously declares the following view:*

The locating of faults can be facilitated in many cases by looping procedures in modems. These loops allow local or remote measurements, analogue/or digital, to be carried out optionally by the Administrations and/or users concerned.

---

[1] In this Recommendation the terms "data terminal equipment" and "data circuit-terminating equipment" are indicated by DTE and DCE respectively.

## 2.    *Definition of the loops*

Four loops are defined (numbered 1 to 4) and their locations as seen from DTE A are shown in Figure 1/V.54. A symmetrical set of four loops could exist as seen from DTE B.

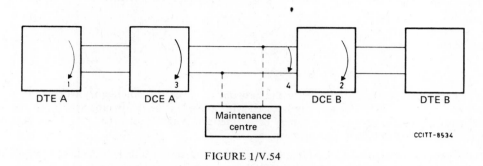

FIGURE 1/V.54

### 2.1    *Loop 1*

This loop is used as a basic test on the operation of the DTE, by returning transmitted signals to the DTE for checking. The loop should be set up inside the DTE as close as possible to the interface.

While the DTE is in the Loop 1 test condition:

—  transmitted data (circuit 103) is connected to received data (circuit 104) within the DTE;

—  circuit 108/1 or 108/2 must be in the same condition as it was before the test;

—  circuit 105 must be in the OFF condition;

—  circuit 125 should continue to be monitored by the DTE so that an incoming call can be given priority over a routine loop test.

Interchange circuit 103 as presented to the DCE must be in the binary 1 condition.

The conditions of the other interchange circuits are not specified but they should if possible permit normal working. The transmitter timing information, in particular, if it comes from the DCE, will continue to be sent [see Recommendation V.24, IV.6c)].

*Note.* —  When circuits 108 and 105 are not used by the DTE (for certain applications on leased lines, for example) the DCE will not be informed of the test condition. This is considered acceptable provided that the remote station is not disturbed.

### 2.2    *Loop 3*

This is a local loop established in analogue mode as close as possible to the line to check the satisfactory working of the DCE. The loop should include the maximum number of circuits used in normal working (in particular the signal conversion function, if possible) which may in some cases necessitate the inclusion of devices for attenuating signals, for example.

The establishment of the loop presents no difficulty when using a 4-wire line, except in certain cases in which parts of the line equalization system are removed from service.

For certain 2-wire lines the loop may be obtained by simple unbalance of the hybrid transformer.

While the DCE is in the loop 3 test condition:

—  the transmission line is suitably terminated, as required by national regulations;

—  all interchange circuits are operated normally, except for the clamping options involving circuits 104, 105 and 109 in case of 2-wire lines;

— circuit 125 should continue to be monitored by the DTE so that an incoming call can be given priority over a routine loop test, after abandoning the loop 3 condition;

— **no signal is transmitted to line on the data channel.**

*Note.* — In certain switched networks the loop 3 procedure may clear the connection due to national regulations. During the loop 3 condition, however, the DCE must not be switched to the line, if not already connected.

2.3     *Loop 2*

Loop 2 is designed to permit station A or the network to check the satisfactory working of the line (or part of the line) and of the DCE B. It can only be used with a duplex DCE; the application to the backward channel is left for further study. Pseudo loop 2 may be defined for a half-duplex DCE and will be specified in the Recommendation relating to the DCE concerned.

The establishment of the loop will be effective when the control is applied, regardless of the condition of circuit 108 presented by the DTE associated with the DCE in which the loop is set up.

While the DCE B is in the loop 2 test condition:

— circuit 104 is connected to the circuit 103 within the DCE;

— the signal on circuit 109 is used to activate circuit 105;

— the signal element timing signal on circuit 115 is connected to circuit 113 (where these circuits are provided); (the signal levels for the above may either be that of the interchange circuits or of the logic level used inside the DCE)

— circuit 107 is maintained in the OFF condition;

— circuit 115 continues to function;

— circuit 104 is maintained in the binary 1 condition.

2.4     *Loop 4*

This loop arrangement is only considered in the case of 4-wire lines. Loop 4 is designed for the maintenance of lines by Administrations using analogue-type measurements. When receiving and transmitting pairs are connected in tandem, such a connection cannot be measured as a data circuit (conformity to a line characteristic curve, for example).

In the loop position the two pairs are disconnected from the DCE and are connected to each other through a symmetrical attenuator designed to prevent any oscillation of the circut (the loop, therefore, does not include any of the amplifiers and/or distortion correctors used in the DCE). The value of the attenuator will be fixed by each Administration, however the minimum attenuation in the loop formed by virtual switching point — subscriber — virtual switching point [2] should be of the order of 6 dB for stability reasons.

Loop 4 may be established inside the DCE or in a separate unit.

**When loop 4 is inside the DCE, and while in the test condition, it presents circuits 107 and 109 to the DTE in the OFF condition and circuit 142 is in the ON condition. When loop 4 is in a separate unit, these conditions are desirable, but not mandatory.**

3.     *Loop control*

Two (non-exclusive) types of control might be possible on the DCE:

— manual control by a switch on the equipment;

— automatic control through the DCE-DTE interface.

---

[2] Virtual switching point is the reference point defined by Recommendation G.111 and can be used for leased circuits as well.

Manual control would always have priority over automatic control and in particular it should be capable of returning the DCE to the normal operating condition.

Interchange circuit 142 shall be used to inform the DTE of a loop condition in the local DCE, even in the case of manual control (but see Note 3 to Table 1/V.54). To avoid ambiguity in interpretation of circuit 142 only one loop should be established at any one time in the DCE.

## 3.1    Manual control

TABLE 1/V.54 — Interface signalling for manual control of loops

| Loop | Control switch on | Signal to DTE A | | | Signal to DTE B | | Notes |
|------|------|------|------|---|------|------|------|
| | | Circuit 107 | Circuit 142 | | Circuit 107 | Circuit 142 | |
| 2 | DCE B | * | * | | OFF | ON | Note 1 |
| 3 | DCE A | ON | ON | | * | * | Note 2 |
| 4 | DCE B | * | * | | OFF | ON | Note 3 |

* not applicable

*Note 1.* – Data station A is in the normal operating condition. The loop is established by a switch on DCE B.

*Note 2.* – DTE B is not concerned with this test. The condition of circuit 107 will be determined by that of circuit 108. The normal case is considered in the table.

**Note 3.—"When loop 4 is in a unit separate from the DCE, the signals to DTE B are desirable but not mandatory due to the difficulty of implementation."**

The conditions represented by ON in Table 1/V.54 may also activate a visual indicator on the DCE.

### 3.2   Automatic control through the DTE/DCE interface

(*Note.* The control of remote loopback for multipoint circuits is the subject for further study.)

Automatic control through the interface is achieved by using circuits 140, 141 and 142 as defined in Recommendation V.24. Circuit 140 is used to control loop 2 and circuit 141 is used to control loop 3. The turning ON of circuit 142 indicates the test mode is established. If circuit 107 is ON, the associated terminal is concerned and subsequent data transmitted on circuit 103 will be looped back on circuit 104. If circuit 107 is OFF, the associated terminal is not concerned.

*Note 1.*—Automatic control of loop 4 is considered of no use either locally or in the remote station and therefore is not provided.

*Note 2.*—Existing systems or equipments which are using another method of test loop selection, such as one test control circuit in conjunction with circuit 103, are not concerned with this recommendation.

### Interface signalling for automatic control of loops

| Loop | Control signals from DTE A | | Signals to DTE A | | Signals to DTE B | | Notes |
|------|------|------|------|------|------|------|------|
| | Circuit 140 | Circuit 141 | Circuit 107 | Circuit 142 | Circuit 107 | Circuit 142 | |
| 2 | ON | OFF | ON | ON | OFF | ON | Note 1 |
| 3 | OFF | ON | ON | ON | * | * | Note 2 |

* not applicable

*Note 1.*—There is a risk of head-on collision of controls from the two ends.

*Note 2.*—In DCE A, the condition of circuit 107 will be determined by the condition of circuit 108. The normal case is considered in the table. DTE B is not concerned with this test.

Normally circuit 103 can only be used to transmit data or the test sequence, so long as the conditions of circuits 106, 140, 141 and 142 are as follows:

| Circuit 103 | Circuit 106 | Circuit 140 | Circuit 141 | Circuit 142 |
|---|---|---|---|---|
| Data | ON | OFF | OFF | OFF |
| Loop 2 test sequence | ON | ON | OFF | ON |
| Loop 3 test sequence | ON | OFF | ON | ON |

THE INTERNATIONAL TELEGRAPH AND TELEPHONE CONSULTATIVE COMMITTEE

# CCITT

# SIXTH PLENARY ASSEMBLY

GENEVA, 27 SEPTEMBER - 8 OCTOBER 1976

## ORANGE BOOK

## VOLUME VIII.2

## PUBLIC DATA NETWORKS

Published by the
INTERNATIONAL TELECOMMUNICATION UNION
GENEVA, 1977

ISBN 92-61-00461-X

# PART II

## Series X Recommendations

# DATA TRANSMISSION OVER PUBLIC DATA NETWORKS

# SECTION 1

## SERVICES AND FACILITIES

**Recommendation X.1**

### INTERNATIONAL USER CLASSES OF SERVICE IN PUBLIC DATA NETWORKS

*(Geneva, 1972, amended at Geneva, 1976)*

The establishment in various countries of public networks for data transmission creates a need to standardize user data signalling rates, terminal operating modes, address selection and call progress signals to facilitate international interworking.

Recommendations in Series V already standardize data signalling rates for synchronous data transmission in the general telephone network and modulation rates for modems. These rates are, however, not necessarily the most suitable for public networks devoted entirely to data transmission and this leads to the requirement for an additional Recommendation.

The CCITT,

*bearing in mind,*

a) the desirability of providing sufficient data signalling rates to meet users' needs;

b) the requirement to optimize terminal, transmission and switching costs to provide an overall economic service to the user;

c) the particular operating modes of users' data terminals;

d) the users' need to transfer information consisting of any bit sequence and of any number of bits up to a certain amount;

e) the interaction between users' requirements, technical limitations and tariff structure,

*unanimously declares the view*

that users' data transmission requirements via public data networks may best be served by defined user classes of service.

These user classes of service are shown in Table 1/X.1; they cater for three particular types of users' data terminals, namely for:

— terminals operating in the start-stop mode (as typified by teleprinters used for message transfer in classes 1 and 2,

— terminals operating in a synchronous mode in classes 3 to 7,

— terminals operating in the packet mode in classes 8 to 11.

TABLE 1/X.1 – **User classes of service for public data networks**

*a)* Terminal operating mode – start-stop (See Notes 1-7)

| User class of service | Data signalling rate and code structure | Address selection and call progress signals |
|---|---|---|
| 1 | 300 bit/s, 11* units/character, start-stop | 300 bit/s, International Alphabet No. 5 (11 units/character) |
| 2 | 50-200 bit/s, 7.5-11* units/character, start-stop | 200 bit/s, International Alphabet No. 5 (11 units/character) |

\* Usage in accordance with Recommendation X.4.

*b)* Terminal operating mode – synchronous (See Note 7)

| User class of service | Data signalling rate | Address selection and call progress signals |
|---|---|---|
| 3 | 600 bit/s | 600 bit/s, International Alphabet No. 5 |
| 4 | 2400 bit/s | 2400 bit/s, International Alphabet No. 5 |
| 5 | 4800 bit/s | 4800 bit/s, International Alphabet No. 5 |
| 6 | 9600 bit/s | 9600 bit/s, International Alphabet No. 5 |
| 7 | 48000 bit/s | 48000 bit/s, International Alphabet No. 5 |

*c)* Terminal operating mode – packet (See Note 7)

| User class of service | Data signalling rate | Address selection and call progress signals |
|---|---|---|
| 8 | 2400 bit/s | 2400 bit/s |
| 9 | 4800 bit/s | 4800 bit/s |
| 10 | 9600 bit/s | 9600 bit/s |
| 11 | 48000 bit/s | 48000 bit/s |

See Recommendation X.25 for user packet format

*Note 1.* — There is no user class of service for the data signalling rate of 50 bit/s, the transmission mode of 7.5 units/character start-stop and address selection and call progress signals at 50 bit/s, International Telegraph Alphabet No. 2. However, several Administrations have indicated that their telex service (50-baud, International Telegraph Alphabet No. 2) will be provided as one of the many services carried by their public data network.

*Note 2.* — Although it is recognized that start-stop data terminals operating on a character-by-character basis will continue to exist for a long time, it is expected that their long-range development direction is towards the use of synchronous mode of transmission at the DCE/DTE interface.

*Note 3.* — Taking account of the existence of data terminal equipments operating in the start-stop mode at a data signalling rate of 300 bit/s and with a 10 unit/character code structure, some Administrations have indicated that their public data networks will accommodate such terminals. Other Administrations, however, have indicated that they cannot guarantee acceptable transmission if such terminals are connected to their networks. The implications of admitting such terminals in class 1 is for further study.

Note 4. — Class 2 will provide, in the data transfer phase, for operation at the following data signalling rates and code structures:

$$
\begin{array}{lll}
50 & \text{bit/s} & (\ 7.5 \ \text{units/character}) \\
100 & \text{bit/s} & (\ 7.5 \ \text{units/character}) \\
110 & \text{bit/s} & (11 \quad \text{units/character}) \\
134.5 & \text{bit/s} & (\ 9 \quad \text{units/character}) \\
200 & \text{bit/s} & (11 \quad \text{units/character})
\end{array}
$$

Address selection and call progress signals would be at 200 bit/s, International Alphabet No. 5 (11 units/character) as indicated in Table 1a)/X.1.

Note 5. — Some Administrations have indicated that for certain of the data signalling rates listed in Note 4 above they will permit users in class 2 to operate the same signalling rate and code structure for both data transfer and address selection and to receive call progress signals at these signalling rates and code structures. Where International Alphabet No. 5 is used for the address selection and call progress signals the appropriate parts of Recommendation X.20 or X.20 bis shall apply. Procedures to be used with other alphabets is a matter for the discretion of individual Administrations.

Note 6. — For data terminals in user class of service 2, it should be noted that some public data networks may not be able to prevent two terminals working at different data signalling rates and code structures from being connected together by means of a circuit switched connection.

Note 7. — In a packet switched data transmission service it should be noted that terminals in different user classes of service may be connected together by means of a packet switched connection.

**Recommendation X.2**

## INTERNATIONAL USER FACILITIES IN PUBLIC DATA NETWORKS

*(Geneva, 1972, amended at Geneva, 1976)*

The CCITT,

*bearing in mind,*

a)   the international user classes of service indicated in Recommendation X.1;

b)   the need to standardize user facilities in public data networks which should be made available on an international basis;

c)   the need to standardize additional user facilities which may be provided by Administrations and which may be available on an international basis;

d)   the impact which these user facilities could have on tariff structures.

*unanimously declares the view that,*

1.    the user facilities should be standardized for each of the user classes of service indicated in Recommendation X.1 for each of the following:

i)    circuit switched data transmission services;

ii)   packet switched data transmission services;

iii)  leased circuit data transmission services.

2.    the user facilities to be available on an international basis are as indicated in Table 1/X.2. Some of the user facilities are available on a per call basis and others may be assigned for an agreed contractual period. In all cases the user has the option of requesting a given user facility.

## TABLE 1/X.2 – International user facilities in public data networks

*a)* Circuit switched service (See Notes 1 and 2)

| User facility | User classes of service | | Definition reference |
|---|---|---|---|
| | 1-2 | 3-7 | |
| *Optional user facilities assigned for an agreed contractual period* | | | 52.44 |
| Direct call | E | E | 52.49 |
| Closed user group | E | E | 52.48 |
| Closed user group with outgoing access | A | A | 52.481 |
| Calling line identification | A | A | 53.625 |
| Called line identification | A | A | 53.62 |
| *Optional user facilities per call* | | | 52.44 |
| Abbreviated address calling | A | A | 52.50 |
| Multi-address calling | A | A | 52.51 |

*b)* Packet switched service (See Notes 1-4)

| User facility | User classes of service | | | | Definition reference |
|---|---|---|---|---|---|
| | 1-7 | 8-11 | | | |
| | See Note 3 | DG | VC | PVC | |
| *Optional user facilities assigned for an agreed contractual period* | | | | | 52.44 |
| Permanent virtual circuit (PVC) | E | – | – | E | 53.545 |
| Closed user group | E | A | E | – | 52.48 |
| Closed user group with outgoing access | A | A | A | – | 52.481 |
| Packet assembly and/or disassembly | E | – | – | – | 52.462 |
| *Optional user facilities per call* | | | | | 52.44 |
| Abbreviated address calling | A | See Note 4 | | – | 52.50 |
| Datagram (DG) | See Note 4 | A | – | – | |
| Virtual call (VC) | E | – | E | – | 53.54 |

*c)* Leased circuit service (See Note 1)

| User facility | User classes of service | | Definition reference |
|---|---|---|---|
| | 1-2 | 3-7 | |
| Point-to-point | E | E | 37.17. |

E = an essential user facility to be made available internationally.
A = an additional user facility which may be available in certain public data networks and may also be available internationally.
DG = applicable when the datagram facility is being used.
VC = applicable when the virtual call facility is being used.
PVC = applicable when the permanent virtual circuit is being used for certain user classes of service.
– = not applicable.

*Note 1.* — It is assumed that terminals have an interface to only one of the three data transmission services identified.

*Note 2.* — The subject of interworking between data terminals having interfaces to different data transmission services is for further study.

*Note 3.* — The interface signalling scheme for terminals in these user classes of service having an interface to the packet switched data transmission service is for further study.

*Note 4.* — For further study.

**Provisional Recommendation X.3**  *(Geneva, 1977)*

## PACKET ASSEMBLY/DISASSEMBLY FACILITY (PAD)
## IN A PUBLIC DATA NETWORK

### Contents

Preface

*Preface*

The establishment in various countries of public data networks providing packet-switched data transmission services creates a need to produce standards to facilitate access from the public telephone network, circuit- switched public data networks and leased circuits.

The CCITT,

*considering*,

a)   that Recommendations X.1 and X.2 define the user classes of service and user facilities in public data networks, Recommendation X.95 defines network parameters, Recommendation X.96 defines call progress signals, Recommendation X.29 defines the procedures for a packet-mode DTE to access the PAD, Recommendation X.28 defines the DTE/DCE interface for a start-stop mode DTE accessing the PAD;

b)   that the logical control links for packet-switched data transmission services are defined in Recommendation X.92, and that in particular Recommendation X.92 allows for the incorporation of a PAD;

c)   the urgent need to allow interworking between a start-stop mode DTE on a public-switched telephone network, a public switched data network or a leased circuit, and a packet-mode DTE using the virtual call facility of the packet-switched data service;

d)   that DTEs operating in the start-stop mode will send and receive network control information and user information in the form of characters;

e)   that DTEs operating in the packet-mode will send and receive network control information and user information in the form of packets in accordance with Recommendation X.25;

f)   that the packet-mode DTE shall not be obliged to use the control procedures for PAD functions, but that some packet-mode DTEs may wish to control specific functions of the PAD;

*unanimously declares the view*

i)   that the functions performed by and operational characteristics of the PAD for the start-stop mode DTE are described below in 1. *Description of the basic functions and user selectable functions of the PAD*;

**Provisional Recomendation X.3**

ii) that the operation of the PAD for the start-stop mode DTE should depend on the possible values of internal variables known as PAD parameters which are described below in 2. *Characteristics of PAD parameters*;

iii) that the PAD parameters for the start-stop mode DTE and their possible values should be those which are listed below in 3. *List of PAD parameters and possible values*;

iv) that the PAD features described in 1., 2. and 3. below could be expanded by future studies to allow interworking with non-packet mode DTEs other than start-stop mode DTEs.

## 1. *Description of the basic functions and user selectable functions of the PAD*

1.1 The PAD performs a number of functions and exhibits operational characteristics. Some of the functions allow either or both the start-stop mode DTE and the packet-mode DTE to configure the PAD so that its operation is adapted to the start-stop mode DTE characteristics, and possibly to the application.

1.2 The operation of the PAD depends on the values of the set of internal variables called PAD parameters. This set of parameters exists for each start-stop mode DTE independently. The current value of each PAD parameter defines the operational characteristics of its related function.

### 1.3 *Basic functions of the PAD*

These basic functions include:
- assembly of characters into packets destined for the packet- mode DTE;
- disassembly of the user data field of packets destined for the start-stop mode DTE;
- handling of virtual call set-up and clearing, resetting and interrupt procedures;
- generation of service signals;
- a mechanism for forwarding packets when the proper conditions exist, e.g. a packet is full or an idle timer expires;
- a mechanism for transmitting data characters, including start, stop and parity elements as appropriate, to the start-stop mode DTE;
- a mechanism for handling a *break signal* from the start-stop mode DTE.

### 1.4 *User selectable functions which may be provided by the PAD*

These functions concern:
- management of the procedure between the start-stop mode DTE and the PAD;
- management of the assembly and disassembly of packets;
- a limited number of additional functions related to the operational characteristics of the start-stop mode DTE.

The method for the control of these functions is specified in Recommendation X.28 for start-stop mode DTE and in Recommendation X.29 for the packet-mode DTE.

#### 1.4.1 *PAD recall by escaping from the data transfer state*

This function allows the start-stop mode DTE to initiate an escape from the *data transfer state* of a virtual call in order to send *PAD command signals*.

**Provisional Recomendation X.3**

### 1.4.2 Echo

This function provides for all characters received from the start-stop mode DTE to be transmitted to the start-stop mode DTE as well as being interpreted by the PAD.

### 1.4.3 Recognition of data forwarding signals

This function allows the PAD to recognize defined character(s) or the *break signal* received from the start-stop mode DTE as an indication to complete the assembly and forward a packet as defined in Recommendation X.25.

### 1.4.4 Selection of idle timer delay

This function allows the PAD to terminate the assembly of a packet and to forward it in the event that the interval between successive characters received from the start-stop mode DTE exceeds a selected value.

### 1.4.5 Ancillary device control

This function allows for flow-control between the PAD and the start-stop mode DTE. The PAD indicates whether it is ready or not to accept characters from the start-stop mode DTE by transmitting special characters. These characters are those which in International Alphabet No. 5(IA5) are used to switch an ancillary transmitting device on and off.

### 1.4.6 Suppression of PAD service signals

This function provides for the suppression of all *PAD service signals* to the start-stop mode DTE.

### 1.4.7 Selection of operation of PAD on receipt of the break signal

This function allows the selection of the operation of the PAD after the receipt of a *break signal* from the start-stop mode DTE.

### 1.4.8 Discard output

This function provides for a PAD to discard the contents of a user data field in packets rather than disassembling and transmitting these to the start-stop mode DTE.

### 1.4.9 Padding after carriage return

This function provides for the automatic insertion by the PAD of padding characters in the character string transmitted to the start-stop mode DTE after the occurrence of a *carriage return* character. This allows for the printing mechanism of the start-stop mode DTE to perform the carriage return function correctly.

### 1.4.10 Line folding

This function provides for the automatic insertion by the PAD of appropriate format effectors in the character string transmitted to the start-stop mode DTE. The predetermined maximum number of graphic characters per line may be set.

### 1.4.11 Flow control of the PAD by the start-stop mode DTE

This function allows for flow-control between the start-stop mode DTE and the PAD. The start-stop mode DTE indicates whether it is ready or not to accept characters from the PAD by transmitting special characters. These characters are those which in IA5 are used to switch an ancillary transmitting device on and off.

## 2. Characteristics of PAD parameters

2.1 In this Recommendation parameters are identified by decimal reference numbers.

**Provisional Recomendation X.3**

2.2    In this Recommendation the possible values of the parameters are represented by decimal numbers.

2.3    Specific procedures, described in Provisional Recommendations X.28 and X.29 are available for initializing, reading and changing the values of PAD parameters.

### 2.4    *Determination of the values of PAD parameters*

### 2.4.1    *Initial values of PAD parameters*

On initialization the initial value of each PAD parameter is set according to a predetermined set of values called a "standard profile".

### 2.4.2    *Current values of PAD parameters*

The current values of PAD parameters at a given time are the values resulting from possible modifications by the start-stop mode DTE and/or the packet-mode DTE.

### 3.    *List of PAD parameters and possible values*

### 3.1    *PAD recall by escaping from the data transfer state*

*Reference 1*

The parameter will have the following selectable values:

not possible                                       — represented by decimal 0 ;

possible                                           — represented by decimal 1.

### 3.2    *Echo*

*Reference 2*

The parameter will have the following selectable values:

no echo                                            — represented by decimal 0;

echo                                               — represented by decimal 1.

### 3.3    *Selection of data forwarding signal*

*Reference 3*

The parameter will have the following selectable values:

no data forwarding signal                          — represented by decimal 0;

carriage return                                    — represented by decimal 2 ;

all characters in columns 0 and 1 and
character 7/15 (DEL) of International
Alphabet No. 5                                     — represented by decimal 126.

Forwarding by other characters, or sets of characters, and the appropriate decimal representation is for further study.

### 3.4    *Selection of idle timer delay*

*Reference 4*

The parameter will have the following selectable values:

any number from 0 to 255                           — represented by the respective decimal
                                                   number.

The value 0 will indicate that no data forwarding on time-out is required; a value between 1 and 255 will indicate the value of the delay in twentieths of a second.

**Provisional Recomendation X.3**

3.5    *Ancillary device control*

*Reference 5*

The parameter will have the following selectable values:

no use of X-ON (DC1) and X-OFF (DC3)          — represented by decimal 0;

use of X-ON and X-OFF          — represented by decimal 1.

3.6    *Suppression of PAD service signals*

*Reference 6*

The parameter will have the following selectable values:

no service signals are transmitted to the start-
stop mode DTE          — represented by decimal 0;

service signals are transmitted          — represented by decimal 1.

3.7    *Selection of operation of PAD on receipt of break signal from the start-stop mode DTE*

*Reference 7*

This parameter is represented by the following encoding of basic functions, each having a decimal value as shown below:

nothing          — represented by decimal 0;

send to packet mode DTE an *interrupt packet*          — represented by decimal 1;

reset          — represented by decimal 2;

send to packet-mode DTE an *indication of
break PAD message*          — represented by decimal 4;

*escape from data transfer state*          — represented by decimal 8;

discard output to start-stop mode DTE          — represented by decimal 16.

Only the following functions and combinations of functions will be selectable: 0, 1, 2, 8 and 21 (1+4+16). The use of other values is for further study.

*Note:* The decimal representation of individual values of this parameter allows coding to represent a single function or combination of functions.

3.8    *Discard output*

*Reference 8*

The parameter will have the following selectable values:

normal data delivery to the start-stop mode
DTE          — represented by decimal 0;

discard output to the start-stop mode DTE          — represented by decimal 1.

3.9    *Padding after carriage return*

*Reference 9*

The parameter will have the following selectable values:

any number from 0 to 7          — represented by the respective decimal number.

**Provisional Recomendation X.3**

A value between 0 and 7 will indicate the number of padding characters to be generated by the PAD after a carriage return character is transmitted to the start-stop mode DTE.

When parameter 9 is 0, PAD service signals will contain a number of padding characters according to the data rate of the start-stop mode DTE.

## 3.10  *Line folding*

*Reference 10*

The parameter will have the following selectable values:

| no line folding | — represented by decimal 0; |
| any value between 1 and 255 characters per line | — represented by the respective decimal number. |

## 3.11  *Binary speed*

This parameter is a read-only parameter and cannot be changed by either of the DTEs. It enables the packet-mode DTE to access a characteristic of the start-stop mode DTE which is known by the PAD.

*Reference 11*

The parameter will have the following values:

| 50 bit/s | — represented by decimal 10; |
| 100 bit/s | — represented by decimal 9; |
| 110 bit/s | — represented by decimal 0; |
| 134.5 bit/s | — represented by decimal 1; |
| 200 bit/s | — represented by decimal 8; |
| 300 bit/s | — represented by decimal 2. |

Only values 0, 2 and 8 are for use at this time. The provision of other speeds and the appropriate decimal representation is for further study.

## 3.12  *Flow control of the PAD by the start-stop mode DTE*

*Reference 12*

The parameter will have the following selectable values:

| no use of X-ON (DC1) and X-OFF (DC3) for flow control | — represented by decimal 0; |
| use of X-ON and X-OFF | — represented by decimal 1. |

### GENERAL STRUCTURE OF SIGNALS OF INTERNATIONAL ALPHABET No. 5 CODE FOR DATA TRANSMISSION OVER PUBLIC DATA NETWORKS [1]

*(Geneva, 1976)*

The CCITT,

I. *considering, firstly,*

the agreement between the International Organization for Standardization (ISO) and the CCITT on the main characteristics of a seven-unit alphabet (International Alphabet No. 5) to be used for data transmission and for telecommunications requirements that cannot be met by the existing five-unit International Telegraph Alphabet No. 2;

the interest, both to the users and to the telecommunication services, of an agreement concerning the chronological order of transmission of bits in serial working,

*declares the view*

that the agreed rank number of the unit in the alphabetical table of combinations should correspond to the chronological order of transmission in serial working on telecommunication circuits;

that, when this rank in the combination represents the order of the bit in binary numbering, the bits should be transmitted in serial working with the low order bit first;

that the numerical meaning corresponding to each information unit considered in isolation is that of the digit:

0 for a unit corresponding to condition A (travail = space), and

1 for a unit corresponding to condition Z (repos = mark),

in accordance with the definitions of these conditions for a two-condition transmission system;

II. *considering, moreover,*

that it is often desirable, in data and messages transmission, to add an extra "parity" unit to allow for the detection of errors in received signals;

the possibility offered by this addition for the detection of faults in terminal equipment;

the need to reserve the possibility of making this addition during the transmission itself, after the seven information units proper have been sent,

---

[1] See Recommendation V.4 for data transmission over public telephone networks.

*declares the view*

that signals of International Alphabet No. 5 code for data and messages transmission should in general include an additional "parity" unit;

that the rank of this unit and, hence, the chronological order of the transmission in serial working should be the eighth of the combination thus completed;

### III. *considering*

that, in start-stop systems working with electromechanical equipment, the margin of such equipment and the reliability of the connection are considerably increased by the use of a stop element corresponding to the duration of two-unit intervals of the modulation;

that for start-stop systems using International Alphabet No. 5 at modulation rates of 200 and 300 bauds, Recommendations X.1 and X.31 specify that transmit devices should use a stop element lasting at least two units;

that the previously expressed preference for a two-unit stop element arises from a transmission point of view where anisochronous public data networks are concerned,

*declares the view*

that in start-stop systems using combinations of International Alphabet No. 5 normally followed by a parity unit, the first information unit of the transmitted combination should be preceded by a start element corresponding to condition A (space);

that the duration of this start element should be one-unit interval for the modulation rate under consideration, at transmitter output;

that the combination of seven information units, normally completed by its parity unit, should be followed by a stop element corresponding to condition Z (mark);

that for public anisochronous data networks, data terminal equipment using International Alphabet No. 5 should comply with Recommendations X.1 and X.31 and use a stop element lasting at least two units;

that the start-stop receivers should be capable of correctly receiving start-stop signals from a source which appears to have a nominal cycle of 10 units (i.e., with a nominal one-unit stop element). However, for certain electromechanical equipment the receivers may only be capable of correctly receiving signals when the stop element is not reduced below one unit (even in the presence of distortion);

### IV. *considering, finally,*

that the direction of the parity unit can only be that of the even parity on the perforated tapes, particularly owing to the possibility of deletion (combination 7/15 of the alphabet) which causes a hole to appear in all tracks;

that, on the other hand, the odd parity is considered essential in the equipment which depends on transitions in the signals to maintain synchronism [in cases where combination 1/6 (SYNC) of the alphabet does not permit of an economical solution],

*declares the view*

that the parity unit of the signal should correspond to the even parity in links or connections operated on the principle of the start-stop system.

that this parity should be odd on links or connections using end-to-end synchronous operation,

that arrangements should be made when necessary to reverse the direction of the parity unit at the input and output of the synchronous equipment connected either to apparatus working on the start-stop principle or receiving characters on perforated tape.

# SECTION 2

## DATA TERMINAL EQUIPMENT AND INTERFACES

**Recommendation X.20**

### INTERFACE BETWEEN DATA TERMINAL EQUIPMENT (DTE) AND DATA CIRCUIT-TERMINATING EQUIPMENT (DCE) FOR START-STOP TRANSMISSION SERVICES ON PUBLIC DATA NETWORKS

*(Geneva, 1972, amended at Geneva, 1976)*

### CONTENTS

1. *Scope*

1.1　　This Recommendation applies to the interface between data terminal equipment (DTE) and data circuit-terminating equipment (DCE) for start-stop transmission services on public data network. In particular it is applicable to:

　　*a)*　circuit-switched service,

　　*b)*　lease circuit service, either point-to-point or multipoint connections.

The application in packet-switched service is for further study.

This Recommendation refers to other Recommendations which define the electrical and functional characteristics of the interchange circuits.

1.2　　In the case of switched data network service it additionally defines the procedural characteristics of the interface in terms of a call control procedure for start-stop transmission in user classes of service 1 and 2 of Recommendation X.1.

1.3　　The DCE of these user classes of service provides all conversions required between the DTE/DCE interface and the line. It also serves for the provision of the essential and optional user facilities according to Recommendation X.2. Nationally assigned user facilities may also be included.

1.4　　The information content of the specified selection signal format and the call progress signal format is not included in this Recommendation. It is the subject of other Recommendations.

2. *Physical characteristics*

2.1　*Interchange circuits*

　　A list of the interchange circuits concerned is presented in Table 1/X.20. Definitions of these interchange circuits are given in Recommendation X.24.

TABLE 1/X.20

| Interchange circuit | Interchange circuit name | Direction | |
| --- | --- | --- | --- |
| | | to DCE | from DCE |
| G (see Note) | Signal ground or common return | | |
| Ga | DTE common return | X | |
| Gb | DCE common return | | X |
| T | Transmit | X | |
| R | Receive | | X |

　　*Note.* – This conductor may be used to reduce environmental signal interference at the interface. In case of shielded interconnecting cable, the additional connection considerations are part of Recommendation X.24.

2.2　*Electrical characteristics*

　　The electrical characteristics of the interchange circuits at the DCE side of the interface will comply with Recommendation X.26.

　　The electrical characteristics at the DTE side of the interface may be applied according to Recommendations X.26, X.27 (without cable termination in the load), or Recommendation V.28.

2.3    *Mechanical characteristics*

Refer to ISO DIS 4903 *(15-pin DTE/DCE interface connector and pin assignment)* for mechanical arrangements.

3.    *Call control procedures*

Figure 1/X.20 — Sequence of signals at the interface, shows the procedures at the interface during call establishment, the data transfer phase, and the clearing of a call including unsuccessful calls. Figure 2/X.20 shows the state diagrams, which define the logical relationships of events at the interface. (Annex 1 defines the symbols used for the state diagrams.)

3.1    *Call establishment*

3.1.1    *Ready* (state 1)

Circuits T and R show binary 0.

3.1.2    *Call request* (state 2)

Circuit T is changed to binary 1.

3.1.3    *Proceed to select* (state 3)

Circuit R is changed to binary 1.

3.1.4    *Selection signals* (state 4)

The selection characters are transmitted on circuit T. In the case of direct calls, no *Selection signals* are transmitted.

3.1.5    *Incoming call* (state 5)

Circuit R is changed to binary 1.

3.1.6    *Call accepted* (state 6)

Circuit T is changed to binary 1 not later than 600 ms after the incoming call is recognized. 10-100 ms thereafter, the DTE transmits the call control character ACK.

3.1.7    *Called and Calling line identification* (states 7A and 7B)

When provided, the *Called line identification* (state 7A) will be transmitted by the DCE to the calling DTE after all *Call progress* signals, if any.

When provided, the *Calling line identification* (state 7B) will be transmitted by the DCE to the called DTE after *Call accepted* has been sent by the DTE.

3.1.8    *Connected* (state 8)

The call control character ACK is transferred on circuit R. The event *Connected* begins nominally at the middle of the first stop unit of the call control character.

3.1.9    *Ready for data* (state 9)

After receipt of *Connected* the DTE shall be ready for the reception of data.

Twenty milliseconds after the beginning of *Connected*, the DTE may commence with the transmission of data.

3.2     *Data transfer* (state 10)

The events during *Data transfer* are in the responsibility of the DTE.

3.3     *Clearing*

3.3.1   *DTE Clear request* (state 11)

Circuit T is changed to binary 0 for more than 210 ms and shall not be reversed to binary 1 before *DCE Ready*.

3.3.2   *DCE Clear confirmation* (state 12)

Within 6 seconds after the beginning of *DTE Clear request* circuit R is changed to binary 0 for more than 210 ms and will not be reversed to binary 1 before *DCE Ready*.

3.3.3   *DCE Clearing* (state 13)

Circuit R is changed to binary 0 for more than 210 ms and will not be reversed to binary 1 before *DCE Ready*.

3.3.4   *DTE Clear confirmation* (state 14)

Within 210-490 ms after the beginning of *DCE Clearing*, circuit T is changed to binary 0 for more than 210 ms and shall not be reversed to binary 1 before *DCE Ready*.

3.3.5   *DTE Ready* (state 15)

Within 210-490 ms after the beginning of *DCE Clear confirmation* or < 490 ms after the beginning of *DTE Clear confirmation*, respectively, the DTE shall be ready to accept an *Incoming call*.

3.3.6   *DCE Ready* (state 1)

490 ms after the beginning of DCE or DTE *Clear confirmation*, respectively, the DCE is ready to accept a new *Call request*.

3.3.7   *Clear collision*

In case *DTE Clear request* and *DCE Clearing* occur at the same instant or during an overlapping time of 210 ms, the DTE shall proceed in its clearing procedure.

3.4     *Unsuccessful call*

3.4.1   *Unsuccessful call request*

If a *Call request* is not successful, the DCE may indicate the reasons by means of call progress signals. In any case, *DCE Clearing* will be performed.

3.4.2   *Call not accepted* (state 17)

If an *Incoming call* cannot be accepted, circuit T has to be changed to binary 1 not later than 600 ms after the beginning of *Incoming call*. 10-100 ms thereafter the DCE shall transmit the call control character NAK and then *DTE Clear request*.

### 3.5 Fault conditions

#### 3.5.1 No Proceed to select

If the *Proceed to select* signal has not been received on circuit R within 6 seconds after the beginning of *Call request*, the DTE shall perform *DTE Clear request*.

#### 3.5.2 No Selection signal

If no *Selection signal* is transmitted within 6 seconds from the beginning of *Proceed to select* or the preceding *Selection signal*, *DCE Clearing* will be performed.

#### 3.5.3 No Connected signal

If the call control character for *Connected* has not been received on circuit R within 60 seconds after

a)   the end of the selection signal sequence, or

b)   the end of the call control character for *Call accepted*

the DTE shall perform *DTE Clear request*.

*Note.* — It should be noted that the setting up time for connections may vary. The value of 60 seconds is a maximum time-out which, if it expires, should lead to the call attempt being abandoned by the calling DTE.

#### 3.5.4 Circuit T in circuit failure state

If circuit T is in a power-off or an open circuit condition, the DCE interprets this condition as binary 0.

#### 3.5.5 Circuit R in circuit failure state

If circuit R is in a power-off or an open circuit condition, the DTE interprets this condition as binary 0.

### 3.6 Call collision (state 18)

A *Call collision* is detected by the DCE when it receives *Call request* in response to an *Incoming call*. The DCE may either accept the *Call request* or may perform *DCE Clearing*.

### 4. Call control formats

### 4.1 Format of selection sequence (see Annex 2)

A selection sequence shall consist of facility request block or address block, or both.

#### 4.1.1 Facility request block

A *Facility request* block shall consist of one or more *Facility request* signals.

Multiple *Facility request* signals shall be separated by character 2/12 (",").

End of a *Facility request* block shall be indicated by character 2/13 (" – ").

#### 4.1.2 Address

An *Address block* shall consist of one or more address or abbreviated address signals.

Multiple *Address* or multiple *Abbreviated address signals* shall be separated by character 2/12 (",").

Start of *Abbreviated address signals* shall be indicated by a prefix character 2/14 (".").

4.1.3    *End of selection sequence*

The end of selection sequence shall be indicated by character 2/11 (" + ").

4.2    *Format of a Call progress block*

A *Call progress block* shall consist of one or more *Call progress* signals. Each *Call progress signal* need not be repeated.

Each *Call progress signal* is preceded by characters 0/13 (CR = carriage return) and 0/10 (LF = line feed).

End of *Call progress block* will be indicated by character 2/11 (" + ").

4.3    *Format of called and calling line identification*

A *Calling line identification signal* and a *Called line identification block* shall be preceded by characters 0/13 (CR = carriage return), 0/10 (LF = line feed) and 2/10 ("*").

A *Called line identification block* shall consist of one or more *Called line identification signals*.

Multiple *Called line identification signals* shall be separated by characters 0/13 (CR) and 0/10 (LF).

The end of a *Calling line identification signal* and a *Called line identification block* shall be indicated by character 2/11 (" + ").

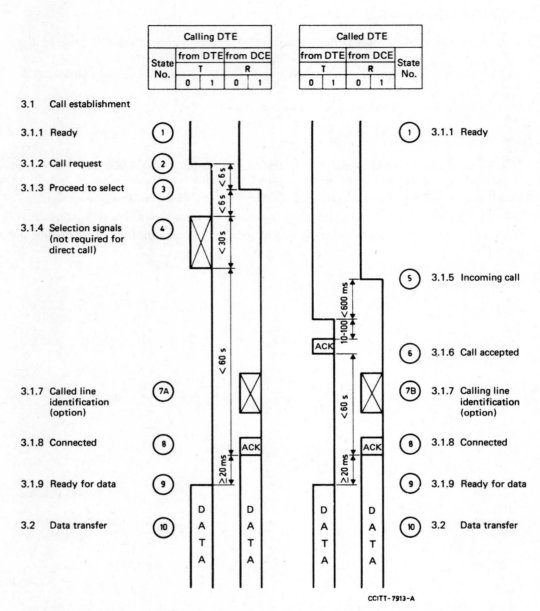

FIGURE 1a/X.20 — **Sequence of signals at the interface** (call establishment/data transfer)

204

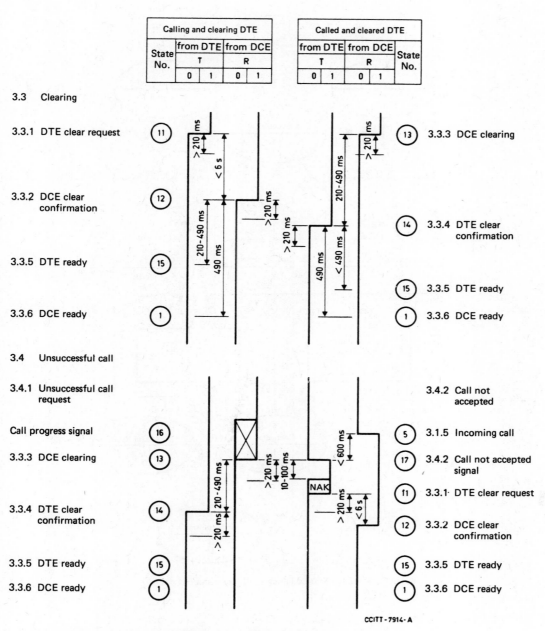

*Note.* – The numbering refers to 3. *Call control procedures* in the text.

FIGURE 1b/X.20 – **Sequence of signals at the interface (clearing/unsuccessful call)**

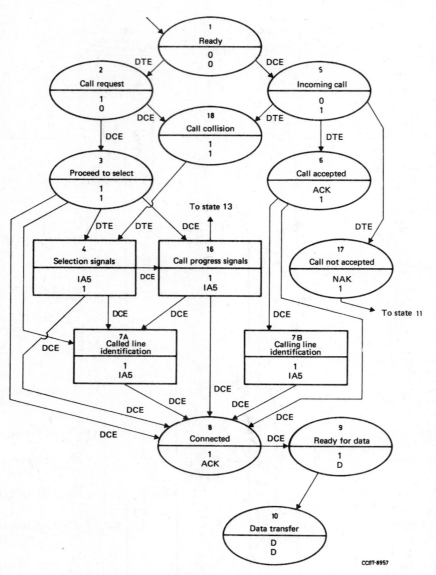

FIGURE 2a/X.20 – **State diagram** (call establishment phase)

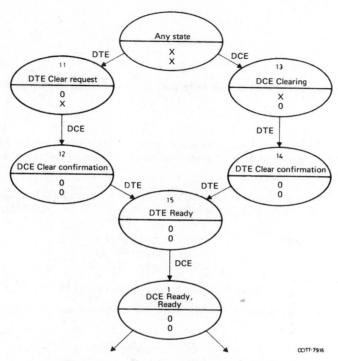

FIGURE 2b/X.20 — **State diagram** (clearing phase)

ANNEX 1

(to Recommendation X.20)

**Definition of symbols used in the state diagrams**

Each state is represented by an ellipse wherein the state name and number is indicated, together with the signals on the interchange circuits which represent that state.

Each state transition is represented by an arrow and the equipment responsible for the transition (DTE or DCE) is indicated beside that arrow. A rectangle represents a sequence of states and transitions.

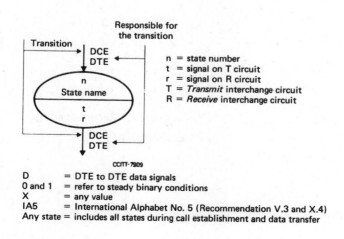

D          = DTE to DTE data signals
0 and 1   = refer to steady binary conditions
X          = any value
IA5        = International Alphabet No. 5 (Recommendation V.3 and X.4)
Any state = includes all states during call establishment and data transfer

## ANNEX 2

### (to Recommendation X.20)

### Formats of selection, call progress and line identification signals

The following description uses Backus Normal Form as the formalism for syntactic description, for example, as in ISO Recommendation R 1538 [1]. A vertical line " | " separates alternatives.

<Selection Sequence> : : = <Facility Request Block>  <–>  <Address Block>  <+> |
<Facility Request Block>  <–>  <+> | <Address Block>  <+>

<Facility Request Block> : : = <Facility Request Signal>
<Facility Request Block>  <,>
<Facility Request Signal>

<Address Block> : : = <Address Signal> | <Address Block>  <,>  <Address Signal>

<Address> : : = <Full Address Signal> | <.>  <Abbreviated Address Signal>

<Call Progress Block> : : = <CR> <LF> <Call Progress Signal>  <+> | <Call Progress Signal>
<Call Progress Block>

<Calling Line Identification> : : = <CR> <LF> <*> <Calling Line Identification Signal>  <+>

<Called Line Identification> : : = <CR> <LF> <*> <Called Line Identification Block>  <+>

<Called Line Identification Block> : : = <Called Line Identification Signal > | <Called Line
Identification Block> <CR> <LF>  <Called Line
Identification Signal>

<CR> : : = IA 5   character  0/13

<LF> : : = IA 5   character  0/10

<*>  : : = IA 5   character  2/10

<+>  : : = IA 5   character  2/11

<,>  : : = IA 5   character  2/12

<–>  : : = IA 5   character  2/13

<.>  : : = IA 5   character  2/14

| | |
|---|---|
| <Facility Request Signal> | |
| <Full Address Signal> | |
| <Abbreviated Address Signal> | Subject of another Recommendation |
| <Call Progress Signal> | (not yet available) |
| <Calling Line Identification Signal> | |
| <Called Line Identification Signal> | |

---

[1] ISO Recommendation R 1538: Programming Language "ALGOL", March 1972.

**Recommendation X.20** *bis*

## V.21-COMPATIBLE INTERFACE BETWEEN DATA TERMINAL EQUIPMENT (DTE) AND DATA CIRCUIT-TERMINATING EQUIPMENT (DCE) FOR START-STOP TRANSMISSION SERVICES ON PUBLIC DATA NETWORKS

*(Geneva, 1976)*

*Introduction*

Many DTEs are in use which are equipped with interfaces recommended for DCEs on telephone-type networks. For an interim period public data networks should also provide such interfaces in order to enable the connection of existing DTEs to these networks.

1.    *Scope*

This Recommendation applies to the interface between DTE designed for interfacing to modems according to Recommendation V.21 and DCE on public data networks.

The operation is limited to start-stop transmission at data signalling rates and character structures specified in user classes of service 1 and 2 of Recommendation X.1.

The application comprises:

*a)*   circuit switched service,

*b)*   leased circuit service.

The application in packet switched service is for further study.

2.    *Interchange circuits*

2.1   *Functional characteristics*

The functional characteristics of the interchange circuits concerned (see Table 1/X.20 *bis*) comply with Recommendation V.24.

TABLE  1/X.20 *bis*

| Interchange circuit | |
|---|---|
| **Number** | **Designation** |
| 102 | Signal ground or common return |
| 103 | Transmitted data |
| 104 | Received data |
| 106 | Ready for sending |
| 107 | Data set ready |
| 108/1 [a] | Connect data set to line |
| 108/2 [b] | Data terminal ready |
| 109 | Data channel received line signal detector |
| 125 [c] | Calling indicator |

[a] Used in case of automatic control of the direct call facility.
[b] Used in case of switched data network service.
[c] Not provided in leased circuit service.

2.2     *Electrical characteristics*

The electrical characteristics of the interchange circuits comply with Recommendation V.28, using the 25-pin interface connector and pin assignments in ISO DIS 2110.2.

3.      *Operational requirements*

3.1     *Operation of interchange circuit 107*

The DCE switches circuit 107 to ON after establishment of the data circuit.

The circuit 107 is switched to OFF only as a response to the OFF condition of the circuit 108.

3.2     *Operation of interchange circuits 109 and 106*

The DCE switches circuit 109 to ON together with circuit 107. Circuit 106 is put to ON 20 to 40 ms after the appearance of the ON condition on circuit 107.

The circuits 109 and 106 are switched to OFF either when circuit 108 is switched to OFF or when circuit 108 is ON and the DCE indicates a *Clear request* from the network (DCE Clearing).

3.3     *Operation of interchange circuit 125*

Circuit 125 will be switched OFF when circuit 107 comes ON.

3.4     *Unaccepted call*

If the ON condition of circuit 125 is not answered by an ON condition on circuit 108 within 450 ms, the incoming call will be rejected by the DCE.

3.5     *Multipoint operation*

As the circuits 106 and 109 are always in the ON condition, the transmission disciplines must be determined by end-to-end control procedures of the DTEs.

3.6     *Direct call facility*

Manual control: The direct call is initiated when the direct call button at the DCE is pressed and circuit 108/2 is ON.

Automatic control: The direct call is initiated when the DTE switches the circuit 108/1 to ON.

**Recommendation X.21**

## GENERAL PURPOSE INTERFACE BETWEEN DATA TERMINAL EQUIPMENT (DTE) AND DATA CIRCUIT-TERMINATING EQUIPMENT (DCE) FOR SYNCHRONOUS OPERATION ON PUBLIC DATA NETWORKS

*(Geneva, 1972, amended at Geneva, 1976)*

CONTENTS

*Preface*

The CCITT,

*considering*

a) that Recommendations X.1 and X.2 define the services and facilities to be provided by a public data network;

b) that it is desirable for the characteristics of the interface between the DTE and DCE in a public data network to be standardized.

*unanimously declares the view*

that the interface between the DTE and DCE in public data networks for user classes of service employing synchronous mode of transmission should be as defined in this Recommendation.

211

1.    *Scope*

1.1    This Recommendation defines the physical characteristics and control procedures for a general purpose interface between DTE and DCE for user classes of service as defined in Recommendation X.1 employing a synchronous mode of transmission across the interface.

1.2    The information content of *Selection, Call progress* and *Line identification* signals is not included in this Recommendation. It is the subject of other Recommendations.

1.3    The operation of the interface when the data circuit interconnects with Recommendation X.21 *bis* DTEs is described in Annex 6.

2.    *Interchange circuits*

Definitions of the interchange circuits concerned (see Table 1/X.21) are given in Recommendation X.24.

TABLE 1/X.21

| Interchange circuit | Name | Direction | | Remarks |
|---|---|---|---|---|
| | | to DCE | from DCE | |
| G | Signal ground or common return | | | see Note 1 |
| Ga | DTE common return | X | | |
| T | Transmit | X | | |
| R | Receive | | X | |
| C | Control | X | | |
| I | Indication | | X | |
| S | Signal element timing | | X | see Note 2 |
| B | Byte timing | | X | see Note 3 |

*Note 1.* — This conductor may be used to reduce environmental signal interference at the interface. In the case of shielded interconnecting cable, the additional connection considerations are part of Recommendation X.24.

*Note 2.* — Continuous isochronous transmission will be provided.

*Note 3.* — May be provided as an optional additional facility (see 4. below).

3.    *Physical characteristics*

3.1    *Electrical characteristics*

3.1.1    *Data signalling rates of 9600 bit/s and below*

The electrical characteristics of the interchange circuits at the DCE side of the interface will comply with Recommendation X.27 (without cable termination in the load). The electrical characteristics at the DTE side of the interface may be applied according to Recommendation X.27 (without cable termination in the load), or Recommendation X.26. The B-leads of circuits R, I, S and B, if provided, must be connected to the individual differential inputs and not connected together. (See ISO DIS 4903.)

3.1.2    *Data signalling rates above 9600 bit/s*

The electrical characteristics of the interchange circuits at both the DCE side and the DTE side of the interface will comply with Recommendation X.27 (with implementation of the cable termination in the load).

212

3.2    *Mechanical characteristics*

Refer to ISO DIS 4903 *(15-pin DTE/DCE interface connector and pin assignment)* for mechanical arrangements.


4.     *Character alignment*


4.1    *Call establishment and other call control phases*

For the interchange of information [2] between the DTE and the DCE for call control purposes, it is necessary to establish correct alignment of characters. Each sequence of call control characters to and from the DCE shall be preceded by two or more contiguous 1/6 (SYN) characters.

4.1.1    Certain Administrations will require the DTE to align call control characters transmitted from the DTE to either SYN characters delivered to the DTE or to signals on the byte timing interchange circuit.

Administrations who require this alignment shall provide the byte timing interchange circuit but its use and termination by the DTE shall not be mandatory.

4.1.2    Certain Administrations will permit call control characters to be transmitted from the DTE, independently of the SYN characters delivered to the DTE.

4.1.3    Additionally, for an intermediate period (see Note) Administrations will provide connection to the public data network of DTEs operating as described in 4.1.2 above.

*Note.* — The intermediate period would be determined by customer demand and other relevant factors as interpreted by individual Administrations.


4.2    *Data phase*

For the interchange of information between one DTE and another DTE after the call has been established, the DTEs will be responsible for establishing their own alignment.

(The byte timing interchange circuit, when implemented, may be utilized by the DTEs for mutual character alignment.)


5.     *Interface procedures and timing of events*

Annex 1 shows the state diagrams which give the definition of logical relationships of events at the interface. (Annex 4 defines the symbols used for the state diagrams).

Annex 2 shows examples of timing diagrams of the signals at the interface. Those diagrams show only the most frequent sequences of signals. All timing diagrams are derived from the state diagrams.

Annex 3 provides the tables of DTE time limits and DCE time-outs which define the timing relationships of events at the interface.

All call control characters are selected from International Alphabet No. 5 according to Recommendations V.3 and X.4 and have odd parity.

Steady binary conditions 0 and 1 on circuit T or R together with the associated condition on circuit C or I persist for at least 15 bits.

ON and OFF conditions for circuit C *(Control)* and I *(Indication)* respectively refer to continuous ON (binary 0) and continuous OFF (binary 1) conditions.

In this Recommendation signals on interchange circuits T, C, R and I are designated by t, c, r and i, respectively.

---

[2] See Definition 53.029 *Information, Green Book*, Volume VIII, page 33.

**5.1** *Quiescent states for circuit switched service* (see Annex 1, Figure 2/X.21)

**5.1.1** *Ready* (state 1)

Both DTE and DCE shall signify their readiness for a new call by sending *Ready* signals, that is, t = 1, c = OFF, r = 1, i = OFF.

**5.1.2** *DTE Uncontrolled not ready* (state 21)

*DTE Uncontrolled not ready* indicates that the DTE is unable to accept incoming calls, generally because of abnormal operating conditions, and will inhibit the connection of incoming calls. This state is signalled by t = 0, c = OFF.

**5.1.3** *DTE Controlled not ready* (state 14)

*DTE Controlled not ready* indicates that, although the DTE is operational, it is temporarily unable to accept incoming calls.

The state is signalled by t = 01 . . . . (alternate bits are binary 0 and binary 1), c = OFF.

In some networks no action will be taken when DTE *Controlled not ready* is signalled from the DTE. The consequences of this for the caller should be studied further.

*Note 1.* — The *Controlled not ready* state is normally entered from the *Ready* state. Other states from which *Controlled not ready* may be entered are left for further study.

*Note 2.* — Possible use and specification of a family of *Controlled not ready* signals is left for further study.

**5.1.4** *DCE not ready* (state 18)

*DCE not ready* indicates that no service is available and will be signalled whenever possible during network fault conditions and when test loops are activated. This state is signalled by steady binary condition r = 0, i = OFF.

**5.2** *Call establishment and clearing for circuit switched service* (see Annex 1, Figures 1/X.21 and 2/X.21)

**5.2.1** *Call request* (state 2)

The calling DTE shall indicate a request for a call by signalling steady binary condition t = 0, c = ON, provided that it was previously signalling *DTE Ready*.

The change of state from *Ready* (t = 1, c = OFF) to *Call request* (t = 0, c = ON) shall be such that the transition to t = 0 occurs within the same bit interval as the transition to c = ON.

**5.2.2** *Proceed to select* (state 3)

When the network is prepared to receive selection information the DCE will transmit continuously character 2/11 (" + ") preceded by 2 or more contiguous characters 1/6 ("SYN") on the R circuit with i = OFF.

*Proceed to select* is maintained until end of selection signal or *DTE Waiting* is received.

The *Proceed to select* signal shall start within 3 seconds of the *Call request* being sent.

### 5.2.3    *Selection signals* (state 4)

The *Selection signals* shall be transmitted by the DTE on the T circuit with c = ON and shall be preceded by two or more contiguous 1/6 ("SYN") characters with c = ON.

The format of *Selection signals* is defined in 6.1 below.

The information content of *Selection signals* is the subject of another Recommendation.

Selection shall start within 6 seconds of *Proceed to select* being received and shall be completed within 36 seconds.

The maximum permissible interval between individual selection characters is 6 seconds.

The period, if any, between individual selection characters shall be filled by character 1/6 ("SYN") with c = ON.

### 5.2.4    *DTE Waiting* (state 5)

The period following transmission of the end of selection signal during which no information is transmitted from the DTE will be signalled by steady binary condition t = 1, c = ON. (See also 5.2.16 below.)

### 5.2.5    *Incoming call* (state 8)

The DCE will indicate an incoming call by continuous transmission of character 0/7 ("BEL") preceded by two or more contiguous 1/6 ("SYN") characters on the R circuit with i = OFF.

### 5.2.6    *Call accepted* (state 9)

The DTE shall accept the incoming call as soon as possible by signalling the steady state binary condition t = 1, c = ON.

1)    The DCE will return to the *Ready* state if the incoming call is not accepted within 500 milliseconds. (See Note below.)

     or, where manual answering is permitted,

2)    the DCE will return to the *Ready* state if the incoming call is not accepted within 60 seconds.

*Note.* — This time-out limit is subject to further study with the objective to reduce the maximum response time in the future to 100 milliseconds.

### 5.2.7    *DCE Waiting* (state 6)

The period following receipt of the end of selection signal or receipt of *Call accepted* by the DCE and during which no information is transmitted from the DCE shall be filled by the transmission of two or more contiguous 1/6 ("SYN") characters on the R circuit with i = OFF.

The transition from this state to state 11 or 12 need not be on a SYN character boundary.

### 5.2.8    *Call progress signals* (state 7) (Refer to Annex 5)

The format of *Call progress signals* is defined in 6.2 below.

The *Call progress signals* will be transmitted by the DCE to the calling DTE on the R circuit with i = OFF.

In some cases there may be no *Call progress signals*.

In some cases there may be several *Call progress blocks*, in that case the period between these blocks will be filled by *DCE Waiting* (state 6B). A *Call progress block* will be preceded by 2 or more contiguous 1/6 ("SYN") characters.

*Call progress signals* will be transmitted by the DCE within 20 seconds of the end of selection signal being sent by the DTE.

The information content of *Call progress signals* is the subject of another Recommendation.

### 5.2.9   *Line identification*  (states 10 and 10 *bis*) (Refer to Annex 5)

*Calling* and *Called line identification* is an optional additional facility.

*Calling* and *Called line identification* will be transmitted by the DCE on the R circuit with i = OFF.

When provided the *Called line identification* (state 10) will be transmitted by the DCE to the calling DTE after all *Call progress signals*, if any.

When provided the *Calling line identification* (state 10 *bis*) will be transmitted by the DCE to the called DTE after *Call accepted* has been sent by the DTE.

The period after Calling or *Called line identification* during which no information is transmitted from the DCE will be filled by the DCE Waiting condition (state 6 C).

The format of *Calling* and *Called line identification* is defined in 6.3 below.

The information content of *Calling* and *Called identification* is the subject of another Recommendation.

### 5.2.10   *Connection in progress*  (state 11)

While the connection process is in progress the DCE will indicate *Connection in progress* (state 11) by signalling r = 1, i = OFF.

In some circumstances *Connection in progress* (state 11) may be bypassed.

### 5.2.11   *Ready for data*  (state 12)

When the connection is available for data transfer between both DTEs, the DCE will indicate *Ready for data* (state 12) by signalling r = 1, i = ON.

1)  *Ready for data* will be indicated by the DCE to the calling DTE within 2 seconds of the last *Call progress signals* being received by the DTE or within 20 seconds of the end of selection signal being signalled by the DTE,

    or, when manual answering is permitted at the called DTE,

2)  *Ready for data* will be indicated by the DCE to the calling DTE within 60 seconds of the appropriate *Call progress signal* being received or within 80 seconds of the end of selection signal being received.

    It will be indicated to the called DTE within 2 seconds of *Call accepted* being signalled by the DTE.

    Subsequent procedures are described in 5.4 below, *Data transfer*.

### 5.2.12   *Clearing by the DTE*  (states 16, 17, 21)

Either a calling or a called DTE may clear a call at any time. The DTE should indicate clearing by signalling the steady binary condition t = 0, c = OFF, *DTE Clear request* (state 16).

Within 2 seconds the DCE will signal the steady binary condition r = 0, i = OFF, *DCE Clear confirmation* (state 17), followed by the steady binary condition r = 1, i = OFF, *DCE ready* (state 21). In some networks where the coding r = 1, i = OFF, is used only for the indication of the *DCE Ready* the *DCE Clear confirmation* state may be bypassed.

The DTE should respond to *DCE Ready* within 100 milliseconds by signalling t = 1, c = OFF, *Ready* (state 1).

### 5.2.13    *Clearing by the DCE* (states 19, 20, 21)

The DCE will indicate clearing to the DTE by signalling the steady binary condition r = 0, i = OFF, *DCE Clear indication* (state 19).

The DTE should signify *Clear confirmation* (state 20) by signalling the steady binary condition t = 0, c = OFF, within 100 milliseconds. The DCE will signal r = 1, i = OFF, *DCE Ready* (state 21) within 2 seconds of receiving *DTE Clear confirmation*.

The DTE should respond to *DCE Ready* within 100 milliseconds by signalling t = 1, c = OFF, *Ready* (state 1).

### 5.2.14    *Unsuccessful call*

If the required connection cannot be established the DCE will indicate this and the reason to the calling DTE by means of a *Call progress signal*. Afterwards the DCE will signal *DCE Clear indication* (state 19).

### 5.2.15    *Call collision* (state 15)

A *Call collision* is detected by a DTE when it receives *Incoming Call* in response to *Call request*.

It is detected by a DCE when it receives *Call request* in response to *Incoming call*.

When a *Call collision* is detected by the DCE, the DCE will indicate *Proceed to select* (state 3) and cancel the *Incoming call*.

### 5.2.16    *Direct call*

For the direct call facility *Selection Signals* (state 4) are bypassed; *Proceed to select* (state 3) is followed by *DTE Waiting* (state 5) within 6 seconds.

### 5.3      *Procedures for leased circuit services*

### 5.3.1    *Leased circuit data transmission service point-to-point operation*

#### 5.3.1.1    *DCE Not ready*

As in 5.1.4 above.

#### 5.3.1.2    *Operation of interchange circuits*

When c = ON:

1)    i = ON at the distant interface,

2)    data transmitted on circuit T are delivered at the distant interface on circuit R.

When c = OFF, the DTE must signal t = 1:

1)    i = OFF at the distant interface,

2)    r = 1 at the distant interface.

### 5.3.2    *Demand leased circuit service — point-to-point operation*

A demand leased circuit service is not yet defined as an international service, and consequently does not appear in Recommendation X.2. However, in view of possible national provision of such a service and its study for introduction internationally the following information is provided for guidance.

217

A possible procedure is described for the establishment of data transfer phase for a demand leased circuit service. However, *DCE Not ready* should always be as indicated below. (See Annex 1, Figure 3/X.21 and Annex 2, Figure 5/X.21.)

### 5.3.2.1    *Ready* (state 1)

Both DTE and DCE shall signify their readiness to establish the data transfer phase by signalling t = 1, c = OFF, r = 1, i = OFF respectively.

### 5.3.2.2    *DTE Uncontrolled not ready*

The DTE shall indicate it is unable to enter data transfer phase by signalling t = 0, c = OFF.

### 5.3.2.3    *DCE Not ready*

As in 5.1.4 above.

### 5.3.2.4    *Request data transfer* (state 11A)

The originating DTE signals t = 1, c = ON.

### 5.3.2.5    *Data transfer requested* (state 11B)

The DCE indicates data transfer requested by signalling r = 1, i = ON.

### 5.3.2.6    *Ready for data* (state 12)

The DTE responds to the data transfer requested signal by signalling t = 1, c = ON. The DCE indicates to the originating DTE that it is ready for data by signalling, r = 1, c = ON. Subsequent procedures are described in 5.4 below, Data Transfer.

### 5.3.2.7    The procedures for the termination of the data transfer phase require further study.

### 5.3.3    *Leased circuit services  —  multipoint operation*

For further study.

## 5.4    *Data transfer* (state 13)

Transmission of *Ready for data* (state 12) indicates to the DTE that data transmission and reception can commence.

*DTE Clear request* or *DCE Clear indication* (state 16 or 19) indicates the end of *Data transfer* (state 13).

All bits sent by a DTE after receiving *Ready for data* and before sending *DTE Clear request* will be delivered to the corresponding DTE after that corresponding DTE has received *Ready for data* and before it has received *DCE Clear indication* (provided that the corresponding DTE does not take the initiative of clearing).

All bits received by a DTE after receiving *Ready for data* and before receiving *DCE Clear indication* or receiving *DCE Clear confirmation* have been sent by the corresponding DTE. Some of them may have been sent before that corresponding DTE has received *Ready for data*; those bits are steady 1.

During *Data transfer* (state 13) any bit sequence may be sent by either DTE.

The action to be taken when circuit C is turned OFF during *Data transfer* (state 13), while the DTE does not signal t = 0, is a subject of further study.

**6.**     *Selection, Call progress and Line identification formats* (also see Annex 5)

**6.1**     *Format of selection sequence*

*Selection sequence* shall consist of a *Facility request block* or *Address block* or both.

**6.1.1**     *Facility request block*

A *Facility request block* shall consist of one or more *Facility request signals*.

Multiple *Facility request signals* shall be separated by character 2/12 (","). 

End of a *Facility request block* shall be indicated by character 2/13 ("-").

**6.1.2**     *Address*

An *Address block* shall consist of one or more *Address* or *Abbreviated address signals*.

Multiple *Address* or multiple *Abbreviated address signals* shall be separated by character 2/12 (",")₊

Start of *Abbreviated address signals* shall be indicated by a prefix character 2/14 (".").

**6.1.3**     *End of selection sequence*

The end of selection sequence shall be indicated by character 2/11 (" + ").

**6.2**     *Format of a Call progress block*

A *Call progress block* shall consist of one or more *Call progress signals*.

Each *Call progress signal* need not be repeated.

Multiple *Call progress signals* shall be separated by character 2/12 (",").

End of *Call progress block* shall be indicated by character 2/11 (" + ").

**6.3**     *Format of called and calling line identification*

*Calling line identification signal* and *Called line identification block* shall be preceded by character 2/10 ("*").

A *Called line identification block* shall consist of one or more *Called line identification signals*.

Multiple *Called line identification signals* shall be separated by character 2/12 (",").

End of *Calling line identification signal* and *Called line identification block* shall be indicated by character 2/11 (" + ").

**7.**     *Failure detection and isolation*

**7.1**     *Indeterminate condition of interchange circuits*

If the DTE is unable to determine the binary state of the R, I, or, if provided, B circuit it should interpret r = 0, i = OFF, b = OFF respectively.

If the DCE is unable to determine the binary state of the T or C circuit it will interpret t = 0, c = OFF respectively.

## 7.2    *DCE fault conditions*

If the DCE is unable to provide service (e.g. loss of alignment or of incoming line signal) for a period longer than a fixed duration it will indicate *DCE Not ready* by signalling r = 0, i = OFF. (see 5.1.4 above.) The value of that duration has to be defined after further study. Prior to this *DCE Not ready* signal, garbled signals may be delivered to the DTE.

## 7.3    *Test loops*

### 7.3.1    *Test of the DTE*

In order to assist the test of the DTE, and specifically of the interconnecting cable, a loop is provided in the DCE which causes signals on the T circuit to be presented on the R circuit and signals on the C circuit to be presented on the I circuit.

The loop should be near the DCE/DTE interface. The DCE drivers and terminators may be included in the loop. The precise implementation of the loop within the DCE is a national option.

Manual control should be provided on the DCE for activation of the loop.

Consideration of automatic control is an item for further study.

### 7.3.2    *Network maintenance*

For network maintenance purposes a loop is implemented in the DCE.

That loop may be controlled manually on the DCE or automatically from the network. That is a national option, as is the method used for the automatic control of the loop when implemented.

In circuit switched service, the loop may be activated without the knowledge and agreement by the customer for periods which do not exceed one second. The loop should not be activated when the DTE is engaged in a call.

In the leased circuit service, the loop should not be activated before the customer has been informed of it. Some Administrations may activate the loop when an abnormal condition is detected in the network without first informing the customer.

When the test is in progress the DCE will signal r = 0, i = OFF.

In case of a collision between *Call request* and the activation of the loop, the loop activation command will have priority.

Garbled signals may be delivered to the DTE on the R and I circuits prior to the closing of the loop.

## 7.4    *Tolerance of the signal element timing signal under fault condition and when test loop is closed*

The signal element timing signal is delivered to the DTE on the S circuit whenever possible.

In particular it is delivered to the DTE when one of the loops described in 7.3 above is activated or when the DCE loses alignment or incoming line signal.

The tolerance of the signal element timing will be ± 1%.

## ANNEX 1
### (to Recommendation X.21)

### Interface signalling state diagrams

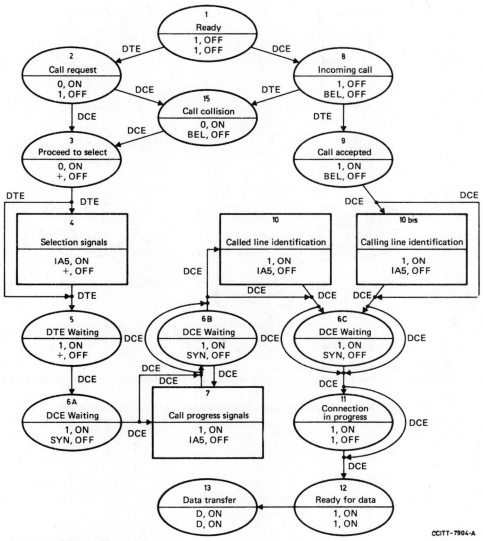

CCITT-7904-A

*Legend:* See Annex 4.

*Note 1.* – As indicated in Figure 2/X.21 the DCE may enter state 19 from any state and the DTE may enter state 16 from any state.

*Note 2.* – States 6A, 6B and 6C are presented in Figure 1/X.21 for convenience. They are all functionally equivalent and are referred to in the text as state 6.

FIGURE 1/X.21 – **Call establishment phase for circuit switched service**

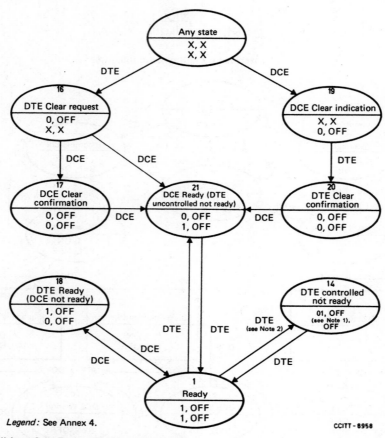

Legend: See Annex 4.

CCITT-8958

Note 1. — The condition of the R circuit is for further study.
Note 2. — The DTE may be able to enter the *controlled not ready* state from states other than *ready*, this is for further study.

FIGURE 2/X.21 — **Clearing phase and quiescent states for circuit switched service**

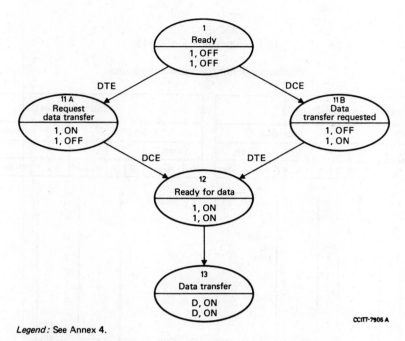

Legend: See Annex 4.

Note. – The procedures for termination of the *Data transfer* phase require further study.

FIGURE 3/X.21 – **Establishment of Data transfer phase for demand leased circuit service, point-to-point operation**

## ANNEX 2
(to Recommendation X.21)

### Interface signalling sequence diagrams

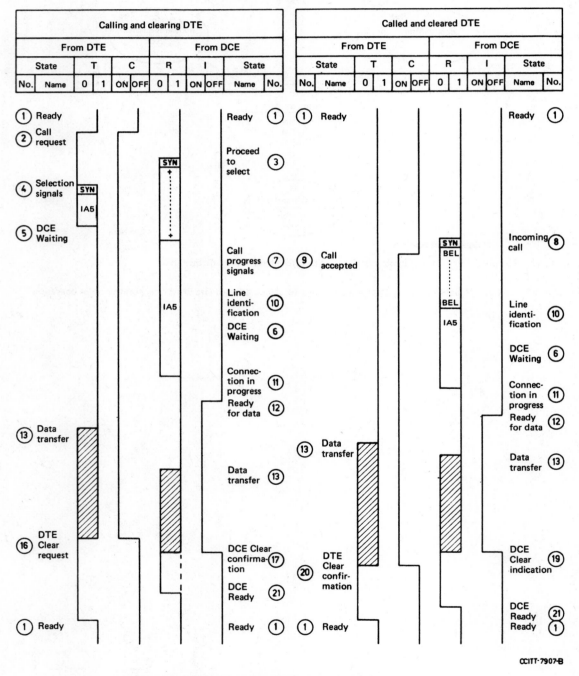

FIGURE 4/X.21 − Successful call and clear for circuit switched service
(Example of sequence of events at the interface)

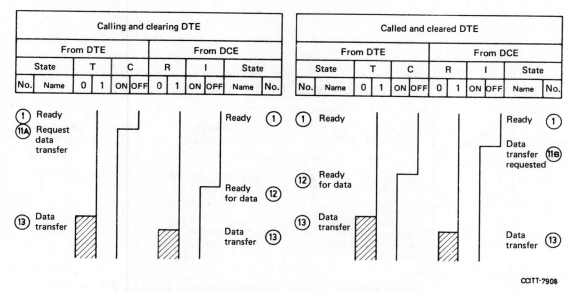

FIGURE 5/X.21 – Call establishment for demand leased circuit service ·

ANNEX 3

(to Recommendation X.21)

**DTE time-limits and DCE time-outs**

**DTE time-limits**

Under certain circumstances this Recommendation requires the DCE to respond to a signal from the DTE within a stated maximum time. If any of these maximum times is exceeded, the DTE should abandon the call. To maximize efficiency, the DTE should incorporate time-limit to send the appropriate signal under the defined circumstances summarized in Table 2/X.21. The time limits given in the first column are the maximum times allowed for the DCE to response and are consequently the lower limits of times a DTE must allow for proper network operation. A time-out longer than the minimum time shown may optionally be used in the DTE; for example, all DTE time-outs could have one single value equal to or greater than the longest time limit shown in this table. However, the use of a longer time will result in reduced efficiency of network utilization.

TABLE 2/X.21 – **DTE Time limits**

| Time limit | Time-limit number | Started by | Normally terminated by | Preferred action to be taken when time expires |
|---|---|---|---|---|
| 3 s | T1 | Signalling of *Call request* (state 2) | Reception of *Proceed to select* (state 3) | DTE signals *DTE Ready* (state 1) |
| 20 s | T2 | Signalling of end of selection or *DTE Waiting* (direct call) (state 4 or 5) | Reception of *Call progress signals* or *Ready for data* (states 7, 10, 11, 12) | DTE signals *DTE Clear request* (state 16) |
| 2 s | T3A | Reception of *Call progress signals* (state 7) | Reception of *Ready for data* or *DCE Clear indication* (state 12 or 19) | DTE signals *DTE Clear request* (state 16) |
| 60 s | T3B (see Note) | Reception of applicable *Call progress signals* (state 7) | Reception of *Ready for data* or *DCE Clear indication* (state 12 or 19) | DTE signals *DTE Clear request* (state 16) |
| 2 s | T4 | Change of state to *DTE Call accepted* (state 9) | Reception of *Ready for data* or *DCE Clear indication* (state 12 or 19) | DTE signals *DTE Clear request* (state 16) |
| 2 s | T5 | Change of state to *DTE Clear request* (state 16) | Change of state to *DCE Ready* (state 21) | DTE regards the DCE as *Not ready* and signals *DTE Ready* (state 18) |
| 2 s | T6 | Change of state to *DTE Clear confirmation* (state 20) | Reception of *DCE Ready* (state 21) | DTE regards the DCE as *Not ready* and signals *Ready* (state 18) |

*Note.* – 60 s (T3B) applies for manual answering DTEs.

226

## DCE time-outs

Under certain circumstances this Recommendation requires the DTE to respond to a signal from the DCE within a stated maximum time. If any of these maximum times is exceeded, a time-out in the DCE will initiate the actions summarized in Table 3/X.21. These constraints must be taken into account in the DTE design. The time-outs given in the first column of the table are the minimum time-out values used in the DCE for the appropriate DTE response and are consequently the maximum times available to the DTE for response to the indicated DCE action.

### TABLE 3/X.21 – DCE Time-outs

| Time-out | Time-out number | Started by | Normally terminated by | Action to be taken when time-out expires |
|---|---|---|---|---|
| 36 s | T11 (See Note 1) | DCE signalling of *Proceed to select* (state 3) | DCE reception of end of selection signal (state 5) | DCE will signal *DCE Clear indication* (state 19) or transmit appropriate call progress signal followed by *DCE Clear indication* |
| 6 s | T12 | DCE signalling of *Proceed to select* (state 3) | DCE reception of first selection character (state 4) or *DTE Waiting* (state 5) | |
| 6 s | T13 (See Note 1) | DCE reception of nth selection character (state 4) | DCE reception of (n+1)th selection character or end of selection signal (state 4) | |
| 500 ms (See Note 2) | T14A | DCE signalling of incoming call (state 8) | Change of state to *Call accepted* (state 9) | The DTE is noted as not answering. The DCE will signal *Ready* (state 1) |
| 60 s (See Note 3) | T14B | | | |
| 100 ms | T15 | Change of state to *DCE Clear indication* (state 19) | Change of state to *DTE Clear confirmation* (state 20) | DCE will signal *DCE Ready* and mark *DTE uncontrolled not ready* (state 21) |
| 100 ms | T16 | Change of state to *DCE Ready* (state 21) | Change of state to *Ready* (state 1) | DCE will mark the DTE as *Uncontrolled not ready* (state 21) |

*Note 1.* – T11 and T13 does not apply in case of a direct call.
*Note 2.* – This time-out value is a subject for further study with the objective of reducing it in the future to 100 ms.
*Note 3.* – T14B will be provided when manual answering DTEs are allowed.

227

## ANNEX 4

### (to Recommendation X.21)

### Definition of symbols used in the state diagrams

Each state is represented by an ellipse wherein the state name and number is indicated, together with the signals on the four interchange circuits which represent that state.

Each state transition is represented by an arrow and the equipment responsible for the transition (DTE or DCE) is indicated beside that arrow.

A rectangle represents a sequence of states and transitions.

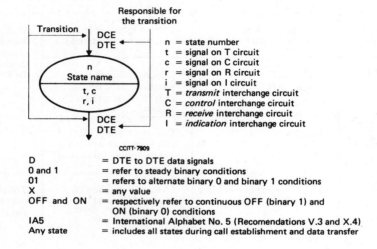

| | |
|---|---|
| n | = state number |
| t | = signal on T circuit |
| c | = signal on C circuit |
| r | = signal on R circuit |
| i | = signal on I circuit |
| T | = *transmit* interchange circuit |
| C | = *control* interchange circuit |
| R | = *receive* interchange circuit |
| I | = *indication* interchange circuit |

CCITT-7909

| | |
|---|---|
| D | = DTE to DTE data signals |
| 0 and 1 | = refer to steady binary conditions |
| 01 | = refers to alternate binary 0 and binary 1 conditions |
| X | = any value |
| OFF and ON | = respectively refer to continuous OFF (binary 1) and ON (binary 0) conditions |
| IA5 | = International Alphabet No. 5 (Recommendations V.3 and X.4) |
| Any state | = includes all states during call establishment and data transfer |

## ANNEX 5

### (to Recommendation X.21)

### Formats of Selection, Call progress, and line identification signals

The following description uses Backus Normal Form as the formalism for syntactic description, for example as in ISO Recommendation R 1538 [3]. A vertical line " | " separates alternatives.

<Selection Sequence> : : = <Facility Request Block> <–> <Address Block> <+> |
<Facility Request Block> <–> <+> | <Address Block> <+>

<Facility Request Block> : : = <Facility Request Signal> | <Facility Request Block> <,> <Facility Request Signal>

<Address Block> : : = <Address Signal> | <Address Block> <,> <Address Signal>

<Address> : : = <Full Address Signal> | <.> <Abbreviated Address Signal>

---

[3] ISO Recommendation R 1538: Programming Language "ALGOL", March 1972.

228

<*Call Progress Block*> : : = <Call Progress Signal> < + > | <Call Progress Signal> < , > <Call Progress Block>

<*Calling Line Identification*> : : = <*> <Calling Line Identification Signal> < + >

<*Called Line Identification*> : : = <*> <Called Line Identification Block> < + >

<Called Line Identification Block> : : = <Called Line Identification Signal> | <Called Line Identification Block> < , > <Called Line Identification Signal>

<*> : : = IA 5 character 2/10

< + > : : = IA 5 character 2/11

< , > : : = IA 5 character 2/12

< − > : : = IA 5 character 2/13

< . > : : = IA 5 character 2/14

<Facility Request Signal>

<Full Address Signal>

<Abbreviated Address Signal>          Subject of another Recommendation

<Call Progress Signal>               (not yet available)

<Calling Line Identification Signal>

<Called Line Identification Signal>

ANNEX 6

(to Recommendation X.21)

**Interworking between DTEs conforming to Recommendations X.21 and X.21 *bis***

Interworking between V-Series DTEs connected to a public data network according to Recommendation X.21 *bis* at one end and Recommendation X.21 DTEs at the other end should always be possible for DTEs not requiring half-duplex transmission at the Recommendation X.21 *bis* interface, i.e. V-Series DTEs which do not require circuit 109 to be OFF before transmission can commence.

In addition to this, certain Administrations may also provide facilities allowing DTEs to use half-duplex transmission according to the figure below.

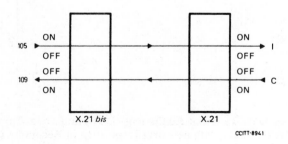

Those Administrations not providing this facility shall cause the Recommendation X.21 DCE to signal r = 1, i = ON when the Recommendation X.21 *bis* DTE signals circuit 105 OFF. This will permit half-duplex operation for those DTEs that do not require circuit 109 to be OFF before signalling circuit 105 ON.

**Recommendation X.21** *bis*

## USE ON PUBLIC DATA NETWORKS OF DATA TERMINAL EQUIPMENTS (DTEs) WHICH ARE DESIGNED FOR INTERFACING TO SYNCHRONOUS V-SERIES MODEMS

*(Geneva, 1976)*

CONTENTS

*Preface*

The CCITT

  *considering* that

  a)    the interface between Data Terminal Equipment (DTE) and Data Circuit-terminating Equipment (DCE) for synchronous operation on public data networks is specified in Recommendation X.21;

  b)    several Administrations are also planning to provide as an interim measure the connection to public data networks of synchronous DTEs which are designed for interfacing to synchronous V-Series modems;

  c)    the international standardization of the implementation of such connection, if provided, is of great importance, *unanimously declares the following view:*

1.    The connection of DTEs with V-Series-type interface to public data network may allow for:

i)    leased circuit service,

ii)   direct call facility,

iii)  address call facility.

2.    This Recommendation defines operational modes and optional features. It is left to the discretion of each Administration to decide which optional features to implement. When implemented they should be in accordance with this Recommendation.

The Recommendation specifies the operation at the interface when the data circuit interconnects V-Series DTEs. Interworking between V-Series DTEs and X.21 DTEs is described in Annex 1.

## 1.    *The use of V-series DTEs for leased circuit service*

### 1.1    *General*

V-Series DTEs utilizing the leased circuit service in public data networks is discussed in the following.

The data signalling rates are those defined in Recommendation X.1 for user classes of service employing synchronous mode of transmission across the interface.

### 1.2    *Use of interchange circuits*

The electrical characteristics of the interchange circuits at the DCE side of the interface comply with Recommendation V.28 using the 25-pin connector and pin allocation standardized by ISO (DIS 2110.2).

The electrical characteristics of the interchange circuits at the DTE side of the interface may be applied according to Recommendation V.28 or Recommendation X.26 as specified in the ISO standard for the assignments of the 37-pin interface connector (ISO DIS 4902).

For applications of the data signalling rate of 48 kbit/s, indications concerning the connector and electrical characteristics at both the DCE side and the DTE side of the interface are given in the ISO standard for the assignment of the 34-pin interface connector (ISO IS 2593) and in Recommendation V.35 respectively.

Table 1/X.21 *bis* shows the use of interchange circuits for the leased circuit service.

TABLE 1/X.21 *bis*

| V.24 interchange circuit No. | Designation |
|---|---|
| 102 | Signal ground or common return |
| 103 | Transmitted data |
| 104 | Received data |
| 105 | Request to send |
| 106 | Ready for sending |
| 107 | Data set ready (See Note 1) |
| 108.1 | Connect data set to line (See Notes 2 and 3) |
| 109 | Data channel received line signal detector |
| 114 | Transmitter signal element timing (DCE) (See Note 4) |
| 115 | Receiver signal element timing (DCE) (See Note 4) |
| 142 | Test indicator (DCE) (See Note 5) |

All these circuit functions are in accordance with Recommendation V.24 and the appropriate modem Recommendations (see also 1.2.1 below).

*Note 1.* — Circuit 107 shall go OFF only in cases of DCE power-off (normally the indeterminate state is interpreted as OFF), loss of service (see paragraph 3.2 below) or when circuit 108.1, when implemented, is turned OFF.

*Note 2.* — Not required for V.35 compatible interface.

*Note 3.* — The DCE interprets the ON condition on circuit 108.1, when implemented, as an indication that the DTE is operational. If circuit 108.1 is not provided the DCE will consider the lack of circuit 108.1 as ON condition. The DCE turns circuit 107 ON while circuit 108.1, if present, is ON and the circuit connection is available.

*Note 4.* — The DCE shall provide the DTE with transmitter and receiver signal element timings, this is done by feeding circuits 114 and 115 with the same timing signal from the DCE.

*Note 5.* — This circuit is used to indicate to the DTE the test mode status of the DCE. The ON condition indicates that the DCE is in a test mode, precluding transmission of data to a remote DTE. The OFF condition indicates that the DCE is in the non-test mode with no test in progress.

### 1.2.1 *Operational requirements*

i) *Half-duplex facility*

When the half-duplex facility is provided for the DTE, in order to take care of those DTEs which do require circuit 109 to be OFF before circuit 105 can be turned ON, circuit 105 shall be able to control circuit 109 at the other end, so that circuit 109 OFF can be used as an indication to the DTE that it can turn circuit 105 ON.

*Note.* — Attention is drawn to the fact that, although circuit 105 can control circuit 109 at the other end, in case of the half-duplex facility, the detection of a line signal should be replaced by some other control mechanism.

ii) *Response times*

The response time of the OFF to ON transition of circuit 106 as a response to circuit 105 OFF to ON should provisionally be between 30 and 50 ms for the 600 bit/s user rate, and 10 to 20 ms for the higher user rate.

iii) *Clamping*

The following conditions apply:

— In the event of line failure (e.g. channel out of service, loss of alignment) the DCE shall clamp circuit 104 to steady binary 1 condition and circuit 109 to OFF condition.

— In all applications the DCE shall hold circuit 104 in binary 1 condition, when circuit 109 is in the OFF condition.

— In addition, when the half-duplex facility is provided, the DCE shall hold circuit 104 in the binary 1 condition and circuit 109 in the OFF condition when circuit 105 is in the ON condition.

iv) *Timing arrangements*

Timing signals on circuits 114 and 115 should always be maintained when the DCE is capable of generating them, disregarding the conditions of the other circuits. Circuit 114 and 115 should be held by the DCE in the OFF condition when the DCE is unable to generate the timing information.

2.      *The use of V-Series DTEs for direct call and address call facilities*

2.1     *General*

V-Series DTEs utilizing the direct call or the address facility in public data networks is discussed in the following.

The data signalling rates are those defined in Recommendation X.1 for user classes of service employing synchronous mode of transmission across the interface.

2.2     *Use of interchange circuits*

The electrical characteristics of the interchange circuits at the DCE side of the interface comply with Recommendation V.28 using the 25-pin connector and pin allocation standardized by ISO (DIS 2110.2). The electrical characteristics of the interchange circuits at the DTE side of the interface may be applied according to Recommendation V.28 or Recommendation X.26 as specified in the ISO standard for the assignments of the 37-pin interface connector (ISO DIS 4902).

For applications of the data signalling rate of 48 kbit/s, indications concerning the connector and electrical characteristics at both the DCE side and the DTE side of the interface are given in the ISO standard for the assignment of the 34-pin interface connector (ISO IS 2593) and in Recommendation V.35 respectively.

For further definitions of the interchange circuits than outlined below, refer to Recommendation V.24 and the appropriate V-Series modem Recommendations.

2.2.1   *Call establishment and disconnection phases*

The following interchange circuits should be used for control signalling in the call establishment and disconnection phases:

*Circuit 102  —  Signal ground or common return*

*Circuit 107  —  Data set ready*

This circuit is used to indicate the following operational functions.

| Condition of circuit 107 | Function in the network |
|:---:|:---|
| ON<br>OFF<br>OFF | Ready for data<br>DCE Clear indication (See 2.2.1.1)<br>DCE Clear confirmation |

*Circuit 108.1 Connect data set to line*

This circuit is used alternatively to circuit 108.2. The following operational functions should be indicated.

| Condition of circuit 108.1 | Function in the network |
|:---:|:---|
| ON<br>ON<br>OFF<br>OFF | Call request<br>Call accepted<br>DTE Clear request<br>DTE Clear confirmation (See 2.2.1.1) |

A DTE receiving an incoming call should turn circuit 108.1 ON to indicate *Call accepted* within 500 ms, otherwise the call will be cleared.

*Circuit 108.2 Data terminal ready*

This circuit is used alternatively to circuit 108.1. The following operational functions should be indicated.

| Condition of circuit 108.2 | Function in the network |
|---|---|
| ON | Call accepted |
| OFF | DTE Clear request |
| OFF | DTE Clear confirmation (See 2.2.1.1) |

When a DTE with circuit 108.2 OFF receives an incoming call the DTE should turn circuit 108.2 from OFF to ON within 500 ms to indicate *Call accepted*, otherwise the call will be cleared.

Optionally where a DTE does not provide circuit 108.1 or 108.2, normally the *Call accepted* signal would be generated within the DCE as an answer to incoming call received from the network. However it is possible to signal to the network *DTE Controlled not ready* by a manual action on the DCE.

It is not mandatory for the DTE to turn circuit 108.2 OFF to give clear confirmation. Optionally in the case of DTEs not providing circuit 108 or unable to use circuit 108.2 for disconnection the *Clear confirmation* signal would be generated within the DCE as an answer to *Clear indication* received from the network.

*Circuit 114 — Transmitter signal element timing (DCE)*

*Circuit 115 — Receiver signal element timing (DCE)*

The DCE shall provide the DTE with transmitter and receiver element timings. This is done by feeding circuits 114 and 115 with the same timing signal from the DCE.

*Circuit 125 — Calling indicator*

The ON condition indicates Incoming Call. The circuit will be turned OFF in conjunction with circuit 107 turned ON or when DCE *Clear indication* is received.

*Circuit 142 — Test indicator*

*Direction: from DCE*

This circuit is used to indicate to the DTE the test-mode status of the DCE.

2.2.1.1    *Operational requirements*

i)    *DCE Clear indication*

*DCE Clear indication* to the DTE, when implemented, is signalled to the DTE by turning circuit 107 OFF. *DTE Clear confirmation* should be given by the DTE within 50 ms after *DCE Clear indication* is received on circuit 107.

However, not all DTEs allow circuit 107 to be turned OFF if circuit 108 has not been turned OFF previously. In this case the Administration will not insist that the DCE will turn 107 OFF to indicate to the DTE that the connection has been cleared down, unless circuit 108 has been turned OFF. The implications of the latter ase are for further study.

ii)    *Line identification*

*Calling* and *Called line identification* signals cannot be handled by V-Series DTEs.

iii)  *Call progress signals*

*Call progress signals* cannot be handled by V-Series DTEs. If automatic address calling is provided in accordance with Recommendation V.25 the reception of negative *Call progress signals* will be indicated to the DTE on circuit 205.

2.2.2   *Data transfer phase*

The interchange circuits shown in Table 2/X.21 *bis* should be used in the data transfer phase.

TABLE  2/X.21 *bis*

| V.24 Interchange circuit No. | Designation |
|---|---|
| 102 | Signal ground or common return |
| 103 | Transmitted data |
| 104 | Received data |
| 105 | Request to send |
| 106 | Ready for sending |
| 109 | Data channel received line signal detector |
| 114 | Transmitter signal element timing (DCE) (See Note 1) |
| 115 | Receiver signal element timing (DCE) (See Note 1) |

235

All the circuit functions are in accordance with Recommendation V.24 and the appropriate modem Recommendations.

*Note 1.* —  The DCE shall provide the DTE with transmitter and receiver element timings. This is done by feeding circuits 114 and 115 with the same timing signal from the DCE.

2.2.2.1   *Operational requirement*

i)  *Half-duplex facility*

When the half-duplex facility is provided for the DTE, circuit 105 shall be able to control circuit 109 at the other end, so that circuit 109 OFF can be used as an indication to the DTE that it can turn circuit 105 ON.

ii)  *Response times*

The response time of the OFF to ON transition of circuit 106 as a response to circuit 105 OFF to ON should provisionally be between 30 and 50 ms for the 600 bit/s user rate, and 10 to 20 ms for the higher user rate.

iii)  *Clamping*

The following conditions shall apply:

—  In the event of line failure (e.g. channel out of service, loss of alignment) the DCE shall clamp circuit 104 to steady binary 1 condition and circuit 109 to OFF condition.

—  In all applications the DCE shall hold circuit 104 in binary 1 condition, when circuit 109 is in the OFF condition.

—  In addition, when the half-duplex facility is provided, the DCE shall hold circuit 104 in the binary 1 condition and circuit 109 in the OFF condition when circuit 105 is in the ON condition.

iv) *Timing arrangements*

Timing signals on circuits 114 and 115 should always be maintained when the DCE is capable of generating them, disregarding the conditions of the other circuits. Circuit 114 and 115 should be held by the DCE in the OFF condition when the DCE is unable to generate the timing information.

Continuous isochronous operation should be used.

2.3    *Operational modes*

2.3.1    *Direct call facility*

The following operational modes may be provided for:

i)    Automatic direct call and automatic disconnection from DTE. Circuit 108.1 should be used.

ii)    Manual direct call from DCE and automatic disconnection from DTE. Circuit 108.2 should be used.

iii)    Manual direct call and manual disconnection from DCE. For DTEs not providing circuit 108 or unable to use circuit 108.2 for disconnection.

Only automatic call answering controlled by circuit 108.1 or 108.2 when provided, or automatically within the DCE itself should be implemented. However in the last case it is possible to signal to the network *DTE Controlled not ready* by a manual action on the DCE.

*Note.* – Consideration of manual answering and the implications of manual *DTE Clear confirmation* are for further study.

2.3.2    *Address call facility*

The following operational modes may be provided for:

i)    Manual address calling from DCE and automatic disconnection from DTE. Circuit 108.2 should be used.

ii)    Manual address calling and manual disconnection from DCE. For DTEs not providing circuit 108.1 or 108.2 or unable to use circuit 108.2 for disconnection.

Only automatic answering controlled by circuit 108.2 when provided, or automatically within the DCE itself should be implemented. However in the last case it is possible to signal to the network *DTE Controlled not ready* by a manual action on the DCE.

iii)    Automatic address calling and automatic disconnection from DTE if provided, should use the 200 series interchange circuits and the V.25 relevant procedures. The spare control characters on the digit signal circuits 206-209 may be used for special purposes (e.g. start of prefix, prefix separator, start of address) during the selection sequence.

3.    *Failure detection and isolation*

3.1    *Indeterminate conditions on interchange circuits*

If the DTE or DCE is unable to determine the condition of circuits 105, 107, 108.1 or 108.2 and possibly circuits 103 and 104 as specified in the relevant electrical interfaces specifications it shall interpret this as OFF condition or binary 1 (circuits 103 and 104).

3.2    *DCE fault conditions*

If the DCE is unable to provide service (e.g. loss of alignment or of incoming line signal) for a period longer than a fixed duration it will turn circuit 107 to the OFF condition. The value of this duration has to be defined after further study.

Moreover, as soon as the DCE detects this condition it turns circuit 109 in the OFF condition and circuit 104 in the binary 1 condition.

## 3.3    Test loops

### 3.3.1    Test of the DTE

In order to assist the test of the DTE, and specifically of the interconnecting cable, a loop is provided in the DCE where the signals transmitted on circuits 103 and 105 are presented on circuits 104 and 109 respectively. Circuit 106 should follow circuit 105 with or without the usual delay. Circuit 107 should be ON. Additionally, circuit 142 should be ON whilst the loop is activated.

In circuit switched service, as described in 2. above, any existing connection with a remote station should be cleared by the DCE when the loop is activated.

The loop should be near the DCE/DTE interface. The DCE drivers and terminators may be included in the loop. The precise implementation of the loop within the DCE is a national option. Manual control should be provided on the DCE for activation of the loop.

### 3.3.2    Network maintenance

For network maintenance purposes a loop is implemented in the DCE: signals incoming from the network towards circuits 104 and 109 are diverted from these circuits and looped back to the network in place of signals from circuits 103 and 105 respectively.

The loop may be controlled manually on the DCE or automatically by the network. The control of the loop and the method used for the automatic control, when implemented, is a national option.

In the circuit switched service the loop should not be activated when the DTE is engaged in a call. The loop may be activated by the network without the knowledge and agreement of the DTE for periods which do not exceed one second.

In the leased circuit service the loop should not be activated before the customer has been informed of it. Some Administrations may activate the loop when abnormal conditions are detected in the network without first informing the customer.

When the test is in progress the DCE will turn circuits 107 and 109 in the OFF condition, circuit 104 in the binary 1 condition and circuit 142 in the ON condition.

In case of a collision between a *Call request* and the activation of the loop, the loop activation command will have priority.

## 3.4    Tolerance of the signal element timing signal under fault conditions

The signal element timing signal is delivered to the DTE on circuits 114 and 115 whenever possible.

In particular, it is delivered to the DTE when one of the loops described in 3.3 above is activated or when the DCE loses alignment or incoming line signal. The tolerance of the signal element timing will be ± 1%.

ANNEX

(to Recommendation X.21 *bis*)

**Interworking between DTEs conforming to Recommendations X.21 and X.21 bis**

Interworking between V-Series DTEs connected to a public data network according to Recommendation X.21 *bis* at one end and X.21 DTEs at the other end should always be possible for DTEs not requiring half-duplex transmission at the X.21 *bis* interface, i.e. V-Series DTEs which do not require circuit 109 to be OFF before transmission can commence.

In addition to this, certain Administrations may also provide facilities allowing DTEs to use half-duplex transmission according to the figure below.

Those Administrations not providing this facility shall cause the X.21 DCE to signal r = 1, i = ON when the X.21 *bis* DTE signals circuit 105 OFF. This will permit half-duplex operation for those DTEs that do not require circuit 109 to be OFF before signalling circuit 105 ON.

**Recommendation X.24**

### LIST OF DEFINITIONS FOR INTERCHANGE CIRCUITS BETWEEN DATA TERMINAL EQUIPMENT (DTE) AND DATA CIRCUIT-TERMINATING EQUIPMENT (DCE) ON PUBLIC DATA NETWORKS

*(Geneva, 1976)*

238

CONTENTS

Introduction

*Introduction*

The CCITT

*considering* that

a)   the interface between DTE and DCE on public data neworks requires, in addition to the electrical and functional characteristics of the interchange circuits, the definition of procedural characteristics for call control functions and selection of the facilities according to Recommendation X.2;

b)   the functions of the circuits defined in Recommendation V.24 are based on the requirements of data transmission over the general telephone network and are not appropriate for use at DTE/DCE interfaces in public data networks;

*expresses the view* that

a new Recommendation to include the list of definitions of interchange circuits for use in public data networks is required.

1.    *Scope*

1.1    This Recommendation applies to the functions of the interchange circuits provided at the interface between DTE and DCE of data networks for the transfer of binary data, call control signals and timing signals.

For any type of practical equipment, a selection will be made from the range of interchange circuits defined in this Recommendation, as appropriate. The actual interchange circuits to be used in a particular DCE for a user class of service according to Recommendation X.1 and defined user facilities according to Recommendation X.2, are those indicated in the relevant Recommendation for the procedural characteristics of the interface, e.g., Recommendation X.20 or X.21.

To enable a standard DTE to be developed, the use and termination by the DTE of certain circuits even when implemented in the DCE are not mandatory. This is covered by the individual interface Recommendations.

The interchange circuits defined for the transfer of binary data are also used for the exchange of call control signals.

The electrical characteristics of the interchange circuits are detailed in the appropriate Recommendation for electrical characteristics of interchange circuits. The application of those characteristics for a particular DCE is specified in the Recommendation for the procedural characteristics of the interface.

1.2    The range of interchange circuits defined in this Recommendation is applicable to the range of services which could be offered on a public data network, e.g., circuit switching services (synchronous and start/stop), telex service, packet switching services, message registration and retransmission service and facsimile service.

2.    *Line of demarcation*

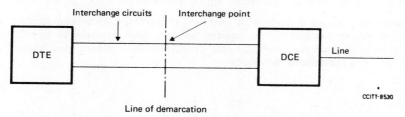

FIGURE 1/X.24 – **General illustration of interface equipment layout**

The interface between DTE and DCE is located at a connector which is the interchange point between these two classes of equipment shown in Figure 1/X.24.

2.1    The connector will not necessarily be physically attached to the DCE and may be mounted in a fixed position near the DTE. The female part of the connector belongs to the DCE.

2.2    An interconnecting cable will normally be provided together with the DTE. The cable length is limited by electrical parameters specified in the appropriate Recommendations for the electrical characteristics of the interchange circuits.

3.    *Definition of interchange circuits*

A list of the data network series interchange circuits is presented in tabular form in Table 1/X.24.

*Circuit G  –  Signal ground or common return*

This conductor establishes the signal common reference potential for unbalanced double-current interchange circuits with electrical characteristics according to Recommendation V.28. In case of interchange circuits according to Recommendations X.26 and X.27, it interconnects the zero volt reference points of a generator and a receiver to reduce environmental signal interference, if required.

TABLE 1/X.24 – **Data network interchange circuits**

| Interchange circuit designation | Interchange circuit name | Data | | Control | | Timing | |
|---|---|---|---|---|---|---|---|
| | | From DCE | To DCE | From DCE | To DCE | From DCE | To DCE |
| G | Signal ground or common return | | | | | | |
| Ga | DTE common return | | | | X | | |
| Gb | DCE common return | | | X | | | |
| T | Transmit | | X | | X | | |
| R | Receive | X | | X | | | |
| C | Control | | | | X | | |
| I | Indication | | | X | | | |
| S | Signal element timing | | | | | X | |
| B | Byte timing | | | | | X | |

Within the DCE, this conductor shall be brought to one point, protective ground or earth, by means of a metallic strap within the equipment. This metallic strap can be connected or removed at installation, as may be required to minimize the introduction of noise into electronic circuitry or to meet applicable regulations.

*Note.* – Where a shielded interconnecting cable is used at the interface, the shield may be connected either to circuit G, or to protective ground in accordance with national regulations. Protective ground may be further connected to external grounds as required by applicable electrical safety regulations.

For unbalanced interchange circuits with electrical characteristics in accordance with Recommendation X.26, two common-return conductors are required, one for each direction of signalling, each conductor being connected to ground only on the generator side of the interface. Where used, these shall be designated circuits Ga and Gb, and they are defined as follows:

*Circuit Ga – DTE common return*

This conductor is connected to the DTE circuit common and is used as the reference potential for the unbalanced X.26 type interchange circuit receivers within the DCE.

*Circuit Gb – DCE common return*

This conductor is connected to the DCE circuit common and is used as the reference potential for the unbalanced X.26 type interchange circuit receivers within the DTE.

*Circuit T – Transmit*

*Direction:* To DCE

The binary signals originated by the DTE to be transmitted during the data transfer phase via the data circuit to one or more remote DTEs are transferred on this circuit to the DCE.

This circuit also transfers the call control signals originated by the DTE, to be transmitted to the DCE in the call establishment and other call control phases as specified by the relevant Recommendations for the procedural characteristic of the interface.

The DCE monitors this circuit for detection of electrical circuit fault conditions, according to the specifications of the electrical characteristics of the interface. A circuit fault is to be interpreted by the DCE as defined in the Recommendation for the procedural characteristics of the interface.

240

*Circuit R — Receive*

*Direction:* From DCE

The binary signals sent by the DCE as received during the data transfer phase from a remote DTE, are transferred on this circuit to the DTE.

This circuit also transfers the call control signals sent by the DCE as received during the call establishment and other call control phases as specified by the relevant Recommendations for the procedural characteristics of the interface.

The DTE monitors this circuit for detection of electrical circuit fault conditions, according to the specifications of the electrical characteristics of the interface. A circuit fault is to be interpreted by the DTE as defined in the Recommendation for the procedural characteristics of the interface.

*Circuit C — Control*

*Direction:* To DCE

Signals on this circuit control the DCE for a particular signalling process.

Representation of a control signal requires additional coding of circuit T-*Transmit* as specified in the relevant Recommendation for the procedural characteristics of the interface. During the data phase, this circuit shall remain ON. During the call control phases, the condition of this circuit shall be as specified in the relevant Recommendation for the procedural characteristics of the interface.

*Note.* — After appropriate selection of special user facilities (not yet defined), it might be required to change the ON condition after entering the data phase in accordance with the regulations for the use of these facilities. This subject is for further study.

The DCE monitors this circuit for detection of electrical circuit fault conditions, according to the specifications of the electrical characteristics of the interface. A circuit fault is to be interpreted by the DCE as defined in the Recommendation for the procedural characteristics of the interface.

*Circuit I — Indication*

*Direction:* From DCE

Signals on this circuit indicate to the DTE the state of the call control process.

Representation of a control signal requires additional coding of circuit R-*Receive*, as specified in the relevant Recommendation for the procedural characteristics of the interface. The ON condition of this circuit signifies that signals on circuit R contain information from the distant DTE. The OFF condition signifies a control signalling condition which is defined by the bit sequence on circuit R as specified by the procedural characteristics of the interface.

The DTE monitors this circuit for detection of electrical circuit fault conditions, according to the specifications of the electrical characteristics of the interface. A circuit fault is to be interpreted by the DTE as defined in the Recommendation for the procedural characteristics of the interface.

*Note.* — For use with special user facilities (not yet defined) it might be required to use the OFF condition after entering the data transfer phase in accordance with the regulations for the use of these facilities. This subject is for further study.

*Circuit S — Signal element timing*

*Direction:* From DCE·

Signals on this circuit provide the DTE with signal element timing information. The condition of this circuit shall be ON and OFF for nominally equal periods of time. However, for burst isochronous operations, longer periods of OFF condition may be permitted equal to an integer odd number of the nominal period of the ON condition as specified by the relevant procedural characteristics of the interface.

241

The DTE shall present a binary signal on circuit T-*Transmit* and a condition on circuit C-*Control*, in which the transitions nominally occur at the time of the transitions from OFF to ON condition of this circuit.

The DCE presents a binary signal on circuit R-*Receive* and a condition on circuit I-*Indication* in which the transitions nominally occur at the time of the transitions from OFF to ON condition of this circuit.

The DCE shall transfer signal element timing information on this circuit across the interface at all times that the timing source is capable of generating this information.

*Circuit B — Byte timing (see Note 2)*

*Direction:* From DCE

Signals on this circuit provide the DTE with 8-bit byte timing information. The condition of this circuit shall be OFF for nominally the period of the ON condition of circuit S-*Signal element timing* which indicates the last bit of an 8-bit byte and shall be ON at all other times within the period of the 8-bit byte.

During the call control phases, the call control characters and steady state conditions used for all information transfers between the DCE and the DTE, in either direction, shall be correctly aligned to the signals of circuit B.

The DTE shall present the beginning of the first bit of each call control character on circuit T-*Transmit* nominally at the time of the OFF to ON transition of circuit S which follows the OFF to ON transition of circuit B.

A change of condition of circuit C-*Control* may occur at any OFF to ON transition of circuit S, but it will be sampled in the DCE at the time of the OFF to ON transition of circuit B, i.e., for evaluation of the following call control character on circuit T.

The centre of the last bit of each call control character will be presented by the DCE on circuit R-*Receive* nominally at the time of the OFF to ON transition of circuit B.

A change of condition of circuit I-*Indication* will occur nominally at the OFF to ON transition of circuit S which follows the OFF to ON transition of circuit B.

The DCE shall transfer byte timing information on this circuit across the interface at all times that the timing source is capable of generating this information.

*Note 1.* — During the data transfer phase, DTEs communicating by means of an 8-bit code may utilize the byte timing information for mutual character alignment.

It is a prerequisite for the provision of this feature that character alignment is preserved after the call has entered the data transfer phase and that the alignment obtained at one interface is synchronized to the alignment at the other interface. (This is only possible on some connections.)

Furthermore, where this feature is available, a change of condition on circuit C as defined above may result in an equivalent change in the relative alignment on circuit I at the distant interface.

*Note 2.* — In some Recommendations for the procedural characteristics of the interface (e.g., X.21), the use and termination of this circuit by the DTE is not mandatory even when implemented in the DCE.

**Provisional Recommendation X.25**

### INTERFACE BETWEEN DATA TERMINAL EQUIPMENT (DTE) AND DATA CIRCUIT-TERMINATING EQUIPMENT (DCE) FOR TERMINALS OPERATING IN THE PACKET MODE ON PUBLIC DATA NETWORKS

*(Geneva, 1976, amended at Geneva, 1977)*

The establishment in various countries of public data networks providing packet-switched data transmission services creates a need to produce standards to facilitate international interworking.

The CCITT,

*considering*

a)  that Recommendation X.1 includes specific user classes of service for data terminal equipments operating in the packet mode, Recommendation X.2 defines user facilities, Recommendations X.21 and X.21 *bis* define DTE/DCE interface characteristics, Recommendation X.95 defines network parameters, and Recommendation X.96 defines call progress signals;

b)  that the logical control links for packet-switched data transmission services are defined in Recommendation X.92;

c)  the need for defining an international recommendation for the exchange between DTE and DCE of control information for the use of packet-switched data transmission services;

d)  that data terminal equipments operating in the packet mode will send and receive network call control information and user information in the form of packets;

e)  that the necessary elements for an interface recommendation should be defined independently as:

*Level 1* — The physical, electrical, functional and procedural characteristics to establish, maintain and disconnect the physical link between the DTE and the DCE.

*Level 2* — The link access procedure for data interchange across the link between the DTE and the DCE.

*Level 3* — The packet format and control procedures for the exchange of packets containing control information and user data between the DTE and the DCE.

f)  that certain data terminal equipments operating in the packet mode will use a packet interleaved, synchronous data circuit;

g)  the desirability of being able to use a single data circuit to a DSE for all user facilities;

*unanimously declares the view*

that for data terminal equipments operating in the packet mode:

1.  The physical, electrical, functional and procedural characteristics to establish, maintain and disconnect the physical link between the DTE and the DCE should be as specified in 1. below *Level 1 DTE/DCE interface characteristics*, see below.

2.  The link access procedure for data interchange across the link between the DTE and the DCE should be as specified in 2. below *Link access procedure across the DTE/DCE interface (level 2)*, see below.

**Provisional Recomendation X.25**

3. The packet level control procedures for the exchange of call control information and user data at the DTE/DCE interface should be as specified in 3. below *Description of the packet level DTE/DCE interface for virtual call and permanent virtual circuit facilities (level 3)*, see below.

4. The format for packets exchanged between the DTE and the DCE should be as specified in 4. below *Packet formats for virtual call and permanent virtual circuit facilities*, see below.

5. Procedures and formats for optional user facilities should be as specified in 5. below *Procedures and formats for optional user facilities to be studied for virtual call and permanent virtual circuit facilities*, see below.

*Note.* — It is for further study whether alternative procedures for levels 2 and 3 of Recommendation X.25 would be advantageously introduced for cases where packet-switched facilities are accessed through circuit-switched connections.

## INDEX TO RECOMMENDATION X.25

## 1. LEVEL 1 DTE/DCE INTERFACE CHARACTERISTICS

The DTE/DCE interface characteristics defined as the level 1 element shall be in accordance with Recommendation X.21 for user classes of service 8 to 11. For an interim period, some Administrations may offer a DTE/DCE interface at this level in accordance with Recommendation X.21 *bis*, for user classes of service 8 to 11.

*Note.* — The level 1 DTE/DCE interface also includes the failure detection and isolation procedures as defined in Recommendation X.21 or X.21 *bis*.

## 2. LINK ACCESS PROCEDURE ACROSS THE DTE/DCE INTERFACE (LEVEL 2)

### 2.1 *Scope and field of application*

2.1.1 The Link Access Procedure (LAP) described hereunder is used for data interchange between a DCE and a DTE operating in user classes of service 8 to 11 as indicated in Recommendation X.1.

2.1.2 The procedure uses the principle and terminology of the High Level Data Link Control (HDLC) procedure specified by the International Organization for Standardization (ISO).

*Note.* — Reference ISO documents IS 3309, IS 4335 plus approved amendments (TC 97/SC 6/N 1300 and 1445), and DP 6256.

2.1.3 The transmission facility is duplex.

2.1.4 DCE compatibility of operation with the ISO balanced class of procedure (Class BA, options 2, 8, 11 and 12) is achieved using the provisions found under the headings annotated as "applicable to LAPB" in this Recommendation.

A DTE may continue to use the provisions found under the headings annotated as "applicable to LAP" in this Recommendation, but for new DTE implementations, LAPB should be preferred.

*Note.* — Other possible applications for further study are, for example:

— two-way alternate, asynchronous response mode

— two-way simultaneous, normal response mode

— two-way alternate, normal response mode

— the need for and use of option 11.

## 2.2 *Frame structure*

2.2.1 All transmissions are in frames conforming to one of the formats of Table 1/X.25. The flag preceding the address field is defined as the opening flag.

TABLE 1/X.25 — **Frame formats**

| Bit order of transmission | 12345678 | 12345678 | 12345678 | 16 to 1 | 12345678 |
|---|---|---|---|---|---|
| | Flag | Address | Control | FCS | Flag |
| | F<br>01111110 | A<br>8-bits | C<br>8-bits | FCS<br>16-bits | F<br>01111110 |

FCS = frame checking sequence

| Bit order of transmission | 12345678 | 12345678 | 12345678 | | 16 to 1 | 12345678 |
|---|---|---|---|---|---|---|
| | Flag | Address | Control | Information | FCS | Flag |
| | F<br>01111110 | A<br>8-bits | C<br>8-bits | I<br>N-bits | FCS<br>16-bits | F<br>01111110 |

FCS = frame checking sequence

## 2.2.2 *Flag sequence*

All frames shall start and end with the flag sequence consisting of one 0 followed by six contiguous 1s and one 0. A single flag may be used as both the closing flag for one frame and the opening flag for the next frame.

## 2.2.3 *Address field*

The address field shall consist of one octet. The coding of the address field is described in 2.4.2 below.

**Provisional Recomendation X.25**

### 2.2.4 Control field

The control field shall consist of one octet. The content of this field is described in 2.3.2 below.

*Note.* — The use of the extended control field is a subject for further study.

### 2.2.5 Information field

The information field of a frame is unrestricted with respect to code or grouping of bits except for the packet formats specified in 3. below.

See 2.3.4.10 and 2.4.11.3 below with regard to the maximum information field length.

### 2.2.6 Transparency

The DTE or DCE, when transmitting, shall examine the frame content between the two flag sequences including the address, control, information and FCS sequences and shall insert a 0 bit after all sequences of 5 contiguous 1 bits (including the last 5 bits of the FCS) to ensure that a flag sequence is not simulated. The DTE or DCE, when receiving, shall examine the frame content and shall discard any 0 bit which directly follows 5 contiguous 1 bits.

### 2.2.7 Frame checking sequence (FCS)

The FCS shall be a 16-bit sequence. It shall be the 1s complement of the sum (modulo 2) of:

1. the remainder of $x^k(x^{15} + x^{14} + x^{13} + \ldots + x^2 + x + 1)$ divided (modulo 2) by the generator polynomial $x^{16} + x^{12} + x^5 + 1$, where k is the number of bits in the frame existing between, but not including, the final bit of the opening flag and the first bit of the FCS, excluding bits inserted for transparency, and

2. the remainder after multiplication by $x^{16}$ and then division (modulo 2) by the generator polynomial $x^{16} + x^{12} + x^5 + 1$ of the content of the frame, existing between but not including, the final bit of the opening flag and the first bit of the FCS, excluding bits inserted for transparency.

As a typical implementation, at the transmitter, the initial remainder of the division is preset to all 1s and is then modified by division by the generator polynomial (as described above) on the address, control and information fields; the 1s complement of the resulting remainder is transmitted as the 16-bit FCS sequence.

At the receiver, the initial remainder is preset to all 1s, and the serial incoming protected bits and the FCS when divided by the generator polynomial will result in a remainder of 0001110100001111 ($x^{15}$ through $x^0$, respectively) in the absence of transmission errors.

### 2.2.8 Order of bit transmission

Addresses, commands, responses and sequence numbers shall be transmitted with the low order bit first (for example the first bit of the sequence number that is transmitted shall have the weight $2^0$).

The order of transmitting bits within the information field is not specified under 2. of this Recommendation. The FCS shall be transmitted to the line commencing with the coefficient of the highest term.

*Note.* — The low order bit is defined as bit 1, as depicted in Tables 1/X.25 to 4/X.25.

### 2.2.9 Invalid frames

A frame not properly bounded by two flags, or having fewer than 32 bits between flags, is an invalid frame.

## 2.2.10 *Frame abortion*

Aborting a frame is performed by transmitting at least seven contiguous 1s (with no inserted 0s).

## 2.2.11 *Interframe time fill*

Interframe time fill is accomplished by transmitting contiguous flags between frames.

## 2.2.12 *Link channel states*

### 2.2.12.1 *Active channel state*

A channel is in an active condition when the DTE or DCE is actively transmitting a frame, an abortion sequence or interframe time fill.

### 2.2.12.2 *Idle channel state*

A channel is defined to be in an idle condition when a contiguous 1s state is detected that persists for at least 15 bit times.

*Note 1.* — The action to be taken upon detection of the idle channel state is a subject for further study.

*Note 2.* — A link channel as defined here is the means of transmission for one direction.

## 2.3 *Elements of procedure*

2.3.1 The elements of procedure are defined in terms of actions that occur on receipt of commands at a DTE or DCE.

The elements of procedure specified below contain a selection of commands and responses relevant to the link and system configuration described in 2.1 above.

A procedure is derived from these elements of procedure and is described in 2.4 below. Together 2.2 and 2.3 form the general requirements for the proper management of the access link.

## 2.3.2 *Control field formats and state variables*

### 2.3.2.1 *Control field formats*

The control field contains a command or a response, and sequence numbers where applicable.

Three types of control field formats (see Table 2/X.25) are used to perform numbered information transfer (I frames), numbered supervisory functions (S frames) and unnumbered control functions (U frames).

TABLE 2/X.25 – **Control field formats**

| Control field bits | 1 | 2 | 3 | 4 | 5 | 6 | 7 | 8 |
|---|---|---|---|---|---|---|---|---|
| I frame | 0 | N(S) | | | P/F | N(R) | | |
| S frame | 1 | 0 | S | | P/F | N(R) | | |
| U frame | 1 | 1 | M | | P/F | M | | |

N(S) = transmitter send sequence number (Bit 2 = low order bit)
N(R) = transmitter receive sequence number (Bit 6 = low order bit)
S = supervisory function bits
M = modifier function bits
P/F = poll bit when issued as a command, final bit when issued as a response. (1 = Poll/Final)

**Provisional Recomendation X.25**

*Information transfer format − I*

The I format is used to perform an information transfer. The functions of N(S), N(R) and P/F are independent; i.e. each I frame has an N(S), an N(R) which may or may not acknowledge additional frames received by the DTE or DCE, and a P/F bit.

*Supervisory format − S*

The S format is used to perform link supervisory control functions such as acknowledge I frames, request retransmission of I frames, and to request a temporary suspension of transmission of I frames.

*Unnumbered format − U*

The U format is used to provide additional link control functions. This format contains no sequence numbers. The encoding of the unnumbered commands is as defined in Table 3/X.25.

### 2.3.2.2 *Control field parameters*

The various parameters associated with the control field formats are described below.

### 2.3.2.3 *Modulus*

Each I frame is sequentially numbered and may have the value 0 through modulus minus one (where "modulus" is the modulus of the sequence numbers). The modulus equals 8 and the sequence numbers cycle through the entire range.

### 2.3.2.4 *Frame variables and sequence numbers*

### 2.3.2.4.1 *Send state variable V(S)*

The send state variable denotes the sequence number of the next in-sequence I frame to be transmitted. The send state variable can take on the value 0 through modulus minus one. The value of the send state variable is incremented by one with each successive I frame transmission, but at the DCE cannot exceed N(R) of the last received I or S frame by more than the maximum number of outstanding I frames (k). The value of k is defined in 2.4.11.4 below.

### 2.3.2.4.2 *Send sequence number N(S)*

Only I frames contain N(S), the send sequence number of transmitted frames. Prior to transmission of an in-sequence I frame, the value of N(S) is updated to equal the value of the send state variable.

### 2.3.2.4.3 *Receive state variable V(R)*

The receive state variable denotes the sequence number of the next in-sequence I frame to be received. This receive state variable can take on the values 0 through modulus minus one. The value of the receive state variable is incremented by the receipt of an error free, in-sequence I frame whose send sequence number N(S) equals the receive state variable.

### 2.3.2.4.4 *Receive sequence number N(R)*

All I frames and S frames contain N(R), the expected sequence number of the next received I frame. Prior to transmission of a frame of the above types, the value of N(R) is updated to equal the current value of the receive state variable. N(R) indicates that the DTE or DCE transmitting the N(R) has correctly received all I frames numbered up to N(R) − 1.

### 2.3.3 *Functions of the poll/final bit*

The poll/final (P/F) bit serves a function in both command frames and response frames. In command frames the P/F bit is referred to as the P bit. In response frames it is referred to as the F bit.

The use of the P/F bit is described in 2.4.3 below.

**Provisional Recomendation X.25**

## 2.3.4 *Commands and responses*

The following commands and responses will be used by either the DTE or DCE and are represented in Table 3/X.25.

TABLE 3/X.25 – Commands and responses

| Format | Commands | Responses | Encoding | | | | | | | |
|---|---|---|---|---|---|---|---|---|---|---|
| | | | 1 | 2 | 3 | 4 | 5 | 6 | 7 | 8 |
| Information transfer | I     (information) | | 0 | N(S) | | | P | N(R) | | |
| Supervisory | RR   (receive ready) <br> RNR (receive not ready) <br> REJ   (reject) | RR   (receive ready) <br> RNR (receive not ready) <br> REJ   (reject) | 1<br>1<br>1 | 0<br>0<br>0 | 0<br>1<br>0 | 0<br>0<br>1 | P/F<br>P/F<br>P/F | N(R)<br>N(R)<br>N(R) | | |
| Unnumbered | SARM (set asynchronous response mode) | DM (disconnected mode) | 1 | 1 | 1 | 1 | P/F | 0 | 0 | 0 |
| | SABM (set asynchronous balanced mode) | | 1 | 1 | 1 | 1 | P | 1 | 0 | 0 |
| | DISC (disconnect) | | 1 | 1 | 0 | 0 | P | 0 | 1 | 0 |
| | | UA (unnumbered acknowledgement) | 1 | 1 | 0 | 0 | F | 1 | 1 | 0 |
| | | CMDR (command reject) <br> FRMR (frame reject) | 1 | 1 | 1 | 0 | F | 0 | 0 | 1 |

*Note 1.* – The need for, and use of, additional commands and responses are for further study.

*Note 2.* – RR, RNR, and REJ commands are not transmitted by the DCE, but can be received from the DTE. DTEs do not have to implement both SARM and SABM. DTEs which do not implement SABM do not have to implement DM.

The commands and responses are as follows:

### 2.3.4.1 *Information (I) command*

The function of the information (I) command is to transfer across a data link sequentially numbered frames containing an information field.

### 2.3.4.2 *Receive ready (RR)*

The receive ready (RR) supervisory frame is used by the DTE or DCE to:

1) indicate it is ready to receive an I frame;

2) acknowledge previously received I frames numbered up to N(R) — 1.

RR may be used to clear a busy condition that was initiated by the transmission of RNR. The RR command with the P bit set to 1 may be used by the DTE to ask for the status of the DCE.

**Provisional Recomendation X.25**

### 2.3.4.3 *Reject (REJ)*

The reject (REJ) supervisory frame is used by the DTE or DCE to request retransmission of I frames starting with the frame numbered N(R). I frames numbered N(R) − 1 and below are acknowledged. Additional I frames pending initial transmission may be transmitted following the retransmitted I frame(s).

Only one REJ exception condition for a given direction of information transfer may be established at any time. The REJ exception condition is cleared (reset) upon the receipt of an I frame with an N(S) equal to the N(R) of the REJ.

### 2.3.4.4 *Receive not ready (RNR)*

The receive not ready (RNR) supervisory frame is used by the DTE or DCE to indicate a busy condition; i.e., temporary inability to accept additional incoming I frames. I frames numbered up to N(R) − 1 are acknowledged. I frame N(R) and subsequent I frames received, if any, are not acknowledged; the acceptance status of these I frames will be indicated in subsequent exchanges.

An indication that the busy condition has cleared is communicated by the transmission of a UA, RR, REJ or SABM. The RNR command with the P bit set to 1 may be used by the DTE to ask for the status of the DCE.

### 2.3.4.5 *Set asynchronous response mode (SARM) command*

The SARM unnumbered command is used to place the addressed DTE or DCE in the asynchronous response mode (ARM) information transfer phase.

No information field is permitted with the SARM command. A DTE or DCE confirms acceptance of SARM by the transmission at the first opportunity of a UA response. Upon acceptance of this command the DTE or DCE receive state variable V(R) is set to 0.

Previously transmitted I frames that are unacknowledged when this command is actioned remain unacknowledged.

### 2.3.4.6 *Set asynchronous balanced mode (SABM) command*

The SABM unnumbered command is used to place the addressed DTE or DCE in the asynchronous balanced mode (ABM) information transfer phase.

No information field is permitted with the SABM command. A DTE or DCE confirms acceptance of SABM by the transmission at the first opportunity of a UA response. Upon acceptance of this command the DTE or DCE send state variable V(S) and receive state variable V(R) are set to 0.

Previously transmitted I frames that are unacknowledged when this command is actioned remain unacknowledged.

### 2.3.4.7 *Disconnect (DISC) command*

The DISC unnumbered command is used to terminate the mode previously set. It is used to inform the DTE or DCE receiving the DISC that the DTE or DCE sending the DISC is suspending operation. No information field is permitted with the DISC command. Prior to actioning the command, the DTE or DCE receiving the DISC confirms the acceptance of DISC by the transmission of a UA response. The DTE or DCE sending the DISC enters the disconnected phase when it receives the acknowledging UA response.

Previously transmitted I frames that are unacknowledged when this command is actioned remain unacknowledged.

### 2.3.4.8 *Unnumbered acknowledge (UA) response*

The UA unnumbered response is used by the DTE or DCE to acknowledge the receipt and acceptance of the U format commands. Received U format commands are not actioned until the UA response is transmitted. The UA response is transmitted as directed by the received U format command. No information field is permitted with the UA response.

250

### 2.3.4.9  *Disconnected mode (DM) response*

The DM response is used to report a status where the DTE or DCE is logically disconnected from the link, and is in the disconnected phase. The DM response is sent in this phase to request a set mode command, or, if sent in response to the reception of a set mode command, to inform the DTE or DCE that the DCE or DTE, respectively, is still in disconnected phase and cannot action the set mode command. No information field is permitted with the DM response.

A DTE or DCE in a disconnected phase will monitor received commands, and will react to SABM as outlined in 2.4.5 below and will respond DM to any other command received with the P bit set to 1.

### 2.3.4.10  *Command reject (CMDR) response*
### *Frame reject (FRMR) response*

The CMDR (FRMR) response is used by the DTE or DCE to report an error condition not recoverable by retransmission of the identical frame; i.e., one of the following conditions, which results from the receipt of a frame without FCS error:

1.  the receipt of a command or response that is invalid or not implemented;

2.  the receipt of an I frame with an information field which exceeds the maximum established length;

3.  the receipt of an invalid N(R).

An invalid N(R) is defined as one which points to an I frame which has previously been transmitted and acknowledged or to an I frame which has not been transmitted and is not the next sequential I frame pending transmission.

An information field which immediately follows the control field, and consists of 3 octets, is returned with this response and provides the reason for the CMDR (FRMR) response. This format is given in Table 4/X.25.

TABLE 4/X.25 – **CMDR (FRMR) field format**

Information field bits

| 1 | 2 | 3 | 4 | 5 | 6 | 7 | 8 | 9 | 10 | 11 | 12 | 13 | 14 | 15 | 16 | 17 | 18 | 19 | 20 | 21 | 22 | 23 | 24 |
|---|---|---|---|---|---|---|---|---|----|----|----|----|----|----|----|----|----|----|----|----|----|----|----|
| Rejected frame control field | | | | | | | | 0 | V(S) | | | (see Note) | V(R) | | | W | X | Y | Z | 0 | 0 | 0 | 0 |

– Rejected frame control field is the control field of the received frame which caused the command (frame) reject.
– V(S) is the current send state variable value at the DTE or DCE reporting the rejection condition (Bit 10 = low order bit).
– V(R) is the current receive state variable at the DTE or DCE reporting the rejection condition (Bit 14 = low order bit).
– W set to 1 indicates that the control field received and returned in bits 1 through 8 was invalid.
– X set to 1 indicates that the control field received and returned in bits 1 through 8 was considered invalid because the frame contained an information field which is not permitted with this command. Bit W must be set to 1 in conjunction with this bit.
– Y set to 1 indicates that the information field received exceeded the maximum established capacity of the DTE or DCE reporting the rejection condition. This bit is mutually exclusive with bit W above.
– Z set to 1 indicates that the control field received and returned in bits 1 through 8 contained an invalid N(R). This bit is mutually exclusive with bit W above.

*Note.* – Bits 9, 13, 21 to 24 shall be set to 0 for CMDR. For FRMR, bits 9, 21 to 24 shall be set to 0. Bit 13 shall be set to 1 if the frame rejected was a response, and set to 0 if the frame rejected was a command.

**Provisional Recomendation X.25**

### 2.3.5 *Exception condition reporting and recovery*

The error recovery procedures which are available to effect recovery following the detection/occurrence of an exception condition at the link level are described below. Exception conditions described are those situations which may occur as the result of transmission errors, DTE or DCE malfunction or operational situations.

#### 2.3.5.1 *Busy condition*

The busy condition results when a DTE or DCE is temporarily unable to continue to receive I frames due to internal constraints, e.g., receive buffering limitations. In this case an RNR frame is transmitted from the busy DCE or DTE. I frames pending transmission may be transmitted from the busy DTE or DCE prior to or following the RNR. Clearing of the busy condition is indicated as described in 2.3.4.4 above.

#### 2.3.5.2 *N(S) sequence error*

The information field of all I frames whose N(S) does not equal the receive state variable V(R) will be discarded.

An N(S) sequence exception condition occurs in the receiver when an I frame received error-free (no FCS error) contains an N(S) which is not equal to the receive state variable at the receiver. The receiver does not acknowledge (increment its receive state variable) the I frame causing the sequence error, or any I frame which may follow, until an I frame with the correct N(S) is received.

A DTE or DCE which receives one or more I frames having sequence errors but otherwise error-free shall accept the control information contained in the N(R) field and the P bit to perform link control functions; e.g., to receive acknowledgement of previously transmitted I frames and to cause the DTE or DCE to respond (P bit set to 1). Therefore, the retransmitted I frame may contain an N(R) field and P bit that are updated from, and therefore different from, that contained in the originally transmitted I frame.

#### 2.3.5.3 *REJ recovery*

The REJ is used to initiate an exception recovery (retransmission) following the detection of a sequence error.

Only one "sent REJ" exception condition from a DTE or DCE is established at a time. A sent REJ exception condition is cleared when the requested I frame is received.

A DTE or DCE receiving REJ initiates sequential (re-)transmission of I frames starting with the I frame indicated by the N(R) obtained in the REJ frame.

#### 2.3.5.4 *Time-out recovery*

If a DTE or DCE, due to a transmission error, does not receive (or receives and discards) a single I frame or the last I frame in a sequence of I frames, it will not detect an out-of-sequence exception condition and therefore will not transmit REJ. The DTE or DCE which transmitted the unacknowledged I frame(s) shall, following the completion of a system specified time-out period (see 2.4.11.1 below), take appropriate recovery action to determine at which I frame retransmission must begin.

#### 2.3.5.5 *FCS error*

Any frame received with an FCS error is not accepted by the receiver. The frame is discarded and no action is taken as the result of that frame.

#### 2.3.5.6 *Rejection condition*

A rejection condition is established upon the receipt of an error free frame which contains an invalid command/response in the control field, an invalid frame format, an invalid N(R) or an information field which exceeds the maximum information field length which can be accommodated.

At the DCE, this exception is reported by a CMDR (FRMR) response for appropriate DTE action. Once a DCE has established a CMDR (FRMR) exception, no additional I frames are accepted, except for examination of the P bit and the N(R) field until the condition is reset by the DTE. The CMDR (FRMR) response is repeated at each opportunity until recovery is effected by the DTE.

## 2.4 *Description of the procedure*

### 2.4.1 *Procedure to set the mode variable B*

The DCE will maintain an internal mode variable B, which it will set as follows:

— to 1, upon acceptance of an SABM command from the DTE

— to 0, upon acceptance of an SARM command from the DTE.

Changes to the mode variable B by the DTE should occur only when the link has been disconnected as described in 2.4.4.3 or 2.4.5.3 below.

Should a DCE malfunction occur, the internal mode variable B will upon restoration of operation, but prior to link set-up by the DTE, be initially set to 1.

Whenever B is 1, the DCE will use the LAPB link set-up and disconnection procedure and is said to be in the LAPB (balanced) mode.

Whenever B is 0, the DCE will use the LAP link set-up and disconnection procedure, and is said to be in the LAP mode.

The following are applicable to both LAP and LAPB modes: 2.4.2, 2.4.3, 2.4.6, 2.4.11.

The following are applicable only to the LAP mode: 2.4.4, 2.4.7, 2.4.8.

The following are applicable only to the LAPB mode: 2.4.5, 2.4.9, 2.4.10.

### 2.4.2 *Procedure for addressing (applicable to both LAP and LAPB)*

Frames containing commands transferred from the DCE to the DTE will contain the address A.

Frames containing responses transferred from the DTE to the DCE shall contain the address A.

Frames containing commands transferred from the DTE to the DCE shall contain the address B.

Frames containing responses transferred from the DCE to the DTE will contain the address B.

A and B addresses are coded as follows:

| Address | 1 2 3 4 5 6 7 8 |
|---------|-----------------|
| A | 1 1 0 0 0 0 0 0 |
| B | 1 0 0 0 0 0 0 0 |

*Note.* — The action to be taken by the DCE following the receipt of an address other than A or B is for further study.

### 2.4.3 *Procedure for the use of the poll/final bit (applicable to both LAP and LAPB)*

The DCE receiving a SARM, SABM, DISC, supervisory command or an I frame with the poll bit set to 1, will set the final bit to 1 in the next response frame it transmits.

The response frame returned by the DCE to a SARM, SABM or DISC command with the poll bit set to 1 will be a UA (or DM) response with the final bit set to 1. The response frame returned by the DCE to an I frame with the poll bit set to 1 will be an RR, REJ or RNR response in supervisory format with the final bit set to 1.

The response frame returned by the DCE to a supervisory command frame with the poll bit set to 1 will be an RR or RNR response with the final bit set to 1.

The P bit may be used by the DCE in conjunction with the timer recovery condition (see 2.4.6.8 below).

*Note.* — Other use of the P bit by the DCE is a subject for further study.

### 2.4.4 *Procedure for link set-up (applicable to LAP)*

#### 2.4.4.1 *Link set-up*

The DCE will indicate that it is able to set up the link by transmitting contiguous flags (active channel state).

The DTE shall indicate a request for setting up the link by transmitting a SARM command to the DCE.

Whenever receiving a SARM command, the DCE will return a UA response to the DTE and set its receive state variable $V(R)$ to 0.

Should the DCE wish to indicate a request for setting up the link, or when receiving from the DTE a first SARM command as a request for setting up the link, the DCE will transmit a SARM command to the DTE and start Timer T1 (see 2.4.11.1 below). The DTE will confirm the reception of the SARM command by transmitting a UA response.

When receiving the UA response the DCE will set its send state variable to 0 and stop its Timer T1. If Timer T1 runs out before the UA response is received by the DCE, the DCE will retransmit a SARM command and restart Timer T1.

After transmission of SARM N2 times by the DCE, appropriate recovery action will be initiated.

The value of N2 is defined in 2.4.11.2 below.

#### 2.4.4.2 *Information transfer phase*

After having both received a UA response to a SARM command transmitted to the DTE and transmitted a UA response to a SARM command received from the DTE, the DCE will accept and transmit I and S frames according to the procedures described in 2.4.6 below.

When receiving a SARM command, the DCE will conform to the resetting procedure described in 2.4.7 below. The DTE may also receive a SARM command according to this resetting procedure.

#### 2.4.4.3 *Link disconnection*

During the information transfer phase the DTE shall indicate a request for disconnecting the link by transmitting a DISC command to the DCE.

Whenever receiving a DISC command, the DCE will return a UA response to the DTE.

During an information transfer phase, should the DCE wish to indicate a request for disconnecting the link, or when receiving from the DTE a first DISC command as a request for disconnecting the link, the DCE will transmit a DISC command to the DTE and start Timer T1 (2.4.11.1 below). The DTE will confirm reception of the DISC command by returning a UA response. After transmitting a SARM command, the DCE will not transmit a DISC command until a UA response is received for this SARM command or until Timer T1 runs out.

When receiving a UA response to the DISC command, the DCE will stop its Timer T1. If Timer T1 runs out before a UA response is received by the DCE, the DCE will transmit a DISC command and restart Timer T1. After transmission of DISC N2 times by the DCE, appropriate recovery action will be initiated. The value of N2 is defined in 2.4.11.2 below.

### 2.4.5 *Procedures for link set-up (applicable to LAPB)*

The DCE will indicate that it is able to set up the link by transmitting contiguous flags (active channel state).

### 2.4.5.1 *Link set-up*

Whenever receiving an SABM command, the DCE will return a UA response to the DTE and set both its send and receive state variable V(S) and V(R) to 0.

*Note.* — The possible use of the SABM command by the DCE is a subject for further study.

### 2.4.5.2 *Information transfer phase*

After having transmitted the UA response to an SABM command, the DCE will accept and transmit I and S frames according to the procedures described in 2.4.6 below.

When receiving an SABM command while in the information transfer phase, the DCE will conform to the resetting procedure described in 2.4.9 below.

### 2.4.5.3 *Link disconnection*

During the information transfer phase, the DTE shall indicate disconnecting of the link by transmitting a DISC command to the DCE.

When receiving a DISC command, the DCE will return a UA response to the DTE and enter the disconnected phase.

*Note.* — The possible use of the DISC command by the DCE is a subject for further study.

### 2.4.5.4 *Disconnected phase*

2.4.5.4.1 After having received a DISC command from the DTE and returned a UA response to the DTE, the DCE will enter the disconnected phase.

In the disconnected phase, the DCE will react to the receipt of an SABM command as described in 2.4.5.1 above and will transmit a DM response in answer to a received DISC command.

When receiving any other command frame with the poll bit set to 1, the DCE will transmit a DM response with the final bit set to 1.

Other frames received in the disconnected phase will be ignored by the DCE.

2.4.5.4.2 The DCE may also enter the disconnected phase after detecting error conditions as listed in 2.4.10 below, or exceptionally after recovery from an internal temporary malfunction. In these cases, the DCE will transmit DM and start its Timer T1 (see 2.4.11.1 below). If Timer T1 runs out before the reception of an SABM or DISC command from the DTE, the DCE will retransmit the DM response and restart Timer T1.

After transmission of the DM response N2 times by the DCE, the DCE will remain in the disconnected phase and will no longer transmit the DM response on time-out.

## 2.4.6 *Procedures for information transfer (applicable to both LAP and LAPB)*

The procedures which apply to the transmission of I frames in each direction during the information transfer phase are described below.

In the following, "number one higher" is in reference to a continuously repeated sequence series, i.e. 7 is one higher than 6 and 0 is one higher than 7 for modulo eight series.

### 2.4.6.1 *Sending I frames*

When the DCE has an I frame to transmit (i.e. an I frame not already transmitted, or having to be retransmitted as described in 2.4.6.5 below), it will transmit it with an N(S) equal to its current send state variable V(S), and an N(R) equal to its current receive state variable V(R). At the end of the transmission of the I frame, it will increment its send state variable V(S) by one.

If the Timer T1 is not running at the instant of transmission of an I frame, it will be started.

**Provisional Recomendation X.25**

If the send state variable V(S) is equal to the last value of N(R) received plus k (where k is the maximum number of outstanding I frames — see 2.4.11.4 below) the DCE will not transmit any I frames, but may retransmit an I frame as described in 2.4.6.5 or 2.4.6.8 below.

*Note.* — In order to ensure security of information transfer, the DTE should not transmit any I frame if its send state variable V(S) is equal to the last value of N(R) it has received from the DCE plus 7.

When the DCE is in the busy or frame/command rejection condition it may still transmit I frames, provided that the DTE is not busy itself. If the frame rejection condition was caused by receipt at the DCE of an invalid N(R) (see 2.4.10.1 below), the DCE will stop transmitting I frames.

### 2.4.6.2    *Receiving an I frame*

2.4.6.2.1 When the DCE is not in a busy condition and receives with the correct FCS an I frame whose send sequence number is equal to the DCE receive state variable V(R), the DCE will accept the information field of this frame, increment by one its receive state variable V(R), and act as follows:

i)   If an I frame is available for transmission by the DCE, it may act as in 2.4.6.1 above and acknowledge the received I frame by setting N(R) in the control field of the next transmitted I frame to the value of the DCE receive state variable V(R). The DCE may also acknowledge the received I frame by transmitting an RR with the N(R) equal to the value of the DCE receive state variable V(R).

ii)  If no I frame is available for transmission by the DCE, it will transmit an RR with the N(R) equal to the value of the DCE receive state variable V(R).

2.4.6.2.2 When the DCE is in a busy condition, it may ignore the information contained in any received I frame.

### 2.4.6.3    *Incorrect reception of frames*

When the DCE receives a frame with an incorrect FCS, this frame will be discarded.

When the DCE receives an I frame whose FCS is correct, but whose send sequence number is incorrect, i.e., not equal to the current DCE receive state variable V(R), it will discard the information content of the frame and transmit an REJ response with the N(R) set to one higher than the N(S) of the last correctly received I frame. The DCE will then discard the information content of all I frames until the expected I frame is correctly received. When receiving the expected I frame, the DCE will then acknowledge the frame as described in 2.4.6.2 above. The DCE will use the N(R) and P bit indications in the discarded I frames.

### 2.4.6.4    *Receiving acknowledgement*

When correctly receiving an I or S frame (RR, RNR or REJ), even in the busy or command/frame rejection condition, the DCE will consider the N(R) contained in this frame as an acknowledgement for all the I frames it has transmitted with an N(S) up to the received N(R) minus one. The DCE will reset the Timer T1.

If there are outstanding I frames still unacknowledged, it will restart the Timer T1. If the Timer then runs out, the DCE will follow the retransmission procedure (in 2.4.6.5 and 2.4.6.8 below) in respect of the unacknowledged I frames.

### 2.4.6.5    *Receiving reject*

When receiving an REJ, the DCE will set its send state variable V(S) to the N(R) received in the REJ control field. It will transmit the corresponding I frame as soon as it is available or retransmit it. Retransmission will conform to the following:

i)   If the DCE is transmitting a supervisory or unnumbered command or response when it receives the REJ, it will complete that transmission before commencing transmission of the requested I frame.

ii) If the DCE is transmitting an I frame when the REJ is received, it may abort the I frame and commence transmission of the requested I frame immediately after abortion.

iii) If the DCE is not transmitting any frame when the REJ is received, it will commence transmission of the requested I frame immediately.

In all cases, if other unacknowledged I frames had already been transmitted following the one indicated in the REJ, then those I frames will be retransmitted by the DCE following the retransmission of the requested I frame.

If the REJ frame was received from the DTE as a command with the P bit set to 1, the DCE will transmit an RR or RNR response with the F bit set to 1 before transmitting or retransmitting the corresponding I frame.

### 2.4.6.6  *Receiving RNR*

After receiving an RNR, the DCE may transmit the I frame with the send sequence number equal to the N(R) indicated in the RNR. If the Timer T1 runs out after the reception of RNR, the DCE will follow the procedure described in 2.4.6.8 below. In any case the DCE will not transmit any other I frames before receiving an RR or REJ.

### 2.4.6.7  *DCE busy condition*

When the DCE enters a busy condition, it will transmit an RNR response at the earliest opportunity. While in the busy condition, the DCE will accept and process S frames and return an RNR response with the F bit set to 1 if it receives an S or I command frame with the P bit set to 1. To clear the busy condition, the DCE will transmit either an REJ response or an RR response with N(R) set to the current receive state variable V(R) depending on whether or not it discarded information fields of correctly received I frames.

### 2.4.6.8  *Waiting acknowledgement*

The DCE maintains an internal retransmission count variable which is set to 0 when the DCE receives a UA or RNR, or when the DCE correctly receives an I or S frame with the N(R) higher than the last received N(R) (actually acknowledging some outstanding I frames).

If the Timer T1 runs out, the DCE will enter the timer recovery condition, add one to its retransmission count variable and set an internal variable $x$ to the current value of its send state variable.

The DCE will restart Timer T1, set its send state variable to the last N(R) received from the DTE and retransmit the corresponding I frame with the poll bit set to 1.

The timer recovery condition is cleared when the DCE receives a valid S frame from the DTE, with the final bit set to 1.

If, while in the timer recovery condition, the DCE correctly receives a frame with an N(R) within the range from its current send state variable to $x$ included, on or before clearing the timer recovery condition, it will not enter a rejection condition, and will set its send state variable to the received N(R).

If the retransmission count variable is equal to N2, the DCE initiates a resetting procedure for the direction of transmission from the DCE as described in 2.4.7.3 or 2.4.9.3 below. N2 is a system parameter (see 2.4.11.2 below).

*Note.* — Although the DCE will implement the internal variable $x$, other mechanisms do exist that achieve the identical functions. Therefore, the internal variable $x$ is not necessarily implemented in the DTE.

## 2.4.7  *Procedures for resetting (applicable to LAP)*

2.4.7.1   The resetting procedure is used to reinitialize one direction of information transmission, according to the procedure described below. The resetting procedure only applies during the information transfer phase.

2.4.7.2   The DTE will indicate a resetting of the information transmission from the DTE by transmitting an SARM command to the DCE. When receiving an SARM command, the DCE will return, at the earliest opportunity, a UA response to the DTE and set its receive state variable V(R) to 0. This also indicates a clearance of the DCE busy condition, if present.

2.4.7.3    The DCE will indicate a resetting of the information transmission from the DCE by transmitting an SARM command to the DTE and will start Timer T1 (see 2.4.11.1 below). The DTE will confirm reception of the SARM command by returning a UA response to the DCE. When receiving this UA response to the SARM command, the DCE will set its send state variable to 0 and stop its Timer T1. If Timer T1 runs out before the UA response is received by the DCE, the DCE will retransmit an SARM command and restart Timer T1. After transmission of SARM N2 times, appropriate recovery action will be initiated. The value of N2 is defined in 2.4.11.2 below.

The DCE will not act on any received response frame which arrives before the UA response to the SARM command. The value of N(R) contained in any correctly received I command frames arriving before the UA response will also be ignored.

2.4.7.4    When receiving a CMDR response from the DTE, the DCE will initiate a resetting of the information transmission from the DCE as described in 2.4.7.3 above.

2.4.7.5    If the DCE transmits a CMDR response, it enters the command rejection condition. This command rejection condition is cleared when the DCE receives an SARM or DISC command. Any other command received while in the command rejection condition will cause the DCE to retransmit this CMDR response. The coding of the CMDR response will be as described in 2.3.4.10 above.

### 2.4.8    *Rejection conditions (applicable to LAP)*

#### 2.4.8.1    *Rejection conditions causing a resetting of the transmission of information from the DCE*

The DCE will initiate a resetting procedure as described in 2.4.7.3 above when receiving a frame with the correct FCS, with the address A (coded 1 1 0 0 0 0 0 0) and with one of the following conditions:

— the frame type is unknown as one of the responses used;

— the information field is invalid;

— the N(R) contained in the control field is invalid;

— the response contains an F bit set to 1 except during a timer recovery condition as described in 2.4.6.8 above.

A valid N(R) must be within the range from the lowest send sequence number N(S) of the still-unacknowledged frame(s) to the current DCE send state variable included, even if the DCE is in a rejection condition, but not if the DCE is in the timer recovery condition (see 2.4.6.8 above).

#### 2.4.8.2    *Rejection conditions causing the DCE to request a resetting of the transmission of information from the DTE*

The DCE will enter the command rejection condition as described in 2.4.7.5 above when receiving a frame with the correct FCS, with the address B (coded 1 0 0 0 0 0 0 0) and with one of the following conditions:

— the frame type is unknown as one of the commands used;

— the information field is invalid.

### 2.4.9    *Procedures for resetting (applicable to LAPB)*

2.4.9.1    The resetting procedures are used to initialize both directions of information transmission according to the procedure described below. The resetting procedures only apply during the information transfer phase.

2.4.9.2    The DTE shall indicate a resetting by transmitting an SABM command. After receiving an SABM command, the DCE will return, at the earliest opportunity, a UA response to the DTE, and reset its send and receive state variables V(S) and V(R) to 0. This also clears a DCE and/or DTE busy condition, if present.

2.4.9.3    Under certain rejection conditions listed in 2.4.6.8 and 2.4.10.2, the DCE may ask the DTE to reset the link by transmitting a DM response.

After transmitting a DM response, the DCE will enter the disconnected phase as described in 2.4.5.4.2 above.

2.4.9.4    Under certain rejection conditions listed in 2.4.10.1 below, the DCE may ask the DTE to reset the link by transmitting a FRMR response.

After transmitting a FRMR response, the DCE will enter the frame rejection condition. The frame rejection condition is cleared when the DCE receives an SABM or DISC command or DM response. Any other command received while in the frame rejection condition will cause the DCE to retransmit the FRMR response with the same information field as originally transmitted.

### 2.4.10  *Rejection conditions (applicable to LAPB)*

2.4.10.1    The DCE will initiate a resetting procedure as described in 2.4.9.4 above, when receiving, during the information transfer phase, a frame with the correct FCS, with the address A or B, and with one of the following conditions:

— the frame is unknown as a command or as a response;

— the information field is invalid;

— the N(R) contained in the control field is invalid as described in 2.4.8.1 above.

The coding of the information field of the FRMR response which is transmitted is given in 2.3.4.10 above. Bit 13 of this information field is set to 0 if the address of the rejected frame is B. It is set to 1 if the address is A.

2.4.10.2    The DCE will initiate a resetting procedure as described in 2.4.9.3 above when receiving during the information transfer phase a frame with the correct FCS, with address A, and with one of the following conditions:

— it is a UA or DM response;

— it contains a final bit set to 1, except during a timer recovery condition as described in 2.4.6.8.

### 2.4.11  *List of system parameters (applicable to both LAP and LAPB)*

The system parameters are as follows:

#### 2.4.11.1  *Timer T1*

The period of the Timer T1, at the end of which retransmission of a frame may be initiated according to the procedures described in 2.4.4 to 2.4.6 above, is a system parameter agreed for a period of time with the Administration.

The proper operation of the procedure requires that Timer T1 be greater than the maximum time between transmission of frames (SARM, SABM, DM, DISC or I) and the reception of the corresponding frame returned as an answer to this frame (UA, DM or acknowledging frame). Therefore, the DTE should not delay the response or acknowledging frame returned to the above frames by more than a value T2 less than T1, where T2 is a system parameter.

The DCE will not delay the response or acknowledging frame returned to a command by more than T2.

**Provisional Recomendation  X.25**

### 2.4.11.2  *Maximum number of transmissions N2*

The value of the maximum number N2 of transmission and retransmissions of a frame following the running out of Timer T1 is a subject for further study.

### 2.4.11.3  *Maximum number of bits in an I frame N1*

The maximum number of bits in an I frame is a system parameter which depends upon the maximum length of the information fields transferred across the DTE/DCE interface.

### 2.4.11.4  *Maximum number of outstanding I frames k*

The maximum number (k) of sequentially numbered I frames that the DCE may have outstanding (i.e. unacknowledged) at any given time is a system parameter which can never exceed seven. It shall be agreed for a period of time with the Administration.

*Note.* — As a result of the further study proposed in 2.2.4 above, the permissible maximum number of outstanding I frames may be increased.

## 3.  DESCRIPTION OF THE PACKET LEVEL DTE/DCE INTERFACE FOR VIRTUAL CALL AND PERMANENT VIRTUAL CIRCUIT FACILITIES (LEVEL 3)

This section of the Recommendation relates to the transfer of packets at the DTE/DCE interface. The procedures apply to packets which are successfully transferred across the DTE/DCE interface.

Each packet to be transferred across the DTE/DCE interface shall be contained within the link level information field which will delimit its length, and only one packet shall be contained in the information field.

*Note.* — Possible insertion of more than one packet in the link level information field is for further study.

To enable simultaneous virtual calls and/or permanent virtual circuits, logical channels are used. Each virtual call or permanent virtual circuit is assigned a logical channel group number (less than or equal to 15) and a logical channel number (less than or equal to 255). For virtual calls a logical channel group number and a logical channel number are assigned during the call set-up phase. For permanent virtual circuits a logical channel group number and a logical channel number are assigned in agreement with the Administration at the time of subscription to the service. The range of logical channels used for virtual calls is agreed with the Administration at the time of subscription to the service.

### 3.1  *Procedures for virtual calls*

Annex 1, Figures 15/X.25, 16/X.25 and 17/X.25 shows the state diagrams which give a definition of events at the packet level DTE/DCE interface for each logical channel used for virtual calls (Annex 2 defines the symbols used for the state diagrams).

Annex 3 gives details of the action taken by the DCE on receipt of packets in each state shown in Annex 1. Details of the actions which should be taken by the DTE are for further study.

Packet formats are given in 4. below.

### 3.1.1  *Ready state*

If there is no call in existence, a logical channel is in the *Ready* state.

### 3.1.2  *Call request packet*

The calling DTE shall indicate a call request by transferring a *call request* packet across the DTE/DCE interface. The logical channel selected by the DTE is then in the *DTE Waiting* state (p2). The *call request* packet includes the called DTE address. (See 3.6 below).

*Note.* – A DTE address may be a DTE network address, an abbreviated address or any other DTE identification agreed for a period of time between the DTE and the DCE.

### 3.1.3  *Incoming call packet*

The DCE will indicate that there is an incoming call by transferring across the DTE/DCE interface an *incoming call* packet. This places the logical channel in the *DCE Waiting* state (p3) (see 3.6 below).

The incoming call packet will use the logical channel in the *Ready* state with the lowest number. The incoming call packet includes the calling DTE address.

*Note.* – A DTE address may be a DTE network address, an abbreviated address or any other DTE identification agreed for a period of time between the DTE and the DCE.

### 3.1.4  *Call accepted packet*

The called DTE shall indicate its acceptance of the call by the transferring across the DTE/DCE interface a *call accepted* packet specifying the same logical channel as that of the *incoming call* packet. This places the specified logical channel in the *Data transfer* state (p4).

### 3.1.5  *Call connected packet*

The receipt of a *call connected* packet by the calling DTE specifying the same logical channel as that specified in the *call request* packet indicates that the call has been established and the logical channel is then in the *Data transfer* state (p4).

### 3.1.6  *Call collision*

*Call collision* occurs when a DTE and a DCE simultaneously transfer a *call request* packet and an *incoming call* packet specifying the same logical channel. The DCE will proceed with the call request and cancel the incoming call.

### 3.1.7  *Clearing by the DTE*

The DTE may indicate clearing by transferring across the DTE/DCE interface a *clear request* packet. The logical channel is then in the *DTE Clear request* state (p6). When the DCE is prepared to free the logical channel, the DCE will transfer across the DTE/DCE interface a *DCE clear confirmation* packet specifying the logical channel. The logical channel is now in the *Ready* state (p1) (see 3.6 below).

It is possible that subsequent to transferring a *clear request* packet the DTE will receive other types of packet, dependent on the state of the logical channel, before receiving a *DCE clear confirmation* packet.

### 3.1.8  *Clearing by the DCE*

The DCE will indicate clearing by transferring across the DTE/DCE interface a *clear indication* packet. The logical channel is then in the *DCE Clear indication* state (p7). The DTE shall respond by transferring across the DTE/DCE interface a *DTE clear confirmation* packet. The logical channel is now in the *Ready* state (p1) (see 3.6 below).

**Provisional Recomendation  X.25**

### 3.1.9   *Clear collision*

Clear collision occurs when a DTE and a DCE simultaneously transfer a *clear request* packet and a *clear indication* packet specifying the same logical channel. The DCE will consider that the clearing is completed and will not transfer a *DCE clear confirmation* packet.

### 3.1.10   *Unsuccessful call*

If a call cannot be established, the DCE will transfer a *clear indication* packet specifying the logical channel indicated in the *call request* packet.

### 3.1.11   *Call progress signals*

The DCE will be capable of transferring to the DTE *clearing call progress signals* as specified in Recommendation X.96.

*Clearing call progress signals* will be carried in *clear indication* packets which will terminate the call to which the packet refers. The method of coding *clear indication* packets containing call progress signals is detailed in 4. below.

### 3.1.12   *Data transfer phase*

The procedures for the control of packets between DTE and DCE while in the *Data transfer* state are contained in 3.3 below.

### 3.2   *Procedures for permanent virtual circuits*

Annex 1, Figures 16/X.25 and 17/X.25 shows the state diagrams which give a definition of events at the packet level DTE/DCE interface for logical channels assigned for permanent virtual circuits (Annex 2 defines the symbols used in the state diagrams). Annex 3, Tables 11/X.25 and 12/X.25 give details of the action taken by the DCE on receipt of packets in each state shown in Annex 1, Figures 16/X.25 and 17/X.25. Details of the action which should be taken by the DTE are for further study.

For permanent virtual circuits there is no call set-up phase and the logical channel is permanently in the data transfer state (p4).

### 3.3   *Procedures for data and interrupt transfer*

The data transfer procedure described in the following applies independently to each logical channel existing at the DTE/DCE interface.

### 3.3.1   *States for data transfer in virtual calls*

*Data, interrupt, flow control* and *reset* packets may be transmitted and received by a DTE in the *Data transfer* state of a logical channel at the DTE/DCE interface. In this state, the flow control and reset procedures described in 3.4 below apply to data transmission on that logical channel to and from the DTE. *Data, interrupt, flow control* and *reset* packets transmitted by a DTE will be ignored by the DCE when the logical channel is in the *DCE Clear indication* state.

*Data, interrupt, flow control* and *reset* packets may also be received by a DTE when the interface is in the *DTE Clear request* state.

### 3.3.2 *Numbering of data packets*

Each *data* packet transmitted at the DTE/DCE interface for each direction of transmission in a virtual call or permanent virtual circuit is sequentially numbered.

The sequence numbering scheme of the packets is performed modulo 8. The packet sequence numbers cycle through the entire range 0 to 7. As an additional facility, some Administrations will provide a sequence numbering scheme for packets being performed modulo 128. In this case, packet sequence numbers cycle through the entire range 0 to 127.

Only *data* packets contain this sequence number called the *packet send sequence number* P(S).

The first data packet to be transmitted across the DTE/DCE interface for a given direction of data transmission after the virtual call or permanent virtual circuit has been established, has a *packet send sequence number* equal to 0.

### 3.3.3 *Data field length*

In the absence of an optional user facility, the maximum data field length applying to a given DTE is fixed at subscription time and is common to all logical channels at the DTE/DCE interface (however, see 5.1.3 below). The maximum data field length of data packets is 128 octets. Additionally some Administrations may support other maximum data field lengths from the following list of powers of two, i.e. 16, 32, 64, 256, 512 and 1024 octets or exceptionally 255 octets.

The data field of *data* packets transmitted by a DTE may contain any number of bits up to the agreed maximum.

If a DTE wishes to indicate a sequence of more than one packet, it uses a *More data* mark as defined below:

In a full *data* packet a DTE may indicate that more data is to follow with a mark called *More data*. This indication has the effect that such a packet may be combined with the subsequent *data* packet within the network.

Two categories of *data* packets are defined:

Category 1    { *a)* Packets which do not have the local maximum data field length.
                 { *b)* Packets having the local maximum data field length and no *More data* mark.

Category 2    Packets having the local maximum data field length and a *More data* mark.

Category 1 packets will not be combined with subsequent packets.

A complete packet sequence is defined as being composed of either a single category 1 packet or consecutive category 2 packets and a category 1 packet. The sequence shall not be preceded by a category 2 packet.

When transmitted by a source DTE, a complete packet sequence is always delivered to the destination DTE as a complete packet sequence.

*Note.* – A national network which provides a single maximum data field length need not take any action as a result of the *More data* mark.

### 3.3.4 *Data qualifier*

A packet sequence may be on one of two levels. If a DTE wishes to transmit data on more than one level, it uses an indicator called *Data qualifier*.

When only one level of data is being transmitted on a logical channel, the *Data qualifier* is always set to zero. If two levels of data are being transmitted, the *Data qualifier* in all packets of a complete packet sequence are all set to the same value, either zero or one. Within a complete packet sequence, different levels of data may not be intermixed. Packets are numbered consecutively regardless of their data level.

**Provisional Recomendation X.25**

### 3.3.5    *Interrupt procedure*

The interrupt procedure allows a DTE to transmit data to the remote DTE, without following the flow control procedure applying to data packets (see 3.4 below). The interrupt procedure can only apply in the *Flow control ready* state (d1) within the *Data transfer* state (p4).

The interrupt procedure has no effect on the transfer and flow control procedures applying to the *data* packets on the virtual call or permanent virtual circuit.

To transmit an interrupt, a DTE transfers across the DTE/DCE interface a *DTE interrupt* packet. The DCE will then confirm the receipt of the interrupt by transferring a *DCE interrupt confirmation* packet. The DCE will ignore further *DTE interrupt* packets until the first one is confirmed with a *DCE interrupt confirmation* packet.

The DCE indicates an interrupt from the remote DTE by transferring across the DTE/DCE interface a *DCE interrupt* packet containing the same data field as in the *DTE interrupt* packet transmitted by the remote DTE. The DTE will confirm the receipt of the *DCE interrupt* packet by transferring a *DTE interrupt confirmation* packet.

### 3.4    *Procedures for flow control*

The following only applies to the data transfer phase and specifies the procedures covering flow control of *data* packets and reset on each logical channel used for a virtual call or a permanent virtual circuit.

### 3.4.1    *Procedure for flow control*

At the DTE/DCE interface of a logical channel used for a virtual call or permanent virtual circuit, the transmission of *data* packets is controlled separately for each direction and is based on authorizations from the receiver.

#### 3.4.1.1    *Window description*

At the DTE/DCE interface of a logical channel used for a virtual call or permanent virtual circuit and for each direction of data transmission, a window is defined as the ordered set of W consecutive *packet send sequence numbers* P(S) of the *data* packets authorized to cross the interface.

The lowest sequence number in the window is referred to as the lower window edge. When a virtual call or permanent virtual circuit at the DTE/DCE interface has just been established, the window related to each direction of data transmission has a lower window edge equal to 0.

The *packet send sequence number* of the first *data* packet not authorized to cross the interface is the value of the lower window edge plus W (modulo 8, or 128 when extended).

In the absence of an optional user facility, the window size W for each direction of data transmission at a DTE/DCE interface is common to all the logical channels and agreed for a period of time with the Administration for the DTE. The value of W does not exceed 7, or 127 when extended (see 5.1.2 below).

#### 3.4.1.2    *Flow control principles*

A number modulo 8 or 128 when extended, referred to as a *packet receive sequence number* P(R), conveys across the DTE/DCE interface information from the receiver for the transmission of *data* packets. When transmittd across the DTE/DCE interface, a P(R) becomes the lower window edge. In this way, additional *data* packets may be authorized by the receiver to cross the DTE/DCE interface.

When the *sequence number* P(S) of the next *data* packet to be transmitted by the DTE is within the window, the DCE will accept this *data* packet. When the *sequence number* P(S) of the next *data* packet to be transmitted by the DTE is outside of the window, the DCE will consider the receipt of this *data* packet from the DTE as a procedure error and will reset the virtual call or permanent virtual circuit. The DTE should follow the same procedure.

**Provisional Recomendation X.25**

When the *sequence number* P(S) of the next *data* packet to be transmitted by the DCE is within the window, the DCE is authorized to transmit this *data* packet to the DTE. When the *sequence number* P(S) of the next *data* packet to be transmitted by the DCE is outside of the window, the DCE shall not transmit a data packet to the DTE.

The *packet receive sequence number*, P(R), is conveyed in *data*, *Receive ready* (RR) and *Receive not ready* (RNR) packets.

The value of a P(R) received by the DCE must be within the range from the last P(R) received by the DCE to the *packet send sequence number* of the next *data* packet to be transmitted by the DCE. Otherwise, the DCE will consider the receipt of this P(R) as a procedure error and will reset the virtual call or permanent virtual circuit. The DTE should follow the same procedure.

The only universal significance of a P(R) value is a local updating of the window across the packet level interface, but the P(R) value may be used within some Administrations' networks to convey an end-to-end acknowledgement.

### 3.4.1.3    DTE and DCE Receive ready (RR) packets

*RR* packets are used by the DTE or DCE to indicate that it is ready to receive the W *data* packets within the window starting with P(R), where P(R) is indicated in the *RR* packet.

### 3.4.1.4    DTE and DCE Receive not ready (RNR) packets

*RNR* packets are used by the DTE or DCE to indicate a temporary inability to accept additional *data* packets for a given virtual call or permanent virtual circuit. A DTE or DCE receiving an *RNR* packet shall stop transmitting *data* packets on the indicated logical channel.

The receive not ready situation indicated by the transmission of an *RNR* packet is cleared by the transmission in the same direction of an *RR* packet or by a reset procedure being initiated.

### 3.4.2    Procedure for reset

The reset procedure is used to reinitialize the virtual call or permanent virtual circuit and in so doing removes in each direction all *data* and *interrupt* packets which may be in the network. The reset procedure can only apply in the *Data transfer* state of the DTE/DCE interface. In any other state of the DTE/DCE interface, the reset procedure is abandoned.

For flow control, there are three states d1, d2 and d3 within the *Data transfer* state (p4). They are *Flow control ready* (d1), DTE *Reset request* (d2), and *DCE Reset indication* (d3) as shown in the state diagram in Annex 1, Figure 16/X.25. When entering state p4 the logical channel is placed in state d1. Annex 3, Table 11/X.25 specifies actions taken by the DCE on the receipt of packets from the DTE.

When a virtual call or permanent virtual circuit at the DTE/DCE interface has just been reset, the window related to each direction of data transmission has a lower window edge equal to 0, and the numbering of subsequent data packets to cross the DTE/DCE interface for that direction of data transmission shall start from 0.

### 3.4.2.1    Reset request packet

The DTE shall indicate a request for reset by transmitting a *reset request* packet specifying the logical channel. This places the logical channel in the *DTE Reset request* state (d2) (see 3.6 below).

### 3.4.2.2    Reset indication packet

The DCE shall indicate a reset by transmitting to the DTE a *reset indication* packet specifying the logical channel and the reason for the resetting. This places the logical channel in the *DCE Reset indication* state (d3). In this state, the DCE will ignore *data*, *interrupt*, *RR* and *RNR* packets.(see 3.6 below).

### 3.4.2.3    Reset collision

Reset collision can occur when a DTE and a DCE simultaneously transmit a *reset request* packet, and a *reset indication* packet. In such a case, the second one of both those packets crossing the interface is considered as a reset confirmation. This places the logical channel in the *Flow control ready* state (d1).

**Provisional Recomendation X.25**

### 3.4.2.4   Reset confirmation packets

When the logical channel is in the *DTE Reset request* state, the DCE will confirm reset by transmitting to the DTE a *DCE reset confirmation* packet. This places the logical channel in the *Flow control ready* state (d1).

When the logical channel is in the *DCE Reset indication* state, the DTE will confirm reset by transmitting to the DCE a *DTE Reset confirmation* packet. This places the logical channel in the *Flow control* state (d1) (see 3.6 below).

## 3.5   Procedure for restart

The restart procedure is used to simultaneously clear all the virtual calls and reset all the permanent virtual circuits at the DTE/DCE interface.

Annex 1, Figure 17/X.25 gives the state diagram which defines the logical relationships of events related to the restart procedure.

Annex 3, Table 12/X.25 specifies actions taken by the DCE on the receipt of packets from the DTE. Details of the action which should be taken by the DTE are for further study.

### 3.5.1   Restart by the DTE

The DTE may at any time request a restart by transferring across the DTE/DCE interface a *restart request* packet. The interface for each logical channel is then in the *DTE Restart request* state.

The DCE will confirm the restart by transferring a *DCE Restart confirmation* packet placing the logical channels used for virtual calls in the *Ready* state (p1), and the logical channels used for permanent virtual circuits in the *Flow control ready* state (d1) (see 3.6 below).

### 3.5.2   Restart by the DCE

The DCE may indicate a restart by transferring across the DTE/DCE interface a *restart indication* packet. The interface for each logical channel is then in the *DCE Restart indication* state. In this state of the DTE/DCE interface, the DCE will ignore *data, interrupt, call set-up* and *clearing, flow control* and *reset* packets.

The DTE will confirm the restart by transferring a *DTE Restart confirmation* packet placing the logical channels used for virtual calls in the *Ready* state (p1), and the logical channels used for permanent virtual circuits in the *Flow control ready* state (d1) (see 3.6 below).

### 3.5.3   Restart collision

Restart collision can occur when a DTE and a DCE simultaneously transfer a *restart request* and a *restart indication* packet. Under this circumstance, the DCE will consider that the restart is completed and will not expect a *DTE Restart confirmation* packet, neither will it transfer a *DCE Restart confirmation* packet.

## 3.6   List of system parameters

The system parameters applying under 3. are for further study. This study should include considerations of both time-outs and numbers of re-tries.

# 4. PACKET FORMATS FOR VIRTUAL CALL AND PERMANENT VIRTUAL CIRCUIT FACILITIES

## 4.1  *General*

The possible extension of packet formats by the addition of new fields is for further study.

Bits of an octet are numbered 8 to 1 where bit 1 is the low order bit and is transmitted first. Octets of a packet are consecutively numbered starting from 1 and are transmitted in this order.

### 4.1.1  *General format identifier*

The general format identifier field is a four-bit binary coded field which is provided to indicate the general format of the rest of the header. The general format identifier field is located in bit positions 8, 7, 6 and 5 of octet 1, and bit 5 is the low order bit (see Table 5/X.25).

Two of the sixteen possible codes are used to identify the formats for the DTE/DCE interface defined herein, which provide for virtual call and permanent virtual circuit facilities. Two other codes are used to identify the similar formats in the case where the sequence numbering scheme of *data* packets is performed modulo 128. Other codes of the general format identifier are unassigned.

*Note.* — It is envisaged that unassigned codes could identify alternative packet formats associated with other facilities or simplified access procedures, for example, datagram facility or single access DTE procedures.

TABLE 5/X.25 – **General format identifier**

| General format identifier | | Octet 1 Bits | | | |
|---|---|---|---|---|---|
| | | 8 | 7 | 6 | 5 |
| Data packets | Sequence numbering scheme modulo 8 | X | 0 | 0 | 1 |
| | Sequence numbering scheme modulo 128 | X | 0 | 1 | 0 |
| Call set-up and clearing, flow control interrupt, reset and restart packets | Sequence numbering scheme modulo 8 | 0 | 0 | 0 | 1 |
| | Sequence numbering scheme modulo 128 | 0 | 0 | 1 | 0 |

*Note.* — A bit which is indicated as "X" may be set to either 0 or 1.

### 4.1.2  *Logical channel group number*

The logical channel group number appears in every packet except in restart packets (see 4.5 below) in bit positions 4, 3, 2 and 1 of octet 1. This field is binary coded and bit 1 is the low order bit of the logical channel group number.

### 4.1.3  *Logical channel number*

The logical channel number appears in every packet except in *restart* packets (see 4.5 below) in all bit positions of octet 2. This field is binary coded and bit 1 is the low order bit of the logical channel number.

**Provisional  Recomendation  X.25**

### 4.1.4 *Packet type identifier*

Each packet shall be identified in the octet 3 of the packet according to Table 6/X.25.

TABLE 6/X.25 -- **Packet type identifier**

| Packet type | | Octet 3 Bits | | | | | | | |
| From DCE to DTE | From DTE to DCE | 8 | 7 | 6 | 5 | 4 | 3 | 2 | 1 |
|---|---|---|---|---|---|---|---|---|---|
| *Call set-up and clearing* | | | | | | | | | |
| Incoming call | Call request | 0 | 0 | 0 | 0 | 1 | 0 | 1 | 1 |
| Call connected | Call accepted | 0 | 0 | 0 | 0 | 1 | 1 | 1 | 1 |
| Clear indication | Clear request | 0 | 0 | 0 | 1 | 0 | 0 | 1 | 1 |
| DCE clear confirmation | DTE clear confirmation | 0 | 0 | 0 | 1 | 0 | 1 | 1 | 1 |
| *Data and interrupt* | | | | | | | | | |
| DCE Data | DTE Data | X | X | X | X | X | X | X | 0 |
| DCE Interrupt | DTE Interrupt | 0 | 0 | 1 | 0 | 0 | 0 | 1 | 1 |
| DCE Interrupt confirmation | DTE Interrupt confirmation | 0 | 0 | 1 | 0 | 0 | 1 | 1 | 1 |
| *Flow control and reset* | | | | | | | | | |
| DCE RR | DTE RR | X | X | X | 0 | 0 | 0 | 0 | 1 |
| DCE RNR | DTE RNR | X | X | X | 0 | 0 | 1 | 0 | 1 |
| | DTE REJ | X | X | X | 0 | 1 | 0 | 0 | 1 |
| Reset indication | Reset request | 0 | 0 | 0 | 1 | 1 | 0 | 1 | 1 |
| DCE Reset confirmation | DTE Reset confirmation | 0 | 0 | 0 | 1 | 1 | 1 | 1 | 1 |
| *Restart* | | | | | | | | | |
| Restart indication | Restart request | 1 | 1 | 1 | 1 | 1 | 0 | 1 | 1 |
| DCE Restart confirmation | DTE Restart confirmation | 1 | 1 | 1 | 1 | 1 | 1 | 1 | 1 |

*Note.* -- A bit which is indicated as "X" may be set to either 0 or 1.

### 4.2 *Call set-up and clearing packets*

### 4.2.1 *Call request and incoming call packets*

Figure 1/X.25 illustrates the format of *call request* and *incoming call* packets.

*Address lengths field*

Octet 4 consists of field length indicators for the called and calling DTE addresses. Bits 4, 3, 2 and 1 indicate the length of the called DTE address in semi-octets. Bits 8, 7, 6 and 5 indicate the length of the calling DTE address in semi-octets. Each address length indicator is binary coded and bit 1 or 5 is the low order bit of the indicator.

**Provisional Recomendation X.25**

*Address field*

Octet 5 and the following octets consist of the called DTE address when present, then the calling DTE address when present.

Each digit of an address is coded in a semi-octet in binary coded decimal with bit 5 or 1 being the low order bit of the digit.

Starting from the high order digit, the address is coded in octet 5 and consecutive octets with two digits per octet. In each octet, the higher order digit is coded in bits 8, 7, 6 and 5.

The address field shall be rounded up to an integral number of octets by inserting zeros in bits 4, 3, 2 and 1 of the last octet of the field when necessary.

*Note.* — This field may be used for optional addressing facilities such as abbreviated addressing. The optional addressing facilities employed as well as the coding of those facilities is for further study.

*Facility length field*

Bits 6, 5, 4, 3, 2 and 1 of the octet following the address field indicate the length of the facility field in octets. The facility length indicator is binary coded and bit 1 is the low order bit of the indicator.

Bits 8 and 7 of this octet are unassigned and set to 0.

*Facility field*

The facility field is present only when the DTE is using an optional user facility requiring some indication in the *call request* and *incoming call* packets.

The coding of this facility field is defined in 5. below.

The facility field contains an integral number of octets. The actual maximum length of this field depends on the facilities which are offered by the network. However, this maximum does not exceed 62 octets.

*Call user data field*

Following the facility field, user data may be present. The call user data field has a maximum length of 16 octets. The contents of the field are passed unchanged to the called DTE.

4.2.2    *Call accepted and Call connected packets*

Figure 2/X.25 illustrates the format of *call accepted* and *call connected* packets.

4.2.3    *Clear request and Clear indication packets*

Figure 3/X.25 illustrates the format of *clear request* and *clear indication* packets.

*Clearing cause field*

Octet 4 is the clearing cause field and contains the reason for the clearing of the call.

The coding of the clearing cause field in *clear indication* packets is given in Table 7/X.25.

The bits of the cause field in *clear request* packets are set to 0.

**Provisional Recomendation X.25**

TABLE 7/X.25 – Coding of clearing cause field in clear indication packet

| | Bits | | | | | | | |
|---|---|---|---|---|---|---|---|---|
| | 8 | 7 | 6 | 5 | 4 | 3 | 2 | 1 |
| DTE Clearing | 0 | 0 | 0 | 0 | 0 | 0 | 0 | 0 |
| Number busy | 0 | 0 | 0 | 0 | 0 | 0 | 0 | 1 |
| Out of order | 0 | 0 | 0 | 0 | 1 | 0 | 0 | 1 |
| Remote procedure error | 0 | 0 | 0 | 1 | 0 | 0 | 0 | 1 |
| Number refuses reverse charging | 0 | 0 | 0 | 1 | 1 | 0 | 0 | 1 |
| Invalid call | 0 | 0 | 0 | 0 | 0 | 0 | 1 | 1 |
| Access barred | 0 | 0 | 0 | 0 | 1 | 0 | 1 | 1 |
| Local procedure error | 0 | 0 | 0 | 1 | 0 | 0 | 1 | 1 |
| Network congestion | 0 | 0 | 0 | 0 | 0 | 1 | 0 | 1 |
| Not obtainable | 0 | 0 | 0 | 0 | 1 | 1 | 0 | 1 |

### 4.2.4  DTE and DCE Clear confirmation packets

Figure 4/X.25 illustrates the format of the *DTE* and *DCE Clear confirmation* packets.

### 4.3  Data and Interrupt packets

### 4.3.1  DTE and DCE data packets

Figure 5/X.25 illustrates the format of the *DTE* and *DCE data* packets.

*Note.* – In the event that the sequence numbering scheme is performed modulo 128 (see 3.3.1.2 above), this format is effected.

#### Data qualifier

Bit 8 in octet 1 is used for *Data qualifier*.

#### Packet receive sequence number

Bits 8, 7 and 6 of octet 3 are used for indicating the *packet receive sequence number* P(R). P(R) is binary coded and bit 6 is the low order bit.

#### More data indication

Bit 5 in octet 3 is used for *More data* indication: 0 for no more data and 1 for more data.

#### Packet send sequence number

Bits 4, 3 and 2 of octet 3 are used for indicating the *packet send sequence number* P(S). P(S) is binary coded and bit 2 is the low order bit.

#### User data field

Bits following octet 3 contain user data.

270

### 4.3.2 *DTE and DCE Interrupt packets*

Figure 6/X.25 illustrates the format of the *DTE* and *DCE interrupt* packets.

*Interrupt user data field*

Octet 4 contains user data.

### 4.3.3 *DTE and DCE interrupt confirmation packets*

Figure 7/X.25 illustrates the format of the *DTE* and *DCE interrupt confirmation* packet.

### 4.4 *Flow control and reset packets*

### 4.4.1 *DTE and DCE receive ready (RR) packets*

Figure 8/X.25 illustrates the format of the *DTE* and *DCE RR* packets.

*Note.* – In the event that the sequence numbering scheme is performed modulo 128 (see 3.3.1.2 above), this format is effected.

*Packet receive sequence number*

Bits 8, 7 and 6 of octet 3 are used for indicating the *packet receive sequence number* P(R). P(R) is binary coded and bit 6 is the low order bit.

### 4.4.2 *DTE and DCE receive not ready (RNR) packets*

Figure 9/X.25 illustrates the format of the *DTE* and *DCE RNR* packets.

*Note.* – In the event that the sequence numbering scheme is performed modulo 128 (see 3.3.1.2 above), this format is effected.

*Packet receive sequence number*

Bits 8, 7 and 6 of the octet 3 are used for indicating the *packet receive sequence number* P(R). P(R) is binary coded and bit 6 is the low order bit.

### 4.4.3 *Reset request and reset indication packets*

Figure 10/X.25 illustrates the format of the *reset request* and *reset indication* packets.

*Resetting cause field*

Octet 4 is the resetting cause field and contains the reason for the reset. The coding of the resetting cause field in a *reset indication* packet is given in Table 8/X.25.

*Diagnostic code*

Octet 5, is the diagnostic code and contains additional information on the reason for the reset when the cause field indicates a local procedure error.

The bits of the diagnostic code field in a *reset request* packet are unassigned and set to 0.

The bits of the diagnostic code field in a *reset indication* packet are all set to 0 when no specific reason for the reset is supplied. Other values are not specified at this time.

TABLE 8/X.25 – Coding of resetting cause field in reset indication packet

| | Bits | | | | | | | |
|---|---|---|---|---|---|---|---|---|
| | 8 | 7 | 6 | 5 | 4 | 3 | 2 | 1 |
| DTE Reset | 0 | 0 | 0 | 0 | 0 | 0 | 0 | 0 |
| Out of order | 0 | 0 | 0 | 0 | 0 | 0 | 0 | 1 |
| Remote procedure error | 0 | 0 | 0 | 0 | 0 | 0 | 1 | 1 |
| Local procedure error | 0 | 0 | 0 | 0 | 0 | 1 | 0 | 1 |
| Network congestion | 0 | 0 | 0 | 0 | 0 | 1 | 1 | 1 |

*Note.* – The bits of the resetting cause field in a *reset request* packet are set to 0.

### 4.4.4    DTE and DCE Reset confirmation packets

Figure 11/X.25 illustrates the format of the *DTE* and *DCE reset confirmation* packets.

### 4.5    Restart packets

### 4.5.1    Restart request and restart indication packets

Figure 12/X.25 illustrates the format of the *restart request* and *restart indication* packets. Bits 4, 3, 2 and 1 of the first octet and all bits of the second octet are set to 0.

*Restarting cause field*

Octet 4 is the restarting cause field and contains the reason for the restart.

The coding of the restarting cause field in the *restart indication* packets is given in Table 9/X.25.

The bits of the restarting cause field in *restart request* packets are set to 0.

TABLE 9/X.25 – Coding of restarting cause field in restart indication packets

| | Bits | | | | | | | |
|---|---|---|---|---|---|---|---|---|
| | 8 | 7 | 6 | 5 | 4 | 3 | 2 | 1 |
| Local procedure error | 0 | 0 | 0 | 0 | 0 | 0 | 0 | 1 |
| Network congestion | 0 | 0 | 0 | 0 | 0 | 0 | 1 | 1 |

### 4.5.2    DTE and DCE restart confirmation packets

Figure 13/X.25 illustrates the format of the *DTE* and *DCE restart confirmation* packets. Bits 4, 3, 2 and 1 of the first octet and all bits of the second octet are set to 0.

**Provisional  Recomendation  X.25**

## 4.6   *Packets required for optional user facilities*

### 4.6.1   *DTE reject (REJ) packets*

Figure 14/X.25 illustrates the format of the *DTE REJ* packets, used in conjunction with the Packet Retransmission Facility described in 5. below.

*Note.* – In the event that the sequence numbering scheme is performed modulo 128 (see 3.3.1.2 above), this format is effected.

*Packet receive sequence number*

Bits 8, 7 and 6 of octet 3 are used for indicating the *packet receive sequence number* P(R). P(R) is binary coded and bit 6 is the low order bit.

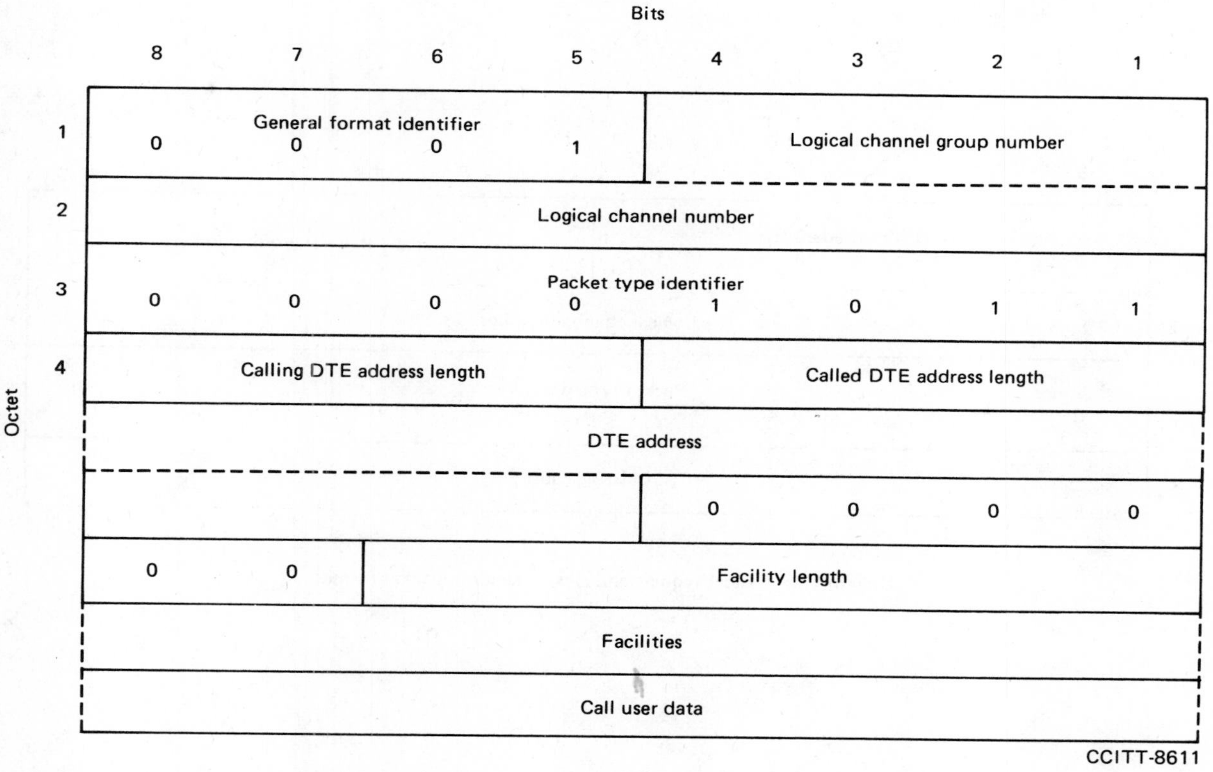

FIGURE 1/X.25 – **Call request and incoming call packet format**

*Note.* – The Figure assumes that a single address is present, consisting of an odd number of digits, and that the call user data field contains an integral number of octets.

Bits

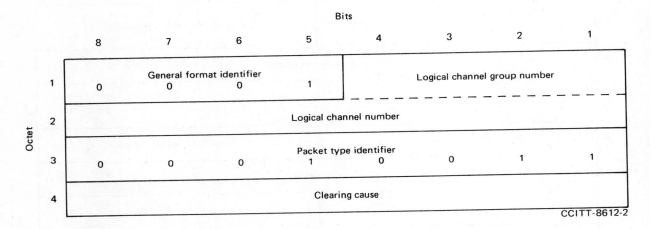

FIGURE 2/X.25 – Call accepted and call connected packet format

FIGURE 3/X.25  Clear request and clear indication packet format

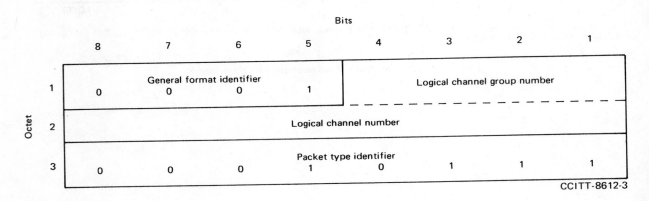

FIGURE 4/X.25 – DTE and DCE clear confirmation packet format

**Provisional Recomendation X.25**

40

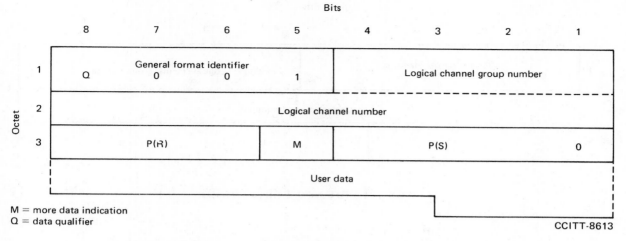

M = more data indication
Q = data qualifier

CCITT-8613

*Note.* — The Figure assumes that the user data field does not contain an integral number of octets.

FIGURE 5/X.25 — **DTE and DCE data packet format**

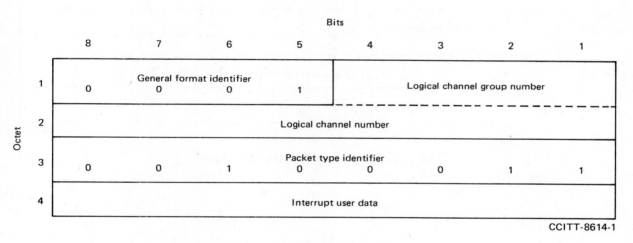

CCITT-8614-1

FIGURE 6/X.25 — **DTE and DCE interrupt packet format**

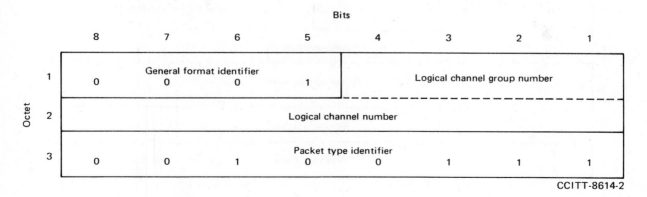

CCITT-8614-2

FIGURE 7/X.25 — **DTE and DCE interrupt confirmation packet format**

**Provisional Recomendation X.25**

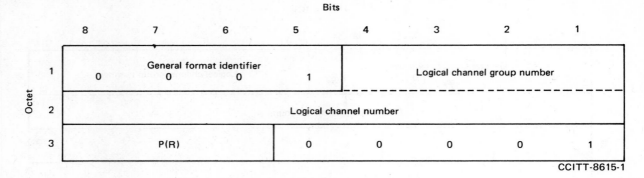

FIGURE 8/X.25 — DTE and DCE RR packet format

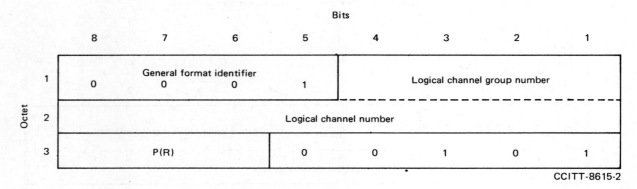

FIGURE 9/X.25 — DTE and DCE RNR packet format

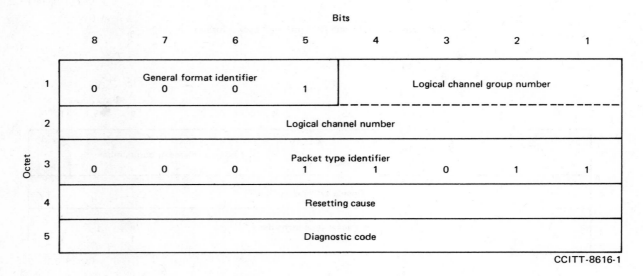

FIGURE 10/X.25 — Reset request and reset indication packet format

**Provisional Recomendation X.25**

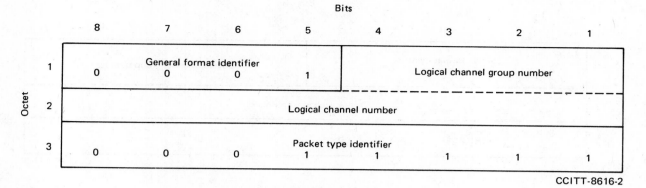

FIGURE 11/X.25 – **DTE and DCE reset confirmation packet format**

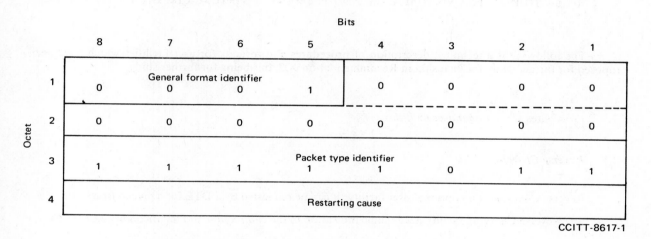

FIGURE 12/X.25 – **Restart request and restart indication packet format**

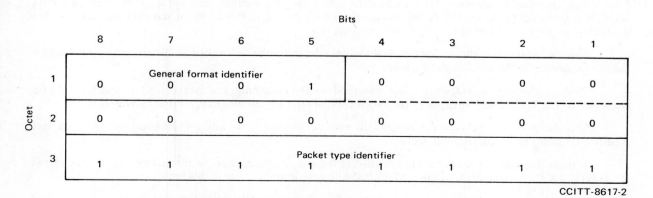

FIGURE 13/X.25 **DTE and DCE restart confirmation packet format**

**Provisional Recomendation X.25**

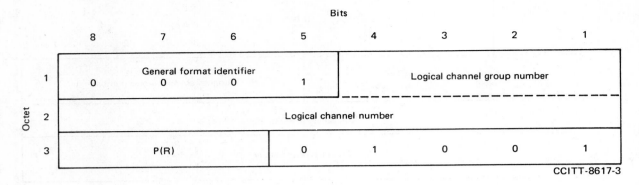

FIGURE 14/X.25 – **DTE REJ packet format**

# 5. PROCEDURES AND FORMATS FOR OPTIONAL USER FACILITIES
## TO BE STUDIED FOR VIRTUAL CALL AND PERMANENT VIRTUAL CIRCUIT FACILITIES

The following is a technical description of procedures and formats for user facilities which have been proposed for further study for inclusion in Recommendation X.2, this being for further study.

### 5.1 *Procedures for optional user facilities*

### 5.1.1 *Reverse Charging*

Reverse Charging is an optional user facility; it can be requested by a DTE for a given virtual call.

The Reverse Charging facility needs some indication in the *call request* and the *incoming call* packets.

### 5.1.2 *Flow Control Parameters Selection*

Flow Control Parameters Selection is an optional user facility agreed for a period of time; it can be used by a DTE for virtual calls.

This user facility allows a DTE to operate with a specific window size and/or maximum data field length depending on the throughput class requested for each direction, that is, the actual data transfer rate that the DTE does not have the need to exceed for this virtual call.

Specific throughput classes for a virtual call may be indicated to a called DTE using the optional user facility parameters in the *incoming call* packet.

If those throughput classes are not specified in the *incoming call* packet, the throughput classes considered at the DTE/DCE interface of the called DTE are the highest attainable one at this interface.

Specific throughput classes for virtual call may be selected by a calling DTE using the optional user facility parameters in the *call request* packet.

If those throughput classes are not specified in the *call request* packet, the throughput classes considered at the DTE/DCE interface of the calling DTE are the highest attainable one at this interface.

For each throughput class, a correspondence agreed with the Administration, for the DTE, for a period of time specifies either the window size, the maximum data field length, or both.

**Provisional Recomendation X.25**

### 5.1.3  *Reverse Charging Acceptance*

Reverse Charging Acceptance is an optional user facility agreed for a period of time.

This user facility, if subscribed to, authorizes the DCE to transmit to the DTE incoming calls which request the Reverse Charging facility. In the absence of this facility, the DCE will not transmit to the DTE incoming calls which request the Reverse Charging facility.

### 5.1.4  *One-way Logical Channel*

One-way Logical Channel is an optional user facility agreed for a period of time.

This is a user facility which limits the use of a logical channel to either incoming or outgoing calls (incoming or outgoing access on logical channels).

### 5.1.5  *Packet Retransmission*

Packet Retransmission is an optional user facility agreed for a period of time.

This user facility allows a DTE to request retransmission of one or several consecutive *data* packets from the DCE (up to the window size) by transferring across the DTE/DCE interface a *DTE reject* packet specifying a logical channel number and a *sequence number* P(R).

When receiving a *DTE reject* packet, the DCE initiates on the specified logical channel retransmission of the *data* packets whose *packet send sequence numbers* start from P(R), where P(R) is indicated in the *DTE reject* packet. Until the DCE transfers across the DTE/DCE interface a *DCE data* packet with a *packet send sequence number* equal to the P(R) indicated in the *DTE reject* packet, the DCE will consider the receipt of another *DTE reject* packet as a procedure error and reset the virtual call or permanent virtual circuit.

Additional *data* packets pending initial transmission may follow the retransmitted *data* packet(s).

A DTE receive not ready situation indicated by the transmission of *RNR* packet is cleared by the transmission of a *DTE reject* packet.

The conditions under which the DCE ignores a *DTE reject* packet, or considers it as a procedure error, are those described for flow control packets in 2. and 3. above.

### 5.2  *Formats for Optional User Facilities*

### 5.2.1  *General*

The facility field is present only when a DTE is using an optional user facility requiring some indication in the call request packet or the incoming call packet.

The facility field contains a sequence of octet pairs. The first octet of each is a facility code field to indicate the facility or facilities requested and the second octet is the facility parameter field.

If any given user facility requires more than a single octet parameter field, several facility codes will be assigned to it. In the facility field, the octet pairs related to the same user facility are placed consecutively in the order specified for the parameters.

The facility code is binary coded. The coding of the facility parameter field is dependent on the facility being requested.

**Provisional Recomendation X.25**

### 5.2.2 *Coding of facility field for particular facilities*

#### 5.2.2.1 *Coding of Reverse Charging facility*

The coding of facility code field and parameters for Reverse Charging is the same in code request and incoming call packets.

*Facility code field*

The coding of the facility code field for Reverse Charging is:

| | |
|---|---|
| bit: | 8 7 6 5 4 3 2 1 |
| code: | 0 0 0 0 0 0 0 1 |

*Facility parameter field*

The coding of the facility parameter field is as follows:

bit 1 = 0 for Reverse Charging not requested

bit 1 = 1 for Reverse Charging requested

*Note.* — Bits 8, 7, 6, 5, 4, 3 and 2 could be used for user facilities other than for Reverse Charging; if not, they are set to 0.

#### 5.2.2.2 *Coding of Flow Control Parameters Selection facility*

The inclusion of facility code and parameter fields for Flow Control Parameters Selection is optional.

The coding in incoming call packets and in call request packets is the same.

*Facility code field*

The facility code field for Flow Control Parameters Selection is coded:

| | |
|---|---|
| bit: | 8 7 6 5 4 3 2 1 |
| code: | 0 0 0 0 0 0 1 0 |

*Facility parameter field*

The throughput class for transmission from the calling DTE is indicated in bits 4, 3, 2, and 1. The throughput class for transmission from the called DTE is indicated in bits 8, 7, 6 and 5.

The four bits indicating each throughput class are binary coded and express the logarithm base 2 of the number of octets per second defining the throughput class. Bit 1 or 5 is the low order bit of each throughput class indicator.

46

ANNEX 1
(to Recommendation X.25)

**Packet level DTE/DCE interface state diagram for a logical channel**

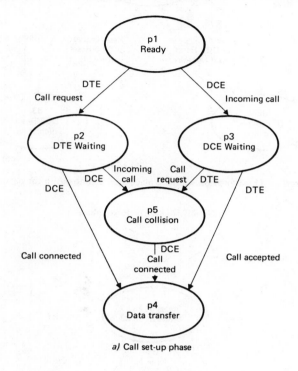

*a)* Call set-up phase

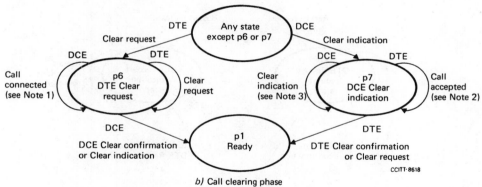

*b)* Call clearing phase

CCITT-8618

*Note 1.* – This transition is possible only if the previous state was *DTE Waiting* (p2).
*Note 2.* – This transition is possible only if the previous state was *DCE Waiting* (p3).
*Note 3.* – This transition will take place after a time-out in the network.

FIGURE 15/X.25

281

**Provisional Recomendation X.25**

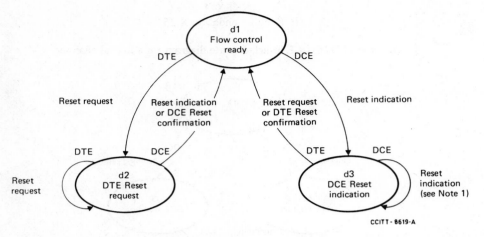

*Note 1.* – This transition will take place after a time-out in the network.

FIGURE 16/X.25 – **Reset phase**

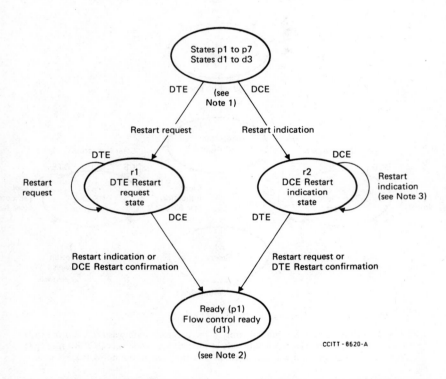

*Note 1.* – States p1 to p7 for virtual calls or states d1 to d3 for permanent virtual circuits.
*Note 2.* – State p1 for virtual calls or state d1 for permanent virtual circuits.
*Note 3.* – This transition will take place after a time-out in the network.

FIGURE 17/X.25 – **Restart phase**

# ANNEX 2
## (to Recommendation X.25)

### Symbol definition of the state diagrams

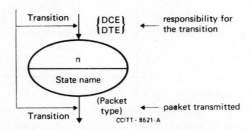

*Note 1.* – Each state is represented by an ellipse wherein the state name and number are indicated.

*Note 2.* – Each state transition is represented by an arrow. The responsibility for the transition (DTE or DCE) and the packet it has successfully transmitted are indicated beside that arrow.

# ANNEX 3
## (to Recommendation X.25)

**TABLE 10/X.25 – Action taken by the DCE on receipt of packets in a given state of the packet level DTE/DCE interface: call set-up and clearing**

| Packet from the DTE \ State of the interface | Ready p1 | DTE Waiting p2 | DCE Waiting p3 | Data transfer p4 | Call collision p5 | DTE Clear request p6 | DCE Clear indication p7 |
|---|---|---|---|---|---|---|---|
| Call request | NORMAL | ERROR | NORMAL | ERROR | ERROR | ERROR | ERROR |
| Call accepted | ERROR | ERROR | NORMAL | ERROR | ERROR | ERROR | NORMAL |
| Clear request | NORMAL | NORMAL | NORMAL | NORMAL | NORMAL | NORMAL | NORMAL |
| DTE Clear confirmation | ERROR | ERROR | ERROR | ERROR | ERROR | ERROR | NORMAL |
| Data, interrupt, reset or flow control | ERROR | ERROR | ERROR | See Table 11/X.25 | ERROR | ERROR | NORMAL |

NORMAL: the action taken by the DCE follows the normal procedures as defined in 3. of the text.

ERROR: the DCE indicates a clearing by transmitting to the DTE a *clear indication* packet, with an indication of Local Procedure Error. If connected through the virtual call, the distant DTE is also informed of the clearing by a *clear indication* packet, with an indication of Remote Procedure Error.

TABLE 11/X.25 — Action taken by the DCE on receipt of packets in a given state
of the packet level DTE/DCE interface: flow control and data transfer

| State of the interface / Packet from the DTE | Data transfer (p4) | | |
|---|---|---|---|
| | Flow control ready (d1) | DTE Reset request (d2) | DCE Reset indication (d3) |
| Reset request | NORMAL | NORMAL | NORMAL |
| DTE Reset confirmation | FLOW CONTROL ERROR | FLOW CONTROL ERROR | NORMAL |
| Data, interrupt or flow control | NORMAL | FLOW CONTROL ERROR | NORMAL |

NORMAL:   the action taken by the DCE follows the normal procedures as defined in 3. of the text.

FLOW CONTROL ERROR:   the DCE indicates a reset by transmitting to the DTE a *reset indication* packet, with an indication of Local Procedure Error. The distant DTE is also informed of the reset by a *reset indication* packet, with an indication of Remote Procedure Error.

TABLE 12/X.25 — Action taken by the DCE on receipt of packets in a given state
of the packet level DTE/DCE interface: restart

| State of the interface / Packet from the DTE | Any state p1 to p7 and d1 to d3 | DTE restart request state | DCE restart indication state |
|---|---|---|---|
| Restart request | NORMAL | NORMAL | NORMAL |
| DTE restart confirmation | ERROR | ERROR | NORMAL |
| Data, interrupt, call set-up and clearing, flow control or reset | See Note | ERROR | NORMAL |

NORMAL:   the action taken by the DCE follows the normal procedures as defined in 3. of the text.

ERROR:   the DCE indicates a restarting by transmitting to the DTE a *restart indication* packet with an indication of Local Procedure Error.

*Note.* — See Table 10/X.25 for call set-up and clearing. See Table 11/X.25 for data, interrupt, flow control and reset.

**Provisional  Recomendation  X.25**

**Recommendation X.26**

## ELECTRICAL CHARACTERISTICS FOR UNBALANCED DOUBLE-CURRENT INTERCHANGE CIRCUITS FOR GENERAL USE WITH INTEGRATED CIRCUIT EQUIPMENT IN THE FIELD OF DATA COMMUNICATIONS

(For the text of this Recommendation, see Recommendation V.10 in Volume VIII.1 for which Study Group XVII is responsible.)

**Recommendation X.27**

## ELECTRICAL CHARACTERISTICS FOR BALANCED DOUBLE-CURRENT INTERCHANGE CIRCUITS FOR GENERAL USE WITH INTEGRATED CIRCUIT EQUIPMENT IN THE FIELD OF DATA COMMUNICATIONS

(For the text of this Recommendation, see Recommendation V.11 in Volume VIII.1 for which Study Group XVII is responsible.)

**Provisional Recommendation X.28**   *(Geneva, 1977)*

## DTE/DCE INTERFACE FOR A START-STOP MODE DATA TERMINAL EQUIPMENT ACCESSING THE PACKET ASSEMBLY/DISASSEMBLY FACILITY (PAD) IN A PUBLIC DATA NETWORK SITUATED IN THE SAME COUNTRY

**Contents**

Preface

The establishment in various countries of public data networks providing packet-switched data transmission services creates a need to produce standards to facilitate access from the public telephone network, circuit- switched public data networks and leased circuits.

*Preface*

The CCITT,

*considering*

a)   that Recommendations X.1 and X.2 define user classes of service and user facilities provided by a public data network, Recommendation X.95 defines network parameters and Recommendation X.96 defines call progress signals,

b)   that Provisional Recommendation X.29 defines procedures for a packet mode DTE to control the PAD,

c)   that the logical control links for packet-switched data transmission services are defined in Recommendation X.92,

d)   the need for defining an international recommendation for the exchange of control information between a start-stop mode DTE and a PAD,

e)   that DTEs operating in the start-stop mode will send and receive network call control information and user information in the form of characters according to Recommendation V.3,

f)   that the necessary elements for an interface recommendation should be defined independently as:

1.      Procedures for the establishment of a national access information path between a start-stop mode DTE and a PAD.

2.      Procedures for character interchange and service initialization between a start-stop mode DTE and a PAD.

3.      Procedures for the exchange of control information between a start-stop mode DTE and a PAD.

4.      Procedures for the exchange of user data between a start-stop mode DTE and a PAD,

**Provisional Recommendation X.28**

*unanimously declares the view*

that start-stop mode DTE accessing the PAD should operate in accordance with this Recommendation.

1. *Procedures for the establishment of a national access information path between a start-stop mode DTE and a PAD*

1.1 *Access via a public switched telephone network or leased lines with V-Series interfaces*

1.1.1 *DTE/DCE interface*

The access information path will be provided by the use of modems standardized for use in the public switched telephone network or leased line operating at rates up to 300 bit/s (nominally at 200 bit/s), in accordance with Recommendation V.21.

The particular interchange circuits provided, and their operation, shall be in accordance with 8. of Recommendation V.21 and clamping of circuit 104 shall be implemented in accordance with 8.e)ii) of Recommendation V.21. The modem shall be set up for channel operation in accordance with 7.a) of Recommendation V.21.

*Note.* — The interface requirements for data signalling rates greater than 300 bit/s are for further study.

1.1.2 *Electrical characteristics*

The electrical characteristics of the DTE/DCE interface shall be in accordance with Recommendation V.28.

1.1.3 *Procedure for setting up and disconnecting the access information path*

1.1.3.1 *Setting up the access information path by the DTE*

The access information path shall be established in accordance with 6. of Recommendation V.25 for a manual data station calling an automatic answering station.

The mechanism for echo suppressor disablement may not be implemented in some national networks where the access information path does not include echo suppressors.

Subsequent to the completion of the above, both the DTE and the DCE shall transmit binary 1 on circuits 103 and 104.

1.1.3.2 *Disconnecting the access information path by the DTE*

The access information path shall be disconnected by:

i) reversion of the data circuit to the voice mode, or

ii) the DTE turning circuit 108/1 or 108/2 OFF for a period greater than Z. The value of Z is for further study.

1.1.3.3 *Setting up the access information path by the PAD*

The procedure for the PAD to establish an access information path is for further study.

1.1.3.4 *Disconnecting the access information path by the PAD*

Disconnection by the PAD will be indicated by the DCE turning circuit 106 and 109 OFF, while circuit 108 is ON.

*Note.* — Access information path clear indication to the DTE is not signalled by circuit 107 OFF. Not all DTEs allow circuit 107 to be turned OFF f circuit 108 has not been turned OFF previously.

**Provisional Recommendation X.28**

**1.2**    *Access via a public switched data network or via leased lines with X-Series interfaces*

**1.2.1**    *DTE/DCE interface designed for start-stop transmission services on public data networks (Recommendation X.20)*

### 1.2.1.1    *Physical characteristics*

The physical characteristics of the DTE/DCE interface are defined in 2. of Recommendation X.20.

### 1.2.1.2    *Procedures for setting up and disconnecting the access information path (call control)*

The procedures and formats for call control of the public switched data network are described in 3. and 4. of Recommendation X.20.

**1.2.2**    *DTE/DCE interface designed for operation on telephone type networks (Recommendation X.20bis)*

In the case of DTEs with interfaces designed for operation on telephone type networks (V-Series interfaces), the access information path will be established by the use of DCEs standardized for start-stop transmission services on public data networks according to Recommendation X.20*bis* (V.21-compatible interface).

### 1.2.2.1    *Characteristics*

The characteristics of the interchange circuits are described in 2. of Recommendation X.20*bis*.

### 1.2.2.2    *Operational requirements*

The requirements for the operation of the interchange circuits 106, 107, 108, 109 and 125 are described in 3. of Recommendation X.20*bis*.

### 1.2.2.3    *Operational requirements for disconnecting the access information path by the DTE*

The access information path shall be disconnected either

*manually* by depressing the clearing key of the DCE

or

*automatically* by the DTE turning OFF circuit 108/1 or 108/2 for a period longer than 210 ms.

### 1.2.2.4    *Indication of disconnection by the PAD*

Disconnection by the PAD, i.e. DCE clearing, will be indicated by the DCE by turning OFF circuits 106 and 109. The DTE should then perform clear confirmation by turning OFF circuit 108.

### 1.2.2.5    *Setting up the access information path by the PAD*

The procedure for the PAD to establish an access information path is for further study.

### 1.2.2.6    *Operational constraints for maintaining the access information path during information transfer*

Some DTEs are equipped with a break key to allow signalling without loss of character transparency. The user should be advised that the transmission of a *break signal* longer than 200 ms may cause clearing in a public switched data network. Therefore, the transmission of a *break signal* in either direction should either be avoided or the timer of the circuit generating a *break signal* should be adjusted to a signal length being considerably shorter than 200 ms.

**2.**    *Procedures for character interchange and service initialization between a start-stop mode DTE and a PAD*

**2.1**    *Format of characters interchanged*

**2.1.1**    The start-stop mode DTE shall generate and be capable of receiving characters in accordance with International Alphabet No. 5 as described in Recommendation V.3. The general structure of characters shall be in accordance with Recommendation X.4.

**Provisional Recommendation X.28**

**2.1.2**   For an interim period, PADs provided by some Administrations act as described below.

The PAD will transmit and expect to receive 8-bit characters.

Whenever the PAD has to interpret a received character for a specific action different from or additional to the transfer of this data character to the packet-mode DTE, it will only inspect the first seven bits and will ignore the eighth bit (the last bit preceding the stop element). Whenever the PAD generates characters (e.g. service signals), they will be transmitted by the PAD with even parity.

The PAD will accept characters which have a single stop element and will transmit characters with at least two stop elements if the start-stop mode DTE is operating at 110 bit/s. If the DTE is operating at 200 bit/s or 300 bit/s the PAD will transmit and accept characters with a single unit stop element.

## 2.2   *Procedures for initialization*

The references to states in the following procedures correspond to the state diagrams, see Figures 1/X.28 and 2/X.28.

### 2.2.1   *Active link (state 1)*

After the access information path has been established, the start-stop mode DTE and PAD exchange binary 1 signals across the start-stop mode DTE/DCE interface and the interface is in the *active link state*.

### 2.2.2   *Service request (state 2)*

If the interface is in the *active link state,* the DTE shall transmit a sequence of characters to indicate *service request* and to initialize the PAD. The *service request signal* enables the PAD to detect the data rate and code used by the DTE and to select the *initial standard profile* of the PAD. The parameters of *initial standard profiles* are summarized in Table 1/X.28.

The format of the *service request signal* to be transmitted by the DTE is given in 3.5.18 below.

The information content of the *service request signal* is a subject for further study.

### 2.2.3   *DTE waiting (state 3A)*

Following the transmission of the *service request signal* the DTE shall transmit binary 1 and the interface will be in the *DTE waiting state*.

### 2.2.4   *Service ready (state 4)*

After receiving a *service request signal* the PAD will transmit a *PAD identification PAD service signal* at the data rate of the DTE, unless the value of parameter 6 is 0.

When the value of parameter 6 is 0, the interface will directly enter the *PAD waiting state* following receipt of a valid *service request signal*.

### 2.2.5   *Fault condition*

If a valid *service request signal* is not received by the PAD within Y seconds after the transmission of binary 1, it will perform PAD clearing by disconnecting the access information path.

The value of Y is for further study.

*Note.* — Some networks may allow states 2 to 4 to be bypassed. In these cases the condition described under 2.2.5 does not apply.

3.    *Procedures for the exchange of control information between a start-stop mode DTE and a PAD*

3.1    *General*

3.1.1    *PAD command signals and PAD service signals*

The operation of the PAD depends on the current values of internal PAD variables which are known as PAD parameters. Initially PAD parameter values depend on the profile selected at the time of sending a *service request signal*. The values of *initial standard profiles* are summarized in Table 1/X.28.

*PAD command signals* (direction DTE to PAD) are provided for:

a)    the establishment and clearing of a virtual call (see 3.2 below);

b)    the selection of a set of preset values of PAD parameters known as a *standard profile* (see 3.3.1 below);

c)    the selection of individual PAD parameter values (see 3.3.2 below);

d)    requesting the current values of PAD parameters to be transmitted by the PAD to the DTE (see 3.4 below).

*PAD service signals* (direction PAD to DTE) are provided to:

a)    transmit *call progress signals* to the calling DTE;

b)    acknowledge *PAD command signals*;

c)    transmit information regarding the operation of the PAD to the start-stop mode DTE.

The formats of *PAD command signals* and *PAD service signals* are given in 3.5 below.

The information content of *PAD command signals* and *PAD service signals* are summarized in the Annex.

3.1.2    *Break signal*

The *break signal* is provided to allow the start-stop mode DTE to signal to the PAD without loss of character transparency. The *break signal* can also be transmitted by the PAD to the start-stop mode DTE.

The *break signal* is defined as the transmission of binary 0 for more than 150 ms. The maximum permitted duration shall depend upon the type of access information path used (see, for example, 1.2.2.6 above).

A *break signal* shall be separated from any following start-stop character or other *break signal* by the transmission of binary 1 for greater than 100 ms.

3.2    *Procedures for virtual call control*

Figure 1/X.28 (Sequence of events at the interface) shows the procedures at the DCE/DTE interface during call establishment, data transfer and call clearing. Figure 2/X.28 shows the state diagram.

3.2.1    *Call establishment*

3.2.1.1    *Pad waiting (state 5)*

Following the transmission of a *PAD service signal* the interface will be in the *PAD waiting state* unless a virtual call is established or is being established. During the *PAD waiting state* the PAD will transmit binary 1.

*Note.* — In certain networks the *active link state* will lead directly to the *PAD waiting state* (see Note in 2.2 above).

3.2.1.2    *Network user identification (NUI)*

When used for security, billing and network management purposes, the network user shall transmit a *network user identification signal*. Some Administrations may not implement a *network user identification signal*.

**Provisional Recommendation X.28**

The format and information content of the *network user identification signal* are for further study.

When *network user identification* is not used and the calling DTE is not identified by other means, the reverse charging facility will be used.

### 3.2.1.3  *PAD command (state 6)*

The DTE may transmit a *PAD command signal* when the interface is in the *PAD waiting state* (state 5) and enters the *PAD command state* at the start of a *PAD command signal*.

The DTE may also transmit *PAD command signals* after escaping from the *data transfer state* of a virtual call (see 4.9.1 below).

If parameter 6 is set to 1, following the receipt of a *PAD command signal*, the PAD will ignore all characters received from the DTE until the associated *PAD service signal* or sequence of *PAD service signals* has been transmitted to the DTE by the PAD.

The formats of *PAD command signals* are given in 3.5 below.

The DTE may request the establishment of a virtual call by transmitting a *selection PAD command signal*. The information content of the *selection PAD command signal* is the subject of future recommendations.

### 3.2.1.4  *DTE waiting (state 3B)*

Following the transmission of a *PAD command signal* the DTE will transmit binary 1 and the interface will be in the *DTE waiting state*.

### 3.2.1.5  *Connection-in-progress (state 7)*

If parameter 6 is set to 1, on receipt of a valid *selection PAD command signal* the PAD will transmit an *acknowledgement PAD service signal* followed by binary 1 and will enter the *connection-in-progress state*. The interface will enter the *PAD service signal state* as necessary and the PAD will transmit the *connected PAD service signal* or a *clear indication PAD service signal* to the DTE. During this period the PAD will not accept any *PAD command signals*.

If the value of parameter 6 is 0, the PAD will not transmit *PAD service signals* to the start-stop mode DTE. Following the receipt of a valid *selection PAD command signal*, the interface shall remain in the *connection-in-progress state* until the virtual call has been established.

### 3.2.1.6  *PAD service signals (state 8)*

Following receipt by the DTE of a *PAD service signal* or a sequence of *PAD service signals* (in the case of call set-up) in response to a previously transmitted *PAD command signal*, the interface will be in either:

a)  a *PAD waiting state* (state 5) if no virtual call is in progress, or

b)  a *data transfer state* (state 9) if a virtual call is in progress.

Any *PAD service signal* arising from events within the packet network will not be transmitted until any *PAD service signal* outstanding from a previously received *PAD command signal* has been transmitted.

*PAD service signals* will not be transmitted if the value of parameter 6 is set to 0.

The format of *PAD service signals* is defined in 3.5 below.

A summary of *PAD service signals* is given in the Annex.

### 3.2.1.7  *Incoming calls*

The procedures for signalling an incoming call and the start-stop mode DTE responses are for further study.

**Provisional Recommendation X.28**

### 3.2.2 Clearing

#### 3.2.2.1 Clearing by the start-stop mode DTE

a) When parameter 6 is set to 1, DTE clearing shall be indicated by either:

    i) transmitting a *clear request PAD command signal* after escaping from the *data transfer state* during a virtual call (see 4.9 below). The format of a *clear request PAD command signal* is given in 3.5.8 below. The PAD will transmit a *clear confirmation PAD service signal*. The format of the *clear confirmation PAD service signal* is given in 3.5.9 below. Following the transmission of the *clear confirmation PAD service signal*, the interface will be in the *PAD waiting state* and the DTE will be allowed to make a follow-on call;

    or

    ii) disconnecting the access information path.

b) When parameter 6 is set to 0, DTE clearing shall be indicated by disconnecting the access information path.

#### 3.2.2.2 PAD clearing

a) When parameter 6 is set to 1, PAD clearing may be indicated by:

    i) transmitting a *clear indication PAD service signal*. The format of a *clear indication PAD service signal* is given in 3.5.19 below. After transmitting a *clear indication PAD service signal*, the interface will be in the *PAD waiting state*. The DTE shall stop sending data on receipt of a *clear indication PAD service signal* and shall transmit binary 1;

    or

    ii) disconnecting the access information path.

b) When parameter 6 is set to 0, the PAD will not clear the access information path and the interface will enter the *PAD waiting state (state 3B)*.

### 3.2.3 Unsuccessful calls

If a call is unsuccessful for any reason, the PAD will indicate the reason to the start-stop mode DTE by means of *PAD service signals*. If parameter 6 is set to 0, a *clear indication PAD service signal* is not transmitted.

After transmission of the *PAD service signals* the PAD will be in the *PAD waiting state*.

#### 3.2.3.1 Fault conditions

a) *Failure to receive a PAD command signal*

If the first character of a *PAD command signal* is not received within X of the interface entering the *PAD waiting state*, the PAD will perform PAD clearing in accordance with 3.2.2.2 above. The value of X is for further study.

If following the first character of a *PAD command signal* a complete *PAD command signal* is not received within 30s, the PAD will transmit an *error PAD service signal*, if parameter 6 is set to 1, indicating that an error has occurred (see 3.5.21 below) and the interface will return to the *PAD waiting state*.

If the PAD received an unrecognized *PAD command signal*, it will transmit an *error PAD service signal*, if parameter 6 is set to 1, indicating that an error has occurred and the interface will return to the *PAD waiting state*.

b) *Failure to establish a virtual call*

If the interface enters the *PAD waiting state* more than N times after receiving a *service request signal* without a virtual call being established, the PAD will perform PAD clearing in accordance with 3.2.2.2 above. The value of N is for further study.

**Provisional Recommendation X.28**

c) *Invalid clear request PAD command signal*

If the PAD receives a *clear request PAD command signal* while the interface is in the *PAD waiting state*, the PAD will transmit a *clear-in-error PAD service signal* and the interface returns to the *PAD waiting state*. The format of the *clear-in-error PAD service signal* is given in 3.5.10 below.

### 3.2.3.2   *Failure of the access information path*

If the access information path is disconnected for any reason, the call attempt or virtual call will be cleared by the PAD.

### 3.2.4   *Data transfer*

The procedures for data transfer are given in 4. below.

### 3.3   *Procedures for setting the values of PAD parameters*

These procedures may be used before the *selection PAD command signal* is sent and also after escaping from the *data transfer state*.

### 3.3.1   *Selection of a standard profile by the start-stop mode DTE*

The start-stop mode DTE may select a set of defined values of PAD parameters known as a *standard profile* [see 3.1.1b) above] by sending the *profile selection PAD command signal* which includes a profile identifier. This procedure is additional to the selecting of an *initial standard profile* by transmitting the *service request signal*.

The format of the *profile selection PAD command signal* is given in 3.5.5 below.

A list of the parameter values associated with the *transparent* and *simple standard profiles* is given in Table 1/X.28. Other *standard profiles,* their corresponding parameter values and their identifiers are subjects for further study.

When parameter 6 is set to 1, the PAD will acknowledge the *profile selection PAD command signal* by sending an *acknowledgement PAD service signal* to the start-stop mode DTE.

The format of the *acknowledgement PAD service signal* is defined in 3.5.3 below.

### 3.3.2   *Procedures for setting or changing one or several parameters by the start-stop mode DTE*

The start-stop mode DTE may change the values of one or several parameters by sending a *set or set and read PAD command signal* including the parameter reference(s) and value(s). The format of *PAD command signals* is defined in 3.5 below.

The PAD will respond to a valid *set and read PAD command signal* by transmitting a *parameter value PAD service signal* as described in 3.4 below, showing the newly set parameter values. The PAD will respond to a valid *set PAD command signal* by transmitting an *acknowledgement PAD service signal*. If at least one of the requested PAD parameters is invalid, the PAD will send a *parameter value PAD service signal* to the start-stop mode DTE to identify the invalid parameters. In this case the valid parameters will be accepted and invoked. Valid parameter references and values are given in Recommendation X.3.

The format of the *parameter value PAD service signal* is defined in 3.5.16 below.

### 3.4   *Procedure for reading the values of one or several parameters by the start-stop mode DTE*

These procedures may be used when parameter 6 is set to 1, before the *selection PAD command signal* is sent and also after escaping from the *data transfer state*.

**Provisional Recommendation   X.28**

The start-stop mode DTE may enquire about the current values of one or several PAD parameters by sending the *read PAD command signal* and the references of the required parameters. The format of the *read PAD command signal* is defined in 3.5.4 below.

The PAD will respond by sending a *parameter value PAD service signal* containing the requested parameter values. The format of the *parameter value PAD service signal* is defined in 3.5.16 below.

## 3.5  Formats of PAD command signals and PAD service signals

All characters in columns 2 to 7 of International Alphabet No. 5, excluding the characters 2/0 (SP), 7/15 (DEL) and 2/11 (+) will be recognized by the PAD as forming part of a *PAD command signal*. All other characters are not considered as part of a *PAD command signal* and those characters will be ignored. The *PAD command signal* delimiter is not part of the command.

All *PAD command signals* shall be terminated with the *PAD command signal delimiter*.

*PAD service signals*, other than the *acknowledgement PAD service signal*, will commence with and be followed by the *format effector*.

### 3.5.1  Format of the PAD command signal delimiter

The character 0/13 (CR) or character 2/11 (+) may be sent as a delimiter.

### 3.5.2  Format of the format effector

The characters 0/13 (CR), 0/10 (LF) will be sent by the PAD followed by, when parameter 9 is set to 0, two padding characters if the start-stop mode DTE operates at a data rate of 110 bit/s and four padding characters if the start-stop mode DTE operates at 200 bit/s or 300 bit/s.

If parameter 9 is not set to 0, then the number of padding characters transmitted after the character 0/10 (LF) will be equal to the current value of that parameter.

The format of the padding characters is given in 3.5.22 below.

### 3.5.3  Format of the acknowledgement PAD service signal

The *format effector* will be sent.

### 3.5.4  Format of read PAD command signal

The characters 5/0 (P) 4/1 (A) 5/2 (R) 3/15 (?) shall be sent followed by the decimal reference of the parameter to be read.

*Note.* — IA5 characters will be sent to represent both the parameter reference and parameter value, e.g. decimal value 12 would be sent as characters 3/1 (1) and 3/2 (2).

If no parameter reference number is indicated in the *read PAD command signal* then it applies implicitly to all the parameters.

When more than one parameter is required to be read by sending the *read PAD command signal*, the character 2/12 (,) shall be sent between the decimal references of the parameters.

Example: PAR?1,3,5

### 3.5.5  Format of standard profile selection PAD command signal

The characters 5/0 (P) 5/2 (R) 4/15 (O) 4/6 (F) shall be sent followed by a profile identifier which is for further study.

### 3.5.6 *Format of set PAD command signal and the set and read PAD command signal*

The characters 5/3 (S) 4/5 (E) 5/4 (T) shall be sent, followed by the decimal reference of the parameter to be set, followed by character 3/10 (" :") and the value for the *set PAD command signal*. The characters 5/3 (S) 4/5 (E) 5/4 (T) 3/15 (?) shall be sent, followed by the decimal reference of the parameter to be set and read, followed by character 3/10 (:) and the value for the *set and read PAD command signal*.

If more than one parameter is to be set or set and read by the *PAD command signal*, the character 2/12 (,) shall be sent between a parameter value and the next parameter reference.

Example: SET2 :0,3 :2,9 :4

### 3.5.7 *Format of the reset PAD service signal*

The characters 5/2 (R) 4/5 (E) 5/3 (S) 4/5 (E) 5/4 (T) will be sent, followed by the character 2/0 (SP), followed by one of the following:

    3.5.7.1    The characters 4/4 (D) 5/4 (T) 4/5 (E).

    3.5.7.2    The characters 4/5 (E) 5/2 (R) 5/2 (R).

    3.5.7.3    The characters 4/14 (N) 4/3 (C).

### 3.5.8 *Format of the clear request PAD command signal*

The characters 4/3 (C) 4/12 (L) 5/2 (R) shall be sent.

### 3.5.9 *Format of the clear confirmation PAD service signal*

The characters 4/3 (C) 4/12 (L) 5/2 (R) will be sent, followed by the character 2/0 (SP), followed by the characters 4/3 (C) 4/15 (0) 4/14 (N) 4/6 (F).

### 3.5.10 *Format of the clear-in-error PAD service signal*

The characters 4/3 (C) 4/12 (L) 5/2 (R) will be sent, followed by the character 2/0 (SP), followed by the characters 4/5 (E) 5/2 (R) 5/2 (R).

### 3.5.11 *Format of the status PAD command signal*

The characters 5/3 (S) 5/4 (T) 4/1 (A) 5/4 (T) will be sent.

### 3.5.12 *Format of the status engaged and status free PAD service signals*

The characters 4/5 (E) 4/14 (N) 4/7 (G) 4/1 (A) 4/7 (G) 4/5 (E) 4/4 (D) will be sent for status engaged. The characters 4/6 (F) 5/2 (R) 4/5 (E) 4/5 (E) will be sent for status free.

### 3.5.13 *Format of the reset PAD command signal*

The characters 5/2 (R) 4/5 (E) 5/3 (S) 4/5 (E) 5/4 (T) will be sent.

### 3.5.14 *Format of the interrupt PAD command signal*

The characters 4/9 (I) 4/14 (N) 5/4 (T) will be sent.

### 3.5.15 *Format of the interrupt and discard output PAD command signal*

The characters 4/9 (I) 4/14 (N) 5/4 (T) 4/4 (D) will be sent.

**Provisional Recommendation X.28**

### 3.5.16 *Format of the parameter value PAD service signal*

The characters 5/0 (P) 4/1 (A) 5/2 (R) will be sent, followed by the decimal reference of the parameter, followed by the character 3/10 (:) and the appropriate parameter value. If the requested parameter's reference is invalid, the characters 4/9 (I) 4/14 (N) 5/6 (V) will be sent in place of the appropriate parameter value.

If more than one parameter value is contained in the *parameter value PAD service signal*, the character 2/12 (,) will be sent between a parameter value and the next parameter reference.

Example: PAR2:1,3:2,64:INV

### 3.5.17 *Format of the selection PAD command signal*

A *selection PAD command signal* shall, in the following order, consist of a facility request block, or an address block, or both, optionally followed by a call user data field.

#### 3.5.17.1 *Format of facility request block*

Characters representing the facility request code shall be sent. When more than one facility request code is to be sent, the character 2/16 (,) is sent to separate the facility request codes. The character 2/13 (−) shall be sent at the end of facility request block.

The format of the facility request code is for further study.

#### 3.5.17.2 *Format of address block*

Characters representing a full address or an abbreviated address shall be sent. When an abbreviated address is sent, it shall be prefixed by character 2/14 (.). When more than one address, either full address or abbreviated address, is sent, the character 2/12 (,) is sent as a separator.

The format of the full address and the abbreviated address is for further study.

#### 3.5.17.3 *Format of call user data field*

The character 5/0 (P) or the character 4/4 (D) shall be sent, followed by up to 12 characters of user data. Some networks may not make this field available to the user.

### 3.5.18 *Format of service request signal*

The format is for further study.

### 3.5.19 *Format of clear indication PAD service signals*

The characters 4/3 (C) 4/12 (L) 5/2 (R) 2/0 (SP) shall be sent, followed by one of the following character sequences:

3.5.19.1   The characters 4/15 (O) 4/3 (C) 4/3 (C).

3.5.19.2   The characters 4/14 (N) 4/3 (C).

3.5.19.3   The characters 4/9 (I) 4/14 (N) 5/6 (V).

3.5.19.4   The characters 4/14 (N) 4/1 (A).

3.5.19.5   The characters 4/5 (E) 5/2 (R) 5/2 (R).

3.5.19.6   The characters 5/2 (R) 5/0 (P) 4/5 (E).

3.5.19.7   The characters 4/14 (N) 5/0 (P).

3.5.l9.8   The characters 4/4 (D) 4/5 (E) 5/2 (R).

3.5.19.9   The characters 5/0 (P) 4/1 (A) 4/4 (D).

*Note.* − The coding of these *PAD service signals* is provisional and may be the subject of another recommendation.

**Provisional Recommendation X.28**

### 3.5.20  *Format of the PAD identification PAD service signal*

The characters that will comprise this *PAD service signal* will be network dependent, but would probably indicate the PAD identity and port identity.

### 3.5.21  *Format of the error PAD service signal*

The characters 4/5 (E) 5/2 (R) 5/2 (R) will be sent, followed by other characters which are for further study.

### 3.5.22  *Format of padding chararcters*

The padding character will be 0/0 (NUL) or the equivalent duration of binary 1 according to the particular network.

### 3.5.23  *Format of the connected PAD service signal*

The character 4/3 (C) 4/15 (O) 4/13 (M) or, as a network option, the character 0/6 (ACK), will be sent.

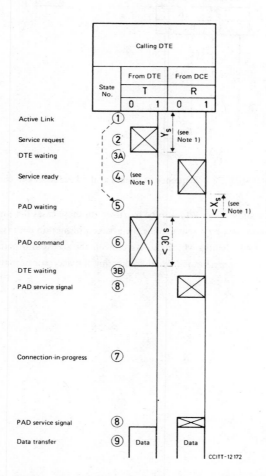

FIGURE 1a/X.28 – **Sequence of events at the interface : call establishment**

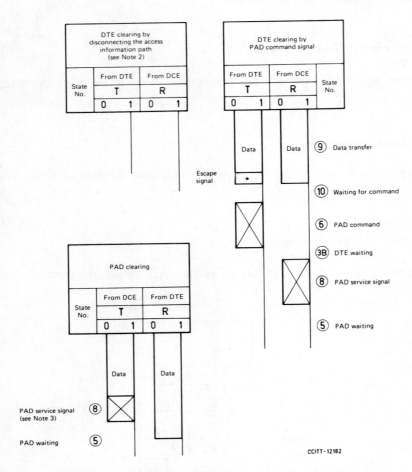

FIGURE 1b/X.28 – **Sequence of events at the interface: call clearing**

*Note 1.* – Some networks may allow states 2 to 4 to be bypassed. In these cases the time-outs X and Y do not apply.

*Note 2.* – The DTE clear is performed by disconnecting the access information path (see 1. of this Recommendation).

The response from the DCE is PAD clearing which also disconnects the access information path.

*Note 3.* – PAD clearing may also be performed by disconnecting the access information path (see 1. of this Recommendation).

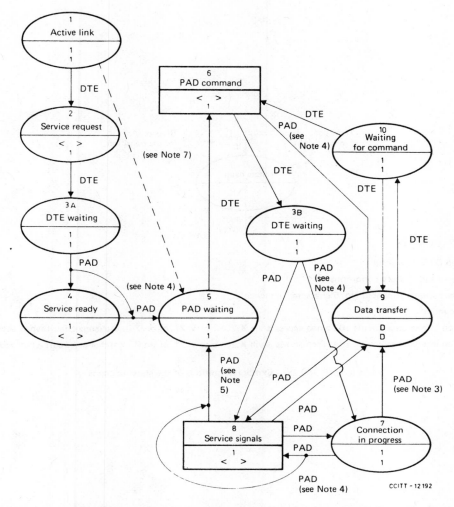

Note 1. – States 3A and 3B are represented in Figure 2/X.28 for convenience. They are functionally equivalent.

Note 2. – State 8 is used to represent a state during which all *PAD service signals* are transmitted.

Note 3. – When parameter 6 is set to 0, the change in state occurs when the PAD receives an indication that the virtual call to the packet-mode DTE has been established.

Note 4. – When parameter 6 is set to 0, states 4 and 8 are bypassed.

Note 5. – The PAD will permit entry to the *PAD waiting state* N times before performing PAD clearing.

Note 6. – Under certain circumstances DTE clearing is performed by disconnecting the access information path (see 1. of this Recommendation).

Note 7. – Some networks may allow states 2 to 4 to be bypassed.

Note 8. – See Figure 3/X.28 for the symbol definitions of the state diagram.

FIGURE 2/X.28 – **State diagram of call establishment and call clearing by PAD command and service signals**

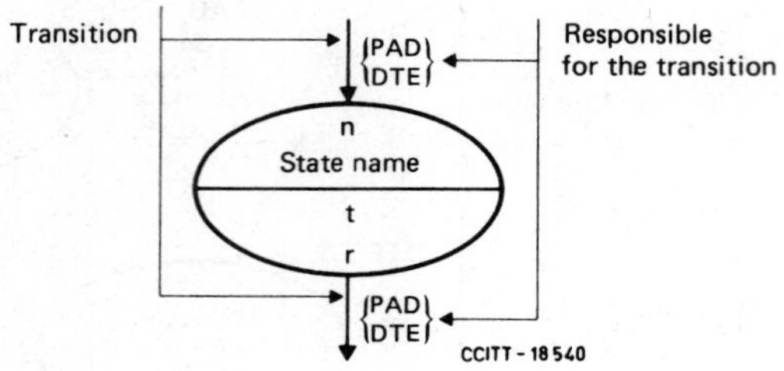

D:       DTE to DTE data signal

0 and 1:    refer to steady binary conditions

< >:      an International Alphabet No. 5 character sequence

n:       state number

t:       value on interchange circuit 103 when access is via X.20*bis* or V.21; or on T interchange circuit when access is via X.20

r:       value on interchange circuit 104 when access is via X.20*bis* or V.21; or on R interchange circuit when access is via X.20

FIGURE 3/X.28 – **Symbol definitions of the state diagrams**

65

## TABLE 1/X.28 – PAD parameter settings

The parameter references and values relate to Recommendation X.3

| Parameter reference number | Parameter description | Parameter setting for the various profiles | | |
|---|---|---|---|---|
| | | Transparent profile | Simple profile | Values available in other (standard) profiles which are for further study (see Note 1) |
| 1 | PAD recall by escaping from the *data transfert state* | Set to *not possible* (value 0) | Set to *possible* (value 1) | Selectable, either *not possible* (value 0), or *possible* (value 1) |
| 2 | Echo | Set to *no echo* (value 0) | Set to *no echo* (value 0) | Selectable, either *no echo* (value 0), or *echo* (value 1) |
| 3 | Selection of *data forwarding signal* | Set to *no data forwarding signal* (value 0) | Set to *all characters in columns 0 and 1 and character 7/15 (DEL) of International Alphabet No. 5* (value 126) | Selectable, either *no data forwarding signal* (value 0) or *carriage return* (value 2) or *all characters in columns 0 and 1 and character 7/15 (DEL) of International Alphabet No. 5* (value 126) |
| 4 | Selection of *idle timer delay* | Set to *one second* (value 20) | Set to *no time out* (value 0) | Selectable, any time between *1 and 255* twentieths of a second, in increments of a twentieth of a second; or *no time out* (value 0) |
| 5 | *Ancillary device control* | Set to *no use of X-ON and X-OFF* (value 0) | Set to *use of X-ON and X-OFF* (value 1) | Selectable, either *no use of X-ON and X-OFF* (value 0), or *use of X-ON and X-OFF* (value 1) |
| 6 | Suppression of *PAD service signals* | Set to *no service signals sent to the start-stop mode DTE* (value 0) | Set to *service signals are sent* (value 1) | Selectable, either *no service signals sent to the start-stop mode DTE* (value 0) or *service signals are sent* (value 1) |
| 7 | Selection of operation of PAD on receipt of *break signal* from the start-stop mode DTE | Set to *reset* (value 2) | Set to *reset* (value 2) | Selectable, either *nothing* (value 0), or *interrupt* (value 1), or *reset* (value 2), or *escape from data transfer state* (value 8), or *discard output, interrupt and indication of break* (value 21) |
| 8 | *Discard output* | Set to *normal data delivery* (value 0) | Set to *normal data delivery* (value 0) | Set to *discard output* (value 1) by PAD and subsequently to *normal data delivery* (value 0) by the packet mode terminal |
| 9 | Padding after carriage return (CR) | Set to *no padding after CR* (value 0) | Set to *no padding after CR* (value 0) | Selectable, any number between *1 and 7* padding characters inserted after each CR character, or *no padding after CR* (value 0) |
| 10 | Line folding | Set to *no line folding* (value 0) | Set to *no line folding* (value 0) | Selectable, any number between *1 and 255* of characters per line, or *no line folding* (value 0) |

301

**Provisional Recommendation X.28**

TABLE 1/X.28 (concluded)

| Parameter reference number | Parameter description | Parameter setting for the various profiles | | |
|---|---|---|---|---|
| | | Transparent profile | Simple profile | Values available in other (standard) profiles which are for further study (see Note 1) |
| 11 | Binary speed of start-stop mode DTE | Set as a result of automatic speed detection as either *110 bit/s* (value 0), or *200 bit/s* (value 8), or *300 bit/s* (value 2) | Set as a result of automatic speed detection as either *110 bit/s* (value 0), or *200 bit/s* (value 8), or *300 bit/s* (value 2) | Set as a result of automatic speed detection as either *110 bit/s* (value 0), or *200 bit/s* (value 8), or *300 bit/s* (value 2) |
| 12 | Flow control of the PAD by the start-stop mode DTE | Set to *no use of X-ON and X-OFF* (value 0) | Set to *use of X-ON and X-OFF* (value 1) | Selectable, either *no use of X-ON* d *X-OFF* (value 0) or *use of X-ON and X-OFF* (value 1) |

*Note 1.* – When using any other profile besides the *transparent standard profile* and *simple standard profile*, the default value of any parameter is the value specified for the profile selected by the *service request signal*.

4. *Procedures for the exchange of user data between a start-stop mode DTE and a PAD*

The procedures described apply during the *data transfer state* for a start-stop mode DTE.

4.1 *Data transfer state*

After receipt of the *connected PAD service signal*, the start-stop mode DTE shall be in the *data transfer state* and will remain in that state, unless it escapes as described in 4.9 below, until the virtual call is cleared by the PAD or by the start-stop mode DTE as described in 3.2.2 above.

If parameter 1 is set to 0, during the *data transfer state* any character sequence may be transmitted by the start-stop mode DTE for delivery to the packet mode DTE. If parameter 1 is set to 1, the character 1/0 (DLE) can only be transferred to the packet-mode DTE by following the procedure described in 4.9.1.1 below.

The values of other parameters may affect the characters which may be transferred during the data phase.

4.2 *Data from the start-stop mode DTE received by the PAD*

Characters received from the start-stop mode DTE are defined as consisting of all the bits received between, but not including, the start and stop bits.

**Provisional Recommendation X.28**

## 4.3 Delivery of user data to the start-stop mode DTE

Data received by the PAD for delivery to the start-stop mode DTE will be transmitted to the start-stop mode DTE at the data signalling rate appropriate to the start-stop mode DTE. Start and stop bits will be added to the data characters in accordance with Recommendation X.4.

## 4.4 Packet forwarding conditions

A packet will be forwarded whenever more than enough data has been received from the start-stop mode DTE to fill a packet after the last packet was forwarded.

In addition, the start-stop mode DTE may indicate to the PAD that a packet should be forwarded, subject to flow control, whenever it performs any one, or more of the following:

4.4.1 Allows the idle timer delay period (see Parameter 4 in Table 1/X.28), after the transmission of the previous character to the PAD, to elapse without sending a character.

If, due to flow control constraints, the packet cannot be forwarded, characters from the start-stop mode DTE will continue to be added to the packet until flow control permits the packet to be forwarded or the packet becomes full. The start-stop mode DTE will be advised (see 4.6 below) if this latter condition occurs.

4.4.2 Transmits one of the data forwarding characters (see Parameter 3 in Table 1/X.28). The character will be included in the data field of the packet it delimits before the packet is forwarded.

4.4.3 Transmits the *break signal* when parameter 7 is set to any value except 0.

4.4.4 Transmits the first character of a *PAD command signal* after the interface has entered a *waiting for command state* as described in 4.9.1 below.

## 4.5 Procedure for the PAD to indicate to the start-stop mode DTE, by means of a PAD service signal, a temporary inability to accept additional information

The procedure to enable the PAD to indicate a temporary inability to receive additional characters and to subsequently indicate that characters will be accepted, using PAD service signals, is for further study.

## 4.6 Procedures for ancillary device control

If parameter 5 is set to 1, the following procedure applies:

The PAD will send the *X-ON character* to the DTE as soon as the interface enters the *data transfer state*. The character 1/1 (DC1) will be transmitted by the PAD.

The PAD will send the *X-OFF character* to the start-stop mode DTE when it is incapable of receiving more than M characters from the ancillary device at the start-stop mode DTE and another character is received from that DTE. The PAD will also send the *X-OFF character* before the interface leaves the *data transfer state*. The character 1/3 (DC3) will be transmitted by the PAD.

When the PAD is again able to receive at least $M + 1$ characters from the start-stop mode DTE, it will send the *X-ON character* to that DTE.

The value of M is for further study.

## 4.7 Procedures for reset

### 4.7.1 Start-stop mode DTE sending a reset PAD command signal

The start-stop mode DTE shall send a *reset PAD command signal* to the PAD when it wishes to reset the virtual call to the packet mode DTE.

a) The *break signal* (see 3.1.2 above) will be recognized by the PAD as a *reset PAD command signal* if parameter 7 is set to 2.

**Provisional Recommendation X.28**

b) Alternatively the start-stop mode DTE may request reset by escaping from the *data transfer state* and sending a *reset PAD command signal* according to the procedure of 4.9.2.3 below.

### 4.7.2 *Sending a reset PAD service signal to the start-stop mode DTE*

If the virtual call is reset by the packet mode DTE, or by the network, the PAD will send a *reset PAD service signal*, if the value of parameter 6 is set to 1, to the start-stop mode DTE. The *PAD service signal* will indicate the cause of the reset.

The following reset causes will be indicated to the start-stop mode DTE.

a) the remote DTE has reset the virtual call. The format is given in 3.5.7.1 above;

b) a local procedure error has occurred. The format is given in 3.5.7.2 above;

c) network congestion has occurred. The format is given in 3.5.7.3 above.

### 4.8 *Procedure for indication of break*

The PAD will inform the start-stop mode DTE that an incoming *indication of break PAD message* has been received from the packet mode DTE (see Recommendation X.29) by sending the *break signal (see 3.1.2 above).*

### 4.9 *Escape from the data transfer state*

4.9.1 During the *data transfer state*, the start-stop mode DTE may escape from that state by transmitting an *escape signal* to the PAD. On detection of the *escape signal,* the interface will enter the *waiting for command state*. On entering the *waiting for command state*, delivery of any data characters to the start-stop mode DTE will be delayed until the interface returns to the *data transfer state*.

If parameter 1 is set to 1, the PAD will recognize the character 1/0 (DLE) as the *escape signal* from the start-stop mode DTE.

On receipt of the next character from the start-stop mode DTE, the PAD will act in accordance with one of the following conditions:

4.9.1.1 If the character is the character 1/0 (DLE) the interface will immediately return to the *data transfer state*. This character will be treated as user data.

4.9.1.2 If the character received is the *PAD command signal delimiter* [characters 2/11 (+) or 0/13 (CR)] the PAD will not transfer it and the interface will return to the *data transfer state*.

4.9.1.3 If the character received is in columns 2 to 7 of International Alphabet No. 5, excluding the characters 2/0 (SP), 2/11 (+) [see 4.9.1.2 above] and 7/15 (DEL), the interface will enter the *PAD command state*. Characters 2/0 (SP) and 7/15 (DEL) will be ignored. Entering the *PAD command state* is a data forwarding condition and data will be sent to the packet mode DTE as described in 4.4 above.

If the complete *PAD command signal* is not received within 30s of entering the *PAD command state*, or an invalid *PAD command signal* is received, the PAD will transmit a *PAD service signal*, when parameter 6 is set to 1, indicating that an error has occurred. Following transmission of the *PAD service signal* the interface will be in the *data transfer state*.

If a valid *PAD command signal* is received the interface will, if parameter 6 is set to 1, subsequently enter the *PAD service signal state*, and on transmission of the last character of the *PAD service signal* will enter the *PAD waiting state* or the *data transfer state* as appropriate. If parameter 6 is set to 0, the interface will enter the *PAD waiting state* or the *data transfer state*, as appropriate, following the transmission of a *PAD command signal*.

If parameter 7 is set to 8, the *break signal* may be used as the *escape signal* from the *data transfer state*, allowing for escape from the *data transfer state* without loss of character transparency.

**Provisional Recommendation X.28**

4.9.2   The ability to escape from the *data transfer state* allows a start-stop mode DTE to use the following *PAD command signals* and procedures:

### 4.9.2.1   *Clearing*

The procedure for clearing of the virtual call by the start-stop mode DTE sending a *clear request PAD command signal* is described in Section 3.2.2.1a)i).

### 4.9.2.2   *Request for status of the virtual call*

The start-stop mode DTE may, if parameter 6 is set to 1, enquire whether a virtual call is existing by sending the *status PAD command signal* to the PAD. The PAD will respond by sending the *status engaged* or *status free PAD service signal* to the DTE. The format of the *PAD command signal* and *PAD service signal* is given above in 3.5.11·and 3.5.12 respectively.

### 4.9.2.3   *Reset*

The start-stop mode DTE may request a resetting of the virtual call to the packet mode DTE by sending a *reset PAD command signal* to the PAD. The format of the *reset PAD command signal* is given in 3.5.13 above.

The PAD will acknowledge the *PAD command signal*, if parameter 6 is set to 1, by transmitting the *acknowledgement PAD service signal*.

### 4.9.2.4   *Interrupt*

The start-stop mode DTE may request that an interrupt be sent to the packet-mode DTE by sending an *interrupt PAD command signal* to the PAD. The format of the *interrupt PAD command signal* is given in 3.5.14 above.

The PAD will acknowledge the *PAD command signal*, if parameter 6 is set to 1, by transmitting the *acknowledgement PAD service signal*.

*Note.* — The use of the *interrupt and discard PAD command signal* is for further study.

### 4.9.2.5   *Setting, setting and reading, and reading PAD parameter values after having entered the data transfer state*

The start-stop mode DTE shall be able to send the following *PAD command signals* to set, set and read, and read PAD parameter values:

a)   *profile selection PAD command signal;*

b)   *set PAD command signal;*

c)   *set and read PAD command signal;*

d)   *read PAD command signal.*

The procedures for sending the above *PAD command signals* are described in 3.3 and 3.4 above.

## 4.10   *Echo*

If parameter 2 is set to 1, the following procedures will apply:

4.10.1   In the *data transfer state*, received characters will be echoed to the start-stop mode DTE prior to the transmission of data characters waiting for delivery at the time.

In the case where the PAD cannot handle and ignores a data character coming from the start-stop mode DTE, because of flow control constraints, the PAD will not echo the character.

4.10.2   In the *PAD waiting state*, all characters are echoed.

4.10.3   In the *PAD command state*, characters in *PAD command signals* are echoed, except the characters following the character P in a *selection PAD command signal*, which are not echoed.

4.10.4   In the *connection-in-progress state* and the *PAD service signal state*, characters are not echoed.

4.11  *Selection of the procedure on receipt of the break signal from the start-stop mode DTE*

The start-stop mode DTE, by means of parameter 7, will be able to select which procedure the PAD will perform when it receives the *break signal* from the start-stop mode DTE. The DTE may select any one of the following:

4.11.1  Take no action.

4.11.2  Send an *interrupt packet* to the packet mode DTE (see Recommendation X.29).

4.11.3  Reset the virtual call to the packet mode DTE (see Recommendation X.29).

4.11.4  Discard all data received for delivery to the start-stop mode DTE and send an *interrupt packet* to the packet mode DTE followed by an *indication of break PAD message* (see Recommendation X.29).

4.11.5  Escape from the *data transfer state* and enter the *waiting for command state*.

Other procedures which may be selected by the start-stop mode DTE are for further study.

4.12  *Selection of padding characters to be inserted after the character 0/13 (CR)*

The start-stop mode DTE, by means of parameter 9, will be able to select the number of padding characters that will be inserted after each character 0/13 (CR) transmitted or echoed to it. The value selected will also apply to the number of padding characters transmitted after the character 0/10 (LF) of the *format effector* as described in 3.5.2 above.

Other padding sequences and other padding rules are for further study.

4.13  *Selection of line folding*

The start-stop mode DTE, by means of parameter 10, will be able to select line folding and specify the maximum number (L) of graphic characters that the PAD may send as a single line to the start-stop mode DTE.

When line folding is requested, the PAD will maintain a count (C) which is incremented by 1 each time a graphic character is sent to the start-stop mode DTE.

The graphic characters are those shown in columns 2 to 7 of International Alphabet No. 5, excluding the character 7/15 (DEL).

If the value of C is equal to the value of L, the PAD will transmit to the start-stop mode DTE a format effector (see 3.5.2 above) and set the value of C to O.

The PAD will set the value of C to 0 when the PAD transmits the character 0/13 (CR) to the start-stop mode DTE.

When echo is provided, the PAD will increment the value of C by 1 each time a character is echoed.

Line folding also applies to *PAD service signals*.

4.14  *Selection of start-stop DTE flow control*

The start-stop mode DTE, by means of parameter 12, will be able to select to use or not flow control. When the use of flow control is selected it shall be achieved by use of the X-ON (DC1) and X-OFF (DC3) characters.

On entering the data transfer phase the PAD will be in the X-ON condition.

**Provisional Recommendation X.28**

ANNEX

(to Recommendation X.28)

**PAD command signals and PAD service signals**

TABLE 1 – **PAD command signals**

| PAD command signal format | Function | PAD service signal sent in response |
|---|---|---|
| STAT | To request status information regarding a virtual call connected to the DTE | FREE or ENGAGED |
| CLR | To clear down a virtual call | CLR CONF or CLR ERR (in the case of local procedure error) |
| PAR? List of parameter references | To request the current values of specified parameters | PAR [list of parameter references with their current values] |
| SET? List of parameter references and corresponding values | To request changing or setting of the current values of the specified parameters | PAR [list of parameter references with their current values] |
| PROF [identifier] | To give to PAD parameters a standard set of values | Acknowledgement |
| RESET | To reset the virtual call or permanent virtual circuit | Acknowledgement |
| INT or INTD | To transmit an *interrupt packet* to the packet mode DTE | Acknowledgement |
| SET List of parameters with requested values | To set or change parameter values | Acknowledgement |
| *Selection PAD command signal* | To set-up a virtual call | Acknowledgement |

TABLE 2 – **PAD service signals**

| Format of the PAD service signal | | Explanation |
|---|---|---|
| RESET | DTE | Indication that the remote DTE has reset the virtual call or permanent virtual circuit |
| | ERR | Indication of a reset of a virtual call or permanent virtual circuit due to a local procedure error |
| | NC | Indication of a reset of a virtual call or permanent virtual circuit due to network congestion |
| COM | – | Indication of call connected |
| CLR | See Table 3 | Indication of clearing |
| *PAD identification PAD service signal* | The characters to be sent are network dependent and are for further study | |
| ERROR | ERR | Identification that a *PAD command signal* is in error |

**Provisional Recommendation X.28**

TABLE 3 — Clear indication PAD service signals

| Clear indication PAD service signal | Possible mnemonics (see Note) | Explanation |
|---|---|---|
| Number busy | OCC | The called number is fully engaged in other calls |
| Network congestion | NC | Congestion conditions within the network temporarily prevent the requested virtual call from being established |
| Invalid call | INV | Facility invalid requested |
| Access barred | NA | The calling DTE is not permitted to obtain the connection to the called DTE. Incompatible closed user group would be an example |
| Local procedure error | ERR | The call is cleared because of a local procedure error |
| Remote procedure error | RPE | The call is cleared because of a remote procedure error |
| Not obtainable | NP | The called number is not assigned or is no longer assigned |
| Out of order | DER | The called number is out of order |
| Clearing after invitation | PAD | The PAD has cleared the call following the receipt of an invitation to clear from the packet mode DTE |

*Note.* — The final coding of *clear indication PAD service signals* may be the subject of another Recommendation.

**Provisional Recommendation X.28**

**Provisional Recommendation X.29**   *(Geneva, 1977)*

## PROCEDURES FOR THE EXCHANGE OF CONTROL INFORMATION AND USER DATA BETWEEN A PACKET MODE DTE AND A PACKET ASSEMBLY/DISASSEMBLY FACILITY (PAD)

**Contents**

Preface

The establishment in various countries of public data networks providing packet-switched data transmission services creates a need to produce standards to facilitate international interworking.

310

*Preface*

The CCITT,

*considering*

a)   that Recommendations X.1 and X.2 define the user classes of service and facilities in a public data network, Recommendation X.95 defines network parameters and Recommendation X.96 defines call progress signals;

b)   that Recommendation X.25 defines the interface between the DTE and the DCE for DTEs operating in the packet mode in public data networks;

c)   that Recommendations X.20, X.20 *bis*, X.21 and X.21 *bis* define the DTE/DCE interfaces in a public data network;

d)   that Recommendation X.3 defines the PAD in a public data network;

e)   that Recommendation X.28 defines the DTE/DCE interface for a start-stop mode DTE accessing the PAD in a public data network;

f)   the need to allow interworking between a packet mode DTE and a non-packet mode DTE in the packet-switched transmission service;

g)   the urgent need to allow interworking between a start-stop mode DTE in a public switched telephone network, public switched data network or a leased line and a packet mode DTE using the virtual call facility of the packet-switched transmission service;

h)   that the packet mode DTE shall not be obliged to use the control procedures for PAD functions, but that some packet mode DTEs may wish to control specific functions of the PAD.

**Provisional Recommendation X.29**

*unanimously declares the view* that:

1. the Recommendation X.29 procedures shall apply to the Recommendation X.25 interface between the DCE and the packet mode DTE;

2. the procedures be as specified below in 1. *Procedures for the exchange of PAD control information and user data*;

3. the manner in which user data is transferred between the packet mode DTE and the PAD be as specified below in 2. *User data transfer*;

4. the procedures for the control of the PAD via PAD messages be as specified below in 3. *Procedures for the use of PAD messages*;

5. the formats of the data fields which are transferable on a virtual call be as specified below in 4. *Formats*.

*Note.* — The following items are for further study:

— the possibility of a packet mode DTE establishing a virtual call to a non-packet mode DTE;

— the use of the permanent virtual circuit facility;

— interworking between non-packet mode DTEs in the packet-switched data transmission service;

— interworking between DTEs having interfaces to different data transmission services;

— operation of non-packet mode DTEs in other than start-stop mode.

## 1. *Procedures for the exchange of control information and user data*

1.1    The exchange of control information and user data between a PAD and a packet mode DTE is performed by using user data fields defined in Recommendation X.25.

1.2    The Appendix briefly describes some of the characteristics of virtual calls, as well as of the Recommendation X.25 interface as related to the PAD representation of a start-stop mode DTE to a packet mode DTE.

### 1.3 *Call user data*

The call user data field of *incoming call* packets to the packet mode DTE is comprised of two fields (see the Appendix):

a)    the protocol identifier field, and

b)    the call data field.

The protocol identifier field is used for protocol identification purposes and the call data field contains user data received by the PAD from the start-stop mode DTE during the call establishment phase.

### 1.4 *User sequences*

1.4.1    User sequences are used to exchange user data between the PAD and the packet mode DTE.

1.4.2    User sequences are conveyed in the user data fields of complete *data* packet sequences with Q = 0 (see the Appendix), and in both directions on a virtual call.

1.4.3    There will be only one user sequence in a complete *data* packet sequence.

**Provisional Recommendation X.29**

### 1.5    *PAD messages*

1.5.1    *PAD messages* are used to exchange control information between the PAD and the packet mode DTE. A *PAD message* consists of a control identifier field and a message code field possibly followed by a parameter field (see 4.4 below).

1.5.2    *PAD messages* are conveyed in the user data fields of complete *data* packet sequences with $Q = 1$ (see the Appendix) and in both directions on a virtual call.

1.5.3    There will be only one *PAD message* in a complete *data* packet sequence.

1.5.4    The PAD will take into consideration a *PAD message* transmitted by a packet mode DTE only when it has been completely received.

1.5.5    In the case where a parameter reference (see 3. below) appears more than once in a *PAD message*, only the last appearance is taken into account.

1.5.6    In the direction from the PAD to the packet mode DTE, the data fields of *interrupt* packets (see the Appendix) are set to 0. The use of other values is for further study. In the direction from the packet mode DTE to the PAD, the PAD ignores the data fields of *interrupt* packets received. Other procedures are for further study.

### 1.5.7    *PAD messages from a packet mode DTE*

1.5.7.1    A packet mode DTE may request the PAD to set and/or read the current values of PAD parameters by sending a *PAD message* specifying the PAD parameters to be set and/or read. If PAD parameters are to be changed, the *PAD message* specifies the new values in the *PAD message* parameter field (see 3. below).

1.5.7.2    A packet mode DTE may send an *indication of break PAD message* to the PAD; a *break signal* will be sent to the start-stop mode DTE by the PAD (see Recommendation X.28). No parameter field is required.

1.5.7.3    The packet mode DTE may send an *invitation to clear PAD message*. This procedure is defined in 3. below. No parameter field is required.

### 1.5.8    *PAD messages to the packet-mode DTE*

1.5.8.1    On receipt of a *read* or *set and read PAD message*, the PAD will indicate, in a *parameter indication PAD message*, the state of some or all of the PAD parameters. This *PAD message* contains the control identifier field and a message code field followed by a parameter field (see 3. below).

1.5.8.2    When requested to do so by the start-stop mode DTE, the PAD will transmit an *indication of break PAD message* to the packet mode DTE consisting of a control identifier field, a message code field and a parameter field as defined in 3. below.

1.5.8.3    The PAD will transmit an *error PAD message* to the packet mode DTE in response to the receipt of a *PAD message* from the packet mode DTE which the PAD does not recognize as a valid *PAD message*. The parameter field of the *error PAD message* will define the type of error detected (see 3. below).

### 2.    *User data transfer*

2.1    Packets will be forwarded from the PAD when a *set*, *read*, or *set and read PAD message* is received from a packet mode DTE, or under any of the other data forwarding conditions provided by the PAD (see Recommendation X.28).

2.2    The occurrence of a data forwarding condition will not cause the PAD to transmit empty data packets.

**Provisional Recommendation X.29**

# 3. Procedures for the use of PAD messages

## 3.1 Procedures for reading, setting, and reading and setting of PAD parameters

3.1.1   A packet mode DTE may change and read the current values of PAD parameters by transmitting to the PAD a *set*, *read*, or *set and read PAD message*.

3.1.2   When the PAD receives a *set*, *read* or *set and read PAD message*, any data previously transmitted by the packet mode DTE will be delivered to the start-stop mode DTE before taking action on the *PAD message*. The PAD will also consider the arrival of such a *PAD message* as a data forwarding condition in the direction PAD to packet mode DTE.

3.1.3   The PAD will respond to a valid *read* or *set and read PAD message* by returning to the packet mode DTE a *parameter indication PAD message* with a parameter field containing a list of the current values, after any necessary modification, of the PAD parameters to which the message from the packet mode DTE referred.

3.1.4   The PAD will not return a *parameter indication PAD message* in response to a valid *set PAD message* received from the packet mode DTE.

3.1.5   Table 1/X.29 specifies the PAD's response to *set*, *set and read*, and *read PAD messages*.

TABLE 1/X.29 – **PAD messages transmitted by the PAD in response to set, set and read,
and read PAD messages from the packet mode DTE**

| *PAD message* received by the PAD | | Action upon PAD parameters | Corresponding *parameter indication PAD message* transmitted to the packet mode DTE |
|---|---|---|---|
| Type | Parameter field | | |
| Set | None | Reset all parameters to their initial values (corresponding to the initial profile) | None |
| | List of selected parameters with the desired values | Set the selected parameters to the given values: <br> a) if no error is encountered <br> b) if the PAD fails to modify the values of some parameters | a) None <br> b) List of these invalid parameters with the error bit set |
| Set and read | None | Reset all parameters to their initial values (corresponding to the initial profile) | List all parameters, and their initial values |
| | List of selected parameters with the desired values | Set the selected parameters to the given values | List of these parameters with their new current values with the error bit set, as appropriate |
| Read | None | None | List of all parameters with their current values |
| | List of selected parameters | None | List of these parameters with their current values |

**Provisional Recommendation X.29**

## 3.2 *Procedures for inviting the PAD to clear*

3.2.1 The *invitation to clear PAD message* is used to request the PAD to clear the virtual call, after transmission to the start-stop mode DTE of all previously transmitted data.

*Note.* — The *clear indication* packet, which is transmitted to the packet mode DTE after delivery of the last data character to the start-stop mode DTE, will have a clearing cause field set to DTE clearing.

## 3.3 *Interrupt and discard procedures*

3.3.1 If parameter 7 is set to 21, the PAD will transmit an *interrupt* packet followed by an *indication of break PAD message* to indicate to the packet mode DTE that the PAD, at the request of the start-stop mode DTE, is discarding the user sequences received from the packet mode DTE. The *PAD message* will contain an indication in its parameter field that parameter 8 (see Recommendation X.3) has been set to 1 (*discard output*).

3.3.2 Before resuming data transmission to the PAD, the packet mode DTE shall respond to the *indication of break PAD message* by transmitting a *set* or *set and read PAD message*, indicating that parameter 8 should be set to 0 (*normal data delivery*).

## 3.4 *Procedure for resets*

The procedures defined in Recommendation X.25 are used. The effect of the resetting procedure on the value of PAD parameter 8 is to reset its value to 0 (normal data delivery). The current values of PAD parameters 1 to 7 and 9 to 12 are not affected.

## 3.5 *Error handling procedures by the PAD*

3.5.1 If the PAD receives a *set, read* or *set and read PAD message* from a packet mode DTE containing an invalid reference to a PAD parameter, the parameter field within the *parameter indication PAD message* returned to the packet mode DTE will contain an indication of that fact. The remaining valid references to PAD parameters are processed by the PAD.

Possible reasons for an invalid access to a PAD parameter are:

a) the parameter reference does not exist;

b) the parameter reference corresponds to an additional user facility which is not available;

c) the parameter is a read-only one: (set/set and read only);

d) the requested value is invalid: (set/set and read only).

3.5.2 The PAD will transmit an *error PAD message* containing the message code of an invalid *PAD message* received from the packet mode DTE under the following conditions:

a) if the PAD receives an unrecognizable message code;

b) if the parameter field following a recognizable message code is incorrect or incompatible with the message code;

c) if the parameter field following a recognizable message code has an invalid format.

3.5.3 The PAD will transmit an *error PAD message* if a *PAD message* containing less than 8 bits is received from the packet mode DTE.

**Provisional Recommendation X.29**

## 4. *Formats*

### 4.1 *Introduction*

Bits of an octet are numbered 8 to 1 where bit 1 is the low order bit and is transmitted first. Octets of the call user data, of user sequences, of *PAD messages* and of interrupt user data are consecutively numbered starting from 1 and are transmitted in this order.

### 4.2 *Call user data format* (see Figure 1/X.29)

| | 8 | 7 | 6 | 5 | 4 | 3 | 2 | 1 | |
|---|---|---|---|---|---|---|---|---|---|
| Octet 1 | 0 | 0 | 0 | 0 | 0 | 0 | 0 | 1 | Protocol identifier |
| 2 | 0 | 0 | 0 | 0 | 0 | 0 | 0 | 0 | |
| 3 | 0 | 0 | 0 | 0 | 0 | 0 | 0 | 0 | |
| 4 | 0 | 0 | 0 | 0 | 0 | 0 | 0 | 0 | |
| ⋮ | | | | | | | | | Call data |
| N | | | | | | | | | |

4 ⩽ N ⩽ 16 octets

FIGURE 1/X.29 – **Call user data field format**

#### 4.2.1 *Protocol identifier format*

The protocol identifier field consists of four octets.

The first octet is coded as follows:

| | | |
|---|---|---|
| bits 8 and 7 | = 00 | for CCITT use |
| | = 01 | for national use |
| | = 10 | reserved for international user bodies |
| | = 11 | for DTE-DTE use |

When bits 8 and 7 are equal to 00, bits 6 to 1 are equal to 000001 for indicating PAD messages relating to the packer assembly/disassembly facility for the start-stop mode DTE. Other coding of bits 6 to 1 is reserved for future standardization by the CCITT. The use of octets 2, 3 and 4 is reserved and all bits are set to 0. Octets 2, 3 and 4 are reserved as a future mechanism for providing the called PAD or packet mode DTE with additional information pertinent to the calling party.

#### 4.2.2 *Call data format*

Octets of the call data field will contain the user characters received by the PAD from the start-stop mode DTE during the call establishment phase. The coding of these octets is similar to that of user sequences (see 4.3 below). The call data field is limited to 12 octets (see Figure 1/X.29).

### 4.3 *User sequence format*

4.3.1 The order of bit transmission to the packet mode DTE is the same as the order that bits are received from the start-stop mode DTE. The order of bit transmission to the start-stop mode DTE is the same as the order that bits are received from the packet mode DTE.

4.3.2 No maximum is specified for the length of a user sequence.

### 4.4 *Control message format*

4.4.1 The control identifier field (bits 8, 7, 6, 5 of octet 1) of all *control messages* is used to identify the facility, such as PAD, to be controlled. The control identifier field coding for a PAD for a start-stop mode DTE is 0000. Other codings of the control identifier field are reserved for future standardization.

*Note.* — The possiblity of extending the control identifier field is for further study.

4.4.2 When the control identifier field (see 4.4.1 above) is set to 0000, bits 4, 3, 2, 1 of octet 1 are defined as a message code field. The message code field is used to identify specific types of *PAD messages*, as given in Table 2/X.29.

TABLE 2/X.29 – Type and coding of octet 1 of PAD messages

| Type | Message code | | | |
|------|:---:|:---:|:---:|:---:|
| | Bits 4 | 3 | 2 | 1 |
| Set PAD message | 0 | 0 | 1 | 0 |
| Read PAD message | 0 | 1 | 0 | 0 |
| Set and read PAD message | 0 | 1 | 1 | 0 |
| Parameter indication PAD message | 0 | 0 | 0 | 0 |
| Invitation to clear PAD message | 0 | 0 | 0 | 1 |
| Indication of break PAD message | 0 | 0 | 1 | 1 |
| Error PAD message | 0 | 1 | 0 | 1 |

*Note.* — The possibility of extending the message code field is for further study.

4.4.3 All *PAD messages* consist of a control identifier field (bits 8, 7, 6, 5 of octet 1 equal to 0000) and a message code (bits 4, 3, 2, 1 of octet 1).

*Set, read, set and read* and *parameter indication PAD messages* consist of octet 1 which may be followed by one or more parameter fields. Each parameter field consists of a parameter reference octet and a parameter value octet.

The parameter value octets of the *read PAD message* contain the value 0.

The *error PAD message* consists of octet 1 and one or two octets, giving the reason for the error.

The *indication of break PAD message* consists of octet 1 and one parameter field.

The *invitation to clear PAD message* consists of octet 1 only.

**Provisional Recommendation X.29**

4.4.4   The maximum length is for further study.

4.4.5   *Parameter field for set, read, set and read, and parameter indication PAD messages* (see Figure 2/X.29)

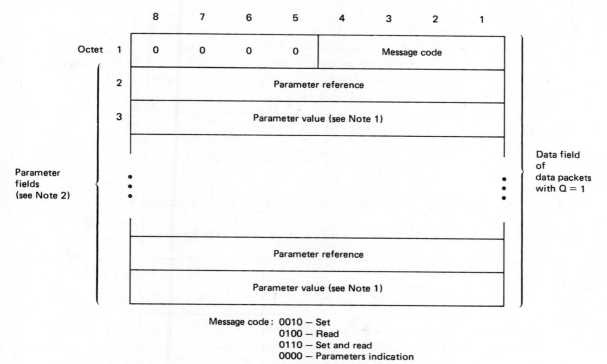

Message code : 0010 — Set
0100 — Read
0110 — Set and read
0000 — Parameters indication

*Note 1.* – These octets contain all 0s in read PAD messages.

*Note 2.* – Parameter fields need not be present (see Table 1/X.29).

FIGURE 2/X.29 – Set, read, set and read, and parameter indication PAD message format

The parameter field of these *PAD messages*, when present, will consist of successive parts of reference fields and value fields. Each one of these fields will be one octet long.

4.4.5.1   A reference field consists of a parameter reference, identified as a decimal number in Recommendation X.3, and is binary coded in bits 7 to 1, where bit 1 is the low order bit. Reference fields need not be ordered by increasing parameter reference numbers.

4.4.5.2   In *PAD messages* received by the PAD, bit 8 of each octet will be ignored. In *parameter indication PAD messages*, bit 8 set to 1 will indicate an invalid access to the referred parameter as described in 3.5 above.

4.4.5.3   A parameter value field consists of a value of the parameter reference, identified as a decimal number in Recommendation X.3, and is binary coded in bits 8 to 1, where bit 1 is the low order bit. Value fields in *read PAD messages* are coded as all binary 0s. In *set* and *set and read PAD messages*, they will indicate the requested values of parameters. In *parameter indication PAD messages*, they will indicate the current values of PAD parameters, after modification if any. If bit 8 (error bit) is set to 1 in the preceding octet (i.e. reference field), they will be set to 0.

*Note.* – The coding 1111111 (decimal 127) in bits 7 to 1 of the reference field will be used for the extension of this field. Such coding will indicate that there is another octet following. The following octet is coded with the parameter reference of Recommendation X.3 minus 127.

**Provisional Recommendation X.29**

**4.4.6** *Parameter field for error PAD messages* (see Figure 3/X.29)

|  | 8 | 7 | 6 | 5 | 4 | 3 | 2 | 1 |
|---|---|---|---|---|---|---|---|---|
| Octet 1 | 0 | 0 | 0 | 0 | 0 | 1 | 0 | 1 |
| 2 | Error type (see Table 3/X.29) | | | | | | | |
| 3 | Invalid message code (see Note) | | | | | | | |

Data field
of data packets
with Q = 1

*Note.* – Does not occur for error type 00000000.

FIGURE 3/X.29 – **Error PAD message format**

**4.4.6.1** Octet 2 of the *error PAD message* will be coded as shown in Table 3/X.29.

TABLE 3/X.29 – **Coding and meaning of octet 2 of error PAD messages**

| Case | Meaning | Coding |
|---|---|---|
| | | Bits    8 7 6 5 4 3 2 1 |
| a | Received *PAD message* contained less than eight bits | 0 0 0 0 0 0 0 0 |
| b | Unrecognized message code in received *PAD message* | 0 0 0 0 0 0 0 1 |
| c | Parameter field format of received *PAD message* was incorrect or incompatible with message code | 0 0 0 0 0 0 1 0 |
| d | Received *PAD message* did not contain an integral number of octets | 0 0 0 0 0 0 1 1 |

**4.4.6.2** In cases b, c and d in Table 3/X.29, octet 3 of an *error PAD message* will contain the message code of the received *PAD message*.

**4.4.7** *Parameter field for indication of break PAD messages* (see Figure 4/X.29)

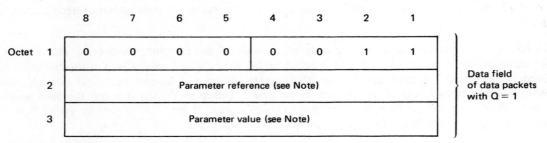

|  | 8 | 7 | 6 | 5 | 4 | 3 | 2 | 1 |
|---|---|---|---|---|---|---|---|---|
| Octet 1 | 0 | 0 | 0 | 0 | 0 | 0 | 1 | 1 |
| 2 | Parameter reference (see Note) | | | | | | | |
| 3 | Parameter value (see Note) | | | | | | | |

Data field
of data packets
with Q = 1

*Note.* – PAD to packet mode DTE only.

FIGURE 4/X.29 – **Indication of break PAD message format**

**Provisional Recommendation X.29**

4.4.7.1    When transmitted by the packet mode DTE, this *PAD message* will contain no parameter field.

4.4.7.2    When transmitted by the PAD, the parameter field will contain two octets (i.e. one reference field and one value field) and will be coded as follows: the reference field will be coded 00001000 (indicating parameter 8) and the value field will be coded 00000001 (indicating decimal 1).

4.4.8    *Parameter field for invitation to clear PAD message* (see Figure 5/X.29)

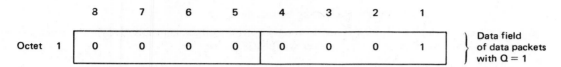

FIGURE 5/X.29 – Invitation to clear PAD message format

This *PAD message* will contain no parameter field.

## APPENDIX

## (to Recommendation X.29)

**Characteristics of virtual calls and Recommendation X.25
as related to the PAD representation of
a start-stop mode DTE to a packet mode DTE**

1.    *General interface characteristics*

1.1    The physical, electrical, functional and procedural characteristics to establish, maintain and disconnect the physical link between the DTE and the DCE will be in accordance with the level 1 procedures of Recommendation X.25.

1.2    The link access procedure for data interchange across the link between the DTE and the DCE will be in accordance with the level 2 procedures of Recommendation X.25.

1.3    The packet format and control procedures for the exchange of packets containing control information and user data between the DTE and the DCE will be in accordance with the level 3 procedures of Recommendation X.25.

2.    *Interface procedures for virtual call control*

2.1    Incoming calls are indicated to the packet mode DTE as specified in 3.1.3 of Recommendation X.25. Any use of optional user facilities are indicated in accordance with 5. of Recommendation X.25.

2.2    The throughput classes used for the virtual call will correspond to the data rates used by the start-stop mode DTE.

2.3    The PAD and the packet mode DTE will use the clearing procedures specified in 3.1.7, 3.1.8 and 3.1.9 of Recommendation X.25.

3.    *Interface procedures for data transfer*

3.1    Data transfer on a virtual call can only take place in the data transfer state and when flow control permits (see 3.4 of Recommendation X.25). The same is true for the transfer of *interrupt* packets (see 3.3 of Recommendation X.25).

**Provisional Recommendation X.29**

3.2    *Interrupt* packets transmitted by the packet mode DTE will be confirmed by the PAD following the procedures in Recommendation X.25 (see 3.3.5 of Recommendation X.25).

3.3    The reset procedure may be used by the packet mode DTE or the PAD, to re-initialize the virtual call and will conform to the procedures described in 3.4.2 of Recommendation X.25.

3.4    A reset of the virtual call originated by the packet mode DTE or due to network congestion may be indicated by the PAD to the start-stop mode DTE.

3.5    A reset procedure initiated by the PAD may be due either to:

   a)    the receipt at the PAD of a request to reset from the non-packet mode DTE. The resetting cause contained in the reset indication packet will be *DTE reset*; or

   b)    a PAD or network failure.

### 4.    *Virtual call characteristics*

### 4.1    *Resetting*

4.1.1    There may be a loss of data characters in any case of reset, as stated in Recommendation X.25. Characters generated by either of the DTEs prior to the reset indication or confirmation will not be delivered to the other DTE after the reset indication or confirmation.

### 4.2    *Interrupt transfer*

4.2.1    An *interrupt* packet is always delivered at or before the point in the data packet stream at which it was generated.

### 4.3    *Call clearing*

Data transmitted immediately before a *clear request* packet is sent may be overtaken within the network by the *clear request* packet and subsequently be destroyed.

# SECTION 4

## NETWORK PARAMETERS AND HYPOTHETICAL REFERENCE CONNECTIONS

**Recommendation X.92**

### HYPOTHETICAL REFERENCE CONNECTIONS FOR PUBLIC SYNCHRONOUS DATA NETWORKS

*(Geneva, 1976)*

The CCITT

*bearing in mind*

a)   the international user classes of service indicated in Recommendation X.1;

b)   the overall user-to-user performance objectives;

c)   the need to standardize the procedures for use over public synchronous data networks;

d)   in the case of packet switching, the need to standardize several procedural levels;

*unanimously recommends*

the use of the five hypothetical reference connections contained in this Recommendation.

1.     The five hypothetical reference connections set down in the present Recommendation (see Figure 1/X.92) are intended for assessing the overall customer-to-customer performance objectives, for determining some data characteristics requirements of the various items in the connections and for setting limits to the impairments these items may introduce.

These hypothetical reference connections should be used for circuit switched services, packet switched services and leased line services in public synchronous data networks.

Other hypothetical reference connections may be set up in the future after experience of the design of synchronous public data networks has been gained.

2.     The hypothetical reference connections of Figure 1/X.92 are intended for the user data signalling rates as recommended in Recommendation X.1.

Between points X and Y, transmission takes place over 64 kbit/s digital paths. Such paths may include digital sections using modems over analogue facilities.

It should be assumed that the signalling for the circuit switched data-call control follows the same route as the data connection.

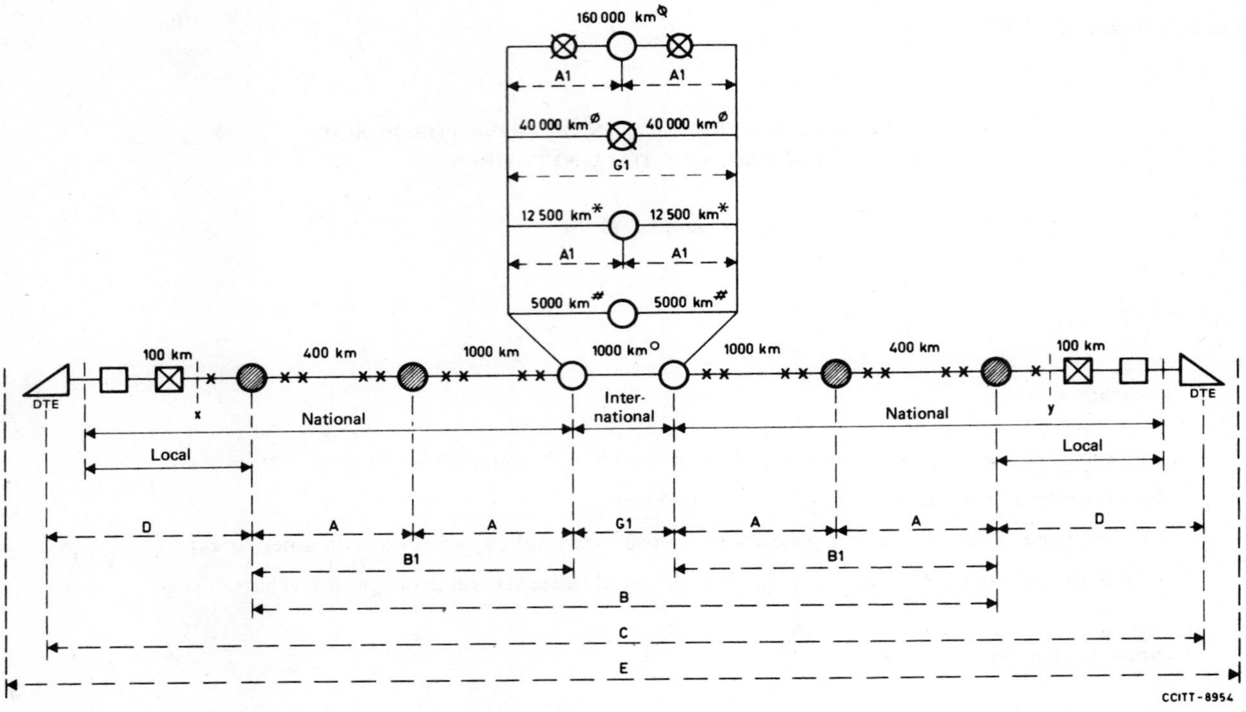

FIGURE 1/X.92 – Hypothetical reference connections for public synchronous data networks

3.    The legend to the symbols used in Figure 1/X.92 is as follows:

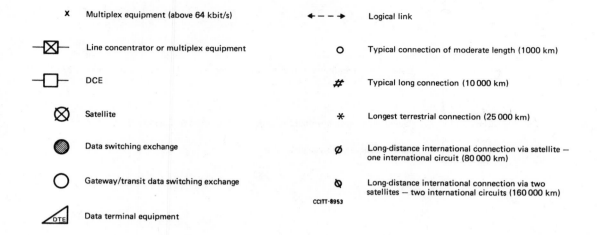

4.    a)   The logical links to be considered in the case of packet switching are indicated on Figure 1/X.92 by the dotted lines. The legend is as follows:

Link A    = data link between two adjacent data switching exchanges in a national network

Link A1   = data link between two adjacent gateway data switching exchanges in an international connection

Link B    = data link between a source DSE and a destination DSE

Link B1   = data link between a local DSE and a gateway DSE

Link G1   = data link between a source gateway DSE and a destination gateway DSE in an international connecton

Link C    = data link between source DTE and destination DTE

Link D    = data link between source DTE and the source local DSE or the data link between destination DTE and destination local DSE

Link E    = data link between communicating processes

    b)   To allow for the incorporation of packet assembly/disassembly facilities, the variants to logical Link D, shown in Figure 2/X.92, are recognized.

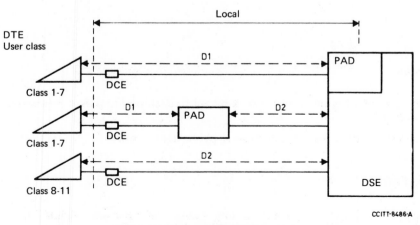

FIGURE 2/X.92 – **Variants of logical link D**

The legend to the symbols used is as follows:

Class 1-7 — Data terminal equipment in user classes of service 1-7

Class 8-11 — Packet mode data terminal equipment in user classes of service 8-11

PAD — Packet assembly/disassembly equipment

CCITT-8487

Link $D_1$ = data link between a data terminal equipment in user class of service 1-7 and a packet assembly/disassembly equipment

Link $D_2$ = data link between a data terminal equipment in user class of service 8-11 or a packet assembly/disassembly equipment and a local data switching exchange.

*Note 1* — A user may see two different types of logical interfaces with the network (Link $D_1$ and $D_2$).

*Note 2.* — Link $D_2$ could provide an interface for a single access terminal as well as for a multiple access terminal.

**Recommendation X.95**

## NETWORK PARAMETERS IN PUBLIC DATA NETWORKS

*(Geneva, 1976)*

The CCITT

*bearing in mind*

a) the need to standardize the basic network parameters relevant to public data networks to facilitate international interworking;

b) the need to standardize the user interfaces to public data networks;

*unanimously declares the view*

that the network parameters inherent in any public data network should be as indicated in Tables 1/X.95 and 2/X.95.

TABLE 1/X.95 — Network parameters for circuit switched and leased circuit services

| Network parameter | User classes of service | | | Definition reference |
|---|---|---|---|---|
| | 1 | 2 | 3-7 | |
| Timing from the network<br>*a)* Bit timing<br>*b)* Byte timing (8-bit) | N<br>N | N<br>N | P<br>S | —<br>— |
| Bit sequence independence | N | N | P | 52.52 |
| Data signalling rate transparency | N | S | N | 53.362 |
| Symmetrical duplex channels | P | P | P | 53.765 |
| Time out | P | P | P | 52.605 |

TABLE 2/X.25 — Network parameters for packet-switched services

| Network parameter | User classes of service | | | Definition reference |
|---|---|---|---|---|
| | 1-2 | 3-7 | 8-11 | |
| Timing from the network<br>*a)* Bit timing<br>*b)* Byte timing (8-bit) | N<br>N | P<br>S | P<br>S | —<br>— |
| Packet format | N | N | P | 53.55 |
| Transmit flow control | see<br>Note 1 | see<br>Note 1 | P | 53.635 |
| Interworking between data terminals in different user classes of service | P | P | P | 52.471 |
| Error-correcting system | see<br>Note 2 | see<br>Note 2 | P | 52.34 |

P = inherent network parameter
S = network parameter inherent in some public data networks
N = parameter not applicable

*Note 1.* — Awaiting further study.
*Note 2.* — Not applicable over the data circuit between the data terminal equipment and the packet assembly/disassembly equipment.

**Recommendation X.96**

### CALL PROGRESS SIGNALS IN PUBLIC DATA NETWORKS

*(Geneva, 1976)*

The CCITT,

*bearing in mind*

that the establishment of public data networks for data transmission in various countries and the subsequent international interconnection of these networks creates the possibility that, in certain circumstances, a switched connection requested by a user will not be established to the called number,

*unanimously declares the view*

that *Call progress signals* should be returned to the caller to indicate the circumstances which have prevented the connection being established to a called number;

that *Call progress signals* should be returned to the caller to indicate in some circumstances the progress made towards establishing the connection requested.

The circumstances giving rise to *Call progress signals* in public data networks are under study within the CCITT and Tables 1/X.96 and 2/X.96 summarize the outcome of the study to date.

TABLE 1/X.96 – Circuit switched data transmission services
(see notes 1-3)

| Call progress signal | Brief description of circumstances |
|---|---|
| Selection signal procedure error | The selection signals received did not conform to the specified procedure. |
| Selection signal transmission error | A transmission error was detected in the selection signals by the first DSE. |
| Invalid call | Facility request invalid. |
| Access barred | The calling DTE is not permitted to obtain a connection to the called DTE. Incompatible closed user group or incoming calls barred are examples. |
| Not obtainable | The called number is not assigned, or is no longer assigned or there is an incompatible user class of service. |
| Number busy | The called number is engaged in another call. |
| Out of order | The called number is out of order (DTE "uncontrolled" not ready). Possible reasons include: 1) DTE not functioning; 2) mains power off to DTE/DCE; 3) line fault between DSE and DCE. |
| Changed number | The called number has recently been assigned a new number. |
| Call the information service | The called number is temporarily unobtainable, call the information service for details. |
| Network congestion | The establishment of the connection has been prevented due to: 1) temporary congestion conditions; 2) temporary fault conditions, e.g. expiry of a time-out. |
| Terminal called (see Note 2) | The incoming call was signalled to the DTE and call acceptance is awaited. |
| Controlled not ready (see Note 3) | The called DTE is in the *Controlled not ready* state. |

TABLE 2/X.96 — **Packet switched data transmission service — Virtual calls only**
(see notes 1 and 4-8)

| Call progress signal | Brief description of circumstances |
|---|---|
| Access barred | The calling DTE is not permitted to obtain the connection to the called DTE. Incompatible closed user group would be an example. |
| Not obtainable | The called DTE is not assigned, or is no longer assigned. |
| Number busy | The called DTE is engaged in other calls on all its logical channels and cannot accept another call. |
| Invalid call | Facility request invalid. |
| Local procedure error | The call is cleared because of a local procedure error. |
| Remote procedure error | The call is cleared because of a remote procedure error. |
| Out of order | The call number is out of order. Possible reasons include: 1) DTE not functioning; 2) Mains power off to DTE/DCE; 3) Line fault between DSE and DCE; 4) Access link not functioning; 5) Link access procedure level not in operation. |
| Network congestion | Congestion conditions within the network temporarily prevent the requested virtual call from being established. |

*Notes*

1. Call progress signal format shall be in accordance with Recommendation X.20 or X.20 *bis* for user classes of service 1 and 2 and Recommendation X.21 for user classes of service 3 to 7.
2. This is a "positive" *Call progress signal* and is not accompanied by a *Clearing signal*. The international implications of "manual answering" are for further study.
3. The international implications concerning the use of *Controlled not ready* are for further study.
4. Call progress signal format and codes shall be in accordance with Recommendation X.25 for user classes of service 8-11.
5. Whilst *Call progress signals* may only be transmitted by the network to the DTE during the call set-up phase, in the packet switched service certain other service or advisory signals may be transmitted by the network to the DTE during the data transfer phase and the call clearing phase of a virtual call. The possible signals are for further study.
6. The call progress signals *Terminal called* and *Controlled not ready* are for further study in relation to packet switched data transmission services.
7. The fact that a DTE is also out of order when the link access procedure level is not operating correctly is a subject for further study.
8. The *Call progress signals* (if any) for the Datagram facility is a subject for further study.

# Part 2

# ISO Data Communications Standards

A prime example of successful international cooperation, the International Organization for Standardization (ISO) is a voluntary, non-treaty group devoted to achieving world-wide agreement on International Standards. ISO was established in 1947 and now has a membership of approximately 82 countries. Over 574 new or revised standards have been completed for a very wide range of applications by about 150 technical committees. These committees cover a large variety of subjects including such diverse areas as screw threads, nuclear energy, data processing, and many other advancing technologies.

The ISO's membership is made up of the prinicpal standardization body of each represented nation. The USA member body is the American National Standards Institute (ANSI). Participating ("P") members actively contribute to the work of the technical committees and have the power to vote for or against the approval of developed standards. Correspondent ("O") members do not vote but act as observers to keep themselves informed of current standardization activities. The administration of each technical committee and subcommittee is assigned to a "P" member body which is charged with seeing that the work gets done.

Technical Committee 97 is responsible for Computers and Information Processing. Their scope is:

"Standardization in the areas of computers and information processing systems and peripheral equipment and devices and media related thereto."

Under TC 97 there are 15' subcommittees which develop Draft Proposals in their specific areas of expertise. The work in the data communications field is carried on by Subcommittee 6 which consists of 17 "P" members and 11 "O" members. Their scope is:

"To define the system functions, procedures, and parameters necessary for the transfer of data between data systems over communications networks. To effect liaison with CCITT and CCIR and to prepare proposals for their consideration and for inclusion in CCITT and CCIR Recommendations as appropriate. To prepare International Standards and/or Technical Reports relating to those aspects of data communications for which ISO is responsible."

ANSI serves as the Secretariat to administer the activities of both TC 97 and SC 6.

Because of the broad scope of SC 6's work, there are three Working Groups to deal with specific technical areas. Working Group 1 is responsible for procedures and formats for data transfer. Working Group 2 deals with the interfaces to, and operation with, public data networks in direct cooperation with CCITT Study Group VII. Working Group 3 maintains liaison with CCITT Study Group XVII in relation to interfacing with V-series interfaces. It also develops International Standards for interface connectors and pin assignments. Each of the three Working Groups meets twice a year on its own and presents its results to the annual meeting of SC 6 for approval.

The ISO International Standards in this section were developed by TC 97/SC 6 Working Groups with the exception of ISO 646 and ISO 2022 which were the responsibility of TC 97/SC 2, Character Sets and Coding.

New work that is advancing toward completion in SC 6 includes the bit-oriented High-level Data Link Control (HDLC) procedures. The first in the HDLC series has been published as ISO 3309 which specifies the frame structure. To follow soon is 4335 (presently a Draft International Standard) which specifies the elements of procedure for HDLC. Also in the status of Draft International Standards being balloted by TC 97 for approval are 6159 and 6256 which specify classes of procedure for HDLC. Other Draft International Standards related to data communications that are in earlier stages of balloting include:

2110 (revised)—25-pin DTE/DCE interface connector and pin assignments. (applicable to CCITT (V24/V28) interfaces)

4902—37-pin and 9-pin DTE/DCE interface connectors and· pin assignments. (applicable to CCITT V.24/V.10/V.11 interfaces )

4903—15-pin DTE/DCE interface connector and pin assignments. (applicable to CCITT X-series interfaces )

The newest work under study in ISO/TC97/SC6 and the newly formed Subcommittee 16 is the establishment of a common architecture for all data processing and data communications standards. Successful completion of an agreed architecture will establish orderly relationships of present and future work while facilitating the completeness of technical areas covered.

| NUMBER | TITLE | RELATED CROSS REFERENCE STANDARDS |
|---|---|---|
| ISO 646-1973 | 7-bit coded character set for information processing interchange | CCITT V.3; ANSI A.4 |
| ISO 1155-1973 | Information processing–use of longitudinal parity to detect errors in information messages | CCITT V.4, X.4; ANSI X3.15; X3.16; FED-STD-1010, 1011 |
| ISO 1177-1973 | Information processing–character structure for start/stop and synchronous transmission | CCITT V.4, X.4; ANSI X3.1, X3.16; FED-STD-1010, 1011 |
| ISO 1745-1975 | Information processing–basis mode control procedures for data communications systems | ANSI X3.38 |
| ISO 2022-1973 | Code extension techniques for use with ISO 7-bit coded character sets | ANSI X3.41 |
| ISO 2110-1972 | Data communication–Data terminal and data communication equipment–interchange circuits–assignment of connector pin numbers | EIA RS-232C |
| ISO 2111-1972 | Data communication–basic mode control procedures–code-independent information transfer | None |
| ISO 2593-1973 | Connector pin allocations for use with high speed data terminal equipment | None |
| ISO 2628-1973 | Basic mode control procedures–complements | None |
| ISO 2629-1973 | Basic mode control procedures–conversational information message transfer | None |
| ISO 3309-1976 | Data communication–high-level data link control procedures–frame structure | CCITT X.25 LAP; ANSI DRAFT BSR X3.66 |

# INTERNATIONAL STANDARD  646

INTERNATIONAL ORGANIZATION FOR STANDARDIZATION ·МЕЖДУНАРОДНАЯ ОРГАНИЗАЦИЯ ПО СТАНДАРТИЗАЦИИ ·ORGANISATION INTERNATIONALE DE NORMALISATION

# 7-bit coded character set for information processing interchange

**First edition — 1973-07-01**

331

UDC 681.3.042 : 003.62

Ref. No. ISO 646-1973 (E)

**Descriptors** : data processing, data transmission, character sets, coding, information interchange.

## FOREWORD

ISO (the International Organization for Standardization) is a worldwide federation of national standards institutes (ISO Member Bodies). The work of developing International Standards is carried out through ISO Technical Committees. Every Member Body interested in a subject for which a Technical Committee has been set up has the right to be represented on that Committee. International organizations, governmental and non-governmental, in liaison with ISO, also take part in the work.

Draft International Standards adopted by the Technical Committees are circulated to the Member Bodies for approval before their acceptance as International Standards by the ISO Council.

International Standard ISO 646 was drawn up by Technical Committee ISO/TC 97, *Computers and information processing*, and circulated to the Member Bodies in April 1972.

It has been approved by the Member Bodies of the following countries :

| | | |
|---|---|---|
| Belgium | Ireland | Sweden |
| Brazil | Italy | Switzerland |
| Canada | Japan | Thailand |
| Czechoslovakia | Netherlands | United Kingdom |
| Denmark | Portugal | U.S.A. |
| Egypt, Arab Rep. of | Romania | U.S.S.R. |
| France | South Africa, Rep. of | |
| Germany | Spain | |

No Member Body expressed disapproval of the document.

332

# 7-bit coded character set for information processing interchange

## 1 SCOPE AND FIELD OF APPLICATION

**1.1** This International Standard contains a set of 128 characters (control characters and graphic characters such as letters, digits and symbols) with their coded representation. Most of these characters are mandatory and unchangeable, but provision is made for some flexibility to accommodate special national and other requirements.

**1.2** The need for graphics and controls in data processing and in data transmission has been taken into account in determining this character set.

**1.3** This International Standard consists of a general table with a number of options, notes, a legend and explanatory notes. It also contains a specific international reference version, guidance on the exercise of the options to define specific national versions and application oriented versions.

**1.4** This character set is primarily intended for the interchange of information among data processing systems and associated equipment, and within message transmission systems.

**1.5** This character set is applicable to all latin alphabets.

**1.6** This character set includes facilities for extension where its 128 characters are insufficient for particular applications.

**1.7** The definitions of some control characters in this International Standard assume that data associated with them is to be processed serially in a forward direction. Their effect when included in strings of data which are processed other than serially in a forward direction or included in data formatted for fixed record processing may have undesirable effects or may require additional special treatment to ensure that the control characters have their desired effect.

## 2 IMPLEMENTATION

**2.1** This character set should be regarded as a basic alphabet in an abstract sense. Its practical use requires definitions of its implementation in various media. For example, this could include punched tapes, punched cards, magnetic tapes and transmission channels, thus permitting interchange of data to take place either indirectly by means of an intermediate recording in a physical medium, or by local electrical connection of various units (such as input and output devices and computers) or by means of data transmission equipment.

**2.2** The implementation of this coded character set in physical media and for transmission, taking into account the need for error checking, is the subject of other ISO publications. (See Appendix Y.)

## 3 BASIC CODE TABLE

TABLE 1 — Basic code table

| b7 →<br>b6 →<br>b5 →<br>b4 b3 b2 b1 | column →<br>↓ row | 0<br>0<br>0<br>0 | 0<br>0<br>1<br>1 | 0<br>1<br>0<br>2 | 0<br>1<br>1<br>3 | 1<br>0<br>0<br>4 | 1<br>0<br>1<br>5 | 1<br>1<br>0<br>6 | 1<br>1<br>1<br>7 |
|---|---|---|---|---|---|---|---|---|---|
| 0 0 0 0 | 0 | NUL | TC7 (DLE) | SP | 0 | ③ | P | ` ④ | p |
| 0 0 0 1 | 1 | TC1 (SOH) | DC1 | ! | 1 | A | Q | a | q |
| 0 0 1 0 | 2 | TC2 (STX) | DC2 | " ⑥ | 2 | B | R | b | r |
| 0 0 1 1 | 3 | TC3 (ETX) | DC3 | £(#) ② | 3 | C | S | c | s |
| 0 1 0 0 | 4 | TC4 (EOT) | DC4 | $(¤) ② | 4 | D | T | d | t |
| 0 1 0 1 | 5 | TC5 (ENQ) | TC8 (NAK) | % | 5 | E | U | e | u |
| 0 1 1 0 | 6 | TC6 (ACK) | TC9 (SYN) | & | 6 | F | V | f | v |
| 0 1 1 1 | 7 | BEL | TC10 (ETB) | ' ⑥ | 7 | G | W | g | w |
| 1 0 0 0 | 8 | FE0 (BS) | CAN | ( | 8 | H | X | h | x |
| 1 0 0 1 | 9 | FE1 (HT) | EM | ) | 9 | I | Y | i | y |
| 1 0 1 0 | 10 | FE2 (LF)① | SUB | * | : | J | Z | j | z |
| 1 0 1 1 | 11 | FE3 (VT)① | ESC | + | ; | K | ③ | k | ③ |
| 1 1 0 0 | 12 | FE4 (FF)① | IS4 (FS) | , ⑥ | < | L | ③ | l | ③ |
| 1 1 0 1 | 13 | FE5 (CR)① | IS3 (GS) | - | = | M | ③ | m | ③ |
| 1 1 1 0 | 14 | SO | IS2 (RS) | . | > | N | ^ ④⑤ | n | ‾ ④⑤ |
| 1 1 1 1 | 15 | SI | IS1 (US) | / | ? | O | _ | o | DEL |

334

## NOTES ABOUT TABLE 1

① The format effectors are intended for equipment in which horizontal and vertical movements are effected separately. If equipment requires the action of CARRIAGE RETURN to be combined with a vertical movement, the format effector for that vertical movement may be used to effect the combined movement. For example, if NEW LINE (symbol NL, equivalent to CR + LF) is required, $FE_2$ shall be used to represent it. This substitution requires agreement between the sender and the recipient of the data.

The use of these combined functions may be restricted for international transmission on general switched telecommunication networks (telegraph and telephone networks).

② The symbol £ is assigned to position 2/3 and the symbol $ is assigned to position 2/4. In a situation where there is no requirement for the symbol £ the symbol # (number sign) may be used in position 2/3. Where there is no requirement for the symbol $ the symbol ¤ (currency sign) may be used in position 2/4. The chosen allocations of symbols to these positions for international information interchange shall be agreed between the interested parties. It should be noted that, unless otherwise agreed between sender and recipient, the symbols £, $ or ¤ do not designate the currency of a specific country.

③ National use positions. The allocations of characters to these positions lies within the responsibility of national standardization bodies. These positions are primarily intended for alphabet extensions. If they are not required for that purpose, they may be used for symbols.

④ Positions 5/14, 6/0 and 7/14 are provided for the symbols UPWARD ARROW HEAD, GRAVE ACCENT and OVERLINE. However, these positions may be used for other graphical characters when it is necessary to have 8, 9 or 10 positions for national use.

⑤ Position 7/14 is used for the graphic character ‾ (OVERLINE), the graphical representation of which may vary according to national use to represent ~(TILDE) or another diacritical sign provided that there is no risk of confusion with another graphic character included in the table.

⑥ The graphic characters in positions 2/2, 2/7, 2/12 and 5/14 have respectively the significance of QUOTATION MARK, APOSTROPHE, COMMA and UPWARD ARROW HEAD; however, these characters take on the significance of the diacritical signs DIAERESIS, ACUTE ACCENT, CEDILLA and CIRCUMFLEX ACCENT when they are preceded or followed by the BACKSPACE character (0/8).

## 4  LEGEND

### 4.1  Control characters

| Abbrevi-ation | Note | Meaning | Position in the code table |
|---|---|---|---|
| ACK | | Acknowledge | 0/6 |
| BEL | | Bell | 0/7 |
| BS | | Backspace | 0/8 |
| CAN | | Cancel | 1/8 |
| CR | 1 | Carriage Return | 0/13 |
| DC | | Device control | — |
| DEL | | Delete | 7/15 |
| DLE | | Data Link Escape | 1/0 |
| EM | | End of Medium | 1/9 |
| ENQ | | Enquiry | 0/5 |
| EOT | | End of Transmission | 0/4 |
| ESC | | Escape | 1/11 |
| ETB | | End of Transmission Block | 1/7 |
| ETX | | End of Text | 0/3 |
| FE | | Format Effector | — |
| FF | 1 | Form Feed | 0/12 |
| FS | | File Separator | 1/12 |
| GS | | Group Separator | 1/13 |
| HT | | Horizontal Tabulation | 0/9 |
| IS | | Information Separator | — |
| LF | 1 | Line Feed | 0/10 |
| NAK | | Negative Acknowledge | 1/5 |
| NUL | | Null | 0/0 |
| RS | | Record Separator | 1/14 |
| SI | | Shift-In | 0/15 |
| SO | | Shift-Out | 0/14 |
| SOH | | Start of Heading | 0/1 |
| SP | | Space (see 7.2) | 2/0 |
| STX | | Start of Text | 0/2 |
| SUB | | Substitute Character | 1/10 |
| SYN | | Synchronous Idle | 1/6 |
| TC | | Transmission Control | — |
| US | | Unit Separator | 1/15 |
| VT | 1 | Vertical Tabulation | 0/11 |

### 4.2  Graphic characters

| Graphic | Note | Name | Position in the code table |
|---|---|---|---|
| (space) | | Space (see 7.2) | 2/0 |
| ! | | Exclamation mark | 2/1 |
| " | 6 | Quotation mark, Diaeresis | 2/2 |
| £ | 2 | Pound sign | 2/3 |
| # | 2 | Number sign | 2/3 |
| $ | 2 | Dollar sign | 2/4 |
| ¤ | 2 | Currency sign | 2/4 |
| % | | Percent sign | 2/5 |
| & | | Ampersand | 2/6 |
| ' | 6 | Apostrophe, acute accent | 2/7 |
| ( | | Left parenthesis | 2/8 |
| ) | | Right parenthesis | 2/9 |
| * | | Asterisk | 2/10 |
| + | | Plus sign | 2/11 |
| , | 6 | Comma, Cedilla | 2/12 |
| — | | Hyphen, Minus sign | 2/13 |
| . | | Full stop (period) | 2/14 |
| / | | Solidus | 2/15 |
| : | | Colon | 3/10 |
| ; | | Semi-colon | 3/11 |
| < | | Less than sign | 3/12 |
| = | | Equals sign | 3/13 |
| > | | Greater than sign | 3/14 |
| ? | | Question mark | 3/15 |
| ^ | 4, 6 | Upward arrow head, Circumflex accent | 5/14 |
| _ | | Underline | 5/15 |
| ` | 4 | Grave accent | 6/0 |
| ~ | 4, 5 | Overline, Tilde | 7/14 |

336

# 5 EXPLANATORY NOTES

## 5.1 Numbering of the positions in Table 1

Within any one character the bits are identified by $b_7$, $b_6 \ldots b_1$, where $b_7$ is the highest order, or most significant bit, and $b_1$ is the lowest order, or least significant bit.

If desired, these may be given a numerical significance in the binary system, thus :

| Bit identification : | $b_7$ | $b_6$ | $b_5$ | $b_4$ | $b_3$ | $b_2$ | $b_1$ |
|---|---|---|---|---|---|---|---|
| Significance : | 64 | 32 | 16 | 8 | 4 | 2 | 1 |

In the table the columns and rows are identified by numbers written in binary and decimal notations.

Any one position in the table may be identified either by its bit pattern, or by its column and row numbers. For instance, the position containing the digit 1 may be identified :

— by its bit pattern in order of decreasing significance, i.e. 011 0001;

— by its column and row numbers, i.e. 3/1.

The column number is derived from bits $b_7$, $b_6$ and $b_5$ giving them weights of 4, 2 and 1 respectively. The row number is derived from bits $b_4$, $b_3$, $b_2$ and $b_1$ giving them weights of 8, 4, 2 and 1 respectively.

## 5.2 Diacritical signs

In the 7-bit character set, some printing symbols may be designed to permit their use for the composition of accented letters when necessary for general interchange of information. A sequence of three characters, comprising a letter, "backspace" and one of these symbols, is needed for this composition; the symbol is then regarded as a diacritical sign. It should be noted that these symbols take on their diacritical significance when they are preceded or followed by one "backspace" character; for example the symbol corresponding to the code combination 2/7 normally has the significance of "apostrophe", but becomes the diacritical sign "acute accent" when it precedes or follows a "backspace" character.

In order to increase efficiency, it is possible to introduce accented letters (as single characters) in the positions marked by Note ③ in the code table. According to national requirements, these positions may contain special diacritical signs.

## 5.3 Names, meanings and fonts of graphic characters

This International Standard assigns at least one name to denote each of the graphic characters displayed in Tables 1 and 2. The names chosen to denote graphic characters are intended to reflect their customary meanings. However, this International Standard does not define and does not restrict the meanings of graphic characters. In addition, it does not specify a particular style or font design for the graphic characters.

Under the provision of Note ③ of Table 1, graphic characters which are different from the characters of the international reference version may be assigned to the national use positions. When such assignments are made, the graphic characters shall have distinct forms and be given distinctive names which are not in conflict with any of the forms or the names of any of the graphic characters in the international reference version.

## 5.4 Uniqueness of character allocation

A character allocated to a position in Table 1 may not be placed elsewhere in the table. For example, in the case of position 2/3 the character not used cannot be placed elsewhere. In particular the POUND sign (£) can never be represented by the bit combination of position 2/4.

# 6 VERSIONS OF TABLE 1

## 6.1 General

6.1.1 In order to use Table 1 for information interchange, it is necessary to exercise the options left open, i.e. those affected by Notes ② to ⑤. A single character must be allocated to each of the positions for which this freedom exists or it must be declared to be unused. A code table completed in this way is called a "version".

6.1.2 The notes to Table 1, the explanatory notes and the legend apply in full to any version.

## 6.2 International reference version

This version is available for use when there is no requirement to use a national or an application-oriented version. In international information processing interchange the international reference version (Table 2) is assumed unless a particular agreement exists between sender and recipient of the data. The following characters are allocated to the optional positions of Table 1 :

| # | Number sign | 2/3 |
|---|---|---|
| ¤ | Currency sign | 2/4 |
| @ | Commercial at | 4/0 |
| [ | Left square bracket | 5/11 |
| \ | Reverse solidus | 5/12 |
| ] | Right square bracket | 5/13 |
| { | eft curly bracket | 7/11 |
| | | Vertical line | 7/12 |
| } | Right curly bracket | 7/13 |

It should be noted that no substitution is allowed when using the international reference version.

337

TABLE 2 — International reference version

| b7 | 0 | 0 | 0 | 0 | 1 | 1 | 1 | 1 |
|---|---|---|---|---|---|---|---|---|
| b6 | 0 | 0 | 1 | 1 | 0 | 0 | 1 | 1 |
| b5 | 0 | 1 | 0 | 1 | 0 | 1 | 0 | 1 |
| b4 b3 b2 b1 / row / column | 0 | 1 | 2 | 3 | 4 | 5 | 6 | 7 |
| 0 0 0 0 — 0 | NUL | TC7 (DLE) | SP | 0 | @ | P | ` | p |
| 0 0 0 1 — 1 | TC1 (SOH) | DC1 | ! | 1 | A | Q | a | q |
| 0 0 1 0 — 2 | TC2 (STX) | DC2 | " | 2 | B | R | b | r |
| 0 0 1 1 — 3 | TC3 (ETX) | DC3 | # | 3 | C | S | c | s |
| 0 1 0 0 — 4 | TC4 (EOT) | DC4 | ¤ | 4 | D | T | d | t |
| 0 1 0 1 — 5 | TC5 (ENQ) | TC8 (NAK) | % | 5 | E | U | e | u |
| 0 1 1 0 — 6 | TC6 (ACK) | TC9 (SYN) | & | 6 | F | V | f | v |
| 0 1 1 1 — 7 | BEL | TC10 (ETB) | ' | 7 | G | W | g | w |
| 1 0 0 0 — 8 | FE0 (BS) | CAN | ( | 8 | H | X | h | x |
| 1 0 0 1 — 9 | FE1 (HT) | EM | ) | 9 | I | Y | i | y |
| 1 0 1 0 — 10 | FE2 (LF) | SUB | * | : | J | Z | j | z |
| 1 0 1 1 — 11 | FE3 (VT) | ESC | + | ; | K | [ | k | { |
| 1 1 0 0 — 12 | FE4 (FF) | IS4 (FS) | , | < | L | \ | l | | |
| 1 1 0 1 — 13 | FE5 (CR) | IS3 (GS) | - | = | M | ] | m | } |
| 1 1 1 0 — 14 | SO | IS2 (RS) | . | > | N | ^ | n | ‾ |
| 1 1 1 1 — 15 | SI | IS1 (US) | / | ? | O | _ | o | DEL |

338

## 6.3 National versions

**6.3.1** The responsibility for defining national versions lies with the national standardization bodies. These bodies shall exercise the options available and make the required selection.

**6.3.2** If so required, more than one national version can be defined within a country. The different versions shall be separately identified. In particular when for a given national use position, for example, 5/12 or 6/0, alternative characters are required, two different versions shall be identified, even if they differ only by this single character.

**6.3.3** If there is in a country no special demand for specific characters, it is strongly recommended that the characters of the international reference version be allocated to the same national use positions.

## 6.4 Application-oriented versions

Within national or international industries, organizations or professional groups, application-oriented versions can be used. They require precise agreement among the interested parties, who will have to exercise the options available and to make the required selection.

## 7 FUNCTIONAL CHARACTERISTICS RELATED TO CONTROL CHARACTERS

Some definitions in this section are stated in general terms and more explicit definitions of use may be needed for specific implementation of the code table on recording media or on transmission channels. These more explicit definitions and the use of these characters are the subject of other ISO Publications.

## 7.1 General designations of control characters

The general designation of control characters involves a specific class name followed by a subscript number.

They are defined as follows :

**TC   Transmission control characters**

Control characters intended to control or facilitate transmission of information over telecommunication networks.

The use of the TC characters on the general telecommunication networks is the subject of other ISO publications.

The transmission control characters are :

ACK, DLE, ENQ, EOT, ETB, ETX, NAK, SOH, STX and SYN.

**FE   Format effectors**

Control characters mainly intended for the control of the layout and positioning of information on printing and/or display devices. In the definitions of specific format effectors, any reference to printing devices should be interpreted as including display devices.

The definitions of format effectors use the following concept :

a)  a page is composed of a number of lines of characters;

b)  the characters forming a line occupy a number of positions called character positions;

c)  the active position is that character position in which the character about to be processed would appear, if it were to be printed. The active position normally advances one character position at a time.

The format effector characters are : BS, CR, FF, HT, LF and VT (see also Note 1 ).

**DC   Device control characters**

Control characters for the control of a local or remote ancillary device (or devices) connected to a data processing and/or telecommunication system. These control characters are not intended to control telecommunication systems; this should be achieved by the use of TCs.

Certain preferred uses of the individual DCs are given in 7.2.

**IS   Information separators**

Control characters that are used to separate and qualify data logically. There are four such characters. They may be used either in hierarchical order or non-hierarchically; in the latter case their specific meanings depend on their applications.

When they are used hierarchically, the ascending order is :

US, RS, GS, FS.

In this case, data normally delimited by a particular separator cannot be split by a higher order separator but will be considered as delimited by any higher order separator.

## 7.2 Specific control characters

Individual members of the classes of controls are sometimes referred to by their abbreviated class name and a subscript number (for example, $TC_5$) and sometimes by a specific name indicative of their use (for example, ENQ).

Different but related meanings may be associated with some of the control characters but in an interchange of data this normally requires agreement between the sender and the recipient.

**ACK  Acknowledge**

A transmission control character transmitted by a receiver as an affirmative response to the sender.

**BEL   Bell**

A control character that is used when there is a need to call for attention; it may control alarm or attention devices.

**BS   Backspace**

A format effector which moves the active position one character position backwards on the same line.

**CAN   Cancel**

A character, or the first character of a sequence, indicating that the data preceding it is in error. As a result, this data is to be ignored. The specific meaning of this character must be defined for each application and/or between sender and recipient.

**CR   Carriage return**

A format effector which moves the active position to the first character position on the same line.

340

**Device controls**

**$DC_1$**   A device control character which is primarily intended for turning on or starting an ancillary device. If it is not required for this purpose, it may be used to restore a device to the basic mode of operation (see also $DC_2$ and $DC_3$), or for any other device control function not provided by other DCs.

**$DC_2$**   A device control character which is primarily intended for turning on or starting an ancillary device. If it is not required for this purpose, it may be used to set a device to a special mode of operation (in which case $DC_1$ is used to restore the device to the basic mode), or for any other device control function not provided by other DCs.

**$DC_3$**   A device control character which is primarily intended for turning off or stopping an ancillary device. This function may be a secondary level stop, for example wait, pause, stand-by or halt (in which case $DC_1$ is used to restore normal operation). If it is not required for this purpose, it may be used for any other ancillary device control function not provided by other DCs.

**$DC_4$**   A device control character which is primarily intended for turning off, stopping or interrupting an ancillary device. If it is not required for this purpose, it may be used for any other device control function not provided by other DCs.

Examples of use of the device controls :

1) One switching

|  |  |
|---|---|
| on—$DC_2$ | off—$DC_4$ |

2) Two independent switchings

| | | |
|---|---|---|
| first one | on—$DC_2$ | off—$DC_4$ |
| second one | on—$DC_1$ | off—$DC_3$ |

3) Two dependent switchings

| | | |
|---|---|---|
| general | on—$DC_2$ | off—$DC_4$ |
| particular | on—$DC_1$ | off—$DC_3$ |

4) Input and output switching

| | | |
|---|---|---|
| output | on—$DC_2$ | off—$DC_4$ |
| input | on—$DC_1$ | off—$DC_3$ |

**DEL   Delete**

A character used primarily to erase or obliterate an erroneous or unwanted character in punched tape. DEL characters may also serve to accomplish media-fill or time-fill. They may be inserted into or removed from a stream of data without affecting the information content of that stream but then the addition or removal of these characters may affect the information layout and/or the control of equipment.

**DLE   Data link escape**

A transmission control character which will change the meaning of a limited number of contiguously following characters. It is used exclusively to provide supplementary data transmission control functions. Only graphic characters and transmission control characters can be used in DLE sequences.

**EM   End of medium**

A control character that may be used to identify the physical end of a medium, or the end of the used portion of a medium, or the end of the wanted portion of data recorded on a medium. The position of this character does not necessarily correspond to the physical end of the medium.

**ENQ   Enquiry**

A transmission control character used as a request for a response from a remote station — the response may include station identification and/or station status. When a "Who are you" function is required on the general switched transmission network, the first use of ENQ after the connection is established shall have the meaning "Who are you" (station identification). Subsequent use of ENQ may, or may not, include the function "Who are you", as determined by agreement.

**EOT   End of transmission**

A transmission control character used to indicate the conclusion of the transmission of one or more texts.

**ESC   Escape**

A control character which is used to provide additional control functions. It alters the meaning of a limited number of contiguously following bit combinations. The use of this character is specified in ISO 2022.

**ETB   End of transmission block**

A transmission control character used to indicate the end of a transmission block of data where data is divided into such blocks for transmission purposes.

**ETX   End of text**

A transmission control character which terminates a text.

**FF   Form feed**

A format effector which advances the active position to the same character position on a pre-determined line of the next form or page.

**HT   Horizontal tabulation**

A format effector which advances the active position to the next pre-determined character position on the same line.

**Information separators**

**IS₁**   (US) A control character used to separate and qualify data logically; its specific meaning has to be defined for each application. If this character is used in hierarchical order as specified in the general definition of IS, it delimits a data item called a UNIT.

**IS₂**   (RS) A control character used to separate and qualify data logically; its specific meaning has to be defined for each application. If this character is used in hierarchical order as specified in the general definition of IS, it delimits a data item called a RECORD.

**IS₃**   (GS) A control character used to separate and qualify data logically; its specific meaning has to be defined for each application. If this character is used in hierarchical order as specified in the general definition of IS, it delimits a data item called a GROUP.

**IS₄**   (FS) A control character used to separate and qualify data logically; its specific meaning has to be defined for each application. If this character is used in hierarchical order as specified in the general definition of IS, it delimits a data item called a FILE.

**LF   Line feed**

A format effector which advances the active position to the same character position of the next line.

**NAK   Negative acknowledge**

A transmission control character transmitted by a receiver as a negative response to the sender.

**NUL   Null**

A control character used to accomplish media-fill or time-fill. NUL characters may be inserted into or removed from a stream of data without affecting the information content of that stream; but then the addition or removal of these characters may affect the information layout and/or the control of equipment.

**SI   Shift-in**

A control character which is used in conjunction with SHIFT-OUT and ESCAPE to extend the graphic character set of the code. It may reinstate the standard meanings of the bit combinations which follow it. The effect of this character when using code extension techniques is described in ISO 2022.

**SO   Shift-out**

A control character which is used in conjunction with SHIFT-IN and ESCAPE to extend the graphic character set of the code. It may alter the meaning of the bit combinations of columns 2 to 7 which follow it until a SHIFT-IN character is reached. However, the characters SPACE (2/0) and DELETE (7/15) are unaffected by SHIFT-OUT. The effect of this character when using code extension techniques is described in ISO 2022.

**SOH   Start of heading**

A transmission control character used as the first character of a heading of an information message.

**SP   Space**

A character which advances the active position one character position on the same line.

This character is also regarded as a non-printing graphic.

**STX   Start of text**

A transmission control character which precedes a text and which is used to terminate a heading.

**SUB   Substitute character**

A control character used in the place of a character that has been found to be invalid or in error. SUB is intended to be introduced by automatic means.

**SYN  Synchronous idle**

A transmission control character used by a synchronous transmission system in the absence of any othor character (idle condition) to provide a signal from which synchronism may be achieved or retained between data terminal equipment.

**VT  Vertical tabulation**

A format effector which advances the active position to the same character position on the next pre-determined line.

## APPENDIX Y

## RELEVANT ISO PUBLICATIONS

ISO/R 961, *Implementation of the 6 and 7-bit coded character sets on 7 track 12,7 mm (1/2 in) magnetic tape.*

ISO/R 962, *Implementation of the 7-bit coded character set on 9 track 12,7 mm (1/2 in) magnetic tape.*

ISO/R 963, *Guide for the definition of 4-bit character sets derived from the ISO 7-bit coded character set for information processing interchange.*

ISO/R 1113, *Representation of 6 and 7-bit coded character sets on punched tape.*

ISO/R 1155, *The use of longitudinal parity to detect errors in information messages.*

ISO/R 1177, *Character structure for start/stop and synchronous transmission.*

ISO/R 1679, *Representation of ISO 7-bit coded character set on 12-row punched cards.*

ISO/R 1745, *Basic mode control procedures for data communication systems.*

ISO 2022, *Code extension techniques for use with the ISO 7-bit coded character set.*

ISO 2047, *Graphical representations for the control characters of the ISO 7-bit coded character set.*

## APPENDIX Z

ISO Recommendation R 646-1967 contained a table for a 6-bit coded character set. This character set has been deleted from the present International Standard and is no more part of the standard. The corresponding table with the relevant Notes is reproduced in this Appendix for information only.

| Bits | $b_6$ $b_5$ $b_4$ $b_3$ $b_2$ $b_1$ | Column / Row | 0 (0 0) | 0 (0 1) | 1 (1 0) | 1 (1 1) |
|------|------|------|------|------|------|
| | | | 0 | 1 | 2 | 3 |
| | 0 0 0 0 | 0 | SP | 0 | NUL | P |
| | 0 0 0 1 | 1 | $F_1$ (HT) | 1 | A | Q |
| | 0 0 1 0 | 2 | $F_2$ (LF) ① | 2 | B | R |
| | 0 0 1 1 | 3 | $F_3$ (VT) | 3 | C | S |
| | 0 1 0 0 | 4 | $F_4$ (FF) | 4 | D | T |
| | 0 1 0 1 | 5 | $F_5$ (CR) ① | 5 | E | U |
| | 0 1 1 0 | 6 | SO | 6 | F | V |
| | 0 1 1 1 | 7 | SI | 7 | G | W |
| | 1 0 0 0 | 8 | ( | 8 | H | X |
| | 1 0 0 1 | 9 | ) | 9 | I | Y |
| | 1 0 1 0 | 10 | * | : ⑧ | J | Z |
| | 1 0 1 1 | 11 | + | ; ⑧ | K | ([) ③ |
| | 1 1 0 0 | 12 | , | <   \$ ② | L | (£) ②③ |
| | 1 1 0 1 | 13 | — | = ⑨ % | M | (]) ③ |
| | 1 1 1 0 | 14 | . | > & | N | ESC |
| | 1 1 1 1 | 15 | / | ' | O | DEL |

①     The controls CR and LF are intended for printer equipment which requires separate combinations to return the cariage and to feed a line.

For equipment which uses a single control for a combined carriage return and line feed operation,

    — in the 6-bit set table, the function $F_2$ will have the meaning of "New Line" (NL), $F_5$ will then have the meaning of "Back-space" (BS);

    — in the 7-bit set table, the function $FE_2$ will have the meaning of "New Line" (NL).

These substitutions require agreement between the sender and the recipeint of the data.

The use of this function "NL" is not allowed for international transmission on general telecommunication networks (Telex and Telephone networks).

②     For international information interchange, \$ and £ symbols do not designate the currency of a given country. The use of these symbols combined with other graphic symbols to designate national currencies may be the subject of other ISO Recommendations.

③     Reserved for National Use. These positions are primarily intended for alphabetic extensions. If they are not required for that purpose, they may be used for symbols and a recommended choice is shown in parenthesis in some cases.

Some restrictions are placed on the use of these characters on the general telecommunication networks for international transmission.

⑧     If 10 and 11 as single characters are needed (for example, for Sterling currency subdivision), they should take the place of "colon" (:) and "semi-colon" (;) respectively. These substitutions require agreement between the sender and the recipient of the data.

On the general telecommunication networks, the characters "colon" and "semi-colon" are the only ones authorized for international transmission.

⑨     Either of the two sets of three symbols shown in these positions in the table may be chosen; the interpretation of the corresponding combinations requires agreement between the sender and the recipient of the data.

# INTERNATIONAL STANDARD  1155

INTERNATIONAL ORGANIZATION FOR STANDARDIZATION ·МЕЖДУНАРОДНАЯ ОРГАНИЗАЦИЯ ПО СТАНДАРТИЗАЦИИ·ORGANISATION INTERNATIONALE DE NORMALISATION

# Information processing — Use of longitudinal parity to detect errors in information messages

First edition — 1973-10-01

345

UDC 681.3.042.4

Ref. No. ISO 1155-1973 (E)

**Descriptors** : data processing, error detection codes, parity check.

## FOREWORD

ISO (the International Organization for Standardization) is a worldwide federation of national standards institutes (ISO Member Bodies). The work of developing International Standards is carried out through ISO Technical Committees. Every Member Body interested in a subject for which a Technical Committee has been set up has the right to be represented on that Committee. International organizations, governmental and non-governmental, in liaison with ISO, also take part in the work.

Draft International Standards adopted by the Technical Committees are circulated to the Member Bodies for approval before their acceptance as International Standards by the ISO Council.

Prior to 1972, the results of the work of the Technical Committees were published as ISO Recommendations; these documents are now in the process of being transformed into International Standards. As part of this process, Technical Committee ISO/TC 97, *Computers and information processing,* has reviewed ISO Recommendation R 1155-1969 and found it technically suitable for transformation. International Standard ISO 1155 therefore replaces ISO Recommendation R 1155-1969, which was approved by the Member Bodies of the following countries :

| | | |
|---|---|---|
| Australia | Germany | Romania |
| Belgium | Greece | Spain |
| Brazil | Israel | Sweden |
| Canada | Italy | Switzerland |
| Czechoslovakia | Japan | Thailand |
| Denmark | New Zealand | United Kingdom |
| Egypt, Arab Rep. of | Peru | U.S.A. |
| France | Poland | U.S.S.R. |

No Member Body expressed disapproval of the Recommendation.

# Information processing — Use of longitudinal parity to detect errors in information messages

## 0 INTRODUCTION

In data communication systems the information formats and the redundancy in the data to be transmitted differ widely from one application to another. It is therefore clear that a number of classes of error protection systems may be required.

This International Standard defines one method of error detection which satisfies a wide range of applications. It consists of accompanying the data block or text by one checking character (in addition to character parity) and it is often referred to as the "Longitudinal Parity Method".

## 1 SCOPE AND FIELD OF APPLICATION

This International Standard specifies a method for detecting errors in information messages by attaching one block check character to the transmitted information block (or text) and checking this character when it is received. The method of correcting errors when they are detected is specified in ISO...[1]

The method is applicable to systems which use the 7-bit coded character set which is the subject of ISO 646, *7-bit coded character set for information processing interchange,* and the basic mode of implementing this 7-bit code in data communication systems, which is the subject of ISO 1745, *Information processing — Basic mode control procedures for data communication systems*[2].

The rules for generating the character parity bits, according to ISO 1177, *Information processing — Character structure for start/stop and synchronous transmission,* are that the character parity sense shall be odd in synchronous systems and even in asynchronous systems.

## 2 RULES FOR GENERATING THE LONGITUDINAL PARITY BLOCK CHECK CHARACTER

### 2.1 Block check character

**2.1.1** The block check character shall be composed of 7 bits plus a parity bit.

**2.1.2** Each of the first 7 bits of the block check character shall be the modulo 2 binary sum of every element in the same bit 1 to bit 7 column of the successive characters of the transmitted block.

**2.1.3** The longitudinal parity of each column of the block, including the block check character, shall be even.

**2.1.4** The sense of the parity bit of the block check character shall be the same as for the information characters (odd for synchronous transmission, even for asynchronous transmission).

### 2.2 Summation

**2.2.1** The summation to obtain the block check character shall be started by the first appearance of either SOH (Start of Heading) or STX (Start of Text).

**2.2.2** The starting character shall not be included in the summation.

**2.2.3** If an STX character appears after the summation has been started by SOH, then the STX character shall be included in the summation as if it were a text character.

**2.3** With the exception of SYN (Synchronous Idle), all the characters which are transmitted after the start of the block check summation shall be included in the summation, including the ETB (End of Transmission/Block) or ETX (End of Text) control character which signals that the next following character is the block check character.

**2.4** No character, SYN or otherwise, shall be inserted between the ETB or ETX character and the block check character.

---

1) At present at the stage of draft proposal.

2) At present at the stage of draft. (Revision of ISO/R 1745.)

# INTERNATIONAL STANDARD  1177

INTERNATIONAL ORGANIZATION FOR STANDARDIZATION ·МЕЖДУНАРОДНАЯ ОРГАНИЗАЦИЯ ПО СТАНДАРТИЗАЦИИ·ORGANISATION INTERNATIONALE DE NORMALISATION

# Information processing — Character structure for start/stop and synchronous transmission

348

**First edition — 1973-11-01**

# FOREWORD

ISO (the International Organization for Standardization) is a worldwide federation of national standards institutes (ISO Member Bodies). The work of developing International Standards is carried out through ISO Technical Committees. Every Member Body interested in a subject for which a Technical Committee has been set up has the right to be represented on that Committee. International organizations, governmental and non-governmental, in liaison with ISO, also take part in the work.

Draft International Standards adopted by the Technical Committees are circulated to the Member Bodies for approval before their acceptance as International Standards by the ISO Council.

Prior to 1972, the results of the work of the Technical Committees were published as ISO Recommendations; these documents are now in the process of being transformed into International Standards. As part of this process, Technical Committee ISO/TC 97, *Computers and information processing*, has reviewed ISO Recommendation R 1177-1970 and found it technically suitable for transformation. International Standard ISO 1177 therefore replaces ISO Recommendation R 1177-1970, which was approved by the Member Bodies of the following countries :

| | | |
|---|---|---|
| Australia | Greece | Sweden |
| Belgium | Israel | Switzerland |
| Brazil | Italy | Thailand |
| Canada | Japan | Turkey |
| Czechoslovakia | New Zealand | United Kingdom |
| Denmark | Poland | U.S.A. |
| France | Romania | U.S.S.R. |
| Germany | Spain | |

No Member Body expressed disapproval of the Recommendation.

# Information processing — Character structure for start/stop and synchronous transmission

## 1 SCOPE AND FIELD OF APPLICATION

This International Standard specifies the character structure to be used for serial-by-bit start/stop and synchronous data transmissison systems using the 7-bit coded character set which is the subject of ISO 646, *7-bit coded character set for information processing interchange*[1].

It applies to the information transfer through the interface standardized by the CCITT and the IEC/ISO between the data terminal equipment and data communications equipment as defined in CCITT Recommendation V 24, and the relevant CCITT modem recommendations.

## 2 BIT SEQUENCING — START/STOP AND SYNCHRONOUS OPERATION

In serial working data transmission systems, the chronological order of transmission of the information bits shall correspond to the bit identification $b_1$ to $b_7$ as defined in the 7-bit code table of ISO 646 with least significant bit transmitted first.

When the rank in the combination represents the order of the bit in binary numbering, the bits shall be transmitted in serial, working with the low order bit first.

The numerical meaning corresponding to each information bit considered in isolation is that of the digit :

0 for a unit corresponding to condition A (Travail = Space), and

1 for a unit corresponding to condition Z (Repos = Mark),

in accordance with the definitions of these conditions for two-condition transmission systems.

## 3 PARITY BIT — START/STOP AND SYNCHRONOUS OPERATION

A parity bit is added to every character and is located in the eight position, $b_8$, and is therefore transmitted after the seven signifiant bits for the character.

## 4 PARITY SENSE — START/STOP AND SYNCHRONOUS OPERATION

For asynchronous systems, the parity bit is chosen in such a way that the number of "ONE" bits is even in the sequence of eight bits thus formed.

For synchronous systems, the parity bit is chosen in such a way that the number of "ONE" bits is odd in the sequence of eight bits thus formed.

## 5 CHARACTER FRAMING

### 5.1 Start/Stop operation

In start/stop systems using the 7-bit coded character set (see ISO 646), ten or eleven unit elements shall be used per character.

The first information bit of the transmitted coded combinations shall be preceded by a start element corresponding to condition A (Travail = Space). The duration of this start element shall be one unit interval at the data signalling rate at the transmitting interface.

---

1) This character set is also standardized by CCITT : International Telegraphic Alphabet No. 5 Recommendation V 3.

The combination of seven information elements completed by its parity element shall be followed by a stop element corresponding to condition Z (Repos = Mark).

For systems using the 7-bit coded character set over the general switched telephone and telegraph networks with electromechanical data terminal equipment operating at modulation rates up to and including 200 bauds, the stop element duration at the transmitter shall be TWO unit intervals at the data signalling rate of the transmitter.

In other cases the use of a stop element with a duration of ONE unit interval is preferable. However, this is subject to mutual agreement between the administrations and/or recognized private operating agencies concerned.

Similar situations when a ONE unit interval stop element can be used may apply to leased circuits.

The start/stop receivers should be capable of correctly receiving start/stop signals comprising a single-unit stop element, whose duration will be reduced by a time interval equal to the deviation corresponding to the degree of gross start/stop distortion permitted at the receiver input. However, for electromechanical eqipment which must use a two-unit stop element (11-unit alphabet) with a modulation rate of 200 bauds or less, receivers should be capable of correctly receiving signals with a stop element reduced to one unit.

The time between the end of the stop element of a character and the beginning of the start element of the next character may be of any duration; the polarity of the signal during this time is the same as that of the stop element.

## 5.2 Synchronous operation

In synchronous systems using the 7-bit coded character set (see ISO 646), eight bits per character shall be used : the seven information bits followed by the parity bit.

The time between the end of the last bit of a character and the beginning of the first bit of the next character shall be zero time or a multiple of the unit interval at the data signalling rate of the transmitter. When character synchronism must be maintained, this time interval shall be zero or a multiple of the character interval. Where necessary, parity sense should be maintained.

# INTERNATIONAL STANDARD  ISO 1745

INTERNATIONAL ORGANIZATION FOR STANDARDIZATION ·МЕЖДУНАРОДНАЯ ОРГАНИЗАЦИЯ ПО СТАНДАРТИЗАЦИИ ·ORGANISATION INTERNATIONALE DE NORMALISATION

# Information processing — Basic mode control procedures for data communication systems

*Traitement de l'information — Procédures de commande pour transmission de données en mode de base*

**First edition — 1975-02-01**

352

UDC 681.327.18.01

Ref. No. ISO 1745-1975 (E)

**Descriptors :** data processing, data transmission, control procedures, control characters.

# FOREWORD

ISO (the International Organization for Standardization) is a worldwide federation of national standards institutes (ISO Member Bodies). The work of developing International Standards is carried out through ISO Technical Committees. Every Member Body interested in a subject for which a Technical Committee has been set up has the right to be represented on that Committee. International organizations, governmental and non-governmental, in liaison with ISO, also take part in the work.

Draft International Standards adopted by the Technical Committees are circulated to the Member Bodies for approval before their acceptance as International Standards by the ISO Council.

International Standard ISO 1745 was drawn up by Technical Committee ISO/TC 97, *Computers and information processing*, and circulated to the Member Bodies in May 1973.

It has been approved by the Member Bodies of the following countries :

| | | |
|---|---|---|
| Australia | Japan | Switzerland |
| Brazil | Netherlands | Thailand |
| Canada | New Zealand | Turkey |
| Czechoslovakia | Poland | United Kingdom |
| Egypt, Arab Rep. of | Romania | U.S.A. |
| France | South Africa, Rep. of | U.S.S.R. |
| Hungary | Spain | Yugoslavia |
| Italy | Sweden | |

The Member Bodies of the following countries expressed disapproval of the document on technical grounds :

Bulgaria
Germany

This International Standard cancels and replaces ISO Recommendation R 1745-1971, of which it constitutes a technical revision.

# CONTENTS

354

# Information processing — Basic mode control procedures for data communication systems

## 0 INTRODUCTION

### 0.1 General

A data communication system may be considered as the ensemble of the terminal installations and the interconnecting network that permits information to be exchanged.

A data link concept is identifiable when considering terminal installations connected to the same network, operating at the same speed, in the same code. Whenever actions on the respective transmission control characters take place, a separation of data links is constituted. Typical examples where this applies are : store and forward switching centres, concentrators, intermediate reformatting and speed-change devices.

The information transfer in a data link is monitored by data link control procedures where some characters, selected within a code set, are given particular meanings according to the transmission phase and are used for various purposes such as to delineate information, to reverse the direction of transmission, to ask questions, to answer, etc.

The data link control procedures are categorized in classes which are referred to as modes of operation. The present considerations relate to one class called "basic mode", which is defined as follows :

In the basic mode all the necessary transmission control information (for example, message framing and supervisory instructions) passing from one station to another is carried over the link by discrete control characters selected from the ten transmission control characters which are defined in the ISO/CCITT 7-bit code (ISO 646). The information exchanges are carried out in the alternate mode on standard communication facilities. The control of the data link is not affected by any characters other than the ten transmission control characters. Other codes than the ISO/CCITT code may therefore be transmitted provided that they do not contain any of the ten transmission control characters in either heading or text. Sequences of transmission control character combinations such as DLE.XXX are not permitted, with the one exception DLE.EOT which is defined as "Disconnect".

Extensions to the basic mode are contained in the following International Standards :

ISO 2111, *Basic mode control procedures — Code independent information transfer;*

ISO 2628, *Basic mode control procedures — Complements;*

ISO 2629, *Basic mode control procedures — Conversational information message transfer;*

and also in annexes B and C of this International Standard.

The following considerations have been taken into account in developing the rules for the basic mode :

The rules are based on the assumption that one of the stations in each connection would be either a computer or a device capable of handling automatically an exchange of information. The rules are designed to allow the complexity of operation to be increased from a basic level by adding options. These options are designed so that any number of stations can still communicate even though they normally operate at different levels of complexity.

It is desirable to reduce optional features in this International Standard to a minimum, but still retain a balance between an economic solution for the "low cost systems" and extendability for encompassing more complex systems. The rules may be difficult to implement in very simple systems involving low cost devices and human control. On the other hand, in complex high speed computer links, the rules may seriously restrict the throughput of information. These two cases are regarded as the upper and lower fringes of the present International Standard and may be the subject of future International Standards.

With the above considerations, typical limitations of basic mode control procedures are :

— restriction of efficiency by the time delay which is due to the alternate mode of operation;

— single link operation only.

## 0.2 Communication phases

The table below shows the various possible phases and sub-phases of a data communication.

Phases 1 and 5, which relate to the establishment and clearing of connections over the general switched network, are under the responsibility of the CCITT and are therefore not covered by this International Standard.

In each phase, one of the stations directs the operation and is responsible for the continuity of the communication. The other station or stations only react to the actions of the responsible station.

The transmission control characters which are shown alongside the various sub-phases are those which are involved in the basic mode of operation.

EOT is shown in parentheses in Phases 2 and 3 because its use within the phases initiates a changeover to Phase 4.

## 1 SCOPE AND FIELD OF APPLICATION

### 1.1 General

This International Standard specifies the method of implementation of the ISO/CCITT 7-bit coded character set[1] for information interchange on data transmission channels. It also defines the formats of the transmitted messages and the supervisory sequences which are part of the transmission control procedures. It covers the majority of existing data transmission systems and data link configurations used in conjunction with data processing systems.

These control procedures deal with transmission over one link at a time and do not describe the operation of data links in "tandem". They relate to the class of control procedures which is known as the basic mode and apply at the interface between data communication equipment and data terminal equipment.

**Table of phases**

| Phase | | | Function: Action | Function: Reaction | Station's name: Responsible | Station's name: Responsive | Transmission control characters used in basic mode: Forward | Transmission control characters used in basic mode: Backward | Notes |
|---|---|---|---|---|---|---|---|---|---|
| 1 | Establishment of connection over general network | a) Switching | | | | | | | CCITT Responsibility |
| | | b) Identification | | | | | | | |
| 2 | Establishment of data link | a) Switching | Call | Answer | Calling | Called | | | Not covered at present |
| | | b) Polling | Poll | Reply | Control | Tributary | (EOT), ENQ | (EOT) | |
| | | c) Selecting | Select | Reply | Master | Slave | (EOT), ENQ | ACK, NAK | |
| 3 | Information transfer | | Transfer | Supervision | Master | Slave | SOH, STX ETB, ETX, (EOT) | ACK, NAK, (EOT) | |
| 4 | Termination | a) Return to neutral state | Terminate Interrupt | | Master | Slave | EOT | EOT | |
| | | b) Return to control station | Terminate Interrupt | | Master | Slave | EOT | EOT | |
| | | c) Disconnect | Disconnect Disconnect | | Master | Slave | DLE, EOT | DLE, EOT | |
| 5 | Clearing of connection | | | | | | | | CCITT Responsibility |

---

1) See ISO 646. CCITT : Alphabet No. 5.

It is accepted that, in their present form, the control procedures are a framework upon which a system can be built and that, before the successful interconnection of equipment from different supplies can be ensured, it will be necessary to define additional details, such as :

— structure of prefixes or addresses when used;

— "time-out" procedures and the recovery procedures which follow the various time-out conditions (see ISO 2628).

This International Standard must be considered in conjunction with the following ISO publications :

1) ISO 1177, *Information processing — Character structure for start/stop and synchronous transmission;*

2) ISO 1155, *Information processing — Use of longitudinal parity to detect errors in information messages.*

### 1.2 Assumptions

1) The information to be transmitted will normally be coded in accordance with the 7-bit ISO/CCITT code.

2) All transmission control functions will be performed by the use of ten specific transmission control characters which are defined in this code as TC 1 to TC 10.

3) No recommendation is made regarding

— the technique used (hardware or software);

— the part of the terminal installation where the information messages and supervisory sequences are generated and recognized.

4) Transmission may be at any data transfer rate, either serial or parallel and either start/stop or synchronous.

5) Responses to an information message or a supervisory sequence may be either by turn around of the channel or by using another channel.

6) The basic mode control procedures are applicable to systems of varied complexity based on either-way transmission using :

a) One-way transfer of information with alternate supervision on the same channel.

b) One-way transfer of information with alternate supervision on a separate channel.

c) Alternate two-way transfer of information with alternate supervision on the same channel.

d) Alternate two-way transfer of information with alternate supervision on separate channels.

7) The following cases will be the subject of further study :

a) One-way transfer of information with simultaneous supervision.

b) Alternate two-way transfer of information with simultaneous supervision.

c) Two-way simultaneous transfer of information with alternate supervision.

d) Two-way simultaneous transfer of information with simultaneous supervision.

## 2 DEFINITIONS OF THE TRANSMISSION CONTROL CHARACTERS

The basic definitions of the ten transmission control characters, as taken from ISO 646, are listed below (see clause 5 for description of use).

**(TC1) SOH Start of heading**

A transmission control character used as the first character of a heading of an information message.

**(TC2) STX Start of text**

A transmission control character which precedes a text and which is used to terminate a heading.

**(TC3) ETX End of text**

A transmission control character which terminates a text.

**(TC4) EOT End of transmission**

A transmission control character used to indicate the conclusion of the transmission of one or more texts.

**(TC5) ENQ Inquiry**

A transmission control character used as a request for a response from a remote station — the response may include station identification and/or station status. When a "Who are you" function is required on the general switched transmission network, the first use of ENQ after the connection is established shall have the meaning "Who are you" (station identification). Subsequent use of ENQ may, or may not, include the function "Who are you", as determined by agreement.

**(TC6) ACK Acknowledge**

A transmission control character transmitted by a receiver as an affirmative response to the sender.

**(TC7) DLE Data link escape**

A transmission control character which will change the meaning of a limited number of contiguously following characters. It is used exclusively to provide supplementary data transmission control functions. Only graphic characters and transmission control characters can be used in DLE sequences.

(TC8) **NAK   Negative acknowledge**

A transmission control character transmitted by a receiver as a negative response to the sender.

(TC9) **SYN   Synchronous idle**

A transmission control character used by a synchronous transmission system in the absence of any other character (idle condition) to provide a signal from which synchronism may be achieved or retained between data terminal equipment.

(TC10) **ETB   End of transmission block**

A transmission control character used to indicate the end of a transmission block of data where data is divided into such blocks for transmission purposes.

## 3   MESSAGE FORMATS

The various possible messages are categorized as follows :

— information messages;

— forward supervisory sequences;

— backward supervisory sequences.

### 3.1   General rules

Every transferred sequence of characters contains at least one transmission control character. These are used either to define the nature of the information contained in a sequence of data or to convey supervisory functions.

— They must not be considered information. Therefore, they must not be transmitted as part of the text or heading of an information message with the exception of SYN which may be inserted as required but which must not be regarded as information.

— When used singly or at the end of a message or sequence, they invite the station receiving them to take action.

#### a)   *Information messages*

Information messages consist of a text which can be preceded by a heading; the heading is delivered with the text. Routing indication, for intermediate points in particular, must be in the heading. Other auxiliary information may be either in the heading or in the text.

SOH, STX, ETB and ETX are used as information framing characters. They cannot be sent singly.

Information messages, or information blocks, may be accompanied by a block checking character in accordance with ISO 1155. The use of this block checking character, shown in parentheses, is optional and therefore subject to prior agreement.

#### b)   *Supervisory sequences*

All supervisory sequences except DLE.EOT are composed of either a single transmission control character or a single transmission control character preceded by one or several graphics.

In some of the following supervisory sequences the meaning of the character or characters which precede the transmission control character is defined (for example Polling address). In others, it is simply shown as a prefix which may include one or more of the following :

— identity information;

— address information;

— status information;

— any other qualifier as necessary (for example response number).

The use of these prefixes and their description is subject to prior agreement. They may be standardized at a later date.

EOT, ENQ, ACK and NAK are used for supervision. They can never appear contiguously.

The prefix must not contain more than 15 characters.

### 3.2   Information messages

a)   S
     T – – TEXT – – T   (B / C / C)
     X             X

b)   S
     T – – TEXT – – T   (B / C / C)
     X             B

(See note 2, below)

c)   S                 S              E (B / C / C)
     O – – HEADING – – T – – TEXT – – T
     H                 X              X

d)   S                 S              E (B / C / C)
     O – – HEADING – – T – – TEXT – – T
     H                 X              B

(See note 2, below)

e)   S                       E (B / C / C)
     O – – HEADING – – T
     H                       B

(See note 2, below)

NOTES

1   Fillers may be inserted in the heading and the text (for example SYN).

2   In formats b), d) and e) above which end with ETB, some continuation is required.

3   All the above messages can be aborted by terminating them at any point with EOT. Future study may lead to the specifying of another method for aborting which allows the continuation of the communication.

### 3.3 Forward supervisory sequences

a) Polling

```
                          E
        Polling address   N
                          Q
```

(See note below)

b) Selecting

1) Station selection

```
                          E
        Selecting address N
                          Q
```

(See note below)

If a reply is not required, ENQ is not used and the selecting sequence is immediately followed by the information message.

2) Identification and status

```
                  E
        (Prefix)  N
                  Q
```

3) Out of neutral

```
                  E
        (Prefix)  N
                  Q
```

c) Return to control station — Return to neutral state

```
                  E
        (Prefix)  O
                  T
```

(See note below)

d) Disconnect

```
                  D   E
        (Prefix)  L   O
                  E   T
```

NOTE — Polling sequences are always preceded by EOT except in systems involving Phase 1 where the omission of EOT is optional. Selecting sequences may also be preceded by EOT.

Some systems may not be able to tolerate a polling or selecting sequence immediately following EOT. In such cases it may be necessary to ensure a short delay between the EOT and the address by using, for example, a number of "filler" characters.

### 3.4 Backward supervisory sequences

a) Positive reply to :

— an information message

— selecting

```
                  A
        (Prefix)  C
                  K
```

b) Negative reply to :

— an information message

```
                  N
        (Prefix)  A
                  K
```

c) Negative reply to :

— a polling supervisory sequence

```
                  E
        (Prefix)  O
                  T
```

d) Negative reply to :

— a selecting supervisory sequence

```
                  N
        (Prefix)  A
                  K
```

e) Request for :

— an interruption

```
                  E
        (Prefix)  O
                  T
```

— a return of responsibility to the control station

```
                  E
        (Prefix)  O
                  T
```

— return to neutral state

```
                  E
        (Prefix)  O
                  T
```

(See note 2, below)

f) Disconnect

```
                  D   E
        (Prefix)  L   O
                  E   T
```

NOTES

1 The procedures for the cases of "no reply" are covered in 4.3.

2 Future study may lead to replacing the interruption by EOT with another method.

## 4 DESCRIPTION OF PHASES

The operational procedures of a complete system can be constructed from the following separate phases and sub-phases :

**Phase 1**[1] Establishment of connection over the general network

a) Switching

b) Identification

---

1) This phase is under the responsibility of the CCITT.

**Phase 2** Establishment of data link

    a)   Switching

    b)   Polling

    c)   Selecting

**Phase 3** Information transfer

**Phase 4** Termination

    a)   Return to neutral state

    b)   Return to control station

    c)   Disconnect

**Phase 5**[1] Clearing of connection

### 4.1  Phase linkage

Figure 1 represents the various phases of a communication which are linked (thick lines) to achieve one transmission, or information transfer, in the most general case encompassed by the basic mode control procedures.

The sequence of events for such a communication would be as follows :

**Phase 1** a), b) Establishment of connection over the general switched network

    Here the connection is established by the telecommunications administration and this is likely to be divided into two sub-phases : "Switching" and "Identification". They will both be under the responsibility of the administration.

    Unless otherwise stipulated by the administration, once this phase is achieved, the calling station takes on the responsibility for the communication and acts as master station or control station.

    Means for signalling the completion of Phase 1 will be defined with reference to Recommendations on interfaces (for example, CCITT-V24, Circuit 107).

**Phase 2** a), b), c) Establishment of data link

    After establishing the Connection on the general network it is required to establish the data link. This procedure may involve some private line Switching performed by a private Switching exchange or a line concentrator before polling and selecting.

    The "Polling" procedure, carried out by the control station, invites a tributary station to transmit any message it may have.

    This procedure transfers the responsibility of the communication to the polled station, which takes the status of master station.

    The "Selecting" procedure, carried out by the so-designated master station, invites in turn another station to get ready to receive an information message.

    This procedure gives to the selected station the status of slave station.

**Phase 3**  Information transfer

    Assuming the slave station(s) has accepted to receive the information message, the master station commences its transmission.

    During this phase there are no changes of station status or responsibility.

**Phase 4** a), b), c)  Termination

    When the information message has been transmitted and satisfactorily received by the slave station(s), the master station sends EOT to announce to the control station that its transmission requirement has temporarily ceased. By doing so, the master station relinquishes its master status and returns the responsibility of the communication to the control station.

    If there are no further transmission requirements, the control station, by sending DLE.EOT, releases the possibly involved private switching equipment.

**Phase 5**  Clearing of connection (General network)

    The Disconnect function (DLE.EOT) of the termination phase will initiate the clearing of the connection over the general switched network. The procedure for so doing is under the responsibility of the administration.

As a matter of fact, in most systems, several data link establishments and several information transfers take place in sequence within a communication.

This is illustrated by the phase linkage arrows marked PL 1, 2,...,6. An example of such a multiple communication could be :

    Phase 1 a), b) — We reach a multistation link via the general network;

    Phase 2 c) — We try to select station X;

    Phase 4 a) — Station X refuses to receive;

    Phase 2 b) — We poll station Y;

    Phase 3 — Station Y transmits information to us;

    Phase 4 — Station Y terminates its transmission;

    Phase 4 c) — We decide to disconnect;

    Phase 5 — The general network is cleared.

All the permitted phase linkages are shown on the phase diagrams in 4.2, along with more detailed descriptions of the phases and sub-phases.

In some systems, not all the phases or sub-phases shown on the phase linkage diagram will be required. Examples are illustrated by different by-passes :

**By-pass 1 (BP1)**

This by-pass applies to systems composed entirely of leased or private circuits not connected to the general switched network.

---

1)  This phase is under the responsibility of the CCITT.

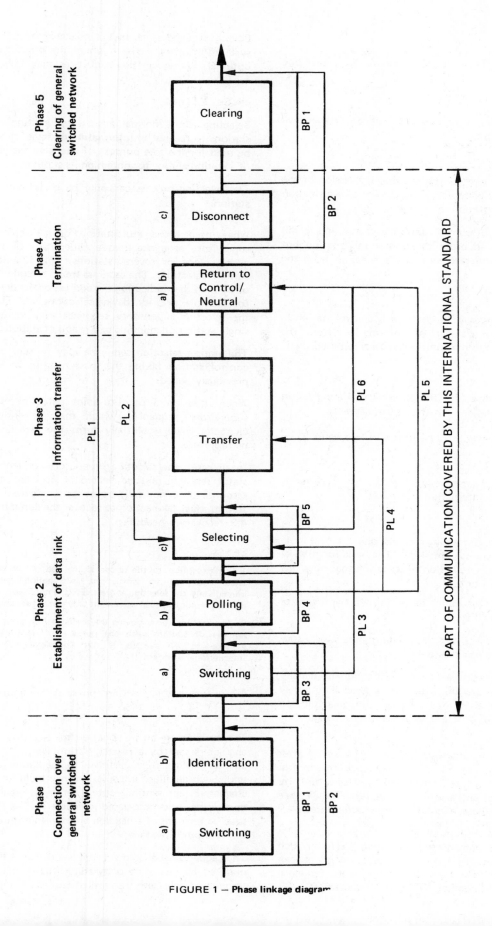

FIGURE 1 — Phase linkage diagram

**By-pass 2 (BP2)**

This by-pass applies to systems which do not involve line switching.

**By-pass 3 (BP3)**

This by-pass applies to systems that do not involve private line switching.

**By-pass 4 (BP4)**

In systems with control station, the suppression of the polling sub-phase allows only the sending of information from the control station to the others.

In systems without control station, each station can still select the other in order to transfer information but cannot poll in order to request transmission from the other end.

**By-pass 5 (BP5)**

This by-pass applies to systems with control station in which only the transfer of information from the tributary stations to the control station is required.

**4.2 Phase diagrams**

The detailed procedures for the phases 2 to 4 are given in the following text and illustrated by flow diagrams.

**4.2.1** *Establishment of data link* (Phase 2) (see figure 2)

a) SWITCHING

A private switching process may be used. This, however, is not described in this International Standard.

b) POLLING

Polling is the process of inviting stations, one at a time in an orderly fashion, to transmit messages. The basic function of polling is to prevent contention by ensuring that only one station transmits at a time.

The polling process can only be performed by a control station, following EOT.

When a station receives its appropriate polling supervisory sequence, it becomes the master station.

Each polling supervisory sequence must uniquely identify one station on the data link. However, a given station may be assigned more than one address (for example, to distinguish between different precedences of originating traffic).

If no response or a non-valid response is received after transmission of a polling supervisory sequence, the control station must clear the possibly established data link by sending a termination supervisory sequence (EOT, see 4.2.3, termination phase, and 4.3, recovery procedures).

Some systems may not be able to tolerate a polling or selecting sequence immediately following EOT. In such cases it may be necessary to assure a short delay between the EOT character and the polling or selecting sequence using, for example, a number of "filler" characters.

Polling sequences are always preceded by EOT except in systems involving Phase 1 where the omission of EOT is optional. Some selecting sequences may be preceded by EOT.

c) SELECTING

Selecting is the normal process for inviting one or more stations to receive an information message. However, it can be used for the sole purpose of checking the identification of a station and/or for obtaining its status.

The selecting process can only be performed by a master station.

When used on a multistation data link, the selecting supervisory sequence uniquely identifies, by means of its address, one or several stations. This function is called station selection. The address may include information other than that indicating the address of the desired station, for example, priority, device selections, etc. The address of the selecting supervisory sequence may identify either a single station on the link or a group of stations on the link.

The station selection sequence may be sent either by the control station taking the master status or by a station previously polled.

When used on a point-to-point data link, the selecting supervisory sequence, which may be limited to the character ENQ, essentially turns the data link out of its neutral state.

In all cases, the selecting supervisory sequence calls for a status reply from the selected or opposed station. If no response or a non-valid response is received, the master station must take action to recover the communication (see 4.3, recovery procedures).

NOTES

1  An exception to the above is the use of the so-called "Fast select" method in which the information message follows immediately the selecting address (ENQ is not used). The use of this method requires special agreement between sender and recipient.

2  The method for achieving the sequential or group selection of a number of stations with the purpose of transmitting the same message to all stations is not completely defined by this International Standard.

**4.2.2** *Information transfer* (phase 3) (see figure 3)

a) HEADING

The heading of an information message is a sequence of characters sent by a master station which constitutes the auxiliary information pertinent to the communication of a text. Such auxiliary information may include, for instance, characters representing routing, priority, security, message numbering, and associated characters. The definition of specific portions of a heading is not within the scope of this International Standard.

The sequence of characters which constitutes the heading is prefixed by the start of heading (SOH) character and is terminated only with the start of text (STX) character.

362

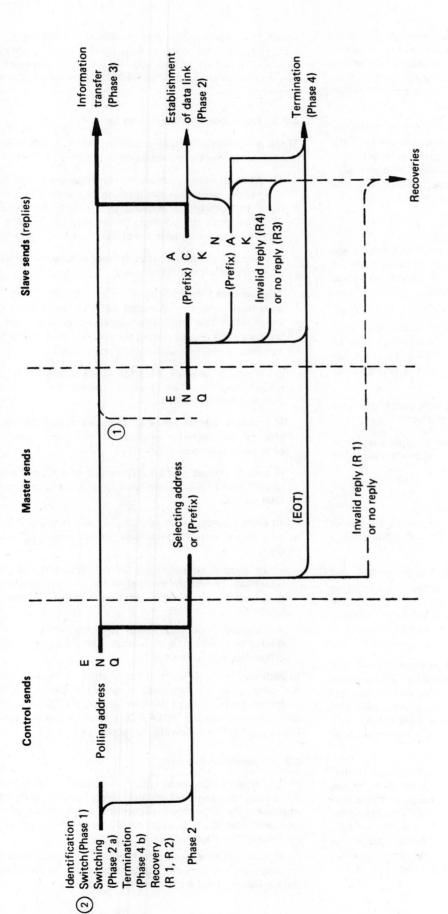

① For those systems where the information message follows the selecting address (Fast Select path), ENQ is not used.

② In most cases, the transmission of EOT is necessary after the connection has been established and before a polling or selecting supervisory sequence is transmitted.

FIGURE 2 — Phase 2 — Establishment of data link

The heading is not a "stand alone" message but must always be immediately followed by a text and is applicable only to that text. Any arrangement for association of one heading with more than one text can only be made by prior agreement between the affected parties and is not within the scope of this International Standard.

The heading may be sub-divided into more than one information block by terminating each such block with the end of transmission block (ETB) character and by prefixing the next following portion of the heading with the SOH character. The block check character (BCC), if used, immediately follows the ETB character (see ISO 1155).

### b) TEXT

The text portion of an information message contains the information that is to be transmitted as an entity from the sender to the recipient(s). A text is always embodied between the STX and ETX characters, and is always transmitted by a master station.

The text may be sub-divided into more than one information block. Each of the blocks is terminated by an end of block control character (ETB, or ETX if it is the last block of text). The following portion of the text must be prefixed with the STX character.

If block checking is employed, the block check character immediately follows the ETB or ETX character.

The master station stops sending after each text or block has been sent, and normally does not resume transmission until the reply has been received.

### c) REPLIES

Replies are used by the slave station to inform the master station of the status of the slave station and of the validity of the received message.

If the information message, or block, was acceptable, and the slave station is ready to receive, the slave station replies by transmitting the acknowledge character (ACK). The master station then transmits the next information message or block. If the master station has no more to transmit, it passes to termination phase.

If the information message or block is unacceptable, the slave station responds with the negative acknowledge character (NAK). This negative acknowledgement indicates to the master station that the data was unacceptable and also that the slave station is ready to receive. The next block of data transmitted is normally a retransmission of the previous information message or block.

If, during the information transfer, the slave station becomes unable to receive further, this station waits for the end of the transmitted information message or block and then responds by EOT. This EOT shall be interpreted by the other station(s) as a request for interruption and, according to the type of system used, either returns the responsibility of the communication to the control station or returns the data link to neutral state (see 4.2.3, termination phase).

Up to 15 characters may precede the final character of the reply sequence to convey information of a qualifying nature. The nature of this information is not a subject of this International Standard (see 3.1b), supervisory sequences formats).

**4.2.3** *Termination* (Phase 4) (see figure 4).

There are essentially three situations when a station may elect to terminate the transmission in progress :

a) When a station refuses the establishment of a data link, either because it has nothing to transmit (negative reply to polling), or because it is unable to transmit (negative reply to polling), or because it is unable to receive (negative reply to selecting).

b) When the master station has successfully transmitted all of the data it desires to send :

The master station then transmits the end of transmission control character (EOT), indicating to the slave station that the master station has no more data to transmit. The master station thus relinquishes its right to transmit (unless it is also the control station).

c) When an unusual situation arises where either the master or the slave station desires to stop the transmission in progress :

If a master station sends EOT at any time other than after a terminated transmission, the transmission in progress is said to be aborted.

If a slave station sends EOT instead of a normal reply (ACK, NAK), the transmission in progress is said to be interrupted.

In all circumstances, such as a), b), c), the sending of EOT by any of the stations terminates the transmission, that is to say :

— in systems comprising a control station, returns the responsibility and the control of the communication to that control station; this function is called "Return to Control";

— in point-to-point systems, without control station, returns the data link to neutral state; this function is called "Return to Neutral".

In addition, in all the cases of termination, if the clearing of the connection (private and/or general network) is the intended consequence of the termination, the "Disconnect" supervisory sequence (DLE.EOT) shall be sent, either instead of EOT or following the receipt of an EOT.

### 4.3 Recovery procedures

It has been recognized that a number of recovery procedures are required to deal with various abnormal situations. Some recovery procedures are outlined in the following and their linkage to the appropriate phase is outlined in the diagrams in 4.2.

In all cases, after appropriate time-out periods, it shall be the responsibility of either the control station or the master station (never of a slave station) to take action.

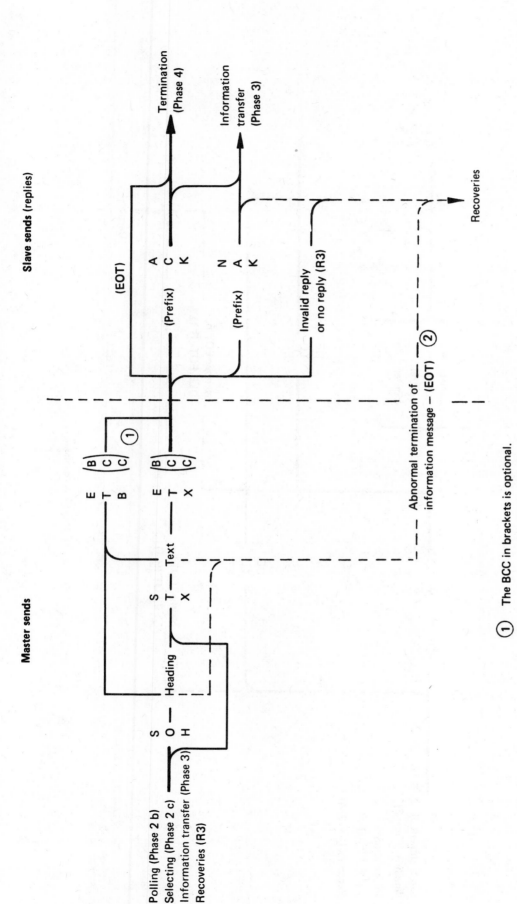

FIGURE 3 — **Phase 3 — Information transfer**

① The BCC in brackets is optional.

② EOT can abort the transfer at any point in the transfer phase.

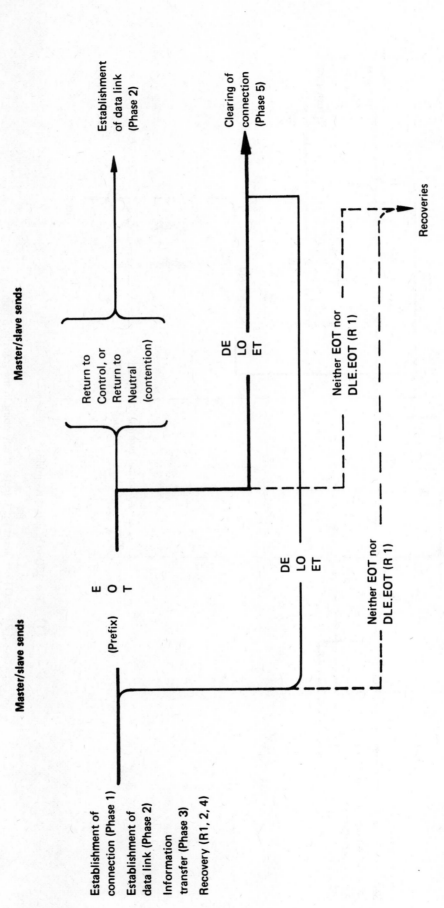

FIGURE 4 — **Phase 4** — Termination

**4.3.1** *Recovery procedures by control station*

**R1** — In the case of :

— no reply or invalid reply to a polling supervisory sequence, or

— invalid, or absence of termination supervisory sequence,

the control station must transmit EOT.

**R2** — In the case of repeated unsuccessful polling of one, several or all stations, the control station should, for example, set an alarm or report to the operator. Subsequent manual or automatic continuation is system-dependent and falls outside the scope of this International Standard.

**4.3.2** *Recovery procedures by master station*

**R3** — In the case of :

— no reply or invalid reply to an information message, or

— no reply or invalid reply to a selecting supervisory sequence,

the master station must repeat the previous transmission.

This procedure can lead to duplication of blocks. A possible alternative is :

In the case of no reply or an invalid reply to an information message, ENQ can be transmitted by the master station as a request to the slave station to repeat its previous response (ACK or NAK). This procedure is preferably used in conjunction with a response numbering scheme to ensure that blocks are neither added nor deleted.

**R4** — In the case of :

— repeated unsuccessful transmission of an information message, or

— repeated unsuccessful transmission of a selecting supervisory sequence, or

— repeated negative replies (NAK) to an information message,

the master station should set an alarm or report to the operator and go to the termination phase. Subsequent manual or automatic continuation is system-dependent and falls outside the scope of this International Standard.

## 5 DESCRIPTION OF USE OF THE TRANSMISSION CONTROL CHARACTERS

In the following, the definition of the ten transmission control characters already given in clause 2 is recalled and the functional description of their use is summarized. See also 3.1.

### SOH — Start of heading — TC 1

*Definition*

A transmission control character used as the first character of a heading of an information message.

*Description of use*

1) SOH is transmitted by a master station.

2) If a heading is used it must be preceded by SOH.

3) If a heading is sub-divided into transmission blocks, each block must be preceded again by SOH.

4) If block checking is used, see ISO 1155.

5) SOH is not permitted in text.

### STX — Start of text — TC 2

*Definition*

A transmission control character which precedes a text and which is used to terminate a heading.

*Description of use*

1) STX is transmitted by the master station.

2) A text must be preceded by STX.

3) If a heading is used, STX in the message indicates the end of the heading.

4) If the text is sub-divided into transmission blocks, each block must be preceded again by STX.

5) If block checking is used, see ISO 1155.

6) STX is not permitted in the heading.

### ETX — End of text — TC 3

*Definition*

A transmission control character which terminates a text.

*Description of use*

1) ETX is only transmitted by the master station.

2) ETX indicates the end of the text of each information message.

3) If an information message is sub-divided into transmission blocks, ETX is used to terminate the last block.

4) ETX calls for a reply from the slave station.

5) If block checking is used, ETX signals that the next following character is a block check character. See ISO 1155.

6) ETX is not permitted in text or heading.

### EOT — End of transmission — TC 4

*Definition*

A transmission control character used to indicate the conclusion of the transmission of one or more texts.

*Description of use*

1) EOT may be transmitted by a control, master, or slave station.

2) The control station transmits EOT to condition all tributary stations to anticipate the reception of a forward supervisory sequence.

3) The master station, in a system with control station, transmits EOT to relinquish its right to transmit in favour of the control station.

4) The master station, in a system without control station, transmits EOT to indicate either the end or the aborting of a transmission and resets the master and slave stations to the neutral state.

5) The slave station transmits EOT to indicate its inability to receive further information messages. This is an abnormal reply or interruption and leads into termination phase.

6) EOT is not permitted in text or heading.

NOTE — Depending upon the system characteristics and configuration, EOT transmitted by a tributary station (master or slave) may reset some or all tributary stations.

### ENQ — Enquiry — TC 5

*Definition*

A transmission control character used as a request for a response from a remote station — the response may include station identification and/or station status. When a "Who are you" function is required on the general switched transmission network, the first use of ENQ after the connection is established shall have the meaning "Who are you" (station identification). Subsequent use of ENQ may, or may not, include the function "Who are you", as determined by agreement.

*Description of use*

1) ENQ is transmitted by the control station during polling and by the master station during selecting.

2) In the polling sub-phase, ENQ is used to indicate the end of a polling address.

3) In the selecting sub-phase, ENQ is used to indicate the end of a selecting address or prefix when a reply is required from the slave station.

More specifically, ENQ can in this sub-phase

— terminate a station selection sequence;

— request identification and/or status;

— turn a data link out of neutral state.

4) ENQ is not permitted in text or heading.

### ACK — Acknowledge — TC 6

*Definition*

A transmission control character transmitted by a receiver as an affirmative response to the sender.

*Description of use*

1) ACK is transmitted only by a slave station as an affirmative reply to a master station.

2) When supplementary information is included in the reply (for example, station identification or status information) it is prefixed to ACK.

3) In the selecting sub-phase, ACK is transmitted as a reply to a selecting supervisory sequence to indicate that the slave station is ready to receive.

4) In the information transfer phase, ACK indicates that the last transmitted information message or block was received correctly, and that the slave station is ready to receive the next one.

5) ACK is not permitted in text or heading.

### DLE — Data link escape — TC 7

*Definition*

A transmission control character which will change the meaning of a limited number of contiguously following characters. It is used exclusively to provide supplementary data transmission control functions. Only graphics and transmission control characters can be used in DLE sequences.

*Description of use*

1) DLE immediately followed by EOT is transmitted by a master or a slave station to "disconnect", that is, to initiate the clearing of the connection over private and/or general switched network.

2) Other uses of DLE require prior agreement until additional DLE sequences are defined in future International Standards for modes other than the basic mode.

### NAK — Negative acknowledge — TC 8

*Definition*

A transmission control character transmitted by a receiver as a negative response to the sender.

*Description of use*

1) NAK is transmitted only by a slave station as a negative reply to the master station.

2) When supplementary information is included in the reply (for example, station identification or status information) it is prefixed to NAK.

3) The slave station transmits NAK after receipt of a selecting supervisory sequence to indicate its inability to receive an information message.

4) In the information transfer phase, NAK indicates that the last transmitted information message or block was not received correctly, and the slave station is ready to receive the same one.

5) NAK is not permitted in text or heading.

**SYN  — Synchronous idle — TC 9**

*Definition*

A transmission control character used by a synchronous transmission system in the absence of any other character (idle condition) to provide a signal from which synchronism may be achieved or retained between terminal equipments.

*Description of use*

1) SYN may be transmitted by a control, master, or slave station.

2) SYN is used to achieve character synchronization in synchronous data communication systems. At least two SYN characters must be transmitted prior to the transmission of any information message or supervisory sequence.

NOTE — It is assumed that the receiving station requires two consecutive SYN characters to reliably achieve character synchronization.

3) SYN can be used as "time-fill" to maintain, for instance, synchronization during periods when no other characters are available for transmission. When used as "time-fill", SYN may be added at any point in a character sequence, except :

a) between ETX or ETB and the block check character when block checking is implemented;

b) within DLE sequences.

4) SYN is generally removed at the receiving terminal installation.

5) If block checking is used, see ISO 1155.

**ETB  — End of transmission block — TC 10**

*Definition*

A transmission control character used to indicate the end of a transmission block of data where data is divided into such blocks for transmission purposes.

*Description of use*

1) ETB is transmitted only by the master station.

2) If an information message is sub-divided into transmission blocks, ETB is used to terminate each block, with the exception of the final one.

3) ETB calls for a reply from the slave station.

4) If block checking is used, ETB indicates that the next following character is a block check character (BCC). See ISO 1155.

## ANNEX A
(not part of the standard)

## DEFINITIONS

**A.1** The following definitions have been included for use with this International Standard since these definitions are not in accordance with those given in the corresponding sections of the ISO Information Processing Vocabulary (ISO 2382) :

**coded character set :** A finite set of characters arranged in a specified order according to given rules and conventions.

**filler :** A character that is used as a time or space fill when a block of a specified size is required and there are insufficient heading and/or text characters for this purpose.

**heading :** A sequence of characters preceding the text of a message. It enables the receiving station to handle the text(s).

**information block :** A sequence of characters of fixed or variable length which is a subdivision of an information message formed for the purpose of meeting transmission requirements.

**transmission control characters** (TCC) **:** Characters used either to define the nature of the information contained in a sequence of data characters or to convey supervisory instructions. They must not be transmitted as part of the text or heading.

**A.2** The following definitions have been drawn in part from the corresponding sections of the ISO Information Processing Vocabulary (ISO 2382).

**address :** A sequence of characters to select or poll another station.

**asynchronous transmission**[1] **:** A transmission process such that between any two significant instants in the same group (block or character) there is always an integral number of unit intervals. Between two significant intervals located in different groups there is not always an integral number of unit intervals.

**backward supervision :** The use of supervisory sequences sent from the slave to the master station.

**block check**[1] **:** A system of error control based on the check that some preset rules for the formation of blocks are observed.

**both-way :** A mode of operation on two channels so that transmission may occur simultaneously. One of the channels is equipped for transmission in one direction while the other is equipped for transmission in the opposite direction.

**channel**[1] **:** A means of one-way transmission. Several channels may share a common path as in carrier systems; in this case, each channel is allotted a particular frequency band which is reserved to it.

**code transparent transmission :** A transmission process which is capable of handling any character set or binary arrangement.

**connection :** The established path between two or more terminal installations. It is a permanent connection when it is established without using switching facilities, and a temporary connection when it is established by using switching facilities. It may consist of one or more channels in tandem.

**contention :** A condition arising on a communication channel when two or more stations try to transmit at the same time.

**control station :** The station on a network which supervises the procedures such as polling, selecting and recovery. It is also responsible for establishing order on the line in the event of contention, or any other abnormal situation, arising between any stations on the network.

**data communication system :** One or more data links each of which may be operating in the same or a different mode.

**data link :** An ensemble of terminal installations and the interconnecting network operating in a particular mode that permits information to be exchanged between terminal installations.

**data terminal equipment**[1] **:** Equipment comprising the data source, the data link or both.

**either-way :** A mode of operation of a channel to permit the transmission of signals in either direction. These transmissions cannot take place simultaneously.

**error :** Any received character or sequence of characters that does not conform to those transmitted.

370

---

1) Definition taken from CCITT yellow book and appropriate supplements.

**error control or error protection :** A procedure for detecting and reducing the effects of errors generated during the process of data transmission.

**forward supervision :** Use of supervisory sequences sent from the master to the slave station.

**identifier :** A sequence of one or more characters transmitted by a station in order to identify itself.

**information message :** A sequence of characters conveying the test. It may also convey supplementary information forming a heading.

**invalid reception :** A character or sequence of characters that has not been recognized in accordance with the expected character or sequence.

**master station :** The station which, at a given instant, has the right to select and to transmit an information message to a slave station and the responsibility of ensuring the information transfer. There should be only one master station on a data link at one time.

**multi-point network :** A configuration in which a connection is established between more than two terminal installations. The connection may include switching facilities.

**network :** The ensemble of equipment through which connections are made between terminal installations. These equipments operate in real time and do not introduce store and forward delays.

**one-way :** A mode of operation of a channel for one-way transmission of signals in a preassigned direction.

**parity bit :** A bit associated with a character or block for the purpose of checking the absence of error within that character or block. This is chosen to make the modulo 2 sum of the bits (including the parity bit) in the character or block a "0" or a "1" as required.

**parallel transmission[1] :** The simultaneous transmission of a certain number of signal elements constituting the same telegraph or data signal.

**point-to-point connection :** A configuration in which a connection is established between two, and only two, terminal installations. The connection may include switching facilities.

**polling :** The process of inviting another station to become a master station.

**prefix :** A sequence of characters (other than TCCs) used in a supervisory sequence to define or qualify the meaning of the supervisory sequence.

**query :** The process by which a master station asks a slave station to identify itself and to give its status.

**recovery procedure :** A process by which a responsible station within the network attempts to resolve either conflicting, or erroneous conditions arising in the communication process. The control or master station is responsible for this procedure.

**route :** The selected path between master station and slave station for the purpose of information transfer.

**selecting :** The process of inviting a station to receive.

**serial transmission[1] :** Transmission, at successive intervals, of signal elements constituting the same telegraph or data signal. The sequential elements may be transmitted with or without interruption provided that they are not transmitted simultaneously.

**slave station :** A station which, at a given instant, is intended to receive an information message from a master station.

**start element** (in a start/stop system) : Binary element serving to prepare the terminal installation for the reception of a character.

**start/stop transmission[1] :** Asynchronous transmission in which a group of code elements corresponding to a character signal is preceded by a start signal which serves to prepare the receiving mechanism for the reception and registration of a character and is followed by a stop signal which serves to bring the receiving mechanism to rest in preparation for the reception of the next character.

**station :** See terminal installation.

**status :** The capability at a given instant of a station to receive or transmit. It is indicated by a sequence of one or more characters which may be an acknowledgement of the previous data exchange.

---

1) Definition taken from CCITT yellow book and appropriate supplements.

371

**stop element** (in a start/stop system) : Binary element serving to bring the terminal installation to rest.

**supervisory sequence :** A sequence of one or more characters used for transmission control, whose structure does not follow the formatting rules applied to the information message.

**switching**[1] **:** Operations involved in interconnecting circuits in order to establish a temporary communication between two or more stations.

**synchronous transmission**[1] **:** A transmission process such that between any two significant instants there is always an integral number of unit intervals.

**terminal installation** (for data transmission) **or station**[1] **:** Installation comprising

— the data terminal equipment,

— the signal conversion equipment,

— and any intermediate equipment.

NOTE — In some instances the data terminal equipment may be connected directly to a data processing machine or may be a part of it.

**text :** A sequence of characters forming part of a transmission, which is transmitted as an entity to the ultimate destination and which contains the information to be conveyed.

**tributary station :** Any station on a non-switched multi-point network, other than the control station.

---

1) Definition taken from CCITT yellow book and appropriate supplements.

## ANNEX B
### (not part of the standard)

## EXTENSIONS OF TRANSMISSION CONTROL FUNCTIONS USING DLE SEQUENCES

This annex describes a standard method for providing additional transmission control functions through the definition of Extension Sequences using DLE.

The character DLE is used as a prefix to one or more additional ISO 7-bit characters to form a sequence to represent a communication control function not directly represented by a single control character.

Preference should be given to two-character sequences.

Two-character sequences consist of DLE followed by one "final" character. When Extension Sequences formed by more than two characters are required, the Extention Sequence begins with DLE, continues with any number of "intermediate" characters and end with one "final" character.

The intermediate characters are those in column 2 of the ISO 7-bit code table.

The final characters are the TC's and those in columns 3, 4, 5, 6 and 7, with the exception of DEL $\{7/15\}$ of the ISO 7-bit code table, i.e. the alphabetics, the numerics and several of the special graphics.

The final characters in columns 4, 5, 6 are for private use. The TC's and the characters in columns 3 and 7 with the exception of DEL $\{7/15\}$ are reserved for future international standardization.

## ANNEX C
### (not part of the standard)

## ALTERNATIVE POSITIVE ACKNOWLEDGEMENT OPTION

The following basic rules are recommended for alternate response numbering, if used :

1)   Use of DLE 3/0 instead of ACK in phase 2.

2)   Use of DLE 3/0 alternating with DLE 3/1 instead of ACK in phase 3, starting with DLE 3/1 as the acknowledgement for the first information block.

3)   Use of DLE 3/0 or DLE 3/1, as appropriate, in response to ENQ in phase 3.

# INTERNATIONAL STANDARD

INTERNATIONAL ORGANIZATION FOR STANDARDIZATION ·МЕЖДУНАРОДНАЯ ОРГАНИЗАЦИЯ ПО СТАНДАРТИЗАЦИИ·ORGANISATION INTERNATIONALE DE NORMALISATION

# Code extension techniques for use with the ISO 7-bit coded character set

374

**First edition — 1973-07-01**

UDC 681.3.042

Ref. No. ISO 2022-1973 (E)

**Descriptors** : data processing, data transmission, character sets, coding, extensions.

## FOREWORD

ISO (the International Organization for Standardization) is a worldwide federation of national standards institutes (ISO Member Bodies). The work of developing International Standards is carried out through ISO Technical Committees. Every Member Body interested in a subject for which a Technical Committee has been set up has the right to be represented on that Committee. International organizations, governmental and non-governmental, in liaison with ISO, also take part in the work.

Draft International Standards adopted by the Technical Committees are circulated to the Member Bodies for approval before their acceptance as International Standards by the ISO Council.

International Standard ISO 2022 was drawn up by Technical Committee ISO/TC 97, *Computers and information processing,* and circulated to the Member Bodies in June 1972.

It has been approved by the Member Bodies of the following countries :

| | | |
|---|---|---|
| Bulgaria | Italy | Thailand |
| Czechoslovakia | Japan | United Kingdom |
| Denmark | Netherlands | U.S.A. |
| France | Romania | U.S.S.R. |
| Germany | Sweden | |
| .Ireland | Switzerland | |

The Member Bodies of the following countries expressed disapproval of the document on technical grounds :

Australia
Belgium
South Africa, Rep. of

Printed in Switzerland

# CONTENTS

376

# Code extension techniques for use with the ISO 7-bit coded character set

## 1 SCOPE

**1.1** This International Standard specifies methods of extending the 7-bit code, remaining in a 7-bit environment or increasing to an 8-bit environment. These techniques are described in four inter-related sections dealing respectively with :

  a) the extension of the 7-bit code remaining in a 7-bit environment;

  b) the structure of a family of 8-bit codes;

  c) the extension of an 8-bit code remaining in an 8-bit environment;

  d) the relationship between the 7-bit code and an 8-bit code.

**1.2** While the 7-bit code of ISO 646 is the agreed code for information interchange, an 8-bit code as described in this International Standard is provided for information interchange within an 8-bit environment.

**1.3** It is not the intention of this International Standard that all instances of its application accommodate all of its provisions. However, it is intended that, when code extension techniques are used, the applicable parts of this International Standard are to be followed.

When two systems with different levels of implementation of code extension techniques are required to communicate with one another they will do so using the code extension techniques they have in common.

**1.4** Code extension techniques are classified and some classes are given a structure in this International Standard. Other assignments of bit combinations associated with the designation of the classes are made in accordance with ISO 2375. Specific assignments of bit combinations to relate individual codes with their invocation or designation are also to be made in accordance with that International Standard.

**1.5** Code extension techniques are designed to be used for data to be processed serially in a forward direction. Use of these techniques in strings of data which are processed other than serially in a forward direction or included in data formatted for fixed record processing may have undesirable results or may require additional special treatment to ensure correct interpretation.

## 2 FIELD OF APPLICATION

The 7-bit code of ISO 646 allows, through its different versions, the representation of up to 128 characters. Additionally, that document allows the representation of other graphics by the combination of two graphic characters with the back space control. In some instances the code of ISO 646 lacks sufficient controls or graphics to satisfy the needs of an application.

These needs may be satisfied by means of code extension which is the subject of this International Standard.

This International Standard presents a review of the salient structure of the 7-bit code and then builds upon that structure to describe various means of extending the control and graphic sets of the code. It also describes structures and techniques to construct and formalize codes related to the 7-bit code. These related codes are structured so as to allow application dependent usage without preventing the interchangeability of data employing them. This document describes :

a) the structure of the 7-bit code;

b) extension of the 7-bit code, remaining in a 7-bit environment and making use of code extension techniques;

c) increasing the number of bits to 8, yet retaining a structure compatible with the 7-bit structure;

d) increasing the number of bits to 8 and applying similar code extension techniques.

In order to use identical techniques in each of the above cases, and to facilitate conversion between them, standard rules are necessary. This has the advantage of :

a) reducing the risk of conflict between systems required to inter-operate;

b) permitting provision for code extension in the design of systems;

c) providing standardized methods of calling into use agreed sets of characters;

d) allowing the interchange of data between 7-bit and 8-bit environments, etc.

This International Standard also describes the structure of families of codes which are related to the code of ISO 646 by their structure.

## 3  REFERENCES

ISO 646, *7-bit coded character set for information processing interchange.*

ISO 2375, *Data processing — Procedure for registration of escape sequences.*[1]

## 4  DEFINITIONS AND NOTATION

### 4.1  Definitions

For the purpose of this International Standard, the following definitions apply :

**4.1.1  character** : A member of a set of elements that is used for the organisation, control or representation of data.

**4.1.2  code; coded character set** : A set of unambiguous rules that establish a character set and the one-to-one relationship between the characters of the set and their bit combinations.

**4.1.3  bit combination** : An ordered set of bits that represents a character.

**4.1.4  code table** : A table showing the character corresponding to each bit combination in a code.

**4.1.5  position** : An item in a code table identified by its column and row coordinates.

**4.1.6  byte** : A bit string that is operated upon as a unit and whose size is independent of redundancy or framing techniques.

**4.1.7  control function** : An action that affects the recording, processing, transmission or interpretation of data.

**4.1.8  control character** : A character whose occurrence in a particular context initiates, modifies or stops a control function.

**4.1.9  graphic character** : A character, other than a control character, that has a visual representation normally handwritten, printed or displayed.

**4.1.10  code extension** : Techniques for the encoding of characters that are not included in the character set of a given code.

**4.1.11  escape sequence** : A bit string that is used for control purposes in code extension procedures and that consists of two or more bit combinations. The first of these combinations corresponds to the escape character.

**4.1.12  final character** : The character whose bit combination terminates an escape sequence.

**4.1.13  intermediate character** : A character whose bit combination occurs between the escape character and the final character in an escape sequence consisting of more than two bit combinations.

**4.1.14  to designate** : To identify a set of characters that are to be represented, in some cases immediately and in others on the occurrence of a further control function, in a prescribed manner.

---

**4.1.15  to invoke** : To cause a designated set of characters to be represented by the prescribed bit combinations whenever those bit combinations occur, until an appropriate code extension function occurs.

**4.1.16  to represent** :

1)  To use a prescribed bit combination with the meaning of a character in a set of characters that has been designated and invoked.

2)  To use an escape sequence with the meaning of an additional control character.

**4.1.17  environment** : The characteristic that identifies the number of bits used to represent a character in a data processing or data communication system or in part of such a system.

**4.1.18  national version** : A code of 128 characters that is identical to the 7-bit coded character set of ISO 646 except in those positions in which ISO 646 makes provision for the assignment of alternative graphics and in those positions conforms to the requirements of ISO 646.

**4.2  Notation**

In this International Standard the following notations are used :

For the bits
of a 7-bit
combination :          $b_7$    $b_6$    $b_5$    $b_4$    $b_3$    $b_2$    $b_1$

For the bits
of an 8-bit
combination : $a_8$    $a_7$    $a_6$    $a_5$    $a_4$    $a_3$    $a_2$    $a_1$

Bit weight
for column
and row
reference :     $2^3$   $2^2$   $2^1$   $2^0$     $2^3$   $2^2$   $2^1$   $2^0$
                      $\underbrace{\hspace{3cm}}$        $\underbrace{\hspace{3cm}}$
                              column                              row

A bit combination is sometimes referred to by the column and row numbers of its position in the code table. The column number is the decimal equivalent of bits $b_7 - b_5$ (or $a_8 - a_5$) and the row number is the decimal equivalent of bits $b_4 - b_1$ (or $a_4 - a_1$), giving to these bits the weights shown above.

In representing the decimal equivalents, the convention is to append a leading zero to the column number for 8-bit columns 00 to 09. As an example the position of the "space" character in the 7-bit code table is 2/0; the position of the same character in an 8-bit code table is 02/0.

Character mnemonics such as SO, ESC and column/row numbers such as 0/5 and 1/7 are shown underlined to emphasize the fact that they stand for one bit-combination only.

# 5  EXTENSION OF THE 7-BIT CODE REMAINING IN A 7-BIT ENVIRONMENT

## 5.1  Introduction

### 5.1.1  *The structure of the 7-bit code*

The 7-bit code table which is the basis of code extension techniques for use with the 7-bit coded character set of ISO 646 consists of areas for an ordered set of control characters and graphic characters grouped as follows :

1)  the area for a set of 32 control characters allocated to columns 0 and 1;

2)  the space character in position 2/0 which may be regarded either as a control character or a non-printing graphic character;

3)  the area for a set of 94 graphic characters allocated to columns 2 to 7;

4)  the delete character in position 7/15.

This is shown in Figure 1.

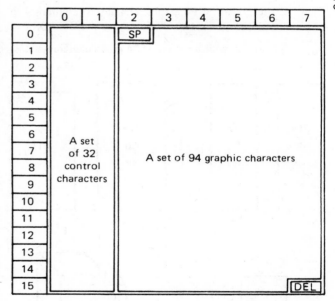

FIGURE 1 – **The structure of the 7-bit code**

### 5.1.2  *Extension by substitution*

In many cases the provisions of ISO 646 will satisfy the requirements of an application. Other applications will be satisfied by the use of a similarly structured code in which some of the characters of ISO 646 are substituted by other characters. Such substitution may be regarded as a replacement of the control set and/or the graphic set according as new controls and/or graphic characters are required.

**5.1.3** *Extension by increasing the repertoire of characters*

This International Standard provides for additional characters to the 128 provided by the structure of the 7-bit code in the following ways :

1) additional single controls;

2) additional sets of 32 control characters;

3) additional sets of 94 graphic characters;

4) additional sets of more than 94 graphic characters each represented by more than one byte.

**5.1.4** *The elements of code extension*

Many applications will require combinations of the above code extension facilities. The elements of code extension are shown in Figure 2, where the names of elements are defined as follows :

— C0 set : a set of 32 control characters (columns 0 and 1);

— C1 set : an additional set of 32 control characters;

— G0 set : a set of 94 graphic characters (columns 2 to 7) (a multiple byte set also functions as the G0 set);

— G1 set : an additional set of 94 graphic characters.

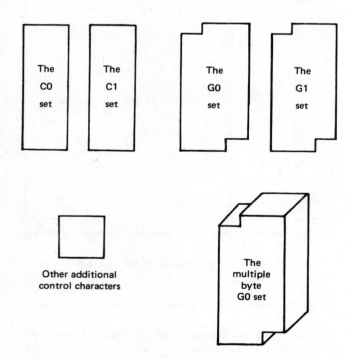

FIGURE 2 — **The elements of code extension**

NOTE — It is intended that a set of control characters and a set of graphic characters which are permitted by ISO 646, if they are used, are assigned to the C0 set and the G0 set respectively.

**5.1.5** *Compatibility*

For purpose of interchange there are identified various levels of compatibility which may be preserved when applying extension facilities. The following three such levels are distinguished in this International Standard :

1) a version permitted by ISO 646.

2) a compatible variant.

A compatible variant is defined as a set which is compatible with the ISO 646 inasmuch as

— columns 0 and 1 contain only control characters;

— columns 2 to 7 are used for graphic characters only (apart from DEL);

— the ten transmission control characters and NUL, SO, SI, CAN, SUB, ESC, SP and DEL remain unaltered in their meanings and in their positions in the code table;

— graphics of ISO 646 are not moved to other positions (a non-latin alphabet containing graphics which are also included in the latin alphabet is not subject to this rule).

3) other sets structured as in 5.1.1 above. To be able to provide the facilities of code extension of this International Standard, the escape and/or the shift-out and shift-in characters must remain unaltered in their meanings and their positions in the code table.

**5.1.6** *Code extension characters*

In the 7-bit code, the following characters are provided for the purpose of code extension :

— the escape character              ESC

— the shift-out character           SO

— the shift-in character             SI

— the data link escape character    DLE

This International Standard does not describe the use of the data link escape character which is reserved for the provision of additional transmission controls; the use of this character is specified in other ISO publications.

**5.2 Extension of the graphic set by means of the shift-out and shift-in characters**

**5.2.1** *Use of shift-out and shift-in*

The shift-out character SO and the shift-in character SI are used exclusively for extension of the graphics.

The character SO invokes an additional set of 94 graphics : the G1 set. This set replaces the graphic characters of the G0 set. Graphic characters need not be assigned to all the positions of the additional set, nor, except as specified below, need all the graphic characters of the additional set be different from the graphic characters of the G0 set.

The character SI invokes the graphic characters of the G0 set that are to replace the graphic characters of the additional set.

The meanings of the following bit combinations are not affected by the occurrence of SO and SI :

   1) those corresponding to the control characters in columns 0 and 1 and position 7/15;

   2) the one corresponding to the space character in position 2/0;

   3) those included in any escape sequence.

The space character occurs only at position 2/0; it shall not be assigned to any position in the alternative graphic set. These provisions do not preclude the assignment to positions in any graphic set of characters equivalent to spaces of size other than that of the space assigned to position 2/0.

At the beginning of any information interchange the shift status shall be defined by SI or SO. When in the shift-in status SI has no effect, and when in the shift-out status SO has no effect.

### 5.2.2 *Unique shift-out set*

Some applications require the use of only one additional set of 94 graphic characters. In such a case, that unique set is invoked by each use of SO. The set is identified either by an appropriate ESC sequence as described in 5.3.7 or by agreement between the interchanging parties.

### 5.2.3 *Multiple shift-out sets*

If two or more additional graphic sets are required to coexist in a system, the set to be used next is designated by the appropriate ESC sequence. That set is then invoked by the use of SO.

The use of SI re-invokes the graphics of the G0 set last designated but does not affect the identity of the designated G1 set. An additional set may be invoked any number of times by successive use of SO until it is superseded by another G1 set designated by another escape sequence.

It is not necessary to revert to the G0 set by use of SI before changing from one G1 set to another by means of a further escape sequence. When the system is in the shift-out state, the use of such a further escape sequence leaves the shift status unaltered, and the additional set is invoked.

Figure 3 is a schematic representation of the above.

In some devices or systems there may be a requirement to re-establish the shift-in state before designating a new shift-out set by means of an escape sequence. This can be achieved by inserting SI before the escape sequence which designates the subsequent shift-out set. The use of such a procedure is subject to agreement between the interchanging parties.

### 5.3 Code extension by means of escape sequences

### 5.3.1 *Purposes of escape sequences*

Escape sequences provide single or sets of control functions other than for transmission control. Escape sequences are also used to designate sets of graphics, different uses of some or all of the 7-bit code combinations, and coded character sets with a number of bits other than 7.

Thus escape sequences are required to provide, for example :

— a single control character not already in the code;

— a set of control characters not already in the code;

— a set of graphic characters not already in the code;

— a code structure different from the code.

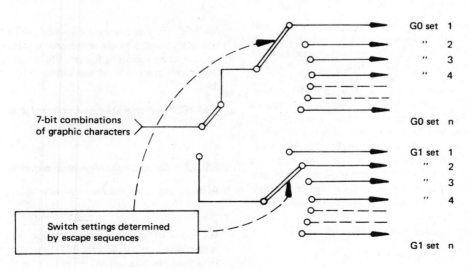

FIGURE 3

### 5.3.2　*Structure of escape sequences*

An escape sequence consists of two or more 7-bit combinations. The first is always the bit combination of ESC and the last is always one of the Final characters. An escape sequence may also contain any number of 7-bit combinations representing Intermediate characters.

The meaning of an escape sequence is determined by the 7-bit combination representing its Intermediate character(s), if any, and by the 7-bit combination representing its Final character.

WARNING : Although in this International Standard, escape sequences are described in terms of characters or of positions of the code table, the meaning of an escape sequence is determined only by its bit combinations and it is unaffected by any meaning previously assigned to these bit combinations taken individually.

Intermediate characters are the 16 characters of column 2 of the 7-bit code table.

NOTE — In this International Standard, any one of these 16 intermediate characters is denoted by the symbol : (I).

Final characters are the 79 characters of columns 3 to 7 of the 7-bit code table excluding position 7/15.

NOTE — In this International Standard, any one of these 79 final characters is denoted by the symbol : (F).

Prohibited characters are the control characters in columns 0 and 1 and the character in position 7/15.

The 33 prohibited characters shall not be used as either intermediate or final characters to construct an escape sequence.

As these prohibited characters may appear in an ESC sequence in error, it may be necessary within an application to provide methods of identifying such a situation and of recovering from it but this is not covered by this International Standard.

### 5.3.3　*Categories of escape sequence*

The use of escape sequences is specified in this International Standard. However, escape sequences with final characters from column 3 are reserved for private use subject to the categorization outlined below.

WARNING : The implementors of any private escape sequence described as such in this document are alerted to the fact that other implementors may give different meanings to the same escape sequence or may use different escape sequences to mean the same thing. Furthermore, such meanings may subsequently be assigned to standardized escape sequences. Interchanging parties are warned that the use of such private escape sequences may reduce their capability to interchange data subsequently.

### 5.3.3.1　*Two-character escape sequences*

A two-character escape sequence takes the form :

ESC (F)

Such escape sequences are used to represent single additional control characters.

The 79 two-character sequences are split into three types, depending on the Final character, as shown in Figure 4.

|   | 0 | 1 | 2 | 3 | 4 | 5 | 6 | 7 |
|---|---|---|---|---|---|---|---|---|
| 0 | | | | | | | | |
| 1 | | | | | | | | |
| 2 | | | | | | | | |
| 3 | | | | | | | | |
| 4 | | | | | | | | |
| 5 | | | | | | | | |
| 6 | | | | | | | | |
| 7 | | | | $F_p$ | $F_e$ | | $F_s$ | |
| 8 | | | | | | | | |
| 9 | | | | | | | | |
| 10 | | | | | | | | |
| 11 | | | | | | | | |
| 12 | | | | | | | | |
| 13 | | | | | | | | |
| 14 | | | | | | | | |
| 15 | | | | | | | | |

FIGURE 4

An ESC ($F_s$) sequence represents, depending on the Final character used, a single additional standardized control character. 31 Final characters of columns 6 and 7 are provided for this purpose.

An ESC ($F_e$) sequence represents, depending on the Final character used, an individual control character of an additional standardized set of 32 control characters (see 5.3.6). The 32 Final characters of columns 4 and 5 are provided for this purpose. Some applications require the use of only one such additional set. In this case, the set is identified either by the appropriate ESC sequence, as described in 5.3.6, or by agreement between the interchanging parties. If more than one additional set of controls are required to coexist in a system, the set to be used next is designated and invoked by the appropriate ESC sequence.

An ESC ($F_p$) sequence represents, depending on the Final character used, a single additional control character without standardized meaning for private use as required, subject to the prior agreement of the sender and the recipient of the data.

The 16 Final characters of column 3 are provided for this purpose.

### 5.3.3.2　*Three-character escape sequences*

A three-character escape sequence takes the form :

ESC (I) (F)

All types of three-character escape sequences are grouped into classes, according to their purpose, by means of their Intermediate characters, as shown in 5.3.4 to 5.3.11 (see Table 1, page 10).

These sequences are split into two types according to their Final character as shown in Figure 5.

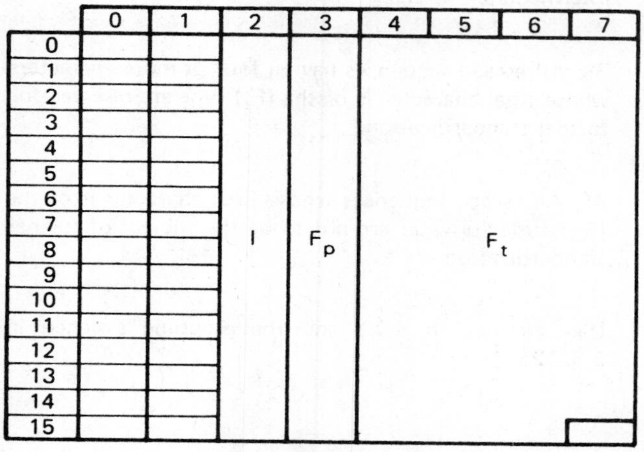

FIGURE 5

ESC (I) (F$_t$) sequences are used for standardized purposes. 63 F$_t$ characters of columns 4 to 7 are provided for this purpose.

ESC (I) (F$_p$) sequences are reserved for private use. 16 F$_p$ characters of column 3 are provided for this purpose.

### 5.3.4 Single additional control characters

ESC 2/3 (F) represents a single additional control character depending on the final character used.

### 5.3.5 Sets of 32 control characters for columns 0 and 1

ESC 2/1 (F) designates and invokes the C0 set of 32 control characters for representation by the bit combinations of columns 0 and 1.

The ten transmission control characters, when included in a C0 set, shall retain their meanings and their positions in the code table. No other transmission control characters may be included in a C0 set.

To reduce the risk of conflict in the interchange of data, this set should have the following characteristics :

— inclusion of the ten transmission control characters;

— inclusion of the characters NUL, SO, SI, CAN, SUB, and ESC with their meanings and their positions in the 7-bit code table unaltered.

Consideration should be given to the effect that changing the meaning of control characters can have on equipment when interchanging data. For example the bit combination corresponding to HT will have the effect of "horizontal tabulation" to a system designed to respond to this control character.

### 5.3.6 Sets of 32 control characters for representation by ESC F$_e$

ESC 2/2 (F) designates and invokes the C1 set of 32 control characters without affecting the C0 set.

Individual control characters of such a set are represented by means of ESC (F$_e$) sequences instead of a single bit-combination. A C1 set shall not include transmission control characters.

### 5.3.7 Sets of 94 graphic characters

ESC 2/8 (F) and ESC 2/12 (F) designate sets of 94 graphic characters which will be used as the G0 set. The designated set is invoked by SI.

ESC 2/9 (F) and ESC 2/13 (F) designate sets of 94 graphic characters which will be used as the G1 set. The designated set is invoked by SO.

NOTE — Two groups of graphic sets are mentioned above. The groups together make up a single repertory of graphic sets which may be designated either as a G0 or as a G1 set. No significance is attached to the groupings other than that their existence allows more such graphic sets to be defined within the scope of three character escape sequences as defined in this International Standard. There are therefore 126 (2 × 63) such sets possible without requiring further extension.

### 5.3.8 Codes which require special interpretation

ESC 2/5 (F) designates and invokes a code that requires special interpretation, such as :

— a code with a number of bits other than 7, excluding those 8-bit codes structured in accordance with this International Standard.

— a 7-bit code whose characteristics differ from those in this International Standard.

The final character assignments are such that within the F$_t$ and F$_p$ groups the following classification occurs :

| Final in column | Broad categorization |
|---|---|
| 3 | a private code with any number of bits |
| 4 | a code of less than 7 bits |
| 5 | a code of 7 bits |
| 6 | a code of 8 bits |
| 7 | a code of more than 8 bits |

### 5.3.9 Sets of graphics with multiple byte representation

ESC 2/4 (F) designates sets of graphic characters that are represented by two or more bytes each corresponding to a bit combination in columns 2 to 7, apart from positions 2/0 and 7/15. The designated set is invoked by SI and is therefore regarded as a G0 set. Within such a set, each graphic character is represented by the same number of bytes, as shown in Figure 6 below :

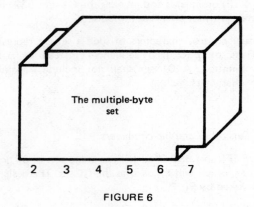

The multiple-byte set

2    3    4    5    6    7

FIGURE 6

### 5.3.10 *Announcement of extension facilities*

ESC 2/0 (F) announces the extension facilities used in conjunction with data which follow. The use of these sequences is specified in section 8.

### 5.3.11 *Three-character escape sequences without assigned meanings*

The escape sequences ESC 2/6 (F), ESC 2/7 (F), ESC 2/10 (F), ESC 2/11 (F), ESC 2/14 (F), and ESC 2/15 (F) have not been assigned meanings and are reserved for further standardization.

### 5.3.12 *Escape sequences having four or more characters*

Escape sequences having four or more characters will be interpreted according to the following :

1) The first intermediate character will indicate the class of usage identical with three character escape sequences above.

2) The second and any additional intermediate characters will be associated with the final character to permit additional entities within the class defined by the intermediate character.

3) All escape sequences having four or more characters whose final character is of the $(F_t)$ type are reserved for further standardization.

4) All escape sequences whose final character is of the $(F_p)$ type (private) are not to be the subject of further standardization.

(See warning in 5.3.3 and the grouping assigned in 5.3.3.2.)

### 5.4 Omission of escape sequences

If the interchanging parties have agreed upon a single G0 set, a single G1 set, a single C0 set and a single C1 set (or on as many of these sets as are to be used), they may also agree to omit the use of escape sequences to designate or invoke them. Interchanging parties are warned however that such agreements may reduce their capability to interchange data subsequently.

### 5.5 Pictorial and tabular representations

Figure 7 summarises, in a schematic form, the standard means of code extension within a 7-bit environment.

Table 1 summarises the assignment of intermediate characters in escape sequences.

384

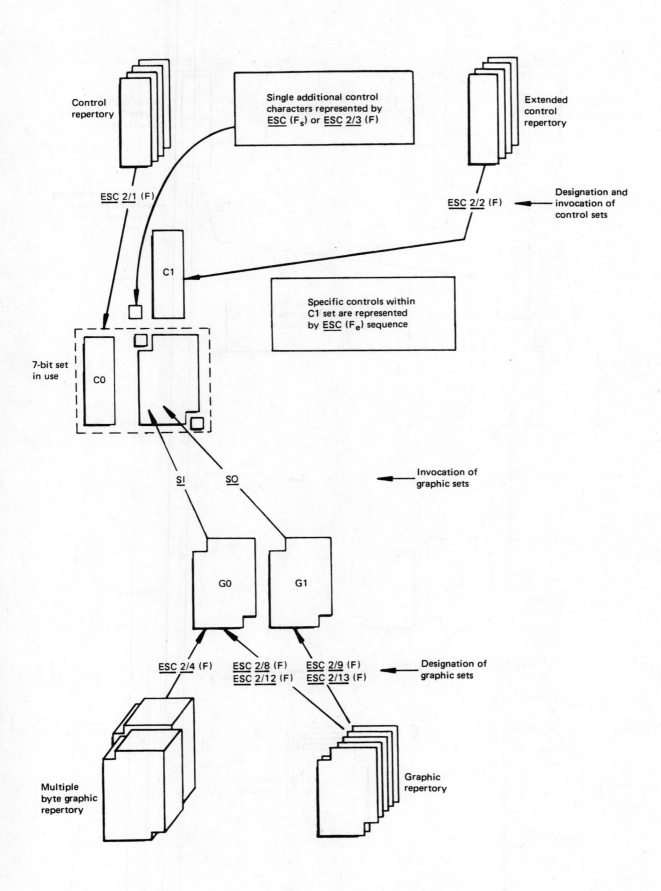

FIGURE 7 — Code extension in a 7-bit environment

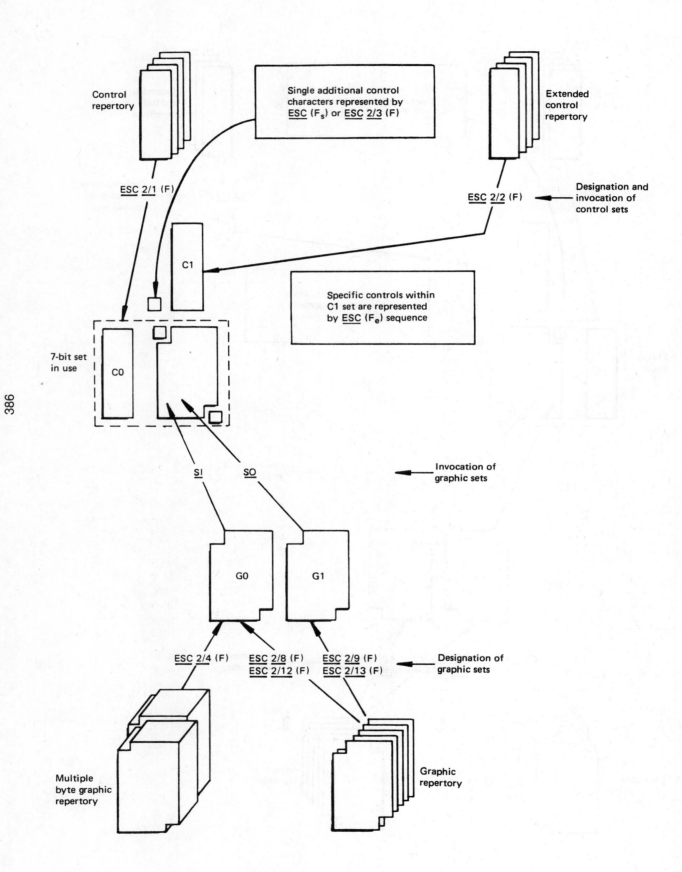

FIGURE 7 — **Code extension in a 7-bit environment**

Columns 00 to 07 of this array contain 128 character positions which are in one-to-one correspondence with the characters of the 7-bit set. Their coded representation is the same as in the 7-bit environment with the addition of an eighth, most significant bit, which is ZERO.

Columns 08 to 15 of this array contain a further 128 code positions; the eighth bit of their coded representations is ONE.

Columns 08 and 09 are provided for control characters and columns 10 to 15 for graphic characters, subject to the exception of positions 10/0 and 15/15 described below.

The control characters in columns 08 and 09 of an 8-bit code shall not include transmission control characters. Provision of data transmission capability for 8-bit codes includes the use of the data link escape characters and is the subject of other ISO publications.

## 6.2 The family concept

In order to cope with the different needs of the various industries, fields of application or systems, this International Standard defines the concept of a family of 8-bit codes as follows :

  1) a set of 32 additional control characters can be selected for columns 08 and 09;

  2) a set of 94 additional graphic characters can be selected for columns 10 to 15 (excluding positions 10/0 and 15/15).

There are standard techniques for identifying selections of sets of controls and graphics for 8-bit codes. These techniques are described below.

## 7  THE USE OF CODE EXTENSION IN AN 8-BIT CODE

The techniques of extending an 8-bit code described in this International Standard have been purposely made compatible with those used to extend the 7-bit code.

The escape character is used in an 8-bit code in exactly the same way as in the 7-bit code to construct ESC sequences. The meanings of these sequences are not altered in an 8-bit code. All characters in columns 08 -15 are excluded from assignment in escape sequences and any occurrences of them in an escape sequence are error conditions for which no standard recovery procedures are prescribed in this International Standard.

### 7.1  Definition of an 8-bit code

As described in section 6, the code table can be considered as having four main parts :

  — the C0 control set;

  — the G0 graphic set;

  — the C1 control set;

  — the G1 graphic set;

The remainder of the code table consists of positions 02/0 (SP), 07/15 (DEL), 10/0 and 15/15.

The C0 and the G0 sets are designated, and invoked by the same escape sequences as in the 7-bit environment (see 5.3.5 and 5.3.7).

The C1 set of control characters is designated and invoked by means of an escape sequence as in the 7-bit environment (see 5.3.6). These control characters are represented by the bit combinations of columns 08 and 09.

The G1 set of graphic characters is designated and invoked by means of an escape sequence as in the 7-bit environment (see 5.3.7). These graphic characters are represented by the bit combination of columns 10 to 15.

### 7.2  Code extension by means of escape sequences

Once the 8-bit code is established in accordance with 7.1, code extension means are available, making use of escape sequences as described therein and below.

#### 7.2.1  Two-character escape sequences

Two-character escape sequences have the same structure as in the 7-bit environment (see 5.3.2).

ESC ($F_s$) sequences represent single additional controls with the same meaning they have in the 7-bit environment.

The use of ESC ($F_e$) sequence in an 8-bit environment is contrary to the intention of this International Standard but should they occur their meaning is the same as in the 7-bit environment.

#### 7.2.2  Three-character escape sequences

Three-character escape sequences have the same structure and meaning as in the 7-bit environment (see 5.3).

#### 7.2.3  Escape sequences with four or more characters

These escape sequences have the same structure and meaning as in the 7-bit environment (see 5.3).

### 7.3  Pictorial representation

Figure 9 summarises in a schematic form a standard means of extension available in an 8-bit environment.

388

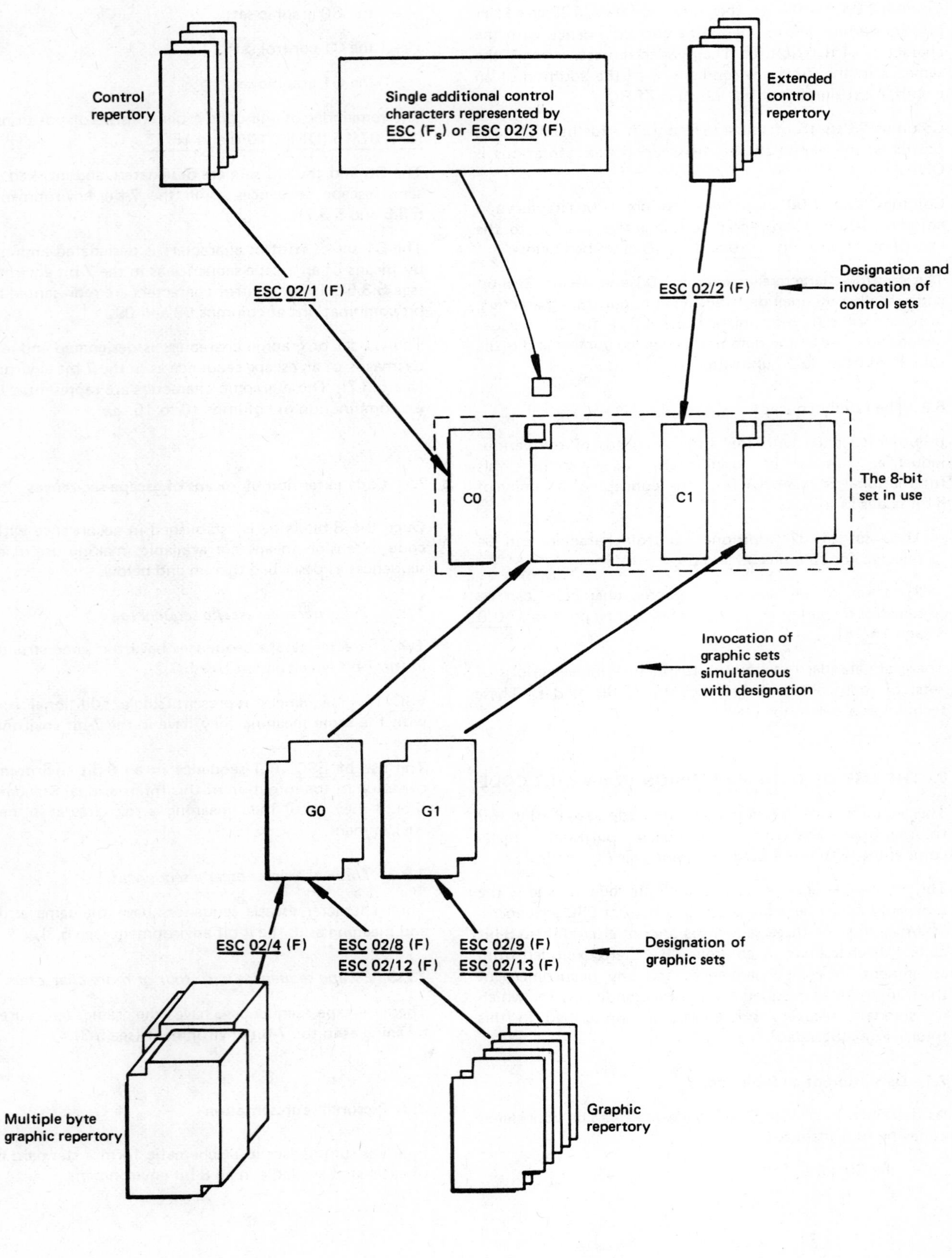

**Control repertory**

**Single additional control characters represented by ESC (Fs) or ESC 02/3 (F)**

**Extended control repertory**

ESC 02/1 (F)

ESC 02/2 (F) ← **Designation and invocation of control sets**

C0      C1     **The 8-bit set in use**

← **Invocation of graphic sets simultaneous with designation**

G0     G1

ESC 02/4 (F)    ESC 02/8 (F)    ESC 02/9 (F)    ← **Designation of graphic sets**
             ESC 02/12 (F)   ESC 02/13 (F)

**Multiple byte graphic repertory**

**Graphic repertory**

FIGURE 9 — **Code extension in a 8-bit environment**

## 8 ANNOUNCEMENT OF EXTENSION FACILITIES USED

The class of three character escape sequences ESC 2/0 (F) is used in data interchange to announce the code extension facilities utilized in the data which follow. Subject to agreement between the interchanging parties, such an announcing sequence may be omitted. The final character of the announcing sequence indicates the facilities used for representing graphic sets in 7 and 8-bit environments, and the number of bits used as follows :

| Final characters | Facilities utilized |
|---|---|
| 4/1 | The G0 set only shall be used. The escape sequence which designates this set also invokes it into columns 2 to 7. SI and SO shall not be used. In an 8-bit environment, columns 10 to 15 are not used. |
| 4/2 | The G0 and G1 sets shall be used. In both 7 and 8-bit environments SI invokes G0 into columns 2 to 7 and SO invokes G1 into columns 2 to 7. In an 8-bit environment columns 10 to 15 are not used. |
| 4/3 | The G0 and G1 sets shall be used in an 8-bit environment only. The designating escape sequences also invoke the G0 and G1 sets into columns 2 to 7 and columns 10 to 15 respectively. SI and SO shall not be used. |
| 4/4 | The G0 and G1 sets shall be used. In a 7-bit environment, SI invokes G0 into columns 2 to 7 and SO invokes G1 into columns 2 to 7. In an 8-bit environment, the designating escape sequences also invoke the G0 and G1 sets into columns 2 to 7 and 10 to 15 respectively; SI and SO shall not be used. |

(A pictorial representation of these cases is shown in Figure 10.)

NOTE — In a 7-bit environment, data announced by a sequence ESC 2/0 4/4 have the same form as data announced by a sequence ESC 2/0 4/2.

Both are provided for those interchange situations in which it is agreed to differentiate between 7-bit and 8-bit originated data, in the 7-bit environment.

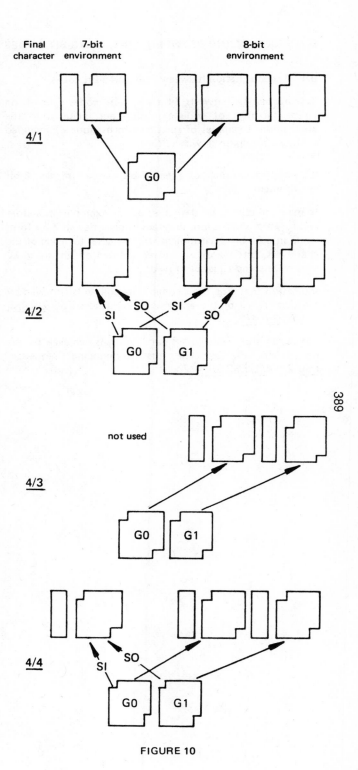

FIGURE 10

389

## 9  RELATIONSHIP BETWEEN 7-BIT AND 8-BIT CODES

### 9.1  Transformation between 7-bit and 8-bit codes

Transformation between 7-bit and 8-bit codes depends on which facilities of code extension are included in the application. Examples of such transformations are included in the Appendix in flowchart form.

### 9.2  Representation of the 7-bit code in an 8-bit environment

It may sometimes be desirable, as for example, in a store and forward application, to retain information in 7-bit form while in an 8-bit environment. In this case, for each of the characters, bits $b_7$ to $b_1$ are represented by bits $a_7$ to $a_1$ respectively and $a_8$ is set to zero.

Indication that true 8-bit coded data follow is achieved by the occurrence of the announcing sequences ESC 2/0 4/3, or ESC 2/0 4/4.

Indication that 7-bit coded data follow is achieved by the occurrence of one of the announcing sequences ESC 2/0 4/1, ESC 2/0 4/2.

### 9.3  Representation of positions 10/0 and 15/15 in a 7-bit environment

No meaning is assigned to positions 10/0 and 15/15 in this International Standard. If there is a requirement to represent these positions in a 7-bit environment, a private escape sequence shall be used.

## 10  SPECIFIC MEANING OF ESCAPE SEQUENCES

**10.1**  The meanings of individual escape sequences are not specified in this International Standard. Instead, their meanings will be specified using the procedures established by ISO 2375. That International Standard is to be followed in preparing and maintaining a register of escape sequences and their meanings. These registration procedures do not apply to escape sequences reserved for private use.

**10.2**  Furthermore, when required, the classes of 3-character escape sequences which are not defined in this International Standard will be allocated by the ISO Coding Committee (ISO/TC 97/SC 2) using the same procedures mentioned in 10.1.

## APPENDIX

The implementation of the extension of the 7-bit code, described in this International Standard, depends on the practical need of the implementor or user of the code extension. Remaining in accordance with this International Standard, they may differ in

— including all or parts of the control sets and graphic sets;

— identifying the sets by ESC-sequences or having an agreement;

— dropping or not the superfluous invokers;

— examining or not the structure of an ESC-sequence;

— handling the character DEL, etc.

In this Appendix there are shown a few of the possible implementations which may help the user to understand this International Standard. A summary of the figures shown in this Appendix is given in Table 2. There are three types of figures : Structure, Interpretation and Transformation.

Table 3 shows the symbols used in the flow charts and the associated meaning. Table 4 shows the items used in the flow charts, their meaning and the possible values.

### Z.1  STRUCTURE

The pictorial representations of the structure of the 7-bit and 8-bit code extension have all the same format as follows :

— at the top of the diagram are the repertory of single controls and sets of controls;

— below that are the escape sequences that are used for the designation and invocation of the control sets as the C0 and C1 set;

— at the bottom of the diagram is the repertory of graphic sets;

— above that are the escape sequences used for designation of the graphic set as the G0 or G1 set;

— above that are the shift characters that are used for the invocation of the G0 or G1 set;

— at the centre all these elements appear as a single form, the 7-bit set or 8-bit set in use.

### Z.2  INTERPRETATION

The interpretation of data containing 7-bit and/or 8-bit code extension is shown by means of flowcharts. The interpretation is separate from the transformation.

### Z.3  TRANSFORMATION

The transformation of the data from 7-bit to 8-bit and from 8-bit to 7-bit is shown by means of flowcharts. The transformation is separate from the interpretation.

TABLE 2 — Summary of the given examples

| Included Sets | Bits | Structure | Interpret-ation | Transformation | |
|---|---|---|---|---|---|
| | | | | 7 to 8 | 8 to 7 |
| C0 C1 | 7 | Figure 11 | Figure 14 | Figure 16 | Figure 17 |
| | 8 | Figure 12 | Figure 15 | | |
| G0 G1 | 7 and 8 | Figure 13 | | | |

TABLE 3 — Meaning of the flowchart symbols

| Symbol | Name | Meaning |
|---|---|---|
| Input next Character / End | Input | To fetch the next character from the input character string and go to the end if there is no further character available. |
| Process / Error End | Process | To execute the defined process. There may be an error end because of lack of more characters. |
| Item / Value | Decision | To examine the item as having the specified value. |
| Value → Item | Preparation | To set the item to the specified value. |
| Output something | Output | To take the last character (output character) or the specified character(s) (output 1/11, 4/14) or the manipulated character (output, $a_8$ = dropped) to the output character string. |
| Sub-Routine | Predefined process | To execute the appropriate predefined process. |

392

TABLE 4 — Items used in the flowcharts

| Item | Meaning | Possible values in 7 bits | in 8 bits |
|---|---|---|---|
| Character | The bit pattern of the current character | 0/0 to 7/15 | 00/0 to 15/15 |
| Column | The column number of the current character | 0 to 7 | 00 to 15 |
| Status 7 | Shift-status of the 7-bit code | SI SO undefined | |
| Bits | Number of the valid bits of the character | | 7 8 undefined |

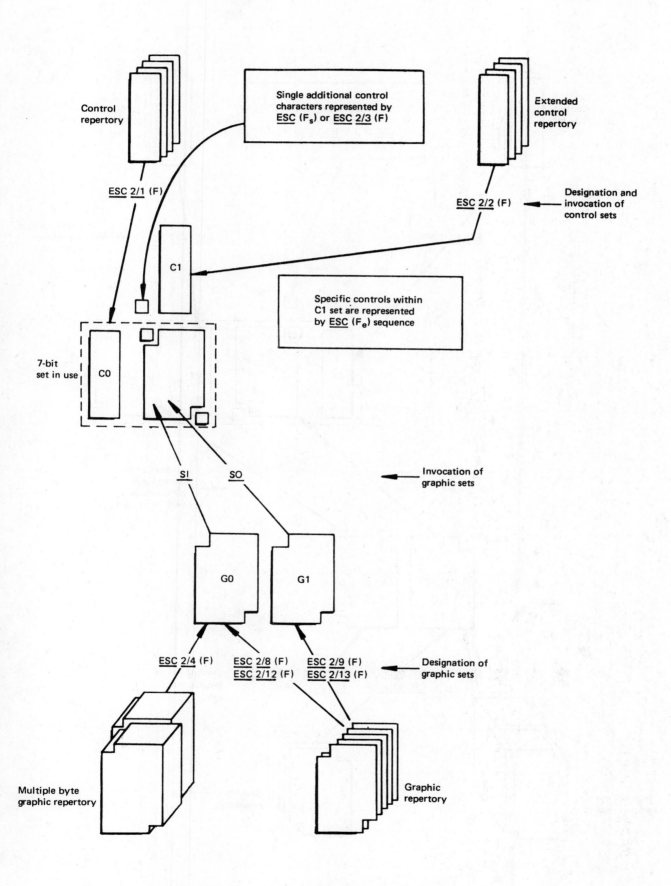

FIGURE 11 — 7-bit code extension structure using C0, C1, G0 and G1 sets

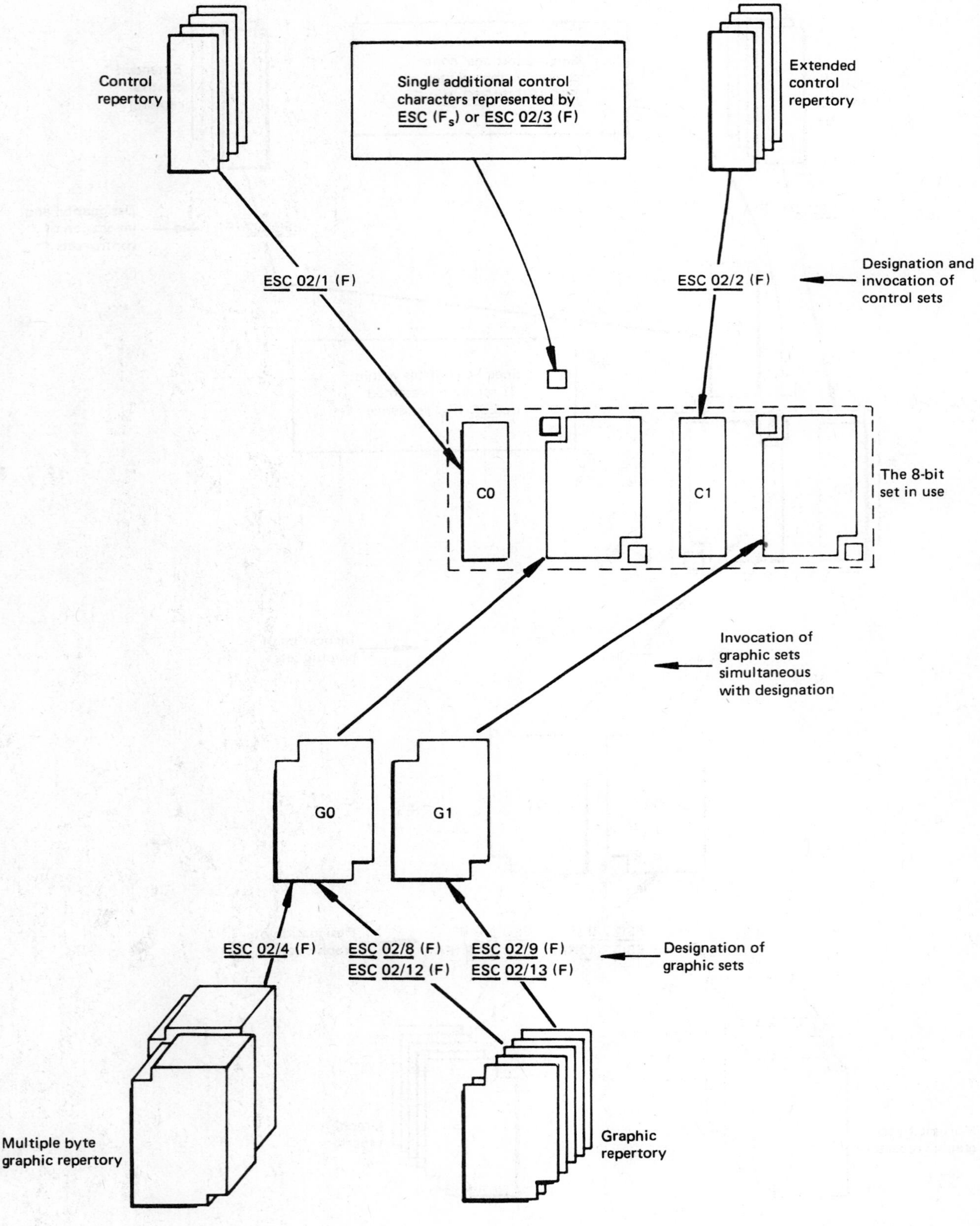

FIGURE 12 — 8-bit code extension structure using C0, C1, G0 and G1 sets

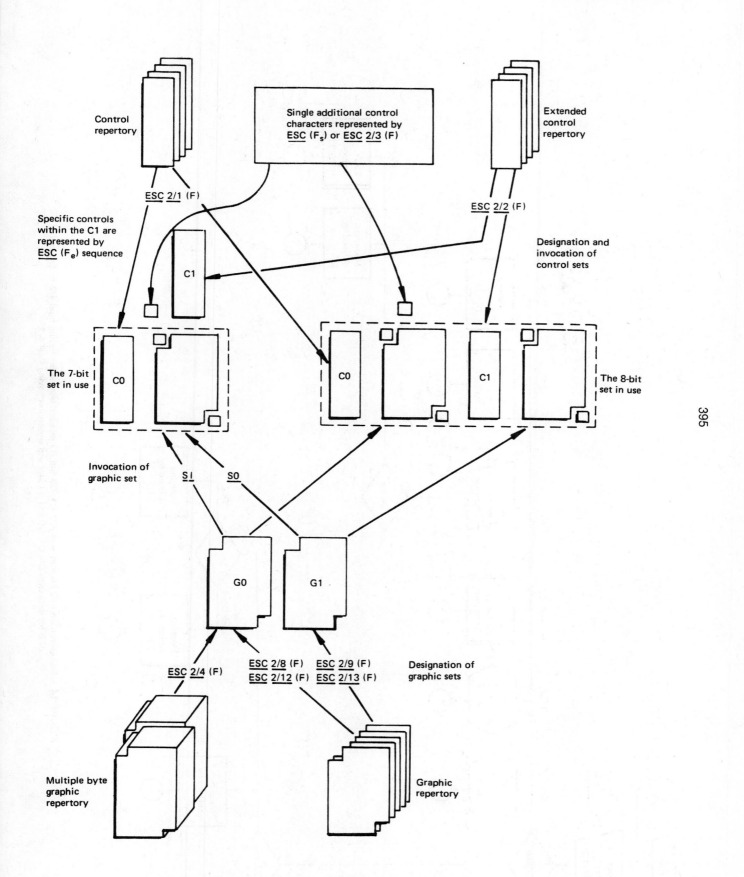

FIGURE 13 — **Composite 7-bit and 8-bit code extension structure using C0, C1, G0 and G1 sets**

396

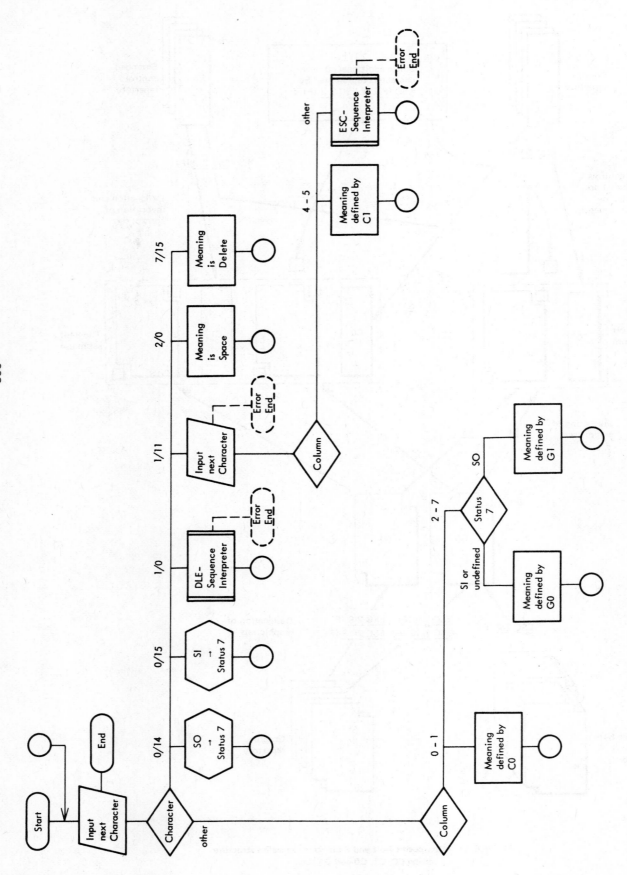

FIGURE 14 — Interpretation of 7-bit data with C0, C1, G0 and G1 sets and announcer having final 4/2 or 4/4 and
of 8-bit data with C0, C1, G0 and G1 sets and announcer having final 4/2

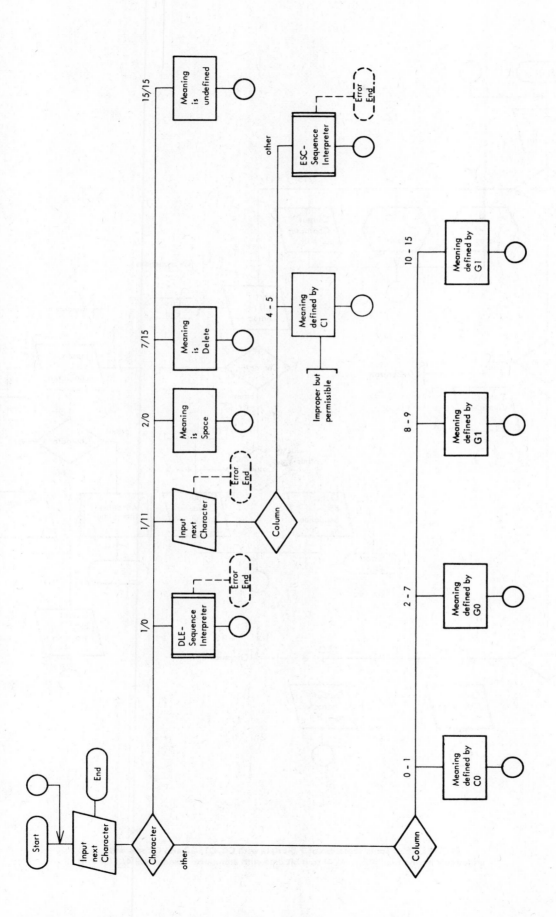

FIGURE 15 — Interpretation of 8-bit data with C0, C1, G0 and G1 sets and announcer having final 4/3 or 4/4

398

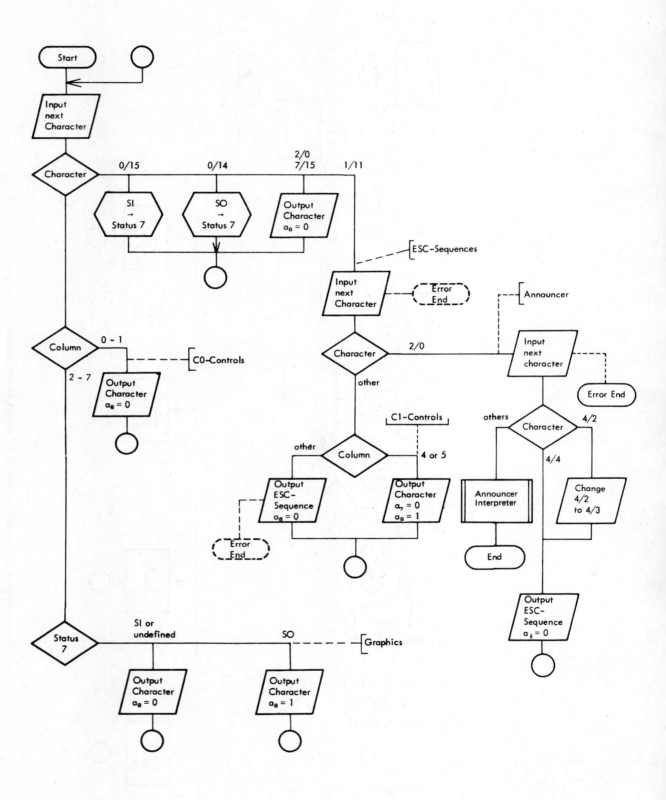

FIGURE 16 — Transformation of 7-bit data with C0, C1, G0 and G1 sets and
announcer having final 4/2 (or 4/4) to 8-bit data with announcer having final 4/3 (or 4/4)

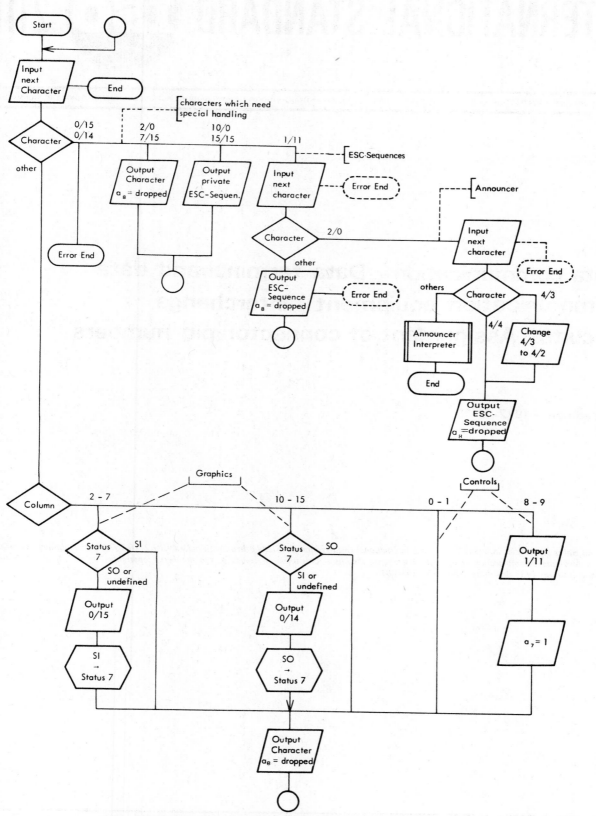

FIGURE 17 — Transformation of 8-bit data with C0, C1, G0 and G1 sets and
announcer having final 4/3 (or 4/4) to 7-bit data with announcer having final 4/2 (or 4/4)

# INTERNATIONAL STANDARD ISO 2110

INTERNATIONAL ORGANIZATION FOR STANDARDIZATION ·МЕЖДУНАРОДНАЯ ОРГАНИЗАЦИЯ ПО СТАНДАРТИЗАЦИИ·ORGANISATION INTERNATIONALE DE NORMALISATION

## Data communication – Data terminal and data communication equipment – Interchange circuits – Assignment of connector pin numbers

400

First edition – 1972-02-15

UDC 681.327.2 : 621.316.541

Ref. No. ISO 2110-1972 (E)

Descriptors : connectors,  data transmission,  numbering,  pins.

## FOREWORD

ISO (the International Organization for Standardization) is a worldwide federation of national standards institutes (ISO Member Bodies). The work of developing International Standards is carried out through ISO Technical Committees. Every Member Body interested in a subject for which a Technical Committee has been set up has the right to be represented on that Committee. International organizations, governmental and non-governmental, in liaison with ISO, also take part in the work.

Draft International Standards adopted by the Technical Committees are circulated to the Member Bodies for approval before their acceptance as International Standards by the ISO Council.

International Standard ISO 2110 was drawn up by Technical Committee ISO/TC 97, *Computers and information processing.*

It was approved in June 1971 by the Member Bodies of the following countries:

| | | |
|---|---|---|
| Belgium | Germany | South Africa, Rep. of |
| Brazil | Greece | Sweden |
| Canada | Italy | Switzerland |
| Czechoslovakia | Japan | Thailand |
| Denmark | Netherlands | United Kingdom |
| Egypt, Arab Rep. of | New Zealand | U.S.A. |
| France | Portugal | U.S.S.R. |

No Member Body expressed disapproval of the document.

# Data communication – Data terminal and data communication equipment – Interchange circuits – Assignment of connector pin numbers

## 1 SCOPE AND FIELD OF APPLICATION

This International Standard specifies the assignment of connector pin numbers at the interface between data terminal equipment and data communication equipment either modems or automatic calling equipment where CCITT - V 24 is applicable.

## 2 REFERENCES

CCITT Recommendation V 24, *Function and electrical characteristics of circuits at the interface between data terminal equipment and data communication equipment.*

CCITT Recommendation V 21, *200 baud modem standardized for use in the general switched telephone network.*

CCITT Recommendation V 23, *600 and 1200 baud modem standardized for use in the general switched telephone network.*

CCITT Recommendation V 26, *2400 bit/second modem for use on 4-wire based point-to-point circuits.*

CCITT Recommendation V 25, *Automatic calling and/or answering equipment of the general switched telephone network.*

CCITT Recommendation V 11, *Automatic calling and/or answering on the telex network.*

## 3 COMPLIANCE WITH THIS INTERNATIONAL STANDARD

Existing operational installations need not comply.

For new installations consisting of both new and old equipment, incompatibilities shall be resolved by modifying that equipment which does not comply with this International Standard by strapping arrangements or other suitable means.

All new equipment should comply with this International Standard within 12 months of its publication.

## 4 CONNECTOR

Although the connector still remains to be the subject of a Recommendation by Technical Committee 48 of the International Electrotechnical Commission (IEC), it is generally accepted to be a 25 pin connector with separate connectors provided for the data communication equipment (or telex service) and for the automatic calling equipment interfaces. The male connector (plug) is associated with the data terminal equipment and the female connector (socket) is associated with the data communication equipment.

## 5 ASSIGNMENT OF PIN NUMBERS

Pin assignment is given in Table 1 and is recommended for the following equipments:

A. Modem complying with CCITT - V 21
B. Modem complying with CCITT - V 23
C. Modem complying with CCITT - V 26
D. Future data communication equipment
E. Telex
F. Other telegraph services
L. Automatic calling over telephone networks complying with CCITT - V 25
M. Automatic calling over telex networks complying with CCITT - V 11

The descriptions of the interchange circuits are given in Table 2 for reference.

402

TABLE 1

| Pin number | Circuit numbers | | | | | | | |
|---|---|---|---|---|---|---|---|---|
| | Modem interfaces | | | | Telegraph interfaces | | Automatic calling interfaces | |
| | Modem | Modem | Modem | Future | Telex | Other | Telephone | Telex |
| | A | B | C | D | E | F | L | M |
| 1 | 101 | 101 | 101 | 101 | 101 | 101 | 212 | 212 |
| 2 | 103 | 103 | 103 | 103 | 103 | 103 | 211 | 211 |
| 3 | 104 | 104 | 104 | 104 | 104 | 104 | 205 | 205 |
| 4 | 105 | 105 | 105 | 105 | N | F | 202 | 202 |
| 5 | 106 | 106 | 106 | 106 | 106 | 106 | 210 | 210 |
| 6 | 107 | 107 | 107 | 107 | 107 | 107 | 213 | 213 |
| 7 | 102 | 102 | 102 | 102 | 102 | 102 | 201 | 201 |
| 8 | 109 | 109 | 109 | 109 | 109 | 109 | F | F |
| 9 | N | N | N | N | N | N | N | N |
| 10 | N | N | N | N | N | N | N | N |
| 11 | 126 | N | N | F | N | N | F | F |
| 12 | F | 122 | 122 | 122 | F | F | F | F |
| 13 | F | 121 | 121 | 121 | F | F | 204 | 204 |
| 14 | F | 118 | 118 | 118 | F | F | 206 | 206 |
| 15 | F | see Note | 114 | 114 | F | F | 207 | 207 |
| 16 | F | 119 | 119 | 119 | F | F | 208 | 208 |
| 17 | F | see Note | 115 | 115 | F | F | 209 | 209 |
| 18 | F | F | F | F | 132 | F | F | F |
| 19 | F | 120 | 120 | 120 | F | F | F | F |
| 20 | 108/1 108/2 | 108/1 108/2 | 108/1 108/2 | 108/1 108/2 | 108/2 | 108/2 | F | F |
| 21 | F | F | F | F | F | F | F | F |
| 22 | 125 | 125 | 125 | 125 | 125 | 125 | 203 | 203 |
| 23 | N | 111 | F | 111 | N | N | F | F |
| 24 | N | N | N | N | N | N | N | N |
| 25 | F | F | F | F | F | F | F | F |

N = Pin number permanently reserved for national use.
F = Pin number reserved for future International Standard and should not be used for national use.

NOTE — Where signal timing is provided in the data communication equipment complying with CCITT - V 23, pin No. 15 will be used for circuit 114 and pin No. 17 will be used for circuit 115.

TABLE 2

| Circuit No. | Description |
|---|---|
| 101 | Protective ground or earth |
| 102 | Signal ground or common return |
| 103 | Transmitted data |
| 104 | Received data |
| 105 | Request to send |
| 106 | Ready for sending |
| 107 | Data set ready |
| 108/1 | Connect data set to line |
| 108/2 | Data terminal ready |
| 109 | Data channel received line signal detector |
| 111 | Data signalling rate selector (Data terminal equipment source) |
| 114 | Transmitter signal element timing (Data communication equipment source) |
| 115 | Receiver signal element timing (Data communication equipment source) |
| 118 | Transmitted backward channel data |
| 119 | Received backward channel data |
| 120 | Transmit backward channel line signal |
| 121 | Backward channel ready |
| 122 | Backward channel received line signal detector |
| 125 | Calling indicator |
| 126 | Select transmit frequency |
| 132 | Return to non-data mode |
| 201 | Signal ground or common return |
| 202 | Call request |
| 203 | Data line occupied |
| 204 | Distant station connected |
| 205 | Abandon call |
| 206 | Digit signal $(2^0)$ |
| 207 | Digit signal $(2^1)$ |
| 208 | Digit signal $(2^2)$ |
| 209 | Digit signal $(2^3)$ |
| 210 | Present next digit |
| 211 | Digit present |
| 212 | Protective ground or earth |
| 213 | Power indication |

403

# INTERNATIONAL STANDARD

**ISO** 2111

INTERNATIONAL ORGANIZATION FOR STANDARDIZATION · МЕЖДУНАРОДНАЯ ОРГАНИЗАЦИЯ ПО СТАНДАРТИЗАЦИИ · ORGANISATION INTERNATIONALE DE NORMALISATION

# Data communication – Basic mode control procedures – Code independent information transfer

**First edition – 1972-02-15**

404

UDC 681.327.18

Ref. No. ISO 2111-1972 (E)

**Descriptors** : control procedures, data link escape, data transmission.

## FOREWORD

ISO (the International Organization for Standardization) is a worldwide federation of national standards institutes (ISO Member Bodies). The work of developing International Standards is carried out through ISO Technical Committees. Every Member Body interested in a subject for which a Technical Committee has been set up has the right to be represented on that Committee. International organizations, governmental and non-governmental, in liaison with ISO, also take part in the work.

Draft International Standards adopted by the Technical Committees are circulated to the Member Bodies for approval before their acceptance as International Standards by the ISO Council.

International Standard ISO 2111 was drawn up by Technical Committee ISO/TC 97, *Computers and information processing.*

It was approved in June 1971 by the Member Bodies of the following countries:

| | | |
|---|---|---|
| Belgium | Greece | Sweden |
| Brazil | Italy | Switzerland |
| Canada | Japan | Thailand |
| Czechoslovakia | Netherlands | United Kingdom |
| Egypt, Arab Rep. of | New Zealand | U.S.S.R. |
| France | Portugal | |
| Germany | South Africa, Rep. of | |

The Member Body of the following country expressed disapproval of the document on technical grounds:

U.S.A.

# Data communication– Basic mode control procedures – Code independent information transfer

## 0 INTRODUCTION

A data communication system may be considered as the set of the terminal installations and the interconnecting network that permits information to be exchanged.

A data link comprises terminal installations connected to the same network, operating at the same speed, in the same code. Any "store and forward" delay or intermediate data processing really separates data links. Any system is constituted of one or several data links.

The information transfer in a data link is monitored by data link control procedures where some characters, selected within a coded character set, are given particular meanings according to the transmission phase and are used for various purposes such as to delineate information, to reverse the direction of transmission, to ask questions, to answer, etc.

The data link control procedures are categorized in classes which are referred to as modes of operation. The basic mode is defined in the Introduction to ISO/R 1745, fifth paragraph.

This International Standard defines the means by which systems can transfer information without coding restrictions. This is achieved by use of the DLE character, and is called code independent transmission.

The procedures described allow for code dependent information messages to alternate with code independent information messages.

## 1 SCOPE AND FIELD OF APPLICATION

**1.1** This International Standard defines the means by which a data communication system operating according to the basic mode procedures defined in ISO/R 1745 can transfer information messages without code restrictions.

**1.2** This International Standard extends Phase 3 (Information transfer) as defined in ISO/R 1745. It also describes other uses of the DLE character than that described in ISO/R 1745. Phase 2 (Establishment of data link) and Phase 4 (Termination) are not affected by this International Standard.

## 2 REFERENCES

ISO/R 646, *6 and 7 bit coded character sets for information processing interchange.*

ISO/R 1177, *Character structure for start/stop and synchronous transmission.*

ISO/R 1745, *Basic mode control procedures for data communication systems.*

## 3 FORMATING RULES

### 3.1 Initiation of code independent text

The sequence "DLE.STX" shall be used to initiate a code independent text.

### 3.2 Termination of code independent text

The sequence "DLE.ETB" or "DLE.ETX" shall be used to terminate a code independent block or text respectively.

### 3.3 Filling

When filling is necessary within the information message, the sequence "DLE.SYN" shall be used in lieu of the single character "SYN". Filling sequences shall not be inserted between any two characters forming a DLE sequence.

### 3.4 Character parity

The characters forming the DLE sequences shall carry the character parity used by the data transmission system through which the code independent text is being transferred. When the system is asynchronous, the character parity shall be even; when the system is synchronous, it shall be odd. (See ISO/R 1177.)

## 4 PRESENTATION OF DATA

**4.1** Texts will be presented in octets, or 8 bit characters (for example, 7 bit plus parity bit), a sequence expressed in any 8 bit code, packed numerics, etc. If a binary data stream split into groups of 8 bits is used, bit padding by an agreed method may be required (to complete the last octet).

**4.2** All 8 bit combinations are acceptable in the original text.

**4.3** For each occurrence of the 8 bit combination corresponding to "DLE" an additional "DLE" shall be inserted for transmission adjacent to it.

**4.4** "DLE" characters which are used to form DLE sequences for transmission control (for example, DLE.STX, DLE.SYN, DLE.ETB) shall not be doubled.

## 5 RECEPTION OF DATA

Received data shall be inspected for DLE sequences and the following independent rules observed:

**5.1** "DLE.STX" shall be interpreted as the initiator of the code independent text.

**5.2** When a double DLE sequence occurs, one "DLE" shall be suppressed and the other shall be regarded as data. The data following shall be inspected for new DLE sequences.

**5.3** "DLE.ETB" or "DLE.ETX", when not immediately preceded by an odd number of "DLE" characters, shall be interpreted as the terminators of the code independent block or text.

**5.4** Unless immediately preceded by an odd number of "DLE" characters, the "DLE.SYN" sequence will normally be discarded.

## 6 HEADING

If a heading is required it may be transmitted

   a) as a separate information message in the 7 bit code of ISO/R 646, in conformity with ISO/R 1745; or

   b) as part of the code independent information message transmission; or

   c) as a code dependent, or independent, heading which can be prefixed to the code independent text by the use of the sequence "DLE.SOH" as the initiating sequence. The rules for code independent text generally apply. However, special consideration must be given to the handling of text function.

## 7 ERROR PROTECTION

**7.1** Since the use of character parity cannot be guaranteed for error checking within the data link a block check sequence (BCS) shall be used.

**7.2** The first appearance of either "SOH", "DLE.SOH", "DLE.STX" shall initiate the calculation of the BCS.

**7.3** The initiating sequence shall not be included in the BCS calculation.

**7.4** The filling sequence "DLE.SYN" shall not be included in the BCS calculation.

**7.5** The first "DLE" in each two-character DLE sequence (DLE.DLE, DLE.ETB, etc.), as received from the data communication equipment, shall not be included in the BCS calculation.

**7.6** The BCS shall follow immediately after the terminating sequence.

**7.7** The form which the BCS takes is given in section 8.

## 8 BLOCK CHECK SEQUENCE (BCS)

The BCS shall conform to the following rules;

**8.1** It shall be a 16 bit sequence (two octets).

**8.2** The BCS is the remainder after the division (modulo 2) of the information bits to be protected arranged in serial form as they will be transferred to the data communication equipment, by the generating polynomial $(x^{16} + x^{12} + x^5 + 1)$.

**8.3** The BCS is transmitted to the line commencing with the highest order bits.

**8.4** At the receiver the serial incoming protected data and the BCS when divided by the generating polynomial will result in a zero remainder in the absence of transmission errors.

NOTE — If future applications show that more bits are needed to provide adequate protection, the number of bits comprising the BCS shall be $8n$.

# INTERNATIONAL STANDARD  2593

INTERNATIONAL ORGANIZATION FOR STANDARDIZATION ·МЕЖДУНАРОДНАЯ ОРГАНИЗАЦИЯ ПО СТАНДАРТИЗАЦИИ·ORGANISATION INTERNATIONALE DE NORMALISATION

# Connector pin allocations for use with high-speed data terminal equipment

408

**First edition — 1973-03-01**

UDC 681.14 : 621.316.3

Ref. No. ISO 2593-1973 (E)

**Descriptors** : data processing, data transmission, electric connectors, connector pins, layout, numbering.

# FOREWORD

ISO (the International Organization for Standardization) is a worldwide federation of national standards institutes (ISO Member Bodies). The work of developing International Standards is carried out through ISO Technical Committees. Every Member Body interested in a subject for which a Technical Committee has been set up has the right to be represented on that Committee. International organizations, governmental and non-governmental, in liaison with ISO, also take part in the work.

Draft International Standards adopted by the Technical Committees are circulated to the Member Bodies for approval before their acceptance as International Standards by the ISO Council.

International Standard ISO 2593 was drawn up by Technical Committee ISO/TC 97, *Computers and information processing.*

It was approved in July 1972 by the Member Bodies of the following countries :

| | | |
|---|---|---|
| Australia | Italy | Spain |
| Belgium | Japan | Sweden |
| Canada | Netherlands | Switzerland |
| Czechoslovakia | New Zealand | Thailand |
| Denmark | Poland | United Kingdom |
| Egypt, Arab Rep. of | Portugal | U.S.A. |
| France | Romania | |
| Germany | South Africa, Rep. of | |

The Member Body of the following country expressed disapproval of the document on technical grounds :

U.S.S.R.

# Connector pin allocations for use with high-speed data terminal equipment

## 0  INTRODUCTION

This International Standard was prepared as a complement to the CCITT Recommendation V 35 and as a guide to other data communication equipment, and to data terminal equipment which operates at a data signalling rate greater than approximately 20 000 bits per second.

## 1  SCOPE AND FIELD OF APPLICATION

This International Standard provides a correspondence between the interface circuit numbers used in CCITT Recommendation V 35, and the pin numbers of the connector used on the data communication equipment and the data terminal equipment.

NOTE — This International Standard defines the connector by a military standard as no international standard exists at this time.

410

## 2  REFERENCES

CCITT Recommendation V 35, *Modem for 48 kilobits per second.*

MIL Specification MIL-C-28748, *Connector, electrical rectangular, crimp type, removable contact, for rack and panel and other application.*

NOTE — Available from Navy Publications and Form Center, 5801 Tabor Ave, PHILA, PA 19120, U.S.A.

## 3  CONNECTOR PIN ALLOCATIONS

The data terminal equipment shall be terminated on a 34 pin connector conforming to MIL Specification No. MIL-C-28748.

The pin allocation shall be as follows :

| Pin | Function | CCITT circuit No. | Direction |
|-----|----------|-------------------|-----------|
| A | Protective ground or earth | 101 | common |
| B | Signal ground or common return | 102 | common |
| C | Request to send | 105 | from DTE |
| D | Ready for sending | 106 | to DTE |
| E | Data set ready | 107 | to DTE |
| F | Data channel received line signal detector | 109 | to DTE |
| H | Connect data set to line | 108/1 | from DTE |
|   | Data terminal ready | 108/2 | from DTE |
| J | Calling indicator | 125 | to DTE |
| K | $F_1$ | — | — |
| L | $F_2$ | — | — |
| M | $F_1$ | — | — |
| N | $F_2$ | — | — |
| R | Received data A-wire | 104 | to DTE |
| T | Received data B-wire | 104 | to DTE |
| V | Receiver signal element timing A-wire | 115 | to DTE |
| X | Receiver signal element timing B-wire | 115 | to DTE |
| Y | Transmitter signal element timing A-wire | 114 | to DTE |
| AA | Transmitter signal element timing B-wire | 114 | to DTE |
| P | Transmitted data A-wire | 103 | from DTE |
| S | Transmitted data B-wire | 103 | from DTE |
| U | Transmitter signal element timing A-wire | 113 | from DTE |
| Z | $F_3$ | — | — |
| W | Transmitter signal element timing B-wire | 113 | from DTE |
| BB | $F_3$ | — | — |
| CC | $F_4$ | — | — |
| DD | $F_5$ | — | — |
| EE | $F_4$ | — | — |
| FF | $F_5$ | — | — |
| HH | $N_1$ | — | — |
| JJ | $N_2$ | — | — |
| KK | $N_1$ | — | — |
| LL | $N_2$ | — | — |
| MM | F | — | — |
| NN | F | — | — |

N = Pin number permanently reserved for national use.

F = Pin number reserved for future International Standard and should not be used for national use.

Subscripts indicate pins which may be associated to form pairs; for example, $F_1$ on pins K and M form a pair.

NOTES

1  Pins HH, JJ and KK are used in the United Kingdom for transmiter clock control, alternate use transmit and alternate use receive respectively.

2  The Figure opposite illustrates the physical layout of the connector pins.

⊖ A ⊖ B
⊖ C ⊖ D ⊖ F
⊖ H ⊖ E ⊖ L
⊖ M ⊖ J ⊖ R
⊖ S ⊖ K ⊖ V
⊖ W ⊖ N ⊖ Z
⊖ AA ⊖ P ⊖ DD
⊖ EE ⊖ T ⊖ JJ
⊖ KK ⊖ U ⊖ NN

FIGURE — 34 Pin connector layout

412

# INTERNATIONAL STANDARD  2628

INTERNATIONAL ORGANIZATION FOR STANDARDIZATION •МЕЖДУНАРОДНАЯ ОРГАНИЗАЦИЯ ПО СТАНДАРТИЗАЦИИ •ORGANISATION INTERNATIONALE DE NORMALISATION

# Basic mode control procedures — Complements

**First edition — 1973-06-01**

413

UDC 681.14

Ref. No. ISO 2628-1973 (E)

**Descriptors :** data processing, data transmission, control procedures, character sets, codes.

## FOREWORD

ISO (the International Organization for Standardization) is a worldwide federation of national standards institutes (ISO Member Bodies). The work of developing International Standards is carried out through ISO Technical Committees. Every Member Body interested in a subject for which a Technical Committee has been set up has the right to be represented on that Committee. International organizations, governmental and non-governmental, in liaison with ISO, also take part in the work.

Draft International Standards adopted by the Technical Committees are circulated to the Member Bodies for approval before their acceptance as International Standards by the ISO Council.

International Standard ISO 2628 was drawn up by Technical Committee ISO/TC 97, *Computers and information processing,* and circulated to the Member Bodies in May 1972.

It has been approved by the Member Bodies of the following countries :

| | | |
|---|---|---|
| Australia | Ireland | Spain |
| Belgium | Italy | Sweden |
| Canada | Japan | Switzerland |
| Czechoslovakia | Netherlands | Thailand |
| Egypt, Arab Rep. of | New Zealand | United Kingdom |
| France | Portugal | U.S.A. |
| Germany | South Africa, Rep. of | U.S.S.R. |

No Member Body expressed disapproval of the document.

414

# Basic mode control procedures — Complements

## 0  INTRODUCTION

A data communication system may be considered as the set of the terminal installations and the interconnecting network that permits information to be exchanged.

A data link comprises terminal installations connected to the same network, operating at the same speed, in the same code. Any "store and forward" delay or intermediate data processing separates data links. Any system is constituted of one or several data links.

The information transfer in a data link is monitored by data link control procedures where some characters, selected within a coded character set, are given particular meanings according to the transmission phase and are used for various purposes such as to delineate information, to reverse the direction of transmission, to ask questions, to answer, etc.

This International Standard defines complements to the basic mode and its extensions :

1)  Recovery procedures

    —  System guidelines are given for the use of timers, counters, etc.;

2)  Abort and interrupt procedures

    —  Defines abort procedures which are always initiated by the master station, and interrupt procedures which are always initiated by the slave station;

3)  Multiple station selection

    —  Gives means whereby a master station may select more than one slave station so that all the selected slave stations receive the same transmission at the same time.

## 1  SCOPE AND FIELD OF APPLICATION

This International Standard extends the digital basic mode control procedures as defined in ISO/R 1745 and ISO 2111, to allow the following features :

1)  Recovery procedures;

2)  Abort and interrupt procedures;

3)  Multiple station selection.

Those systems which conform to ISO/R 1745 do not necessarily have to include the functions described in this

International Standard. However, those systems implementing the functions described in this International Standard and conforming to ISO/R 1745 and ISO 2111, must follow these recommendations.

## 2  REFERENCES

ISO/R 1745, *Basic mode control procedures for data communication systems*. (At present under revision.)

ISO 2111, *Data communication — Basic mode control procedures — Code independent information transfer.*

CCITT Recommendation V24, *Function and electrical characteristics of circuits at the interface between data terminal equipment and data communication equipment.*

## 3  RECOVERY PROCEDURES

### 3.1  General

These recovery procedures are system guidelines which should be used by all stations, as applicable. However, it is recognized that the detailed method of station mechanization, absolute value of timers, etc., may vary with applications and communication facilities.

In some cases, these recovery procedures can only detect the error condition and then notify the operator or the processor program, or both. In more sophisticated cases, automatic recovery is partially or completely possible. In other cases, only operators can perform the recovery procedures. Operator recovery procedures are not part of this International Standard. However, the operator may do such things as retry *n* more times, establish voice communication to the distant station in order to determine trouble, etc.

For a good system, the functions of timers A, B, and C defined below, must be utilized. The value of the timer may vary over a wide range depending upon whether they are implemented via hardware, software, or human operator.

It is recognized that in some systems additional timers may be required for such purposes as aiding synchronization procedures, added reliability, etc.

### 3.2  Timers and counters

Timers are primarily used as aids in recovery procedures when recognition of specific control characters does not occur. The action taken following a time-out is specified in general terms to provide system protection. The absolute values of the timers are dependent upon such things as manual use, non-manual data entry, speed of transmission, type of data source/sink, etc.

Counting is primarily used as an aid in determining what recovery alternative is applicable in each error condition. The number of consecutive negative or invalid replies and the number of consecutive attempts to recover using one recovery procedure before an alternative is chosen depends upon the network configuration, quality of the channel, and application.

### 3.2.1 Timer A (No-response timer)

Where implemented : control station, master station, or both.

Purpose : protection against an invalid response or no response.

Start : after transmitting any ending character where a reply is expected; for example ENQ, ETB, ETX, DLE ETB, DLE ETX.

Stop : upon receipt of a valid reply from the communication line; for example ACK, NAK, STX, EOT, DLE STX.

When time-out occurs :

1) — retransmit same information (up to $n$ times)[1], or

— transmit different information; for example ENQ, different polling/selection sequence;

2) transmit EOT, when station abort procedures are used;

3) notify operator or processor program, or both;

4) return to non-transparent mode, if applicable.

### 3.2.2 Timer B (Receive timer)

Where implemented : slave station.

Purpose : protection against non-recognition of any block terminating character, for example ETB, ETX, ENQ, DLE ETB or DLE ETX received from the communication line.

Start :

1) receipt of SOH, STX (if not preceded by SOH), DLE SOH, DLE STX or other opening characters or sequence as required.

2) this timer may be restarted to permit receipt of variable length blocks.

Stop : upon receipt of a valid terminating character or sequence; for example ETB, ETX, ENQ, DLE ETB, DLE ETX.

When time-out occurs :

1) remain in slave status and initiate search for character synchronization in synchronous systems;

2) prepare to receive another transmission;

3) notify operator or processor program or both, and discard the incomplete block;

4) return to non-transparent mode, if applicable.

NOTE — For maximum system efficiency, the duration of the no-response timer (Timer A) should be short and the receive timer (Timer B) should time-out before the no-response timer.

### 3.2.3 Timer C (No-activity timer for switched lines)

Where implemented : all stations.

Purpose : facilitates disconnection procedures of the communication line if data transmission stops due to not recognizing DLE EOT, or due to remote station or communication facility problems.

Start or restart :

1) upon receipt of indication of circuit connection; for example receipt of ON condition of circuit 107 (data set ready[2]) or circuit 125 (calling indicator[2]) and circuit 108.2 (data terminal ready[2]).

2) upon receipt or transmission of any character in asynchronous systems or the synchronizing sequence in synchronous systems.

Stop :

1) upon receipt or transmission of DLE EOT, or

2) loss of circuit 107 (data set ready[2])

When time-out occurs :

1) disconnect communication circuit;

2) notify operator or processor program, or both;

3) return to control mode, if applicable;

4) return to non-transparent mode, if applicable.

### 3.2.4 Timer D (No-activity timer for non-switched lines)

Where implemented : control station

Purpose : serves as a "no-activity" time-out for all stations in a system.

Start or restart : upon receipt or transmission of any character in asynchronous systems or after the synchronizing sequence in synchronous systems.

Stop : upon receipt or transmission of EOT

---

1) Retransmission of a data block can result in duplication of a block at the receiving location if a block numbering or other protective scheme is not used.

2) CCITT — V 24 designation.

When time-out occurs :

1) notify operator or processor program, or both;

2) return to control mode, if applicable;

3) return to non-transparent mode, if applicable.

## 3.3 Recovery procedures

Some recovery procedures are outlined in the following with their linkage to the appropriate phase diagrams in 4.2 of ISO/R 1745 and to the timers A, B and C described in this International Standard.

In all cases, after the appropriate time-out periods, it shall be the final responsibility of either the control station or the master station to take action.

### 3.3.1 *Recovery procedures by control station*

R1 — In the case of :

1) invalid or absence of termination supervisory sequence detected by time-out of either timer A or timer C, the control station must transmit EOT or DLE EOT whichever is appropriate;

2) invalid or no response to a polling/selection sequence detected by time-out of timer A, the control station may transmit the same or a different polling/selection sequence following the transmission of an EOT and/or notify operator or processor program, or both.

R2 — In the case of :

repeated unsuccessful polling of one, several or all stations, the control station should notify operator or processor program, or both.

### 3.3.2 *Recovery procedures by master station*

R3 — In the case of :

1) invalid or no response to a selecting supervisory sequence detected by the time-out of timer A, the master station may

    a) terminate by transmitting EOT;

    b) transmit same or another selecting supervisory sequence (up to *n* times);

    c) notify operator or processor program, or both.

2) invalid or no response to information message detected by time-out of timer A the master station may

    a) repeat the previous transmission (up to *n* times). This procedure can lead to duplication of blocks;

    b) transmit prefix ENQ (up to *n* times) which requests the slave station to repeat its previous response (ACK or NAK). This procedure can lead to loss of blocks unless used in conjunction with a response numbering scheme to ensure that blocks are neither added nor deleted.

R4 — In the case of :

1) repeated negative replies (NAK) or invalid or no responses to a selection supervisory sequence, the master station should notify the operator or processor program, or both;

2) repeated negative replies (NAK) or failure to receive a valid reply for an information block, the master station may transmit an EOT (if master station abort is used) and/or notify the operator or processor program, or both.

### 3.3.3 *Recovery procedures by a slave station*

Recovery procedures by a slave station are explained by the functions of timer B (see 3.2.2).

## 4 ABORT AND INTERRUPT PROCEDURES

### 4.1 General

Abort procedures are always initiated by the master station wishing either

1) to stop transmitting a block of information before its normal end (ETB or ETX) but without returning to control mode or neutral; or

2) to stop transmitting at any time during the information transfer phase and then return to control mode or neutral.

Interrupt procedures are always initiated by the slave station which desires to stop receiving either instantaneously or within a short period of time.

### 4.2 Abort procedures

#### 4.2.1 *Block abort*

##### 4.2.1.1 DESCRIPTION

The master station decides to terminate a block in an unusual way so that the slave station rejects this block. There is no return to control mode or neutral and the master station resumes transmission to the same slave station.

##### 4.2.1.2 PROCEDURE

When the master station decides to abort a block, it terminates it immediately with ENQ (DLE ENQ if applicable). The slave station replies with NAK which is the only valid acknowledgement in this case. The master station then resumes transmission beginning with STX (or SOH). If the reply from the slave station is invalid, or if there is no reply, the normal recovery procedures may apply (*n* retries, time-out). (See Figure 1.)

NOTE — As examples, block abort may be used in the following cases :

— the master station determines that invalid data have been sent;

for example, errors are detected at the buffer storage level, or when reading data from their media, or by the source (operator).

— with fixed length blocks when, due to transmission, programming or operator errors, the block being transmitted overflows normal length.

— when the master station determines that the block being transmitted will not be accepted by the slave station.

### 4.2.2 *Station abort*

#### 4.2.2.1 DESCRIPTION

The master station is sending a message and decides, either while a block of information is being sent or between two blocks of information, to stop transmitting and return to control mode or neutral.

#### 4.2.2.2 PROCEDURE

a)  While a block is being sent

When the master station decides to abort a transmission, it immediately sends the transmission control character ENQ (DLE ENQ). The slave station detects this unusual termination with ENQ (instead of ETB or ETX) and then replies with NAK which is the only valid reply in this case. After receiving NAK, the master station sends EOT and the communication link returns to control mode or neutral.

When there is no answer or an invalid answer, the normal recovery procedures may apply (*n* retries, time-out). (See Figure 2.)

b)  Between two blocks of information

The master station terminates the block being transmitted in the usual way. The usual answer of the slave station is ACK. The master station then sends EOT and the communication link goes back to control mode or neutral.

If the answer is NAK or if there is no answer or an invalid one, the master station may or may not decide to use the normal recovery procedures (*n* retries, time-out) before transmitting EOT with the resulting return to control mode or neutral. (See Figure 3.)

NOTES

1  In switched line applications, DLE EOT may be used in place of EOT.

2  Examples of use : when it is intended to disconnect the line.

Master station abort may be used in the following cases :

— Master station detects its own malfunction, or a malfunction of the transmitting media.

— Master station detects a failure in the slave station or in the link (persisting NAK, or invalid reply, or absence of reply) or the master station detects that the slave station is no longer in a position to receive.

— Master station is notified that the transmission media are urgently required for another purpose.

### 4.3 Interrupt procedures

### 4.3.1 *Block interrupt*

#### 4.3.1.1 DESCRIPTION

The slave station, at the end of a message or of an information block, is no longer in a position to receive and wishes the master station to cease transmission immediately.

#### 4.3.1.2 PROCEDURE

The slave station replies EOT instead of its normal reply. EOT indicates a negative acknowledgement of the last received block and the conclusion of the current transmission. The communication link returns to control mode or neutral. (See Figure 4.)

NOTE — The transmission systems fall into one of the following classes :

1  Control station is also master station.

2  Control station is also slave station.

3  Control station is neither master nor slave but is monitoring only the transmissions of the master station.

4  Control station, being neither master nor slave, is monitoring all data exchange within the system.

The block interrupt procedure, as described in 4.3.1.2 above, can only be used in classes 1, 2 and 4. As regards class 3. the control station is not aware of the EOT sent by the slave station and there is no way to return to control mode or neutral other than through recovery procedures (control station time-out, for instance).

For this reason, the use of block interrupt procedure is not recommended for systems falling in class 3 above. Concerning systems falling in classes 1, 2 and 4, block interrupt is not recommended for frequent utilization; its use should be reserved for emergency situations.

### 4.3.2 *Station interrupt*

#### 4.3.2.1 DESCRIPTION

Station interrupt is the means whereby a slave station can request the master station to stop transmitting as soon as possible.

#### 4.3.2.2 PROCEDURE

Station interrupt is accomplished by the slave transmitting the control sequence DLE < instead of the normal positive acknowledgement. This reply has a double meaning :

1)  it includes the positive acknowledgement which would have been normally sent;

2)  it means a request from the slave station to have the current transmission terminated at the earliest possible time (by the master station sending EOT). However, the master station may not stop transmitting immediately

and may, for instance, continue to transmit so that its buffers are cleared and readily available for further transmissions. The point where the master station effectively stops is system dependent.

NOTES

1   Example of use. The control station being also the slave station may want to interrupt so as to be able to urgently poll or select another tributary station.

2   Recovery procedures. The possibility of a station interrupt sequence being garbled by line disturbances should be considered. In particular, if backward supervision numbering should be established, this may lead either to use the same numbering scheme for the station interrupt sequence as for the supervisory sequence, or to impose other rules to preserve the correct information blocks sequencing.

For instance, it should not be permitted to continuously send the station interrupt sequence.

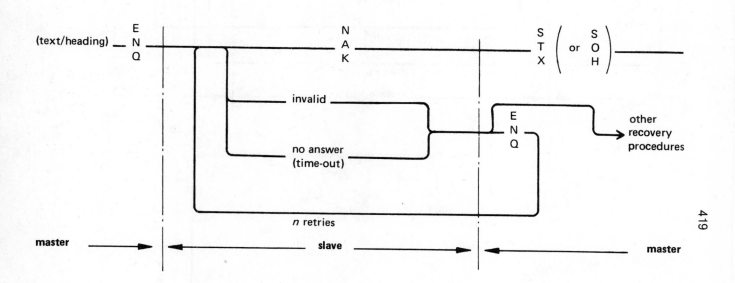

FIGURE 1 — **Block abort**

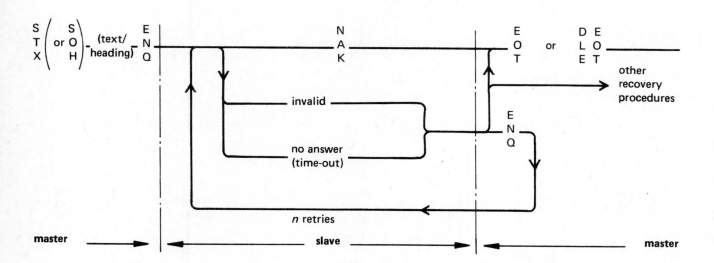

FIGURE 2 — **Station abort (while a block is being sent)**

419

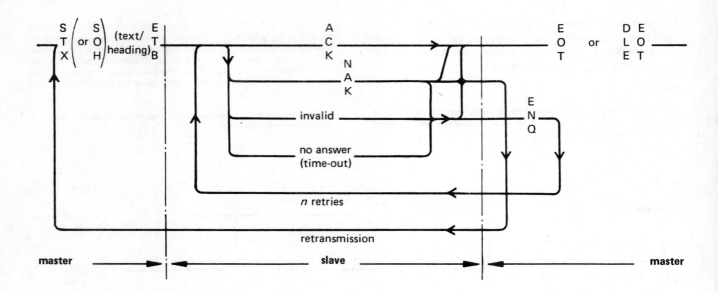

FIGURE 3 — **Master station abort (between two information blocks)**

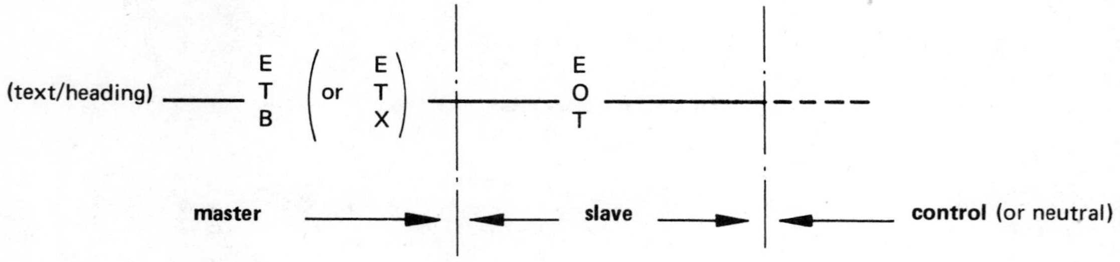

420

FIGURE 4 — **Block interrupt**

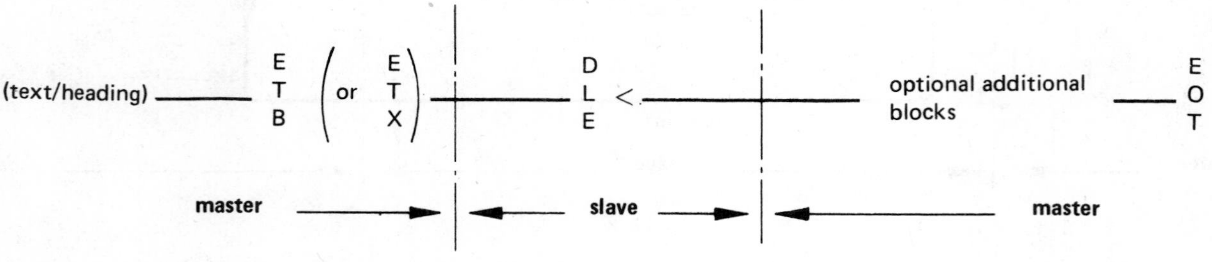

FIGURE 5 — **Station interrupt**

## 5 MULTIPLE STATION SELECTION

### 5.1 General

**5.1.1** Multiple selection is a means whereby a master station may select more than one slave station so that all the selected slave stations receive the same transmission at the same time.

**5.1.2** The procedures for multiple selection are not covered by the basic mode control procedures, hence they are considered to be an extension to them.

**5.1.3** When a system is designed to work both with and without the multiple selection procedure, some means must be provided for the master station to notify the slave stations which procedure is to be entered. For example, by assigning two different addresses to each station having both facilites.

### 5.2 Selection sub-phase

Three methods providing various degrees of protection are proposed for the selection sub-phase. They are listed in the order of decreasing degree of protection.

**5.2.1** *Sequential selection with individual replies from the selected stations.*

```
Master │ Sel.        E         Sel.        E         Sel.        E              S
       │ Address 1   N         Address 2   N         Address n   N              T   Data
       │             Q                     Q                     Q              X
Slaves │              \Address 1  A         \Address 2  A         \Address n  A
       │                         C                     C                     C
       │                         K                     K                     K
```

**5.2.2** *Group selection with reply from one designated station*, for example, the most distant one, or strategically located, or any station indicated within the selection sequence.

```
        Master      │ Group Select   E                        S
                    │ Address X      N                        T   Data
                    │                Q                  A   / X
        Designated  │                 \Address X       C /
        slave       │                                  K
```

**5.2.3** *Group selection with fast select.*

```
            Master  │     Group Select     S
                    │     Address          T    Data
                    │                      X
```

### 5.3 Information transfer phase

Three methods providing various degrees of protection are proposed for the information transfer phase. They are listed in the order of decreasing degree of protection.

**5.3.1** *Information transfer with individual replies from the slave station*

After each transmission block, the master station sends a delivery verification supervisory sequence, consisting of a prefix identifying a single slave station, followed by ENQ. Only tributary stations having slave status should respond to delivery verification supervisory sequences.

```
Master │ E   /B\    Address 1   E              Address n   E            /
       │ T   (C)                N                          N           /
       │ X   \C/                Q                          Q          A
Slaves │                         \(Address 1)  A            \(Address n)  C
       │                                       C                       K
       │                                       K
```

**5.3.2** *Information transfer with reply from one designated station,* for example the most distant, or strategically located.

|  | | |
|---|---|---|
| **Master** | E     ⎛B⎞ | |
|  | T     ⎜C⎟ | A |
|  | X     ⎝C⎠ (Address *X*) | C |
| **Designated slave** | | K |

*X* being the address of the designated station.

**5.3.3** *No reply*

Although the "no reply" case is not considered by the basic mode, it is recognized that it may conveniently be used for general announcement (for example conference) and the broadcasting of messages of a "clear text" type.

**5.4 Relations between selection procedures and information transfer procedures**

Although the adoption of one of the three selection procedures does not preclude the adoption of any one of the three procedures for information transfer, it is recognized that some pairings would not be realistic. Straightforward pairings could be, for example, 2.1 with 3.1, 2.2 with 3.2, and 2.3 with 3.3.

# INTERNATIONAL STANDARD  ISO 2629

INTERNATIONAL ORGANIZATION FOR STANDARDIZATION ·МЕЖДУНАРОДНАЯ ОРГАНИЗАЦИЯ ПО СТАНДАРТИЗАЦИИ·ORGANISATION INTERNATIONALE DE NORMALISATION

# Basic mode control procedures — Conversational information message transfer

**First edition — 1973-02-15**

423

**UDC 681.14**

**Ref. No. ISO 2629-1973 (E)**

**Descriptors** : data processing, data transmission, control procedures.

## FOREWORD

ISO (the International Organization for Standardization) is a worldwide federation of national standards institutes (ISO Member Bodies). The work of developing International Standards is carried out through ISO Technical Committees. Every Member Body interested in a subject for which a Technical Committee has been set up has the right to be represented on that Committee. International organizations, governmental and non-governmental, in liaison with ISO, also take part in the work.

Draft International Standards adopted by the Technical Committees are circulated to the Member Bodies for approval before their acceptance as International Standards by the ISO Council.

International Standard ISO 2629 was drawn up by Technical Committee ISO/TC 97, *Computers and information processing.*

It was approved in September 1972 by the Member Bodies of the following countries :

| | | |
|---|---|---|
| Australia | Germany | South Africa, Rep. of |
| Belgium | Ireland | Spain |
| Brazil | Italy | Sweden |
| Canada | Japan | Switzerland |
| Czechoslovakia | Netherlands | Thailand |
| Denmark | New Zealand | United Kingdom |
| Egypt, Arab Rep. of | Portugal | U.S.A. |
| France | Romania | U.S.S.R. |

No Member Body expressed disapproval of the document.

424

# Basic mode control procedures — Conversational information message transfer

## 0 INTRODUCTION

This International Standard defines an addition to the basic mode control procedures for data communication systems and allows the reversal of information transfer while remaining in Phase 3 (information transfer) of the basic mode.

Although applicable to many types of terminals, this type of operation is particularly adaptable to inquiry/response systems.

In some systems, the security of the data link operation may be obtained by the use of block checking and ACK-NAK supervisory sequences.

To preserve a high degree of security, it is recognized that numbering schemes (forward and/or backward) can also be used.

In other systems utilizing less sophisticated terminals, supervisory information and control information may be contained within the message and/or handled by operator procedures. These systems are not covered by this International Standard.

## 1 SCOPE AND FIELD OF APPLICATION

**1.1** This International Standard defines the means by which a data communication system operating according to the basic mode control procedures defined in ISO/R 1745 can interchange information messages in a fast conversational manner and where the operator plays a significant role in the operation of the terminal.

**1.2** This International Standard extends Phase 3 (information transfer) as defined in ISO/R 1745, to allow two stations connected by a data link to reverse their master/slave status, thereby reversing the direction of the information transfer, without leaving Phase 3.

**1.3** During one conversation process considered here, only two stations are involved at one time. Conversation with any other station requires termination of the existing data link and establishment of another data link.

**1.4** This procedure applies to the following system configurations : point-to-point, centralized multipoint.

## 2 REFERENCE

ISO/R 1745, *Basic mode control procedures for data communication systems.*

## 3 CONVERSATION RULES (See Figure)

**3.1** The positive acknowledgement can be replaced by the transfer of an information message in the opposite direction.

**3.2** The information message can be sent in lieu of the positive acknowledgement only when the received message is terminated by ETX (or DLE.ETX); not ETB (DLE.ETB).

**3.3** The opening character of the returned message (i.e. STX, SOH or DLE.STX, DLE.SOH) shall be considered as having the additional meaning of positive acknowledgement.

**3.4** If there is no information message to be transferred in the opposite direction after the received message is correctly received, the positive acknowledgement shall be used (see Note 2).

**3.5** An information message shall never be sent instead of a negative acknowledgement, except as indicated in Note 2.

**3.6** Termination is initiated by the station with link responsibility.

### NOTES

1 Typical applications require one reversal of transmission only. If multiple reversals occur, the error protection capabilities may be degraded. If no forward numbering scheme is used, consecutive reversals may lead to loss or duplication of messages or to conflicting situations (for example master/slave decision). In these cases, operator intervention may be required.

2 In systems which use backward supervisory sequences, ACK and NAK are used for the acknowledgements. In systems where ACK and NAK are not used, acknowledgement information may be included in the reply messages.

## 4 RECOVERY PROCEDURES

In the case of no reply or an invalid reply to an information message, the information message should be repeated. This procedure can lead to a duplication of blocks. Alternatively, ENQ can be transmitted to request the replying station to repeat its previous response or message.

This implies that the station that has reversed its status to master must be prepared to receive ENQ in reply to an information message requesting it to repeat that message.

In either case a numbering scheme is needed to ensure that blocks are neither added nor deleted.

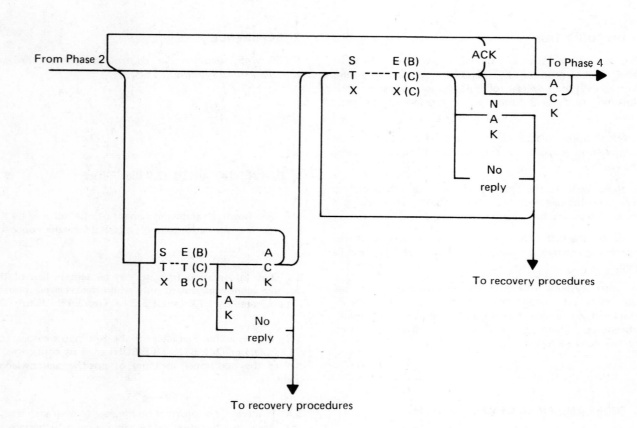

FIGURE

# INTERNATIONAL STANDARD  3309

INTERNATIONAL ORGANIZATION FOR STANDARDIZATION • МЕЖДУНАРОДНАЯ ОРГАНИЗАЦИЯ ПО СТАНДАРТИЗАЦИИ • ORGANISATION INTERNATIONALE DE NORMALISATION

# Data communication — High-level data link control procedures — Frame structure

*Téléinformatique — Procédures de commande de chaînon à haut niveau — Structure de trame*

**First edition — 1976-04-01**

427

UDC 681.3 : 621.391

Ref. No. ISO 3309-1976 (E)

**Descriptors :** data processing, teleprocessing, data transfer, control procedures, high-level data link control.

## FOREWORD

ISO (the International Organization for Standardization) is a worldwide federation of national standards institutes (ISO Member Bodies). The work of developing International Standards is carried out through ISO Technical Committees. Every Member Body interested in a subject for which a Technical Committee has been set up has the right to be represented on that Committee. International organizations, governmental and non-governmental, in liaison with ISO, also take part in the work.

Draft International Standards adopted by the Technical Committees are circulated to the Member Bodies for approval before their acceptance as International Standards by the ISO Council.

International Standard ISO 3309 was drawn up by Technical Committee ISO/TC 97, *Computers and information processing,* and circulated to the Member Bodies in June 1975.

It has been approved by the Member Bodies of the following countries :

| | | |
|---|---|---|
| Australia | Hungary | Sweden |
| Belgium | Italy | Switzerland |
| Brazil | Japan | Turkey |
| Canada | Netherlands | United Kingdom |
| Czechoslovakia | New Zealand | U.S.A. |
| Egypt, Arab Rep. of | Romania | U.S.S.R. |
| France | South Africa, Rep. of | Yugoslavia |
| Germany | Spain | |

No Member Body expressed disapproval of the document.

428

# Data communication — High-level data link control procedures — Frame structure

## 0 INTRODUCTION

This document is one of a series of International Standards, to be used for implementation of various applications with synchronous transmission facilities.

## 1 SCOPE AND FIELD OF APPLICATION

This International Standard defines in detail the frame structure for data communication systems using bit-oriented high-level data link control (HDLC) procedures. It defines the relative positions of the various components of the basic frame and the bit combination for the frame delimiting sequence (Flag). The bit escaping mechanism which is used to achieve bit pattern independence within the frame is also defined. The document also specifies the frame checking sequence (FCS). No details of the address or control field allocations are included, other than address extension outlined in clause 4.

## 2 FRAME STRUCTURE

In HDLC, all transmissions are in frames, and each frame conforms to the following format :

| Flag | Address | Control | Information | FCS | Flag |
|------|---------|---------|-------------|-----|------|
| 01111110 | 8 bits | 8 bits | * | 16 bits | 01111110 |

*  An unspecified number of bits which in some cases may be a multiple of a particular character size, for example an octet.

where

   Flag = flag sequence

   Address = secondary station address field

   Control = control field

   Information = information field

   FCS = frame checking sequence

Frames containing only supervisory control sequences form a special case where there is no information field. The format for these frames shall be :

| Flag | Address | Control | FCS | Flag |
|------|---------|---------|-----|------|
| 01111110 | 8 bits | 8 bits | 16 bits | 01111110 |

## 3 ELEMENTS OF THE FRAME

### 3.1 Flag sequence

All frames shall start and end with the flag sequence. All stations which are attached to the data link shall continuously hunt for this sequence. Thus, the flag is used for frame synchronization. A single flag may be used as both the closing flag for one frame and the opening flag for the next frame.

### 3.2 Address field

The address shall in all cases identify the secondary or secondaries which are involved in the particular frame interchange.

### 3.3 Control field

The control field contains commands or responses, and sequence numbers. The control field shall be used by the primary to command the addressed secondary to perform a particular operation. It shall be used by the secondary to respond to the primary.

### 3.4 Information field

Information may be any sequence of bits. In most cases it will be linked to a convenient character structure, for example octets, but if required, it may be an unspecified number of bits and unrelated to a character structure.

### 3.5 Transparency

The transmitter shall examine the frame content between the two flag sequences including the address, control and FCS sequences and shall insert a "0" bit after all sequences of 5 contiguous "1" bits (including the last 5 bits of the FCS) to ensure that a flag sequence is not simulated. The receiver shall examine the frame content and shall discard any "0" bit which directly follows 5 contiguous "1" bits.

### 3.6 Frame checking sequence (FCS)

The FCS shall be a 16-bit sequence. It shall be the ones complement of the sum (modulo 2) of :

1) the remainder of $x^k$ $(x^{15} + x^{14} + x^{13} + \ldots + x^2 + x + 1)$ divided (modulo 2) by the generator polynomial $x^{16} + x^{12} + x^5 + 1$, where $k$ is the number of bits in the frame existing between, but not including, the final bit of the opening flag and the first bit of the FCS, excluding bits inserted for transparency, and

2) the remainder after multiplication by $x^{16}$ and then division (modulo 2) by the generator polynomial $x^{16} + x^{12} + x^5 + 1$ of the content of the frame, existing between, but not including, the final bit of the opening flag and the first bit of the FCS, excluding bits inserted for transparency.

As a typical implementation, at the transmitter, the initial remainder of the division is preset to all ones and is then modified by division by the generator polynomial (as described above) on the Address, Control and Information fields; the ones complement of the resulting remainder is transmitted as the 16-bit FCS sequence.

At the receiver, the initial remainder is preset to all ones and the serial incoming protected bits and the FCS when divided by the generator polynomial will result in a remainder of $0001110100001111$ ($x^{15}$ through $x^0$, respectively) in the absence of transmission errors.

NOTES

1 If future applications show that a higher degree of protection is needed, the number of bits of the FCS shall be increased by octets.

2 See the annex for explanatory notes on implementation of the FCS.

### 3.7 Order of bit transmission

Addresses, commands, responses, and sequence numbers shall be transmitted low-order bit first (for example the first bit of the sequence number that is transmitted shall have the weight $2^0$).

The order of transmitting bits within the information field is not specified by this International Standard.

The FCS shall be transmitted to the line commencing with the coefficient of the highest term.

### 3.8 Inter-frame time fill

Inter-frame time fill shall be accomplished by transmitting either contiguous flags or a minimum of seven contiguous "1" bits or a combination of both.

A selection of the inter-frame time fill methods depends on systems requirement.

### 3.9 Invalid frame

An invalid frame is defined as one that is not properly bounded by two flags or one which is too short (for example shorter than 32 bits between flags). Invalid frames shall be ignored. Thus, a frame which ends with an all "1" bit sequence equal to or greater than seven bits in duration shall be ignored.

As an example, one method of aborting a frame would be to transmit 8 contiguous "1" bits.

## 4 EXTENSIONS

### 4.1 Extended address field

Normally a single octet address shall be used and all 256 combinations shall be available.

However, by prior agreement the address range can be extended by reserving the first transmitted bit (low order) of each address octet which would then be set to binary zero to indicate that the following octet is an extension of the basic address. The format of the extended octet(s) shall be the same as that of the basic octet. Thus, the address field may be recursively extended.

When extensions are used, the presence of a binary "1" in the first transmitted bit of the basic address octet signals that only one address octet is being used. The use of address extension thus restricts the range of single octet addresses to 128.

### 4.2 Extended control field

The control field may be extended by one or more octets. The extension method and the bit patterns for the commands and responses will be defined in a separate International Standard.

ANNEX

## EXPLANATORY NOTES ON IMPLEMENTATION OF THE FRAME CHECKING SEQUENCE

(Not part of the standard)

In order to permit the use of existing devices that are arranged to use a zero preset register, the following implementation may be used.

At the transmitter, generate the FCS sequence in the following manner while transmitting the elements of the frame unaltered onto the line :

a)  preset the FCS register to zeros;

b)  invert the first 16 bits (following the opening flag) before shifting into the FCS register;

c)  shift the remaining fields of the frame into the FCS register uninverted;

d)  invert the contents of the FCS register (remainder) and shift onto the line as the FCS sequence.

At the receiver, operate the FCS checking register in the following manner while receiving (and storing) unaltered the elements of the frame as·received from the line :

a)  preset the FCS register to zeros;

b)  invert the first 16 bits (following the opening flag) before shifting them into the FCS checking register;

c)  shift the remaining elements of the frame, up to the beginning of the FCS, into the checking register uninverted;

d)  invert the FCS sequence before shifting into the checking register. In the absence of errors, the FCS register will contain all zeros after the FCS is shifted in.

In the above, inversion of the first 16 bits is equivalent to a ones preset and inversion of the FCS at the receiver causes the registers to go to the all zeros state.

The transmitter or the receiver can independently use the ones preset or the first 16-bit inversion. Also, the receiver can choose not to invert the FCS, in which case it must check for the unique non-zero remainder specified in 3.6.

It must be realized that inversion of the FCS by the receiver requires a 16-bit storage delay before shifting message bits into the register. The receiver cannot anticipate the beginning of the FCS. Such storage, however, will normally take place naturally as the FCS checking function will need to differentiate the FCS from data anyway, and it will thus withhold 16 bits from the next function at all times.

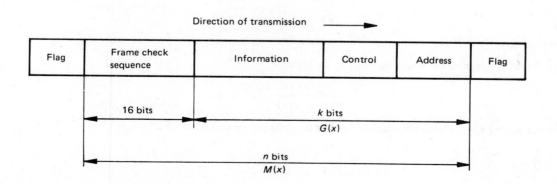

The procedure for using the FCS assumes the following :

1) The $k$ bits of data which are being checked by the FCS can be represented by a polynomial $G(x)$.

*Example :* $G(x) = x^5 + x^3 + 1$ represents 101001.

2) The Address field, Control field and Information field (if it exists in the message) are represented by the polynomial $G(x)$.

3) For the purpose of generating the FCS, the first bit following the opening flag is the most significant bit of $G(x)$ regardless of the actual representation of the Address, Control and Information fields.

4) There exists a generator polynomial $P(x)$ of degree 16, having the form $P(x) = x^{16} + x^{12} + x^5 + 1$.

The FCS is defined as a ones complement of a remainder, $R(x)$, obtained from the modulo 2 division of

$$x^{16}G(x) + x^k (x^{15} + x^{14} + x^{13} + x^{12} + x^{11} + x^{10} + x^9 + x^8 + x^7 + x^6 + x^5 + x^4 + x^3 + x^2 + x^1 + 1)$$

by the generator polynomial $P(x)$.

$$\frac{x^{16}G(x) + x^k (x^{15} + x^{14} + \ldots + x + 1)}{P(x)} = Q(x) + \frac{R(x)}{P(x)} \quad \overline{\text{FCS}}$$

The multiplication of $G(x)$ by $x^{16}$ corresponds to shifting the message $G(x)$ 16 places and thus providing the space of 16 bits for the FCS.

The addition of $x^k (x^{15} + x^{14} + \ldots + x + 1)$ to $x^{16}G(x)$ (equivalent to inverting the first 16 bits of $x^{16}G(x)$) corresponds to initializing the initial remainder to a value of all "ones". This addition is provided for the purpose of protection against the obliteration of leading flags, which may be non-detectable if the initial remainder is zero. The complementing of $R(x)$, by the transmitter, at the completion of the division ensures that the received, error-free message will result in a unique, non-zero remainder at the receiver. The non-zero remainder provides protection against potential non-detectability of the obliteration of trailing flags.

At the transmitter the FCS is added to the $x^{16}G(x)$ and results in the total message $M(x)$ of length $n$, where $M(x) = x^{16}G(x) + \text{FCS}$.

At the receiver, the incoming $M(x)$ is multiplied by $x^{16}$, added to $x^n (x^{15} + x^{14} + \ldots + x + 1)$ and divided by $P(x)$.

$$\frac{x^{16} [x^{16}G(x) + \text{FCS}] + x^n (x^{15} + x^{14} + \ldots + x + 1)}{P(x)} = Qr(x) + \frac{Rr(x)}{P(x)}$$

If the transmission is error free, the remainder $Rr(x)$ will be "0001110100001111" ($x^{15}$ through $x^0$).

$Rr(x)$ is the remainder of the division : $\dfrac{x^{16}L(x)}{P(x)}$

where $L(x) = x^{15} + x^{14} + \ldots + x + 1$. This can be shown by establishing that all other terms of the numerator of the receiver division are divisible by $P(x)$.

Note that $\text{FCS} = \overline{R(x)} = L(x) + R(x)$. (Adding $L(x)$ without carry to a polynomial of its same length is equivalent to a bit-by-bit inversion of the polynomial.)

The equation above, for the FCS receiver residual, is used in the following to show that inverting the FCS at the receiver returns the checking register to zero. This equation is

$$\frac{x^{16}L(x)}{P(x)} = Q(x) + \frac{Rr(x)}{P(x)}$$

where $L(x)$ is as before and $Rr(x)$ is the residual contents of the FCS register.

If another $x^{16}L(x)$ is added to the above numerator, the result is

$$\frac{x^{16}L(x) + x^{16}L(x)}{P(x)} = 0$$

Physically, this second $x^{16}L(x)$ quantity is added to the bit stream by inverting the FCS.

# Part 3

# ANSI Data Communications Standards

In the USA, the principal standards development body is the American National Standards Institute (ANSI). ANSI was formed in 1918 and evolved through several forms until 1969, when it adopted its current structure and name. ANSI is a non-profit, non-governmental organization supported by over 1,000 trade organizations, professional societies, and companies. It serves as the national clearing house and coordinating activity for voluntary standards and is the USA's member body to ISO.

The majority of the standards produced by ANSI come from the work of about 300 Standards Committees that work in association with Technical Committees and Task Groups. Other ANSI standards evolve out of the work of associated groups such as the Electronic Industries Association (EIA).

The ANSI procedures require that all members of Technical Committees and Task Groups serve as technically qualified, participating individuals rather than as company or organizational representatives. At this level of activity, membership is open to anyone with a bona fide interest and a willingness to participate. At the Standards Committee level, however, members are categorized by the type of organization they represent. These categories are "consumers," "producers," and "general interest." The membership of each Standards Committee must be distributed in such a way as to ensure that no one category has a majority of the total vote. This provides a balance of influence which, in turn, ensures that an agreed standard truly represents a national consensus.

Standards Committee X3, Computers and Information Processing, was established in 1960 to deal with this emerging technology. X3 is sponsored and administered by the Computer and Business Equipment Manufacturers Association (CBEMA) which also acts as X3's secretariat.

There are some 25 Technical Committees—each dealing with an assigned technical area within the overall scope of X3. One of the larger Technical Committees is X3S3, Data Communications. X3S3 was established in 1961 and now has seven Task Groups that are mandated to accomplish the specific technical work of developing proposed American National Standards. These Task Groups are identified as:

X3S31—Planning
X3S32—Glossary
X3S33—Transmission Formats
X3S34—Control Procedures
X3S35—System Performance
X3S36—Signaling Speeds
X3S37—Public Data Networks

Nine of the American National Standards reprinted in this section have evolved out of the work of X3S3 and its Task Groups. The other two, X3.4 and X3.41, were developed by Technical Committee X3L2 and are important references for data communication applications.

In the advance stages of processing are two draft American National Standards which are being balloted. The first is BSR X3.66, Advanced Data Communication Control Procedures (ADCCP), which is similar to ISO HDLC and FED-STD-1003. The other draft is BSR X3.69, General Purpose Interface between Data Terminal Equipment and Data Circuit-terminating Equipment for Synchronous Operation on Public Data Networks. Additional work under development includes: code independent message-heading formats, performance of data communication systems, and an interface for packet-switched public data networks providing both datagram and virtual circuit services.

| NUMBER | TITLE | RELATED CROSS REFERENCE STANDARDS |
|---|---|---|
| X3.1 | Synchronous signaling rates for data transmission | CCITT V.5, V.6; EIA RS 269B; FED-STD-1013 |
| X3.4 | Code of information interchange | CCITT V.3; ISO 646 |
| X3.15 | Bit sequencing of the American national standard code for information interchange in serial-by-bit data transmission | CCITT V.4, X.4; ISO 1155, 1177; FED-STD 1010 |
| X3.16 | Character structure and character parity sense for serial-by-bit data communication information interchange | CCITT V.4, X.4; ISO 1155, 1177, FED-STD-1011 |
| X3.24 | Signal quality at interface between data processing technical equipment for synchronous data transmission | EIA RS-334 |
| X3.25 | Character structure and character parity sense for parallel-by-bit communication in American national standard code for information interchange | FED-STD-1012 |
| X3.28 | Procedures for the use of the communication control characters for American national standard code for information interchange in specific data communication links | ISO 1745 |
| X3.36 | Synchronous high-speed data signaling rates between data terminal equipment and data communication equipment | FED-STD-1001 |
| X3.41 | Code extension techniques for use with 7-bit coded character set of American national standard code for information interchange | ISO 2022 |
| X3.44 | Determination of performance of data communication systems | None |
| X3.57 | Structure for formatting message headings for information interchange using the American national standard code for information interchange for data communication system control | None |

# American National Standard

## synchronous signaling rates
## for data transmission

american national standards institute, inc.
1430 broadway, new york, new york 10018

This standard has been adopted for Federal Government use.

Details concerning the use of this standard within the Federal Government are contained in FIPS PUB 22-1, SYNCHRONOUS SIGNALING RATES BETWEEN DATA TERMINAL AND DATA COMMUNICATION EQUIPMENT. For a complete list of the publications available in the FEDERAL INFORMATION PROCESSING STANDARDS Series, write to the Office of Technical Information and Publications, National Bureau of Standards, Washington, D.C. 20234.

ANSI®
X3.1-1976
Revision of
X3.1-1969

# American National Standard
# Synchronous Signaling Rates
# for Data Transmission

437

Secretariat

**Computer and Business Equipment Manufacturers Association**

Approved August 14, 1976

**American National Standards Institute, Inc**

## American National Standard

An American National Standard implies a consensus of those substantially concerned with its scope and provisions. An American National Standard is intended as a guide to aid the manufacturer, the consumer, and the general public. The existence of an American National Standard does not in any respect preclude anyone, whether he has approved the standard or not, from manufacturing, marketing, purchasing, or using products, processes, or procedures not conforming to the standard. American National Standards are subject to periodic review and users are cautioned to obtain the latest editions.

**CAUTION NOTICE**: This American National Standard may be revised or withdrawn at any time. The procedures of the American National Standards Institute require that action be taken to reaffirm, revise, or withdraw this standard no later than five years from the date of publication. Purchasers of American National Standards may receive current information on all standards by calling or writing the American National Standards Institute.

438

Published by

**American National Standards Institute**
**1430 Broadway, New York, New York  10018**

Printed in the United States of America

P7M177/3

# Foreword

(This Foreword is not a part of American National Standard Synchronous Signaling Rates for Data Transmission, X3.1-1976.)

This standard provides information on the transfer of data between data processing equipment and data communication equipment that transmits data over media commonly referred to as voice band. Signaling rates for use at greater than voice bandwidth are prescribed by American National Standard Synchronous High-Speed Data Signaling Rates between Data Terminal Equipment and Data Communication Equipment, X3.36-1976.

This standard is a revision of X3.1-1969, to which two changes have been made: Recognition of the interim standard serial signaling rate of 2000 bits per second has been withdrawn, and preferred standard parallel signaling rates are specified. X3.1-1969 was a consolidation and revision of still earlier American National Standards: X3.1-1962, Signaling Speeds for Data Transmission, and X3.13-1966, Parallel Signaling Speeds for Data Transmission. Other issued and future standards define and will define additional electrical parameters vital to the connection of and transfer of data between data terminal equipment and data communication equipment.

This standard was developed by Subcommittee X3S3 in coordination with Electronic Industries Association (EIA) Engineering Committee TR-30 and is technically identical to EIA RS-269-B, January 1976, Synchronous Signaling Rates for Data Transmission. The following considerations were taken into account: existing standards — even those partially covering the field, present state-of-the-art and design trends for future data transmission equipment, and the possible future work of the subcommittee in the areas of speeds and other parameters pertinent to the performance of data transmission systems.

Suggestions for improvement of this standard will be welcome. They should be sent to the American National Standards Institute, 1430 Broadway, New York, N.Y. 10018.

This standard was processed and approved for submittal to ANSI by American National Standards Committee on Computers and Information Processing, X3. Committee approval of the standard does not necessarily imply that all committee members voted for its approval. At the time it approved this standard, the X3 Committee had the following members:

J. F. Auwaerter, Chairman
R. M. Brown, Vice-Chairman
W. F. Hanrahan, Secretary

| Organization Represented | Name of Representative |
| --- | --- |
| Addressograph Multigraph Corporation | A. C. Brown |
|  | D. Anderson (Alt) |
| Air Transport Association | F. C. White |
|  | C. Hart (Alt) |
| American Library Association | J. R. Rizzolo |
|  | J. C. Kountz (Alt) |
|  | M. S. Malinconico (Alt) |
| American Nuclear Society | D. R. Vondy |
|  | M. K. Butler (Alt) |
| American Society for Information Science | M. C. Kepplinger |
| Association for Computing Machinery | P. Skelly |
|  | J. A. N. Lee (Alt) |
|  | N. E. Thiess (Alt) |
| Association for Educational Data Systems | A. K. Swanson |
| Association of American Railroads | R. A. Petrash |
| Association of Computer Programmers and Analysts | L. A. Ruh |
|  | M. A. Morris, Jr (Alt) |
| Association of Data Processing Service Organizations | J. B. Christiansen |
| Burroughs Corporation | E. Lohse |
|  | J. S. Foley (Alt) |
|  | J. F. Kalbach (Alt) |
| California Computer Products, Inc | R. C. Derby |
| Computer Industry Association | N. J. Ream |
|  | A. G. W. Biddle (Alt) |

| *Organization Represented* | *Name of Representative* |
|---|---|
| Control Data Corporation | C. E. Cooper |
| | G. I. Williams (Alt) |
| Data Processing Management Association | A. E. Dubnow |
| | D. W. Sanford (Alt) |
| Datapoint Corporation | V. D. Poor |
| | H. W. Swanson (Alt) |
| Digital Equipment Corporation | P. White |
| | A. R. Kent (Alt) |
| Edison Electric Institute | S. P. Shrivastava |
| | J. L. Weiser (Alt) |
| General Services Administration | D. L. Shoemaker |
| | M. L. Burris (Alt) |
| GUIDE International | T. E. Wiese |
| | B. R. Nelson (Alt) |
| | D. Stanford (Alt) |
| Honeywell Information Systems, Inc. | T. J. McNamara |
| | E. H. Clamons (Alt) |
| Institute of Electrical and Electronics Engineers, Communication Society | (Representation Vacant) |
| Institute of Electrical and Electronics Engineers, Computer Society | R. L. Curtis |
| | T. Feng (Alt) |
| International Business Machines Corporation | L. Robinson |
| | W. F. McClelland (Alt) |
| ITEL Corporation | R. A. Whitcomb |
| | W. E. Topercer (Alt) |
| Joint Users Group | T. E. Wiese |
| | L. Rodgers (Alt) |
| Life Office Management Association | R. Ricketts |
| | J. F. Foley (Alt) |
| Litton Industries | I. Danowitz |
| National Association of State Information Systems | G. H. Roehm |
| | J. L. Lewis (Alt) |
| | G. I. Theis (Alt) |
| National Bureau of Standards | H. S. White, Jr |
| | R. E. Rountree (Alt) |
| National Communications Systems | M. L. Cain |
| | G. W. White (Alt) |
| National Machine Tool Builders' Association | O. A. Rodriques |
| NCR Corporation | R. J. Mindlin |
| | A. R. Daniels (Alt) |
| | T. W. Kern (Alt) |
| Olivetti Corporation of America | E. J. Almquist |
| Printing Industries of America | N. Scharpf |
| | E. Masten (Alt) |
| Recognition Equipment, Inc. | H. F. Schantz |
| | W. E. Viering (Alt) |
| Sanders Associates, Inc. | A. L. Goldstein |
| | T. H. Buchert (Alt) |
| Scientific Apparatus Makers Association | A. Savitsky |
| | J. E. French (Alt) |
| SHARE Inc | T. B. Steel, Jr |
| | E. Brubaker (Alt) |
| Society of Certified Data Processors | T. M. Kurihara |
| | A. E. Dubnow (Alt) |
| Telephone Group | V. N. Vaughan, Jr |
| | S. M. Garland (Alt) |
| | J. C. Nelson (Alt) |
| 3M Company | R. C. Smith |
| UNIVAC, Division of Sperry Rand Corporation | M. W. Bass |
| | C. D. Card (Alt) |
| U.S. Department of Defense | W. L. McGreer |
| | W. C. Rinehuls (Alt) |
| | W. B. Robertson (Alt) |
| U.S. Department of Health, Education and Welfare | W. R. McPherson, Jr |
| | M. A. Evans (Alt) |
| VIM (CDC 6000 Users Group) | S. W. White |
| | M. R. Speers (Alt) |
| Xerox Corporation | J. L. Wheeler |
| | A. R. Machell (Alt) |

440

Subcommittee X3S3 on Data Communications, which developed this standard, had the following members:

G. C. Schutz, Chairman
J. L. Wheeler, Vice-Chairman
S. M. Harris, Secretary

P. A. Arneth
J. R. Aschenbrenner
P. C. Baker
M. W. Baty
M. J. Bedford
R. C. Boepple
W. Brown
M. T. Bryngil
M. L. Cain
D. E. Carlson
G. E. Clark, Jr
J. W. Conard
H. J. Crowley
J. L. Dempsey
W. F. Emmons
H. C. Folts
G. O. Hansen
T. L. Hewitt
L. G. Kappel

P. W. Kiesling, Jr
C. C. Kleckner
W. E. Landis
D. S. Lindsay
G. J. McAllister
R. C. Matlack
B. L. Meyer
O. C. Miles
R. T. Moore
L. S. Nidus
N. F. Priebe
S. J. Raiter
S. R. Rosenblum
D. L. Shoemaker
J. M. Skaug
N. E. Snow
E. R. Stephan
E. E. Udo-Ema
G. W. White

Task Group X3S36 on Digital Data Signaling Rates, which had technical reponsibility for the development of this standard, had the following members:

H. J. Crowley, Chairman

N. Kramer
S. Lechter
W. Lyons
S. Schreiner

R. J. Smith
J. L. Wheeler
C. E. Young

# Contents

442

# American National Standard Synchronous Signaling Rates for Data Transmission

## 1. Scope and Purpose

**1.1 Scope.** This standard provides a group of specific signaling rates for synchronous serial or parallel binary data transmission. These rates exist on the received data and transmitted data circuits of the interface between data terminal equipment and data communication equipment that operate over nominal 4-kHz voice bandwidth channels.

**1.2 Purpose.** This standard results from indications by data equipment manufacturers and suppliers of data communication channels that the establishment of specific signaling rates is important to ensure compatibility between communication channels and data terminal equipment of data communication systems.

## 2. Standard Signaling Rates

**2.1** The standard serial signaling rates shall be $600 \times N$ bits per second, where $N$ may be any positive integer from 1 through 16.

**2.1.1** The preferred standard signaling rates shall be 600, 1200, 2400, 4800, 7200, and 9600 bits per second.

**2.1.2** For those applications requiring synchronous operation below 600 bits per second, the standard signaling rates shall be 75, 150, and 300 bits per second.

**2.2** The standard parallel signaling rates for equipment designed to operate with up to 8 parallel data bits per character shall be $75 \times N$ characters per second, where $N$ may be any positive integer from 1 through 16.

**2.2.1** The preferred standard parallel signaling rates shall be 75, 150, 300, 600, 900, and 1200 characters per second.

## 3. Rate Tolerances

The serial signaling rates defined herein shall conform to the tolerance defined in American National Standard Signal Quality at Interface between Data Processing Terminal Equipment and Synchronous Data Communication Equipment for Serial Data Transmission, X3.24-1968 (EIA RS-334-1967).

The tolerance on parallel signaling rates is not defined at this time.

## 4. Unit Element Duration

A synchronous signal train at the interface between the data communication equipment and the data terminal equipment after synchronization is established shall consist of a sequence of marking and spacing signals whose durations are all nominally integral multiples of the unit interval. The unit interval duration in seconds is the reciprocal of the modulation rate in bauds.

# Appendix

(This Appendix is not a part of American National Standard Synchronous Signaling Rates for Data Transmission, X3.1-1976, but is included for information purposes only.)

## Application

### A1. Error Control Devices

Where error control or similar devices that change the signaling rates by a fixed ratio are inserted between the data terminal source/sink equipment and the data communication channel equipment, the $600 \times N$ standard signaling rates apply at the interface between the data terminal equipment and the data communication equipment, as shown in Fig. A1.

### A2. Data Rate Converters

When data rate converters that result in a nonstandard signaling rate at the data processing equipment junction with the rate converter are used, the converter should furnish the clocking for the data processing equipment.

### A3. Parallel Data Transfer

For parallel data transfer at the interface, the standard rates for up to 8 bits per character are as specified in 2.2. If eight channels are supplied, but fewer than eight used for data, the excess bit channels should be run in an idling condition or by using no-data bits.

444

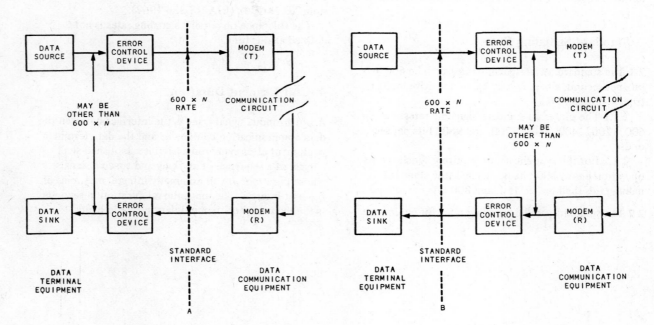

**Fig. A1**
**Interface between Data Terminal Equipment and**
**Data Communication Equipment**

8

# American National Standard

## code for information interchange

american national standards institute, inc.
1430 broadway, new york, new york 10018

ANSI®
X3.4-1977
Revision of
X3.4-1968

# American National Standard
# Code for Information Interchange

446

Secretariat

**Computer and Business Equipment Manufacturers Association**

Approved June 9, 1977

**American National Standards Institute, Inc**

Published by

**American National Standards Institute**
**1430 Broadway, New York, New York 10018**

# Foreword

(This Foreword is not a part of the American National Standard Code for Information Interchange, X3.4-1977).

This American National Standard presents the standard coded character set to be used for information interchange among information processing systems, communications systems, and associated equipment.

Other standards prescribe the means of implementing this standard in media, such as perforated tape, punched cards, magnetic tape, magnetic tape cassettes and cartridges, and optical character recognition. Further standards deal with error control, data communication formats, keyboards, graphic representation of control characters, code extension techniques, and media labels and file structures.

The 7-bit coded character set was developed from a careful review of past work in the field and after a comprehensive program of original research and code design was completed. Careful consideration has been given to the several conflicting code set requirements, and their resolution is reflected in the standard code.

This standard is a revision of X3.4-1968, which was developed in parallel with its international counterpart, ISO 646-1973. This current revision retains the same technical relationship to ISO 646-1973 as the earlier edition. However, some definitions have been changed to adopt more customary U.S. terminology and to reduce ambiguity. Several previously permitted dualities for specific graphics have been eliminated, based on evolving practice in the United States. The relationship with the American National Standard Code Extension Techniques for Use with the 7-Bit Coded Character Set of American National Standard Code for Information Interchange, ANSI X3.41-1974, has also been included.

Suggestions for improvement of this standard will be welcome. They should be sent to the American National Standards Institute, 1430 Broadway, New York, N.Y. 10018.

This standard was processed and approved for submittal to ANSI by American National Standards Committee on Computers and Information Processing, X3. Committee approval of the standard does not necessarily mean that all committee members voted for its approval. At the time it approved this standard, the X3 Committee had the following members:

J. F. Auwaerter, Chairman
R. M. Brown, Vice-Chairman
W. F. Hanrahan, Secretary

| *Organization Represented* | *Name of Representative* |
|---|---|
| Addressograph Multigraph Corporation | A. C. Brown |
| | D. D. Anderson (Alt) |
| Air Transport Association | F. C. White |
| | C. Hart (Alt) |
| American Bankers Association | M. E. McMahon |
| | A. Miller (Alt) |
| American Library Association | J. R. Rizzolo |
| | J. C. Kountz (Alt) |
| | M. S. Malinconico (Alt) |
| American Nuclear Society | D. R. Vondy |
| | M. K. Butler (Alt) |
| American Society for Information Science | M. C. Kepplinger |
| Association for Computing Machinery | P. Skelly |
| | J. A. N. Lee (Alt) |
| | H. E. Thiess (Alt) |
| Association for Educational Data Systems | A. K. Swanson |
| Association of American Railroads | R. A. Petrash |
| Association of Computer Programmers and Analysts | L. A. Ruh |
| | M. A. Morris, Jr (Alt) |
| Association of Data Processing Service Organizations | J. B. Christiansen |
| Burroughs Corporation | E. Lohse |
| | J. S. Foley (Alt) |
| | J. F. Kalbach (Alt) |

| *Organization Represented* | *Name of Representative* |
|---|---|
| Computer Industry Association | N. J. Ream |
| | A. G. W. Biddle (Alt) |
| Control Data Corporation | C. E. Cooper |
| | G. I. Williams (Alt) |
| Data Processing Management Association | A. E. Dubnow |
| | D. W. Sanford (Alt) |
| Datapoint Corporation | V. D. Poor |
| | H. W. Swanson (Alt) |
| Digital Equipment Corporation | P. W. White |
| | A. R. Kent (Alt) |
| Edison Electric Institute | S. P. Shrivastava |
| | J. L. Weiser (Alt) |
| General Services Administration | D. L. Shoemaker |
| | M. L. Burris (Alt) |
| GUIDE International | T. E. Wiese |
| | B. R. Nelson (Alt) |
| | D. Stanford (Alt) |
| Honeywell Information Systems, Inc. | T. J. McNamara |
| | E. H. Clamons (Alt) |
| Institute of Electrical and Electronics Engineers, Communications Society | (Representation Vacant) |
| Institute of Electrical and Electronics Engineers, Computer Society | R. L. Curtis |
| | T. Feng (Alt) |
| International Business Machines Corporation | L. Robinson |
| | W. F. McClelland (Alt) |
| Joint Users Group | T. E. Wiese |
| | L. Rodgers (Alt) |
| Life Office Management Association | R. E. Ricketts |
| | J. F. Foley (Alt) |
| | E. J. Jowdy (Alt) |
| Litton Industries | I. Danowitz |
| National Association of State Information Systems | G. H. Roehm |
| | J. L. Lewis (Alt) |
| | G. I. Theis (Alt) |
| National Bureau of Standards | H. S. White, Jr |
| | R. E. Rountree (Alt) |
| National Communications System | M. L. Cain |
| | G. W. White (Alt) |
| National Machine Tool Builders' Association | O. A. Rodriques |
| NCR Corporation | R. J. Mindlin |
| | A. R. Daniels (Alt) |
| | T. W. Kern (Alt) |
| Olivetti Corporation of America | E. J. Almquist |
| Pitney Bowes, Inc. | R. H. Fenn |
| | E. T. Warzecha (Alt) |
| Printing Industries of America | N. Scharpf |
| | E. Masten (Alt) |
| Recognition Equipment, Inc. | H. F. Schantz |
| | W. E. Viering (Alt) |
| Sanders Associates, Inc. | A. L. Goldstein |
| | T. H. Buchert (Alt) |
| Scientific Apparatus Makers Association | A. Savitsky |
| | J. E. French (Alt) |
| SHARE Inc | T. B. Steel, Jr |
| | E. Brubaker (Alt) |
| Society of Certified Data Processors | T. M. Kurihara |
| | A. E. Dubnow (Alt) |
| Telephone Group | V. N. Vaughan, Jr |
| | J. C. Nelson (Alt) |
| | P. G. Wray (Alt) |
| 3M Company | R. C. Smith |
| UNIVAC, Division of Sperry Rand Corporation | M. W. Bass |
| | C. D. Card (Alt) |
| U.S. Department of Defense | W. L. McGreer |
| | W. B. Rinehuls (Alt) |
| | W. B. Robertson (Alt) |
| U.S. Department of Health, Education and Welfare | W. R. McPherson, Jr |
| | M. A. Evans (Alt) |
| VIM (CDC 6000 Users Group) | S. W. White |
| | M. R. Speers (Alt) |
| Xerox Corporation | J. L. Wheeler |
| | A. R. Machell (Alt) |

Subcommittee X3L2 on Character Sets and Codes, which developed this standard, had the following members:

C. D. Card, Chairman

J. B. Booth
H. Bullard
C. C. Chandler*
B. C. Duncan
T. F. Fitzsimons
S. M. Garland
W. L. Hafner
T. N. Hastings

M. F. Hill
T. O. Holtey
W. F. Huf
H. F. Ickes
W. F. Keenan
T. W. Kern
J. L. Little
R. D. Prigge

---

*Deceased.

# Contents

# American National Standard
# Code for Information Interchange

## 1. Scope

This coded character set is to be used for the general interchange of information among information processing systems, communications systems, and associated equipment.

## 2. Standard Code

| b4 | b3 | b2 | b1 | COLUMN ↓ / ROW ↓ | 000 / 0 | 001 / 1 | 010 / 2 | 011 / 3 | 100 / 4 | 101 / 5 | 110 / 6 | 111 / 7 |
|---|---|---|---|---|---|---|---|---|---|---|---|---|
| 0 | 0 | 0 | 0 | 0 | NUL | DLE | SP | 0 | @ | P | ` | p |
| 0 | 0 | 0 | 1 | 1 | SOH | DC1 | ! | 1 | A | Q | a | q |
| 0 | 0 | 1 | 0 | 2 | STX | DC2 | " | 2 | B | R | b | r |
| 0 | 0 | 1 | 1 | 3 | ETX | DC3 | # | 3 | C | S | c | s |
| 0 | 1 | 0 | 0 | 4 | EOT | DC4 | $ | 4 | D | T | d | t |
| 0 | 1 | 0 | 1 | 5 | ENQ | NAK | % | 5 | E | U | e | u |
| 0 | 1 | 1 | 0 | 6 | ACK | SYN | & | 6 | F | V | f | v |
| 0 | 1 | 1 | 1 | 7 | BEL | ETB | ' | 7 | G | W | g | w |
| 1 | 0 | 0 | 0 | 8 | BS | CAN | ( | 8 | H | X | h | x |
| 1 | 0 | 0 | 1 | 9 | HT | EM | ) | 9 | I | Y | i | y |
| 1 | 0 | 1 | 0 | 10 | LF | SUB | * | : | J | Z | j | z |
| 1 | 0 | 1 | 1 | 11 | VT | ESC | + | ; | K | [ | k | { |
| 1 | 1 | 0 | 0 | 12 | FF | FS | , | < | L | \ | l | \| |
| 1 | 1 | 0 | 1 | 13 | CR | GS | – | = | M | ] | m | } |
| 1 | 1 | 1 | 0 | 14 | SO | RS | . | > | N | ^ | n | ~ |
| 1 | 1 | 1 | 1 | 15 | SI | US | / | ? | O | ___ | o | DEL |

8

## 3. Character Representation and Code Identification

The standard 7-bit character representation, with $b_7$ the high-order bit and $b_1$ the low-order bit, is shown below:

*Example:* The bit representation for the character "K," positioned in column 4, row 11, is

$$b_7 \quad b_6 \quad b_5 \quad b_4 \quad b_3 \quad b_2 \quad b_1$$
$$1 \quad \quad 0 \quad \quad 0 \quad \quad 1 \quad \quad 0 \quad \quad 1 \quad \quad 1$$

The code table for the character "K" may also be represented by the notation "column 4, row 11" or alternatively as "4/11." The decimal equivalent of the binary number formed by bits $b_7$, $b_6$, and $b_5$, collectively, forms the column number, and the decimal equivalent of the binary number formed by bits $b_4$, $b_3$, $b_2$, and $b_1$, collectively, forms the row number.

The standard code may be identified by the use of of the notation ASCII.

The notation ASCII (pronounced 'as-₁key) should ordinarily be taken to mean the code prescribed by the latest edition of this standard. To explicitly designate a particular (perhaps prior) edition, the last two digits of the year of issue may be appended, as, "ASCII 68" or "ASCII 77."

## 4. Legend

### 4.1 Control Characters

| Col/Row | Mnemonic and Meaning[1] | | Col/Row | Mnemonic and Meaning[1] | |
|---------|------|------|---------|------|------|
| 0/0 | NUL | Null | 1/0 | DLE | Data Link Escape (CC) |
| 0/1 | SOH | Start of Heading (CC) | 1/1 | DC1 | Device Control 1 |
| 0/2 | STX | Start of Text (CC) | 1/2 | DC2 | Device Control 2 |
| 0/3 | ETX | End of Text (CC) | 1/3 | DC3 | Device Control 3 |
| 0/4 | EOT | End of Transmission (CC) | 1/4 | DC4 | Device Control 4 |
| 0/5 | ENQ | Enquiry (CC) | 1/5 | NAK | Negative Acknowledge (CC) |
| 0/6 | ACK | Acknowledge (CC) | 1/6 | SYN | Synchronous Idle (CC) |
| 0/7 | BEL | Bell | 1/7 | ETB | End of Transmission Block (CC) |
| 0/8 | BS | Backspace (FE) | 1/8 | CAN | Cancel |
| 0/9 | HT | Horizontal Tabulation (FE) | 1/9 | EM | End of Medium |
| 0/10 | LF | Line Feed (FE) | 1/10 | SUB | Substitute |
| 0/11 | VT | Vertical Tabulation (FE) | 1/11 | ESC | Escape |
| 0/12 | FF | Form Feed (FE) | 1/12 | FS | File Separator (IS) |
| 0/13 | CR | Carriage Return (FE) | 1/13 | GS | Group Separator (IS) |
| 0/14 | SO | Shift Out | 1/14 | RS | Record Separator (IS) |
| 0/15 | SI | Shift In | 1/15 | US | Unit Separator (IS) |
| | | | 7/15 | DEL | Delete |

[1] (CC) Communication Control; (FE) Format Effector; (IS) Information Separator

## 4.2 Graphic Characters

| Column/Row | Symbol | Name |
|---|---|---|
| 2/0 | SP | Space (Normally Nonprinting) |
| 2/1 | ! | Exclamation Point |
| 2/2 | " | Quotation Marks (Diaeresis)[2] |
| 2/3 | # | Number Sign[3] |
| 2/4 | $ | Dollar Sign |
| 2/5 | % | Percent Sign |
| 2/6 | & | Ampersand |
| 2/7 | ' | Apostrophe (Closing Single Quotation Mark; Acute Accent)[2] |
| 2/8 | ( | Opening Parenthesis |
| 2/9 | ) | Closing Parenthesis |
| 2/10 | * | Asterisk |
| 2/11 | + | Plus |
| 2/12 | , | Comma (Cedilla)[2] |
| 2/13 | - | Hyphen (Minus) |
| 2/14 | . | Period (Decimal Point) |
| 2/15 | / | Slant |
| 3/0 to 3/9 | 0 . . . 9 | Digits 0 through 9 |
| 3/10 | : | Colon |
| 3/11 | ; | Semicolon |
| 3/12 | < | Less Than |
| 3/13 | = | Equals |
| 3/14 | > | Greater Than |
| 3/15 | ? | Question Mark |
| 4/0 | @ | Commercial At[3] |
| 4/1 to 5/10 | A . . . Z | Uppercase Latin Letters A through Z |
| 5/11 | [ | Opening Bracket[3] |
| 5/12 | \ | Reverse Slant[3] |
| 5/13 | ] | Closing Bracket[3] |
| 5/14 | ^ | Circumflex[3] |
| 5/15 | _ | Underline |
| 6/0 | ` | Opening Single Quotation Mark (Grave Accent)[2,3] |
| 6/1 to 7/10 | a . . . z | Lowercase Latin letters a through z |
| 7/11 | { | Opening Brace[3] |
| 7/12 | \| | Vertical Line[3] |
| 7/13 | } | Closing Brace[3] |
| 7/14 | ~ | Tilde[2,3] |

454

[2] The use of the symbols in 2/2, 2/7, 2/12, 5/14, 6/0, and 7/14 as diacritical marks is described in A5.2 of Appendix A.

[3] These characters should not be used in international interchange without determining that there is agreement between sender and recipient. (See Appendix B5.)

# 5. Definitions

## 5.1 General

**Control Character.** A character whose occurrence in a particular context initiates, modifies, or stops an action that affects the recording, processing, transmission, or interpretation of data.

**Graphic Character.** A character, other than a control character, that has a visual representation normally handwritten, printed, or displayed.

**(CC) Communication Control.** A control character intended to control or facilitate transmission of information over communication networks.

**(FE) Format Effector.** A control character that controls the layout or positioning of information in printing or display devices.

**(IS) Information Separator.** A control character that is used to separate and qualify information in a logical sense. There is a group of four such characters, which are to be used in a hierarchical order.

## 5.2 Control Characters

**0/0 NUL (Null).** A control character used to accomplish media fill or time fill. Null characters may be inserted into or removed from a stream of data without affecting the information content of that stream. However, the addition or removal of these characters may affect the information layout or the control of equipment.

**0/1 SOH (Start of Heading).** A communication control character used as the first character of a heading of an information message.

**0/2 STX (Start of Text).** A communication control character that precedes a text and is used to terminate a heading.

**0/3 EXT (End of Text).** A communication control character that terminates a text.

**0/4 EOT (End of Transmission).** A communication control character used to indicate the conclusion of a transmission, which may have contained one or more texts and any associated headings.

**0/5 ENQ (Enquiry).** A communication control character used in data communication systems as a request for a reponse from a remote station. It may be used as a "Who Are You" (WRU) to obtain identification, or may be used to obtain station status, or both.

**0/6 ACK (Acknowledge).** A communication control character transmitted by a receiver as an affirmative reponse to a sender.

**0/7 BEL (Bell).** A control character for use when there is a need to call for attention. It may control alarm or attention devices.

**0/8 BS (Backspace).** A one-active-position format effector that moves the position backward on the same line.

**0/9 HT (Horizontal Tabulation).** A format effector that advances the active position to the next predetermined character position on the same line.

**0/10 LF (Line Feed).** A format effector that advances the active position to the same character position on the next line. Where appropriate, this character may have the meaning "New Line" (NL), a format effector that advances the active position to the first character position on the next line. Use of the NL convention requires agreement between sender and recipient of data.

**0/11 VT (Vertical Tabulation).** A format effector that advances the active position to the same character position on the next predetermined line. When agreed upon between the interchange parties, VT may advance the active position to the first character position on the next predetermined line.

**0/12 FF (Form Feed).** A format effector that advances the active position to the same character position on a predetermined line of the next form or page. When agreed upon between the interchange parties, FF may advance the active position to the first character position on a predetermined line of the next form or page.

**0/13 CR (Carriage Return (Return)).** A format effector that moves the active position to the first character position on the same line.

**0/14 SO (Shift Out).** A control character that is used in conjunction with Shift In to extend the graphic character set. It may alter the meaning of the bit combinations of columns 2 to 7 that follow it until a Shift In character is reached. However, the characters Space (2/0) and Delete (7/15) are unaffected. The effect of this character is described in American National Standard Code Extension Techniques for Use with the 7-Bit Coded Character Set of American National Standard Code for Information Interchange, ANSI X3.41-1974.

**0/15 SI (Shift In).** A control character that is used in conjunction with Shift Out to extend the graphic set. It may reinstate the standard meanings of the bit combinations which follow it. The effect of this character is described in American National Standard X3.41-1974.

**1/0 DLE (Data Link Escape).** A communication control character that will change the meaning of a limited number of contiguously following characters. It is used exclusively to provide supplementary data transmission control functions. Appropriate sequences are defined in American National Standard Procedures for the Use of the Communication Control Characters of American National Standard Code for Information Interchange in Specified Data Communication Links, ANSI X3.28-1976.

**1/1–1/4 DC1, DC2, DC3, DC4 (Device Controls).** Control characters for the control of ancillary devices associated with data processing or telecommunication systems, more especially switching devices "on" or "off." (If a single "stop" control is required to interrupt or turn off ancillary devices, DC4 is the preferred assignment.)

**1/5 NAK (Negative Acknowledge).** A communication control character transmitted by a receiver as a negative response to the sender.

**1/6 SYN (Synchronous Idle).** A communication control character used by a synchronous transmission system in the absence of any other character to provide a signal from which synchronism may be achieved or retained.

**1/7 ETB (End of Transmission Block).** A communication control character used to indicate the end of a block of data for communication purposes. ETB is used for blocking data where the block structure is not necessarily related to the processing format.

**1/8 CAN (Cancel).** A control character used to indicate that the data with which it is sent are in error or are to be disregarded. The specific meaning of this character must be defined for each application.

**1/9 EM (End of Medium).** A control character that may be used to identify the physical end of a medium, the end of the used portion of a medium, or the end of the wanted portion of data recorded on a medium. The position of this character does not necessarily correspond to the physical end of the medium.

**1/10 SUB (Substitute).** A control character that may be substituted for a character that is determined to be invalid or in error.

**1/11 ESC (Escape).** A control character intended to provide supplementary characters (code extension). The Escape character itself is a prefix affecting the interpretation of a limited number of contiguous bit patterns. The effect of this character is described in American National Standard X3.41-1974.

**1/12–1/15 FS (File Separator), GS (Group Separator), RS (Record Separator), and US (Unit Separator).** These information separators may be used with data in optional fashion, except that their hierarchical relationship shall be: FS as the most inclusive, then GS, then RS, and US as least inclusive. (The content and length of a file, group, record, or unit are not specified.)

**7/15 DEL (Delete).** A character used primarily to erase or obliterate an erroneous or unwanted character in punched tape. DEL characters may also serve to accomplish media fill or time fill. They may be inserted into or removed from a stream of data without affecting the information content of that stream. However, the addition or removal of these characters may affect the information layout or the control of equipment, or both.

### 5.3 Graphic Characters

NOTE: No specific meaning is prescribed for any of the graphics in the code table except that which is understood by the users. Furthermore, this standard does not specify a type style for the printing or display of the various graphic characters. In specific applications it may be desirable to employ distinctive styling of individual graphics to facilitate their use for specific purposes.

**2/10 SP (Space).** A graphic character that is usually represented by a blank site in a series of graphics. The space character, though not a control character, has a function equivalent to that of a format effector that causes the active position to move one position forward without producing the printing or display of any graphic. Similarly, the space character may have a function equivalent to that of an information separator.

## 6. General Considerations

**6.1** This standard does not define the means by which the coded set is to be recorded in any physical medium, nor does it include any redundancy or define techniques for error control. Further, this standard does not define data communication character structure, data communication formats, code extension techniques, or graphic representation of control characters.

**6.2** Deviations from the standard may create serious difficulties in information interchange and should be used only with full cognizance of the parties involved.

**6.3** The relative sequence of any two characters, when used as a basis for collation, is defined by their binary values. The Null character (position 0/0) will be ranked low and the Delete character (position 7/15) will be

456

ranked high. Other collating sequences may be used by prior agreement between interchanging parties.

**6.4** The representation of this 7-bit code in an 8-bit environment is specified in American National Standard X3.41-1974.

**6.5** The Appendixes to this standard contain additional information on the design and use of this code.

## 7. Related Standards

### 7.1 American National Standards

Perforated Tape Code for Information Interchange, ANSI X3.6-1965 (R1973)

Recorded Magnetic Tape for Information Interchange (200 CPI, NRZI), ANSI X3.14-1973

Bit Sequencing of the American National Standard Code for Information Interchange in Serial-by-Bit Data Transmission, ANSI X3.15-1976

Character Structure and Character Parity Sense for Serial-by-Bit Data Communication in the American National Standard Code for Information Interchange, ANSI X3.16-1976

Character Set and Print Quality for Optical Character Recognition (OCR-A), ANSI X3.17-1977

Recorded Magnetic Tape for Information Interchange (800 CPI, NRZI), ANSI 3.22-1973

Character Structure and Character Parity Sense for Parallel-by-Bit Communication in the American National Standard Code for Information Interchange, ANSI X3.25-1976

Hollerith Punched Card Code, ANSI X3.26-1970

Magnetic Tape Labels and File Structure for Information Interchange, ANSI X3.27-1977

Procedures for the Use of the Communication Control Characters of American National Standard Code for Information Interchange in Specified Data Communication Links, ANSI X3.28-1976

Graphic Representation of the Control Characters of American National Standard Code for Information Interchange, ANSI X3.32-1973

Recorded Magnetic Tape for Information Interchange (1600 CPI, PE), ANSI X3.39-1973

Code Extension Techniques for Use with the 7-Bit Coded Character Set of American National Standard Code for Information Interchange, ANSI X3.41-1974

Specifications for Character Set for Handprinting, ANSI X3.45-1974

Magnetic Tape Cassettes for Information Interchange (3.810-mm [0.015-in] Tape at 32 bpmm [800 bpi], PE), ANSI X3.48-1977

Character Set for Optical Character Recognition (OCR-B), ANSI X3.49-1975

Recorded Magnetic Tape for Information Interchange (6250 CPI, Group-Coded Recording), ANSI X3.54-1976

Recorded Magnetic Tape Cartridge for Information Interchange (4 Track, 0.250 in [6.30 mm], 1600 bpi [63 bpmm], PE); ANSI X3.56-1977

Structure for Formulating Message Headings for Information Interchange Using the American National Standard Code for Information Interchange for Data Communication System Control, ANSI X3.57-1977

Alphanumeric Keyboard Arrangements Accommodating the Character Sets of American National Standard Code for Information Interchange and American National Standard Character Set for Optical Character Recognition, ANSI X4.14-1971

### 7.2 International Standards[4]

7- Bit Coded Character Set for Information Processing Interchange, ISO 646-1973

Information Processing — Basic Mode Control Procedures for Data Communication Systems, ISO 1745-1975

Code Extension Techniques for Use with the ISO 7-Bit Coded Character Set, ISO 2022-1973

---

[4]Publications of the International Organization for Standardization are available from the American National Standards Institute, 1430 Broadway, New York, N.Y. 10018.

## Appendix A
## Design Considerations for the Coded Character Set

### A1. Introduction

The standard coded character set is intended for the interchange of information among information processing systems, communication systems, and associated equipment.

### A2. Considerations Affecting the Code

There were many considerations that determined the set size, set structure, character selection, and character placement of the code. Among these were (not listed in order of priority):

(1) Need for adequate number of graphic symbols

(2) Need for adequate number of device controls, format effectors, communication controls, and information separators

(3) Desire for a nonambiguous code, that is, one in which every code combination has a unique interpretation

(4) Physical requirements of media and facilities

(5) Error control considerations

(6) Special interpretation of the all-zeros and all-ones characters

(7) Ease in the identification of classes of characters

(8) Data manipulation requirements

(9) Collating conventions
  (a) Logical
  (b) Historical

(10) Keyboard conventions
  (a) Logical
  (b) Historical

(11) Other set sizes

(12) International considerations

(13) Programming Languages

(14) Existing coded character sets

### A3. Set Size

A 7-bit set is the minimum size that will meet the re-

quirements for graphics and controls in applications involving general information interchange.

### A4. Set Structure

**A4.1** In discussing the set structure it is convenient to divide the set into eight columns and sixteen rows as indicated in this standard.

**A4.2** It was considered essential to have a dense subset that contained only graphics. For ease of identification this graphic subset was placed in six contiguous columns.

**A4.3** The first two columns were chosen for the controls for the following reasons:

(1) The character NUL by its definition has the location 0/0 in the code table. NUL is broadly considered a control character.

(2) The representations in column 7 were felt to be most susceptible to simulation by a particular class of transmission error — one that occurs during an idle condition on asynchronous systems.

(3) To permit the considerations of graphic subset structure described in Section A6 to be satisfied, the two columns of controls had to be adjacent.

**A4.4** The character set was structured to enable the easy identification of classes of graphics and controls.

### A5. Choice of Graphics

**A5.1** Included in the set are the digits 0 through 9, uppercase and lowercase Latin letters A through Z, and those punctuation, mathematical, and business symbols considered most useful. The set includes a number of characters commonly encountered in programming languages. In particular, all the COBOL and FORTRAN graphics are included.

**A5.2** In order to permit the representation of languages other than English, two diacritical (or accent) marks have been included, and provision has been made for the use of four punctuation symbols alterna-

458

13

## Table A1
### Punctuation and Diacritical Marks

| Col/<br>Row | Code Table<br>Symbol | Use | |
|---|---|---|---|
| | | Punctuation | Diacritical |
| 2/2 | " | Quotation marks | Diaeresis |
| 2/7 | ' | Apostrophe | Acute accent |
| 2/12 | , | Comma | Cedilla |
| 5/14 | ^ | (None) | Circumflex |
| 6/0 | ` | Opening single<br>quotation mark | Grave accent |
| 7/14 | ~ | (None) | Tilde |

tively as diacritical marks in conjunction with backspace. The pairing of these punctuation symbols with their corresponding diacritical marks was done to facilitate the design of a typeface that would be acceptable for both uses.

These arrangements are given in Table A1.

## A6. Graphic Subset Structure

**A6.1** The basic structure of the dense graphic subset was influenced by logical collating considerations, the requirements of simply related 6-bit sets, and the needs of typewriter-like devices. For information processing it is desirable that the characters be arranged in such a way as to minimize both the operating time and the hardware components required for ordering and sequencing operations. This requires that the relative order of characters, within classes, be such that a simple comparison of the binary codes will result in information being ordered in a desired sequence.

**A6.2** Conventional usage requires that SP (Space) be ahead of any other symbol in a collatable set. This permits a name such as "JOHNS" to collate ahead of a name such as "JOHNSON." The requirement that punctuation symbols such as *comma* also collate ahead of the alphabet ("JOHNS, A" should also collate before "JOHNSON") establishes the special symbol locations, including SP, in the first column of the graphic subset.

**A6.3** To simplify the design of typewriter-like devices, it is desirable there be only a common 1-bit difference between characters to be paired on keytops. This, together with the requirements for a contiguous alphabet and the collating requirements outlined above, resulted in the placement of the alphabet in the last four columns of the graphic subset and the placement of the digits in the second column of the graphic subset.

**A6.4** It is expected that devices having the capability of printing only 64 graphic symbols will continue to be important. It may be desirable to arrange these devices to print one symbol for the bit pattern of both uppercase and lowercase of a given alphabetic letter. To facilitate this, there should be a single-bit difference between the uppercase and lowercase representations of any given letter. Combined with the requirement that a given case of the alphabet be contiguous, this dictated the assignment of the alphabet, as shown in columns 4 through 7.

**A6.5** To minimize ambiguity caused by the use of a 64-graphic device as described above, it is desirable, to the degree possible, that each character in column 6 or 7 differ little in significance from the corresponding character in column 4 or 5. In certain cases this was not possible.

**A6.6** The assignment of *reverse slant* and *vertical line*, the *brackets* and *braces,* and the *circumflex* and *tilde* were made with a view to the considerations of A6.5.

**A6.7** The resultant structure of "specials" (S), "digits" (D), and "alphabetics" (A) does not conform to the most prevalent collating convention (S-A-D) because of other code requirements.

**A6.8** The need for a simple transformation from the set sequence to the prevalent collating convention was recognized, and it dictated the placement of some of the "specials" within the set. Specifically, those special symbols, namely, *ampersand* (&), *asterisk* (*), *comma* (,). *hyphen* (-), *period* (.), and *slant* (/), that are most often used as identifiers for ordering information and normally collate ahead of both the alphabet and the digits were not placed in the column containing the digits. Thus the entire numeric column could be rotated via a relatively simple transformation to a position higher than the alphabet. The sequence of the aforementioned "specials" were also established to the extent practical to conform to the prevalent collating convention.

**A6.9** The need for a useful 4-bit numeric subset also played a role in the placement of characters. Such a 4-bit subset, including the digits and the symbols *asterisk, plus* (+), *comma, hyphen, period,* and *slant,* can easily be derived from the code.

**A6.10** Considerations of other domestic code sets, including the Department of Defense former standard 8-bit data transmission code ("Fieldata"-1961), as well as international requirements, played an important role in deliberations that resulted in the code. The selection and grouping of the symbols *dollar sign* ($), *percent sign* (%), *ampersand* (&), *apostrophe* ('), *Less than* (<), *equals* (=), and *greater than* (>) facilitate contraction to either a business or scientific 6-bit subset. The position of these symbols, and of the symbols *comma, hyphen, period,* and *slant,* facilitates achievement of commonly accepted pairing on a keyboard. The historic pairing of question mark and slant is preserved and the *less than* and *greater than* symbols, which have comparatively low usage, are paired with period and comma, so that in dual-case keyboard devices where it is desired to have period and comma in both cases, the *less than* and *greater than* symbols are the ones displaced. Provision was made for the accommodation of alphabets containing more than 26 letters and for 6-bit contraction by the location of low-usage characters in the area following the alphabet.

## A7. Control Subset Content and Structure

**A7.1** The control characters included in the set are those required for the control of terminal devices, input and output devices, format, or communication transmission and switching, and are general enough to justify inclusion in a standard set.

**A7.2** Many control characters may be considered to fall into the following categories:
 (1) Communication Controls
 (2) Format Effectors
 (3) Device Controls
 (4) Information Separators
 To the extent that was practical, controls of each category were grouped in the code table.

**A7.3** The information separators (FS, GS, RS, US) identify boundaries of various elements of information, but differ from punctuation in that they are primarily intended to be machine sensible. They were arranged in accordance with an expected hierarchical use, and the lower end of the hierarchy is contiguous in binary order with SP (Space), which is sometimes used as a machine-sensible separator. Subject to this hierarchy, the exact nature of their use within data is not specified.

**A7.4** The character SYN (Synchronous Idle) was located so that its binary pattern in serial transmission was unambiguous as to character framing, and also to optimize certain other aspects of its communication usage.

**A7.5** ACK (Acknowledge) and NAK (Negative Acknowledge) were located so as to gain the maximum

practical protection against mutation of one into the other by transmission errors.

**A7.6** The function "New Line" (NL) was associated with LF (rather than with CR or with a separate character) to provide the most useful combinations of functions through the use of only two character positions, and to allow the use of a common end of line format for printers having separate CR-LF functions as well as for printers having a combined (that is, NL) function. This sequence would be CR-LF, producing the same result on printers of both classes, and would be useful during conversion of a system from one method of operation to the other.

**A7.7** This standard is enhanced by the inclusion of the option "New Line" which conforms to traditional electric typewriter practice. Data processing keyboard implementors are cautioned of a potential confusion between the use of the terms "Return" and "New Line." The large key on the right side of keyboards has often been marked "Return," although it sometimes accomplishes a "New Line" function, rather than a "Return" function (which according to this standard only has a horizontal motion).

**A7.8** The function "Vertical Tabulation" (VT) and "Form Feed" (FF) are defined to advance the active position to the same character position of the subsequent line similarly to "Line Feed." By agreement, these functions may also return the active position to the first character position on the subsequent line. The practice of preceding either of these characters with "CR" may be useful during conversion, as with "LF."

## A8. Collating Sequence

This supplements the consideration of collating sequence in Section A6.

It is recognized that the collating sequence defined in 6.3 of this standard cannot be used in many specific applications that define their own sequence. In some applications, groups of characters may be assigned exactly equal collating weight to preserve an initial ordering. In other applications a different sequence may be desired to meet the needs of the particular application. Nonetheless, it was deemed essential to define a standard sequence and standard results for comparisons of two items of data, to serve the needs of many applications.

The standard sequence will facilitate, but will not provide directly by means of simple sorting, the ordering of items customarily found in (1) algebraically signed data, in which the largest positive value is high and the largest negative value is low, and (2) complex alphabetic listings, such as those found in telephone directories, library catalogs, or dictionaries. However,

general use of these standard collating sequences and standard comparison evaluations will facilitate the transfer of programs and the general interchange of data among various computer systems.

# Appendix B
# Notes on Application

## B1. Introduction

**B1.1** The standard code was developed to provide for information interchange among information processing systems, communications systems, and associated equipment. In a system consisting of equipment with several local or native codes, maximum flexibility will be achieved if each of the native codes is translated to the standard code whenever information interchange is desired.

**B1.2** Within any particular equipment, system, or community of activity, it may be necessary to substitute characters. For example, some systems may require special graphic symbols, and some devices may require special control codes. (Design efforts on the standard code included consideration of these types of adaptations.) So-called "secular sets" produced by such substitutions, though not conforming to this standard, may nonetheless be consonant with it if substitutions are made in accordance with the guidelines of Section B2.

**B1.3** In recognition of these requirements for control and graphic characters, additional national and international standardization efforts beyond those provided in this standard are in progress to extend the 7-bit code. The techniques for this extension are provided by International Standard ISO 2022-1973, Code Extension Techniques for Use with the ISO 7-Bit Coded Character Set, and by American National Standard X3.41-1974. Standards for additional graphic character sets and for con-

trol character sets for displays and enhanced printing devices are currently under development.

## B2. Character Substitutions

**B2.1** Any character substitution will result in a coded character set that does not conform to this standard.

**B2.2** It is recommended that when a character is substituted in the code table for a standard character, the standard character should not be reassigned elsewhere in the table. Such a substitute character, once assigned, should not be subsequently reassigned elsewhere.

**B2.3** It is recommended that graphic substitutions be made only in the graphic area and control substitutions only in the control area. Any substitution involving a control should be made only with full cognizance of all possible operational effects.

**B2.4** It should be noted that this standard specifies, for each position of the code table, the information represented by the character and not necessarily the precise action taken by the recipient when the character is received. In the case of graphics, considerable variation in the actual shape printed or displayed may be appropriate to different units, systems, or fields of application. In the case of controls, the action performed is dependent upon the use for which the particular system is intended, the application to which it is being put, and a number of conventions established by the user or designer — some system-wide and some unique to a particular unit.

461

**B2.5** Typical examples of diversity in execution not necessarily contrary to this standard are:

(1) A number of graphic symbols, other than those used in the code table, are used for *ampersand* in various type styles; still other symbols may be more appropriate to electronic display devices. The use of such alternate symbols does not in itself constitute deviation from the standard as long as *ampersand* is the concept associated with the character. Note that this does not necessarily restrict the use of such an alternate symbol design to mean "*and*"; in any type style *ampersand* may, of course, be used with arbitrary meaning.

(2) A card punch in one application may "skip" when the character HT (Horizontal Tabulation: used as skip) is presented to it; in another application the character HT may be recorded in the card without further action.

## B3. Interoperation of "LF" and "NL" ASCII Equipment

Several bit pattern sequences in ASCII will cause a device receiving these sequences to move its active position to the first (leftmost) column and also move the active position down one row. Some of these sequences are:

Using "Line Feed" convention:
    CR LF
    CR CR LF
    CR CR LF DEL

Using "New Line" option:
    NL
    CR NL
    CR CR NL
    CR CR NL DEL

where DEL (Delete) is merely a "time waster" to accommodate mechanical devices. The functions involving only horizontal motion are shown preceding those involving vertical motion for the reason that mechanical devices may require more time to accomplish the horizontal motion.

Interoperation of equipment conforming to the ASCII control conventions of CR (Carriage Return) and LF (Line Feed) with equipment conforming to the optional control NL (New Line) in position 0/10 can be assured if the operational sequences CR NL or CR CR NL or CR NL DEL or CR CR NL DEL are always used to move the active position to the first position of the next line. The sequence CR NL sent from an "option" device will be received by a "conventional" device as CR LF, and the reaction will be the desired

one. Likewise, the sequence CR LF sent from a "conventional" device will be received as CR NL on an "option" device, and reaction will be as desired.

## B4. Related Larger and Smaller Sets

Consideration has been given to the relationship between the standard set and sets of other sizes. A number of straightforward logical transformations are possible, which result in a variety of sets related to the standard. None of the transformed sets is recognized by this standard, except through the related standards concerning code extension.

## B5. International Considerations

**B5.1 General.** This standard conforms to the recommendations of the International Organization for Standardization (ISO), and the International Telegraph and Telephone Consultative Committee (CCITT)[5] for a 7-bit code. It includes all the character assignments specified by those bodies for international standardization. Their recommendations, however, permit national standardization by the various countries in seven code table positions. Also, the characters in three additional positions have been designed as "supplementary" use positions, which are replaceable by national characters in only those countries having an extraordinary requirement in this regard.

The ten National Use positions and their assignments in this standard are as follows:

| Column/Row | Character (U.S.) |
|---|---|
| National use: | |
| 4/0 | @ |
| 5/11 | [ |
| 5/12 | \ |
| 5/13 | ] |
| 7/11 | { |
| 7/12 | | |
| 7/13 | } |
| Supplementary national use: | |
| 5/14 | ^ |
| 6/0 | ` |
| 7/14 | ~ |

---

[5] An international body that establishes standards and conventions for international telecommunications, especially where the public telegraph and telephone services are governmentally owned and operated. Their recommendations are often embodied in the regulations applying to such services.

462

17

In international interchange of information these ten characters should not be used except where it is determined that there is agreement between sender and recipient. In such an interchange, where accented letters are to be formed via combinations of graphics and backspace, users are cautioned that other standards prescribe the syntax of such constructs rigidly.

In addition, in other countries, the number sign (#) (in position 2/3) may be replaced by "£", and the dollar sign ($) (in position 2/4) may be replaced by the currency symbol ( ¤ ).

### B5.2 International Reference Version.

The related standard, ISO 646-1973, describes an "International Reference Version" (IRV). The IRV is the default case of ISO 646-1973 in the lack of need of a defined national version. The CCITT permits international interchange using the IRV.

The graphic characters of this standard (ASCII) are consistent with the IRV with the exception of the following two positions:

| Column/Row | Character (U.S.) | Character (IRV) |
|---|---|---|
| 2/4 | $ | (currency sign) |
| 7/14 | ~ | (overline) |

### B5.3 International Terminology Differences.

Where practical, this standard has adopted terminology in use in the English version of associated international standards. Variations from this are due to the demands of common American usage and the desire for consistency with prior versions of this standard.

## B6. Communication Considerations

Certain control characters are designated as "Communication Controls." They are:

SOH   (Start of Heading)
STX   (Start of Text)
ETX   (End of Text)
EOT   (End of Transmission)
ENQ   (Enquiry)
ACK   (Acknowledge)
DLE   (Data Link Escape)
NAK   (Negative Acknowledge)
SYN   (Synchronous Idle)
ETB   (End of Transmission Block)

These may be used by communication systems for their internal signaling or for the exchange of information relating to the control of the communication system between that system and its end terminals. Some such systems may impose restrictions on the use of these communication control characters by the end terminals. For example, the use of some of them may be completely prohibited; others may be restricted to use in conformity with the formats and procedures required by the communication system for its operation.

463

# Appendix C
# Original Criteria

## C1. Introduction

**C1.1** This Appendix contains the original criteria upon which the design of the code was based. Not all criteria have been entirely satisfied. Some are conflicting, and the characteristics of the set represent accepted compromises of these divergent criteria.

**C1.2** The criteria were drawn from communication, processing, and media recording aspects of information interchange.

## C2. Criteria

**C2.1** Every character of the code set shall be represented by the same number of bits.

**C2.2** The standard set shall be so structured as to facilitate derivation of logically related larger or smaller sets.

**C2.3** In a code of $n$ bits, all possible $2^n$ patterns of ones and zeros will be permitted and considered valid.

**C2.4** The number of bits, $n$, shall be sufficient to provide for the alphabetic and numeric characters, commonly used punctuation marks, and other special symbols, along with those control characters required for interchange of information.

**C2.5** The digits 0 through 9 shall be included within a 4-bit subset.

**C2.6** The digits 0 through 9 shall be so represented that the four low-order bits shall be the binary-coded-decimal form of the particular digit that the code represents.

**C2.7** The interspersion of control characters among the graphic characters shall be avoided. The characters devoted to controls shall be easily separable from those devoted to graphics.

**C2.8** Within the standard set, each character shall stand by itself and not depend on surrounding characters for interpretation.

**C2.9** An entire case of the Latin alphabet (A through Z) shall be included within a 5-bit subset. Consideration shall be given to the need for more than 26 characters in some alphabets.

**C2.10** The letters of each case of the alphabet shall be assigned, in conventional order (A through Z), to successive, increasing binary representations.

**C2.11** Suitable control characters required for communication and information processing shall be included.

**C2.12** Escape functions that provide for departures from the standard set shall be incorporated.

**C2.13** A simple binary comparison shall be sufficient to determine the order within each class of characters. (In this regard, the special graphics, the digits, and the alphabet are each defined as distinct classes.) Simple binary rules do not necessarily apply between classes when ordering information.

**C2.14** Space must collate ahead of all other graphics.

**C2.15** Special symbols used in the ordering of information must collate ahead of both the alphabet and the digits.

**C2.16** Insofar as possible, the special symbols shall be grouped according to their functions; for example, punctuation and mathematical symbols. Further, the set shall be so organized that the simplest possible test shall be adequate to distinguish and identify the basic alphabetic, numeric, and special symbol subsets.

**C2.17** Special symbols shall be placed in the set so as to simplify their generation by typewriters and similar keyboard devices. This criterion means, in effect, that the codes for pairs of characters that normally appear on the same keytops on a typewriter shall differ only in a common single-bit position.

**C2.18** The set shall contain the graphic characters of the principal programming languages.

**C2.19** The codes for all control characters shall contain a common, easily recognizable bit pattern.

**C2.20** The Null (0000000) and Delete (1111111) characters shall be provided.

**Appendix D**

**Revision Criteria and Guidelines**

**D1. Introduction**

**D1.1** This Appendix has been added to assist users of the standard. The criteria used in coming to a given revision are briefly stated in this Appendix. Also included are guidelines now in use as well as recommended guidelines for successive revisions (at the mandatory 5-year intervals).

**D1.2** The criteria listed here have been adopted from many sources. Primarily they have come from users of the 1968 edition of this standard or from suggestions by others involved in international standards.

## D2. The 1977 Revision

**D2.1 General.** The 1968 edition was reviewed and revised to bring terminology into more consistent U.S. practice.

**D2.2 Graphics.** The primary considerations in revision of the graphics area of the standard were:

(1) Elimination of formerly recognized dualities (positions 2/1, 5/14). These allowed stylization of these characters to reflect their possible usage as logical OR and NOT, respectively. Evolving practice has shown these to be unnecessary.

(2) Elimination of a formerly recognized duality (position 2/3), which allowed substitution of the "Pound Sterling" symbol. Evolving practice has shown this to be unnecessary.

(3) Clarification of conflict between graphic shape and description (position 7/12).

**D2.3 Controls.** The primary considerations in revision of the controls area of the standard were:

(1) To adopt definitions consistent with associated standards (positions 0/14, 0/15, and 1/11).

(2) To make no change of substance to the communications controls without the explicit request of the group responsible for data communication procedures.

(3) To accommodate the definitions of the controls of ISO 646-1973 wherever domestic conflicts did not exist.

(4) To clarify definitions where use application had given indications of need for clarification.

(5) To recognize evolving practice aimed at providing an "optional implicit CR" function with all three vertical movement Format Effectors (positions 0/10, 0/11, 0/12).

## D3. Succeeding Revisions

**D3.1 General.** A review will be made for domestic and international consistency of use.

**D3.2 Controls.** A study will be made in the following areas:

(1) Communications Controls
(2) Information Separators
(3) Format Effectors
(4) Device Controls

Each group will be evaluated from the viewpoint of current usage, suitability of the definitions, possible improvement of definitions, potential replacement by more desirable functions, or any combinations thereof.

**D3.3 Graphics.** A study will be made concerning the following positions:

(1) 5/12
(2) 5/14
(3) 6/0
(4) 7/14

If characters of more frequent usage emerge in U.S. practice, these positions will be evaluated for potential replacement. This is entirely consistent with the cautions described in Appendix B, wherein the national use considerations were introduced.

## Appendix E

## Terminology

This Appendix is intended to clarify the sense in which certain terms are used.

**Active Position.** That character position in which the character about to be processed would appear. The active position normally advances one character position at a time.

**Bit.** Contraction of "binary digit."

**Bit Pattern.** The binary representation of a character.

**Character.** A member of a coded character set; the binary representation of such a member and its graphic symbol or control function.

**Code.** A system of discrete representations of a set of symbols and functions.

# American National Standard

for bit sequencing of the
american national standard code for
information interchange in serial-by-bit
data transmission

466

X3.15-1976

american national standards institute, inc.
1430 broadway, new york, new york 10018

This standard has been adopted for Federal Government use.

Details concerning the use of this standard within the Federal Government are contained in FIPS PUB 16-1, BIT SEQUENCING OF THE CODE FOR INFORMATION INTERCHANGE IN SERIAL-BY-BIT DATA TRANSMISSION. For a complete list of the publications available in the FEDERAL INFORMATION PROCESSING STANDARDS Series, write to the Office of Technical Information and Publications, National Bureau of Standards, Washington, D.C. 20234.

ANSI®
X3.15-1976
Revision of
X3.15-1966

# American National Standard for Bit Sequencing of the American National Standard Code for Information Interchange in Serial-by-Bit Data Transmission

468

Secretariat

**Computer and Business Equipment Manufacturers Association**

Approved June 25, 1976

**American National Standards Institute, Inc**

# American National Standard

An American National Standard implies a consensus of those substantially concerned with its scope and provisions. An American National Standard is intended as a guide to aid the manufacturer, the consumer, and the general public. The existence of an American National Standard does not in any respect preclude anyone, whether he has approved the standard or not, from manufacturing, marketing, purchasing, or using products, processes, or procedures not conforming to the standard. American National Standards are subject to periodic review and users are cautioned to obtain the latest editions.

CAUTION NOTICE: This American National Standard may be revised or withdrawn at any time. The procedures of the American National Standards Institute require that action be taken to reaffirm, revise, or withdraw this standard no later than five years from the date of publication. Purchasers of American National Standards may receive current information on all standards by calling or writing the American National Standards Institute.

469

Published by

**American National Standards Institute**
**1430 Broadway, New York, New York 10018**

Printed in the United States of America

(This Foreword is not a part of American National Standard for Bit Sequencing of the American National Standard Code for Information Interchange in Serial-by-Bit Data Transmission, X3.15-1976.)

This standard specifies means of bit sequencing for the transmittal of the American National Standard Code for Information Interchange, X3.4-1968 (ASCII) and the codes invoked when applying the American National Standard Code Extension Techniques for Use with the 7-Bit Coded Character Set of American National Standard Code for Information Interchange, X3.41-1974, for serial-by-bit, serial-by-character data transmission at the interface between data processing terminal equipment and data communication equipment. It is a revision of American National Standard for Bit Sequencing of the American National Standard Code for Information Interchange in Serial-by-Bit Data Transmission, X3.15-1966.

This standard is one of a series developed by Task Group X3S3.3 on Data Communication Formats under the coordination of the X3S3 Subcommittee on Data Communications of American National Standards Committee on Computers and Information Processing, X3. Task Group X3S3.3, which was organized late in 1962 and held its first meeting in January 1963, is charged with the responsibility for standardizing character format, data transmission of characters within a hierarchy of groupings (that is, words, blocks, messages, etc) including group error control, and the order or sequence of bits within characters (including parity).

Other standards prescribe the character structure, the sense (odd or even) of parity bits, the formats for parallel-by-bit, serial-by-character data transmission, and other parameters vital to the transmission of information between the types of equipment previously mentioned.

Suggestions for improvement of this standard will be welcome. They should be sent to the American National Standards Institute, 1430 Broadway, New York, N.Y. 10018.

This standard was processed and approved for submittal to ANSI by American National Standards Committee on Computers and Information Processing, X3. Committee approval of the standard does not necessarily imply that all committee members voted for its approval. At the time it approved this standard, the X3 Committee had the following members:

J. F. Auwaerter, Chairman
R. M. Brown, Vice-Chairman
W. F. Hanrahan, Secretary

| Organization Represented | Name of Representative |
| --- | --- |
| Addressograph Multigraph Corporation | A. C. Brown |
| | D. Anderson (Alt) |
| Air Transport Association | F. C. White |
| | C. Hart (Alt) |
| American Library Association | J. R. Rizzolo |
| | J. C. Kountz (Alt) |
| | M. S. Malinconico (Alt) |
| American Nuclear Society | D. R. Vondy |
| | M. K. Butler (Alt) |
| American Society for Information Science | M. C. Kepplinger |
| Association for Computing Machinery | P. Skelly |
| | J. A. N. Lee (Alt) |
| | H. E. Thiess (Alt) |
| Association for Educational Data Systems | A. K. Swanson |
| Association of American Railroads | R. A. Petrash |
| Association of Computer Programmers and Analysts | L. A. Ruh |
| | M. A. Morris, Jr (Alt) |
| Association of Data Processing Service Organizations | J. B. Christiansen |
| Burroughs Corporation | E. Lohse |
| | J. S. Foley (Alt) |
| | J. F. Kalbach (Alt) |
| Computer Industry Association | N. J. Ream |
| | A. G. W. Biddle (Alt) |
| Control Data Corporation | C. E. Cooper |
| | G. I. Williams (Alt) |

| Organization Represented | Name of Representative |
|---|---|
| Data Processing Management Association | A. E. Dubnow |
| | D. W. Sanford (Alt) |
| Datapoint Corporation | V. D. Poor |
| | H. W. Swanson (Alt) |
| Digital Equipment Corporation | P. White |
| | A. R. Kent (Alt) |
| Edison Electric Institute | S. P. Shrivastava |
| | J. L. Weiser (Alt) |
| General Services Administration | D. L. Shoemaker |
| | M. L. Burris (Alt) |
| GUIDE International | T. E. Wiese |
| | B. R. Nelson (Alt) |
| | D. Stanford (Alt) |
| Honeywell Information Systems, Inc | T. J. McNamara |
| | E. H. Clamons (Alt) |
| Institute of Electrical and Electronics Engineers, Communications Society | (Representation Vacant) |
| Institute of Electrical and Electronics Engineers, Computer Society | R. L. Curtis |
| | T. Feng (Alt) |
| International Business Machines Corporation | L. Robinson |
| | W. F. McClelland (Alt) |
| Joint Users Group | T. E. Wiese |
| | L. Rodgers (Alt) |
| Life Office Management Association | R. Ricketts |
| | J. F. Foley (Alt) |
| | E. J. Jowdy (Alt) |
| Litton Industries | I. Danowitz |
| National Association of State Information Systems | G. H. Roehm |
| | J. L. Lewis (Alt) |
| | G. I. Theis (Alt) |
| National Bureau of Standards | H. S. White, Jr |
| | R. E. Rountree (Alt) |
| National Communications System | M. L. Cain |
| | G. W. White (Alt) |
| National Machine Tool Builders' Association | O. A. Rodriques |
| NCR Corporation | R. J. Mindlin |
| | A. R. Daniels (Alt) |
| | T. W. Kern (Alt) |
| Olivetti Corporation of America | E. J. Almquist |
| Pitney Bowes, Inc | R. H. Fenn |
| | E. T. Warzecha (Alt) |
| Printing Industries of America | N. Scharpf |
| | E. Masten (Alt) |
| Recognition Equipment, Inc | H. F. Schantz |
| | W. E. Viering (Alt) |
| Sanders Associates, Inc | A. L. Goldstein |
| | T. H. Buchert (Alt) |
| Scientific Apparatus Makers Association | A. Savitsky |
| | J. E. French (Alt) |
| SHARE Inc | T. B. Steel, Jr |
| | E. Brubaker (Alt) |
| Society of Certified Data Processors | T. M. Kurihara |
| | A. E. Dubnow (Alt) |
| Telephone Group | V. N. Vaughan, Jr |
| | J. C. Nelson (Alt) |
| | P. G. Wray (Alt) |
| 3M Company | R. C. Smith |
| UNIVAC, Division of Sperry Rand Corporation | M. W. Bass |
| | C. D. Card (Alt) |
| U.S. Department of Defense | W. L. McGreer |
| | W. B. Rinehuls (Alt) |
| | W. B. Robertson (Alt) |
| U.S. Department of Health, Education and Welfare | W. R. McPherson, Jr |
| | M. A. Evans (Alt) |
| VIM (CDC 6000 Users Group) | S. W. White |
| | M. R. Speers (Alt) |
| Xerox Corporation | J. L. Wheeler |
| | A. R. Machell (Alt) |

Subcommittee X3S3 on Data Communications, which coordinated the development of this standard, had the following members:

G. C. Schutz, Chairman
J. L. Wheeler, Vice-Chairman
S. M. Harris, Secretary

P. A. Arneth
J. R. Aschenbrenner
P. C. Baker
M. W. Baty
M. J. Bedford
R. C. Boepple
W. Brown
M. T. Bryngil
M. L. Cain
D. E. Carlson
G. E. Clark, Jr
J. W. Conard
H. J. Crowley
J. L. Dempsey
W. F. Emmons
H. C. Folts
G. O. Hansen
T. L. Hewitt
L. G. Kappel

P. W. Kiesling, Jr
C. C. Kleckner
W. E. Landis
D. S. Lindsay
G. J. McAllister
R. C. Matlack
B. L. Meyer
O. C. Miles
R. T. Moore
L. S. Nidus
N. F. Priebe
S. J. Raiter
S. R. Rosenblum
D. L. Shoemaker
J. M. Skaug
N. E. Snow
E. R. Stephan
E. E. Udo-Ema
G. W. White

Task Group X3S3.3 on Data Communication Formats, which developed this standard, had the following members:

W. F. Emmons, Chairman

J. V. Fayer
J. G. Griffis
M. E. McMahon
L. Nidus

E. Novgorodoff
G. W. White
C. E. Young

Other persons who contributed to the development of this standard were:

G. A. Barletta
E. Berezin
J. B. Booth

M. E. Cook
J. L. Little
J. Lowe

O. C. Miles
J. J. O'Donnell
J. R. Sink

## Contents

472

# American National Standard
## for Bit Sequencing of the American National Standard Code for Information Interchange in Serial-by-Bit Data Transmission

## 1. Scope

**1.1** This standard specifies the bit sequencing of the American National Standard Code for Information Interchange, X3.4-1968 (ASCII) and the codes invoked when applying the American National Standard Code Extension Techniques for Use with the 7-Bit Coded Character Set of American National Standard Code for Information Interchange, X3.41-1974, for serial-by-bit, serial-by-character data transmission.

**1.2** This standard applies to general information interchange in the ASCII and extended ASCII, at the interface between data processing terminal equipment (such as data processors, data media input-output devices, and office machines) and data communication equipment (such as data sets and modems).

## 2. Standard Bit Sequence

The bit sequence for an ASCII character shall be least significant bit first to most significant bit — in terms of the 7-bit ASCII nomenclature (American National Standard Code for Information Interchange, X3.4-1968) $b_1$ through $b_7$ in ascending (consecutive) order, or in terms of the 8-bit nomenclature (American National Standard Code Extension Techniques for Use with the 7-Bit Coded Character set of American National Standard Code for Information Interchange, X3.41-1974) $a_1$ through $a_8$ in ascending (consecutive) order.

## 3. Character Parity

In a 7-bit environment a character parity bit, if transmitted, shall follow the most significant bit, $b_7$, of the character to which it applies. No character parity bit is provided in an 8-bit environment.

## 4. Qualifications

**4.1** This standard does not specify that a character parity bit shall or shall not be transmitted, nor does it specify the total number of bits per character, the bit rate, the character rate, or the transmission technique.

**4.2** This standard does not apply to parallel-by-bit, serial-by-character data transmission.

## 5. Revision of American National Standards Referred to in This Document

When the following American National Standards referred to in this document are superseded by a revision approved by the American National Standards Institute, Inc, the revision shall apply:

American National Standard Code for Information Interchange, X3.4-1968 (ASCII)

American National Standard Code Extension Techniques for Use with the 7-Bit Coded Character Set of American National Standard Code for Information Interchange, X3.41-1974

6

# Appendix

(This Appendix is not a part of American National Standard for Bit Sequencing of the American National Standard Code for Information Interchange in Serial-by-Bit Data Transmission, X3.15-1976, but is included for information purposes only.)

## Criteria, Considerations, and Conclusions

### A1. Introduction

This Appendix contains criteria pertinent to the selection of the bit sequencing in serial-by-bit data transmission. Not all of the listed criteria have been entirely satisfied. Some are conflicting, and the selected bit sequence was based upon a detailed analysis and weighing of these divergent criteria. The final choice of transmission bit sequence represents an acceptable compromise.

### A2. Specific Criteria

NOTE: The following criteria are not mutually consistent and are not listed in order of importance.

**A2.1** The transmission bit sequence should be in consecutive numerical order (ascending or descending), in terms of ASCII nomenclature.

**A2.2** The transmission bit sequence should minimize the amount and complexity of existing and future hardware.

**A2.3** The transmission bit sequence should be selected to minimize adverse consequences of equipment or system malfunction.

**A2.4** The transmission of a binary bit stream should not be precluded.

**A2.5** The transmission of encrypted material should not be precluded.

**A2.6** There should be a correspondence between media track (channel or row) designation, ASCII bit number, and transmission bit sequence, in order to minimize training and reduce confusion of operating, maintenance, and engineering personnel.

**A2.7** The transmission bit sequence should allow a logical extension to supersets of ASCII.

**2.8** The transmission bit sequence of any subset or perset of ASCII should provide that any designated bit be immutable in its position in the transmission sequence as well as in its logical order and media representation.

**A2.9** The character parity bit should be positioned to allow it to be generated "on the fly," following the data bits.

**A2.10** The transmission bit sequence should allow maximum design flexibility in future systems utilizing ASCII.

### A3. Considerations

**A3.1** Considerations of the various possible bit sequences for serial-by-bit, serial-by-character data transmission produced the following two choices for further study:
 (1) Low-order bit first to high-order bit
 (2) High-order bit first to low-order bit
The basic structure of ASCII — that is, the separation of graphics and controls, and the location of the symbols, digits, and alphabet within the graphic portion — favored a consecutive bit sequence.

**A3.2** There is unanimity of opinion that the character parity bit, where included on a per-character basis, should appear last in the bit sequence so that this bit can be generated "on the fly."

**A3.3** These two bit-sequencing choices, A3.1(1) and A3.1(2), were then subjected to an exhaustive investigation to determine their influence on data interchange from the following points of view:
 (1) Flexibility of hardware design
 (2) Hardware efficiency
 (3) Ease of maintenance
 (4) Contraction of ASCII to subsets
 (5) Expansion of ASCII to supersets
 (6) Error consequences

**A3.4** With the low-order bit transmitted first, the first data pulse can correspond to ASCII bit $b_1$, the second to bit $b_2$, etc. Thus "third" will mean the third data pulse as well as bit $b_3$. It can also mean the third track (or channel or row) in media. This simple, orderly relationship between media track number, pulse number, and bit designation number is desirable in the maintenance of communication equipment, especially in discussions between remote technicians and between technicians and engineers. This arrangement is as follows:

| Media track: | 1 | 2 | 3 | 4 | 5 | 6 | 7 | 8 |
|---|---|---|---|---|---|---|---|---|
| ASCII bit: | $b_1$ | $b_2$ | $b_3$ | $b_4$ | $b_5$ | $b_6$ | $b_7$ | P |
| Data pulse number: | 1 | 2 | 3 | 4 | 5 | 6 | 7 | 8 |

where P is the character parity bit.

**A3.5** The 7 bits of ASCII are designated $b_1$ through $b_7$ in increasing order of significance. Thus, additional high-order bits may be added and designated $b_8$, $b_9$, etc, in an orderly manner. It is desirable, from the transmission standpoint, to have a code "open-ended" at the end of a character bit sequence rather than at the start. Compatibility between equipments using different size sets is less difficult, since each numbered bit always appears in a given position with respect to the start of the character sequence.

**A3.5.1** Terminals using a subset of ASCII need only operate with the appropriate numbers of bits at the beginning of the character.

**A3.5.2** Terminals transmitting supersets of ASCII need only append the additional ($b_8$, $b_9$, etc) bits to the ASCII.

**A3.5.3** In all cases, the character parity bit then maintains its defined position as the last bit of the character to be transmitted.

**A3.6** Low-order bit first is in agreement with recently established standards, such as bit sequencing of ASCII in Military Standard MIL-STD-188 C, Military Communication System Technical Standards, Nov 24, 1969.[1]

**A3.7** High-order bit first allows future systems and

---

[1] Available from Naval Publications and Forms Center, 5801 Tabor Avenue, Philadelphia, Pa. 19120.

hardware to be designed to take advantage of the unique organization of ASCII into control character columns and graphic columns; this could, for example, reduce the bit storage requirement in simple input-output (I/O) printers and control mechanisms.

**A3.8** If the transmission sequence is high-order bit first, the implementation of ASCII into present-day, 6-bit character based processors and into upper-lower I/O typewriters can be simplified, resulting in hardware reduction.

**A3.9** The only criterion pertinent to character parity is that it can be capable of generation "on the fly." This implies that it be transmitted later than all data bits. This does not contradict any of the criteria for the sequence of data bits.

**A3.10** The error rate of a system is not dependent on the choice of bit sequencing. The consequence of an error is influenced by the choice of bit sequence; however, no definite conclusions could be drawn as to which bit sequence resulted in the least harmful error condition.

## A4. Conclusions

**A4.1** The question of bit sequence resolves itself into a choice between the following two specific interests:

(1) Low-order bit first offers the advantage of convenient expansion to 8-bit supersets, allows direct correspondence between the ASCII bit designators in media and the transmitted bit sequence, eases maintenance and training, and is consistent with international and military standards.

(2) High-order bit first offers the possibility of more economical implementation of ASCII by present-day 6-bit processors and by simple I/O printers and typewriters.

**A4.2** The conclusion reached is that the known advantages of low-order bit first outweigh the possible advantages of high-order bit first.

# American National Standard

character structure and character parity
sense for serial-by-bit data communication
in the american national standard code
for information interchange

476

X3.16-1976

american national standards institute, inc.
1430 broadway, new york, new york 10018

This standard has been adopted for Federal Government use.

Details concerning the use of this standard within the Federal Government are contained in FIPS PUB 17-1, CHARACTER STRUCTURE AND CHARACTER PARITY SENSE FOR SERIAL-BY-BIT DATA COMMUNICATION IN THE CODE FOR INFORMATION INTERCHANGE. For a complete list of the publications available in the FEDERAL INFORMATION PROCESSING STANDARDS Series, write to the Office of Technical Information and Publications, National Bureau of Standards, Washington, D.C. 20234.

477

ANSI®
X3.16-1976
Revision of
X3.16-1966

# American National Standard
# Character Structure and Character Parity Sense
# for Serial-by-Bit Data Communication in the
# American National Standard Code for
# Information Interchange

478

Secretariat

**Computer and Business Equipment Manufacturers Association**

Approved June 25, 1976

American National Standards Institute, Inc

## American National Standard

An American National Standard implies a consensus of those substantially concerned with its scope and provisions. An American National Standard is intended as a guide to aid the manufacturer, the consumer, and the general public. The existence of an American National Standard does not in any respect preclude anyone, whether he has approved the standard or not, from manufacturing, marketing, purchasing, or using products, processes, or procedures not conforming to the standard. American National Standards are subject to periodic review and users are cautioned to obtain the latest editions.

**CAUTION NOTICE**: This American National Standard may be revised or withdrawn at any time. The procedures of the American National Standards Institute require that action be taken to reaffirm, revise, or withdraw this standard no later than five years from the date of publication. Purchasers of American National Standards may receive current information on all standards by calling or writing the American National Standards Institute.

479

Published by

**American National Standards Institute**
**1430 Broadway, New York, New York  10018**

Printed in the United States of America

A7M1176/3

# Foreword

(This Foreword is not a part of American National Standard Character Structure and Character Parity Sense for Serial-by-Bit Data Communication in the American National Standard Code for Information Interchange, X3.16-1976.)

This standard defines character structure for both synchronous and asynchronous transmission modes, since arguments on the nature of character structure are closely related to the method by which transmission is accomplished. The standard applies to general information interchange at the interface between data processing terminal equipment and data communication equipment. It is a revision of American National Standard Character Structure and Character Parity Sense for Serial-by-Bit Data Communication in the American National Standard Code for Information Interchange, X3.16-1966.

This standard is one of a series developed by Task Group X3S3.3 on Data Communication Formats under the coordination of the X3S3 Subcommittee on Data Communications of American National Standards Committee on Computers and Information Processing, X3. Task Group X3S3.3, which was organized late in 1962 and held its first meeting in January 1963, is charged with the responsibility for standardizing character format, data transmission of characters within a hierarchy of groupings (that is, words, blocks, messages, etc) including group error control, and the order or sequence of bits within characters (including parity).

Other standards provide specifications for bit sequencing, formats for parallel-by-bit, serial-by-character data transmission, and other parameters vital to the transmission of information between the types of equipment previously mentioned.

Suggestions for improvement of this standard will be welcome. They should be sent to the American National Standards Institute, 1430 Broadway, New York, N.Y. 10018.

This standard was processed and approved for submittal to ANSI by American National Standards Committee on Computers and Information Processing, X3. Committee approval of the standard does not necessarily imply that all committee members voted for its approval. At the time it approved this standard, the X3 Committee had the following members:

J. F. Auwaerter, Chairman
R. M. Brown, Vice-Chairman
W. F. Hanrahan, Secretary

| Organization Represented | Name of Representative |
|---|---|
| Addressograph Multigraph Corporation | A. C. Brown |
| | D. Anderson (Alt) |
| Air Transport Association | F. C. White |
| | C. Hart (Alt) |
| American Library Association | J. R. Rizzolo |
| | J. C. Kountz (Alt) |
| | M. S. Malinconico (Alt) |
| American Nuclear Society | D. R. Vondy |
| | M. K. Butler (Alt) |
| American Society for Information Science | M. C. Kepplinger |
| Association for Computing Machinery | P. Skelly |
| | J. A. N. Lee (Alt) |
| | H. E. Thiess (Alt) |
| Association for Educational Data Systems | A. K. Swanson |
| Association of American Railroads | R. A. Petrash |
| Association of Computer Programmers and Analysts | L. A. Ruh |
| | M. A. Morris, Jr (Alt) |
| Association of Data Processing Service Organizations | J. B. Christiansen |
| Burroughs Corporation | E. Lohse |
| | J. S. Foley (Alt) |
| | J. F. Kalbach (Alt) |
| Computer Industry Association | N. J. Ream |
| | A. G. W. Biddle (Alt) |
| Control Data Corporation | C. E. Cooper |
| | G. I. Williams (Alt) |

| Organization Represented | Name of Representative |
|---|---|
| Data Processing Management Association | A. E. Dubnow |
| | D. W. Sanford (Alt) |
| Datapoint Corporation | V. D. Poor |
| | H. W. Swanson (Alt) |
| Digital Equipment Corporation | P. White |
| | A. R. Kent (Alt) |
| Edison Electric Institute | S. P. Shrivastava |
| | J. L. Weiser (Alt) |
| General Services Administration | D. L. Shoemaker |
| | M. L. Burris (Alt) |
| GUIDE International | T. E. Wiese |
| | B. R. Nelson (Alt) |
| | D. Stanford (Alt) |
| Honeywell Information Systems, Inc | T. J. McNamara |
| | E. H. Clamons (Alt) |
| Institute of Electrical and Electronics Engineers, Communications Society | (Representation Vacant) |
| Institute of Electrical and Electronics Engineers, Computer Society | R. L. Curtis |
| International Business Machines Corporation | L. Robinson |
| | W. F. McClelland (Alt) |
| | T. Feng (Alt) |
| Joint Users Group | T. E. Wiese |
| | L. Rodgers (Alt) |
| Life Office Management Association | R. Ricketts |
| | J. F. Foley (Alt) |
| | E. J. Jowdy (Alt) |
| Litton Industries | I. Danowitz |
| National Association of State Information Systems | G. H. Roehm |
| | J. L. Lewis (Alt) |
| | G. I. Theis (Alt) |
| National Bureau of Standards | H. S. White, Jr |
| | R. E. Rountree (Alt) |
| National Communications System | M. L. Cain |
| | G. W. White (Alt) |
| National Machine Tool Builders' Association | O. A. Rodriques |
| NCR Corporation | R. J. Mindlin |
| | A. R. Daniels (Alt) |
| | T. W. Kern (Alt) |
| Olivetti Corporation of America | E. J. Almquist |
| Pitney Bowes, Inc | R. H. Fenn |
| | E. T. Warzecha (Alt) |
| Printing Industries of America | N. Scharpf |
| | E. Masten (Alt) |
| Recognition Equipment, Inc | H. F. Schantz |
| | W. E. Viering (Alt) |
| Sanders Associates, Inc | A. L. Goldstein |
| | T. H. Buchert (Alt) |
| Scientific Apparatus Makers Association | A. Savitsky |
| | J. E. French (Alt) |
| SHARE Inc | T. B. Steel, Jr |
| | E. Brubaker (Alt) |
| Society of Certified Data Processors | T. M. Kurihara |
| | A. E. Dubnow (Alt) |
| Telephone Group | V. N. Vaughan, Jr |
| | J. C. Nelson (Alt) |
| | P. G. Wray (Alt) |
| | R. C. Smith |
| 3M Company | M. W. Bass |
| UNIVAC, Division of Sperry Rand Corporation | C. D. Card (Alt) |
| U.S. Department of Defense | W. L. McGreer |
| | W. B. Rinehuls (Alt) |
| | W. B. Robertson (Alt) |
| U.S. Department of Health, Education and Welfare | W. R. McPherson, Jr |
| | M. A. Evans (Alt) |
| VIM (CDC 6000 Users Group) | S. W. White |
| | M. R. Speers (Alt) |
| Xerox Corporation | J. L. Wheeler |
| | A. R. Machell (Alt) |

Subcommittee X3S3 on Data Communications, which coordinated the development of this standard, had the following members:

G. C. Schutz, Chairman
J. L. Wheeler, Vice-Chairman
S. M. Harris, Secretary

| | |
|---|---|
| P. A. Arneth | P. W. Kiesling, Jr |
| J. R. Aschenbrenner | C. C. Kleckner |
| P. C. Baker | W. E. Landis |
| M. W. Baty | D. S. Lindsay |
| M. J. Bedford | G. J. McAllister |
| R. C. Boepple | R. C. Matlack |
| W. Brown | B. L. Meyer |
| M. T. Bryngil | O. C. Miles |
| M. L. Cain | R. T. Moore |
| D. E. Carlson | L. S. Nidus |
| G. E. Clark, Jr | N. F. Priebe |
| J. W. Conard | S. J. Raiter |
| H. J. Crowley | S. R. Rosenblum |
| J. L. Dempsey | D. L. Shoemaker |
| W. F. Emmons | J. M. Skaug |
| H. C. Folts | N. E. Snow |
| G. O. Hansen | E. R. Stephan |
| T. L. Hewitt | E. E. Udo-Ema |
| L. G. Kappel | G. W. White |

Task Group X3S3.3 on Data Communication Formats, which developed this standard, had the following members:

W. F. Emmons, Chairman

| | |
|---|---|
| J. V. Fayer | E. Novgorodoff |
| J. G. Griffis | G. W. White |
| M. E. McMahon | C. E. Young |
| L. Nidus | |

Other persons who contributed to the development of this standard were:

| | |
|---|---|
| G. A. Barletta | J. Lowe |
| E. Berezin | O. C. Miles |
| J. B. Booth | J. J. O'Donnell |
| M. E. Cook | J. R. Sink |
| J. L. Little | |

# Contents

# American National Standard
# Character Structure and Character Parity Sense
# for Serial-by-Bit Data Communication in the
# American National Standard Code for
# Information Interchange

## 1. Scope

1.1 This standard specifies the character structure and sense of character parity for serial-by-bit, serial-by-character synchronous and asynchronous data communication in the American National Standard Code for Information Interchange, X3.4-1968 (ASCII), and the codes invoked when applying the American National Standard Code Extension Techniques for Use with the 7-Bit Coded Character Set of American National Standard Code for Information Interchange, X3.41-1974.

1.2 This standard applies to general information interchange at the interface between data processing terminal equipment (such as data processors, data media input-output devices, and office machines) and data communication equipment (such as data sets and modems).

## 2. Synchronous Data Communication

### 2.1 Seven-Bit Environment

**2.1.1 Standard Character Structure.** The character structure for synchronous data communication shall consist of 8 bits (7 ASCII bits, $b_1$ through $b_7$, plus 1 character parity bit) having equal time intervals. See Fig. 1(a).

**2.1.2 Standard Sense of Character Parity.** The sense of character parity for synchronous data communication shall be *odd* over the 8 bits (7 ASCII bits and 1 character parity bit), that is, an odd number of "1" (marking) bits per character.

### 2.2 Eight-Bit Environment

**2.2.1 Standard Character Structure.** The character structure for synchronous data communication shall consists of 8 bits ($a_1$ through $a_8$) having equal time intervals. See Fig. 1(b).

**2.2.2 Standard Sense of Character Parity.** There is no parity bit in the 8-bit environment.

## 3. Asynchronous Data Communication

### 3.1 Seven-Bit Environment

**3.1.1 Standard Character Structure.** The character structure for asynchronous data communication shall consist of 10 signal elements having equal time intervals: one "0" (spacing) start element, 7 ASCII bits ($b_1$ through $b_7$), one character parity bit, and one "1" (marking) stop element. The intercharacter interval (the time interval between the end of a stop element and the beginning of the next start element) may be of any length, and is of the same sense as the stop element, that is, "1" (marking). See Fig. 2(a).

**3.1.2 Standard Sense of Character Parity.** The sense of character parity for asynchronous data communication shall be *even* over the 8 bits (7 ASCII bits and one character parity bit), that is, an even number of "1" (marking) bits per character.

### 3.2 Eight-Bit Environment

**3.2.1 Standard Character Structure.** The character structure for asynchronous data communication shall consist of 10 signal elements having equal time intervals: one "0" (spacing) start element, 8 bits ($a_1$ through $a_8$), and one "1" (marking) stop element. The intercharacter interval (the time interval between the end of a stop element and the beginning of the next start element) may be of any length, and is of the same sense as the stop element, that is, "1" (marking). See Fig. 2(b).

**3.2.2 Standard Sense of Character Parity.** There is no parity bit in the 8-bit environment.

## 4. Qualifications

4.1 Some configurations of communication facilities

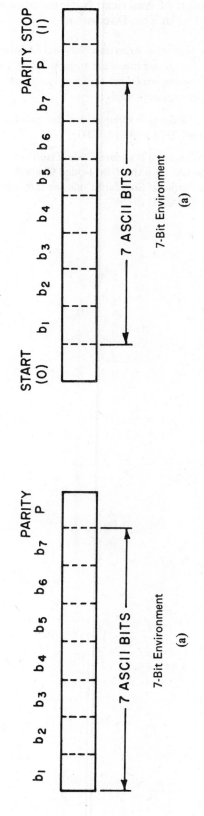

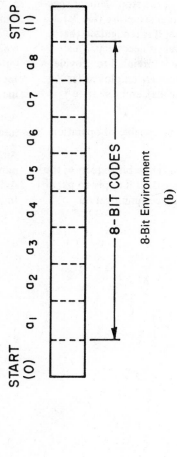

NOTE: The order of transmission is from left to right.

Fig. 2

Character Structure for Asynchronous Data Communication

NOTE: The order of transmission is from left to right.

Fig. 1

Character Structure for Synchronous Data Communication

485

cannot operate satisfactorily with the stop element specified in the asynchronous character structure (see 3.1.1 and 3.2.1). Where this is the case, it is recognized that a stop element of two time intervals is necessary. This exception to character structure is intended to provide relief where character regenerators are employed (as on long-haul, multistation networks), and its use requires prior agreement.

**4.2** Receiving equipment must be capable of operation with no intercharacter interval.

**4.3** This standard does not specify the bit sequence, the bit rate, or the character rate, nor does it apply to parallel-by-bit, serial-by-character data communication.

## 5. Revision of American National Standards Referred to in This Document

When the following American National Standards referred to in this document are superseded by a revision approved by the American National Standards Institute, Inc, the revision shall apply:

American National Standard Code for Information Interchange, X3.4-1968 (ASCII)

American National Standard Code Extension Techniques for Use with the 7-Bit Coded Character Set of American National Standard Code for Information Interchange, X3.41-1974

## Appendix A
## Criteria

### A1. Introduction

**A1.1** This Appendix contains the criteria upon which the character structure and the character parity sense were based. Not all criteria have been entirely satisfied. Some of these criteria conflict with others, and the character structure and the character parity sense specified represent accepted compromises of these divergent criteria.

**A1.2** The criteria were drawn primarily from communication aspects of information interchange; processing and media recording aspects of information interchange were also considered.

### A2. Specific Criteria: Character Structure

NOTE: The following criteria are not mutually consistent and are not listed in order of importance.

**A2.1** One hundred twenty-eight characters should be uniquely specified.

**A2.2** A single character structure should be specified regardless of the transmission facility, speed, or modulation technique.

**A2.3** No ASCII character should require special treatment.

**A2.4** The highest possible character rate should be provided.

**A2.5** Each character should contain a single character parity bit.

**A2.6** Character structure should cause minimum confusion to maintenance and operating personnel.

**A2.7** Simple and economic means of error checking should be possible.

**A2.8** In asynchronous communication the start and stop signal elements should always be of the same duration as the data bits.

**A2.9** In asynchronous communication the character structure should contain one start and one stop signal element.

**A2.10** In asynchronous communication the character structure should contain more than one stop signal element.

### A3. Specific Criteria: Sense of Character Parity

NOTE: The following criteria are not mutually consistent and are not listed in order of importance.

**A3.1** No ASCII character should require special treatment.

**A3.2** There should be no restrictions on sequences of characters (for example, Successive Nulls, Syncs, Spaces, Zeros, or Deletes).

**A3.3** All characters in the ASCII should have the same parity sense (odd or even).

**A3.4** The sense (odd or even) of the character parity bit should minimize hardware complexity.

**A3.5** Maximum compatibility should be provided with the parity sense requirements of the various media.

**A3.6** The sense of the character parity should be the same regardless of the data transmission technique (for example, synchronous or asynchronous), the transmission facility (for example, bandwidth or distortion), speed, or modulation technique.

**A3.7** Equipment complexity should be minimized when alternately handling other 8-bit codes or a binary bit stream.

**A3.8** The character parity sense should cause minimum confusion to maintenance and operating personnel.

487

10

## Appendix B
### Design Considerations

## B1. Introduction

**B1.1** System factors considered in this standard are transmission efficiency, reliability, error control, media requirements, equipment complexity, maintenance confusion, and transition to and from alternate non-ASCII codes or binary bit stream data.

**B1.2** A single character structure could not be specified for both asynchronous (start-stop) and synchronous data communication. Therefore two character structures, one for synchronous and one for asynchronous communication, are specified.

## B2. Character Structure for Synchronous Data Communication

**B2.1** Synchronous data communication can be achieved by transmitting only the 7 information bits for each ASCII character. However, the Null and Delete characters of ASCII would provide no signal element transitions, and successive Null or Delete characters would cause self-clocking synchronous systems to lose synchronization. To overcome this difficulty, either the Null and Delete characters require special handling, or an eighth signal element must be added to each character to assure at least one signal transition within every ASCII character. The latter method was determined to require less equipment complexity and was therefore selected.

**B2.2** Two choices were available for the use of the eighth signal element that assures at least one transition per character:

(1) A bit always opposite in value to a specified information bit

(2) An odd character parity bit

The odd character parity bit was selected because it also provides a basic method of error checking.

**B2.3** All 7 ASCII bits and the odd character parity bit were specified to be of unit time interval in order to simplify timing equipment in transmitters, receivers, and regenerators.

## B3. Character Structure for Asynchronous Data Communication

**B3.1** Asynchronous binary data communication requires that synchronization be derived from the signal elements of each character.

**B3.2** A straightforward technique for enabling character synchronization is to append a start signal element and a stop signal element to each ASCII character. Although there is no synchronization requirement for signal elements, a single character parity bit is included to permit a simple error check.

**B3.3** The start signal element, the 7 ASCII bits, the character parity bit, and the stop signal element are of the same duration to simplify the requirements for timing equipment in transmitters and receivers.

**B3.4** Two choices of character parity sense in asynchronous data communication were possible:

(1) *Even character parity sense* permits consistent character parity sense when using perforated tape or edge-punched documents; facilitates the use of the Delete character as a "timing" character in electromechanical control devices; facilitates the use of an arbitrary spacing interval to turn on motors in asynchronous electromechanical devices; and facilitates disconnection of electromechanical line-switched asynchronous terminals upon receipt of a timed spacing interval.

(2) *Odd character parity sense* permits consistent character parity sense for all data transmission; avoids parity sense inversion when interchanging data between synchronous and asynchronous systems; simplifies maintenance techniques and documentation; and facilitates the handling of parity sense in mixed (synchronous and asynchronous) systems.

**B3.5** The question of asynchronous character parity sense resolves itself into a choice between two specific alternatives: even parity sense results in more economical punched media electromechanical terminals; odd parity sense is consistent with synchronous data transmission, and results in hardware savings in mixed (synchronous and asynchronous) systems.

The conclusion reached is that the economic advantage of even parity sense in electromechanical asynchronous equipment outweighs the advantage that consistent parity sense provides in mixed data communication systems.

488

11

# ANSI
# X3.24-1968
### Approved September 27, 1968
## AMERICAN NATIONAL STANDARDS INSTITUTE

**EIA RS-334**

# EIA STANDARD

*Signal Quality at Interface Between Data Processing Terminal Equipment and Synchronous Data Communication Equipment for Serial Data Transmission*

## RS-334

489

*March 1967*

*Engineering Department*

# ELECTRONIC INDUSTRIES ASSOCIATION

## NOTICE

EIA engineering standards are designed to serve the public interest through eliminating mis-understandings between manufacturers and purchasers, facilitating interchangeability and improvement of products, and assisting the purchaser in selecting and obtaining with minimum delay the proper product for his particular need. Existence of such standards shall not in any respect preclude any member or non-member of EIA from manufacturing or selling products not conforming to such standards, nor shall the existence of such standards preclude their voluntary use by those other than EIA members whether the standard is to be used either domestically or internationally.

Recommended standards are adopted by EIA without regard to whether or not their adoption may involve patents on articles, materials, or processes. By such action, EIA does not assume any liability to any patent owner, nor does it assume any obligation whatever to parties adopting the recommended standards.

Published by

# ELECTRONIC INDUSTRIES ASSOCIATION

### Engineering Department

### 2001 Eye Street, N. W., Washington, D. C. 20006

490

## Price $2.90

Printed in U.S.A.

# SIGNAL QUALITY AT INTERFACE BETWEEN DATA PROCESSING TERMINAL EQUIPMENT AND SYNCHRONOUS DATA COMMUNICATION EQUIPMENT FOR SERIAL DATA TRANSMISSION

*(From Standards Proposal No. 921 formulated under the cognizance of EIA Subcommittee TR-30.1 on Signal Quality)*

## 1. SCOPE

1.1   This standard is applicable to the exchange of serial binary data signals and timing signals across the interface between data processing terminal equipment and synchronous data communication equipment, as defined in EIA Standard RS-232-B.   The data communication equipment is considered to be synchronous if the timing signal circuits are required at the transmitting terminal or the receiving terminal, or both.

1.2   This standard is of particular importance when the equipments in a system are furnished by different organizations.   It does not attempt to indicate what action, if any, is to be taken if the limits are not met, but it is intended to provide a basis for agreement between the parties involved.

1.3   This standard does not describe any requirements for error performance, either for a complete system or any system components.   It should not be assumed that compliance with these standards will produce error rates that are acceptable in any particular application.

1.4   Any equipment which is represented as complying with this standard shall meet the applicable specifications within the ranges of those factors which are described as appropriate for the normal operation of the equipment, such as primary power voltage and frequency, ambient temperature, and humidity.

## 2. DEFINITIONS

2.1   Usage of terminology in this standard is consistent with that in EIA Standard RS-232-B.

2.2   Terms related to signal distortion are taken to have the significance defined in the International Telecommunication Union List of Definitions of Essential Telecommunication Terms, Part 1, General Terms, Telephony, Telegraph, 1961, and Supplement 1 thereto.

2.3   For discussion of isochronous and individual peak distortions, used in this standard, see Section 6, Explanatory Notes.

## 3. INTERFACE AT TRANSMITTING TERMINAL

### 3.1 TIMING SIGNALS

The following standards apply to synchronous systems using either Circuit DA (Transmitter Signal Element Timing, Data Processing Terminal Equipment Source), or Circuit DB (Transmitter Signal Element Timing, Data Communication Equipment Source).   (See Figure 1)

#### 3.1.1 DISTORTION

The degree of peak individual distortion of the significant transitions of the timing signal shall not exceed 0.5 per cent, at the modulation rate of the data signal.   The duty cycle shall be 50 plus or minus 10 per cent.

#### 3.1.2 FREQUENCY

The frequency of the timing signals shall not deviate more than 0.01 per cent from its assigned value.

### 3.2 TRANSMITTED DATA SIGNAL

#### 3.2.1 DATA TRANSITIONS NOT SIGNIFICANT FOR TIMING

For systems in which the individual transitions on circuit BA (Transmitted Data) are not significant in determining the timing of the modulations of the signal transmitted by the data communication equipment, no distortion limit is specified, provided that the relationship between the transitions on Circuit BA and the transitions on either Circuit DA or DB is met as specified in Section 3.3.

491

### 3.2.2 DATA TRANSITIONS SIGNIFICANT FOR TIMING

For systems in which the individual transitions on Circuit BA (Transmitted Data) are significant in determining the timing of the modulations of the signal transmitted by the data communication equipment, the degree of isochronous distortion on Circuit BA shall not exceed 4 per cent.

## 3.3 RELATIONSHIPS BETWEEN TIMING SIGNALS AND DATA SIGNALS

The following relationships apply to the case described in paragraph 3.2.1. (See Figure 1)

### 3.3.1 TIMING SIGNALS PROVIDED BY DATA PROCESSING TERMINAL EQUIPMENT

In synchronous systems using Circuit DA (Transmitter Signal Element Timing, Data Processing Terminal Equipment Source) the interval between any transition on Circuit BA (Transmitted Data) and any ON to OFF transition on Circuit DA, shall not be less than 25 per cent of the nominal unit interval of the data signal.

### 3.3.2 TIMING SIGNALS PROVIDED BY DATA COMMUNICATION EQUIPMENT

In synchronous systems using Circuit DB (Transmitter Signal Element Timing, Data Communication Equipment Source), the interval between any transition on Circuit BA (Transmitted Data) and the corresponding OFF to ON transition on Circuit DB, shall not exceed 25 per cent of the nominal unit interval of the data signal.

## 4. INTERFACE AT RECEIVING TERMINAL

### 4.1 TIMING SIGNALS

The following standards apply to synchronous systems using Circuit DD (Receiver Signal Element Timing, Data Communication Equipment Source). Since the use of Circuit DC (Receiver Signal Element Timing, Data Processing Terminal Equipment Source) is rare, no standards for it are established at this time. (See Figure 2)

### 4.1.1 DISTORTION

The degree of peak individual distortion of the ON to OFF transitions of the timing signals on Circuit DD (Receiver Signal Element Timing, Data Communication Equipment Source) shall not exceed 5 per cent, at the modulation rate of the data signal. The duty cycle shall be 50 plus or minus 10 per cent.

### 4.2 RECEIVED DATA SIGNAL

#### 4.2.1 DISTORTION

The degree of isochronous distortion of the signals on Circuit BB (Received Data) shall not exceed 10 per cent.

### 4.3 RELATIONSHIP BETWEEN TIMING SIGNALS AND DATA SIGNALS

#### 4.3.1 TIMING SIGNALS PROVIDED BY DATA COMMUNICATION EQUIPMENT

The interval between any transition on Circuit BB (Received Data) and any ON to OFF transition on Circuit DD (Receiver Signal Element Timing, Data Communication Equipment Source) shall not be less than 25 per cent of the nominal unit interval of the data signal. (See Figure 2)

## 5. STANDARD DISTORTION MEASUREMENT

5.1 Standard distortion measurement and interface voltage measurement shall be made on the particular signal interchange circuit of interest while the circuit is loaded with a Standard Test Load.

### 5.2 STANDARD TEST LOAD

The Standard Test Load shall consist of a 3000-ohm resistor shunted by 2500 picofarad capacitance and connected from the signal interchange circuit under test to Circuit AB (Signal Ground).

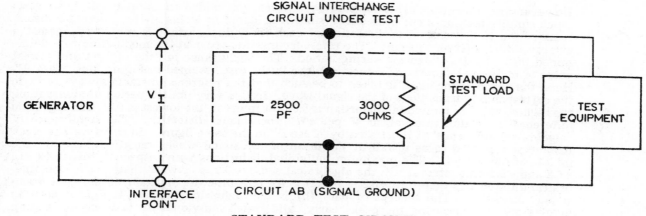

STANDARD TEST CIRCUIT

## 5.3 INTERFACE VOLTAGE MEASUREMENT

With the standard test load, the generator shall produce a voltage $V_I$ across the test load of not less than ± 5.0 volts.

## 5.4 STANDARD DISTORTION MEASUREMENT THRESHOLD

The standard distortion measurement shall be made using a +3 volt and a −3 volt threshold to determine the occurrence of signal transitions.

5.4.1 A mark-to-space or an OFF to ON transition shall be taken to occur at the instant $V_I$ crosses +3 volts on a positive-going transition.

5.4.2 A space-to-mark or an ON to OFF transition shall be taken to occur at the instant $V_I$ crosses −3 volts on a negative-going transition.

## 6. EXPLANATORY NOTES

### 6.1 DEGREE OF ISOCHRONOUS DISTORTION

The following definition is from Supplement 1 to the ITU List, definition 33.07:

*Degree of Isochronous Distortion*

(a) Ratio to the unit interval of the maximum measured difference, irrespective of sign, between the actual and the theoretical intervals separating any two significant instants of modulation (or of restitution), these instants being not necessarily consecutive.

(b) Algebraical difference between the highest and lowest value of individual distortion affecting the significant instants of an isochronous modulation. (This difference is independent of the choice of the reference ideal instant.)

The degree of distortion (of an isochronous modulation or restitution) is usually expressed as a percentage.

*Note:* The result of the measurement should be completed by an indication of the period, usually limited, of the observation. For a prolonged modulation (or restitution ) it will be appropriate to consider the probability that an assigned value of the degree of distortion will be exceeded.

With respect to part (a) of the definition, the following interpretations are made:

(1) The Term "theoretical interval" (i.e., unit interval) is taken to be the reciprocal of the modulation (or restitution) rate, as determined throughout a stated measuring period.

(2) The term "maximum" requires that the intervals between all possible pairs of significant instants (i.e., transitions), that occur during the period of observation have been considered. This appears to involve making a very large number of measurements simultaneously.

However, the alternate definition, part (b), which was added in Supplement 1, suggests much simpler measurement procedures, since it is stated in terms of the individual distortions of the transitions. The distortion of each individual instant is measured against a corresponding reference instant. The reference instants occur at the modulation rate determined throughout a stated measuring period. The significance of "the highest and lowest value of individual distortion" is most obvious in the case assumed in Figure 3 as "Case A". Here, the phase of the clock is taken to be such that the reference instants always precede the corresponding transitions of the signal wave. In this very limited period of observation, the highest value of individual distortion is 0.6 unit interval; the lowest is 0.2. The algebraic difference is 0.4 unit interval, or 40 per cent isochronous distortion. The significance of "algebraical difference" is illustrated by "Case B" in the same figure. In this case, the clock instants are phased so as to fall at the mean of the signal wave transitions, (i.e., in the middle of the spread). The maximum individual distortions with respect to this clock are +0.2 and −0.2 unit interval. In the algebraical sense, −0.2 is "lower" than +0.2. The algebraical difference is therefore +0.2 − (−0.2) = 0.4 unit interval, or again, 40 per cent isochronous distortion. The concept of part (b) of the definition in Case B thus is seen to involve merely measuring the peak "early" and "late" individual distortions, and adding them to find the degree of isochronous distortion.

## 6.2 DEGREE OF PEAK INDIVIDUAL DISTORTION

For the purpose of this Standard, the following definition is used:

*Degree of Peak Individual Distortion*
The maximum individual distortion, irrespective of sign, of all significant instants occurring during a particular measuring period.

This term is defined and used in this Standard in stating the timing signal requirements, instead of isochronous distortion, for the following reasons:

(1) Only alternate transitions are considered (either "ON" to "OFF", or "OFF" to "ON", but not both).

(2) The reference is a unit interval, not of the timing signal, but of the data signal.

Figure 4 illustrates the case when it is the "ON" to "OFF" transitions that are significant. Individual distortions are measured with respect to a clock operating at the modulation rate of the associated data signal, and phased to occur at the mean of the significant timing transitions. (The "OFF" to "ON" transitions are shown dotted to indicate that they are not considered in the distortion measurement.) In the example, the peak individual distortion is (−)0.2 of a data signal unit interval, or 20 per cent, "irrespective of sign".

## PARAGRAPH 3.1.1 — TRANSMITTER SIGNAL ELEMENT TIMING SIGNALS, DISTORTION

(A): FOR
PARAGRAPH 3.3.1

CIRCUIT DA

(TIMING SIGNALS
PROVIDED BY
DATA PROCESSING
TERMINAL
EQUIPMENT)

(B): FOR
PARAGRAPH 3.3.2

CIRCUIT DB

(TIMING SIGNALS
PROVIDED BY
DATA COMMUNICATION
EQUIPMENT)

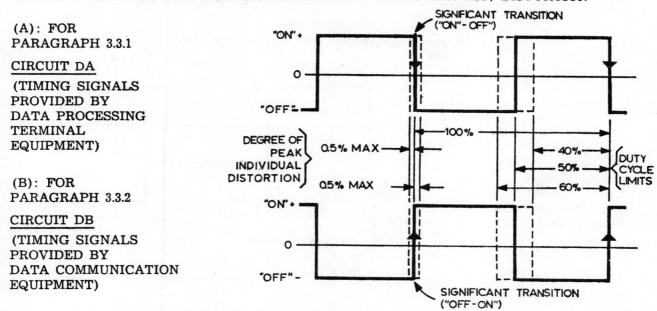

## PARAGRAPH 3.3.1 — RELATIONSHIP BETWEEN TIMING SIGNALS AND DATA SIGNALS— TIMING SIGNALS PROVIDED BY DATA PROCESSING TERMINAL EQUIPMENT.

CIRCUIT DA

TRANSMITTER
SIGNAL ELEMENT
TIMING, DATA
PROCESSING
TERMINAL
EQUIPMENT
SOURCE

CIRCUIT BA

TRANSMITTED
DATA

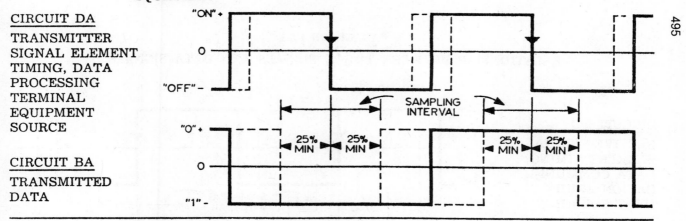

495

## PARAGRAPH 3.3.2 — RELATIONSHIP BETWEEN TIMING SIGNALS AND DATA SIGNALS— TIMING SIGNALS PROVIDED BY DATA COMMUNICATION EQUIPMENT.

CIRCUIT DB

TRANSMITTER
SIGNAL ELEMENT
TIMING, DATA
COMMUNICATION
EQUIPMENT
SOURCE

CIRCUIT BA

TRANSMITTED
DATA

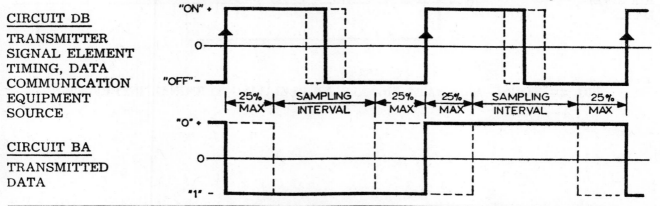

FIG. 1—TIMING AND DATA SIGNALS AT TRANSMITTING TERMINAL INTERFACE

**PARAGRAPH 4.1.1**

## MAXIMUM DISTORTION OF TIMING SIGNALS

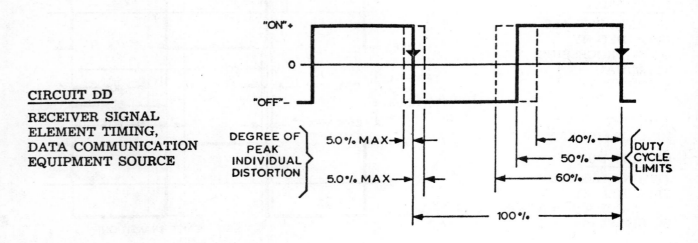

CIRCUIT DD

RECEIVER SIGNAL
ELEMENT TIMING,
DATA COMMUNICATION
EQUIPMENT SOURCE

496

**PARAGRAPH 4.3.1**

## RELATIONSHIP BETWEEN TIMING SIGNALS AND DATA SIGNALS

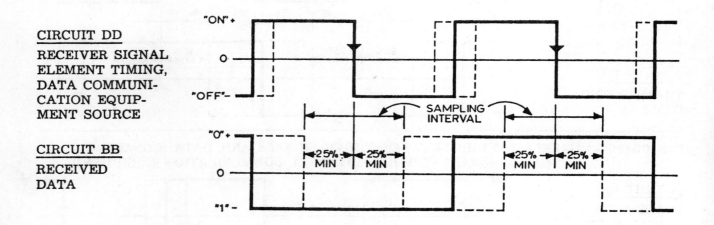

CIRCUIT DD

RECEIVER SIGNAL
ELEMENT TIMING,
DATA COMMUNI-
CATION EQUIP-
MENT SOURCE

CIRCUIT BB

RECEIVED
DATA

**FIG. 2—TIMING AND DATA SIGNALS AT RECEIVING TERMINAL INTERFACE**

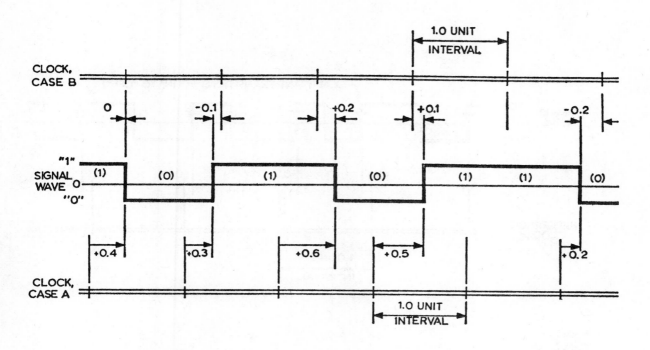

*CASE A:* REFERENCE CLOCK NOT IN PHASE WITH MEAN TRANSITIONS OF SIGNAL WAVE

| ALGEBRAIC HIGHEST INDIVIDUAL DISTORTION | +0.6 (UNIT INTERVAL) |
| ALGEBRAIC LOWEST INDIVIDUAL DISTORTION | +0.2 |
| | |
| ALGEBRAIC DIFFERENCE | 0.4 (UNIT INTERVAL) |
| DEGREE OF ISOCHRONOUS DISTORTION | 40% |

*CASE B:* REFERENCE CLOCK IN PHASE WITH MEAN TRANSITIONS OF SIGNAL WAVE

| ALGEBRAIC HIGHEST INDIVIDUAL DISTORTION | +0.2 (UNIT INTERVAL) |
| ALGEBRAIC LOWEST INDIVIDUAL DISTORTION | −0.2 |
| | |
| ALGEBRAIC DIFFERENCE | 0.4 (UNIT INTERVAL) |
| DEGREE OF ISOCHRONOUS DISTORTION | 40% |

FINDING DEGREE OF ISOCHRONOUS DISTORTION
BY ITU DEFINITION 33.07, PART (b)

**FIGURE 3**

*NOTE: This figure is explanatory only. See text for specific signal requirements.*

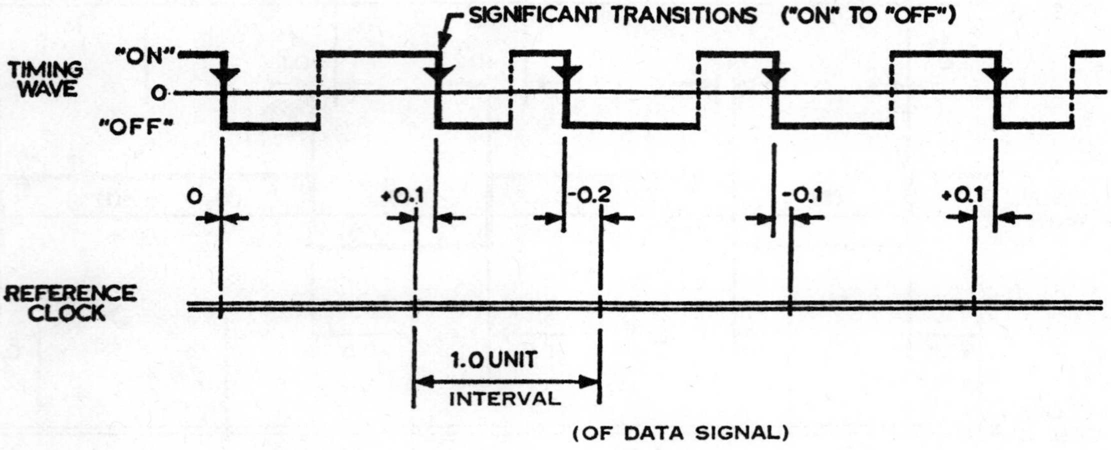

MAXIMUM INDIVIDUAL DISTORTION OF SIGNIFICANT TRANSITIONS
(REFERENCE CLOCK IN PHASE WITH MEAN OF SIGNIFICANT TRANSITIONS):

0.2 DATA SIGNAL UNIT INTERVAL
DEGREE OF PEAK INDIVIDUAL DISTORTION: 20%

FINDING DEGREE OF PEAK INDIVIDUAL DISTORTION
FOR CASE OF "ON" TO "OFF" SIGNIFICANT TRANSITIONS

FIGURE 4

NOTE: *This figure is explanatory only. See text for specific signal requirements.*

# BACKGROUND INFORMATION

1. Background information is given in "EIA Industrial Electronics Bulletin No. 5, March 1965, Tutorial Paper on Signal Quality at a Digital Interface."

2. Interface requirements for non-synchronous data communication equipment differ significantly from those for synchronous data communication equipment, and therefore will be covered separately.

3. This standard is concerned only with data and timing signals at the specified digital interfaces of synchronous data communication equipment.

4. This standard is intended to allow equipment designers as much latitude as practicable in obtaining an economic balance between the cost of precision timing sources, the long-term stability of critical components, the time between equipment failures, and maintenance costs.

5. Definition 6.2 was developed because no existing definition of this type of distortion was listed in the ITU List of Definitions, and it was considered desirable to use it in writing the standards.

6. RS-232-B appears to have an ambiguity which would limit the accuracy of measurements. The ambiguity is evident upon consideration of the following paragraphs of that document:

> "4.5 The terminating impedance at the receiving end of an interchange circuit shall have a dc resistance of not less than 3000 ohms or more than 7000 ohms, and the voltage in open-circuited condition shall not exceed two volts. The effective shunt capacitance at the receiving end of a signal interchange circuit, measured at the interchange point and including any connecting cable, shall not exceed 2500 picofarads."

> "4.7 For both data circuits and timing signal circuits (Circuit BA, BB, DA, DB, DC and DD), neither the rise time nor the fall time, through the six volt range in which the signal condition is not defined, shall exceed three per cent of the nominal duration of a signal element. The circuitry used to generate a signal voltage on an interchange circuit shall meet this specification with any receiving termination which complies with Section 4.5".

> "4.8 Equivalent Circuit. The following conditions shall apply:

> \*    \*    \*    \*    \*

> (8) The circuitry that receives signals from an interchange circuit shall recognize the binary signal when $V_I$ equals $\pm 3$ volts. (See Section 5.4.) The signal condition is not defined when $V_I$ is in the range between $+3$ and $-3$ volts."

That is, the signal, when undergoing a transition, may be in an undefined condition for as long as three per cent of a unit element duration, and for a given source, is influenced by the impedance of the load.

Paragraph 5 of the Standard defines a set of standard conditions for making distortion measurements at the digital interface, which will eliminate the ambiguity of the range in which the signal condition is not defined. The specified "Standard Test Load" represents the most restrictive condition according to RS-232-B, Paragraph 4.5. The signal level ambiguity is resolved by specifying that the transitions shall be taken to occur when the signal first emerges from the undefined six volt range, at the $+3$ volt and $-3$ volt levels according to the sense of the transition.

499

# American National Standard

character structure and character parity
sense for parallel-by-bit data communication
in the american national standard code
for information interchange

Adopted for Use by
the Federal Government

FIPS PUB 18-1
See Notice on Inside
Front Cover

american national standards institute, inc.
1430 broadway, new york, new york 10018

This standard has been adopted for Federal Government use.

Details concerning the use of this standard within the Federal Government are contained in FIPS PUB 18-1, CHARACTER STRUCTURE AND CHARACTER PARITY SENSE FOR PARALLEL-BY-BIT DATA COMMUNICATION IN THE CODE FOR INFORMATION INTERCHANGE. For a complete list of the publications available in the FEDERAL INFORMATION PROCESSING STANDARDS Series, write to the Office of Technical Information and Publications, National Bureau of Standards, Washington, D.C. 20234.

ANSI®
X3.25-1976
Revision of
X3.25-1968

American National Standard
Character Structure and Character Parity Sense for
Parallel-by-Bit Data Communication in the
American National Standard Code for
Information Interchange

502

Secretariat

**Computer and Business Equipment Manufacturers Association**

Approved June 25, 1976

**American National Standards Institute, Inc**

## American National Standard

An American National Standard implies a consensus of those substantially concerned with its scope and provisions. An American National Standard is intended as a guide to aid the manufacturer, the consumer, and the general public. The existence of an American National Standard does not in any respect preclude anyone, whether he has approved the standard or not, from manufacturing, marketing, purchasing, or using products, processes, or procedures not conforming to the standard. American National Standards are subject to periodic review and users are cautioned to obtain the latest editions.

**CAUTION NOTICE:** This American National Standard may be revised or withdrawn at any time. The procedures of the American National Standards Institute require that action be taken to reaffirm, revise, or withdraw this standard no later than five years from the date of publication. Purchasers of American National Standards may receive current information on all standards by calling or writing the American National Standards Institute.

503

Published by

**American National Standards Institute**
**1430 Broadway, New York, New York 10018**

Printed in the United States of America

A7M1176/3

**Foreword**

(This Foreword is not a part of American National Standard Character Structure and Character Parity Sense for Parallel-by-Bit Data Communication in the American National Standard Code for Information Interchange, X3.25-1976.)

This standard specifies the character structure and sense of character parity for parallel-by-bit, serial-by-character data communication in the American National Standard Code for Information Interchange, X3.4-1968 (ASCII), at the interface between data processing terminal equipment and data communication equipment. It also specifies the codes invoked when applying the American National Standard Code Extension Techniques for Use with the 7-Bit Coded Character Set of American National Standard Code for Information Interchange, X3.41-1974.

This standard is one of a series developed by Task Group X3S3.3 on Data Communication Formats under the coordination of the X3S3 Subcommittee on Data Communications of American National Standards Committee on Computers and Information Processing, X3. Task Group X3S3.3, which was organized late in 1962 and held its first meeting in January 1963, is charged with the responsibility for standardizing character format, data transmission of characters within a hierarchy of groupings (that is, words, blocks, messages, etc) including group error control, and the order or sequence of bits within characters (including parity).

Other standards provide specifications for bit sequencing, formats for serial-by-bit, serial-by-character data transmission, and other parameters vital to the transmission of information between the types of equipment previously mentioned.

Suggestions for improvement of this standard will be welcome. They should be sent to the American National Standards Institute, 1430 Broadway, New York, N.Y. 10018.

This standard was processed and approved for submittal to ANSI by American National Standards Committee on Computers and Information Processing, X3. Committee approval of the standard does not necessarily imply that all committee members voted for its approval. At the time it approved this standard, the X3 Committee had the following members:

J. F. Auwaerter, Chairman
R. M. Brown, Vice-Chairman
W. F. Hanrahan, Secretary

| Organization Represented | Name of Representative |
|---|---|
| Addressograph Multigraph Corporation | A. C. Brown |
| | D. Anderson (Alt) |
| Air Transport Association | F. C. White |
| | C. Hart (Alt) |
| American Library Association | J. R. Rizzolo |
| | J. C. Kountz (Alt) |
| | M. S. Malinconico (Alt) |
| American Nuclear Society | D. R. Vondy |
| | M. K. Butler (Alt) |
| American Society for Information Science | M. C. Kepplinger |
| Association for Computing Machinery | P. Skelly |
| | J. A. N. Lee (Alt) |
| | H. E. Thiess (Alt) |
| Association for Educational Data Systems | A. K. Swanson |
| Association of American Railroads | R. A. Petrash |
| Association of Computer Programmers and Analysts | L. A. Ruh |
| | M. A. Morris, Jr (Alt) |
| Association of Data Processing Service Organizations | J. B. Christiansen |
| Burroughs Corporation | E. Lohse |
| | J. S. Foley (Alt) |
| | J. F. Kalbach (Alt) |
| Computer Industry Association | N. J. Ream |
| | A. G. W. Biddle (Alt) |
| Control Data Corporation | C. E. Cooper |
| | G. I. Williams (Alt) |

504

| Organization Represented | Name of Representative |
|---|---|
| Data Processing Management Association | A. E. Dubnow |
| | D. W. Sanford (Alt) |
| Datapoint Corporation | V. D. Poor |
| | H. W. Swanson (Alt) |
| Digital Equipment Corporation | P. White |
| | A. R. Kent (Alt) |
| Edison Electric Institute | S. P. Shrivastava |
| | J. L. Weiser (Alt) |
| General Services Administration | D. L. Shoemaker |
| | M. L. Burris (Alt) |
| GUIDE International | T. E. Wiese |
| | B. R. Nelson (Alt) |
| | D. Stanford (Alt) |
| Honeywell Information Systems, Inc | T. J. McNamara |
| | E. H. Clamons (Alt) |
| Institute of Electrical and Electronics Engineers, Communications Society | (Representation Vacant) |
| Institute of Electrical and Electronics Engineers, Computer Society | R. L. Curtis |
| | T. Feng (Alt) |
| International Business Machines Corporation | L. Robinson |
| | W. F. McClelland (Alt) |
| Joint Users Group | T. E. Wiese |
| | L. Rodgers (Alt) |
| Life Office Management Association | R. Ricketts |
| | J. F. Foley (Alt) |
| | E. J. Jowdy (Alt) |
| Litton Industries | I. Danowitz |
| National Association of State Information Systems | G. H. Roehm |
| | J. L. Lewis (Alt) |
| | G. I. Theis (Alt) |
| National Bureau of Standards | H. S. White, Jr |
| | R. E. Rountree (Alt) |
| National Communications System | M. L. Cain |
| | G. W. White (Alt) |
| National Machine Tool Builders' Association | O. A. Rodriques |
| NCR Corporation | R. J. Mindlin |
| | A. R. Daniels (Alt) |
| | T. W. Kern (Alt) |
| Olivetti Corporation of America | E. J. Almquist |
| Pitney Bowes, Inc | R. H. Fenn |
| | E. T. Warzecha (Alt) |
| Printing Industries of America | N. Scharpf |
| | E. Masten (Alt) |
| Recognition Equipment, Inc | H. F. Schantz |
| | W. E. Viering (Alt) |
| Sanders Associates, Inc | A. L. Goldstein |
| | T. H. Buchert (Alt) |
| Scientific Apparatus Makers Association | A. Savitsky |
| | J. E. French (Alt) |
| SHARE Inc | T. B. Steel, Jr. |
| | E. Brubaker (Alt) |
| Society of Certified Data Processors | T. M. Kurihara |
| | A. E. Dubnow (Alt) |
| Telephone Group | V. N. Vaughan, Jr |
| | J. C. Nelson (Alt) |
| | P. G. Wray (Alt) |
| 3M Company | R. C. Smith |
| UNIVAC, Division of Sperry Rand Corporation | M. W. Bass |
| | C. D. Card (Alt) |
| U.S. Department of Defense | W. L. McGreer |
| | W. B. Rinehuls (Alt) |
| | W. B. Robertson (Alt) |
| U.S. Department of Health, Education and Welfare | W. R. McPherson, Jr |
| | M. A. Evans (Alt) |
| VIM (CDC 6000 Users Group) | S. W. White |
| | M. R. Speers (Alt) |
| Xerox Corporation | J. L. Wheeler |
| | A. R. Machell (Alt) |

Subcommittee X3S3 on Data Communications, which coordinated the development of this standard, had the following members:

G. C. Schutz, Chairman
J. L. Wheeler, Vice-Chairman
S. M. Harris, Secretary

| | | |
|---|---|---|
| P. A. Arneth | J. L. Dempsey | B. L. Meyer |
| J. R. Aschenbrenner | W. F. Emmons | O. C. Miles |
| P. C. Baker | H. C. Folts | R. T. Moore |
| M. W. Baty | G. O. Hansen | L. S. Nidus |
| M. J. Bedford | T. L. Hewitt | N. F. Priebe |
| R. C. Boepple | L. G. Kappel | S. J. Raiter |
| W. Brown | P. W. Kiesling, Jr | S. R. Rosenblum |
| M. T. Bryngil | C. C. Kleckner | D. L. Shoemaker |
| M. L. Cain | W. E. Landis | J. M. Skaug |
| D. E. Carlson | D. S. Lindsay | N. E. Snow |
| G. E. Clark, Jr | G. J. McAllister | E. R. Stephan |
| J. W. Conard | R. C. Matlack | E. E. Udo-Ema |
| H. J. Crowley | | G. W. White |

Task Group X3S3.3 on Data Communication Formats, which developed this standard, had the following members:

W. F. Emmons, Chairman

| | | |
|---|---|---|
| J. V. Fayer | M. E. McMahon | G. W. White |
| J. G. Griffis | L. Nidus | C. E. Young |
| | E. Novgorodoff | |

Other persons who contributed to the development of this standard were:

| | | |
|---|---|---|
| J. H. Bardon | E. Berezin | J. L. Little |
| G. A. Barletta | J. B. Booth | J. J. O'Donnell |

506

# Contents

# American National Standard
# Character Structure and Character Parity Sense for
# Parallel-by-Bit Data Communication in the
# American National Standard Code for
# Information Interchange

## 1. Scope

**1.1** This standard specifies the character structure and sense of character parity for parallel-by-bit, serial-by-character data communication in the American National Standard Code for Information Interchange, X3.4-1968 (ASCII), and the codes invoked when applying the American National Standard Code Extension Techniques for Use with the 7-Bit Coded Character Set of American National Standard Code for Information Interchange, X3.41-1974.

**1.2** This standard applies to general information interchange at the interface between data processing terminal equipment (such as data processors, data media input-output devices, and office machines) and data communication equipment (such as data sets and modems).

## 2. Standard Character Structure

**2.1 Seven-Bit Environment.** The character structure shall consist of 8 bits, that is, 7 ASCII bits, $b_1$ through $b_7$, plus 1 character parity bit.

**2.2 Eight-Bit Environment.** The character structure shall consist of 8 bits, $a_1$ through $a_8$.

## 3. Standard Bit-to-Channel Relationship

**3.1 Seven-Bit Environment.** The 7 ASCII bits ($b_1$ through $b_7$) plus the character parity bit (P) shall be assigned to an ordered series of channel designators as follows: $b_1$ to the lowest designator and in ascending order, with P to the highest designator.

| ASCII bit: | $b_1$ | $b_2$ | $b_3$ | $b_4$ | $b_5$ | $b_6$ | $b_7$ | P |
|---|---|---|---|---|---|---|---|---|
| Channel: | 1 | 2 | 3 | 4 | 5 | 6 | 7 | 8 |

**3.2 Eight-Bit Environment.** The 8 bits ($a_1$ through $a_8$)

shall be assigned to an ordered series of channel designators as follows: $a_1$ to the lowest designator and in ascending order, with $a_8$ to the highest designator.

| 8-bit codes: | $a_1$ | $a_2$ | $a_3$ | $a_4$ | $a_5$ | $a_6$ | $a_7$ | $a_8$ |
|---|---|---|---|---|---|---|---|---|
| Channel: | 1 | 2 | 3 | 4 | 5 | 6 | 7 | 8 |

## 4. Standard Sense of Character Parity

**4.1 Seven-Bit Environment**

    **4.1.1** Where the transmission system is of the type where character timing is not separately signaled, the sense of the character parity shall be *odd*; that is, the parity bit for each character shall be such that there are an odd number of "1" (marking) bits in the character.

    **4.1.2** Where the transmission system is of the type providing character timing information by means of a separate timing channel, the sense of the character parity shall be *even*; that is, the parity bit for each character shall be such that there are an even number of "1" (marking) bits in the character.

**4.2 Eight-Bit Environment.** There is no parity bit in the 8-bit environment.

## 5. Revision of American National Standards Referred to in This Document

When the following American National Standards referred to in this document are superseded by a revision approved by the American National Standards Institute, Inc, the revision shall apply:

American National Standard Code for Information Interchange, X3.4-1968 (ASCII)

American National Standard Code Extension Techniques for Use with the 7-Bit Coded Character Set of American National Standard Code for Information Interchange, X3.41-1974

# Appendixes

(These Appendixes are not a part of American National Standard Character Structure and Character Parity Sense for Parallel-by-Bit Data Communication in the American National Standard Code for Information Interchange, X3.25-1976, but are included for information purposes only.)

## Appendix A
## Criteria

### A1. Introduction

**A1.1** This Appendix contains the criteria upon which the character structure, bit-to-channel relationship, and the character parity sense were based. Not all criteria have been entirely satisfied. Some of these criteria conflict with others, and the character structure and the character parity sense specified represent compromises among these divergent criteria.

**A1.2** The criteria were drawn from communication aspects of information interchange as well as processing and media recording aspects of information interchange.

### A2. Specific Criteria: Character Structure and Bit-to-Channel Relationship

NOTE: The following criteria are not mutually consistent and are not listed in order of importance.

**A2.1** One hundred twenty-eight characters should be uniquely specified.

**A2.2** A single character structure should be specified regardless of the transmission facility, speed, or modulation technique.

**A2.3** No ASCII character should require special treatment.

**A2.4** The highest possible character rate should be provided.

**A2.5** The character structure for parallel-by-bit communication should be consistent with that for serial-by-bit communication.

**A2.6** Each character should contain a single character parity bit.

**A2.7** The character structure should cause minimum confusion to maintenance and operating personnel.

**A2.8** Simple and economic means of error checking should be possible.

**A2.9** The character structure should be independent of the transmission technique (for example, fixed or random character rates).

**A2.10** The character structure should be independent of the existence of a timing channel.

### A3. Specific Criteria: Sense of Character Parity

NOTE: The following criteria are not mutually consistent and are not listed in order of importance.

**A3.1** No ASCII character should require special treatment.

**A3.2** There should be no restrictions on sequences of characters (for example, successive Null, Sync, Space, Zero, or Delete characters).

**A3.3** All characters in the ASCII should have the same parity sense (odd or even).

**A3.4** The sense (odd or even) of the character parity bit should minimize hardware complexity.

**A3.5** Maximum compatibility should be provided with the parity sense requirements of the various media.

**A3.6** The sense of the character parity should be the same regardless of the data transmission technique, transmission facility, speed, or modulation technique.

**A3.7** Equipment complexity should be minimized when alternately handling other codes or random binary data.

**A3.8** The character parity sense should cause minimum confusion to maintenance and operating personnel.

**A3.9** The character parity sense should not preclude any transmission techniques.

508

7

**Appendix B**
**Design Considerations**

## B1. Introduction

System factors considered in this standard are transmission efficiency, reliability error control, media requirements, equipment complexity, maintenance confusion, and transition to and from alternate non-ASCII codes or random binary data.

## B2. Character Structure and Bit-to-Channel Relationship

**B2.1** The overriding consideration affecting the choice of character structure is compatibility with serial-by-bit data communication in order to minimize confusion. An 8-bit character structure (7 ASCII bits and parity) satisfies this requirement as well as all other criteria.

**B2.2** The overriding consideration affecting the assignment of the bits to the channels is the need for a simple and orderly relationship. This requirement and all other criteria are satisfied by a $b_1$ to channel 1, $b_2$ to channel 2, etc, relationship.

## B3. Sense of Character Parity

The choice of character parity sense, odd or even, involves a choice between clearly conflicting sets of criteria. Consistency with American National Standard Character Structure and Character Parity Sense for Serial-by-Bit Data Communication in the American National Standard Code for Information Interchange, X3.16-1976, cannot resolve this choice since that standard specifies even parity sense for asynchronous transmission and odd parity sense for synchronous transmission.

Essentially, the remaining criteria present a choice between:

(1) Media consistency with punched paper tape in such a way as to facilitate the use of the same equipment for handling binary data and other codes (argument for even parity)

(2) Providing the basis for recovery of character timing (that is, detecting character presence) from the data alone (argument for odd parity)

Since it is envisioned that one of these two considerations will become controlling in a distinct system, the standard sense of parity was chosen as different for the two types of systems based on presence or absence of a separate timing channel.

# American National Standard

### procedures for the use of the communication control characters of american national standard code for information interchange in specified data communication links

510

X3.28-1976

american national standards institute, inc.
1430 broadway, new york, new york 10018

ANSI
X3.28-1976
Revision of
X3.28-1971

# American National Standard
# Procedures for the Use of the Communication
# Control Characters of American National Standard
# Code for Information Interchange in Specified
# Data Communication Links

511

Secretariat

**Computer and Business Equipment Manufacturers Association**

Approved December 17, 1975

**American National Standards Institute, Inc**

## American National Standard

An American National Standard implies a consensus of those substantially concerned with its scope and provisions. An American National Standard is intended as a guide to aid the manufacturer, the consumer, and the general public. The existence of an American National Standard does not in any respect preclude anyone, whether he has approved the standard or not, from manufacturing, marketing, purchasing, or using products, processes, or procedures not conforming to the standard. American National Standards are subject to periodic review and users are cautioned to obtain the latest editions.

**CAUTION NOTICE**: This American National Standard may be revised or withdrawn at any time. The procedures of the American National Standards Institute require that action be taken to reaffirm, revise, or withdraw this standard no later than five years from the date of publication. Purchasers of American National Standards may receive current information on all standards by calling or writing the American National Standards Institute.

512

Published by

**American National Standards Institute**
**1430 Broadway, New York, New York  10018**

Printed in the United States of America

Q1½M876/1050

# Foreword

(This Foreword is not a part of American National Standard Procedures for the Use of the Communication Control Characters of American National Standard Code for Information Interchange in Specified Data Communication Links, X3.28-1976.)

This standard presents control procedures for specified data communication links that employ the communication control characters of American National Standard Code for Information Interchange, X3.4-1968 (ASCII). Other applicable standards include American National Standard for Bit Sequencing of the American National Standard Code for Information Interchange in Serial-by-Bit Data Transmission, X3.15-1976, and American National Standard Character Structure and Character Parity Sense for Serial-by-Bit Data Communication in the American National Standard Code for Information Interchange, X3.16-1976. Additional standards, which prescribe the signaling speed, message format, and other parameters vital to the communication of information in a data system, are being developed. It is expected that additional standards or future revisions of this standard, or both, will increase the number and types of data communication links and systems for which standard control procedures are defined.

This standard supersedes American National Standard Procedures for the Use of the Communication Control Characters of American National Standard Code for Information Interchange in Specified Data Communication Links, X3.28-1971. In general, the functional properties of the categories specified in American National Standard X3.28-1971 have not been altered; in some cases, however, additional requirements are specified, particularly in the areas of aborts, interrupts, and recovery procedures. In addition to editorial changes, the major substantive additions include twelve new subcategories (2.7, 2.8, 3.1, A4, C1, C2, D1, E1, E2, E3, F1, and F2) and new sections pertaining to aborts, interrupts, timers, and link block numbering.

In the development of this standard, both historical and present data communication practices were considered. In addition, consideration was given to the requirements for future systems and to the applicable standards being developed internationally.

Suggestions for improvement of this standard will be welcome. They should be sent to the American National Standards Institute, 1430 Broadway, New York, N.Y. 10018.

This standard was processed and approved for submittal to ANSI by American National Standards Committee on Computers and Information Processing, X3. Committee approval of the standard does not necessarily imply that all committee members voted for its approval. At the time it approved this standard, the X3 Committee had the following members:

J. F. Auwaerter, Chairman
V. E. Henriques, Vice-Chairman
R. M. Brown, Secretary

| Organization Represented | Name of Representative |
| --- | --- |
| Addressograph Multigraph Corporation | A. C. Brown |
| | R. H. Trenkamp (Alt) |
| Air Transport Association | F. C. White |
| | C. Hart (Alt) |
| American Bankers Association | M. E. McMahon |
| | A. Miller (Alt) |
| American Library Association | J. R. Rizzolo |
| | J. C. Kountz (Alt) |
| | M. S. Malinconico (Alt) |
| American Nuclear Society | D. R. Vondy |
| | M. K. Butler (Alt) |
| American Society for Information Science | M. C. Kepplinger |
| American Society of Mechanical Engineers | I. Berman |
| | R. T. Woythal (Alt) |
| Association for Computer Programmers and Analysts | T. G. Grieb |
| Association for Computing Machinery | P. Skelly |
| | J. A. N. Lee (Alt) |
| | H. Theiss (Alt) |

| Organization Represented | Name of Representative |
|---|---|
| Association for Educational Data Systems | A. K. Swanson |
| Association of American Railroads | R. A. Petrash |
| Association of Data Processing Service Organizations | J. B. Christiansen |
| Burroughs Corporation | E. Lohse |
| | J. F. Kalbach (Alt) |
| Computer Industry Association | N. Ream |
| | A. G. W. Biddle (Alt) |
| Control Data Corporation | C. E. Cooper |
| | G. I. Williams (Alt) |
| Data Processing Management Association | A. E. Dubnow |
| | D. W. Sanford (Alt) |
| Digital Equipment Corporation | P. White |
| | E. Corell (Alt) |
| Edison Electric Institute | J. W. Fish |
| | K. C. Adgate (Alt) |
| General Electric Company | R. R. Hench |
| | J. K. Snell (Alt) |
| General Services Administration | D. L. Shoemaker |
| | M. L. Burris (Alt) |
| GUIDE International | T. E. Wiese |
| | B. R. Nelson (Alt) |
| | D. Stanford (Alt) |
| Honeywell Information Systems, Inc | T. J. MacNamara |
| | E. H. Clamons (Alt) |
| Institute of Electrical and Electronics Engineers, Communications Society | G. E. Friend |
| Institute of Electrical and Electronics Engineers, Computer Society | G. C. Schutz |
| | T. Feng (Alt) |
| International Business Machines Corporation | L. Robinson |
| | W. F. McClelland (Alt) |
| Joint Users Group | T. E. Wiese |
| | L. Rodgers (Alt) |
| Life Office Management Association | A. J. Tufts |
| Litton Industries | I. Danowitz |
| National Association of State Information Systems | G. H. Roehm |
| | C. Vorlander (Alt) |
| National Bureau of Standards | H. S. White, Jr |
| | R. E. Rountree (Alt) |
| National Machine Tool Builders' Association | O. A. Rodriques |
| NCR Corporation | R. J. Mindlin |
| | A. R. Daniels (Alt) |
| | T. W. Kern (Alt) |
| Olivetti Corporation of America | E. J. Almquist |
| Pitney Bowes, Inc | R. H. Fenn |
| | E. T. Warzecha (Alt) |
| Printing Industries of America | N. Scharpf |
| | E. Masten (Alt) |
| Recognition Equipment, Inc | H. F. Schantz |
| | W. E. Viering (Alt) |
| Sanders Data Systems, Inc | W. Bernstein |
| | T. H. Buchert (Alt) |
| Scientific Apparatus Makers Association | A. Savitsky |
| | J. E. French (Alt) |
| SHARE Inc | T. B. Steel, Jr |
| | R. H. Wahlen (Alt) |
| Society of Certified Data Processors | A. Taylor |
| | A. E. Dubnow (Alt) |
| Sperry UNIVAC | M. W. Bass |
| | C. D. Card (Alt) |
| Systems Engineering Laboratories, Inc | A. E. Roberts |
| | M. Olson (Alt) |
| Telephone Group | V. N. Vaughan, Jr |
| | J. C. Nelson (Alt) |
| | P. G. Wray (Alt) |
| 3M Company | R. C. Smith |
| U.S. Department of Defense | W. L. McGreer |
| | W. B. Rinehuls (Alt) |
| | W. B. Robertson (Alt) |
| Xerox Corporation | J. L. Wheeler |

Subcommittee X3S3 on Data Communications, which developed this standard, had the following members:

G. C. Schutz, Chairman
A. H. Stillman,* Former Chairman
J. L. Wheeler, Vice-Chairman
S. M. Harris, Secretary

| | | |
|---|---|---|
| J. R. Aschenbrenner | D. L. Hanna | R. T. Moore |
| M. W. Baty | D. R. Hearsum | P. E. Muench |
| M. J. Bedford | T. L. Hewitt | L. T. O'Connor |
| R. C. Boepple | W. M. Hornish | N. F. Priebe |
| W. R. Brown | L. G. Kappel | R. R. Propp |
| M. T. Bryngil | P. W. Kiesling, Jr | S. R. Rosenblum |
| M. L. Cain | C. C. Kleckner | J. M. Skaug |
| G. E. Clark | W. E. Landis | H. H. Smith |
| J. W. Conard | D. S. Lindsay | N. E. Snow |
| H. J. Crowley | R. C. Matlack | E. R. Stephan |
| R. Dawson | B. L. Meyer | A. F. Taibl |
| J. L. Dempsey | R. F. Meyer | C. P. Van Lidth de Juede |
| W. F. Emmons | O. C. Miles | G. W. White |

*Deceased.

Task Group X3S34 on Control Procedures, which had technical responsibility for the development of this standard, had the following members:

D. E. Carlson, Chairman
R. F. Meyer, Former Chairman

| | | |
|---|---|---|
| M. W. Baty | J. G. Griffis | D. E. Rogness |
| E. J. Brown | W. E. Hahn | S. R. Rosenblum |
| R. H. G. Chan | J. F. Henchy | P. D. Simpson |
| G. E. Clark | R. E. Huettner | L. T. Snapko |
| M. E. Cook | D. R. Hughes | R. C. Tannenbaum |
| D. D'Andrea | B. Hyman | B. Tymann |
| R. R. Donecker | D. C. Johnson | R. A. Varekamp |
| W. F. Emmons | J. L. Lindinger | G. W. White |
| T. F. Fitzsimons | N. F. Priebe | G. Zoolakis |

Others who contributed to the development of this standard are:

| | | |
|---|---|---|
| J. R. Aschenbrenner | B. J. Griffith* | A. Reszka |
| M. J. Bedford | D. Heiser | J. A. Schwartz |
| E. Berezin | R. Kerker | D. R. Spicer |
| A. B. J. Cuccio | D. A. Kerr | L. A. Tate |
| R. B. Gibson | R. A. Kitchener | K. Terhorst |

*Deceased.

# Contents

516

# American National Standard Procedures for the Use of the Communication Control Characters of American National Standard Code for Information Interchange in Specified Data Communication Links

## 1. Introduction

**1.1 Purpose.** This standard is intended for systems that are controlled through the use of the ten communication control characters provided in American National Standard Code for Information Interchange, X3.4-1968 (ASCII). It offers categories of data communication control procedures for the following application needs and link configurations:

(1) One-way-only message transfer
(2) Two-way alternate message transfer
(3) Two-way simultaneous message transfer
(4) Conversational message interchange
(5) Links with and without error protection
(6) Circuit-switched data communication links
(7) Nonswitched data communication links
(8) Point-to-point communication links
(9) Multipoint communication links

The standard categories of communication control procedures are outlined in Section 2.

**1.2 Scope.** This standard specifies a variety of data communication control procedures to meet the needs of data communication links having different configurations or complexity. Covered in each category are standard control procedures to perform the communication functions necessary for (1) the establishment and termination of transmission between stations on a link and (2) the transfer of messages between stations. Later revisions to this standard may include additional categories.

**1.3 Application.** The standard categories of data communication control procedures specified herein are applicable to a data communication system on a per-link basis. This standard does include, however, guidance on the application of these control procedures to multiple-link data communication systems to augment the primary definition of the control procedures on a link-by-link basis. Compliance with this standard requires operation in one or more categories. The specific categories supported must be stated.

In addition to the procedures contained in this standard, other features of the total system, as, for example, those in the following list, must be considered to ensure data interchange:

(1) Required signaling capabilities (that is, speed, error rate, etc) for communication channels
(2) Signals necessary to control a data set (modem)
(3) The subject of protective time-outs
(4) The need for additional error recovery procedures
(5) The need for additional security of data, polling and selective addresses, control signals, and reply sequences
(6) The possible use of more than one category on a single link
(7) Requirements for compatibility between stations implementing different categories

Such additional features may be the subjects of additional standards or may be left to the designers of a particular system.

Future standards will provide for the following types of procedures: interactive, code independent, and additional multipoint. Other characteristics will include emphasis on high-reliability links and systems, and on data network operations. Additional message-oriented categories will be added as needed.

Data communication systems span a wide group of applications, products, and system configurations. The application of this standard to a particular system requires careful analysis of the requirements of the system and of the procedures provided in this standard. Additions to the procedures provided in this standard may be required.

## 2. Standard Categories of Communication Control Procedures

**2.1 General.** This standard provides for a number of categories of data communication control procedures. Each category consists of two subcategories:

(1) Establishment and termination (see Table 1)
(2) Message transfer (see Table 2)

13

**Table 1**
**Establishment and Termination Subcategories**

| Type of System | Salient Features | Subcategory Designation |
|---|---|---|
| *One-Way-Only Systems* | | |
| Nonswitched multipoint | Master status permanently assigned<br>Single or group selection without replies | 1.1 |
| *Point-to-Point Two-Way Alternate Systems* | | |
| Switched | No identification<br>Calling station has master status initially<br>Terminate<br>Mandatory disconnect | 2.1 |
| Switched | Station identification<br>Calling station has master status initially<br>Terminate<br>Mandatory disconnect | 2.2 |
| Nonswitched | Contention<br>Replies<br>Terminate | 2.3 |
| *Multipoint Two-Way Alternate Systems* | | |
| Centralized operation | Polling<br>Selection (single slave)<br>Control—tributary communication only<br>Return to control on termination | 2.4 |
| Centralized operation | Polling<br>Selection or fast selection (single slave)<br>Control—tributary communication only<br>Return to control on termination | 2.5 |
| Noncentralized operation | Polling<br>Selection (single slave)<br>Tributary—tributary communication permitted<br>Return to control on termination | 2.6 |
| Centralized operation | Polling<br>Selection (multiple slave)<br>Control—tributary communication only<br>Return to control on termination<br>Delivery verification | 2.7 |
| Noncentralized operation | Polling<br>Selection (multiple slave)<br>Tributary—tributary communication permitted<br>Return to control on termination<br>Delivery verification | 2.8 |
| *Point-to-Point Two-Way Simultaneous Systems* | | |
| Switched | Station identification<br>Both stations have concurrent master and slave status<br>Mandatory disconnect | 3.1 |

523

14

Table 2
Message Transfer Subcategories

| Type of System | Salient Features | Subcategory Designation |
|---|---|---|
| Message-oriented | Without replies<br>Without longitudinal checking | A1 |
| Message-oriented | Without replies<br>With longitudinal checking | A2 |
| Message-oriented | With replies<br>Without longitudinal checking | A3 |
| Message-oriented | With replies<br>With longitudinal checking | A4 |
| Message-associated blocking | With longitudinal checking<br>Retransmission of unacceptable blocks<br>Single-character acknowledgment | B1 |
| Message-associated blocking | With longitudinal checking<br>Retransmission of unacceptable blocks<br>Alternating acknowledgments | B2 |
| Message-independent blocking | With longitudinal checking<br>Retransmission of unacceptable blocks<br>Alternating acknowledgments<br>Noncontinuous operation<br>Nontransparent heading and text | C1 |
| Message-independent blocking | With longitudinal checking<br>Retransmission of unacceptable blocks<br>Modulo-8 numbering of blocks and acknowledgments<br>Continuous operation<br>Nontransparent heading and text | C2 |
| Message-independent blocking | With cyclic redundancy checking<br>Retransmission of unacceptable blocks<br>Alternating acknowledgments<br>Noncontinuous operation<br>Transparent heading and text | D1 |
| Conversational | Without blocking<br>Without longitudinal checking | E1 |
| Conversational | With blocking<br>With longitudinal checking | E2 |
| Conversational | With blocking<br>With longitudinal checking<br>With batch transmission capabilities | E3 |
| Message-associated blocking for two-way simultaneous transmission | With longitudinal checking<br>Retransmission of unacceptable blocks<br>Alternating acknowledgments<br>Embedded responses | F1 |
| Message-independent blocking for two-way simultaneous transmission | With longitudinal checking<br>Retransmission of unacceptable blocks<br>Modulo-8 numbering of blocks and acknowledgments<br>Continuous operation<br>Nontransparent heading and text<br>Embedded response | F2 |

524

The subcategories describing establishment and termination procedures are given numeric designations, and the subcategories describing message transfer procedures are given alphanumeric designations. A category is specified by selecting one numeric and one alphanumeric subcategory of control procedures. Thus category 2.1/B2 consists of subcategory 2.1 establishment and termination procedures and subcategory B2 message transfer procedure.

**2.2 Standard Categories.** Establishment and termination subcategory 1.1 may only be used with message transfer subcategories A1 and A2. Establishment and termination subcategory 3.1 may only be used with message transfer subcategories F1 and F2. Other combinations of establishment and termination subcategories and message transfer subcategories are permitted, subject to the detailed provisions contained in the subcategory descriptions (Section 5). Sections 3 and 4 contain generalized discussions of the communication control characters and their usage, but Sections 5 and 6 are to be consulted for the specific usage prescribed for each subcategory.

# 3. General

## 3.1 Communication Control Characters

**3.1.1** The ten communication control characters provided in the coded character set of ASCII and additional communication controls provided through the use of DLE sequences are used to perform the control functions specified in the categories. The character structure and parity sense in these characters and sequences are in accordance with American National Standard Character Structure and Character Parity Sense for Serial-by-Bit Data Communication in the American National Standard Code for Information Interchange, X3.16-1976, and American National Standard for Bit Sequencing of the American National Standard Code for Information Interchange in Serial-by-Bit Data Transmission, X3.15-1976.

**3.1.2** The ten communication control characters and any additional communication controls provided through the use of DLE sequences must not be used as message characters in the heading or text, except as prescribed in 6.10, which describes transparent transmission procedures.

**3.1.3** The communication control characters SOH, STX, and ETX and the control character sequences TSOH, TSTX, and TETX are defined for use as message delimiters and are, therefore, integral parts of a message. Accordingly, these message delimiters must not be modified or deleted by the link control pro-

cedures; that is, these message delimiter characters must be propagated with the message through successive links in a multilink system.

**3.1.4** The communication control characters SOH and STX and the control character sequences SOTB, TSOH, and TSTX are further defined, along with ETB and the sequence TETB, for use as control signals for transmission blocking of messages. Transmission blocking is defined as a link control function, and where SOH, STX, ETB, SOTB, TSOH, TSTX, and TETB are used only for transmission blocking, they may be deleted or modified by the link control procedures. However, the ability to propagate transmission blocking information with the message is recognized as a requirement for systems in which the originator must prescribe blocking formats for the destination station(s). This standard does not describe means for the propagation of this information.

**3.1.5** The communication control characters EOT, ENQ, ACK, NAK, and SYN and the control character sequences DEOT, ACK*N*, TDLE, WACK, SOSS, and TSYN are defined for use on each link independently of their use on other links and, therefore, are not required to be propagated through successive links in a multiple-link system. In multiple-link systems, however, the functional properties of these control characters may be required to be passed through the system. This standard does not describe means for the propagation of this information.

**3.1.6** The character DLE is used exclusively to represent additional communication control functions. Depending upon the particular functions defined for these additional controls, they may be used on either an intralink or an interlink basis.

**3.2 Code Extension Sequences Using DLE.** This section specifies a standard method of providing additional communication control functions through the definition of code extension sequences using DLE.

**3.2.1** The character DLE is used as a prefix to one or more additional ASCII characters to form a sequence representing a communication control function not directly represented by a single control character.

**3.2.2** A code extension sequence begins with DLE, continues with any number of "intermediate" characters (see 3.2.3), and ends with one "final" character (see 3.2.4). Two-character sequences consist of DLE followed by one "final" character.

**3.2.3** The intermediate characters are those in column 2 of the ASCII code table (see Appendix E), that is, SPACE and certain special graphics.

**3.2.4** The final characters are the communication controls in columns 0 and 1 and all those in columns 3, 4, 5, 6, and 7 of the ASCII code table, that is, the communication control characters, the alphabetics, the numerics, and special graphics, except as noted in 3.2.5.

3.2.5 The following characters are reserved as final characters of DLE sequences for use in future extensions of this standard:

(1) Two-character sequences: communication controls and all characters in columns 3 and 7

(2) Three or more character sequences: communication controls and all characters in column 3

The DEL character in column 7 shall never be assigned as part of a DLE sequence.

3.3 Prefixes. In general, supervisory sequences consist of a prefix followed by a communication control character. The prefix may consist of up to 15 characters, other than communication control characters, in order to convey additional information (for example, identity or status). Note, however, that EOT must never have a prefix. Stations implementing a subcategory of control procedures in which replies are used must accept, without confusion, prefix characters to the control character. Interpretation of the prefix, however, must be by prior agreement between the users. A standard interpretation of prefixes is not covered by this document. A prefix must not change the meaning of the associated control character.

3.4 Aborts and Interrupts. At times during data interchange, the sending station may desire to end a block or transmission in an unusual manner such that the receiving station disregards that portion of the block or transmission that has been received. These procedures are called *aborts*.

At other times during data interchange, a receiving station may wish to cause the sending station to stop sending, either temporarily (in order to permit the receiving station to send) or permanently. These procedures are called *interrupts*.

Mandatory use of aborts and interrupts is indicated under category-dependent usage. In subcategories where abort and interrupt procedures are *not* mandatory, systems employing such functions must implement them in the manner prescribed in the following sections. However, in order to remain standard, these systems must be capable of operation without those functions.

3.4.1 Block Abort Procedures

3.4.1.1 Description. The sending station, in the process of sending a block, but before the end of the block, decides to end the block in an unusual manner such that the receiving station(s) will discard the block. Such a procedure is called a *block abort*.

NOTE: If the abort resulted from a fault, the user is cautioned that under some conditions retransmission of the aborted block may not be possible or, if possible, may not correct the

fault. In these cases message cancellation or termination may be performed. This standard does not define procedures for message cancellation.

3.4.1.2 Application

3.4.1.2.1 Block abort may be used by a sending station when, in the process of sending data, there occurs an indication that the data being sent may not be valid. Such an indication would result from a buffer parity check, a card reader check, or a tape read check, or include error indications from the source or media from which the data are being received or read prior to transmission, etc.

3.4.1.2.2 Block abort may be used by a sending station when, in the process of sending a fixed length block, it is determined that the block exceeds the allowed length. Such a condition, if allowed to continue, might overflow a receiving station's buffer or media. The condition could occur as the result of errors in the transmitting mechanism or software, or by the operator.

3.4.1.2.3 Block abort may be used in the message transfer state to cause a temporary text delay after receipt of the previous acknowledgment if the sending station is not capable of transmitting the text of the next transmission block before the predetermined time-out period. The reasons for such a delay might be the unavailability of buffer space or that the speed of the input device is considerably slower than the transmission speed and a full block has not yet been read from the media.

3.4.1.3 Nontransparent Mode Procedure. Block abort is accomplished by the sending station's ending the block (at any time) with the communication control character ENQ. The sending station then halts transmission and waits for a reply. The receiving station(s) detects that the block was ended with ENQ rather than with a normal ending character (ETB or ETX), discards that portion of the block that had been received, and sends a NAK response (with optional prefix) to the sending station and remains in the receive condition.

Following receipt of the NAK response, the sending station will normally reinitiate the transmission with the same or a new block in accordance with the rules of the message transfer subcategory in use and the system discipline being employed. Alternatively, the sending station may choose to initiate an appropriate termination or recovery procedure. Note that when the application requires that the data be transmitted in strict sequence, recovery procedures must be employed if the aborted block cannot be successfully retransmitted.

In the case of a NAK response that is not received, the sending station will time out (expiration of timer

A–see 3.5.2.1) and then will normally send an ENQ character to request a retransmission of the last response. This additional ENQ will help to ensure that the receiving station(s) detects the block abort action. Upon receipt of the NAK response, the sending station will proceed as indicated in the preceding paragraph. Alternatively to sending this additional ENQ, the sending station may choose to reinitiate transmission with the same or a new block, or it may choose to initiate an appropriate termination or recovery procedure. The specific choice of operation will generally be a function of both the message transfer subcategory in use and the system discipline being employed.

In conversational subcategories, where both the initiator and the responder each function alternately as a sending and a receiving station, block abort is accomplished and, with the exception of subcategory E1, responded to in the same way as other nontransparent mode procedures. In subcategory E1, the receiving station discards the portion of the message that has been received and remains in the receive condition without sending a response.

**3.4.1.4 Transparent Mode Procedure.** The transparent mode procedure for block abort is identical to the nontransparent mode procedure of 3.4.1.3 except that the sending station transmits DLE ENQ instead of ENQ. The receiving station(s) thus detects that the block ends with DLE ENQ instead of with the normal ending DLE ETB or DLE ETX. Since DLE ENQ returns the data link to a nontransparent condition, NAK (with optional prefix) remains the only valid response to a block that has been aborted by this procedure.

**3.4.1.5 Category-Dependent Usage**

**3.4.1.5.1** The nontransparent block abort procedure is used in subcategories A3, A4, B1, B2, C1, C2, E3, F1, and F2.

**3.4.1.5.2** The transparent block abort procedure is used in subcategory D1.

**3.4.2 Sending Station Abort Procedures**

**3.4.2.1 Description.** The sending station, in the process of sending several blocks per text, decides to prematurely terminate transmission to the particular station at the end of a block and after receipt of the proper acknowledgment reply. Such a procedure is called a *sending station abort*.

**3.4.2.2 Application**

**3.4.2.2.1** Sending station abort procedures may be used by a sending station when, in the process of sending multiple blocks per text, it determines that it should prematurely terminate transmission to the particular receiving station. Such a determination would be caused by the processor's not receiving the remaining blocks in time from a peripheral device or another communication line; the need for a higher priority mes-

sage to be transmitted to the same or another station; the temporary inability of the sending station to continue transmission; etc.

**3.4.2.2.2** Sending station abort procedures may be used following block abort procedures to accomplish a transmission abort condition; that is, the sending station prematurely terminates the transmission within a transmission block (or text).

**3.4.2.3 Procedures.** Sending station abort procedures are accomplished by the sending station's completing the transmission of a block, for example, ETB, DLE ETB, ENQ, DLE ENQ. Then, upon receipt of the proper acknowledgment reply (ACK, NAK, etc) or a timer-A time-out, the sending station transmits EOT to terminate the transmission to the particular receiving station. The receiving station detects this sending station abort procedure by recognizing receipt of EOT following ETB, ENQ, DLE ETB, or DLE ENQ instead of ETX or DLE ETX. Receipt or transmission of EOT resets all stations on the link to the control phase in a nontransparent mode. These sending station abort procedures may be followed by any procedure that is normal for the control phase. In a multipoint system the control station should regain control and poll/select a station. The disposition of the partial text blocks (which were received without error) is dependent upon the application and system usage.

**3.4.2.4 Category-Dependent Usage.** Sending station abort procedures are used in the following subcategories: A3, A4, B1, B2, C1, C2, D1, E3, F1, and F2.

**3.4.3 Termination Interrupt Procedures**

**3.4.3.1 Description.** The receiving station, after receipt of a message or transmission block, causes the sending station to cease transmission. Such a procedure is called *termination interrupt*.

**3.4.3.2 Application.** Termination interrupt may be used by the receiving station to cause the transmission to be interrupted because it is not in condition to receive. Causes for such inability to receive could include the running out of paper or of cards, a card jam, a hardware malfunction, etc.

**3.4.3.3 Procedures.** Termination interrupt procedures are accomplished by the receiving station's transmitting EOT in lieu of one of its normal responses. This response indicates a negative acknowledgment of the last transmission and the conclusion of a transmission. The EOT resets all stations on the link to the control phase and may be followed by any procedure that is normal for the control phase. In a multipoint system the control station regains control and polls/selects a station.

**3.4.3.4 Category-Dependent Usage.** Termination interrupt procedures are used in the following subcategories: A3, A4, B1, B2, C1, C2, D1, E3, F1, and F2.

### 3.4.4 Reverse Interrupt Procedures (RINT)

**3.4.4.1 Description.** A receiving station may request the sending station to terminate the transmission in progress prematurely in order to facilitate a reversal in the direction of data transfer. Such a procedure is called *reverse interrupt.*

**3.4.4.2 Application**

**3.4.4.2.1** Reverse interrupt procedures may be used by a receiving station to interrupt its receiving of a "batch" message stream so that it may transmit a priority message or messages to the original sending station.

**3.4.4.2.2** In a multipoint environment, the control station, acting as a receiving station, may want to interrupt in order to poll or select another tributary station.

**3.4.4.3 Procedures.** Reverse interrupt procedures may be used by a receiving station only after reception of a block with a valid block check character (BCC) or cyclic redundancy check (CRC). Reverse interrupt procedures are accomplished by the receiving station's transmitting the control character sequence DLE < in lieu of the normal affirmative acknowledgment. This reply is interpreted as an affirmative reply to the last transmission, and it signals a request by the receiving station that the sending station terminate the transmission sequence in progress as soon as the sending station is in such status that it can receive a message without destroying or losing information that may have previously been stored in buffers.

When alternating or modulo-8 acknowledgments are being used, the DLE < sequence may not be repeated by the receiving station to successive transmission blocks without transmitting intervening affirmative acknowledgments (ACK0, ACK1, etc). However, successive DLE < sequences may be transmitted for a given block in response to ENQ's being received for retry purposes.

Upon receipt of DLE <, the sending station should terminate the transmission by transmitting EOT after it has completed transmitting all data that would prevent it from receiving a message. The number of transmission blocks to be transmitted prior to termination is variable and dependent upon station design.

The receipt of DLE < as a response to a sending station's ENQ should be treated as a repeated response if the last valid response received was ACK0 or ACK1. The sending station should continue by transmitting the next block or EOT. If the last valid response was DLE <, the sending station must assume that the last transmitted block was garbled. The sending station should retransmit the previous block.

**3.4.4.4 Category-Dependent Usage.** Reverse interrupt is used in the following subcategories: B2, C1, C2, and D1.

### 3.4.5 Temporary Interrupt Procedure—WACK (Optional)

**3.4.5.1 Description.** After receipt of a message or transmission block, the receiving station affirmatively acknowledges the information received and also indicates that it is temporarily unable to receive additional data by transmitting the WACK control function. The sending station must then delay transmission of any subsequent blocks or messages to that station until it has determined that the receiving station is once again ready to receive. Such a procedure is called *temporary interrupt.*

NOTE: This function is intended for receiving devices that are unable to accept data continuously, that is, devices that become temporarily unable to accept a contiguous data block or message immediately. WACK may also be used as a response to a selection or request for identification supervisory sequence, or to a bid for master status.

**3.4.5.2 Application.** Temporary interrupt may be used by a receiving station to delay the receipt of additional data until it has the opportunity to clear a buffer that will be required for reception of the next block or message.

Temporary interrupt may also be used by a receiving station to provide a temporary time delay during the establishment control procedure (that is, before receiving data) for such terminal functions as motor start-up or accessing I/O units.

**3.4.5.3 Procedures.** Temporary interrupt is accomplished by the receiving station's transmitting the WACK control character sequence consisting of a DLE character followed by a semicolon character. It is transmitted in lieu of the appropriate affirmative acknowledgment (ACK or ACK*N*). The WACK response indicates that the receiving station received the last transmission block or message correctly, but that it is temporarily unable to accept additional data.

Upon receipt of WACK, the sending station may release the information just transmitted from storage and prepare for transmission of the next block or message. However, before actually initiating such transmission, the sending station must first determine the state of readiness of the receiving station by transmitting the control character ENQ to the receiving station.

The receiving station shall respond to ENQ with the WACK sequence if it is still unable to receive. The sending station may then repeat the ENQ query. The receiving station shall continue to transmit the WACK sequence in response to ENQ until it is ready to receive additional data, at which time it shall respond with the appropriate ACK or ACK*N*. Upon receipt of the appropriate ACK or ACK*N*, the sending station shall resume sending additional data.

If no response to an ENQ is received during the ENQ/WACK interchange, the sending station shall exit to an appropriate recovery procedure.

If an invalid response to an ENQ is received during the ENQ/WACK interchange, the sending station shall transmit ENQ *N* (*N* ≥ 0) times in order to solicit a valid response. After *N* unsuccessful attempts, the sending station shall exit to an appropriate recovery procedure.

Alternatively, upon receipt of the WACK sequence from the receiving station, the sending station may, as appropriate:

(1) Initiate a normal termination procedure (if a complete message ending with ETX or DLE ETX has been transmitted)

(2) Initiate a sending station abort function

Temporary interrupt, when used as a response to a selection or identification supervisory sequence, or to a bid for master status, provides a time delay only and contains no indication as to the acceptance or rejection of a previous transmission block or message.

**3.4.5.4 Category-Dependent Usage.** Temporary interrupt (WACK) is optionally applicable in the following subcategories: A3, A4, B1, B2, C1, D1, E3, F1, 2.2. through 2.8, and 3.1.

## 3.5 Recovery Procedures

**3.5.1 General.** Recovery procedures are system guidelines that should be used by all stations, as applicable. However, it is recognized that the detailed method of station mechanization, absolute value of timers, etc, may vary with applications and communication facilities.

In some cases these recovery procedures can only detect the error condition and then notify the operator or the processor program, or both. In more sophisticated cases automatic recovery is partially or completely possible. In still other cases only operators can perform the recovery procedures. Operator recovery procedures are not part of this standard.

**3.5.2 Timers.** Timers are primarily used to indicate when recognition of specific control characters does not occur within specified periods. It is to be noted that the timers specified in this section are functional only and do not necessarily imply a specific implementation. The action taken following a time-out is specified in general terms to provide system protection. The absolute values of the timers are dependent upon system properties (for example, manual versus nonmanual data entry, type of data source/sink, or communication facilities) and are therefore not defined in this standard.

**3.5.2.1 Timer A (Response Timer)**

**3.5.2.1.1 Purpose.** Timer A is used by a sending station to protect against an invalid response or no response.

**3.5.2.1.2 Start.** Timer A is started after the transmission of any ending character where a reply is expected, for example, ENQ, ETB, ETX, DLE ETB, or DLE ETX.

**3.5.2.1.3 Stop.** Timer A is stopped upon receipt of a valid reply from the communication line, for example, ACK, ACK0, ACK1, NAK, or EOT, or upon receipt of STX in conversational subcategories E1, E2, or E3.

**3.5.2.1.4 Time-Out.** When time-out occurs:
(1) Either:
  (a) Retransmit same information (up to *N* times)[1]
  (b) Transmit different information, for example, ENQ or different polling/selection sequence
(2) Transmit EOT when sending station abort procedures are used
(3) Notify operator or processor program, or both
(4) Return to nontransparent mode, if applicable

**3.5.2.2 Timer B (Receive Timer)**

**3.5.2.2.1 Purpose.** Timer B is used by a receiving station to protect against nonrecognition of end-of-block or end-of-text, for example, ETB or ETX, by a receiving station.

**3.5.2.2.2 Start.** Timer B is started upon receipt of start-of-block or start-of-text, for example, SOH or STX (if not preceded by SOH), or DLE SOH or DLE STX (if not preceded by DLE SOH).

**3.5.2.2.3 Restart.** Timer B may be restarted while data continue to be received to permit receipt of variable length blocks.

**3.5.2.2.4 Stop.** Timer B is stopped upon receipt of a valid terminating character or sequence, for example, ETB, ETX, ENQ, DLE ETB, or DLE ETX.

**3.5.2.2.5 Time-Out.** When time-out occurs:
(1) Initiate search for character synchronization in synchronous systems
(2) Prepare to receive another transmission
(3) Notify operator or processor program, or both, and discard incomplete block
(4) Return to nontransparent mode, if applicable

NOTE: For maximum system efficiency, the duration of the response timer (A) should be short, and the receive timer (B) should time-out before the response timer (A).

**3.5.2.3 Timer C (Gross Timer)**

**3.5.2.3.1 Purpose.** Timer C facilitates disconnection procedures of the switched communication line if

---

[1] Retransmission of a data block can result in duplication of a block at the receiving location if a block numbering or other protective scheme is not used.

data transmission stops because DLE EOT is not recognized or because of remote station or communication facility problems.

**3.5.2.3.2 Start.** Timer C is started upon initiation or a receipt of a call, for example, upon receipt of CE ("Ring Indicator")[2] signal from called station's data set going ON and the station's having CD ("Data Terminal Ready")[2] ON.

**3.5.2.3.3 Restart.** Timer C is restarted upon receipt or transmission of any character in asynchronous systems or of the synchronizing sequence in synchronous systems.

**3.5.2.3.4 Stop.** Timer C is stopped either:

(1) Upon receipt or transmission of DLE EOT

(2) When line disconnection occurs, for example, loss of CC ("Data Set Ready")[2]

**3.5.2.3.5 Time-Out.** When time-out occurs:

(1) Disconnect communication circuit

(2) Notify operator or processor program, or both

(3) Return to nontransparent mode, if applicable

**3.5.2.4 Timer D (No-Activity Timer)**

**3.5.2.4.1 Purpose.** Timer D serves as a "no-activity" time-out for all stations on a multistation line.

**3.5.2.4.2 Start or Restart.** Timer D is started or restarted upon receipt or transmission of any character in asynchronous systems or of the synchronizing sequence in synchronous systems.

**3.5.2.4.3 Stop.** Timer D is stopped upon receipt or transmission of EOT/DLE EOT.

**3.5.2.4.4 Time-Out.** When time-out occurs:

(1) Notify operator or processor program, or both

(2) Return to control mode, if applicable

(3) Return to nontransparent mode, if applicable

**3.5.3 Category-Dependent Recovery Procedures**

**3.5.3.1 Subcategories 2.2 through 2.8, and 3.1.** The recovery procedures consist of notifying the operator, processor program, etc.[3]

**3.5.3.2 Subcategories A3 and A4.** For a viable system the functions of timers A, B, and C must be utilized. The value of a timer may vary, depending upon whether it is implemented by hardware, software, or human operator.

The recovery procedures [see Fig. 13 (11) and 14 (11)] following an invalid or no reply may consist of retransmitting the previous message up to *N* times. Then, if there is still no reply, the operator or the processor program, or both, should be notified.

The recovery procedures following no reply after transmittal of prefix ENQ *N* times consist of notifying the operator or processor program, or both.[3]

**3.5.3.3 Subcategories B1, B2, C1, C2, D1, F1, and F2.** The recovery procedures following receipt of NAK after *N* unsuccessful retransmissions consist of notifying the operator or the processor program, or both.[3] Similar recovery procedures are also applicable in the following cases: receipt of invalid or no response after *N* retransmissions, or receipt of invalid or no responses after transmitting prefix ENQ to solicit last ACK/NAK reply.

Timers A, B, and either C or D are required.

**3.5.3.4 Subcategory E1.** The timer-A function is required at the initiator location. The timing may be implemented by hardware or software, or performed by the operator. Also, timer-C and -D functions are required on switched or multipoint networks, respectively.

The recovery procedures consist of notifying the operator or processor program, or both.

**3.5.3.5 Subcategories E2 and E3.** The timers A, B, and either C or D are required. The timing may be implemented by hardware or software, or performed by the operator.

The recovery procedures consist of notifying the operator or the processor program, or both.

**3.6 Synchronization**

**3.6.1 General.** The categories defined in this standard are intended for use where the transmission of characters may be either synchronous or asynchronous. This standard does not specify the means necessary to achieve bit synchronization.

In addition to the control procedures defined in the categories, the character synchronization procedures specified in 3.6.2 are applicable to synchronous systems.

**3.6.2 Synchronous Procedures.** After a period in which no characters have been transmitted on a channel, and prior to the transmission of any other character, a minimum of two SYN characters shall be transmitted.

This standard does not specify the number of SYN characters that a receiver is required to recognize in order to achieve character synchronization.

This standard does not specify procedures for detecting the loss or reestablishment of character synchronization.

When either the SYN character or the DLE SYN character sequence is used as a communication "time-fill" character during a transmission, SYN or DLE SYN may be arbitrarily added at any point in the transmission except:

(1) In DLE sequence

(2) Between ETB, ETX, DLE ETB, and DLE ETX,

---

[2] As designated in EIA RS-232-C-August 1969, Interface between Data Terminal Equipment and Data Communication Equipment Employing Serial Binary Data Interchange.

[3] The operator may retry *N* more times, establish voice communication to the distant station in order to determine trouble, etc.

and the next following block check character (BCC) or cyclic redundancy check (CRC)

DLE SYN may only be used in message transfer subcategory D1 incorporating transparency.

Although deletion of SYN or DLE SYN by a receiving station is not mandatory, removal of added SYN characters or DLE SYN character sequences may be required to avoid disrupting other control procedures (for example, ESC control sequence or fixed-length blocks).

**3.7 Link Block Number (BLK).** The link block number (BLK) is mandatory within message transfer subcategories C2 and F2. When BLK is used, it immediately follows each start-of-block delimiter for all transmission blocks. The BLK character is a single ASCII numeric character that varies from zero through seven. The first transmission block, after completion of the call establishment control procedures, shall be assigned the numeric character one (1). Subsequent transmission blocks will use sequential numbers (2, 3, 4, 5, 6, 7, 0, 1, 2, etc).

The BLK character is added by the transmitting station and is functionally deleted by the receiving station. It is not considered as part of the end-to-end heading or text. In order to obtain good system reliability, BCC must be correct before a decision regarding the validity of the BLK character is made.

If the received transmission block is accepted (that is, contains the proper BLK character, satisfies other accuracy control criteria, etc) and the receiving station is ready to receive another block, it sends an optional prefix followed by ACK*N*.

If the received BLK character is lower than expected, the receiving station discards the received transmission block and sends an optional prefix followed by ACK*N* when the receiving station is ready to receive another block.

If the received BLK character is higher than expected, the receiving station discards the received transmission block and sends an optional prefix followed by NAK when the receiving station is ready to receive another block.

The BLK counter is reset to "one" upon a timer-C or -D time-out or upon transmission or reception of EOT or DLE EOT.

NOTE: In subcategories where BLK is not mandatory, systems employing such a function should implement it in the manner prescribed above. Such implementation may be considered an extension to the standard subcategories.

**3.8 Link Interface.** The concept of link interface is one of a functional rather than a physical interface. Therefore, the link interface within a data terminal may be implemented with the same physical hardware as that which operates on heading or text (for instance, data processing equipment). It is cautioned, therefore, that the characters of an escape sequence that appear in heading or text may become separated by communication blocking characters (ETB BCC, STX, etc).

**3.9 Secondary Channels for Supervisory Use.** A secondary channel is a lower speed backward channel that can be used to provide additional functions of the type described in 3.9.1 through 3.9.5. The use of such functions is optional for message transfer subcategories A1 through A4, B1, B2, C1, D1, and E1 through E3.

**3.9.1 Information Path Assurance Function.** The information path assurance function may be used to prevent a transmitting station from transmitting to an open circuit or to an inoperable or nonexistent receiving station, particularly when a subcategory without replies is being used.

**3.9.1.1 Procedure.** The transmitting station must not transmit data until receipt of an ON condition on the following two data set circuits: "Clear-to-Send"[4] *and* "Secondary Received Line Signal Detector."[5] The receiving station places an ON condition on its "Secondary Request to Send"[6] only when it is ready to receive data.

**3.9.2 Information Path Failure Detection.** The information path failure detection function may be used to detect that the line has become disconnected or that the receiving station has become inoperable, particularly when a subcategory without replies is being used.

**3.9.2.1 Procedure.** An information path failure is indicated to the station transmitting information on the primary channel by an OFF condition on the "Secondary Received Line Signal Detector"[7] until the failure is corrected.

**3.9.3 "Break" Function.** The "break" function is used by a receiving station to notify the transmitting station to stop transmitting (usually immediately).

**3.9.3.1 Procedure.** When in the process of receiving data, the receiving station notifies the transmitting station to stop transmitting by placing an OFF condition

---

[4] Designated in accordance with EIA RS-232-C-August 1969.

[5] Designated in accordance with EIA RS-232-C-August 1969. (NOTE: On some commercially available data sets the function of this circuit is performed by the "Supervisory Received Data" circuit.)

[6] Designated in accordance with EIA RS-232-C-August 1969. (NOTE: On some commercially available data sets the function of this circuit is performed by the "Supervisory Transmitted Data" circuit.)

[7] Designated in accordance with EIA RS-232-C-August 1969. (NOTE: On some commercially available data sets the function of this circuit is performed by the "Supervisory Received Data" circuit.)

on the "Secondary Request to Send"[8] (receiving station) for a fixed time period. Upon detecting this OFF condition, the transmitter enters a recovery procedure.

**3.9.4 Timing.** The absolute values for the time that an ON or OFF condition is applied or detected, or both, depend upon such things as line speed (character time) and application.

**3.9.5 ON/OFF Condition.** The ON and OFF conditions are as defined in EIA RS-232-C-August 1969, Interface between Data Terminal Equipment and Data Communication Equipment Employing Serial Binary Data Interchange.

## 4. Communication Control Characters

**4.1 ASCII Communication Control Characters.** The definitions of the ASCII communication control characters are taken from American National Standard Code for Information Interchange, X3.4-1968 (ASCII).

**4.1.1 SOH (Start of Heading)**

**4.1.1.1 Definition.** SOH is a communication control character used at the beginning of a sequence of characters that constitute a machine-sensible address or routing information. Such a sequence is referred to as the "heading." An STX character has the effect of terminating a heading.

**4.1.1.2 Function.** SOH delimits the start of a message heading.

**4.1.1.3 Category-Dependent Usage**

**4.1.1.3.1** In subcategories B1, B2, and F1, if the heading is subdivided into transmission blocks by the use of ETB, SOH delimits the start of each block that continues transmission of the heading.

**4.1.1.3.2** In subcategories A2, A4, B1, B2, and F1, SOH is used in block check procedures, as described in 4.3.

**4.1.2 STX (Start of Text)**

**4.1.2.1 Definition.** STX is a communication control character that precedes a sequence of characters that is to be treated as an entity and entirely transmitted through to the ultimate destination. Such a sequence is referred to as "text." STX may be used to terminate a sequence of characters (heading) started by SOH.

**4.1.2.2 Function**

**4.1.2.2.1** STX delimits the start of a message text.

**4.1.2.2.2** If a heading precedes the text, STX delimits the end of the message heading.

**4.1.2.3 Category-Dependent Usage**

**4.1.2.3.1** In subcategories A2, A4, B1, B2, E2, E3, and F1, STX is used in block check procedures, as described in 4.3.

**4.1.2.3.2** In subcategories B1, B2, E2, E3, and F1, if the text is subdivided into transmission blocks by the use of ETB, STX delimits the start of each block that continues transmission of the text.

**4.1.2.3.3** In conversational subcategories E1, E2, and E3, the receipt of STX implies acknowledgment of the previous message sent.

**4.1.3 ETX (End of Text)**

**4.1.3.1 Definition.** ETX is a communication control character used to terminate a sequence of characters started with STX and transmitted as an entity.

**4.1.3.2 Function.** ETX delimits the end of a message text.

**4.1.3.3 Category-Dependent Usage**

**4.1.3.3.1** In subcategories A2, A4, B1, B2, E2, E3, and F1, ETX is used in block check procedures, as described in 4.3.

**4.1.3.3.2** In subcategories A2, A4, B1, B2, E2, E3, and F1, ETX signals that the character immediately following is BCC.

**4.1.3.3.3** In subcategories A3, A4, B1, B2, E1, E2, E3, and F1, ETX calls for a reply from the station receiving the message.

**4.1.3.3.4** In subcategories B1, B2, E2, E3, and F1, if the text is subdivided into transmission blocks by the use of ETB, ETX delimits the end of the last transmission block of the message.

**4.1.3.3.5** In subcategory E1, ETX calls for a message reply in response to a message received. In subcategory E2, ETX calls for a NAK or a message reply as a response. In subcategory E3, ETX calls for a message reply or ACK or NAK as a response to a message.

**4.1.4 EOT (End of Transmission)**

**4.1.4.1 Definition.** EOT is a communication control character used to indicate the conclusion of a transmission that may have contained one or more texts and any associated headings.

**4.1.4.2 Function**

**4.1.4.2.1** EOT cancels any previous master/slave or initiator/responder assignment.

**4.1.4.2.2** EOT must never have a prefix.

**4.1.4.2.3** EOT is sent by a master (initiator) station after the completion of the message transfer phase in order to effect a normal termination of the transmission.

**4.1.4.2.4** EOT is sent by a master (initiator) station prior to the completion of the message transfer phase in order to effect a sending station abort function.

---

[8] Designated in accordance with EIA RS-232-C-August 1969. (NOTE: On some commercially available data sets the function of this circuit is performed by the "Supervisory Transmitted Data" circuit.)

**4.1.4.2.5** EOT is sent by a slave (responder) station in place of one of its normal responses in order to effect a termination interrupt function.

**4.1.4.3 Category-Dependent Usage**

**4.1.4.3.1** When used as the indicator of a normal transmission termination, EOT has identical usage in all establishment and termination subcategories; however, the control procedures that follow EOT are different in different subcategories.

**4.1.4.3.2** In subcategories A3, A4, B1, B2, C1, C2, D1, E3, F1, and F2, EOT is used by the master (initiator) station in sending station abort procedures, as described in 3.4.2.

**4.1.4.3.3** In subcategories A3, A4, B1, B2, C1, C2, D1, E3, F1, and F2, EOT is used by the slave (responder) station in termination interrupt procedures, as described in 3.4.3.

**4.1.5 ETB (End of Transmission Block)**

**4.1.5.1 Definition.** ETB is a communication control character used to indicate the end of a block of data for communication purposes. ETB is used for blocking data where the block structure is not necessarily related to the processing format.

**4.1.5.2 Function.** ETB delimits the end of a transmission block (heading or text).

**4.1.5.3 Category-Dependent Usage**

**4.1.5.3.1** In subcategories B1, B2, C1, C2, E2, E3, F1, and F2, ETB is used in block check procedures, as described in 4.3.

**4.1.5.3.2** In subcategories B1, B2, C1, C2, E2, E3, F1, and F2, ETB signals that the character immediately following is a BCC.

**4.1.5.3.3** In subcategories B1, B2, C1, C2, E2, E3, F1, and F2, ETB calls for a reply from the station receiving the transmission block.

**4.1.6 ENQ (Enquiry)**

**4.1.6.1 Definition.** ENQ is a communication control character used in data communication systems as a request for a response from a remote station. It may be used to obtain station status, or as a "who are you?" (WRU) to obtain identification, or both.

**4.1.6.2 Function.** ENQ is used to solicit a response from a station.

**4.1.6.3 Category-Dependent Usage**

**4.1.6.3.1** In subcategories 2.2 through 2.8, and 3.1, ENQ calls for a reply indicating the slave station's readiness to receive.

**4.1.6.3.2** In subcategory 2.2 and 3.1, ENQ calls for the slave station to transmit its identity as a prefix to the reply.

**4.1.6.3.3** In subcategories 2.4 through 2.8, ENQ is the last character of a polling supervisory sequence.

**4.1.6.3.4** In subcategories 2.4 through 2.8, ENQ is the last character of a selection supervisory sequence.

**4.1.6.3.5** In subcategory 2.3 (contention system, ENQ is used to bid for master status.

**4.1.6.3.6** In subcategories A3, A4, B1, B2, C1, C2, D1, F1, and F2, ENQ calls for a retransmission of the last reply (ACK/NAK).

**4.1.6.3.7** In subcategories 2.7 and 2.8, ENQ calls for a delivery verification to be sent by the addressed slave station.

**4.1.6.3.8** In subcategories A3, A4, B1, B2, C1, C2, E3, F1, and F2, ENQ is used by the master (initiator) station in block abort procedures, as described in 3.4.1.

**4.1.7 ACK (Acknowledgment)**

**4.1.7.1 Definition.** ACK is a communication control character transmitted by a receiver as an affirmative response to a sender.

**4.1.7.2 Function.** ACK is transmitted by a slave (responder) station as an affirmative reply. Supplementary information may be prefixed to ACK.

**4.1.7.3 Category-Dependent Usage**

**4.1.7.3.1** In subcategories 2.2 through 2.8, and 3.1, ACK indicates that the slave (responder) station is ready to receive.

**4.1.7.3.2** In subcategories A3, A4, B1, E2, and E3, ACK indicates that the last message or transmission block was accepted and that the slave (responder) station is ready to receive.

**4.1.8 NAK (Negative Acknowledge)**

**4.1.8.1 Definition.** NAK is a communication control character transmitted by a receiver as a negative response to the sender.

**4.1.8.2 Function.** NAK is transmitted by a slave (responder) station as a negative reply. Supplementary information may be prefixed to NAK.

**4.1.8.3 Category-Dependent Usage**

**4.1.8.3.1** In subcategories 2.2 through 2.8, and 3.1, NAK indicates that the selected station is not ready to receive.

**4.1.8.3.2.** In subcategories A3, A4, B1, B2, C1, C2, D1, E2, E3, F1, and F2, NAK indicates that the last message or transmission block was not accepted and that the slave (responder) station is ready to receive. (In some subcategories the station may receive a retransmission of the unaccepted block.)

**4.1.9 SYN (Synchronous Idle)**

**4.1.9.1 Definition.** SYN is a communication control character used by a *synchronous* transmission system in the absence of any other character to provide a signal from which synchronism may be achieved or retained.

**4.1.9.2 Function**

**4.1.9.2.1** SYN is used to achieve and maintain character synchronism in synchronous communication systems.

**4.1.9.2.2** SYN is used as a communication "time-fill" character during periods in a transmission when no other characters are available to send.

NOTE: Since it is expected that many systems will be arranged to add or delete SYN characters during the process of communication, the SYN character should not be used for time-fill or media-fill functions that are to be conveyed through the system. Although deletion of SYN characters by a receiving station is not mandatory, deletion may be a system requirement to avoid conflict with other control procedures.

**4.1.9.3 Category-Dependent Usage.** In subcategories A2, A4, B1, B2, C1, C2, E2, E3, F1, and F2, SYN is used in block check procedures, as described in 4.3.

### 4.1.10 DLE (Data Link Escape)

**4.1.10.1 Definition.** DLE is a communication control character that will change the meaning of a limited number of contiguously following characters. It is used exclusively to provide supplementary controls in data communication networks.

**4.1.10.2 Function.** Additional control functions are represented by a contiguous sequence of characters, the first character of which is always DLE (see 3.2).

**4.1.10.3 Category-Dependent Usage.** DLE sequences are used to achieve:

(1) Mandatory disconnect in subcategories 2.1, 2.2, and 3.1

(2) Alternating acknowledgments in subcategories B2, C1, D1, and F1

(3) Numbered acknowledgments in subcategories C2 and F2

(4) The start-of-block delimiter in subcategories C1, C2, and F2

(5) Transparent block delimiting sequences (TSOH, TSTX, TETX, and TETB) for subcategory D1

(6) Transparent synchronous idle in subcategory D1

(7) Transparent DLE in data in subcategory D1

(8) The WACK function option in subcategories A3, A4, B1, B2, C1, D1, E3, F1, 2.2 through 2.8, and 3.1

(9) Start of supervisory sequence delimiter in subcategories F1 and F2

(10) The block abort function in subcategory D1

(11) The reverse interrupt function in subcategories B2, C1, C2, and D1

### 4.2 Control Sequences Represented by Code Extension

#### 4.2.1 DEOT (Mandatory Disconnect)

**4.2.1.1 Representation.** DEOT is represented by the character sequence: DLE followed by EOT.

**4.2.1.2 Definition.** DEOT is a communication control sequence used to signal that a disconnect of a circuit-switched communication channel must be initiated.

**Table 3**
**Representation of ACK$N$**

| Abbreviation | Representation |
|---|---|
| ACK0 | DLE followed by 0 |
| ACK1 | DLE followed by 1 |
| ACK2 | DLE followed by 2 |
| ACK3 | DLE followed by 3 |
| ACK4 | DLE followed by 4 |
| ACK5 | DLE followed by 5 |
| ACK6 | DLE followed by 6 |
| ACK7 | DLE followed by 7 |

**4.2.1.3 Function**

**4.2.1.3.1** DEOT indicates the end of a transmission and signals the mandatory disconnect of a circuit-switched communication channel.

**4.2.1.3.2** A master station relinquishes its right to transmit by sending DEOT.

**4.2.1.3.3** The receipt of DEOT cancels any previous selection of a slave station.

**4.2.1.3.4** DEOT must never have a prefix.

**4.2.1.4 Category-Dependent Usage**

**4.2.1.4.1** DEOT is used in subcategories 2.1 through 2.3, and 3.1. When used in subcategory 2.3, DEOT has no additional meaning from EOT.

**4.2.1.4.2** In recovery procedures within subcategories 2.1, 2.2, and 3.1, DEOT may be used to initiate a disconnect of a communication channel.

#### 4.2.2 ACK$N$ (Acknowledgment $N$)

**4.2.2.1 Representation.** ACK$N$ is represented as shown in Table 3.

**4.2.2.2 Definition.** ACK$N$ is a set of communication control sequences transmitted by a receiver as affirmative responses to a sender. These are used in lieu of ACK to number affirmative replies.

**4.2.2.3 Function.** ACK$N$ is transmitted by a slave station as a numbered affirmative reply. Supplementary information may be prefixed to ACK$N$.

**4.2.2.4 Category-Dependent Usage**

**4.2.2.4.1** In subcategories 2.2 through 2.6, when used in conjunction with message transfer subcategory B2, C1, C2, or D1, ACK0 must be used in lieu of ACK (in response to a selection or identification supervisory sequence, or to a bid for master status). ACK0 must also be employed with subcategory 3.1

**4.2.2.4.2** In subcategories B2, C1, D1, and F1, ACK0 and ACK1 must be used in lieu of ACK. ACK1 is used as the affirmative reply to the first block. The affirmative reply to subsequent blocks alternates between ACK0 and ACK1.

**4.2.2.4.3** In subcategory C2 and F2, ACK$N$ must be used in lieu of ACK. ACK1 is used as the affirmative reply to the first block, ACK2 to the second block, through ACK7, restarting with ACK0.

### 4.2.3 SOTB (Start of Transmission Block)

**4.2.3.1 Representation.** SOTB is represented by the character sequence: DLE followed by an equals (=) character.

**4.2.3.2 Definition.** SOTB is a communication control sequence that denotes the start of a block of data for communication purposes. SOTB is used for blocking data where the block structure is not necessarily related to the processing format.

**4.2.3.3 Function.** SOTB delimits the start of a transmission block.

**4.2.3.4 Category-Dependent Usage**

**4.2.3.4.1** In subcategories C1, C2, and F2, SOTB is used in block check procedures, as described in 4.3.

**4.2.3.4.2** In subcategories C2 and F2, SOTB signals that the character immediately following is a block number (BLK).

### 4.2.4 TSOH (Transparent Start of Heading)

**4.2.4.1 Representation.** TSOH is represented by the character sequence: DLE followed by SOH.

**4.2.4.2 Definition.** TSOH is a communication control sequence used at the beginning of a sequence of characters that constitutes address or routing information for a message of transparent text. Such a sequence is referred to as the "heading." A DLE STX sequence has the effect of terminating such a heading.

**4.2.4.3 Function**

**4.2.4.3.1** TSOH delimits the start of a message heading for a block of transparent heading or transparent heading and text.

**4.2.4.3.2** TSOH places the data link in a condition to handle a transparent heading.

**4.2.4.4 Category-Dependent Usage**

**4.2.4.4.1** In subcategory D1, if the heading is subdivided into transmission blocks by the use of DLE ETB, DLE SOH delimits the start of each block that continues transmission of the heading.

**4.2.4.4.2** In subcategory D1, DLE SOH is used in block check procedures as described in 4.3.

### 4.2.5 TSTX (Transparent Start of Text)

**4.2.5.1 Representation.** TSTX is represented by the character sequence: DLE followed by STX.

**4.2.5.2 Definition.** TSTX is a communication control sequence that precedes a block of transparent text and that may be used to terminate a transparent heading and initiate a block of transparent text.

**4.2.5.3 Function**

**4.2.5.3.1** TSTX delimits the start of a block of transparent text.

**4.2.5.3.2** If a heading precedes it, TSTX delimits the end of the transparent heading.

**4.2.5.3.3** TSTX places the data link in a condition to handle transparent text.

**4.2.5.4 Category-Dependent Usage**

**4.2.5.4.1** In subcategory D1, TSTX is used in block check procedures, as described in 4.3.

**4.2.5.4.2** In subcategory D1, if the text is subdivided into transmission blocks by the use of TETB, TSTX delimits the start of each transparent block that continues transmission of a text.

### 4.2.6 TETX (Transparent End of Text)

**4.2.6.1 Representation.** TETX is represented by the character sequence: DLE followed by ETX.

**4.2.6.2 Definition.** TETX is a communication control sequence that terminates the last block of a transparent text.

**4.2.6.3 Function**

**4.2.6.3.1** TETX delimits the end of a transparent message text.

**4.2.6.3.2** TETX returns the data link to a nontransparent condition.

**4.2.6.4 Category-Dependent Usage**

**4.2.6.4.1** In subcategory D1, TETX is used in block check procedures, as described in 4.3.

**4.2.6.4.2** In subcategory D1, TETX signals that the two characters immediately following constitute a CRC.

**4.2.6.4.3** In subcategory D1, TETX calls for a reply from the slave station receiving the block.

**4.2.6.4.4** In subcategory D1, if the text is subdivided into transmission blocks by the use of TETB, TETX delimits the end of the last transmission block of the message.

### 4.2.7 TETB (Transparent End of Block)

**4.2.7.1 Representation.** TETB is represented by the character sequence: DLE followed by ETB.

**4.2.7.2 Definition.** TETB is a communication control sequence used to indicate the end of a block of transparent data for communication purposes. TETB is used for the blocking of transparent data where the block structure is not necessarily related to the processing format.

**4.2.7.3 Function.** TETB delimits the end of a transparent transmission block (heading or text) that is not the last block of a message.

**4.2.7.4 Category-Dependent Usage**

**4.2.7.4.1** In subcategory D1, TETB is used in block check procedures, as described in 4.3.

**4.2.7.4.2** In subcategory D1, TETB signals that the two characters immediately following constitute a CRC.

**4.2.7.4.3** In subcategory D1, TETB calls for a reply from the slave station receiving the block.

**4.2.7.4.4** In subcategory D1, TETB returns the data link to a nontransparent condition.

### 4.2.8 TSYN (Transparent Synchronous Idle)

**4.2.8.1 Representation.** TSYN is represented by the character sequence: DLE followed by SYN.

**4.2.8.2 Definition.** TSYN is a communication control sequence used by a synchronous transmission system in the absence of any other character to provide a signal from which synchronism may be maintained while sending and receiving transparent data.

**4.2.8.3 Function**

**4.2.8.3.1** TSYN is used to maintain character synchronization within transparent data in synchronous communication systems.

**4.2.8.3.2** TSYN is used as a communication "time-fill" during periods in a transparent transmission when no other characters are available to send.

NOTE: Since it is expected that many systems will be arranged to add or delete DLE SYN characters during the process of communication, the DLE SYN character sequence should not be used for time-fill or media-fill functions that are to be conveyed through the system. Although deletion of DLE SYN characters by a receiving station is not mandatory, deletion may be a system requirement to avoid conflict with other control procedures.

**4.2.8.4 Category-Dependent Usage.** In subcategory D1, TSYN is used in block check procedures as described in 4.3.

### 4.2.9 TDLE (Data DLE in Transparent Data)

**4.2.9.1 Representation.** TDLE is represented by the character sequence: DLE followed by DLE.

**4.2.9.2 Definition.** TDLE is a sequence that indicates that a DLE character bit pattern is being transmitted as data within a transparent text.

**4.2.9.3 Function.** TDLE is used within transparent text whenever a data DLE character occurs in the transparent data.

**4.2.9.4 Category-Dependent Usage**

**4.2.9.4.1** In subcategory D1, TDLE is used to make a communication link transparent to the transmission of a data DLE character within a transparent header or text.

**4.2.9.4.2** In subcategory D1, TDLE is substituted for a data DLE character within a transparent header or text by a transmitting station.

**4.2.9.4.3** In subcategory D1, DLE is substituted for a TDLE character sequence within a transparent header or text by a receiving station.

### 4.2.10 WACK (Wait after Positive Acknowledgment)

**4.2.10.1 Representation.** WACK is represented by the character sequence: DLE followed by a semicolon (;) character.

**4.2.10.2 Definition.** WACK is a communication control sequence transmitted by a receiver as an alternative affirmative response to a sender and also as an indication that the receiver is temporarily unable to receive data.

**4.2.10.3 Function**

**4.2.10.3.1** WACK may be used as a positive acknowledgment to indicate that the block or message was received properly but that the receiving station is subsequently temporarily unable to receive. Supplementary information may be prefixed to WACK.

**4.2.10.3.2** WACK may be used as a positive acknowledgment to a selection or identification supervisory sequence, or to a bid for master status, to indicate that the receiver is temporarily unable to receive. Supplementary information may be prefixed to WACK.

**4.2.10.4 Category-Dependent Usage.** Subject to the procedures of 3.4.5.3, WACK is an option that may be used as indicated in 4.2.10.4.1 through 4.2.10.4.6.

**4.2.10.4.1** In subcategories A3, A4, B1, and E3, WACK may be used in lieu of ACK as a positive response to information received.

**4.2.10.4.2** In subcategories B2, C1, D1, and F1, WACK may be used in lieu of ACK0 or ACK1 (as appropriate) as a positive response to information received.

**4.2.10.4.3** In subcategories 2.7 and 2.8, WACK may be used in lieu of ACK as a positive response to a delivery verification sequence.

**4.2.10.4.4** In subcategories 2.2 and 3.1, WACK may be used in lieu of ACK or ACK0 (as appropriate) as a positive response to an identification supervisory sequence.

**4.2.10.4.5** In subcategory 2.3, WACK may be used in lieu of ACK or ACK0 (as appropriate) as a positive response to a bid for master status.

**4.2.10.4.6** In subcategories 2.4 through 2.8, WACK may be used in lieu of ACK or ACK0 (as appropriate) as a positive response to a selection supervisory sequence.

### 4.2.11 SOSS (Start of Supervisory Sequence)

**4.2.11.1 Representation.** SOSS is represented by the character sequence: DLE followed by a colon (:) character.

**4.2.11.2 Definition.** SOSS is a communication control function used to indicate the beginning of a supervisory sequence.

**4.2.11.3 Function**

**4.2.11.3.1** SOSS delimits the beginning of a supervisory sequence.

**4.2.11.3.2** SOSS temporarily halts any BCC (or CRC) block checking summation that may be in progress until the supervisory sequence is terminated.

**4.2.11.4 Category-Dependent Usage.** In subcategories F1, F2, and 3.1, SOSS is used to facilitate the embedding of slave responses within the header or text of a message in a two-way simultaneous transmission environment.

#### 4.2.12 TENQ (Transparent Block Abort)

**4.2.12.1 Representation.** TENQ is represented by the character sequence: DLE followed by ENQ.

**4.2.12.2 Definition.** TENQ is a communication control sequence sent by the master station to signify that the transparent text in the process of being transmitted is being aborted.

**4.2.12.3 Function**

4.2.12.3.1 TENQ abnormally terminates the transmission of a block of transparent text.

4.2.12.3.2 TENQ returns the data link to a non-transparent condition.

4.2.12.3.3 TENQ terminates the CRC block check summation and procedure.

4.2.12.3.4 TENQ causes the receiving slave station to discard the block.

**4.2.12.4 Category-Dependent Usage.** In subcategory D1, TENQ aborts the transmission of a block in progress and calls for a NAK reply from the slave station receiving the block.

#### 4.2.13 RINT (Reverse Interrupt)

**4.2.13.1 Representation.** RINT is represented by the character sequence: DLE followed by a "less-than" ($<$) character.

**4.2.13.2 Definition.** RINT is a communication control sequence sent by a slave (responder) station, in lieu of an affirmative response, as a request for premature termination of the current transmission by the master (initiator) station in order to facilitate a reversal in the direction of data transfer.

**4.2.13.3 Function**

4.2.13.3.1 RINT serves as an affirmative acknowledgment for the transmission block or message just received.

4.2.13.3.2 RINT requests that the master (initiator) station terminate its transmission so that the slave (responder) station can become master (initiator) and initiate a transmission of its own.

**4.2.13.4 Category-Dependent Usage.** In subcategories B2, C1, C2, and D1, RINT may be used in lieu of ACK*N* responses after receipt of alternate transmission blocks to request the master (initiator) station to terminate its transmission.

### 4.3 Block Checking

#### 4.3.1 BCC (Block Check Character)

**4.3.1.1 Definition.** BCC is a character added at the end of a message or transmission block to facilitate error detection.

4.3.1.1.1 The BCC is generated by taking a binary sum independently (without carry) on each of the seven individual levels of the transmitted code (b1 through b7).

4.3.1.1.2 In each code level the number of "one"

bits (including any in the BCC) is caused to be even. Thus, the sense of longitudinal parity is said to be even.

4.3.1.1.3 The correct value of the character parity bit of the BCC is defined as that which makes the sense of character parity the same as for text characters.

**4.3.1.2 Function**

4.3.1.2.1 The summation to obtain the block check character is started at the first appearance in a message or transmission block of either SOH, STX, or SOTB. The SOH or STX character or the SOTB sequence is not included in the summation. STX may appear once within a message or transmission block started with SOH. In this case the BCC summation is started with SOH, and STX is included in the summation.

4.3.1.2.2 All characters, except SYN, transmitted after the start of the BCC summation are included in the summation, through and including the control character that signals that the next following character is BCC.

4.3.1.2.3 The BCC is transmitted as the next character following the transmission of each end-of-text or end-of-block delimiter (ETX or ETB). SYN must not be transmitted between ETX or ETB and BCC.

**4.3.1.3 Category-Dependent Usage.** The block checking procedures are defined identically for all subcategories that use BCC.

NOTE: Since the BCC summation may produce any of the 128 code combinations, receiving stations must avoid erroneous interpretation of BCC.

#### 4.3.2 CRC (Cyclic Redundancy Check)

**4.3.2.1 Definition.** CRC designates the bits added at the end of a message or transmission block to facilitate error control. The degree of the generator polynomial may be $16 + 8n$ (where $n$ is a nonnegative integer greater than or equal to zero). If the degree of the generator polynomial is 16, then at least

$$CRC-16 = x^{16} + x^{15} + x^2 + 1$$

and

$$CCITT = x^{16} + x^{12} + x^5 + 1$$

must be available.

**4.3.2.2 Function**

4.3.2.2.1 The CRC generation is started by the first appearance in a message or transmission block of either TSOH or TSTX.

4.3.2.2.2 The CRC generation is stopped by the appearance of TETB, TETX, or TENQ.

4.3.2.2.3 Table 4 indicates which control characters are included in the CRC generation. All characters

**Table 4
Control Sequences — Inclusion in CRC Accumulation**

| Character of Sequence | Included in CRC Accumulation | |
| --- | --- | --- |
| | Yes | No |
| TSYN | – | DLE SYN |
| TSOH | – | DLE SOH |
| TSTX* | – | DLE STX |
| TSTX† | STX | DLE |
| TETB | ETB | DLE |
| TETX | ETX | DLE |
| TDLE | DLE (one) | DLE (one) |

*If not preceded within the same block by transparent heading information.

†If preceded within the same block by transparent heading information.

not shown in the table are included in the CRC generation.

**4.3.2.3 Usage.** Those bits of a transmission block that are included in the CRC (information characters plus included control characters) correspond to the coefficients of a message polynomial having terms from $x^{(n-1)}$ down to $x^{16}$ (where $n$ equals the total number of included bits in a block or sequence). This polynomial is divided modulo 2 by the generating polynomial. The check bits correspond to the coefficients of the terms from $x^{15}$ to $x^0$ in the remainder polynomial found at the completion of this division. The complete block, consisting of the information bits followed by the check bits, corresponds to the coefficients of a polynomial that is integrally divisible in modulo-2 fashion by the generating polynomial.

At the transmitter the information bits are subjected to an encoding process equivalent to a division by the generator polynomial. The resulting remainder is transmitted to the line immediately after the information bits, as delimited by TETB or TETX, commencing with the highest order bits.

At the receiver the incoming block is subjected to a decoding process equivalent to a division by the generator polynomial, which, in the absence of errors, will result in a zero remainder. If the division results in other than a zero remainder, errors are indicated.

The preceding processes may conveniently be carried out by a 16-stage cyclic shift register with appropriate feedback gates (see Appendix B, B3.5), which is set to the all 0 position before starting to process each block; at the receiver, after the processing of a block, the all 0 condition indicates error-free reception.

**4.3.2.4 Category-Dependent Usage.** These CRC procedures are defined for use in transparent message transfer subcategory D1.

NOTE: Since the CRC summation may produce any of the code combinations, receiving stations must avoid erroneous interpretation of CRC as a control sequence.

## 5. Subcategories of Establishment and Termination Control Procedures

**5.1 Introduction.** This section describes a set of subcategories of establishment and termination control procedures. Each subcategory description defines how the communication control characters or character sequences are to be used in that particular establishment and termination control procedure to perform the functions required. Such functions may include polling, selection, contention, replies, termination, and disconnect. Narrative descriptions of these functions and their use can be found in the appendixes.

Each subcategory description includes a flowchart that depicts the establishment and termination control procedures. Primary paths are shown by heavy lines. Prefixes are shown as either mandatory or optional with the optional prefixes given in parentheses. The functions and operations depicted in the flowcharts are referenced in the narrative description of each subcategory by a bracketed figure citation that includes the specific reference number given in the chart.

For the purposes of this document, "nonswitched" means that no procedures are included for terminal identification or for disconnecting a communication circuit. Subcategories intended for use on nonswitched circuits may be used on switched circuits provided the required additional control functions are performed manually or by some other means external to the communication control procedures. Also, subcategories intended for use on switched circuits may be used on nonswitched circuits provided there is prior agreement regarding which station is the master station and which station is the slave station initially.

When operating the conversational subcategories (E1, E2, and E3) on a nonswitched line with point-to-point operation, an establishment procedure is not employed for some applications. The initiator may proceed directly to the message transfer procedure as shown in subcategory 2.1.

**5.2 Subcategory 1.1: One-Way Only, Nonswitched Multipoint**

**5.2.1 General Description.** Subcategory 1.1 is primarily applicable to a link consisting of a master station and one or more receive-only stations on a nonswitched multipoint link. Fig. 1 is a flowchart of subcategory 1.1. A selection function is included to designate slave station(s) to receive a transmission. A terminate function is provided at the end of the transmission to negate the slave status of those station(s) that have been selected. The system operates one-way only; therefore, there are no replies.

**5.2.2 Establishment.** Prior to the establishment of

transmission [Fig. 1 (1)], tributary stations monitor the link for their assigned selection supervisory sequence.

To select a slave station, the master station sends a selection supervisory sequence [Fig. 1 (2)], consisting of a prefix. When a tributary station detects its assigned selection supervisory sequence, it assumes slave status.

The prefix part of the selection supervisory sequence may include selection of a single station or a group of stations. The selection function may be repeated [Fig. 1 (3)] as many times as necessary to select all of the desired slave stations. There are no replies from slave stations.

**5.2.3 Message Transfer.** Only message transfer subcategories A1 and A2 may be employed with establishment and termination subcategory 1.1.

**5.2.4 Termination.** The master station transmits EOT [Fig. 1 (4)] to release selected stations from their slave status in preparation for the next transmission.

## 5.3 Subcategory 2.1: Two-Way Alternate, Switched Point-to-Point without Establishment (Identification) Procedure

**5.3.1 General Description.** Subcategory 2.1 is applicable to a link that has been made via a circuit-switched network. Initially, the calling station is the master station and the called station is the slave station. No identification procedure is provided; the master station proceeds directly to message transfer.

The master (calling) station may relinquish its master status to the called station immediately after call establishment or following message transfer.

Messages may be interchanged between the two stations until at some point the master station sends the

mandatory disconnect sequence to initiate termination of the circuit connection.

Fig. 2 is a flowchart of subcategory 2.1. The circuit connection procedure [Fig. 2 (1)] and the circuit disconnection procedure [Fig. 2 (9)] shown are not subjects of this standard but are indicated for reference.

**5.3.2 Establishment.** No establishment procedures are employed; the master station proceeds directly [Fig. 2 (2)] to message transfer. Alternatively, the calling station may relinquish its master status to the called station [Fig. 2 (3), (4), (5)] to permit the called station to send the first message.

**5.3.3 Message Transfer.** Message transfer subcategories A1 through A4, B1, B2, C1, C2, D1, and E1 through E3 may be employed with establishment and termination subcategory 2.1.

**5.3.4 Termination.** The termination procedure includes:

(1) A terminate function, which defines the end of a transmission and effects an interchange of master/slave status for the next transmission

(2) A mandatory disconnect function, which initiates the release of the circuit connection

**5.3.4.1 Terminate.** Upon completion of the message transfer operation, the master station may bypass the terminate function [Fig. 2 (7)] and proceed directly to the mandatory disconnect function [Fig. 2 (8)]. Alternatively, the master station may send EOT [Fig. 2 (5)]. The master station then assumes slave status. The slave station, upon receiving EOT, assumes master status.

If the new master station has a message for the new slave station, it may proceed [Fig. 2 (6)] to the message transfer procedure.

If the new master station has no message for the new

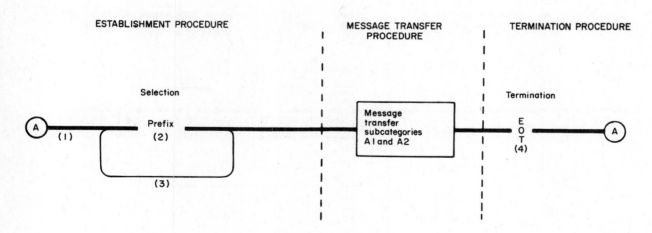

**Fig. 1**
**Subcategory 1.1: One-Way Only, Nonswitched Multipoint**

slave station, it initiates [Fig. 2 (8)] the mandatory disconnect function.

**5.3.4.2 Disconnect.** The master station transmits the mandatory disconnect sequence DEOT [Fig. 2 (8)] and then initiates a local disconnect of the switched circuit [Fig. 2 (9)]. When DEOT is received, the slave station also initiates a local disconnect of the switched circuit.

The master station can abort the transmission at any time by sending the DEOT.

**5.3.5 Aborts, Interrupts, and Recovery Procedures.** Refer to 3.4 and 3.5.

### 5.4 Subcategory 2.2: Two-Way Alternate, Switched Point-to-Point with Establishment (Identification) Procedure

**5.4.1 General Description.** Subcategory 2.2 is applicable to a link that has been established via a circuit-switched network. Initially, the calling station is the master station and the called station is the slave station.

To ensure that a proper circuit connection has been made, the stations go through an identification procedure. After proper identification, the master station begins transmission.

After transmission from the master (calling) station to the slave (called) station is completed, the calling station may relinquish master status to the called station. If the new master station has traffic to send, it may perform the identification procedure or it may proceed with the transmission of message data.

Messages may be interchanged between the two stations until at some point the master station sends the mandatory disconnect sequence and then initiates termination of the circuit connection.

Fig. 3 is a flowchart of subcategory 2.2. The circuit connection procedure [Fig. 3 (1)] and the circuit disconnection procedure [Fig. 3 (15)] shown are not subjects of this standard but are indicated for reference.

**5.4.2 Establishment.** An identification procedure is required to ensure connection to the proper station. The master station requests the identity of the slave station by sending an identification supervisory sequence consisting of an optional prefix followed by ENQ [Fig. 3 (2)]. This prefix may include master station identity, device selection, etc. This identification procedure serves to obtain the identity and status of the slave station.

When the slave station detects ENQ, it sends one of two replies:

(1) If the slave station is ready to receive, it sends a prefix that includes its identity followed by an affirmative reply [Fig. 3 (4)].[9] Upon receipt of a valid reply, the master station proceeds with message transfer. Alternatively, after receiving ACK [Fig. 3 (4)],[9] the calling station can immediately ask the called station to transmit first [Fig. 3 (17)].

(2) If the slave station is not ready to receive, it sends a prefix that includes its identity followed by NAK [Fig. 3 (5)]. Upon detecting the NAK, the master station may try again [Fig. 3 (7)], may proceed to the mandatory disconnect function [Fig. 3 (8), (13)], may exit to a recovery procedure [Fig. 3 (8), (14)], or may relinquish master status by sending EOT [Fig. 3 (18), (9)].

If the master station receives an invalid reply or no

---

[9] When used with message transfer subcategory B2, C1, C2, or D1, the affirmative reply is ACK0 in lieu of ACK.

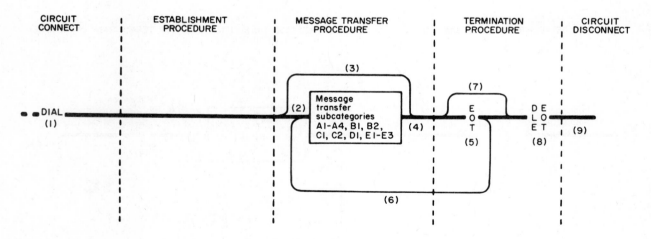

**Fig. 2**
**Subcategory 2.1: Two-Way Alternate, Switched Point-to-Point**

540

reply [Fig. 3 (6)], it may send the identification supervisory sequence again [Fig. 3 (7)], or proceed to either the recovery procedures [Fig. 3 (8), (14)] or to the mandatory disconnect function [Fig. 3 (8), (13)]. $N$ retries ($N \geqslant 0$) may be made to verify the identity and status of the slave station.

In some systems the initial inquiry from the master station may be implied [Fig. 3 (3)] by the initial seizure of the switched circuit. The called station sends a reply [Fig. 3 (4)]. A prefix to the reply identifies the called station and may contain other information.

**5.4.3 Message Transfer.** Message transfer subcategories A1 through A4, B1, B2, C1, C2, D1, and E1 through E3 may be employed with establishment and termination subcategory 2.2.

**5.4.4 Termination.** The termination procedure includes:

(1) A terminate function, which defines the end of a transmission and effects an interchange of master/slave status for the next transmission

(2) A mandatory disconnect function, which initiates the release of the circuit connection

**5.4.4.1 Terminate.** Upon completion of the identification or message transfer operation, the master station may bypass the terminate function [Fig. 3 (16)] and proceed directly to the mandatory disconnect function [Fig. 3 (13)]. Alternatively, the master station may send EOT [Fig. 3 (9)]. The master station

then assumes slave status. Upon receiving EOT, the slave station assumes master status.

If the new master station has a message for the new slave station, it initiates [Fig. 3 (10)] the identification function [Fig. 3 (2)] described above to determine the readiness of the new slave station to receive. Alternatively, it may proceed directly [Fig. 3 (11)] to the message transfer procedure.

If the new master station has no message for the new slave station, it initiates [Fig. 3 (12)] the mandatory disconnect function [Fig. 3 (13)].

**5.4.4.2 Disconnect.** The master station transmits the mandatory disconnect sequence DEOT [Fig. 3 (13)] and initiates a local disconnect of the switched circuit [Fig. 3 (15)]. When DEOT is received, the slave station also initiates a local disconnect of the switched circuit.

The master station can abort the transmission at any time by sending DEOT.

**5.4.5 Aborts, Interrupts, and Recovery Procedures.** Refer to 3.4 and 3.5.

**5.5 Subcategory 2.3: Two-Way Alternate, Nonswitched Point-to-Point**

**5.5.1 General Description.** Subcategory 2.3 is applicable to systems in which two stations are on a nonswitched point-to-point link and where both stations may contend for master status and either seize it if the other station is not also attempting to seize it. A differ-

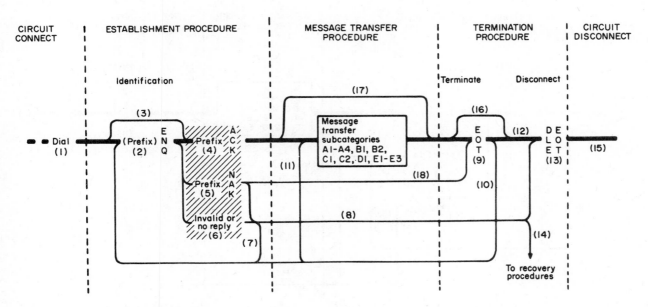

NOTE: Crosshatched area is slave response.

**Fig. 3**
**Subcategory 2.2: Two-Way Alternate, Switched Point-to-Point with Identification**

ent time-out is employed by each station to stagger reattempts to attain master status if an initial attempt has not broken contention. Fig. 4 is a flowchart of subcategory 2.3.

**5.5.2 Establishment.** Prior to the establishment of transmission [Fig. 4 (1)], neither station has master status, but either or both stations may bid for master status.

A station wishing to transmit a message bids for master status by sending ENQ without a prefix [Fig. 4 (2)]. There may occur cases where both stations bid for master status simultaneously. Recovery from this condition is system dependent and is not covered herein (see Appendix B, B2.4).

A station that has not sent ENQ but that has received ENQ takes the following action:

(1) Inhibits the sending of ENQ to bid for master status

(2) If ready to receive, assumes slave status and sends an optional prefix followed by the appropriate affirmative reply [Fig. 4 (3)] [9]

(3) If not ready to receive, sends an optional prefix followed by NAK [Fig. 4 (4)]

Upon receipt of an affirmative reply, the bidding sta-

tion assumes master status and proceeds with message transfer.

Upon receipt of NAK, the bidding station reinitiates a bid for master status [Fig. 4 (2)]. The station reinitiates its bid $M$ $(M \geqslant 0)$ times [Fig. 4 (5)] or exits to a recovery procedure [Fig. 4 (6), (10)].

In the case of an invalid or no reply [Fig. 4 (7)] to ENQ, the bidding station reinitiates its bid for master status [Fig. 4 (2)]. The station reinitiates its bid $N$ $(N \geqslant 0)$ times [Fig. 4 (8)]. After the defined number of unsuccessful bids [Fig. 4 (9)], the station exits to a recovery procedure [Fig. 4 (10)].

**5.5.3 Message Transfer.** Message transfer subcategories A1 through A4, B1, B2, C1, C2, D1, and E1 through E3 may be employed with establishment and termination subcategory 2.3.

**5.5.4 Termination.** The master station transmits EOT [Fig. 4 (11)] to indicate it has no more data to transmit. EOT negates the master/slave relationship that existed for the last transmission and returns the station to contention [Fig. 4 (1)].

Use of DEOT in subcategory 2.3 is permitted to facilitate use of terminal equipment capable of operating on switched circuits. In this subcategory, however, DEOT has the same meaning as EOT.

**5.5.5 Aborts, Interrupts, and Recovery Procedures.** Refer to 3.4 and 3.5.

---

[9] When used with message transfer subcategory B2, C1, C2, or D1, the affirmative reply is ACK0 in lieu of ACK.

542

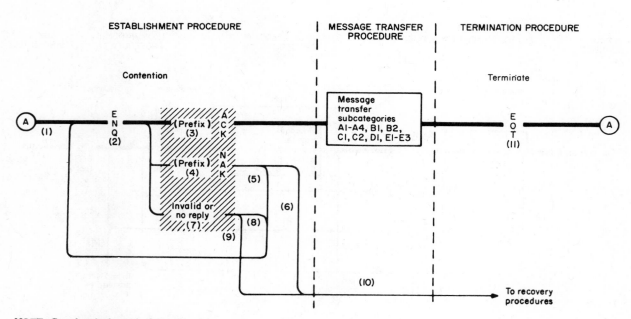

NOTE: Crosshatched area is slave response.

**Fig. 4**
**Subcategory 2.3: Two-Way Alternate, Nonswitched Point-to-Point**

## 5.6 Subcategory 2.4: Two-Way Alternate, Nonswitched Multipoint with Centralized Operation

**5.6.1 General Description.** Subcategory 2.4 is applicable to systems where one or more tributary stations are on a nonswitched multipoint link and all messages are delivered via a control station. One station is designated as the control station. The control station may assume master status and deliver messages to a single selected slave station, or it may assign master status to any one of the tributary stations and then assume the slave status itself to accept the transmission. Tributary-to-tributary transmission is not permitted. The control station monitors the link at all times to ensure orderly operation. Fig. 5 is a flowchart of subcategory 2.4.

**5.6.2 Establishment.** Prior to the establishment of transmission [Fig. 5 (1)], the control station is designated as the master station, and none of the tributary stations have slave status. The control station may either:

(1) Poll [Fig. 5 (2)] to assign master status to one of the tributary stations

(2) Select a slave station [Fig. 5 (3), (7)] to receive a transmission

**5.6.2.1 Polling.** The control station sends a polling supervisory sequence [Fig. 5 (2)]. The prefix identifies a single tributary station and may also include other information. ENQ defines the end of the polling supervisory sequence.

A tributary station detecting its assigned polling supervisory sequence assumes master status and responds in one of two ways:

(1) If the station has a message to send, it initiates message transfer [Fig. 5 (4)]. The control station assumes slave status.

(2) If the station has no message to send, it sends EOT [Fig. 5 (5), (14)], terminating its master status. Master status reverts to the control station, which may then poll or select the next station.

If the control station receives an invalid or no response [Fig. 5 (6)], it terminates by sending EOT [Fig. 5 (14)] prior to resuming polling or selection.

**5.6.2.2 Selection.** The control station sends a selection supervisory sequence [Fig. 5 (7)]. The prefix selects a single tributary station and may also include other information. ENQ defines the end of the selection supervisory sequence.

A tributary station detecting its assigned selection

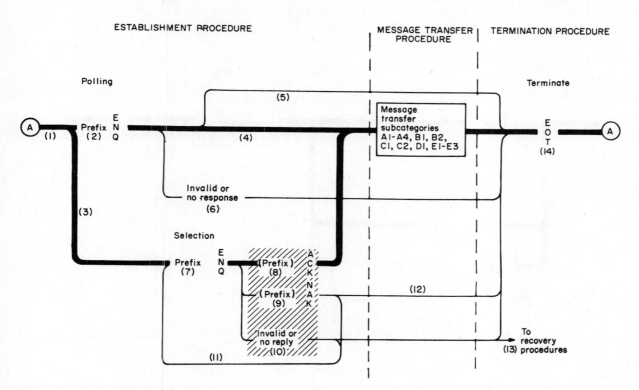

NOTE: Crosshatched area is slave response.

**Fig. 5**
**Subcategory 2.4: Two-Way Alternate, Nonswitched Multipoint with Centralized Operation**

supervisory sequence assumes slave status and sends one of two replies:

(1) If the station is ready to receive, it sends an optional prefix followed by the appropriate affirmative reply [Fig. 5 (8)].[9] Upon detecting this reply, the master station proceeds with message transfer.

(2) If the station is not ready to receive, it sends an optional prefix followed by NAK [Fig. 5 (9)]. Upon detecting NAK, the master station may either select the same station [Fig. 5 (11), (7)] or initiate a terminate function [Fig. 5 (12), (14)].

If the master station receives an invalid or no reply [Fig. 5 (10)], it may select the same station [Fig. 5 (11), (7)], exit to a recovery procedure [Fig. 5 (13)], or initiate a terminate function [Fig. 5 (14)]. $N$ retries ($N \geqslant 0$) may be made to select a station.

**5.6.3 Message Transfer.** Message transfer subcategories A1 through A4, B1, B2, C1, C2, D1, and E1 through E3 may be employed with establishment and termination subcategory 2.4.

**5.6.4 Termination.** The master station transmits EOT [Fig. 5 (14)] to indicate that it has no more messages to transmit. EOT negates the master/slave status of

---

[9] When used with message transfer subcategory B2, C1, C2, or D1, the affirmative reply is ACK0 in lieu of ACK.

both stations and returns master status to the control station.

NOTE: In some network configurations, tributary stations may be unable to detect transmissions from another tributary station. For this reason it may be necessary for the control station to send EOT prior to proceeding.

**5.6.5 Aborts, Interrupts, and Recovery Procedures.** Refer to 3.4 and 3.5.

**5.7 Subcategory 2.5: Two-Way Alternate, Nonswitched Multipoint with Centralized Operation and Fast Select**

**5.7.1 General Description.** Subcategory 2.5 is applicable to systems where one or more tributary stations are on a nonswitched multipoint link and all messages are delivered via a control station. One station is designated as the control station. The control station may assume master status and deliver messages to a single selected slave station, or it may assign master status to any one of the tributary stations and then assume the slave status itself to accept the transmission. Tributary-to-tributary transmission is not permitted. The control station monitors the link at all times to ensure orderly operation. This subcategory also provides the additional function of fast selection of one or more stations. Fig. 6 is a flowchart of subcategory 2.5.

544

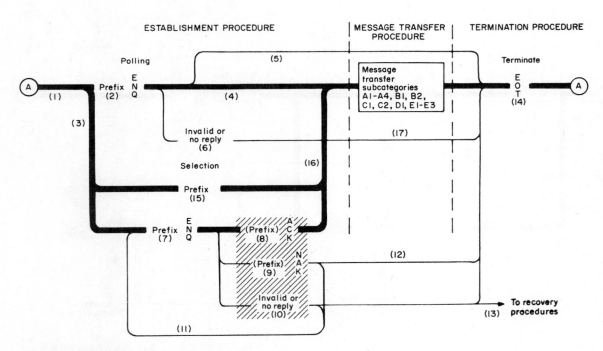

NOTE: Crosshatched area is slave response.

**Fig. 6**
**Subcategory 2.5: Two-Way Alternate, Nonswitched Multipoint with Centralized Operation and Fast Select**

**5.7.2 Establishment.** Prior to the establishment of transmission [Fig. 6 (1)], the control station is designated as the master station and none of the tributary stations have slave status. The control station may either:

(1) Poll [Fig. 6 (2)] to assign master status to one of the tributary stations

(2) Select a slave station [Fig. 6 (3), (7), or (3), (15)] to receive a transmission

**5.7.2.1 Polling.** The control station sends a polling supervisory sequence [Fig. 6 (2)]. The prefix identifies a single tributary station and may also include other information. ENQ defines the end of the polling supervisory sequence.

A tributary station detecting its assigned polling supervisory sequence assumes master status and responds in one of two ways:

(1) If the station has a message to send, it initiates message transfer [Fig. 6 (4)]. The control station assumes slave status.

(2) If the station has no message to send, it sends EOT [Fig. 6 (5), (14)], terminating its master status. Master status reverts to the control station, which may then poll or select the next station.

If the control station receives an invalid or no reply [Fig. 6 (6)], it terminates by sending EOT [Fig. 6 (17), (14)] prior to resuming polling or selection.

**5.7.2.2 Selection with Response.** The control station sends a selection supervisory sequence [Fig. 6 (7)]. The prefix selects a single tributary station and may also include other information. ENQ defines the end of the selection supervisory sequence.

A tributary station detecting its assigned selection supervisory sequence assumes slave status and sends one of two replies:

(1) If the station is ready to receive, it sends an optional prefix followed by the appropriate affirmative reply [Fig. 6 (8)].[9] Upon detecting this reply, the master station proceeds with message transfer.

(2) If the station is not ready to receive, it sends an optional prefix followed by NAK [Fig. 6 (9)]. Upon detecting NAK, the master station may either select the same station [Fig. 6 (11), (7)] or initiate a terminate function [Fig. 6 (12), (14)].

If the master station receives an invalid or no reply [Fig. 6 (10)], it may select the same station [Fig. 6 (11), (7)], exit to a recovery procedure [Fig. 6 (13)], or initiate a terminate function [Fig. 6 (14)]. $N$ retries ($N \geqslant 0$) may be made to select a station.

**5.7.2.3 Fast Select.** Alternatively, the master (control) station may send a selection supervisory sequence [Fig. 6 (15)] without an ENQ. This selects tributary station(s) and may include other information. The master station proceeds [Fig. 6 (16)] directly to a message transfer operation. When more than one station is selected [Fig. 6 (15)], only message transfer subcategories A1 and A2 may be used.

**5.7.3 Message Transfer.** Message transfer subcategories A1 through A4, B1, B2, C1, C2, D1, and E1 through E3 may be employed with establishment and termination subcategory 2.5 unless more than one station is selected without response.

**5.7.4 Termination.** The master station transmits EOT [Fig. 6 (14)] to indicate that it has no more messages to transmit. EOT negates the master/slave status of both stations and returns master status to the control station.

NOTE: In some network configurations, tributary stations may be unable to detect transmissions from another tributary station. For this reason, it may be necessary for the control station to send EOT prior to proceeding.

**5.7.5 Aborts, Interrupts, and Recovery Procedures.** Refer to 3.4 and 3.5.

## 5.8 Subcategory 2.6: Two-Way Alternate, Nonswitched Multipoint with Noncentralized Operation

**5.8.1 General Description.** Subcategory 2.6 is applicable to a system where a number of tributary stations are on a nonswitched multipoint link. One station is designated as the control station. The control station may assume master status and deliver messages to a single selected slave station, or it may assign master status to any one of the tributary stations for the purpose of delivering messages to a single selected slave station. A tributary station having master status may select and deliver messages to another tributary station or the control station. The control station monitors the link at all times and takes the necessary action to ensure orderly operation. Fig. 7 is a flowchart of subcategory 2.6.

**5.8.2 Establishment.** Prior to the establishment of transmission [Fig. 7 (1)], the control station is designated as master and none of the tributary stations have slave status. The control station may either:

(1) Poll [Fig. 7 (2)] to assign master status to one of the tributary stations

(2) Select a slave station [Fig. 7 (3), (4)] to receive a transmission

**5.8.2.1 Polling.** The control station sends a polling supervisory sequence [Fig. 7 (2)]. The prefix identifies a single tributary station and may also include other information. ENQ defines the end of the polling supervisory sequence.

A tributary station detecting its assigned polling su-

---

[9] When used with message transfer subcategory B2, C1, C2, or D1, the affirmative reply is ACK0 in lieu of ACK.

pervisory sequence assumes master status and responds in one of two ways:

(1) If the station has message(s) to send, it initiates a selection supervisory sequence [Fig. 7 (4)].

(2) If the station has no message(s) to send, it sends EOT [Fig. 7 (5), (14)], terminating its master status. Master status reverts to the control station.

If the control station detects an invalid or no reply [Fig. 7 (6)], it terminates by sending EOT [Fig. 7 (10), (14)] prior to resuming polling or selection.

**5.8.2.2 Selection.** The master station sends a selection supervisory sequence [Fig. 7 (4)]. The prefix selects a single tributary station and may also include other information. ENQ defines the end of the selection supervisory sequence.

A station detecting its assigned selection supervisory sequence assumes slave status and sends one of two replies:

(1) If the station is ready to receive, it sends an optional prefix followed by the appropriate affirmative reply [Fig. 7 (7)].[9] Upon detecting this reply, the master station proceeds with message transfer.

(2) If the station is not ready to receive, it sends an optional prefix followed by NAK [Fig. 7 (8)]. Upon

---

[9] When used with message transfer subcategory B2, C1, C2, or D1, the affirmative reply is ACK0 in lieu of ACK.

detecting NAK, the master station may either select the same station [Fig. 7 (11), (4)] or initiate a terminate function [Fig. 7 (12), (14)].

If the master station receives an invalid or no reply [Fig. 7 (9)], it may select the same station [Fig. 7 (11), (4)], exit to a recovery procedure [Fig. 7 (13)], or initiate a terminate function [Fig. 7 (10), (14)]. $N$ ($N \geqslant 0$) retries may be made to select a station.

**5.8.3 Message Transfer.** Message transfer subcategories A1 through A4, B1, B2, C1, C2, and D1 may be employed with establishment and termination subcategory 2.6.

**5.8.4 Termination.** The master station transmits EOT [Fig. 7 (14)] to indicate that it has no more messages to transmit. EOT negates the master/slave status of both stations and returns master status to the control station.

**5.8.5 Aborts, Interrupts, and Recovery Procedures.** Refer to 3.4 and 3.5.

**5.9 Subcategory 2.7: Two-Way Alternate, Nonswitched Multipoint with Centralized Operation and Multiple Slave Transmission**

**5.9.1 General Description.** Subcategory 2.7 is applicable to systems where one or more tributary stations are on a nonswitched multipoint link and all messages

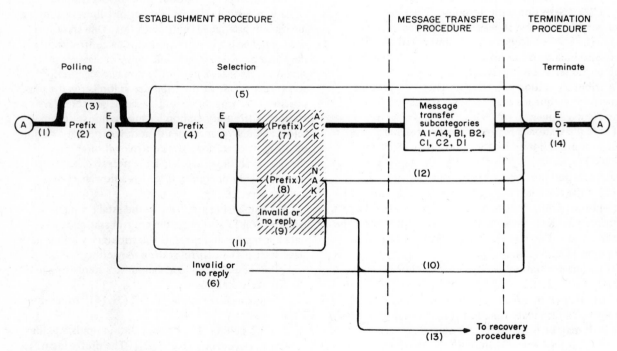

NOTE: Crosshatched area is slave response.

**Fig. 7**
**Subcategory 2.6: Two-Way Alternate, Nonswitched Multipoint with Noncentralized Operation**

37

are delivered via a control station. One station is designated as the control station. The control station may assume master status and deliver messages to one or more individually selected slave stations, or it may assign master status to any one of the tributary stations and then assume slave status itself to accept the transmission. Tributary-to-tributary transmission is not permitted. The control station monitors the link at all times to ensure orderly operation.

The control station may select a multiplicity of slave stations to receive a transmission. In addition, a delivery verification function is provided so that at the end of a transmission to a multiplicity of slave stations, the control station can receive a reply from each selected slave station.

Fig. 8 is a flowchart of subcategory 2.7.

**5.9.2 Establishment.** Prior to the establishment of transmission [Fig. 8 (1)], the control station is designated as the master station and none of the tributary stations have slave status. The control station may either:

(1) Poll [Fig. 8 (2)] to assign master status to one of the tributary stations

(2) Select a slave station [Fig. 8 (3), (7)] to receive a transmission

**5.9.2.1 Polling.** The control station sends a polling supervisory sequence [Fig. 8 (2)]. The prefix identifies a single tributary station and may also include other information. ENQ defines the end of the polling supervisory sequence.

A tributary station detecting its assigned polling supervisory sequence assumes master status and responds in one of two ways:

(1) If the station has a message to send, it initiates message transfer [Fig. 8 (4)]. The control station assumes slave status.

(2) If the station has no message to send, it sends EOT [Fig. 8 (5), (24), (22)] terminating its master status. Master status reverts to the control station.

If the control station receives an invalid or no response [Fig. 8 (6)], it terminates by sending EOT [Fig. 8 (24), (22)] prior to resuming polling or selection.

**5.9.2.2 Selection.** The control station sends a selection supervisory sequence [Fig. 8 (7)]. The prefix selects a single tributary station and may also include

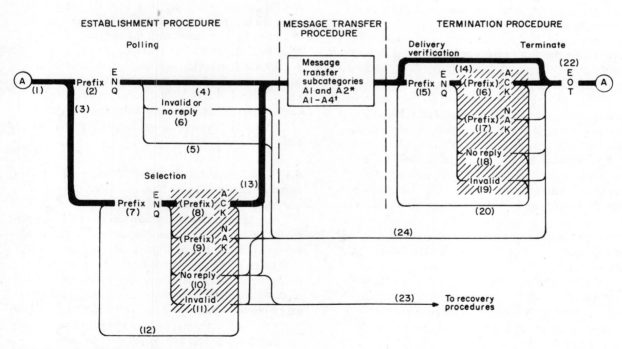

NOTE: Crosshatched area is slave response.

*Control station sending to tributary station(s).

†Tributary station sending to control station.

**Fig. 8**
**Subcategory 2.7: Two-Way Alternate, Nonswitched Multipoint with Centralized Operation and Multiple Slave Transmission**

other information. ENQ defines the end of the selection supervisory sequence.

A tributary station detecting its assigned selection supervisory sequence assumes slave status and sends one of two replies:

(1) If the station is ready to receive, it sends an optional prefix followed by ACK [Fig. 8 (8)]. Upon detecting this reply, the master station either selects another tributary station [Fig. 8 (12), (7)] or proceeds with message transfer [Fig. 8 (13)].

(2) If the station is not ready to receive, it sends an optional prefix followed by NAK [Fig. 8 (9)] and thereby relinquishes slave status. If the master station receives NAK or no reply [Fig. 8 (9) or (10)], it may select another or the same tributary station [Fig. 8 (12), (7)] or terminate [Fig. 8 (24), (22)]. $N$ retries ($N \geq 0$) may be made to select a station for which NAK or no response has been received.

Alternatively, if one or more tributary stations have been selected and have properly responded with ACK, the master station may proceed with an appropriate message transfer [Fig. 8 (13)]. In this case, it must be noted that all tributary stations with either an outstanding unresolved NAK or no response will require subsequent selection and retransmission to avoid loss of traffic.

If the master station receives an invalid response to a selection supervisory sequence [Fig. 8 (11)], it may terminate by sending EOT [Fig. 8 (24), (22)]. Selected slave stations relinquish slave status.

Alternatively, for both invalid or no reply the master station may exit to a recovery procedure [Fig. 8 (23)].

Upon completion of a successful selection function, the master station proceeds with message transfer.

**5.9.3 Message Transfer.** Message transfer subcategories A1 and A2 may be employed with establishment and termination subcategory 2.7 for deliveries by the control station acting as a master station. Message transfer subcategories A1 through A4 may be employed with this establishment and termination subcategory for transmission by a tributary station with master status.

**5.9.4 Termination.** The termination procedure consists of:

(1) A delivery verification function

(2) A terminate function to return the link to the idle condition

**5.9.4.1 Delivery Verification.** Two conditions are described. In the case of a tributary station with master status sending to the control station, verification of delivery, if employed, is an inherent function of the **message transfer subcategory** used. Hence, after message transfer, the master station proceeds [Fig. 8 (14)] to the termination function [Fig. 8 (22)].

In the case of the control station acting as a master station, a delivery verification function is employed. The master station sends a delivery verification supervisory sequence [Fig. 8 (15)] consisting of a prefix followed by ENQ. The prefix identifies a single slave station. The control character ENQ defines the end of the delivery verification supervisory sequence. Only tributary stations having slave status should respond to delivery verification supervisory sequences.

A slave station detecting its assigned delivery verification supervisory sequence sends one of two replies:

(1) If the slave station receives the message properly, it sends an optional prefix followed by ACK [Fig. 8 (16)]. Upon detecting ACK, the master station sends a delivery verification supervisory sequence for another slave station [Fig. 8 (20)] or proceeds with the terminate function [Fig. 8 (22)].

(2) If the slave station does not receive the message properly, it sends an optional prefix followed by NAK [Fig. 8 (17)]. Upon detecting NAK, the master station sends the delivery verification supervisory sequence for another slave station [Fig. 8 (20)] or proceeds with the terminate function [Fig. 8 (22)].

If the master station receives either no reply [Fig. 8 (18)] or an invalid reply [Fig. 8 (19)], it may either request a reply from the same or another slave station [Fig. 8 (20)] or proceed with the terminate function [Fig. 8 (22)].

**5.9.4.2 Terminate.** To terminate, the master station transmits EOT [Fig. 8 (22)]. EOT negates the master or slave status of all tributary stations and returns master status to the control station.

**5.9.5 Aborts, Interrupts, and Recovery Procedures.** Refer to 3.4 and 3.5.

**5.10 Subcategory 2.8: Two-Way Alternate, Non-switched Multipoint with Noncentralized Operation and Multiple Slave Transmission**

**5.10.1 General Description.** Subcategory 2.8 is applicable to systems where a number of tributary stations are on a nonswitched multipoint link. One station is designated as the control station. The control station may assume master status and deliver messages to one or more individually selected slave stations, or it may assign master status to any one of the tributary stations so that the tributary station may transmit messages to one or more individually selected slave stations. A tributary station having master status may select and deliver messages to other tributary stations or to the control station. The control station monitors the link at all times to ensure orderly operation.

Both the tributary stations and the control station may individually select a multiplicity of slave stations to receive a transmission. In addition, a delivery verification function is provided so that at the end of a transmission to a multiplicity of slave stations, the master station can receive a reply from each selected slave station.

Fig. 9 is a flowchart of subcategory 2.8.

**5.10.2 Establishment.** Prior to the establishment of transmission [Fig. 9 (1)], the control station is designated as master and none of the tributary stations have slave status. The control station may either:

(1) Poll [Fig. 9 (2)] to assign master status to one of the tributary stations

(2) Select a slave station [Fig. 9 (3), (4)] to receive a transmission

**5.10.2.1 Polling.** The control station sends a polling supervisory sequence [Fig. 9 (2)]. The prefix identifies a single tributary station and may also include other information. ENQ defines the end of the polling supervisory sequence.

A tributary station detecting its assigned polling supervisory sequence assumes master status and responds in one of two ways:

(1) If the station has a message to send, it initiates a selection supervisory sequence [Fig. 9 (4)].

(2) If the station has no message to send, it sends EOT [Fig. 9 (5), (24), (22)] terminating its master status. Master status reverts to the control station.

If the control station detects an invalid or no reply [Fig. 9 (6)], it terminates by sending EOT [Fig. 9 (24), (22)] prior to resuming polling or selection.

**5.10.2.2 Selection.** The master station sends a selection supervisory sequence [Fig. 9 (4)]. The prefix selects a single tributary station and may also include other information. ENQ defines the end of the selection supervisory sequence.

A station detecting its assigned selection supervisory sequence assumes slave status and sends one of two replies:

(1) If the station is ready to receive, it sends an optional prefix followed by ACK [Fig. 9 (8)]. Upon detecting this reply, the master station either selects another station [Fig. 9 (12), (4)] or proceeds with message transfer [Fig. 9 (13)].

(2) If the station is not ready to receive, it sends an optional prefix followed by NAK [Fig. 9 (9)] and thereby relinquishes slave status. If the master station receives NAK or no reply [Fig. 9 (9) or (10)], it may select another or the same tributary station [Fig. 9 (12), (4)] or terminate [Fig. 9 (24), (22)]. $N$ retries ($N \geq 0$) may be made to select a station for which NAK or no response has been received.

Alternatively, if one or more tributary stations have been selected and have properly responded with ACK, the master station may proceed with an appropriate message transfer [Fig. 9 (13)]. In this case, it must be noted that all tributary stations with either an outstanding unresolved NAK or no response will require subsequent selection and retransmission to avoid loss of traffic.

549

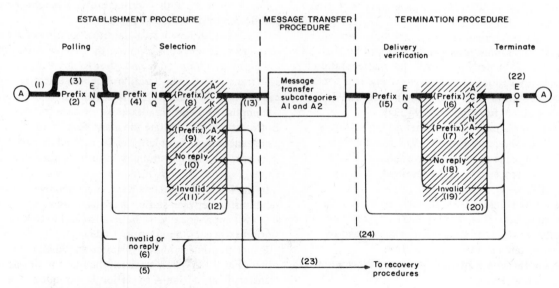

NOTE: Crosshatched area is slave response.

**Fig. 9**
**Subcategory 2.8: Two-Way Alternate, Nonswitched Multipoint with Noncentralized Operation and Multiple Slave Transmission**

If the master station receives an invalid response [Fig. 9 (11)], it may terminate by sending EOT [Fig. 9 (24), (22)], and any selected slave stations relinquish their slave status.

Alternatively, for both invalid or no reply the master station may exit to a recovery procedure [Fig. 9 (23)].

Upon completion of a successful selection function, the master station proceeds with message transfer.

**5.10.3 Message Transfer.** Message transfer subcategories A1 and A2 may be employed with establishment and termination subcategory 2.8.

**5.10.4 Termination.** The termination procedure consists of:

(1) A delivery verification function

(2) A terminate function to return the link to the idle condition

**5.10.4.1 Delivery Verification.** Upon completion of message transfer, the master station sends a delivery verification supervisory sequence [Fig. 9 (15)] consisting of a prefix followed by ENQ. The prefix identifies a single slave station. The control character ENQ defines the end of the delivery verification supervisory sequence. Only stations having slave status should respond to delivery verification supervisory sequences.

A slave station detecting its assigned delivery verification supervisory sequence sends one of two replies:

(1) If the slave station receives the message properly, it sends an optional prefix followed by ACK [Fig. 9 (16)]. Upon detecting ACK, the master station sends a delivery verification supervisory sequence for another slave station [Fig. 9 (20), (15)] or proceeds with the terminate function [Fig. 9 (22)].

(2) If the slave station does not receive the message properly, it sends an optional prefix followed by NAK [Fig. 9 (17)]. Upon detecting NAK, the master station sends a delivery verification supervisory sequence for another slave station [Fig. 9 (20), (15)] or proceeds with the terminate function [Fig. 9 (22)].

If the master station receives either no reply [Fig. 9 (18)] or an invalid reply [Fig. 9 (19)], it requests a reply from the same or another slave station [Fig. 9 (20), (15)] or proceeds with the terminate function [Fig. 9 (22)].

**5.10.4.2 Terminate.** To terminate, the master station transmits EOT [Fig. 9 (22)]. EOT negates the master or slave status of all tributary stations and returns master status to the control station.

**5.10.5 Aborts, Interrupts, and Recovery Procedures.** Refer to 3.4 and 3.5.

**5.11 Subcategory 3.1: Two-Way Simultaneous, Switched Point-to-Point**

**5.11.1 General Description.** Subcategory 3.1 is applicable to systems with two stations on a switched point-to-point full-duplex link where both stations may simultaneously transmit information to each other. To accomplish such operation, each station must have the capability of maintaining concurrent master and slave status, that is, master status on its transmit side and slave status on its receive side. The link is established for two-way simultaneous transmissions by (1) either station calling the other to establish the connection and (2) both stations executing an identification procedure. The transfer of messages in either direction is time and subject matter independent of message transfers in the other direction.

Establishment and termination subcategory 3.1 may only be used with message transfer subcategories F1 and F2, which permit slave responses (supervisory sequences) to be transmitted either between (interleaved) or within (embedded) messages or transmission blocks in accordance with specified rules. Such systems, constructed for two-way simultaneous operations, require that outstanding slave response obligations be satisfied at the earliest possible time (interleaved or embedded) consistent with the message transfer subcategory mode of operation and before the initiation of any new master-originated function. In no case may a slave response be embedded in a master-originated supervisory sequence. Subject to these and other specified message transfer subcategory constraints, either station may initiate and proceed with the transfer of a message or a transmission block independent of the transfers being initiated by or received from the other station.

Fig. 10 includes both a configuration block diagram and flowchart of subcategory 3.1.

**5.11.2 Establishment.** An identification procedure is required to ensure connection to the proper station following circuit connection and prior to initiating message transfer in either direction. The identification procedure must be executed by both stations in order to accomplish two-way simultaneous message transfer. Both stations may initiate the identification procedure simultaneously if both have traffic to send; normally, however, the calling station executes the procedure first.

In the following, all statements made relative to the relationship between station A (master) and station B (slave) apply equally to the relationship between station B (master) and station A (slave).

Either station establishes the circuit by dialing the other [Fig. 10 (1)]. To establish the link for message transfers from station A to station B, A requests the identity of B by sending an identification supervisory sequence [Fig. 10 (2)] consisting of SOSS (a DLE character followed by a colon character), an optional prefix, and ENQ. (Note that at this time, immediately following circuit establishment, station B cannot

550

have any other outstanding slave response for station A since no traffic has been exchanged.) Upon detecting ENQ, station B sends one of two replies:

(1) If ready to receive [Fig. 10 (3)], station B sends SOSS, a prefix that includes its identity, followed by ACK0. This establishes the link for message transfers from station A to station B. Upon detecting ACK0, station A proceeds with message transfer.

(2) If not ready to receive [Fig. 10 (4)], station B sends SOSS, a prefix that includes its identity, followed by NAK. Station A, upon detecting NAK may try again [Fig. 10 (12), (2)], proceed to termination [Fig. 10 (12), (6)], or exit to a recovery procedure [Fig. 10 (13)].

If station A receives an invalid reply or no reply [Fig. 10 (5)], it may send an identification supervisory sequence again [Fig. 10 (12), (2)], proceed to termination [Fig. 10 (12), (6)], or exit to a recovery procedure [Fig. 10 (13)]. $N$ retries ($N \geqslant 0$) may be made to verify the identity of the other station.

Establishment of the link for message transfers from station B to station A (accomplished in the same manner as described above for transfer from A to B) may be initiated by station B at any time following circuit establishment as long as station B does not have an outstanding slave response.

**5.11.3 Message Transfer.** Message transfer subcategories F1 and F2, designed especially for two-way simultaneous transmission environments, provide for slave responses to be transmitted either between (interleaved) or within (embedded) messages or transmission blocks.

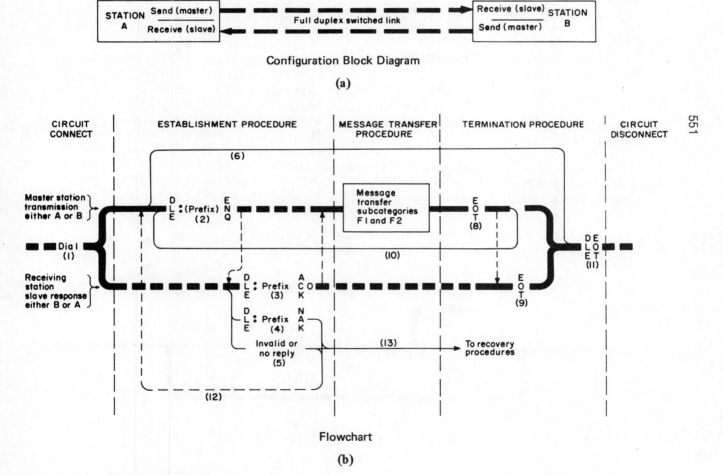

Configuration Block Diagram

**(a)**

Flowchart

**(b)**

Legend: — — — Sequence indication

**Fig. 10**
**Subcategory 3.1: Two-Way Simultaneous, Switched Point-to-Point**

42

**5.11.4 Termination.** If the link has not been established for message transfer in either direction (that is, neither station has successfully completed the identification procedure), either station may proceed to initiate disconnect by transmitting DLE EOT [Fig. 10 (6), (11)].

If the link has been established for message transfers in either direction (station A to B, station B to A, or both), then both stations must, under normal conditions, transmit (and receive) EOT [Fig. 10 (8), (9)] prior to either station's initiating disconnect by transmitting DLE EOT [Fig. 10 (11)].

If the link has been established for message transfers in both directions, the sending of EOT by either station signals the end of the message transfers in that direction. To resume message transfers after sending EOT (but prior to receiving EOT), the link must be reestablished [Fig. 10 (10)] by the station that initiated termination.

EOT may only be transmitted by a station after all outstanding slave responses have been received or otherwise accounted for.

**5.11.5 Aborts, Interrupts, and Recovery Procedures.** Refer to 3.4 and 3.5.

# 6. Subcategories of Message Transfer Control Procedures

**6.1 Introduction.** This section describes a set of subcategories of message transfer control procedures. Each subcategory defines the control procedures for framing messages and transmission blocks and their associated replies under normal conditions.

Procedures for use under abnormal conditions are defined in 3.4, Aborts and Interrupts; 3.5, Recovery Procedures; and 4.2.1, DEOT (Mandatory Disconnect).

Each subcategory description includes a flowchart that depicts the message transfer control procedures. Primary paths are shown by heavy lines. Prefixes are shown as either mandatory or optional with the optional prefixes given in parentheses. The functions and operations depicted in the flowcharts are referenced in the narrative description of each subcategory by a bracketed figure citation that includes the specific reference number given in the chart.

**6.2 Subcategory A1: Message-Oriented, without Replies and without Longitudinal Checking**

**6.2.1 General Description.** Subcategory A1 uses only one direction of transmission and is applicable in systems where the master station sends a message or series of messages, which may or may not include headings, to designated slave stations. Fig. 11 is a flowchart of subcategory A1.

**6.2.2 Message Transfer.** Message transfer is initiated by the master station after an appropriate establishment procedure [Fig. 11 (1)]. If the message has a heading, the master station begins the transmission with SOH [Fig. 11 (2)]. If the message has no heading [Fig. 11 (3)], the master station begins the transmission with STX [Fig. 11 (4)]. The entire message is sent, terminating with ETX [Fig. 11 (5)]. The master station may repeat this operation [Fig. 11 (6)] until all messages for the selected slave station(s) have been

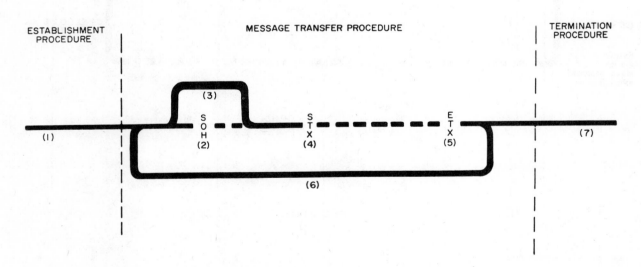

**Fig. 11**
**Subcategory A1: Message-Oriented, without Replies and without Longitudinal Checking**

552

43

sent. The master station then proceeds with the termination procedure [Fig. 11 (7)] that is part of the establishment and termination subcategory employed.

### 6.3 Subcategory A2: Message-Oriented, without Replies and with Longitudinal Checking

**6.3.1 General Description.** Subcategory A2 is applicable in systems where the master station sends a message or series of messages, which may or may not include headings, to designated slave station(s). Fig. 12 is a flowchart of subcategory A2.

**6.3.2 Message Transfer.** Message transfer is initiated by the master station after an appropriate establishment procedure [Fig. 12 (1)]. If the message has a heading, the master station begins the transmission with SOH [Fig. 12 (2)]. If the message has no heading [Fig. 12 (3)], the master station begins the transmission with STX [Fig. 12 (4)]. The entire message is sent, terminating with ETX [Fig. 12 (5)]. The ETX character is immediately followed by a block check character (BCC). The master station may repeat this operation [Fig. 12 (6)] until all messages for the selected slave station(s) have been sent. The master station then proceeds with the termination procedure [Fig. 12 (7)] that is part of the establishment and termination subcategory employed.

### 6.4 Subcategory A3: Message-Oriented, with Replies and without Longitudinal Checking

**6.4.1 General Description.** Subcategory A3 is applicable in systems where the master station sends a message, which may or may not include a heading, to a single slave station. It then waits for a reply from the slave station. Upon receipt of an affirmative reply, the master station may proceed with the transmission of

another message. A negative reply from the slave station does not require immediate retransmission of the message that was not accepted. Fig. 13 is a flowchart of subcategory A3.

**6.4.2 Message Transfer.** Message transfer is initiated by the master station after an appropriate establishment procedure [Fig. 13 (1)]. If the message has a heading, the master station begins the transmission with SOH [Fig. 13 (2)]. If the message has no heading [Fig. 13 (3)], the master station begins the transmission with STX [Fig. 13 (4)]. The entire message is sent, terminating with ETX [Fig. 13 (5)]. The master station then waits for a reply from the slave station.

**6.4.3 Replies.** Upon detecting ETX, the slave station sends one of two replies:

(1) An optional prefix followed by ACK [Fig. 13 (6)] if the message was accepted and the slave station is ready to receive.

(2) An optional prefix followed by NAK [Fig. 13 (8)] if the message was not accepted and the slave station is ready to receive.

The master station, detecting the ACK or NAK reply, may initiate the transmission of another message to this slave station [Fig. 13 (7)] or may proceed to the termination procedure [Fig. 13 (12)]. Following a NAK reply, the next transmission need not be a retransmission of the message that was not accepted.

If an invalid or no reply [Fig. 13 (9)] is received, the master station may send a reply-request supervisory sequence [Fig. 13 (10)] consisting of an optional prefix followed by ENQ, or the master station may exit to a recovery procedure [Fig. 13 (11)].

Upon receipt of a reply-request supervisory sequence, the slave station repeats its last reply. $N$ retries ($N \geqslant 0$)

553

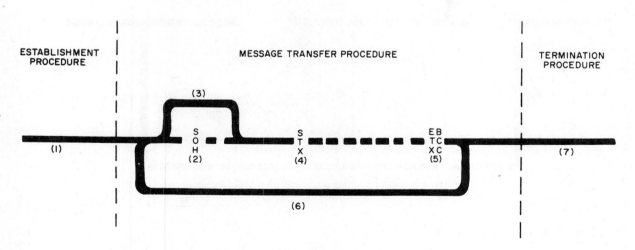

**Fig. 12**
**Subcategory A2: Message-Oriented, without Replies and with Longitudinal Checking**

may be made to get a valid reply. If a valid reply is not received after $N$ retries, the master station exits [Fig. 13 (11)] to a recovery procedure.

NOTE: The use of a reply-request supervisory sequence [Fig. 13 (10)] may result in lost or duplicate messages. Loss of messages may be avoided by always retransmitting the previous message, or by sending only a single message for each link establishment. Although these precautions do not eliminate the possible receipt of duplicate messages, the reply-request supervisory sequence permits the detection of their occurrence. The use of a numbering scheme will aid in detecting loss or duplication of messages.

## 6.5 Subcategory A4: Message-Oriented, with Replies and with Longitudinal Checking

**6.5.1 General Description.** Subcategory A4 is applicable in systems where the master station sends a message, which may or may not include a heading, to a single slave station. The master sends each message to the slave station and waits for a reply. If the reply indicates that the message was accepted, the master station may send another message, or it may terminate. A negative reply from the slave station does not require immediate retransmission of the message that was not accepted. Fig. 14 is a flowchart of subcategory A4.

**6.5.2 Message Transfer.** Message transfer is initiated by the master station after an appropriate establishment procedure [Fig. 14 (1)]. If the message has a heading, the master station begins the transmission with SOH [Fig. 14 (2)]. If the message has no heading [Fig. 14 (3)], the master station begins the transmission with STX [Fig. 14 (4)]. The entire message is sent, terminating with ETX [Fig. 14 (5)]. The ETX character is immediately followed by a block check character (BCC). The master station then waits for a reply from the slave station.

**6.5.3 Replies.** The slave station, upon detecting ETX followed by the BCC, sends one of two replies:

(1) If the message was accepted and the slave station is ready to receive another message, it sends an optional prefix followed by ACK [Fig. 14 (6)]. Upon detecting ACK, the master station may either transmit the next message [Fig. 14 (7)] or initiate termination [Fig. 14 (12)].

(2) If the message was not accepted and the slave station is ready to receive another message, it sends an optional prefix followed by NAK [Fig. 14 (8)]. Upon detecting NAK, the master station may either transmit another message [Fig. 14 (7)] or initiate termination [Fig. 14 (12)]. Following the NAK reply, the next message transmitted need not be a retransmission of the message that was not accepted.

If the master station receives an invalid or no reply to a message [Fig. 14 (9)], it may send a reply-request supervisory sequence [Fig. 14 (10)] consisting of an

554

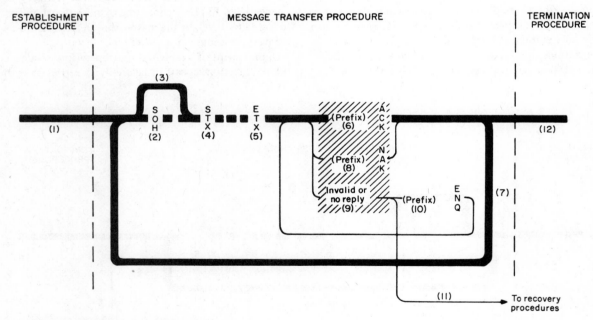

NOTE: Crosshatched area is slave response.

**Fig. 13**
**Subcategory A3: Message-Oriented, with Replies and without Longitudinal Checking**

45

optional prefix followed by ENQ, or it may exit to a recovery procedure [Fig. 14 (11)].

Upon receipt of a reply-request supervisory sequence, the slave station repeats its last reply. $N$ retries ($N \geqslant 0$) may be made to get a valid reply. If a valid reply is not received after $N$ retries, the master station exits [Fig. 14 (11)] to a recovery procedure.

NOTE: The use of a reply-request supervisory sequence [Fig. 14 (10)] may result in the loss or duplication of messages. Loss of messages may be avoided by always retransmitting the previous message or by sending only a single message for each link establishment. Although these precautions do not eliminate the possible receipt of duplicate messages, the reply-request supervisory sequence permits the detection of such an occurrence. The use of a numbering scheme will aid in detecting loss or duplication of messages.

### 6.6 Subcategory B1: Message-Associated Blocking, with Longitudinal Checking and Single Acknowledgment

**6.6.1 General Description.** Subcategory B1 is applicable primarily in systems where messages may be subdivided into blocks. A transmission block may be a complete message or a portion of a message. The master station sends each transmission block to the slave station and waits for a reply. If the reply indicates that the block was accepted, the master station

may send another block, or it may terminate. If the reply is negative, the master station immediately retransmits the block that was not accepted. Fig. 15 is a flowchart of subcategory B1.

**6.6.2 Transmission Blocks.** The transmission of blocks is initiated by the master station after an appropriate establishment procedure [Fig. 15 (1)]. If the message has a heading, the master station begins the transmission with SOH [Fig. 15 (2)]. If the message has no heading [Fig. 15 (3)], the master station begins the transmission with STX [Fig. 15 (4)].

An intermediate block that continues a heading [Fig. 15 (7), (2)] is started with SOH. An intermediate block that either begins or continues a text [Fig. 15 (7), (3), (4)] is started with STX. If the last information character of a heading ends on a block boundary (ended by ETB), the subsequent block may be started by either SOH or STX. Note that in such a case the receiver must be able to handle both situations.

A block that ends at an intermediate point within the message is ended with ETB [Fig. 15 (5)]. A block that ends at the end of a message is ended with ETX [Fig. 15 (6)]. The ETB or ETX character is immediately followed by a block check character (BCC). After the ETB or ETX and BCC are sent, the master station waits for a reply.

555

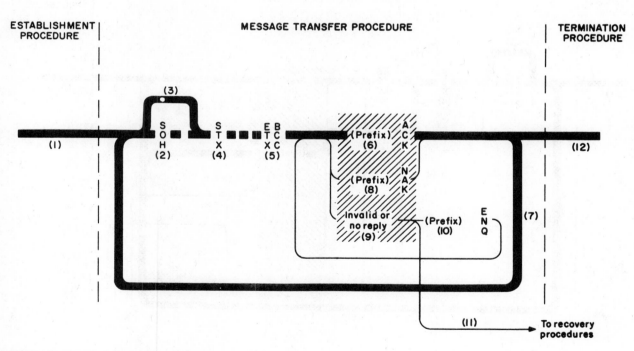

NOTE: Crosshatched area is slave response.

**Fig. 14**
**Subcategory A4: Message-Oriented, with Replies and with Longitudinal Checking**

**6.6.3 Replies.** The slave station, upon detecting the ETB or ETX followed by the BCC, sends one of two replies:

(1) If the transmission block was accepted and the slave station is ready to receive another block, it sends an optional prefix followed by ACK [Fig. 15 (8)]. Upon detecting ACK, the master station may either transmit the next block [Fig. 15 (7)], or initiate termination [Fig. 15 (9)] if the last block ended in ETX BCC [Fig. 15 (6)].

(2) If the transmission block was not accepted and the slave station is ready to receive another block, it sends an optional prefix followed by NAK [Fig. 15 (10)]. Upon detecting NAK, the master station initiates retransmission of the last transmission block [Fig. 15 (11), (7)]. $N$ retransmissions ($N \geqslant 0$) may be made, after which the master station exits [Fig. 15 (12)] to a recovery procedure.

If the master station receives an invalid or no reply to a transmission block [Fig. 15 (13)], it may send a reply-request supervisory sequence [Fig. 15 (14)] consisting of an optional prefix followed by ENQ, or it may exit to a recovery procedure [Fig. 15 (15), (12)].

Upon receipt of a reply-request supervisory sequence,

the slave station repeats its last reply. $N$ retries ($N \geqslant 0$) may be made to get a valid reply. If a valid reply is not received after $N$ retries, the master station exits [Fig. 15 (15), (12)] to a recovery procedure.

NOTE: The use of a reply-request supervisory sequence [Fig. 15 (14)] may result in the loss or duplication of transmission blocks. Loss of transmission blocks may be avoided by always retransmitting the previous block. Although this precaution does not eliminate the possible receipt of duplicate blocks, the reply-request supervisory sequence permits the detection of such an occurrence. The use of a numbering scheme will aid in detecting loss or duplication of blocks.

## 6.7 Subcategory B2: Message-Associated Blocking, with Longitudinal Checking and Alternating Acknowledgments

**6.7.1 General Description.** Subcategory B2 is applicable primarily in systems where messages may be subdivided into blocks. A transmission block may be a complete message or a portion of a message. The master station sends each transmission block to the slave station and waits for a reply.

Alternating acknowledgments are used in lieu of ACK. If the reply indicates that the block was accepted, the master station may send another block, or

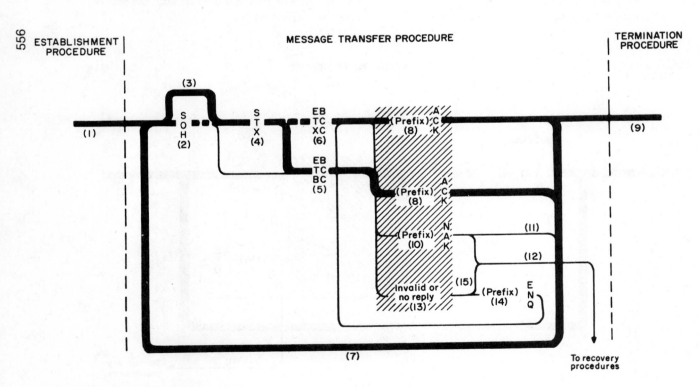

NOTE: Crosshatched area is slave response.

**Fig. 15**
**Subcategory B1: Message-Associated Blocking, with Longitudinal Checking and Single Acknowledgment**

it may terminate. If the reply is negative, the master station immediately retransmits the block that was not accepted.

Fig. 16 is a flowchart of subcategory B2.

**6.7.2 Transmission Blocks.** The transmission of blocks is initiated by the master station after an appropriate establishment procedure [Fig. 16 (1)]. If the message has a heading, the master station begins the transmission with SOH [Fig. 16 (2)]. If the message has no heading [Fig. 16 (3)], the master station begins the transmission with STX [Fig. 16 (4)].

An intermediate block that continues a heading [Fig. 16 (7), (2)] is started with SOH. An intermediate block that either begins or continues a text [Fig. 16 (7), (3), (4)] is started with STX. If the last information character of a heading ends on a block boundary (ended by ETB), the subsequent block may be started by either SOH or STX. Note that in such a case the receiver must be able to handle both situations.

A block that ends at an intermediate point within the message is ended with ETB [Fig. 16 (5)]. A block that ends at the end of a message is ended with ETX [Fig. 16 (6)]. The ETB or ETX character is immediately followed by a block check character (BCC). After the ETB or ETX and BCC are sent, the master station waits for a reply.

**6.7.3 Replies.** The slave station, upon detecting the ETB or ETX followed by the BCC, sends one of two replies:

(1) If the transmission block was accepted and the slave station is ready to receive another block, it sends an optional prefix followed by the appropriate ACK$N$ [Fig. 16 (8)]. ACK1 is used as the reply to the first block received. ACK0 is used as the reply to the second block received. Subsequent positive acknowledgments alternate between ACK1 and ACK0. Upon detecting the appropriate ACK$N$, the master station may either transmit the next block [Fig. 16 (7)], or initiate termination [Fig. 16 (9)] if the last block ended in ETX BCC [Fig. 16 (6)].

(2) If the transmission block was not accepted and the slave station is ready to receive another block, it sends an optional prefix followed by NAK [Fig. 16 (10)]. Upon detecting NAK, the master station initiates retransmission of the last transmission block [Fig. 16 (11), (7)]. $N$ retransmissions ($N \geq 0$) may be made, after which the master station exits [Fig. 16 (12)] to a recovery procedure.

The use of NAK does not alter the sequence of alternating acknowledgments. The same affirmative reply (ACK0 or ACK1) is used for a successful retransmission as would have been used if the previous trans-

557

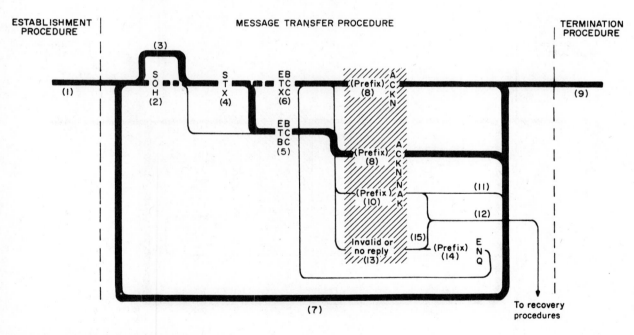

NOTE: Crosshatched area is slave response.

**Fig. 16**
**Subcategory B2: Message-Associated Blocking, with Longitudinal Checking and Alternating Acknowledgments**

mission of the unaccepted block had been successful.

If the alternating reply indicates that the slave station missed the outstanding block (receipt of ACK1 instead of ACK0, or vice versa), the master station initiates retransmission of that block as if the slave station had returned a NAK [Fig. 16 (11)].

If the master station receives an invalid or no reply to a transmission block [Fig. 16 (13)], it may send a reply-request supervisory sequence [Fig. 16 (14)] consisting of an optional prefix followed by ENQ, or it may exit to a recovery procedure [Fig. 16 (15), (12)].

Upon receipt of a reply-request supervisory sequence, the slave station repeats its last reply. $N$ retries ($N \geqslant 0$) may be made to get a valid reply. If a valid reply is not received after $N$ retries, the master station exits [Fig. 16 (15), (12)] to a recovery procedure.

### 6.8 Subcategory C1: Message-Independent Blocking—Noncontinuous Operation with Longitudinal Checking and Alternating Acknowledgments

**6.8.1 General Description.** Subcategory C1 is applicable to systems where nontransparent message data are blocked independently of the message format. A block may contain partial, complete, or multiple messages. The master station sends each transmission block to a single slave station and waits for a reply. If the reply indicates that the block was accepted, the

master station may send another block, or it may terminate. If the reply is negative, the master station immediately retransmits the block that was not accepted. Fig. 17 is a flowchart of subcategory C1.

**6.8.2 Transmission Blocks.** The transmission of blocks is initiated by the master station after an appropriate establishment procedure [Fig. 17 (1)]. Transmission blocks always begin with the SOTB (DLE =) sequence [Fig. 17 (2)] and end with ETB [Fig. 17 (3)] followed immediately by the block check character (BCC). After the ETB and BCC are sent, the master station waits for a reply. The ETB character is included in the block check summation, but the SOTB sequence is not. Any character in the text other than SYN is included in the summation. If STX, SOH, or ETX occurs in text, this character is also included.

**6.8.3 Replies.** The slave station, upon detecting an ETB followed by the BCC, sends one of the following replies:

(1) If the transmission block was accepted and the slave station is ready to receive another block, it sends an optional prefix followed by the appropriate alternating acknowledgment [Fig. 17 (8)]. ACK1 is used as a reply to the first block received. ACK0 is used as a reply to the second block received. Subsequent positive acknowledgments alternate between ACK1 and ACK0. Upon detecting the appropriate alternating acknowl-

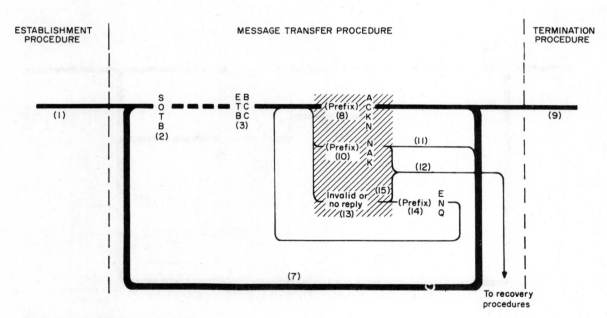

NOTE: Crosshatched area is slave response.

**Fig. 17**
**Subcategory C1: Message-Independent Blocking—Noncontinuous Operation with Longitudinal Checking and Alternating Acknowledgments**

edgment, the master station may either transmit the next block [Fig. 17 (7)] or initiate termination [Fig. 17 (9)].

(2) If the transmission block was not accepted and the slave station is ready to receive another block, it sends an optional prefix followed by NAK [Fig. 17 (10)]. Upon detecting NAK, the master station initiates retransmission of the last transmission block [Fig. 17 (11), (7)]. $N$ retransmissions ($N \geqslant 0$) may be made, after which the master station exits [Fig. 17 (12)] to a recovery procedure.

The use of NAK does not alter the sequence of alternating acknowledgments. The same affirmative reply (ACK0 or ACK1) is used for a successful retransmission as would have been used if the previous transmission of the unaccepted block had been successful.

If the alternating reply indicates that the slave station missed the outstanding block (receipt of ACK1 instead of ACK0, or vice versa), the master station initiates retransmission of that block as if the slave station had returned a NAK [Fig. 17 (11)].

If the master station receives an invalid or no reply to a transmission block [Fig. 17 (13)], it may send a reply-request supervisory sequence [Fig. 17 (14)] consisting of an optional prefix followed by ENQ.

Upon receipt of a reply-request supervisory sequence, the slave station repeats its last reply. $N$ retries ($N \geqslant 0$) may be made to get a valid reply. If a valid reply is not received after $N$ retries [Fig. 17 (15)], the master station exits [Fig. 17 (12)] to a recovery procedure.

## 6.9 Subcategory C2: Message-Independent Blocking—Continuous Operation with Longitudinal Checking and Modulo-8 Numbering of Blocks and Acknowledgments

**6.9.1 General Description.** Subcategory C2 is applicable to systems where nontransparent message data are blocked independently of the message format and where the master station sends blocks to a single slave station continuously without waiting for a reply. By means of a backward channel, replies are sent while the master station is sending subsequent blocks. By the use of modulo-8 numbering of blocks and replies, the master station may get as many as seven blocks ahead of the received replies before stopping to wait for the replies to catch up. If a negative reply is received, the master station starts retransmission with the block following the last block for which the proper affirmative acknowledgment was received. Fig. 18 is a flowchart of subcategory C2.

**6.9.2 Transmission Blocks.** Message transfer is initiated by the master station after an appropriate establishment procedure [Fig. 18 (1)]. Transmission blocks always begin with the SOTB sequence (DLE =) followed by BLK [Fig. 18 (2)] and end with ETB [Fig.

18 (3)] followed immediately by a block check character (BCC). BLK refers to the set of eight block numbers (see 3.7). Block numbering begins with BLK1 and progresses through BLK 7 before recycling with BLK0.

The ETB character and BLK are included in the block check summation, but the SOTB sequence is not. Any character in the text other than SYN is included in the summation. If STX, SOH, or ETX occurs in text, this character is also included.

The master station may send blocks continuously [Fig. 18 (4), (5)] as long as it does not get more than seven blocks ahead of the replies.

**6.9.3 Replies.** Upon detecting an ETB (followed immediately by a BCC), the slave station sends one of two replies on the backward channel:

(1) If the block is acceptable and the slave station is ready to receive subsequent blocks, it sends:

(a) An optional prefix followed by the appropriate affirmative reply [Fig. 18 (6)] if the block number is either as expected or lower than expected. The appropriate affirmative reply is ACK1 for the first block received, ACK2 for the second block received, etc, through ACK7 and then restarting with ACK0 [Fig. 18 (6)]. The slave station should discard blocks containing block numbers lower than expected.

(b) An optional prefix followed by NAK [Fig. 18 (7)] if the block number received is higher than expected. The slave station does not reply to subsequent blocks received, whether they are in error or not, until the proper block number is detected or a reply-request supervisory sequence is received.

(2) If the block was not acceptable and the slave station is ready to receive another block, it sends an optional prefix followed by NAK [Fig. 18 (7)]. The slave station does not reply to subsequent blocks received, whether they are in error or not, until the proper block number is detected or a reply-request supervisory sequence is received.

As long as appropriate affirmative replies are detected, in the proper sequence, the master station continues transmission [Fig. 18 (4), (5)]. After all transmission blocks have been sent, the master station continues to monitor replies until all blocks have been affirmatively acknowledged by the slave station. The master station then initiates termination [Fig. 18 (4), (13)].

Upon detecting a NAK reply, the master station stops transmission at the end of the block and initiates retransmission starting with the block following the last block for which the proper affirmative acknowledgment was received. $N$ retransmissions ($N \geqslant 0$) may be made, after which the master station may either exit [Fig. 18 (9)] to a recovery procedure [Fig. 18 (14)] or initiate termination [Fig. 18 (13)].

The use of NAK does not alter the sequence of acknowledgments. The slave station uses the same affirmative reply (ACK0, ACK1, etc) for a successful retransmission as it would have used if the previous transmission of that block had been successful.

If the master station detects an invalid or no reply to a transmission block [Fig. 18 (10)], it halts transmission at the end of a block being transmitted and sends a reply-request supervisory sequence [Fig. 18 (11)].

Upon receipt of a reply-request supervisory sequence, the slave station sends the affirmative reply for the last block accepted. The master station then resumes transmission with the proper transmission block. $N$ retries ($N \geqslant 0$) may be made to get a valid reply. If a valid reply is not received after $N$ attempts, the master station may either exit [Fig. 18 (12)] to a recovery procedure [Fig. 18 (14)] or initiate termination [Fig. 18 (13)].

The block counter is reset to "one" upon a timer-C time-out or upon transmission/reception of EOT/ DEOT.

## 6.10 Subcategory D1: Message-Independent Blocking, with Cyclic Checking, Alternating Acknowledgments, and Transparent Heading and Text

**6.10.1 General Description.** Subcategory D1 is applicable where blocking may be performed independent

of messages or where the text of a message can contain all 256 combinations[10] of an 8-bit character (transparency), or both. Thus a block may contain partial, complete, or multiple messages. These messages may be either basic or transparent. The master station sends each transmission block to a single slave station and waits for a reply. If the reply indicates that the block was accepted, the master station may send another block or terminate. If the reply is negative, the master station transmits the block that was not accepted prior to the transmission of any other blocks. Fig. 19 is a flowchart of subcategory D1.

**6.10.2 Transmission Blocks.** The transmission of blocks is initiated by the master station after an appropriate establishment procedure [Fig. 19 (1)]. If the message has a heading, the master begins the transmission with DLE SOH [Fig. 19 (2)]. If the message has no heading [Fig. 19 (3)], the master station begins the transmission with DLE STX [Fig. 19 (4)].

An intermediate block that continues a heading [Fig. 19 (7), (2)] is started with DLE SOH. An inter-

[10] NOTE: There may be cases in which the data communication equipment or network places restrictions on long sequences or certain bit patterns; these patterns may be avoided by judicious insertion of DLE SYN sequences.

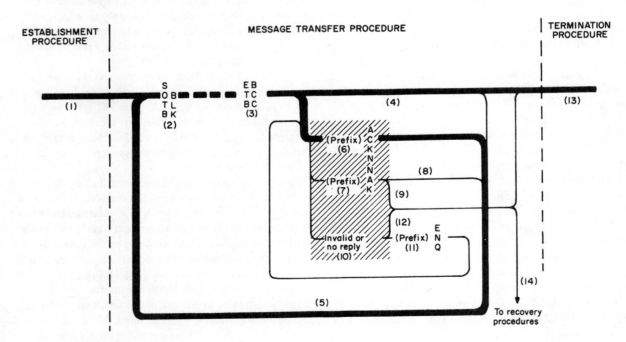

560

NOTE: Crosshatched area is slave response.

Fig. 18

Subcategory C2: Message-Independent Blocking — Continuous Operation with Longitudinal Checking and Modulo-8 Numbering of Blocks and Acknowledgments

mediate block that either begins or continues a text [Fig. 19 (7), (3), (4)] is started with DLE STX. If the last information byte of a transparent heading ends on a block boundary (ended by DLE ETB), the subsequent block may be started by either DLE SOH or DLE STX. Note that in such a case the receiver must be able to handle both situations.

A block that ends at an intermediate point within the message is ended with DLE ETB [Fig. 19 (5)]. A block that ends at the end of a message is ended with DLE ETX [Fig. 19 (6)]. The ETB or ETX character is immediately followed by a cyclic redundancy check (CRC). After the ETB or ETX and CRC are sent, the master station waits for a reply.

Since any of the 256 combinations of 8 bits may occur in transparent data, the stations must be able to distinguish between control characters and data. The control characters that can be recognized must always be character sequences started by DLE. When a DLE appears from the source of transparent data, the master inserts an additional DLE so that the data transmitted on the line will contain two successive DLE characters (TDLE).

Slave stations must scan the incoming data for DLE characters. On the first appearance of a DLE, the character immediately following determines the action to be taken:

(1) If the character immediately following is also a DLE, indicating a data character, the slave station discards one DLE character. The character immediately succeeding is not considered part of any control character sequence. The station continues receiving data.

(2) If the character immediately following is either ENQ, ETB, or ETX, the slave station leaves transparent mode and performs the action designated by the control character (completion of a time-out also causes the slave station to leave transparent mode).

(3) If the character immediately following is SYN, the slave treats this as an "idle" character sequence and discards both characters.

(4) If the character immediately following is other than ENQ, ETB, ETX, DLE, or SYN, it is to be considered invalid unless other control code extension sequences are defined in the system.

Refer to 4.3 for information regarding the cyclic check summation.

6.10.3 Replies. The slave station, upon detecting the TETB or TETX followed by the CRC, sends one of two replies:

(1) If the transmission block was accepted and the slave station is ready to receive another block, it sends an optional prefix followed by the appropriate ACK$N$ [Fig. 19 (8)]. ACK1 is used as the reply to the first block received. ACK0 is used as the reply to the second block received. Subsequent positive acknowledgments

561

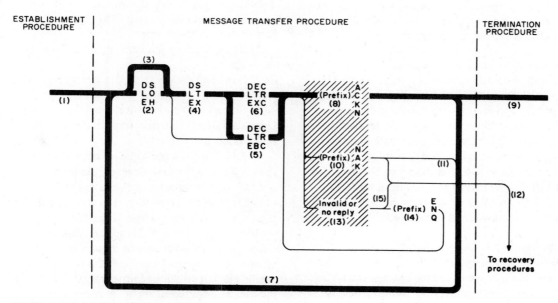

NOTE: Crosshatched area is slave response.

**Fig. 19**
**Subcategory D1: Message-Independent Blocking, with Cyclic Checking, Alternating Acknowledgments, and Transparent Heading and Text**

alternate between ACK1 and ACK0. Upon detecting the appropriate ACK$N$, the master station may either transmit the next block [Fig. 19 (7)], or initiate termination [Fig. 19 (9)] if the last block ended in TETX CRC [Fig. 19 (6)].

(2) If the transmission block was not accepted and the slave station is ready to receive another block, it sends an optional prefix followed by NAK [Fig. 19 (10)]. Upon detecting NAK, the master station initiates retransmission of the last transmission block [Fig. 19 (11), (7)]. $N$ retransmissions ($N \geqslant 0$) may be made, after which the master station exits [Fig. 19 (12)] to a recovery procedure.

The use of NAK does not alter the sequence of alternating acknowledgments. The same affirmative reply (ACK0 or ACK1) is used for a successful retransmission as would have been used if the previous transmission of the unaccepted block had been successful.

If the alternating reply indicates that the slave station missed the outstanding block (receipt of ACK1 instead of ACK0, or vice versa), the master station initiates retransmission of that block as if the slave station had returned a NAK [Fig. 19 (11)].

If the master station receives an invalid or no reply to a transmission block [Fig. 19 (13)], it may send a reply-request supervisory sequence [Fig. 19 (14)] consisting of an optional prefix followed by ENQ, or it may exit to a recovery procedure [Fig. 19 (15), (12)].

Upon receipt of a reply-request supervisory sequence, the slave station repeats its last reply. $N$ retries ($N \geqslant 0$) may be made to get a valid reply. If a valid reply is not received after $N$ retries, the master station exits [Fig. 19 (15), (12)] to a recovery procedure.

**6.10.4 Character Structure of Control Function.** The communication control functions in subcategory D1 use the ASCII characters with character structure and parity sense in accordance with American National Standard Character Structure and Character Parity Sense for Serial-by-Bit Data Communication in the American National Standard Code for Information Interchange, X3.16-1976, and American National Standard Bit Sequencing of the American National Standard Code for Information Interchange in Serial-by-Bit Data Transmission, X3.15-1976.

## 6.11 Subcategory E1: Conversational

**6.11.1 General Description.** Subcategory E1 provides for alternating exchange of related messages and is intended for one or more contiguous inquiry/response exchanges between two parties (one of which is operator attended) in single-link systems.

Subcategory E1 is applicable in systems where the initiating station sends a message (inquiry) to a single responding station. The initiator then waits for a re-

sponse message, which may contain line control procedures information. Operator procedures may play a large part in the timing and accuracy control functions.

Fig. 20 is a flowchart of subcategory E1.

**6.11.2 Message Transfer.** Message transfer is started by the initiator after an appropriate establishment procedure [Fig. 20 (1)]. The initiator begins the transmission with STX [Fig. 20 (2)]. The entire message (inquiry text) is then sent, terminating with ETX [Fig. 20 (4)]. The initiator then waits for a reply from the responder.

**6.11.3 Replies**

**6.11.3.1** Upon detecting ETX, the responder evaluates the received message and sends one of two responses:

(1) A response message [Fig. 20 (5), (6), (7)] if the inquiry message was acceptable

(2) A "request repeat inquiry" message [Fig. 20 (8), (9), (10)] if the inquiry message was not acceptable

After the initiator receives an acceptable response message [Fig. 20 (5), (6), (7)], it can send another inquiry message [Fig. 20 (17), (13)], or it can proceed to the termination procedure [Fig. 20 (18)].

**6.11.3.1.1** If the initiator receives a "request repeat inquiry" message [Fig. 20 (8), (9), (10)], or if an invalid reply or no reply [Fig. 20 (11)] is detected, the initiator should retransmit the same inquiry message [Fig. 20 (13), (2), (3), (4)]. If either a "request repeat inquiry" message is received or no response is detected after $N$ retries, the initiator should then go to the recovery procedures [Fig. 20 (19)], which may consist of such things as:

(1) Retransmitting the inquiry message up to $M$ more times

(2) Notifying the operator, processor program, or both

(3) Going to the termination procedure
Operator intervention may be desirable at either or both stations.

**6.11.3.1.2** If a transmission error [Fig. 20 (12)] is detected in the responder's reply, the initiator should send a "request repeat reply" message [Fig. 20 (14), (15), (16)] to the responder, who should then retransmit its last response message. If the retransmitted response message is acceptable to the initiator, the initiator can either send another inquiry message or terminate in the normal manner. If the retransmitted message is again in error, the initiator should repeat sending the "request repeat reply" message (as applicable) up to $M$ times before exiting to a recovery procedure [Fig. 20 (19)], which can be the same as given in 6.11.3.1.1.

**6.11.4 Termination Responsibility.** The initiator has the responsibility for the termination procedures on both single-station and multistation lines. However, the control station can also perform the termination procedures on a multistation line.

**6.11.5 Timers.** The timer-A function (see 3.5.2.1) is required at the initiator location. The timing may be implemented in hardware or software, or performed by the operator.

## 6.12 Subcategory E2: Conversational, with Blocking and Block Checking

**6.12.1 General Description.** Subcategory E2 provides primarily for alternating exchange of related messages and is intended for one or more contiguous inquiry/ response exchanges between two stations (one of which is normally operator attended) in single-link systems. Additional provisions include message-associated blocking and block checking. Fig. 21 is a flowchart of subcategory E2.

**6.12.2 Message Transfer.** Message transfer is started by the initiator after an appropriate establishment procedure [Fig. 21 (1)]. The initiator begins the transmission with STX [Fig. 21 (2) or (3)]. A block that ends at an intermediate point within the message is ended with ETB [Fig. 21 (5)]. The last block of a message is ended with ETX [Fig. 21 (4)]. The ETB or ETX character is immediately followed by a block check charac-

ter (BCC). After the ETB or ETX and BCC are sent, the initiator waits for a reply from the responder.

**6.12.3 Replies**

**6.12.3.1** Upon detecting ETB [Fig. 21 (5)], the responder sends one of two replies:

(1) If the received block was accepted and the responder is ready to receive another block, it sends an optional prefix followed by ACK [Fig. 21 (6)]. Upon detecting ACK, the initiating station may send another intermediate block [Fig. 21 (7), (22), (3), (5)], or it may send the last block of the message [Fig. 21 (7), (22), (2), (4)].

(2) If the received block was not accepted and the responder is ready to receive a block, it sends an optional prefix followed by NAK [Fig. 21 (8)]. Upon detecting NAK, the initiating station should retransmit the unaccepted block [Fig. 21 (7), (22), (3), (5)]. The initiating station may retransmit $N$ times ($N \geqslant 0$), and if $N$ retrials are unsuccessful, it exits [Fig. 21 (9)] to a recovery procedure, which may consist of such things as:

(a) Trying $M$ additional retransmissions

(b) Notifying the operator or processor program, or both

(c) Going to the termination procedure

Operator intervention may be desirable at each station.

563

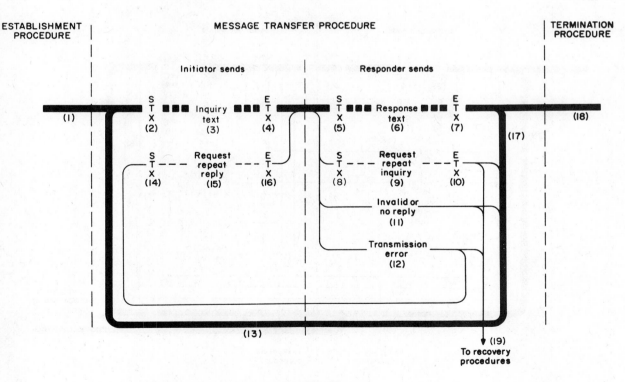

**Fig. 20**
**Subcategory E1: Conversational**

**6.12.3.1.1** If the initiating station receives an invalid or no reply [Fig. 21 (10)], it should retransmit the block [Fig. 21 (7), (22), (3), (5)]. $N$ ($N \geqslant 0$) retransmissions may be made, after which time the initiating station exits [Fig. 21 (9)] to a recovery procedure, which can be the same as given in 6.12.3.1 (2).

**6.12.3.2** Upon detecting ETX [Fig. 21 (4)], the responder evaluates the received message and then sends one of two replies:

(1) If the received message (inquiry) was accepted, the responder transmits a response message [Fig. 21 (11), (13), (15), or (11), (12), (14) as applicable]. The initiator then sends an optional prefix followed by ACK or NAK as a reply for ETB [Fig. 21 (15), (16), and (17) or (20)]. The responder can continue to send blocks and receive ACK/NAK replies [Fig. 21 (18), (11), (13), (15), (16), and (17) or (20)] until it has transmitted the last block ending in ETX BCC [Fig. 21 (14)]. If the last block was accepted by the initiator, the initiator proceeds to another inquiry transmission [Fig. 21 (28)] or to termination [Fig. 21 (21), (27)]. If the last block was not accepted, the initiator replies with NAK [Fig. 21 (20)], which may result in

the retransmission of the block [Fig. 21 (18), (11), (12), (14)] or the execution of a recovery procedure [Fig. 21 (25)].

(2) If the received block was not accepted and the responder is ready to receive another block, it sends NAK [Fig. 21 (8)]. The initiator should then retransmit the block [Fig. 21 (7), (22) and (3), (5) or (2), (4)].

The recovery procedure for handling multiple retransmissions due to receipt of NAK, invalid reply, or no reply are the same as described in 6.12.3.1.

NOTE: The recovery procedure described in 6.12.3 may result in the loss or duplication of blocks. The use of a numbering scheme can aid in detecting loss or duplication of blocks.

**6.12.4 Termination Responsibility.** The initiator has the responsibility for the termination procedures on both single-station and multistation lines. However the control station can also perform the termination procedures on a multistation line after a timer-D time-out.

**6.12.5 Timers.** Timers A, B, C, and D, are required as explained in 3.5.2.

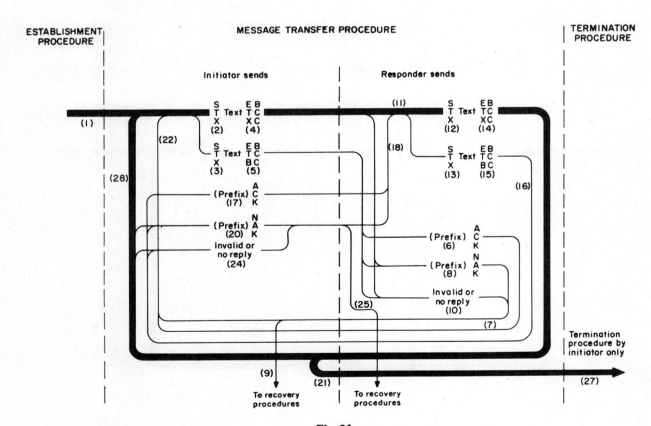

**Fig. 21**
**Subcategory E2: Conversational, with Blocking and Block Checking**

## 6.13 Subcategory E3: Conversational, with Blocking, Longitudinal Checking, and Batch or Alternate Message Transfer

**6.13.1 General Description.** Subcategory E3 primarily provides for alternating exchange of related messages and is intended for one or more contiguous inquiry/response exchanges between two stations (one of which is normally operator attended) in single-link systems. Additional provisions include batch operation, message-associated blocking, and block checking. Fig. 22 is a flowchart of subcategory E3.

**6.13.2 Message Transfer.** Message transfer is started by the initiator after an appropriate establishment procedure [Fig. 22 (1)]. The initiator begins the transmission with STX [Fig. 22 (2) or (3)]. A block that ends at an intermediate point within the message is ended with ETB [Fig. 22 (5)]. The last block of a message is ended with ETX [Fig. 22 (4)]. The ETB or ETX character is immediately followed by a block check character (BCC). After the ETB or ETX and BCC are sent, the initiator waits for a reply from the responder.

**6.13.3 Replies**

**6.13.3.1** Upon detecting ETB [Fig. 22 (5)], the responder sends one of two replies:

(1) If the received block was accepted and the responder is ready to receive another block, it sends an optional prefix followed by ACK [Fig. 22 (6)]. Upon detecting ACK, the initiating station may send another intermediate block [Fig. 22 (7), (22), (3), (5)], or it may send the last block of the message [Fig. 22 (7), (22), (2), (4)].

(2) If the received block was not accepted and the responder is ready to receive a block, it sends an optional prefix followed by NAK [Fig. 22 (8)]. Upon detecting NAK, the initiating station should retransmit the unaccepted block [Fig. 22 (7), (22), (3), (5)]. The initiating station may retransmit $N$ times ($N \geqslant 0$), and if the $N$ retrials are unsuccessful, it exits [Fig. 22 (9)] to a recovery procedure, which may consist of such things as:

(a) Trying $M$ additional retransmissions

(b) Notifying the operator or processor program, or both

(c) Going to the termination procedure

Operator intervention may be desirable at each station.

**6.13.3.1.1** If the initiating station receives an invalid or no reply [Fig. 22 (10)], it should retransmit the block [Fig. 22 (7), (22), (3), (5)]. $N$ ($N \geqslant 0$) re-

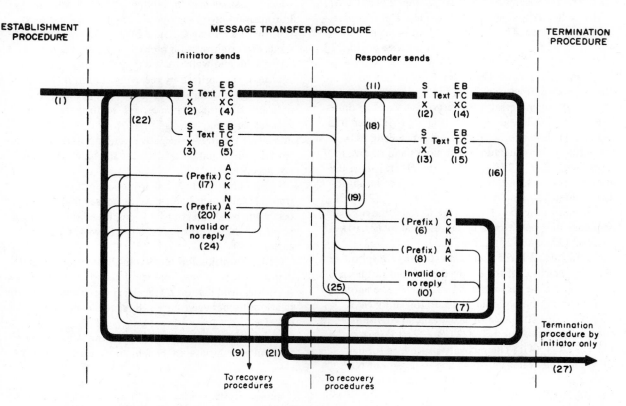

**Fig. 22**
**Subcategory E3: Conversational, with Blocking, Longitudinal Checking, and Batch or Alternate Message Transfer**

transmissions may be made, after which time the initiating station exits [Fig. 22 (9)] to a recovery procedure, which can be the same as given in 6.13.3.1 (2).

6.13.3.2 Upon detecting ETX [Fig. 22 (4)], the responder evaluates the received message and then sends one of three replies, as given in 6.13.3.2.1 through 6.13.3.2.3.

6.13.3.2.1 Inquiry/Response Operation. If the received message (inquiry) was accepted and the responder has a response message for the initiator, the response message is then transmitted by the responder [Fig. 22 (11), (13), (15), or (11), (12), (14), as applicable]. The initiator then sends a message (in response to ETX only) or an optional prefix followed by ACK or NAK, depending upon whether the received message was acceptable. These procedures, along with the recovery procedures, were explained in 6.13.3.1 (2). The responder can continue to send blocks and receive ACK/NAK replies [Fig. 22 (11), (13), (15), (16), (17), (18), (11), (13), (15), etc, that is, batch operation from responder] until it has transmitted the last block ending in ETX BCC. Additional texts can be sent by the responder upon receipt of ACK from the initiator. After the responder has received ACK for the last block [Fig. 22 (17)], the responder sends ACK [Fig. 22 (19), (6)], which indicates that all of the responder's blocks have been transmitted. The initiator then initiates termination [Fig. 22 (7), (21), (27)] or a new message [Fig. 22 (7), (22), (2), (4), or (7), (22), (3), (5)].

6.13.3.2.2 Batch Operation from Initiation. If the received message is accepted by the responder and there is no reply message, the responder sends an optional prefix followed by ACK [Fig. 22 (6)] when it is ready to receive another block. After receiving ACK, the initiator can send additional blocks and receive replies [Fig. 22 (7), (22), (3), (5), (6), (7), etc] until the initiator receives ACK for its last block. The initiator then initiates termination [Fig. 22 (7), (21), (27)]. If the initiator only has one message to send, the normal sequence of events would be represented by (2), (4), (6), (7), (21), and (27) in Fig. 22.

6.13.3.2.3 Error Condition (Inquiry/Response or Batch Operation). If the received block has not been accepted and the responder is ready to receive another block, it sends an optional prefix followed by NAK [Fig. 22 (8)]. The initiator should then retransmit the unaccepted block [Fig. 22 (7), (22), (3), (5), or (7), (22), (2), (4)]. The recovery procedure for handling multiple retransmissions due to receipt of NAK, invalid or no replies, or both, is the same as defined in 6.13.3.1 (2).

NOTE: The recovery procedures described in 6.13.3 may result in the loss or duplication of blocks. The use of a numbering scheme can aid in detecting loss or duplication of blocks.

6.13.4 Termination Responsibility. The initiator has the responsibility for termination procedures on both point-to-point and multistation lines. However, the control station can also perform the termination procedures on a multistation line after a timer-D time-out.

6.13.5 Timers. Timers A, B, and C or D are required as explained in 3.5.2.

## 6.14 Subcategory F1: Message-Associated Blocking with Longitudinal Checking and Alternating Acknowledgments for Two-Way Simultaneous Operation with Embedded Responses

6.14.1 General Description. Message transfer subcategory F1 may only be used with establishment and termination subcategory 3.1.[11] It provides for the embedding (as well as the interleaving) of slave responses for systems operating in a two-way simultaneous data transmission environment employing full-duplex communication facilities. In such environment each station must have the capability of maintaining concurrent master and slave status, that is, master status on its transmit side and slave status on its receive side.

Subcategory F1 is primarily applicable in systems in which messages may be subdivided into blocks. A transmission block may be a complete message or a portion of a message. Sending station A (master side) sends each transmission block to receiving station B (slave side) and waits for a reply; the reply, a slave response, is transmitted by receiving station B (master side) to sending station A (slave side). If the reply indicates the block was accepted, sending station A (master side) may send another block or it may terminate. If the reply is negative, sending station A (master side) immediately retransmits the block that was not accepted.

Fig. 23 illustrates typical sequences of messages, transmission blocks, and slave responses that could occur when subcategory F1 is employed on a point-to-point full-duplex link. Fig. 24 is a flowchart of subcategory F1.

6.14.2 Transmission Blocks. In this section, all statements made relative to the relationship between station A (sender) and station B (receiver) apply equally to the relationship between station B (sender) and station A (receiver).

The transmission of blocks by station A is initiated after an appropriate establishment procedure [Fig. 24

[11] Message transfer subcategory F1 may also be used on a nonswitched, dedicated, full-duplex link to accomplish two-way simultaneous message transfers, in which case, establishment/termination procedures, such as those prescribed for subcategory 3.1, may be required.

(1)].[12] At this time, transmission may or may not already be in progress from B to A. As long as station A does not have an outstanding slave response obligation that has yet to be satisfied, it may initiate transmission of its first message.

If the message has a heading, station A begins the transmission with SOH [Fig. 24 (2)]. If the message has no heading [Fig. 24 (3)], station A begins the transmission with STX [Fig. 24 (4)].

An intermediate block that continues a heading [Fig. 24 (7), (2)] is started with SOH. An intermediate block that either begins or continues a text [Fig. 24 (7), (3), (4)] is started with STX. If the last information character of a heading ends on a block boundary (ended by ETB), the subsequent block may be started by either SOH or STX. Note that in such a case the receiver must be able to handle both situations.

---

[12] On a nonswitched, dedicated, full-duplex link where both stations idle in the "ready-to-receive" state, no establishment procedures are required.

A block that ends at an intermediate point within a message is ended with ETB [Fig. 24 (5)]. A block that ends at the end of a message is concluded with ETX [Fig. 24 (6)]. The ETB or ETX character is immediately followed by a block check character (BCC). After the ETB or ETX and BCC are sent, station A waits for a reply.

Upon detecting ETB or ETX followed by BCC from station A, station B may interrupt its transmission of a message or block in order to embed (insert) its slave response pertaining to the transmission received from station A. The rules governing the embedding of slave responses are given in 6.14.2.1 and 6.14.2.2.

6.14.2.1 Slave responses must not be embedded:

(1) Within another supervisory sequence

(2) Within any DLE sequences

(3) Between a start-of-block indicator and a following link block number

(4) Between an ETB or ETX used to delineate the end of a transmission block and the following block check character

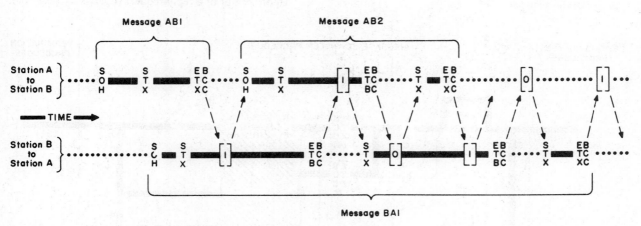

567

**Legend**

... { Asynchronous—Idle line
Synchronous—The transmission of any number of SYN characters (2 to *N*)

▬ The transmission of any number of data characters (0 to *M*)

[1] An affirmative slave response L: (prefix) L1 (with D/E prefix and D/E)

[2] An affirmative slave response L: (prefix) L0 (with D/E prefix and D/E)

**Fig. 23**
**Subcategory F1: Typical Control Character and Slave Response Sequences for Message-Associated Blocking with Longitudinal Checking and Alternating Acknowledgments for Two-Way Simultaneous Operation with Embedded Responses**

**6.14.2.2** The error check summation in progress must be halted for the duration of the embedded slave response and then resumed so that the slave response is not included in the block check.

**6.14.3 Replies.** Station B, upon detecting ETB or ETX followed by BCC, sends one of two replies:

(1) If the transmission block was accepted and station B is ready to receive another block, it sends SOSS, an optional prefix, and the appropriate acknowledgment [Fig. 28 (8)]. ACK1 is used as the affirmative reply to the first block received. ACK0 is used as the affirmative reply to the second block received. Subsequent affirmative acknowledgments alternate between ACK1 and ACK0. Upon detecting the appropriate affirmative acknowledgment, station A, being free of any slave response obligation, may either begin transmission of the next block [Fig. 24 (7)], or initiate termination [Fig. 24 (9)] [13] if the last block ended in ETX BCC [Fig. 24 (6)].

(2) If the transmission block was not accepted and

---

[13] On a nonswitched, dedicated, full-duplex link where both stations idle in the "ready-to-receive" state, no termination procedures are required.

station B is ready to receive another block, it sends SOSS, an optional prefix, and NAK [Fig. 24 (10)]. Upon detecting NAK, station A, being free of any slave response obligation, initiates retransmission of the last transmission block [Fig. 24 (11), (7)]. $N$ retransmissions ($N \geqslant 0$) may be made, after which station A (master) exits [Fig. 24 (12)] to a recovery procedure.

If the alternating reply indicates that station B missed the outstanding block (receipt of ACK1 instead of ACK0, or vice versa), station A initiates retransmission of that block as if the slave station had returned a NAK [Fig. 24 (11)].

The use of NAK does not alter the sequence of alternating acknowledgments. The same affirmative reply (ACK0 or ACK1) is used for a successful retransmission as would have been used if the previous transmission of the block in error had been successful.

If station A receives an invalid or no reply to a transmission block [Fig. 24 (13)], it may send a reply-request supervisory sequence [Fig. 24 (14)] consisting of SOSS, an optional prefix, and ENQ, or it may exit to a recovery procedure [Fig. 24 (15), (12)].

Upon receipt of a reply-request supervisory sequence,

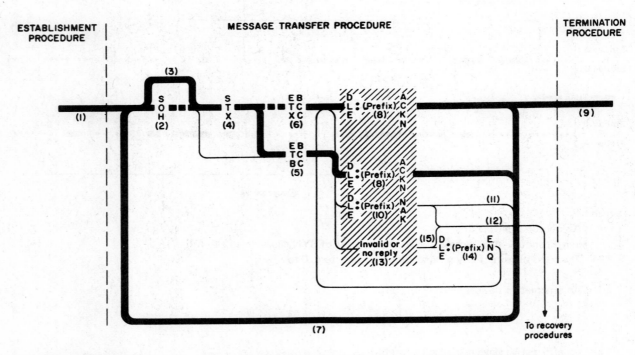

**Fig. 24**
**Subcategory F1: Message-Associated Blocking with Longitudinal Checking and Alternating Acknowledgments for Two-Way Simultaneous Operation with Embedded Responses**

station B repeats its last reply at the earliest possible time (embedded or interleaved). $N$ retries ($N \geqslant 0$) may be made to get a valid reply. If a valid reply is not received after $N$ retries, station A exits [Fig. 24 (15), (12)] to a recovery procedure.

NOTE: The use of a reply-request supervisory sequence [Fig. 24 (14)] may result in the loss or duplication of transmission blocks. Loss of transmission blocks may be avoided by always retransmitting the previous block. Although this precaution does not eliminate the possible receipt of duplicate blocks, the reply-request supervisory sequence permits the detection of such an occurrence. The use of a numbering scheme will aid in detecting loss or duplication of blocks.

## 6.15 Subcategory F2: Message-Independent Blocking—Continuous Operation with Longitudinal Checking and Modulo-8 Numbering of Blocks and Acknowledgments for Two-Way Simultaneous Operation with Embedded Responses

**6.15.1 General Description.** Message transfer subcategory F2 may only be used with establishment and termination subcategory 3.1.[14] It is similar to subcategory C2, except that it provides for the embedding (as well as the interleaving) of slave responses for systems operating in a two-way simultaneous data transmission environment employing full-duplex communication facilities. In such environment each station must have the capability of maintaining concurrent master and slave status, that is, master status on its transmit side and slave status on its receive side.

Subcategory F2 is applicable to systems where nontransparent message data are blocked independently of the message format. A block may contain partial, complete, or multiple messages. Sending station A (master side) may send blocks continuously to receiving station B (slave side) without waiting for a reply; replies, or slave responses, are transmitted by receiving station B (master side) while sending station A (master side) is sending subsequent blocks. By use of modulo-8 numbering of blocks and replies, sending station A (master side) may get as many as seven blocks ahead of the received replies before stopping to wait for the replies to catch up. If a negative reply is received, sending station A (master side) starts retransmission with the block following the last block for which the proper affirmative acknowledgment was received.

Fig. 25 illustrates typical sequences of transmission blocks and slave responses that could occur when

subcategory F2 is employed on a point-to-point full-duplex link. Fig. 26 is a flowchart of subcategory F2.

**6.15.2 Transmission Blocks.** In this section, all statements made relative to the relationship between station A (sender) and station B (receiver) apply equally to the relationship between station B (sender) and station A (receiver).

The transmission of blocks by station A is initiated after an appropriate establishment procedure [Fig. 26 (1)].[15] At this time, transmission may or may not already be in progress from B to A. As long as station A does not have an outstanding slave response obligation that has yet to be satisfied, it may initiate transmission of its first block.

Transmission blocks always begin with SOTB (DLE followed by =) followed by BLK [Fig. 26 (2)] and end with ETB [Fig. 26 (3)] followed immediately by a block check character (BCC).[16] BLK refers to the set of eight block numbers (see 3.7). Block numbering begins with BLK1 and progresses through BLK7 and BLK0 before recycling with BLK1.

Upon detecting ETB followed by BCC from station A, station B may interrupt its transmission of a block or message in order to embed (insert) a slave response pertaining to the transmission received from station A. The rules governing the embedding of slave responses are given in 6.15.2.1 and 6.15.2.2.

**6.15.2.1** Slave responses must not be embedded:
(1) Within another supervisory sequence
(2) Within any DLE sequences
(3) Between a start-of-block indicator and a following link-block number
(4) Between ETB and the following block check character

**6.15.2.2** The error check summation in progress must be halted for the duration of the embedded slave response and then resumed so that the slave response is not included in the block check.

Station A may send blocks continuously [Fig. 26 (4), (5)] as long as it does not get more than seven blocks ahead of the replies.

**6.15.3 Replies.** Upon detecting ETB followed immediately by BCC, station B sends one of two replies at the earliest possible time (embedded or interleaved):

---

[14] Message transfer subcategory F2 may also be used on a nonswitched, dedicated, full-duplex link to accomplish two-way simultaneous message transfers, in which case establishment/termination procedures, such as those prescribed for subcategory 3.1, may be required.

[15] On a nonswitched, dedicated, full-duplex link where both stations idle in the "ready-to-receive" state, no establishment procedures are required.

[16] The BLK number and the ETB character are included in the block check character, but the SOTB sequence is not. Any characters in the text other than SYN and embedded slave response sequences from A to B are included in the summation; for example, if STX, SOH, or ETX occur in text these characters are also included.

(1) If the block is acceptable and station B is ready to receive subsequent blocks, it replies in either of the two following ways:

(a) If BLK is either as expected or is lower than expected, station B sends SOSS, an optional prefix, and the appropriate affirmative reply. The affirmative reply [Fig. 26 (6)] is ACK1 for the first block received, ACK2 for the second block received, etc, through ACK7, ACK0, restarting with ACK1 [Fig. 26 (6)]. Station B should discard blocks containing block numbers lower than expected.

(b) If BLK is higher than expected, station B sends SOSS and an optional prefix followed by NAK [Fig. 26 (7)]. Station B does not reply to subsequent blocks

received, whether they are in error, or not, until the proper block number is detected or a reply-request supervisory sequence [Fig. 26 (11)] is received.

(2) If the block was not acceptable and station B is ready to receive another block, it sends SOSS and an optional prefix followed by NAK [Fig. 26 (7)]. Station B does not reply to subsequent blocks received, whether they are in error, or not, until the proper block number is detected or a reply-request supervisory sequence [Fig. 26 (11)] is received.

As long as appropriate affirmative replies, in the proper sequence, are detected, station A continues transmission [Fig. 26 (4), (5)]. After all transmission blocks have been sent, station A (slave side) continues

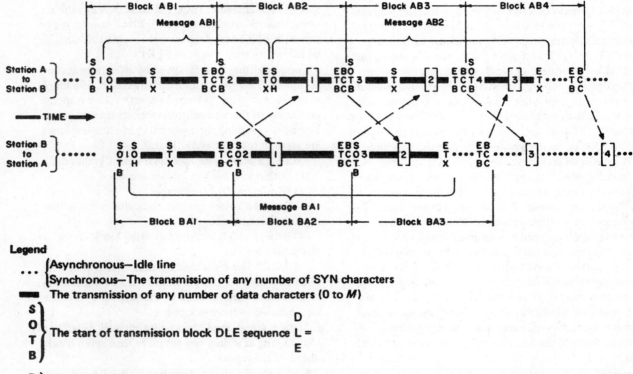

570

**Legend**

··· {Asynchronous—Idle line
Synchronous—The transmission of any number of SYN characters

▬ The transmission of any number of data characters (0 to *M*)

S
O        The start of transmission block DLE sequence L =         D
T                                                               E
B

B
L    The link block number transmitted as a single ASCII numeric character in the sequence 1, 2, 3, 4, 5, 6,
K    7, 0, 1, etc

[*N*]  A numbered affirmative slave response L: (prefix) L *N*, where *N* (0 to 7) corresponds to the number of
                                          D          D
                                          E          E
       the block being acknowledged

**Fig. 25**
**Subcategory F2: Typical Control Character and Slave Response Sequence for Message-Independent**
**Blocking—Continuous Operation with Longitudinal Checking and Modulo-8 Numbering of Blocks**
**and Acknowledgments for Two-Way Simultaneous Operation with Embedded Responses**

to monitor replies until all blocks have been affirmatively acknowledged by station B (master side). Station A then initiates termination [Fig. 26 (4), (13)].[17]

Upon detecting a NAK reply, station A stops transmission at the end of the block and initiates retransmission starting with the block following the last block for which the proper affirmative acknowledgment was received. $N$ retransmissions ($N \geq 0$) may be made, after which station A may either exit [Fig. 26 (9)] to a recovery procedure [Fig. 26 (14)] or initiate termination [Fig. 26 (13)].[17]

The use of NAK does not alter the sequence of acknowledgments. Station B uses the same affirmative

reply (ACK0, ACK1, etc) for a successful retransmission as would have been used if the previous transmission of that block had been successful.

If station A detects an invalid or no reply to a transmission block [Fig. 26 (10)], it halts transmission at the end of the block being transmitted and sends a reply-request supervisory sequence [Fig. 26 (11)].

Upon receipt of a reply-request supervisory sequence, station B sends the affirmative reply for the last block accepted. Station A then resumes transmission with the proper transmission block. $N$ retries ($N \geq 0$) may be made to get a valid reply. If a valid reply is not received after $N$ attempts, station A may either exit [Fig. 26 (12)] to a recovery procedure [Fig. 26 (14)] or initiate termination [Fig. 26 (4), (13)].

The block counter is reset to "one" upon a timer-C time-out or upon transmission/reception of EOT.

---

[17]On a nonswitched, dedicated, full-duplex link where both stations idle in the "ready-to-receive" state, no termination procedures are required.

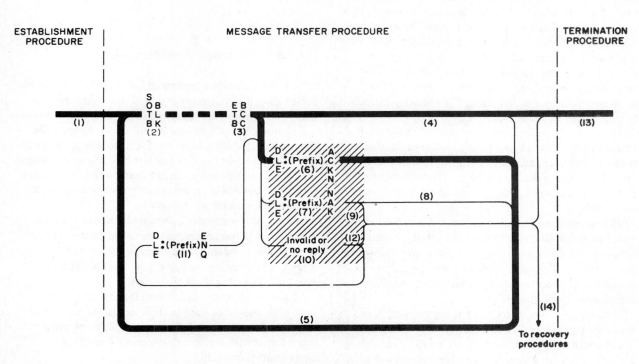

Fig. 26
Subcategory F2: Message-Independent Blocking—Continuous Operation with Longitudinal Checking and Modulo-8 Numbering of Blocks and Acknowledgments for Two-Way Simultaneous Operation with Embedded Responses

## Appendix A

## Design Considerations for Data Communication Control Procedures

### A1. General Design Considerations

The problem of standardizing data communication control procedures is complicated by the diversity of needs of different types of communication systems. A standard covering data communication control procedures must meet the needs of the numerous types of communication systems of significance. At the same time, however, optionality in the standard must be controlled so that the advantages of a standard can accrue to its users.

Three approaches to the specification of data communication control procedures were considered during the formulation of this standard. They differ principally with regard to subject level:

(1) Control Character Level. A standard at the control character level would specify only the communication control characters and their functions. The standard control characters could be used by system designers to implement communication systems in accordance with their needs.

(2) Communication-Function Level. A standard at the communication-function level would specify a standard set of control procedures for each of several communication functions. The standard would not specify particular associations of communication functions to form an overall operational discipline.

(3) Category Level. A standard at the category level would specify sets (called categories) of data communication control procedures for use on a data communication link. Hence, a standard at this level would not only specify control procedures for each communication function, but would also specify standard associations of functions to form categories.

This standard specifies standard categories of data communication control procedures. A relatively large number of categories have been included to meet the needs of many different types of data communication systems. Additional categories will be added to the standard as the need for them arises.

Section 4 of this standard defines the functional properties of the communication control signals. The section is included primarily to provide guidance to systems designers implementing functions not covered in the categories. Such implementations are viewed as being outside the scope of this standard, but not necessarily as constituting a violation of this standard.

### A2. Specific Design Considerations

**A2.1 Code Extension Sequences Using DLE.** The criteria for dividing the characters of ASCII into "intermediate" and "final" groups are discussed in the Proposed American National Standard Code Extension Techniques for Seven and Eight-Bit Codes. For DLE sequences the partition selected is the same as that for ESC sequences except that the communication control characters from columns 0 and 1 of the ASCII coded character set (see Appendix E) may be used as final characters.

In 3.2.5 of this standard, certain characters of ASCII are reserved as final characters for use in future extensions to this standard. The criteria in A2.1.1 and A2.1.2 are recognized as being applicable to future assignments involving these characters.

**A2.1.1** No character in columns 0 and 1 of the ASCII code table should be used in DLE sequences (of any length) unless the intended control function obtained is consistent with the functional properties of the single control character.

**A2.1.2** The characters in A2.1.2.1 through A2.1.2.5 must not be assigned for use in DLE sequences of any length. However, in some systems SUB, DEL, and CAN might appear in the reception of DLE sequences as a result of error, error control, or recovery procedures. The standard interpretation of such sequences is not specified.

**A2.1.2.1** NUL is excluded because of the hazards associated with the lack of clearly established conventions for its use.

**A2.1.2.2** CAN is excluded since its purpose is to "cancel" a portion of the data and possibly eventually to "cancel" a DLE sequence.

**A2.1.2.3** SUB is similarly excluded because it may be used to replace a character determined to be in error and may thus unpredictably appear in a DLE sequence as a result of this process.

**A2.1.2.4** DEL is excluded because it may unpredictably appear as a result of correction of operation errors in perforated tape. Also, some portions of a system may "delete" this character from the data stream.

**A2.1.2.5** The ESC character is excluded from DLE sequences to avoid potential interference with the control logic of data stations.

## Appendix B
## Description of Communication Functions

### B1. Introduction

The categories described in this standard were developed by associating various communication functions in different ways. Hence the control procedures of a category are the aggregate of the control procedures for each of the communication functions in the category. Since they constitute the basis for the categories, a brief description of the communication functions is included for reference purposes.

### B2. Establishing Communication

**B2.1 General.** This standard does not prescribe the means for establishing a channel (signaling path) between two or more stations. However, after a channel has been established, procedures to identify and control the status of connected stations are specified. That is, the control procedures are used to designate which station is to be the master station and which station(s) is (are) to be the slave station(s).

The control procedures used in establishing a communication link not only make use of communicaton control characters but of other characters as well. All characters can be interpreted unambiguously since the control functions are not performed during message transmission.

**B2.2 Polling.** In polling, the control station transfers master status to a tributary station. A station is polled

so that it may transmit a message, but a station need not have a message to send to be polled.

The control station sends a polling supervisory sequence consisting of a prefix following by ENQ. The prefix identifies a single tributary station and may also include other information. ENQ defines the end of the polling supervisory sequence.

When a station is polled, it may respond in either of three ways: (1) If the station has a message to send, it commences the transmission of the message. In such case, the message is deemed to be directed to the control station. (2) If the message is for one or more tributary stations on the link, the polled station may precede the message transmission with the selection of the appropriate slave station(s). (3) If the polled station does not have a message to send, it should relinquish master status by terminating the transmission. In this case, the control station may either poll another tributary station or it may select a tributary station to receive a message.

Use of EOT to initiate a polling supervisory sequence is permitted in some systems.

**B2.3 Identification.** An identification procedure is used (1) to determine the readiness of a slave station to receive a message transmission, (2) to determine the identity of the station(s), and (3) to exchange supplementary information concerning the status of the station(s).

An identification procedure may only be initiated by a station that has master status or, in contention systems, a station that wishes to obtain master status. The identification procedure consists of the transmission of an ENQ character (in noncontention systems, ENQ may be preceded by a prefix). Upon receipt of the ENQ character (with or without a prefix), the receiving station should condition itself to receive a message and should send a reply. If the station receiving ENQ is ready to receive a message, it should reply with ACK (possibly preceded by a prefix to identify itself and, perhaps, to convey supplementary information). If it is not ready to receive a message, it should reply with NAK (possibly preceded by a prefix to identify itself and, perhaps, to convey supplementary information). If establishment and termination subcategory 2.2 or 3.1 is being used, a prefix that includes the identity of the called station is mandatory.

**B2.4 Contention.** In contention systems the reply by the receiving station either confirms or denies the other station's bid for master status. If both stations bid for the line, either station or both may receive the other's ENQ character. Recovery from this condition is systems dependent. One way of resolving this situation requires designating one station as primary and the other as secondary. If the primary station receives ENQ, it restarts its time-out to provide time for the secondary station's reply to the primary station's ENQ. If a normal reply is received before the time-out expires, the primary station follows normal procedure (for example, if ACK0 is received, transmit a message). If a second ENQ character is received, the primary station should follow systems recovery procedure (for example, retransmit ENQ $N$ times, awaiting a reply each time in an effort to break contention).

In some systems it may be desirable (or necessary) to establish the master/slave relationship by convention in order to avoid contention situations. This is exemplified by circuit-switched, point-to-point systems.

**B2.5 Selection.** In selection, a tributary station is nominated to be a slave station. Selection is used primarily on multipoint data communication links and may only be performed by a station with master status. The selection process (1) determines the readiness of a slave station to receive a message transmission, (2) provides for the identification of the slave station, and (3) provides for the exchange of supplementary information concerning the status of the station(s).

The selection procedure consists of the transmission of a selection supervisory sequence by the master station and the transmission of a reply to this sequence by the slave station. A selection supervisory sequence consists of a selection prefix (to identify the station

being selected and, perhaps, to convey supplementary information) followed by an ENQ character. (If a polling function is provided, the polling and selection prefixes should be different.) A station receiving its selection supervisory sequence may reply in either of two ways:

(1) If the selected station is ready to receive a message, it should reply with ACK (possibly preceded by a prefix to identify and, perhaps, to convey supplementary information).

(2) If the selected station is not ready to receive a message, it should reply with NAK (possibly preceded by a prefix to identify itself and, perhaps, to convey supplementary information).

In some data communication systems, replies are not used (subcategories 1.1 and 2.5). In these systems either the master/slave relationship may be fixed (in which case the selection function is not used), or a form of selection in which no reply is expected may be used.

Use of EOT to initiate a selection supervisory sequence is permitted in some systems.

## B3. Message Transmission

**B3.1 Introduction.** The transmission of messages on a data communication link may be separated into two functions: (1) the framing of messages or transmission blocks and (2) the use of replies to messages or transmission blocks.

**B3.2 Framing of Messages and Transmission Blocks.** A message or transmission block is a sequence of characters, transmitted as a unit, for which a reply is normally expected (except in message transfer subcategories A1 and A2). The message or transmission block is both preceded and followed by communication control characters that serve as delimiters.

Messages or transmission blocks that include a heading begin with SOH or TSOH. Messages or transmission blocks that do not include a heading begin with STX or TSTX. If a transmission block includes both heading and text, STX or TSTX is used to delimit the end of the heading and the beginning of the text. Messages end with ETX or TETX. When a message is subdivided, ETB or TETB terminates all but the last block of the message.

Messages may or may not be followed by a block check character (BCC) or cyclic redundancy check (CRC). Transmission blocks must always be followed by a BCC or CRC.

The master station must wait after transmitting each message or block for a reply, except in subcategories

C2 and F2. When a master station receives an affirmative reply, it may proceed with the transmission of the next message or transmission block, or it may terminate the transmission.

If a master station receives a negative reply to a message or transmission block, it may either retransmit the unacceptable message or block, or it may send the next message or block, depending upon the system discipline in use. However, in conversational message transfers, complete messages serve as replies. Affirmative or negative response is either implied by transmission of STX (as part of a message or a block) or explicitly transmitted in subcategory E2 or E3.

### B3.3 Replies to Messages and Transmission Blocks.

When a slave[18] station receives the control character that delimits the end of a message or block, it should accept the character immediately following as a block check character (if longitudinal checking is used) and transmit a reply for the message or block. This reply:

(1) Indicates the acceptability of the received data.

(2) Indicates that the slave station is ready to receive the next message or transmission block.

(3) May contain the identity of the slave station and, possibly, supplementary information concerning the status of the slave station. The slave station should transmit the reply as soon as possible after the receipt of the message or transmission block.

There are two valid replies. If the message or block is acceptable to the slave station, an affirmative reply (for example, ACK or ACK$N$) should be transmitted. If the message or block is unacceptable, a negative

reply (NAK) should be transmitted. Either of these replies may have a prefix to identify the slave station or convey supplementary information, or both.

The criteria for acceptance of a message or transmission block by a slave station are not covered by this standard. In general, the criteria include consideration of the parity of the received characters and the validity of the block character; however, additional factors pertaining to the received data or the operational state of the station may be considered.

### B3.4 Invalid or No Reply.

If a reply to a message or transmission block is not received, the master station may, after an appropriate time interval, request a reply from the slave station, or exit to a recovery procedure. The request for a reply should have the form of a reply-request supervisory sequence. If a valid reply is not received in response to this request, the master station may repeat the request up to $N$ times ($N \geqslant 0$). If, after $N$ requests, a valid reply is not received, the master station should exit to a recovery procedure. It is possible to lose or duplicate blocks unless a numbering scheme is used.

### B3.5 Checking Using CRC

**B3.5.1 Encoding.** Fig. B1 shows an arrangement for encoding using a shift register and the CRC-16 polynomial. To encode, the storage stages are set to zero, gates A, B, and E are enabled, gate C is inhibited, and $K$ service and information bits are clocked into the input. They will appear simultaneously at the output.

After the bits have been entered, gates A and B are inhibited and gate C enabled, and the register is clocked a further 16 counts. During these counts the required check bits will appear in succession at the output. When characters not included in the CRC check are encountered in the bit stream, gate B is enabled and gate E is inhibited.

---

[18] In conversational message transfer, the master/slave relationship is not defined, since either initiator or responder may transmit text or replies. A condition or status similar to master status may reverse each time a complete message is transmitted by either initiator or responder.

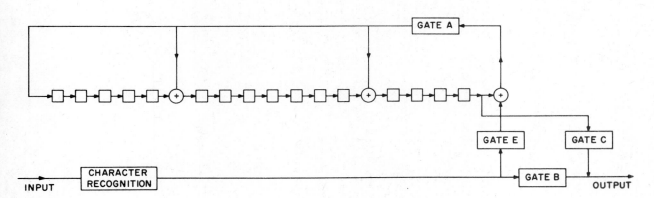

**Fig. B1**
**Encoder**

**B3.5.2 Decoding.** Fig. B2 shows an arrangement for decoding using the shift register on the CRC-16 polynomial. To decode, gates A, B, and E are enabled, and the storage stages are set to zero.

The $K$ information bits are then clocked into the input and after $K$ counts gate B is inhibited; the 16 check bits are then clocked into the input and the contents of the storage stages are examined. For an error-free block the contents will be zero. A nonzero content indicates an erroneous block. When characters not included in the CRC check are encountered in the bit stream, gate B is enabled and gate E is inhibited.

## B4. Termination

**B4.1 Introduction.** Termination includes two control functions: (1) terminate and (2) mandatory disconnect.

**B4.2 Terminate.** The terminate function consists of the transmission, by a master station, of the EOT character. Upon transmission of EOT the master station relinquishes its master status, and the selection of all slave stations is canceled.

The terminate function is performed by a master station to decline master status or to relinquish use of the link upon completing a transmission.

In multipoint systems the transmission of EOT transfers master status to the control station.

**B4.3 Mandatory Disconnect.** The mandatory disconnect function consists of the transmission, by a master station, of the character sequence DLE followed by EOT (DEOT). Upon transmission of DEOT the master station relinquishes its master status, and the selection of the slave station is canceled.

In circuit-switched point-to-point systems, DEOT initiates the mandatory disconnect of the switched circuit.

In nonswitched systems DEOT has no additional meaning from EOT.

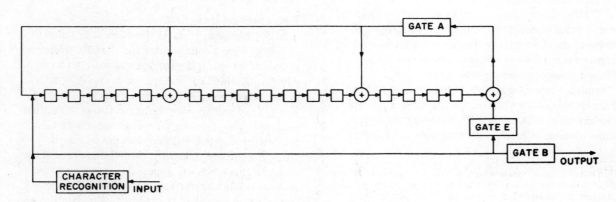

**Fig. B2**
**Decoder**

## Appendix C
## Future Work

At the time of approval of this standard, the following topics were under consideration by Task Group X3S34:

(1) An investigation of bit-oriented advanced data communication control procedures to provide a new generation of functional capabilities — such as code independence, interactive operation, and network configuration independence — accompanied by improved reliability and efficiency. It is not intended that these bit-oriented procedures obsolete the ASCII-oriented procedures specified in this standard.

(2) An investigation of the requirements for network control procedures.

These topics represent efforts to broaden the scope and field of application of control procedures beyond what is defined in this standard.

## Appendix D
## Glossary

The following technical terms are used in this standard. For a more comprehensive vocabulary, including both data-processing and communication terms, reference is made to American National Standard Vocabulary for Information Processing, X3.12-1970.

Parentheses are used to indicate words that are sometimes deleted in text when they are implied by the context. Italicization of terms in the definitions indicates that the terms have been defined elsewhere in the glossary.

**American National Standard Code for Information Interchange—ASCII.** The set of 128 control and data characters defined as a standard data communications code in American National Standard Code for Information Interchange, X3.4-1968.

**Block, Transmission.** See *(Transmission) Block.*

**Centralized Operation.** A control discipline for *multipoint data communication links* by which *transmission* may be between the *control station* and *tributary station* but not between *tributary stations.*

**Circuit.** See *(Data Communication) Circuit.*

**Circuit Switched.** See *(Switched Circuit).*

**Code Transparent Data Link Control Procedure.** A procedure for transmitting all combinations of coded or binary data for which link control (for example, framing) is exercised by *control characters* (from a fixed coded character set) residing between character unit boundaries. Block length is a multiple of integral numbers of character units (proposed American National Standard implementations consider only variable undeclared block length).

**Communication Circuit.** See *(Data Communication) Circuit.*

**Communication Control Procedures.** See *(Data Communication) Control Procedures.*

**Communication Facility.** The communication medium used to provide *communication circuits* (for example, cable, radio, open-wire).

**Communication Link.** See *(Data Communication) Link.*

**Contention.** An operational condition on a *data communication link* when two stations try to transmit at the same time and no *station* is designated a *master station.*

**Continuous Operation.** A type of *transmission* through which the *master station* need not stop for a *reply* after transmitting each *message* or *transmission block.*

**Control Character.** See *(Data Communication) Control Character.*

**Control Procedures.** See *(Data Communication) Control Procedures.*

**Control Station.** The *station* on a *data communication link* with the overall responsibility for the orderly operation of the *link.* A *control station* has the responsibility for initiating *recovery procedures* in the event of abnormal conditions on the *link.* The designation of a particular *station* as a *control station* is not affected by the *control procedures* of this standard.

**Conversational Procedure.** Message transfer procedures for two *stations,* by means of which the direction of information transfer can alternate between *messages* without recourse to separate link establishment/termination procedures.

**Data Communication Circuit.** A means of both-way communication between two points, comprising associated "forward" and "backward" channels.

**(Data Communication) Control Character.** Within ASCII, a functional character intended to control or facilitate *transmission* of information over communication networks. There are ten such characters: EOT, SOH, STX, ETX, ACK, NAK, ENQ, ETB, SYN, DLE.

**(Data Communication) Control Procedures.** The means used to control the orderly communication of information between *stations* on a *data communication link.*

**(Data Communication) Link.** The logical association of two or more *stations* interconnected by the same *data communication circuit,* including the communication control capability of the interconnected *stations.*

**(Data Communication) Station.** The independently controllable configuration of logical elements, from or to which *messages* are transmitted on a *data communication link.* It includes those elements (data communication equipment, intermediate terminal equipment) that control the message flow on the *link* via *data communication control procedures.*

**Data Link.** See *(Data Communication) Link.*

**Delimiters.** *Control characters* used to define the extent of a particular sequence of characters.

**Disconnect.** The disassociation or release of a *switched circuit* between two *stations.*

**Heading.** See *(Message) Heading.*

**Identification, Station.** See *Station Identification.*

**Initiator.** The *station* that initiates message exchange in a conversational subcategory. The initiator is so

designated by virtue of receiving master status in the establishment procedure.

**Link.** See *(Data Communication) Link.*

**Link, Multipoint.** See *Multipoint Link.*

**Link, Point-to-Point.** See *Point-to-Point Link.*

**Master Station.** A *station* that has control of the *data communication link* at a given instant. The assignment of master status to a given *station* is temporary and is controlled by the procedures set forth in the categories described in this standard. Master status is normally conferred upon a *station* so that it may transmit a *message,* but a *station* need not have a *message* to send to be nominated master.

**Message.** A sequence of characters arranged for the purpose of conveying information from an originator to one or more destinations (or addresses). It contains the supplementary information in a *heading.*

**(Message) Heading.** The part of a *message* containing all components preceding the *text.* A message heading is preceded by an SOH character or TSOH sequence and followed by an STX character or TSTX sequence.

**(Message) Text.** The part of a *message* beginning with the first character following an STX character or TSTX sequence and followed by an ETX character or TETX sequence.

**Multipoint Link.** A *data communication link* connecting two or more *stations.*

**Noncentralized Operation.** A control discipline for *multipoint data communication links* by which *transmission* may be between *tributary stations* or between the *control station* and *tributary station(s).*

**One-Way-Only Transmission.** A type of *transmission* through which *message* or *transmission blocks* are sent in one direction only, with no return transmission.

**Operation, Centralized.** See *Centralized Operation.*

**Operation, Continuous.** See *Continuous Operation.*

**Operation, Noncentralized.** See *Noncentralized Operation.*

**Point-to-Point Link.** A *data communication link* connecting only two *stations.*

**Polling.** A technique for inviting a *station* to transmit *messages* at a given time. One *station* is designated as a *control station* to invite *tributary station(s)* to transmit.

**Polling Supervisory Sequence.** A *supervisory sequence* that performs a polling function.

**Prefix.** A sequence of characters (other than *communication control characters*) used in a *supervisory sequence* to define or qualify the meaning of the *supervisory sequence*.

**Recovery Procedure.** Control procedures used to restore normal operation to a *data communication link* after unusual (abnormal) events have occurred.

**Reply.** A *supervisory sequence* by which a *slave station* informs the *master station* of its operational condition or status, or the status of a received *message* or *block*.

**Reply-Request Supervisory Sequence.** A *supervisory sequence* used by a *master station* to request a *reply* from a *slave station*.

**Responder.** The station that is responsible for supplying reply *messages* to the *initiator*.

**Selection.** A technique for assignment of slave status to *station(s)* on a *data communication link*.

**Selection Supervisory Sequence.** A *supervisory sequence* that performs a selection function.

**Sequence, Supervisory.** See *Supervisory Sequence*.

**Slave Station.** A *station* that has been selected to receive a *transmission* from the *master station*. The assignment of slave status is (1) temporary, (2) under the control of the *master station*, and (3) continuous for the duration of a *transmission*.

**Station.** See *(Data Communication) Station*.

**Station, Control.** See *Control Station*.

**Station Identification.** A sequence of characters used to identify a *station*.

**Station, Master.** See *Master Station*.

**Station, Slave.** See *Slave Station*.

**Station, Tributary.** See *Tributary Station*.

**Supervisory Sequence.** A sequence of *communication control characters* and possibly other characters that performs a defined control function. A supervisory se-

quence may contain a *prefix* together with any required *delimiters*.

**Supervisory Sequence, Polling.** See *Polling Supervisory Sequence*.

**Supervisory Sequence, Reply-Request.** See *Reply-Request Supervisory Sequence*.

**Supervisory Sequence, Selection.** See *Selection Supervisory Sequence*.

**Switched Circuit.** A *circuit* that may be set up and cleared at the request of one or more of the connected *stations*.

**Synchronization.** The establishment and maintenance of a desired timing relationship between a transmitted and a received signal.

**Text.** See *(Message) Text*.

**Transmission.** The entirety of the data transmitted between the *master station* and *slave station(s)* for the period of their uninterrupted assignment to such status.

**(Transmission) Block.** A group of characters transmitted as a unit. A transmission block may contain a *message*, a portion of a *message,* or combinations thereof.

**Transmission, Two-Way Alternate.** See *Two-Way Alternate Transmission*.

**Transmission, Two-Way Simultaneous.** See *Two-Way Simultaneous Transmission*.

**Tributary Station.** A *station* on a *data communication link* that is not a *control station*.

**Two-Way Alternate Transmission.** A type of *transmission* through which *messages* or *transmission blocks* may be sent in either one direction or the other, but not both directions simultaneously.

**Two-Way Simultaneous Transmission.** A type of *transmission* through which *messages* or *transmission blocks* are sent in both directions simultaneously.

579

## Appendix E

## American National Standard Code for Information Interchange (ASCII)

Fig. E1 in this appendix shows the coded character set used for the general interchange of information among information processing systems, communication systems, and associated equipment.

| Bits | | | | | | 0 0 0 | 0 0 1 | 0 1 0 | 0 1 1 | 1 0 0 | 1 0 1 | 1 1 0 | 1 1 1 |
|---|---|---|---|---|---|---|---|---|---|---|---|---|---|
| $b_4$ | $b_3$ | $b_2$ | $b_1$ | ROW / COLUMN | | 0 | 1 | 2 | 3 | 4 | 5 | 6 | 7 |
| 0 | 0 | 0 | 0 | 0 | | NUL | DLE | SP | 0 | @ | P | ` | p |
| 0 | 0 | 0 | 1 | 1 | | SOH | DC1 | ! | 1 | A | Q | a | q |
| 0 | 0 | 1 | 0 | 2 | | STX | DC2 | " | 2 | B | R | b | r |
| 0 | 0 | 1 | 1 | 3 | | ETX | DC3 | # | 3 | C | S | c | s |
| 0 | 1 | 0 | 0 | 4 | | EOT | DC4 | $ | 4 | D | T | d | t |
| 0 | 1 | 0 | 1 | 5 | | ENQ | NAK | % | 5 | E | U | e | u |
| 0 | 1 | 1 | 0 | 6 | | ACK | SYN | & | 6 | F | V | f | v |
| 0 | 1 | 1 | 1 | 7 | | BEL | ETB | ' | 7 | G | W | g | w |
| 1 | 0 | 0 | 0 | 8 | | BS | CAN | ( | 8 | H | X | h | x |
| 1 | 0 | 0 | 1 | 9 | | HT | EM | ) | 9 | I | Y | i | y |
| 1 | 0 | 1 | 0 | 10 | | LF | SUB | * | : | J | Z | j | z |
| 1 | 0 | 1 | 1 | 11 | | VT | ESC | + | ; | K | [ | k | { |
| 1 | 1 | 0 | 0 | 12 | | FF | FS | , | < | L | \ | l | \| |
| 1 | 1 | 0 | 1 | 13 | | CR | GS | – | = | M | ] | m | } |
| 1 | 1 | 1 | 0 | 14 | | SO | RS | . | > | N | ^ | n | ~ |
| 1 | 1 | 1 | 1 | 15 | | SI | US | / | ? | O | _ | o | DEL |

Reprinted from American National Standard Code for Information Interchange, X3.4-1968.

**Fig. E1**
**ASCII Coded Character Set**

580

71

# American National Standard

synchronous high-speed
data signaling rates
between data terminal equipment
and data communication equipment

581

X3.36-1975

american national standards institute, inc.
1430 broadway, new york, new york 10018

This standard has been adopted for Federal Government use.

Details concerning its use within the Federal Government are contained in FIPS 37, SYNCHRONOUS HIGH SPEED DATA SIGNALING RATES BETWEEN DATA TERMINAL EQUIPMENT AND DATA COMMUNICATION EQUIPMENT. For a complete list of the publications available in the FEDERAL INFORMATION PROCESSING STANDARDS Series, write to the Office of Technical Information and Publications, National Bureau of Standards, Washington, D.C. 20234.

582

# American National Standard
# Synchronous High-Speed
# Data Signaling Rates
# between Data Terminal Equipment
# and Data Communication Equipment

Secretariat

**Computer and Business Equipment Manufacturers Association**

Approved April 2, 1975

**American National Standards Institute, Inc**

# American National Standard

An American National Standard implies a consensus of those substantially concerned with its scope and provisions. An American National Standard is intended as a guide to aid the manufacturer, the consumer, and the general public. The existence of an American National Standard does not in any respect preclude anyone, whether he has approved the standard or not, from manufacturing, marketing, purchasing, or using products, processes, or procedures not conforming to the standard. American National Standards are subject to periodic review and users are cautioned to obtain the latest editions.

**CAUTION NOTICE:** This American National Standard may be revised or withdrawn at any time. The procedures of the American National Standards Institute require that action be taken to reaffirm, revise, or withdraw this standard no later than five years from the date of publication. Purchasers of American National Standards may receive current information on all standards by calling or writing the American National Standards Institute.

584

Published by

**American National Standards Institute**
**1430 Broadway, New York, New York 10018**

# Foreword

(This Foreword is not a part of American National Standard Synchronous High-Speed Signaling Rates between Data Terminal Equipment and Data Communication Equipment, X3.36-1975.)

This American National Standard specifies a series of standard and preferred signaling rates to be employed in information systems. The standard applies to data communication systems that utilize synchronous data rates higher than those commonly used in analog voice bandwidth channels. It is expected that future revisions of this standard will identify additional standard or preferred rates as more experience is gained with high data rate systems.

American National Standard Synchronous Signaling Rates for Data Transmission, X3.1-1969, established a set of standard rates for nominal 4-kHz voice bandwidth channels. Other standards have been adopted, or are being developed, that prescribe the character structure, bit sequencing, message formats, and other parameters vital to the communication of information in a data communication system.

In the development of this standard, both historical and present data communications practices were considered. In addition, consideration was given to the requirements for future systems and to the applicable standards being developed internationally.

Suggestions for improvement of this standard will be welcome. They should be sent to the American National Standards Institute, 1430 Broadway, New York, N.Y. 10018.

This standard was processed and approved for submittal to ANSI by American National Standards Committee on Computers and Information Processing, X3. Committee approval of the standard does not necessarily imply that all committee members voted for its approval. At the time it approved this standard, the X3 Committee had the following members:

J. F. Auwaerter, Chairman
V. E. Henriques, Vice-Chairman
R. M. Brown, Secretary

| Organization Represented | Name of Representative |
|---|---|
| Addressograph Multigraph Corporation | A. C. Brown |
| | D. S. Bates (Alt) |
| Air Transport Association | F. C. White |
| American Bankers Association | M. E. McMahon |
| | J. Booth (Alt) |
| American Gas Association | J. A. Pinnola |
| American Institute of Certified Public Accountants | N. Zakin |
| | P. B. Goodstat (Alt) |
| | C. A. Phillips (Alt) |
| | F. Schiff (Alt) |
| American Library Association | J. R. Rizzolo |
| | J. C. Kountz (Alt) |
| | M. S. Malinconico (Alt) |
| American Newspaper Publishers Association | W. D. Rinehart |
| American Nuclear Society | D. R. Vondy |
| | M. K. Butler (Alt) |
| American Society for Information Science | S. Furth |
| American Society of Mechanical Engineers | R. W. Rau |
| | R. T. Woythal (Alt) |
| Association for Computing Machinery | J. A. N. Lee |
| | P. Skelly (Alt) |
| | H. Thiess (Alt) |
| Association for Educational Data Systems | C. Wilkes |
| Association for Systems Management | A. H. Vaughan |
| Association of American Railroads | R. A. Petrash |
| Association of Computer Programmers and Analysts | T. G. Grieb |
| | G. Thomas (Alt) |
| Burroughs Corporation | E. Lohse |
| | J. F. Kalbach (Alt) |
| Control Data Corporation | S. F. Buckland |
| | C. E. Cooper (Alt) |
| Data Processing Management Association | A. E. Dubnow |
| | D. W. Sanford (Alt) |

The following members of Task Group X3S3.6 on Digital Data Signaling Rates were the principal contributors to the development of this standard:

H. J. Crowley, Chairman

| | |
|---|---|
| N. Kramer | R. J. Smith |
| S. Lechter | J. L. Wheeler |
| W. Lyons | C. E. Young |
| S. Schreiner | |

Task Group X3S3.6 worked in cooperation with Electronic Industries Association Subcommittee TR30 on Data Transmission in the development of this standard.

Others who contributed to the development of this standard were the following:

| | |
|---|---|
| W. W. Baty | L. T. O'Connor |
| W. Brown | N. Priebe |
| R. H. Bickler | S. Rosenblum |
| M. L. Cain | P. D. Simpson |
| R. Hearsum | H. H. Smith |
| G. R. Hopping | N. E. Snow |
| W. M. Hornish | A. H. Stillman* |
| L. G. Kappel | A. F. Taibl |
| R. F. Meyer | I. Velinov |
| O. C. Miles | G. W. White |

*Former Chairman, deceased.

# Contents

# American National Standard Synchronous High-Speed Data Signaling Rates between Data Terminal Equipment and Data Communication Equipment

## 1. Scope and Purpose

This standard provides a group of specific signaling rates for synchronous high-speed serial data transfer. These rates exist on the received data and transmitted data circuits of the interface between data terminal equipment and data communication equipment that operate over high-speed channels. This standard does not specify rates at the output of data communication equipment (that is, on the communication circuit shown in Fig. 1). Rates on communication circuits may be either higher (because, for example, information required for communication system control has been added) or lower (because modulation of a higher order than binary has been used).

## 2. Standard Signaling Rates

2.1 The standard signaling rates used for data transfer at rates above 9600 bits per second shall be selected integral multiples of 8000 bits per second.

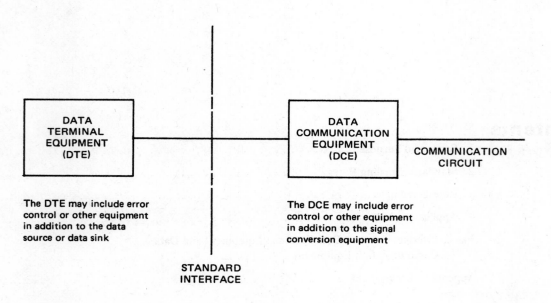

**Fig. 1**
**Interface between Data Terminal Equipment and Data Communication Equipment**

**2.2** The selected standard signaling rates shall be:

16 kilobits per second
56 kilobits per second
1.344 megabits per second
1.544 megabits per second*

**2.3** For those applications where established practice requires the use of the internationally recommended rate for wide-band transmission, a rate of 48 kilobits per second is recognized.

## 3. Rate Tolerances

The serial signaling rates defined herein are nominal values. The tolerance of these rates is under further

___

*This rate will be subject to coding restrictions in many applications involving public networks, as noted in the Appendix.

study and will be specified in a future revision of this standard.

## 4. Application Notes

**4.1** Where data rate converters, such as error control or similar devices which change the signaling rates by a fixed ratio, are inserted between the data terminal source/sink equipment (DTE) and the data communication equipment (DCE), they may be considered as part of the DTE or as part of the DCE as shown in Fig. 1. Whether they are considered as part of the DTE or as part of the DCE, the interface rates shall conform to this standard.

**4.2** When data rate converters are used which result in a nonstandard signaling rate at the data processing equipment junction with the rate converter, the clocking for the data processing equipment, if required, shall be furnished by the data rate converter.

# Appendix

(This Appendix is not a part of American National Standard Synchronous High-Speed Data Signaling Rates between Data Terminal Equipment and Data Communication Equipment, X3.36-1975, but is included for information purposes only.)

## Background

Within the past several years, there has been a tremendous growth in time-division multiplexing of voice channels within the communication networks of the carriers. To date, this multiplexing has been applied primarily on the interoffice trunks of the telephone systems in order to concentrate more channels on a given size cable plant. The digital system will soon be extended to the long-distance trunks because of its many technical and economic advantages. Although there are several different digital modulation schemes that can be used, the pulse code modulation (PCM) scheme is the one being widely implemented. Many thousands of channel miles of PCM carrier service are being installed each year, not only in the United States but throughout the world.

At the present time, the PCM terminals are arranged to accept a conventional analog telephone channel at the input and deliver a conventional analog signal at the output. Between the input and output terminals, the signal is digital and can be propagated only over a line equipped with digital repeaters. Within limits, it is, therefore, feasible to use a particular PCM channel for data transmission with a conventional modem designed for 4-kHz voice channels. For wide-band transmission using PCM channels, however, it is not feasible to employ a wide-band modem that would be suitable for a frequency division multiplex (FDM) group bandwidth channel. The basic reason for this is that there is no wide-band analog channel available in the PCM system. In the PCM multiplexer, the 24 input analog channels are combined into a single digital bit stream at the output. At the next level of multiplexing, four of these streams are again multiplexed into another high-speed bit stream. Therefore, it can be seen that there is no wide-band analog channel in the PCM system that would be equivalent to the group or supergroup channels of an FDM system.

In the United States, the PCM telephone transmission system uses a pulse rate on the line of 1544 kilobits per second (kb/s). The bit stream consists of contiguous 193-bit frames recurring at the rate of 8000 frames per second. Each frame contains one framing bit plus twenty-four channels, each channel containing eight bits. In each channel, one of the eight bits is required to satisfy transmission line constraints and can be used for signaling by the carrier. In some systems, the eighth bit is used for this purpose in every frame; other systems use it for this purpose during every sixth frame, while allowing it to be used for voice during the other frames. With either type of system, the seven-bit bytes always available for speech transmission at 8000 times per second can be used to transmit 56 000 bits per second of synchronous data over the same types of transmission channels that are used for PCM voice transmission. Therefore, it is expected that the preferred synchronous data rate of 56 kb/s will be most widely employed in carrier networks.

In the multichannel case, the line rate is 1544 kb/s

but, again, in the U.S. PCM telephone transmission system this rate is usually not available to the user due to the reservation of the twenty-four bits for satisfying transmission line constraints and for synchronization. For most applications, the user will have access to only 1344 kb/s.

In certain other systems it may be possible to utilize the entire 193 bits for synchronous data transmission at 1544 kb/s, provided other means are used to control the channel supervision and synchronization. In many cases involving the U.S. PCM telephone system, the PCM channel will not be available on a through basis but will be routed through secondary multiplexers which must have the synchronizing and signaling bits to maintain proper operation. For these reasons, the 1544 kb/s rate will often be subject to certain coding restraints by the carriers before it can be used as an interface rate on the public networks.

The only standard for wide-band data rates now in existence was adopted by the International Telegraph and Telephone Consultative Committee (CCITT) in 1968. This standard specifies a rate of 48 kb/s for synchronous operation with an option of 40.8 kb/s on an exception basis. This standard also includes the specifications for a modem to operate at this rate. Although there is at present little commercial use of this rate in the United States, it is expected to be widely used in other parts of the world over FDM systems. The adoption of this speed as a standard in the United States would simplify interoperation with non–United States systems. On the other hand, 48 kb/s is sufficiently below the full data handling capacity of the group channel as to miss the most economical usage of FDM facilities, and it uses only 75%–85% of the capacity of a PCM channel. The 48 kb/s must be recognized for those who are involved in operations extending beyond the United States, but it is not a preferred rate for use in the United States at this time.

Widespread application of data rates between 9.6 and 48 kb/s has not as yet taken place, but there has been considerable experimental application of a number of different rates in this range by government agencies. A rate of 16 kb/s has been chosen as the best estimate of an intermediate rate in this range at the present time, with the expectation that additional rates will probably be added in the future as more experience is gained in different applications.

# American National Standard

## code extension techniques
## for use with the 7-bit coded character set
## of american national standard code
## for information interchange

**ans**

american national standards institute, inc.
1430 broadway, new york, new york 10018

ANSI
X3.41-1974

# American National Standard
# Code Extension Techniques
# for Use with the 7-Bit Coded Character Set
# of American National Standard Code
# for Information Interchange

592

Secretariat

**Computer and Business Equipment Manufacturers Association**

Approved May 14, 1974

**American National Standards Institute, Inc**

## American National Standard

An American National Standard implies a consensus of those substantially concerned with its scope and provisions. An American National Standard is intended as a guide to aid the manufacturer, the consumer, and the general public. The existence of an American National Standard does not in any respect preclude anyone, whether he has approved the standard or not, from manufacturing, marketing, purchasing, or using products, processes, or procedures not conforming to the standard. American National Standards are subject to periodic review and users are cautioned to obtain the latest editions.

**CAUTION NOTICE**: This American National Standard may be revised or withdrawn at any time. The procedures of the American National Standards Institute require that action be taken to reaffirm, revise, or withdraw this standard no later than five years from the date of publication. Purchasers of American National Standards may receive current information on all standards by calling or writing the American National Standards Institute.

Published by

**American National Standards Institute**
**1430 Broadway, New York, New York 10018**

Printed in the United States of America

H1½M175/6

# Foreword

(This Foreword is not a part of American National Standard Code Extension Techniques for Use with the 7-Bit Coded Character Set of American National Standard Code for Information Interchange, X3.41-1974.)

American National Standard Code for Information Interchange (ASCII), X3.4-1968, provides coded representation for a set of graphics and control characters having general utility in information interchange. In some applications, it may be desirable to augment the standard repertory of characters with additional graphics or control functions.

ASCII includes characters intended to facilitate the representation of such additional graphics or control function by a process known as code extension. Although the basic nature of code extension limits the degree to which it may be standardized, there are advantages to adherence to certain standard rules of procedure. These advantages include minimized risk of conflict between systems required to interoperate, and the possibility of including advance provision for code extension in the design of general-purpose data handling systems.

A need has also developed for 8-bit codes for general information interchange in which ASCII is a subset. This standard provides a structure which will accommodate this need.

This standard was developed after extensive study of various potential applications and of trends expected in system design.

Suggestions for improvement of this standard will be welcome. They should be sent to the American National Standards Institute, 1430 Broadway, New York, N.Y. 10018.

This standard was processed and approved for submittal to ANSI by American National Standards Committee on Computers and Information Processing, X3. Committee approval of the standard does not necessarily imply that all committee members voted for its approval. At the time it approved this standard, the X3 Committee had the following members:

J. F. Auwaerter, Chairman
V. E. Henriques, Vice-Chairman
Robert M. Brown, Secretary

| Organization Represented | Name of Representative |
|---|---|
| Addressograph Multigraph Corporation | A. C. Brown |
| | D. S. Bates (Alt) |
| Air Transport Association | F. C. White |
| American Bankers Association | M. E. McMahon |
| | John B. Booth (Alt) |
| American Institute of Certified Public Accountants | N. Zakin |
| | P. B. Goodstat (Alt) |
| | C. A. Phillips (Alt) |
| | F. Schiff (Alt) |
| American Library Association | J. R. Rizzolo |
| | J. C. Kountz (Alt) |
| | M. S. Malinconico (Alt) |
| American Newspaper Publishers Association | W. D. Rinehart |
| American Nuclear Society | D. R. Vondy |
| | M. K. Butler (Alt) |
| American Society of Mechanical Engineers | R. W. Rau |
| | R. T. Woythal (Alt) |
| Association for Computing Machinery | P. Skelly |
| | J. A. N. Lee (Alt) |
| | L. Revens (Alt) |
| | H. Thiess (Alt) |
| Association for Educational Data Systems | C. Wilkes |
| Association for Systems Management | A. H. Vaughan |
| Association of American Railroads | R. A. Petrash |
| Association of Computer Programmers and Analysts | T. G. Grieb |
| | G. Thomas (Alt) |
| Association of Data Processing Service Organizations | J. B. Christiansen |
| Burroughs Corporation | E. Lohse |
| | J. F. Kalbach (Alt) |
| Control Data Corporation | S. F. Buckland |
| | C. E. Cooper (Alt) |
| Data Processing Management Association | A. E. Dubnow |
| | D. W. Sanford (Alt) |

# Contents

# American National Standard
# Code Extension Techniques
# for Use with the 7-Bit Coded Character Set
# of American National Standard Code
# for Information Interchange

## 1. Introduction

In the establishment of American National Standard Code for Information Interchange (ASCII), X3.4-1968, a fundamental decision had to be made as to the size of the code. In making such a decision there is usually a conscious effort to avoid the most obvious problems with a code that is either too large or too small. Should the number of characters included be too small, many users will find their needs not accommodated and will be forced to adopt "parochial" codes for their applications. Should the number of characters be too large, many users will find the code disproportionately costly to implement, or untenably inefficient in transmission or storage, and will again be driven to the use of some other code. Thus, either extreme in code sizing will reduce the generality of the code, defeating the very purpose of standardization in this field.

The 7-bit size (128 characters) adopted for ASCII is thought to be near optimum at present with respect to the foregoing considerations. Nevertheless, there will be numerous applications with requirements that are not accommodated by a code of this size, or at least not by the specific characters assigned within it. Many of these applications can be served by the use of the standard code augmented by standard code extension procedures. Through such an approach, the user may be able to implement much of his system with standard hardware and software. More significantly, he will thereby be able to retain compatibility with other systems for the interchange of that information which can adequately be directly represented by the standard code.

## 2. General

### 2.1 Scope

2.1.1 This standard specifies methods of extending the 7-bit code remaining in a 7-bit environment or increasing to an 8-bit environment. The description of techniques is contained in four interrelated sections dealing respectively with:

(1) The extension of the 7-bit code remaining in a 7-bit environment

(2) The structure of a family of 8-bit codes

(3) The extension of an 8-bit code remaining in an 8-bit environment

(4) The relationship between the 7-bit code and an 8-bit code

2.1.2 While ASCII is the agreed-upon code for information interchange, an 8-bit code as described in this standard is provided for information interchange within an 8-bit environment.

2.1.3 It is not the intention of this standard that all instances of its application accommodate all of its provisions. However, it is intended that, when code extension techniques are used, the applicable parts of this standard are to be followed.

When two systems with different implementations of code extension techniques are required to communicate with one another, they shall do so using only the code extension techniques they have in common.

2.1.4 Code extension techniques are classified, and some classes are given a structure in this standard. Other assignments of bit combinations associated with the designation of the classes will be made in accordance with the procedures for the registration of characters and character sets given in a related standard currently under development. Specific assignments of bit combinations to relate individual codes with their invocation or designation will also be made in accordance with that standard.

2.1.5 Code extension techniques are designed to be used for data to be processed serially in a forward direction. Use of these techniques in strings of data that are processed other than serially in a forward direction or included in data formatted for fixed record processing may have undesirable results.

598

7

## 2.2 Purpose

2.2.1 American National Standard Code for Information Interchange (ASCII), X3.4-1968, specifies the representation of 128 characters. Additionally, it allows the representation of several other graphics by the combination of two graphic characters with the Backspace control character. In many instances, ASCII lacks either controls or graphics to sufficiently satisfy the needs of an application. These needs may be satisfied by means of code extension, which is the subject of this standard.

This standard is intended to present a review of the salient structure of ASCII and then build upon that structure to describe various means of extending the control and graphic set of the code. It also describes structures and techniques to construct or formalize codes related to ASCII. These related codes are structured so as to allow application-dependent usage without preventing the interchangeability of their data.

2.2.2 The standard describes the following:

(1) The structure of ASCII

(2) Extension of the 7-bit code, remaining in a 7-bit environment, and making use of code extension techniques

(3) Increasing the number of bits to 8, yet retaining a structure compatible with the 7-bit structure

(4) Increasing the number of bits to 8 and applying similar code extension techniques

2.2.3 In order to use identical techniques in each of the cases mentioned in 2.2.2, and to facilitate conversion between them, standard rules are necessary. This has the advantage of:

(1) Reducing the risk of conflict between systems required to interoperate

(2) Permitting provision for code extension in the design of systems

(3) Providing standardized methods of calling into use agreed-upon sets of characters

(4) Allowing the interchange of data between 7-bit and 8-bit environments

2.2.4 This standard also describes the structure of families of codes that are related to ASCII by their structure.

## 2.3 Application

Characters and character sequences conforming to this standard are used in ASCII or ASCII-related information interchange where additional control or graphic characters, or both, not in ASCII must be provided.

## 2.4 Conformance

Recorded or transmitted data are in conformance with this standard if all those bit patterns contained in the data stream have exactly, and only, the meanings specified in this standard and in a related standard currently under development.

WARNING: Products may provide or process bit patterns, code extension procedures, classes of Escape sequences, and code relationships not specified in this standard. Products may have the capability to record or transmit data that contain contiguous bits or bytes, the meaning of which is not intended as those specified in this standard.

A user application program may cause nonconforming bit patterns to be transmitted in interchange data or require the use of bit patterns, code extension procedures, classes of Escape sequences, and code relationships not as specified in this standard for further processing of the application data.

The use of such product features or user applications may render the data in nonconformance with this standard.

## 3. Definitions

In this standard, the following definitions shall apply:

**bit combination.** An ordered set of bits that represents a character.

**byte.** A bit string that is operated upon as a unit and whose size is independent of redundancy or framing techniques.

**character.** A member of a set of elements that is used for the organization, control, or representation of data.

**code, coded character set.** A set of unambiguous rules that establish a character set and the one-to-one relationships between the characters of the set and their bit combinations.

**code extension.** Techniques for the encoding of characters that are not included in the character set of a given code.

**code table.** A table showing the character corresponding to each bit combination in a code.

**control character.** A character whose occurrence in a particular context initiates, modifies, or stops a control function.

**control function.** An action that affects the recording, processing, transmission, or interpretation of data.

**designate.** To identify a set of characters that are to be represented, in some cases immediately and in others on the occurrence of a further control function, in a prescribed manner.

**environment.** The characteristic that identifies the number of bits used to represent a character in a data processing or data communication system or in part of such a system.

**Escape sequence.** A bit string that is used for control purposes in code extension procedures and that consists of two or more bit combinations, of which the first is the bit combination corresponding to the Escape character.

**Final character.** The character whose bit combination terminates an Escape sequence.

**graphic character.** A character, other than a control character, that has a visual representation normally handwritten, printed, or displayed.

**Intermediate character.** A character whose bit combination occurs between the Escape character and the Final character in an Escape sequence, consisting of more than two bit combinations.

**invoke.** To cause a designated set of characters to be represented by the prescribed bit combinations whenever those bit combinations occur, until an appropriate code extension function occurs.

**position.** An item in a code table identified by its column and row coordinates.

**represent.** To use a prescribed bit combination with the meaning of a character in a set of characters that has been designated and invoked. To use an Escape sequence with the meaning of an additional control character.

## 4. Notational Conventions

**4.1 General.** The primary concept of code extension is to allow the meaning associated with a bit pattern to be changed in a discrete and orderly way. As a result, some shorthand notations are introduced to disassociate bit patterns and classes of bit patterns from the meanings in any one code set. A number of notational conventions have been established in this document to simplify the expressions for code extension.

A bit combination is sometimes referred to by the column and row numbers of its position in the code table. The column number is the sum of the decimal equivalent of bits $b_7-b_5$ (or $a_8-a_5$), and the row number is the sum of the decimal equivalent of bits $b_4-b_1$ (or $a_4-a_1$), giving to these bits the weights given in 4.2.

In representing the decimal equivalents, the convention is to append a leading zero to the column number for 8-bit columns 00–09. As an example, the position of the Space character in the 7-bit code table is 2/0; the position of the same character in an 8-bit code table is 02/0.

**4.2 Character Representation.** In this standard the following notations are used:

| | | | | | | | |
|---|---|---|---|---|---|---|---|
| The bits of a 7-bit combination: | $b_7$ | $b_6$ | $b_5$ | $b_4$ | $b_3$ | $b_2$ | $b_1$ |
| The bits of an 8-bit combination: | $a_8$ | $a_7$ | $a_6$ | $a_5$ | $a_4$ | $a_3$ | $a_2$ | $a_1$ |

| | Column | | | | Row | | | |
|---|---|---|---|---|---|---|---|---|
| Bit weight for column and row reference: | $2^3$ | $2^2$ | $2^1$ | $2^0$ | $2^3$ | $2^2$ | $2^1$ | $2^0$ |

**4.3 Character Notations.** Characters may be represented as follows:

| | | |
|---|---|---|
| Mnemonics: | ENQ | ETB |
| Denotation: (Location in code table [column/row]): | 0/5 | 1/7 |
| Bit pattern (byte): | 0000101 | 0010111 |

Character mnemonics such as SO and ESC, and column/row numbers such as 0/5 and 1/7, are as shown to emphasize the fact that they stand for one bit combination only.

## 5. Extension of the 7-Bit Code Remaining in a 7-Bit Environment

**5.1 Introduction**

**5.1.1 The Structure of the 7-Bit Code.** The 7-bit code table of ASCII consists of an ordered set of controls and graphic characters grouped (see Fig. 1) as follows:

(1) The set of thirty-two control characters allocated to columns 0 and 1 (hereafter referred to as the C0 set)

(2) The Space character in position 2/0, which may be regarded either as a control character or a nonprinting graphic character

(3) The set of ninety-four graphic characters allocated to columns 2–7 (hereafter referred to as the G0 set)

(4) The Delete character in position 7/15

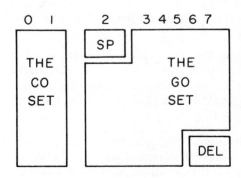

**Fig. 1**
**The Structure of the 7-Bit Code**

**5.1.2 Extension by Substitution.** In many cases the provisions of ASCII will satisfy the requirements of an application. Other applications, however, will be satisfied by the use of a similarly structured code in which some of the characters of ASCII are substituted by other characters. Such a substitution may be regarded as a replacement of the C0 set or the G0 set, or both, accordingly as new controls or graphics, or both, are required (see Appendix B of American National Standard X3.4-1968).

**5.1.3 Extension by Increasing the Repertory of Characters.** This standard provides for the addition of characters to the 128 provided by the structure of the 7-bit code in the following ways:

(1) Additional single controls

(2) Additional sets of thirty-two control characters (referred to as C1 sets)

(3) Additional sets of ninety-four graphic characters (referred to as G1 sets)

(4) Sets of more than ninety-four graphic characters, each represented by more than one byte (these sets function as G0 sets)

Many applications will require combinations of the aforementioned facilities. These facilities are shown in Fig. 2.

**5.1.4 Code Extension Characters.** In the 7-bit code, the following characters are provided for the purpose of code extension:

| | |
|---|---|
| Escape character: | ESC |
| Shift-Out character: | SO |
| Shift-In character: | SI |
| Data Link Escape character: | DLE |

This standard does not describe the use of the Data Link Escape character, which is reserved for the provision of additional transmission controls; the use of this character is specified in American National Standard

Procedures for the Use of the Communication Control Characters of American National Standard Code for Information Interchange in Specified Data Communication Links, X3.28-1971.

**5.1.5 Compatibility.** For purposes of interchange, various levels of compatibility that may be preserved among extension facilities are identified as follows:

(1) A set that is compatible with ASCII inasmuch as:

(a) Columns 0 and 1 contain only control characters

(b) Columns 2–7 are used for graphic characters only (apart from DEL)

(c) The ten transmission control characters and NUL, SO, SI, CAN, SUB, ESC, and DEL remain unaltered in their meanings and in their positions in the code table

(d) Graphics of ASCII are not moved to other positions (A non-Latin alphabet containing graphics that are also included in the Latin alphabet is not subject to this rule.)

(2) Other sets structured as in 5.1.1

To be able to provide the facilities of code extension given in this standard, the Escape, or the Shift-Out and Shift-In characters, or both, shall remain unaltered in their meanings and their positions in the code table.

**5.2 Extension of the Graphic Set by Means of the Shift-Out and Shift-In Characters**

**5.2.1 Use of SO and SI.** The Shift-Out character SO and the Shift-In character SI are used exclusively for extension of the graphics.

The character SO invokes an alternative set of ninety-four graphics, the G1 set. This set replaces the graphic characters of the G0 set. Graphic characters need not be assigned to all the positions of the alternative set, nor need all the graphic characters of the alternative set be different from the graphic characters of the G0 set.

The character SI invokes the graphic characters of the G0 set that are to replace the graphic characters of the alternative set.

The meanings of the following bit combinations are not affected by the occurrence of SO and SI:

(1) Those corresponding to the control characters in columns 0 and 1 and the Delete character in position 7/15

(2) The one corresponding to the Space character in position 2/0

(3) Those included in any Escape sequence

The Space character occurs only at position 2/0; it shall not be assigned to any position in the alternative graphic set. These provisions do not preclude the assignment to positions in any graphic set of characters equivalent to spaces of size other than that of the Space assigned to position 2/0, as, for example, half-space.

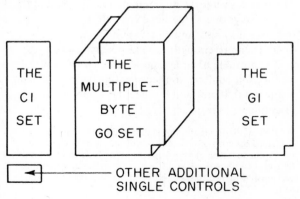

**Fig. 2**
**Structure for the Extended 7-Bit Code**

601

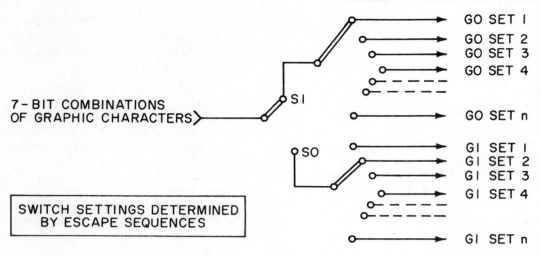

**Fig. 3**
**Schematic Illustration of the Shift-Out and Shift-In Concept**

At the beginning of any information interchange, the shift status should be defined by SI or SO. When in the Shift-In state, SI has no effect; and when in the Shift-Out state, SO has no effect.

**5.2.2 Unique Shift-Out Set.** Some applications require the use of only one alternative set of ninety-four graphic characters. In such a case, that unique set is invoked by each use of SO. The set is identified either by an appropriate ESC sequence as described in 5.3.7 or by agreement between or among the interchanging parties.

**5.2.3 Multiple Shift-Out Sets.** If two or more alternative graphic sets are required to coexist in a system, the set to be used next is designated by the appropriate ESC sequence. That set can then be invoked by the use of SO.

The use of SI reinvokes the graphics of the G0 set last designated, but does not affect the identity of the designated alternative set. The alternative set may be invoked any number of times by successive use of SO until it is superseded by another alternative set designated by another Escape sequence.

It is not necessary to revert to the G0 set by use of SI before changing from one alternative set to another by means of a further Escape sequence. When the system is in the Shift-Out state, the use of such a further Escape sequence leaves the shift status unaltered, and the alternative set is invoked.

A schematic representation of the aforementioned is given in Fig. 3.

WARNING: In some devices or systems there may be a requirement to reestablish the Shift-In state before designating a new Shift-Out set by means of an Escape sequence. This can be achieved by inserting SI before the Escape sequence which designates the subsequent Shift-Out set. Such a requirement shall be agreed upon between or among any interchanging parties.

**5.3 Code Extension by Means of Escape Sequences**

**5.3.1 Purposes of Escape Sequences.** Escape sequences provide single or sets of control functions other than for transmission control. Escape sequences are also used to designate sets of graphics, different uses of some or all of the 7-bit code combinations, and coded character sets with a number of bits other than seven.

Thus, Escape sequences are required to provide, for example:

(1) A single control character not already in the code

(2) A set of control characters not already in the code

(3) A set of graphic characters not already in the code

(4) A code of different structure

**5.3.2 Structure of Escape Sequences.** An Escape sequence consists of two or more 7-bit combinations. The first is always the bit combination of ESC and the last is always one of the Final characters. An Escape sequence may also contain any number of 7-bit combinations representing Intermediate characters.

The meaning of an Escape sequence is determined by the 7-bit combination representing its Intermediate character(s), if any, and by the 7-bit combination representing its Final character.

WARNING: Although in this standard Escape sequences are described in terms of characters or of positions in the code tables, the meaning of an Escape sequence is determined only by its bit combinations, and it is unaffected by any meaning previously assigned to these bit combinations taken individually.

Intermediate characters are the sixteen characters of column 2 of the 7-bit code table.

NOTE: In this standard, any one of these sixteen Intermediate characters is denoted by the symbol "I."

Final characters are the seventy-nine characters of columns 3–7 of the 7-bit code table excluding position 7/15.

NOTE: In this standard, any one of these seventy-nine Final characters is denoted by the symbol "F."

Prohibited characters are the control characters in columns 0 and 1 and the character in position 7/15.

The thirty-three prohibited characters shall not be used as either Intermediate or Final characters to construct an Escape sequence.

As these prohibited characters may appear in an ESC sequence in error, it may be necessary within an application to provide methods of identifying such a situation and of recovering from it, but this is not covered in this standard (see Section A1 of Appendix A).

**5.3.3 Categories of Escape Sequences.** The categories of Escape sequences are specified in this standard. However, Escape sequences with Final characters from column 3 are reserved for private use subject to the categorization outlined in 5.3.3.1 and 5.3.3.2.

WARNING: The implementors of any private Escape sequence described as such in this document are alerted to the fact that other implementors may give different meanings to the same Escape sequence or may use different Escape sequences to mean the same thing. Furthermore, such meanings may subsequently be assigned to standardized Escape sequences. Interchanging parties are warned, therefore, that such agreement may reduce their capability to interchange subsequently.

**5.3.3.1 Two-Character Escape Sequences.** A two-character Escape sequence takes the form ESC F. Such Escape sequences are used to represent single additional characters. The seventy-nine two-character Escape sequences are split into three types, depending on the Final character, as shown in Fig. 4.

An ESC Fs sequence represents, depending on the Final character used, a single additional standardized control character. Thirty-one Final characters of columns 6 and 7 are provided for this purpose.

An ESC Fe sequence represents, depending on the Final character used, an individual control character of an additional standardized set of thirty-two control characters (see 5.3.6). The thirty-two Final characters of columns 4 and 5 are provided for this purpose. Some applications require the use of only one such additional set. In this case, the set is identified either by the appropriate ESC sequence, as described in 5.3.6, or by agreement between or among the interchanging parties. If more than one additional set of controls are required to coexist in a system, the set to be used next is designated and invoked by the appropriate ESC sequence.

An ESC Fp sequence represents, depending on the Final character used, a single additional character without standardized meaning, for private use as required, subject to the prior agreement of the sender and the recipient of the data. The sixteen Final characters of column 3 are provided for this purpose.

**5.3.3.2 Three-Character Escape Sequences.** A three-character Escape sequence takes the form ESC I F.

All types of three-character Escape sequences are grouped into categories according to their purpose, by means of their Intermediate characters, as shown in 5.3.4 through 5.3.10 (see Table 1).

These sequences are split into two types according to their Final character as shown in Fig. 5.

ESC I Ft sequences are used for standardized purposes. The sixty-three Ft characters of columns 4–7 are provided for this purpose.

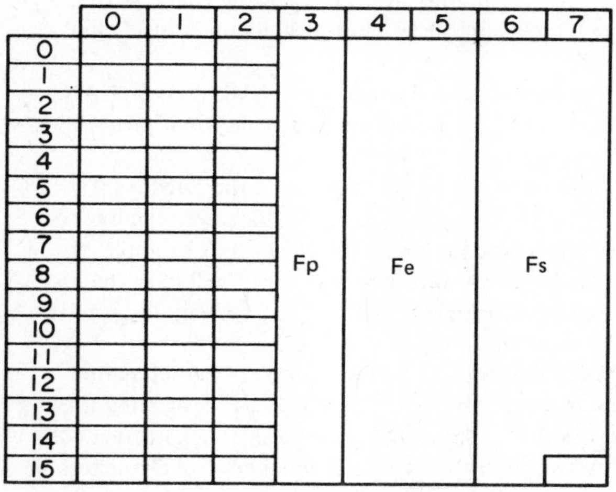

**Fig. 4**
**Portions of the Code Table Used in Two-Character Escape Sequences**

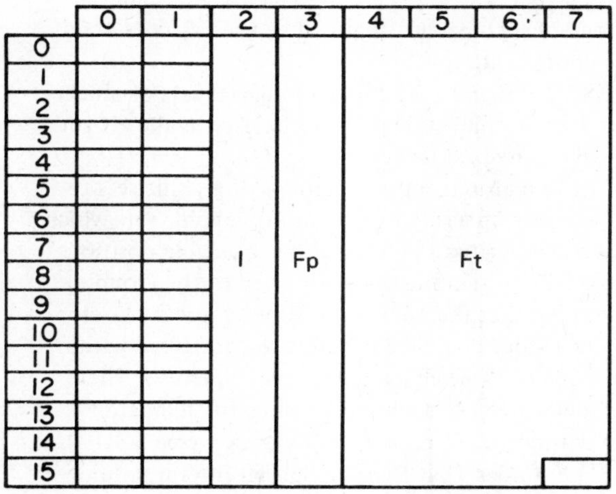

**Fig. 5**
**Portions of the Code Table Used in Three-Character Escape Sequences**

ESC I Fp sequences are reserved for private use. The sixteen Fp characters of column 3 are provided for this purpose.

**5.3.4 Single Additional Characters.** ESC 2/3 F represents, depending on the Final character used, a single additional character.

**5.3.5 Sets of Thirty-Two Control Characters for Columns 0 and 1.** ESC 2/1 F designates and invokes the C0 set of thirty-two control characters for representation by the bit combinations of columns 0 and 1.

The ten transmission control characters, when included in a C0 set, shall retain their meanings and their positions in the code table. No other transmission control characters shall be included in a C0 set.

To reduce the risk of conflict in the interchange of data, this set should have the following characteristics:

(1) Inclusion of the ten transmission control characters

(2) Inclusion of the characters NUL, SO, SI, CAN, SUB, and ESC with their meanings and their position in the code table unaltered

Consideration should be given to the effect that changing the meaning of control characters can have on equipment when interchanging data. For example, the bit combination corresponding to HT will have the effect of "horizontal tabulation" to a system designed to respond to this control character.

**5.3.6 Sets of Thirty-Two Control Characters for Representation by ESC Fe.** ESC 2/2 F designates and invokes the C1 set of thirty-two control characters without affecting the C0 set. Individual control characters of such a set are represented by means of ESC Fe sequences rather than single-character bit combinations. A C1 set shall not include transmission control characters.

**5.3.7 Sets of Ninety-Four Graphic Characters.** ESC 2/8 F and ESC 2/12 F designate sets of ninety-four graphic characters that will be used as the G0 set. The designated set is invoked by SI.

ESC 2/9 F and ESC 2/13 F designate sets of ninety-four graphic characters that will be used as the G1 set. The designated set is invoked by SO.

The two aforementioned groups of graphic sets together make up a single repertory of graphic sets which may be designated to either of two available positions, G0 or G1. No significance is attached to the groupings other than that their existence allows more such sets of ninety-four graphic characters to be defined within the scope of three-character Escape sequences. There are, therefore, 126 such sets possible for standardization without requiring further extension (see 5.3.12).

**5.3.8 Codes That Require Special Interpretation.** ESC 2/5 F designates and invokes a code that requires special interpretation, such as:

(1) A code with a number of bits other than seven,

excluding those 8-bit codes structured in accordance with this standard

(2) A 7-bit code whose characteristics differ from those in this standard

The Final character assignments are such that, within the Ft and Fp groups (columns 3–7), the following classification occurs:

| Final in Column | Broad Categorization |
|---|---|
| 3 | Private code with any number of bits |
| 4 | Code of less than 7 bits |
| 5 | Code of 7 bits |
| 6 | Code of 8 bits |
| 7 | Code of more than 8 bits |

**5.3.9 Sets of Graphics with Multiple-Byte Representation.** ESC 2/4 F designates sets of graphic characters that are represented by two or more bytes, each corresponding to a bit combination in columns 2–7, apart from positions 2/0 and 7/15. The designated set is invoked by SI and is therefore regarded as a G0 set. Within such a set, each graphic character is represented by the same number of bytes as shown in Fig. 6.

**5.3.10 Announcement of Extension Facilities.** ESC 2/0 F announces the extension facilities used in conjunction with data that follow. The use of these sequences is specified in Section 8.

**5.3.11 Three-Character Escape Sequences without Assigned Meanings.** The three-character Escape sequences ESC 2/6 F, ESC 2/7 F, ESC 2/10 F, ESC 2/11 F, ESC 2/14 F, and ESC 2/15 F have not been assigned meanings and are reserved for future standardization.

**5.3.12 Escape Sequences Having Four or More Characters.** Escape sequences having four or more characters shall be interpreted according to the following:

(1) The first Intermediate character shall indicate the category of usage identical with three-character Escape sequences in 5.3.11

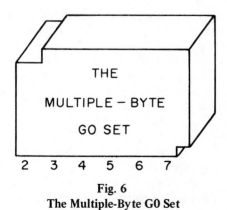

THE
MULTIPLE – BYTE
GO SET

2  3  4  5  6  7

**Fig. 6**
**The Multiple-Byte G0 Set**

604

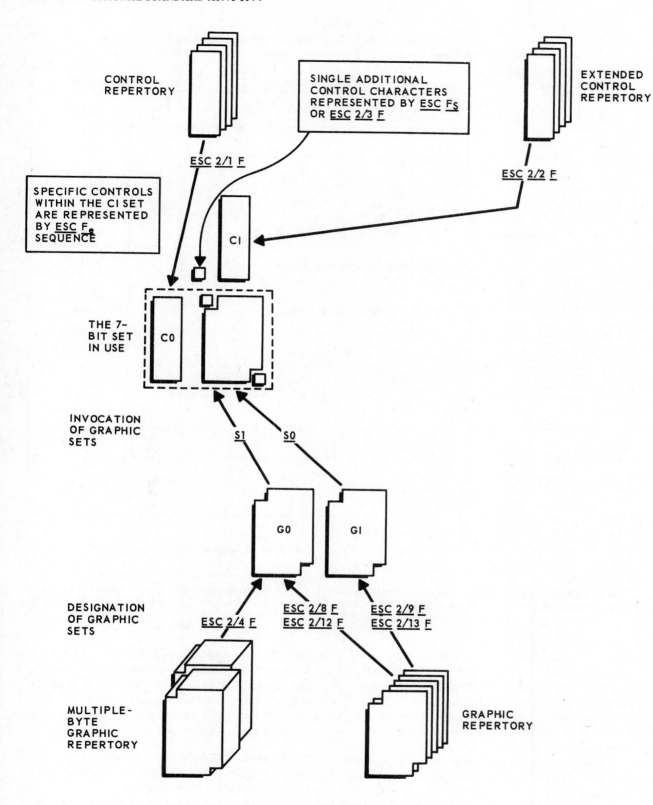

Fig. 7
7-Bit Code Extension Structure Using C0, C1, G0, and G1 Sets

Table 1
Summary of Assignments of Three-Character Escape Sequences

| Column/ Row | Bits of Intermediate $b_7\, b_6\, ..\, ..\qquad b_1$ | Category | Grouping | Subsection |
|---|---|---|---|---|
| 2/0 | 0 1 0 0 0 0 0 | Announcers | | 5.3.10 |
| 2/1 | 0 1 0 0 0 0 1 | CONTROLS | C0 set | 5.3.5 |
| 2/2 | 0 1 0 0 0 1 0 | | C1 set | 5.3.6 |
| 2/3 | 0 1 0 0 0 1 1 | Single characters | | 5.3.4 |
| 2/4 | 0 1 0 0 1 0 0 | GRAPHICS | Multiple-byte sets | 5.3.9 |
| 2/5 | 0 1 0 0 1 0 1 | Codes requiring special interpretation | | 5.3.8 |
| 2/6 | 0 1 0 0 1 1 0 | Reserved for future standardization | | 5.3.11 |
| 2/7 | 0 1 0 0 1 1 1 | | | |
| 2/8 | 0 1 0 1 0 0 0 | GRAPHICS* | G0 set | 5.3.7 |
| 2/9 | 0 1 0 1 0 0 1 | | G1 set | |
| 2/10 | 0 1 0 1 0 1 0 | | Reserved for future standardization | 5.3.11 |
| 2/11 | 0 1 0 1 0 1 1 | | | |
| 2/12 | 0 1 0 1 1 0 0 | | G0 set | 5.3.7 |
| 2/13 | 0 1 0 1 1 0 1 | | G1 set | |
| 2/14 | 0 1 0 1 1 1 0 | | Reserved for future standardization | 5.3.11 |
| 2/15 | 0 1 0 1 1 1 1 | | | |

*There is a single repertory of sets of ninety-four graphic characters. Any member of the repertory may be designated as either a G0 or G1 set. Four designating Escape sequences, two for G0 and two for G1, are provided for designating members of the repertory.

(2) The second and any additional Intermediate characters shall be associated with the Final character to permit additional entities within the category defined by the first Intermediate character

(3) All Escape sequences having four or more characters whose final character is of the Ft type are reserved for standardization

(4) All Escape sequences whose final character is of the Fp type (private) are not to be the subject of standardization

**5.3.13 Omission of Escape Sequences.** If the interchanging parties have agreed upon a single G0 set, a single G1 set, a single C0 set, and a single C1 set (or on as many of these sets as are to be used), they may also agree to omit the use of Escape sequences to designate or invoke them. Interchanging parties are warned, however, that such agreements may reduce their capability to interchange data subsequently.

**5.4 Pictorial and Tabular Representations.** Fig. 7 summarizes in pictorial form the standard means of code extension within a 7-bit environment.

Table 1 summarizes in tabular form the assignment of Intermediate characters in Escape sequences.

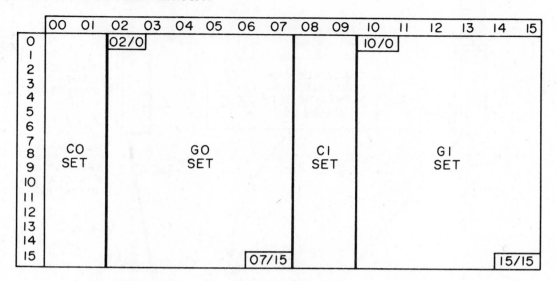

**Fig. 8**
**The 8-Bit Code Table**

## 6. Structure of a Family of 8-Bit Codes

**6.1 General.** The family of 8-bit codes specified in this standard is obtained by the addition of one bit to each of the bit combinations of the 7-bit code, thus producing a set of 256 8-bit combinations. The 128 characters of the 7-bit code, the set as defined under 5.1, form a defined and integral part of an 8-bit code that is structured in accordance with this document. The 128 additional bit combinations, whose 8th bit is 1, are available for future assignment.

**6.2 The 8-Bit Code Table.** A 16 × 16 array of columns numbered 00–15 and rows numbered 0–15 contains 256 code positions (see Fig. 8). The 8-bit code table consists of an ordered set of controls and graphic characters grouped as follows (see Fig. 8):

(1) A set of thirty-two control characters allocated to columns 00 and 01 (C0 set)

(2) A set of ninety-four graphic characters allocated to columns 02–07, subject to the exception of positions 02/0 and 07/15 (G0 set)

(3) The Space character in position 02/0, which may be regarded either as a control character or a nonprinting graphic character

(4) A set of thirty-two control characters allocated to columns 08 and 09 (C1 set)

(5) A set of ninety-four graphic characters allocated to columns 10–15, subject to the exception of positions 10/0 and 15/15

The control characters in columns 08 and 09 of an 8-bit code shall not include transmission control char-

acters. Provision of data transmission capability for 8-bit codes includes the use of the Data Link Escape character and is covered in American National Standard Procedures for the Use of the Communication Control Characters of American National Standard Code for Information Interchange in Specified Data Communication Links, X3.28-1971.

**6.3 The Family Concept.** In order to cope with the different needs of the various industries, fields of application, or systems, this standard defines the concept of a family of 8-bit codes as follows:

(1) A set of thirty-two additional control characters can be selected for columns 08 and 09

(2) A set of ninety-four additional graphic characters can be selected for columns 10–15 (excluding positions 10/0 and 15/15)

There are standard techniques for identifying selections of sets of controls and graphics for 8-bit codes. These techniques are described in Sections 7, 8, and 9.

## 7. The Use of Code Extension in an 8-Bit Code

**7.1 General.** The techniques of extending an 8-bit code described in this standard have been purposely made compatible with those used to extend the 7-bit code.

The Escape character is used in an 8-bit code in exactly the same way as in the 7-bit code to construct ESC sequences. Except as provided in 7.2, the meanings of these sequences are not altered in an 8-bit code. All characters in columns 08–15 are excluded from assign-

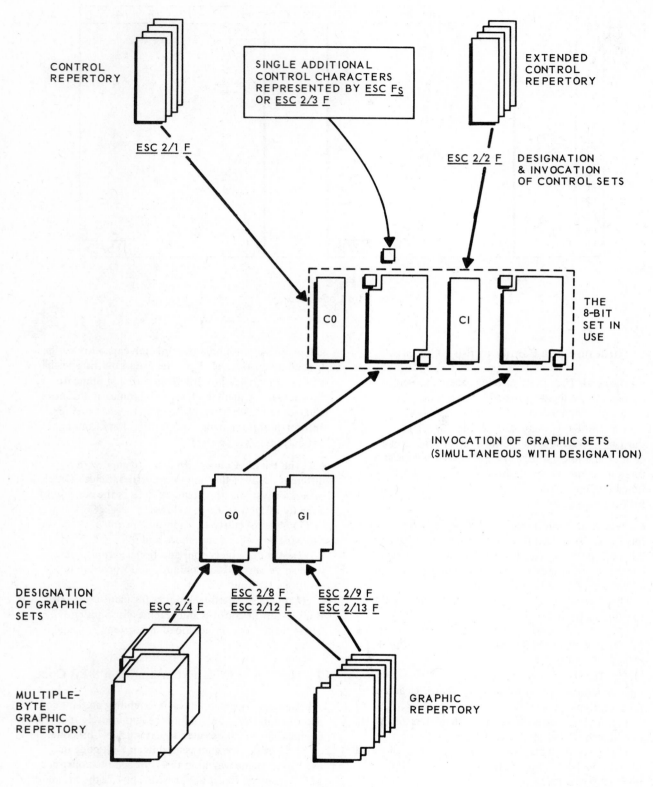

Fig. 9
8-Bit Code Extension Structure Using C0, C1, G0, and G1 Sets

ment in Escape sequences and any occurrences of them in an Escape sequence are error conditions for which no standard recovery procedures are prescribed in this standard.

**7.2 Defining an 8-Bit Code.** As described in Section 6, the code table can be considered as having four main parts: the C0 control set, the G0 graphic set, the C1 control set, and the G1 graphic set.

The remainder of the code table consists of positions 02/0 SP, 07/15 DEL, 10/0, and 15/15.

The C0 and G0 sets are designated and the C0 set is invoked as in the 7-bit environment (see 5.3.5 and 5.3.7).

The C1 set of control characters is designated and invoked by means of an Escape sequence as in the 7-bit environment (see 5.3.6). These control characters are represented by the bit combinations of columns 08 and 09.

The G1 set of graphic characters is designated by means of an Escape sequence as in the 7-bit environment (see 5.3.7). These graphic characters are represented by the bit combinations of columns 10-15.

**7.3 Code Extension by Means of Escape Sequences.** Once the 8-bit code is established in accordance with 7.2, code extension means are available, making use of Escape sequences as described therein and in 7.3.1 through 7.3.3.

**7.3.1 Two-Character Escape Sequences.** Two-character Escape sequences have the same structure as in the 7-bit environment (see 5.3.2).

**7.3.1.1** ESC Fs sequences represent single additional controls with the same meaning they have in the 7-bit environment.

**7.3.1.2** The use of ESC Fe sequence in an 8-bit environment is contrary to the intention of this standard but, should they occur, their meaning is the same as in the 7-bit environment.

**7.3.2 Three-Character Escape Sequences.** Three-character Escape sequences have the same structure as in the 7-bit environment (see 5.3.2).

**7.3.3 Escape Sequences Having Four or More Characters.** These sequences have the same structure and meaning as in the 7-bit environment (see 5.3.2 and 5.4).

**7.4 Pictorial Representation.** Fig. 9 summarizes in pictorial form the standard means of extension available in an 8-bit environment.

# 8. Announcement of Extension Facilities Used

The class of three-character Escape sequences ESC 2/0 F

is used in data interchange to announce the code extension facilities utilized in the data that follow. Subject to agreement between or among the interchanging parties, such an announcing sequence may be omitted. The Final character of the announcing sequence indicates the facilities used for representing graphic sets in 7- and 8-bit environments and the number of bits used as given in Table 2.

# 9. Relationship between 7-Bit and 8-Bit Codes

**9.1 Transformation between 7-Bit and 8-Bit Codes.** Transformation between 7-bit and 8-bit codes depends on which facilities of code extension are included in the application. Examples of such transformations are given in Fig. B7 and B8 in flowchart form.

**9.2 Representation of the 7-Bit Code in an 8-Bit Environment.** It may sometimes be desirable, as for example in a store and forward application, to retain information in 7-bit form while in an 8-bit environment. In this case, for each of the characters, bits $b_7-b_1$ are represented by bits $a_7-a_1$, respectively, and $a_8$ is set to zero.

**Table 2**
**Announcer Sequences and Meanings***

| Final Characters | Facilities Used |
|---|---|
| 4/1 | The G0 set only is to be used. The Escape sequence which designates this set also invokes it into columns 2–7. SI and SO are not to be used. In an 8-bit environment, columns 10–15 are not used. |
| 4/2 | The G0 and G1 sets are to be used. In both 7- and 8-bit environments, SI invokes G0 into columns 2–7 and SO invokes G1 into columns 2–7. In an 8-bit environment, columns 10–15 are not used. |
| 4/3 | The G0 and G1 sets are to be used in an 8-bit environment only. The designating Escape sequences also invoke the G0 and G1 sets into columns 2–7 and 10–15, respectively. SI and SO are not to be used. |
| 4/4 | The G0 and G1 sets are to be used. In a 7-bit environment, SI invokes G0 into columns 2–7 and SO invokes G1 into columns 2–7. In an 8-bit environment, the designating Escape sequences also invoke the G0 and G1 sets into columns 2–7 and 10–15, respectively. SI and SO are not to be used. |

*A pictorial representation of these cases is shown in Fig. 10.

NOTE: In a 7-bit environment, data announced by a sequence ESC 2/0 4/4 have the same form as data announced by a sequence ESC 2/0 4/2. The announcer ESC 2/0 4/4 is provided for those interchange situations in which it is agreed to differentiate between 7-bit and 8-bit originated data in the 7-bit environment.

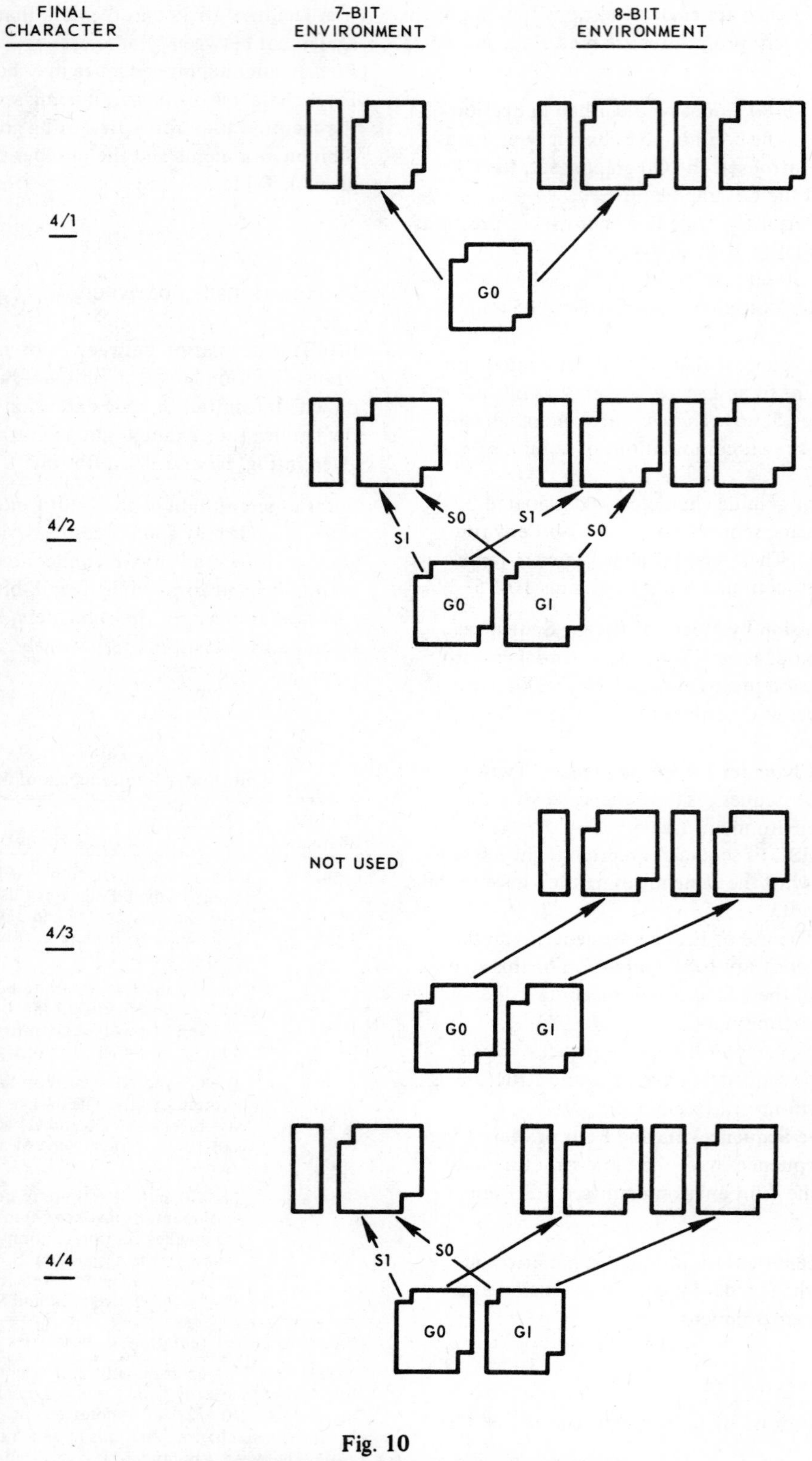

Fig. 10
Pictorial Representation of the Announcer Facilities

Indication that true 8-bit coded data follow is achieved by the occurrence of the announcing sequences ESC 2/0 4/3 or ESC 2/0 4/4.

Indication that 7-bit coded data follow is achieved by the occurrence of one of the announcing sequences ESC 2/0 4/1 or ESC 2/0 4/2.

**9.3 Representation of Positions 10/0 and 15/15 in a 7-Bit Environment.** No meaning is assigned to positions 10/0 and 15/15 in this standard. If there is a requirement to represent these positions in a 7-bit environment, a private Escape sequence shall be used.

## 10. Specific Meaning of Escape Sequences

The meanings of individual Escape sequences are not specified in this standard. Instead, their meanings will be specified using the procedures for the registration of characters and character sets given in a related standard currently under development which are to be followed in preparing and maintaining a register of Escape sequences and their meanings. These registration procedures do not apply to Escape sequences reserved for private use.

## Appendix A

## Implementation Considerations

### A1. Appearance of Control Characters in Escape Sequences

**A1.1** Although this standard precludes the assignment of control characters from columns 0 and 1 in Escape sequences, it does not necessarily preclude the occurrence of control characters in Escape sequences. System designers must be aware that additional standards may have to be considered if the occurrence of control characters is to be taken into account. Of particular concern is American National Standard Procedures for the Use of the Communication Control Characters of American National Standard Code for Information Interchange in Specified Data Communication Links, X3.28-1971. Under the provisions of that standard, communication controls may be interjected into Escape sequences as a part of the communication control procedures.

**A1.2** In interpreting data streams, the communication control characters and communication control sequences formed with the Data Link Escape (DLE) character as well as any associated error check characters (Block Check characters or Cyclic Redundancy Check characters), if present, should not be allowed to affect the meanings of Escape sequences in which they occur. The meaning of the Escape sequences is to be reckoned as if the communication controls had been purged. Of course, this doctrine cannot be applied when the occurrence of a communication control would have the effect of terminating a portion of a message prior to the transmission of the Final character of the affected Escape communication control characters and Data Link Escape sequences and the possibility of their occurring with validity in an Escape sequence at the input to the receiving portion of the system. The occurrence of a communication control character or character sequence formed with DLE which would not validly occur in the data stream should be considered as an error condition. This Appendix does not prescribe the means of recovery from this error condition. For suggested examples of the preceding, see Table A1.

### A2. Interactions When Operating in Multiprogramming and Multiprocessing Environments

The requirements and restrictions that must be satisfied in either a multiprogramming or multiprocessing environment are not increased over those required in a simple batch environment by the introduction of data that include Escape sequences. These requirements and restrictions include the serial examination of the data to ensure that all Escape sequences are interpreted, and recovery procedures to restart after error take into account the serial dependency of the data. Also, that provision is made that no characters in Escape sequences are skipped or encountered in an incorrect sequence.

The last requirement precludes access being per-

**Table A1
Suggested Validity of Communication
Control Characters in Sequences**

| Communication Control Character or Character Sequence | May Validly Occur in an Escape Sequence at a Receiver |
|---|---|
| SOH | Yes |
| STX | Yes |
| ETX | No |
| EOT | No |
| ENQ | No |
| ACK | No |
| DLE | Yes |
| NAK | No |
| ETB | Yes |
| SYN | Yes |
| DLE SOH | Yes |
| DLE STX | Yes |
| DLE ETX | No |
| DLE EOT | No |
| DLE ETB | Yes |
| DLE = | Yes |
| DLE 0 | No |
| DLE 1 | No |
| DLE 2 | No |
| DLE 3 | No |
| DLE 4 | No |
| DLE 5 | No |
| DLE 6 | No |
| DLE 7 | No |
| DLE SYN | Yes |

mitted to stored data at any point where the Escape sequence is not recognized, or to magnetic tape using read backward if the Escape sequences were only provided for read forward.

## A3. Unique Interchange by Prior Agreements

In order that the code represented by a character string can be unambiguously identified, the character string should be preceded by a designating Escape sequence. However, under certain conditions, this Escape sequence can be omitted as illustrated in the following examples:

(1) If only ASCII or an agreed-upon 8-bit code is used, it is not required to designate these codes.

(2) If ASCII or an ASCII variant is used with a single alternative graphic set, it is not required to use designating Escape sequences.

(3) If interchange is only in another code, such interchange is beyond the scope of this standard. Any necessary conventions are subject to the agreement of the parties to the interchange.

However, it is a recommended practice that the designating Escape sequence be included in every interchange in order to preserve the generality of the application and avoid unnecessary limitations to the arena of interchangeability of the data.

## A4. System Startup and Restart

When a link is established or reinitialized, in the absence of any other agreement between or among the exchanging parties, the basic code in effect is assumed to be ASCII (7-bit environment). If the interchange is to be in another code where no prior agreement between or among the exchanging parties exists, appropriate Escape sequences must be used.

## A5. Future Work on Code Extension

This Appendix contains additional code extension techniques that have been discussed. This is one approach to the problem of accommodating more than two pages of graphics, if required to do so.

## A6. Additional Invokers and Announcers

### A6.1 Definitions of Additional Invokers

**A6.1.1 ESI.** ESI is a mnemonic for the control character Extended Shift-In. This control is used in 8-bit extended codes, and its ESC Fe counterpart is used in some 7-bit extended codes to invoke the last designated G2 graphic set.

**A6.1.2 ESO.** ESO is a mnemonic for the control character Extended Shift-Out. This control is used in 8-bit extended codes and its ESC Fe counterpart is used in some 7-bit extended codes to invoke the last designated G3 graphic set.

**A6.2 Categories.** These categories are primarily intended for 8-bit codes, but the facility for their use is available to the application designer or user in 7-bit working where full understanding of the implications exists.

ESC 2/10 F and ESC 2/14 F designate sets of ninety-four graphic characters which will be used as the G2 set. The designated set is invoked by ESI or the Escape sequence ESC 4/15, when appropriate.

**A6.3 Future Announcers.** See Table A2.

### Table A2
### Future Announcers

| Final Character | Facilities Used |
|---|---|
| 4/5 | The G0 set of graphics is to be used. In both 7- and 8-bit environments, the Escape sequence which designates this set also invokes it into columns 2–7. SI and SO are not to be used. The C1 set of controls is to be used also. In an 8-bit environment, individual control characters of the C1 set of controls will be represented by ESC Fe sequences and not by the bit patterns of columns 08 and 09. Thus, in an 8-bit environment, columns 08–15 are not used. |
| 4/6 | The G0 and G1 sets of graphics are to be used. In both 7- and 8-bit environments, SI invokes G0 into columns 2–7 and SO invokes G1 into columns 2–7. The C1 set of controls is to be used also. In an 8-bit environment, individual control characters of the C1 set of controls will be represented by ESC Fe sequences and not by the bit patterns of columns 08 and 09. Thus, in an 8-bit environment, columns 08–15 are not used. |

NOTE: Additional announcers would be required to cover G2 and G3 sets of graphics.

## Appendix B

## Illustration of Implementation

### B1. General

The implementation of the extension of the 7-bit code, described in this standard, depends on the practical need of the implementor or user of the code extension. Implementations may differ in the following aspects and still remain in accordance with this standard:

(1) Including all or parts of the control sets and graphic sets

(2) Identifying the sets by Escape sequences, or having an agreement

(3) Dropping the superfluous invokers or not

(4) Examining the structure of an Escape sequence or not

(5) Handling the character DEL

(6) Combinations of the aforementioned

In this Appendix there are shown a few of the possible implementations that may help the user to a better understanding. A summary of Fig. B1 through B8 is given in Table B1. There are three types of figures: structure, interpretation, and transformation.

Table B2 shows the symbols used in the flowcharts (see Fig. B4 through B8) and their associated meaning. Table B3 shows the items used in the flowcharts, their meaning, and the possible values.

### B2. Structure

The pictorial representations of the structure of the 7-bit and 8-bit code extension (see Fig. B1 through B3)

have all the same format as follows:

(1) At the top of the figure are the repertories of single controls and sets of controls

(2) Below that are the Escape sequences that are used for the designation and invocation of the control sets as the C0 and C1 set

(3) At the bottom of the figure is the repertory of graphic sets

(4) Above that are the Escape sequences used for designation of the graphic set as the G0 or G1 set

(5) Above that are the shift characters that are used for the invocation of the G0 or G1 sets

(6) All this comes together in the center to form the 7-bit set or 8-bit set in use

### B3. Interpretation

The interpretation of data containing 7-bit or 8-bit code extension is shown by means of flowcharts (see Fig. B4 through B6). The interpretation is separate from the transformation.

### B4. Transformation

The transformation of the data from 7-bit to 8-bit and from 8-bit to 7-bit is shown by means of flowcharts (see Fig. B7 and B8). The transformation is separate from the interpretation.

#### Table B1
#### Index to Figures of Appendix B

| Included Sets | Bits | Structure | Interpretation | Transformation 7 to 8 | Transformation 8 to 7 |
|---|---|---|---|---|---|
| C0, C1, G0, G1 | 7 | B1 | B4 | B7 | B8 |
| C0, C1, G0, G1 | 8 | B2 | B5 | – | – |
| C0, C1, G0, G1 | 7 and 8 | B3 | B6 | – | – |

Table B2
Meaning of Flowchart Symbols

| SYMBOL | NAME | MEANING |
|---|---|---|
| INPUT NEXT CHARACTER / END | INPUT | TO FETCH THE NEXT CHARACTER FROM THE INPUT CHARACTER STRING AND GO TO THE END IF THERE IS NO MORE CHARACTER AVAILABLE. |
| PROCESS / ERROR END | PROCESS | TO EXECUTE THE DEFINED PROCESS.  THERE MAY BE AN ERROR END BECAUSE OF LACK OF MORE CHARACTERS. |
| ITEM   VALUE | DECISION | TO EXAMINE THE ITEM OF HAVING THE SPECIFIED VALUE. |
| VALUE ITEM | PREPARATION | TO SET THE ITEM TO THE SPECIFIED VALUE. |
| OUTPUT SOMETHING | OUTPUT | TO TAKE THE LAST CHARACTER (OUTPUT CHARACTER) OR THE SPECIFIED CHARACTER(S) OUTPUT 1/11, 4/14 OR THE MANIPULATED CHARACTER OUTPUT α8  DROPPED TO THE OUTPUT CHARACTER STRING. |
| SUB ROUTINE | PREDEFINED PROCESS | TO EXECUTE THE APPROPRIATE PREDEFINED PROCESS |
| ○ | CONNECTOR | ENTRY FROM, OR EXIT TO, ANOTHER PART OF THE FIGURE. |

Table B3
Items Used in the Flowcharts

| ITEM | MEANING | POSSIBLE VALUES | |
|---|---|---|---|
| | | IN 7 BITS | IN 8 BITS |
| CHARACTER | THE BIT PATTERN OF THE LAST CHARACTER | 0/0 TO 7/15 | 0/0 TO 15/15 |
| COLUMN | THE COLUMN NUMBER OF THE LAST CHARACTER | 0 TO 7 | 0 TO 15 |
| STATUS 7 | SHIFT-STATUS OF THE 7-BIT CODE | S1  S0 UNDEFINED | //////// |
| STATUS LEFT | SHIFT-STATUS OF THE LEFT PART OF THE 8-BIT CODE | //////// | S1 UNDEFINED |
| STATUS RIGHT | SHIFT-STATUS OF THE RIGHT PART OF THE 8-BIT CODE | //////// | S0 UNDEFINED |
| BITS | NUMBER OF THE VALID BITS OF THE CHARACTER | //////// | 7  8 UNDEFINED |

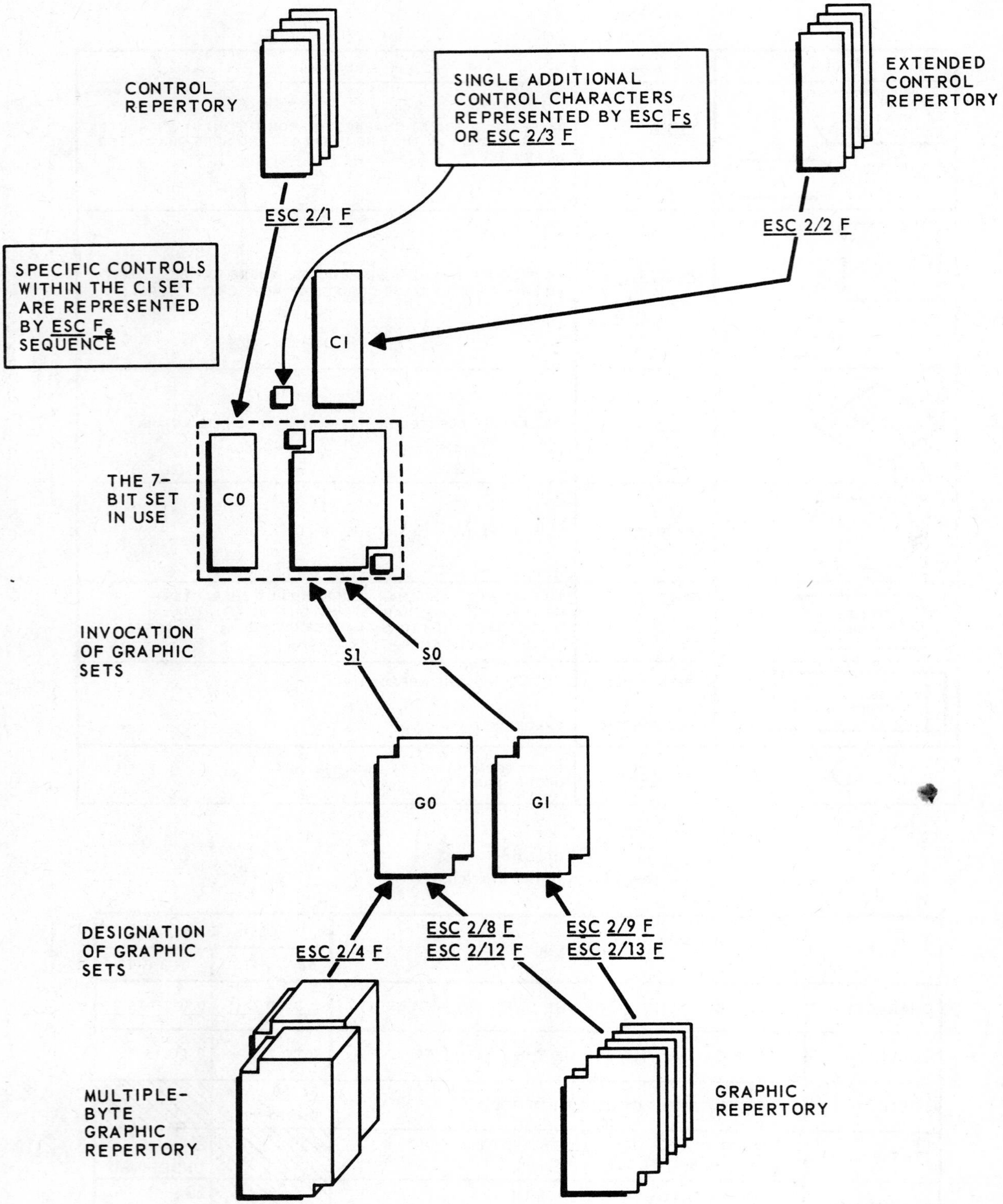

CONTROL
REPERTORY

SINGLE ADDITIONAL
CONTROL CHARACTERS
REPRESENTED BY ESC F$_S$
OR ESC 2/3 F

EXTENDED
CONTROL
REPERTORY

ESC 2/1 F

ESC 2/2 F

SPECIFIC CONTROLS
WITHIN THE CI SET
ARE REPRESENTED
BY ESC F$_e$
SEQUENCE

C1

THE 7-
BIT SET
IN USE

C0

616

INVOCATION
OF GRAPHIC
SETS

S1          S0

G0          G1

DESIGNATION
OF GRAPHIC
SETS

ESC 2/4 F

ESC 2/8 F        ESC 2/9 F
ESC 2/12 F       ESC 2/13 F

MULTIPLE-
BYTE
GRAPHIC
REPERTORY

GRAPHIC
REPERTORY

**Fig. B1**
**7-Bit Code Extension Structure Using C0, C1, G0, and G1 Sets**

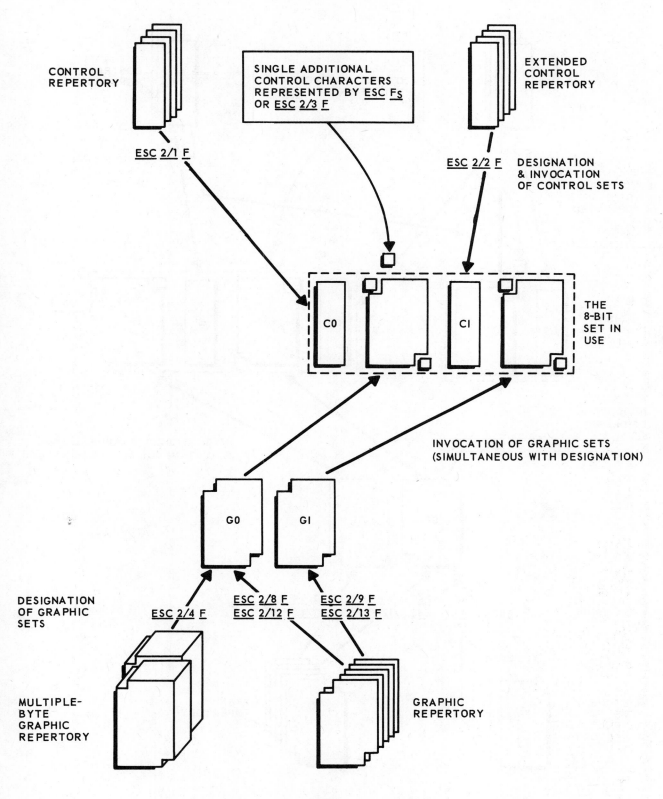

Fig. B2
8-Bit Code Extension Structure Using C0, C1, G0, and G1 Sets

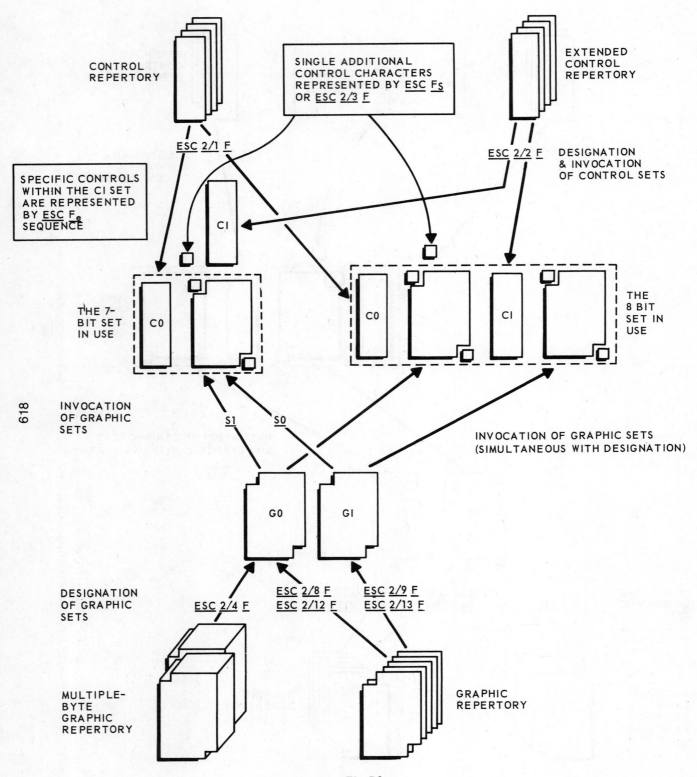

Fig. B3
Composite 7-Bit and 8-Bit Code Extension Structure
Using C0, C1, G0, and G1 Sets

618

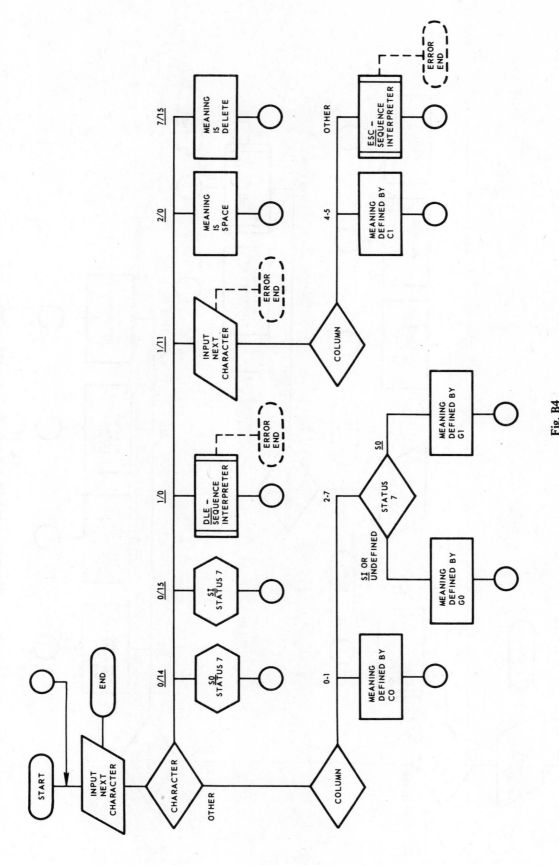

**Fig. B4**

Interpretation of 7-Bit Data with C0, C1, G0, and G1 Sets

620

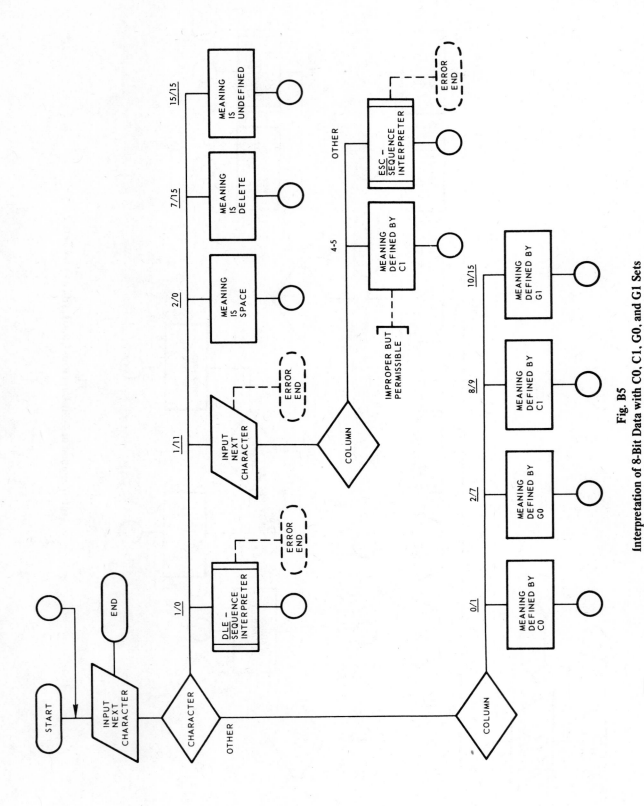

Fig. B5
Interpretation of 8-Bit Data with C0, C1, G0, and G1 Sets

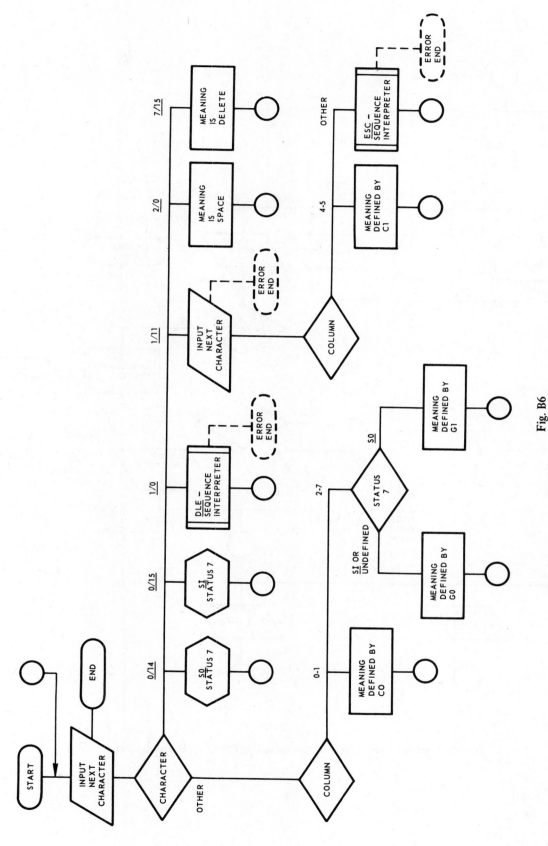

**Fig. B6**

**Interpretation of 7-Bit and 8-Bit Data with C0, C1, G0, and G1 Sets**

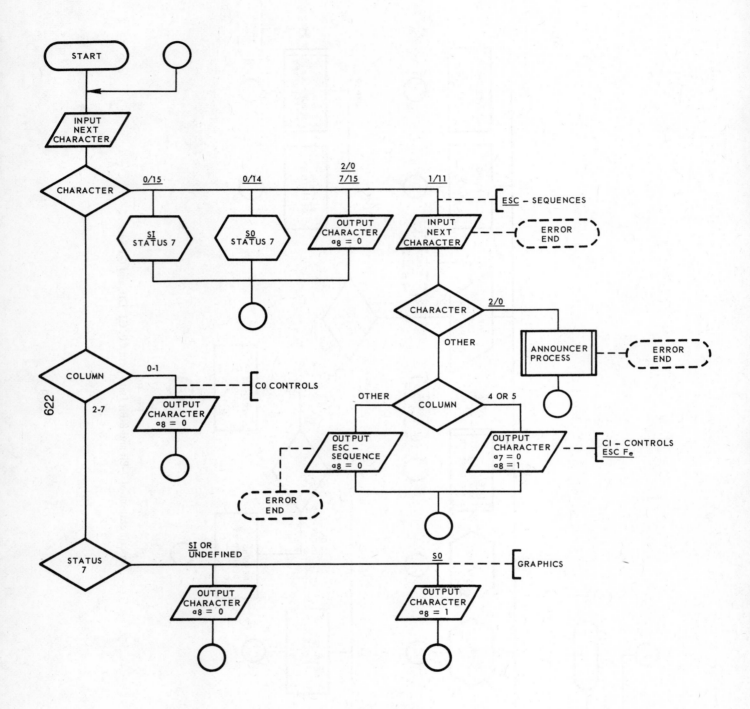

**Fig. B7**
**Transformation from 7 Bits to 8 Bits Including C0, C1, G0, and G1 Sets**

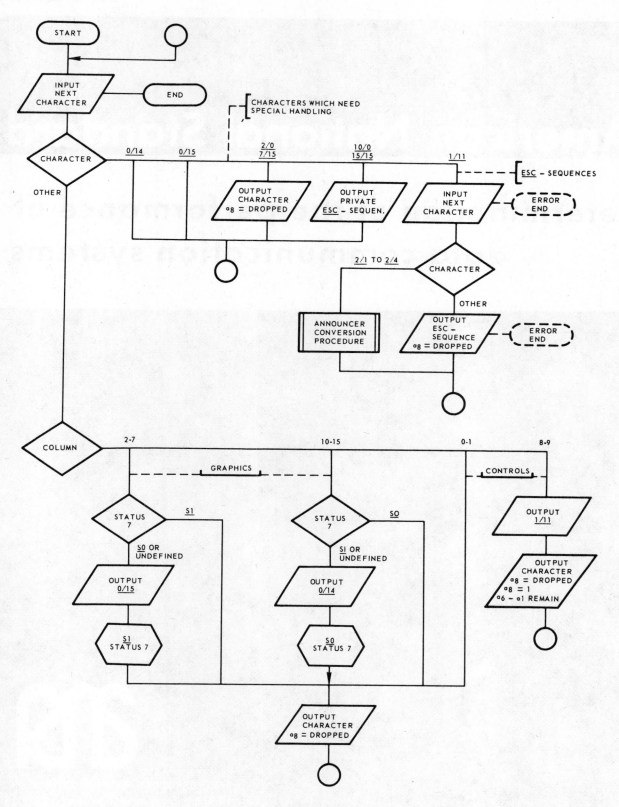

**Fig. B8**
**Transformation from 8 Bits to 7 Bits Including C0, C1, G0, and G1 Sets**

# American National Standard

## determination of the performance of data communication systems

624

X3.44-1974

american national standards institute, inc.
1430 broadway, new york, new york 10018

# American National Standard
# Determination of the Performance of
# Data Communication Systems

Secretariat

**Computer and Business Equipment Manufacturers Association**

Approved August 12, 1974

**American National Standards Institute, Inc**

## American National Standard

An American National Standard implies a consensus of those substantially concerned with its scope and provisions. An American National Standard is intended as a guide to aid the manufacturer, the consumer, and the general public. The existence of an American National Standard does not in any respect preclude anyone, whether he has approved the standard or not, from manufacturing, marketing, purchasing, or using products, processes, or procedures not conforming to the standard. American National Standards are subject to periodic review and users are cautioned to obtain the latest editions.

**CAUTION NOTICE:** This American National Standard may be revised or withdrawn at any time. The procedures of the American National Standards Institute require that action be taken to reaffirm, revise, or withdraw this standard no later than five years from the date of publication. Purchasers of American National Standards may receive current information on all standards by calling or writing the American National Standards Institute.

626

Published by

**American National Standards Institute**
**1430 Broadway, New York, New York 10018**

# Foreword

(This Foreword is not a part of American National Standard Determination of the Performance of Data Communication Systems, X3.44-1974.)

This standard presents the means for determining performance over an Information Path within a data communication system employing the American National Standard Code for Information Interchange, X3.4-1968 (ASCII). It specifically covers the necessary criteria, system description, and the methods for determining performance.

This standard was developed following extensive study of the various performance aspects as well as past, present, and future data communication system configurations and applications. In addition, consideration was given to other standards (national and international) that exist or are being developed for data communication.

This standard was approved as an American National Standard by ANSI on August 12, 1974.

Suggestions for improvement of this standard will be welcome. They should be sent to the American National Standards Institute, 1430 Broadway, New York, N.Y. 10018.

This standard was processed and approved for submittal to ANSI by American National Standards Committee on Computers and Information Processing, X3. Committee approval of the standard does not necessarily imply that all committee members voted for its approval. At the time it approved this standard, the X3 Committee had the following members:

J. F. Auwaerter, Chairman
V. E. Henriques, Vice-Chairman
R. M. Brown, Secretary

| Organization Represented | Name of Representative |
|---|---|
| Addressograph Multigraph Corporation | A. C. Brown |
| | D. S. Bates (Alt) |
| Air Transport Association | F. C. White |
| American Bankers Association | M. E. McMahon |
| | J. Booth (Alt) |
| American Institute of Certified Public Accountants | N. Zakin |
| | P. B. Goodstat (Alt) |
| | C. A. Phillips (Alt) |
| | F. Schiff (Alt) |
| American Library Association | J. R. Rizzolo |
| | J. C. Kountz (Alt) |
| | M. S. Malinconico (Alt) |
| American Newspaper Publishers Association | W. D. Rinehart |
| American Nuclear Society | D. R. Vondy |
| | M. K. Butler (Alt) |
| American Society of Mechanical Engineers | R. W. Rau |
| | R. T. Woythal (Alt) |
| Association for Computing Machinery | P. Skelly |
| | J. A. N. Lee (Alt) |
| | L. Revens (Alt) |
| | H. Thiess (Alt) |
| Association for Educational Data Systems | C. Wilkes |
| Association for Systems Management | A. H. Vaughan |
| Association of American Railroads | R. A. Petrash |
| Association of Computer Programmers and Analysts | T. G. Grieb |
| | G. Thomas (Alt) |
| Association of Data Processing Service Organizations | J. B. Christiansen |
| Burroughs Corporation | E. Lohse |
| | J. F. Kalbach (Alt) |
| Control Data Corporation | S. F. Buckland |
| | C. E. Cooper (Alt) |
| Data Processing Management Association | A. E. Dubnow |
| | D. W. Sanford (Alt) |
| Edison Electric Institute | R. Bushner |
| | J. P. Markey (Alt) |
| Electronic Industries Association | (Representation Vacant) |
| | A. M. Wilson (Alt) |
| General Electric Company | R. R. Hench |
| | J. K. Snell (Alt) |

| Organization Represented | Name of Representative |
|---|---|
| General Services Administration | D. L. Shoemaker |
| | M. W. Burris (Alt) |
| GUIDE International | T. E. Wiese |
| | D. Stanford (Alt) |
| Honeywell Information Systems Inc | T. J. McNamara |
| | E. H. Clamons (Alt) |
| Institute of Electrical and Electronics Engineers, Communications Society | R. Gibbs |
| Institute of Electrical and Electronics Engineers, Computer Society | G. C. Schutz |
| | C. W. Rosenthal (Alt) |
| Insurance Accounting and Statistical Association | W. Bregartner |
| | J. R. Kerber (Alt) |
| International Business Machines Corporation | L. Robinson |
| | W. F. McClelland (Alt) |
| Joint Users Group | T. E. Wiese |
| | L. Rodgers (Alt) |
| Life Office Management Association | B. L. Neff |
| | A. J. Tufts (Alt) |
| Litton Industries | I. Danowitz |
| National Association of State Information Systems | G. H. Roehm |
| | G. Vorlander (Alt) |
| National Bureau of Standards | H. S. White, Jr |
| | J. O. Harrison (Alt) |
| National Cash Register Company | R. J. Mindlin |
| | T. W. Kern (Alt) |
| National Machine Tool Builders' Association | O. A. Rodriques |
| | E. J. Loeffler (Alt) |
| National Retail Merchants Association | I. Solomon |
| Olivetti Corporation of America | E. J. Almquist |
| Pitney-Bowes Inc | D. J. Reyen |
| | B. Lyman (Alt) |
| Printing Industries of America | N. Scharpf |
| | E. Masten (Alt) |
| Scientific Apparatus Makers Association | A. Savitsky |
| | J. French (Alt) |
| SHARE Inc | T. B. Steel, Jr |
| | R. H. Wahlen (Alt) |
| Society of Certified Data Processors | A. Taylor |
| | J. J. Martin (Alt) |
| Telephone Group | V. N. Vaughan, Jr |
| | S. M. Garland (Alt) |
| | J. C. Nelson (Alt) |
| UNIVAC, Division of Sperry Rand Corporation | M. W. Bass |
| | C. D. Card (Alt) |
| U.S. Department of Defense | W. L. McGreer |
| | W. B. Rinehuls (Alt) |
| | W. B. Robertson (Alt) |
| Xerox Corporation | J. L. Wheeler |

Subcommittee X3S3 on Data Communication, which developed this standard, had the following members:

G. C. Schutz, Chairman
A. H. Stillman, Past Chairman*

M. W. Baty
M. J. Bedford
M. Bryngil
M. L. Cain
G. E. Clark
M. E. Cook
H. J. Crowley
V. Dagastino
J. L. Dempsey
D. Hanna
S. M. Harris
R. Hearsum
G. H. Hopping
W. M. Hornish
P. W. Kiesling
R. C. Matlack
W. B. McClelland
B. L. Meyer
R. F. Meyer
O. C. Miles
P. Muench
L. T. O'Connor
N. Priebe
S. Rosenblum
P. D. Simpson
C. P. Van Lidth de Juede
F. W. Warden
J. L. Wheeler
G. W. White

The following members of Task Group X3S3.5 on System Performance were the principal contributors to the development of this standard:

G. J. McAllister, Chairman
W. B. Dickinson, Past Chairman
R. Kerker, Past Chairman

L. M. Borden
R. G. Bounds, Jr
R. E. Huettner
R. T. Moore
R. A. Northrup
S. J. Raiter
G. C. Schutz

629

---

*Deceased.

# Contents

# American National Standard
# Determination of the Performance of
# Data Communication Systems

## 1. Scope

**1.1** This standard establishes:

(1) Descriptions of those elements of a data communication system pertinent to the determination of performance of an Information Path within a data communication system

(2) Criteria to be used in determining performance of an Information Path within a data communication system

(3) Methodology for determining values for these criteria

**1.2** This standard is applicable to any Information Path employing the American National Standard Code for Information Interchange, X3.4-1968 (ASCII), and the following relevant American National Standards pertaining to data communication, wherein the transmission of characters may be synchronous or nonsynchronous:

American National Standard for Bit Sequencing of the American National Standard Code for Information Interchange in Serial-by-Bit Data Transmission, X3.15-1966

American National Standard Character Structure and Character Parity Sense for Serial-by-Bit Data Communication in the American National Standard Code for Information Interchange, X3.16-1966

American National Standard Character Structure and Character Parity Sense for Parallel-by-Bit Data Communication in the American National Standard Code for Information Interchange, X3.25-1968

American National Standard Procedures for the Use of the Communication Control Characters of American National Standard Code for Information Interchange in Specified Data Communication Links, X3.28-1971

**1.3** The objective of this standard is to provide a means of:

(1) Prescribing performance requirements

(2) Determining projected and actual performance

(3) Determining temporal variations in performance

(4) Comparing the performance of an Information Path with the performance of any other Information Path

This standard is specifically not applicable to Information Paths that utilize modes of operation for which American National Standard control procedures have not been developed.

The effects of interaction between or among two or more Information Paths within a data communication system may require the future addition of other criteria to express other aspects of total system performance.

## 2. System Description

**2.1 General.** In this standard, performance is considered with respect to the functional rather than physical aspects of data communication systems. The physical items of equipment that perform these communication functions are identified in Appendix D to assist in the understanding of these functions.

**2.2 Functional Elements**

**2.2.1 General.** The functional terminology set forth herein describes elements of a data communication system for performance measurement purposes. These functional concepts are applied uniformly from system to system, regardless of the physical differences of the systems. Likewise, performance criteria as defined in Section 4 are applied uniformly. The elements of data communication system description that are established by this standard are Information Transfer Channel, Terminal Configuration, and Information Path. (See Fig. 1.)

**2.2.2 Information Transfer Channel.** The Information Transfer Channel functionally accomplishes the unidirectional transfer of information over a circuit from one Terminal Configuration to another within an Information Path and also allows the transmission of sequences in either direction for the purpose of supervision and control. The Information Transfer Channel is not required to recognize, and (as seen at the interfaces with the sending and receiving Terminal Configu-

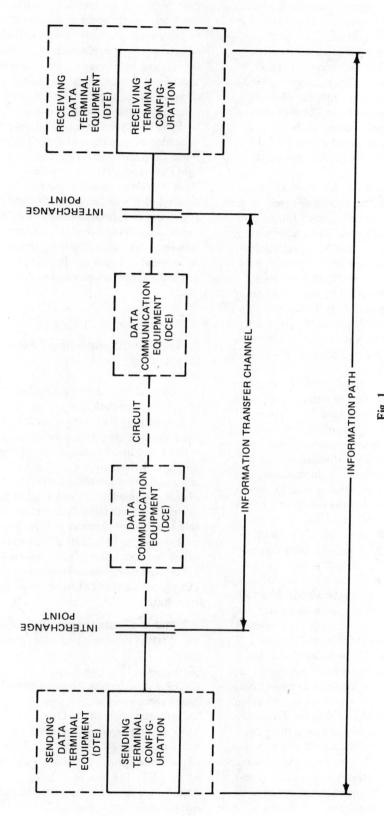

**Fig. 1**
**Block Diagram of an Information Path**

rations) must not alter, the character or message structure or the sequence of the binary data stream passing through it. The Information Transfer Channel is bounded by the interchange points located between the transmitting and receiving data communication equipments and their respective interconnecting data terminal equipments. The physical interconnections necessary to establish an Information Transfer Channel are either permanent by system design (as with a full-period private-line service) or are established temporarily by the operation of the data communication system.

**2.2.3 Terminal Configuration.** A Terminal Configuration comprises the specific functional elements within a data terminal equipment interconnected for the purpose of information transfer over a single Information Path at a given time. A transmitting Terminal Configuration performs all functions required to convert data into a form suitable for transfer to the interchange point and subsequent transmission by the Information Transfer Channel. A receiving Terminal Configuration performs the inverse of these functions. Data-communication-related procedures, such as format construction and conversion, error control, communications control procedures, transcription to and from media/storage, etc, are performed by a Terminal Configuration for the Information Path of which it is a functional part. A Terminal Configuration may be established as a consequence of reaction to communications control procedures, or it may be established by system design.

**2.2.4 Information Path.** The Information Path is the functional route by which data communication is accomplished. It comprises the sending and the receiving Terminal Configurations and the Information Transfer Channel. It encompasses all of the functions performed, including forward and reverse supervision, to allow information transfer in a one-way direction from a single data source to a single data sink over a single Information Transfer Channel.

**2.2.5 Activation of an Information Path.** When all of the five communication phases are utilized (see Section 3), the Information Path is activated and deactivated by the sequential progression of the system through these phases. Phase 1 accomplishes the electrical interconnection of all equipment required to establish the function of the Information Transfer Channel, and Phase 2 completes the logical interconnection of those functions required to establish the Terminal Configurations. Thus at the end of Phase 2 the Information Path is fully functionally defined and activated. Similarly, Phase 4 disestablishes the logical interconnections of the Terminal Configurations, and Phase 5 accomplishes the electrical disconnection of the Information Transfer Channel. Although the activation and

deactivation of the functions ascribed to an Information Path usually occur as described at the beginning of this paragraph, an Information Path is considered to exist between a particular data source and a particular data sink whenever the data communication system is designed in such a way that it will allow the transfer of information between this source and sink. However, connection establishment and link establishment may be fixed by system design as in systems employing a dedicated Information Transfer Channel and fixed Terminal Configuration. In such an instance, the Information Path is activated by the Information Transfer Phase and deactivated by the Link Termination Phase. In systems where the Terminal Configuration is not fixed, the Link Establishment Phase determines the extent of penetration of the Information Path into the sink data terminal equipment; that is, the data sink is established and may be identified by means of the communication control procedures employed during this phase.

# 3. Data Communication Phases and Phase Sequencing

**3.1 General.** In order to determine the performance of an Information Path, it is necessary to distinguish between the time that the subject Information Path is in the process of performing data-communication-related activity and the time when the Information Path is not. When the Information Path is actively engaged in data-communication-related activity, that is, any of the five phases described in 3.2.1 through 3.2.5, it is said to be in the Communication State. At all other times it is said to be in the Quiescent State. While it is in this state, portions of the Information Path may be performing other useful functions. All criteria value determinations are derived from measurements or calculations made during, or as a result of, time spent in the Communication State.

**3.2 Phases of Data Communication.** The Communication State is subdivided into five phases (see 3.2.1 through 3.2.5). All phases need not be employed on an Information Path (see 3.3).

**3.2.1 Connection Establishment Phase ($\phi$1).** This phase represents the time interval required to establish an Information Transfer Channel when switched circuits are employed. When leased or private circuits are employed, and the Information Transfer Channel is always established, this phase is not applicable.

START: The instant of going "OFF-HOOK" at the calling data station in order to start the dialing process; for example, lifting the handset off the cradle in normal dialing, or putting on "ON" condition on interchange circuit CRQ — Call Request (CCITT 202)[1] in automatic calling.

END: The instant when an "ON" condition exists on interchange circuit CB — Clear to Send (CCITT 106) or equivalent function at the calling data station. Alternatively, going "ON-HOOK" terminates Phase 1 and causes Phase 5 to be entered.

**3.2.2 Link Establishment Phase ($\phi$2).** This phase represents the time interval required to establish the Terminal Configurations resulting in activation of an Information Path. An Information Transfer Channel must be established prior to entry into this phase. Phase 2 includes the time required to poll or select, and also includes the time associated with any attempts to retry selection. If contention is employed as the method of link establishment, times associated with retries are included in this phase. On simple point-to-point Information Paths, where no link establishment procedure is required, this phase is not applicable.

START: The end of the Connection Establishment Phase. When Phase 1 is not employed, the Link Establishment Phase begins at the instant of initiation of transfer of communication control characters or prefixes over interchange circuit BA — Transmitted Data (CCITT 103) or when a "MARK-HOLD" is applied to this circuit.

END: The initiation of the transfer of a message or transmission block. This shall be indicated by the initiation of the transfer of a SYN, SOH, or STX character over interchange circuit BA — Transmitted Data (CCITT 103) or by the initiation of the transfer of SYN or "MARK-HOLD" over this circuit for the purpose of time fill. Communication control characters, or prefixes and communication control characters, shall have been transferred over interchange circuit BA between the time that a "MARK-HOLD" is used to determine the start of this phase and the time that it is used to determine the end of this phase. Alternatively, Phase 2 shall be terminated by initiation of the transfer of an EOT character, or a DLE EOT sequence, whichever applies. In synchronous systems, this instant will occur when the transfer of a SYN character, required to precede the EOT or DLE EOT, is initiated over interchange circuit BA — Transmitted Data (CCITT 103).

**3.2.3 Information Transfer Phase ($\phi$3).** This phase represents the time required to transfer all messages associated with a single activation of an Information Path. It also includes: (1) the time required to retransmit messages or transmission blocks; (2) the time required to perform supervision related to these messages

or transmission blocks; (3) time delays associated with device start-up, media loading, or media change, which are incurred after Phase 3 has started but not ended; and (4) any other delay within the Information Path affecting execution of this phase as defined by the formats of the messages and supervisory sequences.

START: The initiation of the transfer of a message or transmission block. This shall be indicated by the initiation of the transmission block. This shall be indicated by the initiation of the transfer of a SYN, SOH, or STX character over interchange circuit BA — Transmitted Data (CCITT 103) or by the initiation of the transfer of SYN or by the application of "MARK-HOLD" over this circuit for the purpose of time fill.

END: The initiation of the transfer of an EOT character or a DLE EOT sequence, whichever applies. In synchronous systems, this instant will occur when the transfer of a SYN character, required to precede the EOT or DLE EOT, is initiated over interchange circuit BA — Transmitted Data (CCITT 103).

**3.2.4 Link Termination Phase ($\phi$4).** This phase represents the time required to disestablish that portion of the Information Path not a part of the Information Transfer Channel, specifically the Terminal Configuration(s). It includes the time required to transmit an EOT character or the time required to transmit a mandatory disconnect sequence (DLE EOT), or both.

START: The initiation of the transfer of an EOT character or a DLE EOT sequence from any preceding phase. In synchronous systems, this instant will occur when the transfer of a SYN character, required to precede the EOT or DLE EOT, is initiated over interchange circuit BA — Transmitted Data (CCITT 103).

END: On switched circuits, the end of Phase 4 interchange circuit CD — Data Terminal Ready (CCITT 108.2) or equivalent is turned OFF unless reentry to Phase 2 is allowed for the purpose of establishing another Terminal Configuration. In this case, the end of Phase 4 occurs at the instant of initiation of transfer of communication control characters or prefixes over interchange circuit BA — Transmitted Data (CCITT 103) or when a "MARK-HOLD" is applied to this circuit. On leased or private circuits, the end of Phase 4 occurs when the control station returns to the Quiescent State or on Information Paths with distributed control, when the originating data station returns to the Quiescent State.

**3.2.5 Connection Clearing Phase ($\phi$5).** This phase represents the time interval required to disestablish an Information Transfer Channel when switched circuits are employed.

START: When interchange circuit CD — Data Terminal Ready (CCITT 108.2) or equivalent is turned OFF at the calling data station.

---

[1] See Section G5 of Appendix G for references related to interchange circuits.

END: The instant when the circuit switching equipment has completed those disconnect functions, such that reentry to the Connection Establishment Phase is possible.

**3.3 Phase Sequencing.** Subject to the methodology provisions of Section 5, any phase sequences permitted by the applicable control procedures may be employed in determining values for the performance criteria defined in Section 4. (See also Appendix C.)

# 4. Performance Criteria

**4.1 General.** Four performance criteria are established by this standard. They are:

(1) Transfer Rate of Information Bits (TRIB)
(2) Transfer Overhead Time (TOT)
(3) Residual Error Rate (RER)
(4) Availability (A)

All four are necessary for the complete determination and expression of performance on any one Information Path of a data communication system.

**4.2 Transfer Rate of Information Bits (TRIB).** The TRIB criterion expresses the ratio of the number of Information Bits accepted by the receiving Terminal Configuration during a single Information Transfer Phase (Phase 3) to the duration of that Information Transfer Phase. TRIB is expressed in bits per second.

**4.2.1 Information Bits.** Information Bits are all bits excluding start-stop elements (if used) and parity bits contained in Information Characters. Information Characters are the characters in the graphic subset of ASCII (American National Standard Code for Information Interchange, X3.4-1968) that are transmitted as message text except when the characters are utilized for the purpose of control or effecting format. (See 4.2.3.)

**4.2.2 Acceptance of Information Bits.** Information Bits are defined to be accepted by the receiving Terminal Configuration if a positive acknowledgment to a transmission block is received by the sending Terminal Configuration. If, because of system design, no such positive acknowledgment is ever generated, Information Bits are assumed to be accepted by the receiving Terminal Configuration at the completion of the Information Transfer Phase.

**4.2.3 Non–Information Characters in Message Text.** ESC sequences used for any control or format purposes, or the multiple (more than two) sequential use of any character for the purpose of indenting, positioning, or ordering other characters in the message text, shall not be counted as Information Characters.

**4.3 Transfer Overhead Time (TOT).** The TOT criterion

is a measure of overhead time distributed over accepted Information Bits. The overhead time is the time required to establish and disestablish an Information Path in order to effect information transfer. TOT is the ratio of the sum of the elapsed times in Phases 1, 2, 4, and 5, which constitute a complete transition through all applicable phases, to the number of Information Bits accepted by the receiving Terminal Configuration during a single Information Transfer Phase.

$$TOT = \frac{\Sigma\phi 1 + \Sigma\phi 2 + \Sigma\phi 4 + \Sigma\phi 5}{\text{Number of Information Bits accepted}}$$

TOT is expressed in seconds per bit.

**4.4 Residual Error Rate (RER).** The RER criterion is a measure of the inaccuracy of the transfer of Information Characters over the Information Path. RER is the ratio of the sum of (1) erroneous Information Characters accepted by the receiving Terminal Configuration and (2) Information Characters transmitted by the sending Terminal Configuration but not delivered to the receiving Terminal Configuration, to the total number of Information Characters contained in the source data. Within 4.4(1), erroneous Information Characters may be either undetected, or detected but uncorrected, within the Information Path.

$$RER = \frac{C_e + C_u + C_d}{C_t}$$

where

$C_e$ = Erroneous Information Characters accepted by the receiving Terminal Configuration

$C_u$ = Information Characters that were transmitted and assumed by the sending Terminal Configuration to be accepted, but were not accepted by the receiving Terminal Configuration (undelivered)

$C_d$ = Information Characters that were accepted in duplicate by the receiving Terminal Configuration, but were not intended for duplication (duplicates)

$C_t$ = Information Characters contained in the source data

RER is expressed in terms of Information Characters in error per Information Character.

**4.5 Availability (A).** This criterion identifies, as a percentage, the portion of a selected time interval during which the Information Path is capable of performing its assigned data communication function. Availability is expressed as follows:

$$A = \frac{\text{Base Time Period (hours)} - \text{Unavailable Time (hours)}}{\text{Base Time Period (hours)}} \times 100\%$$

The Base Time Period is that time when the Information Path is expected to perform its assigned data communication function and the Unavailable Time is that time, within the Base Time Period, when the Information Path is not capable of performing its assigned data communication function.

# 5. Methods for Determining Performance

**5.1 General.** Section 5 identifies the information required to accompany values for criteria, and establishes rules for the determination of values for the criteria, including accuracy requirements and the circumstances that result in invalid determinations.

**5.2 Minimum Information Required to Accompany Criteria Values.** As a minimum, the information outlined in 5.2.1 and 5.2.2 must be provided in conjunction with the values of performance criteria determined in accordance with this standard. This information is required to define the Information Path and its utilization for which performance values have been, or are to be, determined. Measured criteria values must be accompanied by the basic Information Path description (see 5.2.1). Calculated values require, in addition, identification of values assumed for significant parameters used in the calculation (see 5.2.2). Additional information beyond that identified in 5.2.1 and 5.2.2 may be required to accompany individual criterion values. This information is identified in 5.3.

**5.2.1 Basic Information for Measured and Calculated Values**

(1) Terminal Configuration
    (a) Device types
    (b) Data representation
    (c) Data rates
(2) Information Transfer Channel
    (a) Point-to-point/multipoint
    (b) Synchronous/asynchronous
    (c) Simplex/half-duplex/full-duplex
    (d) Switched/leased
    (e) Manual/automatic dial
    (f) Signaling rate
    (g) Data communication equipment type
(3) Procedures
    (a) Data communication phases used and associated times for each
    (b) Supervisory sequences
    (c) Prefixes
    (d) Heading length (characters)
    (e) Text length (characters)
    (f) Block length (characters)
    (g) Error control used

(h) Recovery (retries, exits)
    (i) Control procedure category used, including applicable options

**5.2.2 Additional Information for Calculated Values**
(1) Terminal Configuration
    (a) Significant device delays
    (b) Buffer arrangements
    (c) Buffer capacities
    (d) Buffer load and unload times
    (e) Character error rates
    (f) Timing interrelationships of all delay factors with respect to data communication phases
    (g) Frequency and duration of unavailable time due to external or internal causes
(2) Information Transfer Channel
    (a) Data communication equipment delay: CA "ON" to CB "ON"
    (b) Propagation delay
    (c) Character error rates
    (d) Frequency and duration of unavailable time due to external or internal causes
(3) Procedures
    (a) Method of generating and reacting to non-Information Characters

**5.3 Rules for Determining Values for the Criteria**
**5.3.1 Phase Transitions.** Values of TRIB, TOT, and RER shall be determined only when there has been a complete transition through the Information Transfer and Link Termination Phases on the Information Path under consideration. In addition, the determination of TOT requires consideration of times involved in other phases of data communication as identified in 5.3.7.

**5.3.2 Invalid Determinations.** In the event that abnormal termination occurs during determination of TRIB, TOT, or RER, the values shall be discarded.

**5.3.2.1** An abnormal termination is a system failure that disables the communications capability of the Information Path to the extent that normal automatic or manual recovery procedures are not effective.

An abnormal termination may be indicated by expiration of the timer, which facilitates disconnection procedures if data transmission stops due to not recognizing DLE EOT, or due to problems on the Information Path. Aborts and interrupts are not considered abnormal terminations.

**5.3.2.2** All time increments discarded under 5.3.2 shall accrue to the determination of unavailable time.

**5.3.3 Means of Determination.** Values of the performance criteria may be determined in accordance with this standard by calculation, by measurement, or by a combination of calculation and measurement.

**5.3.4 Development and Presentation of Representative Criteria Values.** The values determined for TRIB,

TOT, and RER are singular values for any single determination, and this standard specifies how such singular values may be determined. However, the frequency of individual value determinations, their distribution over a period of time, and the method of combining singular values of a criterion to form a representative performance criterion value are not a part of this standard. Such considerations are dependent on the utilization of the particular Information Path; anticipated variability of individual criterion values; required statistical significance, including tolerance and confidence level of the representative values; and limitations of the measurement tools to be employed.

**5.3.5 Concurrency of Criteria Value Determinations.** When two or more of the performance criteria are utilized to describe the performance of a particular Information Path, the degree of concurrency of time periods during which the value determinations are made shall be stated. Further, when TRIB and TOT are stated together, the values shall have been determined concurrently.

**5.3.6 TRIB Determination Rules**

**5.3.6.1** A TRIB value shall be determined by establishing the number of Information Bits accepted by the receiving Terminal Configuration during a single transition through the Information Transfer Phase and dividing by the duration of that Information Transfer Phase in seconds.

**5.3.6.2** TRIB shall be determined to an accuracy of ± 1%. Since the number of Information Bits can be determined with absolute accuracy, the duration of the Information Transfer Phase may be determined within an accuracy of ± 1%.

**5.3.6.3** Acceptance of Information Bits is defined to occur when a positive acknowledgment to a transmission block is received. Information Bits shall be counted at the source Terminal Configuration. If, because of system design, no such positive acknowledgment is ever generated, acceptance is assumed at the receiving Terminal Configuration at the completion of the Information Transfer Phase.

**5.3.7 TOT Determination Rules**

**5.3.7.1** TOT values shall be determined only when there has been a complete transition through all of the data communication phases (including the Information Transfer Phase) that are employed by the Information Path under consideration.

**5.3.7.2** The accuracy of the individual phase times from which TOT values are determined shall be ± 1%.

**5.3.7.3** TOT values for an Information Path wherein polling is employed shall be determined from the time increments of Phases 1, 2, 4, and 5 that are parts of a single complete transition through all the phases (including the Information Transfer Phase) employed by the Information Path. In polling applications, entry to the Information Transfer Phase will not occur unless the polled station has message traffic. Phase times associated with entry attempts that are unsuccessful for this reason shall not accrue to TOT values.

**5.3.7.4** TOT values for an Information Path wherein selection or contention is employed shall be determined from the time increments of Phases 1, 2, 4, and 5 that result in a single complete transition through all of the phases (including the Information Transfer Phase) employed by the Information Path and all of the preceding associated phase time increments that were utilized in unsuccessful attempts to enter and complete the Information Transfer Phase. Thus, in selection and contention applications, phase times associated with unsuccessful attempts to enter and complete the Information Transfer Phase shall accrue to TOT values; however, intervening periods between attempts, during which the Information Path is not in one of the phases of data communication, shall be excluded from TOT values.

**5.3.8 RER Determination Rules**

**5.3.8.1** RER shall be determined in accordance with the formula given in 4.4.

**5.3.8.2** The number of Information Characters over which RER is to be determined (the denominator of the RER formula) is a function of the anticipated character error rate and the tolerance and confidence level of the desired RER determination.

**5.3.8.3** The number of erroneous, nondelivered, and duplicated Information Characters (the numerator of the RER formula) shall be determined by intercomparing the Information Characters accepted by the receiving Terminal Configuration with those that were contained in the source data. This intercomparison shall be accomplished prior to the use of the medium containing the accepted Information Characters for any other purpose.

**5.3.8.4** The accuracy of intercomparison shall be absolute. This requires extreme care in the selection of techniques and equipment for comparison so as to validate each candidate error completely.

**5.3.8.5** Human errors (as from an operator striking a wrong type key) do not constitute a part of RER.

**5.3.8.6** It should be noted that the impact of non–Information Characters in error on RER may sometimes be evident and sometimes not. The impact of such errors and the effect on RER are dependent to some extent on the specific Terminal Configurations and control procedures employed. These factors should be examined with respect to the particular Information Path under consideration, and resolution of questions arising from this examination should be achieved jointly by the parties involved.

### 5.3.9 Rules for Determining Availability

**5.3.9.1** Availability shall be determined in accordance with the formula given in 4.5.

**5.3.9.2** Precise terms and conditions relative to the determination of Availability shall be formulated pertinent to the configuration and application of each Information Path. These terms and conditions shall be developed and stated by the user or purchaser in preparing requirements, by the manufacturer or supplier in quoting projected or actual performance, or jointly by the user and the supplier in contracting for a level of Availability for a particular Information Path. Such terms and conditions shall encompass at least the following items.

**5.3.9.2.1** An Availability performance period shall be identified in days (for example, 30 days), and this period shall encompass the base time period over which Availability is to be determined.

**5.3.9.2.2** Within the performance period, increments of time significant to the determination of Availability shall be identified, and their inclusion in, or exclusion from, the base time period shall be prescribed. Examples of time increments to be considered are scheduled maintenance time, parts procurement time, travel time, unscheduled maintenance time, idle time, media replenishment time, and power-off time.

**5.3.9.2.3** Conditions or events that could render the Information Path incapable of performing its assigned data communication function shall be identified, and their impact on unavailable time shall be prescribed. Examples of conditions or events to be considered are equipment failures in the Terminal Configurations or in the Information Transfer Channel, frequent or repeated temporary failures in the Information Path, equipment or power failure external to the Information Path, failures occurring outside of the base time period, media failures (such as tapes or cards), environmental failures (such as air conditioning), and catastrophes (such as fire or flood), and the results of human error.

**5.3.9.2.4** Isolated temporary failures occurring during data communication from which recovery is effected by an automatic or manually initiated retry shall not result in unavailable time for the purpose of determining availability. The evaluation of frequent or repeated temporary failures shall be established in accordance with 5.3.9.2.2 and 5.3.9.2.3.

## Appendix A
## General Considerations

### A1. Introduction

The development of this standard required the consideration of many factors that tend to affect the performance of data communication systems. These studies led to the conviction that system performance should best be described in terms that would in some way reflect: (1) the capacity of the system to handle information, for example, some measure of the rate of information flow or throughput; (2) the delay factors associated with the data communication process; (3) the accuracy or freedom from errors; and (4) the availability of the system to perform its intended function.

It was anticipated that some means could be developed to express each of these attributes of performance in terms that would be universally applicable to the expression of the total performance of any data communication system. However, the interactions among these attributes of performance, the variability in total system configuration and complexity, and the wide range in system operating disciplines forced the adoption of performance criteria that have constraints in their application. These constraints are not trivial, and were not accepted without long and exhaustive deliberation. Principal among them is the limitation of applicability of the standard to a single Information Path employing American National Standards X3.4-1968 and X3.28-1971, and other associated communication standards. Within these constraints many possible criteria were examined, and TRIB, TOT, RER, and Availability were adopted as most suitable for the expression of a manageable set of performance attributes.

The values of these criteria can be calculated and measured. They can be used to specify, evaluate, and compare performance of Information Paths. It is further intended that they be used to evaluate performance on an operating basis and be sufficiently sensitive to deterioration in performance to aid the user in applying corrective measures. It is acknowledged that, for systems composed of more than one Information Path, the criteria selected may be incomplete as total system performance measures. Further, it is anticipated

that extension of standard control procedures into other modes of operation will require reexamination, and perhaps modification, of the criteria selected. These considerations have been identified as the subjects of future work in the data communication performance area.

Relating the five phases of data communications as presented in ISO (International Organization for Standardization) International Standard ISO 1745-1975[2] to American National Standard X3.28-1971 was an essential step in the development of the methodology. The ISO concept was expanded to define clearly the start and end delimiters necessary to establish the time utilized in each phase, and therefore provide the data for determining the values of the criteria.

The concept of relating the methodology to a functional procedure in lieu of a physical description of test methods was adopted to avoid some obvious difficulties. Test point, test messages, make and model of instruments, and exact test procedures will vary considerably among systems, and impose impossible conditions on standardization. It remains for the user to reduce the functional procedures described herein to a physical description of the test methods for the system under consideration.

### A2. System Description

System performance standards cannot be developed and fully used without having an organized way of describing the functional and physical conditions to which performance applies. Accordingly, a system description has been developed to the extent required for performance determination. It is recognized that a more complete system description may be necessary to establish methods for classifying systems. This could simplify and reduce the work required for developing perfor-

---

[2] See Appendix G, Reference Documents.

640

17

mance standards for more complex systems, and remains in the program of future work.

## A3. Relationships among Criteria

TRIB, TOT, RER, and Availability are largely independent criteria of performance, and it is not the intent of the standard to relate them. Further, it is not necessary to determine values for all of the criteria for a particular Information Path. Any one or more of the criteria may be utilized, subject to the requirements of 5.3.5 and the conditions noted in 4.1.

## A4. Methodology

In testing the application of the criteria to representative systems it became apparent that the standardization of methods of determining criteria values was necessary. It was found that methodology was highly dependent upon the control procedures employed in the system. Accordingly, the application of this standard is limited to those systems that are operated in accordance with American National Standard X3.28-1971. When new categories of control procedures are standardized, this document will be updated to fit the new categories.

## Appendix B
## Qualifications and Rationale

### B1. Introduction

This Appendix identifies some of the alternatives considered in the development of this standard, provides additional information to facilitate the understanding and use of specific criteria and concepts, and indicates some areas in which additional work is known to be required.

### B2. Transfer Rate of Information Bits

Much of the work done in the past several years in evaluating the performance or effectiveness of the data communication process has been expressed in terms of efficiency or throughput. Examination of many published papers has disclosed a variety of definitions for these terms and the absence of any commonly accepted meanings across the industry. After several alternatives were suggested and evaluated, the term Transfer Rate of Information Bits (TRIB) was selected and defined as one that would uniquely and unambiguously represent the throughput aspect of data communication. It reflects the effective rate of Information Bit flow in an Information Path during the Information Transfer Phase, and

can be used to represent the throughput aspect of performance for any Information Path that falls within the scope of this standard. Since TRIB is a new term in the industry, it is not handicapped by a variety of previous uses and meanings.

### B3. Transfer Overhead Time

The Transfer Overhead Time (TOT) criterion was selected to represent a measure of the amount of time in the data communication process spent in other than the actual Information Transfer Phase. Expressed in terms of Phase 1, 2, 4, and 5 time (seconds) per Information Bit, TOT provides a means of relating these "overhead" times to the amount of data transferred. The value of TOT is sensitive to system design and application considerations and, in particular, will expose the impact of design or application constraints that would not necessarily be reflected in the other performance criteria. Consider, for example, an application utilizing two alternate system designs and requiring the consecutive transmission of a large number of messages over a switched network. In the first, the design of the sending Terminal Configuration is such that only one message may be transferred per connection. The second system has no such limitations. Whereas TRIB for the two cases might be very nearly the same, TOT for the first system would be significantly larger because the overhead time required for connection, link establishment, termination, and clearing (essentially the same in both systems) would be associated with a much smaller number of Information Bits. Although this may be an extreme example, it should be observed that TOT will provide, in any case, an effective measure of the amount of this type of overhead incurred.

As indicated in 5.2 of this standard, the data communication phases employed and the associated times for each must be indicated as part of the basic Information Path description. This information, in conjunction with a TOT value, will provide additional insight into system design and application considerations relative to overhead versus Information Bit Transfer.

## B4.  Residual Error Rate

The Residual Error Rate (RER) criterion provides a measure of the accuracy of the data communication process. Included in the RER value are only those characters that remain in error after the detection and correction facilities provided within the Information Path have been applied. Errors that occur within the Terminal Configurations or the Information Transfer Channel may become residual errors, depending of course on the effectiveness of the detection and correction facilities. RER is character oriented, and the occurrence of a residual error must be determined by a one-for-one intercomparison of Information Characters recorded in the media of the receiving Terminal Configuration against the Information Characters contained in the source data at the sending Terminal Configuration. Fig. B1 illustrates that there are two basically different types of error occurrence that may result in residual error. If the error occurs in an Information Character (the left side of the figure), and it is either undetected or detected but uncorrected, it becomes a single residual error. If, however, the error occurs in another character (a non–Information Character), the results or impact of that error must be examined further. If intercomparison of sink versus source data discloses discrepancies, then residual error has occurred. For example, undetected errors in communication control characters might result in missing or duplicated records. In this case the contribution to the RER count would be the number of characters in the missing or duplicated records as indicated in the RER formula; see terms $C_u$ and $C_d$ in 4.4 of this standard.

In another example, an undetected error in the LF

(Line Feed) character might result in an entire line being overprinted at the sink. Although it is clear that residual error has occurred, it is not clear whether the residual error count should include one (for the control character that produced the problem) or should include all of the characters in the unreadable line or lines. Since the impact and significance of errors such as this are dependent in many cases on the specific Terminal Configurations and control procedures employed, this standard cannot establish the way in which the impact of these errors is to be reflected in RER. Rather, this standard specifies that, except in the case of undelivered or duplicate records, the effect on RER of undetected errors in non–Information Characters should be examined and resolved for the particular Information Path under consideration by the parties involved.

Rererring again to Fig. B1, if an uncorrected error in a non–Information Character does *not* produce an error in the Information Characters in the sink, then by definition (see 4.4) no residual error is counted. It is acknowledged that this branch of the diagram permits an uncorrected character error (in the true sense, a residual error) to be excluded from the RER value. For example, an uncorrected error in a heading character might result in misrouting or mishandling of the associated message in a subsequent Information Path. Although the probability of occurrence of such an error is very small, it is recognized that more study is required. The impact of this type of error on RER has been identified as the subject of negotiation by the parties involved, and will be further addressed in future work.

One other category of error should be recognized with respect to RER. This category, which might be called "misdelivered characters," includes any characters that are recorded in the sink media while a measurement is in progress and were intended for a different Information Path. Presumably, this situation could happen only in a multipoint network as a result of a control error in an Information Path other than the one being measured. This category of error has also been excluded from RER by definition; again, however, further consideration of this area has been identified as the subject of future work.

Because RER values observed in actual system operation will normally be very low, it may be necessary to monitor the communication of a large volume of messages or blocks in order to establish values of adequate statistical significance. It should also be noted that, because of accuracy requirements, measurement procedures and equipment must be carefully planned. Since such procedures and equipment will be different for different Information Paths and applications, a standard approach cannot be defined. It is the responsibility of the parties involved to establish a suitable measurement plan.

642

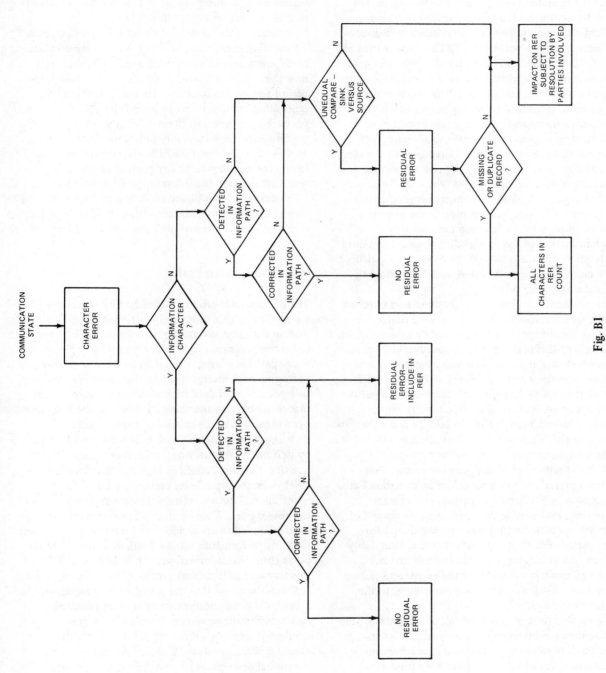

**Fig. B1**
**Types of Error Occurrence**

## B5. Availability

There are many terms currently in use to express various aspects of the ability of a system (or elements of a system) to operate over a period of time without failure, or to recover from a failure. Among these terms are reliability, maintainability, serviceability, availability, mean time between failures (MTBF), mean time to repair (MTTR), etc. There are no universally accepted definitions or meanings of these terms when applied to the data communication process. When such terms are used today in relationships among users and suppliers, specific definitions and rules of measurement must be established appropriate to the particular equipment and applications involved. In this environment the term "Availability" was selected and defined as a basic measure of the portion of time during which an Information Path is capable of performing its assigned data communication function in relation to the time it is expected to perform its function satisfactorily.

Determination of the Availability of an Information Path requires determination of three factors: a performance period, a base time period, and unavailable time during the base time period.

The performance period, usually expressed in terms of calendar days, identifies a specific time period for which Availability is to be determined or demonstrated. The length of this performance period should be based upon consideration of the anticipated frequency and duration of failures and the desired statistical significance of the results. Experience indicates that a performance period of 30 days or more is usually required to arrive at a defendable value of availability and an understanding of the conditions that contribute to that value.

The base time period is within the performance period. It is the sum of all of the time increments that have been prescribed, for a particular Information Path, for inclusion in the base time period. For example, in a particular case the base time period might consist of all the scheduled operating hours during the performance period. For each application of the Availability criterion to a particular Information Path, the standard requires an explicit statement of the events and circumstances that contribute, or do not contribute, to the base time period.

Unavailable time is the sum of all of the times within the base time period when the Information Path is incapable of performing its assigned data communication function, for reasons that have been predefined. For example, for a specific Information Path, it might be prescribed that all downtimes resulting from any failure occurring during the base time period and within the Information Path shall be excluded in unavailable time, and that downtimes resulting from failures exter-

nal to the Information Path, or occurring during other than the base time period, shall not be included in unavailable time. For all cases, the establishment of explicit terms and conditions will require detailed consideration of possible events and circumstances relative to their impact on unavailable time.

Because of the many and varied events and circumstances that may occur, and their varying significance in different applications, this standard cannot specify how they shall be treated in relation to the base time period and unavailable time. Rather, this standard requires that explicit terms and conditions be established for each application of the Availability criterion.

Although the Availability criterion of this standard applies to an Information Path, the concept as stated herein can be adapted to any portion of the Information Path, such as the Information Transfer Channel. This adaptation can be accomplished provided that the three factors of performance period, base time period, and conditions of unavailability are clearly delineated.

## B6. Information Path

The performance criteria established by this standard are applicable to a single Information Path. An Information Path may constitute, in a simple case, an entire data communication system, or it may be only one of many paths in a large system. Methods of combining performance measures for individual Information Paths to represent total data communication system performance are beyond the scope of this standard and have been identified as the subject of future work.

It should be observed that an Information Path is, by definition, unidirectional with regard to information transfer. Further, American National Standard X3.28-1971 does not provide for certain modes of data communication. Thus this standard may not be applicable to those modes. When control procedure standards have been extended to cover additional modes, it is intended that this performance standard will be amended to include them. An Information Path is defined in 2.2.4 of this standard in functional, rather than physical, terms. It should be noted that the use of backup equipment in a Terminal Configuration, or of another switched circuit between the same two Terminal Configurations, would not result in a different Information Path. Thus, even if backup or alternate equipment (of the same functional description) is substituted for the original equipment, the Information Path does not change and performance determinations may be continued. This standard (see 2.2.5) indicates that the extent of penetration of an Information Path into the sink data terminal equipment is determined by the communication control procedures employed during the Link Establishment Phase. This is important because, without this rule, in some cases it may not be entirely clear where the Information Path terminates. Consider, for example, three possible means of transferring data between two magnetic tapes units over an Information Transfer channel.

In the first case the receiving station (one of several on a multipoint network) consists of a transmission control unit (containing a buffer) and a single tape unit. A selection sequence including the receiving station address is sent, and records are subsequently transferred from the sending Terminal Configuration through the receiving buffer directly to the receiving tape unit. In this situation the receiving station address in the selection sequence implicitly identifies the tape unit as the end of the Information Path.

In the second case the receiving data terminal equipment includes a line printer, as well as a tape unit. The selection sequence includes the station address and the address of the tape unit. Here the control procedures in the Link Establishment Phase have explicitly identified the tape unit as part of the Terminal Configuration and the end of the Information Path.

In the third case a central processor is involved. The selection sequence includes only the station address. Incoming records are transferred to the buffer of the transmission control unit and then, based on additional information contained within the header or text of the message, the records are subsequently transferred to a specific area of processor storage and written on tape. In this situation the communication control procedures used in the Link Establishment Phase have implicitly identified the buffer of the control unit as the end of the Information Path. Note that if all incoming records were automatically transferred to the same area of processor storage before the heading or text was examined to determine subsequent message routing, that area of processor storage would be identified as the end of the Information Path.

Another example, involving multiple device selection within Phase 3, should also be considered. An application might call for the selection of a station in Phase 2, followed by the sequential delivery of messages to two or more devices within a single Phase 3, with device selection being accomplished within the message headers. In this situation, again the Information Path is established in Phase 2, and the end of the path is at the last common point where the messages for all of the devices are assembled, probably a control unit buffer. Performance determinations in this case would be made from the source to the common buffer in the receiving stations. Note that TRIB would be determined over all of the messages delivered in a single Information Transfer Phase and that RER would exclude any residual errors injected by the devices themselves, since they are outside of the Information Path.

The intent of this standard is to provide criteria to be used in determining performance of data communication, not data processing. Using the rule discussed here to establish the extent of an Information Path, it is possible to differentiate unambiguously between data communication and data processing.

## B7.  Information Characters and Bits

The performance criteria TRIB and TOT are both expressed in terms of Information Bits, and it is essential that the definition of Information Bits be precise. Several possibilities exist and were studied for identifying Information Bits from among all of those bits transferred across the Information Transfer Channel in the data communication process. The rationale supporting the definition of Information Characters and Bits is presented here.

Since TRIB has been established as a measure of the rate of information transfer once the Information Path has been activated, Information Bits should clearly not include bits transferred in other than Phase 3 (the Information Transfer Phase).

Within Phase 3, bits are transferred in the form of ASCII characters, which make up either control sequences or messages. Since messages represent the information that is to be communicated, and control characters and sequences are provided only as required to permit or facilitate the communication of messages, the definition of Information Characters should be limited to those within a message.

Messages may be composed of heading characters and text characters. Within both heading and text, characters may be identified as control characters or graphic characters. Further, within text, graphics may be used in control (ESC) sequences. The proper selection of Information Characters from within message characters is not obvious, and is subject to different interpretations by users and suppliers of data communication systems.

One alternative considered was to permit the user of this standard to establish which characters from within the message should be counted as Information Characters in his application. This alternative was rejected on the basis that it was not consistent with a standardization effort and would make comparison of performance of different systems quite difficult.

The question of whether or not heading characters should be identified as Information Characters is particularly subjective, and there are good reasons why they should, and why they should not, be included. Resolution is made more difficult by the current absence of standards for heading format and content. This issue has been resolved by the exclusion of all heading characters from the Information Character definition.

Of the remaining characters (those in message text), all control characters and those graphics used in control sequences were excluded because of their control function.

With this exclusion, there remain only graphic characters within message text when not used in control sequences. In applying this definition to examples of actual systems, it was disclosed that the character *Space*, though a graphic, might be used as control for the purpose of effecting format at the sink device. For example, a sequence of *Space* characters might be used instead of the HT (Horizontal Tab) control character. It was agreed that *Space* or any other character, when used in this way as a format effector, should not be counted as an Information Character.

The resulting definition of Information Characters (see 4.2.1) was then utilized to define Information Bits. All bits in Information Characters are included except parity bits and start-stop bits (when used), which are clearly not part of the basic information to be transferred.

## B8. Application Dependencies

**B8.1 General.** Certain aspects of performance criteria determination have not been specified in this standard because of their dependency on the particular configuration and details of application of the specific Information Path under consideration. Each of these aspects is discussed here with reference to the section of the standard wherein the application dependency is identified.

**B8.2 Development and Presentation of Representative Criteria Values.** It is indicated in 5.3.4 that methods of determining singular or individual values for the criteria TRIB, TOT, and RER are a part of this standard, but that means of developing and presenting representative values based on these singular values are not standardized. The reason for this is that the development of representative values is highly application dependent.

A singular value of TRIB, TOT, or RER, resulting from a single determination, may have little significance in terms of the long-range or average performance of the system under consideration. Therefore, a number of singular determinations must be made — the number depending on the anticipated variations in the observed values and the required tolerance and confidence level of the resulting representative value. In determining the time distribution of singular values required, considera-

tion must be given to the different operating conditions that may prevail at different times of the day, week, or month. For example, TRIB, TOT, and RER values may all be influenced by line error rates, which vary hourly in accordance with the density of traffic on a switched network, or from day to day in accordance with prevailing weather conditions.

Once an appropriate number of singular criterion values has been obtained, there are many alternate means of developing and presenting a representative value. For example, the mean value, or perhaps a mean and associated deviation, may be adequate. For some applications it may be meaningful to present, in addition, or instead, a maximum (or minimum) criterion value and associated confidence level. Another alternative would be to present all of the observed singular values as points on a graph. The method used should be selected with regard to the application and requirements of the particular Information Path.

**B8.3 RER Determination.** In 5.3.8.2 it is stated that the number of transmitted characters to be used as a base in determining a singular RER value is not specified in this standard. The number selected in a particular case is a function of the anticipated residual error count and of the required tolerance and confidence level of the resulting RER value. Anticipated residual error count is, in turn, a function of the error characteristics and application details of the particular Information Path under consideration.

It should be observed that once an appropriate number of characters has been selected, many messages may have to be monitored, in view of the very low residual error count normally expected (or tolerated) in a practical system. Further, in developing a representative RER value, the considerations of B8.2 must be applied.

**B8.4 Impact of Non–Information Characters on RER Value.** In 5.3.8.6 and B4 it is indicated that a residual error in a non–Information Character may produce residual errors in the recorded media, such that the resulting contribution to the residual error count is not obvious. Examples of overprinted lines and misrouted messages were cited to illustrate some of the many possibilities. A general solution that would be applicable to all such situations cannot be provided in this standard. Rather, each situation must be examined with regard to its effect and significance in the particular Information Path, and an acceptable method of reflecting such errors in RER must be developed by the parties involved.

**B8.5 Availability Determination.** In 5.3.9.2 and B.5 general guidelines are provided for the determination of Availability, and in these sections it is indicated that precise terms and conditions must be established pertinent to the configuration and application of each Information Path. The terms and conditions required are relative to establishment of a performance period, the constituents of a base time period, and a precise definition of unavailable time. Each of these factors must be reviewed and established in light of the specific configuration and application of the Information Path under consideration.

**B8.6 Resolution of Application Dependencies.** In each of the four situations described in B8.2 through B8.5, knowledge of the specific Information Path, its application, and the required tolerance and confidence limits is required in order to establish precise rules for determination of criteria values. Three different circumstances are recognized in which such rules must be developed in accordance with this standard.

The first circumstance involves the development of performance criteria as a part of general requirements or specifications for an Information Path. In this case it is the responsibility of the developer of these requirements to establish the criteria determination rules.

In the second circumstance a supplier or manufacturer of elements of an Information Path may quote actual or projected performance criteria values. In this case these values must be accompanied by a statement of the rules that governed the determination of the projected or actual performance. Note that the equipment of two or more suppliers may be involved in performance determination for a particular Information Path, in which case the supplier quoting performance values must identify any assumptions made concerning the performance of other elements of the path.

The third, and perhaps most critical, circumstance involves the development of contractual performance requirements for a particular Information Path. In this case the user and supplier(s) must jointly establish determination rules. Where more than two parties are involved, the user or purchaser is responsible to ensure that criteria determination rules are complete and unambiguous.

## B9. Human Interaction within Information Paths

In some Information Paths, and particularly within some Terminal Configurations, an operator may play an essential role in the process of data communication. For example, in a card-oriented terminal, an operator may be required to load blank cards in a card punch during an Information Transfer Phase. In such a situation it is apparent that the time required to accomplish the card loading would be included in the denominator of **TRIB** if the Information Path remained in Phase 3 during this time. In other potential situations it is not entirely clear how operator interaction should be treated with respect to the performance criteria. For example, in a particular Information Path an operator may be expected to examine the printed output and determine whether or not a correction of the message should be called for. Such situations should be anticipated in advance, and agreement should be reached among the parties concerned.

The area of human interaction has also been recognized as requiring further study for a more detailed resolution.

647

## Appendix C

## Data Communication Phases

### C1. Introduction

This Appendix describes the flow of activity in each of the data communication phases. Each phase is discussed relative to the functions required in American National Standard X3.28-1971, as well as those actions required to calculate or measure performance.

### C2. Connection Establishment Phase

This phase represents the time interval required to establish an Information Transfer Channel when switched circuits are employed.

Referring to Fig. C1, the Information Path leaves the Quiescent State, $Q_1$, when the calling data station goes "off-hook," either manually or automatically. The calling data station's data terminal equipment (DTE) must then wait for circuit CB — Clear to Send (CCITT 106) to come "ON," Phase 5 is entered and a retry may be in order. Since the calling data station has a prior knowledge as to the intent of the call, a decision as to the disposition of the elapsed time can be made. If, for example, the calling data station planned to poll another Terminal Configuration, there is no guarantee that the polled Terminal Configuration had data to send, and therefore the elapsed time is discarded. If the calling data station had planned, either explicitly or implicitly, to select or contend for the receiving Terminal Configuration, then the elapsed time represents a delay in transferring data. This time must be charged to Phase 1 for the Information Path under discussion.

When circuit CB comes "ON," the Information Path leaves Phase 1 and enters Phase 2 if connection establishment is explicit. The linkage is to Phase 3 if connection establishment is implicit.

### C3. Link Establishment Phase

This phase represents the time interval required to establish the Terminal Configurations resulting in activation of an Information Path.

Referring to Fig. C2, the Information Path enters Phase 2 either from the Quiescent State, $Q_2$, in the nonswitched case, from Phase 1 in the switched case, or from Phase 4 in either case. This entry occurs when a communication control character (such as SYN or ENQ) or a prefix character is transferred over interchange circuit BA — Transmitted Data (CCITT 103), or when a "MARK-HOLD" is applied to this circuit.

When polling is employed, the sending Terminal Configuration transmits the polling sequence and subsequently the line is turned around ($\tau$). If the polled Terminal Configuration does not have data to send, it transmits a negative reply, and the Information Path leaves Phase 2 and enters Phase 4. If the polled Terminal Configuration does have data to send, then the Information Path enters Phase 3. If the opening communication control character (such as SOH or STX) is *not* received, the Information Path enters Phase 4.

When selection or contention is employed, the sending Terminal Configuration transmits the appropriate sequence. The line is turned around ($\tau$), and the other Terminal Configuration replies positively if it is ready to receive. In the event of a negative or an invalid reply, the sequence is retransmitted $n$ times, after which a termination (Phase 4) or recovery procedure is initiated. The time in Phase 2 is recorded for the purpose of accumulating the total time required to enter Phase 3.

### C4. Information Transfer Phase

This phase represents the time required to transfer all messages associated with a single activation of an Information Path.

Referring to Fig. C3, the Information Path enters Phase 3, either from Phase 2 when link establishment is explicit or from Phase 1 or the Quiescent State, $Q_2$, when link establishment is implicit. The entry occurs when a communication control character (such as SYN, SOH, or STX) is transferred over interchange circuit BA — Transmitted Data, or when a "MARK-HOLD" is applied to this circuit.

After a transmission block is sent, the receiving Terminal Configuration replies positively if the block is correct. If the reply is received correctly, additional blocks may be sent. If the transmission block is received incorrectly, a negative reply is sent. If this reply is received correctly, correction may be attempted by retransmission. After unsuccessful attempts to retransmit the block, an abnormal termination is required. At this

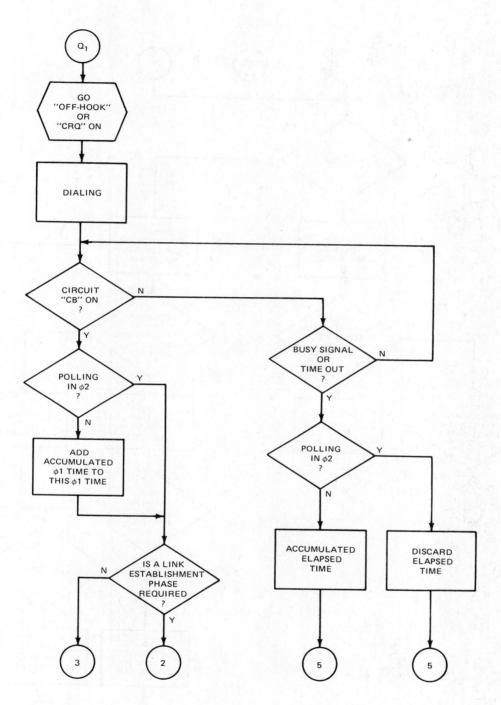

**Legend**

$Q_1$ = Quiescent State entry for switched network

**Fig. C1**
**Connection Establishment Phase (Phase 1)**

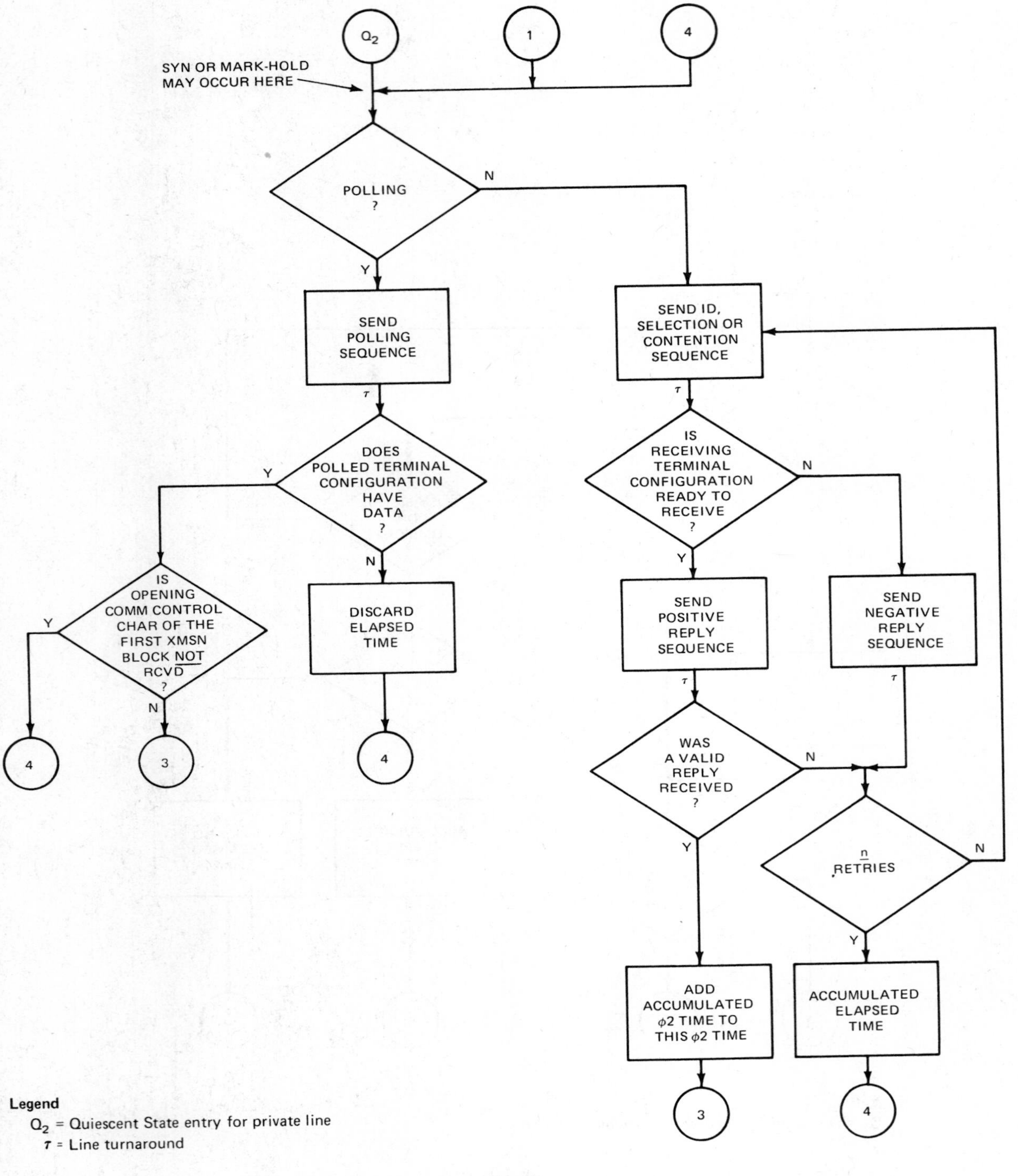

**Fig. C2**
**Link Establishment Phase (Phase 2)**

650

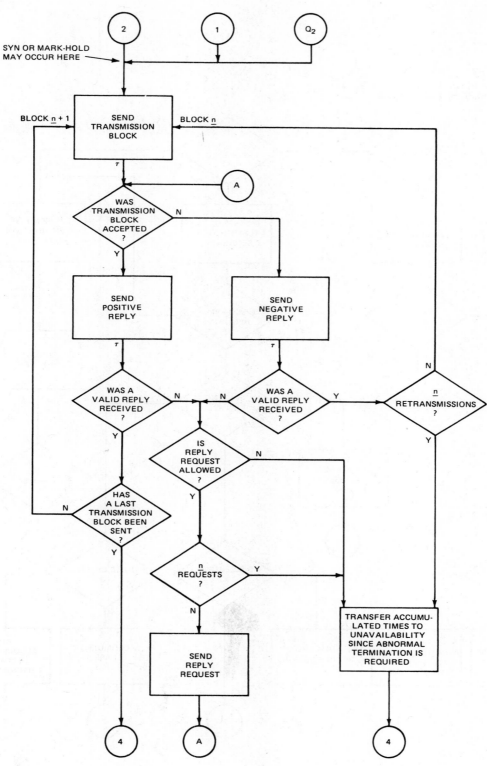

SYN OR MARK-HOLD
MAY OCCUR HERE →

651

**Legend**

$Q_2$ = Quiescent State for private lines
$\tau$ = Line turnaround

**Fig. C3**
**Information Transfer Phase (Phase 3)**

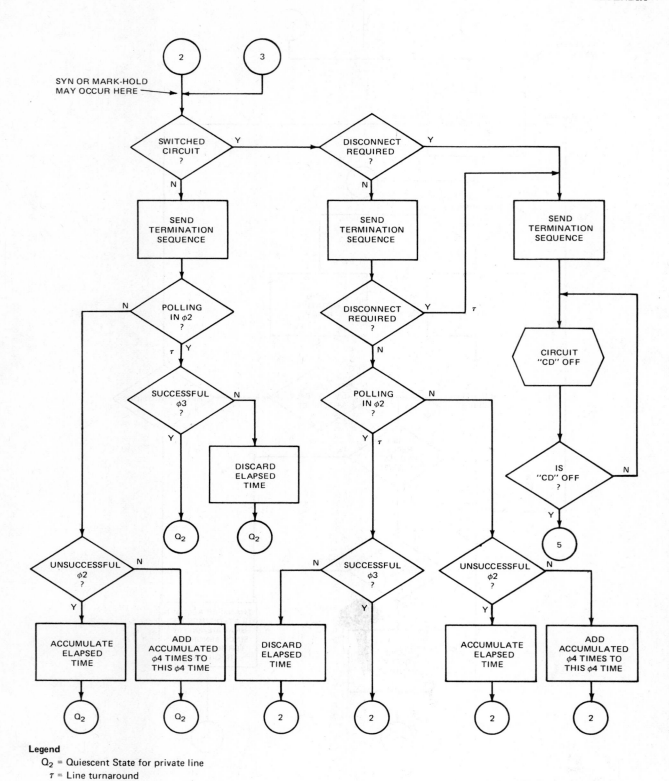

**Legend**

$Q_2$ = Quiescent State for private line

$\tau$ = Line turnaround

**Fig. C4**
**Line Termination Phase (Phase 4)**

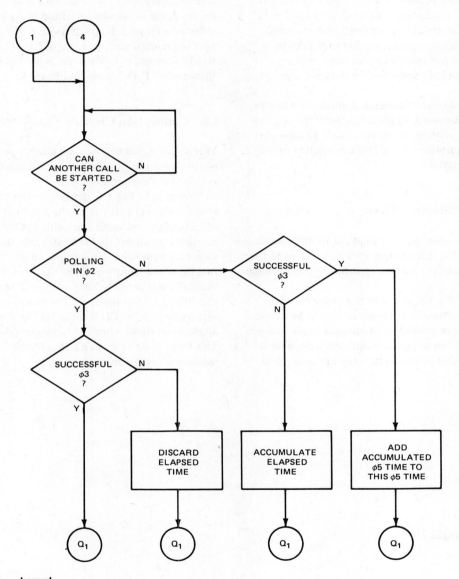

**Legend**

$Q_1$ = Quiescent State for switched network

**Fig. C5**
**Connection Clearing Phase (Phase 5)**

point all measurements that may have been accumulaged for *this* entry into the Communication State are disregarded, except for the determination of Availability. In the event of an invalid or no reply to a transmission block, a reply request (ENQ sequence) may be sent, if the control procedure subcategory allows this sequence. After a specified number of attempts, *n*, to obtain a reply, an abnormal termination is executed, and once again all measurements that may have been accumulated for *this* entry into the Communication State are disregarded except for the determination of Availability.

In the normal course of events, transmission blocks are sent and acknowledged positively until the sending Terminal Configuration has no more data to send. At this point the normal entry to the Termination Phase (Phase 4) is achieved.

## C5. Link Termination Phase

This phase represents the time required to disestablish that portion of the Information Path not related to the Information Transfer Channel, specifically the Terminal Configuration(s).

Referring to Fig. C4, entry to this phase may be from Phase 2 or Phase 3. If termination is to be accomplished on a nonswitched Information Transfer Channel, the termination sequence is sent and a decision is made as to whether or not to turn the line around. If

654

the Phase 2 operation was polling, the line must be turned around before the private line Quiescent State, $Q_2$, is entered. Similarly, on switched networks where disconnect is not required by one end, the termination sequence is sent, and then the decision concerning disconnect is made by the other end. If disconnect is not required, the turnaround decision is made prior to reentry into Phase 2. If the decision is yes, the disconnect sequence is sent, and circuit CD — Data Terminal Ready is turned off. When this is accomplished, the Information Path has entered Phase 5.

## C6. Connection Clearing Phase

This phase represents the time interval required to disestablish an Information Transfer Channel when switched circuits are employed

Referring to Fig. C5, the Information Path enters Phase 5 when circuit CD, or the equivalent, is turned off. Presently, no signal is provided at the data communication equipment that indicates that the circuit switching equipment has completed all disconnect functions and reentry into the Connection Establishment Phase is possible. One method of implementing this function is to enable a fixed timer, which is activated when circuit CD is turned off and provides the appropriate signal when the timing interval has elapsed. This timer is set to a value large enough to allow immediate reentry into Phase 1.

## Appendix D
## System Description

### D1. Rationale for Functional Approach

It was realized quite early in the development effort of this standard that if a performance standard were to be uniformly applicable to a wide variety of data communication systems covered by American National Standards on data communication, a uniform method of describing these various systems would be necessary. Since data communication systems encompass a wide

variety of features and operational capabilities, practical performance criteria and measurement methods cannot be dependent upon the physical characteristics of the many different kinds or makes of equipment used in these systems.

Although terminology has been, and is continuing to be, developed by various groups to identify data communication hardware, this terminology was found to be insufficiently precise, or otherwise undesirable,

31

when attempting to unambiguously identify and describe those points in a complex data communication system between which performance determinations were to be accomplished. It was also realized that although the physical nature of a particular portion of a data communication system can appear quite similar from one time to another, the actual functions being performed by the equipment can be drastically different. Thus it became essential to identify quite carefully which equipment and system functions were truly data communication related and would consequently have to be reflected in any determination of performance made in accordance with this standard. Conversely, it was important that non–communication-related activity, or communication-related activity not pertinent to the particular Information Path under consideration, be excluded from these determinations.

As the development of purely functional concepts proceeded, it became evident that a considerable simplification of communication system descriptive terminology was possible. For example, it was not found to be necessary to describe and understand the detailed physical makeup of an Information Transfer Channel. It was sufficient to understand only that some method existed to deliver the data stream from one interchange point to another. The development of terms to describe the major functional grouping of equipment in a data communication system resulted in the identification of the Information Transfer Channel, the Terminal Configurations, and the resultant Information Path. Existing terminology has been used, without alteration, wherever possible.

## D2. Information Transfer Channel

The function of the Information Transfer Channel is established by the combined functions of the communication circuit and the data communication equipment. The circuit is the physical medium over which data communication takes place. It may be simply a pair of wires, or it may be a complex arrangement of radio and multiplex equipment. A circuit provides the capability for both-way communication. Included in the data communication equipment is all of the hardware necessary to condition the data signals transferred across the interchange point for transmission by the circuit, and vice versa. Electrical activities normally engaged in by the data communication equipment are modulation, demodulation, scrambling, descrambling, encryption, decryption, etc. A data communication equipment must contain at least a signal converter in order to perform the minimum functions assigned to it.

The Information Transfer Channel is bounded by two interchange points, between which the transmission of binary data is possible. An Information Transfer Channel is established by the electrical interconnection of a circuit and two data communication equipments. On switched networks this interconnecting activity takes place during Phase 1, the Connection Establishment Phase. On leased or private networks the interconnecting arrangement of circuits and data communication equipments is usually established permanently by system design, and thus the Information Transfer Channel is also permanently established. Whenever the interconnections between the circuit and the associated data communication equipment are broken, the Information Transfer Channel is no longer capable of transferring data and therefore has been deactivated. On switched networks this disconnecting activity takes place during Phase 5, the Connection Clearing Phase.

## D3. Terminal Configuration

The Terminal Configuration is established by the logical or physical interconnection of a number of communication functions available within a data terminal equipment. Generally, these functions are provided by one or more input/output devices and some kind of buffer/controller hardware to permit the input/output device(s) to interface electrically and operationally with the remainder of the communication system. Thus the electrical activities of a Terminal Configuration are centered about the transcription of data as well as the conditioning (formatting and control) of this data for transmission or reception via the data communication system. (The function(s) performed by a specific combination of the many possible devices define the extent and the characteristics of a Terminal Configuration within the data terminal equipment under consideration.)

One bound of a Terminal Configuration is the interchange point at which the interface with the Information Transfer Channel occurs. This is, of course, the same point at which the data terminal equipment and the data communication equipment interface. The other bound of a Terminal Configuration is determined by the control procedures in use.

A Terminal Configuration is established when all physical or logical interconnections within a data terminal equipment have been accomplished to make possible the transfer of data between the source or sink and the interchange point. This interconnecting activity takes place during Phase 2, the Link Establishment Phase.

A Terminal Configuration is disestablished during Phase 4, the Link Termination Phase, which accomplishes the "disconnection" of the specific devices within the data terminal equipment that comprises the Terminal Configuration under consideration.

## D4. Information Path

An Information Path is functionally activated at the time that an Information Transfer Channel and its associated Terminal Configurations are fully interconnected and activated. Thus an Information Path consists of all of the physical devices and their corresponding communications functions contained within the Information Transfer Channel and the Terminal Configuration.

## Appendix E
## Application Examples

### E1. Introduction

In this Appendix, examples are provided to demonstrate the application of this standard in determining values for the performance criteria TRIB, TOT, and RER. The first example, given in Section E2, illustrates determinations based on measurements. The second example, given in Section E3, involves a more complex system, with values for the criteria based on calculations. The significance of individual determinations is discussed, and the impact of changes in operating parameters, such as message (block) length and retransmission of messages (blocks), is demonstrated.

### E2. Measurement of Criteria

**E2.1 General.** A detailed block diagram of the total system involved in this example is shown in Fig. E1. This is simplified in Fig. E2 to show only the Information Path under consideration. In accordance with 5.2 of this standard, minimum supplementary descriptive information is provided, along with measured values of the criteria, in E2.2 and E2.3.

**E2.2 Terminal Configuration.** The source is a Model _____ paper tape reader, and the sink is a Model _____ paper tape punch, with data rates of 100 characters per second. The data are represented in ASCII. (Other than ASCII representation is permissible, provided that ASCII appears at the Interchange Points.)

**E2.3 Information Transfer Channel.** The Information Transfer Channel employs a point-to-point, half-duplex switched circuit with manual dial-up. The signaling rate is 1200 bits per second, asynchronous. The data communication equipment includes a Model _____ data set and a Model _____ auxiliary set at both source and sink.

**E2.4 Procedures.** The system employs control procedure category 2.4/B2 polling and data communication Phases 1, 2, 3, 4, and 5.

The supervisory sequences are as follows:

(1) Selection: A ENQ (A is one character device address)

(2) Positive acknowledgment: ACK

(3) Positive acknowledgment to selection: I ACK (I is one character device identification)

(4) Negative acknowledgment: NAK

Message length is equal to block length, and each message is assumed to have a heading length of seven characters and a text length of 200 Information Characters. Error control is by character parity and block check. After three unsuccessful retransmission attempts the system exits to termination.

A phase time diagram of the example is shown in Fig. E3 as an aid in understanding the operation of the system and to indicate the time measurements that are necessary to develop measured values of TRIB and TOT. Referring to Fig. E3, Phase 1 begins when the calling station goes "OFF-HOOK" to begin the dialing process and ends when an "ON" condition appears on interchange circuit CB — Clear to Send. At this instant Phase 2 begins. Phase 2 ends and Phase 3 begins with the initiation of transfer of the SOH character over interchange circuit BA. The transition from Phase 3 to Phase 4 occurs at the initiation of the DLE EOT

656

33

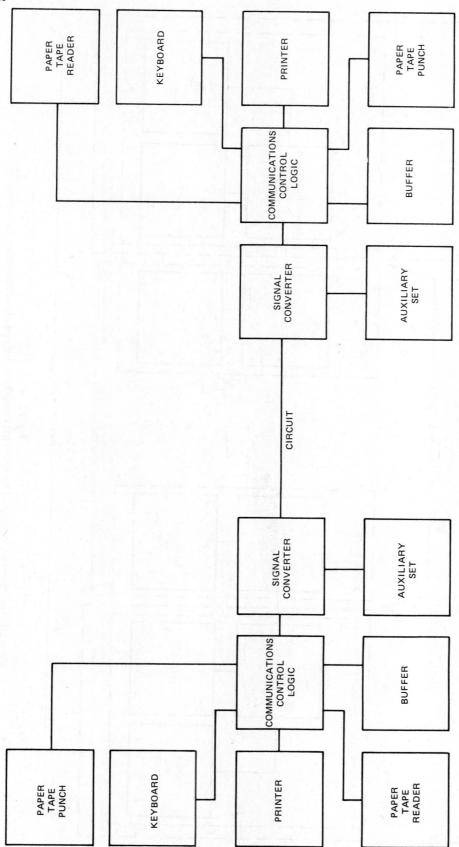

**Fig. E1**
**Detailed System Block Diagram**

34

658

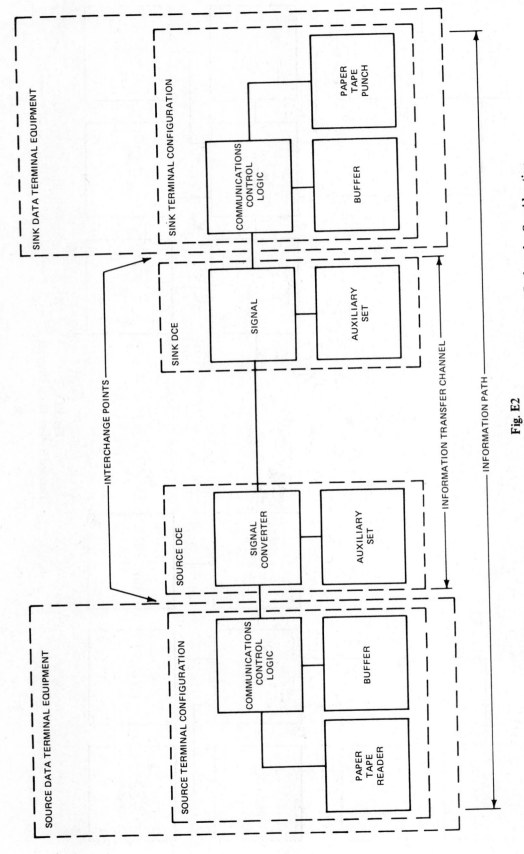

**Fig. E2**

**Block Diagram Showing Portion of System That Constitutes Information Path under Consideration**

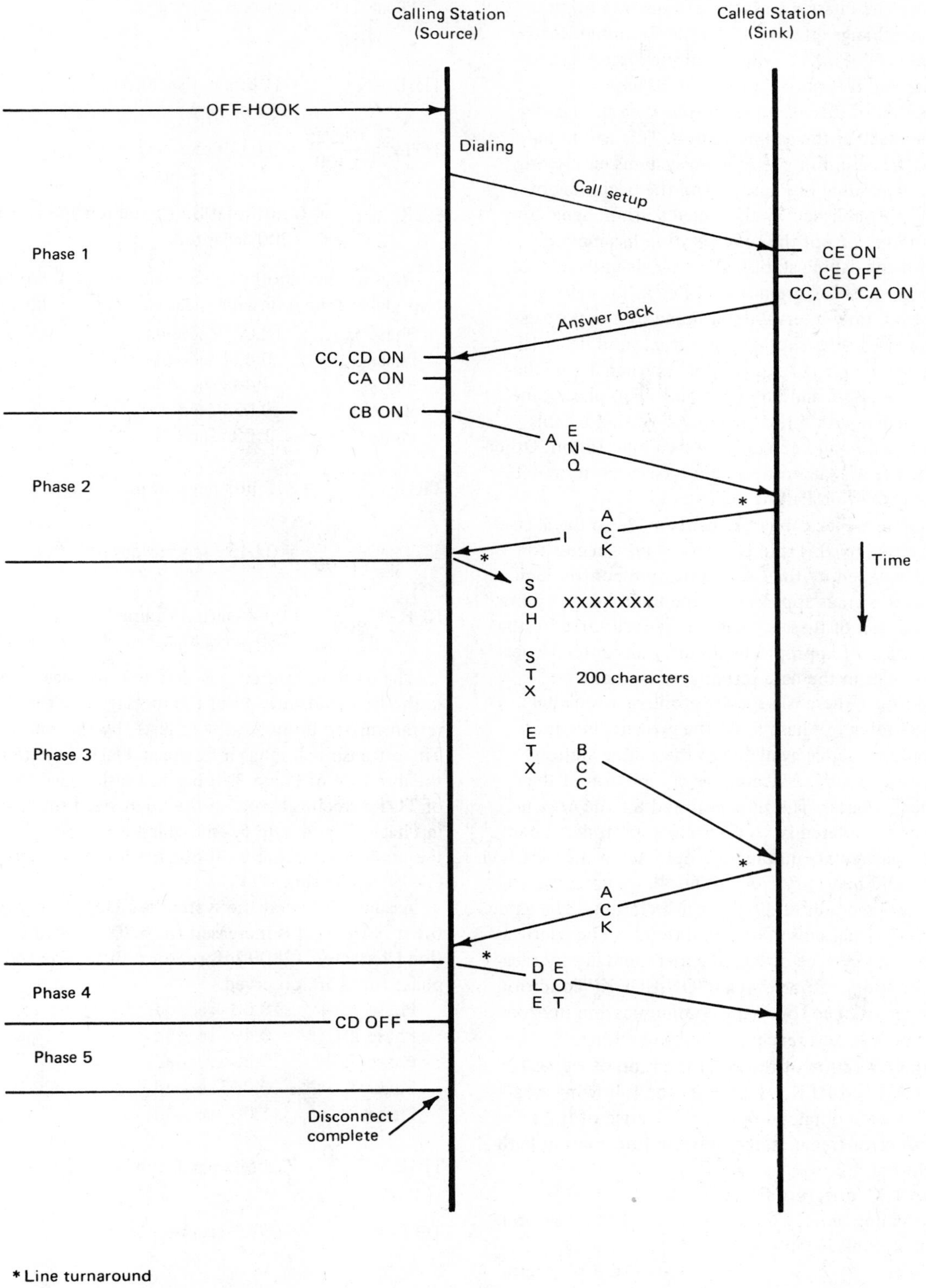

**Fig. E3**
**Phase Time Diagram of Information Path of Fig. E2**

*Line turnaround

659

sequence. The Phase 4 to Phase 5 transition occurs when interchange circuit CD – Data Terminal Ready is turned OFF. Phase 5 ends when the disconnect is complete and it is possible to reenter Phase 1.

Note that in this example all time measurements could be made at the calling station. This will not always be the case. For example, in systems employing polling, the ending of Phase 2 (and the beginning of Phase 3) is established by the called station, depending upon whether or not the called station has message traffic for the calling station. The signals indicative of this phase transition are not available at the calling station for a time interval determined by the propagation delay and echo suppression turnaround delay (if applicable). If this time interval is less than 1% of the respective Phase 2 and Phase 3 times, then phase time measurements within the accuracy specified by this standard could still be made at the calling station. Otherwise, the use of synchronized clocks at each terminal of the Information Path is indicated.

In most instances, measurements made to the accuracy specified by this standard will require access to signals developed by the communication control logic, as well as to signals appearing at the interchange points. The extraction of these signals can be facilitated by the incorporation of appropriate circuitry and interconnection terminals in the data terminal equipment.

In addition, there is a special problem associated with the ending of Phase 5. At the present time there is generally no signal available to the calling station that unambiguously indicates the completion of the disconnect process. The time required for the disconnect to be completed is, to a certain extent, dependent upon the particular equipment employed in the switched network, and may vary from about 40 milliseconds to more than 2 seconds at different installations. The value for any individual subscriber circuit tends to be relatively constant, and may be obtained experimentally by measuring the shortest time that an "ON-HOOK" condition must be maintained following communication in order to obtain consistent reacquisition of dial tone.

As an illustration of the determination of values for TRIB, TOT, and RER assume that the following measurements were obtained during the course of transferring an error-free message over the Information Path shown in Fig. E2.

| | | |
|---|---|---|
| Phase 1, Connection Establishment time: | 18.00 | seconds |
| Phase 2, Link Establishment time: | 0.49 | second |
| Phase 3, Information Transfer time: | 2.22 | seconds |
| Phase 4, Link Termination time: | 0.035 | second |

| | | |
|---|---|---|
| Phase 5, Connection Clearing time: | 1.00 | second |

$$TRIB = \frac{1400}{2.22} = 631 \text{ bits per second}$$

$$TOT = \frac{19.525}{1400} = 0.0139 \text{ second per bit}$$

$$RER = \frac{0}{200} = 0 \text{ Information Characters in error for 200 accepted}$$

Assume that another measurement of the same system yielded the following measurements of phase times:

| | | |
|---|---|---|
| Phase 1: | 18.00 | seconds |
| Phase 2: | 0.49 | second |
| Phase 3: | 4.44 | seconds |
| Phase 4: | 0.035 | second |
| Phase 5: | 1.00 | second |

$$TRIB = \frac{1400}{4.44} = 315 \text{ bits per second}$$

$$TOT = \frac{19.525}{1400} = 0.0139 \text{ second per bit}$$

$$RER = \frac{0}{200} = 0 \text{ Information Characters in error for 200 accepted}$$

The obvious inference is that an error was detected in the first transmission of the message and that it was retransmitted upon receipt of NAK by the source DTE. The retransmission has influenced TRIB by extending the duration of Phase 3. It has not influenced the value of TOT, which is based on the unchanged times spent in Phases 1, 2, 4, and 5, and which is assessed against the unchanged number of information bits *accepted* (1400) at the sink DTE.

Assume now that the system is changed only in that the message text is increased from 200 to 400 Information Characters (2800 Information Bits). The following phase times are observed:

| | | |
|---|---|---|
| Phase 1: | 18.00 | seconds |
| Phase 2: | 0.49 | second |
| Phase 3: | 3.88 | seconds |
| Phase 4: | 0.035 | second |
| Phase 5: | 1.00 | second |

$$TRIB = \frac{2800}{3.88} = 722 \text{ bits per second}$$

$$TOT = \frac{19.525}{2800} = 0.007 \text{ second per bit}$$

$$RER = \frac{0}{400} = 0 \text{ Information Characters in error for 400 accepted}$$

Assume now that message text is retained at 400 Information Characters, but that difficulty is experienced in establishing the connection. Exactly 12 seconds after going "OFF-HOOK," the calling station goes "ON-HOOK," turning interchange circuit CD — Data Terminal Ready OFF as a result of receiving a busy signal. This 12.00 seconds of Phase 1 time and the associated 1.00 second of Phase 5 time are retained and added to the applicable Phase 1 and Phase 5 times, which are measured during a later successful attempt to activate this Information Path and transfer the message (see 5.3.7.4). If this successful attempt resulted in the phase time measurements shown in the preceding list, the TRIB and RER values would be unchanged, but the numerator of TOT would be increased by 13.00 to yield a TOT value of 32.525/2800 = 0.012 second per bit.

There is a clear rationale for determining when to accrue the times associated with unsuccessful attempts to pass through the Information Transfer Phase. In general, if a failure to sequence normally through Phase 3 introduces additional delay to a waiting message, the delay is chargeable. If a failure to sequence normally through Phase 3 results from the fact that no message traffic is available (a frequent condition on some systems employing polling), the times are discarded.

An abnormal termination invalidates any singular determination (see 5.3.2). For instance, if a circuit failure caused an unintended disconnection, all data associated with that determination of TRIB, TOT, or RER would be discarded. The implication of such an event on Availability is in accordance with 5.3.2.2 and 5.3.9.

In the foregoing illustrations the measured values of TRIB and TOT differ greatly with each (singular) determination. TRIB is shown to be sensitive to block length and to the error control procedures employed within the system, even when all other factors remain unchanged. At the same time, TOT is shown to be sensitive to the number of Information Bits accepted during a transition through Phase 3. It is a measure of the time associated with getting into, and out of, the Information Transfer Phase.

Since each measurement results in a singular value, representative values are best expressed in statistical terms, which include the number of measurements involved and define the limits of other parameters influencing the statistics. For example, based on 100 measurements of single message transmissions over the given Information Path and with fixed message length of 200 Information Characters, the average value of TRIB was 625 bits per second; TRIB values less than this average were observed on 1% of the measurements. That is, 100 measurements of TOT with message lengths of from 200 to 400 Information Characters resulted in an average TOT of 0.011 second per bit, with

a standard deviation of 0.005 second per bit.

In the case of RER, systems that experience low error rates may have to be observed for long periods of time, involving many complete transitions through the phases of data communication in order to obtain statistically significant RER data. In such instances, the number of transitions through Phase 3 might usefully be reported, together with the average RER, its standard deviation, and the confidence level of the statistic.

## E3. Calculation of Criteria

**E3.1 General.** As an illustration of determining calculated values for TRIB and TOT, consider a multipoint data communication system, as shown in part in Fig. E4. In this system the control station polls tributary stations and accepts their message traffic. The Information Path for which performance values will be calculated extends from the magnetic tape transport at tributary station A to the magnetic tape transport at the control station.

In accordance with 5.2 of this standard, the following supplementary information is necessary in making the calculations.

**E3.2 Terminal Configuration.** Source and sink are Model _____ magnetic tape transports, with data transfer rates of 36 000 characters per second (tape speed 45 inches per second) and start or stop time delays of 0.005 second. The data are represented in ASCII in accordance with American National Standard Recorded Magnetic Tape for Information Interchange (800 CPI, NRZI), X3.22-1973. Two 1024-character-capacity buffers are employed at both the source and the sink. To maximize core memory utilization at the source (station A), core allocation of buffer space is not made until receipt of a polling sequence from the control station. The tributary station's CPU response to a polling sequence includes the allocation of buffer space and initiation of appropriate data block read instructions to the magnetic tape controller, which result in a delay of 1 millisecond. The first block of data is then read into buffer no. 1 and checked (CRC and LRC) prior to start of transmission. During this interval SYN characters are output to the control station. During the transmission of the first block, the next block of data is read into buffer no. 2 to be available for immediate transmission upon receipt of acknowledgment of the first block. At the control station (sink) acknowledgment is initiated immediately after receipt of the block check character, provided that the remaining half of the double buffer is not filled. If both buffers are filled, SYN characters are transmitted while awaiting the emp-

662

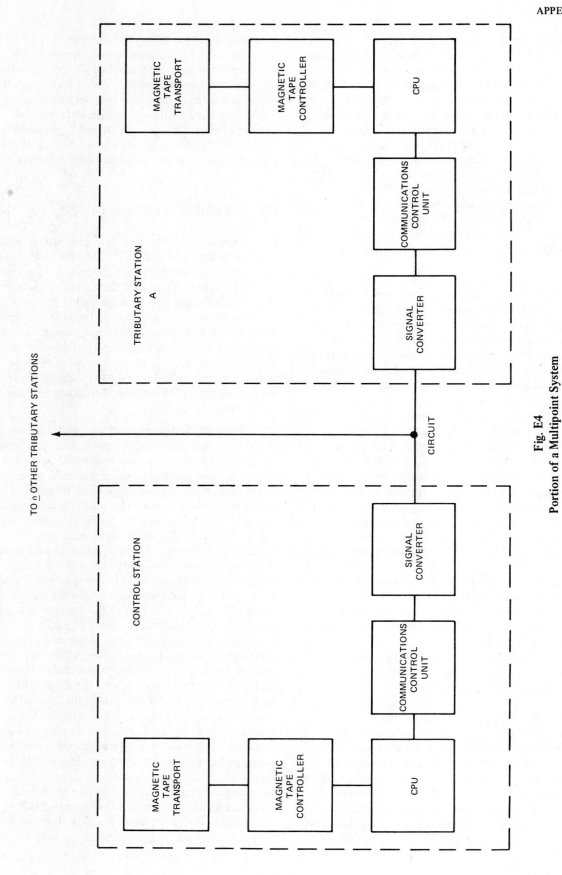

**Fig. E4**
**Portion of a Multipoint System**

tying of one buffer. If this is not accomplished within 0.050 second, an EOT is transmitted as a termination interrupt.

For the purposes of this calculation, error-free operation is assumed, together with 100% Availability.

**E3.3 Information Transfer Channel.** The Information Transfer Channel employs a multipoint, full-duplex dedicated circuit, and operates at the synchronous signaling rate of 40.8 kilobits per second in both directions (character time $T_c = 0.000\ 196$ second), using Model _____ data communication equipments. Propagation delay is assumed to be 0.010 second. Data communication equipment delay, character error rate, and unavailable time are assumed to be zero.

**E3.4 Procedures.** The system employs American National Standard X3.28-1971 category 2.4/B2 polling, and uses communication Phases 2, 3, and 4. The control station sends a polling supervisory sequence with a single-character address prefix (station and device) to each tributary in a predetermined sequence, and all transmissions are preceded by four SYN characters.

Message length is equal to block length and each message is assumed to have a heading length of four characters (excluding SOH) and a text length of 1000 characters (excluding STX, ETX, and BCC), all of which are Information Characters. Error correction is by retransmission of blocks in which character parity or BCC failures are detected. If acknowledgment is not received after three retries, the master station exits to a recovery procedure.

Generation of, and reaction to, non–Information Characters are accomplished by the central processing unit or communications control unit at the control and the tributary stations. Generation and reaction delay times are assumed to be zero.

Given the foregoing descriptive information, Fig. E5 shows calculations of values for TRIB and TOT are based on the assumption that there are two messages (blocks) at tributary station A that are available for transmission at the time of the next polling sequence.

## E4. Effect of Variable Parameters on TRIB

In order to illustrate the effects on TRIB resulting from changes in parameter values, calculations were made, and the results were plotted as shown in Fig. E6, E7, and E8. These calculations were based on the following assumptions, which are required for computing TRIB values.

(1) Terminal Configuration
  (a) Device delays (sink and source): 0 second
  (b) Buffer load and unload times: 0 second (takes

place during reply time)
  (c) Buffer capacity: 2 each 2500-character buffers
  (d) Character error rate: 0.000 000 1
  (e) Data rates (sink and source): 120 characters per second
(2) Information Transfer Channel
  (a) Data communication equipment plus propagation delays: $T = 0.05, 0.25, 0.4$ second
  (b) Character error rates: $E = 0.001, 0.0001, 0.000\ 01$
  (c) Signaling rate: $F = 1200$ bits per second
(3) Procedures
  (a) Character structure: Asynchronous, 10 bits per character; synchronous, 8 bits per character
  (b) Control procedure category: 2.2/B1 (see Fig. E3)
  (c) Number of Information Characters: $M = 50$ to 2000
(4) Additional assumptions
  (a) Multiple message blocks may be transferred during a single Phase 3 transition.
  (b) The number of heading and overhead characters is the same as in the example in Section E2; that is, $H = 7$ and $O = 4$ for asynchronous operation, and $H = 7$ and $O = 8$ for synchronous operation. Likewise, the number of characters in the reply (NAK or Prefix ACK) was increased by 4 SYN characters for the case of synchronous operation.
  (c) An initial 0.5-second period of idle SYN character transmission has been used at the beginning of Phase 3 prior to transmission of SOH.
  (d) A time-out period of 0.5 second has been introduced into the calculations for the case of invalid or no reply and the subsequent reply request (Prefix ENQ) transmission.

Fig. E6 shows the variation of TRIB as a function of the character error rates $E$; Fig. E7 illustrates the change of the TRIB values for different delay times $T$; and Fig. E8 shows the TRIB values for asynchronous and synchronous operations. The fixed values for the parameters $E$, $O$, and $T$ as selected for each of the figures are shown in the upper right side.

The curves in all three figures clearly show the effect of optimum block/message size or, stated in reverse, the reduction in TRIB values if the blocks/messages for one reason or other are smaller or larger than the optimized range. This aspect is rather important when cost/performance trade-off considerations are made (for example, buffer size and buffer arrangement, circuit conditioning, or selection of DCE).

The curves further illustrate clearly the relative impact on TRIB by the variable parameters ($E$, $T$, and Asynchronous/Synchronous).

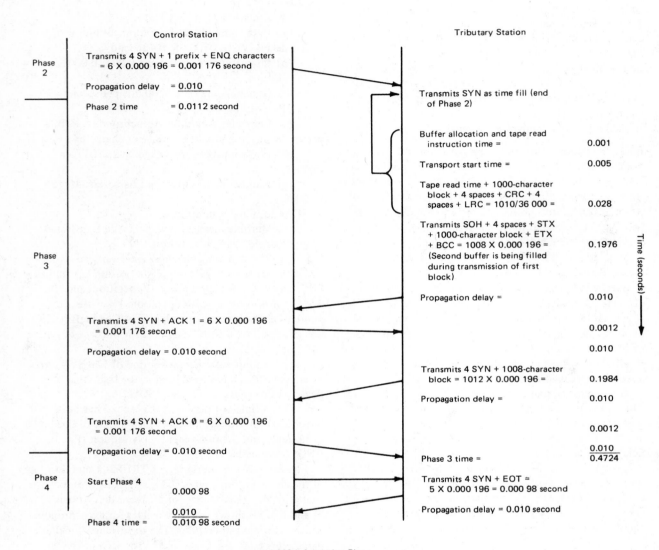

**Control Station**

**Tributary Station**

Phase 2

Transmits 4 SYN + 1 prefix + ENQ characters
= 6 X 0.000 196 = 0.001 176 second

Propagation delay    = 0.010

Phase 2 time    = 0.0112 second

Transmits SYN as time fill (end
of Phase 2)

Buffer allocation and tape read
instruction time =    0.001

Transport start time =    0.005

Tape read time + 1000-character
block + 4 spaces + CRC + 4
spaces + LRC = 1010/36 000 =    0.028

Transmits SOH + 4 spaces + STX
+ 1000-character block + ETX
+ BCC = 1008 X 0.000 196 =    0.1976
(Second buffer is being filled
during transmission of first
block)

Phase 3

Propagation delay =    0.010

Transmits 4 SYN + ACK 1 = 6 X 0.000 196
= 0.001 176 second

0.0012

Propagation delay = 0.010 second

0.010

Transmits 4 SYN + 1008-character
block = 1012 X 0.000 196 =    0.1984

Propagation delay =    0.010

Transmits 4 SYN + ACK Ø = 6 X 0.000 196
= 0.001 176 second

0.0012

Propagation delay = 0.010 second

0.010

Phase 3 time =    0.4724

Phase 4

Start Phase 4

0.000 98

Transmits 4 SYN + EOT =
5 X 0.000 196 = 0.000 98 second

Propagation delay = 0.010 second

0.010

Phase 4 time =    0.010 98 second

Time (seconds) ——►

$$\text{TRIB} = \frac{14\ 000\ \text{Information Bits}}{0.4724\ \text{second}} = 29\ 636\ \text{bits per second}$$

$$\text{TOT} = \frac{0.0222\ \text{second}}{14\ 000\ \text{bits}} = 1.58\ \mu\text{s per bit}$$

**Fig. E5**
**Calculation of TRIB and TOT Values**

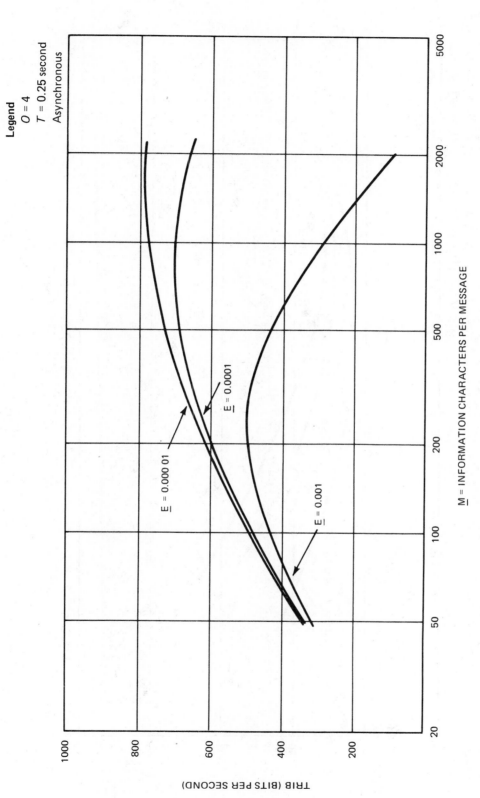

**Legend**
$O = 4$
$T = 0.25$ second
Asynchronous

M = INFORMATION CHARACTERS PER MESSAGE

**Fig. E6**
**Variation of TRIB as a Function of $E$ and $M$**

$\underline{E} = 0.0001$

$\underline{E} = 0.000\ 01$

$\underline{E} = 0.001$

TRIB (BITS PER SECOND)

666

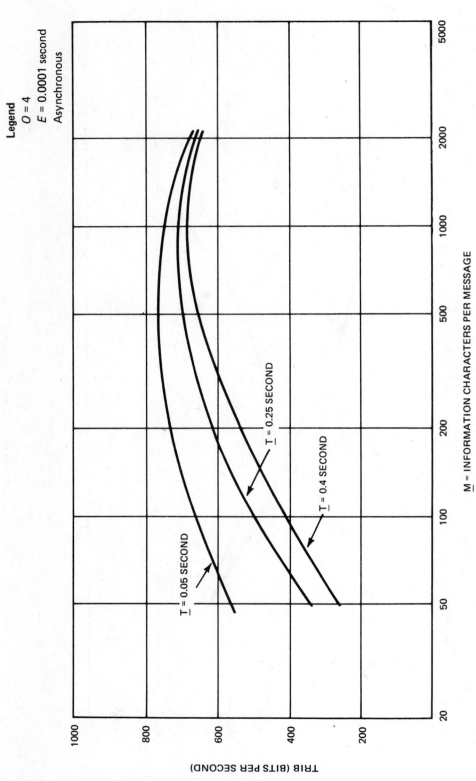

**Legend**
$O = 4$
$E = 0.0001$ second
Asynchronous

$\underline{T} = 0.25$ SECOND

$\underline{T} = 0.4$ SECOND

$\underline{T} = 0.05$ SECOND

$\underline{M}$ = INFORMATION CHARACTERS PER MESSAGE

TRIB (BITS PER SECOND)

**Fig. E7**
**Variation of TRIB as a Function of $T$ and $M$**

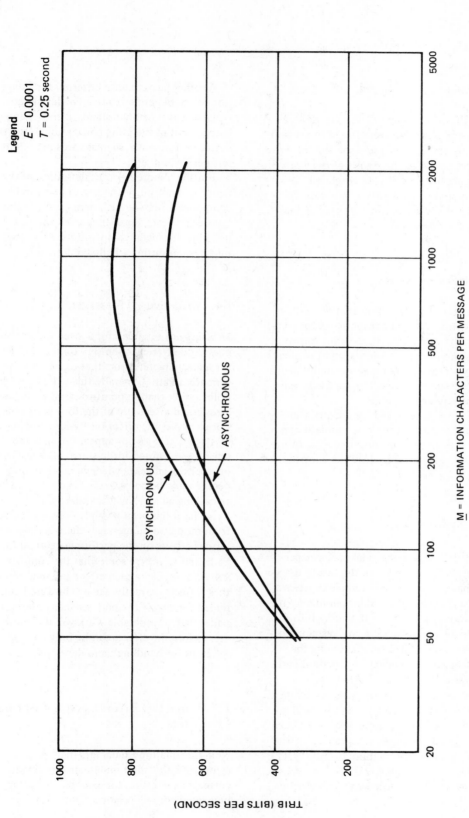

**Fig. E8**

**Variation of TRIB as a Function of _M_ for Synchronous and Asynchronous Operations**

# Appendix F
## Future Work

## F1. Introduction

In the course of the development of this standard, a number of questions arose that were found to warrant additional study. In some cases, the resolution of these questions is dependent upon the results of the work of other standards committees not yet completed. The purpose of this section is to identify and briefly describe those topics that have been recognized as the subject of required future work.

## F2. New Control Procedures

Additional categories of data communication control procedures are being developed as candidates for standardization using the ASCII communication control characters. Transparent, conversational, additional categories of two-way alternate and two-way simultaneous modes are being developed and will probably be completed in the relatively near future.

This standard must be reviewed for adequacy and completeness when applied to the new standard control categories and appropriately amended. The future work cannot be started, however, until the control standard development has been completed.

## F3. Additional Performance Criteria

It is anticipated that certain additional performance criteria may be usefully developed in the future to express the performance attributes of complex systems.

For example, although this standard permits performance determinations over each of the many Information Paths that might be activated in any system, techniques have not been developed for combining the values obtained from the individual Information Paths into a value representative of the total system.

In multilink, store-and-forward message switching systems, it would appear desirable to have some standard method of determining the time delay associated with the transmission of messages. This aspect of performance, called Delay of Information Transfer (DIT), has been the subject of considerable study in the past. Much additional work will be necessary, however, to refine and formalize the concept.

Another performance criterion for future consideration might be termed "effective device rate," and is visualized as possibly a standard conversion factor relating TRIB to the more conventional expressions of device performance, such as cards per minute, characters per second, etc.

Finally, a more rigorous definition of the components of Availability appears to be desirable. These components include such items as mean time between failures (MTBF), mean time to repair (MTTR), and the handling of scheduled downtime and of failures that might occur during scheduled downtime.

## F4. Misdelivered Messages

Messages not intended for a given Information Path may get delivered to the sink on that Information Path through character errors that occur on some other Information Path. Generally this would occur in multipoint rather than in point-to-point arrangements. Such cases could affect any of the four performance criteria. For example, the effect on RER of finding a received message that does not appear on the sending media must be accounted for in some manner. Or, the effect of receiving an acknowledgment when one is not expected must be considered, for example, with respect to causing an abnormal termination.

So far it has been possible only to recognize that from misdelivered messages there is the potential for serious effects on the performance of an Information Path, and to realize somewhat the complexities involved in assessing each effect and in accounting for t these effects correctly when values are being determined for any of the performance criteria. It is anticipated that considerable work will be required in this area before the effects can be fully catalogued and methods for handling them developed.

## F5. Impact of Errors in Non–Information Characters

Non–Information Characters that are accepted with errors that are either undetected or detected and uncorrected can cause, for example, loss of messages, unwanted control responses, unwarranted retransmissions, and other potentially serious performance degradations. It is anticipated that the impact of these errors will be most serious on complex systems. Its evaluation is at least partly dependent upon the development of performance criteria applicable to such systems.

## F6. Human Interaction within an Information Path

An objective of this standard has been to exclude the effects of human interaction within an Information Path, where possible, because of the additional complexity that would result. However, it is recognized that in many cases the human (an operator) may be required to perform a necessary part of the data communication functions to be accomplished by the Information Path. It is not in all cases entirely clear how such human interaction should be accounted for in terms of the various performance criteria. A more detailed examination of the possible implications is required.

## F7. ASCII Code Extension

Efforts are currently under way to add an eighth bit to the ASCII code (CCITT No. 5 alphabet) in order to extend this standard to cover 256 unique structures. This will add 128 characters to the present standard, some of which may be considered as Information Characters and some non–Information Characters. It remains to be determined what impact these new graphic and nonprinting characters will have upon this standard.

It is conceivable that the additional uses found for the new characters will require a completely new definition for "Information Bits." Further, the addition of 32 control characters, such as COMMAND, TEXT, CHARACTER, SIMULTANEOUS SEND RECEIVE, etc, will dictate a careful review of all concepts and methodology for utilizing the current set of criteria with these new characters.

## Appendix G
## Reference Documents

In addition to the American National Standards listed in 1.2 of this standard, the following national and international standards documents, which have direct or indirect impact, have been referred to in the development of this standard.

## G1. American National Standards

American National Standard Synchronous Signaling Rates for Data Transmission, X3.1-1969

American National Standard Vocabulary for Information Processing, X3.12-1970

American National Standard Signal Quality at Interface between Data Processing Terminal Equipment for Synchronous Data Communication Equipment for Serial Data Transmission, X3.24-1968 (EIA RS-334-May 1967)

American National Standard Code Extension Techniques for Use with the 7-Bit Coded Character Set of

American National Standard Code for Information Interchange, X3.41-1974

## G2.  Electronic Industries Association Standards[3]

Interface between Data Terminal Equipment and Data Communication Equipment Employing Serial Binary Data Exchange, EIA RS-232-C-Aug 1969

Standard for Specifying Signal Quality for Transmitting and Receiving Data Processing Terminal Equipments Using Serial Data Transmission at the Interface with Non-Synchronous Data Communication Equipment, EIA RS-363-May 1969

Interface between Data Terminal Equipment and Automatic Calling Equipment for Data Communication, EIA RS-366-Aug 1969

## G3.  European Computer Manufacturers Association Documents[4]

Basic Mode Control Procedures for Data Communication Using the ECMA 7-Bit Code, ECMA-16 (Revised June 1973)

Code Independent Information Transfer (An Extension to the Basic Mode Transmission Control Procedures), ECMA-24 (Dec 1969)

---

[3] Available from the Electronic Industries Association, 2001 Eye Street, NW, Washington, D.C. 20006.

[4] Available from the European Computer Manufacturers Association, Rue du Rhône, 1204 Geneva, Switzerland.

## G4.  ISO Standards[5]

Information Processing — Use of Longitudinal Parity to Detect Errors in Information Messages, ISO 1155-1973

Information Processing — Basic Mode Control Procedures for Data Communication Systems, ISO 1745-1975

## G5.  CCITT Documents[6]

White Book, Vol VIII, 1969. Data Transmissions. Series V Recommendations:

Functions and Electrical Characteristics of Circuits at the Interface between Data Terminal Equipment and Data Communication Equipment, V.24

Automatic Calling and/or Answering on the General Switched Telephone Network, V.25

Organization of the Maintenance of International Telephone-Type Circuits Used for Data Transmission, V.51

Characteristics of Distortion and Error-Rate Measuring Apparatus for Data Transmission, V.52

---

[5] Publications of the International Organization for Standardization are available from the American National Standards Institute, 1430 Broadway, New York, N.Y. 10018.

[6] The International Telegraph and Telephone Consultative Committee (CCITT) is an international body that establishes standards and conventions for international telecommunications, especially where the public telegraph and telephone services are governmentally owned and operated. Their recommendations are often embodied in the regulations applying to such services. (CCITT documents are available from International Telecommunications Union, Place des Nations, 1211 Geneva 20, Switzerland, or the United Nations Book Shop, United Nations Plaza, New York, N.Y. 10017.)

**Appendix H**

**Glossary**

The following technical terms have been introduced in this standard. Compliance with generally accepted usage has been observed with respect to other terms used in this standard.

670

**Availability (Information Path).** The portion of a selected time interval during which the Information Path is capable of performing its assigned data communications function. Availability is expressed as a percentage.

**Base Time Period (Information Path).** The time period, within the performance period, during which the Information Path is expected to perform its assigned data communication function.

**Communication State.** That state when an Information Path is in any of the five communication phases (Phases 1 through 5).

**Connection Clearing Phase (Phase 5, $\phi5$).** The phase that represents the time interval required to disestablish the Information Transfer Channel when switched circuits are employed.

**Connection Establishment Phase (Phase 1, $\phi1$).** The phase that represents the time interval required to establish an Information Transfer Channel when switched circuits are employed.

**Information Bits.** For the purpose of performance determination, all bits excluding start-stop elements (if used) and parity bits contained in Information Characters.

**Information Characters.** For the purpose of performance determination, those characters in the graphic subset of ASCII that are transmitted as text, except when the characters are utilized for the purpose of control or effecting format.

**Information Path.** The Information Path is the functional route by which data communication is accomplished. It comprises the sending and receiving Terminal Configurations and the Information Transfer Channel. It encompasses all of the functions performed including forward and reverse supervision to allow information transfer in a one-way direction from a single data source to a single data sink over a single Information Transfer Channel.

**Information Transfer Channel.** The Information Transfer Channel functionally accomplishes the unidirectional transfer of information from one Terminal Configuration to another within an Information Path and also allows transmission of supervisory sequences in either direction. The Information Transfer Channel is bounded by two interchange points.

**Information Transfer Phase (Phase 3, $\phi3$).** The phase that represents the time interval required to transfer all messages associated with a single activation of an Information Path.

**Link Establishment Phase (Phase 2, $\phi2$).** The phase that represents the time interval required to establish the Terminal Configurations resulting in activation of the Information Path.

**Link Termination Phase (Phase 4, $\phi4$).** The phase that represents the time interval required to disestablish that portion of the Information Path not included in the Information Transfer Channel, specifically the Terminal Configurations.

**Quiescent State.** That state when an Information Path is not in any of the five communication phases.

**Residual Error Rate (RER).** For the purpose of performance determination, the ratio of the sum of (1) erroneous Information Characters accepted by the receiving Terminal Configuration and (2) Information Characters transmitted by the sending Terminal Configuration but not delivered to the receiving Terminal Configuration, to the total number of Information Characters contained in the source data.

**Terminal Configuration.** The specific functional elements within a data terminal equipment interconnected for the purpose of information transfer over a single Information Path at a given time.

**Transfer Overhead Time (TOT).** A measure of the individual communication phase times required to establish and disestablish an Information Path, in order to effect an information transfer.

**Transfer Rate of Information Bits (TRIB).** The ratio of the number of Information Bits accepted by the receiving Terminal Configuration during a single Information Transfer Phase (Phase 3) to the duration of the Information Transfer Phase. TRIB is expressed in bits per second.

**Unavailable Time (Information Path).** The time period, within the Base Time Period, when the Information Path is not capable of performing its assigned data communication function.

# American National Standard

structure for formatting message headings for
information interchange using the american
national standard code for information interchange
for data communication system control

672

X3.57-1977

american national standards institute, inc.
1430 broadway, new york, new york 10018

# American National Standard
# Structure for Formatting Message Headings for Information Interchange Using the American National Standard Code for Information Interchange for Data Communication System Control

673

Secretariat

**Computer and Business Equipment Manufacturers Association**

Approved January 31, 1977

**American National Standards Institute, Inc**

674

Published by

**American National Standards Institute**
**1430 Broadway, New York, New York 10018**

# Foreword

(This Foreword is not a part of American National Standard Structure for Formatting Message Headings for Information Interchange Using the American National Standard Code for Information Interchange for Data Communication System Control, X3.57-1977.)

This standard specifies message heading formats for use in information interchange between systems using the American National Standard Code for Information Interchange, X3.4-1977 (ASCII).

Specified in this standard are the definitions of the message heading items, the sequence of these items, the means of indicating which items are present in the heading, and the use of the ASCII delimiter characters to separate the heading items.

Historical and present practices were considered in the development of this standard.

Other standards prescribe the character structure and character parity sense, the bit sequence, the signaling rates, the data link control procedures, error control, and additional parameters vital to the communication of information between systems.

This standard was approved as an American National Standard by ANSI on January 31, 1977.

Suggestions for improvement of this standard will be welcome. They should be sent to the American National Standards Institute, 1430 Broadway, New York, N.Y. 10018.

This standard was processed and approved for submittal to ANSI by American National Standards Committee on Computers and Information Processing, X3. Committee approval of the standard does not necessarily imply that all committee members voted for its approval. At the time it approved this standard, the X3 Committee had the following members:

J. F. Auwaerter, Chairman
R. M. Brown, Vice-Chairman
W. F. Hanrahan, Secretary

| Organization Represented | Name of Representative |
|---|---|
| Addressograph Multigraph Corporation | A. C. Brown |
| | D. D. Anderson (Alt) |
| Air Transport Association | F. C. White |
| | C. Hart (Alt) |
| American Library Association | J. R. Rizzolo |
| | J. C. Kountz (Alt) |
| | M. S. Malinconico (Alt) |
| American Nuclear Society | M. L. Couchman |
| | M. K. Butler (Alt) |
| | D. R. Vondy (Alt) |
| American Society for Information Science | M. C. Kepplinger |
| Association for Computing Machinery | P. Skelly |
| | J. A. N. Lee (Alt) |
| | H. E. Thiess (Alt) |
| Association for Educational Data Systems | A. K. Swanson |
| Association of American Railroads | R. A. Petrash |
| Association of Computer Programmers and Analysts | L. A. Ruh |
| | M. A. Morris, Jr (Alt) |
| Association of Data Processing Service Organizations | J. B. Christiansen |
| Burroughs Corporation | E. Lohse |
| | J. S. Foley (Alt) |
| | J. F. Kalbach (Alt) |
| California Computer Products, Inc | R. C. Derby |
| Computer and Communications Industry Association | N. J. Ream |
| | A. G. W. Biddle (Alt) |
| Control Data Corporation | C. E. Cooper |
| | G. I. Williams (Alt) |

| Organization Represented | Name of Representative |
|---|---|
| Data Processing Management Association | A. E. Dubnow |
| | D. W. Sanford (Alt) |
| Datapoint Corporation | V. D. Poor |
| | H. W. Swanson (Alt) |
| Digital Equipment Corporation | P. W. White |
| | A. R. Kent (Alt) |
| Edison Electric Institute | S. P. Shrivastava |
| | J. L. Weiser (Alt) |
| General Services Administration | D. L. Shoemaker |
| | M. L. Burris (Alt) |
| GUIDE International | T. E. Wiese |
| | B. R. Nelson (Alt) |
| | D. Stanford (Alt) |
| Honeywell Information Systems, Inc | T. J. McNamara |
| | E. H. Clamons (Alt) |
| Institute of Electrical and Electronics Engineers, Communications Society | (Representation Vacant) |
| Institute of Electrical and Electronics Engineers, Computer Society | R. L. Curtis |
| | T. Feng (Alt) |
| International Business Machines Corporation | L. Robinson |
| | W. F. McClelland (Alt) |
| Joint Users Group | T. E. Wiese |
| | L. Rodgers (Alt) |
| Life Office Management Association | R. E. Ricketts |
| | J. F. Foley (Alt) |
| Litton Industries | I. Danowitz |
| National Association of State Information Systems | G. H. Roehm |
| | J. L. Lewis (Alt) |
| | G. I. Theis (Alt) |
| National Bureau of Standards | H. S. White, Jr |
| | R. E. Rountree (Alt) |
| National Communications Systems | M. L. Cain |
| | G. W. White (Alt) |
| National Machine Tool Builders' Association | O. A. Rodriques |
| NCR Corporation | R. J. Mindlin |
| | A. R. Daniels (Alt) |
| | T. W. Kern (Alt) |
| Olivetti Corporation of America | E. J. Almquist |
| Printing Industries of America | N. Scharpf |
| | E. Masten (Alt) |
| Recognition Equipment, Inc | H. F. Schantz |
| | W. E. Viering (Alt) |
| Sanders Associates, Inc | A. L. Goldstein |
| | T. H. Buchert (Alt) |
| Scientific Apparatus Makers Association | A. Savitsky |
| | J. E. French (Alt) |
| SHARE, Inc | T. B. Steel, Jr |
| | E. Brubaker (Alt) |
| Society of Certified Data Processors | T. M. Kurihara |
| | A. E. Dubnow (Alt) |
| Telephone Group | V. N. Vaughan, Jr |
| | S. M. Garland (Alt) |
| | J. C. Nelson (Alt) |
| 3M Company | R. C. Smith |
| UNIVAC, Division of Sperry Rand Corporation | M. W. Bass |
| | C. D. Card (Alt) |
| U.S. Department of Defense | W. L. McGreer |
| | W. B. Rinehuls (Alt) |
| | W. B. Robertson (Alt) |
| U.S. Department of Health, Education and Welfare | W. R. McPherson, Jr |
| | M. A. Evans (Alt) |
| VIM (CDC 6000 Users Group) | S. W. White |
| | M. R. Speers (Alt) |
| Xerox Corporation | J. L. Wheeler |
| | A. R. Machell (Alt) |

Subcommittee X3S3 on Data Communications, which coordinated the development of this standard, had the following members:

G. C. Schutz, Chairman
J. L. Wheeler, Vice-Chairman
S. M. Harris, Secretary

| | |
|---|---|
| P. A. Arneth | P. W. Kiesling, Jr |
| J. R. Aschenbrenner | C. C. Kleckner |
| P. C. Baker | W. E. Landis |
| M. W. Baty | D. S. Lindsay |
| M. J. Bedford | G. J. McAllister |
| R. C. Boepple | R. C. Matlack |
| W. Brown | B. L. Meyer |
| M. T. Bryngil | O. C. Miles |
| M. L. Cain | R. T. Moore |
| D. E. Carlson | L. S. Nidus |
| G. E. Clark, Jr | J. M. Peacock |
| R. Cleary | N. F. Priebe |
| K. T. Coit | S. J. Raiter |
| J. W. Conard | E. L. Scace |
| H. J. Crowley | P. W. Selvaggi |
| R. Dawson | D. L. Shoemaker |
| J. L. Dempsey | J. M. Skaug |
| W. F. Emmons | N. E. Snow |
| H. C. Folts | E. R. Stephan |
| T. L. Hewitt | E. E. Udo-Ema |
| L. G. Kappel | G. W. White |

Task Group X3S33 on Data Communication Formats, which developed this standard, had the following members:

W. F. Emmons, Chairman

| | |
|---|---|
| J. V. Fayer | E. Novgorodoff |
| J. G. Griffis | G. W. White |
| M. E. McMahon | C. E. Young |
| L. Nidus | |

Others who contributed to the development of this standard are:

| | |
|---|---|
| P. Boehm | J. R. Mineo |
| H. Crumb | P. T. Patterson |
| R. B. Gibson | W. J. Retzbach |
| M. Gottlieb | W. R. Wheeler |
| G. I. Isaacson | J. R. Wynne |

# Contents

678

# American National Standard
# Structure for Formatting Message Headings for Information Interchange Using the American National Standard Code for Information Interchange for Data Communication System Control

## 1. Scope

**1.1** This standard specifies the information items used to construct a message heading and prescribes the sequence of these items. It is intended to satisfy the message heading format requirements for general interchange of information between systems that employ the character set of American National Standard Code for Information Interchange, X3.4-1977 (ASCII), and American National Standard Procedures for the Use of the Communication Control Characters of American National Standard Code for Information Interchange in Specified Data Communication Links, X3.28-1976.

**1.2** The primary consideration in the design of this standard was to provide the capability for interchanging information between systems. The second consideration was to provide a method of obtaining subsets of the general capability for optional use within a given system.

**1.3** Two levels of conformance with this standard are recognized, as specified in Section 5, Conformance.

## 2. Introduction

A message is a sequence of characters arranged for the purpose of conveying information from an originator to one or more destinations (addresses). It contains information (called text) to be conveyed from the composer to the recipient(s), and may, in addition, contain supplementary information (called heading).

The need for message heading information varies widely depending upon the applications considered. In some applications a heading may be required to convey system control information, such as addressing, identification, and status instructions. In other applications, however, no heading information may be required.

This standard defines and positions heading format information items that may be used to perform such functions as handling and delivery of messages, processing of messages, and billing and accounting of messages.

The procedures to perform these functions are not covered in this standard.

## 3. General

**3.1 Message Sections.** A message normally consists of a message heading and a message text. In some cases a heading is not required. The text of a message contains information that the message originator wishes to be conveyed to the message addressee(s). The heading of a message contains supplementary information that may be needed by the communication system or the destination station, or both, to handle the message.

The message heading, if used, is associated with a message text and is applicable only to the text that immediately follows it.

In general, the contents of a message are expected to be provided by the message originator. In handling a message, the communication system may not make alterations to the contents of a message heading except as provided for in this standard. The communication system must deliver the contents of the text of a message to the message addressee(s) without alteration (see Appendix F).

**3.2 Message Framing.** A message is framed by three communication controls (two if a heading is not used). One control delimits the start of the message heading, one delimits the start of the message text, and the third delimits the end of the message text.

Two types of messages are defined: a basic message and a transparent message. In a basic message, any of the 118 noncommunication control characters of

ASCII may be used by the message originator in the preparation of a message heading (subject to the provisions of Section 4, Message Heading Items) or a message text. In a transparent message, any of the ASCII control characters may be present in the message text. The heading of a transparent message, however, is subject to the same 118 noncommunication control character restriction established for basic messages. (For ASCII data link control procedures, see American National Standard X3.28-1976.)

Different framing controls are employed for the two types of messages. For basic messages, the ASCII communication control characters SOH, STX, and ETX are used to represent the start-of-heading, start-of-text, and end-of-text control functions, respectively. For transparent messages, the communication control character sequences DLE SOH, DLE STX, and DLE ETX are used to represent these functions. For consistency, all following descriptions and examples will use the basic message control characters SOH, STX, and ETX. To convert to transparent message control character sequences, substitute DLE SOH for SOH, DLE STX for STX, and DLE ETX for ETX. The two message formats are shown in Fig. 1 and 2.

Message framing controls are considered to be a part of the message.

| S O H | Heading | S T X | Text | E T X |
|---|---|---|---|---|

**Fig. 1**
**Basic Message Format**

| D L E | S O H | Heading | D L E | S T X | Transparent Text | D L E | E T X |
|---|---|---|---|---|---|---|---|

**Fig. 2**
**Transparent Message Format**

| S O H | Heading Address Section | S O R | Heading Reference Section | S T X |
|---|---|---|---|---|

**Fig. 3**
**Basic Message Heading Format**

### 3.3 Message Heading

**3.3.1 General.** A message heading is subdivided into two sections: an address section and an optional reference section.

The address section is started by a start-of-heading control character (SOH) and is ended by a start-of-reference indicator (SOR), which is defined in this standard as the file separator character (FS) or, in the absence of a reference section, by a start-of-text control character (STX). See Fig. 3.

The SOR indicator starts the reference section of the message heading, which continues until a start-of-text control (STX) is reached.

**3.3.2 Message Heading Address Section.** The address section of the heading contains the information necessary for the communication system to handle the message and to route it to its ultimate destination(s). In some cases the address section may also contain information associated with the transmission of the message over a particular data communication link. Link information in the address section may be provided by the originator of the message or by the communication system. When transmitted on a given link, however, a message heading address section should contain only that link information applicable to the link in use.

Information in the address section, other than link information, should be provided by the message origi-

nator. This information may be altered by the communication system (subject to the provisions of 4.3) during the process of delivery of the message to its destination(s).

All messages having a heading will have a heading address section, which, as a minimum, will contain the heading item indicator (HII) (see 4.3.1).

**3.3.3 Message Heading Reference Section.** The reference section of the heading contains communication and processing information that is to be delivered with the message to the destination station(s).

In abnormal message delivery circumstances, the reference section may be used by the communication system to determine the disposition of the message.

In general, the contents of the reference section are to be delivered exactly as provided by the message originator. In certain cases, however, the communication system may add information to the reference section (see 4.4).

## 4. Message Heading Items

**4.1 General.** The function and composition of message heading items are prescribed in 4.2 through 4.4. The communication system is not permitted to delete these

**Table 1**
**Heading Address Section**

| Separator Character | Item | Description | See Section |
|---|---|---|---|
| SOH | | Start-of-heading character | |
| | HII | Heading item indicator | 4.3.1 |
| GS | HA1 | Link message identity/date-time group | 4.3.2 |
| RS | HA2 | Link message status | 4.3.3 |
| GS | HA3 | Privacy/classification | 4.3.4 |
| GS | HA4* | Precedence indicator (per address) | 4.3.5.1 |
| RS | HA5* | Destination address (per address) | 4.3.5.2 |
| US | HA6* | Secondary routing/handling information (per address) | 4.3.5.3 |

*There may be many station addresses or group code addresses, or both, in a message heading. See 4.3.5, 4.3.5.1, 4.3.5.2, and 4.3.5.3 for details.

items except as specified in 4.3 and 4.4. Revision and updating are allowed as specified in 4.3 and 4.4.

The use of any message heading item is optional with the exception of the heading item indicator (HII). Message heading items may have variable lengths. Note that the heading item indicator is a minimum of two characters in length.

Heading items, when included in a message, only pertain to the message of which they are a part.

**4.2 Message Heading Item Separators.** Four ASCII characters have been defined for use as heading item separators. These characters are the file separator (FS), group separator (GS), record separator (RS), and unit separator (US). These characters are used to frame heading items and may not be used within any heading field.

The separator that defines the start of a particular heading item is specified in the subsequent description for each item. See Appendix E for a description of separator use and examples of message headings.

**4.3 Message Heading Address Items.** In addition to the heading item indicator there are six items in the heading address section, numbered HA1 through HA6 to facilitate identification. The sequence of these message heading items is shown in Table 1. (See also Appendix C.)

**4.3.1 HII (Heading Item Indicator).** The heading item indicator is preceded by the control character SOH, and is composed of two ASCII characters that identify those items present in the message heading. This item is present in all messages containing a heading. See Appendix D for description of HII bit significance.

**4.3.1.1 HII Extension.** The heading item indicator may optionally be extended where such extension may be useful to furnish additional information regarding the heading or message content — for example,

to furnish a count field (bits, characters, cards, blocks, number of addresses, etc, in the heading or text) or to indicate block or field lengths.

**4.3.2 HA1 (Link Message Identity/Date-Time Group).** HA1 is a unit of information that, if present, is preceded by the separator GS and may be used to identify a message on a given communication link. This heading item contains only that link message identity applicable to the link in use. The link message identity may include a date-time group to identify the date or time, or both, at which the message transmission occurred on a given communication link.

**4.3.3 HA2 (Link Message Status).** HA2 is a unit of information that, if present, is preceded by the separator RS and may be used to convey status information pertaining to the current transmission on a given communication link. This heading item contains only that link message status information applicable to the link in use.

**4.3.4 HA3 (Privacy/Classification).** HA3 is a unit of information that, if present, is preceded by the separator character GS and is assigned to a message by the message originator. This item indicates the degree of precaution that should be exercised by the system to avoid unauthorized disclosure of the message. When used, the privacy designator is considered to be a permanent part of the message and is to be delivered without alteration by the communication system to all addressed stations.

**4.3.5 Destination Address Items.** The message heading items HA4, HA5, and HA6 are applicable on a per-address basis and, if present, must be contiguous for each address. For multiple address messages, sequences of items HA4, HA5, and HA6 are used within the heading address section.

For multiple address messages where all addresses have the same precedence, HA4 optionally may be

**Table 2**
**Heading Reference Section**

| Separator Character | Item | Description | See Section |
|---|---|---|---|
| FS | | Start-of-reference indicator (SOR) | |
| | | File separator character (FS) | 4.4.1 |
| | HR1 | Reference station identity | 4.4.2 |
| GS | HR2 | Originating station identity | 4.4.3 |
| RS | HR3 | Originating message identity | 4.4.4 |
| US | HR4 | Originating date-time group | 4.4.5 |
| GS | HR5 | Message accounting information | 4.4.6 |
| GS | HR6 | Message status | 4.4.7 |
| STX | | Start-of-text character | |

used only once; in this case, however, it must always precede the first HA5 address field. Note that where a single precedence is used on multiple address messages, care must be taken on interchange links to ensure that the precedence is forwarded with those associated destination addresses requiring action by the receiving station.

**4.3.5.1 HA4 (Precedence Indicator).** HA4 is a unit of information that, if present, is preceded by the separator GS and designates the degree of urgency for the delivery of a message to a particular addressee. A precedence indicator, when used, is applicable only to its associated destination address. It should be inserted in the message heading by the message originator. For a multiple address message with a single precedence, see 4.3.5.

**4.3.5.2 HA5 (Destination Address).** HA5 is a unit of information that, if present, is preceded by the separator character RS and is supplied by the message originator. This item identifies the station or stations to which the message is to be delivered. When transmitted on an interchange link, this message heading item contains only those destination addresses (with any associated precedence indicators and secondary routing information) requiring routing action by the receiving system.

**4.3.5.3 HA6 (Secondary Routing/Handling Information).** HA6 is a unit of information that, if present, is preceded by the separator character US and is supplied by the message originator. The content of HA6 is associated with the immediately preceding destination address and, when present, is used to facilitate the handling of a message after it has arrived at the destination station. This information, for example, may include identification and location of individuals, departments, organizations, or devices. This item may also be used by the communication system when that system has to perform various communication based functions, such as code translation, as part of moving the message through the communication system or delivering it to the addressee.

**4.4 Message Heading Reference Items.** Heading reference items, if used, are a permanent part of the message and are to be delivered to all addressed stations. Unless designated in the item description, no heading reference item may be altered by the communication system. The sequencing of message heading items in the heading reference section is shown in Table 2. (See also Appendix C.)

**4.4.1 SOR (Start-of-Reference Character).** The start-of-reference character (file separator [FS]) delimits the beginning of the heading reference section.

**4.4.2 HR1 (Reference Station Identity).** HR1 is a unit of information that, if present, is preceded by the separator character FS. This item identifies the station that performs communication servicing functions for the message originator. The reference station identity is intended principally for use as a means for a message addressee or the system to perform communication servicing functions (for example, requesting a repeat transmission of the message). In addition, some systems may make use of this information in abnormal circumstances concerning message delivery. The reference station identity is not necessarily the address of the originator of the message.

**4.4.3 HR2 (Originating Station Identity).** HR2 is a unit of information that, if present, is preceded by the separator character GS. This heading item identifies the address of the originating station.

**4.4.4 HR3 (Originating Message Identity).** HR3 is a unit of information that, if present, is preceded by the separator character RS. This heading item distinguishes a message from other messages transmitted from the same originating station. The originating message identity need not uniquely identify the message in the absence of additional information. Additional information that may be needed to uniquely identify the message may be implied by the administrative environment (for example, reference to message 100 might always imply today's message 100) or may be contained in another heading item (for example, a date-time group).

**4.4.5 HR4 (Originating Date-Time Group).** HR4 is a unit of information that, if present, is preceded by the separator character US. This heading item identifies the date or time, or both, that the message was first entered into the communication system. When an originating date-time group is used, it should indicate the date or time, or both, of transmission by the originating station.

**4.4.6 HR5 (Message Accounting Information).** HR5 is a unit of information that, if present, is preceded by the separator character GS. This heading item may consist, for example, of an account or terminal identification, which is to be billed for a message transmission. The information content of HR5 may be modified as necessary by the communication system.

**4.4.7 HR6 (Message Status).** HR6 is a unit of information that, if present, is preceded by the separator character GS. This heading item is added to a message by the message originator or by the communication system to indicate the delivery status of the message. A repeat transmission of a message (suspected duplicate), or a message presumed (by the communication system) to be in error (because of interrupted transmission, for example), may be so designated in the message status item.

## 5. Conformance

**5.1 General.** Two levels of conformance with this standard are recognized. These levels are described in 5.2 and 5.3.

**5.2 Total or Class A Conformance.** Total or Class A conformance with this standard exists when all stipulations of the standard are followed.

**5.3 Partial or Class B Conformance.** Partial or Class B conformance with this standard is recognized because systems planned prior to the publication of the standard may not be able to conform fully, either for economic reasons or because the equipment cannot generate all of the ASCII control characters (for example, the information separators and all of the binary encodings of HII). Therefore, in order to promote the interchange of traffic between these systems, and also to facilitate interfacing with Class A systems, partial conformance will exist when all stipulations of the standard are followed except that:

(1) Use of HII, heading item indicator, is optional and, when used, the first two characters may be redefined.

(2) Use of the standard heading item separators is optional, and other means not defined herein may be used to identify the presence or location, or both, of heading items.

# Appendixes

(These Appendixes are not a part of American National Standard Structure for Formatting Message Headings for Information Interchange Using the American National Standard Code for Information Interchange for Data Communication System Control, X3.57-1977, but are included for information purposes only.)

## Appendix A
## Glossary

The following technical terms are used in this standard. For a more comprehensive vocabulary, including both data-processing and communication terms, reference is made to American National Standard Vocabulary for Information Processing, X3.12-1970.

**Center Switching.** An installation in a data communication system where equipment is used to interconnect communication circuits on a message or circuit switching basis.

**Communication Link.** The physical means of connecting one location to another for the purpose of transmitting and receiving data. (See American National Standard X3.12-1970.)

**Communication Node.** The connection point for two or more data communication links.

**Destination Station.** The program or machine that receives the message over the last link of the data communication system.

**Message Composer.** The person, program, or machine that first writes, generates, or composes the information (message text) to be conveyed to the recipient.

**Message Originator.** The person, program, or machine that writes or generates a message in a format suitable for entry into the data communication system.

**Originated Station.** The location (program or machine) at which the composed message first enters the data communication system.

**Recipient.** The person or program for whom the message is ultimately intended.

## B1. Introduction

**B1.1** This Appendix contains the criteria upon which the message heading format was based. Not all criteria have been entirely satisfied. Some of these criteria conflict with others, and the message heading format specified represents accepted compromises of these divergent criteria.

**B1.2** The criteria were drawn primarily from communication aspects of information interchange; however, processing and media recording aspects of information interchange were considered.

## B2. Specific Criteria

NOTE: The following criteria are not mutually consistent and are not listed in order of importance.

**B2.1** The capability of interchanging information between systems should be provided.

**B2.2** A method of obtaining message heading format subsets should be provided.

**B2.3** Automatic insertion of format control characters should be provided.

**B2.4** A unified method of specifying heading formats for simple and complex systems, independent of characteristics of the data system or the transmission system, should be provided.

**B2.5** Format overhead should be minimized.

**B2.6** The format should cause a minimum of confusion to operating personnel.

**B2.7** A simple and accurate means of generating the format should be possible.

**B2.8** Message heading formats should minimize hardware and software complexity.

**B2.9** The standard should encompass all heading functions.

**B2.10** Equipment complexity should be minimized when converting from one format to another.

**B2.11** Nothing in the format should cause data link control problems.

**B2.12** There should be a simple means of uniquely specifying the format content.

**B2.13** The standard should be structured to facilitate derivation of logically related smaller sets, including no heading at all.

**B2.14** The standard should provide for easy identification of fields within the heading.

**B2.15** Each field should stand by itself and not be dependent upon adjacent or surrounding fields.

**B2.16** Heading items should be grouped according to function performed and frequency of usage.

# Appendix C
## Design Considerations

### C1. General

This standard is intended to satisfy the message heading format requirements for general interchange of information between systems that employ the character set of the American National Standard Code for Information Interchange, X3.4-1977 (ASCII), for data communication system control. (This is illustrated in Fig. C1.) The primary considerations in the design of this standard were, therefore, to provide the capability of interchanging information between systems and to provide a method of obtaining subsets of the general capability for optional use within a given system. Provision has also been made to permit the system to perform message heading format control functions on behalf of stations.

Note that the link between the systems (see Fig. C1) may be point-to-point or multipoint.

### C2. Determination of Position of Message Heading Items

C2.1 The message heading items are arranged into two groups, as described in C2.1.1 and C2.1.2.

C2.1.1 The first group contains information necessary for the communication system to handle/route a message to its destination; this information may be altered by the communication system during the process of delivering the message to its destination. This is called the address section of the heading.

Information items in the address section of the heading may be used for fault recovery across a link (for example, to request retransmission or to trace a message across a link).

C2.1.2 The second group contains information intended to be delivered in its original form with the message to the destination; this information may not be deleted by the communication system and, except for unusual circumstances, the system may not add any information. This is called the reference section of the heading.

Information items in the reference section of the heading may be used to perform message accounting in a communication network or for fault recovery from origin to destination (for example, to trace or request retransmission of a message).

C2.2 Items within each of the two heading sections should be arranged in the order in which they will be used, keeping in mind the frequency of use of each item.

(1) Address section:

(a) The first item (HII) describes in detail the contents of the entire heading; it is the index of the heading.

(b) The link information items (HA1 and HA2) are next since the link is the lowest level of control in the system. Each link operates independently of all other links so is likely to have new information to be placed in these items.

(c) The following item (HA3) is placed ahead of the addressing information since it applies to all destination addresses and therefore need not be repeated for each addressee.

(d) The next three items (HA4, HA5, and HA6) are arranged as a group to permit more than one level of precedence handling and to allow the inclusion of special routing/handling/programming information for each addressee on multiple address messages.

(2) Reference section:

(a) The first item (HR1) identifies the station to be contacted for communication servicing functions (such as requesting a repeat transmission of the message). This was considered to be of primary importance.

(b) The "originating" heading items (HR2, HR3, and HR4) (used to identify the originating station and message) were considered the most frequently used reference section items.

(c) The following item (HR5) was considered to be less frequently used than the previous items.

(d) The final item (HR6) was positioned last to provide a convenient location for the message originator (or the communication system in unusual circumstances) to insert status information.

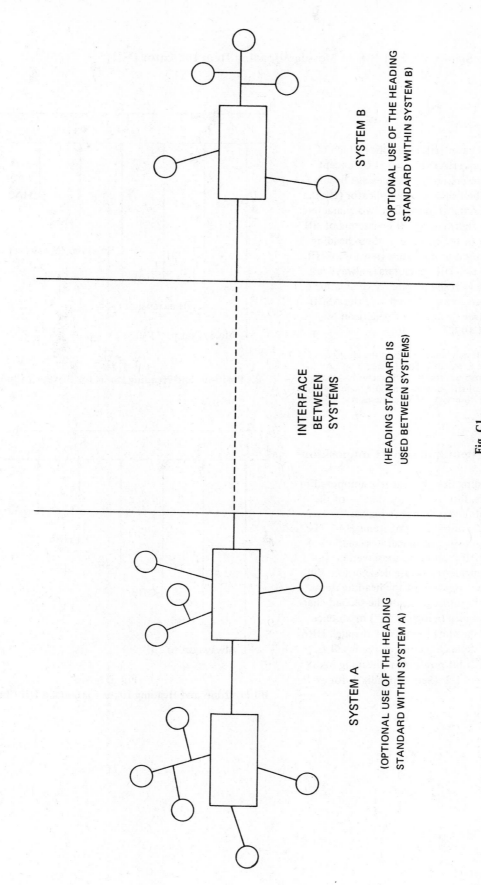

SYSTEM B

(OPTIONAL USE OF THE HEADING
STANDARD WITHIN SYSTEM B)

INTERFACE
BETWEEN
SYSTEMS

(HEADING STANDARD IS
USED BETWEEN SYSTEMS)

SYSTEM A

(OPTIONAL USE OF THE HEADING
STANDARD WITHIN SYSTEM A)

Fig. C1
Links between Systems

## Appendix D

## Description of the Significance of Bits in Message Heading Item Indicator (HII)

### D1. General

The heading item indicator (HII) identifies which of the other heading items (HA1 through HR6, except for HA4 and HA6) are present in the message heading, and whether or not the heading item indicator (HII) is extended, that is, contains more than two characters. Individual bits in the first two ASCII characters of HII indicate the presence or the absence of these heading items, and one bit is used to indicate extension of HII.

Bit $b_7$ of the first two HII characters is always set to "1" in order to prevent the possibility of creating a control character from columns 0 and 1 of the ASCII standard code table (see Section 2 of American National Standard X3.4-1977).

NOTE: Since $b_7 = 1$, HII may include any of the characters contained in columns 4, 5, 6, and 7. Special care may have to be taken in some systems since the DEL character may occur as the second character of HII. Bits $b_1$ through $b_6$ are used to indicate the presence or absence of heading items and whether or not HII is extended.

### D2. HII (Heading Item Indicator) Composition

The unextended heading item indicator is composed of two ASCII characters. Bits $b_1$, $b_2$, $b_3$, and $b_5$ of the first character indicate the presence (a logical "1") or absence (a logical "0") of the heading items HA1, HA2, HA3, and HA5, respectively, in the address section of the heading. Bit $b_4$ is designated as a reserved bit (always set to "0"), available for future designation. Bit $b_6$ is used to indicate extension of the heading item indicator (HII). Bits $b_1$ through $b_6$ of the second character indicate the presence (a logical "1") or absence (a logical "0") of the heading items HR1 through HR6, respectively, in the reference section of the heading. The relationship of HII bit positions to heading items is shown in Fig. D1 and D2 (See Appendix E for examples.)

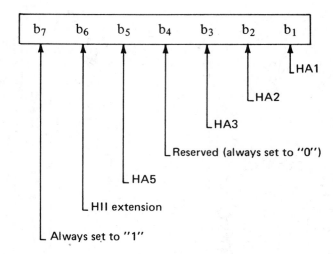

**Fig. D1**
**Bit Positions and Heading Items for First HII Character**

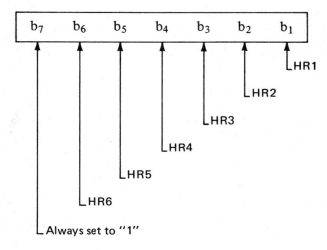

**Fig. D2**
**Bit Positions and Heading Items for Second HII Character**

## Appendix E

## Application of Message Heading Item Separators and Heading Item Indicator (HII)

### E1. General

The ASCII characters used as heading item separators and the rules governing their usage depart, to some extent, from analogous characters and usage rules employed in systems planned before the issuance of this standard. The purpose of this Appendix is to explain the separator character rules and to illustrate how they may be used to form standard message headings of varying types and complexity.

Separator characters are essential in certain types of messages to permit the various heading items and the text to be identified and located easily, especially when these functions are automated. In other types of messages a "fixed" format is used, wherein the size and location of each field in the message is always the same. Still other types of messages contain an indicator (usually at the beginning of the message) that specifies which one of several different "fixed" formats is being used in each message. Typically, systems handling messages with variable field lengths or many optional fields, or both, have a much greater internal system need for separator characters than do systems handling only rigidly formatted messages. The message formatting needs of different systems are so diverse, however, that even a large number of standard rigid formats would be insufficient to satisfy all of the existing message format needs (to say nothing of future needs). It is for this reason that this standard makes use of separator characters to delineate message heading fields.

In several communication systems planned prior to the publication of this standard, through careful and expert planning it was possible to combine the functions of field separation with one or more other functions, such as line feed, carriage return, vertical tabulation, horizontal tabulation, form feed, and space. In these systems no additional characters were needed solely for indicating field separations because some or all of the aforementioned formatting characters served a dual purpose, thus increasing the overall internal system efficiency. Wherever increased efficiency is attained through dual-purpose use of formatting characters, however, restrictions must be made concerning formatting flexibility. For example, if the ASCII LINE FEED character is used to separate message fields, no single field may be longer than one line (unless more complex rules are devised to obviate this re-

striction). To improve a system's efficiency through dual use of formatting characters, the particular rules adopted for field separation and formatting must be custom-designed to fit the particular internal formatting needs of the system. Any set of rules involving dual use of formatting characters as field separators will result in format restrictions that are quite undesirable in some systems, though the same format restrictions might have little or no adverse effect on other systems.

### E2. Message Heading Item Separators

Considerations such as these have led to the conclusion that heading item separators used for *standard* message interchange should not restrict the formatting flexibility within any individual systems. This precludes the dual use of formatting characters as heading item separators only for standard intersystem message exchange. It does not curtail the use of formatting characters for any other desired purpose. It was also determined that printing characters are undesirable for use as heading item separators because they could easily confuse or change the meaning of other adjacent printed characters used in message headings. Also, use of any printing characters as heading item separators would prohibit use of those characters for other purposes in message headings — an unnecessary restriction. The following four ASCII control characters satisfy these criteria, and they were selected for use as message heading item separators:

| UNIT SEPARATOR | US |
|---|---|
| RECORD SEPARATOR | RS |
| GROUP SEPARATOR | GS |
| FILE SEPARATOR | FS |

This standard allows the use of these four ASCII characters as desired in the text of a message, but allows their use only to separate heading items in a standard message heading.

In addition to the distinct advantage of not restricting the formatting flexibility within any system, these characters could be inserted automatically into the message heading on the originator's behalf by a terminal device or a programmed switching center within a given system, prior to transferring the message to a different system. This might be done as part of the routine message processing service provided for the originator in many systems. If, for any reason, use of the standard separator characters within a message heading is found to be undesirable in a given system, and yet that system needs to communicate externally in a standard manner, these characters could be inserted, removed, or translated as necessary at the external system interfaces. If this not feasible, partial conformance may be adopted. This option is described in Section 5 of this standard.

## E3. Heading Item Indicator (HII)

Used in conjunction with the heading separator characters is the heading item indicator (HII). The first two characters of HII must appear immediately after the start-of-heading character (SOH). When HII is extended, the remaining characters and its total length are not specified. One of the fourteen bits in the first two ASCII characters of HII is used to indicate that HII is extended (has more than two characters). The high-order bit of each character is always set at 1 to avoid generating ASCII communication control characters. One bit is a spare, and the other ten bits are used on a one-to-one basis to indicate the presence or absence of ten of the twelve other heading items (HA1, HA2, HA3, HA5 and HR1 through HR6). No bits in HII are dedicated for indicating the presence of HA4 or HA6 because these heading items may be present for only some of the addresses in a multiple address message. Thus, HII avoids ambiguities that could otherwise exist where multiple address messages do not always include HA4 and HA6 for each address. It also facilitates automated heading analysis. The three heading items that may appear more than once in a message (HA4, HA5, and HA6) are uniquely distinguishable through their separators.

Heading item indicator separator characters may be inserted in a message directly by the originator, or they may be inserted on his behalf by the system. The second alternative may be implemented either through use of an appropriate terminal device at the originating station or by other automated means in the originating system. Studies have shown that control information when directly inserted by humans is more apt to contain errors than when automatically inserted. The number

of different characters needed for use in the first two character positions of HII (64 characters, whenever the spare bit is assigned for use) necessitated use of printable ASCII characters. In some cases, delivery of the two HII characters to an addressed station or system may be useful, but in other cases it may not be. In cases where it would not be useful to deliver these characters, an agreement may be made for the transmitting communication system to omit the characters before delivering the message.

## E4. Message Heading Format Examples

**E4.1 General.** The three examples given in E4.2 through E4.4 illustrate the use of the heading item indicator (HII) and the various item separators. Example 1 illustrates a "processor-to-processor" message, which would be handled completely automatically. Examples 2 and 3 include forms control characters to aid human recognition. Example 2 includes each of the message heading items defined in the standard. Example 3 includes an extended heading item indicator (HII), two addresses, their secondary routing information, and the text. All of the ASCII characters in the message headings are shown; the characters in Examples 2 and 3 are positioned as they might be on a hard copy printout. Forms control characters are omitted from the text portion of the examples. Note in Examples 2 and 3 that extra characters may be required at the end of each line for timing purposes to accommodate some unbuffered terminal devices.

**E4.2 Example 1: HII = SB.** When the bits of the first character of HII are 1010011, the ASCII character S, it indicates that items HA1, HA2, and HA5 are present and HA3 is absent, and that HII is not extended.

When the bits of the second character of HII are 1000010, the ASCII character B, it indicates that item HR2 only is present (see Fig. E1).

**E4.3 Example 2: HII = W DEL.** When the bits of the first character of HII are 1010111, the ASCII character W, it indicates that address section items HA1, HA2, HA3, and HA5 are present, and that HII is not extended.

When the bits of the second character of HII are 1111111, the ASCII character DEL, it indicates that reference section items HR1 through HR6 are present (see Fig. E2). Note that the message text in Fig. E2 is shown garbled to illustrate a usage of HR6.

**E4.4 Example 3: HII = p@02.** When the bits of the first character of HII are 1110000, the ASCII character p, it indicates the presence of item HA5, the absence of items HA1, HA2, and HA3, and that HII is extended. The ASCII characters 02 comprise the extension of HII and in this example are used to indicate the number of addresses contained.

When the bits of the second character of HA1 are 1000000, the ASCII character @, it indicates that no items are contained in the reference section; that is, there is no reference section in the heading — and also, therefore, no need to include an SOR indicator, which would precede the reference section. (See Fig. E3.)

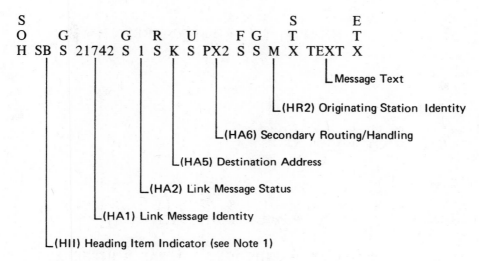

```
S                                           S     E
O     G     G  R  U     F G                 T     T
H SB  S 21742 S 1  S K  S PX2 S S  M  X  TEXT  X
                                          └ Message Text
                                    └ (HR2) Originating Station Identity
                              └ (HA6) Secondary Routing/Handling
                        └ (HA5) Destination Address
                  └ (HA2) Link Message Status
            └ (HA1) Link Message Identity
      └ (HII) Heading Item Indicator (see Note 1)
```

NOTES:
   (1) Bits 1 through 6 in the first two characters following the start-of-heading character indicate which heading items are present. See Appendix D.
   (2) Spaces around nonprinting control characters are added for clarity and readability.

**Fig. E1**
**Diagram for Example 1 (HII = SB)**

```
                S   D
                O     E C L
(HII)           H W L R F       (see Note 1)

                G       S       S           C L
(HA1)           S MSG. P NO. P R3-1/2/3  R F

                R         S           S          C L
(HA2)           S POSSIBLE P DUPLICATE P MESSAGE R F

                G     S       S    C L
(HA3)           S PVCY P CLASS P 1.7 R F

                G       S    C L
(HA4)           S PREC. P 1.3 R F

                R     S S S L
(HA5)           S NYC P P P F

                              U    S    S   S              C L
(HA6)                         S MR. P K. P L. P BROWN(PRES.) R F

                F     C L
(HR1)           S NYC R F

                G     C L
(HR2)           S CHI R F

                R       S      S    C L
(HR3)           S MSG. P NO. P S1 R F

                U     C L
(HR4)           S 1/2/3 R F

                G       S     S    S      S         C L
(HR5)           S BROWN P & P CO. P ACCT. P 3135 R F

                G                      S     S     S              C L
(HR6)           S ***ATTENTION***–––THIS P MSG P WAS P INTERRUPTED R F

                    S               S       S            S     C L
                DURING P TRANSMISSION P FROM P CHI–––REPEAT P COPY R F

                S       C L
                MAY P FOLLOW R F

                S                                               E
                T                                               T
(TEXT)          X TRAVEL REQUEST #39065 REQUIRES YOUR APPROVQZT3 X
```

NOTES:

(1) Bits 1 through 6 in the first two characters following the start-of-heading character indicate which heading items are present. See Appendix D.

(2) Spaces around nonprinting control characters are added for clarity and readability.

**Fig. E2**
**Diagram for Example 2 (HII = W DEL)**

```
         S
         O        C  F
(HII)    H  p@02  R  F     (see Note 1)

         R      S  S  S  L
(HA5)    S CHI  P  P  P  F

                    U H      S      S          C  L
(HA6)               S T MS.  P F.   P SMITH    R  F

         H                      S      S          C  L
         T             ROOM  P  2.3  P (TRAVEL)   R  F

         R      S  S  S  L
(HA5)    S NYC  P  P  P  F

                    U H      S      S          C  L
(HA6)               S T MR.  P M.   P JONES    R  F

         H                      S      S          C  V
         T             ROOM  P  3.5  P (SUPPLY)   R  T
```

(TEXT)   X TRAVEL DEPT.: PLEASE EXPEDITE TRAVEL REQUEST
         #39065 FOR MR. JONES OF OUR SUPPLY DEPT.
                   K. L. BROWN, PRESIDENT

```
         E
         T
         X
```

NOTES:
  (1) Bits 1 through 6 in the first two characters following the start-of-heading character indicate which heading items are present. See Appendix D.
  (2) Spaces around nonprinting control characters are added for clarity and readability.

**Fig. E3**
**Diagram for Example 3 (HII = p@02)**

# Appendix F

## Insertion and Revision of Message Heading Information

### F1. Prior Agreements

As a message traverses a data communication system, the content of the heading portion of the message may be modified from link to link by the system to conform to the agreed-upon message structure for each link. For example, a message originator could, by prior agreement, delegate to a switching center or node of his communication system the responsibility for entering certain heading information items on his behalf such as the heading item indicator, reference station identity, or originating message identity. Such delegation would result in a subsetted message heading format on the link from the message originator to the first switching center or node; however, the switching center would insert the "delegated" information into the message heading prior to forwarding the message to a general information interchange machine or program that expects a standard heading format.

In the same manner the message recipient or destination station may, by prior agreement, delegate to the last switching center or node of the communication system the responsibility for removing certain heading information items on his behalf. For example, the receiving station may delegate to its switching center the function of removing heading items such as the heading item indicator, precedence indicator, and reference station identity, if these items are not needed at the receiving station.

### F2. Revision of Message Heading Information

In general, the less complex a system is, the less complex will be its internal message heading format needs. Some simple point-to-point systems, for example, will require no message heading information. As the number of stations in a community of interest increases, and as the complexity of the communication system increases, the need for more complex message heading information increases. The most stringent message heading requirements exist when entirely different systems having different internal needs find that they must communicate with each other. This standard is primarily designed to resolve this problem. It is recognized that application of this standard internally within some systems may also result in overall benefits accrued from uniformity of procedures, programs, or hardware, or combinations thereof.

In message interchange between two different systems, the exposure to lost, garbled, or misdirected messages can be even more serious than within a single system; moreover, preventing and resolving such problems is more complex in a multisystem environment. For these reasons, capabilities are provided in this standard to permit the routing of undeliverable messages to control points (reference stations) when confusion arises. The capability is also provided for message identification and accountability on both a network and a "per-link" basis. If desired, communication systems using this standard for information interchange may insert link message identity/rate-time group and link message status information into the message heading to minimize the probability of uncorrectable errors occurring. Some systems may wish to log these message heading items at switching centers to permit tracing the exact route taken in the delivery of any given message back to the message originator or, perhaps, from originator to destination station.

Other heading items that may be revised by the communication system, while a message is en route, include message accounting information and message status. These items are intended for delivery to all addresses.

Before forwarding a multiple address message each switching center in the communication system must do one of the following:

(1) Delete message heading items HA4, HA5, and HA6 for which it has delivery responsibility.

(2) By prior agreement, have some other form of identifying these address items, in order to prevent, for example, multiple deliveries of a message to an addressee.

# Part 4

# EIA Data Communications Standards

The Electronic Industries Association (EIA) is a national body that represents the full spectrum of manufacturers in the U.S. electronics industry. EIA was originally founded in 1924 as the Radio Manufacturers Association. Its current membership ranges from the producers of the smallest electronic part to corporations that design and produce the country's most complex systems for defense, space, and industry.

Included along with its other activities of a trade association, the EIA's work in the area of standards has become widely recognized throughout the world. The comprehensive standardization program of the EIA Engineering Department has produced over 400 standards and publications through the efforts of more than 4,000 industry and government reprsentatives who participate on some 225 committees.

Many EIA standards become American National Standards by adoption through ANSI approval and form the basis of USA positions in such international arenas as CCITT, ISO, and the International Electrotechnical Commission (IEC).

In the area of data communications standards, the EIA's activity is centered around Technical Committee TR-30, Data Transmission. The work is divided between two Subcommittees, TR-30.1, Signal Quality, and TR-30.2, Digital Interfaces. Established in 1962, TR-30 developed the famous RS-232 interface that has dominated data communications.

Because of the close relationship of the TR-30 work with that of ANSI Technical Committee X3S3, the two groups maintain close liaison by meeting jointly once every two months. The work of EIA is more hardware-oriented and applicable to data transmission over the telephone and analog networks. On the other hand, the ANSI X3S3 work is more procedurally oriented and directed toward higher level protocols and operation with new digital networks tailored for data communications services.

The most recent effort of note from the EIA is the completion of the new RS-449 interface to supersede the old RS-232. Technology is rapidly advancing and has come a long way since RS-232 was originally developed. New standards are needed to satisfy today's more demanding operational requirements. The RS-422 and RS-423 electrical characteristics were developed by TR-30.1 to be economically implemented in integrated circuit technology while providing an essential transition from RS-232 to the capability to operate over greatly extended distances and at data signaling rates up to 10 Mbit/s. A transition from old RS-232 equipment to new RS-449 equipment is provided through the provision of Industrial Electronics Bulletin No. 12 to avoid forced obsolescence of equipment or costly retrofits for the user community.

EIA work continues to look toward the future to keep pace with technology and to provide an orderly evolution toward future generations of equipment.

| NUMBER | TITLE | RELATED CROSS REFERENCE STANDARDS |
|---|---|---|
| **RS-232C** | Interface between data terminal equipment and data communication equipment employing serial binary data interchange | CCITT V.24, V.28; ISO 2110 |
| **RS-269B** | Synchronous signaling rates for data transmission | CCITT V.5, V.6; ANSI X3.1; FED-STD-1013 |
| **RS-334** | Signal quality at interface between data processing terminal equipment and synchronous data communication equipment for serial data transmission | ANSI X3.24 |
| **RS-363** | Standard for specifying signal quality for transmitting and receiving data processing terminal equipments using serial data transmission at the interface with non-synchronous data communication equipment | None |
| **RS-366** | Interface between data terminal equipment and automatic calling equipment for data communication | CCITT V.25 |
| **RS-404** | Standard for start-stop signal quality betwen terminal equipment and non-synchronous data communication equipment | None |
| **RS-410** | Standard for the electrical characteristics of class A closure interchange circuits | None |
| **RS-422** | Electrical characteristics of balanced voltage digital interface circuits | CCITT V.11 (X.27); FED-STD-1020 |
| **RS-423** | Electrical characteristics of unbalanced voltage digital interface circuits | CCITT V.10 (X.26); FED-STD-1030 |
| **RS-449** | General purpose 37-position and 9-position interface for data terminal equipment and data circuit-terminating equipment employing serial-binary data interchange | CCITT V.24 (with V.10/V.11), V.54; ISO DRAFT 4902 |

# EIA STANDARD

*Interface Between Data
Terminal Equipment and
Data Communication Equipment
Employing Serial Binary
Data Interchange*

## RS-232-C

*August 1969*

*Engineering Department*

# ELECTRONIC INDUSTRIES ASSOCIATION

## NOTICE

EIA engineering standards are designed to serve the public interest through eliminating misunderstandings between manufacturers and purchasers, facilitating interchangeability and improvement of products, and assisting the purchaser in selecting and obtaining with minimum delay the proper product for his particular need.  Existence of such standards shall not in any respect preclude any member or non-member of EIA from manufacturing or selling products not conforming to such standards, nor shall the existence of such standards preclude their voluntary use by those other than EIA members whether the standard is to be used either domestically or internationally.

Recommended standards are adopted by EIA without regard to whether or not their adoption may involve patents on articles, materials, or processes.  By such action, EIA does not assume any liability to any patent owner, nor does it assume any obligation whatever to parties adopting the recommended standards.

Published by

# ELECTRONIC INDUSTRIES ASSOCIATION

### Engineering Department

### 2001 Eye Street, N. W., Washington, D. C. 20006

**Price $** 5.10

# INTERFACE BETWEEN DATA TERMINAL EQUIPMENT
# AND DATA COMMUNICATION EQUIPMENT EMPLOYING
# SERIAL BINARY DATA INTERCHANGE

*(From EIA Standard RS-232-B and Standards Proposal No. 1012 formulated under the cognizance of EIA Subcommittee TR-30.2 on Interface.)*

## INDEX

669

700

# SECTION ONE

## 1. SCOPE

1.1    This standard is applicable to the interconnection of data terminal equipment (DTE) and dat. communication equipment (DCE) employing serial binary* data interchange. It defines:

Section 2 — Electrical Signal Characteristics:—

Electrical characteristics of the interchange signals and associated circuitry.

Section 3 — Interface Mechanical Characteristics:—

Definition of the mechanical characteristics of the interface between the data terminal equipment and the data communication equipment.

Section 4 — Functional Description of Interchange Circuits:—

Functional description of a set of data, timing and control interchange circuits for use at a digital interface between data terminal equipment and data communication equipment.

Section 5 — Standard Interfaces for Selected Communication System Configurations:—

Standard subsets of specific interchange circuits are defined for a specific group of data communication system applications.

In addition, the standard includes:

Section 6 — Recommendations and Explanatory Notes
Section 7 — Glossary of New Terms

1.2    This standard includes thirteen specific interface configurations intended to meet the needs of fifteen defined system applications. These configurations are identified by type, using alphabetic characters A through M. In addition, type Z has been reserved for applications not covered by types A through M, and where the configuration of interchange circuits is to be specified, in each case, by the supplier.

1.3    This standard is applicable for use at data signalling rates in the range from zero to a nominal upper limit of 20,000 bits per second.

1.4    This standard is applicable for the interchange of data, timing and control signals when used in conjunction with electronic equipment, each of which has a single common return (signal ground),

701

---

*See section 6.1.

that can be interconnected at the interface point. It does not apply where electrical isolation between equipment on opposite sides of the interface point is required.

1.5     This standard applies to both synchronous and nonsynchronous serial binary data communication systems.

1.6     This standard applies to all classes of data communication service, including:

1.6.1     Dedicated leased or private line service, either two wire or four wire. Consideration is given to both point-to-point and multipoint operation.

1.6.2     Switched network service, either two-wire or four-wire. Consideration is given to automatic answering of calls; however, this standard does not include all of the interchange circuits required for automatically originating a connection. (See EIA Standard RS-366 "Interface Between Data Terminal Equipment and Automatic Calling Equipment for Data Communication".)

1.7     The data set may include transmitting and receiving signal converters as well as control functions. Other functions, such as pulse regeneration, error control, etc., may or may not be provided. Equipment to provide these additional functions may be included in the data terminal equipment or in the data communication equipment, or it may be implemented as a separate unit interposed between the two.

1.7.1     When such additional functions are provided within the data terminal equipment or the data communication equipment, this interface standard shall apply only to the interchange circuits between the two classes of equipment.

1.7.2     When additional functions are provided in a separate unit inserted between the data terminal equipment and the data communication equipment, this standard shall apply to both sides (the interface with the data terminal equipment and the interface with the data communication equipment — See Section 3.1.1) of such separate unit.

1.8     This standard applies to all of the modes of operation afforded under the system configurations indicated in Section 5, Standard Interfaces for Selected Communication System Configurations.

## SECTION TWO

## 2. ELECTRICAL SIGNAL CHARACTERISTICS

2.1     Figure 2.1, Interchange Equivalent Circuit, shows the electrical parameters which are specified in the subsequent paragraphs of this section. The equivalent circuit shown in Figure 2.1 is applicable to all interchange circuits regardless of the category (data, timing, or control) to which they belong. The equivalent circuit is independent of whether the driver is located in the data communication equipment and the terminator in the data terminal equipment or vice versa.

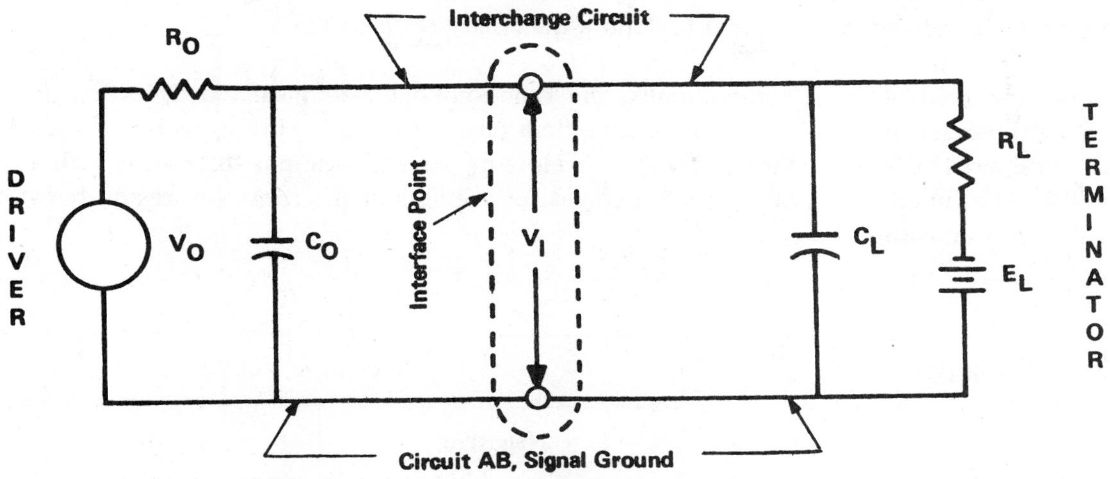

**FIGURE 2.1 - INTERCHANGE EQUIVALENT CIRCUIT**

$V_O$ is the open-circuit driver voltage.

$R_O$ is the driver internal dc resistance.

$C_O$ is the total effective capacitance associated with the driver, measured at the interface point and including any cable to the interface point.

$V_1$ is the voltage at the interface point.

$C_L$ is the total effective capacitance associated with the terminator, measured at the interface point and including any cable to the interface point.

$R_L$ is the terminator load dc resistance.

$E_L$ is the open circuit terminator voltage (bias).

2.2    The driver on an interchange circuit shall be designed to withstand an open circuit, a short circuit between the conductor carrying that interchange circuit in the interconnecting cable and any other conductor in that cable, or any passive non-inductive load connected between that interchange circuit and any other interchange circuit including Circuit AB (Signal Ground), without sustaining damage to itself or its associated equipment. The terminator on an interchange circuit shall be designed to withstand any input signal within the 25 volt limit specified in section 2.6. (see Section 6.6).

2.3    For data interchange circuits, the signal shall be considered in the marking condition when the voltage ($V_1$) on the interchange circuit, measured at the interface point, is more negative than minus three volts with respect to Circuit AB (Signal Ground). The signal shall be considered in the spacing condition when the voltage ($V_1$) is more positive than plus three volts with respect to circuit AB (see 6.3). The region between plus three volts and minus three volts is defined as the transition region. The signal state is not uniquely defined when the voltage ($V_1$) is in this transition region.

During the transmission of data, the marking condition shall be used to denote the binary state ONE and the spacing condition shall be used to denote the binary state ZERO.

For timing and control interchange circuits, the function shall be considered ON when the voltage $(V_1)$ on the interchange circuit is more positive than plus three volts with respect to circuit AB, and shall be considered OFF when the voltage $(V_1)$ is more negative than minus three volts with respect to Circuit AB. The function is not uniquely defined for voltages in the transition region between plus three volts and minus three volts.

| Notation | Interchange Voltage | |
|---|---|---|
| | Negative | Positive |
| Binary State | 1 | 0 |
| Signal Condition | Marking | Spacing |
| Function | OFF | ON |

This specification neither implies nor precludes the use of terminator circuits which utilize hysteresis techniques to enhance their noise immunity; however, the requirements of section 2.5 must also be satisfied.

2.4    The load impedance ($R_L$ and $C_L$) of the terminator side of an interchange circuit shall have a dc resistance ($R_L$) of not less than 3000 Ohms, measured with an applied voltage not greater than 25 volts in magnitude, nor more than 7000 Ohms, measured with an applied voltage of 3 to 25 volts in magnitude. The effective shunt capacitance ($C_L$) of the terminator side of an interchange circuit, measured at the interface point, shall not exceed 2500 picofarads. The reactive component of the load impedance shall not be inductive. The open circuit terminator voltage ($E_L$) shall not exceed 2 volts in magnitude. (See sections 6.4, 6.5, and 6.6).

2.5    The following interchange circuits, where implemented, shall be used to detect either a power-off condition in the equipment connected across the interface, or the disconnection of the interconnecting cable:

Circuit CA        (Request to Send)

Circuit CC        (Data Set Ready)

Circuit CD        (Data Terminal Ready)

Circuit SCA       (Secondary Request to Send)

The power-off source impedance of the driver side of these circuits shall not be less than 300 Ohms, measured with an applied voltage not greater than 2 volts in magnitude referenced to Circuit AB (Signal Ground). The terminator for these circuits shall interpret the power-off condition or the disconnection of the interconnecting cable as an OFF condition.

2.6    The open-circuit driver voltage ($V_O$) with respect to Circuit AB (Signal Ground) on any interchange circuit shall not exceed 25 volts in magnitude. The source impedance ($R_O$ and $C_O$) of the driver side of an interchange circuit including any cable to the interface point is not specified; however, the combination of $V_O$ and $R_O$ shall be selected such that a short circuit between any two conductors (including ground) in the interconnecting cable shall not result in a current in excess of one-half ampere. Additionally, the driver design shall be such that, when the terminator load resistance ($R_L$) is in the range between 3000 Ohms and 7000 Ohms and the terminator open circuit voltage ($E_L$) is zero, the potential ($V_1$) at the interface point shall not be less than 5 volts nor more than 15 volts in magnitude (see section 6.5).

2.7    The characteristics of the interchange signals transmitted across the interface point, exclusive of external interference, shall conform to the limitations specified in this section. These limitations shall be satisfied at the interface point when the interchange circuit is terminated with any receiving circuit which meets the requirements given in Section 2.4. These limitations apply to all interchange signals (Data, Control and Timing) unless otherwise specified.

(1)    All interchange signals entering into the transition region shall proceed through the transition region to the opposite signal state and shall not reenter the transition region until the next significant change of signal condition.

(2)    There shall be no reversal of the direction of voltage change while the signal is in the transition region.

(3)    For Control Interchange Circuits, the time required for the signal to pass through the transition region during a change in state shall not exceed one millisecond.

(4)    For Data and Timing interchange Circuits, the time required for the signal to pass through the transition region shall not exceed one millisecond or four percent of the nominal duration of a signal element on that interchange circuit, whichever is the lesser.

(5)    The maximum instantaneous rate of voltage change shall not exceed 30 volts per microsecond.

## SECTION THREE

## 3. INTERFACE MECHANICAL CHARACTERISTICS

3.1    The interface between the data terminal equipment and data communication equipment is located at a pluggable connector signal interface point between the two equipments. The female connector shall be associated with, but not necessarily physically attached to the data communication equipment and should be mounted in a fixed position near the data terminal equipment. The use of an extension cable on the data communication equipment is permitted. An extension cable with a male connector shall be provided with the data terminal equipment. The use of short cables (each less than approximately 50 feet or 15 meters) is recommended; however, longer cables are permissible, provided that the resulting load capacitance ($C_L$ of Fig. 2.1), measured at the interface point and including the signal terminator, does not exceed 2500 picofarads. (See section 2.4 and 6.5.)

3.1.1   When additional functions are provided in a separate unit inserted between the data terminal equipment and the data communication equipment (See section 1.7), the female connector, as indicated above shall be associated with the side of this unit which interfaces with the data terminal equipment while the extension cable with the male connector shall be provided on the side which interfaces with the data communication equipment.

| Pin Number | Circuit | Description |
|---|---|---|
| 1 | AA | Protective Ground |
| 2 | BA | Transmitted Data |
| 3 | BB | Received Data |
| 4 | CA | Request to Send |
| 5 | CB | Clear to Send |
| 6 | CC | Data Set Ready |
| 7 | AB | Signal Ground (Common Return) |
| 8 | CF | Received Line Signal Detector |
| 9 | — | (Reserved for Data Set Testing) |
| 10 | — | (Reserved for Data Set Testing) |
| 11 | | Unassigned  (See Section 3.2.2) |
| 12 | SCF | Sec. Rec'd. Line Sig. Detector |
| 13 | SCB | Sec. Clear to Send |
| 14 | SBA | Secondary Transmitted Data |
| 15 | DB | Transmission Signal Element Timing (DCE Source) |
| 16 | SBB | Secondary Received Data |
| 17 | DD | Receiver Signal Element Timing (DCE Source) |
| 18 | | Unassigned |
| 19 | SCA | Secondary Request to Send |
| 20 | CD | Data Terminal Ready |
| 21 | CG | Signal Quality Detector |
| 22 | CE | Ring Indicator |
| 23 | CH/CI | Data Signal Rate Selector (DTE/DCE Source) |
| 24 | DA | Transmit Signal Element Timing (DTE Source) |
| 25 | | Unassigned |

Figure 3.1

Interface Connector Pin Assignments

3.2     **Pin Identification**

3.2.1   Pin assignments listed in Figure 3.1 shall be used.

3.2.2   Pin assignments for circuits not specifically defined in section 4 (See section 4.1.1.) are to be made by mutual agreement. Preference should be given to the use of unassigned pins, but in the event that additional pins are required extreme caution should be taken in their selection.

SECTION FOUR

**4. FUNCTIONAL DESCRIPTION OF INTERCHANGE CIRCUITS**

4.1     **General**

This section defines the basic interchange circuits which apply, collectively, to all systems.

4.1.1   Additional interchange circuits not defined herein, or variations in the functions of the defined interchange circuits may be provided by mutual agreement. See sections 3.2.2. and 5.2.

4.2     **Categories**

Interchange circuits between data terminal equipment and data communication equipment fall into four general categories.

Ground or Common Return
Data Circuits
Control Circuits
Timing Circuits

4.2.1   A list of circuits showing category as well as equivalent C.C.I.T.T. identification in accordance with Recommendation V.24 is presented in Figure 4.1.

4.3     **Signal Characteristics, General**

4.3.1   Interchange circuits transferring data signals across the interface point shall hold marking (binary ONE) or spacing (binary ZERO) conditions for the total nominal duration of each signal element.

In synchronous systems using synchronous data communication equipment, distortion tolerances as specified in RS-334[1] shall apply. Acceptable distortion tolerances for data terminal equipment in synchronous and start-stop (i.e. asynchronous) systems using non-synchronous

---

[1] RS-334 "Signal Quality at Interface Between Data Processing Terminal Equipment and Synchronous Data Communication Equipment for Serial Data Transmission" — March 1967.

707

data communication equipment are under consideration for a future companion standard to RS-334.

4.3.2 Interchange circuits transferring timing signals across the interface point shall hold ON and OFF conditions for nominally equal periods of time, consistent with acceptable tolerances as specified in RS-334. During periods when timing information is not provided on a timing interchange circuit, this interchange circuit shall be clamped in the OFF condition.

4.3.3 Tolerances on the relationship between data and associated timing signals shall be in accordance with RS-334.

| Interchange Circuit | C.C.I.T.T. Equivalent | Description | Gnd | Data | | Control | | Timing | |
|---|---|---|---|---|---|---|---|---|---|
| | | | | From DCE | To DCE | From DCE | To DCE | From DCE | To DCE |
| AA | 101 | Protective Ground | X | | | | | | |
| AB | 102 | Signal Ground/Common Return | X | | | | | | |
| BA | 103 | Transmitted Data | | | X | | | | |
| BB | 104 | Received Data | | X | | | | | |
| CA | 105 | Request to Send | | | | | X | | |
| CB | 106 | Clear to Send | | | | X | | | |
| CC | 107 | Data Set Ready | | | | X | | | |
| CD | 108.2 | Data Terminal Ready | | | | | X | | |
| CE | 125 | Ring Indicator | | | | X | | | |
| CF | 109 | Received Line Signal Detector | | | | X | | | |
| CG | 110 | Signal Quality Detector | | | | X | | | |
| CH | 111 | Data Signal Rate Selector (DTE) | | | | | X | | |
| CI | 112 | Data Signal Rate Selector (DCE) | | | | X | | | |
| DA | 113 | Transmitter Signal Element Timing (DTE) | | | | | | | X |
| DB | 114 | Transmitter Signal Element Timing (DCE) | | | | | | X | |
| DD | 115 | Receiver Signal Element Timing (DCE) | | | | | | X | |
| SBA | 118 | Secondary Transmitted Data | | | X | | | | |
| SBB | 119 | Secondary Received Data | | X | | | | | |
| SCA | 120 | Secondary Request to Send | | | | | X | | |
| SCB | 121 | Secondary Clear to Send | | | | X | | | |
| SCF | 122 | Secondary Rec'd Line Signal Detector | | | | X | | | |

**Figure 4.1**

**Interchange Circuits by Category**

## 4.4    Interchange Circuits

*Circuit AA* — Protective Ground (C.C.I.T.T. 101)
Direction:  Not applicable

This conductor shall be electrically bonded to the machine or equipment frame. It may be further connected to external grounds as required by applicable regulations.

*Circuit AB* — Signal Ground or Common Return (C.C.I.T.T. 102)
Direction:  Not applicable

This conductor establishes the common ground reference potential for all interchange circuits except Circuit AA (Protective Ground). Within the data communication equipment, this circuit shall be brought to one point, and it shall be possible to connect this point to Circuit AA by means of a wire strap inside the equipment. This wire strap can be connected or removed at installation, as may be required to meet applicable regulations or to minimize the introduction of noise into electronic circuitry.

*Circuit BA* — Transmitted Data (C.C.I.T.T. 103)
Direction:  TO data communication equipment

Signals on this circuit are generated by the data terminal equipment and are transferred to the local transmitting signal converter for transmission of data to remote data terminal equipment.

The data terminal equipment shall hold Circuit BA (Transmitted Data) in marking condition during intervals between characters or words, and at all times when no data are being transmitted.

In all systems, the data terminal equipment shall not transmit data unless an ON condition is present on all of the following four circuits, where implemented.

1. Circuit CA (Request to Send)
2. Circuit CB (Clear to Send)
3. Circuit CC (Data Set Ready)
4. Circuit CD (Data Terminal Ready)

All data signals that are transmitted across the interface on interchange circuit BA (Transmitted Data) during the time an ON condition is maintained on all of the above four circuits, where implemented, shall be transmitted to the communication channel.

See Section 4.3, for signal characteristics.

*Circuit BB* — Received Data (C.C.I.T.T. 104)
Direction:  FROM data communication equipment

Signals on this circuit are generated by the receiving signal converter in response to data signals received from remote data terminal equipment via the remote transmitting signal converter. Circuit BB

709

(Received Data) shall be held in the binary ONE (Marking) condition at all times when Circuit CF (Received Line Signal Detector) is in the OFF condition.

On a half duplex channel, Circuit BB shall be held in the Binary One (Marking) condition when Circuit CA (Request to Send) is in the ON condition and for a brief interval following the ON to OFF transition of Circuit CA to allow for the completion of transmission (See Circuit BA — Transmitted Data) and the decay of line reflections. See section 4.3 for signal characteristics.

*Circuit CA* — Request to Send (C.C.I.T.T. 105)
Direction:  TO data communication equipment

This circuit is used to condition the local data communication equipment for data transmission and, on a half duplex channel, to control the direction of data transmission of the local data communication equipment.

On one way only channels or duplex channels, the ON condition maintains the data communication equipment in the transmit mode. The OFF condition maintains the data communication equipment in a non-transmit mode.

On a half duplex channel, the ON condition maintains the data communication equipment in the transmit mode and inhibits the receive mode. The OFF condition maintains the data communication equipment in the receive mode.

A transition from OFF to ON instructs the data communication equipment to enter the transmit code (see Section 6.8). The data communication equipment responds by taking such action as may be necessary and indicates completion of such actions by turning ON Circuit CB (Clear to Send), thereby indicating to the data terminal equipment that data may be transferred across the interface point on interchange Circuit BA (Transmitted Data).

A transition from ON to OFF instructs the data communication equipment to complete the transmission of all data which was previously transferred across the interface point on interchange Circuit BA and then assume a non-transmit mode or a receive mode as appropriate. The data communication equipment responds to this instruction by turning OFF Circuit CB (Clear to Send) when it is prepared to again respond to a subsequent ON condition of Circuit CA.

> *NOTE:* A non-transmit mode does not imply that all line signals have been removed from the communication channel. See section 6.8.

When Circuit CA is turned OFF, it shall not be turned ON again until Circuit CB has been turned OFF by the data communication equipment.

An ON condition is required on Circuit CA as well as on Circuit CB, Circuit CC (Data Set Ready) and, where implemented, Circuit CD (Data Terminal Ready) whenever the data terminal equipment transfers data across the interface on interchange Circuit BA.

It is permissible to turn Circuit CA ON at any time when Circuit CB is OFF regardless of the condition of any other interchange circuit.

710

*Circuit CB* – Clear to Send (C.C.I.T.T. 106)
Direction: FROM data communication equipment

Signals on this circuit are generated by the data communication equipment to indicate whether or not the data set is ready to transmit data.

The ON condition together with the ON condition on interchange circuits CA, CC and, where implemented, CD, is an indication to the data terminal equipment that signals presented on Circuit BA (Transmitted Data) will be transmitted to the communication channel.

The OFF condition is an indication to the data terminal equipment that it should not transfer data across the interface on interchange Circuit BA.

The ON condition of Circuit CB is a response to the occurrence of a simultaneous ON condition on Circuits CC (Data Set Ready) and Circuit CA (Request to Send), delayed as may be appropriate to the data communication equipment for establishing a data communication channel (including the removal of the MARK HOLD clamp from the Received Data interchange circuit of the remote data set) to a remote data terminal equipment.

Where Circuit CA (Request to Send) is not implemented in the data communication equipment with transmitting capability, Circuit CA shall be assumed to be in the ON condition at all times, and Circuit CB shall respond accordingly.

711

*Circuit CC* – Data Set Ready (C.C.I.T.T. 107)
Direction: FROM data communication equipment

Signals on this circuit are used to indicate the status of the local data set.

The ON condition on this circuit is presented to indicate that –

a)    the local data communication equipment is connected to a communication channel ("OFF HOOK" in switched service),

AND b)    the local data communication equipment is not in test (local or remote), talk (alternate voice) or dial\* mode, (See section 6.10).

AND c)    the local data communication equipment has completed, where applicable

1.    any timing functions required by the switching system to complete call establishment, and

2.    the transmission of any discreet answer tone, the duration of which is controlled solely by the local data set.

---

\* The data communication equipment is considered to be in the dial mode when circuitry directly associated with the call origination function is connected to the communication channel. These functions include signalling to the central office (dialing) and monitoring the communication channel for call progress or answer back signals.

Where the local data communication equipment does not transmit an answer tone, or where the duration of the answer tone is controlled by some action of the remote data set, the ON condition is presented as soon as all the other above conditions (a, b, and c-1) are satisfied.

This circuit shall be used only to indicate the status of the local data set. The ON condition shall not be interpreted as either an indication that a communication channel has been established to a remote data station or the status of any remote station equipment.

The OFF condition shall appear at all other times and shall be an indication that the data terminal equipment is to disregard signals appearing on any other interchange circuit with the exception of Circuit CE (Ring Indicator). The OFF condition shall not impair the operation of Circuit CE or Circuit CD (Data Terminal Ready).

When the OFF condition occurs during the progress of a call before Circuit CD is turned OFF, the data terminal equipment shall interpret this as a lost or aborted connection and take action to terminate the call. Any subsequent ON condition on Circuit CC is to be considered a new call.

When the data set is used in conjunction with Automatic Calling Equipment, the OFF to ON transition of Circuit CC shall not be interpreted as an indication that the ACE has relinquished control of the communication channel to the data set. Indication of this is given on the appropriate lead in the ACE interface (see EIA Standard RS-366).

> *Note:* Attention is called to the fact that if a data call is interrupted by alternate voice communication, Circuit CC will be in the OFF condition during the time that voice communication is in progress. The transmission or reception of the signals required to condition the communication channel or data communication equipment in response to the ON condition of interchange Circuit CA (Request to Send) of the transmitting data terminal equipment will take place after Circuit CC comes ON, but prior to the ON condition on Circuit CB (Clear to Send) or Circuit CF (Received Line Signal Detector).

*Circuit CD* – Data Terminal Ready (C.C.I.T.T. 108.2)
Direction: To data communication equipment

Signals on this circuit are used to control switching of the data communication equipment to the communication channel. The ON condition prepares the data communication equipment to be connected to the communication channel and maintains the connection established by external means (e.g., manual call origination, manual answering or automatic call origination).

When the station is equipped for automatic answering of received calls and is in the automatic answering mode, connection to the line occurs only in response to a combination of a ringing signal and the ON condition of Circuit CD (Data Terminal Ready): however, the data terminal equipment is normally permitted to present the ON condition on Circuit CD whenever it is ready to transmit or receive data, except as indicated below.

The OFF condition causes the data communication equipment to be removed from the communication channel following the completion of any "in process" transmission. See Circuit BA (Transmitted Data). The OFF condition shall not disable the operation of Circuit CE (Ring Indicator).

712

In switched network applications, when circuit CD is turned OFF, it shall not be turned ON again until Circuit CC (Data Set Ready) is turned OFF by the data communication equipment.

*Circuit CE* — Ring Indicator (C.C.I.T.T. 125)
Direction: FROM data communication equipment

The ON condition of this circuit indicates that a ringing signal is being received on the communication channel.

The ON condition shall appear approximately coincident with the ON segment of the ringing cycle (during rings) on the communication channel.

The OFF condition shall be maintained during the OFF segment of the ringing cycle (between "rings") and at all other times when ringing is not being received. The operation of this circuit shall not be disabled by the OFF condition on Circuit CD (Data Terminal Ready).

*Circuit CF* — Received Line Signal Detector (C.C.I.T.T. 109)
Direction: FROM data communication equipment

The ON condition on this circuit is presented when the data communication equipment is receiving a signal which meets its suitability criteria. These criteria are established by the data communication equipment manufacturer.

The OFF condition indicates that no signal is being received or that the received signal is unsuitable for demodulation.

The OFF condition of Circuit CF (Received Line Signal Detector) shall cause Circuit BB (Received Data) to be clamped to the Binary One (Marking) condition.

The indications on this circuit shall follow the actual onset or loss of signal by appropriate guard delays.

On half duplex channels, Circuit CF is held in the OFF condition whenever Circuit CA (Request to Send) is in the ON condition and for a brief interval of time following the ON to OFF transition of Circuit CA. (See Circuit BB.)

*Circuit CG* — Signal Quality Detector (C.C.I.T.T. 110)
Direction: FROM data communication equipment

Signals on this circuit are used to indicate whether or not there is a high probability of an error in the received data.

An ON condition is maintained whenever there is no reason to believe that an error has occurred.

An OFF condition indicates that there is a high probability of an error. It may, in some instances, be used to call automatically for the retransmission of the previously transmitted data signal. Preferably the response of this circuit shall be such as to permit identification of individual questionable signal elements on Circuit BB (Received Data).

713

*Circuit CH* — Data Signal Rate Selector (DTE Source) (C.C.I.T.T. 111)
Direction: TO data communication equipment

Signals on this circuit are used to select between the two data signalling rates in the case of dual rate synchronous data sets or the two ranges of data signalling rates in the case of dual range non-synchronous data sets.

An ON condition shall select the higher data signalling rate or range of rates.

The rate of timing signals, if included in the interface, shall be controlled by this circuit as may be appropriate.

*Circuit CI* — Data Signal Rate Selector (DCE Source) (C.C.I.T.T. 112)
Direction: FROM data communication equipment

Signals on this circuit are used to select between the two data signalling rates in the case of dual rate synchronous data. sets or the two ranges of data signalling rates in the case of dual range non-synchronous data sets.

An ON condition shall select the higher data signalling rate or range of rates.

The rate of timing signals, if included in the interface, shall be controlled by this circuit as may be appropriate.

*Circuit DA* — Transmitter Signal Element Timing (DTE Source) (C.C.I.T.T. 113)
Direction: TO data communication equipment

Signals on this circuit are used to provide the transmitting signal converter with signal element timing information.

The ON to OFF transition shall nominally indicate the center of each signal element on Circuit BA (Transmitted Data). When Circuit DA is implemented in the DTE, the DTE shall normally provide timing information on this circuit whenever the DTE is in a POWER ON condition. It is permissible for the DTE to withhold timing information on this circuit for short periods provided Circuit CA (Request to Send) is in the OFF condition. (For example, the temporary withholding of timing information may be necessary in performing maintenance tests within the DTE.)

*Circuit DB* — Transmitter Signal Element Timing (DCE Source) (C.C.I.T.T. 114)
Direction: FROM data communication equipment

Signals on this circuit are used to provide the data terminal equipment with signal element timing information. The data terminal equipment shall provide a data signal on Circuit BA (Transmitted Data) in which the transitions between signal elements nominally occur at the time of the transitions from OFF to ON condition of the signal on Circuit DB. When Circuit DB is implemented in the DCE, the DCE shall normally provide timing information on this circuit whenever the DCE is in a POWER ON condition. It is permissible for the DCE to withhold timing information on this circuit for short periods provided Circuit CC (Data Set Ready) is in the OFF condition. (For example, the withholding of timing information may be necessary in performing maintenance tests within the DCE.)

*Circuit DD* – Receiver Signal Element Timing (DCE Source) (C.C.I.T.T. 115)
Direction: FROM data communication equipment.

Signals on this circuit are used to provide the data terminal equipment with received signal element timing information. The transition from ON to OFF condition shall nominally indicate the center of each signal element on Circuit BB (Received Data). Timing information on Circuit DD shall be provided at all times when Circuit CF (Received Line Signal Detector) is in the ON condition. It may, but need not be present following the ON to OFF transition of Circuit CF (See section 4.3.2).

*Circuit SBA* – Secondary Transmitted Data (C.C.I.T.T. 118)
Direction: TO data communication equipment

This circuit is equivalent to Circuit BA (Transmitted Data) except that it is used to transmit data via the secondary channel.

Signals on this circuit are generated by the data terminal equipment and are connected to the local secondary channel transmitting signal converter for transmission of data to remote data terminal equipment.

The data terminal equipment shall hold Circuit SBA (Secondary Transmitted Data) in marking condition during intervals between characters or words and at all times when no data are being transmitted.

In all systems, the data terminal equipment shall not transmit data on the secondary channel unless an ON condition is present on all of the following four circuits, where implemented:

1. Circuit SCA – Secondary Request to Send
2. Circuit SCB – Secondary Clear to Send
3. Circuit CC – Data Set Ready
4. Circuit CD – Data Terminal Ready

All data signals that are transmitted across the interface on interchange Circuit SBA during the time when the above conditions are satisfied shall be transmitted to the communication channel. See Section 4.3.

When the secondary channel is useable only for circuit assurance or to interrupt the flow of data in the primary channel (less than 10 Baud capability), Circuit SBA (Secondary Transmitted Data) is normally not provided, and the channel carrier is turned ON or OFF by means of Circuit SCA (Secondary Request to Send). Carrier OFF is interpreted as an "Interrupt" condition.

*Circuit SBB* – Secondary Received Data (C.C.I.T.T. 119)
Direction: FROM data communication equipment

This circuit is equivalent to Circuit BB (Received Data) except that it is used to receive data on the secondary channel.

When the secondary channel is useable only for circuit assurance or to interrupt the flow of data in the primary channel, Circuit SBB is normally not provided. See interchange Circuit SCF (Secondary Received Line Signal Detector).

*Circuit SCA* — Secondary Request to Send (C.C.I.T.T. 120)
Direction: TO data communication equipment

This circuit is equivalent to Circuit CA (Request to Send) except that it requests the establishment of the secondary channel instead of requesting the establishment of the primary data channel.

Where the secondary channel is used as a backward channel, the ON condition of Circuit CA (Request to Send) shall disable Circuit SCA and it shall not be possible to condition the secondary channel transmitting signal converter to transmit during any time interval when the primary channel transmitting signal converter is so conditioned. Where system considerations dictate that one or the other of the two channels be in transmit mode at all times but never both simultaneously, this can be accomplished by permanently applying an ON condition to Circuit SCA (Secondary Request to Send) and controlling both the primary and secondary channels, in complementary fashion, by means of Circuit CA (Request to Send). Alternatively, in this case, Circuit SCB need not be implemented in the interface.

When the secondary channel is useable only for circuit assurance or to interrupt the flow of data in the primary data channel, Circuit SCA shall serve to turn ON the secondary channel unmodulated carrier. The OFF condition of Circuit SCA shall turn OFF the secondary channel carrier and thereby signal an interrupt condition at the remote end of the communication channel.

*Circuit SCB* — Secondary Clear to Send (C.C.I.T.T. 121)
Direction: FROM data communication equipment

This circuit is equivalent to Circuit CB (Clear to Send), except that it indicates the availability of the secondary channel instead of indicating the availability of the primary channel. This circuit is not provided where the secondary channel is useable only as a circuit assurance or an interrupt channel.

*Circuit SCF* — Secondary Received Line Signal Detector (C.C.I.T.T. 122)
Direction: FROM data communication equipment

This circuit is equivalent to Circuit CF (Received Line Signal Detector) except that it indicates the proper reception of the secondary channel line signal instead of indicating the proper reception of a primary channel received line signal.

Where the secondary channel is useable only as a circuit assurance or an interrupt channel (see Circuit SCA — Secondary Request to Send), Circuit SCF shall be used to indicate the circuit assurance status or to signal the interrupt. The ON condition shall indicate circuit assurance or a non-interrupt condition. The OFF condition shall indicate circuit failure (no assurance) or the interrupt condition.

## SECTION FIVE

## 5. STANDARD INTERFACES FOR SELECTED
## COMMUNICATION SYSTEM CONFIGURATIONS

5.1     This section describes a selected set of data transmission configurations. For each of these configurations a standard set of interchange circuits (defined in section 4) is listed. (See section 6.2.)

5.1.1   Provision is made for additional data transmission configurations not defined herein. Interchange circuits for these applications must be specified separately, for each application, by the supplier.

5.2     Drivers shall be provided for every interchange circuit included in the standard interface. Terminators need not be provided for every interchange circuit included in the standard interface; however, the designer of the equipment which does not provide all of the specified terminators must be aware that any degradation in service due to his disregard of a standard interchange circuit is his responsibility.

In the interest of minimizing the number of different types of equipment, additional interchange circuits may be included in the design of a general unit capable of satisfying the requirements of several different applications.

5.2.1   For a given configuration, interchange circuits which are included in the standard list and for which drivers are provided, but which the manufacturer of equipment at the receiving side of the interface chooses not to use, shall be suitably terminated by means of a dummy load impedance in the equipment which normally provides the terminator. See Section 2.4.

5.2.2   Where interchange circuits not on the standard list are provided for a given configuration, the designer of this equipment must be prepared to find an open circuit on the other side of the interface, and the system shall not suffer degradation of the basic service.

Interference due to unterminated drivers in this category is the responsibility of the designer who includes these drivers.

Terminators shall not interfere with or degrade system performance as a result of open circuited input terminals.

5.3     Circuit configurations for which standard sets of interchange circuits are defined are listed in Figure 5.1.

5.4     The use of Circuit AA (Protective Ground) is optional. Where it is used, attention is called to the applicable Underwriters' regulation applying to wire size and color coding. Where it is not used, other provisions for grounding equipment frames should be made in accordance with good engineering practice.

5.5     The use of Circuit AB (Signal Ground) is mandatory in all systems. See section 1.4.

717

5.6 Secondary channels, where involved in the standard interfaces, are shown as Auxiliary Channels.

5.6.1 Where secondary channels are intended for use as backward channels, Circuit SCA (Secondary Request to Send) shall be interconnected with Circuit CA (Request to Send) within the data communication equipment and need not be brought out to the interface. See Section 4.4, Interchange Circuit SCA (Secondary Request to Send) for detailed information.

5.6.2 Where secondary channels are useable only for circuit assurance or to interrupt the flow of data in the primary channel, they transmit no actual data and depend only on the presence or absence of the secondary channel carrier. For this application only, Circuit SBA (Secondary Transmitted Data), SBB (Secondary Received Data) and SCB (Secondary Clear to Send) are not provided. Circuit SCA (Secondary Request to Send) turns secondary channel carrier ON and OFF as required and Circuit SCF (Secondary Received Line Signal Detector) recognizes its presence or absence. See definitions of Circuits SCA and SCF in Section 4.4 for details.

718

| Data Transmission Configuration | Interface Type |
|---|:---:|
| Transmit Only | A |
| Transmit Only* | B |
| Receive Only | C |
| Half Duplex | D |
| Duplex* | D |
| | |
| Duplex | E |
| Primary Channel Transmit Only * / Secondary Channel Receive Only | F |
| Primary Channel Transmit Only / Secondary Channel Receive Only | H |
| Primary Channel Receive Only / Secondary Channel Transmit Only* | G |
| Primary Channel Receive Only / Secondary Channel Transmit Only | I |
| | |
| Primary Channel Transmit Only* / Half Duplex Secondary Channel | J |
| Primary Channel Receive Only / Half Duplex Secondary Channel | K |
| Half Duplex Primary Channel / Half Duplex Secondary Channel | L |
| Duplex Primary Channel* / Duplex Secondary Channel* | L |
| Duplex Primary Channel / Duplex Secondary Channel | M |
| | |
| Special (Circuits specified by Supplier) | Z |

*Note:* Data Transmission Configurations identified with an asterisk (*) indicate the inclusion of Circuit CA (Request to Send) in a One Way Only (Transmit) or Duplex Configuration where it might ordinarily not be expected, but where it might be used to indicate a non-transmit mode to the data communication equipment to permit it to remove a line signal or to send synchronizing or traning signals as required.

**Figure 5.1**

**Interface Types for Data Transmission Configurations**

720

| Interchange Circuit | Interface Type |  |  |  |  |  |  |  |  |  |  |  |  |  |
|---|---|---|---|---|---|---|---|---|---|---|---|---|---|---|
|  | A | B | C | D | E | F | G | H | I | J | K | L | M | Z |
| AA Protective Ground | – | – | – | – | – | – | – | – | – | – | – | – | – | – |
| AB Signal Ground | x | x | x | x | x | x | x | x | x | x | x | x | x | x |
| BA Transmitted Data | x | x |  | x | x | x | x | x |  | x | x | x | x | ° |
| BB Received Data |  |  | x | x | x | x | x | x | x | x | x | x | x | ° |
| CA Request to Send | x | x |  | x | x | x | x | x |  | x | x | x | x | ° |
| CB Clear to Send | x | x | x | x | x | x | x | x | x | x | x | x | x | ° |
| CC Data Set Ready | x | x | x | x | x | x | x | x | x | x | x | x | x | ° |
| CD Data Terminal Ready | s | s | s | s | s | s | s | s | s | s | s | s | s | ° |
| CE Ring Indicator | s | s | s | s | s | s | s | s | s | s | s | s | s | ° |
| CF Received Line Signal Detector |  |  | x | x | x | x | x | x | x | x | x | x | x | ° |
| CG Signal Quality Detector |  |  |  |  |  |  |  |  |  |  |  |  |  | ° |
| CH/CI Data Signalling Rate Selector (DTE)/(DCE) |  |  |  |  |  |  |  |  |  |  |  |  |  | ° |
| DA/DB Transmitter Sig. Element Timing (DTE)/(DCE) | t | t |  | t | t | t |  | t | t | t | t | t | t | ° |
| DD Receiver Signal Element Timing (DCE) |  |  | t | t | t |  | t | t | t |  | t | t | t | ° |
| SBA Secondary Transmitted Data |  |  |  |  |  |  | x |  | x | x | x | x | x | ° |
| SBB Secondary Received Data |  |  |  |  |  | x | x | x |  | x | x | x | x | ° |
| SCA Secondary Request to Send |  |  |  |  |  |  | x |  | x | x | x | x | x | ° |
| SCB Secondary Clear to Send |  |  |  |  |  |  | x |  | x | x | x | x | x | ° |
| SCF Secondary Received Line Signal Detector |  |  |  |  |  | x | x | x |  | x | x | x | x | ° |

Legend:
° - To be specified by the supplier
– - optional
s - Additional Interchange Circuits required for Switched Service
t - Additional Interchange Circuits required for Synchronous Channel
x - Basic Interchange Circuits, All Systems

Figure 5.2

Standard Interfaces For
Selected Communication Systems Configurations

# SECTION SIX

## 6. RECOMMENDATIONS AND EXPLANATORY NOTES

6.1    The data are to be serialized by the data terminal equipment so that the design of the data communication equipment may be independent of the character length and code used by the data terminal equipment. The data communication equipment shall place no restrictions on the arrangement of the sequence of bits provided by the data terminal equipment.

6.2    The control interchange circuits at the interface point are arranged to permit the alternate use of a higher class of communication service as follows:

A.  Data terminal equipment designed for Transmit-Only or Receive-Only service may also use either Half-duplex or Duplex service.

B.  Data terminal equipment designed for Half-duplex service may also use Duplex service.

6.3    The electrical specifications are intended to provide a two-volt margin in rejecting noise introduced either on interchange circuits or by a difference in reference ground potential across the interface. The equipment designer should maintain this margin of safety on all interchange circuitry.

6.4    To avoid inducing voltage surges on interchange circuits, signals from interchange circuits should not be used to drive inductive devices, such as relay coils. (Note that relay or switch contacts may be used to generate signals on an interchange circuit, with appropriate measures to assure that signals so generated comply with Section 2.7.)

6.5    Alphabetical parenthetical designations are added to the terms used in Sections 2.3, 2.4, and 2.6 to better tie them in with the equivalent circuit of Section 2.1 and stress the point that the 2500 picofarad capacitance ($C_L$) is defined for the receiving end of the interchange circuit and that the capacitance ($C_O$) at the driving end of the interchange circuit, including cable, is not defined. It is the responsibility of the designer to build a circuit capable of driving all of the capacitance in the driver circuitry plus the capacitance in his part of the interconnecting cable (not specified) plus 2500 pF in the load (including the cable on the load side of the interface point).

6.6    The user is reminded that the characteristics of an equivalent load (terminator) circuit used to test for compliance with each of the electrical specifications in section 2 are a function not only of the parameter under test, but also of the tolerance limit to be tested. For example, a driver which just delivers a minimum of 5 Volts into a 7,000 Ohm test load may fail the test if the load is reduced to 3,000 Ohms, whereas, a driver with an output within the 15 Volt limit when driving a 3,000 Ohm load may exceed this limit when driving a 7,000 Ohm load. The 5 Volt tolerance should therefore be tested with a 3,000 Ohm load while the 15 Volt limit should be tested using a 7,000 Ohm load.

6.7    The operation of the transmitting and receiving circuits should minimize the effects of any circuit time constants which would delay the circuit response and introduce time distortion of the signals.

6.8    The turning ON of Circuit CA (Request to Send) does not necessarily imply the turning ON of a line signal on the communication channel. Some data sets might not have a line signal as it is understood in this standard, e.g. the signal can be a modified digital base-band signal.

Conversely, in data sets which do transmit a "line signal", the turning OFF of Circuit CA does not necessarily command the removal of that line signal from the communication channel. On a duplex channel, the data set might autonomously transmit a training signal to hold AGC Circuits or automatic equalizers in adjustment, or to keep timing locked (synchronized) when Circuit CA is OFF.

It is not within the scope of this standard to specify in detail what occurs on the communication channel (line) side of the data communication equipment. Therefore the definition for Circuit CA uses the terminology "assume the transmit mode" intentionally avoiding reference to "carrier" or "line signals".

However, the continued requirement for multipoint systems is recognized. Data sets intended for this type of operation should permit the sharing of a communication channel by more than one data set transmitter and should, when in a non-transmit mode, place no signal on the communication channel which might interfere with the transmission from another data set in the network.

6.9    It is important that, at an answering data station, Circuit CC (Data Set Ready) be turned ON independently of any event which might occur at the remote (calling) data station. This independence permits the use of the OFF to ON transition of Circuit CC to start an "abort timer" in the data terminal equipment. This timer would cause termination of an automatically answered call (by causing Circuit CD to be turned OFF) if other expected events such as Circuit CF ON or proper exchange of data do not occur in a predetermined time interval. Such independence is necessary to assure the starting of the abort timer when an automatically answered incoming call is the result of a wrong number reached from a regular (non data station) telephone instrument.

6.10    Although the method of operation for multi-line automatic calling equipment, RS-366 (when assigned) Interface Type V, has not yet been fully defined, it appears that a situation could arise during call origination where both the DCE and the ACE appear to be idle (at the interface) even though actively engaged in establishing a connection.

One possible solution to this problem requires that circuit CC be turned ON upon completion of dialing to provide continuity of signalling during call origination. When multiline automatic calling equipment is used, Circuit CC would thus turn ON earlier than specified in Section 4.4 herein. This solution is subject to further study; however, data terminal equipment which may, in the future, be used in systems with multi-line automatic calling equipment should not be adversely affected by this early "Data Set Ready" indication.

SECTION SEVEN

7. GLOSSARY OF NEW TERMS

.1   This section defines terms used in this standard which are new or are used in a special sense.

### 1.  Data Transmission Channel

The transmission media and intervening equipment involved in the transfer of information between data terminal equipments  A data transmission channel includes the signal conversion equipment. A data transmission channel may support the transfer of information in one direction only, in either direction alternately, or in both directions simultaneously and the channel is accordingly classified as defined in the following sections. When the ·data communications equipment has more than one speed capability associated with it, for example 1200 baud transmission in one direction and 150 baud transmission in the opposite direction, a channel is defined for each speed capability.

### 2.  Primary Channel

The data transmission channel having the highest signaling rate capability of all the channels sharing a common interface connector. A primary channel may support the transfer of information in one direction only, either direction alternately or both directions simultaneously and is then classified as "one way only", "half duplex" or "duplex" as defined herein.

### 3.  Secondary Channel

The data transmission channel having a lower signaling rate capability than the primary channel in a system in which two channels share a common interface connector. A secondary channel may be either one way only, half duplex or duplex as defined later. Two classes of secondary channels are defined, auxiliary and backward.

### 4.  Auxiliary Channel

A secondary channel whose direction of transmission is independent of the primary channel and is controlled by an appropriate set of secondary control interchange circuits.

### 5.  Backward Channel

A secondary channel whose direction of transmission is constrained to be always opposite to that of the primary channel. The direction of transmission of the backward channel is restricted by the control interchange circuit (Circuit CA — Request to Send) that controls the direction of transmission of the primary channel.

### 6.  One Way Only (Unidirectional) Channel

A primary or secondary channel capable of operation in only one direction. The direction is fixed and cannot be reversed. The term "one way only" used to describe a primary channel does not imply anything about the type of secondary channel or the existence of a secondary channel; similary, the use of the term to describe a secondary channel implies nothing about the type of primary channel present.

### 7. Half Duplex Channel

A primary or secondary channel capable of operating in both directions but not simultaneously. The direction of transmission is reversible. The term half duplex used to describe a primary channel does not imply anything about the type of secondary channel; similarly, the use of the term to describe a secondary channel implies nothing about the type of primary channel present. (Note that as a result of the definitions, both directions of a half duplex channel have the same signaling rate capability.)

### 8. Duplex Channel (Full Duplex Channel)

A primary or secondary channel capable of operating in both directions simultaneously. The term duplex used to describe a primary channel does not imply anything about the type of secondary channel or the existence of a secondary channel; similarly, the use of the term to describe a secondary implies nothing about the type of primary channel present. (Note that a full duplex channel has the same signaling rate capability in both directions. A system with different rates would be considered to be a one way only primary channel in one direction and a one way only secondary channel in the opposite direction.)

### 9. Synchronous Data Transmission Channel

A data channel in which timing information is transferred between the data terminal equipment and the data communication equipment. Transmitter Signal Element Timing signals can be provided by either the data terminal equipment or by the data communication equipment. Receiver Signal Element Timing is normally recovered in and provided by the Data Communication Equipment. A synchronous data channel will not accommodate Start/Stop data signals unless they are transmitted isochronously and timing signals are interchanged at least at the transmitting.station.

### 10. Nonsynchronous Data Transmission Channel

A data channel in which no timing information is transferred between the data terminal equipment and the data communication equipment.

### 11. Dedicated Line

A communications channel which is nonswitched, i.e., which is permanently connected between two or more data stations. These communication channels are also referred to as "leased" or "private"; however, since leased and private switched networks do exist, the term "dedicated" is preferred herein to define a nonswitched connection between two or more stations.

## 12. Interchange Circuit

A circuit between the data terminal equipment and the data communication equipment for the purpose of exchanging data, control or timing signals. Circuit AB (signal ground) is a common reference for all interchange circuits.

## 13. Driver

a. The electronic circuitry or relay contact at the transmitting end (source) of an interchange circuit which transmits binary digital signals to a terminator via an interconnecting cable.

b. The transmitter of a binary digital signal.

## 14. Terminator

a. The electronic circuitry at the receiving end (sink) of an interchange circuit which receives binary digital signals from a driver via an interconnecting cable.

b. The receiver of a binary digital signal.

## 15. Signal Conversion Equipment

Those portions of the data communication equipment which transform (e.g., modulate, shape, etc.) the data signals exchanged across the interface into signals suitable for transmission through the associated communication media or which transform (e.g., demodulate, slice, regenerate, etc.) the received line signals into data signals suitable for presentation to the data terminal equipment.

# APPENDIX I

## INTERFACE CONNECTOR

While no industry standard exists which defines a suitable interface connector, it should be noted that commercial products are available which will perform satisfactorily as electrical connectors for interfaces specified in RS-232C, such as those connectors meeting Military Specification MIL-C-24308 (MS-18275) or equivalent.

It is not intended that the above reference be considered as part of RS-232C or as a standard for the devices to which reference is made.

# EIA STANDARD

*Synchronous Signaling Rates*
*for Data Transmission*

## RS-269-B

*(Revision of RS-269-A)*

*January 1976*

*Engineering Department*

# ELECTRONIC INDUSTRIES ASSOCIATION

## NOTICE

EIA engineering standards are designed to serve the public interest through eliminating misunderstandings between manufacturers and purchasers, facilitating interchangeability and improvement of products, and assisting the purchaser in selecting and obtaining with minimum delay the proper product for his particular need. Existence of such standards shall not in any respect preclude any member or non-member of EIA from manufacturing or selling products not conforming to such standards, nor shall the existence of such standards preclude their voluntary use by those other than EIA members whether the standard is to be used either domestically or internationally.

Recommended standards are adopted by EIA without regard to whether or not their adoption may involve patents on articles, materials, or processes. By such action, EIA does not assume any liability to any patent owner, nor does it assume any obligation whatever to parties adopting the recommended standards.

Published by

# ELECTRONIC INDUSTRIES ASSOCIATION

### Engineering Department

### 2001 Eye Street, N.W., Washington, D. C. 20006

728

**Price:** $2.00

Printed in U.S.A.

## SYNCHRONOUS SIGNALING RATES

## FOR DATA TRANSMISSION

*(From EIA Standard RS-269-A and Standards Proposal No. 1190, formulated under the cognizance of EIA Committee TR-30 on Data Transmission Systems and Equipment.)*

**FOREWORD**

(This Foreword is not part of the EIA Standard on Synchronous Signaling Rates for Data Transmission.)

This standard provides information on the transfer of data between data processing equipment and data communication equipment which transmit the data over media commonly referred to as voice band. Subsequent standards will prescribe signaling rates for use at greater than voice bandwidth.

The rates prescribed herein include the serial speeds and the parallel speeds in EIA RS-269-A. This standard therefore replaces the previous EIA document on Signaling Speeds. Future standards will be issued to define additional electrical parameters which are vital to the interconnection and the transfer of data between the data processing equipment and the data communication equipment.

Committee TR-30 has considered the material contained in this standard most carefully and has co-ordinated its work with ANS Subcommittee X3S3 to the extent that there is complete concurrence. In preparing this standard the following considerations were taken into account: the history of all known efforts in standardization in this field prior to this proposal; existing standards even partially covering this field; the present state-of-the-art and design trends for future data transmission equipments; the possible future work of the committee, both in areas of speeds and other parameters pertinent to the performance of data transmission systems; various considerations of a design nature; and current and re-quired glossary work.

There is an American National Standard which is equivalent to the above-referenced EIA document. AN Standard 3.1–1969 is the same as EIA Standard RS-269-A. ANS Subcommittee X3S3 has sub-mitted this Standards Proposal to Sectional Committee X3 for consideration as an American National Standard. When approved, it will replace the referenced AN Standard.

## 1. SCOPE AND PURPOSE

1.1 This standard provides a group of specific signaling rates for synchronous serial or parallel binary data transmission. These rates exist on the received data and transmitted data circuits at the interface between data terminal equipment and data communication equipments which operate over nominal 4kHz voice bandwidth channels.

729

1.2 This standard arises from indications by data equipment manufacturers and suppliers of data communication channels that the establishment of specific rates is important to insure compatibility between communication channels and data terminal equipments of data communication systems.

## 2. STANDARD SIGNALING RATES

2.1 The standard serial signaling rates shall be 600 times "N" bits per second where N may be any positive integer from 1 through 16.

2.1.1 The preferred standard signaling rates shall be 600, 1200, 2400, 4800, 7200 and 9600 bits per second.

2.1.2 For those applications requiring synchronous operation below 600 bits per second, the standard rates shall be 75, 150 and 300 bits per second.

2.2 The standard parallel signaling rates for equipments designed to operate with up to eight parallel data bits per character shall be 75 times "N" characters per second where N may be any positive integer from 1 through 16.

2.2.1 The preferred standard parallel signaling rates shall be 75, 150, 300, 600, 900 and 1200 characters per second.

## 3. RATE TOLERANCES

The serial signaling rates defined herein shall conform to the tolerance defined in Electronic Industries Association Standard RS-334, "Signal Quality at the Interface Between Data Processing Terminal Equipment and Synchronous Data Communication Equipment for Serial Data Transmission". The tolerance on parallel signaling rates is not defined at this time.

## 4. UNIT ELEMENT DURATION

A synchronous signal train at the interface between the data communication equipment and the data terminal equipment after synchronization is established shall consist of a sequence of marking and spacing signals the duration of which are all nominally integral multiples of the unit interval. The unit interval duration, in seconds, is the reciprocal of the modulation rate in bauds.

# APPENDIX I

## APPLICATION NOTES

*Note 1* – Where error control or similar devices which change the signaling rates by a fixed ratio are inserted between the data terminal source/sink equipment and the data communication channel equipment, the 600 x N standard signaling rates apply at the interface between the data communications equipment and the data terminal equipment as shown in Figure 1.

*Note 2* – When data rate converters are used which result in a non-standard signaling rate at the data processing equipment junction with the rate converter, the clocking for the data processing equipment should be furnished by the data rate converter.

*Note 3* – When using parallel data transfer at the interface with less than 8 bits per character, the standard rates of paragraph 2.2 should be used. If 8 channels are supplied, but less than 8 used for data, the excess bit channels should be run in an idling condition or by using no-data bits.

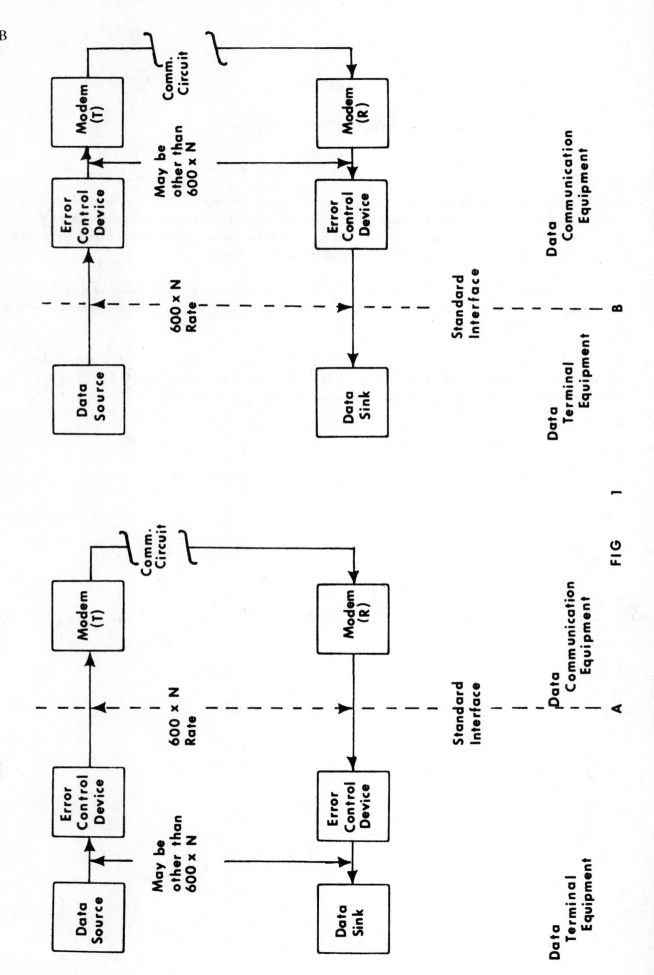

732

FIG 1

# ANSI

## X3.24-1968

Approved September 27, 1968

AMERICAN NATIONAL STANDARDS INSTITUTE

EIA RS-334

# EIA STANDARD

*Signal Quality at Interface
Between Data Processing
Terminal Equipment
and Synchronous Data
Communication Equipment for
Serial Data Transmission*

RS-334

733

*March 1967*

*Engineering Department*

# ELECTRONIC INDUSTRIES ASSOCIATION

## NOTICE

EIA engineering standards are designed to serve the public interest through eliminating misunderstandings between manufacturers and purchasers, facilitating interchangeability and improvement of products, and assisting the purchaser in selecting and obtaining with minimum delay the proper product for his particular need. Existence of such standards shall not in any respect preclude any member or non-member of EIA from manufacturing or selling products not conforming to such standards, nor shall the existence of such standards preclude their voluntary use by those other than EIA members whether the standard is to be used either domestically or internationally.

Recommended standards are adopted by EIA without regard to whether or not their adoption may involve patents on articles, materials, or processes. By such action, EIA does not assume any liability to any patent owner, nor does it assume any obligation whatever to parties adopting the recommended standards.

Published by

# ELECTRONIC INDUSTRIES ASSOCIATION

### Engineering Department

### 2001 Eye Street, N. W., Washington, D. C. 20006

734

## Price $ 2.90

# SIGNAL QUALITY AT INTERFACE BETWEEN DATA PROCESSING TERMINAL EQUIPMENT AND SYNCHRONOUS DATA COMMUNICATION EQUIPMENT FOR SERIAL DATA TRANSMISSION

*(From Standards Proposal No. 921 formulated under the cognizance of EIA Subcommittee TR-30.1 on Signal Quality)*

## 1. SCOPE

1.1   This standard is applicable to the exchange of serial binary data signals and timing signals across the interface between data processing terminal equipment and synchronous data communication equipment, as defined in EIA Standard RS-232-B.  The data communication equipment is considered to be synchronous if the timing signal circuits are required at the transmitting terminal or the receiving terminal, or both.

1.2   This standard is of particular importance when the equipments in a system are furnished by different organizations.  It does not attempt to indicate what action, if any, is to be taken if the limits are not met, but it is intended to provide a basis for agreement between the parties involved.

1.3   This standard does not describe any requirements for error performance, either for a complete system or any system components.  It should not be assumed that compliance with these standards will produce error rates that are acceptable in any particular application.

1.4   Any equipment which is represented as complying with this standard shall meet the applicable specifications within the ranges of those factors which are described as appropriate for the normal operation of the equipment, such as primary power voltage and frequency, ambient temperature, and humidity.

## 2. DEFINITIONS

2.1   Usage of terminology in this standard is consistent with that in EIA Standard RS-232-B.

2.2   Terms related to signal distortion are taken to have the significance defined in the International Telecommunication Union List of Definitions of Essential Telecommunication Terms, Part 1, General Terms, Telephony, Telegraph, 1961, and Supplement 1 thereto.

2.3   For discussion of isochronous and individual peak distortions, used in this standard, see Section 6, Explanatory Notes.

## 3. INTERFACE AT TRANSMITTING TERMINAL

### 3.1 TIMING SIGNALS

The following standards apply to synchronous systems using either Circuit DA (Transmitter Signal Element Timing, Data Processing Terminal Equipment Source), or Circuit DB (Transmitter Signal Element Timing, Data Communication Equipment Source).  (See Figure 1)

#### 3.1.1 DISTORTION

The degree of peak individual distortion of the significant transitions of the timing signal shall not exceed 0.5 per cent, at the modulation rate of the data signal.  The duty cycle shall be 50 plus or minus 10 per cent.

#### 3.1.2 FREQUENCY

The frequency of the timing signals shall not deviate more than 0.01 per cent from its assigned value.

### 3.2 TRANSMITTED DATA SIGNAL

#### 3.2.1 DATA TRANSITIONS NOT SIGNIFICANT FOR TIMING

For systems in which the individual transitions on circuit BA (Transmitted Data) are not significant in determining the timing of the modulations of the signal transmitted by the data communication equipment, no distortion limit is specified, provided that the relationship between the transitions on Circuit BA and the transitions on either Circuit DA or DB is met as specified in Section 3.3.

### 3.2.2 DATA TRANSITIONS SIGNIFICANT FOR TIMING

For systems in which the individual transitions on Circuit BA (Transmitted Data) are significant in determining the timing of the modulations of the signal transmitted by the data communication equipment, the degree of isochronous distortion on Circuit BA shall not exceed 4 per cent.

## 3.3 RELATIONSHIPS BETWEEN TIMING SIGNALS AND DATA SIGNALS

The following relationships apply to the case described in paragraph 3.2.1. (See Figure 1)

### 3.3.1 TIMING SIGNALS PROVIDED BY DATA PROCESSING TERMINAL EQUIPMENT

In synchronous systems using Circuit DA (Transmitter Signal Element Timing, Data Processing Terminal Equipment Source) the interval between any transition on Circuit BA (Transmitted Data) and any ON to OFF transition on Circuit DA, shall not be less than 25 per cent of the nominal unit interval of the data signal.

### 3.3.2 TIMING SIGNALS PROVIDED BY DATA COMMUNICATION EQUIPMENT

In synchronous systems using Circuit DB (Transmitter Signal Element Timing, Data Communication Equipment Source), the interval between any transition on Circuit BA (Transmitted Data) and the corresponding OFF to ON transition on Circuit DB, shall not exceed 25 per cent of the nominal unit interval of the data signal.

## 4. INTERFACE AT RECEIVING TERMINAL

## 4.1 TIMING SIGNALS

The following standards apply to synchronous systems using Circuit DD (Receiver Signal Element Timing, Data Communication Equipment Source). Since the use of Circuit DC (Receiver Signal Element Timing, Data Processing Terminal Equipment Source) is rare, no standards for it are established at this time. (See Figure 2)

### 4.1.1 DISTORTION

The degree of peak individual distortion of the ON to OFF transitions of the timing signals on Circuit DD (Receiver Signal Element Timing, Data Communication Equipment Source) shall not exceed 5 per cent, at the modulation rate of the data signal. The duty cycle shall be 50 plus or minus 10 per cent.

## 4.2 RECEIVED DATA SIGNAL

### 4.2.1 DISTORTION

The degree of isochronous distortion of the signals on Circuit BB (Received Data) shall not exceed 10 per cent.

## 4.3 RELATIONSHIP BETWEEN TIMING SIGNALS AND DATA SIGNALS

### 4.3.1 TIMING SIGNALS PROVIDED BY DATA COMMUNICATION EQUIPMENT

The interval between any transition on Circuit BB (Received Data) and any ON to OFF transition on Circuit DD (Receiver Signal Element Timing, Data Communication Equipment Source) shall not be less than 25 per cent of the nominal unit interval of the data signal. (See Figure 2)

## 5. STANDARD DISTORTION MEASUREMENT

5.1 Standard distortion measurement and interface voltage measurement shall be made on the particular signal interchange circuit of interest while the circuit is loaded with a Standard Test Load.

## 5.2 STANDARD TEST LOAD

The Standard Test Load shall consist of a 3000-ohm resistor shunted by 2500 picofarad capacitance and connected from the signal interchange circuit under test to Circuit AB (Signal Ground).

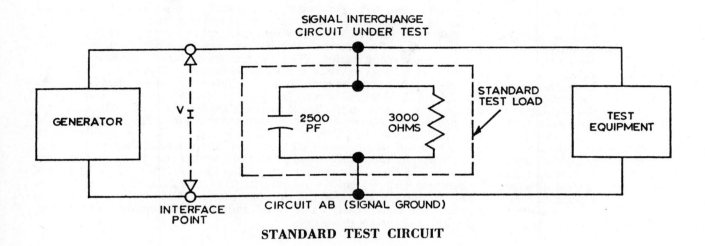

**STANDARD TEST CIRCUIT**

### 5.3 INTERFACE VOLTAGE MEASUREMENT

With the standard test load, the generator shall produce a voltage $V_I$ across the test load of not less than ± 5.0 volts.

### 5.4 STANDARD DISTORTION MEASUREMENT THRESHOLD

The standard distortion measurement shall be made using a +3 volt and a −3 volt threshold to determine the occurrence of signal transitions.

5.4.1   A mark-to-space or an OFF to ON transition shall be taken to occur at the instant $V_I$ crosses +3 volts on a positive-going transition.

5.4.2   A space-to-mark or an ON to OFF transition shall be taken to occur at the instant $V_I$ crosses −3 volts on a negative-going transition.

## 6. EXPLANATORY NOTES

### 6.1 DEGREE OF ISOCHRONOUS DISTORTION

The following definition is from Supplement 1 to the ITU List, definition 33.07:

*Degree of Isochronous Distortion*
(a) Ratio to the unit interval of the maximum measured difference, irrespective of sign, between the actual and the theoretical intervals separating any two significant instants of modulation (or of restitution), these instants being not necessarily consecutive.

(b) Algebraical difference between the highest and lowest value of individual distortion affecting the significant instants of an isochronous modulation. (This difference is independent of the choice of the reference ideal instant.)

The degree of distortion (of an isochronous modulation or restitution) is usually expressed as a percentage.

*Note:* The result of the measurement should be completed by an indication of the period, usually limited, of the observation. For a prolonged modulation (or restitution )it will be appropriate to consider the probability that an assigned value of the degree of distortion will be exceeded.

With respect to part (a) of the definition, the following interpretations are made:

(1) The Term "theoretical interval" (i.e., unit interval) is taken to be the reciprocal of the modulation (or restitution) rate, as determined throughout a stated measuring period.

(2) The term "maximum" requires that the intervals between all possible pairs of significant instants (i.e., transitions), that occur during the period of observation have been considered. This appears to involve making a very large number of measurements simultaneously.

However, the alternate definition, part (b), which was added in Supplement 1, suggests much simpler measurement procedures, since it is stated in terms of the individual distortions of the transitions. The distortion of each individual instant is measured against a corresponding reference instant. The reference instants occur at the modulation rate determined throughout a stated measuring period. The significance of "the highest and lowest value of individual distortion" is most obvious in the case assumed in Figure 3 as "Case A". Here, the phase of the clock is taken to be such that the reference instants always precede the corresponding transitions of the signal wave. In this very limited period of observation, the highest value of individual distortion is 0.6 unit interval; the lowest is 0.2. The algebraic difference is 0.4 unit interval, or 40 per cent isochronous distortion. The significance of "algebraical difference" is illustrated by "Case B" in the same figure. In this case, the clock instants are phased so as to fall at the mean of the signal wave transitions, (i.e., in the middle of the spread). The maximum individual distortions with respect to this clock are +0.2 and −0.2 unit interval. In the algebraical sense, −0.2 is "lower" than +0.2. The algebraical difference is therefore $+0.2 - (-0.2) = 0.4$ unit interval, or again, 40 per cent isochronous distortion. The concept of part (b) of the definition in Case B thus is seen to involve merely measuring the peak "early" and "late" individual distortions, and adding them to find the degree of isochronous distortion.

## 6.2 DEGREE OF PEAK INDIVIDUAL DISTORTION

For the purpose of this Standard, the following definition is used:

*Degree of Peak Individual Distortion*
The maximum individual distortion, irrespective of sign, of all significant instants occurring during a particular measuring period.

This term is defined and used in this Standard in stating the timing signal requirements, instead of isochronous distortion, for the following reasons:

(1) Only alternate transitions are considered (either "ON" to "OFF", or "OFF" to "ON", but not both).

(2) The reference is a unit interval, not of the timing signal, but of the data signal.

Figure 4 illustrates the case when it is the "ON" to "OFF" transitions that are significant. Individual distortions are measured with respect to a clock operating at the modulation rate of the associated data signal, and phased to occur at the mean of the significant timing transitions. (The "OFF" to "ON" transitions are shown dotted to indicate that they are not considered in the distortion measurement.) In the example, the peak individual distortion is (−)0.2 of a data signal unit interval, or 20 per cent, "irrespective of sign".

**PARAGRAPH 3.1.1 — TRANSMITTER SIGNAL ELEMENT TIMING SIGNALS, DISTORTION**

(A): FOR
PARAGRAPH 3.3.1

CIRCUIT DA

(TIMING SIGNALS
PROVIDED BY
DATA PROCESSING
TERMINAL
EQUIPMENT)

(B): FOR
PARAGRAPH 3.3.2

CIRCUIT DB

(TIMING SIGNALS
PROVIDED BY
DATA COMMUNICATION
EQUIPMENT)

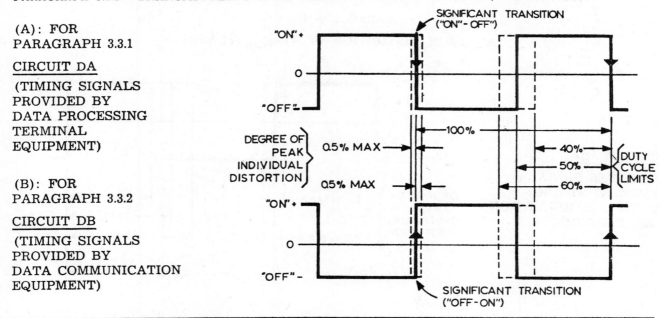

**PARAGRAPH 3.3.1 — RELATIONSHIP BETWEEN TIMING SIGNALS AND DATA SIGNALS — TIMING SIGNALS PROVIDED BY DATA PROCESSING TERMINAL EQUIPMENT.**

CIRCUIT DA

TRANSMITTER
SIGNAL ELEMENT
TIMING, DATA
PROCESSING
TERMINAL
EQUIPMENT
SOURCE

CIRCUIT BA

TRANSMITTED
DATA

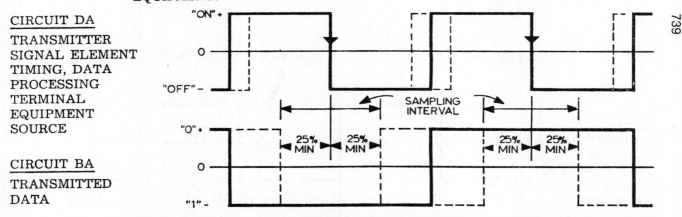

739

**PARAGRAPH 3.3.2 — RELATIONSHIP BETWEEN TIMING SIGNALS AND DATA SIGNALS — TIMING SIGNALS PROVIDED BY DATA COMMUNICATION EQUIPMENT.**

CIRCUIT DB

TRANSMITTER
SIGNAL ELEMENT
TIMING, DATA
COMMUNICATION
EQUIPMENT
SOURCE

CIRCUIT BA

TRANSMITTED
DATA

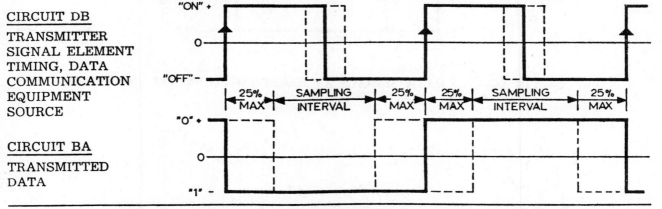

FIG. 1 — TIMING AND DATA SIGNALS AT TRANSMITTING TERMINAL INTERFACE

### PARAGRAPH 4.1.1

## MAXIMUM DISTORTION OF TIMING SIGNALS

CIRCUIT DD

RECEIVER SIGNAL
ELEMENT TIMING,
DATA COMMUNICATION
EQUIPMENT SOURCE

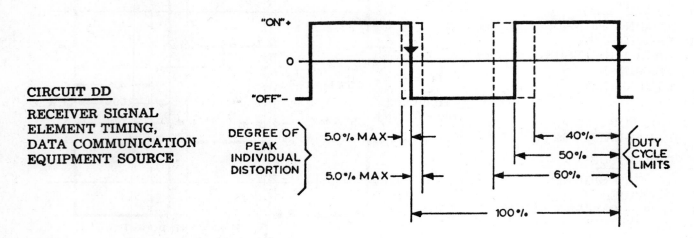

### PARAGRAPH 4.3.1

## RELATIONSHIP BETWEEN TIMING SIGNALS AND DATA SIGNALS

CIRCUIT DD

RECEIVER SIGNAL
ELEMENT TIMING,
DATA COMMUNI-
CATION EQUIP-
MENT SOURCE

CIRCUIT BB

RECEIVED
DATA

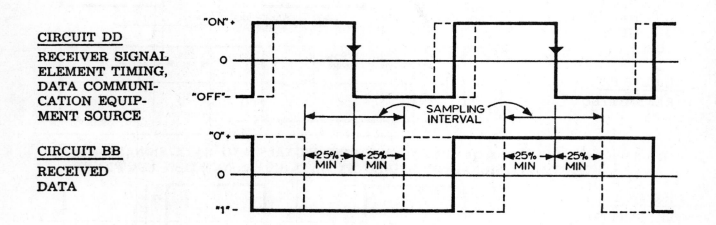

## FIG. 2—TIMING AND DATA SIGNALS AT RECEIVING TERMINAL INTERFACE

740

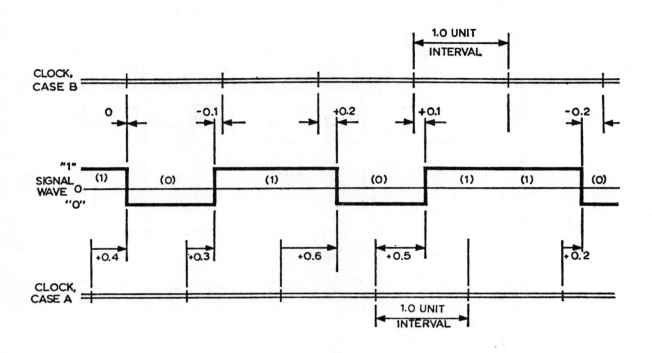

*CASE A:* REFERENCE CLOCK NOT IN PHASE WITH MEAN TRANSITIONS OF SIGNAL WAVE

ALGEBRAIC HIGHEST INDIVIDUAL DISTORTION   +0.6 (UNIT INTERVAL)
ALGEBRAIC LOWEST INDIVIDUAL DISTORTION   +0.2

ALGEBRAIC DIFFERENCE   0.4 (UNIT INTERVAL)
DEGREE OF ISOCHRONOUS DISTORTION   40%

*CASE B:* REFERENCE CLOCK IN PHASE WITH MEAN TRANSITIONS OF SIGNAL WAVE

ALGEBRAIC HIGHEST INDIVIDUAL DISTORTION   +0.2 (UNIT INTERVAL)
ALGEBRAIC LOWEST INDIVIDUAL DISTORTION   −0.2

ALGEBRAIC DIFFERENCE   0.4 (UNIT INTERVAL)
DEGREE OF ISOCHRONOUS DISTORTION   40%

FINDING DEGREE OF ISOCHRONOUS DISTORTION
BY ITU DEFINITION 33.07, PART (b)

**FIGURE 3**

*NOTE: This figure is explanatory only. See text for specific signal requirements.*

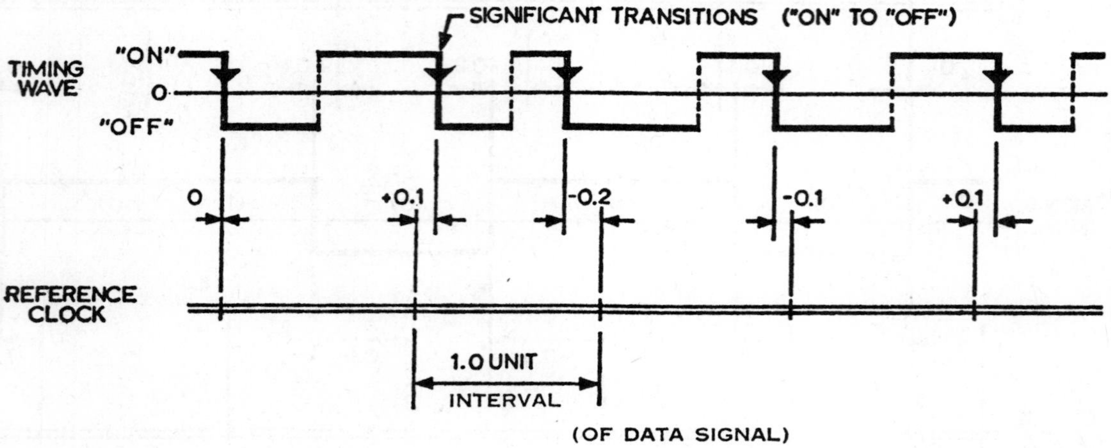

SIGNIFICANT TRANSITIONS ("ON" TO "OFF")

TIMING WAVE "ON" "OFF" 0

REFERENCE CLOCK

0 +0.1 -0.2 -0.1 +0.1

1.0 UNIT INTERVAL

(OF DATA SIGNAL)

MAXIMUM INDIVIDUAL DISTORTION OF SIGNIFICANT TRANSITIONS
(REFERENCE CLOCK IN PHASE WITH MEAN OF SIGNIFICANT TRANSITIONS):

0.2 DATA SIGNAL UNIT INTERVAL
DEGREE OF PEAK INDIVIDUAL DISTORTION: 20%

FINDING DEGREE OF PEAK INDIVIDUAL DISTORTION
FOR CASE OF "ON" TO "OFF" SIGNIFICANT TRANSITIONS

FIGURE 4

NOTE: *This figure is explanatory only. See text for specific signal requirements.*

## BACKGROUND INFORMATION

1. Background information is given in "EIA Industrial Electronics Bulletin No. 5, March 1965, Tutorial Paper on Signal Quality at a Digital Interface."

2. Interface requirements for non-synchronous data communication equipment differ significantly from those for synchronous data communication equipment, and therefore will be covered separately.

3. This standard is concerned only with data and timing signals at the specified digital interfaces of synchronous data communication equipment.

4. This standard is intended to allow equipment designers as much latitude as practicable in obtaining an economic balance between the cost of precision timing sources, the long-term stability of critical components, the time between equipment failures, and maintenance costs.

5. Definition 6.2 was developed because no existing definition of this type of distortion was listed in the ITU List of Definitions, and it was considered desirable to use it in writing the standards.

6. RS-232-B appears to have an ambiguity which would limit the accuracy of measurements. The ambiguity is evident upon consideration of the following paragraphs of that document:

> "4.5   The terminating impedance at the receiving end of an interchange circuit shall have a dc resistance of not less than 3000 ohms or more than 7000 ohms, and the voltage in open-circuited condition shall not exceed two volts. The effective shunt capacitance at the receiving end of a signal interchange circuit, measured at the interchange point and including any connecting cable, shall not exceed 2500 picofarads."

> "4.7   For both data circuits and timing signal circuits (Circuit BA, BB, DA, DB, DC and DD), neither the rise time nor the fall time, through the six volt range in which the signal condition is not defined, shall exceed three per cent of the nominal duration of a signal element. The circuitry used to generate a signal voltage on an interchange circuit shall meet this specification with any receiving termination which complies with Section 4.5".

> "4.8   Equivalent Circuit.   The following conditions shall apply:
>
> \*     \*     \*     \*     \*
>
> (8)   The circuitry that receives signals from an interchange circuit shall recognize the binary signal when $V_I$ equals $\pm 3$ volts. (See Section 5.4.) The signal condition is not defined when $V_I$ is in the range between $+3$ and $-3$ volts."

That is, the signal, when undergoing a transition, may be in an undefined condition for as long as three per cent of a unit element duration, and for a given source, is influenced by the impedance of the load.

Paragraph 5 of the Standard defines a set of standard conditions for making distortion measurements at the digital interface, which will eliminate the ambiguity of the range in which the signal condition is not defined. The specified "Standard Test Load" represents the most restrictive condition according to RS-232-B, Paragraph 4.5. The signal level ambiguity is resolved by specifying that the transitions shall be taken to occur when the signal first emerges from the undefined six volt range, at the $+3$ volt and $-3$ volt levels according to the sense of the transition.

EIA RS-363

744

# EIA STANDARD

*Standard for Specifying
Signal Quality for Transmitting
and Receiving Data Processing
Terminal Equipments Using
Serial Data Transmission
at the Interface With
Non-Synchronous Data
Communication Equipment*

**RS-363**

*May 1969*

*Engineering Department*

## ELECTRONIC INDUSTRIES ASSOCIATION

## NOTICE

EIA engineering standards are designed to serve the public interest through eliminating misunderstandings between manufacturers and purchasers, facilitating interchangeability and improvement of products, and assisting the purchaser in selecting and obtaining with minimum delay the proper product for his particular need. Existence of such standards shall not in any respect preclude any member or non-member of EIA from manufacturing or selling products not conforming to such standards, nor shall the existence of such standards preclude their voluntary use by those other than EIA members whether the standard is to be used either domestically or internationally.

Recommended standards are adopted by EIA without regard to whether or not their adoption may involve patents on articles, materials, or processes. By such action, EIA does not assume any liability to any patent owner, nor does it assume any obligation whatever to parties adopting the recommended standards.

Published by

# ELECTRONIC INDUSTRIES ASSOCIATION

### Engineering Department

### 2001 Eye Street, N.W., Washington, D. C. 20006

**PRICE $4.30**

Printed in U.S.A.

# STANDARD FOR SPECIFYING SIGNAL QUALITY FOR TRANSMITTING AND RECEIVING DATA PROCESSING TERMINAL EQUIPMENTS USING SERIAL DATA TRANSMISSION AT THE INTERFACE WITH NON-SYNCHRONOUS DATA COMMUNICATION EQUIPMENT

*(From Standards Proposal No. 1000, formulated under the cognizance of EIA Subcommittee TR-30.1 on Signal Quality)*

## 1. SCOPE

1.1    This standard is applicable for specifying the quality of serial binary signals exchanged across the interface between synchronous or start-stop (i.e. asynchronous) data processing terminal equipment and non-synchronous data communciations equipment as defined in EIA Standard RS-232B. The data communications equipment is considered to be non-synchronous if the timing signal circuits across the interface are not required at either the transmitting terminal or the receiving terminal.

1.2    This standard is arranged as a guide to define the quality of signals from the Transmitting Data Processing Terminal Equipment and acceptable to the Receiving Data Processing Terminal Equipment and to stipulate those characteristics which must be described and measured to give a meaningful statement concerning signal quality. It does not specify actual values for these characteristics, but provides standard statements into which agreed upon limits may be inserted in place of the alphabetic characters shown. A standard work sheet format is included to assist in preparing a system specification.

1.3    This standard is of particular importance when the equipments in a system are furnished by different organizations. It does not attempt to indicate what action, if any, is to be taken if the limits are not met, but it is intended to provide a basis for agreement between parties involved.

1.4    Any equipment which is represented as complying with limits established on the basis of this standard shall meet the applicable specifications within the ranges of those factors which are described as appropriate for the operation of the equipment, such as primary power voltage, frequency, ambient temperature and humidity.

1.5    The effects on signal quality of the inherent characteristics of the data communication equipment or communication channel are considered beyond the scope of this standard.

1.6    This standard does not describe any requirements for error performance, either for a complete system or any system components. It should not be assumed that compliance with a particular set of limits inserted into the standard statements of signal quality will produce error rates that are acceptable in any particular situation.

## 2. DEFINITIONS

The following definitions apply to the characteristics of signal quality described in this standard and for the purposes of this standard are the only meaning that should be attached to the characteristics described.

2.1 **Synchronous System** — A system in which the sending and receiving data processing terminal equipments are operating continuously at substantially the same frequency, and are maintained in a desired phase relationship by an appropriate means.

2.2 **Start-Stop System** — A system in which each group of code elements corresponding to a character is preceded by a start element which serves to prepare the receiving equipment for the reception and registration of a character, and is followed by a stop element during which the receiving equipment comes to rest in preparation for the reception of the next character.

2.3 **Continuous Start-Stop Operation** — The method of operation in a start-stop system in which the signals representing a series of characters follow one another at the nominal character rate, (e.g. in sending steadily from a perforated paper tape as compared to manual keyboard operation).

Note: Under these conditions, the character intervals will have their nominal durations.

2.4 **Signal Element** — Each of the parts of a digital signal, distinguished from others by its duration, position and sense, or by some of these features only.

2.5 **Unit Interval** — The longest interval of time such that the nominal durations of the signal elements in a synchronous system or the start and information elements in a start-stop system are whole multiples of this interval. (Note: A unit interval is the shortest nominal signal element.)

2.6 **Modulation Rate** — Reciprocal of the unit interval measured in seconds. This rate is expressed in bauds. (Note: As used in this standard, this term relates to the interface signals.)

2.7 **Baud** — The unit of modulation rate. It corresponds to a rate of one unit interval per second. Example: if the duration of the unit interval is 20 milliseconds, the modulation rate is 50 bauds.

2.8 **Character Interval** — In start-stop operation the duration of a character expressed as the total number of unit intervals (including information, error checking and control bits and the start and stop elements) required to transmit any given character in any given communications system.

2.9 **Start Element** — In a character transmitted in a start-stop system, the first element in each character, which serves to prepare the receiving equipment for the reception and registration of the character. The start element is a spacing signal.

2.10 **Start Transition** — In a character transmitted in a start-stop system, the mark-to-space transition at the beginning of the *start* element.

2.11 **Stop Element** — In a character transmitted in a start-stop system, the last element in each character, to which is assigned a minimum duration, during which the receiving equipment is returned to its rest condition in preparation for the reception of the next character. The stop element is a marking signal.

**2.12 Degree of Distortion** at the digital interface for binary signals is a measure of the time displacement of the transitions between signal states from their ideal instants. The degree of distortion is generally expressed as a percentage of the unit interval.

**2.13 Degree of Individual Distortion** of a particular signal transition, is the ratio to the unit interval of the displacement, expressed algebraically, of this transition from its ideal instant. This displacement is considered positive when the transition occurs after its ideal instant (late).

**2.14 Degree of Isochronous Distortion**

a. Ratio to the unit interval of the maximum measured difference, irrespective of sign, between the actual and the theoretical intervals separating any two transitions of modulation (or of restitution), these transitions being not necessarily consecutive.

b. Algebraical difference between the highest and lowest value of individual distortion affecting the transitions of an isochronous modulation. (This difference is independent of the choice of the reference ideal instant.)

The degree of distortion (of an isochronous modulation or restitution) is usually expressed as a percentage.

**2.15 Degree of Start-Stop Distortion** — The ratio to the unit interval of the maximum measured difference, irrespective of sign, between the actual interval and the theoretical interval (the appropriate integral multiple of unit intervals) separating any transition from the start transition preceding it. (See Paragraph 8.2.)

**2.16 Degree of Gross Start-Stop Distortion** — The degree of Start-Stop distortion (2.15) determined using the unit interval which corresponds exactly to the assigned modulation rate of the system. (See Paragraph 8.2)

**2.17 Degree of Synchronous Start-Stop Distortion** — The degree of Start-Stop distortion (2.15) determined using the unit interval which corresponds to the actual mean modulation rate of the signal involved. (See Paragraph 8.2.)

**2.18 Margin** of a receiving equipment. The maximum degree of distortion of the received signal which is compatible with the correct translation of all of the signals which it may possibly receive.

Note: This maximum degree of distortion applies without reference to the form of distortion affecting the signals. In other words, it is the maximum degree of the most unfavorable distortion acceptable, beyond which incorrect translation occurs, which determines the value of the margin. The condition of the measurements of the margin are to be specified in accordance with the requirements of the system.

748

## 3. SYSTEM CHARACTERISTICS

**3.1   Modulation Rate** — On the data interface circuits the nominal modulation rate shall be (A) bauds.

**3.2   Start-Stop Character Interval** — In start-stop operation the nominal character interval shall be (B) unit intervals.

## 4. SIGNAL QUALITY FROM THE TRANSMITTING DATA PROCESSING TERMINAL EQUIPMENT

### 4.1   Distortion

**4.1.1** In a *start-stop* system the signal provided by the transmitting data processing terminal equipment on circuit BA (Transmitted Data) shall have a degree of synchronous start-stop distortion not greater than (C) percent, and a degree of gross start-stop distortion not greater than (D) percent, provided that no signal element shall have a duration of less than (E) percent of a unit interval. Note: The degree of gross start-stop distortion (D) will usually be greater than the degree of synchronous start-stop distortion (C) by an amount determined by the allowable drift of the unit interval timing device used in the data processing terminal equipment.

**4.1.2** In a *synchronous system* the signal provided by the transmitting data processing terminal equipment on circuit BA (Transmitted Data) shall have a modulation rate within $\pm$ (F) percent of the nominal modulation rate, have a rate of change of the modulation rate no greater than (G) percent per (H) (unit of time) and a degree of isochronous distortion no greater than (J) percent.

### 4.2   Character Interval

In *continuous start-stop operation* the signals on circuit BA (Transmitted Data) may have a minimum average character interval which is shorter than the nominal character interval and an occasional character having a still shorter duration called the minimum character interval according to the following requirements:

#### 4.2.1  Minimum Average Character Interval

In *continuous start-stop operation* the interval between successive start transitions on circuit BA (Transmitted Data) averaged over (K) number of consecutive characters shall be no less than the nominal character interval reduced by (L) percent of a unit interval.

#### 4.2.2  Minimum Character Interval

In *continuous start-stop operation* the interval between successive start transitions on circuit BA (Transmitted Data) shall not be less than the nominal character interval reduced by (M) percent of a unit interval.

### 4.3 Continuity of Signal

On circuit BA (Transmitted Data) between successive unit elements of the same sense a continuous state may be maintained. In some equipment, however, a break may occur between successive unit elements of the same sense (such as in that equipment which generates a signal from a contact). These breaks shall not exceed (N) percent of a unit interval in duration and shall occur within the limits of the specified distortion for that equipment. When no breaks are expected (N) is said to be equal to zero.

### 4.4 Contact Bounce

In systems where the signal on circuit BA (Transmitted Data) is generated from a contact, all contact bounce is to be completed within (P) percent of a unit interval. For purposes of this requirement, the period of contact bounce shall be defined for a negative-to-positive transition as beginning when the signal first crosses the +3 volt level and the bounce is terminated when the signal finally remains beyond that level. For a positive-to-negative transition the opposite polarity applies. It is expected that contact bounce is completed within the limits of the specified distortion.

## 5. SIGNAL ACCEPTABLE TO THE RECEIVING DATA PROCESSING TERMINAL EQUIPMENT

### 5.1 Receiving Margin

**5.1.1** In a *start-stop system*, correct operation of the Receiving Data Processing Terminal Equipment shall be possible when the signal on circuit BB (Received Data) generated at the nominal modulation rate has a degree of start-stop distortion which does not exceed (Q) percent (Net Margin), provided that no signal element shall have a duration of less than (R) percent of a unit interval.

**5.1.2** In a *synchronous system* correct operation of the receiving data processing terminal equipment shall be possible when the signals on circuit BB (Received Data) have a modulation rate within ± (S) percent of the nominal modulation rate, have a rate of change of the modulation rate as great as (T) percent for (U) unit of time, and a degree of isochronous distortion as great as (V) percent, provided that no signal element has a duration of less than (W) percent of a unit interval. When measuring the isochronous distortion acceptable to the receiver (Net Margin) the test signal shall have a modulation rate equal to the nominal modulation rate of the system.

### 5.2 Character Interval

In *continuous start-stop operation* the receiving data processing terminal equipment shall respond to signals on circuit BB (Received Data) which have a minimum average character interval which is shorter than the nominal character interval and an occasional character having a still shorter duration called the minimum character interval, according to the following requirements:

### 5.2.1  Minimum Average Character Interval

In *continuous start-stop operation* the receiving data processing terminal equipment shall be prepared to respond to successive start transitions on circuit BB (Received Data) which follow their previous start transitions by a character interval averaged over (K) number of consecutive characters which is no less than the nominal character interval reduced by (X) percent of a unit interval.

### 5.2.2  Minimum Character Interval

In *continuous start-stop operation* when the above average is met, the receiving data processing terminal equipment shall be prepared to respond to a start transition on circuit BB (Received Data) which follows the start transition of the preceding character by an interval which is no less than the nominal character interval reduced by (Y) percent of a unit interval.

### 5.3  Minimum Duration Start Element

In a *start-stop system* when the receiver is in the stop or resting condition, it shall not be required to start reception of a character on a spacing signal element on circuit BB (Received Data) which has a duration of less than (Z) percent of a unit interval.

## 6. REQUIREMENTS FOR MEASUREMENT OF INTERFACE CHARACTERISTICS

Measurements of distortion, voltage, and loading of interface circuits shall meet the following requirements:

### 6.1  Measurement of Interface Driver Characteristics

#### 6.1.1  Use of Standard Test Load

Standard distortion measurement and interface voltage measurement shall be made on the particular signal interchange driver circuit of interest while the circuit is terminated with a standard test load. This standard test load may be the input impedance of the test device or may be an external device but in all cases the total load on the interchange circuit shall meet the following requirement:

#### 6.1.2  Requirement for Standard Test Load

The standard test load shall consist of 3000 ohm resistance shunted by 2500 pF capacitance and connected from the signal interchange circuit under test to circuit AB (Signal Ground) as shown in Figure 1.

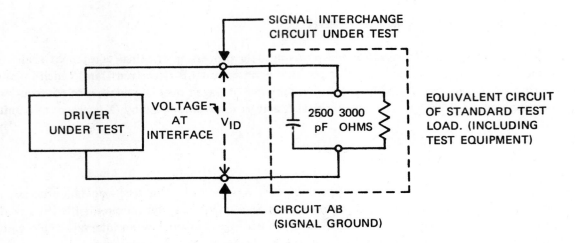

**STANDARD TEST CIRCUIT FOR DRIVER**
**FIGURE 1**

### 6.1.3 Interface Voltage Measurement

The interface voltage measurement shall be made to determine that the driver is producing an output signal $V_{ID}$ of at least $\pm$ 5.0 Volts into the standard test load.

### 6.1.4 The Standard Distortion Measurement Thresholds

The standard distortion measurement shall be made using a +3 Volt and a -3 Volt threshold to determine the occurrence of signal transitions.

**6.1.4.1** A mark-to-space transition shall be taken to occur at the instant $V_{ID}$ crosses +3 Volts on a positive going transition.

**6.1.4.2** A space-to-mark transition shall be taken to occur at the instant $V_{ID}$ crosses -3 Volts on a negative going transition.

### 6.2 Measurement of Interface Terminator Characteristics

Standard measurements of loading and voltage characteristics and of margin on the terminator side of the interface shall be made using the circuit of Figure 2 and shall meet the following requirements:

### 6.2.1 Measurement of Loading Characteristics

The standard loading measurement shall be made to determine that the input resistance of the terminator is not less than 3000 ohms nor more than 7000 ohms and that the input shunt capacitance is not greater than 2500 pF.

### 6.2.2 Voltage Characteristics

The measurement of the voltage response characteristics of the interface terminator circuits shall be made to determine that the terminator will respond correctly to a signal voltage $V_{IT}$ at the interface point of $\pm$ 3 Volts. (It is necessary that the terminator provide some external indication of its response to the received signal.)

### 6.2.3 Margin of Receiving Data Terminal Equipment

The measurement of receiver margin shall be made using a signal $V_{IT}$ of $\pm$ 5.0 Volts when working into the terminator under test. The transitions of the test signal shall be deviated from their ideal instants in such a way as to measure the receiver margin. (See Paragraph 2.18.) The deviations of the transitions in time shall be taken to occur when the signal crosses the $\pm$ 3 Volt thresholds.

> 6.2.3.1 A mark-to-space transition shall be taken to occur at the instant $V_{IT}$ crosses +3 Volts on a positive going transition.

> 6.2.3.2 A space-to-mark transition shall be taken to occur at the instant $V_{IT}$ crosses -3 Volts on a negative going transition.

> 6.2.3.3 It is necessary that the receiving Data Terminal Equipment provide some continuous indication of its response to the distorted test signal to indicate when incorrect translations of the signal occur.

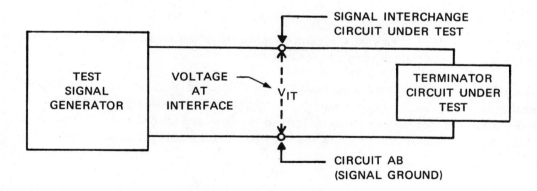

**STANDARD TEST CIRCUIT FOR TERMINATOR**
**FIGURE 2**

## 7. WORK SHEET

The standard format for a work sheet is shown in Table 1, and may be used to prepare a specification for terminals operating within the scope of this standard. Alphabetical references refer to the alphabetical designations used for the various characteristics of signal quality described in Sections 4 and 5.

## 8. EXPLANATORY NOTES

**8.1    Signal Nomenclature.** Figure 3 illustrates a number of definitions of binary digital signals, taking a start-stop character as an example. The figure shows an ideal signal, without distortion, representing the number "9" in the USA Standard Code for Information Interchange, (USASCII) in the standard published as USAS X3.4-1967. In has even parity, a unit start element, and a 2-unit stop element. The code is transmitted with "low order bit" first, followed by the even-parity bit. In the following, the first use of a term defined in Section 2 is followed by its number, in parentheses. Therefore:

   a. The *character interval* (2.8) has a duration of 11 *unit intervals* (2.5) nominal.

   b. The information and parity bits are represented by *signal elements* (2.4).

   c. These have durations of one unit interval each, except that when unit-interval elements ("unit elements") of the same sense are consecutive, they run together to make longer elements having durations of 2, 3 or more unit intervals. A signal element may be of any duration.

   d. The duration of a unit interval (in seconds) is the reciprocal of the *modulation rate* (2.6) the latter being expressed in *bauds* (2.7), and the modulation rate is equal to the number of unit intervals (not "bits", "bauds" or signal elements) per second (as long as the modulation rate is maintained without stopping).

   e. The *start element* (2.9) begins with a mark-to-space transition called the *start transition* (2.10), and its duration is nominally one unit interval.

   f. The *stop element* (2.11) in this example has a nominal duration of 2.0 intervals. It will have this nominal duration during *continuous start-stop operation* (2.3), however, in non-continuous start-stop operation the stop element is considered to be prolonged until the start transition of the following character occurs, however long that may be, as in manual keyboard operation of a teleprinter.

(It is not good practice to refer to this character structure as having "2 stop bits"; the stop element is only one signal element, with a duration of 2 unit intervals, or it is a "2-unit stop element".)

**8.2    Gross Start-Stop Distortion and Synchronous Start-Stop Distortion.** Figure 4 illustrates the relationship between two kinds of "start-stop distortion." Generally *start-stop distortion* (2.15) is measured with reference to the start transition of the character, in terms of the largest per cent of a unit interval by which any transition leads or lags the instant when it should have occurred, if the

754

*degree of distortion* (2.12) were zero. The terms *gross start-stop distortion* (2.16) and *synchronous start-stop distortion* (2.17) point up a useful distinction that is not always appreciated. If all start-stop sending, receiving and distortion measuring equipment in a particular system were controlled by the same clock, only *synchronous* start-stop distortion would be measured. However, if the actual modulation rate of the signal in question differs from the nominal modulation rate (i.e., the rate at which the equipment is intended to work), while the distortion measuring device is controlled by a clock which runs at *exactly* the nominal rate, the measured start-stop distortion will generally be greater because the ideal instants of the signal and of the measuring equipment do not agree. This disagreement increases steadily during each character interval; an otherwise perfectly proportioned signal will show, on the average, the greatest *individual distortion* (2.13), on the last transition in each character. The distortion measured under these conditions is the *gross* start-stop distortion. The existence of the modulation rate discrepancy is readily recognized through the use of measuring equipment which can be made to measure the distortion of any particular transition (e.g., the 5th) within the character. If the rate discrepancy exists, the degree of individual distortion will increase progressively from the earliest transition to the latest transition.

In Figure 4, the distorted start-stop signal to be measured is drawn through the middle of the diagram. After the mark-to-space start transition, there are transitions A, B, C, D, and E, exhibiting various degrees of individual distortion. This could be the character of Figure 3 with distortion. The upper part of the figure illustrates the measurement of gross start-stop distortion when the actual modulation rate of the signal is lower than that used by the measuring equipment. The reference instants numbered from 0 to 10 at the top of the figure are equally spaced at the unit intervals corresponding to the nominal modulation rate, at which the measuring equipment works. Transition A comes 33% of a unit interval after its reference instant; B falls 15% before reference instant No. 2, while E is 50% after reference instant No. 9. Since the latter is the largest value, this determines the *gross* start-stop distortion to be 50%. Actually, in an actual measurement, the values quoted would be the maximum individual distortions for the various transitions during the period of observation.

Now, suppose that the operator slows down the clock signals governing the measuring equipment until he determines that their modulation rate has "synchronized" (hence the name) with the average modulation rate of the signal. This will often be the adjustment which gives the minimum average start-stop distortion during a short period of observation. For the particular character signal illustrated, the new reference instants tend to come closer to the actual transition times than before, so that E is now only 12% late, as shown in the lower part of the figure, and A is only 28% late; which turns out, in the case, to be the largest figure, so that the synchronous start-stop distortion is found to be 28%. It will be noted that C is now 16% early, while D has changed from 5% late to 23% early.

The concept of gross start-stop distortion is useful from the standpoint of routine maintenance of station equipment in switched systems where many transmitting (and receiving) equipment are served over circuits from a given switching or maintenance center. Provided that all of the receivers which might be connected to any of the transmitters meet certain receiving tolerance or margin (2.18) requirements, any receiver should operate correctly if the greatest degree of individual distortion of the signals they receive is not in excess of an established amount, regardless of whether this is due to transmitter misadjustment or modulation rate difference, or both. Measurement of gross start-stop distortion of signals from the transmitting equipment will allow checking that such distortion limits are being met. If a station meets this requirement, it is of no immediate concern whether the measured distortion is due entirely to inaccurate signal formation, or partly or entirely to modulation rate difference. As mentioned above, with some available distortion measuring equipment it is relatively simple to recognize modulation rate error. Going farther, by changing to a synchronous start-stop distortion measurement, it would be possible to analyze the transmitter signal formation to discover which particular transitions may be persistently experiencing the most distortion, as an aid to trouble shooting.

8.3 **Minimum Duration Signal Element.** Figure 4 also illustrates the need, in some cases, to place limits on the minimum duration of signal elements, even though the maximum displacement of transitions may be controlled. The marking element from A to B has a duration of only 52% of a unit interval, (on the upper timing scale). This is due to two simultaneous individual distortions in opposite directions. A comes 33% late while B comes 15% early, on the *same* signal element of this particular character, and thus they consume a total of 48% of the unit interval allotted to it, leaving this element an actual duration of 52%, ignoring any slope of the transition waveforms. Against the lower timing scale, the duration is only 50% of a unit interval.

A similar shortening could occur for a spacing element. With regard to paragraph 5.3, while a valid start element might sometimes be unduly shortened in this way, the main reason for establishing this kind of requirement is to prevent many of the short "hits" or impulse noises on the circuit from starting the receiver through an unnecessary receiving cycle, during which a valid start element might arrive. Such protection could be provided by placing suitable circuitry in the receiver to measure the minimum duration of all signal elements, with provision to inhibit a character receiving cycle from starting on spacing signals shorter than the established minimum duration.

756

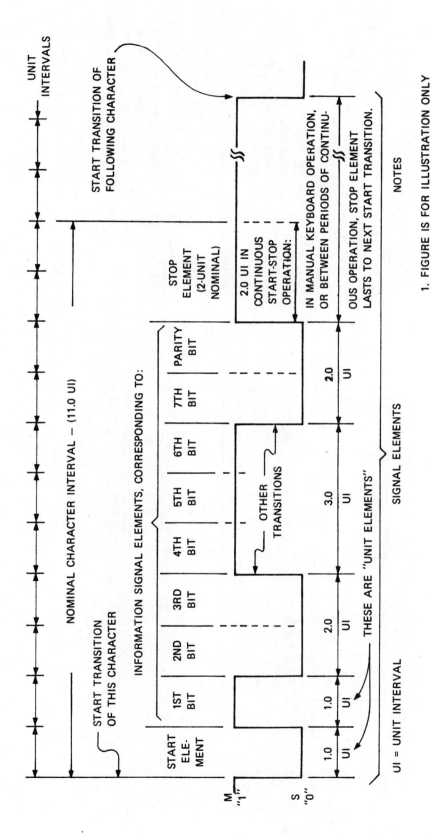

EXAMPLE OF APPLICATION OF TERMS APPLIED
TO CHARACTER IN START-STOP OPERATION

FIGURE 3

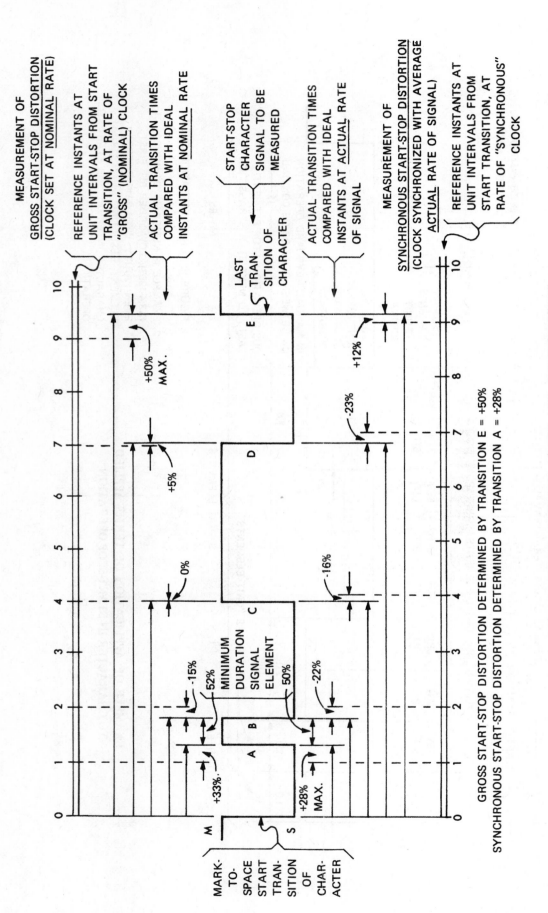

MEASUREMENT OF GROSS START-STOP DISTORTION (CLOCK SET AT NOMINAL RATE)

REFERENCE INSTANTS AT UNIT INTERVALS FROM START TRANSITION, AT RATE OF "GROSS" (NOMINAL) CLOCK

ACTUAL TRANSITION TIMES COMPARED WITH IDEAL INSTANTS AT NOMINAL RATE

START-STOP CHARACTER SIGNAL TO BE MEASURED

ACTUAL TRANSITION TIMES COMPARED WITH IDEAL INSTANTS AT ACTUAL RATE OF SIGNAL

MEASUREMENT OF SYNCHRONOUS START-STOP DISTORTION (CLOCK SYNCHRONIZED WITH AVERAGE ACTUAL RATE OF SIGNAL)

REFERENCE INSTANTS AT UNIT INTERVALS FROM START TRANSITION, AT RATE OF "SYNCHRONOUS" CLOCK

758

MARK-TO-SPACE START TRANSITION OF CHARACTER

LAST TRANSITION OF CHARACTER

MINIMUM DURATION SIGNAL ELEMENT

GROSS START-STOP DISTORTION DETERMINED BY TRANSITION E = +50%
SYNCHRONOUS START-STOP DISTORTION DETERMINED BY TRANSITION A = +28%

RELATIONSHIP BETWEEN GROSS AND SYNCHRONOUS
START-STOP DISTORTION MEASUREMENTS

FIGURE 4

TABLE 1

EIA STANDARD WORK SHEET FOR SPECIFYING SIGNAL QUALITY

| TRANSMITTING TERMINAL | | | | RECEIVING TERMINAL | | | |
|---|---|---|---|---|---|---|---|
| Ref. | Characteristic | Value | Units | Ref. | Characteristic | Value | Units |
| | *Start-Stop Operation* | | | | *Start-Stop Operation* | | |
| 4.1.1 | Synchronous Start-Stop Distortion | C___ | % | 5.1.1 | Net Margin | Q___ | % |
| 4.1.1 | Gross Start-Stop Distortion | D___ | % | 5.1.1 | Minimum Signal Element | R___ | % Unit Int |
| 4.1.1 | Minimum Signal Element | E___ | % Unit Int | 5.2 | Character Interval Requirements | | |
| 4.2 | Character Interval Requirements | | | 5.2.1 | Average: Nominal Reduced by | X___ | % Unit Int |
| 4.2.1 | Average: Nominal Reduced by | L___ | % Unit Int | 5.2.1 | Averaged Over | K___ | Characters |
| 4.2.1 | Averaged Over | K___ | Characters | 5.2.2 | Minimum: Nominal Reduced by | Y___ | % Unit Int |
| 4.2.2 | Minimum: Nominal Reduced by | M___ | % Unit Int | 5.3 | Minimum Duration Start Element | Z___ | % Unit Int |
| | *Synchronous Operation* | | | | *Synchronous Operation* | | |
| | Modulation Rate | | | | Modulation Rate | | |
| 4.1.2 | Maximum Deviation | F___ | % | 5.1.2 | Maximum Deviation | S___ | % |
| 4.1.2 | Maximum Rate of Change $\Big\}$ % Mod. Rate / Time Unit | G__/H__ | % /Time | 5.1.2 | Maximum Rate of Change $\Big\}$ % Mod. Rate / Time Unit | T__/U__ | % /Time |
| 4.1.2 | Isochronous Distortion | J___ | % | 5.1.2 | Net Margin | V___ | % |
| | | | | 5.1.2 | Minimum Duration Signal Element | W___ | % Unit Int |
| | *Characteristics of Signals Generated From Contacts* | | | | *System Characteristics* | | |
| 4.3 | Continuity of Signal Breaks Not To Exceed | N___ | % Unit Int | 3.1 | Modulation Rate | A___ | Bauds |
| 4.4 | Contact Bounce Duration Not To Exceed | P___ | % Unit Int | 3.2 | Start-Stop Character Interval | B___ | Unit Int |

REMARKS: _____

**APPENDIX**

## NUMERICAL VALUES NOT STANDARDIZED

Numerical values for signal quality of typical equipments vary over a large range. To establish a standard specifying actual values for signal quality would require numerous classifications or categories of equipment in order to satisfy the many kinds of equipment and applications that exist today or are anticipated for the future. To illustrate this point, a range of typical performance values are shown in Tables 2 and 3. Here four groups of performance characteristics are shown within this range. While these may be typical of some existing kinds of equipment they are not to be construed as a standard nor does the range necessarily show the actual limit that may be found in practice. It is included to emphasize the conclusion that no useful function could be served by a numerical standard because of the many different sets of numbers required at this time.

For purposes of this illustration start-stop terminals operating at about 150 bauds or less are shown. Such terminals can be roughly divided into two groups. Those generating or detecting signals electronically, and those accomplishing the same function electromechanically. Each of these groups may be further divided depending on the economics of the system involved. Electronic systems may use highly stable crystal oscillators or tuning forks for unit element timers, or less expensive L-C oscillators or R-C relaxation circuits. Electromechanical systems will have characteristics depending upon their design and the precision of parts used. In Table 2, four kinds of transmitting equipment are shown, two electronic and two electromechanical. Transmitters in column I use crystal controlled timers, in column II, L-C oscillators are used for timers. Columns III and IV are electromechanical transmitting terminals with the equipment in column III constructed more precisely than that in column IV. Table 3 shows receive equipment with divisions much like that shown for transmitters; thus column A terminals have crystal controlled selectors, column B uses L-C oscillators, column C is an electromechanical selector, while column D is a less precise electromechanical selector.

The Tables have been shown separate to illustrate the fact that receivers and transmitters of the same type need not be used in a particular system. In fact in systems such as data collection systems where a single receiver may collect data from many transmitters it is likely the most economical solution would be to use transmitters typified by column IV and the receivers typified by column A. By contrast, a broadcast system may be most economically built up by using a single precise transmitter such as in Table 2, column I sending to many receivers having less precise margins such as those in Table 3, column D.

Characteristics shown in the tables are typical of a variety of commercial gear. The reasoning underlying the performance characteristics will now be described. A terminal generating a signal from a crystal oscillator may have a typical unit timer stability of 0.05%. Frequency dividers and associated circuitry add some small amount of jitter to the resultant timing signal. This jitter may amount to as much as 1.0% of a unit element under worst conditions. This is the source of the synchronous start-stop distortion shown in column I. In addition to this, the gross start-stop distortion also depends on the frequency variation of the timing signal. Since this is 0.05%, a character having a length of 10 unit elements would suffer 10 x 0.05% or 0.5% additional distortion due to possible timing error. This is the basis for showing a degree of gross start-stop distortion of 1.5% in column I. Signals controlled by electronic timing generators which are less stable than crystals or tuning forks show more jitter which results in the 5% figure shown in column II. In addition to this, the degree of gross start-stop distortion will be greater than the previous class because of increased frequency drift of the less stable unit interval timer. A timer stability of 0.3% in this illustration produces a degree of gross start-stop distortion of 8%.

The electromechanical generation of signals results in degrees of synchronous start-stop distortion greater than in those equipments above. The principal source of distortion in this equipment is the play in gears and linkages and the hunting of the motor which generates the shaft motion used for signal timing. When the motor is synchronous the timing will have a frequency stability proportional to the stability of the power line frequency. While the power line usually is highly stable with respect to frequency there are those cases where power is generated by small independent or emergency systems displaying considerable frequency variation.

An examination of electromechanical equipment shows a range of performance depending upon the signal generation methods and the degree of precision with which the units are constructed. A typical transmitter shown in column III may have a degree of synchronous start-stop distortion of 8% from the causes mentioned above. Power line frequency variation experienced is generally no worse than 0.75%. In a 10 unit character, this could cause a displacement of the last transition in the character of as much as 7.5% which may be rounded off to 8%, thus, the degree of gross start-stop distortion would be 8% jitter plus 8% drift for a total of 16%.

Where less precise construction is allowed to introduce greater manufacturing economy, the degree of synchronous start-stop distortion may go as high as 12% as shown in column IV. The same consideration on frequency stability would apply to the degree of gross start-stop distortion as for column III. Thus 8% additional for timing error results in 20% as shown in column IV.

The receiver in column A Table 3 has the same kind of timing generator as the transmitter shown in Table 2, column I. The characteristics of the timer responsible for the 1.5% gross start-stop distortion of the transmitter will degrade the perfect margin of the receiver by the same amount. Thus the receiver's margin is 50% less 1.5% or 48.5%. Since sampling may take place at any instant within the ±1.5% range, the minimum signal element that can be detected reliably under such a margin condition is 3% (plus and minus 1.5% from the ideal center of an arriving signal element). However, many receiving systems involving electronic detection of signals will test to ascertain that a signal element is at least 50% of a unit interval in duration before it will act upon that element. Thus two values are shown for minimum signal element. The start element in electronic equipment is usually tested to make sure it is at least 50% of a unit interval in duration before the character receiving cycle is allowed to proceed.

Following similar reasoning for the receiving equipment in column B, the receiving margin is 50% less 8% or 42%. Likewise a minimum signal element of 2 x 8 = 16% would be required. Testing for the minimum signal element will lengthen it to 50% as before.

Electromechanical receiving equipment generally display lower receiving margins because of motor speed variations and tolerances in mechanical selectors. Typical margins are shown in columns C and D, however these are "worse case" conditions and the same equipment operating at the lower speeds may show better performance characteristics. The minimum signal element is determined by the armature operate time of the selector magnet and numbers shown are typical. At lower speeds this number would be smaller. In mechanical selectors the minimum start element and the minimum signal element are usually the same since most equipments do not perform a separate test on the start element duration as do the electronic receiving equipment and this characteristic is again a function of armature operate time.

This appendix has discussed only a small segment of the data transmission field. High speed transmission and synchronous data transmission have not been included and would have resulted in many additional categories tabulated. The numbers that have been listed are illustrative only and must not be construed as a standard or even as guide lines for system performance.

Because of the many categories or groups of equipment that would need to be established and standardized following logical divisions, it becomes apparent that such would be only slightly more useful than no standard at all. Further, since no standard can be established for the communications channel at this time, there seems little usefulness in having numerical standards for terminals which work through unstandardized channels.

A standard for specifying signal quality thus has been created which will be useful to all parties furnishing and maintaining communications systems, and upon which a basis of agreement can be established which is meaningful to all concerned.

762

| GROUP | I | II | III* | IV* |
| CHARACTERISTIC | Crystal | Electronic | Electro-mechanical | Electro-mechanical |
| --- | --- | --- | --- | --- |
| Degree of Synchronous Start-Stop Distortion, Percent | 1.0 | 5 | 8 | 12 |
| Gross Start-Stop Distortion, Percent | 1.5 | 8 | 16 | 20 |

**TABLE 2 – TRANSMITTING TERMINAL CHARACTERISTICS**

| GROUP | A | B | C* | D* |
| CHARACTERISTIC | Crystal | Electronic | Electro-mechanical | Electro-mechanical |
| --- | --- | --- | --- | --- |
| Margin, Percent | 48.5 | 42 | 35 | 25 |
| Minimum Signal Element – Percent Unit Interval | 3 (>50) | 16 (>50) | 30 | 30 |
| Minimum Start Element – Percent Unit Interval | >50 | >50 | 30 | 30 |

**TABLE 3 – RECEIVING TERMINAL CHARACTERISTICS**

*(Both Tables) These characteristics (*) are typical of equipment operating at rates of 100 bauds and below.

# EIA STANDARD

*Interface Between Data
Terminal Equipment and
Automatic Calling Equipment
for Data Communication*

RS-366

*August 1969*

*Engineering Department*

# ELECTRONIC INDUSTRIES ASSOCIATION

## NOTICE

EIA engineering standards are designed to serve the public interest through eliminating misunderstandings between manufacturers and purchasers, facilitating interchangeability and improvement of products, and assisting the purchaser in selecting and obtaining with minimum delay the proper product for his particular need. Existence of such standards shall not in any respect preclude any member or non-member of EIA from manufacturing or selling products not conforming to such standards, nor shall the existence of such standards preclude their voluntary use by those other than EIA members whether the standard is to be used either domestically or internationally.

Recommended standards are adopted by EIA without regard to whether or not their adoption may involve patents on articles, materials, or processes. By such action, EIA does not assume any liability to any patent owner, nor does it assume any obligation whatever to parties adopting the recommended standards.

Published by

# ELECTRONIC INDUSTRIES ASSOCIATION

### Engineering Department

### 2001 Eye Street, N.W., Washington, D.C. 20006

---

**PRICE: $4.50**

# INTERFACE BETWEEN DATA TERMINAL EQUIPMENT AND AUTOMATIC CALLING EQUIPMENT FOR DATA COMMUNICATION

*(From Standards Proposal No. 1011, formulated under the cognizance of EIA TR-30.2 Subcommittee on Interface.)*

## INDEX

766

## SECTION ONE

## 1. SCOPE

1.1     This standard is applicable to the interconnection of data terminal equipment (DTE) and automatic calling equipment (ACE) for data communication.

It defines:

Section 2 — Electrical Signal Characteristics

Electrical characteristics of the interchange signals and associated circuitry.

Section 3 — Interface Mechanical Characteristics

Definition of the mechanical characteristics of the interface between the data terminal equipment and the automatic calling equipment.

Section 4 — Functional descriptions of Interchange Circuits

Functional descriptions of a set of digit and control interchange circuits for use at a digital interface between data terminal equipment and automatic calling equipment.

Section 5 — Standard Interfaces

Standard subsets of specific interchange circuits are defined for the classes of automatic calling equipment defined in Section 1.3.

In addition, the standard includes:

Section 6 — Recommendations and Explanatory Notes

1.2     This standard is applicable for automatic calling equipment used in conjunction with electronic data terminal equipment.

1.3     This standard applies to all classes of automatic calling equipment for data communication, including the following:

1.3.1     Automatic calling where the number(s) to be called are stored in the automatic calling equipment,

1.3.2     Automatic calling where the numbers to be called are passed from the data terminal equipment to the automatic calling equipment,

1.3.3     Automatic calling where the automatic calling equipment may be used both for call origination and for the transmission of data to another data set, and

1.3.4   Automatic calling where one automatic calling equipment is used on one or more communication channels. (See Section 5.9).

1.4   This standard is applicable for the interchange of digit and control signals when used in conjunction with electronic equipments, each of which has a single common return (signal ground) that can be interconnected at the interface point. It does not apply where electrical isolation between equipment on opposite sides of the interface point is required.

1.5   This standard is independent of the technique used by the automatic calling unit to convey information over the communication channel and applies equally to impulse dialing and multi-tone keying.

1.6   Provisions are made for two methods of operation following the transmission of the last digit of the called number. Either or both methods may be provided in a particular automatic calling equipment. (See Section 6.5).

1.6.1   The automatic calling equipment waits for an indication (e.g., a remote data set answer signal) that a connection has been established to a remote data set, and then transfers control of the communication channel to the associated data set. The answer signal detection capability may be provided in the automatic calling unit *or* in the associated data set.

1.6.2   An EON (End Of Number) code combination presented by the data terminal equipment on the digit interchange circuits cause the automatic calling equipment to immediately transfer the communication channel to the associated data set.

1.7   This standard is independent of data rate, transmission mode (synchronous or non-synchronous), modulation technique, error control techniques and line control procedures used in the associated system.

SECTION TWO

2. ELECTRICAL SIGNAL CHARACTERISTICS*

2.1   Figure 2.1, Interchange Equivalent Circuit, shows the electrical parameters which are specified in the subsequent paragraphs of this section. The equivalent circuit shown in Figure 2.1 is applicable to all interchange circuits regardless of the category (digit or control) to which they belong. The equivalent circuit is independent of whether the driver is located in the automatic calling equipment and the terminator in the data terminal equipment or vice versa.

---

* The electrical signal characteristics included here conform to EIA Standard RS-232-C.

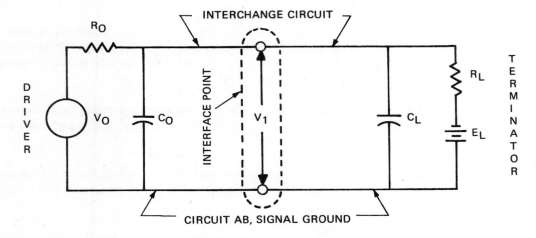

**Figure 2.1 — Interchange Equivalent Circuit**

$V_O$ is the open-circuit driver voltage.

$R_O$ is the driver internal de resistance.

$C_O$ is the total effective capacitance associated with the driver, measured at the interface point and including any cable to the interface point.

$V_1$ is the voltage at the interface point.

$C_L$ is the total effective capacitance associated with the terminator, measured at the interface point and including any cable to the interface point.

$R_L$ is the terminator load de resistance.

$E_L$ is the open-circuit terminator voltage (bias).


2.2     The driver on an interchange circuit shall be designed to withstand an open-circuit, a short-circuit between the conductor carrying that interchange circuit in the inter-connecting cable and any other conductor in that cable, or any passive non-inductive load connected between that interchange circuit and any other interchange circuit, including Circuit AB (Signal Ground), without sustaining damage to itself or its associated equipment. The terminator on an interchange circuit shall be designed to withstand any input signal within the 25 volt limit specified in Section 2.6. (See Section 6.4).

2.3     For the digit signal circuits (NB1, NB2, NB4, and NB8), the signal shall be considered in the binary ONE condition when the voltage ($V_1$) on the interchange circuit measured at the interface point is more negative than minus three volts with respect to Circuit AB (Signal Ground). The signal shall be considered in the binary ZERO condition when the voltage $(V_1)$ is more positive than plus three volts with respect to Circuit AB. (See Section 6.1).

The region between plus three volts and minus three volts is defined as the transition region. The signal state is not uniquely defined when the voltage $(V_1)$ is in this transition region.

For control interchange circuits, the control function shall be considered ON when the voltage $(V_1)$ on the interchange circuit is more positive than plus three volts with respect to circuit AB, and shall be considered OFF when the voltage $(V_1)$ is more negative than minus 3 volts with respect to Circuit AB. The control function is not uniquely defined for voltages in the transition region between plus 3 volts and minus 3 volts.

| Notation | Negative Voltage | Positive Voltage |
|---|---|---|
| Binary State | 1 | 0 |
| Control Function | OFF | ON |

This specification neither implies nor precludes the use of terminator circuits which utilize hysteresis techniques to enhance their noise immunity; however, the requirements of Section 2.5 must also be satisfied.

2.4     The load impedance ($R_L$ and $C_L$) of the terminator side of an interchange circuit shall have a dc resistance ($R_L$) of not less than 3000 ohms, measured with an applied voltage not greater than 25 volts in magnitude, nor more than 7000 ohms, measured with an applied voltage of 3 to 25 volts in magnitude. The effective shunt capacitance ($C_L$) of the terminator side of an interchange circuit, measured at the interface point and including any cable to the interface point, shall not exceed 2500 picofarads. The reactive component of the load impedance shall not be inductive. The open-circuit terminator voltage ($E_L$) shall not exceed 2 volts in magnitude. (See Sections 6.2, 6.3, and 6.4).

2.5     The following interchange circuits, where implemented, shall be used to detect either a power-off condition in the equipment connected across the interface or the disconnection of the inter-connecting cable:

Circuit CRQ (Call Request)

Circuit PWI (Power Indication)

The power-off source impedance of the driver side of these circuits shall not be less than 300 ohms, measured with an applied voltage not greater than 2 volts in magnitude referenced to Circuit AB (Signal Ground). The terminator for these circuits shall interpret the power-off condition or the disconnection of the interconnecting cable as an OFF condition.

2.6     The open-circuit driver voltage ($V_O$) with respect to Circuit AB (Signal Ground) on any interchange circuit shall not exceed 25 volts in magnitude. The source impedance ($R_O$ and $C_O$) of the driver side of an interchange circuit including any cable to the interface point is not specified;

771

however, the combination of $V_O$ and $R_O$ shall be selected such that a short-circuit between any two conductors (including ground) in the interconnecting cable shall not result in a current in excess of one-half ampere. Additionally, the driver design shall be such that, when the terminator load resistance ($R_L$) is in the range between 3000 ohms and 7000 ohms and the terminator open-circuit voltage ($E_L$) is zero, the voltage ($V_1$) at the interface point shall not be less than 5 volts nor more than 15 volts in magnitude. (See Section 6.3).

2.7    The characteristics of the interchange signals transmitted across the interface point, exclusive of external interferences, shall conform to the limitations specified in this section. These limitations shall be satisfied at the interface point when the interchange circuit is terminated with any receiving circuit which meets the requirements given in Section 2.4. These limitations apply to all (Digit and control) interchange signals unless otherwise specified.

(1)    All interchange signals entering into the transition region shall proceed through the transition region to the opposite signal state and shall not re-enter the transition region until the next significant change of signal condition.

(2)    There shall be no reversal of the direction of voltage change while the signal is in the transition region.

(3)    For Control Interchange Circuits, the time required for the signal to pass through the transition region during a change in state shall not exceed one millisecond.

(4)    The maximum instantaneous rate of voltage change shall not exceed 30 volts per microsecond.

(5)    The operation of terminators shall be dependent only on the signal voltage, as specified in Section 2.3, and should, therefore, be insensitive to the rise time, fall time, presence of signal overshoot, etc.

## SECTION THREE

## 3. INTERFACE MECHANICAL CHARACTERISTICS

3.1    The interface between the data terminal equipment and automatic calling equipment is located at a pluggable connector signal interface point between the two equipments. The female connector shall be associated with, but not necessarily physically attached to the automatic calling equipment and should be mounted in a fixed position near the data terminal equipment. The use of an extension cable on the automatic calling equipment is permitted. An extension cable with a male connector shall be provided with the data terminal equipment. The use of short cables (each less than approximately 50 feet or 15 meters) is recommended; however, longer cables are permissible, provided that the resulting load capacitance ($C_L$ of Fig. 2.1), measured at the interface point and including the signal terminator, does not exceed 2500 picofarads. (See Sections 2.4 and 6.3).

3.2    Pin Identification

3.2.1    Pin assignments listed in Figure 3.1 shall be used.

3.2.2    Pin assignments for circuits not specifically defined in Section 4 are to be made by mutual agreement. Preference should be given to the use of unassigned pins, but in the event that additional pins are required, extreme caution should be taken in their selection. (See Section 4.1.1).

## FIGURE 3.1

### INTERFACE CONNECTOR PIN ASSIGNMENTS

| Pin No. | Circuit | Description |
|---------|---------|-------------|
| 1 | AA | Protective Ground |
| 2 | DPR | Digit Present |
| 3 | ACR | Abandon Call and Retry |
| 4 | CRQ | Call Request |
| 5 | PND | Present Next Digit |
| 6 | PWI | Power Indication |
| 7 | AB | Signal Ground |
| 8 | – – | Unassigned |
| 9 | – – | Reserved for automatic calling equipment testing.  These two pins shall not be wired in the data terminal equipment. |
| 10 | – – | |
| 11 | – – | Unassigned |
| 12 | – – | Unassigned |
| 13 | COS | Call Origination Status |
| 14 | NB1 | |
| 15 | NB2 | Digit Signal Circuits |
| 16 | NB4 | |
| 17 | NB8 | |
| 18 | – – | Unassigned |
| 19 | – – | Unassigned |
| 20 | – – | Unassigned |
| 21 | – – | Unassigned |

773

**FIGURE 3.1 (Continued)**

| Pin No. | Circuit | Description |
|---|---|---|
| 22 | DLO | Data Line Occupied |
| 23 | – – | Unassigned |
| 24 | – – | Unassigned |
| 25 | – – | Unassigned |

## SECTION FOUR

## 4. FUNCTIONAL DESCRIPTION OF INTERCHANGE CIRCUITS

### 4.1 General

This section defines the basic interchange circuits which apply, collectively, to all systems.

4.1.1 Additional interchange circuits not defined herein, or variations in the functions of the defined interchange circuits may be provided by mutual agreement. (See Sections 3.2.2 and 5.2).

### 4.2 Categories

Interchange circuits between data terminal equipment and automatic calling equipment fall into three general categories.

Ground or Common Return

Digit Circuits

Control Circuits

4.2.1 A list of circuits showing category as well as equivalent C.C.I.T.T. identification in accordance with C.C.I.T.T. Recommendation V.24 as amended at the IV Plenary Assembly (Mar del Plata, Argentina, Oct. 1968) is shown in Figure 4.1.

### 4.3 Interchange Circuits

**Circuit AA** – Protective Ground                     (CCITT 212)
Direction: Not Applicable

This conductor shall be electrically bonded to the machine or equipment frame. It may be further connected to external grounds as required by applicable regulations.

| Inter-change Circuit | C.C.I.T.T. Equivalent | Description | Gnd. | Digit. to ACE | Control | |
|---|---|---|---|---|---|---|
| | | | | | To ACE | From ACE |
| AA | 212 | Protective Ground | X | | | |
| AB | 201 | Signal Ground | X | | | |
| CRQ | 202 | Call Request | | | X | |
| PWI | 213 | Power Indication | | | | X |
| DLO | 203 | Data Line Occupied | | | | X |
| COS | 204 | Call Origination Status | | | | X |
| ACR | 205 | Abandon Call and Retry | | | | X |
| PND | 210 | Present Next Digit | | | | X |
| DPR | 211 | Digit Present | | | X | |
| NB1 | 206 | Low Order Binary Digit | | X | | |
| NB2 | 207 | Second Order Binary Digit | | X | | |
| NB4 | 208 | Third Order Binary Digit | | X | | |
| NB8 | 209 | High Order Binary Digit | | X | | |

**FIGURE 4.1**

**INTERCHANGE CIRCUITS BY CATEGORY**

*Circuit AB* — Signal Ground or Common Return                    (CCITT 201)
Direction:  Not Applicable

This conductor establishes the common ground reference potential for all interchange circuits except Circuit AA (Protective Ground). Within the automatic calling equipment, this circuit shall be brought to one point, and it shall be possible to connect this point to Circuit AA by means of a wire strap inside the equipment. This wire strap can be connected or removed during Installation, as may be required to meet applicable regulations or to minimize the introduction of noise into electronic circuitry.

*Circuit CRQ* — Call Request                    (CCITT 202)
Direction:  TO automatic calling equipment

Signals on this circuit are generated by the data terminal equipment to request the automatic calling equipment to originate a call.

The ON condition indicates a request to originate a call and must be maintained during call origination, until Circuit COS (Call Origination Status) is turned ON, in order to hold the connection to the communication channel (remain "OFF HOOK"). The call is aborted if Circuit CRQ is turned OFF prior to turning ON Circuit COS.

The OFF condition indicates that the data terminal equipment is not using or has completed a prior use of the automatic calling equipment. To avoid a potential race condition, it is recommended that, if Circuit CRQ is turned OFF prior to Circuit COS coming ON with the intent of aborting a call attempt, Circuit CD (Data Terminal Ready) (See RS-232-C) in the interface of the associated data set should also be turned OFF.

After the automatic calling equipment has turned ON Circuit COS, the data terminal equipment may turn Circuit CRQ OFF without causing a disconnect.

Circuit CRQ must be turned OFF between calls or call attempts and shall not be turned ON unless Circuit DLO (Data Line Occupied) is in the OFF condition.

*Circuit PWI* – Power Indication                                                 (CCITT 213)
Direction:  FROM automatic calling equipment

Signals on this circuit are generated by the automatic calling equipment to indicate whether power is available within the automatic calling equipment.

The ON condition indicates that power is available in the automatic calling equipment.

An inoperative condition resulting from the loss of power should be detected in accordance with the provisions of Section 2.5.

This circuit should not be interpreted to indicate the power status in any other equipment.

*Circuit DLO* – Data Line Occupied                                               (CCITT 203)
Direction:  FROM automatic calling equipment

Signals on this circuit are used to indicate when the communication channel is in use for automatic calling, data communication, voice communication or for testing of the automatic calling or data communication equipment.

The ON condition indicates that the communication channel is in use.

The OFF condition indicates that the data terminal equipment may originate a call provided that Circuit PWI (Power Indication) is ON.

The OFF condition of Circuit DLO shall not be presented until all of the other interchange circuits from the automatic calling equipment are returned to their proper idle condition.

776

*Circuit COS* – Call Origination Status    (CCITT 204)*
Direction:  FROM automatic calling equipment

> *NOTE:* This circuit was called Circuit DSS (Data Set Status) in earlier versions of this document.

Signals on this circuit are generated by the automatic calling equipment to indicate the status of automatic call origination procedures.

The ON condition presented during a call originated by the automatic calling equipment indicates that the automatic calling equipment has completed its call origination functions and that the control of the communication channel has been transferred from Circuit CRQ (Call Request) to Circuit CD (Data Terminal Ready) in the data set interface (See RS-232-C). When Circuit COS is turned ON, the data terminal equipment may turn Circuit CRQ OFF without causing a communication channel disconnect. Disconnection of the channel by the data terminal equipment is then possible only through the associated data set interface.

Once Circuit COS is turned ON, it shall remain ON at least until Circuit CRQ is turned OFF by the data terminal equipment. Circuit COS may come ON at other times, e.g., during an incoming call or a manually originated call, but any ON condition appearing at a time other than during automatic call origination by the automatic calling equipment should be disregarded.

This circuit should not be interpreted to convey information regarding the operational status or state of preparedness of the associated data set.

*Circuit ACR* – Abandon Call and Retry    (CCITT 205)
Direction:  FROM automatic calling equipment

Signals on this circuit are used to indicate the probability of successful completion of the call attempt.

The ON condition, when presented during the process of call origination, indicates that there is a high probability that the connection to a remote data station cannot be successfully established and is a suggestion to the data terminal equipment to abandon the call and to re-initiate the call at a later time. The automatic calling equipment does not determine that the call is to be abandoned. Action required to abandon the call must be initiated by the data terminal equipment.

The OFF condition indicates that there is no reason to believe that the call cannot be successfully completed.

When the answer signal mode of operation indicated in Section 1.6.1 is used, Circuit ACR remains in the OFF condition after Circuit COS (Call Origination Status) is turned ON. When the End of Number mode is used as in Section 1.6.2, Circuit ACR continues to function after Circuit COS is turned ON.

---

\* CCITT Circuit 204 (Distant Station Connected) is defined differently but is used in a similar manner.

777

*Circuit NB1* – Digit Signal Circuit – Low Order Bit                    (CCITT 206)

*Circuit NB2* – Digit Signal Circuit – Second Order Bit                (CCITT 207)

*Circuit NB4* – Digit Signal Circuit – Third Order Bit                 (CCITT 208)

*Circuit NB8* – Digit Signal Circuit – High Order Bit                  (CCITT 209)

Direction: TO automatic calling equipment

Parallel binary signals on these circuits are generated by the data terminal equipment.

The information presented on these interchange circuits may either be transmitted (e.g., digits of the called number) or used locally as a control signal. An important use of these interchange circuits for control purposes is the passing of the EON (end of number) code combination to the automatic calling equipment after the last digit of the number to be called has been passed. In response to EON, the automatic calling equipment transfers the communication channel to the data set immediately without waiting for an answer signal from the called data set. (See Section 1.6.2).

Figure 4.2 defines the character set provided by the sixteen code combinations available.

778

*Circuit PND* – Present Next Digit                                     (CCITT 210)
Direction: FROM automatic calling equipment

Signals on this circuit are generated by the automatic calling equipment to control the presentation of digits on the Digit Signal Circuits.

The ON condition indicates that the automatic calling equipment is ready to accept the next digit indicated on Circuits NB1, NB2, NB4 and NB8 (Digit Signal Circuits).

The OFF condition indicates that the data terminal equipment should turn OFF Circuit DPR (Digit Present) and set the states of the Digit Signal Circuits for the next digit. Circuit PND (Present Next Digit) shall not be changed to the ON condition while Circuit DPR is ON.

Circuit PND may come ON after the data terminal equipment turns Circuit DPR OFF following the presentation of the last code combination on the Digit Signal Circuits.

*Circuit DPR* – Digit Present                                         (CCITT 211)
Direction: TO automatic calling equipment

Signals on this circuit are generated by the data terminal equipment to indicate that the automatic calling equipment may read the code combination presented on the Digit Signal Circuits NB1, NB2, NB4, NB8.

| Digit | Digit Signal Circuit States | | | |
|-------|-----|-----|-----|-----|
|       | NB8 | NB4 | NB2 | NB1 |
| 0 | 0 | 0 | 0 | 0 |
| 1 | 0 | 0 | 0 | 1 |
| 2 | 0 | 0 | 1 | 0 |
| 3 | 0 | 0 | 1 | 1 |
| 4 | 0 | 1 | 0 | 0 |
| 5 | 0 | 1 | 0 | 1 |
| 6 | 0 | 1 | 1 | 0 |
| 7 | 0 | 1 | 1 | 1 |
| 8 | 1 | 0 | 0 | 0 |
| 9 | 1 | 0 | 0 | 1 |
| * | 1 | 0 | 1 | 0 |
| # | 1 | 0 | 1 | 1 |
| EON | 1 | 1 | 0 | 0 |
| Unassigned | 1 | 1 | 0 | 1 (Note 1) |
| Unassigned | 1 | 1 | 1 | 0 |
| Unassigned | 1 | 1 | 1 | 1 |

*Note 1:* Used as the separation control character (SEP) in C.C.I.T.T.
Recommendation V.24.

**FIGURE 4.2**

**DIGIT SIGNAL CHARACTER SET**

779

The OFF to ON transition indicates that the data terminal equipment has set the states of the Digit Signal Circuits for the next digit.

Circuit DPR (Digit Present) must not be turned ON before Circuit PND (Present Next Digit) comes ON. When turned ON, Circuit DPR must remain ON until Circuit PND goes OFF. Circuit DPR may then be turned OFF, and when turned OFF, must be held OFF until Circuit PND comes on again.

*The states of the Digit Signal Circuits must not change when Circuit DPR is in the ON condition.*

After the automatic calling unit has accepted the last digit of the called number (including EON when used) and has turned Circuit PND OFF, Circuit DPR must be turned OFF and held in the OFF condition even though Circuit PND may come ON again.

## SECTION FIVE

## 5. STANDARD INTERFACES

5.1    Section 1.3 of this standard defines four classes of automatic calling equipment for data transmission. For each of these classes, a standard set of interchange circuits (defined in Section 4) is listed.

5.2    Drivers shall be provided for every interchange circuit included in a standard interface. Terminators need not be provided for every interchange circuit included in a standard interface; however, the designer of the equipment which does not provide all of the specified terminators must be aware that any degradation in service due to his disregard of a standard interchange circuit is his responsibility.

In the interest of minimizing the number of different types of equipment, additional interchange circuits may be included in the design of a general unit capable of satisfying the requirements of several different applications.

5.2.1    For a given class of automatic calling equipment interchange circuits which are included in the standard list and for which drivers are provided, but which the manufacturer of equipment at the receiving side of the interface chooses not to use, shall be suitably terminated by means of a dummy load impedance in the equipment which normally provides the terminator. See Section 2.4.

5.2.2    Where interchange circuits which are not included on the standard list for a given configuration are provided, the equipment providing these circuits should not cause any degradation in the basic service due to the lack of a termination for these circuits in the connecting equipment.

Interference due to unterminated drivers in this category is the responsibility of the designer who includes these drivers.

Terminators in this category shall not interfere with or degrade system performance as a result of open circuited input terminals.

5.3     The use of Circuit AA (Protective Ground) is optional. Where it is used, attention is called to the Underwriters' regulations applying to wire size and color coding. Where it is not used, other provisions for grounding equipment frames should be made in accordance with good engineering practice.

5.4     The use of Circuit AB (Signal Ground) is mandatory in *all* systems. See Section 1.4.

5.5     Interface, Type I

5.5.1     This section defines an interface for the class of automatic calling equipment in which one or more numbers to be called are stored in the automatic calling equipment. Where multiple numbers are stored, they are selected in sequence, not under the control of the data terminal equipment, and each remote station is polled in turn. (See Section 1.3.1).

5.5.2     Basic Interchange Circuits Required

Circuit AB     — Signal Ground
Circuit CRQ    — Call Request
Circuit PWI    — Power Indication
Circuit DLO    — Data Line Occupied
Circuit COS    — Call Origination Status
Circuit ACR    — Abandon Call and Retry

5.6     Interface Type II

5.6.1     This section defines an interface for the class of automatic calling equipment in which multiple numbers are stored in the automatic calling equipment and can be selected under the control of the data terminal equipment using a single digit code to identify the set of digits which comprise the number to be called.

5.6.2     Basic Interchange Circuits Required:

Circuit AB     — Signal Ground
Circuit CRQ    — Call Request
Circuit PWI    — Power Indication
Circuit DLO    — Data Line Occupied
Circuit COS    — Call Origination Status
Circuit ACR    — Abandon Call and Retry
Circuits NB1, NB2, NB4, & NB8     — Digit Signal Circuits
Circuit PND    — Present Next Digit
Circuit DPR    — Digit Present

5.7     Interface Type III

5.7.1   This section defines an interface for the class of automatic calling equipment in which the numbers to be called are stored in the data terminal equipment and are passed from the data terminal equipment to the automatic calling equipment over the interface.

5.7.2   Basic Interchange Circuits Required:

Circuit AB        — Signal Ground
Circuit CRQ       — Call Request
Circuit PWI       — Power Indication
Circuit DLO       — Data Line Occupied
Circuit COS       — Call Origination Status
Circuit ACR       — Abandon Call and Retry
Circuit NB1, NB2, NB4 & NB8   — Digit Signal Circuits
Circuit PND       — Present Next Digit
Circuit DPR       — Digit Present

5.8     Interface, Type IV

5.8.1   Where the functions of automatic call origination and data communication are combined into one unit with two separate interface cables, the automatic calling interface will be either a type I or a type II configuration, as required. The data communication interface will require a separate interconnecting cable in accordance with appropriate data terminal equipment/data communication equipment interface standards, e.g., EIA RS-232-C. (See Section 1.3.3).

5.9     Interface, Type V

5.9.1   Multi-line automatic calling equipment designed to accommodate a multiplicity of communication channels is presently under study. One or more interfaces for this type of equipment will be defined in a later edition of this standard. (See Section 1.3.4).


SECTION SIX

6. RECOMMENDATIONS AND EXPLANATORY NOTES

6.1     The electrical specifications are intended to provide a two-volt margin in rejecting noise introduced either on interchange circuits or by a difference in reference ground potential across the interface. The equipment designer should maintain this margin of safety on all interchange circuitry.

6.2     To avoid inducing voltage surges on interchange circuits, signals from interchange circuits should not be used to drive inductive devices, such as relay coils.

(Note that relay or switch contacts may be used to generate signals on an interchange circuit, with appropriate measures to assure that signals so generated comply with Section 2.7).

6.3    Alphabetical parenthetical designations are added to the terms used in Sections 2.3, 2.4, and 2.6 to better tie them in with the equivalent circuit of Section 2.1 and stress the point that the 2500 pF capacitance ($C_L$) is defined for the receiving end of the interchange circuit and that the capacitance ($C_o$) at the driving end of the interchange circuit, including cable, is not defined. It is the responsibility of the designer to build a circuit capable of driving all of the capacitance in the driver circuitry plus the capacitance in his part of the interconnecting cable (not specified) plus 2500 pF in the load (including the cable on the load side of the interface point).

6.4    The user is reminded that the characteristics of an equivalent load (terminator) circuit used to test for compliance with each of the electrical specifications in section 2 are a function not only of the parameter under test, but also of the tolerance limit to be tested. For example, a driver which just delivers a minimum of 5 Volts into a 7,000 Ohm test load may fail the test if the load is reduced to 3,000 Ohms, whereas, a driver with an output within the 15 Volt limit when driving a 3,000 Ohm load may exceed this limit when driving a 7,000 Ohm load. The 5 Volt tolerance should therefore be tested with a 3,000 Ohm load while the 15 Volt limit should be tested using a 7,000 Ohm load.

6.5    As indicated (Figure 6.1), control of the communication channel is transferred between the automatic calling equipment and the associated data set (if used). When the automatic calling equipment is arranged for both methods of operation described in Section 1.6 and it is desired that the automatic calling equipment retain control of the communication channel until the answer signal from the called data set has been detected, the data terminal equipment must not present the EON (End Of Number) code combination before call origination is completed [i.e., prior to Circuit COS – (Call Origination Status) – coming ON].

If the EON method is used, the associated data set is connected to the communication channel while the connection to the called data set is being established. Therefore the data set may produce spurious interface signals caused, for example, by call progress tones.

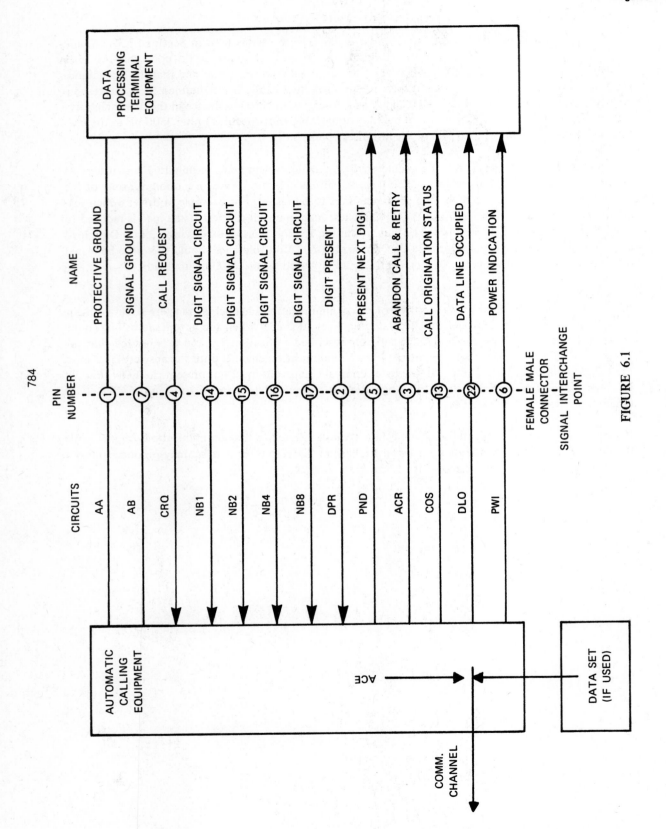

FIGURE 6.1

# APPENDIX I

## INTERFACE CONNECTOR

While no industry standard exists which defines a suitable interface connector, it should be noted that commercial products are available which will perform satisfactorily as electrical connectors for interfaces specified in RS-366, such as those connectors meeting Military Specification MIL-C-24308 (MS-18275) or equivalent.

It is not intended that the above reference be considered as part of RS-366 or as a standard for the device to which reference is made.

# EIA STANDARD

*Standard for Start-Stop Signal Quality Between Data Terminal Equipment and Non-Synchronous Data Communication Equipment*

**RS-404**

*March 1973*

*Engineering Department*

# ELECTRONIC INDUSTRIES ASSOCIATION

**NOTICE**

EIA engineering standards are designed to serve the public interest through eliminating misunderstandings between manufacturers and purchasers, facilitating interchangeability and improvement of products, and assisting the purchaser in selecting and obtaining with minimum delay the proper product for his particular need. Existence of such standards shall not in any respect preclude any member or non-member of EIA from manufacturing or selling products not conforming to such standards.

Recommended standards are adopted by EIA without regard to whether or not their adoption may involve patents on articles, materials, or processes. By such action, EIA does not assume any liability to any patent owner, nor does it assume any obligation whatever to parties adopting the recommended standards.

Published by

# ELECTRONIC INDUSTRIES ASSOCIATION

**Engineering Department**

**2001 Eye Street, N.W., Washington, D. C. 20006**

**PRICE: $4.00**

Printed in U.S.A.

# STANDARD FOR START-STOP SIGNAL QUALITY BETWEEN DATA TERMINAL EQUIPMENT AND NON-SYNCHRONOUS DATA COMMUNICATION EQUIPMENT

*(From Standards Proposal No. 1080, formulated under the cognizance of Subcommittee TR-30.1 on Signal Quality.)*

## 1. SCOPE

1.1   This standard specifies the quality of serial binary signals exchanged across the interface between start-stop (i.e. asynchronous) data terminal equipment (i.e.: processor or teleprinter) and non-synchronous data communications equipment (i.e.: data set or signal converter) as defined in EIA Standard RS-232-C. The data communications equipment is considered to be non-synchronous if the timing signal circuits across the interface are not required at either the transmitting terminal or the receiving terminal.

The scope of this standard is limited to start-stop signals at the interface with non-synchronous data communication equipment, therefore signal quality standards pertaining to synchronous data terminal equipments are not included. For a signal quality standard at the interface with synchronous data communications equipment, see E.I.A. Standard RS-334.

1.2   This standard recognizes the need to have a number of different sets of standard performance requirements because of differing operating conditions which lead to different ways of implementing equipment. A number of performance categories therefore, are defined for transmitting equipment and for receiving equipment, it being the intent that any receiving category may be specified to operate with any transmitting category, the actual selection being dependent upon such factors as channel performance, acceptable error rate, and the economic considerations of the system.

1.3   This standard is of particular importance when the equipments in a system are furnished by different organizations. It does not attempt to indicate what action, if any, is to be taken if the limits are not met, but it is intended to provide a basis for agreement between parties involved.

1.4   Any equipment which is represented as complying with limits established by this standard shall meet the applicable specifications within the ranges of those factors which are described as appropriate for the operation of the equipment, such as primary power voltage and frequency, ambient temperature and humidity.

1.5   This standard does not describe signal quality performance characteristics of the data communication equipment or the communications channel associated with it. Neither does it describe any requirement for error performance of the system or any of its components. Compliance of the terminals with the requirements of a particular standard signal quality performance category should not be construed as establishing an acceptable error rate for the terminals or the system.

## 2. DEFINITIONS

The following definitions apply to the characteristics of signal quality described in this standard and for the purposes of this standard are the only meaning that should be attached to the characteristics described.

2.1   Synchronous System — A system in which the sending and receiving data terminal equipments are operating continuously at substantially the same frequency, and are maintained in a desired phase relationship by an appropriate means.

2.2   Start-Stop System — A system in which each group of code elements corresponding to a character is preceded by a start element which serves to prepare the receiving equipment for the reception and registration of a character, and is followed by a stop element during which the receiving equipment comes to rest in preparation for the reception of the next character.

2.3   Continuous Start-Stop Operation — The method of operation in a start-stop system in which the signals representing a series of characters follow one another at the nominal character rate, (e.g. in sending steadily from a perforated paper tape as compared to manual keyboard operation).

   *NOTE:*  Under these conditions, the character intervals will have their nominal durations.

2.4   Signal Element — Each of the parts of a digital signal, distinguished from others by its duration, position and sense, or by some of these features only.

2.5   Unit Interval — A unit interval is the duration of the shortest nominal signal element. It is the longest interval of time such that the nominal durations of the signal elements in a synchronous system or the start and information elements in a start-stop system are whole multiples of this interval.

2.6   Modulation Rate — Reciprocal of the unit interval measured in seconds. This rate is expressed in bauds. (Note: As used in this standard, this term relates to the interface signals.)

2.7   Baud — The unit of modulation rate. It corresponds to a rate of one unit interval per second. Example: if the duration of the unit interval is 20 milliseconds, the modulation rate is 50 bauds.

2.8   Character Interval — In start-stop operation the duration of a character expressed as the total number of unit intervals (including information, error checking and control bits and the start and stop elements) required to transmit any given character in any given communications system.

2.9   Start Element — In a character transmitted in a start-stop system, the first element in each character, which serves to prepare the receiving equipment for the reception and registration of the character. The start element is a spacing signal.

2.10  Start Transition — In a character transmitted in a start-stop system, the mark-to-space transition at the beginning of the start element.

2.11  Stop Element — In a character transmitted in a start-stop system, the last element in each character, to which is assigned a minimum duration, during which the receiving equipment is returned to

789

its rest condition in preparation for the reception of the next character. The stop element is a marking signal.

2.12 Degree of Distortion at the digital interface for binary signals is a measure of the time displacement of the transitions between signal states from their ideal instants. The degree of distortion is generally expressed as a percentage of the unit interval.

2.12.1 Degree of Individual Distortion of a particular signal transition is the ratio to the unit interval of the displacement, expressed algebraically, of this transition from its ideal instant. This displacement is considered positive when the transition occurs after its ideal instant (late).

2.12.2 Degree of Start-Stop Distortion — The ratio to the unit interval of the maximum measured difference, irrespective of sign, between the actual interval and the theoretical interval (the appropriate integral multiple of unit intervals) separating any transition from the start transition preceding it. (See Paragraph 8.2.)

2.12.2.1 Degree of Gross Start-Stop Distortion — The degree of Start-Stop distortion determined using the unit interval which corresponds exactly to the nominal modulation rate of the system. (See Paragraph 8.2.)

2.12.2.2 Degree of Synchronous Start-Stop Distortion — The degree of Start-Stop distortion determined using the unit interval which corresponds to the actual mean modulation rate of the signal involved. (See Paragraph 8.2.)

*NOTE:* Degree of Gross Start-Stop Distortion differs from the degree of synchronous start-stop distortion only when the actual modulation rate is different from the nominal modulation rate. Under these conditions the degree of gross start-stop distortion will usually be greater than the degree of synchronous start-stop distortion.

2.13 Margin of a receiving equipment. The maximum degree of distortion of the received signal which is compatible with the correct translation of all of the signals which it may possibly receive.

*NOTE:* This maximum degree of distortion applies without reference to the form of distortion affecting the signals. In other words, the value of the margin is the maximum degree of the most unfavorable distortion beyond which incorrect translation occurs.

2.13.1 Synchronous Margin of a start-stop receiver. The maximum degree of start-stop distortion which is acceptable for the correct operation of a receiver using a test signal having an average modulation rate exactly equal to the actual sampling rate of the receiver.

2.13.2 Net Margin of a start-stop receiver. The maximum degree of start-stop distortion which is acceptable for the correct operation of a receiver using a test signal having an average modulation rate equal to the nominal modulation rate.

*NOTE:* Synchronous margin differs from net margin only when the sampling rate of the receiver is different from the nominal modulation rate. Under these conditions the net margin is less than the synchronous margin.

2.13.3 Practical Margin of a start-stop receiver. The maximum degree of start-stop distortion which is acceptable for the correct operation of a receiver using a test signal having an average modulation rate equal to the nominal modulation rate and in which no signal element duration is less than a specified value.

*NOTE:* The Practical Margin is the net margin with an added requirement on the signal element duration.

## 3. SYSTEM CHARACTERISTICS

In application of this standard, the nominal values of modulation rate and character interval are to be specified by the user.

## 4. SIGNAL QUALITY FROM THE TRANSMITTING DATA TERMINAL EQUIPMENT

Start-Stop transmitting data terminal equipment meeting the requirements of this standard shall operate within the specified category of performance requirement, having the signal quality characteristics shown in Table I. The alphabetical designations in the following paragraphs key the parameters to Table I.

Five categories of performance requirements are defined for transmitting data terminal equipment. The signal quality characteristics have been chosen for performance categories I, II, and III to cover the kinds of performance that may be expected from equipment using all electronic signal generation. Performance Categories P1 and P2 are provided for those kinds of terminals deriving their signal timing from the power line frequency. Further comment on the use of performance categories for transmitting terminals will be found in the Explanatory Notes (Section 7).

4.1  Distortion — The signal provided by the transmitting data terminal equipment on circuit BA (Transmitted Data) shall have a degree of synchronous start-stop distortion not greater than (N) percent when specified and a degree of gross start-stop distortion not greater than (P) percent, provided that no signal element shall have a duration of less than (Q) percent of a unit interval.

4.2  Character Interval — In continuous start-stop operation the signals on circuit BA (Transmitted Data) may have a minimum average character interval which is shorter than the nominal character interval and an occasional character having a still shorter duration called the minimum character interval according to the following requirements:

4.2.1  Minimum Average Character Interval — The interval between successive start transitions on circuit BA (Transmitted Data) averaged over (S) number of consecutive characters shall be no less than the nominal character interval reduced by (R) percent of a unit interval.

4.2.2  Minimum Character Interval — The interval between successive start transitions on circuit BA (Transmitted Data) shall not be less than the nominal character interval reduced by (T) percent of a unit interval.

# 5. MARGIN OF RECEIVING DATA TERMINAL EQUIPMENT

Start-Stop receiving data terminal equipment meeting the requirements of this standard shall operate within the specified category of performance requirement having those signal quality characteristics shown in Table I.

Five categories of performance requirements are defined for receiving data terminal equipment. The characteristics of the received signal quality have been chosen for performance categories A, B, and C to cover the kinds of performance that may be expected from equipment using all electronic timing signal generation. Performance categories PA and PB are provided for those kinds of terminals deriving their signal timing from the power line frequency. It is intended that any of these categories of receiving performance requirements may be selected to operate with a transmitter having any performance requirement providing an interconnecting means is chosen that provides suitable system operating characteristics within the requirements and economic limits of the system. Further comment on use of performance categories for receiving data terminals will be found in the Explanatory Notes (Section 7).

5.1   Receiving Margin — In a start-stop system, the receiving data terminal equipment is expected to have a synchronous margin of (U) percent when specified and a practical margin of (V) percent, and is not expected to respond to any signal element having a duration of less than (W) percent of a unit interval.

5.2   Character Interval — In continuous start-stop operation the receiving data terminal equipment shall respond to signals on circuit BB (Received Data) which have a minimum average character interval which is shorter than the nominal character interval and an occasional character having a still shorter duration called the minimum character interval, according to the following requirements:

5.2.1   Minimum Average Character Interval — The receiving data terminal equipment shall be prepared to respond to successive start transitions on circuit BB (Received Data) which follow their previous start transitions by a character interval, averaged over (S) number of consecutive characters, which is no less than the nominal character interval reduced by (X) percent of a unit interval.

5.2.2   Minimum Character Interval — When the above average is met, the receiving data terminal equipment shall be prepared to respond to a start transition on circuit BB (Received Data) which follows the start transition of the preceding character by an interval which is no less than the nominal character interval reduced by (Y) percent of a unit interval.

5.3   Minimum Duration Start Element — In a start-stop system when the receiver is in the stop or resting condition, it shall not be required to start reception of a character on a spacing signal element on circuit BB (Received Data) which has a duration of less than (Z) percent of a unit interval.

# 6. REQUIREMENTS FOR MEASUREMENT OF INTERFACE CHARACTERISTICS

Measurements of distortion, voltage, and loading of interface circuits shall meet the following requirements:

792

6.1  Measurement of Interface Driver Characteristics

6.1.1  Use of Standard Test Load

Standard distortion measurement and interface voltage measurement shall be made on the particular signal interchange driver circuit of interest while the circuit is terminated with a standard test load. This standard test load may be the input impedance of the test device or may be an external device but in all cases the total load on the interchange circuit shall meet the following requirement:

6.1.2  Requirement for Standard Test Load

The standard test load shall consist of 3000 ohm resistance shunted by 2500 pF capacitance and connected from the signal interchange circuit under test to circuit AB (Signal Ground) as·shown in Figure 1.

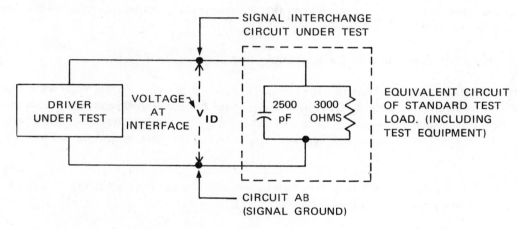

**FIGURE 1 – STANDARD TEST CIRCUIT FOR DRIVER**

6.1.3  Interface Voltage Measurement

The interface voltage measurement shall be made to determine that the driver is producing an output signal $V_{ID}$ of at least ± 5.0 Volts into the standard test load.

6.1.4  The Standard Distortion Measurement Thresholds

The standard distortion measurement shall be made using a +3 Volt and a -3 Volt threshold to determine the occurrence of signal transitions.

6.1.4.1  A mark-to-space transition shall be taken to occur at the instant $V_{ID}$ crosses +3 Volts on a positive going transition.

6.1.4.2  A space-to-mark transition shall be taken to occur at the instant $V_{ID}$ crosses -3 Volts on a negative going transition.

6.2  Measurement of Interface Terminator Characteristics

Standard measurements of loading and voltage characteristics and of margin on the terminator side of the interface shall be made using the circuit of Figure 2 and shall meet the following requirements:

6.2.1  Measurement of Loading Characteristics

The standard loading measurement shall be made to determine that the input resistance of the terminator is not less than 3000 ohms nor more than 7000 ohms and that the input shunt capacitance is not greater than 2500 pF.

6.2.2  Voltage Characteristics

The measurement of the voltage response characteristics of the interface terminator circuits shall be made to determine that the terminator will respond correctly to a signal voltage $V_{IT}$ at the interface point of ± 3 Volts. (It is necessary that the terminator provide some external indication of its response to the received signal.)

6.2.3  Margin of Receiving Data Terminal Equipment

The measurement of receiver margin shall be made using a signal $V_{IT}$ of ± 5.0 Volts when working into the terminator under test. The transitions of the test signal shall be deviated from their ideal instants in such a way as to measure the receiver margin. (See Paragraph 2.13.) The deviations of the transitions in time shall be taken to occur when the signal crosses the ± 3 Volt thresholds.

6.2.3.1  A mark-to-space transition shall be taken to occur at the instant $V_{IT}$ crosses +3 Volts on a positive going transition.

6.2.3.2  A space-to-mark transition shall be taken to occur at the instant $V_{IT}$ crosses -3 Volts on a negative going transition.

6.2.3.3  It is necessary that the receiving data terminal equipment provide some indication of its response to the distorted test signal to allow determination of whether incorrect translations of the signal occur.

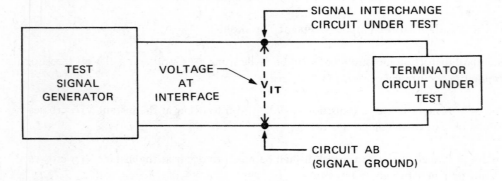

FIGURE  2 – STANDARD  TEST  CIRCUIT  FOR  TERMINATOR

794

6.3 Tolerances of Measuring Equipment

This standard does not specify tolerances of the standard test load, distortion measuring equipment or test signal generator, however, the user is cautioned to consider these tolerances when determining that performance of a particular piece of equipment falls within the requirements of this standard.

## 7. EXPLANATORY NOTES

7.1 Transmitting and receiving performance requirements have been divided into two groups of categories. A group having three divisions each for transmitters and receivers (I, II, III and A, B, C) which are for general use with equipments deriving their signal timing from internal electronic circuitry; and a group having two divisions (P1, P2 and PA, PB) are provided for these equipments deriving their signal timing from the frequency of the power line. The performance requirement assigned to a particular equipment will depend upon its design, speed, mode of operation, environmental factors and other considerations which must be specified along with the appropriate performance requirement. Indeed, a single apparatus in one operating mode may fall into one category requirement, but when operating under the circumstances of a different mode may operate in a different category. Thus it is a part of this standard that the system characteristics must be specified to make the performance requirement meaningful.

7.2 In Table I which establishes the limits of performance requirements for categories, numbers shown are the worst case bounds that should be expected from equipment during its period in service. It is expected that should the signal quality characteristics drop below the limits shown that suitable action be taken by the parties involved.

In particular, category II for transmitting terminals specifying a gross start-stop distortion of 10% and category C for receiving terminals specifying a practical margin of 35% are representative of typical terminals that have been found suitable for general applications, the other categories being given for special purpose applications. When connected to the public telecommunications network, terminals having these characteristics have generally performed satisfactorily. It is beyond the scope of this standard, however, to specify performance of the public telecommunications network since this is subject to tariff regulations. It shall be recognized that the limitation specified in Paragraph 1.5 of this standard apply to these categories as well as all categories listed in Table I.

7.3 The category of performance requirement specified for a particular equipment is expected to describe the performance of that equipment under a particular set of operating conditions. The signal performance limits are not to be exceeded in operation as long as the other specified operating conditions are met. The category thus provides a guide for the system designer and for the maintenance organization for the performance of the equipment. Some kinds of equipment may have different signal quality characteristic when operating in different modes. These may be due to different options selected at the time the equipment is placed in service, or field changes that may be made either by modification or operating controls. Where these changes of mode affect the signal quality it is imperative that such an effect be recognized by expressing the proper category of performance requirement for the mode of the operation selected.

7.4 Terminals deriving their signal timing from the power line present a special situation from the signal quality viewpoint. For this reason, categories P1 and P2 for transmitters and PA and PB for receivers have been established. The power line frequency is generally quite reliable and very close to its nominal value especially at those locations which utilize power from the power grid. Under unusual circumstances, however, the frequency of the power line may vary appreciably from its nominal value. For this reason, both synchronous start-stop distortion and gross start-stop distortion have been specified for transmitter performance, and synchronous margin and net margin for receivers.

> *NOTE:* In this section, no reference is made to the minimum signal element duration, therefore the term "Net Margin" is used instead of "Practical Margin" as used in Table I.

Gross start-stop distortion from the transmitting terminal contains a factor for the timing error present in the timing device used in the transmitter in addition to other causes of inaccurate signal generation. In category P1 or P2 where timing is derived from the power line, the accuracy of the timing device depends upon the accuracy of the frequency of the power line. In rare instances the power line frequency may be incorrect by as much as 0.75%. In a ten unit code, this would result in a displacement of the last transition by about 7.5% (say 8%) which would result in the gross start-stop distortion, being 8% greater than the synchronous start-stop distortion for the same equipment which is the limit shown in Table I. If the equipment is operated from a power line that is always exactly the nominal frequency then the gross start-stop distortion will be the same as the synchronous start-stop distortion. In the receiver, the net margin shown allows for an error in the timing device as well as for other causes of non-accurate reception of the signal code. In receiver performance category PA and PB an allowance has been made for an 0.75% error in the power line frequency. Thus the synchronous margin is 8% higher than the net margin. As a result, if the power line frequency was exactly the nominal value then the net margin would be the same as the synchronous margin. It is important to note that for a transmitter operating in performance category P1, working into a receiver with a PA performance category rating, using the gross start-stop distortion limits and the net margin shown in Table I, a distortion allowance of only 30% − 16% = 14% additional distortion of the signal is allowed for the rest of the system under worst case conditions. If, however, these terminals were running from a perfect power line frequency, the synchronous start-stop distortion and synchronous margin limits would then allow 38% − 8% = 30% additional distortion of the signal allowed for the rest of the system before incorrect reception of the signal would occur. It is also interesting to note that if both transmitter and receiver timing devices deviated in the same direction their timing errors would nullify each other. Thus, if both transmitter and receiver were operating from the same frequency power source (i.e. the power grid) the timing error would cancel even if the power line frequency were incorrect, resulting in 30% distortion allowance for the remainder of the system. If these terminals having category P1 and PA rating did not run from a common frequency source as discussed above, then the latter reasoning cannot apply and a distortion allowance of only 14% additional distortion of the signal is allowed for the rest of the system, but it is quite probable that this would seldom occur. It is important that system designers take cognizance of these facts in choosing the characteristics of system components.

When using devices deriving their signal timing from the power line frequency, it is expected the frequency will be nearly correct most of the time and that deviation will occur only rarely but when it does, total system performance may be expected to be degraded during this time.

# 8. TUTORIAL NOTES

8.1 Signal Nomenclature. Figure 3 illustrates a number of definitions of binary digital signals, taking a start-stop character as an example. The figure shows an ideal signal, without distortion, representing the number "9" in the USA Standard Code for Information Interchange, (USASCII) in the standard published as USAS X3.4-1967. The figure depicts a character with even parity, a unit start element, and a 1-unit stop element. The code is transmitted with "low order bit" first, followed by the even-parity bit. In the following, the first use of a term defined in Section 2 is followed by its number, in parentheses. Therefore:

  a.   The *character interval* (2.8) has a duration of 10.0 *unit intervals* (2.5) nominal.

  b.   The information and parity bits are represented by *signal elements* (2.4).

  c.   These have durations of one unit interval each, except that when unit-interval elements ("unit elements") of the same sense are consecutive, they run together to make longer elements having durations of 2, 3 or more unit intervals.

  d.   The duration of a unit interval (in seconds) is the reciprocal of the *modulation rate* (2.6) the latter being expressed in *bauds* (2.7). The modulation rate is equal to the number of unit intervals per second, (not bits per second, signal elements per second nor bauds per second).

  e.   The *start element* (2.9) begins with a mark-to-space transition called the *start transition* (2.10), and its duration is nominally one unit interval.

  f.   The *stop element* (2.11) in this example has a nominal duration of 1.0 unit intervals. It will have this nominal duration during *continuous start-stop operation* (2.3), however, in non-continuous start-stop operation the stop element is considered to be prolonged until the start transition of the following character occurs, however long that may be, as in manual keyboard operation of a teleprinter.

  (If this character had a stop element 2.0 unit intervals long, it would not be good practice to refer to this character structure as having "2 stop bits"; the stop element is only one signal element, with a duration of 2 unit intervals, or it is a "2-unit stop element".)

8.2 Gross Start-Stop Distortion and Synchronous Start-Stop Distortion

Figure 4 illustrates the relationship between two kinds of "start-stop distortion." Generally *start-stop distortion* (2.12.2) is measured with reference to the start transition of the character, in terms of the largest per cent of a unit interval by which any transition leads or lags the instant when it should have occurred, if the *degree of distortion* (2.12) were zero. The terms *gross start-stop distortion* (2.12.2.1) and *synchronous start-stop distortion* (2.12.2.2) point up a useful distinction that is not always appreciated. If all start-stop sending receiving and distortion measuring equipment in a particular system were controlled by the same clock, only *synchronous* start-stop distortion would be measured. However, if the actual modulation rate of the signal in question differs from the nominal modulation rate (i.e., the rate at which the equipment is intended to work), while the distortion measuring device is controlled by a clock which runs at *exactly* the nominal rate, the measured start-stop distortion will generally be greater because the ideal instants of the signal and of the measuring equipment do not agree. This disagreement increases steadily during each character interval; an otherwise perfectly proportioned signal will show, on the average, the greatest *individual distortion* (2.12.1) on the last transition in each character. The distortion measured under these conditions is the *gross* start-stop distortion. The existence of the modulation rate discrepancy is readily recognized through the use of measuring equipment which can be made to measure the distortion of any particular transition (e.g., the 5th) within the character. If the rate discrepancy exists, the degree of individual distortion will increase progressively from the earliest transition to the latest transition.

In Figure 4, the distorted start-stop signal to be measured is drawn through the middle of the diagram. After the mark-to-space start transition, there are transitions A, B, C, D, and E, exhibiting various degrees of individual distortion. This could be the character of Figure 3 with distortion. The upper part of the figure illustrates the measurement of gross start-stop distortion when the actual modulation rate of the signal is lower than that used by the measuring equipment. The reference instants numbered from 0 to 10 at the top of the figure are equally spaced at the unit intervals corresponding to the nominal modulation rate, at which the measuring equipment works. Transition A comes 33% of a unit interval after its reference instant; B occurs 15% before reference instant No. 2, while E is 50% after reference instant No. 9. Since the latter is the largest value, this determines the *gross* start-stop distortion to be 50%. Actually, in an actual measurement, the values quoted would be the maximum individual distortions for the various transitions during the period of observation.

Now, suppose that the operator slows down the clock signals governing the measuring equipment until he determines that their modulation rate has "synchronized" (hence the name) with the average modulation rate of the signal. This will often be the adjustment which gives the minimum average start-stop distortion during a short period of observation. For the particular character signal illustrated, the new reference instants tend to come closer to the actual transition times than before, so that E is now only 12% late, as shown in the lower part of the figure, and A is only 28% late; which turns out, in the case, to be the largest figure, so that the synchronous start-stop distortion is found to be 28%. It will be noted that C is now 16% early, while D has changed from 5% late to 23% early.

The concept of gross start-stop distortion is useful from the standpoint of routine maintenance of station equipment in switched systems where many transmitting (and receiving) equipment are served over circuits from a given switching or maintenance center. Provided that all of the receivers which might be connected to any of the transmitters meet certain receiving tolerance or *margin* (2.13) requirements, any receiver should operate correctly if the greatest degree of individual distortion of the signals they receive is not in excess of an established amount, regardless of whether this is due to transmitter misadjustment or modulation rate difference, or both. Measurement of gross start-stop distortion of signals from the transmitting equipment will allow checking that such distortion limits are being met. If a station meets this requirement, it is of no immediate concern whether the measured distortion is due entirely to inaccurate signal formation, or partly or entirely to modulation rate difference. As mentioned above, with some available distortion measuring equipment it is relatively simple to recognize modulation rate error. Going farther, by changing to a synchronous start-stop distortion measurement, it would be possible to analyze the transmitter signal formation to discover which particular transitions may be persistently experiencing the most distortion, as an aid to trouble shooting.

8.3 Minimum Duration Signal Element. Figure 4 also illustrates the need, in some cases, to place limits on the minimum duration of signal elements, even though the maximum displacement of transi-

798

tions may be controlled. The marking element from A to B has a duration of only 52% of a unit interval, (on the upper timing scale). This is due to two simultaneous individual distortions in opposite directions. A comes 33% late while B comes 15% early, on the *same* signal element of this particular character, and thus they consume a total of 48% of the unit interval allotted to it, leaving this element an actual duration of 52%, ignoring any slope of the transition waveforms. Against the lower timing scale, the duration is only 50% of a unit interval.

A similar shortening could occur for a spacing element. With regard to paragraph 5.3, while a valid start element might sometimes be unduly shortened in this way, the main reason for establishing this kind of requirement is to prevent many of the short "hits" or impulse noises on the circuit from starting the receiver through an unnecessary receiving cycle, during which a valid start element might arrive. Such protection could be provided by placing suitable circuitry in the receiver to measure the minimum duration of all signal elements, with provision to inhibit a character receiving cycle from starting on spacing signals shorter than the established minimum duration.

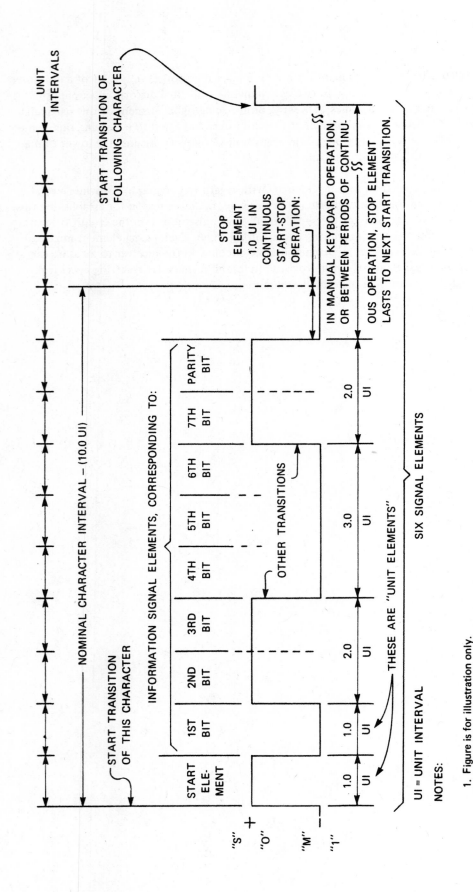

UI = UNIT INTERVAL

NOTES:

1. Figure is for illustration only.

2. Example is numeral "9" in USASCII, (X3.4-1967), low-order bit first, even parity, with 1-unit stop element, resulting in an 10-unit character interval for start-stop operation.

FIGURE 3 – EXAMPLE OF APPLICATION OF TERMS APPLIED
TO CHARACTER IN START-STOP OPERATION

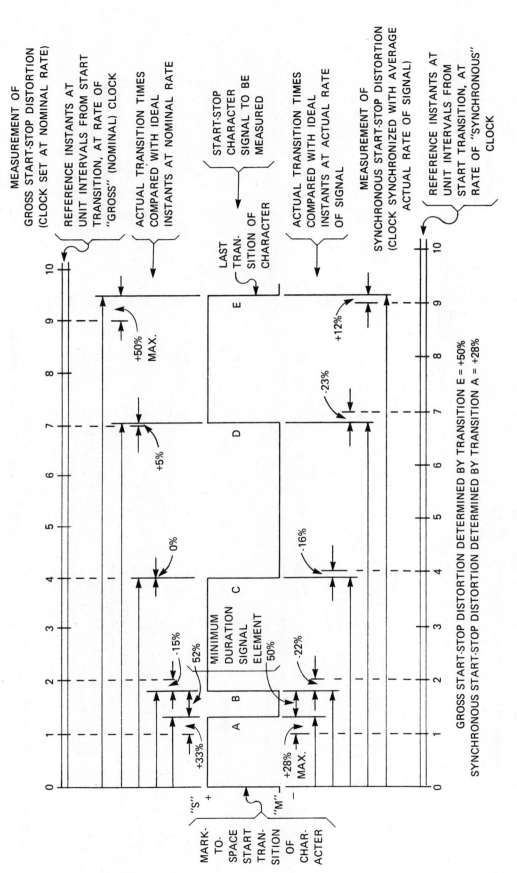

FIGURE 4 – RELATIONSHIP BETWEEN GROSS AND SYNCHRONOUS
START-STOP DISTORTION MEASUREMENTS

801

802

# STANDARD FOR START-STOP SIGNAL QUALITY

## TABLE I

When categorizing performance of data terminal equipment using the table below, modulation rate, character interval, operating mode, and environmental conditions must be specified.

### TRANSMITTING DATA TERMINAL

| PARA. | | DESIG. | UNIT | Performance Category | | | | |
| | | | | Internal Timing | | | Power Line Timing | |
| | | | | I | II# | III | P1 | P2 |
|---|---|---|---|---|---|---|---|---|
| 4.1 | Synchronous Start-Stop Distortion | N | % | – | – | – | 8* | 12* |
| 4.1 | Gross Start-Stop Distortion | P | % | 2 | 10 | 16 | 16* | 20* |
| 4.1 | Minimum Signal Element | Q | % UI | 98 | 90 | 84 | 84 | 84 |
| 4.2 | Character Interval Requirement | | | | | | | |
| 4.2.1 | Average: Nominal Reduced By | R | % UI | 3 | 8 | 10 | 10 | 10 |
| 4.2.1 | Averaged Over | S | Char | 2 | 2 | 2 | 2 | 2 |
| 4.2.2 | Minimum: Nominal Reduced By | T | % UI | 5 | 16 | 20 | 20 | 20 |

*See Explanatory Note 7.4

### RECEIVING DATA TERMINAL

| PARA. | | DESIG. | UNIT | Performance Category | | | | |
| | | | | Internal Timing | | | Power Line Timing | |
| | | | | A | B | C# | PA | PN |
|---|---|---|---|---|---|---|---|---|
| 5.1 | Synchronous Margin | U | % | – | – | – | 38* | 33* |
| 5.1 | Practical Margin | V | % | 45 | 40 | 35 | 30* | 25* |
| 5.1 | Minimum Signal Element | W | % UI | 30 | 30 | 30 | 30 | 34 |
| 5.2 | Character Interval Requirement | | | | | | | |
| 5.2.1 | Average: Nominal Reduced By | X | % UI | 25 | 25 | 25 | 25 | 25 |
| 5.2.1 | Averaged Over | S | Char | 2 | 2 | 2 | 2 | 2 |
| 5.2.2 | Minimum: Nominal Reduced By | Y | % UI | 50 | 50 | 50 | 50 | 50 |
| 5.3 | Minimum Duration Start Element | Z | % UI | 50 | 50 | 50 | 50 | 50 |

*See Explanatory Note 7.4

#It is anticipated the values in these categories will be the accepted CCITT worst case figures.

# EIA STANDARD

*Standard for the Electrical
Characteristics of Class A Closure
Interchange Circuits*

RS-410

APRIL 1974

*Engineering Department*

# ELECTRONIC INDUSTRIES ASSOCIATION

## NOTICE

EIA engineering standards are designed to serve the public interest through eliminating misunderstandings between manufacturers and purchasers, facilitating interchangeability and improvement of products, and assisting the purchaser in selecting and obtaining with minimum delay the proper product for his particular need.  Existence of such standards shall not in any respect preclude any member or non-member of EIA from manufacturing or selling products not conforming to such standards, nor shall the existence of such standards preclude their voluntary use by those other than EIA members whether the standard is to be used either domestically or internationally.

Recommended standards are adopted by EIA without regard to whether or not their adoption may involve patents on articles, materials, or processes.  By such action, EIA does not assume any liability to any patent owner, nor does it assume any obligation whatever to parties adopting the recommended standards.

Published by

# ELECTRONIC INDUSTRIES ASSOCIATION

### Engineering Department

### 2001 Eye Street, N. W., Washington, D. C. 20006

Price **$1.40**

# STANDARD FOR THE ELECTRICAL CHARACTERISTICS
# OF CLASS A CLOSURE INTERCHANGE CIRCUITS

*(From Standards Proposal No. 1107, formulated under the cognizance of Subcommittee TR-30.1 on Signal Quality.)*

## 1. SCOPE

1.1 This standard is applicable, but not limited to interconnection of equipment in voice or data communication services where:

1.1.1 The driver includes a closure or switch (metal contacts or solid-state device) and,

1.1.2 The terminator includes an electromechanical relay winding or a solid-state device and,

1.1.3 Power is always furnished from the terminator side of the interface and,

1.1.4 Information is passed across the interconnection as discrete direct current states.

1.2 Circuits meeting the requirements of this standard are designated as Class A closure interchange circuits. Future standards may define closure interchange circuits having different electrical characteristics. These circuits will be designated by other class symbols.

1.3 This standard is applicable for use in circuits where the nominal duration of either circuit state is not less than 10 milliseconds.

1.4 Three types of circuit configurations are defined: N (negative common), P (positive common) and F (floating circuit). Users of this standard shall specify which type of circuit configuration (N, P or F) is to be used. For example "RS-410 Type N".

1.5 It is intended that within a given type of interchange circuit N, P or F, mechanical closures or the appropriate solid-state device may be used interchangeably. Interconnection of dissimilar types of interchange circuit (N, P, F) is not intended.

1.6 The basic difference between an interchange circuit of the closure type described in this standard and the circuit of RS-232-C is the latter drivers (signal sources) furnish power to the terminator while the closure driver modulates power furnished by the terminator.

## 2. CIRCUIT ARRANGEMENTS

2.1 Three circuit arrangements are provided allowing for either the negative power supply terminal to be common, the positive power supply terminal to be common, or closure return not connected to common such as would be encountered in the use of a "floating contact". In all arrangements, the power supply is electrically part of the terminator circuit. The power supply may be physically located in either terminal, however, if a power supply exists only on the driver side its voltages may be conducted to the terminator through a separate power lead. The load within the terminator may include logic devices, (closures) which under certain conditions may remove the voltage from the interchange lead.

2.1.1 The N arrangement equivalent circuit is shown in Figure 1. The negative terminal of the power supply is circuit common. An NPN type transistor, as well as a metal contact or other device may be used for the closure driver. (Circuit symbols are defined in paragraph 2.2.)

2.1.2 The P arrangement equivalent circuit is shown in Figure 2. The positive terminal of the power supply is circuit common. A PNP type transistor, as well as a metal contact or other device may be used for the closure in the driver.

2.1.3 The F arrangement equivalent circuit is shown in Figure 3. Neither power supply terminal is required to be connected to the closure return and the closure is considered to be floating. It . should be noted that with a type F arrangement, only the driver need be considered floating. The terminator may have circuit logic requirements for closures or other circuit elements between power supply leads and either interchange lead. Also note that any point within the terminator may be connected to common. While a metal contact is a common way of implementing such a floating circuit, this arrangement is intended to also include optoelectronic isolators or other devices that fall within the limits of the specified electrical characteristics.

## 2.2 Definitions of Circuit Elements

2.2.1 Closure: An electromechanical contact, a solid-state device, or an equivalent device with appropriate characteristics.

2.2.2 Load: Any combination of electrical components, e.g.: relay coils, resistors, solid-state devices, contacts, etc.

2.2.3 $R_1$: Total series resistance of the driver circuit including the internal resistance of the closure and any required current limiting resistance to protect the closure from the transient discharge of $C_1$ and $C_2$.

2.2.4 $R_2$: Leakage resistance of the closure.

2.2.5 $C_1$: The total effective capacitance associated with the driver, measured at the interface point and including any cable to the interface point.

2.2.6 $C_2$: The total effective capacitance associated with the terminator, measured at the interface point and including any cable to the interface point.

2.2.7 Common: This circuit establishes the common reference potential for all interchange circuits except protective ground.

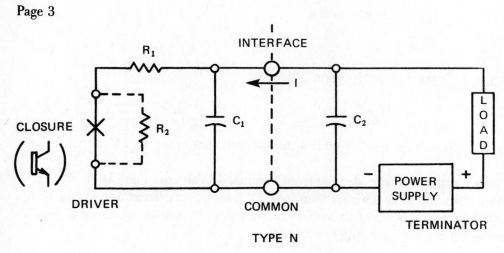

TYPE N

**FIGURE 1**

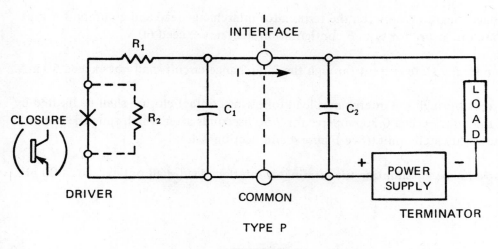

TYPE P

**FIGURE 2**

807

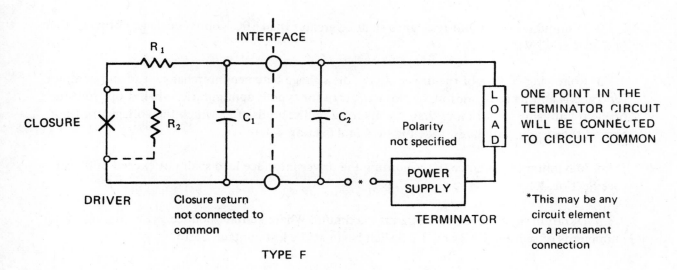

ONE POINT IN THE
TERMINATOR CIRCUIT
WILL BE CONNECTED
TO CIRCUIT COMMON

*This may be any
circuit element
or a permanent
connection

TYPE F

**FIGURE 3**

### 3. INTERFACE ELECTRICAL CHARACTERISTICS

#### 3.1 Power Supply

3.1.1 The D.C. power supply voltage shall be no less than 10 volts nor more than 52 volts.

CAUTION: Use of power supply voltages derived from nominal 48 volt communication or switch gear battery banks may result in voltages higher than the 52V limit being present during equalization charge periods. Appropriate power supply voltage limiting devices may be necessary to prevent component damage when using such supply voltages.

3.1.2 The ripple of the power supply voltage shall not exceed 5% (peak-to-peak) of nominal.

#### 3.2 Terminator

3.2.1 The peak voltage measured between the terminator interchange lead and common for type N and P configuration or closure for type F configuration shall not exceed 60 V.

3.2.2 The maximum steady state current through the interchange circuit shall not exceed 50 mA.

3.2.3 The peak current through the interchange lead following contact closure shall be limited by the terminator and shall not exceed 0.20 amperes for 2 milliseconds after which time the current must not exceed the steady state limit. (See Figure 4 and Section 4.1.)

3.2.4 Maximum capacitance $C_2$ shunting the terminator interface lead shall not exceed 5000 pF. (See Section 4.1.)

#### 3.3 Driver

3.3.1 The maximum closed circuit voltage between the driver interchange lead and common for type N and P configuration or closure return for type F configuration shall be no greater than 2 V with any value of current up to 50 mA through the closure interchange lead.

3.3.2 Minimum open circuit resistance shall be greater than 100 Kohm (Leakage not more than 0.52 mA at 52 V).

3.3.3 Following closure of the driver circuit the voltage between interchange lead and common for Type N and P configuration, or closure return for type F configuration shall drop from the open circuit value to not more than 8 volts for 2 milliseconds after which the voltage must not exceed the steady state limit. (See Figure 4 and Section 4.1.)

3.3.4 Maximum capacitance ($C_1$) shunting the driver interface lead shall not exceed 5000 pF. (See Section 4.1.)

3.3.5 Duration of contact bounce: 2 ms maximum. Where contact bounce exists, timing for requirements of section 3.2.3 and 3.3.3 shall begin at the last contact closure.

## 4. EXPLANATORY NOTES

4.1 The maximum value of capacitance shunting the interface leads has been limited to protect the closure device. Thus the designer should provide a closure (especially when this is a metallic contact), which can dissipate the energy stored in $C_1$ and $C_2$ in parallel when they are charged to the maximum permissible power supply voltage. Current limiting may be introduced in the circuit to provide such protection, but it must not cause a voltage drop in the closure circuit exceeding that value allowed. Likewise, the designer of the terminator must consider the effect this shunt capacitance may have upon any device included in the terminator and provide suitable protection if necessary. Some terminator devices may contain additional capacitance in the terminator circuit in which case this capacitance must be isolated from $C_2$ and current limiting provided in the terminator to meet the requirements of Section 3.3.3 and Figure 4. In Figure 4A the profile of current limit is shown. The initial spike of current (a-b) is that required to discharge $C_2$ and is limited by the driver and its connections, the maximum value of which is not specified but is determined by the driver designer to be within the limits of the driving device. The plateau (b-c) is the maximum current from the terminator permitted during the specified period and must be limited by terminator design. During the remainder of the closure (c-d) the current must be below the steady state value shown.

The voltage limits in Figure 4B provide that before closure up to the maximum supply voltage may exist across the interface. Following closure, the voltage drops (a-b) from the supply voltage to 8 volts or less and must remain (b-c) less than 8 volts or continue to decay for the duration of the transition period allowed in Section 3.2.3 and 3.3.3. At the end of this period, the voltage must have reached the allowable steady state value or less as specified in Section 3.3.1. During the remainder of the contact closure (c-d) the voltage must not exceed the steady state value. Following the closure opening the voltage may rise to as much as the peak allowed in Section 3.2.1 after which it will return (d-e) to its original open circuit value. The time for returning to open circuit value is not specified, however, good practices indicates it should be kept short so that the voltage will return to its steady open circuit value before the next closure operation.

4.2 These interface circuits allow maximum voltages which exceed maximums allowed in RS-232-C and, therefore, the user is cautioned when using this interface to be aware of the hazards of short circuits. Also, the rate of change of voltage on these circuits may be so rapid as to cause interference on high speed digital circuits. Therefore, inclusion of these circuits in the same cable with RS-232-C circuits precludes compliance with RS-232-C Standard.

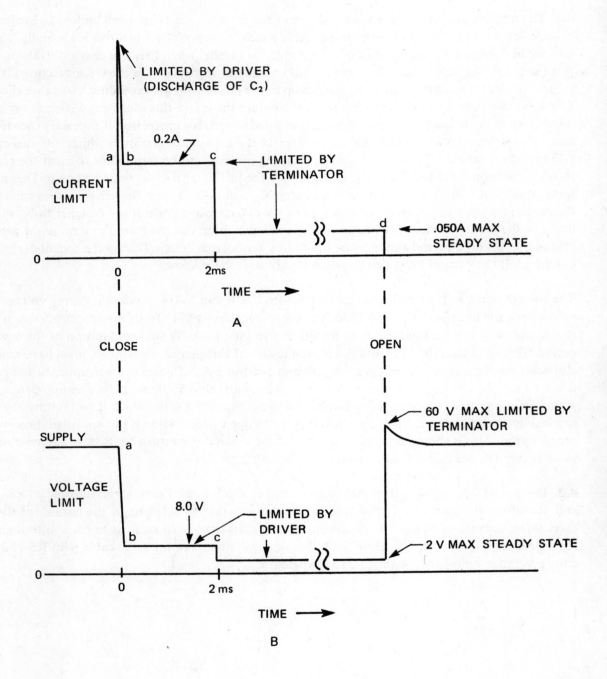

CURRENT AND VOLTAGE LIMITS ON CLOSURE INTERCHANGE CIRCUITS

FIGURE 4

# EIA STANDARD

*Electrical Characteristics*
*of*
*Balanced Voltage Digital*
*Interface Circuits*

**RS-422**

**APRIL 1975**

*Engineering Department*

## ELECTRONIC INDUSTRIES ASSOCIATION

## NOTICE

EIA engineering standards are designed to serve the public interest through eliminating misunderstandings between manufacturers and purchasers, facilitating interchangeability and improvement of products, and assisting the purchaser in selecting and obtaining with minimum delay the proper product for his particular need. Existence of such standards shall not in any respect preclude any member or non-member of EIA from manufacturing or selling products not conforming to such standards, nor shall the existence of such standards preclude their voluntary use by those other than EIA members whether the standard is to be used either domestically or internationally.

Recommended standards are adopted by EIA without regard to whether or not their adoption may involve patents on articles, materials, or processes. By such action, EIA does not assume any liability to any patent owner, nor does it assume any obligation whatever to parties adopting the recommended standards.

Published by

# ELECTRONIC INDUSTRIES ASSOCIATION

### Engineering Department

### 2001 Eye Street, N. W., Washington, D. C. 20006

812

**PRICE:** $4.25

# ELECTRICAL CHARACTERISTICS OF

# BALANCED VOLTAGE DIGITAL INTERFACE CIRCUITS

*(From EIA Standards Proposal No. 1162-A, formulated under the cognizance of EIA Committee TR-30.1 on Signal Quality)*

## 1. SCOPE

This standard specifies the electrical characteristics of the balanced voltage digital interface circuit normally implemented in integrated circuit technology that may be employed when specified for the interchange of serial binary signals between Data Terminal Equipment (DTE) and Data Communications Equipment (DCE) or in any interconnection of binary signals between voice or data equipment.

The interface circuit includes a generator connected by a balanced interconnecting cable to a load consisting of a receiver or receivers and an optional termination resistor. The electrical characteristics of the circuit are specified in terms of required voltage, current, and resistance values obtained from direct measurement of the generator and receiver components. The receiver specification for the interface is electrically identical to that specified for the unbalanced interface circuit in RS-423. The characteristics of the interconnecting cable are specified, and guidance is given with respect to limitations on data modulation rate imposed by the parameters of cable length, balance and termination.

The parameter values specified for the balanced generator and load components of the interface are designed such that balanced interface circuits may be used within the same interconnection as unbalanced interface circuits specified by RS-423. For example, the balanced circuits may be used for data and timing while the unbalanced circuits may be used for low speed control functions.

It is intended that this standard will be referenced by other standards that specify the complete DTE/DCE interface (i.e. protocol, timing, pin assignments, etc.) for applications where the electrical characteristics of a balanced voltage digital circuit are required. Applications are also forseen in other areas using binary signal interchange. This standard does not specify other characteristics of the DTE/DCE interface (such as signal quality and timing, etc.) essential for the interconnected equipment operation.

## 2. CROSS REFERENCE

This standard is one of a series relating to the interconnection of DTE and DCE. Other EIA standards in this series, in addition to RS-423, pertaining to the DTE/DCE and DTE/DTE interface specifications are in existence or in various stages of preparation at the time of publication of this standard. The user is referred to EIA Engineering Headquarters for information on specific standards available.

813

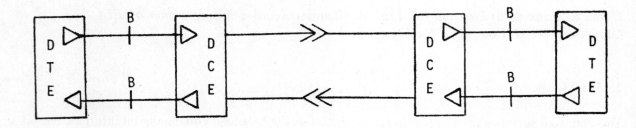

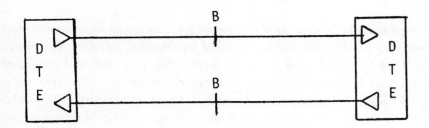

814

Legend:

DTE = Data Terminal Equipment
DCE = Data Communications Equipment
▷ = Interface Generator
─▷ = Interface Load
B
┼ = Balance Interface Circuit
≫ = Telecommunications Channel

FIGURE 3.1

Applications of Balanced Voltage
Digital Interface Circuit

## 3. APPLICABILITY

The provisions of this standard may be applied to the circuits employed at the interface between equipments where the information being conveyed is in the form of binary signals at the dc base-band level. This standard shall be referenced by the specifications and specific interface standards applying these electrical characteristics. Typical points of applicability for this standard are depicted in Figure 3.1.

The balanced voltage digital interface circuit will normally be utilized on data, timing or control where the modulation rate on these circuits is up to 10 megabauds. Balanced voltage digital interface devices meeting the electrical characteristics of this standard need not operate over the entire modulation rate range specified. They may be designed to operate over narrower ranges to more economically satisfy specific applications, particularly at the lower modulation rates.

While the balanced interface is intended for use at the higher modulation rates, it may, in preference to the unbalanced interface circuit, generally be required where any of the following conditions prevail:

a. The interconnecting cable is too long for effective unbalanced operation.

b. The interconnecting cable is exposed to extraneous noise sources that may cause an unwanted voltage in excess of plus or minus one volt measured differentially between the signal conductor and circuit common at the load end of the cable with a 50 ohm resistor substituted for the generator.

c. It is necessary to minimize interference with other signals.

d. Inversion of signals may be required, e.g., plus MARK to minus MARK may be obtained by inverting the cable pair.

While a restriction on maximum cable length is not specified, guidelines are given with respect to conservative operating distances as a function of modulation rate (see Section 7). In general, these conservative values may be greatly exceeded where the installation is engineered to ensure that noise and ground potential values are held within specified limits.

## 4. ELECTRICAL CHARACTERISTICS

The balanced voltage digital interface circuit is shown in Figure 4.1. The circuit consists of three parts: the generator, the balanced interconnecting cable, and the load. The load is comprised of one or more receivers (R) and an optional cable termination resistance $(R_t)$. The electrical characteristics of the generator and receiver are specified in terms of direct electrical measurements while the interconnecting cable is specified in terms of its electrical and physical characteristics.

### 4.1 Generator Characteristics

The generator electrical characteristics are specified in accordance with measurements illustrated in Figures 4.2 and 4.3 and described in paragraphs 4.1.1 through 4.1.5. A generator circuit meeting these requirements results in a low impedance (100 ohms or less) balanced voltage source that will

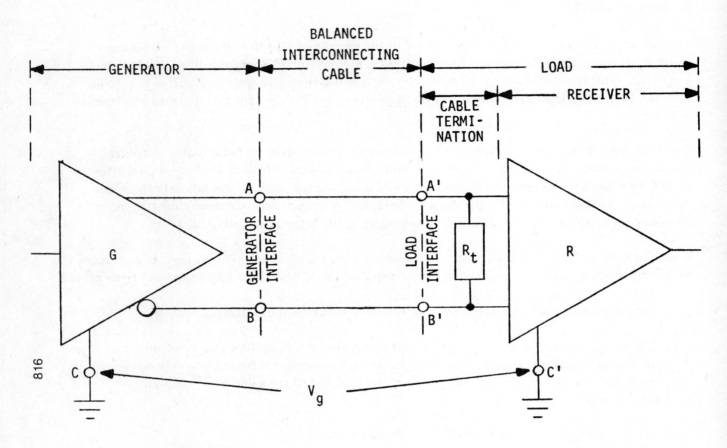

Legend:

$R_t$    =    Optional Cable Termination Resistance

$V_g$    =    Ground Potential Difference

A,B    =    Generator Interface Points

A', B'    =    Load Interface Points

C    =    Generator Circuit Ground

C'    =    Load Circuit Ground

FIGURE 4.1

Balanced Digital Interface Circuit

produce a differential voltage applied to the interconnecting cable in the range of 2 volts to 6 volts. The signalling sense of the voltages appearing across the interconnection cable are defined as follows:

(1) The A terminal of the generator shall be negative with respect to the B terminal for a binary 1 (MARK or OFF) state.

(2) The A terminal of the generator shall be positive with respect to the B terminal for a binary 0 (SPACE or ON) state.

*4.1.1  Open Circuit Measurement (Figure 4.2)* For either binary state, the magnitude of the differential voltage (Vo) measured between the two generator output terminals shall not be more than 6.0 volts; nor shall the magnitude of Voa and Vob measured between the two generator output terminals and generator circuit ground be more than 6.0 volts.

*4.1.2  Test Termination Measurement (Figure 4.2)* With a test load of two resistors, 50 ohm ± 1% each, connected in series between the generator output terminals, the magnitude of the differential voltage $(V_t)$ measured between the two output terminals shall not be less than either 2.0 volts or 50% of the magnitude of Vo whichever is greater. For the opposite binary state the polarity of Vt shall be reversed ($\overline{V}$t). The magnitude of the difference in the magnitude of Vt and $\overline{V}$t shall be less than 0.4 volts. The magnitude of the generator offset voltage Vos measured between the center point of the test load and generator circuit ground shall not be greater than 3.0 volts. The magnitude of the difference in the magnitudes of Vos for one binary state and $\overline{V}$os for the opposite binary state shall be less than 0.4 volts.

*4.1.3  Short- Circuit Measurement (Figure 4.2)* With the generator output terminals short-circuited to generator circuit ground, the magnitudes of the currents flowing through each generator output terminal shall not exceed 150 milliamperes for either binary state.

*4.1.4  Power-off Measurement (Figure 4.2)* Under power-off conditions, the magnitude of the generator output leakage currents (Ixa and Ixb), with voltages ranging between +6.0 and −0.25 volts applied between each output terminal and generator circuit ground, shall not exceed 100 microamperes.

*4.1.5  Output Signal Wave Form (Figure 4.3)* During transitions of the generator output between alternating binary states (one—zero—one—zero, etc.), the differential signal measured across a 100 ohm ± 10% test load connected between the generator output terminals shall be such that the voltage monotonically changes between 0.1 and 0.9 of Vss within 0.1 of the unit interval or 20 nanoseconds, whichever is greater. Thereafter, the signal voltage shall not vary more than 10% of Vss from the steady state value, until the next binary transition occurs, and at no time shall the instantaneous magnitude of Vt or $\overline{V}$t exceed 6 volts, nor be less than 2 volts. Vss is defined as the voltage difference between the two steady state values of the generator output.

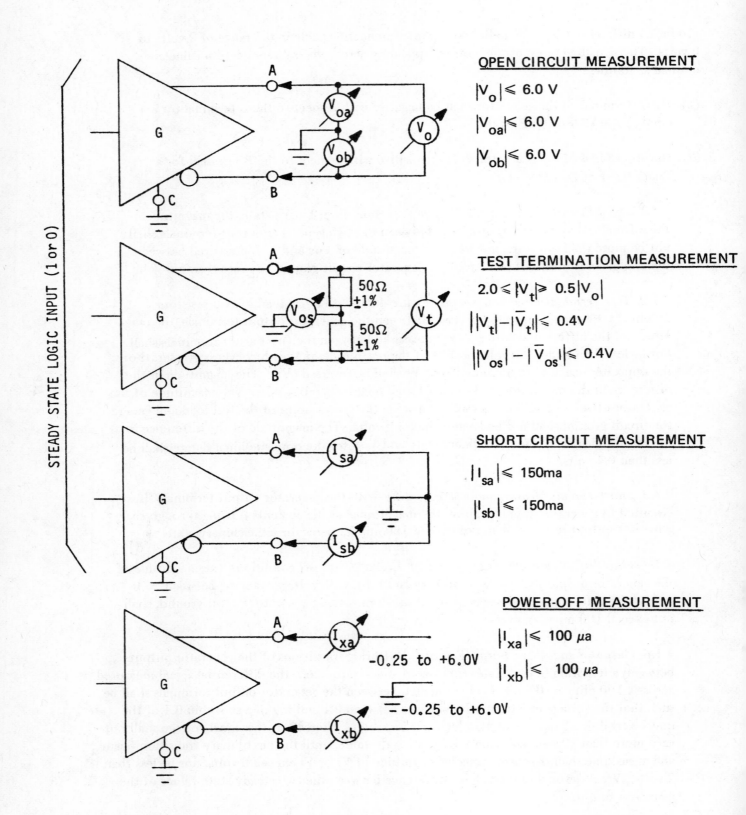

FIGURE 4.2

Generator Parameter Measurements

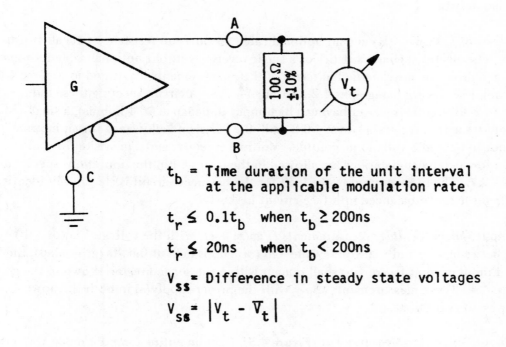

$t_b$ = Time duration of the unit interval at the applicable modulation rate

$t_r \leq 0.1t_b$    when $t_b \geq 200$ns

$t_r \leq 20$ns    when $t_b < 200$ns

$V_{ss}$ = Difference in steady state voltages

$V_{ss} = \left| V_t - \overline{V}_t \right|$

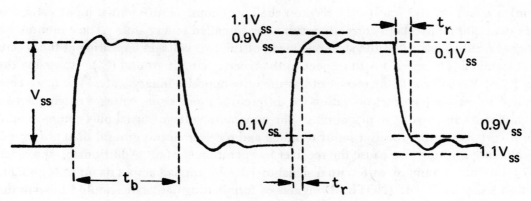

FIGURE 4.3

Generator Output Signal Wave Form

4.2 Load Characteristics

The load consists of a receiver (R) and an optional cable termination resistance (Rt) as shown in Figure 4.1. The electrical characteristics of a single receiver without optional cable termination and fail safe provision are specified in terms of the measurements illustrated in Figures 4.4 through 4.6 and described in paragraphs 4.2.1 through 4.2.5. A circuit meeting these requirements results in a differential receiver having a high input impedance ($\geq$ 4 kohms), a small input threshold transition region between $-0.2$ and $+0.2$ volts, and allowance for an internal bias voltage not to exceed 3 volts in magnitude. Multiple receivers and a provision for fail safe operation for specific applications are allowed in the load within the limitations specified in paragraph 4.2.7. The receiver used in the load for the balanced circuit is electrically identical to that specified for the unbalanced interface circuit in RS-423.

*4.2.1 Input Current-Voltage Measurements (Figure 4.4)* With the voltage Via (or Vib) ranging between $-10$ and $+10$ volts, while Vib (or Via) is held at 0 volts (grounded), the resultant input current Iia (or Iib) shall remain within the shaded region shown in the graph in Figure 4.4. These measurements apply with the power supply(s) in both the power-on and power-off conditions.

*4.2.2 Input Sensitivity Measurement (Figure 4.5)* Over an entire common mode voltage (Vcm) range of $-7$ to $+7$ volts, the receiver shall not require a differential input voltage of more than 200 millivolts to correctly assume the intended binary state. The common mode voltage (Vcm) is defined as the algebraic mean of the two voltages appearing at the receiver input terminals (A' and B') with respect to the receiver circuit ground (C'). Reversing the polarity of Vi shall cause the receiver to assume the opposite binary state. The receiver is required to maintain correct operation for differential input signal voltages ranging between 200 millivolts and 6 volts in magnitude. The maximum voltage (signal plus common mode) present between either receiver input terminal and receiver circuit ground shall not exceed 10 volts in magnitude nor cause the receiver to operationally fail. Additionally, the receiver shall tolerate a maximum differential signal of 12 volts applied across its input terminals without being damaged. (NOTE: Designers of terminating hardware should be aware that slow signal transitions with noise present may give rise to instability or oscillatory conditions in the receiving device, and therefore appropriate techniques should be implemented to prevent such behavior. For example, adequate hysteresis may be incorporated into the receiver to prevent such conditions.)

*4.2.3 Input Balance Measurement (Figure 4.6)* The balance of the receiver input voltage-current characteristics and bias voltages shall be such that the receiver will remain in the intended binary state when a differential voltage (Vi) of 400 millivolts is applied through 500 ohms $\pm$ 1% to each input terminal, as shown in Figure 4.6, and Vcm is varied between $-7$ and $+7$ volts. When the polarity of Vi is reversed, the opposite binary state shall be maintained under the same conditions.

**FIGURE 4.6 Receiver Input Balance Measurement**

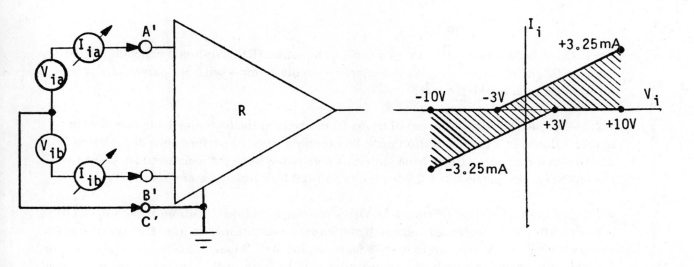

**FIGURE 4.4 Receiver Input Current-Voltage Measurement**

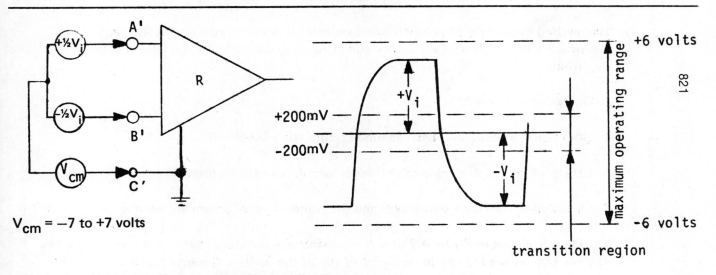

$V_{cm} = -7$ to $+7$ volts

**FIGURE 4.5 Receiver Input Sensitivity Measurement**

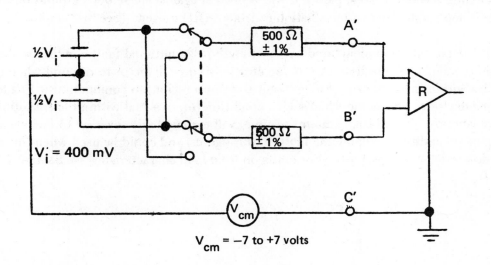

$V_i = 400$ mV

$V_{cm} = -7$ to $+7$ volts

*4.2.4 Cable Termination* The use of a cable termination (Rt) is optional depending upon the specific environment in which the interface circuit is employed. See paragraph 4.2.7 for limit on total load resistance.

*4.2.5 Multiple Receivers* The use of up to 10 receivers in the load may be optionally employed. However, extreme caution must be exercised to avoid performance degradation due to signal reflective effects from stub lines emanating from the load interface point to the receivers. See paragraph 4.2.7 for limits on total load resistance and sensitivity.

*4.2.6 Fail Safe Operation (Figure 4.7)* Other standards and specifications using the electrical characteristics of the balanced voltage digital interface circuit may require that specific interchange leads be made fail safe to certain fault conditions. Where fail safe operation is required by such referencing standards and specifications, a provision shall be incorporated in the load to provide a steady binary condition (either MARK or SPACE as required by the application) to protect against certain fault conditions.

The method of providing fail safe is not standardized, however, the circuit shown in Figure 4.7 will provide a steady binary condition of the receiver for the following fault condition:

a. Generator power off.

b. Both signal wires open (Signal common return still connected).

c. Generator not implemented (Signal leads may or may not be present).

d. Open connector (Both signal leads and the common signal return are open simultaneously).

The fail safe circuit in Figure 4.7 uses two resistors and a voltage source as shown to produce a steady bias on the receiver in the event of any of the faults a. through d. listed above. In normal operation, the low source resistance of the generator will cause the effect of the bias to become negligible on the receiver slicing level. This circuit will not protect against short circuits across the cable pair, nor will it protect against single open ground returns, and is not applicable where a termination resistance (Rt) is used. (See Section 4.2.7)

If the fail safe is implemented by other methods, additional fault conditions may be detected. For example, a threshold region detector (a window detector to respond when the input signal lies within the $-200$ to $+200$ millivolt transition region) in conjunction with a monostable timing device to determine when such a condition (input signal within the transition region) has existed for an abnormal amount of time, will expand fault coverage to include a shorted cable pair (in addition to detecting faults a. through d.) and could be used when the cable termination resistance is present. (For limits on total load characteristics see Section 4.2.7.)

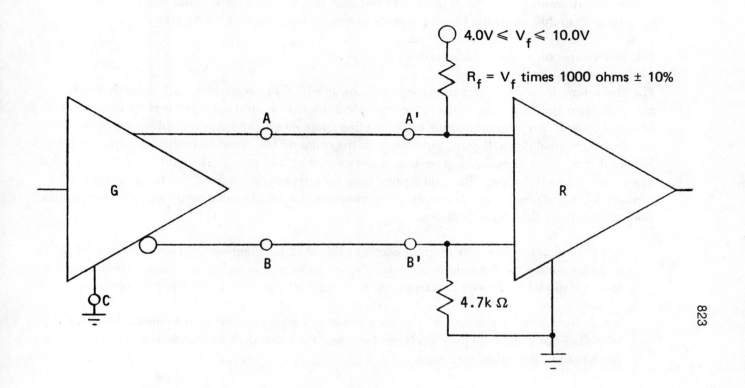

$4.0V \leq V_f \leq 10.0V$

$R_f = V_f$ times 1000 ohms ± 10%

A          A'

G                    R

B          B'

4.7k Ω

C

823

FIGURE 4.7

Example Method of Fail Safe

*4.2.7 Total Load Characteristic Limits* The total load including multiple receivers, fail safe provision, and cable termination shall have a resistance greater than 90 ohms between its input points (A′ and B′, Figure 4.1) and shall not require a differential input voltage of more than 200 millivolts for all receivers to assume the intended binary state.

## 4.3 Interconnecting Cable Characteristics

The characteristics of the interconnecting cable are specified in paragraphs 4.3.1 through 4.3.4, and additional guidance concerning cable characteristics that are not specified is given in Section 7 of this standard. An interconnecting cable meeting these specifications will result in a transmission line with a nominal characteristic impedance on the order of 100 ohms to frequencies greater than 100 kilohertz, and a dc series loop resistance not exceeding 240 ohms. The cable may be composed of twisted or untwisted pair (flat cable) possessing the characteristics described in paragraphs 4.3.1 through 4.3.4 uniformly over its length. Most commonly available cable used for telephone applications should meet these specifications.

*4.3.1 Conductor Size* The interconnecting cable shall be composed of two wires of a 24 AWG or larger conductor for solid or stranded copper wires, or for non-copper conductors, a sufficient size to yield a dc wire resistance not to exceed 30 ohms per 1000 feet per conductor.

*4.3.2 Mutual Pair Capacitance* The capacitance between one wire in the pair to the other wire shall not exceed 20 picofarads per foot, and the value shall be reasonably uniform over the length of the cable.

*4.3.3 Stray Capacitance* The capacitance between one wire in the cable to all others in the cable sheath, with all others connected to ground, shall not exceed 40 picofarads per foot and shall be reasonably uniform for a given conductor over the length of the cable.

*4.3.4 Pair-to-Pair Balanced Crosstalk* The balanced crosstalk from one pair of wires to any other pair in the same cable sheath shall have a minimum value of 40 decibels of attenuation measured at 150 kilohertz.

## 5. ENVIRONMENTAL CONSTRAINTS

A balanced voltage digital interface circuit conforming to this standard will perform satisfactorily at data modulation rates up to 10 megabauds providing that the following operational constraints are simultaneously satisfied·

a. The interconnecting cable length is within that recommended for the applicable modulation rate indicated in Section 7, and the cable is appropriately terminated.

b. The common mode voltage at the receiver is less than 7 volts (peak). The common mode voltage is defined to be any uncompensated combination of generator-receiver ground potential difference, the generator offset voltage (Vos), and longitudinally coupled peak random noise voltage measured between the receiver circuit ground and cable with the generator ends of the cable short-circuited to ground.

## 6. CIRCUIT PROTECTION

Balanced voltage digital interface generator and receiver devices, under either the power-on or power-off condition, complying with this standard shall not be damaged under the following conditions:

a. Generator open circuit.

b. Short-circuit across the balanced interconnecting cable.

c. Short-circuit to any other lead using electrical characteristics complying with this standard and RS-423.

d. Short circuit to ground.

The above faults b. through d. may cause the power dissipation in the interface devices to approach the maximum power dissipation that may be tolerated by a typical integrated circuit (IC) package. The user is therefore cautioned that where multiple generators or receivers are implemented in a single IC package, only one such fault per package may be tolerated at one time without damage occuring.

The user is also cautioned that the generator and receiver devices complying with this standard may be damaged by spurious voltages applied between their input/output terminals and their circuit grounds. In those applications where the interconnecting cable may be inadvertently connected to other circuits or where it may be exposed to a severe electromagnetic environment, protection should be employed as may be specified in a standard yet to be written.

## 7. GUIDELINES

When interconnecting equipment using the electrical interface characteristics specified in this standard, certain consideration should be given to some of the problems that may be encountered due to the interconnecting cable characteristics, cable termination resistance, optional grounding arrangements, and interconnection with interfaces using other electrical characteristics.

### 7.1 Interconnecting Cable

The electrical characteristics of the interconnecting cable are specified in Section 4.3 in this standard. The following section is additional guidance concerning operational constraints imposed by the cable parameters of length and termination resistance.

*7.1.1 Length*  The maximum permissible length of cable separating the generator and load is a function of modulation rate and is influenced by the tolerable signal distortion, the amount of longitudinally coupled noise and ground potential difference introduced between the generator and load circuit grounds as well as by cable balance. Increasing the physical separation and interconnecting cable length between the generator and load interface points, increases exposure to common mode noise, signal distortion, and the effects of cable imbalance. Accordingly, users are advised to restrict cable length to a minimum, consistent with the generator-load physical separation requirements.

825

826

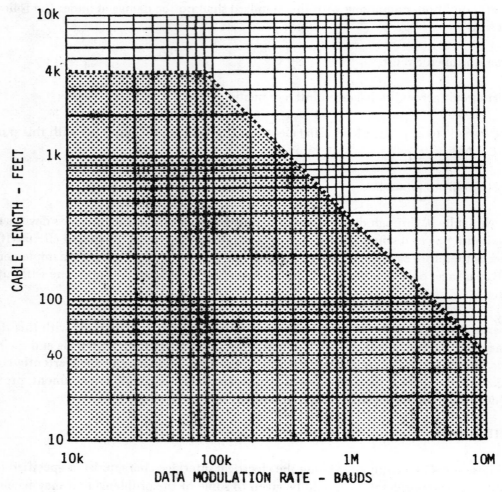

24 AWG TWISTED PAIR CABLE

FIGURE 7.1

Data Modulation Rate Versus
Cable Length for Balanced Interface

The curve of cable length versus modulation rate given in Figure 7.1 may be used as a conservative guide. This curve is based upon empirical data using a 24 gauge twisted-pair telephone cable terminated in a 100 ohm resistive load. The cable length restriction shown by the curve is based upon assumed load signal quality requirements of:

a. Signal rise and fall time equal to, or less than one-half unit interval at the applicable modulation rate.

b. A maximum voltage loss between generator and load of 6dBV.

At the higher modulation rates (0.09 to 10 megabauds) the sloping portion of the curve shows the cable length limitation established by the assumed signal rise and fall time requirements. As the modulation rate is reduced below 90 kilobauds, the cable length has been limited at 4000 feet by the assumed maximum allowable 6dBV signal loss.

The user is cautioned that the curve given in Figure 7.1 does not account for cable imbalance, or common mode noise beyond the limits specified that may be introduced between the generator and load by exceptionally long cables. On the other hand, while signal quality degradation within the bounds of Figure 7.1 will ensure a zero crossing ambiguity of less than .05 unit interval, many applications can tolerate greater timing and amplitude distortion. Thus correspondingly greater cable lengths may be employed than those indicated. Experience has shown that in most practical cases the operating distance at lower modulation rates may be extended to several miles.

*7.1.2 Cable Termination* The characteristic impedance of twisted pair cable is a function of frequency, wire size and type as well as the kind of insulating materials employed. For example, the characteristic impedance of average 24 gauge, copper conductor, plastic insulated twisted pair telephone cable, to a 100 kHz sine wave will be on the order of 100 ohms.

In general, reliable operation of the balanced interface circuit is not particularly sensitive to the presence or absence of the cable termination at lower speeds (below 200 kilobaud) or at any speed where the signal rise time at the load end of the cable is greater than 4 times the one-way propogation delay time of the cable. At other speeds and distances, where signal reflections are of negligible significance, terminating the cable with a resistor ranging in value from 90 to 150 ohms tends to preserve generated signal rise time but at the expense of signal amplitude. At lower modulation rates, where zero crossing ambiguity and signal rise time are not critical, the cable need not be terminated.

## 7.2 Compatibility With Other Interfaces

The electrical characteristics of the balanced voltage digital interface are designed to allow use of both balanced and unbalanced (See RS-423) circuits within the same interconnection cable sheath. For example, the balanced circuits may be used for data and timing while the unbalanced circuits may be used for low speed control functions. The balanced interface circuit is not intended for

828

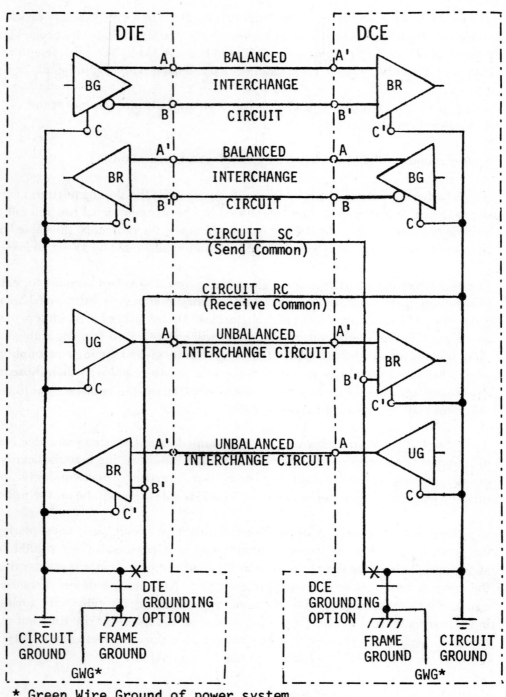

* Green Wire Ground of power system

FIGURE 7.2

Optional Grounding Arrangements

interoperation with other interface electrical characteristics such as RS-232-C, RS-423, MIL-STD-188C, and CCITT Recommendations V.28 and V.35. Under certain conditions with circuits of some of the above interfaces may be possible but may require modification in the interface or within the equipment, therefore, satisfactory operation is not assured and additional provisions not specified herein may be required.

## 7.3 Optional Grounding Arrangements

Proper operation of the interface circuits, whether using balanced, unbalanced, or a combination of electrical characterisitcs, requires the presence of a path between the circuit grounds of the equipments at each end of the interconnection. For example, in a DTE/DCE interface as shown in Figure 7.2, this path may be obtained in a number of ways:

a. Through Earth Ground. In this case, both end equipments have their circuit ground connected to frame ground which in turn is connected to earth ground (e.g., through the third wire (GWG) of the power cord). This is the preferred arrangement when the two earth grounds are at a potential difference of less than four volts.

b. By connecting Circuit SC (Send Common) to DCE circuit ground by means of a wiring option in the DCE. To avoid circulating ground currents, circuit ground must be separated from frame ground in the DCE when this connection is made. Thus, circuit ground for the DCE is obtained from the DTE circuit ground through the interchange circuit SC. The DCE must be capable of withstanding the resulting ground potential differences between its circuit ground and its frame ground. This is the preferred arrangement when the two earth grounds are at a potential difference greater than four volts and the DTE earth ground is the "quieter" of the two earth grounds.

c. By connecting Circuit RC (Receive Common) to DTE circuit ground by means of a wiring option in the DTE. To avoid circulating ground currents, circuit ground must be separated from frame ground in the DTE when this connection is made. Thus, the circuit ground for the DTE is obtained from the DCE circuit ground through interchange Circuit RC. The DTE must be capable of withstanding the resulting ground potential differences between its circuit ground and its frame ground. This is the preferred arrangement when the two earth grounds are at a potential difference greater than four volts and the DCE earth ground is the "quieter" of the two earth grounds.

# EIA STANDARD

*Electrical Characteristics*

*of*

*Unbalanced Voltage Digital*

*Interface Circuits*

**RS-423**

APRIL 1975

*Engineering Department*

# ELECTRONIC INDUSTRIES ASSOCIATION

**NOTICE**

EIA engineering standards are designed to serve the public interest through eliminating misunderstandings between manufacturers and purchasers, facilitating interchangeability and improvement of products, and assisting the purchaser in selecting and obtaining with minimum delay the proper product for his particular need. Existence of such standards shall not in any respect preclude any member or non-member of EIA from manufacturing or selling products not conforming to such standards, nor shall the existence of such standards preclude their voluntary use by those other than EIA members whether the standard is to be used either domestically or internationally.

Recommended standards are adopted by EIA without regard to whether or not their adoption may involve patents on articles, materials, or processes. By such action, EIA does not assume any liability to any patent owner, nor does it assume any obligation whatever to parties adopting the recommended standards.

Published by

# ELECTRONIC INDUSTRIES ASSOCIATION

## Engineering Department

### 2001 Eye Street, N. W., Washington, D. C. 20006

———————

**Price** $4.25

ELECTRICAL CHARACTERISTICS OF

UNBALANCED VOLTAGE DIGITAL INTERFACE CIRCUITS

*(From EIA Standards Proposal No. 1163-A, formulated under the cognizance of EIA Committee TR—30.1 on Signal Quality)*

## 1. SCOPE

This standard specifies the electrical characteristics of the unbalanced voltage digital interface circuit normally implemented in integrated circuit technology that may be employed when specified for the interchange of serial binary signals between Data Terminal Equipment (DTE) and Data Communications Equipment (DCE) or in any interconnection of binary signals between voice or data equipment.

The interface circuit includes a generator connected by an interconnecting cable to a load consisting of a receiver or receivers. The electrical characteristics of the circuit are specified in terms of required voltage, current, and resistance values obtained from direct measurement of the generator and receiver components. The requirements for signal wave shaping, generally necessary to reduce unbalanced circuit near-end crosstalk to adjacent circuits are also described. The receiver specification for the unbalanced interface is electrically identical to that specified for the balanced interface circuit in RS-422. The characteristics of the interconnecting cable are specified, and guidance is given with respect to limitations on data modulation rate imposed by the parameters of cable length and generation of near-end crosstalk.

The parameter values specified for the unbalanced generator and load components of the interface are designed such that unbalanced interface circuits may be used within the same interconnection as balanced interface circuits specified by RS-422. For example, the balanced circuits may be used for data and timing while the unbalanced circuits may be used for low speed control functions. In addition, interoperation may be possible under certain conditions with generators and receivers of other digital interface standards such as EIA RS-232-C and MIL-STD-188C.

It is intended that this standard will be referenced by other standards that specify the complete DTE/DCE interface (i.e. protocol, timing, pin assignments, etc.) for applications where the electrical characteristics of an unbalanced voltage digital circuit are required. Applications are also foreseen in other areas using binary signal interchange. This standard does not specify other characteristics of the DTE/DCE interface (such as signal quality and timing, etc.) essential for the interconnected equipment operation. When this standard is referenced by other standards or specifications, it should be noted that certain options are available. The preparer of those referencing standards and specifications must determine and specify those optional features which are required for that application.

## 2. CROSS REFERENCE

This standard is one of a series relating to the interconnection of DTE and DCE. Other EIA standards in this series, in addition to RS-422, pertaining to the DTE/DCE and DTE/DTE interface specifications are in existence or in various stages of preparation at the time of publication of this standard. The user is referred to EIA Engineering Headquarters for information on specific standards available.

## 3. APPLICABILITY

The provisions of this standard may be applied to the circuits employed at the interface between equipments where the information being conveyed is in the form of binary signals at the dc baseband level. This standard shall be referenced by the specifications and specific interface standards applying these electrical characteristics. Typical points of applicability for this standard are depicted in Figure 3.1.

The unbalanced voltage digital interface circuit will normally be utilized on data, timing or control circuits where the modulation rate on these circuits is up to 100 kilobauds. Unbalanced voltage digital interface devices meeting the electrical characteristics of this standard need not operate over the entire modulation rate range specified. They may be designed to operate over narrower ranges to more economically satisfy specific applications, particularly at the lower modulation rates.

While the unbalanced voltage digital interface circuit is intended for use at lower modulation rates than the balanced circuit, its general use is not recommended where the following conditions prevail:

a. The interconnecting cable is too long for effective unbalanced operation.

b. The interconnecting cable is exposed to extraneous noise sources that may cause an unwanted voltage in excess of plus or minus one volt measured differentially between the signal conductor and circuit common at the load end of the cable with a 50 ohm resistor substituted for the generator.

c. It is necessary to minimize interference with other signals.

d. Inversion of signals may be required, e.g., plus MARK to minus MARK may be obtained by inverting the cable pair.

While a restriction on maximum cable length is not specified, guidelines are given with respect to conservative operating distances as a function of modulation rate (see Section 7). In general, these conservative values may be greatly exceeded where the installation is engineered to ensure that noise and ground potential values are held within specified limits.

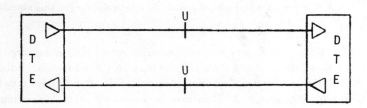

834

LEGEND:

DTE        =    Data Terminal Equipment

DCE        =    Data Communications Equipment

▷          =    Interface Generator

–▷         =    Interface Load

U
+          =    Unbalanced Interface Circuit

≫–         =    Telecommunications Channel

FIGURE 3.1

Applications of Unbalanced Voltage
Digital Interface Circuit

## 4. ELECTRICAL CHARACTERISTICS

The unbalanced voltage digital interface circuit is shown in Figure 4.1. The circuit consists of three parts: the generator, the interconnecting cable, and the load. The load may be comprised of one or more receivers (R). The electrical characteristics of the generator and receiver are specified in terms of direct electrical measurements while the interconnecting cable is specified in terms of its electrical and physical characteristics.

### 4.1 Generator Characteristics

The generator electrical characteristics are specified in accordance with measurments illustrated in Figures 4.2 through 4.5 and described in paragraphs 4.1.1 through 4.1.6. A generator circuit meeting these requirements results in a low impedance (50 ohms or less) unbalanced voltage source that will produce a voltage applied to the interconnecting cable in the range of 4 volts to 6 volts. The signalling sense of the voltages appearing across the interconnecting cable is defined as follows:

(1) The A terminal of the generator shall be negative with respect to the C terminal for a binary 1 (MARK or OFF) state.

(2) The A terminal of the generator shall be positive with respect to the C terminal for a binary 0 (SPACE or ON) state.

*4.1.1 Open Circuit Measurement (Figure 4.2).* For either binary state, the magnitude of the voltage (Vo) measured between the generator output terminal and generator circuit ground shall not be less than 4 volts nor more than 6.0 volts. For the opposite binary state, the polarity of Vo shall be reversed.

*4.1.2 Test Termination Measurement (Figure 4.2).* With a test load of 450 ohm ± 1% connected between the generator output terminal and generator circuit ground, the magnitude of the voltage (Vt) measured between the generator output terminal and generator circuit ground shall not be less than 90% of the magnitude of Vo for either binary state.

*4.1.3 Short-Circuit Measurement (Figure 4.2).* With the generator output terminal short-circuited to generator circuit ground, the magnitude of the current flowing through the generator output terminal shall not exceed 150 milliamperes for either binary state.

*4.1.4 Power-off Measurement (Figure 4.2).* Under power-off conditions, the magnitude of the generator output leakage current (Ix), with voltages ranging between +6.0 and −6.0 volts applied between the generator output terminal and generator circuit ground, shall not exceed 100 microamperes.

*4.1.5 Output Signal Wave Form (Figure 4.3).* During transitions of the generator output between alternating binary states (one—zero—one—zero—, etc.), the signal measured across a

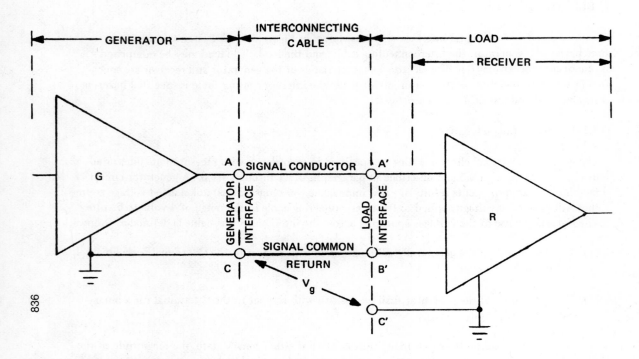

LEGEND:

A, C — Generator Interface

A', B' — Load Interface

C' — Load Circuit Ground

C — Generator Circuit Ground

$V_g$ — Ground Potential Difference

FIGURE 4.1

Unbalanced Digital Interface Circuit

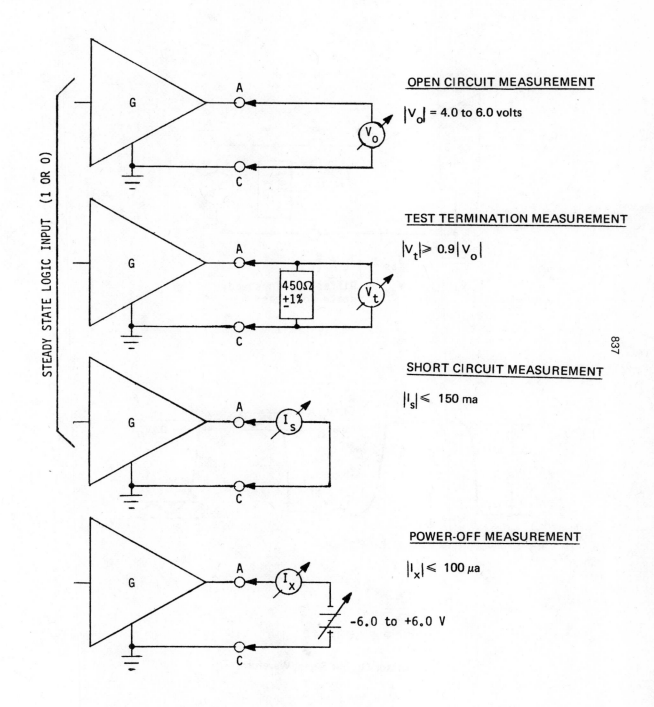

OPEN CIRCUIT MEASUREMENT

$|V_o|$ = 4.0 to 6.0 volts

TEST TERMINATION MEASUREMENT

$|V_t| \geqslant 0.9 |V_o|$

SHORT CIRCUIT MEASUREMENT

$|I_s| \leqslant$ 150 ma

POWER-OFF MEASUREMENT

$|I_x| \leqslant$ 100 μa

-6.0 to +6.0 V

STEADY STATE LOGIC INPUT (1 OR 0)

837

FIGURE 4.2

Generator Parameter Measurements

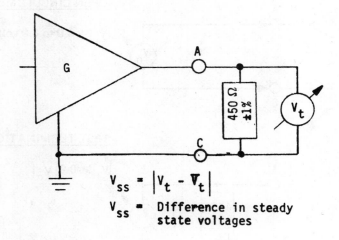

$V_{ss} = \left| V_t - \overline{V}_t \right|$

$V_{ss}$ = Difference in steady
state voltages

838

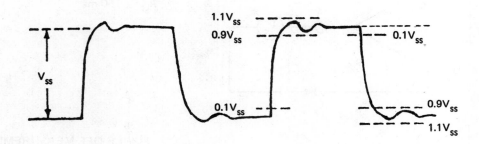

FIGURE 4.3

Generator Output Signal Waveform

450 ohm ± 1% test load connected between the generator output terminal and generator circuit ground shall be such that the voltage monotonically changes between 0.1 and 0.9 of Vss. Thereafter, the signal voltage shall not vary more than 10% of Vss from the steady state value, until the next binary transition occurs, and at no time shall the instantaneous magnitude of Vt or $\overline{V}t$ exceed 6 volts, nor be less than 4 volts. Vss is defined as the voltage difference between the two steady state values of the generator output.

*4.1.6 Wave Shaping (Figure 4.4).* Wave shaping of the generator output signal shall be employed to control the level of interference (near-end crosstalk) that may be coupled to adjacent circuits in an interconnection. The near-end crosstalk is a function of both rise time and cable length (see Section 7.1), and therefore, in establishing the specifications for wave shaping, both of these characteristics have been considered. The rise time tr of the signal shall be controlled to ensure the signal reaches 0.9 Vss between 0.1 and 0.3 of the unit interval (tb) at the maximum modulation rate. Below 1 kilobaud, tr shall be between 100 and 300 microseconds. If a generator is to operate over a range of modulation rates and employs a fixed amount of wave shaping which meets the specification for the maximum modulation rate of the operating range the wave shaping is considered adequate for all lesser modulation rates and equal or lesser cable lengths recommended for the maximum modulation rate in Figure 7.1 even though this may result in a rise time less than 0.1 tb for the actual operating rate. The method of providing the wave shaping is not standardized.

The required wave shaping may be accomplished, for example, either by providing a slow rate control in the generator or by inserting an RC filter at the generator interface point. A combination of these methods may also be employed. An example of the RC filter method is shown in Figure 4.5. Typical values of capacitance Cw, with the value of Rw selected so that Rw plus the output resistance of the generator is approximately 50 ohms, that may be employed with typical twisted pair cable are also given. These values apply for operating distances from zero to that shown by the curve in Figure 7.1 for the applicable modulation rate. Where generators are driving clock leads, the same wave shaping design may be employed as used on the associated data leads, i.e., for clock leads use the period of the clock in lieu of tb to determine rise time.

## 4.2 Load Characteristics

The load consists of one or more receivers (R) as shown in Figure 4.1. The electrical characteristics of a single receiver without fail safe provision are specified in terms of the measurements illustrated in Figures 4.6 through 4.8 and described in paragraphs 4.2.1 through 4.2.4. A circuit meeting these requirements results in a differential receiver having a high input impedance ($\geq$ 4 kohms), a small input threshold transition region between $-$ 0.2 and $+$ 0.2 volts, and allowance for an internal bias voltage not to exceed 3 volts in magnitude. Multiple receivers and a provision for fail safe operation for specific applications are allowed in the load, with the limitations specified in paragraph 4.2.6. The receiver used in the load for the unbalanced circuit is electrically identical to that specified for the balanced interface circuit in RS-422.

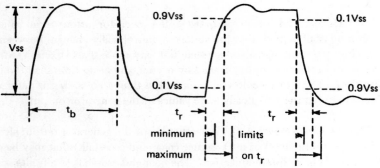

$t_b$ = Time duration of the unit interval at the applicable modulation rate.

$t_r$ = 100$\mu$s to 300$\mu$s when $t_b \geqslant$ 1ms

$t_r$ = 0.1$t_b$ to 0.3$t_b$ when $t_b <$ 1ms

FIGURE 4.4

Signal Wave Shaping Requirement

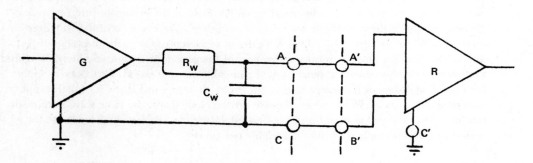

| $C_w$ (microfarads) | Maximum Data Modulation Rate (bauds) | |
|---|---|---|
| 1.0 | 0 | — 2.5k |
| 0.47 | 2.5k | — 5.0k |
| 0.22 | 5.0k | — 10k |
| 0.10 | 10k | — 25k |
| 0.047 | 25k | — 50k |
| 0.022 | 50k | — 100k |

FIGURE 4.5

Example Method for Wave Shaping

840

*4.2.1 Input Current-Voltage Measurements (Figure 4.6).* With the voltage Via or (Vib) ranging between −10 and +10 volts, while Vib (or Via) is held at 0 volts (grounded), the resultant input current Iia (or Iib) shall remain within the shaded region shown in the graph in Figure 4.6. These measurements apply with the power supply(s) in both the power-on and power-off conditions.

*4.2.2 Input Sensitivity Measurement (Figure 4.7).* Over an entire common mode voltage (Vcm) range of −7 to +7 volts, the receiver shall not require a differential input voltage of more than 200 millivolts to correctly assume the intended binary state. The common mode voltage (Vcm) is defined as the algebraic mean of the two voltages appearing at the receiver input terminals (A′ and B′) with respect to the receiver circuit ground (C′). Reversing the polarity of Vi shall cause the receiver to assume the opposite binary state. The receiver is required to maintain correct operation for differential input signal voltages ranging between 200 millivolts and 6 volts in magnitude. The maximum voltage (signal plus common mode) present between either receiver input terminal and receiver circuit ground shall not exceed 10 volts in magnitude nor cause the receiver to operationally fail. Additionally, the receiver shall tolerate a maximum differential signal of 12 volts applied across its input terminals without being damaged. (NOTE: Designers of terminating hardware should be aware that slow signal transitions with noise present may give rise to instability or oscillatory conditions in the receiving device, and therefore appropriate techniques should be implemented to prevent such behavior. For example, adequate hysteresis may be incorporated into the receiver to prevent such conditions.)

*4.2.3 Input Balance Measurement (Figure 4.8).* The balance of the receiver input voltage-current characteristics and bias voltages shall be such that the receiver will remain in the intended binary state when a differential voltage (Vi) of 400 millivolts is applied through 500 ohms ± 1% to each input terminal, as shown in Figure 4.8, and Vcm is varied between −7 and +7 volts. When the polarity of Vi is reversed, the opposite binary state shall be maintained under the same conditions.

*4.2.4 Multiple Receivers.* The use of up to 10 receivers in the load may be optionally employed. However, extreme caution must be exercised to avoid performance degradation due to signal reflective effects from stub lines emanating from the load interface point to the receivers. See paragraph 4.2.6 for limit on total load resistance.

*4.2.5 Fail Safe Operation (Figure 4.9).* Other standards and specifications using the electrical characteristics of the unbalanced interface circuit may require that specific interchange leads be made fail safe to certain fault conditions. Where fail safe operation is required by such referencing standards and specifications, a provision shall be incorporated in the load to provide a steady binary condition (either MARK or SPACE as required by the application) to protect against certain fault conditions.

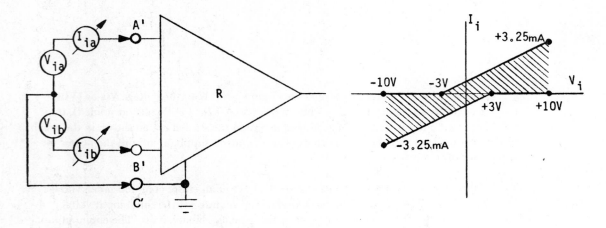

FIGURE 4.6 Receiver Input Current-Voltage Measurement

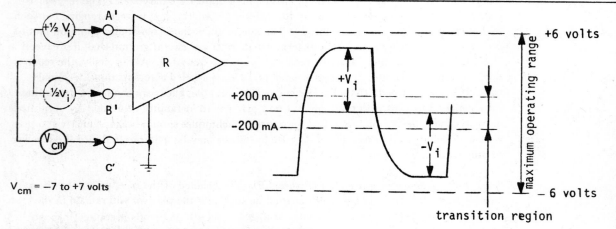

$V_{cm}$ = −7 to +7 volts

FIGURE 4.7 Receiver Input Sensitivity Measurement

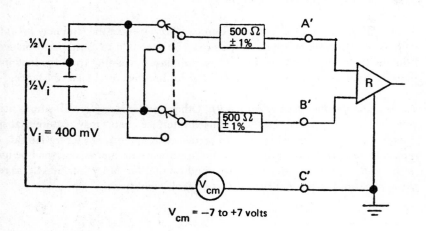

$V_i$ = 400 mV

$V_{cm}$ = −7 to +7 volts

FIGURE 4.8 Receiver Input Balance Measurement

The method of providing fail safe is not standardized, however, the circuit shown in Figure 4.9 as an example, will provide a steady binary condition of a single receiver for the following fault conditions:

a. Generator power off.

b. Open signal lead (signal common return still connected).

c. Generator not implemented (signal lead may or may not be present).

d. Open connector (both signal lead and the signal common return are open simultaneously).

The fail safe circuit shown in Figure 4.9 uses a bias voltage through a resistance to detect faults a. through d. above. Under low source resistance conditions (normal operation), the bias voltage (Vf) has little effect on the receiver operation. Under high source resistance conditions (fail), the bias forces the receiver to the desired binary steady state condition. The capacitor (Cf) is required in the load to reduce any spurious noise as a result of near-end crosstalk to an ineffective level. The suggested values for Cf that will have little effect on normal operation are also given in Figure 4.9. If RC wave shaping as described in paragraph 4.1.6 is employed, an additional 50 ohm resistance Rr should be added in series with the generator output to reduce the effects of reflected waves.

If the fail safe is implemented by other methods, additional fault conditions may be detected. For example, a threshold region detector (a window detector to respond when the input signal lies within the $-200$ to $+200$ millivolt transition region) in conjunction with a monostable timing device to determine when such a condition (input signal within the transition region) has existed for an abnormal amount of time, will expand fault coverage to include a shorted cable pair (in addition to detecting faults a. through d.). For limits on total load characteristics see section 4.2.6.

*4.2.6 Total Load Characteristic Limits.* The total load including multiple receivers and fail safe provision shall have a resistance greater than 400 ohms between its input points (A' and B', Figure 4.1) and shall not require a differential input voltage of more than 200 millivolts for all receivers to assume the intended binary state.

## 4.3 Interconnecting Cable Characteristics

The characteristics of the interconnecting cable are specified in paragraphs 4.3.1 through 4.3.3, and additional guidance concerning cable characteristics that are not specified is given in Section 7 of this standard. An interconnecting cable may be composed of twisted or untwisted pair (flat cable), or unpaired wires possessing the characteristics described in paragraphs 4.3.1 through 4.3.4 uniformly over its length. Most commonly available cable used for telephone applications should meet these specifications. Where twisted pair cable is used with the unbalanced circuit, and the two wires serve as signal conductors for two different interchange circuits, the information flow in both wires shall be in the same direction.

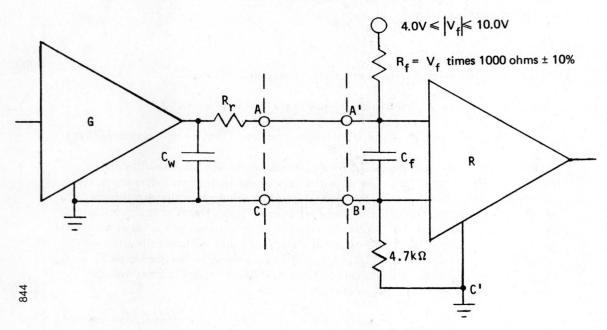

| $C_f$ (microfarads) | Maximum Data Modulation Rate (Bauds) |
|---|---|
| 0.22 | 0–2.5k |
| 0.10 | 2.5k–5.0k |
| 0.05 | 5.0k–10k |
| 0.022 | 10k–25k |
| 0.010 | 25k–50k |
| 0.005 | 50k–100k |

FIGURE 4.9

Example Method of Fail Safe

4.3.1 *Conductor Size.* The interconnecting cable shall be composed of wires of a 24 AWG or larger conductor for solid or stranded copper wires, or for non-copper conductors, a sufficient size to yield a dc wire resistance not to exceed 30 ohms per 1000 feet per conductor.

4.3.2 *Stray Capacitance.* The capacitance between one wire in the cable to all others in the cable sheath, with all others connected to ground, shall not exceed 40 picofarads per foot and shall be reasonably uniform for a given conductor over the length of the cable.

4.3.3 *Mutual Pair Capacitance (applicable only where paired wires are used).* The capacitance between one wire in the pair to the other wire shall not exceed 20 picofarads per foot, and the value shall be reasonably uniform over the length of the cable.

4.4 Signal Common Return (Figure 4.10)

The interconnection between the generator and load interface points shown in Figure 4.1 shall consist of a signal conductor and a signal common return. In order to minimize the effects of ground potential difference (Vg) and longitudinally coupled noise on the signal at the load interface point, the signal common return shall only be connected to circuit ground at the C terminal of the generator interface point. (See section 7.3 for optional signal and protective grounding arrangements.) As shown in Figure 4.10, a separate signal common return shall be used for each signal direction in an interconnection. For example, The B′ terminal of all the receivers in Equipment "W" that interconnect with unbalanced generators in Equipment "E" shall connect to Signal Common Return "W" which is connected to generator circuit ground only in Equipment "E". Signal Common Return "E" is used to interconnect terminal B′ of the receivers in Equipment "E" with the generator circuit ground (C) of the unbalanced generators in Equipment "W".

## 5. ENVIRONMENTAL CONSTRAINTS

An unbalanced voltage digital interface circuit conforming to this standard will perform satisfactorily at data modulation rates ranging between zero and 100 kilobauds providing that the following operational constraints are simultaneously satisfied:

a. The interconnecting cable length is within that recommended for the applicable modulation rate indicated in Section 7.

b. The common mode voltage at the receiver is less than 4 volts (peak). The common mode voltage is defined to be any uncompensated combination of generator-receiver ground potential difference, the generator offset voltage (Vos), and longitudinally coupled peak random noise voltage measured between the receiver circuit ground and cable with the generator ends of the cable short-circuited to ground.

c. The total peak noise between the signal conductor and common return at the load interface point, with a 50 ohm resistor substituted for the generator (and wave shaping network, if used), at the generator interface point is less than the amplitude of the expected received signal minus 0.2 volts.

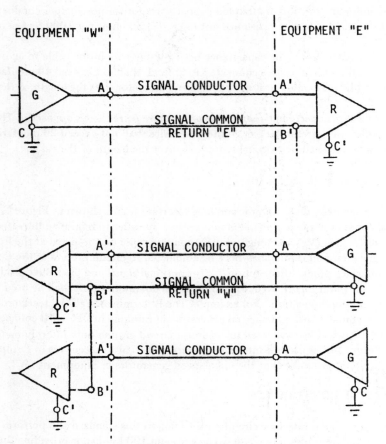

NOTE: See Section 7.3 for optional signal
and Protective grounding arrangements

846

FIGURE 4.10

Interconnection of Signal Common Return

## 6. CIRCUIT PROTECTION

Unbalanced voltage digital interface generator and receiver devices, under either the power-on or power-off condition, complying with this standard shall not be damaged under the following conditions:

a. Generator open circuit.

b. Short-circuit between the signal conductor and common return at either the generator or load interface.

c. Short-circuit to any other lead using electrical characteristics complying with this standard and RS-422 (Balanced Interface).

d. Short circuit to ground.

The above faults b. through d. may cause the power dissipation in the interface devices to approach the maximum power dissipation that may be tolerated by a typical integrated circuit (IC) package. The user is therefore cautioned that where multiple generators or receivers are implemented in a single IC package, only one such fault per package may be tolerated at one time without damage occurring.

The user is also cautioned that the generator and receiver devices complying with this standard may be damaged by spurious voltages applied between their input/output terminals and their circuit grounds. In those applications where the interconnecting cable may be inadvertently connected to other circuits or where it may be exposed to a severe electromagnetic environment, protection should be employed as may be specified in a standard yet to be written.

## 7. GUIDELINES ON INTERCONNECTIONS

When interconnecting equipment using the electrical interface characteristics specified in this standard, certain consideration should be given to some of the problems that may be encountered due to the interconnecting cable length, generation of near-end crosstalk, use of multiple receivers, fail safe operation, and interconnection with interfaces using other electrical characteristics.

### 7.1 Operational Constraints

The maximum operating distance for the unbalanced voltage digital interface is primarily a function of the amount of interference (near-end crosstalk) coupled to adjacent circuits in the equipment interconnection. Additionally, the unbalanced circuit is susceptible to exposure to differential noise resulting from any imbalance between the signal conductor and signal common return at the load interface point. Increasing the physical separation and interconnection cable length between the generator and load interface points increases the exposure to common mode noise and the amount of near-end crosstalk. Accordingly, users are advised to restrict the cable length to a minimum, consistent with the generator-load physical separation requirements.

The curve of cable length versus modulation rate given in Figure 7.1 may be used as a conservative guide. This curve is based upon calculations and empirical data using twisted pair telephone cable with a shunt capacitance of 16 picofarads per foot, a 50 ohm source impedance, a 12 volt peak-to-peak source signal, and allowing a maximum near-end crosstalk of 1 volt peak. The rise time (tr) of the source signal at modulation rates below 900 bauds is 100 microseconds and above 900 bauds is 0.1 unit interval (see Section 4.1.6).

The user is cautioned that the curve given in Figure 7.1 does not account for common mode noise or near-end crosstalk levels beyond the limits specified that may be introduced between the generator and load by exceptionally long cables. On the other hand, while signal quality degradation within the bounds of Figure 7.1 will ensure a zero crossing ambiguity of less than 5%, many applications can tolerate greater timing and amplitude distortion. Thus correspondingly greater cable lengths may be employed than those indicated. The generation of near-end crosstalk can be reduced by using lower source resistances and increasing the amount of wave shaping employed. In practice, the maximum length cable that may be used must be determined on a case by case basis. Experience has shown in most practical cases that the operating distance at lower modulation rates may be extended to several miles.

## 7.2 Compatibility With Other Interfaces

The electrical characteristics of the unbalanced voltage digital interface are designed to allow use of both balanced (See RS-422) and unbalanced circuits within the same interconnection cable sheath. For example, the balanced circuits may be used for data and timing while the unbalanced circuits may be used for low speed control functions. Although the unbalanced electrical characteristics have been designed to permit interoperation, under certain conditions, with other interface electrical characteristics such as RS-232-C, RS-422, MIL-STD-188C, and CCITT Recommendation V.28, satisfactory operation is not assured and additional provisions which are not specified herein may be required.

## 7.3 Optional Grounding Arrangements

Proper operation of the interface circuits, whether using balanced, unbalanced, or a combination of electrical characteristics, requires the presence of a path between the circuit grounds of the equipments at each end of the interconnection. For example, in a DTE/DCE interface as shown in Figure 7.2, this path may be obtained in a number of ways:

a. Through Earth Ground. In this case, both end equipments have their circuit ground connected to frame ground which in turn is connected to earth ground (e.g., through the third wire (GWG) of the power cord). This is the preferred arrangement when the two earth grounds are at a potential difference of less than four volts.

b. By connecting Circuit SC (Send Common) to DCE circuit ground by means of a wiring option in the DCE. To avoid circulating ground currents, circuit ground must be separated from frame

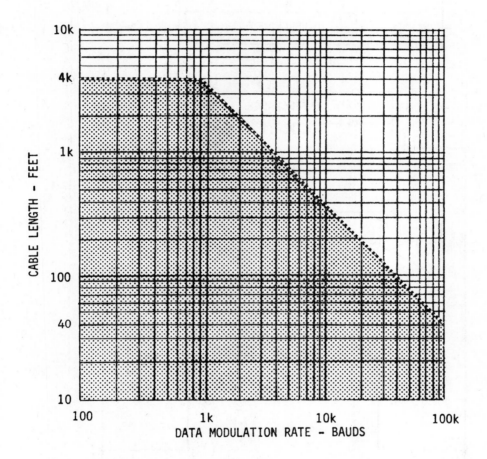

FIGURE 7.1

Data Modulation Rate Versus Cable Length
for Unbalanced Interface

849

850

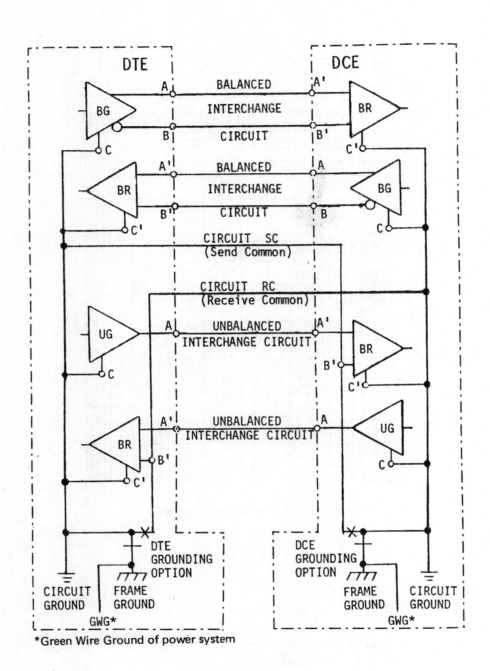

FIGURE 7.2

Optional Grounding Arrangements

ground in the DCE when this connection is made.  Thus, circuit ground for the DCE is obtained from the DTE circuit ground through the interchange circuit SC.  The DCE must be capable of withstanding the resulting ground potential differences between its circuit ground and its frame ground.  This is the preferred arrangement when the two earth grounds are at a potential difference greater than four volts and the DTE earth ground is the "quieter" of the two earth grounds.

c. By connecting Circuit RC (Receive Common) to DTE circuit ground by means of a wiring option in the DTE.  To avoid circulating ground currents, circuit ground must be separated from frame ground in the DTE when this connection is made.  Thus, the circuit ground for the DTE is obtained from the DCE circuit ground through interchange Circuit RC.  The DTE must be capable of withstanding the resulting ground potential differences between its circuit ground and its frame ground.  This is the preferred arrangement when the two earth grounds are at a potential difference greater than four volts and the DCE earth ground is the "quieter" of the two earth grounds.

# EIA STANDARD

GENERAL PURPOSE 37-POSITION AND 9-POSITION
INTERFACE FOR DATA TERMINAL EQUIPMENT
AND DATA CIRCUIT-TERMINATING EQUIPMENT
EMPLOYING SERIAL BINARY DATA INTERCHANGE

# RS-449

NOVEMBER 1977

*Engineering Department*

# ELECTRONIC INDUSTRIES ASSOCIATION

## NOTICE

EIA engineering standards are designed to serve the public interest through eliminating misunderstandings between manufacturers and purchasers, facilitating interchangeability and improvement of products, and assisting the purchaser in selecting and obtaining with minimum delay the proper product for his particular need. Existence of such standards shall not in any respect preclude any member or non-member of EIA from manufacturing or selling products not conforming to such standards, nor shall the existence of such standards preclude their voluntary use by those other than EIA members whether the standard is to be used either domestically or internationally.

Recommended standards are adopted by EIA without regard to whether or not their adoption may involve patents on articles, materials, or processes. By such action, EIA does not assume any liability to any patent owner, nor does it assume any obligation whatever to parties adopting the recommended standards.

EIA Standard RS-449 was developed in close coordination and cooperation with the international standards activities of the C.C.I.T.T (International Telegraph and Telephone Consultative Committee) and ISO (International Organization for Standardization ) and is compatible with C.C.I.T.T. Recommendations V.24, "List of Definitions for Interchange Circuits Between Data Terminal Equipment and Data Circuit-Terminating Equipment", and Recommendation V.54, "Loop Test Devices for Modems" as well as with ISO Draft Proposal DP-4902, "37-Pin and 9-Pin DTE/DCE Interface Connectors and Pin Assignments".

Published by

# ELECTRONIC INDUSTRIES ASSOCIATION

### Engineering Department

### 2001 Eye Street, N. W., Washington, D. C. 20006

PRICE: $9.50

Printed in U.S.A.

# GENERAL PURPOSE 37-POSITION AND 9-POSITION INTERFACE FOR
# DATA TERMINAL EQUIPMENT AND DATA CIRCUIT-TERMINATING EQUIPMENT
# EMPLOYING SERIAL BINARY DATA INTERCHANGE

*(From Standards Proposal No. 1194, formulated under the cognizance*
*of EIA Subcommittee TR-30.2 on Digital Interface.)*

## INDEX

855

856

## LIST OF FIGURES

# GENERAL PURPOSE 37-POSITION AND 9-POSITION INTERFACE FOR
## DATA TERMINAL EQUIPMENT AND DATA CIRCUIT-TERMINATING EQUIPMENT
### EMPLOYING SERIAL BINARY DATA INTERCHANGE

## FOREWORD

(This Foreword provides additional information and does not form an integral part of the EIA Standard specifying the General Purpose 37-Position and 9-Position Interface for Data Terminal Equipment and Data Circuit-Terminating Equipment Employing Serial Binary Data Interchange.)

This Standard, together with EIA Standards RS-422 and RS-423, is intended to gradually replace EIA Standard RS-232-C as the specification for the interface between data terminal equipment (DTE) and data circuit-terminating equipment (DCE) employing serial binary data interchange. With a few additional provisions for interoperability, equipment conforming to this standard can interoperate with equipment designed to RS-232-C. This standard is intended primarily for data applications using analog telecommunications networks.

EIA Standard RS-232-C is in need of replacement in order to specify new electrical characteristics and to define several new interchange circuits. New electrical characteristics are needed to accommodate advances in integrated circuit design, to reduce crosstalk between interchange circuits, to permit greater distances between equipments, and to permit higher data signaling rates. With the expected increase in use of standard electrical interface characteristics between many different kinds of equipment, it is now appropriate to publish the electrical interface characteristics in separate standards. Two electrical interface standards have been published for voltage digital interface circuits:

> EIA Standard RS-422, Electrical Characteristics of Balanced Voltage
> Digital Interface Circuits

> EIA Standard RS-423, Electrical Characteristics of Unbalanced Voltage
> Digital Interface Circuits

With the adoption of EIA Standards RS-422 and RS-423, it became necessary to create a new standard which specifies the remaining characteristics (i.e., the functional and mechanical characteristics) of the interface between data terminal equipment and data circuit-terminating equipment. That is the purpose of this standard.

The basic interchange circuit functional definitions of EIA Standard RS-232-C have been retained in this standard. However, there are a number of significant differences:

a. Application of this standard has been expanded to include signaling rates up to 2,000,000 bits per second.

b.   Ten circuit functions have been defined in this standard which
     were not part of RS-232-C.  These include three circuits for control
     and status of testing functions in the DCE (Circuit LL, Local
     Loopback; Circuit RL, Remote Loopback; and Circuit TM, Test
     Mode), two circuits for control and status of the transfer of the
     DCE to a standby channel (Circuit SS, Select Standby; and Circuit
     SB, Standby Indicator), a circuit to provide an out-of-service func-
     tion under control of the DTE (Circuit IS, Terminal In Service), a
     circuit to provide a new signal function (Circuit NS, New Signal)
     and a circuit for DCE frequency selection (Circuit SF, Select Fre-
     quency).  In addition, two circuits have been defined to provide a
     common reference for each direction of transmission across the
     interface (Circuit SC, Send Common; and Circuit RC, Receive Common).

c.   Three interchange circuits defined in RS-232-C have not been included
     in this standard.  Protective ground (RS-232-C Circuit AA) is not
     included as part of the interface to permit bonding of equipment
     frames, when necessary, to be done in a manner which is in compliance
     with national and local electrical codes.  However, a contact on the
     interface connector is assigned to facilitate the use of shielded inter-
     connecting cable.  The two circuits reserved for data set testing (RS-232-C
     contacts 9 and 10) have not been included in order to minimize the
     size of the interface connector.

d.   Some changes have been made to the circuit function definitions.  For
     example, operation of the Data Set Ready circuit has been changed and
     a new name, Data Mode, has been established due to the inclusion of a separate
     interchange circuit (Test Mode) to indicate a DCE test condition.

e.   A new set of standard interfaces for selected communication system
     configurations has been established.  In order to achieve a greater degree
     of standardization, the option in RS-232-C which permitted the omission
     of the Request to Send interchange circuit for certain transmit only or
     duplex primary channel applications has been eliminated.

f.   A new set of circuit names and mnemonics has been established.  To
     avoid confusion with RS-232-C, all mnemonics in this standard are dif-
     ferent from those used in RS-232-C.  The new mnemonics were chosen
     to be easily related to circuit functions and circuit names.

g.   A different interface connector size and interface connector latching
     arrangement has been specified.  A larger size connector (37-position)
     is specified to accommodate the additional interface leads required for
     the ten newly defined circuit functions and to accommodate balanced
     operation for ten interchange circuits.  In addition, a separate 9-position
     connector is specified to accommodate the secondary channel interchange

circuits. The 37-position and 9-position connectors are from the same connector family as the 25-position connector in general use by equipment conforming to EIA Standard RS-232-C. A connector latching block is specified to permit latching and unlatching of the connectors without the use of a tool. This latching block will also permit the use of screws to fasten together the connectors. The different connectors will also serve as an indication that certain precautions with regard to interface voltage levels, signal risetimes, fail safe circuitry, grounding, etc. must be taken into account before equipment conforming to RS-232-C can be connected to equipment conforming to the new electrical characteristic standards. The connector contact assignments have been chosen to facilitate connection of equipment conforming to this standard to equipment conforming to RS-232-C.

Close attention was given during the development of RS-449 and RS-423 to facilitate an orderly transition from the existing RS-232-C equipment to the next generation without forcing obsolescence or costly retrofits. It will therefore be possible to connect new equipment designed to RS-449 on one side of a interface to equipment designed to RS-232-C on the other side of the interface. Such interconnections can be accomplished with a few additional provisions associated only with the new RS-449 equipment. These provisions are discussed in an EIA Industrial Electronics Bulletin (IE Bulletin No. 12), Application Notes on Interconnection Between Interface Circuits Using RS-449 and RS-232-C.

This standard is designed to be compatible with the specifications of the International Telegraph and Telephone Consultative Committee (C.C.I.T.T.) and the International Organization for Standardization (ISO). However, it should be noted that this standard contains a few specifications which are subjects of further study in C.C.I.T.T. and ISO. These are:

1) Use of interchange circuits Terminal In Service and New Signal.

2) Status of interchange circuits during an equalizer retraining period.

The U.S.A. is actively participating in C.C.I.T.T. and ISO to gain international agreement on these items.

Work is presently underway, in cooperation with C.C.I.T.T. and ISO, to expand the Remote Loopback test function to include testing on multipoint networks. This augmentation will not affect the point-to-point testing capability specified in this document. Work is also underway to augment this standard to cover direct DTE to DTE applications. This augmentation will not affect, in any way, the DTE to DCE operation specified in this document. In addition, work will proceed in cooperation with C.C.I.T.T. toward the development of a more efficient all-balanced interface which minimizes the number of interchange circuits. It is expected that RS-449 will provide the basis for this new work.

860

vii

1. **SCOPE**

1.1 Section Abstracts

This standard is applicable to the interconnection of data terminal equipment (DTE) and data circuit-terminating equipment (DCE) employing serial binary data interchange with control information exchanged on separate control circuits. This standard is intended primarily for data applications using analog telecommunications networks. It defines:

Section 2 - Signal Characteristics

Section 3 - Interface Mechanical Characteristics

Section 4 - Functional Description of Interchange Circuits

Section 5 - Standard Interfaces for Selected Communication System Configurations

In addition, the standard includes:

Section 6 - Recommendations and Explanatory Notes

Section 7 - Glossary of Terms

Figure 1.1 depicts the basic data communications system units.

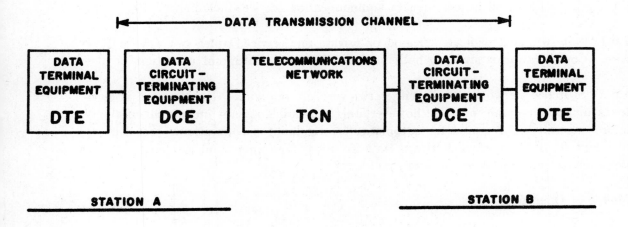

FIGURE 1.1
BASIC DATA COMMUNICATIONS SYSTEM UNITS

## 1.2  Application

This standard applies where equipment on one side of the DTE/DCE interface is intended for connection directly to equipment on the other side without additional technical considerations.  Applications where cable termination, signal waveshaping, interconnecting cable distance, and mechanical configurations of the interface must be tailored to meet specific user needs are not precluded but are beyond the scope of this standard.

## 1.3  Serialization

This standard applies to data communication systems where the data is bit serialized by the DTE and the DCE places no restrictions on the arrangement of the sequence of bits provided by the DTE.

## 1.4  Signaling Rates

This standard is applicable for use at data signaling rates in the range from zero to a nominal upper limit of 2,000,000 bits per second.  Equipment complying with this standard, however, need not operate over this entire data signaling rate range.  They may be designed to operate over a narrower range as appropriate for the specific application.

## 1.5  Synchronous/Nonsynchronous Communication

This standard applies to both synchronous and nonsynchronous serial binary data communication systems.

## 1.6  Classes of Service

This standard applies to all classes of data communication service, including:

1.6.1  Non-switched, dedicated, leased or private line service, either two-wire or four-wire.  Consideration is given to both point-to-point and multipoint operation.

1.6.2  Switched network service, either two-wire or four-wire.  Consideration is given to automatic answering of calls; however, this standard does not include all of the interchange circuits required for automatically originating a connection.  (See EIA Standard RS-366, Interface Between Data Terminal Equipment and Automatic Calling Equipment for Data Communication.)

## 1.7  Modes of Operation

This standard applies to all of the modes of operation afforded under the system configurations indicated in Section 5, Standard Interfaces for Selected Communication System Configurations. Four specific interface configurations, identified as Type SR (Send-Receive), Type SO (Send-Only), Type RO (Receive-Only), and Type DT (Data and Timing only), are defined.

1.8  Allocation of Functions

The DCE may perform transmitting and receiving signal conversion functions as well as control functions.  Other functions, such as pulse regeneration and error control, may or may not be provided.  Equipment to provide additional functions may be included in the DTE or in the DCE, or it may be implemented as a separate unit interposed between the two.

1.8.1  When such additional functions are provided within the DTE or the DCE, this interface standard shall apply only to the interchange circuits between the two classes of equipment.

1.8.2  When additional functions are provided in a separate unit inserted between the DTE and the DCE, this standard shall apply to both sides (the interface with the DTE and the interface with the DCE - see Section 3.2) of such a separate unit.

## 2.  SIGNAL CHARACTERISTICS

2.1  Electrical Characteristics

The electrical characteristics of the interchange circuits are specified in the two following standards:

(1)  RS-422, Electrical Characteristics of Balanced Voltage Digital Interface Circuits

(2)  RS-423, Electrical Characteristics of Unbalanced Voltage Digital Interface Circuits

For the purpose of assigning electrical characteristics, the interchange circuits defined in Section 4.4 are divided into two categories as follows.

2.1.1  *Category I Circuits*

The following ten interchange circuits are classified as Category I circuits:

    Circuit SD   (Send Data),
    Circuit RD   (Receive Data),
    Circuit TT   (Terminal Timing),
    Circuit ST   (Send Timing),
    Circuit RT   (Receive Timing),
    Circuit RS   (Request to Send),
    Circuit CS   (Clear to Send),
    Circuit RR   (Receiver Ready),
    Circuit TR   (Terminal Ready), and
    Circuit DM   (Data Mode).

863

For applications where the signaling rate on the data interchange circuits (Circuits SD and RD) is 20,000 bits per second or less, the individual Category I circuits shall use either the balanced electrical characteristics of RS-422, without the cable termination resistance ($R_t$), or the unbalanced electrical characteristics of RS-423. Guidance for the specific choice of electrical characteristics is given in Section 6.11. Two leads shall be brought out to the interface connector for each Category I circuit as shown in Figure 2.1(a). Thus, each interchange circuit consists of a pair of wires interconnecting a balanced or unbalanced generator and a differential receiver. If RS-423 generators are used, they shall employ waveshaping suitable for operation over an interface cable length of at least 60 meters (200 feet), the maximum cable length allowed for nontailored applications (see Section 6.10).

For applications where the signaling rate on the data interchange circuits (Circuits SD and RD) is above 20,000 bits per second, all Category I circuits shall use the balanced electrical characteristics of RS-422. Use of the optional cable termination resistance ($R_t$) is permitted (see RS-422 for guidance). Each interchange circuit consists of a pair of wires interconnecting a balanced generator and a differential receiver as shown in Figure 2.1(b).

### 2.1.2 *Category II Circuits*

All interchange circuits not classified in Section 2.1.1 as Category I circuits are classified as Category II circuits.

For all applications, the Category II circuits shall use the unbalanced electrical characteristics of RS-423. Each Category II interchange circuit consists of one wire interconnecting an unbalanced generator and a differential receiver as shown in Figure 2.1(c). There are two signal common returns for Category II interchange circuits, one for each direction of transmission. Circuit SC (Send Common) is the common return for all Category II interchange circuits having generators in the DTE. Circuit RC (Receive Common) is the common return for all Category II interchange circuits having generators in the DCE. The RS-423 generators shall employ waveshaping suitable for operation over an interface cable length of at least 60 meters (200 feet), the maximum cable length allowed for nontailored applications (see Section 6.10).

### 2.2 Protective Ground (Frame Ground)

In DTEs and in DCEs, protective ground is a point which is electrically bonded to the equipment frame. It may also be connected to external grounds (e.g., through the third wire of the power cord).

**(a) CATEGORY I CIRCUITS–DATA SIGNALING RATE ≤ 20,000 BITS PER SECOND**

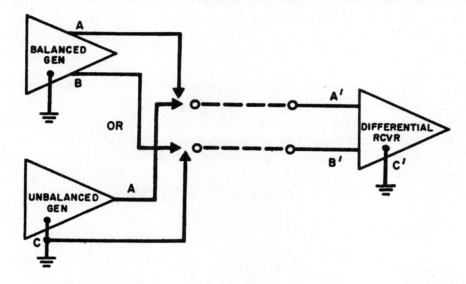

**(b) CATEGORY I CIRCUITS– DATA SIGNALING RATE > 20,000 BITS PER SECOND**

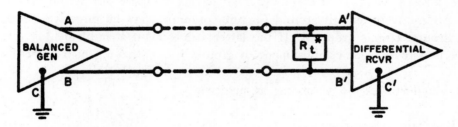

**(c) CATEGORY II CIRCUITS**

**\* OPTIONAL CABLE
TERMINATION RESISTANCE**

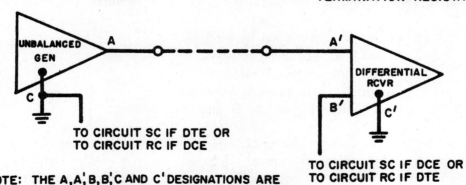

TO CIRCUIT SC IF DTE OR
TO CIRCUIT RC IF DCE

TO CIRCUIT SC IF DCE OR
TO CIRCUIT RC IF DTE

NOTE: THE A, A', B, B', C AND C' DESIGNATIONS ARE
THOSE SPECIFIED IN RS-422 AND RS-423

## FIGURE 2.1
## GENERATOR AND RECEIVER CONNECTIONS AT INTERFACE

865

It should be noted that protective ground (frame ground) is not an interchange circuit in this standard. If bonding of the equipment frames of the DCE and the DTE is necessary, a separate conductor should be used which conforms to the appropriate national or local electrical codes. Attention is called to the applicable Underwriters' Laboratories regulation applying to wire size and color coding.

## 2.3 Shield

In order to facilitate the use of shielded interconnecting cable, interface connector contact number 1 is assigned for this purpose. This will permit the cable associated with the DTE to be composed of tandem connectorized sections with continuity of the shield accomplished by connection through this contact in the connectors. Normally the DCE should make no connection to interface connector contact number 1. It is recognized that for certain electromagnetic interference (EMI) suppression situations, additional provisions may be necessary but are beyond the scope of this standard.

## 2.4 Grounding

Proper operation of the interchange circuits requires the presence of a path between the DTE circuit ground and the DCE circuit ground. This path is obtained by means of interchange Circuit SG, Signal Ground. Both the DCE and the DTE normally should have their circuit ground (circuit common) connected to their protective ground (frame ground) through a resistance of 100 ohms (± 20%) having a power dissipation rating of one-half watt. Alternatively, circuit ground (circuit common) may be directly connected to protective ground (frame ground). If the latter arrangement is implemented, caution should be exercised to prevent establishment of ground loops carrying high currents.

Figure 2.2 illustrates the grounding arrangement.

## 2.5 "Fail Safe" Operation

2.5.1 The receivers for the following interchange circuits:

Circuit IS    (Terminal In Service),
Circuit TR    (Terminal Ready),
Circuit DM    (Data Mode).
Circuit RS    (Request to Send), and
Circuit SRS   (Secondary Request to Send)

shall be used to detect a power-off condition in the equipment connected across the interface and the disconnection of the interconnecting cable. Detection of either of these conditions shall be interpreted as an OFF condition of the interchange circuit.

866

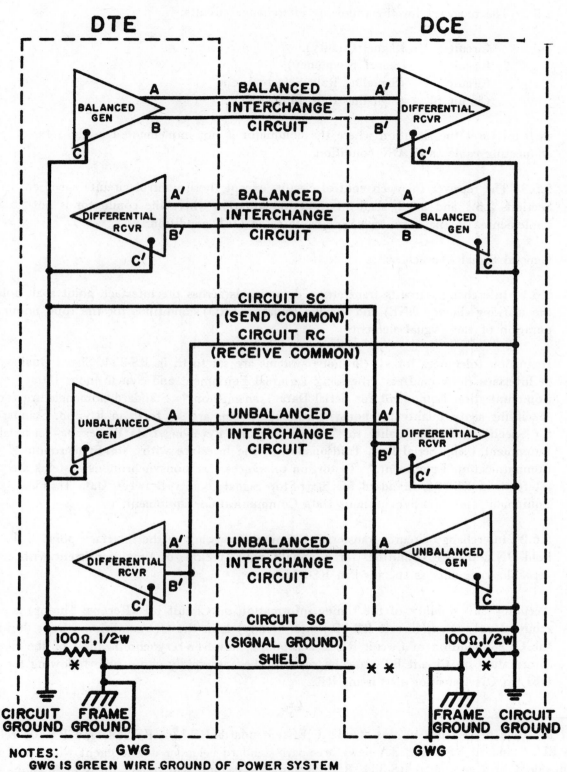

FIGURE 2.2

GROUNDING ARRANGEMENTS

NOTES:
GWG IS GREEN WIRE GROUND OF POWER SYSTEM
✗SEE SECTION 2.4 FOR ALTERNATIVE ARRANGEMENTS
✗✗NORMALLY NO CONNECTION TO SHIELD IN DCE (SEE SECTION 2.3)

867

2.5.2  The receivers for the following interchange circuits:

Circuit SQ    (Signal Quality),
Circuit SF    (Select Frequency),
Circuit SR    (Signaling Rate Selector), and
Circuit SI    (Signaling Rate Indicator)

shall interpret the situation where the conductor is not implemented in the interconnecting cable as an ON condition.

2.5.3  The receiver for each control circuit, except those control circuits specified in Sections 2.5.1 and 2.5.2, shall interpret the situation where the conductor is not implemented in the interconnecting cable as an OFF condition.

2.6  General Signal Characteristics

2.6.1  Interchange circuits transferring data signals across the interface point shall hold the marking (binary ONE) and spacing (binary ZERO) conditions for the total nominal duration of each signal element.

Distortion tolerances for synchronous systems are set forth in RS-334, Signal Quality at Interface Between Data Processing Terminal Equipment and Synchronous Data Communication Equipment for Serial Data Transmission.*  Standard nomenclature for specifying signal quality for nonsynchronous systems are set forth in RS-363, Standard for Specifying Signal Quality for Transmitting and Receiving Data Processing Terminal Equipment Using Serial Data Transmission at the Interface with Non-Synchronous Communication Equipment.*  Distortion tolerances for nonsynchronous systems are set forth in RS-404, Standard for Start-Stop Signal Quality Between Data Terminal Equipment and Non-Synchronous Data Communication Equipment.*

2.6.2  Interchange circuits transferring timing signals across the interface point shall hold ON and OFF conditions for nominally equal periods of time, consistent with acceptable tolerances as specified in RS-334.*

Accuracy and stability of the timing information on Circuit RT (Receive Timing) is required only when Circuit RR (Receiver Ready) is in the ON condition.  Drift during the OFF condition of Circuit RR is acceptable; however resynchronization of the timing information on Circuit RT must be accomplished as rapidly as possible following the OFF to ON transition of Circuit RR.

---

*At the time of issuance of this standard, EIA Standards RS-334, RS-363, and RS-404 relate to EIA Standard RS-232-C.  A new three-part standard is under development which will be equivalent to a combined RS-334, RS-363 and RS-404 and will reflect the specifications contained in RS-422, RS-423, and this standard.

It is desirable that the transfer of timing information across the interface be provided during all times that the timing source is capable of generating this information (i.e., it should not be restricted only to periods when actual transmission of data is in progress). During periods when timing information is not provided on a timing interchange circuit, the interchange circuit shall be clamped in the OFF condition.

2.6.3 Tolerances on the relationship between data and associated timing signals shall be in accordance with RS-334.*

# 3. INTERFACE MECHANICAL CHARACTERISTICS

## 3.1 Definition of Mechanical Interface

The point of demarcation between the DTE and the DCE is located at a plugable connector signal interface point between the two equipments which is less than 3 meters (10 feet) from the DCE (see Figure 3.1). This point may consist of one or two connectors. A 37-position connector is specified for all interchange circuits with the exception of secondary channel circuits which are accommodated in a separate 9-position connector. Provision of the 9-position connector is only necessary when the secondary channel capability is implemented in the interface.

The DCE shall be provided with the connector(s) as specified in Section 3.3, having female contacts and a male shell. The connector shall be either physically attached to the DCE or extended by means of a short cable (less than 3 meters or 10 feet). The DTE shall be provided with a cable having the connector(s) as specified in Section 3.3, having male contacts and a female shell. The total length of the cable associated with the DTE shall not exceed 60 meters (200 feet) for nontailored applications (see Section 6.10). The mechanical configuration for connections of the interface cable at points other than the point of demarcation is not specified.

## 3.2 Intermediate Equipment

When additional functions are provided in a separate unit inserted between the DTE and the DCE (see Section 1.8), the connector(s) with female contacts, as indicated above, shall be associated with the side of this unit which interfaces with the DTE while the cable with the connector(s) with male contacts shall be provided on the side which interfaces with the DCE.

## 3.3 Interface Connector

3.3.1 The 37-position and 9-position interface connectors shall meet the intermating dimensions set forth in MIL-C-24308. These dimensions are contained in the following documents from MIL-C-24308:

_____

*See footnote on Page 8.

870

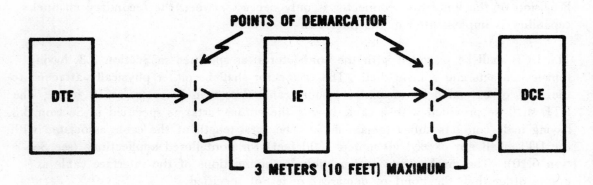

POINT OF DEMARCATION

DTE ⟶ DCE

3 METERS (10 FEET) MAXIMUM

**DTE TO DCE INTERCONNECTION**

POINTS OF DEMARCATION

DTE ⟶ IE ⟶ DCE

3 METERS (10 FEET) MAXIMUM

**DTE TO DCE INTERCONNECTION WITH INTERMEDIATE EQUIPMENT**

⟶    CONNECTOR(S) WITH MALE
CONTACTS AND FEMALE SHELL

⟨    CONNECTOR(S) WITH FEMALE
CONTACTS AND MALE SHELL

IE    INTERMEDIATE EQUIPMENT

**FIGURE 3.1**

**INTERCONNECTION OF EQUIPMENT**

## 37-position connector

| | |
|---|---|
| MS18270-4 | Receptacle Connector |
| MS18271-4 | Plug Connector |
| MS18276-1 | Contact Locations |
| M24308/10-1 | Socket Contact |
| M24308/11-1 | Pin Contact |

## 9-position connector

| | |
|---|---|
| MS18270-1 | Receptacle Connector |
| MS18271-1 | Plug Connector |
| MS18273-1 | Contact Locations |
| M24308/10-1 | Socket Contact |
| M24308/11-1 | Pin Contact |

Only the requirements of MIL-C-24308 pertaining to intermating of connectors must be met.

Figures 3.2-A and 3.2-B illustrate the DTE interface connectors which have male (pin) contacts and a female shell (plug connector). Figures 3.3-A and 3.3-B illustrate the DCE interface connectors which have female (socket) contacts and a male shell (receptacle connector). Contact numbering is also illustrated in these Figures.

3.3.2 Each DCE interface connector shall be equipped with two latching blocks as specified in Figures 3.3-A and 3.3-B. Each DTE interface connector shall be equipped with means for latching to and unlatching from these blocks; preferably without the use of a tool. The means for the DTE connectors to latch to and unlatch from the blocks on the DCE connectors are not standardized. However, this must be accomplished within the envelope specified in Figure 3.4. In addition, the means must be such that the mating and latching and the unlatching and disconnecting of the connectors can be accomplished within the access space available for both of the arrangements illustrated in Figure 3.5. This will permit DCE interface connectors to be mounted with clearances shown for either of the two arrangements in Figure 3.5.

## 3.4 Connector Contact Assignments

3.4.1 Contact assignments listed in Figure 3.6-A for the 37-position connector and in Figure 3.6-B for the 9-position connector shall be used.

3.4.2 Contact assignments for circuits not specifically defined in Section 4 (see Section 4.1) are to be made by mutual agreement. Preference should be given to the use of unassigned contacts, but in the event that additional contacts are required, extreme caution should be taken in their selection.

## DTE CONNECTOR FACE
## 37-POSITION PLUG

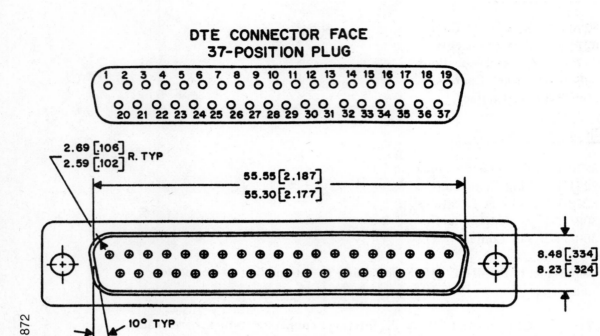

872

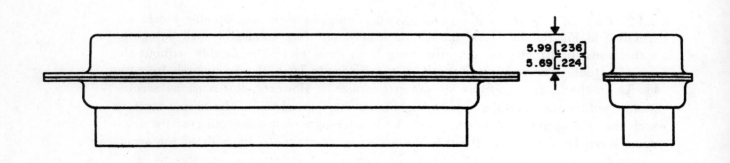

NOTE: MILLIMETERS SPECIFIED;
INCHES IN [ ].
DIMENSIONS IN INCHES
ARE ORIGINAL.

## FIGURE 3.2-A
## 37-POSITION DTE INTERFACE CONNECTOR

DTE CONNECTOR FACE
9-POSITION PLUG

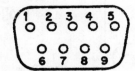

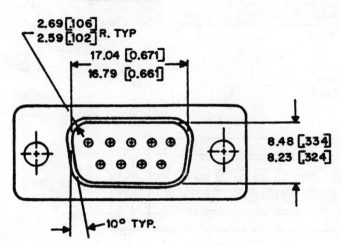

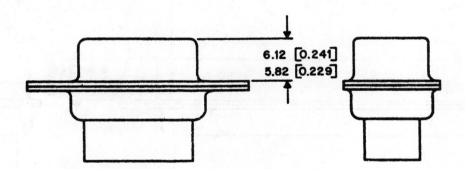

NOTE:  MILLIMETERS SPECIFIED;
INCHES IN [ ].
DIMENSIONS IN INCHES
ARE ORIGINAL.

FIGURE. 3.2-B
9-POSITION DTE INTERFACE CONNECTOR

873

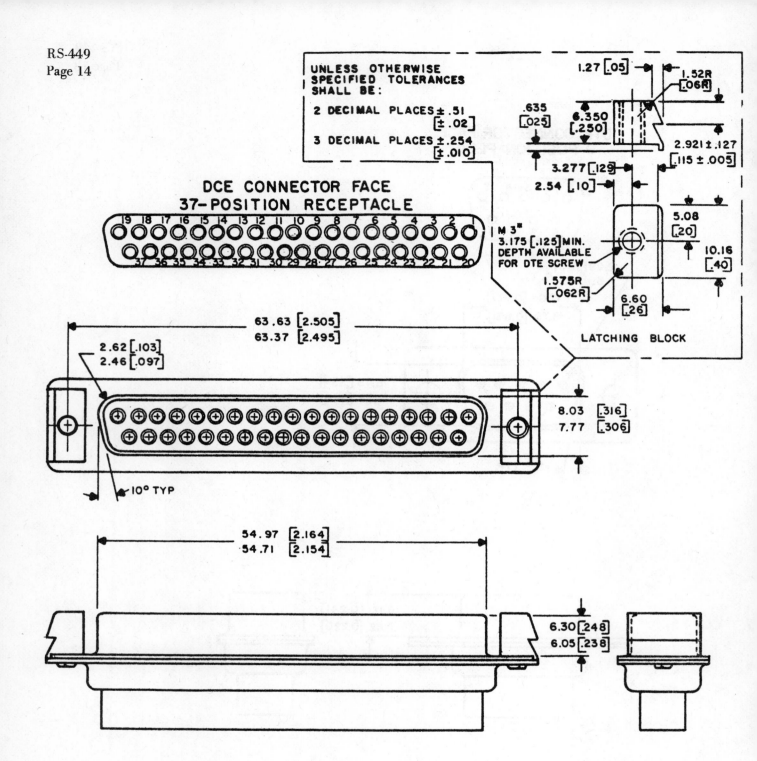

UNLESS OTHERWISE
SPECIFIED TOLERANCES
SHALL BE:

2 DECIMAL PLACES ±.51
[±.02]

3 DECIMAL PLACES ±.254
[±.010]

DCE CONNECTOR FACE
37-POSITION RECEPTACLE

LATCHING BLOCK

*METRIC THREAD PER INTERNATIONAL
STANDARD ISO 261, ISO METRIC SCREW
THREADS—GENERAL PLAN

NOTE: MILLIMETERS SPECIFIED;
INCHES IN [ ].
DIMENSIONS IN INCHES
ARE ORIGINAL.

FIGURE 3.3-A
37-POSITION DCE INTERFACE CONNECTOR

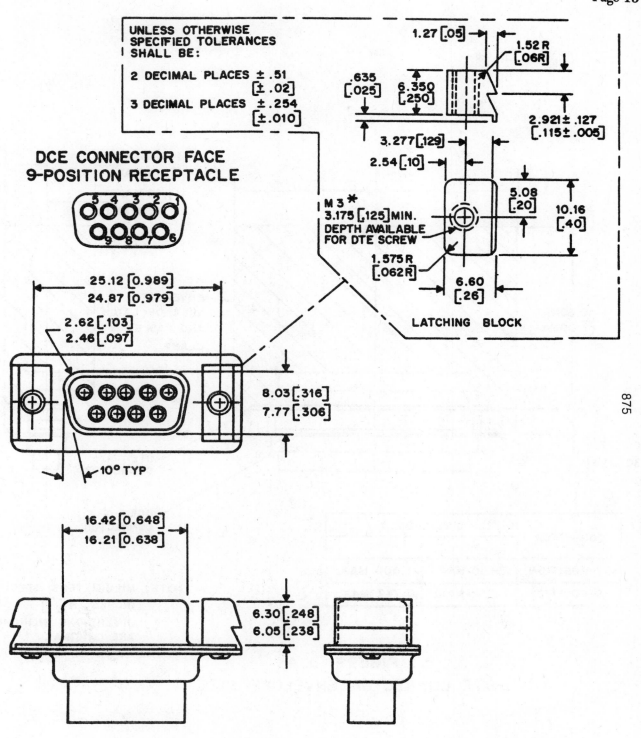

UNLESS OTHERWISE
SPECIFIED TOLERANCES
SHALL BE:

2 DECIMAL PLACES ± .51
[± .02]
3 DECIMAL PLACES ± .254
[±.010]

DCE CONNECTOR FACE
9-POSITION RECEPTACLE

25.12 [0.989]
24.87 [0.979]

2.62 [.103]
2.46 [.097]

8.03 [.316]
7.77 [.306]

10° TYP

1.27 [.05]
1.52 R [.06R]
.635 [.025]
6.350 [.250]
2.921± .127 [.115±.005]
3.277 [.129]
2.54 [.10]
M 3 *
3.175 [.125] MIN. DEPTH AVAILABLE FOR DTE SCREW
1.575 R [.062R]
5.08 [.20]
10.16 [.40]
6.60 [.26]
LATCHING BLOCK

875

16.42 [0.648]
16.21 [0.638]

6.30 [.248]
6.05 [.238]

*METRIC THREAD PER INTERNATIONAL
STANDARD ISO 261, ISO METRIC SCREW
THREADS—GENERAL PLAN

NOTE: MILLIMETERS SPECIFIED;
INCHES IN [ ].
DIMENSIONS IN INCHES
ARE ORIGINAL.

FIGURE 3.3 - B
9-POSITION DCE INTERFACE CONNECTOR

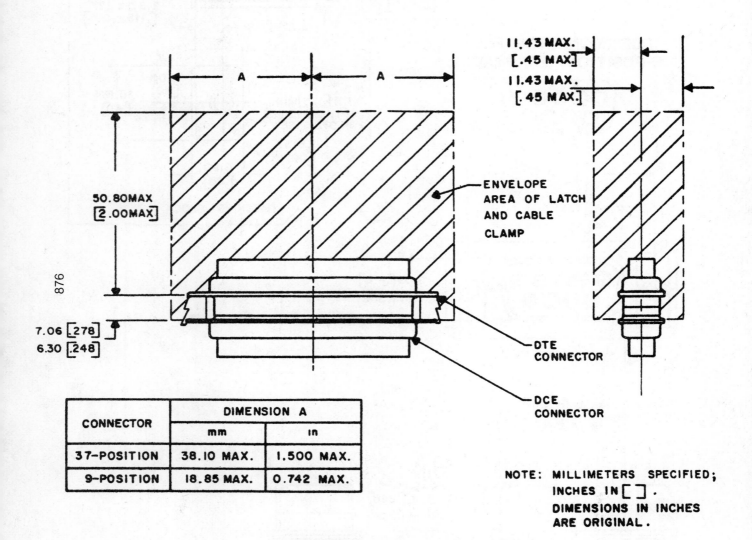

| CONNECTOR | DIMENSION A | |
|---|---|---|
| | mm | in |
| 37-POSITION | 38.10 MAX. | 1.500 MAX. |
| 9-POSITION | 18.85 MAX. | 0.742 MAX. |

NOTE: MILLIMETERS SPECIFIED;
INCHES IN [ ].
DIMENSIONS IN INCHES
ARE ORIGINAL.

FIGURE 3.4
DTE CONNECTOR ENVELOPE SIZE

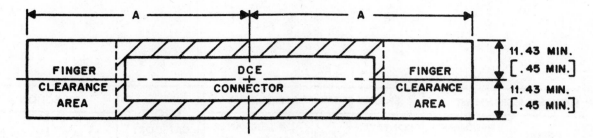

FINGER CLEARANCE AT END

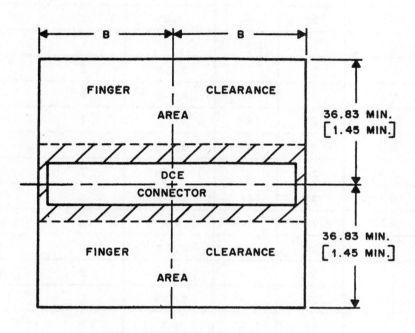

FINGER CLEARANCE AT SIDE

| CONNECTOR | DIMENSION A | | DIMENSION B | |
|---|---|---|---|---|
| | mm | in | mm | in |
| 37-POSITION | 63.50 MIN. | 2.500 MIN. | 38.10 MIN. | 1.500 MIN. |
| 9-POSITION | 44.25 MIN. | 1.742 MIN. | 18.85 MIN. | 0.742 MIN. |

NOTES:
SHADED AREA REPRESENTS MAXIMUM AREA OF DTE CONNECTOR INCLUDING LATCH AND CABLE CLAMP.

FINGER CLEARANCE AREA MAY BE SHARED BY TWO CONNECTORS.

MILLIMETERS SPECIFIED; INCHES IN [ ].
DIMENSIONS IN INCHES ARE ORIGINAL.

FIGURE 3.5
DCE CONNECTOR MOUNTING CLEARANCE

877

| FIRST SEGMENT ASSIGNMENT | | | SECOND SEGMENT ASSIGNMENT | | | CIRCUIT CATEGORY | DIRECTION | |
|---|---|---|---|---|---|---|---|---|
| CONTACT NUMBER | CIRCUIT | INTERCHANGE POINTS* | CONTACT NUMBER | CIRCUIT | INTERCHANGE POINTS* | | TO DCE | FROM DCE |
| 1 | SHIELD | — | | | | | | |
| 2 | SI | A-A' | 20 | RC | C-B' | II | | X |
| 3 | SPARE | | 21 | SPARE | | | | |
| 4 | SD | A-A' | 22 | SD | B/C-B' | I | X | |
| 5 | ST | A-A' | 23 | ST | B/C-B' | I | | X |
| 6 | RD | A-A' | 24 | RD | B/C-B' | I | | X |
| 7 | RS | A-A' | 25 | RS | B/C-B' | I | X | |
| 8 | RT | A-A' | 26 | RT | B/C-B' | I | | X |
| 9 | CS | A-A' | 27 | CS | B/C-B' | I | | X |
| 10 | LL | A-A' | 28 | IS | A-A' | II | X | |
| 11 | DM | A-A' | 29 | DM | B/C-B' | I | | X |
| 12 | TR | A-A' | 30 | TR | B/C-B' | I | X | |
| 13 | RR | A-A' | 31 | RR | B/C-B' | I | | X |
| 14 | RL | A-A' | 32 | SS | A-A' | II | X | |
| 15 | IC | A-A' | 33 | SQ | A-A' | II | | X |
| 16 | SF/SR + | A-A' | 34 | NS | A-A' | II | X | |
| 17 | TT | A-A' | 35 | TT | B/C-B' | I | X | |
| 18 | TM | A-A' | 36 | SB | A-A' | II | | X |
| 19 | SG | C-C' | 37 | SC | C-B' | I | X | |

*SEE FOOTNOTE ON FIGURE 3.6-B

+ CIRCUIT SF AND CIRCUIT SR SHARE THE SAME CONTACT NUMBER

FIGURE 3.6-A ASSIGNMENTS FOR 37-POSITION CONNECTOR

| FIRST SEGMENT ASSIGNMENT | | | SECOND SEGMENT ASSIGNMENT | | | CIRCUIT CATEGORY | DIRECTION | |
|---|---|---|---|---|---|---|---|---|
| CONTACT NUMBER | CIRCUIT | INTERCHANGE POINTS* | CONTACT NUMBER | CIRCUIT | INTERCHANGE POINTS* | | TO DCE | FROM DCE |
| 1 | SHIELD | — | | | | | | |
| 2 | SRR | A-A' | 6 | RC | C-B' | II | | X |
| 3 | SSD | A-A' | 7 | SRS | A-A' | II | X | |
| 4 | SRD | A-A' | 8 | SCS | A-A' | II | | X |
| 5 | SG | C-C' | 9 | SC | C-B' | II | X | |

879

* A, A', B, B', C AND C' INDICATE THE ASSOCIATED GENERATOR-LOAD INTERCHANGE POINTS AS DESIGNATED IN RS-422 AND RS-423. WHERE B/C IS INDICATED, THE B DESIGNATION APPLIES ONLY WHEN A RS-422 BALANCED GENERATOR IS USED AND THE C DESIGNATION APPLIES ONLY WHEN A RS-423 UNBALANCED GENERATOR IS USED (SEE FIGURE 2.1). THE B' LEAD OF CIRCUITS IS, NS, SF/SR, LL, RL, SS, SSD AND SRS IS INTERNALLY CONNECTED WITHIN THE DCE TO CIRCUIT SC. THE B' LEAD OF CIRCUITS IC, SQ, SI, TM, SB, SRD, SCS AND SRR IS INTERNALLY CONNECTED WITHIN THE DTE TO CIRCUIT RC.

FIGURE 3.6-B ASSIGNMENTS FOR 9-POSITION CONNECTOR

3.4.3 Where multipair cable is used, the two conductors of a pair shall be used for one Category I interchange circuit or for two Category II interchange circuits carrying signals in the same direction. The contact assignments listed in Figures 3.6-A and 3.6-B are arranged to facilitate pairing; e.g., acceptable pairs are 2 and 20, 3 and 21, etc., (37-position connector) and 2 and 6, 3 and 7, etc., (9-position connector).

## 4. FUNCTIONAL DESCRIPTION OF INTERCHANGE CIRCUITS

### 4.1 General

This section defines the basic interchange circuits which apply, collectively, to all systems. Additional interchange circuits not defined herein, or variations in the functions of the defined interchange circuits, may be provided by mutual agreement. Careful attention must be paid to the provisions of Section 3.4 and 5.3. Any such addition or variation shall be noted in the manufacturer's specifications.

### 4.2 Classifications of Circuits

Interchange circuits fall into four general classifications:

> Ground or Common Return Circuits,
> Data Circuits,
> Control Circuits, and
> Timing Circuits.

A list of interchange circuits showing circuit mnemonic, circuit name, circuit direction, and circuit type is presented in Figure 4.1.

### 4.3 Equivalency

A list of interchange circuits showing the nearest equivalent RS-232-C and C.C.I.T.T. identification in accordance with Recommendation V.24 is presented in Figure 4.2. It should not be inferred that the circuit definitions given herein are in exact accord with RS-232-C or C.C.I.T.T. Recommendations.

### 4.4 Definitions of Interchange Circuits

#### 4.4.1 *Circuit SG - Signal Ground*

Direction: Not Applicable

This conductor directly connects the DTE circuit ground (circuit common) to the DCE circuit ground (circuit common) to provide a conductive path between the DTE and DCE signal commons (see Section 2.4).

| CIRCUIT MNEMONIC | CIRCUIT NAME | CIRCUIT DIRECTION | CIRCUIT TYPE | |
|---|---|---|---|---|
| SG | SIGNAL GROUND | – | COMMON | |
| SC | SEND COMMON | TO DCE | | |
| RC | RECEIVE COMMON | FROM DCE | | |
| IS | TERMINAL IN SERVICE | TO DCE | CONTROL | |
| IC | INCOMING CALL | FROM DCE | | |
| TR | TERMINAL READY | TO DCE | | |
| DM | DATA MODE | FROM DCE | | |
| SD | SEND DATA | TO DCE | DATA | PRIMARY CHANNEL |
| RD | RECEIVE DATA | FROM DCE | | |
| TT | TERMINAL TIMING | TO DCE | TIMING | |
| ST | SEND TIMING | FROM DCE | | |
| RT | RECEIVE TIMING | FROM DCE | | |
| RS | REQUEST TO SEND | TO DCE | CONTROL | |
| CS | CLEAR TO SEND | FROM DCE | | |
| RR | RECEIVER READY | FROM DCE | | |
| SQ | SIGNAL QUALITY | FROM DCE | | |
| NS | NEW SIGNAL | TO DCE | | |
| SF | SELECT FREQUENCY | TO DCE | | |
| SR | SIGNALING RATE SELECTOR | TO DCE | | |
| SI | SIGNALING RATE INDICATOR | FROM DCE | | |
| SSD | SECONDARY SEND DATA | TO DCE | DATA | SECONDARY CHANNEL |
| SRD | SECONDARY RECEIVE DATA | FROM DCE | | |
| SRS | SECONDARY REQUEST TO SEND | TO DCE | CONTROL | |
| SCS | SECONDARY CLEAR TO SEND | FROM DCE | | |
| SRR | SECONDARY RECEIVER READY | FROM DCE | | |
| LL | LOCAL LOOPBACK | TO DCE | CONTROL | |
| RL | REMOTE LOOPBACK | TO DCE | | |
| TM | TEST MODE | FROM DCE | | |
| SS | SELECT STANDBY | TO DCE | CONTROL | |
| SB | STANDBY INDICATOR | FROM DCE | | |

881

**FIGURE 4.1
INTERCHANGE CIRCUITS**

882

| RS-443 | | RS-232C | | C.C.I.T.T. RECOMMENDATION V.24 | |
|---|---|---|---|---|---|
| SG | SIGNAL GROUND | AB | SIGNAL GROUND | 102 | SIGNAL GROUND |
| SC | SEND COMMON | | | 102a | DTE COMMON |
| RC | RECEIVE COMMON | | | 102b | DCE COMMON |
| IS | TERMINAL IN SERVICE | | | | |
| IC | INCOMING CALL | CE | RING INDICATOR | 125 | CALLING INDICATOR |
| TR | TERMINAL READY | CD | DATA TERMINAL READY | 108/2 | DATA TERMINAL READY |
| DM | DATA MODE | CC | DATA SET READY | 107 | DATA SET READY |
| SD | SEND DATA | BA | TRANSMITTED DATA | 103 | TRANSMITTED DATA |
| RD | RECEIVE DATA | BB | RECEIVED DATA | 104 | RECEIVED DATA |
| TT | TERMINAL TIMING | DA | TRANSMITTER SIGNAL ELEMENT TIMING (DTE SOURCE) | 113 | TRANSMITTER SIGNAL ELEMENT TIMING (DTE SOURCE) |
| ST | SEND TIMING | DB | TRANSMITTER SIGNAL ELEMENT TIMING (DCE SOURCE) | 114 | TRANSMITTER SIGNAL ELEMENT TIMING (DCE SOURCE) |
| RT | RECEIVE TIMING | DD | RECEIVER SIGNAL ELEMENT TIMING | 115 | RECEIVER SIGNAL ELEMENT TIMING (DCE SOURCE) |
| RS | REQUEST TO SEND | CA | REQUEST TO SEND | 105 | REQUEST TO SEND |
| CS | CLEAR TO SEND | CB | CLEAR TO SEND | 106 | READY FOR SENDING |
| RR | RECEIVER READY | CF | RECEIVED LINE SIGNAL DETECTOR | 109 | DATA CHANNEL RECEIVED LINE SIGNAL DETECTOR |
| SQ | SIGNAL QUALITY | CG | SIGNAL QUALITY DETECTOR | 110 | DATA SIGNAL QUALITY DETECTOR |
| NS | NEW SIGNAL | | | | |
| SF | SELECT FREQUENCY | | | 126 | SELECT TRANSMIT FREQUENCY |
| SR | SIGNALING RATE SELECTOR | CH | DATA SIGNAL RATE SELECTOR (DTE SOURCE) | 111 | DATA SIGNALING RATE SELECTOR (DTE SOURCE) |
| SI | SIGNALING RATE INDICATOR | CI | DATA SIGNAL RATE SELECTOR (DCE SOURCE) | 112 | DATA SIGNALING RATE SELECTOR (DCE SOURCE) |
| SSD | SECONDARY SEND DATA | SBA | SECONDARY TRANSMITTED DATA | 118 | TRANSMITTED BACKWARD CHANNEL DATA |
| SRD | SECONDARY RECEIVE DATA | SBB | SECONDARY RECEIVED DATA | 119 | RECEIVED BACKWARD CHANNEL DATA |
| SRS | SECONDARY REQUEST TO SEND | SCA | SECONDARY REQUEST TO SEND | 120 | TRANSMIT BACKWARD CHANNEL LINE SIGNAL |
| SCS | SECONDARY CLEAR TO SEND | SCB | SECONDARY CLEAR TO SEND | 121 | BACKWARD CHANNEL READY |
| SRR | SECONDARY RECEIVER READY | SCF | SECONDARY RECEIVED LINE SIGNAL DETECTOR | 122 | BACKWARD CHANNEL RECEIVED LINE SIGNAL DETECTOR |
| LL | LOCAL LOOPBACK | | | 141 | LOCAL LOOPBACK |
| RL | REMOTE LOOPBACK | | | 140 | REMOTE LOOPBACK |
| TM | TEST MODE | | | 142 | TEST INDICATOR |
| SS | SELECT STANDBY | | | 116 | SELECT STANDBY |
| SB | STANDBY INDICATOR | | | 117 | STANDBY INDICATOR |

FIGURE 4.2
EQUIVALENCY TABLE

### 4.4.2 *Circuit SC - Send Common*

Direction: To DCE

This conductor is connected to the DTE circuit ground (circuit common) and is used at the DCE as a reference potential for Category II interchange circuit receivers.

### 4.4.3 *Circuit RC - Receive Common*

Direction: From DCE

This conductor is connected to the DCE circuit ground (circuit common) and is used at the DTE as a reference potential for Category II interchange circuit receivers.

### 4.4.4 *Circuit IS - Terminal In Service*

Direction: To DCE

Signals on this circuit indicate whether the DTE is available for service.

The ON condition indicates that the DTE is in service and, in switched network applications employing line hunting, allows incoming calls to be connected to the DCE.

The OFF condition signals the DCE that the DTE is not available for service. In switched network applications employing line hunting, the OFF condition causes the DCE to appear busy so that an incoming call will not be connected to the DCE (i.e., the DCE will be skipped over in the line hunting sequence). The OFF condition shall be caused by any DTE out-of-service condition, such as an internal DTE test condition (i.e., a test condition which does not involve the DCE), which the DTE is unable (or unwilling) to interrupt to accept an incoming call.

### 4.4.5 *Circuit IC - Incoming Call*

Direction: From DCE

Signals on this circuit indicate whether an incoming call signal is being received by the DCE.

The ON condition indicates that an incoming call (ringing) signal is being received by the DCE. The ON condition shall appear approximately coincident with the ON segment of the ringing cycle (during rings) on the communication channel.

The OFF condition shall be maintained during the OFF segment of the ringing cycle (between rings) and at all other times when ringing is not being received.

The operation of this circuit shall not be disabled by the OFF condition on Circuit TR (Terminal Ready).

### 4.4.6  *Circuit TR- Terminal Ready*

Direction:    To DCE

Signals on this circuit are used to control switching of the DCE to and from the communication channel.

The ON condition prepares the DCE to be connected to the communication channel and maintains the connection established by external (e.g., manual call origination, manual answering or automatic call origination) or internal (automatic answering) means.

When the station is equipped for automatic answering of received calls and is in the automatic answering mode, connection to the line occurs only in response to a combination of a ringing signal and the ON condition of Circuit TR (Terminal Ready). The DTE is normally permitted to present the ON condition on Circuit TR whenever it is ready to transmit or receive data, except that when Circuit TR is turned OFF, it shall not be turned ON again until Circuit DM (Data Mode) is turned OFF by the DCE.

The OFF condition causes the DCE to be removed from the communication channel following completion of the transfer to the communication channel of all data presented before Circuit TR was turned OFF.  See Circuit SD (Send Data).  The OFF condition shall not disable the operation of Circuit IC   (Incoming Call).

### 4.4.7  *Circuit DM - Data Mode*

Direction:   From DCE

Signals on this circuit indicate the status of the local DCE.

The ON condition indicates that the DCE is in the data transfer mode.  That is:

a)   The local DCE is connected to a communication channel ("OFF HOOK" in switched service),

b)   The local DCE is not in talk (alternative voice) or dial* mode, and

c)   The local DCE has completed the following where applicable:

---

\*    The DCE is considered to be in the dial mode when circuitry directly associated with the call origination function is connected to the communication channel.  These functions include signaling to the central office (dialing) and monitoring the communication channel for call progress or answer back signals.

1. any timing functions required by the switching system to complete call establishment, and

2. the transmission of any discrete answer tone, the duration of which is controlled solely by the local DCE.

Where the local DCE does not transmit an answer tone, or where the duration of the answer tone is controlled by some action of the remote DCE, the ON condition is presented as soon as all the other above conditions (a, b, and c-1) are satisfied.

This circuit shall be used only to indicate the status of the local DCE. The ON condition shall not be interpreted as either an indication that a communication channel has been established to a remote data station or the status of any remote station equipment (see Section 6.3).

The OFF condition shall appear at all other times and shall be an indication that the DTE is to disregard signals appearing on all other interchange circuits with the exception of Circuit IC (Incoming Call), Circuit TM (Test Mode) and Circuit SB (Standby Indicator). The OFF condition shall not impair the operation of Circuit IC. When the OFF condition occurs during the progress of a call before Circuit TR (Terminal Ready) is turned OFF, the DTE shall interpret this as a lost or aborted connection and take action to terminate the call. Any subsequent ON condition on Circuit DM is to be considered a new call.

When the DCE is used in conjunction with Automatic Calling Equipment (ACE), the OFF to ON transition of Circuit DM shall not be interpreted as an indication that the ACE has relinquished control of the communication channel to the DCE. Indication of this is given on the appropriate circuit in the ACE interface (see EIA Standard RS-366).

Note: Attention is called to the fact that if a data call is interrupted by alternate voice communication, Circuit DM will be in the OFF condition during the time that voice communication is in progress. The transmission or reception of the signals required to condition the communication channel or DCE in response to the ON condition of Circuit RS (Request to Send) of the transmitting DTE will take place after Circuit DM comes ON, but prior to the ON condition on Circuit CS (Clear to Send) or Circuit RR (Receiver Ready).

Circuit DM shall be held in the OFF condition for DCE tests where testing is not conducted through the DTE/DCE interface. Circuit DM shall respond normally (i.e., not clamped OFF) for DCE tests where testing is conducted through the DTE/DCE interface.

### 4.4.8  *Circuit SD - Send Data*

Direction:  To DCE

The data signals originated by the DTE, to be transmitted via the data channel to one or more remote data stations, are transferred on this circuit to the DCE.

In all systems, the DTE should hold Circuit SD in the binary ONE (marking) condition unless an ON condition is present on all of the following circuits:

1. Circuit RS  (Request to Send),
2. Circuit CS  (Clear to Send),
3. Circuit DM  (Data Mode),
4. Circuit TR  (Terminal Ready), where implemented, and
5. Circuit IS  (Terminal In Service), where implemented.

The DCE shall disregard all signals appearing on Circuit SD during the time that an OFF condition is present on one or more of the above circuits.

All data signals that are transmitted across the interface on Circuit SD during the time an ON condition is maintained on all of the above circuits shall be transmitted to the communication channel by the DCE.  The term "data signals" includes the binary ONE (marking) condition, reversals, and sequences such as SYN coded characters to maintain timing synchronization.

### 4.4.9  *Circuit RD - Receive Data*

Direction:  From DCE

The data signals generated by the DCE, in response to data channel line signals received from a remote data station, are transferred on this circuit to the DTE.

Circuit RD shall be held in the binary ONE (marking) condition at all times when Circuit RR (Receiver Ready) is in the OFF condition.

On a half duplex channel, Circuit RD shall be held in the binary ONE (marking) condition when Circuit RS (Request to Send) is in the ON condition and for a brief interval following the ON to OFF transition of Circuit RS to allow for the completion of transmission (see Circuit SD - Send Data) and the decay of line reflections.

### 4.4.10 *Circuit TT - Terminal Timing*

Direction:  To DCE

Signals on this circuit provide the DCE with transmit signal element timing information.

The ON to OFF transition shall nominally indicate the center of each signal element on Circuit SD (Send Data).

When Circuit TT is implemented in the DTE, the DTE shall normally provide timing information on this circuit whenever the DTE is in a POWER ON condition. It is permissible for the DTE to withhold timing information on this circuit for short periods provided Circuit RS (Request to Send) is in the OFF condition. (For example, the temporary withholding of timing information may be necessary in performing maintenance tests within the DTE).

### 4.4.11 *Circuit ST - Send Timing*

Direction: From DCE

Signals on this circuit provide the DTE with transmit signal element timing information.

The DTE shall provide a data signal on Circuit SD (Send Data) in which the transitions between signal elements nominally occur at the time of the transitions from OFF to ON condition of the signal on Circuit ST.

The DCE shall normally provide timing information on this circuit whenever the DCE is in a POWER ON condition. It is permissible for the DCE to withhold timing information on this circuit for short periods provided Circuit DM (Data Mode) is in the OFF condition. (For example, the withholding of timing information may be necessary in performing maintenance tests within the DCE.)

### 4.4.12 *Circuit RT - Receive Timing*

Direction: From DCE

Signals on this circuit provide the DTE with receive signal element timing information.

The transition from ON to OFF condition shall nominally indicate the center of each signal element on Circuit RD (Receive Data).

The DCE shall normally provide timing information on this circuit whenever the DCE is in a POWER ON condition (see Section 2.6.2). It is permissible for the DCE to withhold timing information on this circuit for short periods provided Circuit DM (Data Mode) is in the OFF condition. (For example, the withholding of timing information may be necessary in performing maintenance tests within the DCE.)

### 4.4.13 *Circuit RS - Request to Send*

Direction: To DCE

Signals on this circuit control the data channel transmit function of the local DCE and, on a half duplex channel, control the direction of data transmission of the local DCE.

On one-way only channels or duplex channels, the ON condition maintains the DCE in the transmit mode. The OFF condition maintains the DCE in a non-transmit mode.

On a half-duplex channel, the ON condition maintains the DCE in the transmit mode and inhibits the receive mode. The OFF condition maintains the DCE in the receive mode.

A transition from OFF to ON instructs the DCE to enter the transmit mode (see Section 6.2). The DCE responds by taking such action as may be necessary and indicates completion of such actions by turning ON Circuit CS (Clear to Send), thereby indicating to the DTE that data may be transferred across the interface on Circuit SD (Send Data).

A transition from ON to OFF instructs the DCE to complete the transmission of all data which was previously transferred across the interface on Circuit SD and then assume a non-transmit mode or a receive mode as appropriate. The DCE responds to this instruction by turning OFF Circuit CS when it is prepared to again respond to a subsequent ON condition of Circuit RS.

> NOTE: A non-transmit mode does not imply that all line signals
> have been removed from the communication channel (see Section 6.2).

When Circuit RS is turned OFF, it shall not be turned ON again until Circuit CS has been turned OFF by the DCE.

An ON condition is required on Circuit RS as well as on Circuit CS, Circuit DM (Data Mode), Circuit TR (Terminal Ready), where implemented, and Circuit IS (Terminal In Service), where implemented, whenever the DTE transfers data across the interface on Circuit SD.

It is permissible to turn Circuit RS ON at any time when Circuit CS is OFF regardless of the condition of any other interchange circuit.

4.4.14 *Circuit CS - Clear to Send*

Direction: From DCE

Signals on this circuit indicate whether the DCE is conditioned to transmit data on the data channel.

The ON condition together with the ON condition on Circuit RS (Request to Send), Circuit DM (Data Mode), Circuit TR (Terminal Ready), where implemented, and Circuit IS (Terminal In Service), where implemented, is an indication to the DTE that signals presented on Circuit SD (Send Data) will be transmitted to the communication channel.

The OFF condition is an indication to the DTE that it should not transfer data across the interface on Circuit SD since this data will not be transmitted to the line.

The ON condition of Circuit CS is a response to the occurrence of a concurrent ON condition on Circuit DM and Circuit RS, delayed as may be appropriate by the DCE for establishing a data communication channel (including the removal of the mark clamp on Circuit RD of the remote DCE) to a remote DTE.

### 4.4.15 *Circuit RR - Receiver Ready*

Direction: From DCE

Signals on this circuit indicate whether the receiver in the DCE is conditioned to receive data signals from the communication channel, but does not indicate the relative quality of the data signals being received.

Circuit RR is not affected by the condition of an equalizer in a DCE (see Section 6.8).

The ON condition on this circuit is presented when the DCE is receiving a signal which meets its suitability criteria. These criteria are established by the DCE manufacturer.

The OFF condition indicates that no signal is being received or that the received signal does not meet the suitability criteria established by the DCE manufacturer.

The indications on this circuit shall follow the actual onset or loss of signal by appropriate guard delays.

The OFF condition of Circuit RR shall cause Circuit RD (Receive Data) to be clamped to the binary ONE (marking) condition.

On half-duplex channels, Circuit RR is held in the OFF condition whenever Circuit RS (Request to Send) is in the ON condition and for a brief interval of time following the ON to OFF transition of Circuit RS (see Circuit RD).

### 4.4.16 *Circuit SQ - Signal Quality*

Direction: From DCE

Signals on this circuit indicate whether there is a reasonable probability of an error in the receive data (see Section 6.4).

An ON condition is maintained whenever there is no reason to believe that an error has occurred.

An OFF condition indicates that there is a high probability of an error.

The probability of error criterion is established by the DCE manufacturer.

### 4.4.17  *Circuit NS - New Signal*

Direction:  To DCE

Signals on this circuit indicate whether the DCE shall prepare itself to rapidly respond to a new line signal (see Section 6.5).

The ON condition of Circuit NS causes the DCE to place Circuit RR (Receiver Ready) in the OFF condition, place Circuit RD (Receive Data) in the binary ONE (marking) condition, and prepare itself to rapidly respond to the line signal present after Circuit NS is turned OFF.  Once turned ON, Circuit NS may be turned OFF by the DTE any time after Circuit RR is turned OFF by the DCE.  Circuit NS shall be OFF at all other times.

### 4.4.18  *Circuit SF - Select Frequency*

Direction:  To DCE

Signals on this circuit are used to select the transmit and receive frequency bands of a DCE.

The ON condition selects the higher frequency band for transmission to the communication channel and the lower frequency band for reception from the communication channel.

The OFF condition selects the lower frequency band for transmission to the communication channel and the higher frequency band for reception from the communication channel.

### 4.4.19  *Circuit SR - Signaling Rate Selector*

Direction:  To DCE

Signals on this circuit are used to select one of the two data signaling rates of a dual rate synchronous DCE or to select one of the two ranges of data signaling rates of a dual range non-synchronous DCE.

The ON condition selects the higher data signaling rate or range of rates.  The OFF condition selects the lower data signaling rate or range of rates.

The rate of timing signals, if included in the interface, shall be controlled by this circuit as may be appropriate.

### 4.4.20 *Circuit SI - Signaling Rate Indicator*

Direction: From DCE

Signals on this circuit are used to indicate one of the two data signaling rates of a dual rate synchronous DCE or to indicate one of the two ranges of signaling rates of a dual range non-synchronous DCE.

The ON condition indicates the higher data signaling rate or range of rates. The OFF condition indicates the lower data signaling rate or range of rates.

The rate of timing signals, if included in the interface, shall be controlled by this circuit as may be appropriate.

### 4.4.21 *Circuit SSD - Secondary Send Data*

Direction: To DCE

This circuit is equivalent to Circuit SD (Send Data) except that it is used to transmit data via the secondary channel.

Signals on this circuit are generated by the DTE and are connected to the local secondary channel transmitting signal converter for transmission of data to one or more remote data stations.

In all systems, the DTE should hold Circuit SSD in the binary ONE (marking) condition unless an ON condition is present on all of the following circuits:

1. Circuit SRS (Secondary Request to Send),
2. Circuit SCS (Secondary Clear to Send),
3. Circuit DM  (Data Mode),
4. Circuit TR  (Terminal Ready), where implemented, and
5. Circuit IS  (Terminal In Service), where implemented.

The DCE shall disregard all signals appearing on Circuit SSD during the time that an OFF condition is present on one or more of the above circuits.

All data signals that are transmitted across the interface on Circuit SSD during the time when the above condition is satisfied shall be transmitted to the communication channel by the DCE.

When the secondary channel is useable only for circuit assurance or to interrupt the flow of data in the primary channel (less than 10 bits per second capability), Circuit SSD is normally not provided, and the secondary channel carrier is turned ON or OFF by means of Circuit SRS. Carrier OFF is interpreted as an "Interrupt" condition.

### 4.4.22 *Circuit SRD - Secondary Receive Data*

Direction: From DCE

This circuit is equivalent to Circuit RD (Receive Data) except that it is used for data received on the secondary channel.

When the secondary channel is useable only for circuit assurance or to interrupt the flow of data in the primary channel, Circuit SRD is normally not provided. See Circuit SRR (Secondary Receiver Ready).

### 4.4.23 *Circuit SRS - Secondary Request to Send*

Direction: To DCE

This circuit is equivalent to Circuit RS (Request to Send) except that it is used to control the secondary channel transmit function of the DCE.

Where the secondary channel can be used only as a backward channel, the ON condition of Circuit RS shall disable Circuit SRS and it shall not be possible to condition the secondary channel transmitting signal converter to transmit during any time interval when the primary channel transmitting signal converter is so conditioned. Where system considerations dictate that one or the other of the two channels be in transmit mode at all times but never both simultaneously, this can be accomplished by permanently applying an ON condition to Circuit SRS and controlling both the primary and secondary channels, in complementary fashion, by means of Circuit RS. Alternatively, in this case, Circuit SRS need not be implemented in the interface.

When the secondary channel is useable only for circuit assurance or to interrupt the flow of data in the primary data channel, Circuit SRS shall serve to turn ON the secondary channel unmodulated carrier. The OFF condition of Circuit SRS shall turn OFF the secondary channel carrier and thereby signal an interrupt condition at the remote end of the communication channel.

### 4.4.24 *Circuit SCS - Secondary Clear to Send*

Direction: From DCE

This circuit is equivalent to Circuit CS (Clear to Send), except that it indicates whether the DCE is conditioned to transmit data on the secondary channel.

This circuit is not provided where the secondary channel is useable only as a circuit assurance or an interrupt channel.

### 4.4.25 *Circuit SRR - Secondary Receiver Ready*

Direction: From DCE

This circuit is equivalent to Circuit RR (Receiver Ready) except that it indicates whether the secondary channel receiver in the DCE is receiving a proper signal.

Where the secondary channel is useable only as a circuit assurance or an interrupt channel (see Circuit SRS - Secondary Request to Send), Circuit SRR shall be used to indicate the circuit assurance status or to signal the interrupt. The ON condition shall indicate circuit assurance or a non-interrupt condition. The OFF condition shall indicate circuit failure (no assurance) or the interrupt condition.

### 4.4.26 *Circuit LL - Local Loopback*

Direction: To DCE

Signals on this circuit are used to control the LL test condition (see Section 6.6.1) in the local DCE.

The ON condition of Circuit LL causes the DCE to transfer the output of the DCE transmitting signal converter from the communication channel to the receiving signal converter of the same DCE. through such circuitry as may be required for proper operation. After establishing the LL test condition, the DCE turns ON Circuit TM (Test Mode). After Circuit TM is turned ON, the DTE may operate in a duplex mode, exercising all of the circuits in the interface.

The OFF condition of Circuit LL causes the DCE to release the LL test condition.

The LL test condition shall not disable Circuit IC (Incoming Call).

### 4.4.27 *Circuit RL - Remote Loopback*

Direction: To DCE

Signals on this circuit are used to control the RL test condition (see Section 6.6.2) in the remote DCE.

The ON condition of Circuit RL causes the local DCE to signal the establishment of the RL test condition in the remote DCE. After turning ON Circuit RL and detecting an ON condition on Circuit TM (Test Mode), the local DTE may operate in a duplex mode, exercising the circuitry of the local and remote DCE. The OFF condition of Circuit RL causes the DCE to signal the release of the RL test condition.

Test condition RL places the communication system out of service to the DTE associated with the DCE containing the RL loopback. When RL is activated, the DCE containing the RL loopback shall present an OFF condition on Circuit DM (Data Mode) and present an ON condition on Circuit TM. If test condition RL in the remote DCE is activated from the local DCE (by manual means or by means of Circuit RL), the local DCE shall allow Circuit DM to respond normally and shall present an ON condition on Circuit TM.

### 4.4.28  *Circuit TM - Test Mode*

Direction:  From DCE

Signals on this circuit indicate whether the local DCE is in a test condition (see Section 6.6.3).

The ON condition of Circuit TM indicates to the DTE that the DCE has been placed in a test condition. The ON condition of Circuit TM shall be in response to an ON condition of Circuit LL (Local Loopback) or Circuit RL (Remote Loopback) and indicates that the test condition has been established. The ON condition shall also be in response to either local or remote activation by other means of any DCE test condition. Activation of a telecommunications network test condition (e.g., facility loopback) which is known to the DCE shall also cause Circuit TM to assume the ON condition. The OFF condition of Circuit TM indicates that the DCE is not in a test mode and is available for normal service. When testing is conducted through the DTE/DCE interface, Circuit DM (Data Mode) operates in a normal manner. When testing is not conducted through the DTE/DCE interface, Circuit DM is held in the OFF condition.

### 4.4.29  *Circuit SS - Select Standby*

Direction:  To DCE

Signals on this circuit are used to select the normal or standby communication facilities, such as signal converters and communication channels, as provided by the DCE.

The OFF to ON transition instructs the DCE to replace the normal facilities with predetermined standby facilities. The ON condition on this circuit is to be maintained whenever the standby facilities are required for use. The ON to OFF transition instructs the DCE to replace the standby facilities with the normal facilities. The OFF condition on this circuit is to be maintained whenever the standby facilities are not required for use.

### 4.4.30  *Circuit SB - Standby Indicator*

Direction:  From DCE

Signals on this circuit indicate whether the DCE is conditioned to operate with the normal or the standby communication facilities.

An ON condition indicates that the DCE is conditioned to operate in its standby mode with the normal facilities replaced by the predetermined standby facilities. The OFF condition indicates that the DCE is conditioned to operate in its normal mode.

## 5. STANDARD INTERFACES FOR SELECTED COMMUNICATION SYSTEM CONFIGURATIONS

### 5.1 General

This section describes a selected set of data transmission configurations. For each of these configurations a standard set of interchange circuits (defined in Section 4) is listed. (See Section 6.1).

Provision is made for additional data transmission configurations not defined herein. Interchange circuits for these applications must be specified separately, for each application, by the supplier.

### 5.2 Data Transmission Configurations

Data transmission configurations for which standard sets of interchange circuits are defined are as follows:

Type SR   (Send-Receive)
Type SO   (Send-Only)
Type RO   (Receive-Only)
Type DT   (Data and Timing only)

Figure 5.1 lists the interchange circuits to be provided for each data transmission configuration. Figure 5.2 illustrates the interface when the full complement of interchange circuits is employed.

### 5.3 Conditions

For a given interface type, generators and receivers shall be provided for every interchange circuit designated M (mandatory) in Figure 5.1. In addition, generators and receivers shall be provided for interchange circuits designated S, A, and T where the service is switched, switched with answering signaled across the interface, and synchronous, respectively.

All control interchange circuits listed in Figure 4.1 not provided with an operational generator shall be provided with a dummy generator (see Section 6.9). The dummy generator for Circuit IS (Terminal In Service), Circuit SQ (Signal Quality), Circuit SF (Select Frequency), Circuit SR (Signaling Rate Selector) and Circuit SI (Signaling Rate Indicator) shall permanently hold the circuit in the ON condition. The dummy generator for all other control interchange circuits shall permanently hold the circuits in the OFF condition.

| | INTERCHANGE CIRCUIT | CONFIGURATION | | | | NOTES |
|---|---|---|---|---|---|---|
| | | TYPE SR | TYPE SO | TYPE RO | TYPE DT | |
| SG | SIGNAL GROUND | M | M | M | M | |
| SC | SEND COMMON | M | M | M | | |
| RC | RECEIVE COMMON | M | M | M | | |
| IS | TERMINAL IN SERVICE | O | O | O | | |
| IC | INCOMING CALL | A | A | A | | |
| TR | TERMINAL READY | S | S | S | | |
| DM | DATA MODE | M | M | M | | |
| SD | SEND DATA | M | M | | M | |
| RD | RECEIVE DATA | M | | M | M | |
| TT | TERMINAL TIMING | O | O | | O | |
| ST | SEND TIMING | T | T | | T | |
| RT | RECEIVE TIMING | T | | T | T | |
| RS | REQUEST TO SEND | M | M | | | |
| CS | CLEAR TO SEND | M | M | | | |
| RR | RECEIVER READY | M | | M | | |
| SQ | SIGNAL QUALITY | O | | O | | |
| NS | NEW SIGNAL | O | | O | | |
| SF | SELECT FREQUENCY | O | O | O | | |
| SR | SIGNALING RATE SELECTOR | O | O | O | | |
| SI | SIGNALING RATE INDICATOR | O | O | O | | |
| SSD | SECONDARY SEND DATA | O | O | O | | a,d |
| SRD | SECONDARY RECEIVE DATA | O | O | O | | b,d |
| SRS | SECONDARY REQUEST TO SEND | O | O | O | | a,c |
| SCS | SECONDARY CLEAR TO SEND | O | O | O | | a,d |
| SRR | SECONDARY RECEIVER READY | O | O | O | | b |
| LL | LOCAL LOOPBACK | O | | | | |
| RL | REMOTE LOOPBACK | O | | | | |
| TM | TEST MODE | M | M | M | | |
| SS | SELECT STANDBY | O | O | O | | |
| SB | STANDBY INDICATOR | O | O | O | | |

M = MANDATORY INTERCHANGE CIRCUITS FOR A GIVEN CONFIGURATION.
S = ADDITIONAL INTERCHANGE CIRCUIT REQUIRED FOR SWITCHED SERVICE.
A = ADDITIONAL INTERCHANGE CIRCUIT REQUIRED FOR SWITCHED SERVICE WITH ANSWERING
    SIGNALED ACROSS THE INTERFACE.
T = ADDITIONAL INTERCHANGE CIRCUITS REQUIRED FOR SYNCHRONOUS PRIMARY CHANNEL.
O = OPTIONAL INTERCHANGE CIRCUITS.

NOTES

a = UNNECESSARY IF SECONDARY CHANNEL IS RECEIVE ONLY.
b = UNNECESSARY IF SECONDARY CHANNEL IS TRANSMIT ONLY.
c = UNNECESSARY IF SECONDARY CHANNEL IS A BACKWARD CHANNEL.
d = UNNECESSARY IF SECONDARY CHANNEL IS USABLE ONLY FOR CIRCUIT ASSURANCE OR TO INTERRUPT
    THE FLOW OF DATA IN THE PRIMARY CHANNEL.

FIGURE 5.1

STANDARD INTERFACES FOR SELECTED COMMUNICATION SYSTEM CONFIGURATIONS

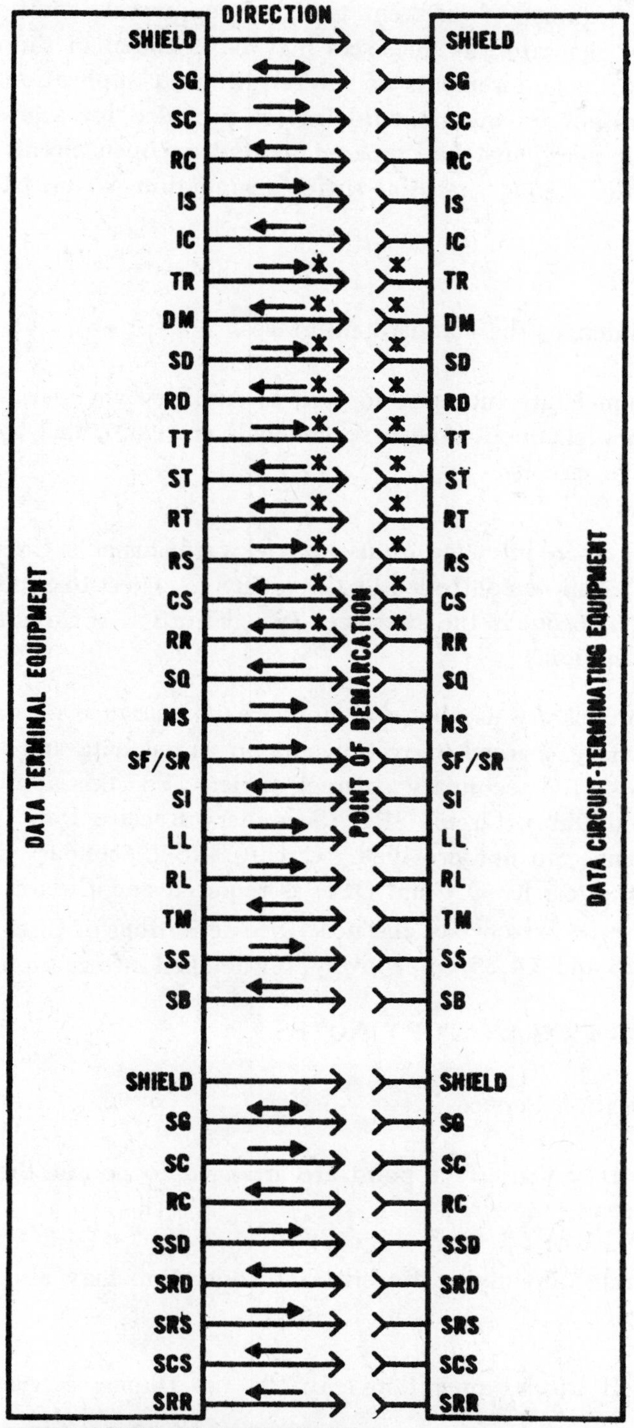

**FIGURE 5.2**
**DTE TO DCE INTERCONNECTION**

✴ **TWO CONNECTOR CONTACTS**

In the interest of minimizing the number of different types of equipment, additional interchange circuits having operational generators or receivers may be included in the design of a general unit capable of satisfying the requirements of several different applications. Where operational generators or receivers not on the standard list are provided for a given configuration, the designer of this equipment must be prepared to find an open circuit on the other side of the interface, and the system shall not suffer degradation of the basic service.

## 5.4 Secondary Channels

Secondary channels may be included in the standard interfaces.

5.4.1 Where secondary channels are intended for use as auxiliary channels, their operation is independent of the primary channel. Auxiliary channels are controlled by the set of secondary control interchange circuits.

5.4.2 Where secondary channels are intended for use as backward channels, Circuit SRS (Secondary Request to Send) shall be controlled by Circuit RS (Request to Send) within the DCE and need not be brought out to the interface. (See definition of Circuit SRS in Section 4.4.23 for detailed information.)

5.4.3 Where secondary channels are useable only for circuit assurance or to interrupt the flow of data in the primary channel, they transmit no actual data and depend only on the presence or absence of the secondary channel carrier. For this application only, Circuit SSD (Secondary Send Data), Circuit SRD (Secondary Receive Data) and Circuit SCS (Secondary Clear to Send) are not provided. Circuit SRS (Secondary Request to Send) turns secondary channel carrier ON and OFF as required and Circuit SRR (Secondary Receiver Ready) indicates its presence or absence. (See definitions of Circuit SRS and Circuit SRR in Sections 4.4.23 and 4.4.25, respectively, for detailed information.)

## 6. RECOMMENDATIONS AND EXPLANATORY NOTES

### 6.1 Alternate Use of Communication Service

The control interchange circuits at the interface point are arranged to permit the alternate use of communication service as follows:

a. A DTE designed for Transmit-Only or Receive-Only operation may also use either Half Duplex or Duplex service.

b. A DTE designed for Half Duplex operation may also use Duplex service.

### 6.2 Line Signals

The turning ON of Circuit RS (Request to Send) does not necessarily imply the turning ON of a line signal on the communication channel. Some DCEs might not have a line signal as it is understood in this standard, e.g., the signal can be a modified digital base-band signal.

Conversely, in DCEs which do not transmit a "line signal", the turning OFF of Circuit RS does not necessarily command the removal of that line signal from the communication channel. On a duplex channel, the DCE might autonomously transmit a training signal to hold AGC circuits or automatic equalizers in adjustment, or to keep timing locked (synchronized) when Circuit RS is OFF.

It is not within the scope of this standard to specify in detail what occurs on the communication channel (line) side of the DCE. Therefore, the definition for Circuit RS uses the terminology "assume the transmit mode" intentionally avoiding reference to "carrier" or "line signals".

However, the requirement for multipoint systems is recognized. DCEs intended for this type of operation should permit the sharing of a communication channel by more than one transmitter and should, when in a non-transmit mode, place no signal on the communication channel which might interfere with the transmission from another DCE in the network.

6.3  Use of Circuit DM - Data Mode

It is important that, at an answering data station, Circuit DM be turned ON independently of any event which might occur at the remote (calling) data station. This independence permits the use of the OFF to ON transition of Circuit DM to start an "abort timer" in the DTE. This timer would cause termination of an automatically answered call (by causing Circuit TR (Terminal Ready) to be turned OFF) if other expected events such as Circuit RR (Receiver Ready) ON or proper exchange of data do not occur in a predetermined time interval. Such independence is necessary to assure the starting of the abort timer when an automatically answered incoming call is the result of a wrong number reached from a regular (non-data station) telephone instrument.

6.4  Use of Circuit SQ - Signal Quality

Circuit SQ is intended to indicate whether or not there is a high probability of error in the receive data. For example, this circuit may indicate the relative quality of the receive data based on the recovery of carrier, recovery of bit timing, or distortion of the eye pattern in the demodulator. Signal quality also may be based on the state of the equalizer for those DCEs equipped with equalizers (see Section 6.8). This circuit may also respond quickly to loss of the received line signal whereas the OFF indication on Circuit RR (Receiver Ready) and the mark clamp on Circuit RD (Receive Data) may be delayed to hold over momentary signal dropouts.

6.5  Use of Circuit NS - New Signal

Circuit NS is intended for use at the control station of multipoint polling systems where the remote DCEs operate in switched carrier mode. The incoming signal to the control station generally appears as a series of short message bursts transmitted by each remote station, in turn, as it responds to the polling message from the control station. In order to permit rapid accommodation to signals from several remote points appearing in quick succession, the control DTE indicates to the DCE when a new signal is about to begin by turning Circuit NS ON for a brief interval.

For synchronous systems, the clock timing on the incoming messages varies from message to message because the remote DCEs are in no way synchronized to each other. If the time interval between messages is too short, the clock holdover after the end of one message may preclude rapid synchronization on the following message. Use of Circuit NS allows the control DTE to reset the DCE receiver timing recovery circuit enabling it to respond much more quickly to the line signal present after Circuit NS is turned off.

For nonsynchronous systems where the time interval between messages is so short that the carrier detector circuit at the control DCE does not detect a loss of received carrier, Circuit RR (Receiver Ready) will not change from ON to OFF to ON. In this situation, Circuit RD (Receive Data) is always active and will probably deliver some spurious signals to the control DTE. Since the DTE has no information from Circuit RR as to when the new line signal is present, it may have difficulty ignoring the spurious signals while properly responding to the actual data information in the next message. Use of Circuit NS allows the control DTE to reset the DCE receiver carrier detection circuit which turns Circuit RR OFF and clamps Circuit RD marking.

Procedures for operation of Circuit NS are system-dependent. Appropriate procedures must be used in the DTE to insure that Circuit NS is not turned OFF until the old line signal has actually disappeared from the communication channel.

6.6  Use of Circuits for Testing

A group of three interchange circuits are defined to permit fault isolation testing to be done under control of the DTE. The three circuits are:

> Circuit LL (Local Loopback),
> Circuit RL (Remote Loopback) , and
> Circuit TM (Test Mode).

The test control (Circuit LL and Circuit RL) and status (Circuit TM) circuits are considered a desirable step toward a uniform methodology for fault isolation. These circuits will assist the user of DTEs and DCEs in identifying the defective system unit.

Figure 6.1 illustrates the local loopback (LL) and remote loopback(RL) tests as seen from the local DTE. A symmetrical set of loopback tests could exist as seen from the remote DTE.

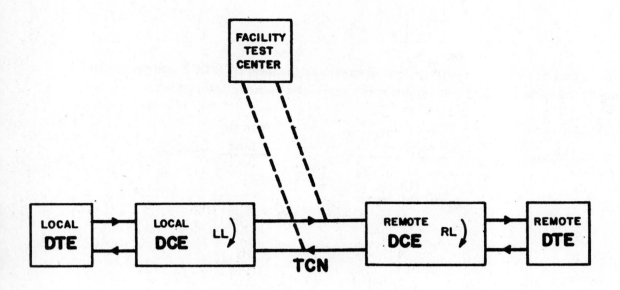

**FIGURE 6.1
ILLUSTRATIVE SYSTEM**

### 6.6.1 *Local Loopback (LL Test)* *

This test condition provides a means whereby a DTE may check the functioning of the DTE/ DCE interface and the transmit and receive sections of the local DCE. The local DCE also may be tested with a test set in place of the DTE. In the LL test, the output of the transmitting section of the DCE is returned to the receiving section of the DCE, through such circuitry as may be required for proper operation. In many DCEs the transmitted signal is not suitable for direct connection to the receiver. In such cases it is preferable to include appropriate signal shaping or conversion in the loop-around circuitry so that all elements used in normal operation are checked in the test condition.

In the LL test, the communication channel is electrically disconnected from the signal processing circuits of the DCE and terminated as appropriate. For switched network applications, the DCE should terminate the call (i.e., go ON HOOK) before establishing the LL test condition. If the DCE is arranged for automatic answering of received calls (see Circuit TR, Terminal Ready), this capability is disabled by the DCE during test LL. However, Circuit IC (Incoming Call) is not disabled during the LL test condition. Circuit IC continues to indicate whether an incoming call (ringing) signal is being received by the DCE during a LL test condition. This will permit the DTE to interrupt the LL test and answer the incoming call, if desired. However, in applications where DCEs are served by a multiline hunting group with line busy-out capability, the DCE should cause a busy indication to be given to the line hunting equipment during test LL. In this case, Circuit IC will be in the OFF condition during test LL.

---

\* Test condition LL is equivalent to C.C.I.T.T. test loop 3.

The condition of various interchange control circuits at the local DTE/DCE interface during a local loopback (LL) test condition is summarized in the following table:

| Interchange Circuit | Switched Network | Private Line |
|---|---|---|
| IS | ON | - |
| IC | ACTIVE OR OFF | - |
| TR | ON | - |
| DM | ON | ON |
| LL | ON | ON |
| RL | ·OFF | OFF |
| TM | ON | ON |

### 6.6.2 *Remote Loopback (RL Test)*\*

This test condition provides a means whereby a DTE or a facility test center may check the transmission path up to and through the remote DCE to the DTE interface and the similar return transmission path. In this test, Circuit SD (Send Data) and Circuit RD (Receive Data) are disconnected or isolated from the remote DTE at the interface and connected to each other in the remote DCE. In synchronous DCEs, arrangements may be required to provide a suitable transmit clock when the RL test condition is activated, and in some instances buffer storage between Circuit RD and Circuit SD may be required.

Remote control of the RL test at the distant DCE through the local DTE/DCE interface is highly desirable to permit automation of the end-to-end testing of a circuit from a central site. Test control is suitable primarily in point-to-point applications but could be used in multipoint configurations with the addition of an address detection capability in the DCE. This test permits circuit verification without the aid of the distant DTE and is supported by inherent remote loopback capability in many present-day DCEs.

Test RL and test LL cannot be performed simultaneously. Consequently, the ON states of Circuit RL (Remote Loopback) and Circuit LL (Local Loopback) are mutually exclusive.

Circuit RL implies that the remote DCE be signaled from the local DCE to activate the RL test condition. Since such control is effected over the communication channel, appropriate measures must be taken to guard against false operation by data or noise.

Consideration must also be given in the implementation of remote control of the RL test condition when these DCEs are used on a common user switched network so that the DTE is not maintained in an out-of-service condition due to occurrences such as:

1. The DCE being left in the RL test condition when the switched connection is dropped.

2. The DCE being deliberately held in the RL test condition by an incoming call.

---

\* Test condition RL is equivalent to C.C.I.T.T. test loop 2.

The condition of various interchange control circuits at the local and remote DTE/DCE interfaces during a remote loopback (RL) test condition is summarized in the following table:

| Interchange Circuit | Switched Network* | | Private Line | |
|---|---|---|---|---|
| | Local Interface | Remote Interface | Local Interface | Remote Interface |
| IS | ON | ON+ | - | |
| IC | OFF | OFF | - | |
| TR | ON | ON+ | - | |
| DM | ON | OFF | ON | OFF |
| LL | OFF | OFF | OFF | OFF |
| RL | ON | OFF | ON | OFF |
| TM | ON | ON | ON | ON |

*Normal call setup procedures apply to establish the connection.

+Must be held ON by DTE to maintain the connection.

903

### 6.6.3 Test Mode

Circuit TM provides the indication from the DCE to the DTE that the DCE is in a test condition. This circuit:

(1) Provides electrical indication that the DCE is in a test condition either in response to control from the DTE or in response to any other action (e.g., telecommunication network initiated testing or manually controlled DCE testing).

(2) Provides action-reaction type control across the interface which verifies completion of requested action.

(3) Permits Circuit DM (Data Mode) to function normally during test modes where testing is conducted through the DTE/DCE interface and prevents any ambiguity between testing and other modes when Circuit DM is in the OFF condition.

### 6.7 Use of Standby Channel Circuits

Two circuits, one for control and one for status, are defined for alternate channel selection by the DTE. These will allow the DTE to switch a DCE to and from one of two alternate channels. The circuits are:

Circuit SS (Select Standby) and
Circuit SB (Standby Indicator).

The two standby circuits permit automatically controlled channel switching methods. Several uses for these circuits have been identified. One use involves switching between two channels for better reliability (e.g., back-up). A second use involves switching between two channels for different types of service. Where line hunting equipment is used, these circuits may be used to switch to a special test line for remote maintenance testing.

The two standby circuits are an action-reaction pair when the standby request comes from the DTE via Circuit SS. Circuit SB should change state upon completion of the requested action. When the request is in the DCE (e.g., by means of a switch in the DCE), Circuit SB indicates which channel (the normal or the standby) is operative in the DCE.

## 6.8 Training of Equalizers

For the purpose of this discussion consider two DCEs, one designated as DCE "E" (East) having the equalizer which requires training and the other designated DCE "W" (West) transmitting toward DCE "E".

The initial training of the equalizer in DCE "E" is performed during the time interval between the ON condition of Circuit RS (Request to Send) and the ON condition of Circuit CS (Clear to Send) of DCE "W". Thus, the initial training of the equalizer in the receiver of DCE "E" is performed prior to the ON condition of Circuit RR (Receiver Ready) of DCE "E". If the initial training of the equalizer is successful, Circuit SQ (Signal Quality) is placed in the ON condition no later than the OFF to ON transition of Circuit RR. (Note that the state of Circuit SQ is not defined when Circuit RR is in the OFF condition.)

If the initial training of the equalizer is unsuccessful or if the equalizer requires retraining at some later point in time, Circuit SQ of DCE "E" is placed in the OFF condition. If the equalizer is such that it can retrain on the normal incoming line signal, the retraining takes place without any change in state of any of the other control interchange circuits. Data received from the communication channel continues to be presented on Circuit RD (Receive Data) of DCE "E". When the equalizer attains proper adjustment, Circuit SQ of DCE "E" is placed in the ON condition.

However, if the equalizer is such that DCE "W" must send a unique training signal to achieve equalization in DCE "E", the states of certain interchange circuits are controlled during this process. If the normal flow of data in the direction toward DCE "W" must be interrupted in order to command DCE "W" to transmit this unique sequence, then Circuit CS of DCE "E" is held in the OFF condition during the interval that the command signal is being transmitted. In this situation, it is preferable that Circuit SQ of DCE "W" be placed in the OFF condition during the reception of the command signal. Circuit RD of DCE "W" may be clamped to the binary ONE (marking) condition during the reception of the command signal. In the reverse direction, Circuit CS of DCE "W" is held in the OFF condition during the interval that the unique training signal is being transmitted. Circuit RD of DCE "E" may be clamped to the binary ONE (marking) condition during the reception of the unique training signal. When the equalizer attains proper adjustment, Circuit SQ of DCE "E" is placed in the ON condition.

904

It should be noted that the state of Circuit RR is not affected by the condition of the equalizer or the equalizer training process.

6.9 Dummy Generator

At the interface connector(s), it is necessary that a voltage be present, from either an operational generator or a dummy generator, for every control interchange circuit listed below.

| | DTE CONTROL INTERCHANGE CIRCUIT GENERATORS | | DCE CONTROL INTERCHANGE CIRCUIT GENERATORS |
|---|---|---|---|
| IS | Terminal In Service | IC | Incoming Call |
| TR | Terminal Ready | DM | Data Mode |
| RS | Request to Send | CS | Clear to Send |
| NS | New Signal | RR | Receiver Ready |
| SF | Select Frequency | SQ | Signal Quality |
| SR | Signaling Rate Selector | SI | Signaling Rate Indicator |
| SRS | Secondary Request to Send* | SCS | Secondary Clear to Send* |
| LL | Local Loopback | SRR | Secondary Receiver Ready* |
| RL | Remote Loopback | TM | Test Mode |
| SS | Select Standby | SB | Standby Indicator |

A dummy generator shall meet the open-circuit, test termination and short-circuit generator requirements of RS-422 or RS-423, as appropriate. A dummy generator satisfying these requirements may be implemented with a 2-watt resistor of 47 ohms (±5%) connected to a dc source voltage between 4 and 6 volts. If a dummy generator is used on one or more of the following interchange circuits:

Circuit IS   (Terminal In Service),
Circuit TR   (Terminal Ready),
Circuit DM (Data Mode),
Circuit RS (Request to Send), and
Circuit SRS (Secondary Request to Send)*,

it shall also meet the generator power-off requirement of RS-422 or RS-423, as appropriate.

A single dummy generator may be employed to supply the signal to more than one interchange circuit. Therefore, only two dummy generators are required when both ON and OFF circuit conditions are to be provided.

It is not necessary for the interface cable associated with the DTE to provide separate conductors for each interchange circuit requiring a dummy generator. Instead, two conductors may be employed in the interface cable, one connected to a positive voltage dummy generator in the DTE and the other

*Required only if the 9-position connector is implemented at the interface (see Section 3.1).

connected to a negative voltage dummy generator in the DTE. At the interface connector, these two conductors are connected to multiple connector contacts, as appropriate.

### 6.10 Relationship Between Signaling Rate, Signal Risetime and Interface Cable Distance

The relationship between signaling rate and interface cable distance for balanced interchange circuits is specified in RS-422. Using the guidelines of RS-422, operation over 60 meters (200 feet) of cable limits the maximum signaling rate of balanced interchange circuits to 2,000,000 bits per second. Timing signals, which operate at twice the signaling rate of data signals, may be up to 4,000,000 bits per second.

The relationship between signaling rate, signal risetime and interface cable distance for unbalanced interchange circuits is specified in RS-423. Using the guidelines of RS-423, operation over 60 meters (200 feet) of cable limits the maximum signaling rate of unbalanced interchange circuits to 60,000 bits per second with exponential wave shaping or 138,000 bits per second with linear (ramp) wave shaping. It is possible to select one value of wave shaping which will meet the 60 meter (200 foot) requirement of this standard and will permit interoperability with RS-232-C at all bit rates and cable distances covered by RS-232-C (see EIA Industrial Electronics Bulletin No. 12, Application Notes on Interconnection Between Interface Circuits Using RS-449 and RS-232-C).

Operation over cable distances greater than 60 meters (200 feet) can be accomplished, in many cases, but is considered a tailored application. In addition to a reduction in the maximum signaling rate from the values given above, the wave shaping of the RS-423 generators in both the DTE and DCE must be suitable for the longer cable length. Additional guidance is provided in RS-422 and RS-423.

### 6.11 Choice of Electrical Characteristics for Category I Circuits

For applications where the data signaling rate is 20,000 bits per second or less, Section 2.1.1 permits the DTE to use any mix of RS-422 and RS-423 generators for four Category I circuits: Circuit SD (Send Data), Circuit TT (Terminal Timing), Circuit RS (Request to Send) and Circuit TR (Terminal Ready). For the same applications, Section 2.1.1 permits the DCE to use any mix of RS-422 and RS-423 generators for six Category I circuits: Circuit RD (Receive Data), Circuit ST (Send Timing), Circuit RT (Receive Timing), Circuit CS (Clear to Send), Circuit RR (Receiver Ready) and Circuit DM (Data Mode). The following paragraphs present guidance to aid in the selection of RS-422 or RS-423 electrical characteristics for these circuits.

Use of unbalanced RS-423 generators may be preferred for Category I circuits where interoperability with RS-232-C equipment is desired. This will facilitate interoperability by permitting use of a passive adapter to achieve this interconnection as described in EIA Industrial Electronics Bulletin No. 12, Application Notes on Interconnection Between Interface Circuits Using RS-449 and RS-232-C.

Use of unbalanced RS-422 generators and receivers without cable termination resistance $R_t$ may be preferred for Category I circuits to simplify the design of equipment which may operate over a range of data signaling rates extending above and below 20,000 bits per second.

In addition to operating with DCEs defined in this standard, a DTE may be designed to also inter-operate with Public Data Network DCEs (see Appendix). The five interchange circuit functions of the Public Data Network interface can be accomplished with six of the Category I interchange circuits. DTEs connecting to Public Data Network DCEs at data signaling rates of 20,000 bits per second or less may use either balanced RS-422 or unbalanced RS-423 generators but balanced RS-422 generators must be used for data signaling rates above 20,000 bits per second. Since the number of interchange circuits is small, it is practical to consider operation over interface cable distances beyond the 60 meters (200 feet) specification of this standard. In such cases, use of balanced RS-422 operation for the six interchange circuits may be preferred for all data signaling rates in order to minimize the susceptibility of these circuits to disturbances when the interconnecting cable is exposed to extraneous noise sources and to minimize the interference to other signals.

A DTE may also be designed to permit direct DTE to DTE operation. Direct DTE to DTE operation is expected to be accomplished with a minimum number of interchange circuits., In addition to a data circuit in each direction, many applications are expected to require a control circuit in each direction plus, if synchronous, a timing circuit in each direction. All these functions can be accomplished within the set of ten Category I interchange circuits. Since operation over considerable distance is important for direct DTE to DTE operation, use of the balanced RS-422 interchange circuits may be preferred. In addition to permitting longer cable distances for applications having data signaling rates above 3,000 bits per second, balanced operation over these longer distances will minimize the susceptibility of these circuits to disturbances when the interconnecting cable is exposed to extraneous noise sources and will minimize the interference to other signals.

## 7. GLOSSARY OF TERMS

This section defines terms used in this standard which are new or are used in a special sense.

### 7.1 Data Transmission Channel

The transmission media and intervening equipment involved in the transfer of information between DTEs (see Figure 1.1). A data transmission channel includes the signal conversion equipment. A data transmission channel may support the transfer of information in one direction only, in either direction alternately, or in both directions simultaneously and the channel is accordingly classified as defined in the following sections. When the DCE has more than one speed capability associated with it, for example 1200 bits per second transmission in one direction and 150 bits per second transmission in the opposite direction, a channel is defined for each speed capability.

### 7.2 Primary Channel

The data transmission channel having the highest signaling rate capability of all the channels sharing a common interface. A primary channel may support the transfer of information in one direction only. either direction alternately or both directions simultaneously and is then classified as "one way only", "half duplex" or "duplex" as defined herein.

### 7.3  Secondary Channel

The data transmission channel having a lower signaling rate capability than the primary channel in a system in which two channels share a common interface.  A secondary channel may be either one way only, half duplex or duplex as defined later.  Two classes of secondary channels are defined, auxiliary and backward.

### 7.4  Auxiliary Channel

A secondary channel whose direction of transmission is independent of the primary channel and is controlled by an appropriate set of secondary control interchange circuits.

### 7.5  Backward Channel

A secondary channel whose direction of transmission is constrained to be always opposite to that of the primary channel.  The direction of transmission of the backward channel is restricted by the control interchange circuit (Circuit RS - Request to Send) that controls the direction of transmission of the primary channel.

### 7.6  One Way Only (Unidirectional) Channel

A primary or secondary channel capable of operation in only one direction.  The direction is fixed and cannot be reversed.  The term "one way only" used to describe a primary channel does not imply anything about the type of secondary channel or the existence of a secondary channel; similarly, the use of the term to describe a secondary channel implies nothing about the type of primary channel present.

### 7.7  Half-Duplex Channel

A primary or secondary channel capable of operating in both directions but not simultaneously.  The direction of transmission is reversible.  The term half-duplex used to describe a primary channel does not imply anything about the type of secondary channel or the existence of a secondary channel; similarly, the use of the term to describe a secondary channel implies nothing about the type of primary channel present.  (Note that as a result of the definitions, both directions of a half-duplex channel have the same signaling rate capability.)

### 7.8  Duplex Channel (Full-Duplex Channel)

A primary or secondary channel capable of operating in both directions simultaneously.  The term duplex used to describe a primary channel does not imply anything about the type of secondary channel or the existence of a secondary channel; similarly, the use of the term to describe a secondary channel implies nothing about the type of primary channel present.  (Note that a duplex channel has the same signaling rate capability in both directions.  A system with different rates would be considered to be a one way only primary channel in one direction and a one way only secondary channel in the opposite direction.)

## 7.9 Synchronous Data Transmission Channel

A data channel in which timing information is transferred between the DTE and the DCE. Transmit timing signals can be provided by either the DTE or by the DCE. Receive timing is provided by the DCE. A synchronous data channel will not accommodate Start/Stop data signals at the same data signaling rate unless the time intervals separating successive significant instants are whole multiples of a unit interval (i.e., signals are transmitted isochronously) and timing signals are interchanged at least at the transmitting station.

## 7.10 Nonsynchronous Data Transmission Channel

A data channel in which no separate timing information is transferred between the DTE and the DCE.

## 7.11 Dedicated Line

A communication channel which is nonswitched, i.e., which is permanently connected between two or more data stations. These communication channels are also referred to as "leased" or "private"; however, since leased and private switched networks do exist, the term "dedicated" is preferred herein to define a nonswitched connection between two or more stations.

## 7.12 Interchange Circuit

A circuit between the DTE and the DCE for the purpose of exchanging data, control or timing signals or for use as a ground or common return. Category II interchange circuits in the direction FROM DCE use Circuit RC (Receive Common) as the common reference. Category II interchange circuits in the direction TO DCE use Circuit SC (Send Common) as the common reference. Circuit SG (Signal Ground) is used to connect DTE circuit ground (circuit common) to DCE circuit ground (circuit common).

## 7.13 Generator

The electronic circuitry at the transmitting end (source) of an interchange circuit which converts binary digital signals to signals having the required electrical characteristics for transmission over the interchange circuit.

## 7.14 Receiver

The electronic circuitry at the receiving end (sink) of an interchange circuit which converts signals received from the interchange circuit to binary digital signals.

## 7.15 Signal Conversion Equipment

Those portions of the DCE which transform (e.g., modulate, shape, etc.) the data signals exchanged across the interface into signals suitable for transmission through the associated communication media or which transform (e.g., demodulate, slice, regenerate, etc.) the received line signals into data signals suitable for presentation to the DTE.

### APPENDIX

### COMPATIBILITY OF EIA STANDARD RS-449
### WITH THE PUBLIC DATA NETWORK INTERFACE

Work has been completed on a proposed American National Standard which defines the DTE/DCE interface for Public Data Networks (PDN). One characteristic of this interface is that data and control information are sent over the same interface leads resulting in the need for only five basic interchange circuit functions (plus signal ground). Effort has been made to insure the maximum degree of compatibility between the Public Data Network interface and the interface described in this standard. This commonality should make it possible for manufacturers to design a DTE that can meet both standards. The following table lists the interchange circuits which are used in interconnecting an RS-449 DTE to a PDN DCE.

| INTERCHANGE CIRCUIT | RS-422 | RS-423* | CONTACT CONNECTION | | RS-422 | INTERCHANGE CIRCUIT |
|---|---|---|---|---|---|---|
| **RS-449 DTE** | | | | | **PDN DCE** | |
| SEND DATA | SD(A) | SD(A) | 4 | 2 | T(A') | TRANSMIT |
| | SD(B) | SD(C) | 22 | 9 | T(B') | |
| RECEIVE DATA | RD(A') | RD(A') | 6 | 4 | R(A) | RECEIVE |
| | RD(B') | RD(B') | 24 | 11 | R(B) | |
| REQUEST TO SEND | RS(A) | RS(A) | 7 | 3 | C(A') | CONTROL |
| | RS(B) | RS(C) | 25 | 10 | C(B') | |
| RECEIVER READY | RR(A') | RR(A') | 13 | 5 | I(A) | INDICATION |
| | RR(B') | RR(B') | 31 | 12 | I(B) | |
| SEND TIMING† | ST(A') | ST(A') | 5 | 6 | S(A) | SIGNAL ELEMENT |
| | ST(B') | ST(B') | 23 | 13 | S(B) | TIMING |
| RECEIVE TIMING† | RT(A') | RT(A') | 8 | 6 | S(A) | SIGNAL ELEMENT |
| | RT(B') | RT(B') | 26 | 13 | S(B) | TIMING |
| SIGNAL GROUND | SG | SG | 19 | 8 | G | SIGNAL GROUND |

These circuits are functionally similar to one another but have significant procedural differences which must be accommodated. Also needed is the adaptation between the 37-position RS-449 connector and the 15-position PDN connector.

---

\* Permitted only for data signaling rates of 20,000 bits per second or less.

† Cable termination resistance, if used with RS-422 receivers on Circuits ST and RT, shall not be less than 200 ohms to permit parallel connection to Circuit S.

# INDUSTRIAL ELECTRONICS BULLETIN NO. 5

## TUTORIAL PAPER

## ON

## SIGNAL QUALITY AT DIGITAL INTERFACE

911

PRICE: $ 1.50

# ELECTRONIC INDUSTRIES ASSOCIATION

**Engineering Department**

2001 Eye Street, N.W., Washington, D.C.    20006

# FOREWORD

This working paper has been prepared by Electronic Industries Association Subcommittee TR-30.1 to provide tutorial background and definitions of appropriate terms, and to lay a basis for a proposed standard, for the quality of digital signals at the interface between data processing terminal equipment, and data communication equipment, in serial-by-bit operation.

* * *

The following is the membership of TR-30.1, Signal Quality, a Subcommittee of EIA Committee TR-30, Data Transmission Equipment:

H. H. Smith, Chairman
   ITT Communication Systems, Inc.

W. B. McClelland, Secretary
   Teletype Corporation

V. L. Dagostino
   Digitronics Corporation

R. A. Day
   Department of Defense

P. H. DeGroat
   Xerox Corporation

N. H. Kramer, (Liaison with ASA Subcommittee X-3.3)
   Stelma, Inc.

H. C. Likel
   Western Union Telegraph Company

G. J. McAllister, (Alternate for Mr. Schmal)
   Bell Telephone Laboratories, Inc.

Pierre Mertz
   Consultant

R. L. Schmal
   Bell Telephone Laboratories, Inc.

R. J. Smith, (Succeeding Mr. Tate)
   International Business Machines Corporation

L. A. Tate
   International Business Machines Corporation

- 1 -

# TABLE OF CONTENTS

913

## 1. INTRODUCTION

In commercial data transmission, it is typical that the data processing equipment and the data transmission facilities are provided by separate organizations. A significant interface therefore exists at the points where signals are exchanged between these two sets of equipment. For serial-by-bit data transmission, this interface is defined, and standards for electrical parameters are established by EIA Standard RS-232-A. To facilitate determination of the portion of the overall system which needs corrective action when operation is found to be unsatisfactory, it is important that agreement be reached as to the quality of the electrical signals at this interface on the circuits carrying data signals sent and received over the channel, and on timing circuits.

## 2. SCOPE

This discussion is concerned with the quality of the electrical data and timing signals at the interface defined in RS-232-A, with respect to timing and distortion, and to necessary relationships between signals on various circuits. It is deemed that consideration of error rate lies outside the scope of the present effort.

## 3. DATA COMMUNICATION SYSTEM

914

For the purposes of this paper, a data communication system is considered to consist of the following distinct parts:

3.1 The data source, that is, the set of data processing equipment which delivers a digital baseband ("dc") signal over an interface circuit for transmission.

3.2 The data sink, that is, the set of data processing equipment which accepts the digital baseband signal received over an interface circuit, and

3.3 The data transmission system or channel, which accepts the signal from the data source, and reproduces, ideally, the identical signal for delivery to the data sink.

## 4. DATA TRANSMISSION SYSTEM

For the purposes of this discussion, it is assumed that the data transmission system includes a pair of data sets interconnected over a suitable circuit. However, the selection of parameters and magnitudes of performance requirements for single links may be determined with the possibility of tandem operation in mind. The data transmission system has three distinguishable components:

4.1     The sending data set, which includes a signal converter which changes the form of the digital signal to one designed to be compatible with the characteristics of the transmission medium, such as bandwidth, amplitude distortion, envelope delay distortion, and signal-to-noise ratio.

4.2     The receiving data set, including a signal converter which changes the received line signal back into the original baseband digital form.

4.3     The transmission medium. At present, the most typical forms of transmission media are either leased private lines, or switched circuits, both fundamentally designed for handling telephone speech. Some grades of private line facilities are available for digital data transmission which have been modified or selected to have improved characteristics for this purpose. Such circuits may include phase or amplitude equalization, or both, and may be selected to have relatively low noise levels.

4.4     Classes of data transmission systems. Serial-by-bit data transmission systems or channels may be classified as either non-synchronous, or synchronous, depending upon the method of operation of the signal converters. As defined in the following paragraph, a "non-synchronous" channel may be used for synchronous operation.

### 5. NON-SYNCHRONOUS DATA TRANSMISSION SYSTEMS

5.1     Terminology

5.1.1     A non-synchronous data transmission system or channel may be defined as one in which the times at which the transitions of the line signal occur are under the control of the data source. That is, for their operation, the sending and receiving signal converters do not require timing from a unit provided for this purpose, usually referred to as a "clock".

5.1.2     A distinction should be made between two terms used in this paper, which have a similar appearance:

       a.     Non-synchronous data transmission channels have no inherent dependence on timing sources. That is, the timing of signal transitions is controlled entirely by the data source, and the receivers do not introduce any intentional retiming.

       b.     Asynchronous operation is actually synchronous operation over a limited time duration, such as within a single character or block. In telegraphy, this is called "start-stop" operation. Asynchronous operation is defined in MIL-STD-188B (24 February 1964) as follows:

- 4 -

"A transmission process such that between any two
significant instants in the same group*, there is
always an integral number of unit intervals.
Between two significant instants located in different
groups, there is not always an integral number of
unit intervals.

*In data transmission this group is a block or a
character.  In telegraphy this group is a character."

## 5.2  Applications

5.2.1    Asynchronous (start-stop) Operation.  The transmission
channel stands ready to accept the beginning of the start
element at any time, and will reproduce it at the receiving signal converter output with a delay which depends on
the transmission propagation time, with any early or late
displacement which may be caused by distortion or noise.
Subsequent transitions arrive at times similarly related
to the transitions delivered by the source.

5.2.2    Synchronous Operation.  In synchronous operation, the
source and sink operate under control of suitable clocks.
The source will originate signal transitions only at
instants comprising a single series, equally spaced in
time.  This is defined by the ITU (31.29) as "isochronous
modulation (or restitution)" [demodulation] , "in which the
time interval separating any two significant instants
[transitions] is theoretically equal to the unit interval
or to a multiple of this".  Since the transmission channel
is able to accept transitions at any time whatever, provided that its rate capabilities are not exceeded, it is
able to function acceptably as part of a "synchronous"
system, although its operation is not directly controlled
by clocks.

5.2.3    Random Two-Level Operation.  A non-synchronous channel
may be used for the transmission of baseband signals which
have transitions between two states or levels, which occur
essentially at random, the time durations between transitions not being even nominally related to a "unit
interval".  An example is two-level facsimile.  A whole line
of scanning may produce only one or two pairs of transitions,
or the interval between transitions may approach zero when
scanning across very narrow lines.  The latter case would
require a much wider transmission bandwidth than is provided,
with the result that such closely sequential transitions may
be subject to considerable distortion and even suppression.
For this reason, the concepts related to distortion and
signal quality for this type of signal involve a number of
factors which are absent or negligible in the transmission
of what is generally understood as "data", and for this
reason, this type of two-level operation will not be
discussed further in this paper.

- 5 -

## 5.3  Characteristics of Non-Synchronous Channels

5.3.1    Overall Distortion.  Any distortion of element duration as delivered to the channel by the source will be reproduced at the output of the channel, together with any additional distortion caused by the signal converters and the transmission medium.  That is, some of the distortion in the received signal is due to causes on the data processing side of the sending interface, and some due to the data transmission system.

5.3.2    Inherent Distortion.  That part of the distortion due to the data transmission channel is called "inherent distortion".  It is defined by the ITU (33.13) as "degree of inherent distortion (of a channel): Degree of distortion of the restitution [demodulation] when the modulation is effected without distortion".  It adds the following notes:

"1)    By inherent distortion is meant the combination of the different types of distortion caused by the channel (bias, characteristic, etc...)

"2)    This notion may be extended to the constituent elements of a channel, such as a telegraph relay.

"3)    It will be necessary to specify in what conditions the channel is used (type of apparatus, modulation rate, manual or automatic keying, etc...) and to effect the modulation under these conditions.

"In particular should be defined:   the point of entry at which the distortionless modulation is applied;   the terminal point where the distortion is measured."

5.3.3    Modulation Rates.  In general, a non-synchronous channel may be operated at any rate up to a maximum which is limited by distortion or error rate.  It should be noted, however, that the inherent distortion may not necessarily be less at some particular rate than at some higher rate, since some signal converter systems exhibit lower distortion and error rate at certain optimum rates than at higher and lower rates.

5.3.4    Distortion in Start-Stop Operation.  In considering objectives for start-stop distortion, it must be noted that any displacement of the beginning of the start element, whether due to transmitter distortion, or to inherent distortion of the channel, or both, will cause effective distortion of all of the subsequent transitions, in addition to that applicable directly to such subsequent transitions individually.  For this reason, it can be stated that with the same inherent distortion in the channel, the distortion measured in start-stop operation may be double the measured individual distortion.  This matter is discussed in more detail in Section 7.

5.3.5     Distortion in Synchronous Operation.  In synchronous
operation, the time reference used in the receiving
signal converter is not directly related to any single
transition.  The distortion of individual transitions
is reckoned against a series of ideal instants of transition, as determined by a clock.  The receiving clock
may be one which is able to maintain a fixed time relation to the sending clock for considerable periods of
time, or one in which the ideal instants are continually
adjusted in accordance with the average phase of the
received bit stream, with a time lag or "flywheel" effect.
A term applicable to distortion in synchronous operation
is "isochronous distortion".  Isochronous distortion is
discussed in Part B.

5.4   Requirements.  Since the source equipment is free to originate
signal transitions at any time, whether start-stop or isochronous
signals, it is important that the non-synchronous channel reproduce
the timing characteristics of the original signals at its output,
within acceptable limits.  This leads to the establishment of a
limit on the inherent distortion for a non-synchronous data transmission channel, so that it will be reasonably "transparent" in
terms of signal element durations.  From this view, it might seem
that no distortion requirements need be imposed on the data processing terminal equipment, since the designer of the latter could
presumably allow for the maximum permitted channel distortion while
establishing optimum tradeoff between source transmitting distortion,
and sink receiving margin.  However, since it must be assumed that
equipment of different manufacture may be interworked over a
channel, it is desirable to establish requirements on maximum source
transmitter distortion and minimum sink receiving margin.

## 6. SYNCHRONOUS DATA TRANSMISSION SYSTEMS

6.1   A synchronous data transmission system or channel may be defined as
one in which the sending and receiving signal converters operate
with isochronous modulation, at a fixed nominal data signaling rate,
under control of clock(s).  That is, the data source is constrained
to supply successive digital signal elements in a specified time
relation to the sending signal converter, or both are constrained
to operate in such a manner that the required time relation is
realized.  The receiving signal converter is synchronized with the
received bit stream, and the data sink must be prepared to accept
successive signal elements at the same rate.  It is possible that,
in some systems, additional signals may be introduced in the line
signal passing between the sending and receiving data sets, but it
is true that the receiving signal converter must remain in proper
phase relation with the received line signal, either by use of
highly stable clocks at both ends, or by continual readjustments
of the phase or rate of the receiving clock, based on continuous
observation of the line signal.

- 7 -

The signal converters in a synchronous data transmission system may be designed to use a variety of modulation principles, and are not restricted to binary (two-level signal) systems. The isochronous binary nature of the interface signals allows translation into multilevel signals, for example, by the use of "logic" devices, such as the dibit logic used in conversion between binary and quaternary modulations. It also allows the signal to be modulated onto a number of simultaneous carrier currents. Both multilevel and multitone schemes can result in a line signal modulation rate (in bauds) which is less than the data signaling rate (in bits per second).

6.2  Application.  For reasons discussed in the preceding paragraph, a synchronous data transmission system may be used only in connection with data sources and sinks designed to operate in a synchronous manner.

6.3  Characteristics of Synchronous Channel

    6.3.1    Distortion.  In a synchronous channel, generally any distortion of element duration as delivered by the source to the sending signal converter, will have no direct effect on the distortion of the signal delivered to the sink by the receiving converter.  It is only necessary that the source distortion be within limits which allow the sending signal converter to identify the successive binary conditions presented by the source.  If these limits are exceeded, the converter may generate a line signal which does not correctly represent these binary conditions, but the line signal elements representing such conditions will be properly timed.  Likewise, line distortion or impulse noise, and the like, may cause the receiving signal converter to make incorrect decisions, but the binary signal elements which it delivers to the sink will be retimed.

    6.3.2    Inherent Regeneration.  It is shown in the preceding discussion that both the sending and the receiving signal converters in a synchronous channel, as defined, accomplish signal regeneration incidentally to their operation.  That is, in addition to being retransmitted in another signal mode, the signals are retimed.  It is therefore meaningless to attempt to apply the concept of "inherent distortion" to a synchronous data transmission channel.

6.4  Requirements.  Since there is no relation between the distortion at the interface at the sending end of the system, and that at the receiving end of the system, it is logical to set up requirements which will merely insure that the signal distortion is held to limits which are acceptable to the device at the receiving end of the Transmitted Data (Circuit BA of RS-232-A) or Received Data

- 8 -

(Circuit BB) interchange circuits, so long as the signals otherwise conform to the electrical characteristics specified in RS-232-A. Requirements need to be placed also on the distortion of the signals passed over timing circuits between data processing terminal equipment and data sets, and the relationships between the "clock" signals and the data signals with which they are associated.

6.4.1    Transmitting Terminal of Synchronous Channel.    In terms of the RS-232-A interface, two interface circuits are involved: Circuit BA, Transmitted Data, and either Circuit DA, Transmitter Signal Element Timing, (Data Processing Terminal Equipment Source), or Circuit DB, Transmitter Signal Element Timing, (Data Communication Equipment Source).

6.4.1.1    Data Processing Terminal Equipment Timing Source. In this case, the data source equipment presents data signals on the BA circuit, and timing signals on the DA circuit. RS-232-A prescribes that for DA, the waveform shall nominally be ON and OFF for equal periods of time, and a transition from ON to OFF shall nominally indicate the center of each signal element on Circuit BA. The ON to OFF transition is defined to be a positive-to-negative transition. The modulation rate of the timing signals must therefore be exactly twice that of the data signals. In the light of this relation, it appears that two requirements need to be set up for this case. One is that the timing signals shall not exceed a specified degree of isochronous distortion. This will limit the effects due to both bias and irregular transitions on the timing circuit DA. The other requirement applicable to this case is that the center of the data signal elements shall coincide with the ON to OFF transitions on the timing circuit, to a stated degree.

6.4.1.2    Data Communication Equipment Timing Source. In this case, the timing signals are provided by the data set, the timing circuit being designated DB. RS-232-A requires that "the data processing terminal equipment shall provide a data signal on circuit BA, Transmitted Data, in which the transitions between signal elements nominally occur at the time of the transitions from OFF to ON condition of the signal on circuit DB". This will insure that the center of the data signal elements will correspond to the ON to OFF timing transitions. In this case, the requirements are similar to the previous one, except that the transitions on circuit BA should be held to a maximum deviation from the OFF to ON transitions on circuit DB.

920

- 9 -

6.4.2        Receiving Terminal of Synchronous Channel.  RS-232-A
             defines use of two interface circuits for data and
             timing:  Circuit BB, Received Data, and either Circuit
             DC, Receiver Signal Element Timing, (Data Processing
             Terminal Equipment Source), or Circuit DD, Receiver
             Signal Element Timing, (Data Communication Equipment
             Source).

             6.4.2.1    Data Processing Terminal Equipment Timing
                        Source.  In this case, the data processing
                        sink equipment provides timing signals on
                        Circuit DC to the data set.  A transition
                        from ON to OFF shall nominally indicate the
                        center of each signal element on Circuit BB,
                        as specified by RS-232-A.  There appear to
                        be two requirements: first, that the data
                        signal on circuit BB should not exceed a
                        stated degree of distortion, and second,
                        that the ON to OFF transitions on circuit
                        DC should fall within a stated range in the
                        middle of the data signal elements on
                        circuit BB.  It is expected that this type
                        of operation will find rare application.

             6.4.2.2    Data Communication Equipment Timing Source.
                        In this case, the timing signals are pro-
                        vided by the receiving data set, over circuit
                        DD.  These timing signals may be used either
                        to control the reading of the received data
                        signals on Circuit BB, or, in some cases of
                        tandem channel operation, to control the
                        timing of a second sending data set.  It is
                        therefore important to set limits on the
                        permissible distortion of the timing signals.
                        Requirements should also be placed on the
                        timing of the received data signal, for two
                        reasons.  One is to insure that the center
                        of data elements will coincide adequately
                        with the significant (ON to OFF) timing
                        transitions.  The other is to provide adequate
                        timing for receiving data processing equip-
                        ment which ignores the data set timing
                        signals, if any, and performs its functions
                        solely on the basis of the received data
                        signal timing.

## 7. COMPARISON OF DISTORTION MEASUREMENT TECHNIQUES

7.1  Types of Distortion Measurements.  Different measurement principles
     are required in the measurement of the distortion of a continuous
     (isochronous) signal, as employed in synchronous operation, and the
     distortion of a start-stop or asynchronous signal.  To clarify the

- 10 -

difference between the principles of these two measurements, a graphical representation is presented. For this purpose, a particular sequence of fourteen signal elements has been selected, which can represent either fourteen consecutive elements of an isochronous signal, or two consecutive seven-element start-stop characters. The latter are separated by a unit-interval stop element to preserve the timing sequence. A number of the signal transitions are shown with either early or late distortion. It is not implied that such a short sequence should be used for actual measurement purposes, it is for illustration only. In general, a relatively long period of measurement should be used. The identical signal sequence is shown in each part of the figure which follows.

Part 1 illustrates the measurement of the Degree of Peak Individual Distortion. The highest degree of individual distortion appears for the transition designated $D_{11}$.

Part 2 illustrates the measurement of the Degree of Start-Stop Distortion. All measurements of distortion refer to the transition at the beginning of the Start element of the particular character. The greatest distortion in this case would be measured with respect to the transition designated $L_{A6}$. The amount of distortion is the difference between the duration of six unit intervals, and the interval from the Start-A transition to the actual occurrence of $L_{A6}$.

Part 3 illustrates the measurement of the Degree of Isochronous Distortion. The displacements are measured from the associated reference instants (marked from 0 to 13), as in Part 1. The algebraic difference between the earliest and latest individual distortions involves transitions $I_2$ and $I_{11}$.

The following comments may be made:

a. Note that the degree of isochronous distortion and peak individual distortion are derived from the same set of numbers.

b. If the displacement is symmetrical about the ideal instants of modulation, the degree of isochronous distortion is equal to twice the degree of peak individual distortion.

c. In start-stop distortion, each start transition begins a measurement. It is not likely that this transition will be at one extreme of the displacement, with one of the character transitions at the other extreme. For this reason, when comparing start-stop distortion with peak individual distortion, one can only say that the start-stop distortion may be equal to twice the peak individual distortion.

922

- 11 -

The three types of measurement:

1. <u>Degree of Peak Individual Distortion</u>

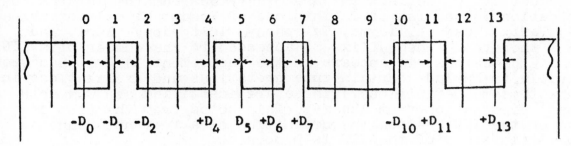

The individual distortions are represented by the numbers $+D_0$ through $+D_{13}$ (provided they exist). The degree of peak individual distortion is given by the largest number of the set regardless of sign, in this case, $D_{11}$. The measurement interval is 14 unit intervals.

2. <u>Degree of Start-Stop Distortion</u>

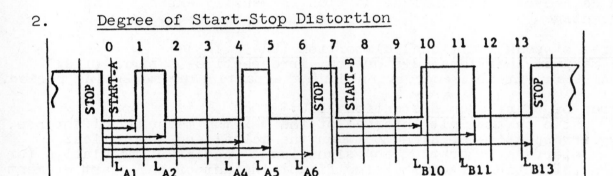

The start-stop distortions are the numbers $L_{A1}-1$, $L_{A2}-2$, through $L_{B13}-6$ (provided they exist). The degree of start-stop distortion is the largest number regardless of sign, in this case, $L_{A6}-6$. The measurement period is two 7-unit-element characters.

3. <u>Degree of Isochronous Distortion</u>

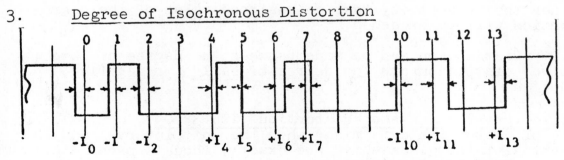

The algebraic difference between the lowest and highest individual distortions is the degree of isochronous distortion, in this case $-I_{11} - (-I_2) = I_{11} + I_2$. The measurement period is 14 unit intervals.

## PART B - DEFINITIONS

The following definitions are applicable to the signals at the digital interface. Most of the definitions are excerpted from the International Telecommunication Union List of Definitions of Essential Telecommunication Terms, Part 1, General Terms, Telephony, Telegraphy, 1961, and Supplement 1 thereto. (The suffix "S" appended to the number of a ITU definition indicates that it appears in the Supplement 1). It has been found desirable to include certain interpretations and explanations in connection with some definitions. These are identified as "Comments" to indicate a distinction from the material quoted from the List of Definitions of the ITU. Also, the definition of a term not found in the List of Definitions of the ITU is proposed.

1. Degree of individual distortion of a particular significant instant (of a modulation or of a restitution). (ITU 33.06)
   Ratio to the unit interval of the displacement, expressed algebraically, of this significant instant from an ideal instant.

   This displacement is considered positive when the significant instant occurs after the ideal instant.

   The degree of individual distortion is usually expressed as a percentage.

2. Degree of peak individual distortion (TR-30.1/3)
   The maximum individual distortion, irrespective of sign, of all significant instants occurring during a particular measuring period.

3. Degree of start-stop distortion (ITU 33.08)
   a) Ratio to the unit interval of the maximum measured difference, irrespective of sign, between the actual and theoretical intervals separating any significant instant of modulation (or restitution) from the significant instant of the start element immediately preceding it.

   b) The highest absolute value of individual distortion affecting the significant instants of a start-stop modulation.

   Note 1. The result of the measurement should be completed by an indication of the period, usually limited, of the observation. For a prolonged modulation (or restitution) it will be appropriate to consider the probability that an assigned value of the degree of distortion will be exceeded.

   Note 2. Distinction can be made between the degree of late (or positive) distortion and the degree of early (or negative) distortion.

4. Degree of gross start-stop distortion (ITU 33.09)
   Degree of distortion determined when the unit interval and the theoretical intervals assumed are exactly those appropriate to the standardized modulation rate.

- 13 -

Note:  The result of the measurement should be completed by an indication of the period, usually limited, of the observation. For a prolonged modulation (or restitution) it will be appropriate to consider the probability that an assigned value of the degree of distortion will be exceeded.

5. Degree of synchronous start-stop distortion (i.e. at the actual mean modulation rate) (ITU 33.10)
Degree of distortion determined when the unit interval and the theoretical intervals assumed are those appropriate to the actual mean rate of modulation (or restitution).

Note 1.  The result of the measurement should be completed by an indication of the period, usually limited, of the observation. For a prolonged modulation (or restitution) it will be appropriate to consider the probability that an assigned value of the degree of distortion will be exceeded.

Note 2.  For the determination of the actual mean modulation rate, account is only taken of those significant instants of modulation (or restitution) which correspond to a change of condition in the same sense as that occurring at the beginning of the start element.

6. Margin   (ITU 34.03S)
Of a telegraph apparatus (or the local end with its termination). The maximum degree of distortion of the circuit at the end of which the apparatus is situated which is compatible with the correct translation of all the signals which it may possibly receive.

Note:  The maximum degree of distortion which results in incorrect translation applies without reference to the form of distortion affecting the signals.  In other words, it is the maximum value of the most unfavorable distortion causing incorrect translation which determines the value of the margin.

7. Margin of start-stop apparatus (or of the local end with its termination) (ITU 34.07S)
The maximum degree of start-stop distortion of the modulation, which it is possible to apply to the apparatus compatible with the correct translation of all the signals which it should be able to receive, whether the signals composing the modulation are transmitted at intervals, or whether they follow one another with the maximum rapidity corresponding to the modulation rate.

8. Net Margin (ITU 34.08S)
The margin represented by the degree of distortion for the margin indicated in 34.07, when the rate of modulation applied to the apparatus is exactly equal to the standard theoretical rate.

9. Synchronous margin (ITU 34.09S)
The margin represented by the degree of distortion for the margin indicated in 34.07, when the mean unit interval of the modulation applied to the apparatus is equal to that which would result from a transmission from the apparatus under examination, assuming it to include a transmitter as well as a receiver.

10.  Degree of Isochronous Distortion (ITU 33.07S)
     a) Ratio to the unit interval of the maximum measured difference,
        irrespective of sign, between the actual and the theoretical
        intervals separating any two significant instants of modulation
        (or of restitution), these instants being not necessarily con-
        secutive.

     b) Algebraic difference between the highest and lowest value of
        individual distortion affecting the significant instants of
        an isochronous modulation.  (This difference is independent
        of the choice of the reference ideal instant.)

     The degree of distortion (of an isochronous modulation or restitu-
     tion) is usually expressed as a percentage.

     Note:  The result of the measurement should be completed by an
     indication of the period, usually limited, of the observation.  For
     a prolonged modulation (or restitution) it will be appropriate to
     consider the probability that an assigned value of the degree of
     distortion will be exceeded.

     Comments
     With respect to part (a) of this definition, the following inter-
     pretations are made:

     1)  The term "maximum" requires that the intervals between all
         possible pairs of significant instants that occur during the
         period of the observation have been considered.

     2)  The term "theoretical interval" is taken to be the reciprocal
         of the modulation (or restitution) rate, as determined through-
         out a stated measuring period.

11.  Displacement (region) Spread (ITU 33.05)
     Time interval at either side of an ideal instant of modulation or
     restitution, in which occur the actual significant instants of the
     modulation or restitution.

12.  Degree of inherent distortion (of a channel) (ITU 33.13)
     Degree of distortion of the restitution when the modulation is
     effected without distortion.

     Notes:
     1)  By inherent distortion is meant the combination of the
         different types of distortion caused by the channel (bias,
         characteristic, etc...).

     2)  This notion may be extended to the constituent elements of
         a channel, such as a telegraph relay.

     3)  It will be necessary to specify in what conditions the channel
         is used (type of apparatus, modulation rate, manual or auto-
         matic keying, etc...) and to effect the modulation under these
         conditions.

926

In particular should be defined:
  the point of entry at which the distortionless modulation is applied;
  the terminal point where the distortion is measured.

13.  Further Comments on Isochronous Distortion

Isochronous distortion is a term used in connection with synchronous operation. It is defined by the ITU in such a way as to treat it as a property of the received bit stream, independent of the timing of any specific receiving device. This is a useful concept, since it permits dealing with the distortion of a signal without having to state or know how the receiving device may or may not adjust its timing in response to bit stream timing. Analysis of the two-part definition shows that both parts define the same thing. However, while part (a) states the distortion in terms of the variations of signal element durations, in (b), it is evident that isochronous distortion is actually the spread of the transitions, if a set of reference or "ideal" instants is established which fall at the center of the spread.

The ITU definition of spread (33.05) is the "time interval at either side of an ideal instant of modulation or restitution, in which occur the actual significant instants of the modulation or restitution". That is, the spread extends from the earliest transition to the latest transition, taken over the observation period.

If the ideal instants fall in the center of the spread, then half of the spread must fall on either side of them. Thus the definition of isochronous distortion gives a figure which is effectively double the distortion of the individual transitions, measured with reference to the mean ideal instant.

This viewpoint may be reflected in the MIL-STD-188B, where distortion is expressed in the form "( )% Marking-0-( )% Spacing." For example, the margin requirement for a certain synchronous receiving device is stated as "49%M-0-49%S". This corresponds to an isochronous distortion of 98 percent.

927

- 16 -

MAY 1971

# INDUSTRIAL ELECTRONICS
# BULLETIN NO. 9

## APPLICATION NOTES

## FOR

## EIA STANDARD RS-232-C

928

# ELECTRONIC INDUSTRIES ASSOCIATION

**Engineering Department**

**2001 Eye Street, N.W., Washington, D.C. 20006**

*PRICE $2.60*

# C O N T E N T S

929

PAGE NO.

930

Application Notes for

EIA Standard RS-232-C

## 1.0 INTRODUCTION

### 1.1 Purpose

This paper reviews methods of operation of DTE (Data Terminal Equipment) and DCE (Data Communication Equipment) which interface according to the provisions of EIA Standard RS-232-C*. It is anticipated that both designers and operators of such equipments will benefit from the summary of service characteristics and transmission facility characteristics provided herein. Also, with particular regard to Sections 4 and 5 of RS-232-C, a coding format is introduced which allows concise graphical description of control circuit sequential states.

The procedures outlined here are not the only possible methods of operation, but they illustrate typical procedures which conform to the provisions of RS-232-C. While all of the thirteen interface types specified in RS-232-C are not covered, examples of the most commonly used configurations are given. Extrapolation of these procedures to other configurations is straightforward.

### 1.2 Relationship to International Standards

This paper relates only to EIA Standard RS-232-C. This document gives the nearest equivalent CCITT identification for the various interchange circuits, shown as (XXX). However, it should not be inferred that the operating methods discussed herein are in exact accord with CCITT recommendations.

### 1.3 Relationship to Automatic Calling Units

The interface between DTE and the ACE (Automatic Calling Equipment) for data communication is specified in EIA Standard RS-366. That standard defines several classes of automatic calling equipment and an associated set of interchange circuits for each class. The electrical characteristics of these circuits are consistent with RS-232-C. It is assumed herein that the functions of automatic calling are separate and apart from the interaction between the DTE and DCE except that when Automatic Calling Equipment is used to establish a channel, its functions will have been completed before the DCE is connected to the line and Circuit CC (Data Set Ready) (107) is turned ON.

* RS-232-C Interface Between Data Terminal Equipment and Data Communication Equipment Employing Serial Binary Data Interchange.

## 1.4    Relationship to Wideband Systems

It is the intent of RS-232-C that it be applicable to data transmission speeds up to about 20,000 bits-per-second.  The characteristics of the RS-232-C electrical interface are not suitable for handling data and timing signals in the megabit range; e.g., increased crosstalk between the normally unbalanced circuits and the reflections caused by not terminating the interchange circuits in a matched impedance.  The control circuits in the RS-232-C interface appear to be satisfactory for use in wideband systems as long as consideration is given to maintaining the proper operating sequences, a problem which is equally important at lower transmission speeds.

## 2.0    GENERAL

There are a number of factors which affect operation (as opposed to performance) of systems using communications channels to transfer information between widely separated data terminal equipments.  These factors are summarized below with regard to facilities, services, and interface types.  A coding format is introduced as an aid in describing control circuit states for these factors.

## 2.1    Facility Characteristics

### 2.1.1    Switched - Dedicated

The switched, common-user channels provided by the common carriers are basically two-wire, point-to-point circuits which are established on request from the stations.  While some sections of the circuit and some of the switches may be four-wire, the sections connecting the stations are two-wire.  Thus, duplex operation on commonly provided switched channels requires that the DCE (modem) be capable of subdividing the channel spectrum to obtain simultaneous two-way data channels.  While some conferencing arrangements have been provided to give multipoint operation on the switched telecommunication network, these have been special cases rather than customary.  The commonly available service has about a 300 to 3000 Hertz passband and transmission characteristics may vary considerably from call to call.  These channels commonly provide alternate voice capability.  Switched private line systems are similar to common-user switched systems, but may provide special features for a given user.

Dedicated lines, on the other hand, may be provided on either a two-wire or a four-wire basis for two-point or multipoint service, as required.  These dedicated lines are used where the volume of traffic justifies their cost, and where some characteristics of switched service (e.g., delay in establishing the connection and variable transmission parameters between successive calls) are not compatible with the data system operations.  The common carriers offer dedicated lines with specified degrees of conditioning which permit higher signaling speeds than generally possible over switched channels.  There are also wider bandwidth channels available, if required.  Alternate voice capability is generally available with dedicated lines.

## 2.1.2   Two-Wire  -  Four-Wire

Four-wire facilities are commonly available only in dedicated lines.  Four-wire facilities provide full bandwidth transmission in each direction without the problems of echoes and slow turn-around of two-wire facilities.

## 2.1.3   Echo Suppressors

Within the switched service networks and on some dedicated circuits there are often separate paths for the two directions of transmission over a portion of the circuit.  In voice communications at least one of the paths is usually idle at any given time.  In order to prevent the person who is talking from hearing a disturbing echo of his voice on long distance calls (circuits with round trip delay greater than 45 msec), the idle receive path is attenuated by a device called an echo suppressor.  As a result, voice or data signals may be transmitted in only one direction at a time, under normal conditions.  If duplex operation is desired, the echo suppressor must be disabled.

In the United States, echo suppressors may be disabled from either end of the circuit by applying a single frequency tone, the level of which is in the range of 0 to 5 dB below the maximum specified data signal level, within the band 2010-2240 Hz for at least 400 milliseconds.  No other signal or tone should be applied elsewhere during this disabling period.  The echo suppressor will remain disabled if, within 100 milliseconds after the disabling tone is removed from the line, signal energy in the 300-3000 Hz band is applied continuously.  Any interruption in the signal over 100 milliseconds in duration may cause the echo suppressor to again become enabled.  Enabling can be assured by a signal interruption greater than 500 milliseconds.

The turn-around time of echo suppressors is a characteristic which concerns the half duplex operation of data modems.  A modem which is operating in a half duplex mode and receiving signals from a distant modem should not begin transmitting at the instant the distant modem ends transmission.  This restriction is caused by the delay in the removal of the high attenuation placed in the receive path by the echo suppressor.  This delay, called turn-around time, is usually about 100 milliseconds.

When duplex data operation is planned on dedicated circuits and there is no need for alternate voice, it may be advisable to request that the circuits be furnished without echo suppressors.

No specific state is illustrated in the attached charts for disabling or turning around echo suppressors.  In half-duplex operation the time required for turn-around should be included in the delay between CA (Request to Send) (105) being turned ON and CB (Clear to Send) (106)coming ON.  In some modems, the data carrier is suitable for disabling echo suppressors, hence a separate disabling tone is not required.

## 2.1.4    Propagation Time

The time required for a data signal to get from the DTE/DCE interface at one end of the circuit to the interface at the other end of the circuit is attributed to the delay in modems, equalizers, filters, and the facilities. The propagation speed of microwave facilities is close to the speed of light; in coaxial cables it may be about nine-tenths the speed of light; and in loaded cable pairs it may be less than one-tenth the speed of light. Since actual circuits may be made up of diverse combinations of facilities and since the actual routing may depart considerably from the airline path between two points, it is not possible to state a general formula which will be accurate in all cases. Some estimates assume a one way propagation delay of one millisecond per hundred miles, which may be near the average for terrestrial facilities, but variations in routes and facility types encountered in alternate routing of long distance switched connections may add as much as 20 milliseconds to the delay. Dedicated lines will not normally vary from call to call, but facility restoral may involve routes considerably different in propagation delay from the original circuit layout. When satellite links are involved, about 300 milliseconds per hop can be anticipated. Where facility propagation is critical the common carrier providing the facility should be consulted for likely values.

In addition to the time required to propagate signals from end to end and to turn around echo suppressors there are the basic problems of detecting the presence of carrier at the receiver, establishing synchronization, and in some cases determining equalizer adjustments before data can be reliably exchanged. Carrier detection may take a few milliseconds to a few hundred milliseconds. Automatic modem equalizers may take a number of seconds to "train" on a new signal. Once "trained", transmission delay in modems typically varies from a few milliseconds in the simpler sets to several hundred milliseconds in more complex sets, which may incorporate error correcting circuitry.

Whether such delays are significant must be determined for each individual system. Systems using to-and-fro or polling-response method of operation may fail to meet the thruput expectations if all the propagation delays are not taken into account.

## 2.1.5    Line Classifications

The channels available from the common carriers in switched services are generally represented as voice grade analog facilities which may have rather wide variations in attenuation distortion and envelope delay distortion from one connection to another. However, even with these variations transmission of data over the switched telecommunications network has proven practical. The approach usually followed in providing service to data stations is to put some administrative limits on the transmission characteristics of the loops (local lines to the serving central office), so that when combined with a reasonable variety of interoffice trunks, acceptable service will result. Specific requirements and maintenance objectives may be obtained from the common carriers.

Page 5

## 2.2  Service Characteristics

The glossary section of RS-232-C defines a number of terms which are of importance to the interface standard.  The following paragraphs are an elaboration of some of these terms and some related factors which affect the characteristics of data services.

### 2.2.1  Duplex - Half Duplex

A half-duplex method of operation is common in many data systems because the nature of the service is a series of action - reaction sequences in which transmission of data in one direction at a time is all that is required.  This coupled with the ability to make maximum use of the available capacity of a two-wire circuit and the ability of terminal equipment to share common equipment between the transmit and receive conditions, favors the use of half-duplex operation. The disadvantages of half-duplex operation lie in the necessity of turning the channel around each time information flow is reversed and the unavailability of the transmitter (receiver) when in the receive (transmit) mode.  The turn-around time of a half-duplex system includes time for turning around echo suppressors and the start-up time of the modem which includes carrier detection and automatic equalization (if used).  To overcome these disadvantages some data terminal equipments use duplex channels, but operate in one direction at a time.

### 2.2.2  Synchronous - Nonsynchronous

In nonsynchronous transmission systems, no timing signals are passed between the DTE and the DCE and all timing information must be derived from transitions of the data.  Nonsynchronous operation is often favored for the simpler, lower cost data systems which typically use frequency shift keying techniques in the modems. Synchronous operation, whether based on a synchronous channel or on terminal-derived timing generally permits higher operating speeds and better performance over a given facility.

Start-up of nonsynchronous modems requires time to detect the presence of carrier, turn ON Circuit CF (Received Line Signal Detector) (109) and remove the MARK HOLD clamp from Circuit BB (Received Data) (104).  Typically, this ranges from 20 ms to 150 ms, depending on the modem design.  In synchronous modems time must also be included to achieve synchronization but this involves other techniques and is not directly related to the start-up time of nonsynchronous modems.  When synchronous data is transmitted through nonsynchronous modems, the DTE must first transmit a synchronizing sequence before attempting to send data.

### 2.2.3  Alternate Voice - Data

Alternate voice capability is normally available on switched network services and is optionally available on dedicated services.  The voice service may be provided as a means of coordinating the operations of the data station attendants, or it may be provided as a primary service during normal business office hours, with transfer to data operation at other times.  Except for some four-wire services, alternate voice data services with over 45 ms round trip delay will generally

935

involve the use of echo suppressors to permit satisfactory voice service. If alternate voice is not a requirement, dedicated circuits may be ordered without echo suppressors.

## 2.2.4    Multipoint - Two Point

Switched common-user channels provided by the common carriers are basically two-wire, point-to-point circuits which are established on request from the stations. While some switched conferencing arrangements have been used to provide multi-point service, these have been special cases rather than customary. Multipoint dedicated line arrangements have found their most common application in those systems where a central location must rapidly exchange a number of short messages with a number of remote stations. Because a given receiver can only accept one transmitter signal at a time, some sort of circuit discipline is required to avoid serious contention problems. Electrically, multipoint circuits can be one way (i.e., broadcast from a central point), half-duplex (signals from any one station are received at all others), or duplex (an outbound channel from a central point to all remote receivers and an inbound channel to be shared by the remote transmitters).

## 2.2.5    Secondary Channels

As defined in RS-232-C, a secondary channel has a lower signaling rate capability than the primary channel and is classed as Auxiliary when its direction is independent of the primary channel, and is classed as Backward when it always operates in the opposite direction to the primary channel. A secondary channel may be used to effect control of the primary channel operations, or it may be used as an independent data channel. Some secondary channels are so limited in speed that they have little data transmission capability and thus are useful only for control and circuit assurance purposes. Others with appreciable speed capability may be used to avoid the necessity for modem turn-around. In this latter use, a nonsymetrical-duplex operation is achieved which may be used to increase thru-put when circuit delay or modem start-up times are long. In circuits which are equipped with echo suppressors, use of a backward secondary channel requires the same disabling procedures as would be required for duplex operation.

## 2.2.6    Originate - Answer

Because a number of DTEs have no need to originate calls, they can be advanta-geously configured as answer-only terminals. In switched offices, lines to groups of such terminals often appear on a hunting arrangement sharing a common directory number so that an incoming call can be directed to the next idle terminal. The advantages from these restrictive configurations are in simplicity of both station hardware and operating procedures.

For terminals which must be able to both originate and answer calls, one can use voice to answer calls and coordinate the direction of transmission. Alterna-tively, the DCEs can incorporate a "hand shaking" routine (see Section 3.1.4), so that when calls are answered automatically the modems will function in the desired mode. In a number of duplex low-speed FSK modems, a convention is used whereby the answering terminal always transmits carrier in the higher of the two frequency bands used for operation. Among any given set of terminals, there must be an established convention so that all calls are handled in the same way.

## 2.2.7    Manual - Automatic

Manual coordination of data calls usually implies voice answer and transfer to the data mode for operation. Terminals can be arranged to automatically answer or originate calls, and when they are so arranged, some additional precautions are necessary to avoid the problem associated with wrong connections or poor transmission which could be identified by the attendants when voice coordination is used.

## 2.3    Interface Types

Section 5 of RS-232-C lists a number of interface types. The types are determined by whether the data terminal transmits, receives, or both, whether the operation is duplex or half duplex, and whether a secondary channel is used. It should be kept in mind that these designations relate to the data transmitted and not to which terminal originates or answers the call. For example, a transmit-only terminal may be arranged for answer-only service whereby it may be called regularly to transmit the data accumulated up to that time.

## 2.4    Control Circuit State Diagrams

The six major control circuits in RS-232 interfaces are CA through CF (105 through 109 plus 125). For convenience in discussion and in drawing diagrams, these six circuits are divided into two groups: the first group CC, CD, CE, (107, 108/2, 125) relates to the alerting and readiness of the equipment to operate, and the second group CA, CB, CF, (105, 106, 109) relates to preparing the communications equipment to transmit and receive data. The ON and OFF states of these two groups in switched service operation have been coded in an octal representation according to the following:

| GROUP A | | | | GROUP B | | |
|---|---|---|---|---|---|---|
| CC (107) Data Set Ready | CD (108/2) Data Terminal Ready | CE (125) Ring Indicator | Octal Code | CA (105) Request to Send | CB (106) Clear to Send | CF (109) Rec'd Line Sig. Detector |
| 0 | 0 | 0 | 0 | 0 | 0 | 0 |
| 0 | 0 | 1 | 1 | 0 | 0 | 1 |
| 0 | 1 | 0 | 2 | 0 | 1 | 0 |
| 0 | 1 | 1 | 3 | 0 | 1 | 1 |
| 1 | 0 | 0 | 4 | 1 | 0 | 0 |
| 1 | 0 | 1 | 5 | 1 | 0 | 1 |
| 1 | 1 | 0 | 6 | 1 | 1 | 0 |
| 1 | 1 | 1 | 7 | 1 | 1 | 1 |
| | Don't Care | | x | | Don't Care | |

1 = ON
0 = OFF

In systems using dedicated lines, the possible range of control circuit conditions is similar to switched service operation. However, CE (Ring Indicator) (125) and CD (Data Terminal Ready) (108/2) are not specified for dedicated lines. This requires some modification of the octal representation of control circuit states in which CE (125) is missing and CD (108/2) is replaced by the condition of the DTE power being ON or OFF according to the following:

| GROUP A | | | | GROUP B | | |
|---|---|---|---|---|---|---|
| CC (107) Data Set Ready | DTE Power | Octal Code | | CA (105) Request to Send | CB (106) Clear to Send | CF (109) Rec'd Line Sig. Detector |
| 0 | 0 | 0 | | 0 | 0 | 0 |
| - | - | 1 | | 0 | 0 | 1 |
| 0 | 1 | 2 | | 0 | 1 | 0 |
| - | - | 3 | | 0 | 1 | 1 |
| 1 | 0 | 4 | | 1 | 0 | 0 |
| - | - | 5 | | 1 | 0 | 1 |
| 1 | 1 | 6 | | 1 | 1 | 0 |
| - | - | 7 | | 1 | 1 | 1 |
| Don't Care | | x | | Don't Care | | |

938

1 = ON
0 = OFF
- = Not Applicable

This method of coding maintains state designations for dedicated line services which parallel the states for switched services.

### 2.4.1    Control Circuit States in Switched Services

To illustrate the various states of control circuits in a system operating over switched channels, an interface of the duplex type (type D in RS-232-C) is assumed (see Chart 1). The control circuit conditions for the originating and for the answering stations are covered separately in Charts 2 and 3. It is assumed in these examples that all six control circuits are implemented in the interface.

From the completely OFF state (0,0), Group A = 0, Group B = 0, there are a number of possible paths to reach the operate state (6,7). The idle state need not be (0,0) but may be any of four conditions (2,0), (0,4), (2,4) or (0,0), depending on whether the DTE has either CD (Data Terminal Ready) (108/2), or CA (Request to Send) (105), or both, or neither, ON when idle. As an aid to following the change of conditions, the circuit which is turned ON when going from one state to the next is indicated alongside the arrow connecting the two states. Circuits turning OFF are indicated by a bar over the designation, e.g., CE. Conditions where the state of all circuits in a given group is immaterial is indicated by an X in place of an octal number. It should be remembered that while the direction of the arrow indicates a favorable direction to advance towards the operating condition, interruptions or anomalies in timing can cause reversals or loops in the diagram and possibly invoke states not shown. Generally, these do not expedite the achievement of the operating condition. Also, there are possible transitory conditions which have little or no meaning (don't care conditions) and thus are not considered in these diagrams.

When data transmission has been completed, the equipment may revert to any of the idle conditions, with the requirement that CD (108/2) be turned OFF and not be turned ON again before CC (107) goes OFF.

Charts 4 and 5 are similar diagrams for a type D Half Duplex interface for a system which switches the direction of data carrier according to needs of the terminals to send data in one direction or the other. A transmit-only or a receive-only station would act in the same manner as a half duplex station except that they would not be reversible. They would operate in state (6,6) or (6,1) respectively.

### 2.4.2 Control Circuit States in Dedicated Services

In systems using dedicated lines, the possible range of control circuit conditions is from (0,0) to (6,7), the same as in switched service operation. However, since the lines are dedicated, the idle state may be any of a number of points on the chart, according to the needs of the terminal devices. While some systems may idle at the (6,7) state, the control circuits do perform useful functions and make for compatible operation of terminals on either switched or dedicated lines. Chart 6 represents a typical duplex system operating over a dedicated line. A half duplex system is shown in Chart 7.

### 3.0 SYSTEM OPERATION

The previously outlined operating sequences show an order of events but do not indicate the times involved or some of the problems which may be encountered. In the following discussion of system operation, some typical numbers are given and some techniques used in present data systems are discussed. It is not intended that engineering values be stated here; they should be obtained from common carriers or equipment manufacturers as appropriate.

### 3.1 General

This section reviews general interface control circuit operations, which apply to a number of system configurations.

### 3.1.1 Alerting

Circuit CE (Ring Indicator) (125) is the basic circuit for alerting data stations of an incoming call. This is primarily used on the switched telecommunication network, where typically the ringing signal on the line is two seconds of 20 Hz voltage applied by the serving central office every six seconds. In auto-answer stations this signal is detected by the DCE which converts the ringing signals to a dc voltage which appears on Circuit CE (125) with approximately two seconds ON (positive) and four seconds OFF (negative). When the DTE is ready .... circuit CD (Data Terminal Ready) (108/2) is ON .... the DCE will go off-hook when it detects the 20 Hz signal. The DCE will then respond with answer tone, data carrier, or silence, according to the system design. DTEs which do not use the alerting signal may keep circuit CD (108/2) ON and ignore indications on Circuit CE (125). Circuit CC (Data Set Ready) (107) may then be used to start operations concerned with answering a call.

With dedicated line services, alerting may or may not be required. A common arrangement on two-point dedicated lines is to use an in-band tone to alert the distant station, expecially where voice is used to coordinate data system operations.

### 3.1.2    Equipment Readiness

Circuits CC (Data Set Ready) (107) and CD (Data Terminal Ready) (108/2) indicate the readiness of the station equipment to operate. Circuit CC (107) is specified as a basic interchange circuit for all types of interfaces. Circuit CD (108/2) is required for switched services.

In some dedicated line DCE's Circuit CC (Data Set Ready) (107) is turned ON when power is on and the DCE is not in the Test Mode; in others Circuit CC (107) comes ON when the DCE starts transmission of its line signal.

Circuit CD (Data Terminal Ready) (108/2) is not specified for dedicated line service because the DCE does not require this information, there being no on-hook/off-hook switching functions and a DTE that is not ready will not attempt to transmit data. Certain DCEs designed to provide a switched network back-up capability in event of failure of the dedicated facility do provide a terminator for Circuit CD (108/2). Where the system does not use alternate switched network capability, Circuit CD (108/2) should be strapped to a permanent ON condition either in the DTE or, if provided, by a strapping option in the DCE. Where the back-up switched network capability is implemented, care should be taken to avoid turning Circuit CD (108/2) OFF when the DCE is connected to the dedicated facility. If this happens, the DCE will go out of data mode and disconnect from the dedicated facility when CD (108/2) goes OFF, but it will not enter data mode when Circuit CD (108/2) is again turned ON since, by definition of Circuit CD (108/2) it can only <u>hold</u> but cannot seize the line. Manual intervention will be required to restore the data transmission channel.

### 3.1.3    Data Channel Readiness

Preparation of the data channel to transmit is controlled by Circuits CA (Request to Send) (105) and CB (Clear to Send) (106). Indication of readiness to receive is given by Circuit CF (Received Line signal Detector) (109).

Circuit CB (106) is required in all transmitting DCEs, just as Circuit CF (109) is required in all receiving DCEs. If Circuit CA (105) is not implemented in the DCE, it is assumed to be ON at all times and Circuit CB (106) must respond appropriately.

The time delay between Circuits CA (105) and CB (106) (Request to Send - Clear to Send) is normally a hardware time-out in the DCE and involves no feedback from the remote data terminal. After this time interval has expired, Circuit CB (106) will come ON even if there is no operating terminal at the other end of the telecommunication channel. In other words, the ON condition on Circuit CB (106) does not guarantee that the remote receiver is "listening" -- circuit assurance cannot be obtained from this interchange circuit.

Nonsynchronous DCEs using frequency shift keying techniques (FSK) transmitting simultaneously in both directions must wait until the receiving data station can recognize the presence of the line signal, turn ON interchange Circuit CF (Received Line Signal Detector) (109) and remove the MARK HOLD clamp from Circuit BB (Received Data) (104). Typically, integrating time of the signal detection circuitry varies from 20 milliseconds to 150 milliseconds, depending on the modem design.

A DTE which transmits an isochronous signal and samples the data using a synchronous clock at the receiving DTE can use either synchronous or nonsynchronous DCEs. Where the DCE is nonsynchronous no timing information is passed between the DCE and the DTE, and all timing between the DTEs is derived from transitions of the data. Where the DCE is synchronous, timing information must be passed between the DTE and the DCE that is transmitting. It is not possible to synchronize the DTE clocks until the data transmission channel is established. With this arrangement, when Circuit CB (Clear to Send) (106) is turned ON by the DCE, the DTE must first transmit bit synchronizing information before attempting to transmit data.

Where a synchronous DCE is used, the ON condition of Circuit CB (Clear to Send) (106) implies that the clock at the remote DCE is synchronized and that the transmitting DTE is free to start transmitting data. Thus, during the time-out interval between the ON condition of Circuit CA (Request to Send) (105) and the ON condition of CB (Clear to Send) (106) the two clocks must achieve synchronization. Where scramblers are used to assure frequent data transitions, their synchronizing time must be taken into consideration in the Circuit CA (105) - Circuit CB (106) delay.

3.1.4    Handshaking

A handshaking technique is used predominantly on the switched telecommunication network. DCEs used on dedicated point-to-point circuits with alternate voice capability may also use this technique. The DCEs transmit signals to each other and perform certain timeouts to establish the data transmission channel and provide circuit assurance prior to turning the channel over to the DTE for data transmission. Where the DCE includes the handshaking function, it guarantees initial circuit assurance, and Circuit CB (Clear to Send) (106) circuit CF (Received Line Signal Detector) (109) must mean exactly what the name implies. Each DCE has communicated with the other end of the telecommunication channel and knows that it is ready to transmit and receive data. In the design of some DCEs this philosophy is carried even further in that the DCE will turn Circuit CB (Clear to Send) (106) OFF if the received line signal disappears from the telecommunication channel. This is based on the logical deduction, "If I can't hear him, he probably can't hear me either. Therefore, stop transmitting."

Certain types of data sets will, after a line signal disappears for a certain length of time, cause the switched network call to be disconnected.

### 3.1.5    Circuit Assurance

An ON condition on interchange Circuit CF (Received Line Signal Detector) (109) normally indicates that the DCE at the other end of the channel is transmitting a line signal, and this provides circuit assurance where data carriers are simultaneously transmitted in both directions.  Where carrier is transmitted in one direction at a time, assurance that the channel is established requires exchange of information in each direction.  They may be performed at the beginning and at the end of each message, or the successful receipt of data may be acknowledged on a block-by-block basis during the message.  Between acknowledgements there is no continuous assurance of circuit continuity in such systems.

In practice there is a delay between the appearance of the line signal on the telecommunication channel and the resultant ON condition on Circuit CF (Received Line Signal Detector) (109).  Similarly, there is a delay between the disappearance of the line signal and the OFF condition of Circuit CF (109), although the turn-off delay is usually a small fraction of the turn-on delay.  However, the DTE may be exposed to spurious signals on Circuit BB (Received Data) (104) during this turnoff interval.

942

### 3.2    Dedicated Service, Two-Point

For a dedicated point-to-point circuit, there are a number of combinations of duplex/half duplex, two-wire/four-wire, and data/voice data service available.

### 3.2.1    Duplex, Four-Wire

This is possibly the simplest of all configurations.  In some systems, the only interchange circuits provided between the DTE and the DCE are Circuits AB (Signal Ground) (102), BA (Transmitted Data) (103) and BB (Received Data) (104). With these three circuits it is possible to communicate; however, there is no information on the status of the associated equipment across the interface nor is there any circuit assurance.  Interface types D or E of EIA RS-232-C (see Chart 1) are recommended for this application.  Chart 6 shows the valid states for a type D interface.  There is no specific idle state for this type of service.  In a type D interface, Circuit CA (Request to Send) (105) has to be ON before Circuit CB (Clear to Send) (106) can be turned ON.  In a type E Interface Circuit CA (Request to Send) (105) is not implemented and Circuit CB (Clear to Send) (106) will be turned ON when the DCE is ready to send data.

Note that use of a DTE having a type D interface with a DCE having a type E interface is not recommended.  The DCE with a type E interface presents a permanent ON condition on Circuit CB (Clear to Send) (106) while the DTE with a type D interface cannot turn Circuit CA (Request to Send) (105) ON during intervals when Circuit CB (Clear to Send) (106) is already ON.  Thus, the DTE can never go to a transmit mode of operation.

## 3.2.2    Duplex, Two-Wire Without Alternate Voice

The general operation of two-wire duplex systems on dedicated lines is fundamentally the same as duplex four-wire systems.  The telecommunications channel must
either be ordered without echo suppressors or they must be disabled for duplex
data transmission.

The available channel can be subdivided into two subchannels; e.g., using
frequency division multiplexing techniques.  Conventionally the subchannel using
the lower band of frequencies is referred to as Channel 1 while the subchannel
using the higher band of frequencies is called Channel 2.  For duplex operation
on a two-wire telecommunications facility it is necessary that one of the two
modems transmit using the frequencies of Channel 2 and receive the frequencies
of Channel 1 while the other must transmit using the frequencies of Channel 1
and receive the frequencies of Channel 2.

Many of the modems available for this type of service are designed for use in
multipoint operation as well as two-point operation and will generally have a
type D interface.

## 3.2.3    Duplex, Two-Wire With Alternate Voice

With alternate voice capability, data transmission channel establishment can
initially be coordinated by voice communication between the data station
attendants.  If signaling capability (ringing) is provided, the DCE may be
arranged for automatic answering.

A subscriber may request that echo suppressors, if any, be removed.  Voice
communication would be somewhat degraded since the speaker would hear his own
voice delayed sufficiently long to create confusion.  On circuits with echo
suppressors, the first DCE to apply a line signal must perform the disabling
function and the other DCE must maintain silence for at least 400 milliseconds
before applying its line signal (see Section 2.1.3).

As in all other duplex cases, once the data transmission channel is set up,
data can be arbitrarily transmitted in either direction or in both directions
simultaneously.

## 3.2.4    Half Duplex With Alternate Voice (Chart 7)

Most half duplex DTEs operate in a manner such that when not transmitting they
are monitoring the data transmission channel for received data.  Since DTEs
designed for half duplex operation may share internal components for transmit
and receive operations, a half duplex DTE may be "blind" when transmitting;
i.e., it cannot receive any data on interchange Circuit BB (Received Data)
(104).  By contrast, a DCE operating on a two-wire half duplex telecommunication channel has both the transmitting and receiving signal converters connected
to the two-wire line and, as a result, the transmitted line signal is impressed
across the receiver input.  EIA Standard RS-232-C requires that for half duplex
operation interchange Circuit CF (Received Line Signal Detector) (109) be

clamped in the OFF Condition and interchange Circuit BB (Received Data) (104) be clamped to MARK HOLD while Circuit CA (Request to Send) (105) is in the ON condition.  This condition is maintained for a brief interval following the ON to OFF transition of Circuit CA (Request to Send) (105) to allow for completion of transmission.

When the DTE turns interchange Circuit CA (Request to Send) (105) OFF, it is no longer in the transmit mode and may immediately revert to the receive mode of operation.  Normally the DTE has just completed a transmission to the remote terminal and is monitoring Circuit BB (Received Data) (104) for a response. The DCE may not have completed the transmission of all of the data which it received from the DTE.  Transmission delays through the DCE may vary from less than a millisecond in a simple FSK modem to several seconds in the more complex equipments, particularly those providing forward acting error control.  If Circuits BB (Received Data) (104) and CF (Received Line Signal Detector) (109) are not clamped or if the clamp is removed immediately when Circuit CA (Request to Send) (105) is turned OFF by the DTE, the DTE would be presented with the last portion of its own transmitted signal, somewhat delayed, on Circuit BB (Received Data) (104).  Because these transitions may appear to be an incomplete response from the remote terminal, the DTE may initiate recovery procedures, go into transmit mode and request a retransmission.

Half duplex DCEs must clamp circuit BB (Received Data) (104) to a MARK HOLD condition and Circuit CF (Received Line Signal Detector) (109) to the OFF condition when Circuit CA (Request to Send) (105) is ON and for a short interval following the ON to OFF transition of Circuit CA (105) until transmission is completed and the echoes subside.

### 3.2.5   Half Duplex Without Alternate Voice (Chart 7)

With this arrangement, both data stations are normally in receive mode awaiting a signal from the other station.  There are no line signals on the telecommunication channel to assure continuity of the channel and the presence of the other data station.  Contention is a problem because both data stations may decide to transmit at the same time.  The DTE intending to transmit may test the line for circuit assurance prior to transmitting blocks of data.  It transmits an inquiry and awaits a response before proceeding.

To increase circuit assurance and eliminate contention, the "ping-pong" method is often used.  The data station which first goes into transmit mode sends a series of idle characters for a fixed period followed by a "turn-around" character which commands the other station to go into transmit mode.  The first data station ceases transmission and awaits a response.  The second data station then sends idles followed by a "turn-around" character.  The two data stations continue to alternately send idle characters to each other to keep the line active until one of them has data ready to send, at which time it substitutes the data for the idle characters.  If the original station had not received any idles after transmitting its turn around character, it would have timed out and repeated the whole procedure.

## 3.3    Dedicated Service, Multipoint

The dedicated multipoint network is a relatively permanent, non-switched communication network serving three or more data stations.  The use of this arrangement is particularly popular in system applications where a central location must rapidly exchange relatively short messages with a multiplicity of remote stations and where the long connect times encountered on a switched telecommunication network would reduce the system efficiency below an acceptable level.

In general, multipoint networks fall into two general categories, "centralized" and "noncentralized".  In the centralized category, the remote stations communicate only with the central or controlling station and traffic between remote stations must be relayed by the central station.  If noncentralized multipoint operation, remote stations can communicate directly with each other; however, because of contention problems one of the stations may be given the authority to control the flow of traffic.

Multipoint networks can be configured as one-way only (broadcast), two-way alternate, or two-way simultaneous.

## 3.3.1    One-Way Sytems

This configuration consists of a master transmitting station and two or more receive-only stations on a dedicated multipoint line.  Such arrangements are often referred to as broadcast systems.

At the transmitting station a Type A interface will suffice for most systems. Under certain circumstances where the DCE is required to transmit idle patterns or training signals during intervals when the DTE is not transmitting data, a Type B interface may be required.  At the receiving terminal a Type C interface is suitable.  The sequencing of these circuits in originating and terminating is quite simple, as shown in Chart 8.  If Circuit CA (Request to Send) (105) is not implemented at the transmitting station, it should be assumed to be in a permanent ON condition.

## 3.3.2    Two-Way Alternate System

This configuration operates in a half duplex mode, but may be implemented on a duplex or half duplex communications channel.  Duplex implementation permits more rapid turnaround of the link between the central station and the selected remote station and offers an increase in system efficiency.

## 3.3.2.1  Centralized Operation

The following discussion assumes centralized operation on a duplex communication channel.  The central station may leave its line signal turned ON at all times because it is the only transmitter on the out-bound circuit (Type E interface). In cases where the DCE transmits training signals, the use of Circuit CA (Request to Send) (105) by the DTE is required.  In this case a type D interface is required. Since all the transmitters of the remote stations share the in-bound telecommunication channel, the use of Circuit CA (Request to Send) (105) is mandatory to permit removal of the data signal from the telecommunication channel when the remote terminal is not transmitting.

The central terminal selects a remote terminal by transmitting the address to which that terminal must respond. The selected terminal transmits a response. There is normally an exchange of additional messages between the two terminals; however, at this point the remote terminal may operate either in half duplex mode and turn Circuit CA (Request to Send) (105) OFF after each transmission, or, it may leave circuit CA (Request to Send) (105) ON, establishing a duplex telecommunication channel between the two terminals until it is commanded by the central terminal to turn Circuit CA (105) OFF and go into standby mode. Note that the central terminal rigidly controls which of the remote terminals may send at any one time.

Chart 6 shows a general status chart for the sequencing of control circuits using the Type D interface without regard to the system application; i.e., this is what the interface can do.

Chart 9 shows Chart 6 adapted to indicate the expected sequencing of these controls at the central terminal. This terminal turns Circuit CA (Request to Send) (105) ON (and leaves it ON) and transmits first.

Chart 10 shows a similar application using a Type E interface which does not provide Circuit CA (Request to Send) (105). The transmitted line signal is turned ON as soon as Circuit CC (Data Set Ready) (107) is turned ON. The ON condition on Circuit CB (Clear to Send) (106) follows. None of this is under the control of the DTE. Line signals can be removed only by turning off power to the DCE.

Chart 11 shows an adaptation of Chart 6 to indicate the sequencing of the Type D interface circuits at the remote terminal on a duplex multipoint network. In this case, the remote terminal does not transmit unless it is interrogated. Circuit CC (Data Set Ready) (107) is ON, Circuit CF (Received Line Signal Detector) (109) comes ON, and a message is received. If the message requires a response by that remote terminal, the DTE turns ON Circuit CA (Request to Send) (105).

In operation on a duplex multipoint network the central DCE and the selected remote DCE may both transmit line signals simultaneously; however, the data flows in only one direction at a time in this type of system, i.e., "two-way alternate". The simultaneous ON condition of both transmitters decreases the circuit turnaround time to the DTE reaction time plus the circuit propagation time. This is a very efficient method of operation.

### 3.3.2.2 Noncentralized Operation

Noncentralized operation on multipoint networks uses a half duplex telecommunication network. With this arrangement, all stations receive everything that is transmitted on the network. This has the advantage that any station on the network can be selected as the "control" station without requiring modification to the network. This characteristic of the network makes direct transmission between remote stations possible, which is the basic purpose of noncentralized operation. The most serious disadvantage of the half duplex multipoint network is the exposure to third party contention. The control terminal selects one of the other terminals and establishes communication with that terminal. If during the transmission of data between the two, the sequence of codes transmitted as part of the data coincides with the address of one of the other terminals, that terminal will interpret this as a command to respond and will start to transmit. The result will be garbled traffic and a serious recovery problem. A defense against this is to use a positive response from the selected terminal to effectively "turn off" all of the other terminals so they will no longer respond to their address until the control terminal restores them by commanding the originally selected terminal to end the current exchange with the control terminal. This is effective unless one of the terminals fails to receive the positive response and consequently does not get "shut off". This defense materially reduces third party contention, but does not completely eliminate it.

Charts 6 and 12 shows the difference in possible states of the <u>same</u> interface used on duplex and half duplex telecommunication channels. Note the combinations listed as "not permitted" on Chart 12. These combinations are permissible on duplex channels, but not on half duplex. On a half duplex telecommunication channel, the operation of the interface is no longer a function of which station is in control and which is the remote station, since all stations, including the control station must turn OFF Circuit CA (Request to Send) (105) when not transmitting data. Referring to Chart 12, states (2,0) and (2,4) become involved only in getting the system turned on initially. State (6,0) is the quiescent state. The control station, when selecting a remote station, proceeds from state (6,0) to states (6,4) and (6,6). It transmits its command and then reverts to the quiescent state, (6,0), by way of state (6,2).

For the remote stations, the conditions move from the quiescent state, (6,0), to the receive state, (6,1), to receive the selection command and then back to the quiescent state, (6,0). The selected remote station then assumes the transmit mode, states 6,0 - 6,4 - 6,6, to respond to the control station, or, in the case of noncentralized operation, to address another remote station. Having completed its transmission, the station then reverts to the quiescent state (6,0) via state (6,2) in the following sequences.

        Transmit Mode:        6,0 - 6,4 - 6,6 - 6,2 - 6,0
        Receive Mode:         6,0 - 6,1 - 6,0


### 3.3.3   Two-Way Simultaneous Systems

The interface operation previously described for multipoint two-way alternate operation using duplex telecommunication channels (Charts 9, 10, and 11) also permits two-way simultaneous operation, where the control station and one remote station at a time exchange data on a duplex basis, or it will permit the control station to receive data from one remote station while simultaneously transmitting to a second remote station. A Type D interface is utilized.

## 3.4    Switched Service

The most commonly available switched channels are two wire, two-point voice grade facilities which may be used to send data between compatible terminals at locations served by the switched telecommunications network.  The modems may be supplied by the common carrier or by the customer.  In the latter case, a data access arrangement may be required between the modem and the telephone line, but the interface with DTE can be the same in either case.

Private switched systems are also available which may include conditioning for improved data performance.  These private switched networks operate in essentially the same manner as common user switched services.

### 3.4.1    One Way Systems

One way systems using the switched network channels are best suited to voice coordinated operation.  Whether used for data collection at a central point or data dissemination to a number of outlying points, it is desirable to know that the proper two points are connected before any data is exchanged.  For those one-way systems where circuit assurance is not a consideration, transmit-only stations may use a Type A or B interface, and receive-only stations may use a Type C interface.  Except for the requirement for Circuit CE (Ring Indicator) (125) and Circuit CD (Data Terminal Ready) (108/2), the operation of these systems after the channel is established is the same as one-way systems on dedicated lines.  The termination of the call follows the procedure of turning CD (Data Terminal Ready) (108/2) OFF and waiting for CC (Data Set Ready) (107) to go OFF.  When manual coordination is used, the call may be terminated by transferring to the voice mode and hanging up the phone.

### 3.4.2    Two Way Alternate Systems

Two-way alternate systems are frequently used on the switched telecommunications network.  Such systems can transmit a full voice bandwidth signal in one direction at a time, turning around as required to acknowledge or to send data in the opposite direction.  Turn around time is controlled by the time required to turn around echo suppressors (100 milliseconds) plus the start up time of the DCE.  A Type D interface is most suitable for this type of system.  These types of systems turn Circuit CA (Request to Send) (105) ON when transmitting, which clamps Circuit BB (Received Data) (104) to MARK.  Coordination of turn-around may be performed by voice, or more efficiently, by control signals in the data stream.  Except for the channel establishment and disconnect sequences, operation of these systems follows the same pattern as two point dedicated line systems.

### 3.4.3    Two Way Simultaneous

Since the basic switched telecommunication network channel is two-wire at the interface, any system which desires to operate in a duplex mode must derive the two channels by band splitting techniques.  For low speed services (up to about 300 bps) this is a commonly used technique, whether the terminals actually use both directions simultaneously or whether they operate in a half duplex manner, switching from one direction to the other with a delay controlled only by the circuit propagation time and the DTE reaction time.  The Type D or Type E interfaces are suitable for this duplex type of service, the difference being that Circuit CA (Request to Send) (105), not being implemented in the Type E interface, is assumed to be ON whenever Circuit CD (Data Terminal Ready) (108/2) is ON.

950

**Column headings (A–Z):**

- A — Transmit Only
- B — Transmit Only*
- C — Receive Only
- D — Duplex* Half Duplex
- E — Duplex
- F — Primary Channel Transmit Only*/Secondary Channel Receive Only
- G — Primary Channel Receive Only/Secondary Channel Transmit Only*
- H — Primary Channel Transmit Only/Secondary Channel Receive Only
- I — Primary Channel Receive Only/Secondary Channel Transmit Only
- J — Primary Channel Transmit Only*/Half Duplex Secondary Channel
- K — Primary Channel Receive Only/ Half Duplex Secondary Channel
- L — Duplex Primary Channel*/Duplex Secondary Channel* Half Duplex Primary Channel/Half Duplex Secondary Channel
- M — Duplex Primary Channel/Duplex Secondary Channel
- Z — Special (Circuits specified by Supplier)

| | INTERCHANGE CIRCUIT | A | B | C | D | E | F | G | H | I | J | K | L | M | Z |
|---|---|---|---|---|---|---|---|---|---|---|---|---|---|---|---|
| AA | Protective Ground | - | - | - | - | - | - | - | - | - | - | - | - | - | - |
| AB | Signal Ground | X | X | X | X | X | X | X | X | X | X | X | X | X | X |
| BA | Transmitted Data | X | X | | X | X | X | | X | | X | | X | X | o |
| BB | Received Data | | | X | X | X | | X | | X | | X | X | X | o |
| CA | Request to Send | | X | | X | | X | | | | X | | X | | o |
| CB | Clear to Send | X | X | | X | X | X | | X | | X | | X | X | o |
| CC | Data Set Ready | X | X | X | X | X | X | X | X | X | X | X | X | X | o |
| CD | Data Terminal Ready | S | S | S | S | S | S | S | S | S | S | S | S | S | o |
| CE | Ring Indicator | S | S | S | S | S | S | S | S | S | S | S | S | S | o |
| CF | Received Line Signal Detector | | | X | X | X | | X | | X | | X | X | X | o |
| CG | Signal Quality Detector | | | | | | | | | | | | | | o |
| CH/CI | Data Signalling Rate Selector (DTE)(DCE) | | | | | | | | | | | | | | o |
| DA/DB | Transmitter Sig. Element Timing (DTE)(DCE) | t | t | | t | t | t | | t | | t | | t | t | o |
| DD | Receiver Signal Element Timing (DCE) | | | t | t | t | | t | | t | | t | t | t | o |
| SBA | Secondary Transmitted Data | | | | | | | X | | X | X | X | X | X | o |
| SBB | Secondary Received Data | | | | | | X | | X | | X | X | X | X | o |
| SCA | Secondary Request to Send | | | | | | | X | | | X | X | X | | o |
| SCB | Secondary Clear to Send | | | | | | | X | | X | X | X | X | X | o |
| SCF | Secondary Received Line Signal Detector | | | | | | X | | X | | X | X | X | X | o |

Legend:  o - To be specified by the supplier
     - - optional
     S - Additional Interchange Circuits required for Switched Service
     t - Additional Interchange Circuits required for Synchronous Channel
     X - Basic Interchange Circuits, All Systems

\* Indicates the inclusion of Circuit CA (Request to Send) in a One Way Only (Transmit) or Duplex Configuration where it might ordinarily not be expected, but where it might be used to indicate a non-transmit mode to the data communication equipment to permit it to remove a line signal or to send synchronizing or training signals as required.

Standard Interfaces for
Selected Communication
Systems Configurations

CHART 1

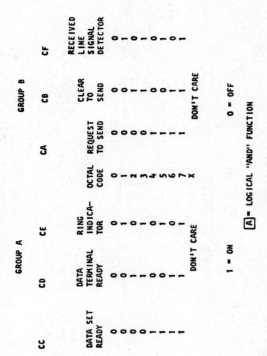

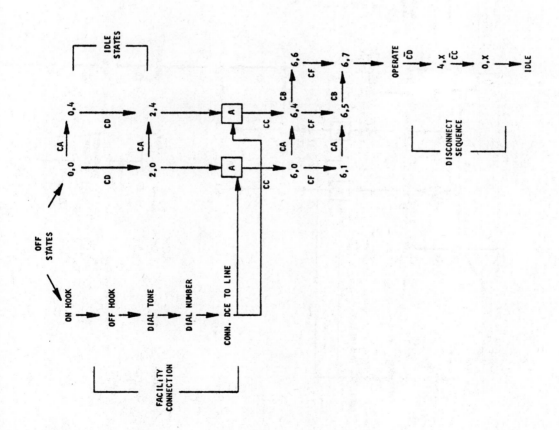

| | GROUP A | | | GROUP B | |
|---|---|---|---|---|---|
| CC | CD | CE | CA | CB | CF |
| DATA SET READY | DATA TERMINAL READY | RING INDICATOR | REQUEST TO SEND | CLEAR TO SEND | RECEIVED LINE SIGNAL DETECTOR |
| 0 | 0 | 0 | 0 | 0 | 0 |
| 0 | 0 | 1 | 0 | 0 | 1 |
| 0 | 1 | 0 | 0 | 0 | 0 |
| 0 | 0 | 1 | 0 | 1 | 1 |
| 1 | 0 | 0 | 1 | 0 | 0 |
| 1 | 0 | 1 | 1 | 0 | 1 |
| 1 | 1 | 0 | 1 | 1 | 0 |
| 1 | 1 | 1 | 1 | 1 | 1 |
| | DON'T CARE | | | DON'T CARE | |

OCTAL CODE: 0 1 2 3 4 5 6 7 X

1 = ON    0 = OFF

Ⓐ = LOGICAL "AND" FUNCTION

ORIGINATING STATION
CONTROL LEAD SEQUENCES
TYPE D DUPLEX INTERFACE
SWITCHED SERVICE

CHART 2

952

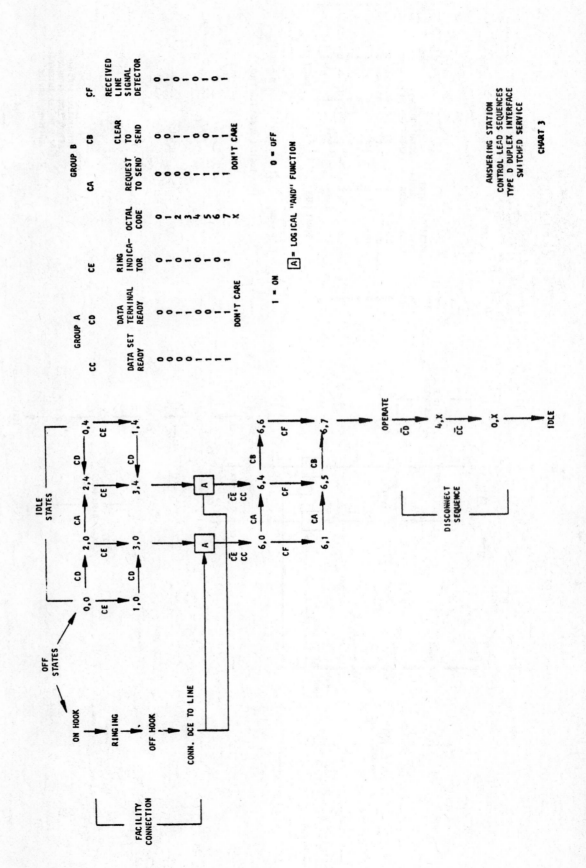

ANSWERING STATION
CONTROL LEAD SEQUENCES
TYPE D DUPLEX INTERFACE
SWITCHED SERVICE

CHART 3

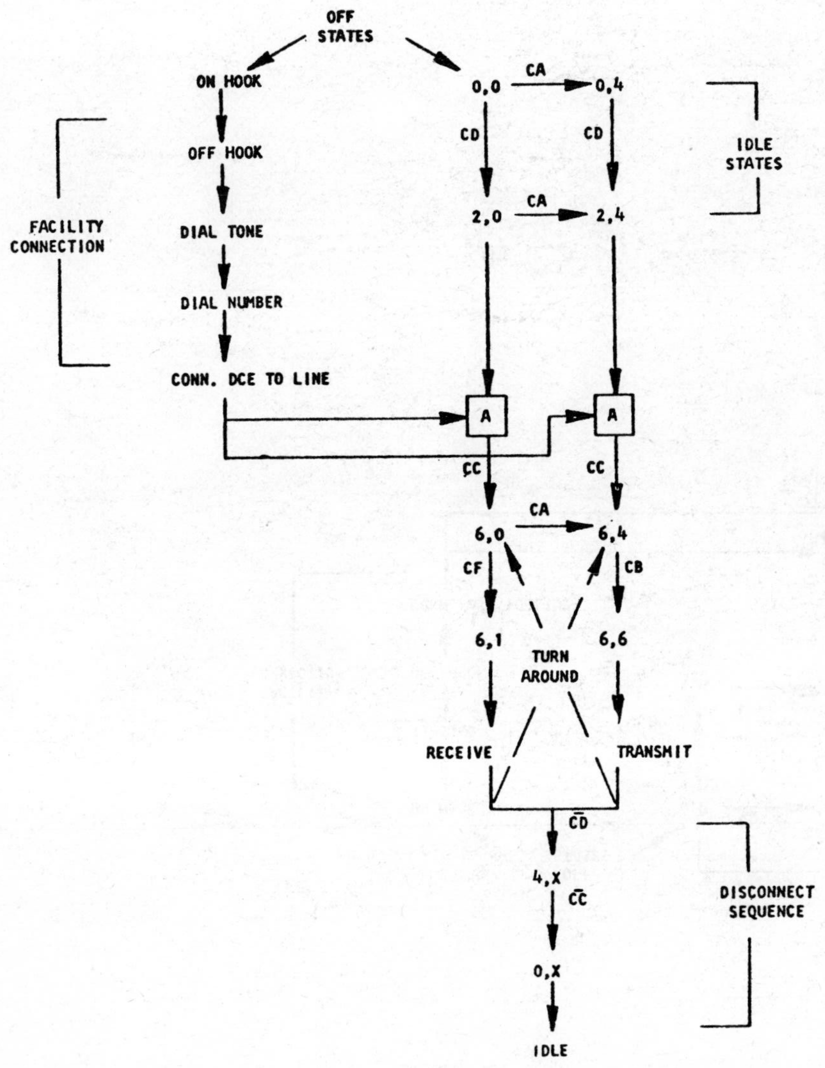


| GROUP A | | | | GROUP B | | |
|---|---|---|---|---|---|---|
| CC | CD | CE | | CA | CB | CF |
| DATA SET READY | DATA TERMINAL READY | RING INDICATOR | OCTAL CODE | REQUEST TO SEND | CLEAR TO SEND | RECEIVED LINE SIGNAL DETECTOR |
| 0 | 0 | 0 | 0 | 0 | 0 | 0 |
| 0 | 0 | 1 | 1 | 0 | 0 | 1 |
| 0 | 1 | 0 | 2 | 0 | 1 | 0 |
| 0 | 1 | 1 | 3 | 0 | 1 | 1 |
| 1 | 0 | 0 | 4 | 1 | 0 | 0 |
| 1 | 0 | 1 | 5 | 1 | 0 | 1 |
| 1 | 1 | 0 | 6 | 1 | 1 | 0 |
| 1 | 1 | 1 | 7 | 1 | 1 | 1 |
| | DON'T CARE | | X | | DON'T CARE | |

1 = ON    0 = OFF

[A] = LOGICAL "AND" FUNCTION

ORIGINATING STATION
CONTROL LEAD SEQUENCES
TYPE D HALF-DUPLEX INTERFACE
SWITCHED SERVICE

CHART 4

954

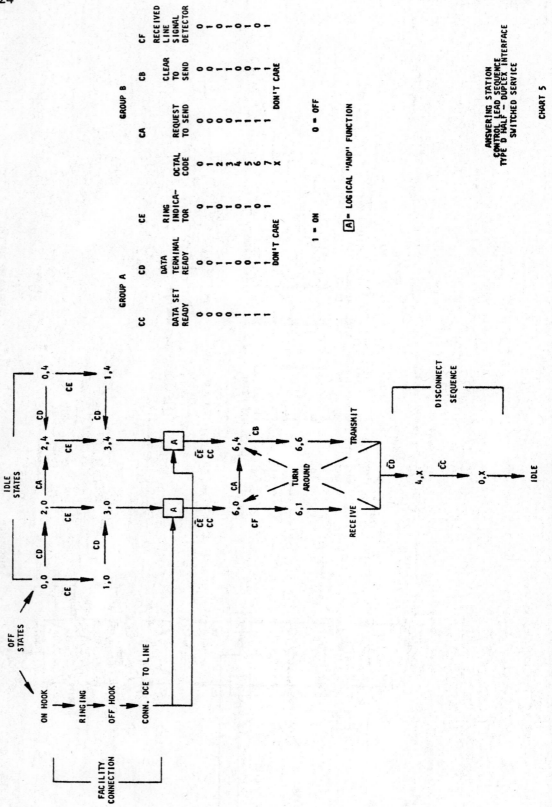

ANSWERING STATION
CONTROL LEAD SEQUENCE
TYPE D HALF — DUPLEX INTERFACE
SWITCHED SERVICE

CHART 5

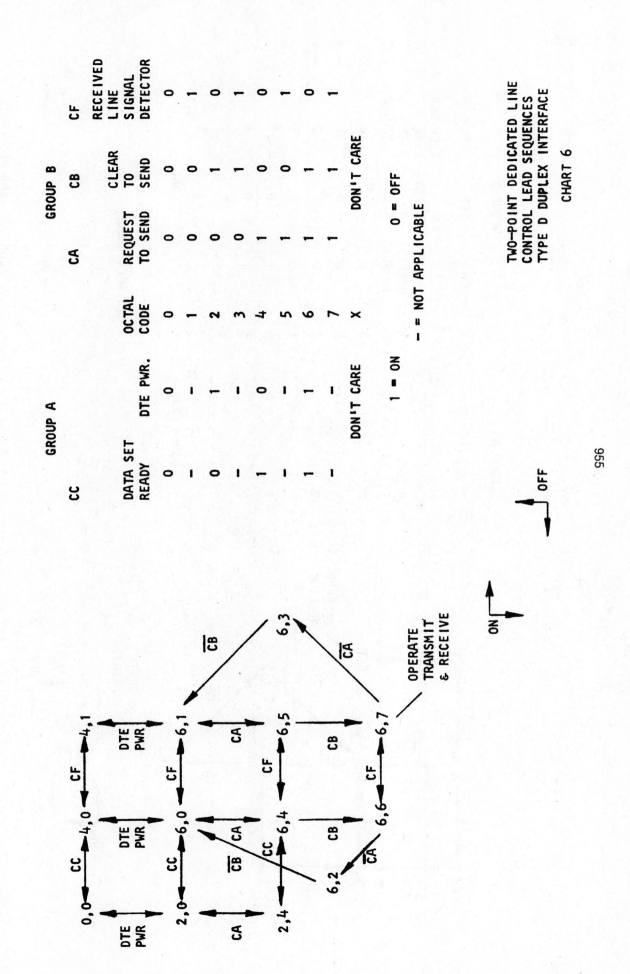

TWO-POINT DEDICATED LINE
CONTROL LEAD SEQUENCES
TYPE D DUPLEX INTERFACE

CHART 6

| | GROUP A | | | GROUP B | | |
|---|---|---|---|---|---|---|
| | CC | | OCTAL CODE | CA | CB | CF |
| | DATA SET READY | DTE PWR. | | REQUEST TO SEND | CLEAR TO SEND | RECEIVED LINE SIGNAL DETECTOR |
| | 0 | 0 | 0 | 0 | 0 | 0 |
| | – | – | 1 | 0 | 0 | 1 |
| | 0 | 1 | 2 | 0 | 1 | 0 |
| | 1 | 0 | 3 | 1 | 1 | 1 |
| | – | 1 | 4 | 1 | 0 | 0 |
| | – | 1 | 5 | 1 | 1 | 1 |
| | – | 1 | 6 | 1 | 1 | 0 |
| | – | – | 7 | | | 1 |
| | – | DON'T CARE | X | DON'T CARE | DON'T CARE | |

1 = ON    0 = OFF

– = NOT APPLICABLE

955

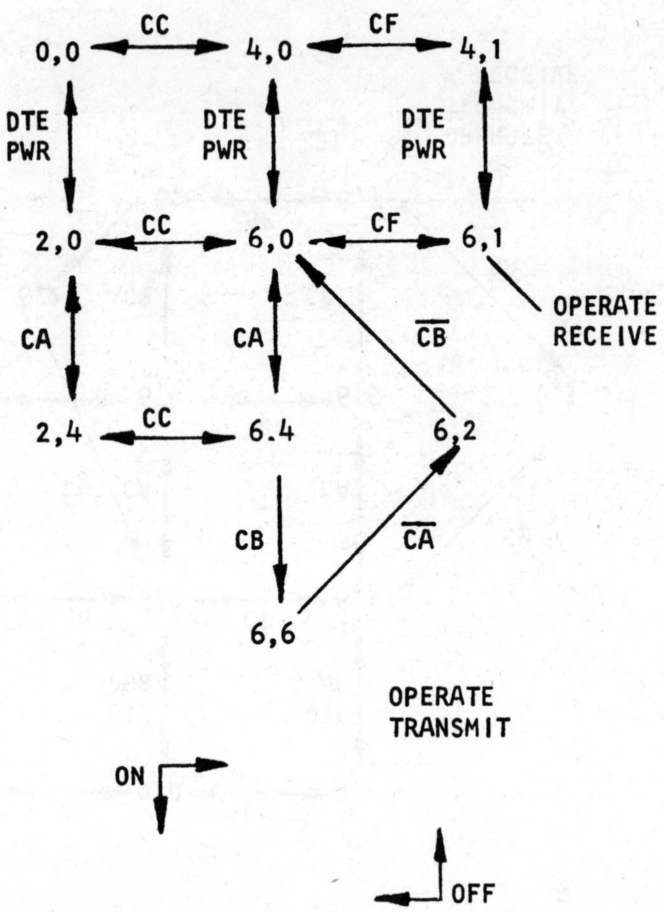

956

| | GROUP A | | | GROUP B | | |
| --- | --- | --- | --- | --- | --- | --- |
| | CC | | | CA | CB | CF |
| | DATA SET READY | DTE PWR. | OCTAL CODE | REQUEST TO SEND | CLEAR TO SEND | RECEIVED LINE SIGNAL DETECTOR |
| | 0 | 0 | 0 | 0 | 0 | 0 |
| | - | - | 1 | 0 | 0 | 1 |
| | 0 | 1 | 2 | 0 | 1 | 0 |
| | - | - | 3 | 0 | 1 | 1 |
| | 1 | 0 | 4 | 1 | 0 | 0 |
| | - | - | 5 | 1 | 0 | 1 |
| | 1 | 1 | 6 | 1 | 1 | 0 |
| | - | - | 7 | 1 | 1 | 1 |
| | DON'T CARE | | X | DON'T CARE | | |

1 = ON    0 = OFF

- = NOT APPLICABLE

CONTROL LEAD SEQUENCES
HALF DUPLEX INTERFACE
TWO-POINT DEDICATED LINE

CHART 7

OPERATE RECEIVE

OPERATE TRANSMIT

ON    OFF

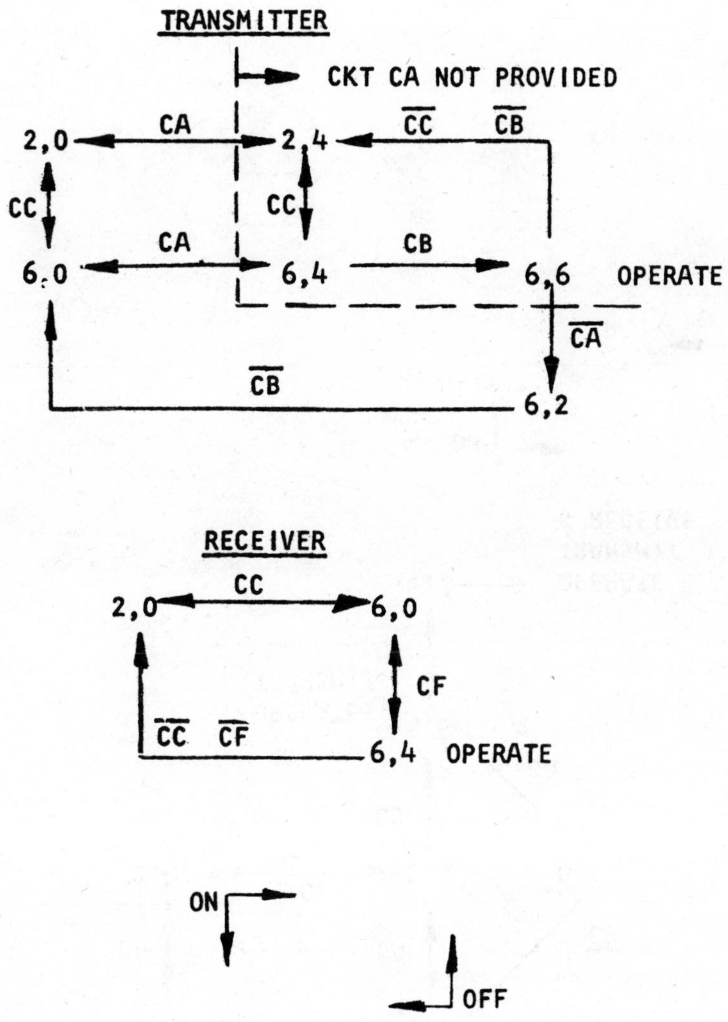

| GROUP A | | | GROUP B | | |
| CC | | | CA | CB | CF |
|---|---|---|---|---|---|
| DATA SET READY | DTE PWR. | OCTAL CODE | REQUEST TO SEND | CLEAR TO SEND | RECEIVED LINE SIGNAL DETECTOR |
| 0 | 0 | 0 | 0 | 0 | 0 |
| 1 | — | 1 | 0 | 0 | 1 |
| 0 | 1 | 2 | 0 | 1 | 0 |
| 1 | — | 3 | 0 | 1 | 1 |
| 1 | 0 | 4 | 1 | 0 | 0 |
| 1 | — | 5 | 1 | 0 | 1 |
| 1 | 1 | 6 | 1 | 1 | 0 |
| 1 | — | 7 | 1 | 1 | 1 |
| DON'T CARE | DON'T CARE | X | DON'T CARE | | |

1 = ON    0 = OFF

— = NOT APPLICABLE

TRANSMITTER

CKT CA NOT PROVIDED

$\overline{CC}$  $\overline{CB}$  6,6 OPERATE  $\overline{CA}$  6,2

2,4  CB

CA  $\overline{CC}$  6,4  CA

2,0  CC  6,0  $\overline{CB}$

RECEIVER

CC  CF  6,4 OPERATE

6,0

2,0  $\overline{CC}$  $\overline{CF}$

ON  OFF

CONTROL LEAD SEQUENCES
ONE-WAY ONLY SERVICE

CHART 8

957

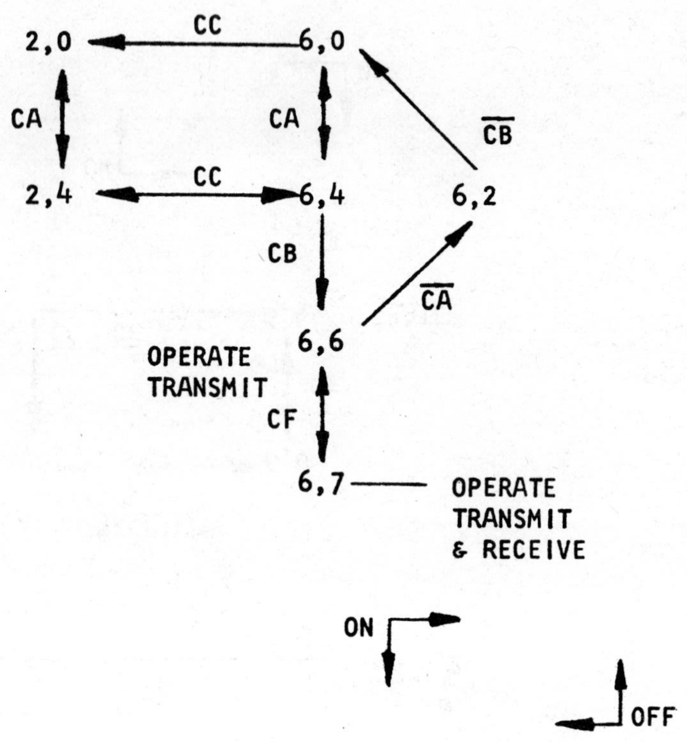

958

| CC DATA SET READY | DTE PWR. | OCTAL CODE | CA REQUEST TO SEND | CB CLEAR TO SEND | CF RECEIVED LINE SIGNAL DETECTOR |
|---|---|---|---|---|---|
| 0 | 0 | 0 | 0 | 0 | 0 |
| - | - | 1 | 0 | 0 | 1 |
| 0 | 1 | 2 | 0 | 1 | 0 |
| - | - | 3 | 0 | 1 | 1 |
| 1 | 0 | 4 | 1 | 0 | 0 |
| - | - | 5 | 1 | 0 | 1 |
| 1 | - | 6 | 1 | 1 | 0 |
| - | - | 7 | 1 | 1 | 1 |
| DON'T CARE | DON'T CARE | X | DON'T CARE | | |

GROUP A — CC, DTE PWR.

GROUP B — CA, CB, CF

1 = ON     0 = OFF

- = NOT APPLICABLE

State diagram:

2,0 — CA — 2,4

6,0 — CA — 6,4 — CB — 6,6 — CF — 6,7 (OPERATE TRANSMIT & RECEIVE)

6,0 — CC — 2,0

6,4 — CC — 2,4

6,0 — $\overline{CB}$ — 6,2

6,2 — $\overline{CA}$ — 6,6

OPERATE TRANSMIT — CF

ON / OFF

TYPE D DUPLEX INTERFACE
MULTIPOINT ~ CONTROL STATION

CHART 9

| | GROUP A | | | OCTAL CODE | GROUP B | | |
|---|---|---|---|---|---|---|---|
| | CC | | CA | | CB | CF |
| | DATA SET READY | DTE PWR. | | | REQUEST TO SEND | CLEAR TO SEND | RECEIVED LINE SIGNAL DETECTOR |
| | 0 | 0 | | 0 | 0 | 0 | 0 |
| | – | 1 | | 1 | 0 | 0 | 1 |
| | 0 | 1 | | 2 | 0 | 1 | 0 |
| | 1 | – | | 3 | 0 | 1 | 1 |
| | – | 0 | | 4 | 1 | 0 | 0 |
| | – | 1 | | 5 | 1 | 0 | 1 |
| | 1 | 1 | | 6 | 1 | 1 | 0 |
| | | | | 7 | 1 | 1 | 1 |
| | DON'T CARE | | | X | DON'T CARE | | |

1 = ON   0 = OFF

– = NOT APPLICABLE

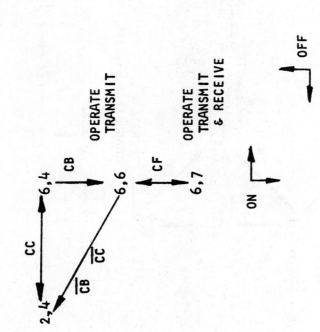

TYPE E DUPLEX INTERFACE
MULTIPOINT — CONTROL STATION

CHART 10

959

960

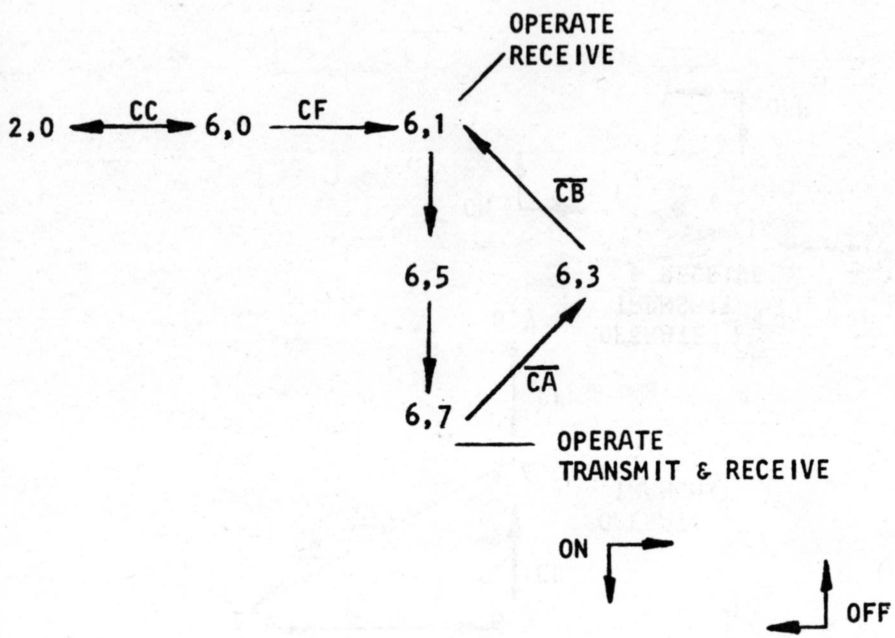

| CC<br>DATA SET READY | DTE PWR. | OCTAL CODE | CA<br>REQUEST TO SEND | CB<br>CLEAR TO SEND | CF<br>RECEIVED LINE SIGNAL DETECTOR |
|---|---|---|---|---|---|
| GROUP A | | | | GROUP B | |
| 0 | 0 | 0 | 0 | 0 | 0 |
| − | − | 1 | 0 | 0 | 1 |
| 0 | − | 2 | 0 | 1 | 0 |
| − | 0 | 3 | 0 | 1 | 1 |
| − | − | 4 | 1 | 0 | 0 |
| − | 1 | 5 | 1 | 0 | 1 |
| − | − | 6 | 1 | 1 | 0 |
| − | − | 7 | 1 | 1 | 1 |
| DON'T CARE | DON'T CARE | X | DON'T CARE | | |

1 = ON     0 = OFF

− = NOT APPLICABLE

**Flow diagram:**

2,0 →(CC) 6,0 —CF→ 6,1 (OPERATE RECEIVE)

6,1 —CB→ 6,3

6,1 → 6,5 → 6,7

6,7 —C̅A̅→ 6,3    OPERATE TRANSMIT & RECEIVE

ON →    OFF ↓

TYPE D DUPLEX INTERFACE
MULTIPOINT — REMOTE STATION
CHART 11

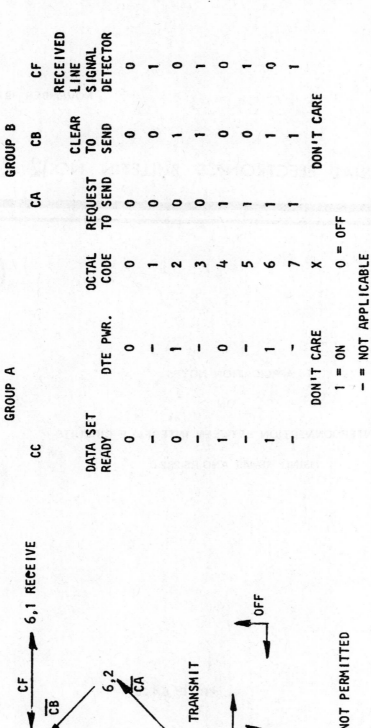

TYPE D HALF DUPLEX INTERFACE
CHART 12

| | GROUP A | | GROUP B | | |
|---|---|---|---|---|---|
| | CC | | CA | CB | CF |
| | DATA SET READY | DTE PWR. | OCTAL CODE | REQUEST TO SEND | CLEAR TO SEND | RECEIVED LINE SIGNAL DETECTOR |
| | 0 | 0 | 0 | 0 | 0 | 0 |
| | 1 | 1 | 1 | 0 | 0 | 1 |
| | 0 | 1 | 2 | 0 | 1 | 0 |
| | 1 | 0 | 3 | 1 | 1 | 1 |
| | — | 1 | 4 | 1 | 0 | 0 |
| | — | — | 5 | 1 | 0 | 1 |
| | — | 1 | 6 | 1 | 1 | 0 |
| | — | — | 7 | 1 | 1 | 1 |
| | | | X | DON'T CARE | DON'T CARE | |

DON'T CARE

1 = ON     0 = OFF

— = NOT APPLICABLE

961

NOTE

STATES X,3; X,5; X,7 ARE NOT PERMITTED

NOVEMBER 1977

# INDUSTRIAL ELECTRONICS BULLETIN NO.12

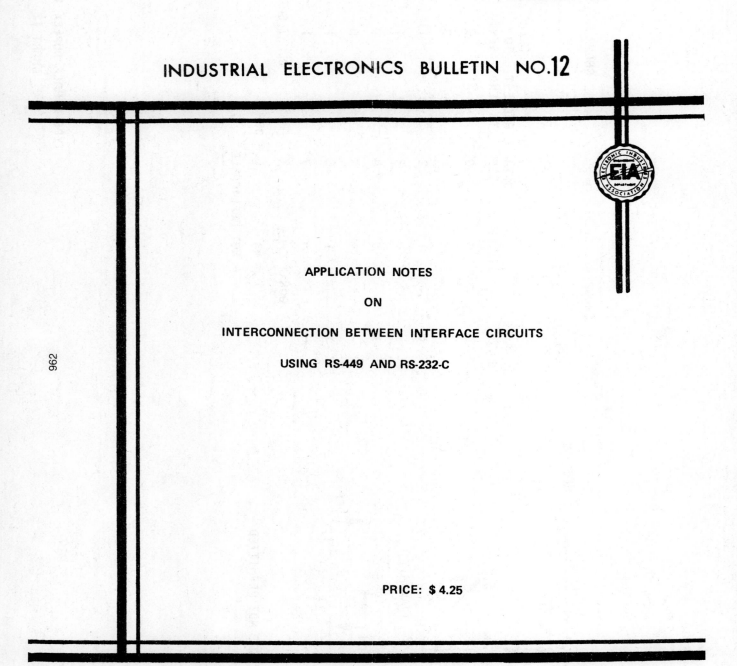

APPLICATION NOTES

ON

INTERCONNECTION BETWEEN INTERFACE CIRCUITS

USING RS-449 AND RS-232-C

962

PRICE: $ 4.25

## ELECTRONIC INDUSTRIES ASSOCIATION

**Engineering Department**

2001 Eye Street, N.W., Washington, D.C.    20006

EIA INDUSTRIAL ELECTRONICS BULLETIN NO. 12

APPLICATION NOTES

ON

INTERCONNECTION BETWEEN INTERFACE CIRCUITS

USING RS-449 AND RS-232-C

Prepared by EIA Subcommittees
TR-30.1 and TR-30.2

NOVEMBER 1977

# TABLE OF CONTENTS

# LIST OF ILLUSTRATIONS

# LIST OF TABLES

964

# APPLICATION NOTES ON INTERCONNECTION
## BETWEEN INTERFACE CIRCUITS USING
### RS-449 AND RS-232-C

## 1.  INTRODUCTION

The new series of digital interface standards, RS-449, RS-422, and RS-423, have been developed to meet the advancing state-of-the-art and greatly enhance the operation between DTEs and DCEs for data communication applications.  Since RS-232C has been the pervasive digital interface standard for a number of years, close attention was given during the development of RS-449, RS-422, and RS-423 to the selection of parameter values that would facilitate an orderly transition from the existing RS-232C equipment to the next generation without forcing obsolescence or costly retrofits.  It will, therefore, be possible to connect new equipment designed to RS-449 on one side of an interface to equipment designed to RS-232C on the other side of the interface.  Such interconnection can be accomplished with a few additional provisions associated only with the new RS-449 equipment, and the performance will be that normally experienced between RS-232C DTEs and DCEs.

This Industrial Electronics Bulletin provides application notes as guidance for implementing the necessary provisions that will allow continued use of existing RS-232C equipment by facilitating a graceful transition to the new equipment using RS-449.

## 2.  REQUIREMENTS

The provisions that are necessary to provide satisfactory operation between RS-232C equipment interconnected with RS-449 equipment fall into three categories - electrical, functional, and mechanical.  The basic requirements are specified in this section putting the onus for adaptation on the equipment designed to RS-449 where it can be made to closely resemble the characteristics of RS-232C.

### 2.1  <u>Electrical Characteristics</u> -

The RS-449 specification divides the interchange circuits into two categories in defining the implementation of the new electrical characteristics.  Category I circuits provide interconnection of either balanced RS-422 or unbalanced RS-423 generators via a pair of wires with a differential receiver.  Category II circuits provide interconnection of only unbalanced RS-423 generators via a single wire to a receiver which uses a common signal return circuit.  Category I applies to primary data, timing, and five selected control circuits while Category II applies to all other circuits.

1

An overlap in values of certain parameters of RS-423 and RS-232C has been established so that a few additional provisions incorporated in the interface circuits using RS-423 will make the necessary adjustments to ensure satisfactory operation with RS-232C circuits. Interface circuits implemented with RS-422 generators will not directly interoperate with RS-232C receivers. Therefore, when operating with RS-232C equipment, RS-423 generators must be used on Category I as well as Category II circuits to facilitate interoperation with RS-232C. An important factor to keep in mind when interworking RS-423 circuits with RS-232C circuits is that the performance is limited to that associated with RS-232C. The following describes the additional considerations that must be taken into account in the design of RS-449 equipment when interoperation with RS-232C equipment is contemplated:

### 2.1.1 Protection

RS-423 specifies that the receivers shall not be damaged by voltages up to 12 volts while RS-232C generators may produce output voltages up to 25 volts. Although many commercially available RS-423 receivers have been designed to withstand and operate properly with the higher RS-232C voltages, protection will be necessary for those receivers which do not have sufficient tolerance. RS-423 generators may also be damaged by the higher RS-232C generator voltages if they are inadvertently interconnected or shorted together. Additional protection for RS-423 generators may only be necessary on circuit SR in an RS-449 DTE or SI in an RS-449 DCE as discussed in section 3.6. Since the short circuit condition between RS-232C and RS-423 generators is purely a fault situation, any further consideration is left to the equipment designer.

### 2.1.2 Load Resistance

RS-423 specifies that individual receivers shall have a resistance of 4 kilohms or greater, while there may be multiple receivers in the load giving a total load resistance of only 400 ohms. RS-232C requires a load resistance of 3 kilohms to 7 kilohms. Therefore, to ensure proper interoperation, the load on each RS-449 interchange circuit must be greater than 3 kilohms. Although RS-423 receivers may have resistance greater than the 7 kilohms maximum specified by RS-232C, proper operation can be expected with RS-232C generators provided the voltage protection consideration of paragraph 2.1.1 have been taken into account.

### 2.1.3 Signal Levels

The generator output signal levels specified in RS-423 and RS-232C have an overlap in the 5 to 6 volt range. Furthermore. RS-423 levels can be as low as 4 volts while RS-232C levels can be as high as 25 volts. The considerations associated with the upper limit levels of RS-232C generators operating with RS-423 receivers have been covered in the previous section on Protection. Although an RS-423

2

generator output between 4 and 5 volts on the lower limit is not within the RS-232C specification, satisfactory operation with RS-232C receivers having a 3 volt transition margin can be expected because of the 15 meter (50 foot) recommended cable length of RS-232C and the low source impedances of RS-423 generators.

### 2.1.4 Risetime, Data Rate, Distance

RS-232C specifies that the risetime for the signal to pass through the +3 volt transition region shall not exceed 4% of the signal element duration. RS-423, on the other hand, generally requires much slower risetimes specified from 10%-90% of the total signal amplitude to reduce cross talk for operation over longer distances. It is possible, however, through proper selection of the waveshaping for a generator in RS-449 equipment to simultaneously meet the requirements of both RS-423 and RS-232C for data rates covered by RS-232C (i.e., up to 20 kbit/s).

RS-423 provides a curve to show the relationships of risetime with modulation rate and cable length. The RS-423 curve has been modified as shown in figure 1.1 to relate the variations which are possible due to the type of waveshaping employed. The revised curve also illustrates improved performance of linear waveshaping as contrasted to exponential waveshaping. It is expected that the more typical implementation will employ linear waveshaping.

The abscissa of figure 1.1 is the risetime of the signal from the RS-423 generator. By reading up to the CABLE LENGTH curve and over to the left hand ordinate scale, the associated maximum cable length can be determined. By reading up to the DATA SIGNALLING RATE curve and over to the right hand ordinate scale, the associated maximum data signalling rate can be determined. Thus, for any specific risetime value both the maximum cable length and maximum data signalling rate can be determined. These values will ensure that the near-end cross-talk levels stay below 1 volt peak.

Figure 1.2 shows the overlap in signal risetime characteristics which will allow interoperation of RS-423 generators with RS-232C receivers. There are two sets of curves which represent selected data signalling rates. One set applies to signals with a linear risetime while the other set applies to signals with an exponential risetime. The right hand limit of overlap of risetime between RS-423 and RS-232C is shown as 2.2 microseconds for linear risetimes and 5 microseconds for exponential risetimes and is based upon the 60 meter (200 foot) requirement of RS-449. This translates to a maximum possible data signalling rate for interoperability of 19200 bit/s or 9600 bit/s, respectively, without need for waveshaping options in the RS-449 equipment. Another influencing factor, shown in figure 1.2, is the RS-423 generator output

3

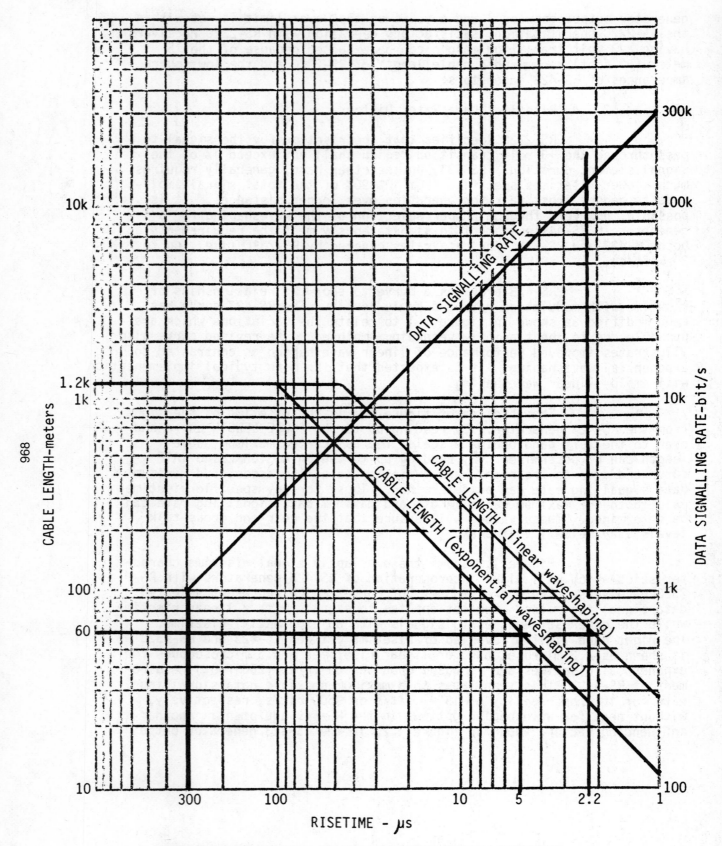

DATA SIGNALLING RATE OR CABLE LENGTH VERSUS RISETIME

<u>FIGURE-1.1</u>

4

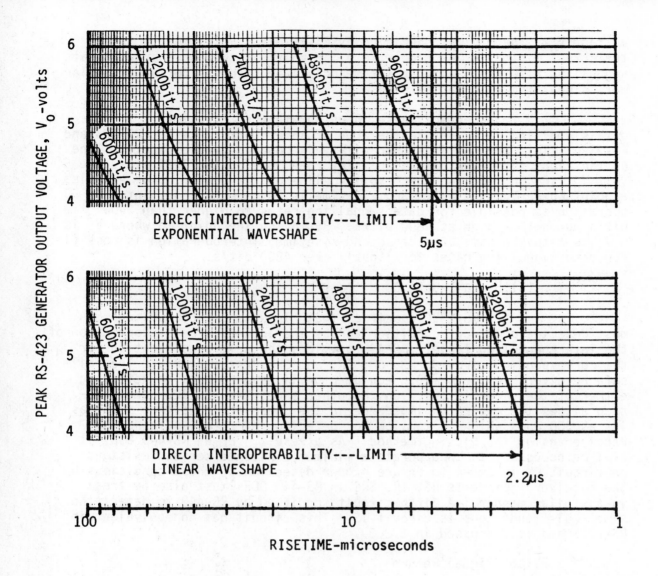

FIGURE 1.2

RISETIME RELATIONSHIP TO RS-449/RS-232C INTEROPERABILITY

5

voltage $V_o$. The minimum specified by RS-423 is 4 volts peak. Should the actual minimum $V_o$ be higher, e.g., 5 volts, either an increased data signalling rate for a fixed risetime or slower risetime for a particular data signalling rate can be used.

In effect these three parameters, risetime, $V_o$, and data signalling rate, define the area of interoperability for RS-423 and RS-232C. This area of interoperability is bounded by the risetime line on the right side, the data signalling rate line on the left side, the 6 volt $V_o$ line on top, and the 4 volt $V_o$ line on the bottom. For example, a point where $V_o$ is 4 volts and risetime is 10 microseconds, inter-operation is possible for data signalling rates to higher than 2400 bit/s but not as high as 4800 bit/s. By moving the point to where $V_o$ is 5 volts with the same 10 microsecond risetime, interoperation is possible for data signalling rates to slightly over 4800 bit/s.

### 2.1.5 Fail-Safe Conditions

RS-232C specifies that circuits CA, CC, CD, and SCA must detect a power-off condition across the interface or disconnection of the interconnecting cable. RS-449 provides the same specification for the corresponding circuits RS, DM, TR, and SRS, respectively. The fail-safe receivers in the RS-232C equipment will have no problem detecting either of these conditions. On the other side, the RS-232C specification allows the generator impedance in the power-off condition to be as low as 300 ohms which is too low for detection by a fail-safe RS-423 receiver using the biasing method. As a result, for an RS-449 DTE, it will be necessary to incorporate a minimum 2 kilohms series resistance on circuit DM in order to ensure proper detection of these conditions by the receiver. Circuits RS, TR, SRS in RS-449 DCEs must also be treated in a similar manner. A fifth circuit specified by RS-449 to detect the same fault conditions is circuit IS. This circuit has no equivalent in RS-232C and is discussed in 2.2.2d.

### 2.1.6 Signal Return

RS-423 requires separate signal return circuits, one for each direction of transmission, in addition to signal ground circuit SG. RS-232C specifies only signal ground circuit AB. It is therefore necessary to connect circuits SG, RC, and SC of the RS-449 equipment with circuit AB of the RS-232C equipment. Additionally, the B' leads appearing at the interface for certain receivers in the RS-449 equipment (TR, SD, TT, and RS for DCEs; DM, RD, ST, RT, CS, and RR for DTEs) must also be connected to circuit AB of the RS-232C equipment.

### 2.2 Functional Characteristics -

2.2.1 The majority of the functional circuits specified by RS-449 correspond directly to circuits of RS-232C. An equivalency table, showing how the circuits from the two standards match, is given

970

6

| | RS-449 | | RS-232C | |
|------|-------------------------------|------|------|----------------------------------------------------|
| | | AA | PROTECTIVE GROUND | |
| SG | SIGNAL GROUND | AB | SIGNAL GROUND | |
| SC | SEND COMMON | | | |
| RC | RECEIVE COMMON | | | |
| IS | TERMINAL IN SERVICE | | | |
| IC | INCOMING CALL | CE | RING INDICATOR | |
| TR* | TERMINAL READY | CD | DATA TERMINAL READY | |
| DM* | DATA MODE | CC | DATA SET READY | |
| SD* | SEND DATA | BA | TRANSMITTED DATA | |
| RD* | RECEIVE DATA | BB | RECEIVED DATA | |
| TT* | TERMINAL TIMING | DA | TRANSMITTER SIGNAL ELEMENT TIMING (DTE SOURCE) | |
| ST* | SEND TIMING | DB | TRANSMITTER SIGNAL ELEMENT TIMING (DCE SOURCE) | |
| RT* | RECEIVE TIMING | DD | RECEIVER SIGNAL ELEMENT TIMING | |
| RS* | REQUEST TO SEND | CA | REQUEST TO SEND | |
| CS* | CLEAR TO SEND | CB | CLEAR TO SEND | |
| RR* | RECEIVER READY | CF | RECEIVED LINE SIGNAL DETECTOR | |
| SQ | SIGNAL QUALITY | CG | SIGNAL QUALITY DETECTOR | |
| NS | NEW SIGNAL | | | |
| SF | SELECT FREQUENCY | | | |
| SR | SIGNALING RATE SELECTOR | CH | DATA SIGNAL RATE SELECTOR (DTE SOURCE) | |
| SI | SIGNALING RATE INDICATOR | CI | DATA SIGNAL RATE SELECTOR (DCE SOURCE) | |
| SSD | SECONDARY SEND DATA | SBA | SECONDARY TRANSMITTED DATA | |
| SRD | SECONDARY RECEIVE DATA | SBB | SECONDARY RECEIVED DATA | |
| SRS | SECONDARY REQUEST TO SEND | SCA | SECONDARY REQUEST TO SEND | |
| SCS | SECONDARY CLEAR TO SEND | SCB | SECONDARY CLEAR TO SEND | |
| SRR | SECONDARY RECEIVER READY | SCF | SECONDARY RECEIVED LINE SIGNAL DETECTOR | |
| LL | LOCAL LOOPBACK | | | |
| RL | REMOTE LOOPBACK | | | |
| TM | TEST MODE | | PINS 9 & 10 TEST FUNCTION | |
| SS | SELECT STANDBY | | | |
| SB | STANDBY INDICATOR | | | |

*Category I Circuits

TABLE 1
EQUIVALENCY TABLE

7

in table 1. There are three circuits in RS-232C which do not have any equivalency in RS-449. The first is protective ground circuit AA which is not consistent with the National Electric Code. It should, therefore, not be connected since protective grounding is normally covered by other equipment provisions. The other circuits were reserved for DCE testing on pins 9 and 10. Absence of any connection to these circuits is of no consequence to normal equipment operation.

2.2.2 There are eight circuits in RS-449 which have no equivalency in RS-232C. These are Local Loopback (LL), Remote Loopback (RL), Test Mode (TM), Terminal in Service (IS), Select Standby (SS), Standby Indicator (SB), New Signal (NS), and Select Frequency (SF). The requirements for satisfactory operation without proper interconnection of these circuits is as follows:

a. Circuits LL and RL - In the case of an RS-449 DCE, the OFF fail-safe condition normally provided will prevent erroneous operation. In the case of an RS-449 DTE requesting a local or remote loopback from an RS-232C DCE, the lack of an ON response on circuit TM should provide sufficient information for the DTE to recognize that such a test mode is either not implemented or not operative.

b. Circuit SS - In the case of an RS-449 DCE, the OFF fail-safe condition normally provided will prevent erroneous operation. In the case of an RS-449 DTE selecting standby from an RS-232C DCE, the lack of an ON response on circuit SB should provide sufficient information for the DTE to recognize that standby is either not implemented or not operative.

c. Circuits TM and SB - These circuits are normally provided with an OFF fail-safe condition in an RS-449 DTE, and since they only provide a response to circuits LL and RL, and circuit SS, respectively, in the ON condition, no impairment to operation is possible.

d. Circuit IS - This circuit, when implemented, must be held continuously in the ON condition for an RS-449 DCE to be operative for switched network service.

e. Circuit NS - This circuit is normally provided with an OFF fail-safe condition in an RS-449 DCE, thus erroneous operation will be prevented. There is no consequence to an RS-449 DTE to the absence of this circuit.

f. Circuit SF - Signals on this circuit are used to select the transmit and receive frequency bands of an RS-449 DCE. Circuit SR, Signalling Rate Selector, is assigned to the same

8

pin as SF since it is expected that only one of the two functions will be necessary for any single application. When adapting to RS-232C, however, additional consideration may be needed as discussed in section 3.6.

2.2.3 There are three circuits in RS-449 which have only near equivalency in RS-232C, thus further consideration for each circuit is necessary as follows:

a. Circuits SC and RC - these circuits provide separate send and receive signal returns in addition to signal ground SG for RS-449, whereas RS-232C only provides a single signal ground, circuit AB. By connecting circuits SG, RC, SC and the B' leads of circuits TR, DM, SD, RD, TT, ST, RT, RS, CS, RR, as appropriate, with circuit AB at the RS-449 side of the interface, satisfactory operation can be expected. (See 2.1.6.)

b. Circuit DM - This circuit is functionally the same as circuit CC in RS-232C except that circuit DM stays ON during test modes in which both the DTE and DCE are involved. Caution must be observed during manually initiated tests using an RS-232C DTE with an RS-449 DCE and an RS-449 DTE with an RS-232C DCE. In the former case, the DTE would have no indication that the DCE might be in a test condition. In the latter case, the DTE could interpret the indication of such a test as voice mode or end of call. In most cases, however, these problems can be overcome by proper operation of the DTE during such tests.

2.3 <u>Mechanical Characteristics</u> -

Although there is no formally standardized connector for RS-232C interfaces, most implementations employ the well-known 25-position connector which has become a de facto standard. RS-449 specifies 37-position and 9-position connectors which belong to the same family of connectors as the 25-position connector used for RS-232C. The reason for the larger 37-position connector for RS-449 is to accommodate additional circuits and the extra contacts required for Category I circuits. The separate 9-position connector is used to accommodate the secondary channel circuits. Using different connectors for RS-449 has the added advantage of preventing inadvertent connection of RS-449 to RS-232C equipment without the further provisions necessary for proper interworking between these different equipments.

The contact assignments for the circuits in RS-449 have been carefully selected to simplify wiring between the 25-position and 37-position connectors. A suggested method of implementation in the following section illustrates how an adapter between an RS-232C and an RS-449 equipment may be configured.

973

9

# 3. SUGGESTED IMPLEMENTATIONS

The actual method of implementation for satisfying the requirements outlined in section 2 is not standardized as it is felt that a number of innovative approaches are possible. Accordingly, it is left to the designer of equipment meeting RS-449 to incorporate the necessary provisions when interconnection with RS-232C equipment is desired as a special feature. It should not be assumed that any equipment meeting RS-449 will interoperate with RS-232C equipment unless a specific reference is made that the requirements of this application note are fulfilled.

An example method of satisfying the requirements set forth herein has been developed. It is presented in this section as guidance for implementing RS-449 where interoperation with existing RS-232C equipment is essential to facilitate an orderly transition to the next equipment generation.

## 3.1 Category I Circuits -

RS-423 generators must be implemented on all Category I circuits.

## 3.2 Protection of RS-423 Receivers -

Although RS-423 specifies that receivers need only withstand 12 volts without being damaged, a number of receiver ICs are available that can withstand and operate properly with the higher voltages which are possible from some RS-232C generators. When the RS-423 receivers in the RS-449 equipment do not have adequate tolerance, however, additional protection will be required. This can be accomplished by the addition of an attenuating L pad in front of the receiver input as shown in figure 2. The L pad with a 2.0 kilohm series resistance and a 3.3 kilohm shunt resistance has an additional effect of appearing as a high impedance source. Therefore, the pad should be no further from the receiver inputs of the RS-449 equipment than 3 cable meters (10 feet) to ensure that near-end cross-talk from adjacent circuits does not reach an unacceptable level (1 volt peak).

## 3.3 Load Resistance -

The total load resistance on each interchange circuit must not be less than 3 kilohms.

## 3.4 Generator Output Signal -

There are two considerations for the generator output signal, waveshape and minimum amplitude as discussed in 2.1.4. With linear waveshaping and a risetime of 2.2 microseconds, interoperation with RS-232C is possible for data signalling rates up to 19.2 kbit/s. Where

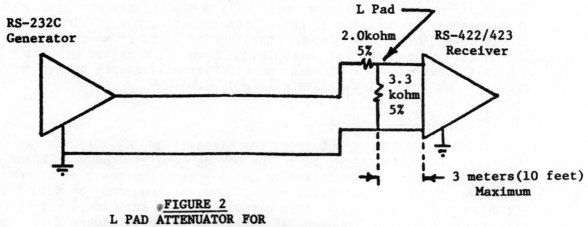

FIGURE 2
L PAD ATTENUATOR FOR
PROTECTION OF RS-423 RECEIVER

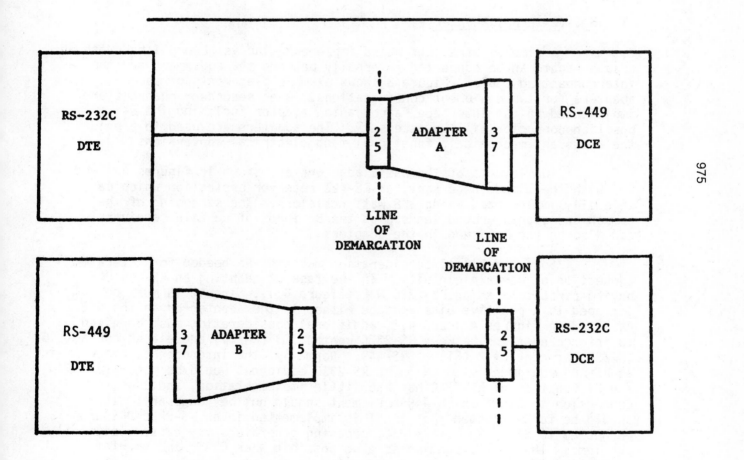

975

FIGURE 3
EXAMPLE INTERCONNECTION CONFIGURATIONS
WITH BASIC ADAPTER

11

lower maximum data signalling rates will apply to interoperation, slower risetimes can be employed as determined from figure 1.2. In the case of linear waveshaping, the minimum amplitude of $V_o$ is of no consequence. In the case of exponential waveshaping, however, a higher minimum $V_o$ of 4.5 volts peak with a 5 microsecond risetime will ensure proper interoperation at 9.6 kbit/s. (See figure 1.2).

### 3.5 Fail-Safe Provisions -

An additional provision for circuits DM, RS, TR, and SRS is necessary for satisfactory detection of the generator power OFF and open cable conditions. As specified in 2.1.5 when using the bias method of fail-safe, a series resistance of 2 kilohms is required for detection of the power-off condition on these circuits. This additional resistance will not be required, however, if the L pad shown in figure 2 is implemented for receiver protection or if other methods are used for fail-safe detection.

### 3.6 External Adapter -

A simple, straightforward implementation is accomplished with an adapter that can be connected externally between the equipment and the interconnecting cable. Figure 3 shows example placements of such adapters for two equipment configurations. When secondary channels are implemented in the interface, a companion adapter including the 9-position connector will be necessary. The adapters are used to provide the necessary mechanical, functional, and electrical conversions.

The layouts of the basic adapters are shown in figures 4.1 and 4.2 with the L pads necessary for RS-423 receiver protection which can be easily implemented using 1/8 watt resistors. The strapping of the common and signal ground leads plus the B' leads of certain receivers can also be accomplished in the adapters.

Some additional consideration may also be needed for interconnection of certain circuits. In the case of adapting an RS-449 DCE having circuit IS to an RS-232C DTE (figure 4.1), circuit IS must be strapped to a positive bias voltage either in the adapter or in the DCE. As noted in figures 4.1 and 4.2, additional consideration must be given to interconnections between RS-232C contact 23 (CH/CI) and either contact 16 (SF/SR) or contact 2 (SI) of RS-449. Normally, the interconnection of SR in an RS-449 equipment with CH in an RS-232C equipment would be more typical. The CI function of RS-232C has had little implementation, and thus, connection to SI of an RS-449 equipment should not be necessary. It should be further noted that if CI is implemented in an RS-232C DCE and connected to SR of an RS-449 DTE, opposing generators may cause electrical problems. The same situation is also possible when CH is implemented in an RS-232C DTE and SI in an RS-449 DCE. An option, therefore, to open such an interconnection may be desirable when such circumstances are

12

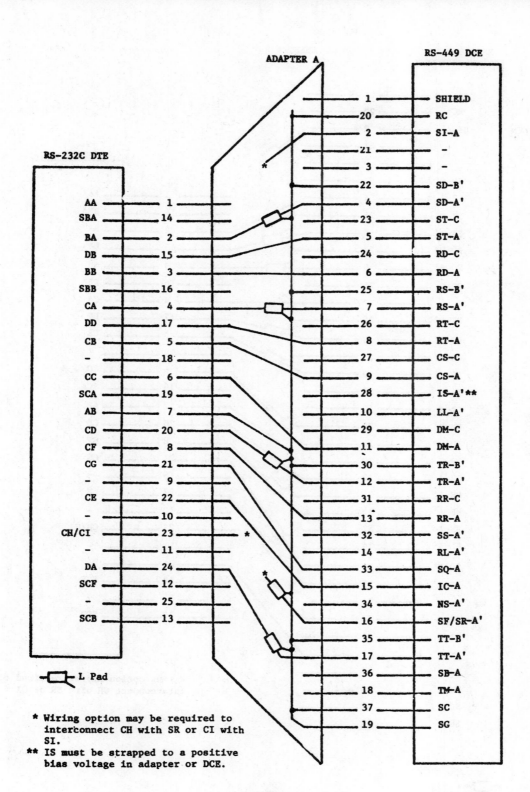

RS-232C DTE

| | |
|---|---|
| AA | 1 |
| SBA | 14 |
| BA | 2 |
| DB | 15 |
| BB | 3 |
| SBB | 16 |
| CA | 4 |
| DD | 17 |
| CB | 5 |
| – | 18 |
| CC | 6 |
| SCA | 19 |
| AB | 7 |
| CD | 20 |
| CF | 8 |
| CG | 21 |
| – | 9 |
| CE | 22 |
| – | 10 |
| CH/CI | 23 |
| – | 11 |
| DA | 24 |
| SCF | 12 |
| – | 25 |
| SCB | 13 |

ADAPTER A

RS-449 DCE

| | |
|---|---|
| 1 | SHIELD |
| 20 | RC |
| 2 | SI-A |
| 21 | – |
| 3 | – |
| 22 | SD-B' |
| 4 | SD-A' |
| 23 | ST-C |
| 5 | ST-A |
| 24 | RD-C |
| 6 | RD-A |
| 25 | RS-B' |
| 7 | RS-A' |
| 26 | RT-C |
| 8 | RT-A |
| 27 | CS-C |
| 9 | CS-A |
| 28 | IS-A'** |
| 10 | LL-A' |
| 29 | DM-C |
| 11 | DM-A |
| 30 | TR-B' |
| 12 | TR-A' |
| 31 | RR-C |
| 13 | RR-A |
| 32 | SS-A' |
| 14 | RL-A' |
| 33 | SQ-A |
| 15 | IC-A |
| 34 | NS-A' |
| 16 | SF/SR-A' |
| 35 | TT-B' |
| 17 | TT-A' |
| 36 | SB-A |
| 18 | TM-A |
| 37 | SC |
| 19 | SG |

⊏▭⊐ L Pad

* Wiring option may be required to
  interconnect CH with SR or CI with
  SI.
** IS must be strapped to a positive
   bias voltage in adapter or DCE.

977

FIGURE 4.1

25/37-POSITION BASIC ADAPTER CONFIGURATION A
(RS-232C DTE TO RS-449 DCE)

13

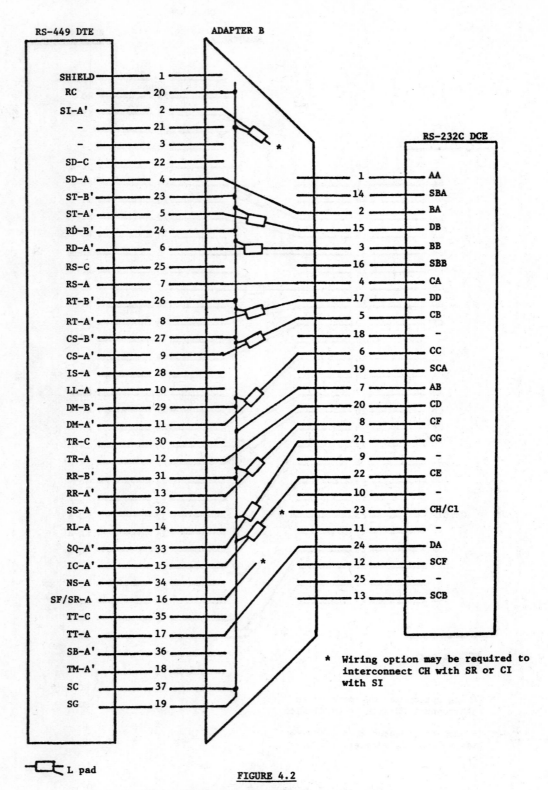

**FIGURE 4.2**

37/25-POSITION BASIC ADAPTER CONFIGURATION B
(RS-449 DTE TO RS-232C DCE)

14

expected. Additional flexibility may be desirable in either type of adapter to provide wiring options which allow alternative interconnection of CH with SR or CI with SI.

When operation of a secondary channel function across the interface is required, adaptation to the 9-position connector will also be necessary. The companion interconnecting configurations for this application are shown in figures 4.3 and 4.4.

As pointed out earlier, the L pads must be located within 3 cable meters (10 feet) of the receivers in the RS-449 equipment to avoid excessive near-end cross-talk. In the case of the RS-232C DTE/RS-449 DCE configuration, there is no problem as RS-449 requires the connector(s) to be on the DCE side of the interconnection. In the other configuration, RS-449 DTE/RS-232C DCE, placement of the adapter at the DCE would not be acceptable. It will therefore be necessary to also implement the 37-position and 9-position connectors at the RS-449 DTE to facilitate proper connection of the adapter. Furthermore, the additional implementation of the connector(s) at the DTE side of the interface would be a good practice for added flexibility in installation where the longer cable lengths are allowed.

3.7  Summary of Suggested Provisions -

a.  25/37-position basic mechanical adapter and, where secondary channel circuits are implemented, 25/9-position companion configuration.

b.  RS-423 generators must be implemented on all Category I circuits.

c.  L pad attenuator, if needed, on appropriate circuits in adapter.

d.  The total load resistance on each interchange circuit must not be less than 3 kilohms.

e.  Waveshaping of RS-423 generators must be fixed at the appropriate value as determined from figure 1.2 for type of waveshaping, and maximum applicable data signalling rate.

f.  Connect signal common return circuits RC and SC, signal ground circuit SG, and the receiver B' leads appearing at the interface from the RS-449 equipment to circuit AB from the RS-232C equipment in the adapter.

g.  For RS-449 DCEs, circuit IS, when implemented, must be strapped to a positive bias voltage in either the adapter or DCE.

979

15

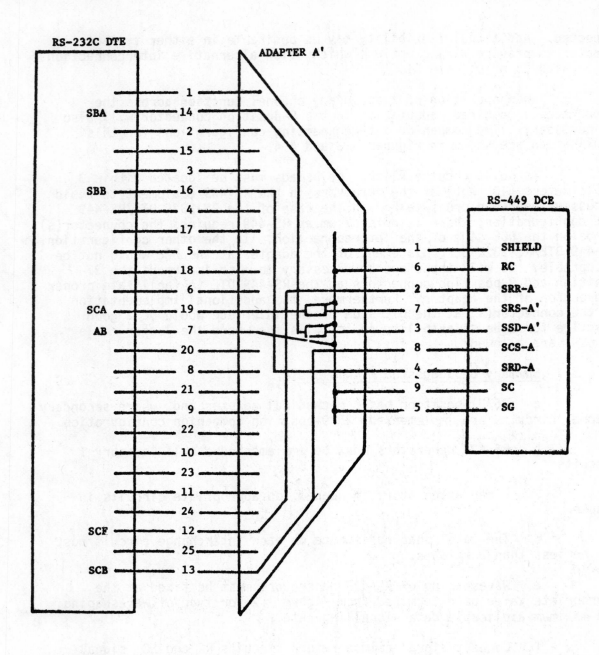

RS-232C DTE

ADAPTER A'

RS-449 DCE

| | |
| --- | --- |
| | 1 |
| SBA | 14 |
| | 2 |
| | 15 |
| | 3 |
| SBB | 16 |
| | 4 |
| | 17 |
| | 5 |
| | 18 |
| | 6 |
| SCA | 19 |
| AB | 7 |
| | 20 |
| | 8 |
| | 21 |
| | 9 |
| | 22 |
| | 10 |
| | 23 |
| | 11 |
| | 24 |
| SCF | 12 |
| | 25 |
| SCB | 13 |

| | |
| --- | --- |
| 1 | SHIELD |
| 6 | RC |
| 2 | SRR-A |
| 7 | SRS-A' |
| 3 | SSD-A' |
| 8 | SCS-A |
| 4 | SRD-A |
| 9 | SC |
| 5 | SG |

980

⊣▭⊢ L pad

FIGURE 4.3

25/9-POSITION OPTIONAL COMPANION ADAPTER CONFIGURATION A'
(RS-232C DTE TO RS-449 DCE
FOR SECONDARY CHANNEL CIRCUITS)

16

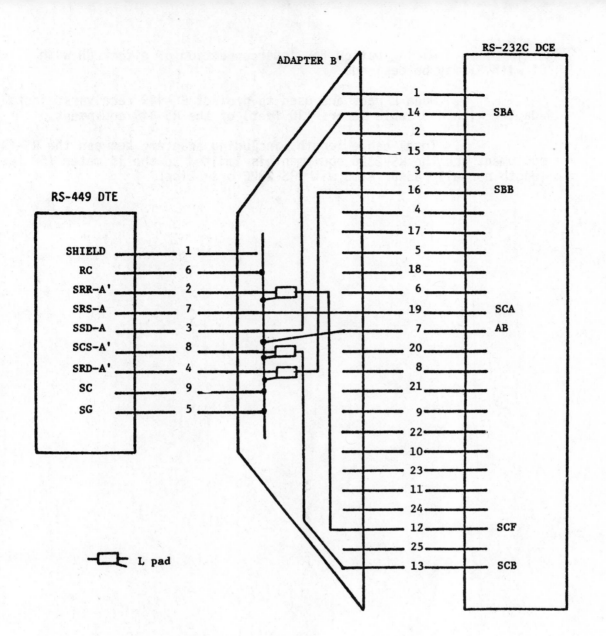

FIGURE 4.4

9/25-POSITION OPTIONAL COMPANION ADAPTER CONFIGURATION B'
(RS-449 DTE TO RS-232C DCE
FOR SECONDARY CHANNEL CIRCUITS)

17

h.  Wiring option for interconnection of either CH with SR or CI with SI may be desired.

i.  When L pads are used to protect RS-449 receivers, install adapter within 3 cable meters (10 feet) of the RS-449 equipment.

j.  Total cable length, including adapter, between the RS-449 equipment and the RS-232C equipment is limited to the 15 meter (50 feet) length normally associated with RS-232C operation.

# Part 5

# Federal Data Communications Standards

The Federal Telecommunications Standards Committee (FTSC) is a Federal Government interagency advisory body established to assist the Manager of the National Communications System in ensuring a timely coordinated movement of the Federal Standardization Program in the area of telecommunications. As outlined in National Communications System Circular 175.1, 23 March 1973, the objectives of the FTSC are:

a. To develop, coordinate, and promulgate the technical and procedural standards required to achieve inter-operability among functionally similar telecommunications networks of the Federal components of the National Communications System.

b. In concert with the National Bureau of Standards, to develop and coordinate technical and procedural standards for data transmission and the computer-telecommunications interface.

c. Increase cohesiveness and effectiveness of the Federal telecommunications community's participation in national/international standards programs and on the Federal Information Processing Standards (FIPS) program.

The principal efforts of the FTSC in the area of data communications have been aimed at establishing mandatory Federal Standards for application to Federal activities. Additional liaison is maintained with the National Bureau of Standards which has overall responsibility for Federal Information Processing Standards (FIPS) for the data processing community. The FIPS data processing effort is complementary to the FTSC's for data communications.

To ensure the commonality of Federal Standards with national and international standards, the FTSC works closely with CCITT, ISO, ANSI, and EIA. The approved standards resulting from the activities of these groups are then either directly applied or adapted for Federal use. The FTSC's policy is to refrain from independently developing unique Federal standards unless there is a clear void to be filled. The objective is to maintain commonality worldwide.

There are three major areas where the FTSC has directly established the development of standards. The first is the development of modem standards (such as FED-STD-1005 and FED-STD-1006) which are compatible with the CCITT's work. There is also a great amount of effort aimed at defining the parameters for the evaluation of system performance. Finally, the newest major effort involves the development of standards for implementing the FIPS 46 Data Encryption Algorithm.

| NUMBER | TITLE | RELATED CROSS REFERENCE STANDARDS |
|---|---|---|
| **FED-STD-1001** | High speed synchronous signaling rates between data terminal equipment and data circuit terminating equipment | ANSI X3.36 |
| **FED-STD-1002** | Time and frequency reference information | None |
| **FED-STD-1003** | Bit oriented data link control procedures | ISO 3309; ANSI DRAFT BSR X3.66 |
| **FED-STD-1005** | 2400 BPS modem | CCITT V.26bis |
| **FED-STD-1006** | 4800 BPS modem | CCITT V.27bis/ter |
| **FED-STD-1010** | ASCII bit sequencing for serial-by-bit transmission | CCITT V.4, X.4; ISO 1155, 1177, ANSI X3.15 |
| **FED-STD-1011** | Character structure for serial-by-bit ASCII transmission | CCITT V.4, X.4; ISO 1155, 1177; ANSI X6 |
| **FED-STD-1012** | Character structure for parallel-by-bit ASCII transmission | ANSI X3.25 |
| **FED-STD-1013** | Data terminal equipment to data circuit-terminating equipment synchronous signaling rates using 4kHz circuits | CCITT V.4, V.5; ANSI X3.1 |
| **FED-STD-1020** | Electrical characteristics of unbalanced voltage digital interface circuits | CCITT V.11 (X.27); EIA RS-422 |
| **FED-STD-1030** | Electrical characteristics of balanced voltage digital interface circuits | CCITT V.10 (X.26); EIA RS-423 |

FEDERAL STANDARD

TELECOMMUNICATIONS:  SYNCHRONOUS HIGH SPEED
DATA SIGNALING RATES BETWEEN DATA TERMINAL EQUIPMENT
AND DATA COMMUNICATION EQUIPMENT

This standard is issued by the General Services Administration
pursuant to the Federal Property and Administrative Services
Act of 1949, as amended, Public Law 89-306 (40 U.S.C. 759 et
seq.), and as implemented by Ex. Order No. 11717, May 9, 1973,
38 F.R. 12315 (1973).  This standard has also been published
as an integral part of Federal Information Processing Standards
(FIPS) Publication 37 dated June 15, 1975, and its application
to both data processing systems and telecommunication systems
is mandatory on all Federal agencies.

1.  Purpose.  The purpose of this standard is to facilitate interoperability
between telecommunication facilities and systems of the Federal Government and
compatibility of these facilities and systems at the computer-communications interface
with data processing equipment (systems) of the Federal Government.

2.  Scope.  This standard establishes signaling rate requirements for data terminal
and data processing equipment which is both (1) employed with synchronous data communication equipment and (2) designed to operate on binary encoded information over
wideband communication channels having greater bandwidth than the nominal 4 kHz bandwidth commonly used in analog voice transmission.  It does not specify any coding or
format restrictions which may be imposed by the telecommunications system on the binary
encoded information emerging from the data terminal or data processing equipment.

3.  Application.  This standard shall be used by all agencies of the Federal
Government.  It is not intended to hasten the obsolescence of equipment currently
existing in the Federal inventory; it shall be used in the planning, design, and
procurement, including lease and purchase, of all new data communication systems.

4.  Applicable documents.  The following document forms a part of this standard to
the extent specified herein:

American National Standards Institute (ANSI) Standard:

X3.36-1975 - Synchronous High Speed Data Signaling Rates Between Data
             Terminal Equipment and Data Communication Equipment

(Applications for copies shall be addressed to the American National Standards
Institute, Inc., 1430 Broadway, New York, N.Y.  Federal agencies may also obtain
copies with FIPS Publication 37 by applying for that document to General Services
Administration, Specifications Sales (3FRSBS), Building 197, Washington Navy Yard
Annex, Washington, D.C. 20407.)

5.  Requirements.

5.1 General.  Signaling rates used for exchange of data between data terminal and
data communication equipment where the data communication equipment is utilizing wide-
band communication channels (greater than 4 kHz bandwidth) shall be in accordance with
ANSI Standard X3.36-1975 with the following exceptions:

a.  The note alluding to certain unspecified coding restrictions on the
    data streams of users operating at 1544 kbit/s is not applicable.

b.  A signaling rate of 64 kbit/s may also be utilized by Federal agencies
    having requirements to interface directly with point-to-point trans-
    mission facilities of foreign communications carriers.

FSC MISC

5.2 <u>Rate tolerance</u>. The signaling rates specified are nominal. The bit rate tolerance shall be consistent with the operational environment in which applied to ensure maintenance of end-to-end synchronism provided that such tolerance complies with other Federal Standards which deal with signal quality and synchronization.

6. <u>Changes</u>. When a Federal agency considers that this standard does not provide for its essential needs, a statement citing inadequacies shall be sent in duplicate to the General Services Administration, Federal Supply Service, Office of Standards and Quality Control, Washington, D.C. 20406, in accordance with provisions of Federal Property Management Regulations 41 C.F.R. 101-29.3. The General Services Administration will determine the appropriate action to be taken and will notify the agency.

7. <u>Conflict with referenced documents</u>. Where the requirements stated in this standard conflict with any requirements in a referenced document, the requirements of this standard shall apply. The nature of the conflict between this standard and a referenced document shall be submitted in duplicate to the General Services Administration, Federal Supply Service, Office of Standards and Quality Control, Washington, D. C. 20406.

8. <u>Notes</u>. Federal Government activities may obtain copies of this standard or other Federal Specifications, Standards, and Handbooks at established distribution points in their agencies. Activities outside the Federal Government may obtain copies of Federal Specifications, Standards, and Handbooks as outlined under General Information in the Index of Federal Specifications and Standards and at the prices indicated in the Index. The Index, which includes cumulative supplements as issued, is for sale on a subscription basis by the Superintendent of Documents, U. S. Government Printing Office, Washington, D.C. 20402. Purchase orders for copies of this standard should be submitted to the General Services Administration, Specifications Sales (3FRSBS), Building 197, Washington Navy Yard Annex, Washington, D.C. 20407. Price 25 cents each.

MILITARY INTERESTS:

CIVIL AGENCY COORDINATING ACTIVITIES:

986

In view of the special nature of this standard, records of coordination with all affected Federal agencies are maintained by the preparing activity.

PREPARING ACTIVITY:

Office of the Manager
National Communications
    System (NCS-TS)
Washington, D.C. 20305

U. S. GOVERNMENT PRINTING OFFICE : 1975 - 585-700/2654

FEDERAL STANDARD

TIME AND FREQUENCY REFERENCE INFORMATION
IN TELECOMMUNICATION SYSTEMS

This standard is issued by the General Services Administration pursuant to the
Federal Property and Administrative Services Act of 1949, as amended.

1. Scope. This standard establishes the requirements for telecommunication facilities and systems of the Federal Government to utilize time and frequency reference information based upon Coordinated Universal Time (UTC).

1.1 Purpose. The purpose of this standard is to facilitate interoperability between telecommunication facilities and systems of the Federal Government.

1.2 Application. This standard shall be used by all Federal agencies where interoperability between Federal Government telecommunication facilities and systems is dependent on time or frequency reference information.

2. Referenced documents. The current issues of the following documents form a part of this standard to the extent specified herein:

General Services Administration, Federal Supply Service:

    41 CFR 101  -  Code of Federal Regulations, Chapter 101
                  Federal Property Management Regulations.

(This GSA publication is for sale by the Superintendent of Documents, U. S. Government Printing Office, Washington, DC 20402.)

National Bureau of Standards (NBS), Department of Commerce:

    NBS Technical Note No. 649  -  The Standards of Time and Frequency in the USA.

(This NBS publication is for sale by the Superintendent of Documents, U. S. Government Printing Office, Washington, DC 20402.)

U. S. Naval Observatory (USNO):

    U. S. Naval Observatory Time Service Announcements:
        Series 1, List of Worldwide VLF and HF Transmissions.
        Series 14, Time Service General Announcements.

(These USNO publications are available from the Superintendent, U. S. Naval Observatory, Washington, DC 20390.)

3. Definitions. As used in this standard, definitions of terms shall be in agreement with their usage in NBS Technical Note No. 649 and in USNO Time and Service Announcements, Series 1 and 14, which provide technical background and list Coordinated Universal Time (UTC) sources.

4. General statement of requirements. The time and frequency reference information utilized in applicable Federal Government telecommunication facilities and systems shall be referenced to (known in terms of) the existing standards of time and frequency maintained by the U. S. Naval Observatory, UTC (USNO), or the National Bureau of Standards, UTC (NBS). For purposes of this standard, coordinated values of UTC will be used to obtain the standard values of time and frequency. The accuracy of this time and frequency reference information with respect to UTC (USNO) or UTC (NBS) shall be commensurate with the individual system design and interface requirements.

5. Changes. When a Federal agency considers that this standard does not provide for its essential needs, a statement citing inadequacies shall be sent in duplicate to the General Services Administration, Federal Supply Service, Washington, DC 20406, in accordance with provisions of Federal Property Management Regulations 41 CFR 101-29.3. The General Services Administration will determine the appropriate action to be taken and will notify the agency.

FSC 5800

FED-STD-1002

MILITARY INTERESTS:                                    CIVIL AGENCY COORDINATING ACTIVITIES:

In view of the special nature of this standard, records of coordination with all affected Federal agencies
are maintained by the preparing activity.

PREPARING ACTIVITY:

Office of the Manager
National Communications System (NCS-TS)
Washington, DC 20305

988

Orders for this publication are to be placed with General Services Administration, acting as an agent
for the Superintendent of Documents.  Single copies of this standard are available at the GSA Business
Service Centers in Boston, New York, Philadelphia, Atlanta, Chicago, Kansas City, MO, Fort Worth, Denver,
San Francisco, Los Angeles, and Seattle.  Additional copies may be purchased for 10 cents each from the
General Services Administration, Specification Sales, Bldg. 197, Washington Navy Yard, Washington,
DC 20407.

2

U. S. Government Printing Office :  1974  0 - 546-583 (2599)

TELECOMMUNICATIONS: SYNCHRONOUS BIT ORIENTED
DATA LINK CONTROL PROCEDURES

This standard is issued by the General Services Administration pursuant to the Federal Property and Administrative Services Act of 1949, as amended. Its application to both data processing and telecommunications systems is mandatory on all Federal agencies to the extend specified herein.

# FOREWORD

The Telecommunication series of
Federal Standards are designed to
facilitate interoperability between
telecommunication facilities and
systems of the Federal Government
and compatibility of these facilities and systems at the computer-
communications interface with data
processing equipment.

990

## 1.0       Description

This standard specifies the frame structure, elements of procedures, and codes of practice (classes of procedure) for data communication systems that transmit synchronous binary data, by electrical or electromagnetic means and which have automatic error detection capabilities. The application of the data link control procedures described in this standard to automatic data processing (ADP) systems is addressed in Federal Information Processing Standard _____.

## 2.0       Application

Federal agencies shall use this standard in the design and procurement of data communications equipment, including associated terminal equipment, using bit oriented data link control procedures. It is not mandatory for new systems whose design was irrevocably committed to the use of other data link control procedures on or before the issue date of this standard. Nor is it applicable to equipment being procured as replacement for, or extensions to, existing systems which use other data link control procedures.

## 3.0       Premises

This standard is defined in terms of the actions that occur on the data link upon receipt of commands at secondary stations and combined stations connected to the data link. To assure standardization (hence compatibility), all features of a stipulated basic class of procedures must apply when equipment is to be operated within the constraints of this standard.

## 4.0       Applicable Documents

With the inclusion of the additional requirements and exceptions listed in paragraph 5.0, this standard adopts in its entirety the provisions of (proposed) American National Standards Standard X3.66 1978. (Copy of BSR X3.66, 14 December 1977, enclosed).

## 5.0       Additional Requirements and Exceptions

References cited at the end of each requirement refer to the most pertinent applicable parts in BSR X3.66 (14 December 1977) standard.

5.1      All systems conforming to this Federal Standard shall implement the 16-bit frame check sequence (FCS) specified in the basic document.  If a higher degree of protection is needed, the length of the FCS shall be increased by multiples of eight bits. When a 32-bit FCS is required, it shall use the generating polynomial

$$X^{32}+X^{26}+X^{23}+X^{22}+X^{16}+X^{12}+X^{11}+X^{10}+X^8+X^7+X^5+X^4+X^2+X^1+1$$

The 32-bit FCS, if used, shall be generated according to the procedures in section 12, of ANSI X3.66 1978, appropriately extended to 32 bits.  That is

$$L(X) = X^{31} + X^{30} + X^{29} \cdots X^2 + X^1 + 1$$

(See paragraph 3.5, section 12, and appendix D of ANSI X3.66 1978).

5.2      Federal networks shall use the extended address format, whether or not optional function 7 (multiple octet addressing) is implemented.  (See paragraphs 4.3.1, 4.3.2, and 4.5 of ANSI X3.66 1978).

5.3      The nonreserved commands and responses shall not be specified or used.  (See paragraphs 7.4.5 and 7.5.5. of ANSI X3.66 1978).

6.0      Changes

When a Federal agency considers that this standard does not provide for its essential needs, a statement citing inadequacies shall be sent in duplicate to the General Services Administration, Federal Supply Service, FMH, Washington, D.C. 20406 (in accordance with provisions of Federal Property Management Regulations 41CFR101-29.3).  The General Services Administration will determine the appropriate action and will notify the agency.

BSR X3.66
X3S34-589
DRAFT 7
14 December 1977

SEVENTH DRAFT


PROPOSED

AMERICAN NATIONAL STANDARD

FOR

ADVANCED DATA COMMUNICATION

CONTROL PROCEDURES (ADCCP)

993

Prepared by

Task Group X3S34

on

Control Procedures

Technical Committee X3S3 on Data Communications

Committee X3 on Computers & Information Processing

Proposed

American National Standard

For Advanced Data Communication

Control Procedures (ADCCP)

Secretariat

Computer and Business Equipment Manufacturers Association

994

Approved _____

AMERICAN NATIONAL STANDARDS INSTITUTE

ABSTRACT Data Communication Control Procedures define the means for exchanging data between business machines (e.g., computers, concentrators and terminals) over communication circuits. The advanced data communication control procedures described in this standard are synchronous, bit oriented (i.e., use bit patterns instead of ASCII characters for control), code independent (i.e., capable of handling any data code or pattern) and interactive (i.e., have relatively high efficiency in an interactive application). Batch operation is handled with efficiency comparable to previous standards. Improvements have also been made with respect to previous standards in the areas of reliability and modularity.

# CONTENTS

996

## 1.0 SCOPE

This standard establishes the procedures to be used on synchronous communication links using ADCCP. This standard does not define any single system and should not be regarded as a specification for a data communications system.

This standard is intended to cover a wide range of applications (e.g., two-way alternate and two-way simultaneous data communication between computers, concentrators and terminals which are normally buffered) and a wide range of data link configurations (e.g., full and half-duplex, multi-point, point-to-point, switched or non-switched).

This standard is defined specifically in terms of the actions that occur on receipt of commands at Secondary stations and Combined stations.

In order to provide a high degree of standardization (and, therefore, of compatibility), any equipment intended to be operated within the constraints of this standard shall implement all features of a stipulated basic class of the procedures.

## 2.0 GENERAL

ADCCP defines a method of data link control in terms of the various combinations of primary link control functions (referred to as Primary station) and secondary link control functions (referred to as a Secondary station) and balanced link control functions (referred to as a Combined station) that make up the control functions and protocols at three types of logical data link control stations:

- Primary station
- Secondary station
- Combined station

In particular, the logical functions and protocols of Secondary stations and Combined stations are specified identically with respect to the action taken and the response frame(s) transmitted as the result of receiving a given command frame(s). The Primary station and Combined station procedures for managing and scheduling the data link, via the transmission of command frames, are the responsibility of the system designer and are not specified.

Since this standard is defined in terms of logical stations it should be noted that a given physical station may be composed of one or more logical stations. For example, a physical station implementation may: 1) have the capability of providing more than one type of logical station capability on a given link at different times (see configurable station Section 2.1.4); 2) have the capability of providing more than one logical station capability on different links at the same time (e.g., a multiplexor that serves several links); 3) house or serve multiple logical stations (e.g., a cluster controller).

## 2.1 Station Types

In ADCCP there are three types of stations: Primary Station; Secondary station; Combined station.

NOTE: As used in this document the word station (by itself) refers to Primary, Secondary and Combined Stations.

## 2.1.1 Primary Station

A Primary station has (only) a primary link control capability. The Primary station transmits command frames (commands) to and receives response frames (responses) from the Secondary station(s) on the link. A Primary station maintains a separate information transmitting ability and/or information receiving

- 2 -

ability with each Secondary station on the link.

## 2.1.2 Secondary Station

A Secondary station has (only) a secondary link control capability. The Secondary station transmits response frames (responses) to and receives command frames (commands) from the Primary station. It maintains one information transmitting ability and/or one information receiving ability with the Primary station.

## 2.1.3 Combined Station

A Combined station has balanced link control capability. The Combined station transmits both command frames (commands) and response frames (responses) to, and receives both commands and responses from, another Combined station. It maintains one information transmitting ability to and one information receiving ability from the other Combined station.

## 2.1.4 Stations Capable of Being Configured

A station is defined as configurable if it has, as the result of mode-setting action, the capability to be, at different times, more than one type of logical station; i.e., Primary station, Secondary station or Combined station.

## 2.2 Logical Data Link Configurations

In ADCCP there are two logical data link configurations:

- Unbalanced configurations which have a Primary station and one or more Secondary stations.

- Balanced configurations which have two Combined stations.

## 2.2.1 Unbalanced Configurations

An Unbalanced configuration has one Primary station and one or more Secondary stations connected to the link. The link may be point-to-point or multipoint, two-way alternate or two-way simultaneous, switched or non-switched. In the Unbalanced configuration the Primary station is responsible for setting each Secondary station in a logical state and mode as appropriate. See Section 6. Additionally, both Primary and Secondary stations are responsible for exchanging data and control information with each other, and initiating the link level error recovery functions defined in this standard. See Figure 2-1.

- 3 -

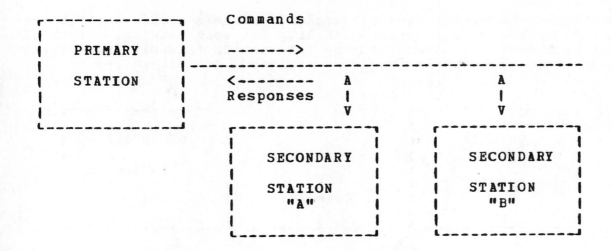

FIGURE 2-1 Unbalanced Configuraticn

## 2.2.2 Balanced Configuration

A Balanced configuration is twc Combined stations connected point-to-point, two-way alternate or twc-way simultaneous, switched or non-switched. Both Combined stations have identical data transfer and link control capability. See Figure 2-2.

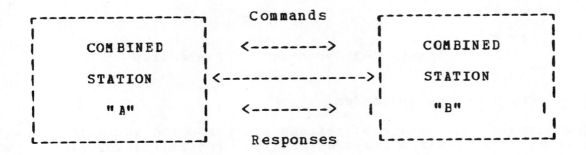

Figure 2-2 Balanced Configuraticn

## 2.2.3 Symmetric Configurations

Two independent point-to-point Unbalanced logical station configurations may be connected in a Symmetric manner and multiplexed on a single link. This configuration may be two-way alternate or two-way simultaneous, switched or non-switched. In this configuration there are two independent Primary

station-to-Secondary station  logical channels where  the Primary
stations have overall responsibility for mode setting.  Each of
the four stations maintains one information transmitting ability
and/or one information receiving ability.  See Figure 2-3.

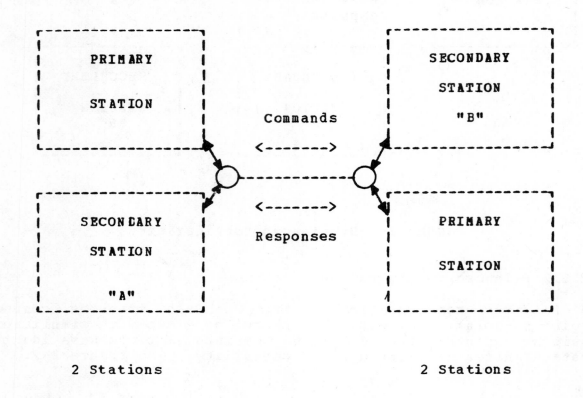

2 Stations                          2 Stations


Figure 2-3 Symmetric Configuration


## 2.3 Logical States and Modes

Communication between two stations is  conducted in three logical
states:   Information Transfer State,  Initialization State,  or
Logically Disconnected State.

### 2.3.1 Information Transfer State (ITS)

While  in  ITS  the Secondary/Combined  station  may transmit  and
receive  information.  Communications  shall  observe  the
constraints of a mode established in a Secondary/Combined station
by the  remote Primary/Combined station.  Each mode specifies a
respond opportunity and  a logical data link  configuration.  See
Section 6.2.

- 5 -

## 2.3.1.1 Normal Response Mode (NRM)

NRM is an Unbalanced configuration operational mode in which the Secondary station may initiate transmission of frames containing information only as the result of receiving explicit permission to do so from the Primary station. After receiving permission, the Secondary station shall initiate a response transmission. The response transmission may consist of one or more frames while maintaining an active channel state. The last frame of the response transmission will be explicitly indicated by the Secondary station. Following the indication of the last frame, the Secondary station will stop transmitting until explicit permission is again received from the Primary station.

## 2.3.1.2 Asynchronous Response Mode (ARM)

ARM is an Unbalanced configuration operational mode in which the Secondary station may initiate transmission without receiving explicit permission from the Primary station. Such an asynchronous transmission may contain single or multiple frames and is used for information field transfer and/or to indicate status changes in the Secondary station (e.g., the number of the next expected frame, transition from a ready to a busy condition or vice versa, occurrance of an exception condition).

## 2.3.1.3 Asynchronous Balanced Mode (ABM)

ABM is a Balanced configuration operational mode in which a Combined station may initiate transmission without receiving permission from the other Combined station. Such an asynchronous transmission may contain single or multiple frames for information transfer and/or to indicate status changes at the transmitting Combined station (e.g., the number of the next expected frame, transition from a ready to a busy condition or vice versa, occurrance of an exception condition).

## 2.3.2 Initialization State (IS)

While in IS communications shall observe the constraints of a system-defined procedure. The system-defined procedure may, for example, cause the Secondary/Combined station's link control to be initialized or regenerated by the remote Primary/Combined station. See Section 6.4.

## 2.3.3 Logically Disconnected State (LDS)

While in LDS the Secondary/Combined station is logically disconnected from the link and is not permitted to transmit or receive information. Communications shall observe the constraints of a mode selected by system definition; each mode specifies a respond opportunity.   See Section 6.3.

2.3.3.1 Normal Disconnected Mode (NDM)

NDM is an Unbalanced configuration  non-operational mode in which
the Secondary station is logically disconnected from the link and
is  not  permitted  to  initiate  or  receive  information.   The
Secondary station may initiate transmission only as the result of
receiving explicit permission to do  so from the Primary station.
After receiving permission, the  Secondary station shall initiate
a single frame transmission indicating its status.

2.3.3.2 Asynchronous Disconnected Mode (ADM)

ADM is  an Unbalanced  or Balanced  configuration non-operational
mode  in  which  the  Secondary/Combined  station  is  logically
disconnected from  the link and is  not permitted to  initiate or
receive information.  A station in  ADM may initiate transmission
without receiving  explicit permission from  the Primary/Combined
station but the  transmission shall be a  single frame indicating
the station status.

# 3.0 FRAME STRUCTURE

In ADCCP, all transmissions are in frames and each frame conforms
to the following structure:

        F, A, C, Info, FCS, F
        Where:
          F = Flag Sequence
          A = Address Field
          C = Control Field
          Info = Information Field
          FCS = Frame Check Sequence

Frames containing only data link control sequences form a special
case where there is no Info field.  The short frame structure is
therefore:

        F, A, C, FCS, F

The elements of the frame are described in the following
paragraphs.  See also Figure 3-1.

## 3.1 Flag Sequence (F)

All frames start and end with the flag sequence.  This sequence
is a zero bit followed by 6 one bits followed by a zero bit
(01111110).  All stations attached to the data link continuously
hunt, on a bit-by-bit basis, for this sequence.  A transmitter
must send only complete eight-bit flag sequences, however the
sequence of 011111101111110 at the receiver is two flag
sequences.  The flag is used for frame synchronization.

In order to achieve transparency the flag sequence is prohibited
from occurring in the Address, Control, Information and FCS
fields via a "zero bit insertion" procedure described in Section
3.7.

The flag sequence which closes a frame may also be the opening
flag sequence for the next frame.  Any number of complete flags
may be used between frames.  See also Appendix B3.1.

## 3.2 Address Field (A)

The Address field contains the link level address of a Secondary
station or a Combined station.  The length of this field (A) is N
octets (N greater than or equal to 1).  The encoding of this
field is described in Section 4.

## 3.3 Control Field (C)

The Control field contains a command or response and may contain sequence numbers. The control field is used by the transmitting Primary/Combined station to instruct the addressed Secondary/Combined station what operation it is to perform. It is also used by the Secondary station or Combined station to respond to the remote Primary station or Combined station. The length of this field (C) is one octet in the case of the basic control field. It is two octets in length in the case of the extended control field. See Section 5.2.2.

Sequence numbers and the formatting of the Control Field are described in Section 5. Commands and responses are functionally described in Section 7.

## 3.4 Information Field (Info)

The Information field may be any number and sequence of bits; the data link procedures are completely transparent. Data contained in the information field is unrestricted with respect to code or grouping of bits. See Appendix B3.4 for examples of typical limitations on maximum length.

## 3.5 Frame Check Sequence (FCS)

All frames include a 16-bit frame check sequence (FCS) just prior to the closing flag for error detection purposes. The contents of the Address, Control and Information fields, excluding the zeros inserted to maintain transparency per Section 3.7, are included in the calculation of the FCS sequence.

The FCS is the remainder of a modulo 2 division process utilizing a generator polynomial as a divisor. The generator polynomial is that used in CCITT Recommendation V.41 and is: $X^{16} + X^{12} + X^5 + 1$.

Section 12 gives a complete description of the FCS generation process and the error checking process which utilizes the FCS. Appendix D gives an example of FCS generation and error detection.

NOTE: If future applications show that a higher degree of protection is needed, the length of the FCS shall be increased by multiples of eight bits. This requires a higher degree generator polynomial the implementation and use of which is outside of this standard.

## 3.6 Abort

Abort is the procedure by which a station in the process of sending a frame, ends the frame in an unusual manner such that the receiving station will ignore the frame.

Aborting a frame is performed by transmitting at least seven, but less than fifteen, contiguous one bits (with no inserted zeros). Receipt of seven contiguous one bits is interpreted as an abort. Receipt of fifteen or more contiguous one bits is interpreted as an abort and Idle Link State. See Section 3.9.

## 3.7 Transparency

ADCCP provides transparency for data coded in the Information field. The occurrence of the flag sequence within a frame is prevented via a "zero bit insertion" procedure as follows:

> The transmitter inserts a zero bit following five contiguous one bits anywhere between the opening flag and the closing flag of the frame. The insertion of the zero bit thus applies to the contents of the Address, Control, Info and FCS fields (including the last 5 bits of the FCS). The receiver continously monitors the received bit stream; upon receiving a zero bit followed by five contiguous one bits, the receiver inspects the following bit: If a zero, the 5 one bits are passed as data and the zero bit deleted. If the sixth bit is a one, the receiver inspects the seventh bit; if it is zero, a flag sequence has been received; if it is a one an abort has been received. See Section 3.6.

## 3.8 Active Link State and Interframe Time Fill

A link is in an ACTIVE state when a Primary station, Secondary station, or Combined station is actively transmitting a frame, an abort sequence, or interframe time fill. When the link is in the active state, the right of the transmitting station to continue transmission is reserved.

Interframe time fill is accomplished by transmitting continuous flags between frames. There is no provision for intraframe time fill. See also Appendix B.

## 3.9 Idle Link State

A link is defined to be in an IDLE state when a continuous ones state is detected that persists for at least 15 bit times.

Idle link time fill is defined to be a continuous one condition on the link.

1007

## 3.10 Invalid Frame

An invalid frame is defined as one that is not properly bounded by two flags (thus an aborted frame is an invalid frame) or one which is too short (i.e., shorter than 32 bits between flags). A Secondary station or Combined station will ignore any invalid frame.

## 3.11 Order of Bit Transmission

Addresses, commands and responses, and sequence numbers are to be transmitted low-order bit first (e.g., the first bit of a sequence number that is transmitted carries the weight $2^0$).

The order of bit transmission for data contained within the information field is application-dependent and not specified.

The order of bit transmission for the FCS is most significant bit first. See Appendix D.

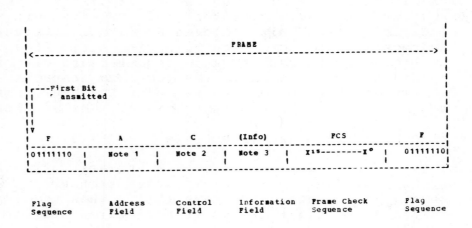

| Flag Sequence | Address Field | Control Field | Information Field | Frame Check Sequence | Flag Sequence |

Note 1. Address Field formats described in Sections 4.3.1 & 4.3.2.

Note 2. Control Field formats described in Section 5.2.

Note 3. Information Field size may be any number of bits.

STRUCTURAL RELATIONSHIP OF DEFINED FIELDS IN ADCCP FORMAT
------------------------------------------------------------
FIGURE 3-1

- 11 -

# 4.0 ADDRESS FIELD

## 4.1 Unbalanced Operation

A unique address is associated with every Secondary station on a link. Additionally, a Secondary station may be capable of accepting frames which utilize a group or global address; however, when such a Secondary station responds, it will utilize its unique address.

The address field in a command frame transmitted by a Primary station contains the address of the (remote) Secondary station. The address field in a response frame transmitted by a Secondary station contains the address of the (local) Secondary station.

## 4.2 Balanced Operation

A unique address is associated with each Combined station on the link. Additionally, a Combined station may be capable of accepting frames which utilize a group or global address; however, when such a Combined station responds it will utilize its unique address.

The address field in a command frame transmitted by a Combined station contains the address of the remote Combined station. The address field in a response frame transmitted by a Combined station contains the address of the local Combined station.

## 4.3 Address Encoding

Two addressing encoding formats are defined for the address field: Basic and Extended. These formats are mutually exclusive for any given Secondary station or Combined station on a link and, therefore, the addressing format must be explicitly specified.

### 4.3.1 Basic Address Format

In basic address format, the address field contains one address, which may be a single Secondary/Combined station address, or a group or global Secondary/Combined station address. This field consists of one octet with the following format:

```
Address Field
Bit Number          1  2  3  4  5  6  7  8

                 ┌─────────────────────────────┐
Address          │  │  │  │  │  │  │  │  │  │   │
                 └─────────────────────────────┘
                    ▲
Least Significant Bit ──┘
First Bit Transmitted ──┘
```

## 4.3.2 Extended Address Format

In extended format, the address field is a sequence of octets
which comprises one address which may be a single
Secondary/Combined station address, cr a group or global
Secondary/Combined station address. When the first bit of an
address octet is "0", the following octet is an extension of the
address field.  See Section 4.5 for exception.  The address field
is terminated by an octet having a "1" in bit position one.
Thus, the address field is recursively extendable.  The format of
the extended address field is:

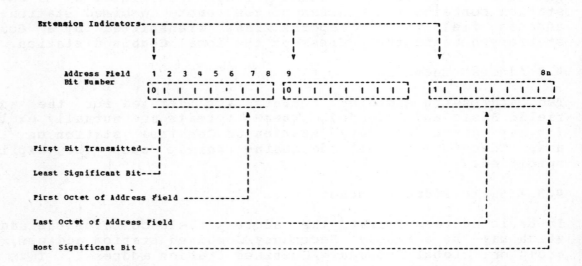

## 4.4 Global Address

The single octet address of eight "1" bits (11111111) is reserved as the global (universal, or broadcast) address in the basic and extended address formats.

The global address is used in situations where a specific Secondary/Combined station address is not known (e.g., switched connection) or is not necessary to the situation (e.g., broadcast transmission).

## 4.5 Null Address

When the first octet of the address field appears as eight "0" bits (00000000) the address is considered to be a null (no station) address and the frame will be ignored.

## 5.0 TRANSMISSION PARAMETERS AND FORMATS

For definitions of the following commonly used terms see Glossary, Appendix A.

- Accept/Acceptance
- Acknowledge
- Action
- Discard
- Invalid

- Implement
- Receive
- Respond Opportunity

### 5.1 Parameters

The various parameters associated with frames are described in the following sections. Figure 5-1 shows the position of parameters within a frame.

### 5.1.1 Modulus

Each Information (I) frame is sequentially numbered and may have the value 0 through modulus - 1 (where modulus is the modulus of the sequence numbers). Modulus equals 8 for the basic control field format and 128 for the extended control field format. The sequence numbers cycle through the entire range.

The maximum number of sequentially numbered I frames that a station may have outstanding (i.e., unacknowledged) at any given time may never exceed one less than the modulus of the sequence numbers. This restriction is to prevent any ambiguity in the association of transmitted I frames with sequence numbers during normal operation and/or error recovery action. The number of outstanding I frames may be further restricted by either the sending or receiving station storage capability; e.g., the number of I frames that can be stored for transmission and/or retransmission in the event of a transmission error.

### 5.1.2 Frame Variables and Sequence Numbers

Every Secondary station in an Information Transfer State maintains a Send Variable on the I frames it transmits to, and a Receive Variable on the I frames it correctly receives from, the Primary station. Every Primary station maintains an individual Send Variable on the I frames it transmits to, and an individual Receive Variable on the I frames it correctly receives from, each Secondary station in an Information Transfer State. Every Combined station in an Information Transfer State maintains one Send Variable on the I frames it transmits to, and one Receive Variable on the I frames it correctly receives from, the remote Combined station.

1012

## 5.1.2.1 Send Variable (S)

Each station capable of transmitting I frames has a Send Variable S which indicates the sequence number of the next I frame to be transmitted. S shall take on the value 0 through modulus-1. S is incremented by one with each completed I frame transmission (i.e., S will not be incremented when an I frame transmission is aborted).

## 5.1.2.2 Send Sequence Number (N(S))

Only I frames contain N(S), the sequence number of the transmitted frame. Prior to transmission cf an I frame, N(S) is set equal to S.

## 5.1.2.3 Receive Variable (R)

Each station capable of receiving I frames shall have a Receive Variable R equal to the expected N(S) contained in the next I frame received. R is incremented by one upon receipt of an error-free I frame whose N(S)=R.

## 5.1.2.4 Receive Sequence Number (N(R))

All I frames and Supervisory (S) frames contain an N(R), the expected sequence number of the next received I frame. Immediately before transmitting or retransmitting an I or S frame, N(R) is set equal to R. N(R) thus indicates that the station transmitting the N(R) has correctly received all I frames numbered up to and including N(R)-1.

## 5.1.3 Poll/Final (P/F) Bit

Poll (P) and Final (F) bits are used for:

1. indicating when a Secondary station may begin and has finished a response transmission under NRM. See Section 6.2.1.

2. checkpointing to determine if error recovery is required. See Section 6.5.2.

3. obtaining a response from a Secondary/Combined station. See Section 6.1.

The P bit set to 1 is used by the Primary/Combined station in command frames to solicit (poll) a response or sequence of response frames from a Secondary station(s) or a Combined station.

The F bit is used by a Secondary station to:

1. indicate in ARM the response frame sent in reply to the receipt of a poll command.

2. in NRM to indicate the final frame transmitted as the result of a previous poll command.

The F bit is used by a Combined station to indicate in ABM the response frame sent in reply to the receipt of a poll command.

See Section 6.1 for further description of the P/F bit operation.

## 5.2 Control Field Formats

The three formats defined for the Control field are used to perform information transfer, basic supervisory control functions, and special or infrequent control functions.

## 5.2.1 Basic Control Field

The basic control field accommodates modulo 8 $N(S)$ and $N(R)$ sequence numbering.

```
                                    <————————Control Field—————————>
First Bit Transmitted      ———————————————+
                                           V

Control Field Bits               1     2     3     4     5     6     7     8
                                 -     -     -     -     -     -     -     -
                            +——————————————————————————————————————————————————+
Information Transfer Format  | 0 |   N(S)     |  P/F  |      N(R)               |
                            |   |            |       |                         |
                            |——————————————————————————————————————————————————|
Supervisory Format          | 1   0 |  S   S|  P/F  |      N(R)               |
                            |       |       |       |                         |
                            |——————————————————————————————————————————————————|
Unnumbered Format           | 1   1 |  M   M|  P/F  | M     M     M           |
                            |       |       |       |                         |
                            +——————————————————————————————————————————————————+
```

Where: $N(S)$ = Transmitting station send sequence number.
(Bit 2 = low order bit)
$N(R)$ = Transmitting station receive sequence number.
(Bit 6 = low order bit)
S = Supervisory function bits
M = Modifier function bits
P/F = Poll bit - Primary/Combined station command frame transmissions.

- 17 -

Final bit - Secondary/Combined station
response frame transmissions.
(1 = Poll/Final)

## 5.2.2 Extended Control Field

The extended control field accommodates module 128 N(S) and N(R) sequence numbering. On long propogation delay links (e.g., satellite transmission) it is desirable for reasons of efficiency to extend the modulus of the sequence numbers N(S) and N(R). The method of extension is defined below.

Control field extension for the three formats is as follows:

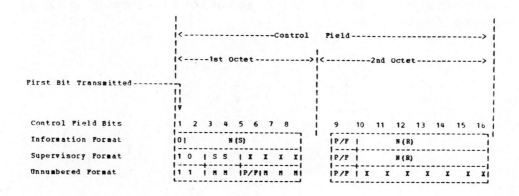

where X bits are reserved and set to "0".

In extended control field format the transmitter sets the P/F bits in bit positions 5 and 9 for Unnumbered format commands and responses. A receiver in extended control field format interprets the P/F bit in bit position 9. A receiver in basic control field format receiving an extended control field format interprets the P/F bit in bit position 5.

## 5.3 Information Transfer Format (I)

The I format is used to perform an information transfer.

The functions of N(S), N(R), and P/F are independent; i.e., each

I frame has an N(S) sequence number, the N(R) sequence number may or may not acknowledge additional I frames at the receiving station, and the P/F bit may or may not be set to "1".

## 5.4 Supervisory Format (S)

The S format is used to perform link supervisory control functions such as acknowledge I frames, request retransmission of I frames, and indicate temporary interruption of capability to receive I/UI frames. The functions of N(R) and P/F are independent.

## 5.5 Unnumbered Format (U)

The U format is used to provide additional link control functions. This format contains no sequence numbers and consequently five "modifier" bit positions are available which allow definition of up to 32 additional command and 32 additional response functions.

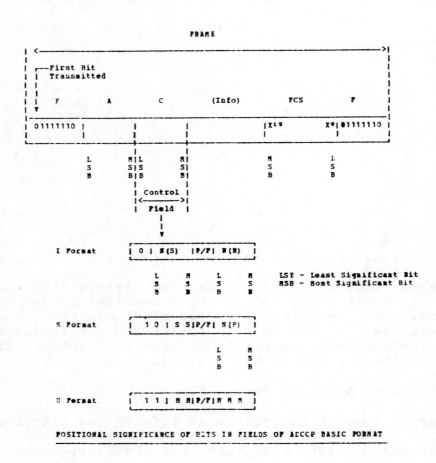

POSITIONAL SIGNIFICANCE OF BITS IN FIELDS OF ADCCP BASIC FORMAT

Figure 5-1

# 6.0 SECONDARY/COMBINED STATION STATES AND MODES

A Secondary/Combined station transmits response frames to a Primary/Combined station based on previous receipt of a command frame. In certain cases, a Secondary/Combined station can also initiate transmission of response frames to a Primary/Combined station. The characteristics of a Secondary/Combined station response are determined by: 1) the type of respond opportunity which exists at the Secondary/Combined station, 2) the current state of the Secondary/Combined station, and 3) the particular mode within the state of the Secondary/Combined station. Secondary/Combined stations do not queue sequential responses for command frames received. A Secondary/Combined station response is predicated on: 1) station status at the time the response is transmitted, 2) an exception condition previously established, or 3) the previous receipt of a command which requires a specific response format.

## 6.1 Poll/Final Bit Usages

The Poll/Final (P/F) bit serves a function in both command frames and response frames. In command frames the P/F bit is referred to as the P bit. In response frames it is referred to as the F bit.

The P bit is used to solicit a response frame with the F bit set to "1" from the Secondary/Combined station at the earliest opportunity. A response frame with the F bit set to "1" also indicates the end of transmission under Normal Respond Opportunity.

For each Primary-Secondary pair on Unbalanced links and each direction on Balanced links only one frame with a P bit set to "1" may be outstanding at a given time. Before a Primary/Combined station can issue another frame with P bit set to "1" it must receive a response frame from the Secondary/Combined station with the F bit set to "1". If no valid response frame is obtained within a system defined time-out, the retransmission of a command with the P bit set to "1" for error recovery purposes is permitted.

## 6.2 Respond Opportunities

## 6.2.1 Normal Respond Opportunity (NRO)

NRO is a Secondary station respond opportunity in which the Secondary station initiates transmission of response frames only as the result of receiving a command frame with the P bit set to "1" or a UP command. See Section 7.4.2.2.

The response transmission may consist of one or more frames while maintaining an Active Link State. In all cases the last frame of the response transmission will have the F bit set to "1". When the response frame with the F bit set to "1" is transmitted the Secondary station will stop transmitting response frames and not initiate any additional transmission of response frames until a subsequent command frame is received with the P bit set to "1" or a UP command is received.

## 6.2.2 Asynchronous Respond Opportunity (ARO)

ARO is a Secondary/Combined station respond opportunity in which the Secondary/Combined station initiates transmission of response frames without regard to the receipt of a command frame with the P bit set to "1". Asynchronous transmission of response frames may be initiated at the first opportunity. In two-way simultaneous (TWS) transmission the opportunity is immediately. In two-way alternate (TWA) transmission the opportunity is the detection of an Idle Link State. An asynchronous transmission may contain multiple frames and is used to initiate information transfer (I/UI) and/or to report status changes in the Secondary/Combined station (e.g., N(R) number change, transition from a ready to a busy condition or vice versa, establishment of an exception condition).

The Secondary/Combined station must transmit a frame with the F bit set to "1" only in response to a received command frame with the P bit set to "1". The F bit is not to be interpreted as the end of transmission by the Secondary/Combined station. Additional response frames with the F bit set to "0" may be transmitted following the response frame which had the F bit set to "1".

In TWS operation a Secondary/Combined station in the process of transmitting when the command frame with the P bit set to "1" is received will set the F bit to "1" in the earliest possible response frame to be transmitted.

When a station has Asynchronous Respond Opportunity, it shall utilize a response time-out function which will cause initiation of appropriate recovery procedures if previously transmitted unsolicited response frames have not been acknowledged within a system-defined time-out period. Since simultaneous contention may occur, in TWA configurations the response timers at each end of the link shall be unequal. In TWA, the interval employed by a Secondary station shall be greater than that employed by the Primary station to permit contention situations to be resolved in favor of the Primary station.

1018

## 6.3 Logically Disconnected State (LDS)

The LDS is provided to prevent a Secondary/Combined station from appearing on the link in a fully operational sense during unusual situations or exception conditions since such operations could cause 1) unintended contention, 2) sequence number mismatch or 3) ambiguity as to the Secondary/Combined station status or mode.

While in LDS the Secondary station, or response capability of a Combined station, is logically disconnected from the data link; i.e., no Information (I), Unnumbered Information (UI) or 8 response frames are transmitted or accepted. The Secondary station capability, or response capability of a Combined station, is limited to 1) accepting one of the mode-setting commands, 2) transmitting a DM or RIM response frame at each respond opportunity and 3) responding to an XID command.

A Secondary/Combined station in LDS, as a minimum capability, must respond DM (Disconnected Mode) to any valid command frame received with the P bit set to "1". A RIM (Request Initialization Mode) response may be transmitted instead of DM; the conditions which cause a Secondary/Combined station to transmit RIM are system defined.

A mode-setting, XID or UP command may be accepted and responded to at the first respond opportunity if the station is capable of accepting and actioning the command. Any other frame received, including a mode setting, XID or UP command that is not accepted, is discarded except for the response requirement described in the paragraph above.

A Secondary/Combined station is system defined as to the condition(s) that cause it to assume one of the two predetermined modes (ADM or NDM).

Examples of possible system-defined conditiors (in addition to receiving a DISC command) which may cause a Secondary/Combined station to enter LDS are:

1. the Secondary/Combined station power is turned on

2. the Secondary/Combined station has a temporary loss of power

3. the Secondary/Combined station link level logic is manually reset

4. the Secondary/Combined station is manually switched from a local (home) condition to a connected-to-the-link condition.

- 22 -

While in LDS a Secondary/Combined station may not establish a Frame Reject exception condition.

### 6.3.1 Modes Within LDS

While in LDS a Secondary/Combined station communicates under the constraints of one of the following two modes.

### 6.3.1.1 Normal Disconnected Mode (NDM)

NDM is a disconnected mode in which the Secondary station is logically disconnected from the data link and follows Normal Respond Opportunity protocol. See Section 6.2.1.

### 6.3.1.2 Asynchronous Disconnected Mode (ADM)

ADM is a disconnected mode in which the Secondary station, or response capability of a Combined station, is logically disconnected from the data link and follows Asynchronous Respond Opportunity protocol. See Section 6.2.2.

### 6.4 Initialization State (IS)

While in IS the Secondary/Combined station may be initialized or regenerated by the remote Primary/Combined station and communicates under the constraints of the following mode.

### 6.4.1 Initialization Mode (IM)

A Secondary/Combined station enters the IM upon sending a UA response, under a system-defined respond opportunity, in reply to the receipt of a Set Initialization Mode (SIM) command. The Secondary/Combined station may request SIM by sending a Request Initialization Mode (RIM) response. While in IM the stations may exchange information in any manner specified for that Secondary/Combined station (e.g., unformatted and unchecked bit streams, UI frames, or I frames); however, in a multipoint configuration care shall be taken to prevent interference with other stations on the link. IM is ended when the Secondary/Combined station receives, actions, and acknowledges a set-mode command (i.e., SNRM, SARM, SABM, SNRME, SARME, SABME, or DISC).

### 6.5 Information Transfer State (ITS)

While in ITS a station is fully operational and capable of transmitting and receiving I, S, and U format frames.

- 23 -

### 6.5.1 Modes Within ITS

While in ITS a Secondary/Combined station communicates under the constraints of one of the following modes. The particular mode utilized is determined by the Primary/Combined station with an appropriate mode-setting command and is entered when the Secondary/Combined station receives, actions, and acknowledges that mode-setting command.

### 6.5.1.1 Normal Response Mode (NRM)

NRM is a Secondary station information transfer mode in which the Secondary station utilizes Normal Respond Opportunity on an unbalanced link configuration. See Sections 2.2.1 and 6.2.1. This mode is selected by a SNRM or SNRME command.

### 6.5.1.2 Asynchronous Response Mode (ARM)

ARM is a Secondary station information transfer mode in which the Secondary station utilizes Asynchronous Respond Opportunity on an unbalanced link configuration. See Sections 2.2.1 and 6.2.2. This mode is selected by a SABM or SABME command.

### 6.5.1.3 Asynchronous Balanced Mode (ABM)

ABM is a Combined station information transfer mode in which the Combined stations utilize Asynchronous Respond Opportunity on a balanced link configuration. See Sections 2.2.2 and 6.2.2. This mode is selected by a SABM or SABME command. A Combined station may transmit command frames at any time; therefore, the ABM definition only describes and applies to the response frame transmitting and command frame receiving capability of the Combined stations.

### 6.5.2 Checkpointing

As the P and F bits are always exchanged as a pair (for every P there is one F, and the next P must not be issued until the previous P has been matched with an F or until the response timer expires), the N(R) contained in a frame with a P or F bit set to "1" can be used to detect I frame sequence errors. This capability can provide early detection of frame sequence errors and indicate the I frame sequence number to begin retransmission. This capability is referred to as checkpointing.

While in ITS the Primary/Combined station shall examine the N(R) contained in any received I or S frame with the F bit set to "1". Appropriate error recovery procedures shall be initiated if this N(R) does not acknowledge all I frames transmitted by the Primary/Combined station prior to and including the last command frame sent with the P bit set to "1". See Section 8.2.1 for

- 24 -

additional qualifying conditions.

Similarly, while in ITS the Secondary station shall examine the N(R) contained in any received I or S frame with the P bit set to "1". Appropriate error recovery procedure shall be initiated if this N(R) does not acknowledge all I frames transmitted by the Secondary station prior to and including the last response frame with the P bit set to "1". See Section 8.2.1 for additional qualifying conditions.

In all cases the N(R) of a correctly received I or S frame shall confirm previously transmitted I frames through N(R)-1.

# 7.0 COMMANDS AND RESPONSES

This standard defines the link control operation in terms of the actions and internal modes of the Secondary/Combined station. The actual link management procedure (i.e., sequence of commands and related responses) is application and link configuration dependent. Consequently, specific Primary/Combined station command sequences are not defined but left to the designer of the Primary/Combined station link control.

Sections 7.1 through 7.3 contain the definition of the set of commands and responses (listed below) for each of the transmission formats.

| Information Transfer Format Commands | Information Transfer Format |
|---|---|
| I - Information | I - Information    Response |

| Supervisory Format Commands | Supervisory Format Responses |
|---|---|
| RR   - Receive Ready | RR - Receive Ready |
| RNR  - Receive Not Ready | RNR - Receive Not Ready |
| REJ  - Reject | REJ - Reject |
| SREJ - Selective Reject | SREJ - Selective Reject |

1023

## Unnumbered Format Commands

### Mode-Setting Commands

SNRM  - Set Normal Response Mode
SARM  - Set Asynchronous Response Mode
SABM  - Set Asynchronous Balanced Mode
SNRME - Set Normal Response Mode
        Extended
SARME - Set Asynchronous
        Response Mode Extended
SABME - Set Asynchronous Balanced Mode
        Extended
SIM   - Set Initialization Mode
DISC  - Disconnect

### Information Transfer Commands

UI - Unnumbered Information
UP - Unnumbered Poll

### Recovery Commands

RSET - Reset

### Miscellaneous Commands

XID - Exchange Identification

### Non-Reserved Commands

4 Encodings

## Unnumbered Format Responses

### Mode-Setting Responses

UA - Unnumbered Acknowledgement
DM - Disconnected Mode

RIM - Request Initialization
      Mode

### Information Transfer Responses

UI - Unnumbered Information

### Recovery Responses

FRMR - Frame Reject

### Miscellaneous Responses

XID - Exchange Identification

RD - Request Disconnect

### Non-Reserved Responses

4 Encodings

## 7.1 Information Transfer Format (I) Command/Response

The function of the Information (I) command/response is to efficiently transfer sequentially numbered frames containing an optional information field.

1024

The encoding of the I command/response control field is:

I Format Command/Response
<---------Control Field--------->

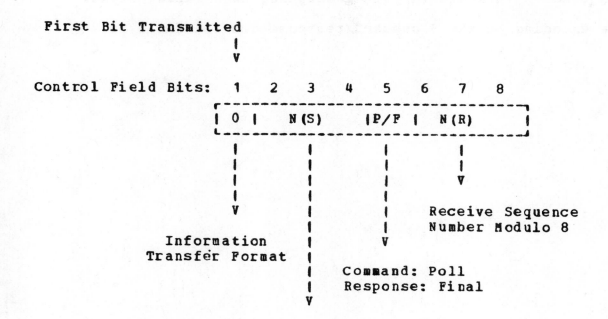

First Bit Transmitted
|
V

Control Field Bits:  1  2  3  4  5  6  7  8

| 0 |   N(S)   | P/F |  N(R)  |

Receive Sequence
Number Modulo 8

Information
Transfer Format

Command: Poll
Response: Final

Send Sequence
Number Modulo 8

For extended control field format see Section 5.2.2.

The I frame control field contains two sequence numbers: N(S), Send Sequence Number, which indicates the sequence number associated with the I frame; N(R) Receive Sequence Number, which indicates the sequence number of the next expected I frame (i.e., I frames numbered up to and including N(R)-1 are accepted.)

An I frame with P/F bit set to "1" may report the end of a station busy condition as specified in Section 8.1.3.

See Sections 6.1, 6.5.2 and 8.2.1 for description of P/F bit operation.

## 7.2 Supervisory Format (S) Commands/Responses

Supervisory (S) commands/responses are used to perform basic
supervisory link control functions such as I frame
acknowledgement, polling, temporary interruption of information
(I/UI) transfer, and error recovery.

Frames with the S format do not contain an information field.
Therefore, a station does not increment its Send Variable (S)
upon the transmission of an S format frame nor does it increment
its Receive Variable (R) upon accepting an S format frame.

The encoding of the S command/response control field is:

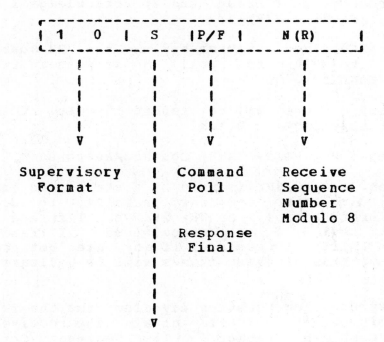

Supervisory Format
|<----Control Field--------------->|

First Bit Transmitted

Control Field Bits:   1   2   3   4   5   6   7   8

|   1    0   |   S   |P/F |   N(R)   |

Supervisory        Command          Receive
Format             Poll             Sequence
                                    Number
                                    Modulo 8
                   Response
                   Final

1027

Commands                          Responses

RR - Receive Reasy          00   RR - Receive Ready
RNR - Receive Not Ready     10   RNR - Receive Not Ready
REJ - Reject                01   REJ - Reject
SREJ - Selective Reject     11   SREJ - Selective Reject

For extended control field format see Section 5.2.2.

An S frame contains an N(R), Receive Sequence Number, which indicates the sequence number of the next expected I frame (i.e., all received I frames numbered up to and including N(R)-1 are accepted). See Sections 6.1, 6.5.2 and 8.2.1 for description of the P/F bit operation.

### 7.2.1 Receive Ready (RR) Command/Response

Receive Ready (RR) is used by a station to: 1) indicate it is ready to receive an I frame and 2) acknowledge I frames numbered up to and including N(R)-1.

The Primary/Combined station may use the RR command with the P bit set to "1" to solicit responses from (poll) a Secondary/Combined.

An RR frame is one way to report the end of a station busy condition. See Section 8.1.3.

### 7.2.2 Receive Not Ready (RNR) Command/Response

Receive Not Ready (RNR) is used by a station to indicate a "busy" condition; i.e. the temporary inability to accept additional incoming information (I or UI) frames. I frames numbered up to and including N(R)-1 are acknowledged. I frame N(R) and any subsequent I frames received, if any, are not acknowledged; the acceptance status of these frames will be indicated in subsequent exchanges.

The Primary/Combined station may also use the RNR command with the P bit set to "1" to obtain the receive status of a Secondary/Combined station. The Secondary/Combined station response will be a frame with the F bit set to "1". See Section 8.1, Busy Condition, for further details on RNR usage.

### 7.2.3 Reject (REJ) Command/Response

Reject (REJ) is used by a station to request retransmission of I frames starting with the frame numbered N(R). I frames numbered N(R)-1 and below are acknowledged. Additional I frames pending initial transmission may be transmitted following the retransmitted I frame(s).

Only one REJ exception condition, from a given station to another station, may be established at any given time; another REJ or SREJ may not be transmitted (i.e., actioned) until the first REJ exception condition has been cleared at the sender.

1028

The REJ exception condition is cleared (reset) upon acceptance of an I frame with an N(S) number equal to the N(R) of the REJ command/response or after a timeout has occurred.

An REJ is one way to report the end of a station busy condition. See Section 8.1.3.

See Section 8.2 for sequence error recovery protocols.

7.2.4 Selective Reject (SREJ) Command/Response

Selective Reject (SREJ) is used by a station to request retransmission of the single I frame numbered N(R). I frames up to and including N(R)-1 are acknowledged.

The SREJ exception condition is cleared (reset) upon acceptance of an I frame with an N(S) number equal to the N(R) of the SREJ command/response.

After a station transmits a SREJ it may not transmit SREJ (except for a SREJ with P or F bit set to "1" and with N(R) equal to the N(R) of the first SREJ; see Section 8.2.3) or REJ for an additional sequence error until the first SREJ error condition has been cleared or a                 timeout has occurred. (To do so would acknowledge as correctly received all I frames up to and including N(R)-1, where N(R) is the sequence number in the second SREJ or REJ.)

I frames that may have been transmitted following the I frame indicated by the SREJ command/response are not retransmitted as the result of receiving an SREJ. Additional I frames pending initial transmission may be transmitted following the retransmission of the specific I frame requested by the SREJ.

An SREJ is one way to report the end of a station busy condition. See Section 8.1.3.

See Section 8.2 for sequence error recovery protocols.

7.3 Unnumbered Format (U) Commands/Responses

Unnumbered (U) commands and responses are used to extend the number of link supervisory functions. U frames do not increment the Send Variable (S) at the transmitting station or increment the Receive Variable (R) at the receiving station. Five "modifier" bits are defined which allow up to 32 additional command functions and 32 additional response functions.

The encoding of the U command/response control field is:

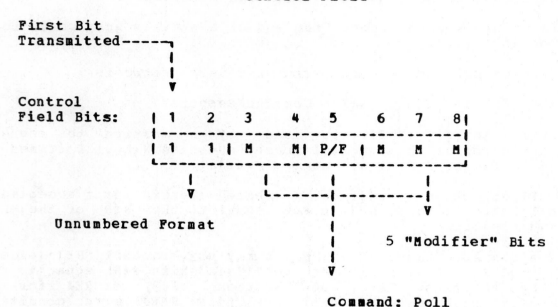

```
                      |                                    |
                      |<------Unnumbered Format------>|
                              Control Field

       First Bit
       Transmitted-----
                       |
                       |
                       V
       Control
       Field Bits:    | 1    2    3    4    5    6    7    8|
                      |---------------------------------------|
                      | 1    1 | M     M| P/F | M    M    M|
                      |---------------------------------------|
                         |           |      |           |
                         V           |      |           |
                                     |------|-----------|
             Unnumbered Format              |           |
                                            |           V
                                            |      5 "Modifier" Bits
                                            |
                                            V

                                    Command: Poll
                                    Response: Final
```

For extended control field format see Section 5.2.2.

See Sections 6.1, 6.5.2 and 8.2.1  for description of the P/F bit operation.


7.4 Unnumbered Format Commands

Unnumbered format commands are grouped  according to the function performed:

- Mode-setting commands:       SNRM, SARM, SABM, SNRME, SARME,
                                    SABME, SIM, DISC

- Information transfer commands: UI, UP

- Recovery commands: RSET

- Miscellaneous commands: XID

- Non-reserved commands: 4 encodings

The following U  format commands are defined;  other commands may be defined  in the  future if  required.  All  bit encodings  not defined are reserved for future standard assignment.

**First Bit Transmitted**

<pre>
             Control Field Bits
    1    2    3    4    5    6    7    8

    1    1    0    0    P    0    0    1    SNRM Command
    1    1    1    1    P    0    0    0    SARM Command
    1    1    1    1    P    1    0    0    SABM Command
    1    1    1    1    P    0    1    1    SNRME Command
    1    1    1    1    P    0    1    0    SARME Command
    1    1    1    1    P    1    1    0    SABME Command
    1    1    1    0    P    0    0    0    SIM Command
    1    1    0    0    P    0    1    0    DISC Command
    1    1    0    0    P    0    0    0    UI Command
    1    1    0    0    P    1    0    0    UP Command
    1    1    1    1    P    0    0    1    RSET Command
    1    1    1    1    P    1    0    1    XID Command
    1    1    0    1    P    0    0    0    Non-Reserved Command
    1    1    0    1    P    0    0    1    Non-Reserved Command
    1    1    0    1    P    0    1    0    Non-Reserved Command
    1    1    0    1    P    0    1    1    Non-Reserved Command
</pre>

For extended control field format see Section 5.2.2.

See Sections 6.1, 6.5.2 and 8.2.1 for description of the P bit operation.

The mode-setting commands, RSET and the XID command form a set of commands which require a specific response from a Secondary/Combined station. The response to a command of this set takes precedence over other responses which may be pending. If more than one command of this set is received prior to a respond opportunity, a single response is transmitted that is referenced to the first such command received; any additional commands of the set are monitored only to detect the next respond opportunity.

NOTE: It is recommended that the Primary/Secondary station transmitting one of the commands in this set provide a respond opportunity for the remote station with each transmitted command, e.g. issue the command with the P bit set to "1".

In the case of TWA operation, following the receipt of one of these U commands, a Secondary/Combined station is restricted to transmitting a single response frame. In the case of TWS operation a Secondary/Combined station which is transmitting concurrent to the receipt of one of these U commands will initiate transmission of a single response frame at the first

- 34 -

respond opportunity. The Secondary/Combined station may continue transmission following return of the response as appropriate to its respond opportunity.

## 7.4.1 Mode-Setting Commands

Mode-setting commands are transmitted by the Primary/Combined station to reset or change the mode of the addressed Secondary/Combined station. Once established a mode remains in effect at a Secondary station until the next mode-setting command is accepted, and at a Combined station until the next mode-setting command is either accepted, or transmitted and acknowledged.

The SNRM, SARM, SABM, SNRME, SARME, SABME, SIM, and DISC commands require the Secondary/Combined station to acknowledge acceptance by responding with a single Unnumbered Acknowledgement (UA) frame at the first respond opportunity. The UA has the F bit set to "1" if the mode-setting command had the P bit set to "1". If other I,S, or U format commands are received following a mode-setting command and prior to a respond opportunity, they are monitored only to determine the respond opportunity.

In the case of the operational mode-setting command (SARM, SNRM, SABM, SARME, SNRME, SABME) the respond opportunity at the Secondary station is determined by the command received (i.e., the mode to which the Secondary/Combined station is directed dictates when the response is transmitted). Unless a response to XID or RSET is pending, a Secondary/Combined station responds as described below to the receipt of a mode-setting command.

1. Upon receipt of a SNRM or SNRME command with the P bit set to "1", the Secondary station responds with a single UA frame with the F bit set to "1"; if the SNRM or SNRME P bit is set to "0" the Secondary station waits until a command frame with the P bit set to "1" is received and then responds with a single UA frame with the F bit set to "1", or until a UP command (with the P bit set to "0") is received and then responds with a single UA frame with the F bit set to "0".

2. Upon receipt of a SARM or SARME command, with or without the P bit set to "1", the Secondary station will transmit a single UA frame:

      a. upon detection of an Idle Link State in TWA operation, or

      b. at the earliest respond opportunity in TWS operation.

The UA frame will have the F bit set to "1" if the command has the P bit set to "1"

3.  Upon receipt of a SABM or SABME command, with or without the P bit set to "1", the Combined station will transmit a single UA frame:

    a. upon detection of an Idle Link state in TWA operation, or

    b. at the earliest respond opportunity in TWS operation.

    The UA frame will have the F bit set to "1" if the command has the P bit set to "1".

In the case of the non-operational mode-setting commands (SIM or DISC) the Secondary/Combined station will respond with a single UA frame at its system-defined respond opportunity; i.e., a given Secondary/Combined station is system defined to always use the normal respond opportunity or the asynchronous respond opportunity for the UA response.

If the Secondary/Combined station can not accept a mode-setting command, it will, at its first respond opportunity, transmit one of the responses, DM, FRMR, RD or RIM, as appropriate, indicating non-acceptance of the command.

NOTE: The protocol defined here requires that the Primary/Combined station restrict the transmission of U commands which require UA responses so that only one such command is outstanding (not acknowledged) to any given Secondary/Combined station at any given time. This eliminates the requirement for the Secondary/Combined station to queue responses and prevents any ambiguity relative to the meaning of the UA response.

7.4.1.1 Set Normal Response Mode (SNRM) Command

The SNRM command is used to place the addressed Secondary station in NRM where all control fields are one octet in length. No information field is permitted with the SNRM command.

Upon acceptance of this command the Secondary station Send and Receive Variables are set to zero. The Secondary station confirms acceptance of SNRM by transmission of a UA in the unextended control field format.

Previously transmitted I frames that are unacknowledged when this command is actioned remain unacknowledged. Transmission of SNRM is one way to report the end of a Primary station busy condition.

See Section 8.1.3.

## 7.4.1.2 Set Asynchronous Response Mode (SARM) Command

The SARM command is used to place the addressed Secondary station in ARM where all control fields are one octet in length. No information field is permitted with the SARM command.

Upon acceptance of this command the Secondary station Send and Receive Variables are set to zero. The Secondary station confirms acceptance of SARM by the transmission of a UA response in the unextended control field format.

Previously transmitted I frames that are unacknowledged when this command is actioned remain unacknowledged. Transmission of SARM is one way to report the end of a Primary station busy condition. See Section 8.1.3.

## 7.4.1.3 Set Asynchronous Balanced Mode (SABM) Command

The SABM command is used to place the addressed Combined station in ABM where all control fields are one octet in length. No information field is permitted with the SABM command.

Upon acceptance of this command the Combined station Send and Receive Variables are set to zero. The Combined station confirms acceptance of SABM by the transmission of a UA response in the unextended control field format.

Previously transmitted I frames that are unacknowledged when this command is actioned remain unacknowledged. Transmission of SABM is one way to report the end of a Combined station busy condition. See Section 8.1.3.

## 7.4.1.4 Set Normal Response Mode Extended (SNRME) Command

The SNRME command is used to place the addressed Secondary station in NRM where all control fields will be two octets in length as defined in Section 5.2.2. No information field is permitted with the SNRME command.

Upon acceptance of this command the Secondary station Send and Receive Variables are set to zero. The Secondary station confirms acceptance of SNRME by transmission of a UA response in the extended control field format.

Previously transmitted I frames that are unacknowledged when this command is actioned remain unacknowledged. Transmission of SARME is one way to report the end of a Primary busy condition. See Section 8.1.3.

## 7.4.1.5 Set Asynchronous Response Mode Extended (SARME) Command

The SARME command is used to place the addressed Secondary station in ARM where all control fields will be two octets in length as defined in Section 5.2.2. No information field is permitted with the SARME command.

Upon acceptance of this command the Secondary station Send and Receive Variables are set to zero. The Secondary station confirms acceptance of SARME by transmission of a UA response in the extended control field format.

Previously transmitted I frames that are unacknowledged when this command is actioned remain unacknowledged. Transmission of SARME is one way to report the end of a Primary busy condition. See Section 8.1.3.

## 7.4.1.6 Set Asynchronous Balanced Mode Extended (SABME) Command

The SABME command is used to place the addressed Combined station in ABM where all control fields will be two octets in length as defined in Section 5.2.2. No information field is permitted with the SABME command.

Upon acceptance of this command the Combined station Send and Receive Variables are set to zero. The Combined station confirms acceptance of SABME by transmission of a UA response in the extended control field format.

Previously transmitted I frames that are unacknowledged when this command is actioned remain unacknowledged. Transmission of SABME is one way to report the end of a Combined station busy condition. See Section 8.1.3.

## 7.4.1.7 Set Initialization Mode (SIM) Command

The SIM command is used to cause the addressed Secondary/Combined station to initiate a station-specified procedure(s) to initialize its link level control functions (e.g., accept a new program or update operational parameters). No information field is permitted with the SIM command.

The Secondary/Combined station confirms acceptance of SIM by transmission of a UA response. The respond opportunity and the control field format of the UA response are system defined.

Previously transmitted I frames that are unacknowledged when this command is actioned remain unacknowledged.

- 38 -

### 7.4.1.8 Disconnect (DISC) Command

The DISC command is used to perform a logical disconnect; i.e., inform the addressed Secondary/Combined station that the transmitting Primary/Combined station is suspending operation with that Secondary/Combined station. In switched networks, this logical disconnect function at the data link level may serve to initiate a physical disconnect operation at the physical interface level; i.e. to go "on-hook" No information field is permitted with the DISC command.

The Secondary/Combined station confirms acceptance of DISC by the transmission of a UA response. The respond opportunity and the control field format of the UA response is system defined. A Secondary/Combined station in ADM or NDM will transmit a DM response upon receiving a DISC command. A RIM response may be transmitted instead of DM under the System defined conditions described in Section 6.3. The respond opportunity and control field format after receipt of DISC is system defined for any given Secondary station. The respond opportunities are defined in Section 6.2.

Previously transmitted I frames that are unacknowledged when this command is actioned remain unacknowledged.

### 7.4.2 Unnumbered Information Transfer Commands

Unnumbered information transfer commands are used to exchange frames containing information.

### 7.4.2.1 Unnumbered Information (UI) Command

The UI command is used to transfer an information field to a Secondary/Combined station or group of Secondary stations without impacting the Send and Receive Variables. The information field is optionally present with the UI command. Reception of the UI frame is not sequence number verified; therefore, the frame may be lost if a link exception occurs during transmission of the UI, or duplicated if an exception occurs during any reply to the UI. Examples of UI frame information are higher level status, operation interruption, temporal data (e.g., time-of-day), or link initialization parameters.

See Appendix B, 7.4.2.1 for additional explanatory information.

### 7.4.2.2 Unnumbered Poll (UP) Command

The UP command is used to solicit response frames from a single Secondary/Combined station (Individual Poll) or from a group of

Secondary stations (Group Poll), by establishing a logical operational condition that exists at each addressed station for one respond opportunity. (In the case of a Group Poll, the mechanism employed to control (schedule) the response transmissions (to avoid simultaneous transmissions) is considered to exist and is not defined in this standard.) Secondary stations receiving UP with a group address will respond in the same manner as when addressed using an individual address. The response frame(s) will contain the sending Secondary/Combined station individual address, plus N(S) and N(R) numbers as required by the particular responses. (The continuity of each Secondary/Combined station N(S) will be maintained.) The UP command does not acknowledge receipt of any response frames that may have been previously transmitted by the Secondary/Combined station. No information field is permitted with the UP command.

A Secondary/Combined station which receives a UP with the P bit set to "1" will respond (at its respond opportunity and consistent with its mode of operation) with a frame which has the F bit set to "1".

A Secondary/Combined station which receives a UP with the P bit set to "0" may or may not respond; responses will have the F bit set to "0" in all response frames. A Secondary/Combined station will respond to a received UP which has the P bit set to "0" when 1) it has an I/UI frame(s) to send, 2) it has accepted but not acknowledged an I frame(s), 3) it has experienced an exception condition or change of status that has not been reported, 4) it has a status to be reported (e.g., DM, FRMR, or optionally an appropriate frame to report a no traffic condition).

## 7.4.3 Unnumbered Recovery Commands

### 7.4.3.1 Reset (RSET) Command

The RSET command is transmitted by a Combined station to reset the Receive State Variable (R) and applicable FRMR conditions in the addressed Combined station. No information field is permitted with the RSET command.

Upon acceptance of this command the station Receive State Variable (R) is set to zero. The Combined station confirms acceptance of RSET by transmission of the UA response while remaining in the previously established operational mode. If the UA is received correctly, the initiating Combined station resets its Send State Variable (S). Previously transmitted I frames that are unacknowledged when this command is actioned remain unacknowledged.

The RSET command will clear all frame rejection conditions except for an invalid N(R) condition in the addressed Combined station. The RSET command may be sent by a Combined station which detects an invalid N(R) instead of reporting such a frame rejection condition via a FRMR response.

### 7.4.4 Miscellaneous Commands

### 7.4.4.1 Exchange Identification (XID) Command

The XID command is used to cause the addressed Secondary/Combined station to report its station identification, and optionally to provide the station identification cf the transmitting Primary/Combined station to the addressed Secondary/Combined station. An information field is optional with the XID command; if present the information field will be the station ID of the Primary/Combined station. The Primary/Combined station may use the global address if the unique address of the Secondary/Combined station is not known. A Secondary/Combined station in any mode receiving an XID command will transmit an XID response unless 1) a UA response is pending, 2) a FRMR condition exists, 3) a RIM condition exists, or 4) the XID command cannot be actioned in a disconnected mode.

### 7.4.5 Non-Reserved Commands

Four non-reserved commands are specified to permit the implementer to define special system-dependent functions that do not have general applicability. Such special system-dependent functions are beyond the scope of this standard.

### 7.5 Unnumbered Format Responses

Unnumbered format responses are grouped according to the function performed:

- Responses to mode-setting and status requests: UA,DM,RIM
- Information transfer responses: UI
- Recovery responses: FRMR
- Miscellaneous responses: XID, RD
- Non-reserved responses: 4 encodings

First Bit Transmitted

```
                    Control Field Bits
      1    2    3    4    5    6    7    8


      1    1    0    0    F    1    1    0     UA Response
      1    1    1    1    F    0    0    0     DM Response
      1    1    1    0    F    0    0    0     RIM Response
      1    1    0    0    F    0    0    0     UI Response
      1    1    1    0    F    0    0    1     FRMR Response
      1    1    1    1    F    1    0    1     XID Response
      1    1    0    0    F    0    1    0     RD Response
      1    1    0    1    F    0    0    0     Non-Reserved Response
      1    1    0    1    F    0    0    1     Non-Reserved Response
      1    1    0    1    F    0    1    0     Non-Reserved Response
      1    1    0    1    F    0    1    1     Non-Reserved Response
```

For extended control field format see Section 5.2.2.

See Sections 6.1, 6.5.2 and 8.2.1 for description of the F bit operation.

7.5.1 Responses to Mode-Setting and Status Requests

The UA, DM and RIM responses are used by the Secondary/Combined station to request transmission of, or to respond to, the mode-setting commands of the Primary/Combined station; DM and RIM are additionally used to indicate Secondary/Combined station status.

7.5.1.1 Unnumbered Acknowledgement (UA) Response

The UA response is used to acknowledge the receipt and acceptance of the SNRM, SARM, SABM, SNRME, SARME, SABME, SIM, DISC, and RSET Unnumbered commands defined in Sections 7.4.1 and 7.4.3. The UA response is transmitted in the basic or the extended control field format as directed by the received Unnumbered command. No information field is permitted with the UA response.

A UA response is one way to report the end of a station busy condition. See Section 8.1.3.

7.5.1.2 Disconnected Mode (DM) Response

The DM response is used to report that the Secondary/Combined station is in the logically disconnected state; i.e., the Secondary/Combined station is, per system definition, in NDM or ADM. See Section 6.3.

The DM response is sent by a Secondary/Combined station in NDM or

- 42 -

ADM to request the remote Primary/Combined station to issue a mode-setting command, or, if sent in response to the reception of a mode-setting command, to inform the addressed Primary/Combined station that the transmitting Secondary/Combined station is still in NDM/ADM and cannot action the mode-setting command. On a switched network where the call is initiated by a Secondary/Combined station DM is sent to request a mode-setting command. On a non-switched line a Secondary/Combined station in ADM may send the DM response at any respond opportunity. No information field is permitted with the DM response.

A Secondary/Combined station in NDM or ADM will monitor received commands (other than those that reset the disconnected mode) only to detect a respond opportunity in order to (re)transmit DM (or RIM if initialization is required); i.e., no I/UI transmissions are exchanged until the disconnected mode is reset by the acceptance of SNRM, SARM, SABM, SNRME, SARME, SABME, OR SIM.

See Appendix C, Example 5.5.


### 7.5.1.3 Request Initialization Mode (RIM) Response

The RIM response is used to request the SIM command. A Secondary/Combined station which has established a RIM condition will monitor any subsequently received commands (other than SIM) only to detect a respond opportunity to (re)transmit RIM or to send DM; i.e., no command transmissions are accepted until the RIM condition is reset by the receipt of SIM. No information field is permitted with the RIM response.


### 7.5.2 Unnumbered Information Transfer Responses

Unnumbered information transfer responses are used to exchange frames containing information.


### 7.5.2.1 Unnumbered Information (UI) Response

The UI response is used to transfer an information field to a Primary/Combined station without impacting the Send and Receive Variables. The information field is optionally present with the UI response. Reception of the UI frame is not sequence-number verified; therefore, the frame may be lost if a link exception condition occurs during transmission of the UI, or duplicated if an exception occurs during any reply to the UI. Examples of UI frame information are higher level status, operation interruption, temporal data, and link initialization parameters.

## 7.5.3 Unnumbered Recovery Responses

Unnumbered recovery responses are used to facilitate the link-level exception condition recovery protocol.

### 7.5.3.1 Frame Reject (FRMR) Response

The FRMR response is used to report an error condition not recoverable by retransmission of the identical frame; i.e., one of the following conditions resulted from the receipt of an error-free frame from the Primary/Combined station:

1. the receipt of a control field that is invalid or not implemented.

2. the receipt of an I/UI frame with an information field which exceeded the maximum established length.

3. the receipt of an invalid N(R) number from the remote Primary/Combined station.

> An invalid N(R) is defined as a number which points to an I frame which has previously been transmitted <u>and</u> acknowledged, or to an I frame which has not been transmitted <u>and</u> is not the next sequential I frame pending transmission.

A Secondary/Combined station in a disconnected mode (NDM or ADM) will not establish a Frame Reject exception condition.

### FRMR Basic Information Field

A basic information field, which immediately follows the basic control field, is returned with this response to provide the reason for the Frame Reject response. The format for the basic information field is:

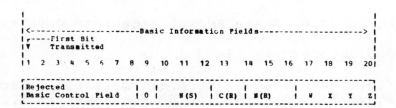

where:

Rejected Control Field is the control field of the received frame which caused the frame reject exception condition.

N(S) is the current Send Variable (S) at the station transmitting the FRMR response.

C/R is set to "1" if the frame which caused the FRMR was a response frame, or is set to "0" if the frame that caused the FRMR was a command frame.

N(R) is the current Receive Variable (R) at the station transmitting the FRMR response.

W set to "1" indicates the control field received and returned in bits 1 through 8 was invalid or not implemented.

X set to "1" indicates the control field received and returned in bits 1 through 8 was considered invalid because the frame contained an information field which is not permitted with this frame. Bit W must be set to "1" in conjunction with this bit.

Y set to "1" indicates the information field received exceeded the maximum established capacity of the Secondary/Combined station.

Z if set to "1" indicates the control field received and returned in bits 1 through 8 contained an invalid N(R) number.

If required, the information field associated with the FRMR may be padded with zero bits so as to end on any convenient, mutually agreed upon character, byte, word or machine-dependent boundary.

FRMR may have bits W,X,Y, and Z all set to zero; however the cause for frame reject shall be as defined in 1, 2 and 3 above.

See also Section 8.4, Frame Reject Exception Condition.

## FRMR Extended Information Field

The format for the extended information field, which immediately follows the extended control field (see Section 5.2.2), returned with the FRMR response is:

```
|                                                                             |
|                                                                             |
|<----------------------------Extended Information---------------------->|
|                                Field Bits                                    |
|First Bit Transmitted                                                        |
| !                                                                           |
| V                                                                           |
| 1                  16  17 18            24  25 26              32 33    36  |
 ---------------------------------------------------------------------------
| Rejected Extended    |  0 |              N(S) |C/R|     N(R)      |  WXYZ   |
| Control Field        |    |                   |   |               |         |
 ---------------------------------------------------------------------------
```

## 7.5.4 Miscellaneous Responses

### 7.5.4.1 Exchange Identification (XID) Response

The XID response is used to reply to an XID command. An
information field containing the identification of the
transmitting Secondary/Combined station is optionally present
with the XID response. A Secondary/Combined station receiving an
XID command will action the XID in any mode unless 1) a UA is
pending, 2) a FRMR condition exists, 3) a RIM condition exists,
or 4) the XID can not be actioned in a disconnected mode.

On switched networks when the Secondary/Combined station is
constrained to send first, it may use the XID response, which may
contain an optional information field, to request an XID
exchange. See Section 10.0, Switched Network Conventions.

### 7.5.4.2 Request Disconnect (RD) Response

The RD response is used to indicate to the remote
Primary/Combined station that the transmitting Secondary/Combined
station wishes to be placed in a logically disconnected mode (NDM
or ADM) by receiving a DISC command. RD may be sent
asynchronously if the Secondary/Combined station is in ARM/ABM,
or if in NRM as a response to a command with the P bit set to "1"
or a UP with the P bit set to "0". See Section 7.4.2.2. A
Secondary/Combined station which has sent an RD response and
receives any non-DISC frame(s) must accept the command frame(s)
if it is able to do so. If the Secondary/Combined station
accepts the non-DISC command frame(s), it follows the normal
ADCCP elements of procedures to respond to the Primary/Combined
station commands. Secondary/Combined station acceptance of
non-DISC frames after having issued an RD response cancels the RD

- 46 -

response. If the Secondary/Combined station still wants to be placed in disconnected mode (NDM or ADM), it must re-issue the RD response. A Secondary/Combined station which cannot accept non-DISC command frames due to internal problems may respond with RD again. No information field is permitted with the RD response.

## 7.5.5 Non-Reserved Responses

Four non-reserved responses are specified to permit the implementer to define special system-dependent functions that do not have general applicability. Such special system-dependent functions are beyond the scope of this standard.

## 8.0 EXCEPTION CONDITION REPORTING AND RECOVERY

This section specifies the procedures to be observed to effect recovery following the detection/occurrence of an exception condition at the link level. Exception conditions described are those situations that may occur as the result of transmission errors, station malfunction or operational situations.

## 8.1 Busy Condition

A busy condition occurs when a station temporarily cannot receive or continue to receive I or UI frames due to internal constraints; e.g., receive buffering limitations. The busy condition is reported by transmission of an RNR frame with the N(R) number of the next I frame that is expected. Traffic pending transmission at the busy station may be transmitted prior to or following the RNR. The continued existence of a busy condition must be reported by retransmission cf RNR at each P/F frame exchange. See Section 8.1.3, Clearing Busy Condition.

## 8.1.1 Secondary/Combined Station Receipt of RNR Command

A Secondary station transmitting TWS in NRM will upon receipt of an RNR command cease transmission at the earliest possible time. The frame in process may be completed or aborted; however, transmission must be terminated with the F bit tc set to "1" (see Example 5.2.1, Appendix C). The Secondary station may resume transmission of I and/or UI frames at the next poll command (an RR, REJ, SREJ, or 1 command frame with the P bit set to "1").

A Secondary/Combined station transmitting TWS in ARM/ABM will, upon receipt of an RNR, cease transmitting I or UI frames at the earliest possible time by completing or aborting the frame in process. If the RNR command frame had the P bit set to "1" the Secondary/Combined station must transmit a frame with the F bit set to "1". See Examples 5.4.1 and 5.4.3 in Appendix C. The Secondary/Combined station must perform a time-out before resuming asynchronous transmission of I or UI frames unless the busy condition is reported as cleared by the remote station.

## 8.1.2 Primary/Combined Station Receipt of RNR Response

Primary/Combined station receipt of an RNR response indicates that the transmitting Secondary/Combined station has a busy condition.

## 8.1.3 Clearing Busy Condition

The busy condition is cleared at the station which transmitted the RNR when the internal constraint ceases.

1045

Clearance of the busy condition is reported to the remote station by transmission of an RR, REJ, SREJ, SARM, SARME, SNRM, SNRME, SABM, SABME, or UA frame (with or without the P/F bit set to "1"). A busy condition is also cleared when a Primary station transmits an I frame with the P bit set to "1", or when a Secondary/Combined station transmits an I frame with the F bit set to "1".

## 8.2 N(S) Sequence Error

An N(S) sequence exception is established in the receiving station when an I frame received error free (no FCS error) contains an N(S) sequence number that is not equal to the Receive Variable (R) at the receiving station. The receiving station does not acknowledge (increment its Receive Variable (R)) the frame causing the sequence error, or any I frames which may follow, until an I frame with the correct N(S) number is received. Unless SREJ is to be used to recover from a given sequence error, the information field of all I frames received whose N(S) does not equal the Receive Variable (R) will be discarded. See Section 8.2.3 for SREJ recovery.

A station which receives one or more I frames having sequence errors, but otherwise error free, will accept the control information contained in the N(R) field and the P/F bit to perform link control functions; e.g., to receive acknowledgement of previously transmitted I frames (via the N(R)), to cause a Secondary/Combined station to respond (P bit set to "1"), and in NRM to detect that the Secondary station will terminate transmission (F bit set to "1"). The retransmitted frame may contain an N(R) and/or P/F bit information that are updated and, therefore, different from that contained in the originally transmitted I frames.

Following the occurrence of a sequence error the following means are available for initiating the retransmission of lost or errored I frames.

## 8.2.1 Checkpoint Recovery

Checkpoint recovery is based on a checkpoint cycle. For a Primary/Combined station a checkpoint cycle begins with the transmission of a frame with a P bit set to "1" and ends with the receipt of a frame with an F bit set to "1" or when the response timer expires. For a Secondary, a check point cycle begins with the transmission of a frame with the F bit set to "1" and ends with the receipt of a frame with a P bit set to "1".

When a Primary/Combined station receives a frame with the F bit set to "1" or when a Secondary station receives a frame with the

- 49 -

P bit set to "1", it initiates retransmission of all unacknowledged I frames with sequence numbers less than the Send Variable (S) at the time the last frame with the P bit set to "1" (Primary/Combined) or frame with the F bit set to "1" (Secondary) was transmitted. Retransmission starts with the lowest numbered unacknowledged I frame. I frames are retransmitted sequentially. New frames may be transmitted if they become available. Such retransmission of I frames is known as checkpoint retransmission.

Note that in Balanced operation either Combined station may initiate a checkpointing cycle independently of the other by the transmission of a frame with the P bit set to "1". Therefore, since two independent checkpointing cycles may be in process simultaneously a Combined station will not initiate checkpoint retransmission upon the receipt of a frame with the P bit set to "1".

To prevent duplicate retransmissions, checkpoint retransmission of a specific I frame (same N(R) in the same numbering cycle) is inhibited for the current checkpoint cycle if during the checkpoint cycle:

   1. A Primary station had previously received and actioned a REJ with the F bit set to "0".

   2. A Secondary station had previously received and actioned a REJ with the P bit set to "0".

   3. A Combined station had previously received and actioned a REJ with the P bit set to "0" or "1" or an F bit set to "0".

If an SREJ with a P/F bit set to "1" is received, checkpoint retransmission is not initiated.

Checkpoint retransmission is also inhibited if an unnumbered format frame with the P/F bit set to "1" is received. This is because there is no N(R) for checkpoint reference.

Finally checkpoint retransmission is inhibited if, after sending a frame with the P/F bit set to "1", a station receives an acknowledgement to that frame before the next checkpoint occurs.

8.2.2 REJ Recovery

The REJ command/response is primarily used to initiate an earlier exception recovery (retransmission) following the detection of a sequence error than is possible by checkpoint recovery; e.g., in two-way simultaneous information transfer if REJ is immediately transmitted upon detection of a sequence error there is no requirement to wait for a frame with P/F bit set to "1".

1047

Only one "sent REJ" exception condition, from a given station to another given station, is established at a time. A "sent REJ" exception is cleared when the requested I frame is received, when a time-out function expires, or when a checkpoint cycle that was initiated concurrent with or following the transmission of REJ is completed. When the station perceives by time-out or by the checkpointing mechanism that the requested I frame will not be received, because either the requested I frame or the REJ was in error/lost, the REJ may be repeated.

A station receiving REJ initiates sequential (re)transmission of I frames starting with the I frame indicated by the N(R) contained in the REJ frame.

If (1) retransmission beginning with a particular frame occurs due to checkpointing (Sections 6.5.2 and 8.2.1), and (2) a REJ is received before a checkpoint cycle completion which would also start retransmission with the same particular frame (as identified by the N(R) in the REJ), the retransmission resulting from the REJ shall be inhibited.

## 8.2.3 SREJ Recovery

The SREJ command/response is primarily used to initiate more efficient error recovery by requesting the retransmission of a single I frame following the detection of a sequence error rather than the retransmission of the I frame requested plus all additional I frames which may have been subsequently transmitted.

Note: To improve transmission efficiency it is recommended that the SREJ command/response be transmitted as the result of the detection of a sequence error where only a single I frame is missing (as determined by receipt of the out-of-sequence N(S)).

When an I frame sequence error is detected, the SREJ is transmitted at the earliest possible time. When a station sends an SREJ with the P bit set to "0" (Primary station), with the F bit set to "0" (Secondary/Combined station), or with the P bit set to 0 or 1 (Combined station), and the "sent SREJ" condition is not cleared when the station is ready to issue the next frame with the P bit (Primary) or the F bit (Secondary/Combined) set to "1", the station sends an SREJ with the same N(R) as the original SREJ with the P/F bit set to "1".

Since a frame sent with the P bit (Primary station) or the F bit (Secondary/Combined station) set to "1" has the potential of causing checkpoint retransmission, a station will not send an SREJ with the same N(R) (same value and same numbering cycle) as that of the previously sent frame with the P bit (Primary station) or the F bit (Secondary/Combined station) set to "1" until the current checkpoint cycle ends.

- 51 -

Only one "sent SREJ" exception condition from a given station to another given station is established at a time. A "sent SREJ" exception condition is cleared when the requested I frame is received, when time-out function expires, or when a checkpoint cycle that was initiated concurrent with or following the transmission of SREJ is completed. When the station perceives by timeout or by the checkpointing mechanism that the requested I frame will not be received, because either the requested I frame or the SREJ was in error/lost, the SREJ may be repeated.

When a station receives and actions an SREJ with the P bit (Secondary station) or F bit (Primary/Combined station) set to "0" or with the P bit set to "0" or "1" (Combined station), it will disable actioning the next SREJ if the SREJ has the P bit (Secondary station) or F bit (Primary/Combined station) set to "1" and has the same N(R) (i.e., has the same value and same numbering cycle) as the original SREJ.

## 8.2.4 Time-out Recovery

In the event a receiving station, due to a transmission error, does not receive (or receives and discards) a single I frame or the last I frame(s) in a sequence of I frames, it will not detect an out-of-sequence exception and, therefore, will not transmit SREJ/REJ. The station which transmitted the unacknowledged I frame(s) shall, following the completion of a system-specified time-out period, take appropriate recovery action to determine the sequence number at which retransmission must begin.

NOTE: It is recommended that a station which has timed out waiting for a response not retransmit all unacknowledged frames immediately. A Secondary station in ARM should, in this time-out case, either retransmit its last single frame or transmit new frames if they are available. A Primary/Combined station may enquire status with a supervisory frame.

To account for possible retransmissions after time-out, a receiving station should not set a SREJ condition when it receives an I frame with an N(S) one less than its Receive Variable (R).

If a station does retransmit all unacknowledged I frames after a time-out, it must be prepared to receive a following REJ frame with an N(R) greater than its Send Variable (S).

## 8.3 FCS Error

Any frame with an FCS error is not accepted by the receiving station and is discarded. At the Secondary/Combined station no action is taken as the result of that frame.

## 8.4 Frame Reject Exception Condition

A frame reject exception condition is established upon the receipt of an error-free frame which contains an invalid or unimplemented control field, an invalid N(R) or an information field which exceeded the maximum established storage capability.

If a frame reject exception condition occurs in a Primary station, or is reported to the Primary station by a FRMR response, recovery action will be taken by the Primary station. This recovery action includes the transmission of an implemented set mode command. Higher level functions may also be included in the recovery.

At the Secondary station this exception condition is reported by transmitting a FRMR response to the Primary station for appropriate action. Once a Secondary station has established a FRMR exception any additional commands (other than those that reset the FRMR exception condition) subsequently received are examined only with regard to the state of the N(R) and the P bit; i.e., only to update the acknowledgement of I frames previously transmitted and to detect a respond opportunity to retransmit FRMR. No additional transmissions are accepted or actioned until the condition is reset by the receipt of an implemented set mode command.

If a frame reject exception condition occurs in a Combined station, the station will either:

1. take recovery action without reporting the condition to the remote Combined station, or

2. report the condition to the remote Combined station with a FRMR response. The remote station will then be expected to take recovery action; if, after waiting an appropriate time, no recovery action appears to have been taken, the Combined station reporting the frame reject exception condition may take recovery action.

Recovery action for Balanced operation includes the transmission of an implemented mode-setting or RSET command, as appropriate. Higher level functions may also be involved in the recovery.

## 8.5 Mode-Setting Contention

A mode-setting contention situation exists when a Combined station issues a mode-setting command and, before receiving an appropriate response (UA or DM), receives a mode-setting command from the remote Combined station. Contention situations shall be resolved in the following manner (see Appendix C, Example 8.5):

1. When the send and receive mode-setting commands are the same, each Combined station shall send an UA response at the earliest respond opportunity. Each Combined station shall either enter the indicated mode immediately or defer entering the indicated mode until receiving an UA response. In the latter case, if the UA response is not received, (1) the mode may be entered when the response timer expires, or (2) the mode-setting command may be reissued.

2. When the mode-setting commands are different, each Combined station shall enter ADM and issue a DM response at the earliest respond opportunity. In the case of DISC contention with a different mode-setting command no further action is required. In the case of contention between SABM and SABME commands, the Combined station sending SABME shall have pricrity in attempting link establishment after the DM responses.

- 54 -

## 9.0 TIME-OUT FUNCTIONS

Time-out functions are used to detect that a required or expected acknowledging action or response has not been received to a previously transmitted frame. Expiration of the time-out function shall initiate appropriate action, e.g., error recovery or reissuance of the P-bit. The duration of time-out functions is system dependent and subject to bilateral agreement.

The following time-out functions represent the minimum requirements, and do not preclude other time-out functions.

### 9.1 Normal Respond Opportunity

The Primary station transmitting a command with the P bit set to "1" or UP with P bit set to "0", anticipates a response and, therefore, starts a time-out function. The time-out function shall be stopped upon receipt of the expected response.

### 9.2 Asynchronous Respond Opportunity

The Primary/Combined station provides a time-out function to determine that a response frame with F bit set to "1" has not been received to a command frame with the P bit set to "1". The time-out function shall be stopped upon receipt of a valid frame with the F bit set to "1".

A Primary/Combined station with no P bit outstanding, and which has transmitted one or more frames for which responses are anticipated, must start a time-out function to detect the no response condition.

The Secondary/Combined station provides a time-out function to determine that a command frame has not been received acknowledging an unsolicited response frame(s).

See Sections 6.2.2 and 8.2.4.

1052

## 10.0 SWITCHED NETWORK CONVENTIONS

Stations connected to a switched communications network may be capable of operation as one station type only (e.g., a Primary station, a Secondary station, or a Combined station); or the station may be configurable as (one at a time) more than one of these types. The capabilities of the called station must be known at the calling station and the calling station must operate accordingly. If the called station is configurable it will:

1. implement the XID command and response, and

2. determine which station type (Primary, Secondary, or Combined) to invoke by recognition of either the remote station address or identification (XID).

The calling or called station will initate the transmission interchange first depending on the characteristics of the transmission network. When initiated by the Secondary station, it sends a single unsolicited Supervisory or Unnumbered response. When initiated by the Primary/Combined station, it sends any appropriate command with an appropriate address.

ASSUMED PRIMARY/SECONDARY/COMBINED ROLES ON SWITCHED NETWORK

| CALLED STATION | CALLING STATION | PRIMARY STATION | SECONDARY STATION | COMBINED STATION | CONFIGURABLE STATION |
|---|---|---|---|---|---|
| PRIMARY STATION | | NA | P / S | NA | P / S |
| SECONDARY STATION | | S / P | NA | NA | S / P |
| COMBINED STATION | | NA | NA | C / C | C / C |
| CONFIGURABLE STATION | | S / P | P / S | C / C | P/S/C / P/S/C |

```
P     = Primary Station
S     = Secondary Station
C     = Combined Station
P/S/C = Primary and/or Secondary and/cr Combined Station
NA    = Not Applicable
```

FIGURE 10-1

1054

- 57 -

# 11.0 CLASSES OF PROCEDURES

All classes of procedures use the two frame formats as defined in Section 3.0 FRAME STRUCTURE. In addition, all procedures assume that the links include Primary and Secondary stations or Combined stations. Primary stations transmit commands (in frames with or without information), and Secondary stations receive the command frames and transmit responses (in frames with or without information). Combined stations transmit and receive commands and responses (in frames with or without information). The Primary/Combined station is responsible for determining which commands to send, within the constraints of the standard.

Procedure differences based on overall system consideration (e.g., network configuration, traffic management, etc.) are accommodated by defining three modes of operation - Asynchronous, Normal and Balanced, and by defining three Classes of Procedures that utilize the capabilities of these modes together with the exception recovery characteristics specified within the standard. Optional Functions are defined to provide additional capabilities. Individual classes implement a prescribed subset of the commands and responses defined in Section 7.0, COMMANDS AND RESPONSES, and include P/F recovery as a minimum capability as defined in Sections 6.1, 6.5.2 and 8.2.1.

## 11.1 Classes of Procedures

The three Classes of Procedures are composed of:

1.  Three types of stations: Primary stations, Secondary stations and Combined stations

2.  Two types of configurations: Unbalanced (for Primary and Secondary stations) and Balanced (for Combined stations).

3.  Two types of respond opportunity: Normal and Asynchronous

| Designation | Class of Procedures Description |
|---|---|
| UA | Unbalanced, Asynchronous Response Mode, Modulo 8 |
| UN | Unbalanced, Normal Response Mode, Modulo 8 |
| BA | Balanced, Asynchronous Balanced Mode, Modulo 8 |

Classes UA and UN can be used on either Unbalanced or Symmetrical configurations. Class BA can be used on Balanced configurations. See Section 2.2.

1055

## 11.1.1 Unbalanced/Symmetrical Configuration

Basic Repertoire of Commands and Responses

| Commands | Responses |
|----------|-----------|
| I        | I         |
| RR       | RR        |
| RNR      | RNR       |
|          | FRMR      |
| *SXXM    | UA        |
| DISC     | DM        |

*SXXM Command is SARM for UA Class
SNRM for UN Class

## 11.1.2 Balanced Configuration

Basic Repertoire of Commands and Responses

| Commands | Responses |
|----------|-----------|
| I        | I         |
| RR       | RR        |
| RNR      | RNR       |
|          | FRMR      |
| SABM     | UA        |
| DISC     | DM        |
| RSET     |           |

## 11.2 Optional Functions

Optional functions are achieved by the addition or deletion of commands and responses or capabilities to those present in any basic Class of Procedures.

| Option | Functional Description | Required Change |
|--------|------------------------|-----------------|
| 1 | Provides the ability to:<br>- exchange identification of stations. See Sections 7.4.4.1 and 7.5.4.1.<br>- request logical disconnection See Section 7.5.4.2. | Add command: XID<br>Add response: XID, RD |
| 2 | Provides the ability for more timely reporting of N(S) sequence errors to improve TWS performance. See Section 7.2.3. | Add command: REJ<br>Add response: REJ |

| 3 | Provides the ability for more efficient recovery from N(S) sequence errors by requesting retransmission of a single I frame.  See Section 7.2.4. | Add command: SREJ<br>Add response: SREJ |
| 4 | Provides the ability to exchange information fields without impacting the Send and Receive Variables.  See Sections 7.4.2.1 and 7.5.2.1. | Add command: UI<br>Add response: UI |
| 5 | Provides the ability to ability to initialize remote stations and the ability to request initialization.<br>See Sections 7.4.1.7 and 7.5.1.3. | Add command: SIM<br>Add response: RIM |
| 6 | Provides the ability to perform unnumbered group polling as well as unnumbered individual polling.<br>See Section 7.4.2.2. | Add command: UP |
| 7 | Provides for greater than single octet addressing.<br>See Section 4.3. | Use extended addressing format in lieu of basic addressing format. |
| 8 | Limits the procedure to allow I frames to be commands only. | Delete response: I |
| 9 | Limits the procedure to allow I frames to be responses only. | Delete command: I |
| 10 | Provides the ability to use extended sequence numbering (modulo 128).<br>See Section 5.2.2. | Use extended control field format in lieu of basic control field format.  Use SXXME in lieu of SXXM. |
| 11 | Removes the ability to reset the Send and Receive variables associated with only one direction of information flow. | Delete command: RSET |

## 11.3 Consistency of Classes of Procedures

Figure 11-1 gives a summary of the basic command/response repertoire of the two Unbalanced and one Balanced Classes of

Procedures, and the commands/responses of the Optional Functions. In the Unbalanced classes the Primary station command repertoire is listed on the left side of each class and the Secondary station response repertoire is listed on the right side. As seen in the figure, the basic repertoire of all Classes of Procedures is identical with the exception of a unique mode-setting command for each class, and the RSET command which is used in the Balanced Class only. This repertoire consistency facilitates the inclusion of multiple Classes of Procedures in a station that is configurable.

## 11.4 Implementation of Classes of Procedure

A station conforms to a given Class of Procedures if it implements the basic repertoire of that class. To implement (see Appendix A definition) a Class of Procedures (or Optional Functions) means:

1. A Primary station has the ability to receive all responses in the Class of Procedures basic repertoire (or Optional Functions).

2. A Secondary station has the ability to receive all commands in the Class of Procedures basic repertoire (or Optional Functions).

3. A Combined station has the ability to receive all commands and responses in the Class of Procedures basic repertoire (or Optional Functions).

## 11.5 Method of Indicating Classes and Optional Functions

Classes of Procedures and the Optional Functions are indicated by specifying the mnemonic designation for the desired Class and the number(s) of the accompanying Optional Functions.

## 11.5.1 Class and Option(s) Examples

Class UN,1,2,6 is the Unbalanced, Normal Response Mode Class of Procedures with the Optional Functions for identification and request disconnect (XID,RD), improved TWS performance (REJ) and unnumbered polling (UP).

Class BA,2,3,10 is the Balanced, Asynchronous Balanced Mode Class of Procedures with the Optional Functions for improved TWS performance (REJ), single frame retransmission (SREJ) and extended sequence numbering (modulo 128).

Class UA,1,5 is the Unbalanced, Asynchronous Response Mode Class of Procedures with the Optional Functions for identification and request disconnect (XID,RD) and initialization (SIM,RIM).

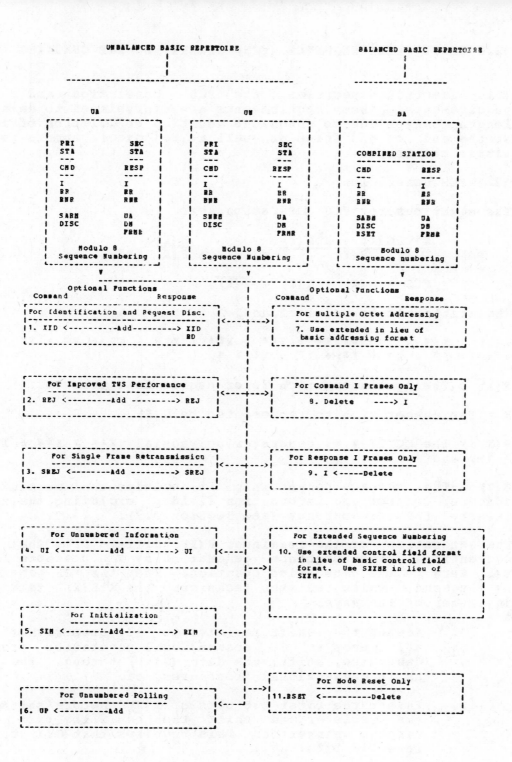

- 62 -

## 12.0 FRAME CHECK SEQUENCE (FCS) GENERATION AND CHECKING

This section specifies the FCS generation and checking requirements. These requirements are formulated to detect frame length changes due to erroneous addition or deletion of zero bits at the end of the frame as well as to detect errors introduced within the frame.

### 12.1 FCS Generation

The equations for FCS generation are:

$$\frac{X^{16} G(X) + X^K L(X)}{P(X)} = Q(X) + \frac{R(X)}{P(X)} \text{ and,}$$

$$FCS = L(X) + R(X) = \overline{R(X)},$$

The arithmetic is modulo 2 and,

$$L(X) = X^{15} + X^{14} + X^{13} + X^{12} + X^{11} + X^{10} + X^9 + X^8 + X^7 + X^6 + X^5 + X^4 + X^3 + X^2 + X^1 + 1,$$

$R(X)$ = The remainder which is of degree less than 16.

$k$ = The number of bits represented by $G(X)$,

$P(X)$ = The CCITT V.41 generator polynomial ($X^{16} + X^{12} + X^5 + 1$), and

$G(X)$ = The message polynomial. It includes the contents of the address, control and information fields, excluding the zero bits inserted for transparency (see Section 3.7).

The generation of the remainder $R(X)$ differs from that used in conventional check sequence generation by the presence of the $X^K L(X)$ term in the generation equation. When the FCS generation is by the usual shift register technique, the $X^K L(X)$ term is added in either of two ways:

1. Preset the shift register to all ones rather than to all zeros as in conventional generation procedures. Otherwise, shift the data ($G(X)$) through the register as in conventional procedures, or,

2. Invert the first 16 bits of $G(X)$ before shifting into the register and shift the remaining part of $G(X)$ through uninverted. This requires that $G(X)$ contain at least 16 bits.

Whether 1 or 2 is used, the shift register contents, after

1060

shifting through G(X), is R(X). These contents are inverted
bit-by-bit and transmitted as the FCS sequence.

The transmitted sequence is always (in algebraic notation):

$$M(X) = X^{16} G(X) + FCS.$$

## 12.2 FCS Checking

The received sequence will be denoted M*(x), which differs from
M(x) if transmission errors are introduced. The checking process
always involves dividing the received sequence by P(x) and
testing the remainder. Direct division, however, does not yield
a unique remainder and it is expected that in most cases the
received sequence will be modified for checking purposes by
addition of terms which will cause the division to yield such a
unique remainder when M*(x) = M(x), i.e., when the frame is error
free.

Two classes of checking equations are given below:

1. $$\frac{X^{r}[M*(X) + X^{k}L(X)]}{P(X)} = Q(X) + \frac{R(X)}{P(X)}$$

In this case the unique remainder is the remainder of
the division $X^{r} \dfrac{L(X)}{P(X)}$

When $r$ = 0 the remainder is L(x) (16 ones).
When $r$ = 16 the remainder is $X^{12} + X^{11} + X^{10} + X^{8} + X^{3} + X^{2} + X + 1$
($X^{15}$ through $X^0$ respectively).

2. $$X^{r}\frac{[M*(X) + (X^{k}+1) L(X)]}{P(X)} = Q(X) + \frac{R(X)}{P(X)}$$

In this case the unique remainder is always zero regardless
of the value of $r$

Shift register implementation of the above equations normally use
$r$ = 16 (pre-multiplication). When this is the case, the added
term $X^{k} L(x)$ in both 1 and 2 is added by either inverting the
first 16 received bits of M*(X) before shifting them through the
checking register or by presetting the register to all 1's and
shifting all of M*(X) through normally. Thus the receiver action
on the leading portion of a frame is the same with either 1 or 2.

The + 1 of the term $(X^{k}+1)L(x)$ of the generation equation of 2 is
added by inverting the FCS. This implies a 16 bit storage delay
by the FCS function at the receiver since the location of the FCS
is not known until the closing flag is received.

- 64 -

**ABORT:** A function invoked by a sending station causing the recipient to discard (and ignore) all bit sequences transmitted by the sender since the preceding FLAG SEQUENCE.

**ACCEPT:** The condition assumed by a STATION upon accepting a correctly RECEIVED FRAME for processing. A station "ACCEPTS" a COMMAND/RESPONSE when the COMMAND/RESPONSE encoded in the CONTROL FIELD of the RECEIVED FRAME is ACTIONED.

**ACKNOWLEDGE:** A STATION "ACKNOWLEDGES" a RECEIVED FRAME when it transmits an appropriate FRAME(S) indicating the RECEIVED FRAME has been ACTIONED.

**ACTION:** A STATION "ACTIONS" a RECEIVED COMMAND/RESPONSE when it performs (or executes) the functions encoded in the CONTROL FIELD of the FRAME.

**ADCCP:** Advanced Data Communication Control Procedures.

**ADDRESS FIELD (A):** The sequence of eight (or any multiple of eight if extended) bits immediately following the opening FLAG of a FRAME identifying the SECONDARY/COMBINED STATION sending a RESPONSE FRAME (or designated to receive a COMMAND FRAME).

**ADDRESS FIELD EXTENSION:** Enlarging the ADDRESS FIELD to include more addressing information.

**COMBINED STATION:** That STATION responsible for performing Balanced LINK LEVEL operations. A COMBINED STATION generates COMMANDS and interprets RESPONSES, and interprets received COMMANDS and generates RESPONSES.

**COMMAND:** The content of the CONTROL FIELD, of a COMMAND FRAME sent by the PRIMARY/COMBINED STATION instructing the addressed SECONDARY/COMBINED STATION to perform some specific LINK LEVEL function.

**COMMAND FRAME:** All FRAMES that are transmitted by the PRIMARY STATION (or by a COMBINED STATION that have the remote/receiving COMBINED STATION address) are referred to as COMMAND FRAMES.

**CONFIGURABLE STATION:** A STATION is CONFIGURABLE if it has as the result of mode-setting action the capability to be, at different times, more than one type of logical station; i.e., PRIMARY STATION, SECONDARY STATION, or COMBINED STATION.

**CONTROL FIELD (C):** The sequence of eight (or sixteen if extended control field) bits immediately following the ADDRESS FIELD of a

A 1

FRAME. The content the CONTROL FIELD is interpreted by the receiving:

1.   SECONDARY STATION, designated by the ADDRESS FIELD, as a COMMAND instructing the performance of some specific function.

2.   PRIMARY STATION, as a RESPONSE from the SECONDARY STATION, designated by the ADDRESS FIELD, to one or more COMMANDS.

3.   COMBINED STATION, 1) as a COMMAND instructing the performance of some specific function, if the ADDRESS FIELD designates the receiving COMBINED STATION, 2) as a RESPONSE to one or more transmitted COMMANDS if the ADDRESS FIELD designates the remote COMBINED STATION.

CONTROL FIELD EXTENSION: Enlarging the CONTROL FIELD to include additional control information.

DATA LINK: An assembly of two or more terminal installations and the interconnecting line operating according to a particular method that permits information to be exchanged; in this context the term "terminal installation" does not include the data source and the data sink.

DISCARD: A STATION may "DISCARD" all or part of a RECEIVED FRAME:

1. A "DISCARDED" FRAME is a RECEIVED FRAME whose control and information fields are not examined or used; i.e., the STATION takes no ACTION on any part of the FRAME.

2. A "RECEIVED" FRAME may have its INFORMATION FIELD (I/UI) "DISCARDED", i.e., the CONTROL FIELD of the FRAME is used but the INFORMATION FIELD is thrown away.

EXCEPTION CONDITION: The condition assumed by a STATION upon receipt of a CONTROL FIELD which it cannot execute due to either a transmission error or an internal processing malfunction.

FLAG SEQUENCE(F): The unique sequence of eight bits(01111110) employed to delimit the opening and closing of a FRAME.

FRAME: The sequence of contiguous bits, bracketed by and including opening and closing FLAG SEQUENCES. A valid FRAME contains at least 32 bits between FLAGS and contains an ADDRESS FIELD, a CONTROL FIELD and a FRAME CHECK SEQUENCE. A FRAME may or may not include an INFORMATION FIELD.

FRAME CHECK SEQUENCE (FCS): The field, immediately preceding the closing FLAG SEQUENCE of a FRAME, containing the bit sequence

A 2

that provides for the detection of transmission errors by the receiving STATION.

HIGH LEVEL: The conceptual level of control or processing logic existing in the hierarchical structure of a STATION that is above the LINK LEVEL and upon which the performance of LINK LEVEL functions are dependent, e.g., device control, buffer allocation, station management, etc.

IMPLEMENT: A COMMAND/RESPONSE is IMPLEMENTED if it is part of the receiving STATION'S repertoire; i.e., the receiving STATION is capable of decoding and ACTIONING the CONTROL FIELD in the RECEIVED COMMAND/RESPONSE.

INFORMATION FIELD (INFO): The sequence of bits, occurring between the last bit of the CONTROL FIELD and the first bit of the FRAME CHECK SEQUENCE. The INFORMATION FIELD contents are not interpreted at the LINK LEVEL.

INTERFRAME TIME FILL: The sequence of bits transmitted between FRAMES. This standard does not provide for time fill within a FRAME.

INVALID: There are three reasons a RECEIVED FRAME may be INVALID:

1. An INVALID FRAME is one that is not properly bounded by two FLAGS (thus an ABORTED FRAME is an INVALID FRAME) or one that is too short (e.g., shorter than 32 bits between FLAGS).

2. An INVALID COMMAND/RESPONSE is a FRAME which has a CONTROL FIELD encoding which is not defined in this standard.

3. An INVALID N(R) is one which points to an I FRAME which has previously been transmitted and acknowledged, or to an I FRAME which has not been transmitted and is not the next sequential I FRAME pending transmission.

LINK LEVEL: The conceptual level of control or processing logic existing in the hierarchical structure of a STATION that is responsible for maintaining control of the DATA LINK. The LINK LEVEL functions provide an interface between the STATION HIGH LEVEL logic and the DATA LINK; these functions include (transmit) bit injection and (receive) bit extraction, ADDRESS/CONTROL FIELD interpretation, COMMAND/RESPONSE generation, transmission and interpretation, and FRAME CHECK SEQUENCE computation and interpretation.

A 3

**PRIMARY STATION:** That STATION responsible for Unbalanced control of the DATA LINK. The PRIMARY STATION generates COMMANDS and interprets RESPONSES. Specific responsibilities assigned to the PRIMARY STATION include:

1. Initialization of (data and control) information interchange

2. Organization and control of data flow

3. Retransmission control

4. All recovery functions at the LINK LEVEL

**RECEIVE:** A STATION "RECEIVES" a COMMAND or RESPONSE FRAME when the incoming bit configuration is bounded by two FLAGS, contains an ADDRESS FIELD recognized by that STATION, and has a correct FCS.

**RESPOND OPPORTUNITY:** The LINK LEVEL logical control condition during which a given SECONDARY/COMBINED STATION may transmit a RESPONSE FRAME(s).

**RESPONSE:** The content of the CONTROL FIELD of a RESPONSE FRAME advising the PRIMARY/COMBINED STATION with respect to the processing by the SECONDARY/COMBINED STATION of one or more COMMAND FRAMES.

**RESPONSE FRAME:** All FRAMES that may be transmitted by a SECONDARY STATION (or by a COMBINED STATION that have the local/transmitting COMBINED STATION address) are referred to as RESPONSE FRAMES.

**SECONDARY STATION:** That STATION responsible for performing Unbalanced LINK LEVEL operations, as instructed by the PRIMARY STATION. A SECONDARY STATION interprets RECEIVED COMMANDS and generates RESPONSES.

**SECONDARY STATUS:** The current condition of a SECONDARY STATION with respect to processing the series of COMMANDS RECEIVED from the PRIMARY STATION.

**STATION:** The word "STATION" unqualified (i.e., not preceeded by PRIMARY, SECONDARY, or COMBINED) applies to all three types of STATIONS: PRIMARY STATION, SECONDARY STATION and COMBINED STATION.

A 4

# APPENDIX B - ADDITIONAL INFORMATION

This appendix provides additional explanatory information to assist in the use of the standard. For ease of reference, the organization of this appendix is identical to that of the body of the standard.

## B.3.4 Frame Structure, Information Field

Although the maximum length of the information field is theoretically unlimited it will be constrained by one or more of the following factors:

1. Error detection capability of the FCS

2. Channel error characteristics and data rates

3. Station buffer sizes and strategies

4. Logical properties of the data

## B.3.1 Flag Sequence, and 3.8 Time Fill

Although this standard permits the closing flag of one frame to be the opening flag of the next frame, it must be recognized that in certain implementations this may result in crisis time problems. Under those conditions, it may be necessary to transmit interframe time fill. The amount of time fill must be determined by prior agreement.

## B.3.9    IDLE Link State

Detection of an IDLE link condition may require the use of a timer or an alternate clock to determine receipt of a continuous one condition for 15 bit times if the link configuration does not provide clock signals in an IDLE condition.

## B.7.4.2.1 Unnumbered Information, UI, Command

A Secondary must respond upon receipt of a UI command frame with the P bit set to "1"; the response shall be any appropriate frame(s), one of which will have the F bit set to "1". A UI command with the P bit set to "0" solicits no response.

## CONTROL FIELD ENCODING

| FORMAT | COMMAND/RESPONSE | COMMAND<br>1 2 3 4 5 6 7 8 | RESPONSE<br>1 2 3 4 5 6 7 8 | INFO FIELD |
|---|---|---|---|---|
| Information | Information   I | 0 N(S) P N(R) | 0 N(S) F N(R) | O |
| Supervisory | Receive Ready   RR | 1 0 0 0 P N(R) | 1 0 0 0 F N(R) | N |
|  | Receive Not Ready   RNR | 1 0 1 0 P N(R) | 1 0 1 0 F N(R) | N |
|  | Reject   REJ | 1 0 0 1 P N(R) | 1 0 0 1 F N(R) | N |
|  | Selective Reject   SREJ | 1 0 1 1 P N(R) | 1 0 1 1 F N(R) | N |

NOTE:   Info Field:   0 – Optional
                       N – Not Allowed
                       R – Required

TABLE B1   COMMAND/RESPONSE SUMMARY

B 2

## CONTROL FIELD ENCODING

| FUNCTION | COMMAND/RESPONSE | | COMMAND 1 2 3 4 5 6 7 8 | RESPONSE 1 2 3 4 5 6 7 8 | INFO FIELD |
|---|---|---|---|---|---|
| Mode Set Commands | Set Normal Response Mode | SNRM | 1 1 0 0 P 0 0 1 | | N |
| | Set Asynchronous Response Mode | SARM | 1 1 1 1 P 0 0 0 | | N |
| | Set Asynchronous Balanced Mode | SABM | 1 1 1 1 P 1 0 0 | | N |
| | Set Asynchronous Response Mode Extended | SARME | 1 1 1 1 P 0 1 0 | | N |
| | Set Normal Response Mode Extended | SNRME | 1 1 1 1 P 0 1 1 | | N |
| | Set Asynchronous Balanced Mode Extended | SABME | 1 1 1 1 P 1 1 0 | | N |
| Mode Set/ Request | Disconnect/ Request Disconnect | DISC | 1 1 0 0 P 0 1 0 | RD 1 1 0 0 F 0 1 0 | N |
| | Set Initialization Mode/ Request Initialization Mode | SIM | 1 1 1 0 P 0 0 0 | RIM 1 1 1 0 F 0 0 0 | N |

TABLE B1 (continued)

## CONTROL FIELD ENCODING

| FUNCTION | COMMAND/RESPONSE | COMMAND 1 2 3 4 5 6 7 8 | | RESPONSE 1 2 3 4 5 6 7 8 | | INFO FIELD |
|---|---|---|---|---|---|---|
| Responses to Mode Set Commands | Unnumbered Acknowledgement | | UA | 1 1 0 0 F 1 1 0 | | N |
| | Disconnected Mode | | DM | 1 1 1 1 F 0 0 0 | | N |
| Information Transfer | Unnumbered Information | UI | 1 1 0 0 P 0 0 0 | UI | 1 1 0 0 F 0 0 0 | O |
| | Unnumbered Poll | UP | 1 1 0 0 P 1 0 0 | | | N |
| | Exchange Identification | XID | 1 1 1 1 P 1 0 1 | XID | 1 1 1 1 F 1 0 1 | O |
| Recovery | Frame Reject Reset | RSET | 1 1 1 P 0 0 1 | FRMR | 1 1 0 F 0 0 1 | R N |
| Non Reserved | NR∅ | | 1 1 0 1 P 0 0 0 | | 1 1 0 1 F 0 0 0 | O |
| | NR1 | | 1 1 0 1 P 0 1 0 | | 1 1 0 1 F 0 1 0 | O |
| | NR2 | | 1 1 0 1 P 0 0 1 | | 1 1 0 1 F 0 0 1 | O |
| | NR3 | | 1 1 0 1 P 0 1 1 | | 1 1 0 1 F 0 1 1 | O |

TABLE B1 (concluded)

# APPENDIX C

## EXAMPLES OF THE USE OF
## COMMANDS AND RESPONSES

The examples in Appendix C are offered for illustrative purposes only and should not be interpreted as establishing any protocol; the exchange of the various command and response frames is limited only by the rules specified in the standard.

The notation used in the Appendix C diagrams is illustrated below:

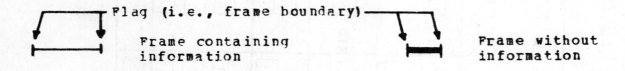

Frame containing information          Frame without information

## UNBALANCED MODE OPERATION

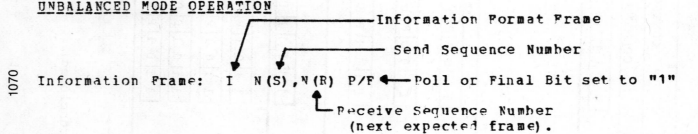

Information Frame: I N(S),N(R) P/F ◄── Poll or Final Bit set to "1"

Information Format Frame

Send Sequence Number

Receive Sequence Number (next expected frame).

Example:    Pri xmits: I2,6P. This denotes a Primary Information format frame with sequence number 2, the next expected frame from the Secondary is sequence number 6 (frames numbered 5 and below are therefore acknkowledged) and the Poll bit is set to "1" (i.e., the Secondary is to initiate transmission with Information format frames if available).

1070

C1

Supervisory command/response

Supervisory Frame:   XXX N(R),P/F ◄——Poll or Final bit set to "1"

——Receive Sequence Number

Example:   Pri xmits: RR2,P.  This  denotes a Receive Ready
           (RR)  command, N(R)=2  (i.e.,  the next  expected
           frame from  the Secondary is sequence  number 2),
           and the Poll bit is set to "1".


Unnumbered command/response

Unnumbered Frame:    YYYY , P/F

↑——Poll or Final Bit set to "1"

Example:   Pri xmits:  SNRM,P.  This  denotes a Set  Normal
           Response Mode  (SNRM) command  with the  Poll bit
           set to "1".


BALANCED MODE OPERATION

Balanced Mode operation notation is identical to that of the
Unbalanced Mode except that a station address must be
indicated in order to designate the  frame as a command of a
response.


Information Frame:   A , I N(S) , N(R)  P/F

└Address:   remote station address indicates
           frame is a command; local station
           address indicates frame is a response.

Example:   Combined xmits: A,I2,6P.   This denotes a command
           Information Format frame with  sequence number 2,
           the next expected frame is  sequence number 6 and
           the Poll bit is set to "1".

1071

C2

Supervisory Frame:    A , XXX N(R)  , P/F

Example:    Combined   xmits:   B,RR2,F.   This  denotes  a
            response Receive Ready (RR) with N(R) = 2 and the
            Final bit set to "1".


Unnumbered Frame:    A , YYYY  , P/F

Example:    Combined  xmits:  A,SABM,P.  This  denotes a  Set
            Asynchronous  Balanced Mode  (SABM) command  with
            the Poll bit set to "1".


NOTE:   Retransmitted Information  Format  frames are  shown
        with a double line: i.e., ▮━━━━━▮

1073

C4

1074

1075

C6

1076

1. Examples of Normal Response Mode (NRM) 2 Way Alternate (TWA) Transmission

1.1 **NRM TWA Without Transmission Errors**

1.1.1 NRM Start-up procedure and Secondary-only information transfer

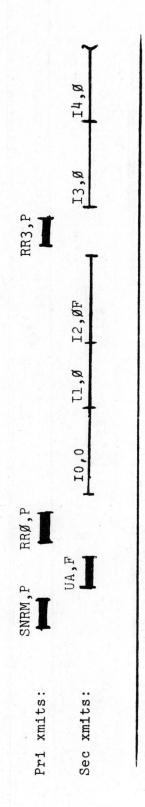

Pri xmits:

Sec xmits:

1.1.2 NRM Start-up procedure and Primary-only information transfer

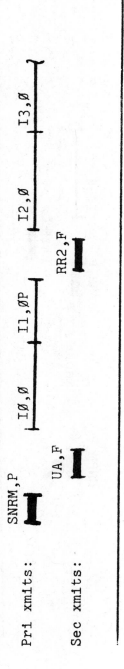

Pri xmits:

Sec xmits:

1.1.3 NRM information transfer by Primary and Secondary

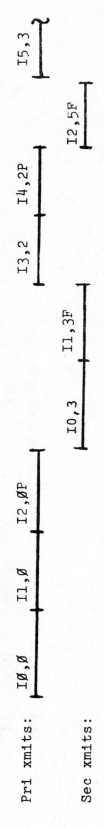

Pri xmits:

Sec xmits:

C8

1.2 **NRM TWA with transmission errors in command frames**

1.2.1 NRM Start-up command error

Pri xmits:  SNRM,P → Timeout → SNRM,P    I0,0

Sec xmits:  UA,F

1.2.2 NRM Primary information frame error

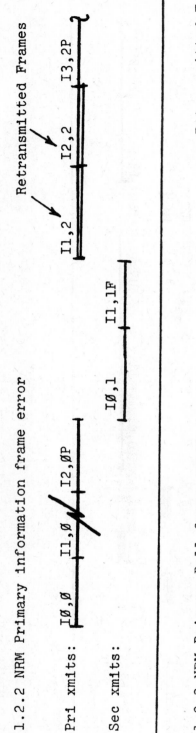

Retransmitted Frames

Pri xmits:  I0,0  I1,0  I2,0P    I1,2  I2,2  I3,2P

Sec xmits:  I0,1  I1,1F

1.2.3 NRM Primary Poll frame error

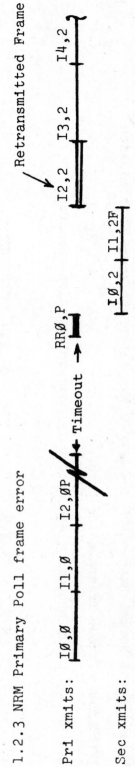

Retransmitted Frame

Pri xmits:  I0,0  I1,0  I2,0P → Timeout → RR0,P    I2,2  I3,2  I4,2

Sec xmits:  I0,2  I1,2F

## 1.3 NRM TWA with transmission errors in response frames

### 1.3.1 NRM Start-up response error

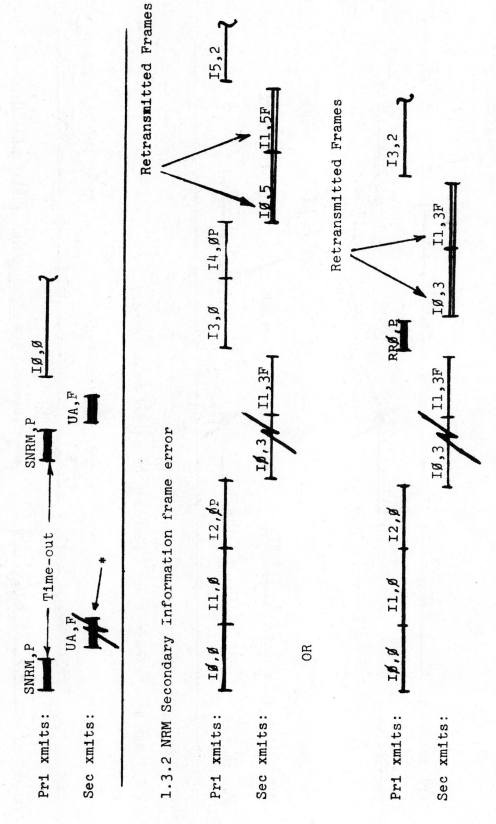

Pri xmits:

Sec xmits:

### 1.3.2 NRM Secondary Information frame error

Pri xmits:

Sec xmits:

OR

Pri xmits:

Sec xmits:

* Idle Link State detection may be used in place of Timeout to initiate Primary transmission.

C10

1080

1.3.3 NRM Secondary "Final" frame error

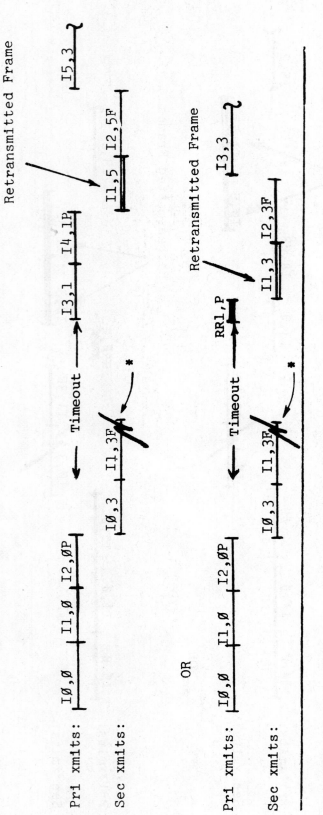

OR

1.4 NRM TWA Command and Response Frame Errors

1.4.1 NRM TWA Primary I and Secondary "Final" I Frame Errors

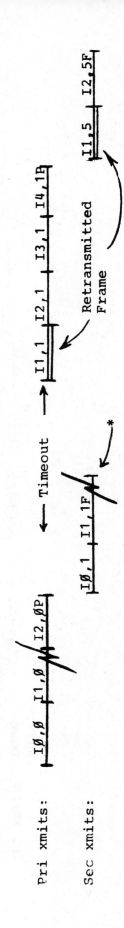

* Idle Link State detection may be used in place of Timeout to initiate Primary Transmission.

C11

2. Examples of Asynchronous Response Mode (ARM) 2-Way Alternate (TWA) Transmission

NOTE: All turnarounds in ARM TWA are by means of Idle Link State detection.

## 2.1 ARM TWA Without Transmission Error

2.1.1 ARM Start-up procedure and Secondary only information transfer

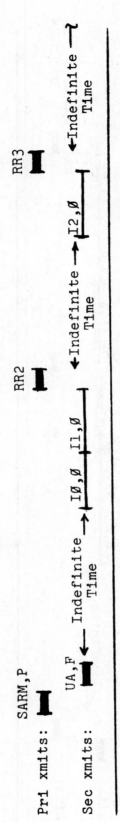

Pri xmits:

Sec xmits:

## 2.1.2 ARM Primary and Secondary information transfer with contention situation

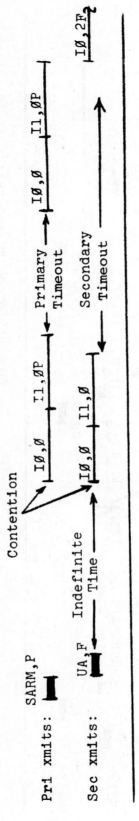

Pri xmits:

Sec xmits:

## 2.2 ARM TWA With Transmission Errors in Command Frames

2.2.1 ARM Start-up command error

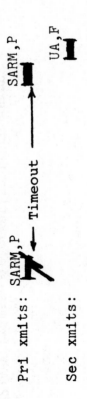

Pri xmits:

Sec xmits:

1081

### 2.2.2 ARM Primary information frame error

### 2.2.3 ARM Primary "Poll" information frame error

### 2.3 ARM TWA With Transmission Errors in Response Frames

### 2.3.1 ARM Start-up

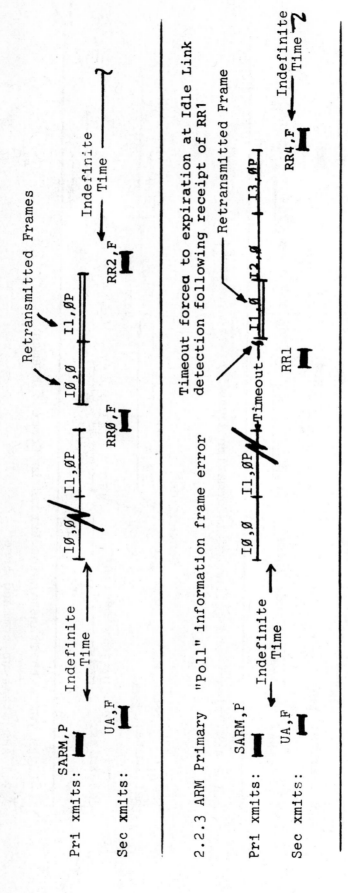

## 2.3.2 ARM Secondary information frame error

Pri xmits: SARM,P

Sec xmits: UA,F

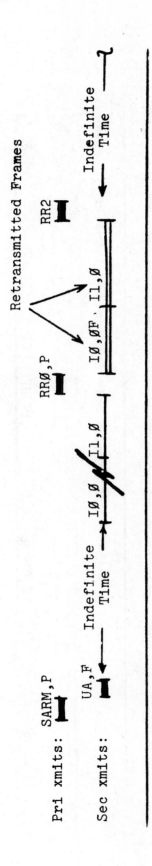

## 2.3.3 ARM Secondary information frame error - no reply received

Pri xmits: SARM,P

Sec xmits: UA,F

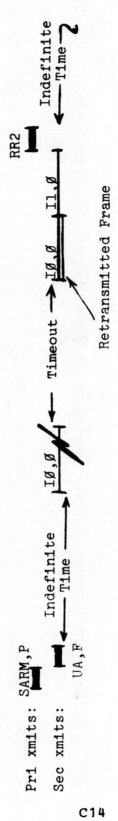

C14

3. Examples of Normal Response Mode (NRM) 2-Way Simultaneous (TWS) Transmission

**3.1 NRM TWS Without Transmission Errors**

3.1.1 NRM Start-up procedure and Secondary only information transfer

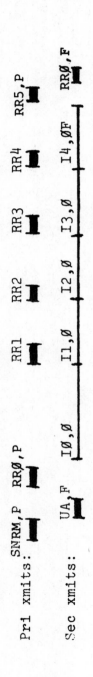

OR (where Primary acknowledgements are returned for several response frames)

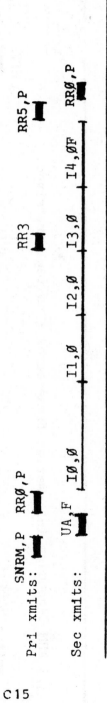

3.1.2 NRM Start-up procedure and Primary only information transfer

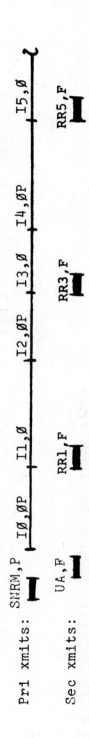

OR (where Primary sets Poll bit to "1" to solicit acknowledgement for several frames)

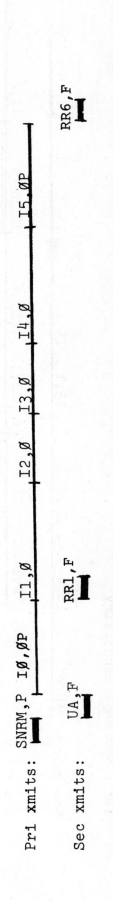

Pri xmits: SNRM,P  I0,0P  I1,0  I2,0  I3,0  I4,0  I5,0P

Sec xmits: UA,F  RR1,F  RR6,F

---

3.1.3 NRM Start-up procedure and Primary/Secondary information transfer

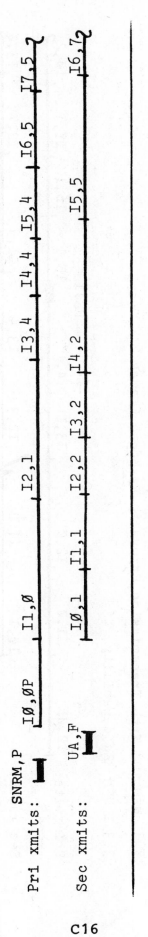

Pri xmits: SNRM,P  I0,0P  I1,0  I2,1  I3,4  I4,4  I5,4  I6,5  I7,5

Sec xmits: UA,F  I0,1  I1,1  I2,2  I3,2  I4,2  I5,5  I6,7

---

3.2 NRM TWS with Transmission Errors in Command Frames

3.2.1 NRM REJ capability

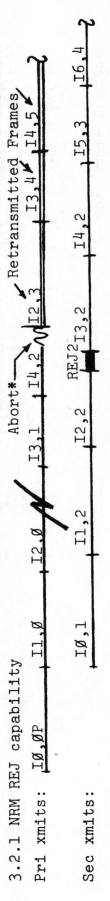

Retransmitted Frames

Abort*

Pri xmits: I0,0P  I1,0  I2,0  I3,1  I4,2  I2,3  I3,4  I4,5  I5,3  I6,4

Sec xmits: I0,1  I1,2  I2,2  REJ2 I3,2  I4,2

* Optional: Frame may be completed or aborted

C16

1085

C16

### 3.2.2 NRM SREJ capability

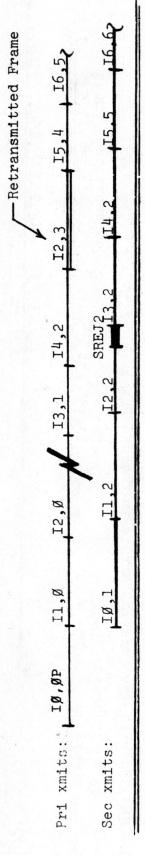

Pri xmits: I0,0P   I1,0   I2,0   I3,1   I4,2   I2,3   I5,4   I6,5

Sec xmits: I0,1   I1,2   I2,2   SREJ2 I3,2   I4,2   I5,5   I6,6

Retransmitted Frame

## 3.3 NRM TWS With Transmission Errors in Response Frames.

### 3.3.1 NRM REJ capability

REJ1

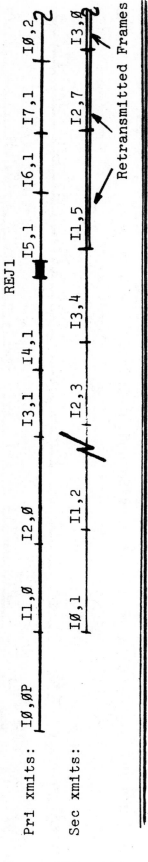

Pri xmits: I0,0P   I1,0   I2,0   I3,1   I4,1   I5,1   I6,1   I7,1   I0,2

Sec xmits: I0,1   I1,2   I2,3   I3,4   I1,5   I2,7   I3,0

Retransmitted Frames

### 3.3.2 NRM SREJ capability

SREJ1

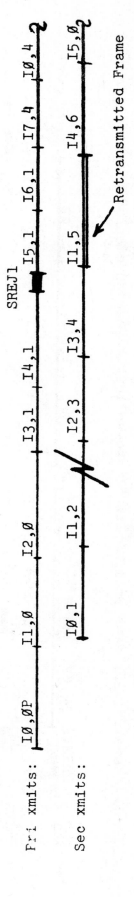

Pri xmits: I0,0P   I1,0   I2,0   I3,1   I4,1   I5,1   I6,1   I7,4   I0,4

Sec xmits: I0,1   I1,2   I2,3   I3,4   I1,5   I4,6   I5,0

Retransmitted Frame

# 4. Examples of Asynchronous Response Mode (ARM) 2-Way Simultaneous (TWS) Transmission

## 4.1 ARM TWS Without Transmission Errors

### 4.1.1 ARM Start-up procedure and intermittent Secondary or Primary information transfer

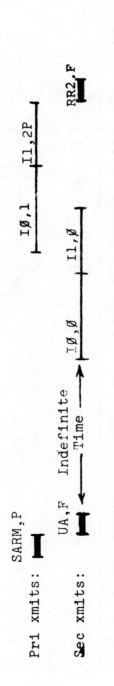

## 4.1.2 ARM Start-up procedure and continuous Primary Secondary information transfer

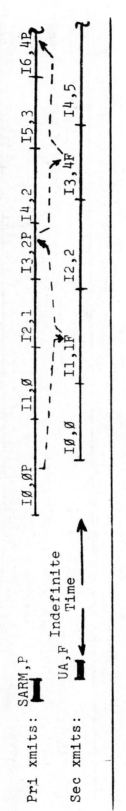

## 4.2 ARM TWS With Transmission Errors in Command Frames

### 4.2.1 ARM Start-up command error

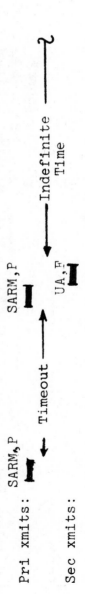

## 4.2.2 ARM REJ capability

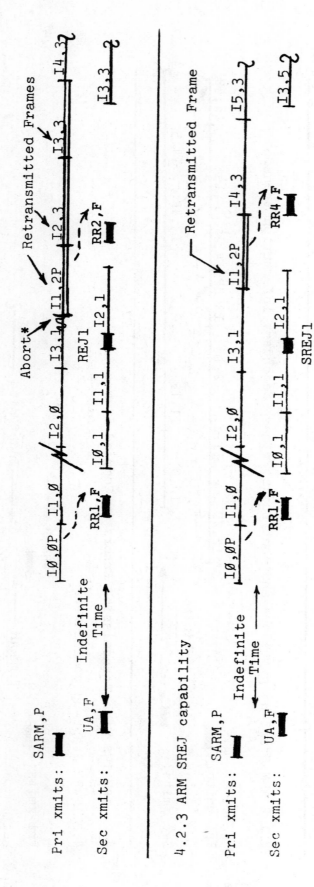

## 4.2.3 ARM SREJ capability

## 4.2.4 ARM P/F bit recovery with transmission error in command frame

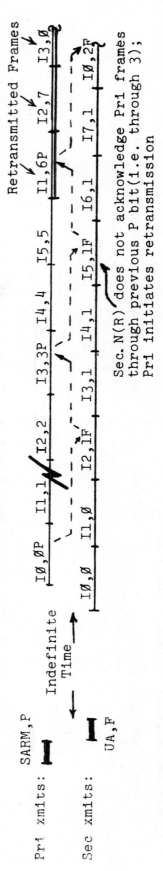

* Optional: Frame may be completed or aborted

# 4.3 ARM TWS With Transmission Errors In Response Frames

## 4.3.1 ARM REJ capability

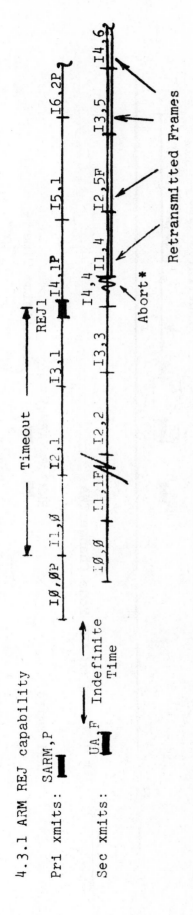

## 4.3.2 ARM SREJ capability

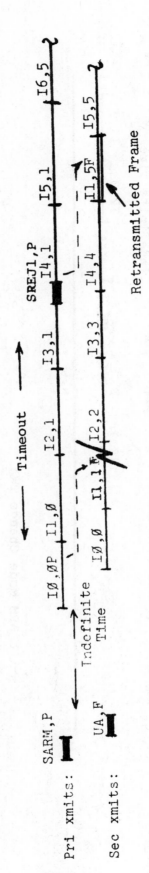

## 4.3.3 ARM P/F bit recovery with transmission error in response frame

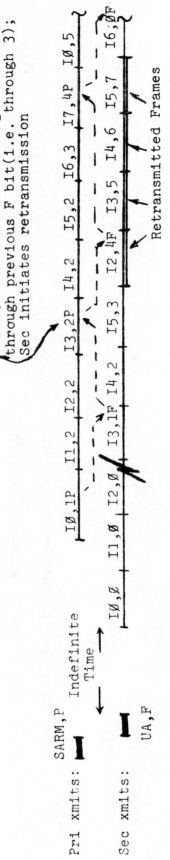

*Optional: Frame may be completed or aborted.

## 5.0 Examples of Changing Control Mode

## 5.1 NRM to ARM 2-Way Alternate (TWA)

### 5.1.1 TWA NRM to ARM Mode Change

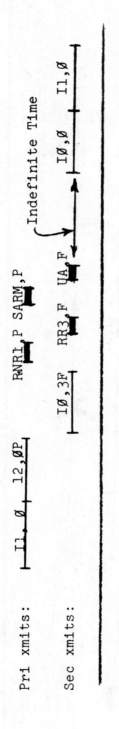

### 5.1.2 NRM to ARM Mode Change TWA

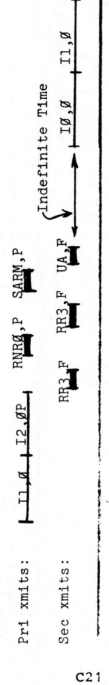

### 5.1.3 NRM to ARM Mode Change TWA

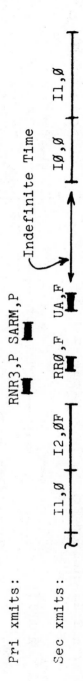

C21

5.2 NRM to ARM 2-Way Simultaneous (TWS)

5.2.1 NRM to ARM Mode Change TWS (Immediate Change)

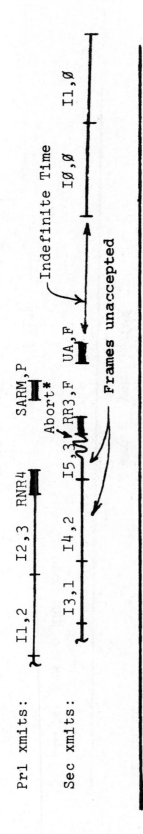

Pri xmits:     I1,2    I2,3    RNR4    SARM,P

Sec xmits:     I3,1    I4,2    I5,3    RR3,F    UA,F    I0,0    I1,0

Abort*

Frames unaccepted

Indefinite Time

5.2.2 NRM to ARM Mode Change TWS (Orderly Change while Pri xmits)

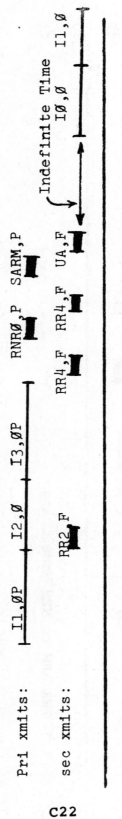

Pri xmits:     I1,0P    I2,0    I3,0P    RNR0,P    SARM,P

sec xmits:     RR2,F    RR4,F    RR4,F    UA,F    I0,0    I1,0

Indefinite Time

5.2.3 NRM to ARM Mode Change TWS (Orderly change while Sec xmits)

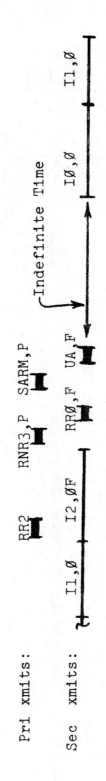

Pri xmits:     RR2    RNR3,P    SARM,P

Sec xmits:     I1,0    I2,0F    RR0,F    UA,F    I0,0    I1,0

Indefinite Time

* Optional: Frame may be completed or aborted

C22

1091

5.3    2-Way Alternate (TWA) ARM to NRM Change

5.3.1  TWA ARM to NRM Mode Change

Pri xmits:

Sec xmits:

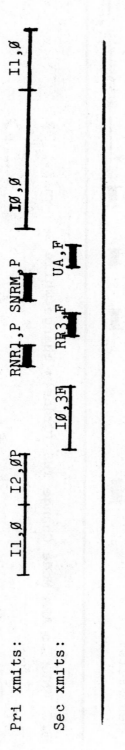

5.3.2  TWA ARM to NRM Mode Change

Pri xmits:

Sec xmits:

5.3.3  TWA ARM to NRM Mode Change

Pri xmits:

Sec xmits:

## 5.4 2-Way Simultaneous (TWS) ARM to NRM Change

### 5.4.1 TWS ARM to NRM Mode Change (Immediate Change)

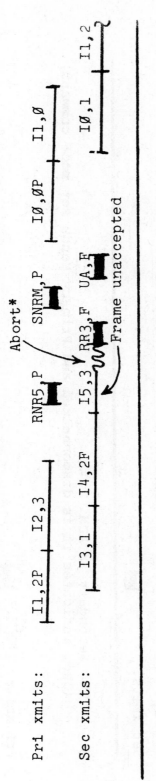

Pri xmits:

Sec xmits:

### 5.4.2 TWS ARM to NRM Mode Change (Orderly Change while Pri xmits)

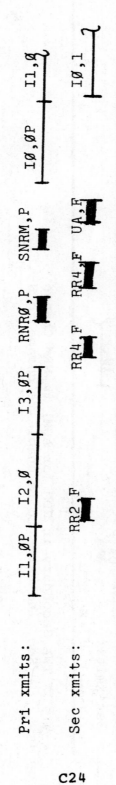

Pri xmits:

Sec xmits:

### 5.4.3 TWS ARM to NRM Mode Change (Orderly change while Sec xmits)

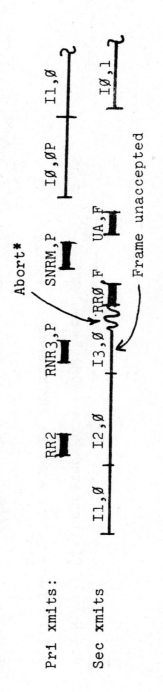

Pri xmits:

Sec xmits

*Optional: Frame may be completed or aborted.

5.5  Normal Disconnected Mode (NDM) Examples

5.5.1  **TWA NDM (or ADM) to ARM Change**

Pri xmits:  SARM,P  $\mid$  I∅.∅  I1.∅P  $\mid$

Sec xmits:  UA,F  $\mid$  I∅.2 ∫

5.5.2  **TWA Secondary in NDM (or ADM) to NRM Change**
       (Sec. indicates it is unable to change to NRM)

Pri xmits:  SNRM,P  $\mid$

Sec xmits:  DM,F  $\mid$

5.5.3  **TWA Secondary in ADM**
       (Sec. indicates it is disconnected and Primary sends set mode command)

Pri xmits: (Poll)  SARM,P  $\mid$

Sec xmits:  DM,F  $\mid$   UA,F  $\mid$

5.5.4  TWA Secondary in NDM (or ADM)
(Sec indicates it is disconnected, and Primary refuses to send set
mode command)

Pri xmits:  (Poll)          DISC,P
                              ⊥

Sec xmits:       DM,F    DM,F
                  ⊥        ⊥

6.0  Examples of End of Operation (General Closing Procedure)

6.1  **NRM TWA**

Pri xmits:   I∅,∅    I1,∅P                RNR2,P        RR2,F

Sec xmits:              I∅,2    I1,2F

6.2  **NRM TWS**

Pri xmits:   I∅,∅P   I1,∅    I2,1    I3,2    RNR3,P        RR4,F

Sec xmits:              I∅,1    I1,1    I2,2F

6.3  **ARM TWA**

Pri xmits:   I∅,∅    I1,∅                RNR2,P    Timeout or Idle Link
                                                    Detection

Sec xmits:                              **RR∅,F**

6.4  **ARM TWS**

Pri xmits:   I∅,∅P   I1,∅    I2,∅    I3,1P            RNR4,P        **RR4,F**

Sec xmits:       I∅,1F   I1,2    I2,3    I3,4F   I4,4
                                                      Frame unaccepted

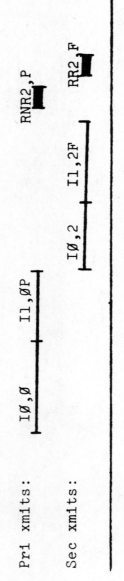
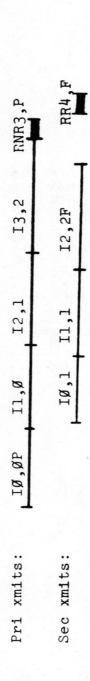
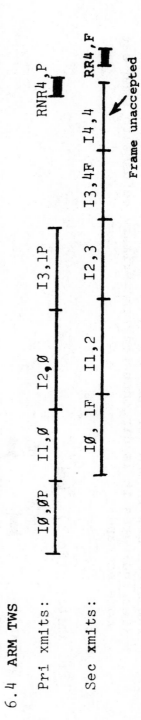

7. Examples of Exception Recovery Procedures

7.1 REJ and Poll/Final Bit Exception Recovery for FDX Operation

7.1.1 NRM – TWS with Information Frame Exception

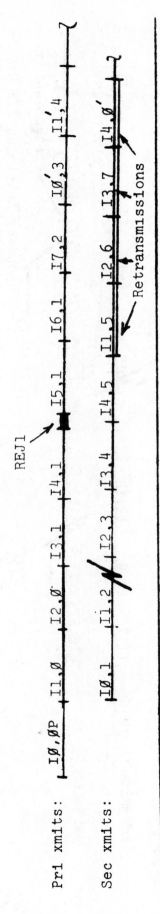

Pri xmits:

Sec xmits:

7.1.2 Example 7.1.1 above except REJ is not received correctly.

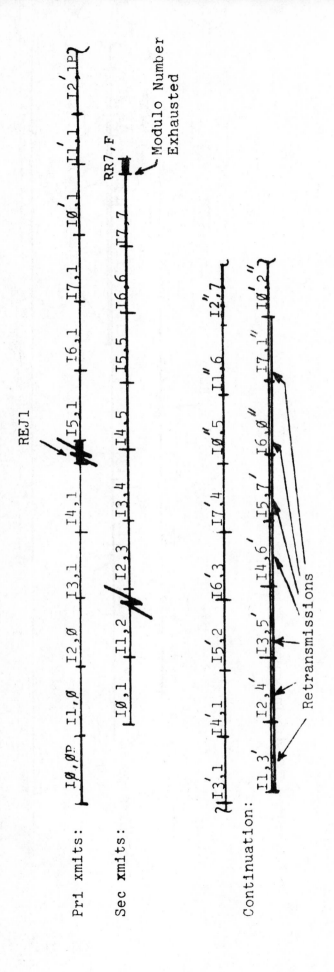

Pri xmits:

Sec xmits:

Continuation:

## 7.1.3 ARM – TWS with Information Frame Exception

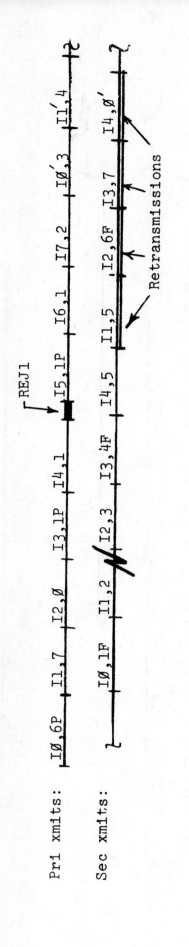

Pri xmits:

Sec xmits:

## 7.1.4 Example 7.1.3 above except REJ is not received correctly

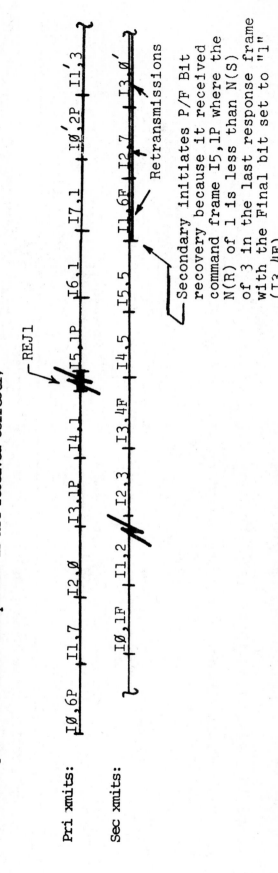

Pri xmits:

Sec xmits:

Retransmissions

Secondary initiates P/F Bit recovery because it received command frame I5,1P where the N(R) of 1 is less than N(S) of 3 in the last response frame with the Final bit set to "1" (I3,4F).

## 7.2 S REJ/REJ Exception Recovery for TWS Operation

### 7.2.1 NRM – TWS with Information Frame Exception

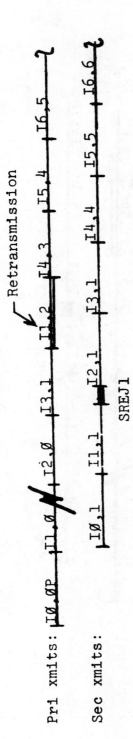

Retransmission

Pri xmits: I0,0P  I1,0  I2,0  I3,1  I4,3  I5,4  I6,5

Sec xmits: I0,1  I1,1  I2,1  I4,4  I5,5  I6,6

SREJ1

Retransmission I1,2

---

### 7.2.2 Example 7.2.1 above except S REJ is not received correctly

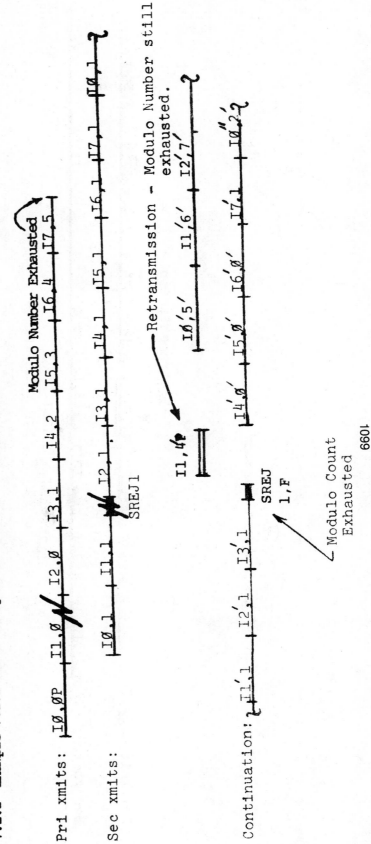

Pri xmits: I0,0P  I1,0  I2,0  I3,1  I4,2  I5,3  I6,4  I7,5

Modulo Number Exhausted  I7,5

Sec xmits: I0,1  I1,1  I2,1  I3,1  I4,1  I5,1  I6,1  I7,1  I0,1

SREJ1

Retransmission – Modulo Number still exhausted.

I1,4P

I0',5'  I1',6'  I2',7'

Continuation: I1',1  I2',1  I3',1

SREJ 1,F

Modulo Count Exhausted

I4',0'  I5',0'  I6',0'  I7',1  I0',2

1099

## 7.2.3 ARM - TWS with I Frame Exception Condition

Pri xmits:

Sec xmits:

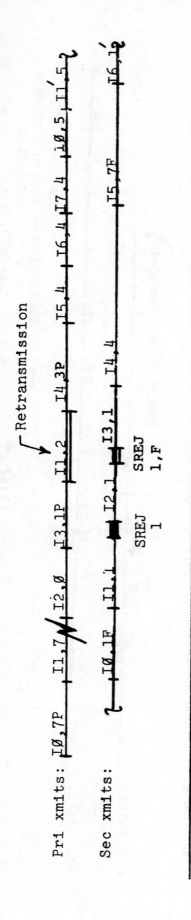

## 7.2.4 Example 7.2.3 above except SREJ is received in error

Pri xmits:

Sec xmits

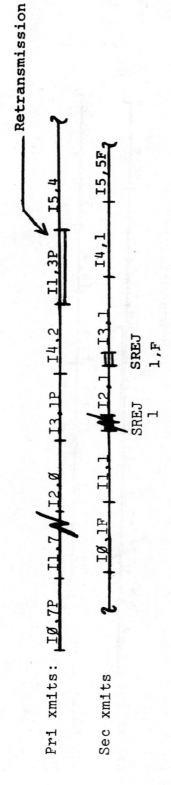

7.2.5 Example 7.2.4 above except two SREJ1 frames received in error.

Pri xmits:

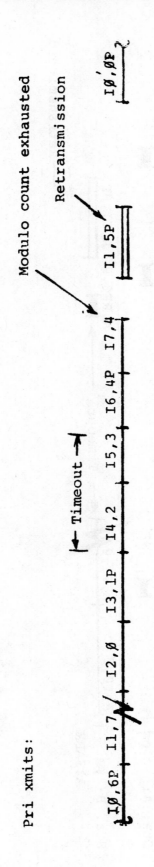

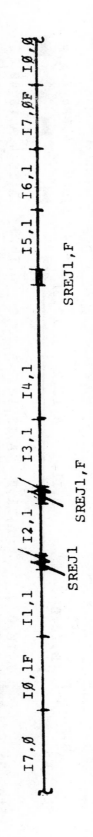

Sec xmits:

8.0 Examples of Balanced Control Operation

8.1 Continuous Information Frames

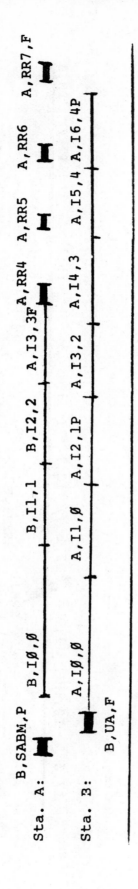

8.2 Discontinuous Information Frames (with error)

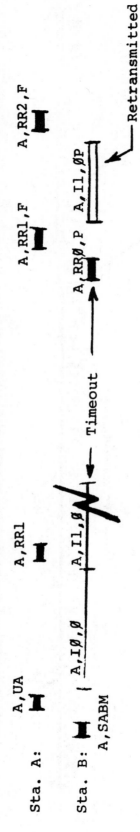

8.3   Simultaneous Mode-Setting Actions   (Contention)

8.3.1   Contention Between SABM and SABM

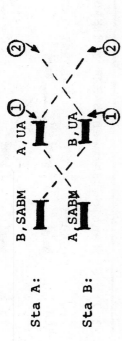

Sta A:

Sta B:

Procedure may be completed at either ① or ② with link available for information transfer.

8.3.2   Contention Between SABM and SABM (Errors)

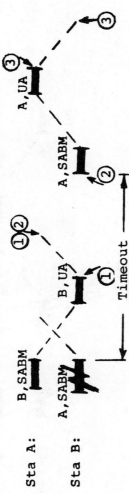

Sta A:

Sta B:

Procedure may be completed at either ①, ② or ③ with link available for information transfer.

8.3.3   Contention Between DISC and DISC

Sta A:

Sta B:

Procedure may be completed at either ① or ② with link in Disconnected Mode.

8.3.4   Contention Between DISC and DISC (Errors)

Sta A:

Sta B:

Procedure may be completed at either ①, ② or ③ with link in Disconnected Mode.

C34

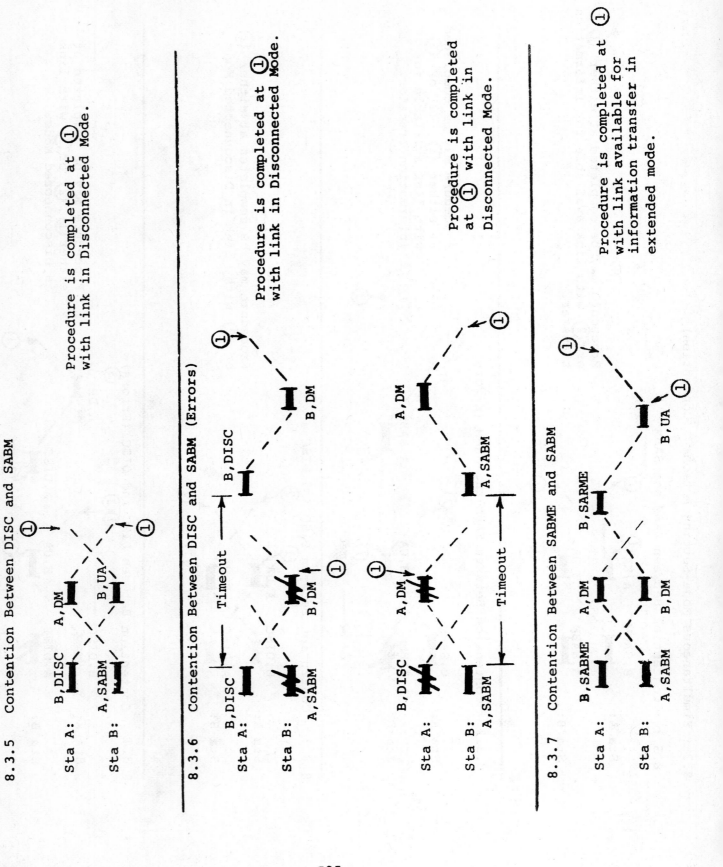

8.3.5  Contention Between DISC and SABM

Sta A:   B,DISC   A,DM

Sta B:   B,UA   A,SABM

Procedure is completed at ① with link in Disconnected Mode.

8.3.6  Contention Between DISC and SABM (Errors)

Sta A:   B,DISC   B,DISC   Timeout

Sta B:   A,SABM   B,DM   B,DM   B,DISC

Procedure is completed at ① with link in Disconnected Mode.

Sta A:   B,DISC   A,DM   A,DM

Sta B:   A,SABM   A,SABM   Timeout

Procedure is completed at ① with link in Disconnected Mode.

8.3.7  Contention Between SABME and SABM

Sta A:   B,SABME   A,DM   B,SARME

Sta B:   A,SABM   B,DM   B,UA

Procedure is completed at ① with link available for information transfer in extended mode.

C35

8.3.8 Contention Between SABME and SABM (Errors)

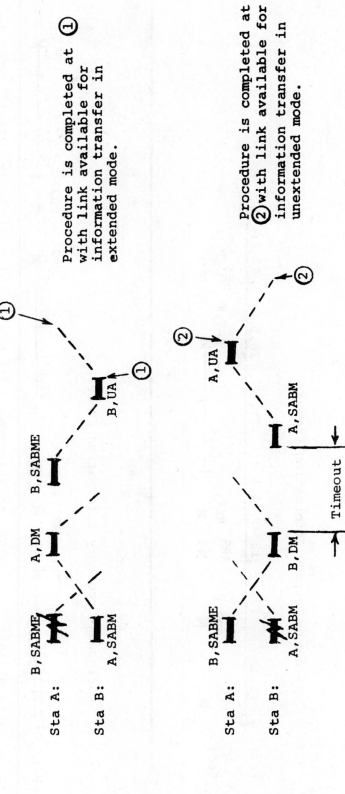

Procedure is completed at ① with link available for information transfer in extended mode.

Procedure is completed at ② with link available for information transfer in unextended mode.

## 9.0 Primary-Secondary ARM Two-Way Simultaneous Point-to-Point Operation

### 9.1 Continuous Information Frames from Primary and Secondary

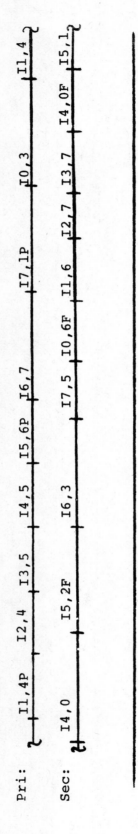

### 9.2 Continuous Primary Information Frames

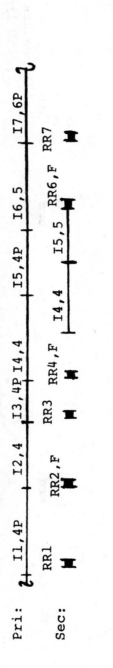

### 9.3 Continuous Secondary Information Frames

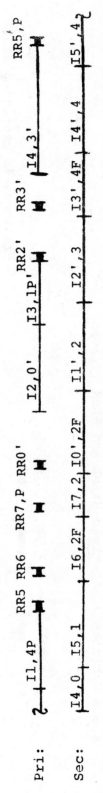

10.0 Symmetrical (Back-to-Back) Primary-Secondary Point-to-Point TWS Operation
Reference Figure 2-3 Configuration

10.1 Secondary B in ARM - Secondary A in NRM Operation

Sta A:

Sta B:

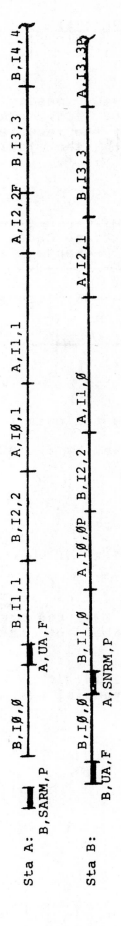

10.2 Use of RNR to restrict Information frames from Secondary Operation

Sta A:

Sta B:

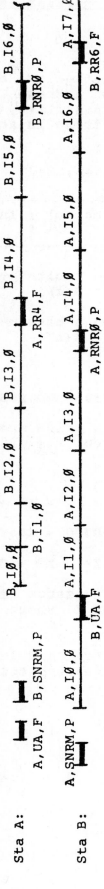

10.3 Secondaries do not transmit Information frames (Optional Function 8) Operation

Sta A:

Sta B:

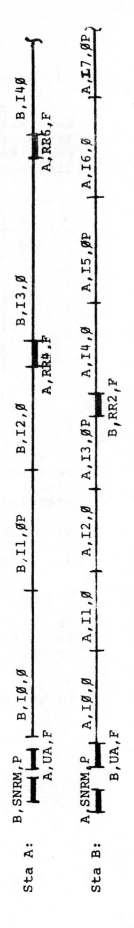

C38

# APPENDIX D - FRAME CHECK SEQUENCE (FCS)

## D1. Description

The transmission integrity of a received message is determined by use of a Frame Check Sequence (FCS). The FCS is generated by a transmitter, inspected by the receiver and positioned within a frame in accordance with the following diagrams:

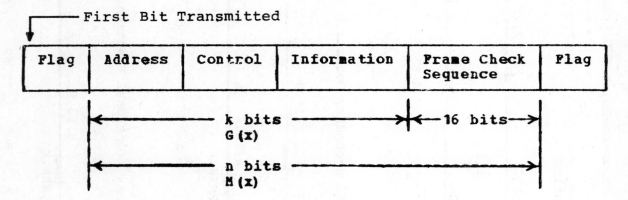

The procedure for using the FCS assumes the following:

1. The k bits of data which are being checked by the FCS can be represented by a polynomial $G(x)$.

Examples:

a.  $G(x) = 10100100 = X^7 + X^5 + X^2 = X^2(X^5+X^3+1)$

b.  $G(x) = 00--010100100 = X^7+X^5+X^2 = X^2(X^5+X^3+1)$

c.  $G(x) = 101001 = X^5 + X^3 + 1$

In general, leading zeros don't change $G(x)$ and trailing zeros add a factor of $X$ where n is the number of trailing zeros.

2. The Address, Control and Information field (if it exists in the message) are represented by the polynomial $G(x)$.

3. For the purpose of generating the FCS, the first bit following the opening flag is the coefficient of the highest degree term of G(x) regardless of the actual representation of the Address, Control and Information fields.

4. There exists a generator polynomial P(x) of degree 16, having the form $P(x) = X^{16} + X^{12} + X^5 + 1$

D.2 Generation and use of FCS

The FCS is defined as a one's complement of a remainder, R(x), obtained from the modulo two division of

$$X^{16} G(x) + X^k (X^{15} + X^{14} + X^{13} + X^{12} + X^{11} + X^{10} + X^9 + X^8 + X^7 + X^6 + X^5 + X^4 + X^3 + X^2 + X^1 + 1)$$

by the generator polynomial P(x).

$$\frac{X^{16} G(x) + X^k (X^{15} + X^{14} - - - - + X + 1)}{P(x)} = Q(x) + \frac{R(x)}{P(x)} \longleftarrow \overline{FCS}$$

The multiplication of G(x) by $X^{16}$ corresponds to shifting the message G(x), 16 places and thus providing the space of 16 bits for the FCS.

The addition of $X^k (X^{15} + X^{14} - - - - - - - - - - + X + 1)$ to $X^{16}$ G(x) is equivalent to inverting the first 16 bits of G(x). It can also be accomplished in a shift register implementation by presetting the register to all "ones" initially. This term is present to detect erroneous addition or deletion of zero bits at the leading end of M(x) due to erroneous flag shifts.

The complementing of R(x), by the transmitter, at the completion of the division insures that the transmitted sequence M(x) has a property which permits the receiver to detect addition or deletion of trailing zeros which may appear as a result of errors.

At the transmitter the FCS is added to the $X^{16}$ G(x) and results in the total message M(x) of length k+16, where M(x) = $X^{16}$ G(x) + FCS.

The receiver can employ one of several detection processes, two of which are discussed here. In the first process, the incoming M(x) (assuming no errors; i.e. M*(x) = M(x)) is

D2

multiplied by $X^{16}$, added to $X^{k+16}$ ($X^{15} + X^{14} - - - - - - + X + 1$) and divided by $P(X)$.

$$\frac{X^{16}[X^{16} G(X) + FCS] + X^{k+16} (X^{15} + X^{14} - - - + X + 1)}{P(x)} =$$

$Qr(x) + Rr(x)/P(x)$

Since the transmission is error free, the remainder $Rr(x)$ will be "0001110100001111" ($X^{15}$ through $X^0$).

Rr(X) is the remainder of the division: $\frac{X^{16} L(x)}{P(x)}$

where $L(x) = X^{15} + X^{14} - - - - + X + 1$. This can be shown by establishing that all other terms of the numerator of the receiver division are divisible by $P(x)$. This will be done below.

Note that $FCS = \overline{R(x)} = L(x) + R(x)$. (Adding $L(x)$ to a polynomial of its same length is equivalent to a bit by bit inversion of the polynomial.)

The receiver division numerator can be rearranged to:

$X^{16}[X^{16} G(x) + X^k L(x) + R(x)] + X^{16} L(x)$

It can be seen that the first term is divisible by $P(x)$ by inspecting the transmitter generation equation, thus the $X^{16}L(x)$ term is the only contributor to $Rr(x)$.

The second process differs from the first in that another term ($X^{16}L(x)$) is added to the numerator of the generation equation. This causes a remainder of zero to be generated if $M^*(x)$ is received error free.

D.3 Implementation

A shift register FCS implementation is described in detail here. It utilizes "ones presetting" at both the sender and the receiver and the receiver does not invert the FCS. The receiver thus checks for the non-zero residual $Rr(x)$ to indicate an error free transmission.

Figure D.1 is an illustration of the implementation. It shows a configuration of storage elements and gates. The addition of $X^k(X^{15}+X^{14}------+X+1)$ to the $X^{16}$ $G(x)$ can be

1110

accomplished by presetting all storage elements to a binary value "1".

The one's complement of R(x) is obtained by the logical bit by bit inversion of the transmitter's R(x).

Figure D.1 shows the implementation of the FCS generation for transmission. The same hardware can also be used for verification of data integrity upon data reception.

Before transmitting data, the storage elements, $X_0 - X_{15}$ are initialized to "ones". The accumulation of the remainder R(x) is begun by enabling the "A" and thereby enabling gates G2 and G3. The data to be transmitted goes out to the receiver via G2 and at the same time the remainder is being calculated with the use of feedback path via G3. Upon completion of transmitting the k bits of data, the "A" is disabled and the stored R(x) is transmitted via G1 and I1 while G2 and G3 are disabled. The I1 provides the necessary inversion of R(x).

At the receiver, before data reception, the storage elements, $X_0 - X_{15}$ are initialized to "ones". The incoming message is then continuously divided by P(x) via G3 ("A" enabled). If the message contained no errors, the storage elements will contain "0001110100001111" ($X^{15}$ through $X^0$) at the end of the M*(x).

Figure D.2 is an example of the receiver and transmitter states during a transmission of a 19 bit G(x) and a 16 bit FCS.

The implementation of the FCS generation and the division by P(x) as described in this Appendix is used as an example only. Other implementations are possible and may be utilized. This standard only requires that the FCS be generated in accordance with the rules of Sections 3.5 and 12.1 and that the checking process involve division by the polynomial P(x). Furthermore, the order of transmission of M(x) is the coefficient of the highest degree term first and thereafter in decreasing order of powers of x, regardless of the actual representation of fields internal to M(x).

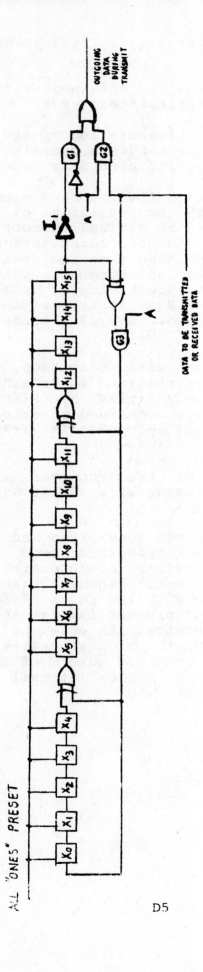

FIGURE D1. FCS IMPLEMENTATION

D5

|  | INPUT<br>TO TX | TX CRC | INPUT<br>TO RX | RX CRC |
|---|---|---|---|---|
| MSB |  | 1111111111111111 |  | 1111111111111111 |
|  | 0 | 1111101111110111 | 0 | 1111101111110111 |
|  | 1 | 0111110111111011 | 1 | 0111110111111011 |
|  | 1 | 0011111011111101 | 1 | 0011111011111101 |
|  | 1 | 0001111101111110 | 1 | 0001111101111110 |
|  | 1 | 1000101110110111 | 1 | 1000101110110111 |
|  | 0 | 1100000111010011 | 0 | 1100000111010011 |
|  | 0 | 1110010011100001 | 0 | 1110010011100001 |
|  | 1 | 0111001001110000 | 1 | 0111001001110000 |
| G(x) | 1 | 1011110100110000 | 1 | 1011110100110000 |
|  | 0 | 0101111010011000 | 0 | 0101111010011000 |
|  | 0 | 0010111101001100 | 0 | 0010111101001100 |
|  | 1 | 1001001110101110 | 1 | 1001001110101110 |
|  | 1 | 1100110111011111 | 1 | 1100110111011111 |
|  | 0 | 1110001011100111 | 0 | 1110001011100111 |
|  | 0 | 1111010101111011 | 0 | 1111010101111011 |
|  | 0 | 1111111010110101 | 0 | 1111111010110101 |
|  | 1 | 0111111101011010 | 1 | 0111111101011010 |
|  | 1 | 1011101110100101 | 1 | 1011101110100101 |
|  | 1 | 0101110111010010 | 1 | 0101110111010010 |
|  |  | 0010111011101001 | 1 | 1010101011100001 |
|  |  | 0001011101110100 | 0 | 1101000101111000 |
|  |  | 0000101110111010 | 1 | 1110110010110100 |
|  |  | 0000010111011101 | 1 | 1111001001010010 |
|  |  | 0000001011101110 | 0 | 0111100100101001 |
|  |  | 0000000101110111 | 1 | 0011110010010100 |
|  |  | 0000000010111011 | 0 | 0001111001001010 |
|  |  | 0000000001011101 | 0 | 0000111100100101 |
|  |  | 0000000000101110 | 0 | 1000001110011010 |
|  |  | 0000000000010111 | 1 | 1100010111000101 |
|  |  | 0000000000001011 | 0 | 1110011011101010 |
|  |  | 0000000000000101 | 0 | 0111001101110101 |
|  |  | 0000000000000010 | 0 | 1011110110110010 |
|  |  | 0000000000000001 | 1 | 1101101011010001 |
|  |  | 0000000000000000 | 0 | 1110100101100000 |
|  |  | 0000000000000000 | 1 | 1111000010111000 |

FIGURE D.2  FCS EXAMPLE

A - Address field
ABM - Asynchronous Balanced Mode
ADCCP - Advanced Data Communication Control Procedures
ADM - Asynchronous Disconnect Mode
ARM - Asynchronous Response Mode
ARO - Asynchronous Respond Opportunity
BA - Balanced, Asynchronous Class
C - Control Field
C - Combined Station (Figure 10-1 only)
CCITT - International Telegraph and Telephone Consultative Committee
Comb - Combined (station)
DISC - Disconnect (Command)
DM - Disconnect Mode (Response)
ECMA - European Computer Manufacturers Association
F - Flag
F bit - Final Bit
FCS - Frame Check Sequence
FRMR - Frame Reject (Response)
I - Information (command, response)
I - Information Format (frame)
I frame - Information Format frame
ID - Identification
IM - Initialization Mode
Info - Information Field
IS - Initialization State
ISO - International Standards Organization
ITS - Information Transfer State
LDS - Logically Disconnected State
LSB - Least Significant Bit
M - Modifier Function Bit
MSB - Most Significant Bit
N - An integer variable
NA - Not Applicable
NDM - Normal Disconnect Mode
N(R) - Receive Sequence Number
NRM - Normal Response Mode
NRO - Normal Respond Opportunity
N(S) - Send Sequence Number
P bit - Poll Bit
P - Primary Station (Figure 10-1 only)
P/F bit - Poll or Final Bit
Pri - Primary (station)
P/S/C - Primary or Secondary or Combined (station)
Pri/Sec - Primary or Secondary (station)
R - Receive Variable
RD - Request Disconnect (Command)
RSET - Reset (Command)
REJ - Reject (Command, Response)
RIM - Request Initialization Mode (Response)

RNR - Receive Not Ready (Command, Response)
RR - Receive Ready (Command, Response)
S - Depending upon usage: - Send Variable
                          - Supervisory Function Bit
                          - Supervisory Format (frame)
S frame - Supervisory Format frame
S - Secondary Station (Figure 10-1 only)
SABM - Set Asynchronous Balanced Mode (Command)
SABME - Set Asynchronous Balanced Mode Extended (Command)
SARM - Set Asynchronous Response Mode (Command)
SARME - Set Asynchronous Response Mode Extended (Command)
Sec - Secondary (station)
SIM - Set Initialization Mode (Command)
SNRM - Set Normal Response Mode (Command)
SNRME - Set Normal Response Mode Extended (Command)
SREJ - Selective Reject (Command, Response)
TO - Timeout
TWA - Two-Way Alternate
TWS - Two-Way Simultaneous
U - Unnumbered Format (frame)
U frame - Unnumbered Format frame
UA - Unnumbered Acknowledgement (Response)
UA - Unbalanced, Asynchronous Class
UI - Unnumbered Information (Command, Response)
UN - Unbalanced, Normal Class
UP - Unnumbered Poll (Command)
XID - Exchange Identification (Command, Response)
W,X,Y,Z - bits in FRMR Status Field

NOTE: The mathematical symbols and abbreviations used in Section 12.0, Frame Check Sequence (FCS) Generation and Checking, and Appendix D, Frame Check Sequence (FCS), are not included above; they are defined as introduced in Section 12 and Appendix D.

FEDERAL STANDARD

TELECOMMUNICATIONS:  CODING AND MODULATION
REQUIREMENTS FOR NONDIVERSITY 2400 BIT/SECOND MODEMS

This standard is issued by the General Services Administration pursuant to the Federal
Property and Administrative Services Act of 1949, as amended.

1.  Scope

1.1  Description.  This standard establishes the coding and modulation requirements for 2400 bit/second
modems owned or leased by the Federal Government for use over analog transmission channels other than
those derived from high-frequency radio facilities.

1.2  Purpose.  This standard is to facilitate interoperability between telecommunication facilities and
systems of the Federal Government.

1.3  Application.  This standard shall be used by all Federal agencies in the design and procurement of
nondiversity 2400 bit/second modems for use with nominal 4 KHz channels derived from either switched
networks or dedicated lines.  Typically, such channels are derived from frequency division multiplex
equipment associated with microwave, cable, and satellite transmission systems.

2.  Requirements

2.1  2400 Bit/second Operation

2.1.1  The transmit carrier frequency shall be 1800 ±1 Hz.

2.1.2  The data stream to be modulated is divided into pairs of consecutive bits (dibits).  Each dibit is
encoded as a phase change of the 1800 Hz carrier relative to the phase of the carrier during transmission
of the immediately preceding dibit as indicated below.

| DIBIT | PHASE CHANGE |
|-------|-------------|
| 00 | + 45° |
| 01 | + 135° |
| 11 | +225° |
| 10 | + 315° |

The phase change is the actual phase shift in the transition region from the end of one signaling element
to the beginning of the following signaling element as illustrated below.

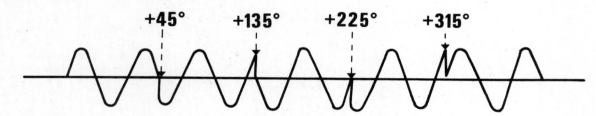

At the demodulator, the dibits are decoded and the bits are reassembled in the correct order.  The left-
hand digit of the dibit is the one occurring first in the data stream as it enters the modulator portion
of the modem.

2.1.3  Synchronization

2.1.3.1  During the interval between Request to Send (RS) indication from an associated Data Terminal
Equipment (DTE) and the modem responding  with a Clear to Send (CS) indication, synchronization signals
will be generated by the transmitting modem(s).  Unless a scrambler is utilized, continuous 225° phase
changes (dibit 11) shall be transmitted for the duration of the synchronization period.

FSC TELE

2.1.3.2  Modems capable of 2-wire, half-duplex operation shall have the ability to delay Clear to Send (CS) indication for a period of at least 185 ms following the receipt of Request to Send (RS) indication. During 2-wire, half-duplex operation using the above delay, modems shall also withhold Receiver Ready (RR) indication after the end of each modem transmission for a period, not to exceed 175 ms, to protect against the effect of line echoes.

2.1.4  The data rate shall be 2400 bit/second $\pm$0.01 percent.

## 2.2  1200 Bit/Second Operation (Optional)

2.2.1  1200 bit/second data rate operation shall be achieved, when available, by encoding binary 0 and 1 bits as $+90^{o}$ and $+270^{o}$ phase changes, respectively.

2.2.2  The data rate shall be 1200 bit/second $\pm$0.01 percent.

2.2.3  All other characteristics shall be as specified for 2400 bit/second operation.

## 2.3  Secondary Channel (Optional)

2.3.1  Secondary channel modulation rate shall not exceed 75 baud.

2.3.2  Characteristic frequencies of the binary 1 (mark) and binary 0 (space) shall be 390 $\pm$1 Hz and 450 $\pm$1 Hz, respectively.

2.3.3  When simultaneous transmission of the primary channel and secondary channel occur in the same direction, the secondary channel shall be 6+0.6dB lower in power level than the primary channel.

## 2.4  Scramber (Optional)

2.4.1  When utilized, digital input signals will be encoded in the modem prior to modulation of the 1800 Hz carrier.  Similarly, demodulated signals will be decoded by an inverse process at the distant modem.

2.4.2  The encoder and decoder shall be capable of being bypassed by switch selection.

## 2.5  General Characteristics

2.5.1  The modem shall be capable of presenting an input and output impedance to the analog line of 600$\pm$ 60 ohms balanced.

2.5.2  The output level of the modulator shall be adjustable from -12dBm to -3dBm in no greater than 1dB steps.  Output level shall not be adjustable by operating personnel.

2.5.3  The demodulator shall have an input sensitivity adjustable to -42+3dBm and -32+3dBm.  When the above-stated input sensitivities are used, the input level dynamic range shall be at least 30dB above the input sensitivity.

2.5.4  A fixed compromise equalizer shall either be incorporated into the receiver or be switchable between the receiver and transmitter.  The characteristics of this equalizer are dependent upon system application.

3.  Changes.  When a Federal agency considers that this standard does not provide for its essential needs, a statement citing inadequacies shall be sent in duplicate to the General Services Administration, Federal Supply Service, Washington, D.C. 20406, in accordance with provisions of Federal Property Management Regulations 41 CFR 101-29.3.  The General Services Administration will determine the appropriate action to be taken and will notify the agency.

Preparing Activity:

National Communications System
Office of Technology and Standards
Washington, D.C.  20305

U. S. GOVERNMENT PRINTING OFFICE : 1977 - 241-237/2164

2

FEDERAL STANDARD

TELECOMMUNICATIONS: CODING AND MODULATION
REQUIREMENTS FOR 4800 BIT/SECOND MODEMS

This standard is issued by the General Services Administration pursuant to the Federal Property and Administrative Services Act of 1949, as amended.

1. Scope

1.1 Description. This standard establishes the coding and modulation requirements for 4800 bit/second modems owned or leased by the Federal Government for use over analog transmission channels.

1.2 Purpose. This standard is to facilitate interoperability between telecommunication facilities and systems of the Federal Government.

1.3 Application. All Federal departments and agencies shall comply with this standard in the design and procurement of 4800 bit/second modems (and equipment containing modems) for use with nominal 4 KHz analog channels. Typically, nominal 4 KHz analog channels are derived from frequency division multiplex equipment associated with microwave, coaxial cable, and satellite transmission systems. Modems described by this standard may also be used on nonmultiplexed transmission systems such as metallic cable facilities.

2. Requirements

2.1 4800 Bit/second Operation

2.1.1 Carrier Frequency. The carrier frequency shall be 1800±1 Hz.

2.1.2 Spectrum. A 50 percent raised cosine energy spectrum shaping shall be equally divided between the modulator and demodulator. The modulator energy spectrum shall be shaped in such a way that, with continuous data ONE's applied to the input of the scrambler, the resulting spectrum shall have a substantially linear phase characteristic over the range of 1100 Hz to 2500 Hz and an energy density at 1000 Hz and 2600 Hz attenuated 3±2 dB with respect to the maximum energy density between 1000 Hz and 2600 Hz.

2.1.3 Data and Modulation Rates. The data rate shall be 4800 bits/ second ±.01 percent, i.e., the modulation (symbol) rate shall be 1600 baud ±.01 percent.

2.1.4 Encoding Data Bits. The data stream to be modulated is divided into groups of three consecutive bits (tribits). Each tribit shall be encoded as a phase change of the 1800 Hz carrier relative to the phase of the carrier during transmission of the immediately preceding signal element as indicated below.

| TRIBIT VALUES | | | PHASE CHANGE |
|:---:|:---:|:---:|:---:|
| 0 | 0 | 1 | 0° |
| 0 | 0 | 0 | 45° |
| 0 | 1 | 0 | 90° |
| 0 | 1 | 1 | 135° |
| 1 | 1 | 1 | 180° |
| 1 | 1 | 0 | 225° |
| 1 | 0 | 0 | 270° |
| 1 | 0 | 1 | 315° |

The phase change is the actual on-line phase shift in the transition region from the end of one signal element to the beginning of the following signal element. At the demodulator, the tribits are decoded and the bits are reassembled in correct order. The left-hand digit of the tribit is the one occurring first in the data stream as it enters the modulator portion of the modem after the scrambler.

FSC TELE

2.2 ·2400 Bit/Second Operation (Optional). 2400 bit/second operation, when utilized, shall follow one of the two options described below.

2.2.1  Option I.

2.2.1.1  Carrier Frequency. The carrier frequency shall be 1800+1 Hz.

2.2.1.2  Spectrum. A minimum of 50 percent raised cosine energy spectrum shaping shall be equally divided between the modulator and demodulator. The modulator energy spectrum shall be shaped in such a way that, with continuous data ONE's applied to the input of the scrambler, the resulting spectrum shall have a substantially linear phase characteristic over the range of 1300 Hz to 2300 Hz and an energy density at 1200 Hz and 2400 Hz attenuated 3+2 dB with respect to the maximum energy density between 1200 Hz and 2400 Hz.

2.2.1.3  Data and Modulation Rates. The data rate shall be 2400 bits/ second +.01 percent, i.e., the modulation (symbol) rate shall be 1200 baud +.01 percent.

2.2.1.4  Encoding Data Bits. For 2400 bit/second operation, the data stream is divided into groups of two bits (dibits). Each dibit shall be encoded as a phase change of the 1800 Hz carrier relative to the phase of the carrier during transmission of the immediately preceding signal element as indicated below.

| DIBIT VALUE | PHASE CHANGE |
|:---:|:---:|
| 0  0 | 0° |
| 0  1 | 90° |
| 1  1 | 180° |
| 1  0 | 270° |

The phase change is the actual on-line phase shift in the transition region from the end of one signal element to the beginning of the following signal element. At the demodulator, the dibits are decoded and reassembled in the correct order. The left-hand digit of the dibit is the one occurring first in the data stream.

2.2.2  Option II. Federal Standard 1005 (2400 bit/second modem) shall be followed completely for 2400 bit/second operation in lieu of this standard.

2.3  Operating Sequences. Modems shall be capable of operation in accordance with the operating sequences described in this section. However, this does not preclude the use of other, additional operating sequences in applications such as fast-polling.

2.3.1  "Turn-On" Sequence

2.3.1.1  During the interval between Request to Send (RS) indication from an associated Data Terminal Equipment (DTE) and the modem responding with a Clear to Send (CS) indication, synchronization signals shall be generated by the transmitting modem(s). These signals are used to establish carrier detection, automatic gain control (AGC) if required, timing synchronization, equalizer convergence, and descrambler synchronization.

2.3.1.2  Two "turn-on" sequences are defined; i.e.,

(a)  A short sequence for operation over well conditioned duplex or one way only channels and for subsequent turn-around operation over half-duplex channels; and

(b)  A longer sequence for operation over less conditioned duplex or one way only channels and for initial establishment of connections over half-duplex channels.

2

During half-duplex operation, sequence (b) is only used after the first OFF to ON transition of the Request to Send (RS) circuit following the OFF to ON transition of the Data Mode (DM) circuit, or at the OFF to ON transition of the DM circuit if the RS circuit is already ON. After every subsequent OFF to ON transition of the RS circuit, sequence (a) is used. The "turn-on" sequences, for both data rates, are divided into three sequences as follows:

| TYPE OF LINE SIGNAL | SEGMENT 1 | SEGMENT 2 | SEGMENT 3 | TOTAL OF SEGMENTS | |
|---|---|---|---|---|---|
| | CONTINUOUS 180° PHASE REVERSALS | 0°-180° CONDITIONING PATTERN | CONTINUOUS SCRAMBLED ONE's | NOMINAL TOTAL "TURN-ON" SEQUENCE TIME | |
| | | | | 4800 b/s | 2400 b/s |
| NUMBER OF SYMBOL INTERVALS (S.I.) | a) 14 S.I. b) 50 S.I. | a) 58 S.I. b) 1074 S.I. | 8 S.I. | a) 50 ms b) 708 ms | a) 67ms b) 943 ms |

2.3.1.3  The composition of segment 1 is continuous 180° phase reversals on line for 14 symbol intervals in the case of sequence (a) and for 50 symbol intervals in the case of sequence (b).

2.3.1.4  Segment 2 is composed of an equalizer conditioning pattern that is a pseudorandom sequence generated by the scrambler. When the pseudorandom sequence contains a zero, 0° phase change is transmitted; when it contains a ONE, 180° phase change is transmitted. Segment 2 begins with the sequence 0°, 180°, 180°, 180°, 180°, 180°, 0°, . . . according to the pseudorandom sequence and continues for 58 symbol intervals in the case of sequence (a) and 1074 symbol intervals in the case of sequence (b). The detailed pseudorandom generation is described in paragraph 2.5.

2.3.1.5  Segment 3 commences transmission according to the encoding described in paragraphs 2.1.4 and 2.2.4 with continuous data ONE's applied to the input of the scrambler. Segment 3 is 8 symbol intervals in duration. At the end of Segment 3, the Clear to Send (CS) circuit is turned ON and data is applied to the input of the scrambler.

2.3.2  Echo Protection.  Modems capable of half-duplex operation shall:

2.3.2.1  Be capable of transmitting, immediately prior to the "turn-on" sequence, an echo protection sequence consisting of 185 to 200 ms of unmodulated carrier followed by 20 to 25 ms of no transmitted energy.

2.3.2.2  When using the echo protection sequence described above, be capable of withholding Receiver Ready (RR) indication after the end of each modem transmission for 150±25 ms to protect against the effect of line echoes.

2.3.2.3  Withhold Receiver Ready (RR) indication during reception of unmodulated carrier.

2.3.3  "Turn-Off" Sequence.  Line signals shall be transmitted after the ON to OFF transition of the Request to Send (RS) circuit as indicated below:

| TYPE OF LINE SIGNAL | SEGMENT A | SEGMENT B | TOTAL OF SEGMENTS |
|---|---|---|---|
| | REMAINING DATA FOLLOWED BY CONTINUOUS SCRAMBLED ONE'S | NO TRANSMITTED ENERGY | TOTAL "TURN-OFF" TIME |
| DURATION | 5ms TO 10ms | 20 TO 22ms | 25 TO 32 ms |

1120

If an OFF to ON transition of the RS circuit occurs during the "turn-off" sequence, it is not taken into account until the end of the "turn-off" sequence.

## 2.4 Secondary Channel (Optional)

2.4.1    Characteristic frequencies of the binary 1 (mark) and binary 0 (space) shall be 390+1 Hz and 450+1 Hz, respectively.

2.4.2    When simultaneous transmission of the primary channel and secondary channel occur in the same direction, the secondary channel shall be 6.0+.6 dB lower in power level than the primary channel.

2.4.3    When the secondary channel is used to transmit data or control information between Data Terminal Equipment (DTE), the modulation rate for such transmission should not exceed 75 baud.  Secondary channel communication between modems themselves, for applications such as fast-polling and diagnostic testing (possibly at modulation rates in excess of 75 baud), is permitted so long as the capability for interoperation with other modems, using the primary channel, is retained.

## 2.5 Scrambler

2.5.1    Scrambler/Descrambler.  A self-synchronizing scrambler/descrambler having the generating polynomial:

$$1 + x^{-6} + x^{-7}$$

with additional guards against repeating patterns of 1, 2, 3, 4, 6, 8, 9, and 12 bits shall be utilized. The purpose of this scrambler is randomization of the data stream in order to maintain convergence of the automatic adaptive equalizer at the demodulator.  Figures 1 and 2 include typical implementations of the scrambler and descrambler.  Logically equivalent circuit designs may be utilized in place of those shown.

2.5.2    Equalizer Training.  Rapid convergence is accomplished by sending only $0^{o}$ phase or $180^{o}$ phase symbols during equalizer training.  For 4800 b/s operation, figure 1 shows circuitry which will generate the proper sequence and figure 3 depicts sequence timing.  When T1 is high, the input to the scrambler is selected; when T2 is high, the first stage of the scrambler is selected, and during the period when T2 is high and C is low, the output is forced high.  When T1 is high and T2 is low, normal operation is restored.

In order to ensure consistent training, the same training pattern must always be sent.  To accomplish this, the data input to the scrambler is mark clamped during training and the first seven stages of the scrambler are loaded with 0011110 (right-hand-most first in time) on the first coincidence of T1 and the signal which will cause the clamp to be removed from the modulator output (generally Request to Send (RS) indication).  This starting point was chosen to ensure a pattern that will provide rapid clock acquisition followed by a pattern providing rapid equalizer convergence.

Eight symbol intervals prior to indicating Clear to Send (CS), the scrambler is switched to normal operation, keeping the scrambler input mark clamped until CS indication is given in order to synchronize the descrambler.

## 2.6 General Characteristics

2.6.1    Impedance.  The modem shall be capable of presenting an input and output impedance to the analog line of 600+60 ohms balanced.

2.6.2    Output Level.  The output level of the modulator shall be adjustable from -12 dBm to -3 dBm in no greater than 1 dB steps.  Output level shall not be adjustable by untrained personnel.

2.6.3    Input Sensitivity.  The demodulator shall have an input sensitivity adjustable to -42+3 dBm and -32+3 dBm.  When the above-stated input sensitivities are used, the input level dynamic range shall be at least 30 dB above the input sensitivity.

2.6.4    Equalizer.  An automatic adaptive equalizer shall be provided.  It shall be in the demodulator.

2.6.5    Digital Interface.  Digital interface characteristics for modems, when applicable, are specified in other Federal standards.  These standards include Federal Standards 1020, 1030, and 1031.

3. <u>Changes</u>. When a Federal agency considers that this standard does not provide for its essential needs, a statement citing inadequacies shall be sent in duplicate to the General Services Administration, Federal Supply Service, Washington, D.C. 20406, in accordance with provisions of Federal Property Management Regulations 41 CFR 101-29.3. The General Services Administration will determine the appropriate action to be taken and will notify the agency.

Preparing Activity:

National Communications System
Office of Technology and Standards
Washington, D.C. 20305

1122

☆U. S. GOVERNMENT PRINTING OFFICE: 1978--261-423/1234

5

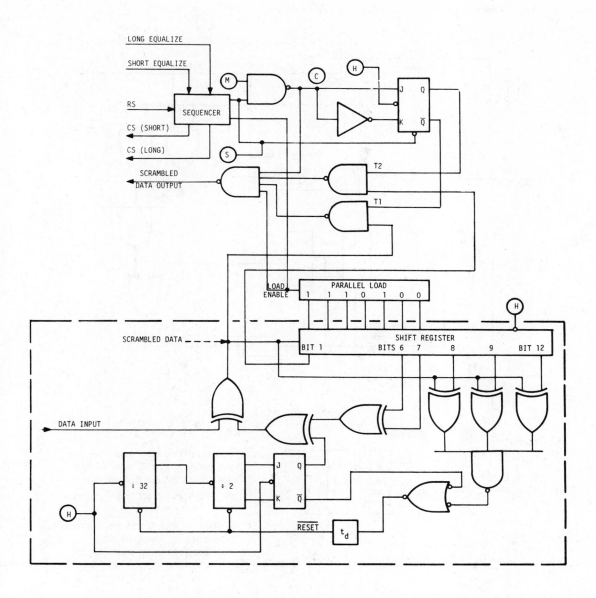

NOTES AND AN EXPLANATION OF SYMBOLS ARE TO BE FOUND
ON FIGURE 2

**FIGURE 1**

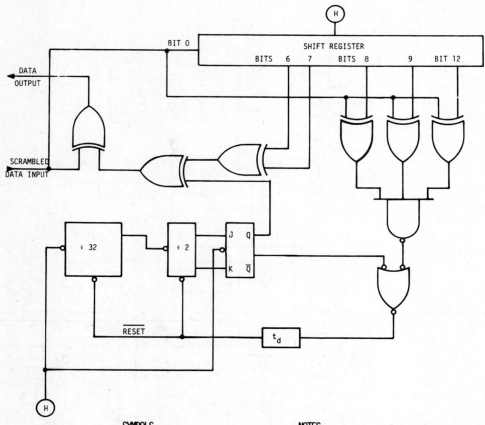

BIT 0

SHIFT REGISTER

BITS 6 7 BITS 8 9 BIT 12

DATA OUTPUT

SCRAMBLED DATA INPUT

÷ 32 ÷ 2 J Q / K Q̄

RESET

$t_d$

H

1124

**SYMBOLS**

| a | b | S1 | S2 | S3 | S4 |
|---|---|----|----|----|----|
| 1 | 0 | 1  | 1  | 0  | 0  |
| 1 | 1 | 0  | 0  | 1  | 0  |
| 0 | 0 | 1  | 0  | 0  | 1  |
| 0 | 1 | 1  | 1  | 0  | 1  |

a
b — S1

a
b — S2

a
b — S3

a — S4

a J Q
t
ā K Q̄

$t_n$ | $t_{n+1}$

J=a
K=ā    $Q_a$

**NOTES:**

(1) H REPRESENTS CLOCK SIGNAL AT THREE TIMES SYMBOL RATE. THE NEGATIVE GOING TRANSITION IS THE ACTIVE TRANSITION

(2) M REPRESENTS CLOCK SIGNAL AT SYMBOL RATE

(3) THERE IS A DELAY TIME DUE TO PHYSICAL CIRCUITS BETWEEN A NEGATIVE GOING TRANSITION OF H AND THE END OF THE "0" STATE REPRESENTED BY $t_d$ ON THE NON-RESET WIRE; THEREFORE THE FIRST COINCIDENCE BETWEEN BIT 0 AND BIT 8, BIT 9, OR BIT 12 IS NOT TAKEN INTO ACCOUNT BY THE COUNTER

(4) DIAGRAMS ARE SHOWN WITH POSITIVE LOGIC

FIGURE 2

7

SYNCHRONIZING SIGNAL SEQUENCE FOR 4800 BITS/S

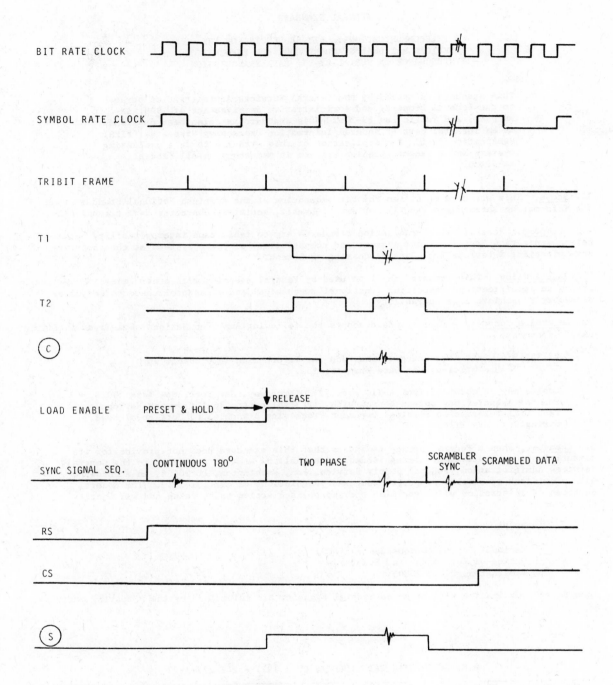

1125

FIGURE 3

FEDERAL STANDARD

TELECOMMUNICATIONS: BIT SEQUENCING OF THE
AMERICAN NATIONAL STANDARD CODE FOR INFORMATION
INTERCHANGE IN SERIAL-BY-BIT DATA TRANSMISSION

This standard is issued by the General Services Administration pursuant to the Federal Property and Administrative Services Act of 1949, as amended, and Public Law 89-306. This standard has also been published as an integral part of Federal Information Processing Standards (FIPS) Publication 16-1. The application of this standard to data processing systems and telecommunication systems is mandatory on all Federal agencies.

1. <u>Scope</u>. This standard specifies the bit sequencing of the American National Standard Code for Information Interchange (ASCII) for serial-by-bit, serial-by-character data transmission.

1.1 <u>Purpose</u>. Federal telecommunication standards are to facilitate interoperability between telecommunication facilities and systems and compatibility of these systems at the computer-communications interface with data processing equipment.

1.2 <u>Application</u>. This standard shall be used by Federal agencies with other Federal Standards or design specifications describing functional, mechanical and procedural characteristics as necessary to achieve interoperability.

2. <u>Reference document</u>. This standard adopts the following American National Standards Institute (ANSI) document:

X3.15-1976  Bit sequencing of the Code for Information Interchange in Serial-by-bit Data Transmission.

Copies may be obtained from ANSI, Inc., 1430 Broadway, New York, New York 10018. Federal Agencies may obtain copies with FIPS Publication 16-1 from the National Technical Information Service, Document Processing Branch, 5285 Port Royal Road, Springfield, VA 22161.

3. <u>Changes</u>. When a Federal agency considers that this standard does not provide for its essential needs, a statement citing inadequacies shall be sent in duplicate to the General Services Administration, Federal Supply Service, FMH, Washington, DC 20406, in accordance with provisions of Federal Property Management Regulations 41 CFR 101-29.3. The General Services Administration will determine the appropriate action to be taken and will notify the agency.

PREPARING ACTIVITY:

National Communications System (NCS)
Office of Technology and Standards
Washington, D.C.  20305

Records of coordination with affected Federal Agencies are maintained by the preparing activity.

U.S. GOVERNMENT PRINTING OFFICE : 1977 - 241-237/2397

FSC   TELE

FEDERAL STANDARD

TELECOMMUNICATIONS:  CHARACTER STRUCTURE AND
CHARACTER PARITY SENSE FOR SERIAL-BY-BIT DATA
COMMUNICATION IN THE AMERICAN NATIONAL STANDARD
CODE FOR INFORMATION INTERCHANGE

This standard is issued by the General Services Administration pursuant
to the Federal Property and Administrative Services Act of 1949, as
amended, and Public Law 89-306.  This standard has also been published
as an integral part of Federal Information Processing Standards (FIPS)
Publication 17-1.  The application of this standard to data processing
systems and telecommuncation systems is mandatory on all Federal
agencies.

1.  Scope.  This standard specifies the character structure and character parity sense for
serial by bit data communication using the American National Standard Code for Information
Interchange (ASCII).

1.1  Purpose.  Federal telecommunication standards are to facilitate interoperability between
telecommunication facilities and systems and compatibility of these systems at the computer-
communications interface with data processing equipment.

1.2  Application.  This standard shall be used by Federal agencies with other Federal Standards
or design specifications describing functional, mechanical and procedural characteristics as
necessary to achieve interoperability.  Systems in existence on the date of this standard which
are designed to a different character parity sense specification are not required to comply
with the parity sense requirement in this standard.

2.  Reference document.  This standard adopts the following American National Standards Institute
(ANSI) document to the extent specified herein:

   X3.16-1976   Character Structure and Character Parity Sense for Serial -by-Bit
                Data Communication in the American National Standard Code for
                Information Interchange.

   Copies may be obtained from ANSI, Inc., 1430 Broadway, New York, New York 10018.
   Federal Agencies may obtain copies with FIPS Publication 17-1 from the National
   Technical Information Service, Document Processing Branch, 5285 Port Royal Road,
   Springfield, VA  22161.

3.  Changes.  When a Federal agency considers that this standard does not provide for its
essential needs, a statement citing inadequacies shall be sent in duplicate to the General
Services Administration, Federal Supply Service, FMH, Washington, DC 20406, in accordance
with provisions of Federal Property Management Regulations 41 CFR 101-29.3.  The General
Services Administration will determine the appropriate action to be taken and will notify
the agency.

   PREPARING ACTIVITY:

   National Communications System (NCS)
   Office of Technology and Standards
   Washington, D.C.  20305

Records of coordination with affected Federal Agencies are maintained by the preparing activity.

U.S. GOVERNMENT PRINTING OFFICE : 1978 - 261-427/2107

This document is available from the General Services Administration (GSA), acting as agent
for the Superintendent of Documents.  A copy for bidding and contracting purposes is available
from GSA Business Services Centers.  Copies are for sale at the GSA, Specification Consumer
Information Distribution Branch, Building 197 (Washington Navy Yard), Washington, DC 20407
for 25cents each.

FSC  TELE

FEDERAL STANDARD

TELECOMMUNICATIONS:  CHARACTER STRUCTURE
AND CHARACTER PARITY SENSE FOR PARALLEL-BY-BIT
DATA COMMUNICATION IN THE AMERICAN NATIONAL
STANDARD CODE FOR INFORMATION INTERCHANGE

This standard is issued by the General Services Administration pursuant to the Federal Property and Administrative Services Act of 1949, as amended, and Public Law 89-306.  This standard has also been published as an integral part of Federal Information Processing Standards (FIPS) Publication 18-1.  The application of this standard to data processing systems and telecommunication systems is mandatory on all Federal agencies.

1.  Scope.  This standard specifies the character structure and character parity sense for parallel-by-bit data communication using the American National Standard Code for Information Interchange (ASCII).

1.1  Purpose.  Federal telecommunication standards are to facilitate interoperability between telecommunication facilities and systems and compatibility of these systems at the computer-communications interface with data processing equipment.

1.2  Application.  This standard is to be used by Federal agencies with other Federal Standards or design specifications describing functional, mechanical and procedural characteristics as necessary to achieve interoperability.  The parity sense required for 8-level perforated paper tape is not a part of this standard.

2.  Reference document.  This standard adopts the following American National Standards Institute (ASNI) document:

X3.25-1976  Character Structure and Character Parity Sense for Parallel-by-Bit Data Communication in the American National Standard Code for Information Interchange.

Copies may be obtained from ANSI, Inc., 1430 Broadway, New York, New York 10018.  Federal Agencies may obtain copies with FIPS Publication 18-1 from the National Technical Information Service, Document Processing Branch, 5285 Port Royal Road, Springfield, VA 22161.

3.  Changes.  When a Federal agency considers that this standard does not provide for its essential needs, a statement citing inadequacies shall be sent in duplicate to the General Services Administration, Federal Supply Service, FMH, Washington, DC 20406, in accordance with provisions of Federal Property Management Regulations 41 CFR 101-29.3.  The General Services Administration will determine the appropriate action to be taken and will notify the agency.

PREPARING ACTIVITY:

National Communications System (NCS)
Office of Technology and Standards
Washington, D.C. 20305

Records of coordination with affected Federal Agencies are maintained by the preparing activity.

U.S. GOVERNMENT PRINTING OFFICE : 1977 - 241-237/2398

This document is available from the General Services Administration (GSA), acting as agent for the Superintendent of Documents.  A copy for bidding and contracting purposes is available from GSA Business Services Centers.  Copies are for sale at the GSA, Specification Consumer Information Distribution Branch, Building 197 (Washington Navy Yard), Washington, DC 20407 for 25cents each.

FSC  TELE

FEDERAL STANDARD

TELECOMMUNICATIONS: SYNCHRONOUS SIGNALING RATES
BETWEEN DATA TERMINAL EQUIPMENT AND DATA CIRUCIT-TERMINATING
EQUIPMENT UTILIZING 4kHz CIRCUITS

This standard is issued by the General Services Administration pursuant
to the Federal Property and Administration Services Act of 1949, as
amended, and Public Law 89-306. This standard has also been published
as an integral part of Federal Information Processing Standards (FIPS)
Publication 22-1. The application of this standard to data processing
systems and telecommunication systems is mandatory on all Federal
agencies.

1. Scope. This standard specifies the standard signaling rates up to 9600 bit/s between
Data Terminal Equipment (DTE) and Data Circuit-Terminating Equipment (DCE) utilizing nominal
4kHz circuits.

1.1 Purpose. Federal telecommunication standards are to facilitate interoperability between
telecommunication facilities and systems and compatibility of these systems at the computer-
communications interface with data processing equipment.

1.2 Application. This standard is to be used by Federal agencies with other Federal Standards
or design specifications describing functional, mechanical and procedural characteristics as
necessary to achieve interoperability. For data rates above 9600 bit/s, whether or not 4kHz
circuits are utilized, FED-STD-1001 applies. Nothing in this standard precludes DTE-DTE
application.

2. Reference document. This standard adopts the following American National Standards Institute
(ANSI) document to the extent specified herein:

X3.1-1976   Synchronous Signaling Rates for Data Transmission

Copies may be obtained from ANSI, Inc., 1430 Broadway, New York, New York 10018.
Federal agencies may obtain copies with FIPS Publication 22-1 from the National
Technical Information Service, Document Processing Branch, 5285 Port Royal Road,
Springfield, VA 22161

3. Changes. When a Federal agency considers that this standard does not provide for its
essential needs, a statement citing inadequacies shall be sent in duplicate to the General
Services Administration, Federal Supply Service, FMH, Washington, DC 20406, in accordance
with provisions of Federal Property Management Regulations 41 CFR 101-29.3. The General
Services Administration will determine the appropirate action to be taken and will notify
the agency.

PREPARING ACTIVITY:

National Communications System (NCS)
Office of Technology and Standards
Washington, D.C.  20305

Records of coordination with affected Federal Agencies are maintained by the preparing activity.

U.S. GOVERNMENT PRINTING OFFICE : 1977 - 241-237/2399

FSC  TELE

FEDERAL STANDARD

## TELECOMMUNICATIONS: ELECTRICAL CHARACTERISTICS OF BALANCED VOLTAGE DIGITAL INTERFACE CIRCUITS

This standard is issued by the General Services Administration pursuant to the Federal Property and Administrative Services Act of 1949, as amended. Its application to both data processing systems and telecommunication systems is mandatory on all Federal agencies to the extent specified herein.

1. <u>Scope</u>. This standard specifies the electrical characteristics of balanced voltage digital interface circuits normally implemented in integrated circuit technology that are to be employed for the interchange of serial binary data, timing, and control signals between voice or data telecommunication equipment where information is being conveyed at the DC baseband level at modulation rates up to 10 megabauds. This standard specifies one of a number of digital interface characteristics, and therefore, must be used in conjunction with other Federal standards or special purpose design specifications giving any functional, mechanical, and procedural characteristics as necessary to achieve full interoperability.

1.1 <u>Purpose</u>. The purpose of this standard is to facilitate interoperability between telecommunication facilities and systems of the Federal Government and compatibility of these facilities and systems at the computer-communications interface with data processing equipment (systems) of the Federal Government.

1.2 <u>Application</u>. This standard shall be used by all Federal departments and agencies in the design and procurement of telecommunication equipment employing balanced voltage digital interface circuits. It is to be used with other applicable Federal standards or design specifications describing functional, mechanical, and procedural characteristics as necessary to achieve compatible interfaces. Exception to the requirement to use this standard is permitted, where equipment compatibility is also dependent on functional and mechanical characteristics, until such time as a Federal standard for these characteristics is published.

2. <u>Applicable documents</u>. The following document forms part of this standard to the extent specified herein:

> Electronic Industries Association (EIA) Standard RS-422, Electrical Characteristics of Balanced Voltage Digital Interface Circuits dated April, 1975.

(Write to Electronic Industries Association, 2001 Eye Street, N.W., Washington, D.C. 20006 for copies of RS-422.)

3. <u>Requirement</u>. The electrical characteristics of balanced voltage digital interface circuits shall conform to EIA Standard RS-422.

4. <u>Changes</u>. When a Federal agency considers that this standard does not provide for its essential needs, a statement citing inadequacies shall be sent in duplicate to the General Services Administration, Federal Supply Service, FMH, Washington, D.C. 20406, in accordance with provisions of Federal Property Management Regulations 41 C.F.R. SS 101-29.3. The General Services Administration will determine the appropriate action to be taken and will notify the agency.

5. <u>Conflict with referenced documents</u>. Where the requirements stated in this standard conflict with any requirements in a referenced document, the requirements of this standard shall apply. The nature of the conflict between this standard and a referenced document shall be submitted in duplicate to the General Services Administration, Federal Supply Service, FMH, Washington, D.C. 20406.

FSC TELE

FED-STD-NO. 1020

<u>PREPARING ACTIVITY:</u>

Office of the Manager
National Communications System (NCS-TS)
Washington, D.C. 20305

Records of coordination with affected Federal Agencies are maintained by the preparing
activity.

U. S. GOVERNMENT PRINTING OFFICE : 1975 - 210-818/2215

2

FEDERAL STANDARD

## TELECOMMUNICATIONS: ELECTRICAL CHARACTERISTICS OF UNBALANCED VOLTAGE DIGITAL INTERFACE CIRCUITS

Authority. This standard is issued by the General Services Administration pursuant to the Federal Property and Administrative Services Act of 1949, as amended. Its application to both data processing systems and telecommunication systems is mandatory on all Federal agencies to the extent specified herein.

1. Scope. This standard specifies the electrical characteristics of unbalanced voltage digital interface circuits normally implemented in integrated circuit technology that are to be employed for the interchange of serial binary data, timing, and control signals between voice or data telecommunication equipment where information is being conveyed at the DC baseband level at modulation rates up to 100 kilobauds. This standard specifies one of a number of digital interface characteristics, and therefore, must be used in conjunction with other Federal standards or special purpose design specifications giving any functional, mechanical, and procedural characteristics as necessary to achieve full interoperability.

1.1 Purpose. The purpose of this standard is to facilitate interoperability between telecommunication facilities and systems of the Federal Government and compatibility of these facilities and systems at the computer-communications interface with data processing equipment (systems) of the Federal Government.

1.2 Application. This standard shall be used by all Federal departments and agencies in the design and procurement of telecommunication equipment employing unbalanced voltage digital interface circuits. It is to be used with other applicable Federal Standards or design specifications describing functional, mechanical, and procedural characteristics as necessary to achieve compatible interfaces. Exception to the requirement to use this standard is permitted, where equipment compatibility is also dependent on functional and mechanical characteristics, until such time as a Federal standard for these characteristics is published.

2. Applicable documents. The following document forms part of this standard to the extent specified herein:

Electronic Industries Association (EIA) Standard RS-423, Electrical Characteristics of Unbalanced Voltage Digital Interface Circuits, dated April, 1975.

(Write to Electronic Industries Association, 2001 Eye Street, N.W., Washington, D.C. 20006 for copies of RS-423.)

3. Requirement. The electrical characteristics of unbalanced voltage digital interface circuits shall conform to EIA Standard RS-423.

4. Changes. When a Federal agency considers that this standard does not provide for its essential needs, a statement citing inadequacies shall be sent in duplicate to the General Services Administration, Federal Supply Service, FMH, Washington, D.C. 20406, in accordance with provisions of Federal Property Management Regulations 41 C.F.R. SS 101-29.3. The General Services Administration will determine the appropriate action to be taken and will notify the agency.

5. Conflict with referenced documents. Where the requirements stated in this standard conflict with any requirements in a referenced document, the requirements of this standard shall apply. The nature of the conflict between this standard and a referenced document shall be submitted in duplicate to the General Services Administration, Federal Supply Service, FMH, Washington, D.C. 20406.

FSC TELE

FED-STD-NO. 1030

MAINTENANCE AGENCY

Office of the Manager
National Communications System (NCS-TS)
Washington, D.C. 20305

Records of coordination with affected Federal Agencies are maintained by the preparing activity.

U. S. GOVERNMENT PRINTING OFFICE : 1975 - 210-818/2216

2